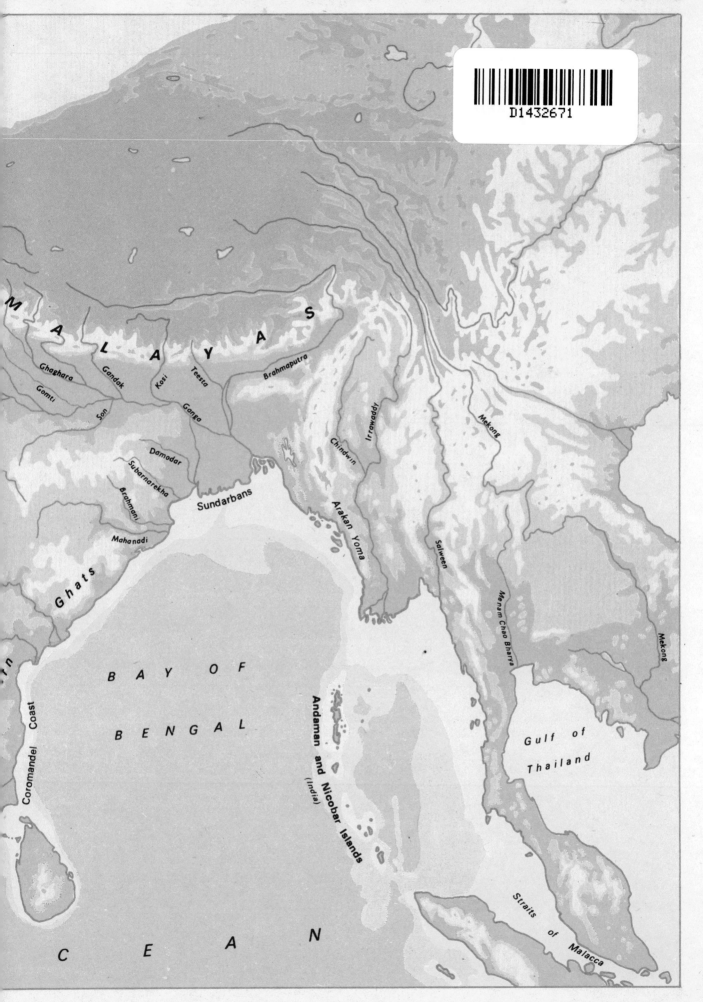

ENCYCLOPEDIA OF
INDIAN NATURAL HISTORY

ENCYCLOPEDIA OF
INDIAN NATURAL HISTORY

Centenary Publication of the
Bombay Natural History Society
1883–1983

General Editor
R. E. HAWKINS

Illustrations Editors
DORIS NORDEN
and for inset plates
BITTU SAHGAL

Published on behalf of the
Bombay Natural History Society
by
OXFORD UNIVERSITY PRESS
DELHI BOMBAY CALCUTTA MADRAS
1986

Oxford University Press, Walton Street, Oxford OX2 6DP

NEW YORK TORONTO
DELHI BOMBAY CALCUTTA MADRAS KARACHI
PETALING JAYA SINGAPORE HONG KONG TOKYO
NAIROBI DAR ES SALAM CAPE TOWN
MELBOURNE AUCKLAND

and associates in
BEIRUT BERLIN IBADAN NICOSIA

Filmset and printed in India
at All India Press, Pondicherry 605 001
and published by R. Dayal, Oxford University Press
YMCA Library Building, Jai Singh Road, New Delhi 110 001.

EDITORS OF SECTIONS

The writing of the Encyclopedia has been undertaken by the following team of editors, whose contributions are seldom signed except when appearing outside their own sections.

AMPHIBIANS

Romulus Whitaker has lived in India since 1951 and has carried out surveys and studies of amphibians and reptiles in Asia, Africa and Papua New Guinea.

With contributions from Ranil Senanayake.

ANATOMY & PHYSIOLOGY

Lata Mehta and A. N. D. Nanavati. Lata Mehta has been Professor of Anatomy at Grant Medical College, Bombay, for the past decade and A. N. D. Nanavati was the medical officer on two Himalayan expeditions, including Everest 1962, and has been the Honorary Secretary of the Bombay Natural History Society since 1974.

ANNELIDS

Renée M. Borges is now studying ecology at the University of Miami and aims to research into evolutionary relationships between certain Old and New World plant and animal communities.

ARACHNIDS

T. V. Subramanyam, Postgraduate in Zoology, 1933, and a keen student of spiders and collector of sea shells for nearly forty years.

BEHAVIOUR

Madhav Gadgil has worked on the ecology and behaviour of social animals ranging from wasps and mynas to elephants and human beings.

With contributions from M. K. Chandrasekaran, Ragavendra Gadagkar (R.G.) and R. Subbaraj.

BIRDS

Sálim Ali has been conducting bird surveys throughout the subcontinent since 1930, and was awarded the Paul Getty Prize for Wildlife Conservation in 1975.

With contributions from Shahid Ali (S.Q.A.), V. C. Ambedkar, B. Biswas & K. K. Tiwari (B.B.& K.K.T.), Renée Borges, Priya Davidar (P.D.), K. S. Dharmakumarsinhji, Zafar Futehally, Madhav Gadgil (M.G.), Prakash Gole (P.G.), R. B. Grubh, H. Himmatsinhji, S. A. Hussain, A. J. T. Johnsingh (A.J.T.J.), P. Kannan (P.K.), Lavkumar Kacher, Reza Khan (R.K.), D. N. Mathew (D.N.M.), A. K. Mukherjee (A.K.M.), R. Narasimha (R.N.), K. K. Neelakantan (K.K.N.), Muhammad Osman (M.O.), D. J. Panday (D.J.P.), N. G. Pillai, Krishna Raju, Ashok Sahni (A.S.), H. Saiduzzafar (H.S.Z.) Shivrajkumar (S.K.), P. T. Thomas, N. S. Tyabji and V. S. Vijayan (V.S.V.).

BOTANY

K. C. Sahni has carried out floristic surveys in the tropical and temperate forests of India (including the Andamans and Nicobars) and Sri Lanka, mostly while posted at the Forest Research Institute, Dehra Dun, where he was Director of Biological Research.

With contributions from K. N. Bahadur (K.N.B.), U. C. Bhattacharyya (U.C.B.), S. S. Bir (S.S.B.), Som Dev Sharma (S.D.), R. K. Gupta (R.K.G.), S. K. Jain (S.K.J.), M. N. Jha (M.N.J.), O. N. Kaul (O.N.K.), V. M. Meher-Homji (V.M.M-H.), K. K. Nanda (K.K.N.), G. S. Paliwal & Usha Rajput (G.S.P. & U.R.), V. Raina (V.R.), A. S. Rao (A.S.R.), M. A. Rau (M.A.R.), Sujan Singh (S.S.), K. Subbaramaiah (K.S.) and J. N. Vohra (J.N.V.).

CLIMATOLOGY

P. R. Pisharoty, now Professor Emeritus at the Physical Research Laboratory, Ahmedabad, was a member of the Indian Meteorological Service whose last post was Director of the Indian Institute of Tropical Meteorology, Poona.

COELENTERATES

T. V. Subrahmanyam

CONSERVATION

Zafer Futehally, former Vice-President of the World Wildlife Fund — India and founder-editor of the *Newsletter for Birdwatchers*.

With contributions from W. E. Alwis, Azra Bhatia, John Blower (J.B.), J. C. Daniel, M. A. Reza Khan (R.K.), T. J. Roberts (T.J.R.), Fred Simmonds and Karna Sakya (K.S.).

CRUSTACEANS

T. V. Subrahmanyam

ECHINODERMS

T. V. Subrahmanyam

EVOLUTION

A. N. D. Nanavati

FISHES

C. V. Kulkarni, former Director of Fisheries in Maharashtra, has long been concerned with the life history and artificial propagation of freshwater fish.

With contributions from B. F. Chhapgar.

GEOGRAPHY

Description of the geography of the subcontinent is beyond the scope of this encyclopedia but contributions of a geographical nature have been made by Shahid Ali, Azra Bhatia (A.S.B.), J. T. M. Gibson (J.T.M.G.), Lavkumar Khacher, M. N. Nigam, D. J. Panday (D.J.P.), Ishwar Prakash (I.P.), Joseph Thomas (J.T.) and Ranjit Tirtha (R.T.).

GEOLOGY

K. V. Subbarao teaches at the Indian Institute of Technology, Powai, His major fields of interest are volcanology and marine geology.

With contributions from R. R. Nair (R.R.N.), C. S. Pitchamuthu (C.S.P.), C. V. R. K. Prasad (C.V.R.K.P.), V. Subramaniam (V.S.), W. D. West (W.D.W.), and R. Vaidyanathan (R.V.).

INSECTS

Kumar D. Ghorpadé has collected and observed insect life in most Indian states since 1968. *Colemania* (an international Journal of Entomology) is being edited and published by him since 1981. He was awarded a postdoctoral fellowship by the Smithsonian Institution, Washington D.C., in 1982.

MAMMALS

T. J. Roberts, author of *Mammals of Pakistan* (1977) based on over thirty years' observation and study of Wildlife in the area.

With contributions from Sálim Ali (S.A.), Satish Bhaskar (S.B.), J. C. Daniel (J.C.D.), G. W. Fulk (G.W.F.), Colin Groves (C.P.L.G.), A.J.T. Johnsingh (A.J.T.J.), M Krishnan (M.K.), W. A. Laurie (W.A.L.), R.C.D. Olivier (R.C.D.O.), Ishwar Prakash (I.P.), Arjan Singh (A.S.), George Schaller (G.B.S.) and R. L. Tilson (R.L.T.).

MICROBES & DISEASE

A. N. D. Nanavati

MOLLUSCS

T. V. Subrahmanyam

REPTILES

Romulus Whitaker

With contributions from Satish Bhaskar (S.B.), T. S. N. Murthy (T.S.N.M.) and Zai Whitaker (Z.W.).

SPONGES

T. V. Subrahmanyam

R. E. HAWKINS was for forty years associated with the Indian branch of the Oxford University Press and has been a member of the Bombay Natural History Society since 1938. DORIS NORDEN studied at the National Academy of Design, New York City, and the New York School of Design, and was formerly Chairman of the Board of Trustees, Madras Snake Park; and BITTU SAHGAL is editor of the magazine *Sanctuary*, founded by him in 1981.

PREFACE

An immense amount of information about India's wildlife has accumulated since the foundation of the Bombay Natural History Society in 1883 and much is recorded in the thousands of pages of its *Journal*. To mark the centenary the Society decided to compile a one-volume encyclopedia of Indian natural history, written by experts in simple language, bringing a small part of this knowledge together in the hope that it would stimulate interest in the teeming plant and animal life of the sub-continent.

Man's fascination with natural history and the living world stems largely from its infinite variety, as well as the fact that we are part of this intricate web of life. An encyclopedia such as this attempts, in the broadest possible way, to satisfy that curiosity and to reveal the vast range and complexity of life forms with all their wonderful adaptations, contributing towards survival. The study of evolution has helped us to understand better how such diverse life forms have come into existence. Evolution also explains, in a logical manner, many of the seemingly strange and weird physical forms and behavioural patterns that we see in nature. The 'Tree of Life' inside the back cover shows relationships between the major divisions of life forms on a time scale. It demonstrates, dramatically, how brief the relative span of time has been since the higher mammals, and in particular man himself, have existed.

Our accumulating knowledge has given us an awesome responsibility. It will determine how we shall change, adapt or even destroy increasing segments of the environment which sustains all of us. A study of evolution shows how many of the earlier branches and twigs of the tree of life have withered away from natural causes. Today, mankind seems able to control the means to damage or cut off some other twigs and branches. The knowledge contained in this encyclopedia will, it is hoped, help to create a better awareness not only of the variety and richness of our wildlife and natural resources but also a consciousness of the interdependence of all life forms and our responsibility to understand, and not upset, irreversibly perhaps, the delicate balance of nature, that will determine our own future.

Headwords in the encyclopedia do not appear in the index, nor obvious derivatives. 'Blister beetles' would not be indexed as one would naturally expect to find them under BEETLES; but one might not expect to find them under BENEFICIAL INSECTS so the index points to that headword only. And the entry 'commensalism, CRABS' does not imply that commensalism is restricted to crabs but only that under CRABS one will find the word used and partially defined. The encyclopedia itself is to be read, and the index only points to appropriate headwords.

R.E.H.

PLATES

DECORATIVE INITIALS AND THE
SIMPLIFIED TREE OF LIFE

The decorative initials devised by Naira Ahmadullah show plants or animals whose common English, Hindi or scientific names are appropriate. To identify the pictures around each letter start at the top right-hand corner or just below it and go round clockwise. Nearly all the plants and animals pictured are described and named in the encyclopedia. A complete list is given below:

A Asoka, Ant, Auger shell, Amaltas, Angler fish, Axis deer

B Bauhinia, Bullfrog, Bulbul, Bummelo, Bamboo, Brown Butterfly, Blackbuck (horns only)

C Cicada, Clouded leopard, Chir pheasant (♀), Crab (Fiddler), Chir pine, Cowry shell

D Dugong, Deccan hemp, Dolphin, Dove (Spotted), Dhub grass, Dove (Red Turtle, ♂)

E Elephant, Earwig, Ebony (*Diospyros ebenum*), Eel, Eagle (Golden).

F Flying Fox, Flamingo, Fungus (bracket), Fungoid Frog, *Ficus benghalensis*, Fern

G Goose (Barheaded), Gulmuhr, Gibbon (hoolock), Gecko, Gourami, Gaur.

H Hibiscus, Hawksbill turtle, Hyena, Honey buzzard, Harp (mollusc)

I Ibis, Ibisbill, *Indoclystus singulare* (antlion larvae and 'trap'), Ibex, Iris

J Jaçana, Jacaranda, Jezebel, Jellyfish, Jackal

K Kingfisher, Kakar, Koel (♀), Kitul, King-crab

L Loris, Lantana, Lynx, Lapwing (Redwattled), Locust

M Mud-skipper, Marking-nut (fruit and leaf), Macaque, Magpie-robin, Mushroom (Morchella), Monitor lizard

N Newt, Nautilus, Nakta duck, Neem, Nilgai (♂)

O Orchid (*Cymbidium sikkimense*), Octopus, Olive shell, *Oligodon*, Owl (Himalayan Wood)

P Python, Pitcher plant (*Nepenthes khasiana*), Partridge (black), Pangolin, Prawn, Puffer fish

Q Quaker babbler, *Quercus dilatata*, Quartz, Quail, Queen's Crape Myrtle

R Roller, Rhinoceros, Rhododendron, Ray (sting), Rhinoceros beetle

S Silk-cotton, Sea Anemone, Sandpiper, Squirrel (Kashmir Flying), Seahorse

T Tailor Bird (and nest), Tiger, Toad, Tortoise (Starred), Tamarind

U *Upupa epops, Uromastix hardwickii*, Urchin (Sea), Upas tree, Urial

V Volute, Vole, *Vanda caerulea*, Viper, Vulture (Bearded)

W Water-lily, Water spider (and 'nest'), Woodpecker (Goldenbacked), Whale (Blue), Weasel (Stripe-backed)

XYZ Yak, Yew, *Zosterops palpebrosa*, Zygopteran, *Xancus pyrum, Xiphias gladius*

The simplified Tree of Life on the back end-paper is reproduced from the Fontana edition of David Attenborough's *Life on Earth*, a book which has been described as the best introduction to natural history ever written. In the simplified tree of life each major group of plants or animals is displayed separately rather than as branching from a central trunk, and the widths of the columns give an indication of the numerical abundance of species — for example, there are about 250,000 species of Flowering Plants at present. The column headed 'Bacteria, Algae, Lichens and Fungi' is still thick at the bottom of the page, and fossils show that it should continue to a time more than 3000 million years ago. Green indicates plant life; yellow very simple animals; blue animals without backbones (invertebrates); and red vertebrate animals.

Note on BNHS

In 1883 eight gentlemen of Bombay formed an association for the purpose of exchanging notes and observations on natural history. In course of time the Bombay Natural History Society acquired a unique national role and international prestige by virtue of its many faunal surveys and collecting expeditions throughout India, resulting in one of the best organized and classified research collections in the country. The Society has published, for 100 years, a journal which is a repository of basic information on the flora and fauna of the subcontinent.

The Society leads the movement for nature conservation in India, and its library and collections continue to attract a large number of professional and amateur scientists, both Indian and foreign, who wish to undertake studies on Indian Wildlife.

The Society's research projects have established field stations all over the country and are continuing the intensive study of nature and natural phenomena of the Indian subcontinent.

The Society has also published books for the general reader, on various aspects of natural history. The best known is the ever popular *The Book of Indian Birds* by Sálim Ali. Books on Indian animals, reptiles, and trees, as well as a series of small illustrated booklets for children, have also been issued under the imprint of the Society.

This Encyclopedia, the Society's most ambitious publishing venture so far, owes its existence to the dedication and perseverance of the General Editor and to a generous grant from the Department of Science and Technology, Government of India, which enabled us to market it at the present low price. It is hoped that the book will answer a long-felt need for a work of reference on natural history suited to the general reader as well as to the scientist in the field.

ACKNOWLEDGEMENTS

We are grateful to the Linnean Society of London for permission to reproduce (in CLASSIFICATION AND NAMING) a drawing made by Linnaeus in the diary he kept during his journey to Lapland; to Benn Technical Books for many drawings made by T. J. Roberts for his book *Mammals of Pakistan* (Ernest Benn, 1977); to the Houghton Mifflin Company, Boston, U.S.A. for two figures (appearing under EARTHQUAKES) from their publication *Investigating the Earth* (1967) and to the Open University, Bletchley, England for another figure reproduced in EARTHQUAKES from Unit 22 of their *Science Foundation Course A100* copyrighted in 1971. The editor of the section on Fishes is particularly indebted to J. R. Norman's *History of Fishes* (Ernest Benn) and to Volume II of the *Oxford Junior Encyclopedia*, from each of which he has taken the liberty of copying several figures. The editor of the section on Insects has based three figures in INSECT IMMATURES on Metcalf & Flint's *Destructive and Useful Insects* (McGraw-Hill).

ACCENTORS or 'HEDGE SPARROWS' are sparrow-sized passerine birds (family Prunellidae) living at high altitudes in the Himalaya, descending in winter to lower levels. They subsist on insects in summer, seeds and berries in winter. The sexes are alike, and both share in building an open-cup nest of leaves and rootlets. 3 to 5 greenish eggs are laid.

AGGRESSION. A cow elephant pushing another one aside at a small water-hole in summer, two deer stags sparring in the rutting season, a female langur roughly pushing aside a squealing infant that wants to suckle, crows mobbing a kite and a rhino charging a man who suddenly blundered into its path: all of this diversity of behaviour will be classified under the rubric of aggression. Its common feature is the expression of intention, or actual accomplishment, of physical injury by one animal to another. But by convention we exclude from this category instances of predation: a cat does not commit aggression against a rat, it preys on it. The two broad functional categories that are then left under aggression involve its use as a device of competition for scarce resources and its use as a device to reduce predation.

In a large number of instances, particularly those involving individuals belonging to the same species, aggression is a ritualized affair. Two male dogs may need nothing more than a bit of growling and sniffing at each other's hindquarters to decide on who struts away as the winner and who withdraws with the tail tucked between his legs. Even if they get into a fight, the winner will not inflict on the loser a grievous injury in spite of an opportunity to do so. By and large animals settle their disputes with much show of their might and with fearsome threats, but without any serious physical injury.

This is because most animal contests would tend to be asymmetrical. One of the contestants may be physically stronger, or may have already established itself in the territory and so on. When such asymmetries exist, one of the contestants stands a greater than half chance of losing the contest. It is evident that under these conditions, it is best for the animals to probe each other's capabilities till they know which one is more likely to lose if the contest were fought to its bitter end. At that juncture, it would be to the advantage of the contestant more likely to lose to withdraw, and not escalate the fight. At the same time, it is advantageous for the winner not to try to press home its advantage too much, for if it tried to do so, it may no longer be advantageous for the potential loser to withdraw. The winner may then run unnecessary risks of injury in such an escalated fight, without any commensurate gain by attempting to force its advantage too much. We therefore expect most animal conflicts to end without serious injury to either party.

Graduated aggressive displays of the rhesus macaque and the night heron. The animals will attack only if neither of the contestants has retreated till the most intense display level is reached.

There are two conditions, however, under which escalated fights do occur. If the two contestants are perfectly evenly matched so that each will judge itself to stand a good chance of success, neither may accept defeat and the conflict may become escalated. A male gaur was in fact killed in such a fight fought to the bitter end in Bandipur Tiger Reserve in 1975. Secondly, the gains of a win may be so high that it may be worth the risk of even a serious injury. Thus in the case of some species of seals the females all congregate for breeding on a small island. The males then fight it out amongst themselves for sexual access over this harem. A male who wins will corner a very large number of females, while a male who loses may get no chance for breeding at all till the next year. Under these circumstances males may continue to fight even if there is only a very small chance of winning, and the fights are long-drawn-out and bloody.

Elephant butting in an aggressive encounter.

Now in the normal ritualized contests the two rivals attempt in various ways to probe each other's strength and determination. Each individual will naturally attempt to convey an exaggerated estimate of its own strength and determination, while attempting to penetrate the exaggerations of the rival. We therefore expect good indicators of strength to evolve as signals, and indicators that can be faked to be lost in the process of evolution. For instance, in the Roe Deer a male's ability for sustained roaring correlates well with his physiological vigour and hence most disputes among the stags are settled in roaring contests. In such contests the two rivals simultaneously start roaring, increasing the tempo with each roar till one of the two gives up and tapers off his roaring and withdraws.

In any contest, there may be asymmetries not only in strength but also in what each contestant stands to lose. Thus when spiders fight over webs, the spider that has already invested in constructing the web stands to lose much more. It may then be willing to risk a more serious injury than the spider which has not invested in the web. The resident may then fight with much greater determination than the intruder. It is therefore necessary for the two contestants to probe each other's determination as well as strength. However, if one of the contestants signals right from the beginning that its determination is weak, it will further

Barasingha stags sparring

© Cyrus Adenwalla

reduce its chance of winning. Hence, none of the contestants will communicate its willingness to give up till the point it actually gives up. This has been beautifully brought out in analysis of contests in certain fish species. The sequence of displays adapted by the contestants cannot be used to predict the ultimate winner till the last. Only in a ground spider has it been possible to characterize the winner in some fashion on the basis of its moves during the contest. The conclusion is that size and residence largely decide the winner. But above and beyond this, the only way in which a winner differs from a loser is in its sequence of moves departing much more from the average pattern than that of a loser. Thus an element of surprise plays a rôle in deciding who wins.

We expect aggressive behaviour of animals to be much more pronounced against unrelated individuals. Thus in the case of lions, the core of the pride is a group of related females. On reaching maturity, males are driven out from such prides. The adult males staying with a pride are then males displaced from another pride. Hence, there is a continual tussle between the floating bachelor males and the adult males attached to a pride. The fights are often bloody and adult males bear innumerable scars and mutilated ears as marks of this struggle. Notably enough, males thrown out of any given pride, which would be siblings or cousins or some such relatives, tend to band together and to fight together to take over a new pride. Once settled in a pride these males fight remarkably little with each other even over copulations with a female in heat.

This does not mean that there is no conflict or aggressive behaviour amongst blood relatives. Indeed there is, even in so apparently well-knit a group as a colony of social wasps. In such a colony there may be just one egg-laying queen, and a number of sterile workers who are her daughters. The workers take care of the brood, which is made up of their brothers and sisters. It is a remarkable fact that in these wasps sisters are more closely related to each other than are mother and daughter, but brother–sister are less closely related than mother–son. It is therefore of genetic advantage to the workers to raise their mother's female eggs, but to raise their own male eggs, the males developing from unfertilized eggs. In certain species a conflict ensues over this with workers laying male eggs which the queen may destroy by eating. The workers also devote more care towards raising their mother's female brood than her male brood.

Aggression within a species may involve not just pairs of individuals, but whole groups, particularly in species which maintain border territories. Thus, among spotted hyenas, border fights may lead to killing of members of the neighbouring groups which are then devoured. Similar pitched battles with large numbers of deaths also occur in many species of ants.

So much for aggression within the species. Potential prey species may also exercise aggression against predators. The best-known example of this is the mobbing of owls and snakes by birds. Often a very large number of small species are involved who chatter around and fly at the predator. Our crows and drongos are also pugnacious and often mob predatory birds and cats. But the attacks appear restricted to driving the predator away and do not lead to any physical injury. Presumably, when the predator's presence is continually being advertised by the mobbing birds, it is difficult for it to make an effective kill, and mobbing may discourage a predator from frequenting a particular area.

Recent years have witnessed a serious controversy over whether aggression is a drive, like hunger or sex. The problem is whether certain animals deliberately seek out situations in which to vent their aggression just as they deliberately seek out prey for eating or mates for mating. For instance, experiments have been conducted in which male fish are kept in aquaria. They have to exert energy to obtain a view of a rival male at which to display aggressively. There is some suggestion that fish will actually do so. This suggests that aggression may in fact be a drive. In general, however, the consensus is that aggression is not a drive in the same fashion that hunger and sex are, and that animals use aggression in particular situations as a device to reach other rewards such as food and sex, but do not seek conflict as an end in itself.

Human aggression has also been a matter of considerable debate in this context. The human species is indeed a species in which aggression including killing of other humans is expressed in a number of contexts. Human aggression is expressed in several contexts—within the family, within a community, amongst different communities. Within the family, the aggression nearly universally involves a certain domination and exploitation of females by males and of children by adults. Serious violence is however most commonly prompted in the context of suspected or real adultery. Within the community considerable aggression arises from the failure of individuals to reciprocate certain acts of altruism. But the most serious aggression relates to sexual jealousy. For instance, most murders in Bushman society are traceable to instances of adultery.

Large-scale homicide however derives from aggression between different communities particularly in disputes over group territories. This is the human parallel of the hyena's killing members of a neighbouring pack. This is known from Stone Age populations such as that of the head-hunters of New Guinea and continues unabated, or perhaps considerably magnified, to this day.

AIR-BREATHING FISHES. A group among the bony fishes known as *Dipnoi*, or *Dipneusti*, the lung-fishes, are the real air-breathing fishes. Another group which are customarily called air-breathing fishes have accessory respiratory organs in addition to the usual gills of the modern bony fishes and since they utilize atmospheric air for oxygenation of their blood with the help of modified vascular structures in their body, they are also called air-breathing fishes.

The first group (*Dipnoi*) is represented by the Australian lung-fish *Neoceratodus*, the African species *Protopterus* and the South American *Lepidosiren*. None of these, or similar Dipnoans, at present occur in India but in the geological past such fishes did exist in some parts of the country. The so-called lung of some of these fishes consists of a single elongate sac, as in *Neoceratodus*, and extends across the entire length of the body; while in *Protopterus* and *Lepidosiren* the sac is double and is very like the bilobed mammalian lung (see Fig.). On the inner wall of this sac are fibrous strands, which extend inwards, divide and subdivide into small cavities or alveoli which are lined by a soft flat layer of tissue which contains numerous blood capillaries. The elongated lung-like structure opens ventrally into the front part of the gullet by a tube known as the pneumatic duct, its opening guarded by a vascular valve (glottis). External

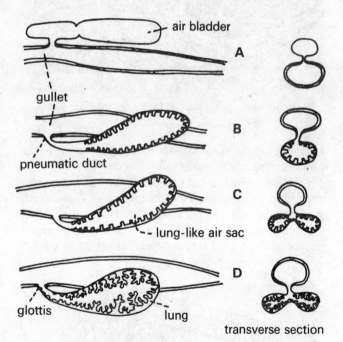

DIAGRAMMATIC REPRESENTATION OF
AIR SAC OF FISHES

A Air bladder of an ordinary carp
B Single lung-like air sac of an Australian lung fish *Neoceratodus*
C Left side of double lung-like air sac of an African lung fish *Protopterus*
D Left side of double-lobed Mammalian lung

nostrils (nares) take air and pass it through a part of the gullet and then through the pneumatic duct into the tubular lung-like structure where oxygenation of the blood passing through the thin-walled blood capillaries is effected as in the land vertebrates where aerial respiration is the rule.

The second group is one which has developed special modifications, known as the accessory respiratory organs, as adaptations to tide over the difficult conditions of life when water becomes deficient in oxygen and the gills are unable to obtain sufficient dissolved oxygen from the water. These are, as the words suggest, additional respiratory organs found in some fishes. Prominent among such fishes are the climbing perch (*Anabas testudineus*), mud-skippers (*Boleophthalmus* sp. and *Periophthalmus* sp.), murrel (*Channa* sp.), shingi (*Heteropnesteus fossilis*), magur (*Clarias*) and cuchia

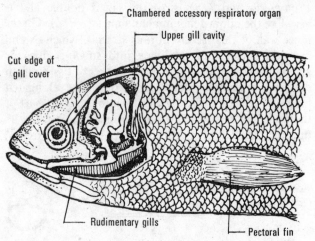

Dissected head of a Climbing Perch, showing accessory respiratory organ × ²/₃

(*Amphinous*). In these fishes, some cavities in the throat and gill chamber region are covered by flat tissue (epithelium) which is richly supplied with blood and are connected to minute blood capillaries. These fishes occasionally come to the surface of the water and gulp a small quantity of air. This air taken by the mouth passes into the aforesaid cavities and comes in contact

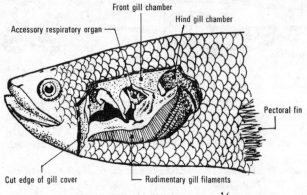

Head of a Snake Head (murrel) × ¹/₃

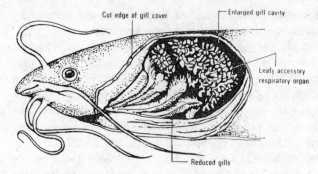

Head of a Catfish (magur) × ²/₃

with the thin epithelial layer containing blood supply. The intake of oxygen and release of carbon dioxide takes place in these cavities, in addition to the normal gill respiration which is considerably reduced in these forms. Acquisition of such accessory respiratory organs enables these fishes to remain out of water for a long time. Murrel and Shingi are reported to live out of water even for 24 hours if kept in moist and cool places.

In magur (*Clarias* sp.), a well known catfish of northeastern India, a leafy structure is developed in a cavity above the gills. In cuchia, a small eel-like fish occurring in the perennial ponds of Bihar and Bengal, a pair of elongated sacs grow from the wall of the throat above the gills and these, with the help of their inner vascular lining (blood-supplied epithelium), perform the function of an accessory respiratory organ.

Another interesting example of accessory respiratory method is reported in the case of loaches (Cobitidae), a group of small slender freshwater fishes, with tiny barbels. These are generally found in ricefields and shallow tanks on the west coast of India and some hill streams in other parts of the country. In these fishes, the posterior or the end part of the intestines serves both assimilative and respiratory functions, but alternately. In some cases the intestinal portion becomes respiratory only in summer when water becomes deficient in oxygen.

In another common riverine fish chital (*Netopterus chitala*) the air-bladder is divided and subdivided into smaller lobes and their inner wall is supplied with blood capillaries, through the thin walls of which oxygenation of blood takes place. In *Megalops cyprinoides*, a cousin

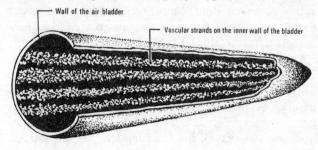

Cut-open air bladder of Indian Tarpon, *Megalops cyprimoides* × ¹/₂

of the American tarpon, locally known around Bombay as *vadas*, strands of blood capillaries projecting on the inner walls of the air-bladder serve as accessory respiratory organs.

Most fishes having accessory respiratory organs are so accustomed to depend on these structures for their respiration that their normal gills are reduced in size and function. Consequently, if these fishes are prevented from coming to the surface of water to gulp air, i.e. they are made to depend on their gills only, they become asphyxiated.

ALGAE. Splashes of green, yellow, brown or black often meet our eyes in roadside ponds and puddles, on tree-trunks, rocks and even buildings. If you put your fingers on them, some appear brittle, some powdery, some spongy and some tough and felt-like. Often we approach them with a broom and a scrubber or a killer chemical. These are the simplest of all pigmented plants, known as algae. The spectrum of colours they splash is due to different types of pigments present in them, such as chlorophyll (green), cartenoid (brown), phycocyanin (blue-green) and phycoerythin (red)— (phyco- means seaweed, or alga). With the help of these pigments they manufacture their own food by PHOTO-SYNTHESIS. Their size shows an enormous range of variation. The simplest of them is a unicellular form like *Chroococcus* commonly found as a yellowish brown powdery crust on tree-trunks, rocks and roofs of buildings, too small to be seen individually by the naked eye. At the other extreme are seaweeds like the giant kelp, 500 m long. There are seaweeds of the genus *Sargassum*, whose berry-like air vessels enable them to float, which drift together in the Atlantic in such masses that the area is called the Sargasso Sea and used to be avoided by sailors. *Sargassum* also occurs abundantly in the Indian Ocean and the Bay of Bengal and bits of it are often washed ashore, in winter when the fronds are shed.

Algae are found almost everywhere where there is water or dampness, being looked upon as environmental pollutants in places like waterworks and recreational areas and as extremely valuable in soils. They occur in a wide range of situations: in hot springs (60–80°C); in salt-water lakes and seas; and even on snow. They also grow in waters polluted with heavy metals like lead, cadmium and mercury. The legendary red snow in some parts of the Himalaya and elsewhere is due to the resting spores of the unicellular alga *Chlamydomonas*. The Red Sea derives its name from the dense growth of the reddish blue-green alga *Trichodesmium*. This also occurs abundantly in the Arabian sea. The green tide often observed on the Indian coast is caused by the alga *Hornellia*. Some algae live in association with fungi to form lichens, some grow epiphytically on

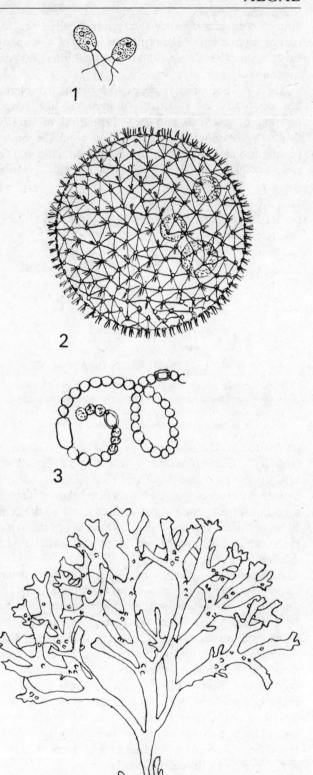

FOUR VARIETIES OF ALGAE
Magnified about × 2000

1 Two *Chlamydomonas* (unicellular). 2 *Volvox* (colonial).
3 *Anabaena* (filamentous). 4 *Dictyota*

animals and plants, some live symbiotically. The red rust of tea and coffee is a typical example of algal parasitism. Algae have been found in animal intestines and human skulls.

Algae do not produce seeds like the higher plants, but reproduce by a variety of methods like simple division of one cell into two formation of asexual spores by developing a thick wall around the protoplasm, and also by sexual fusion between male and female sexual bodies (gametes) which are either similar

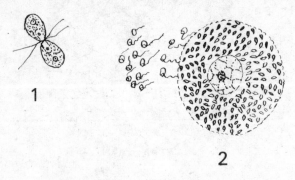

SEXUAL REPRODUCTION OF ALGAE

1 Two motile *Chlamydomonas* have come together by means of their flagella and the contents of their cells mingle (isogamy).
2 One large nonmotile female gamete of *Dictyota* is approached by male gametes (oogamy). About × 1000

or dissimilar. The simplest type of sexual reproduction is seen in *Chlamydomonas* whose two similar motile cells (gametes) come together, their flagella entwine and the contents of one cell moves into another through a small bridge formed at the contact point of the cells. In some forms the egg cell is formed in special structures called oogonia. In red algae the sexual reproduction is extremely complex.

A taxonomist groups algae into 11 classes based largely on their pigment and protosynthetic products and to a lesser extent on their mode of reproduction. In common jargon they are blue-green algae, green algae, red algae, brown algae, cryptomonads, dinoflagellates, diatoms, golden-brown algae, yellow-green algae, euglenoids and stoneworts. *Euglena* is both plant and animal. Like a plant it has chlorophyll and can make its own food, but it ingests organic material as an amoeba does. Diatoms have a silica wall and form an important constituent of marine and fresh-water plankton. Their fossilized silica frames are characteristic of the gravelly earth called kieselguhr.

The blue-green algae are the most primitive ones, their fossil history dating back to precambrian rocks, long before any fossil bones have been found. They resemble bacteria in cell organization with a primitive nucleus. But they are photosynthetic. Their body ranges from single cells to branched or unbranched filaments. They reproduce vegetatively either by simple division or by thin-walled spores formed within the cell (endo-

spores), multicellular segments capable of gliding (hormogones), and thick-walled resting spores (akinetes). They are free-living or occur inside Liverworts and naked-seeded plants like Cycas and Pines (Gymnosperms). They form an important component of lichens. Some of the blue-green algae have the capacity to utilize the nitrogen from the atmosphere and form ammonia. This is called biological nitrogen fixation. The blue-green alga *Spirulina* is rich in proteins (65 percent) and has been safely used as human food in Equatorial Africa. Some noxious blue-green algae like *Microcystis* multiply on the surface of reservoirs and may lead to pollution.

Green algae are grass-green in colour due to the dominance of the chlorophyll pigment. They are either non-motile, or motile single cells; or colonies formed by groups of cells; unbranched or branched filaments, leafy or differentiated into nodes and internodes as in stoneworts. Reproduction ranges from conjugation of exactly similar gametes (isogamy), or of gametes of different size and mobility (anisogamy), or of one gamete smaller and motile and the other larger and non-motile (oogamy), with post-fertilization ramifications. The green algae are generally found in fresh water, although a few are marine.

Brown and red algae are marine and commonly cal-

Sargassum × 1

Seaweeds (*Sargassum, Gracilaria*) growing on the Gujarat coast

led seaweeds. The brown alga or the kelps include the largest seaweeds known. An important representative of this group in the Indian region is *Sargassum* which provides the raw material for our alginate industries.

The Indian coast is also rich in red seaweeds, some of which serve as the raw material for the manufacture of agar-agar, notably *Gelidiella* and *Gracilaria*.

Economic Importance

Many of these algae are of general economic importance in industry, agriculture, pharmaceuticals, medicine and cosmetics. Some seaweeds form a part of the human diet in many countries, particularly in East Asia, and are systematically cultivated in Japan. *Graci-laria* meal has been found to be a good supplement to poultry feed in India. Seaweeds also serve as a good source of organic manures, particularly for vegetable crops in coastal areas.

The red algae serve as a raw material for agar and the brown algae for alginic acid and its derivatives. In India there is an annual indigenous production of 35 tonnes of agar-agar and 400 tonnes of alginate. Agar is used in many industries besides being used in laboratories as a solidifying agent for growing microbes. Alginic acid has a variety of industrial applications, in textiles for sizing and printing, in artificial fibres and also as a stabilizing agent in cosmetics and in dairying and dentistry. Diatomaceous earths (kieselguhr) provide abrasives and are also used in making toothpastes, absorbents, filters and insulating material. The nitrogen-fixing blue-green algae are widely used in India as a fertilizer for rice.

K.S.

ALPINE PLANTS. The alpine region of the Himalaya refers to land above the upper limit of coniferous trees. The tree-line in the Western Himalaya is at about 3600 m. The ascent of trees in the Eastern Himalaya

The Blue Poppy, the Queen of Himalayan flowers (Meconopsis aculeata) × ½

is higher, up to about 4600 m, because of higher humidity and rainfall. This abundant rainfall in the Eastern Himalaya is due to the horseshoe-shaped form assumed by these mountains owing to the bending of the Himalaya in the Arunachal–Assam corner which is responsible for catching the bulk of the monsoon-bearing clouds arising from the Bay of Bengal—the world's largest bay. The zone above the tree-line, which is too cold for tree growth being under snow for six months, is the habitat of most spectacular and gaily coloured flowering herbs and shrubs. The base camps of expeditions to such mountains as Nanga Parbat, Nanda

© K.C. Sahni

Rhododendron campanulatum, *shrub with handsome white flowers tinged with lilac, occurs all along the Himalaya above the tree-line. Leaves rusty hairy beneath to protect them against cold and decrease* water loss caused by transpiration during intense flowering in May. Monal pheasants are sometimes seen in these bushes. Bandarpunch (Monkey's Tail mountain), Garhwal, at 3800 m.

© K.C. Sahni

Primula macrophylla *with brilliant purple flowers. The leaves and stems have a covering of felt to protect them from cold. Flowers in May. Garhwal, 4000 m.*

Devi and Everest are generally fixed near the timber-line and these camping sites abound in Himalayan alpines. The flowers, as brilliant as gems, have been photographed and filmed in colour by mountaineers and botanists for the world to enjoy.

The flora at these lofty heights is characterized by the relative abundance of mosses and lichens encrusting the rocks. The stunted growth of the plants is generally attributed to the retarding action of light, exposure to high-velocity winds, etc. Many alpines show xerophytic adaptation in the form of crowded, narrow, fleshy or very hairy leaves, or leaves with a thick cuticle, for the low temperature retards root absorption, while the conditions of light, low pressure and high winds favour increased transpiration. Vegetative reproduction is common, and many of the flowers are either wind- or self-pollinated. Himalayan alpines in the cold deserts of Ladakh, Sikkim and Tibet are in the form of hard hemispheric cushions as an adaptation against intense cold at night, heat during the day and high-velocity winds.

Above the tree-line, three stages in vegetation are recognized, that is (1) alpine scrub, (2) alpine meadow, and (3) stony desert and perpetual snow.

The *alpine scrub* is in the subalpine zone (FOREST TYPES, No.14). It is mostly a willow-juniper-rhododendron scrub community, with outlying patches of the large-flowered *Rhododendron campanulatum*. *Rhododendron hypenanthum* (small cream flowers), drooping juniper, honeysuckle, etc. comprise the alpine scrub. Ornamental herbs like *Anemone, Fritillaria roylei,* etc. are seen.

© M. A. Rau.

Saussurea gossypiphora. *Flowering head and drooping foliage with woolly hair, like a mountaineer's eiderdown jacket,* × ³/₄ . *Hemkund, Garhwal, 4800 m, Season.*

© K.C. Sahni

Brahma Kamal (Saussurea obvallata) *is not a* kamal *(lotus) but looks like one with large cream-coloured bracts. Flowers in September. Above Harki Dun, 4000 m, Garhwal.* × ¹/₂

Alpine meadows occur over moist localities, in hollows and gentle slopes moistened by snow and glacial streams. The best time to visit these meadows is June when alpines come into bloom. One of their characteristics is their tendency to 'rush into flower' early when the snow melts about May, their short but brilliant flowering period being confined to this time. Several primulas are there, e.g. *Primula macrophylla, P. rosea,* which burst through the snow and reveal their brilliant colours. Common associates are *Iris, Potentilla* with reddish flowers and a galaxy of others. Marsh Marigolds with yellow flowers grow near streams and melting snows. Many alpines are a dazzling blue (gentians and aconites) or sky-blue (blue poppies, *Meconopsis*).

Rhododendrons are an outstanding feature of the alpine Himalaya and dominate the plant life particularly in the E Himalaya where 80 species are known to occur as compared to 5 in the W Himalaya. Every possible habitat is occupied: stream sides, meadows, moun-

tain tops and even the trees at lower altitudes where they grow on them as epiphytes. The most outstanding are *R. campanulatum* (from Kashmir to Arunachal), *R. nivale, R. cinnabarinum, R. thomsonii. R. nivale,* 5 cm tall, is one of the smallest. It grows at elevations up to 5800 m, which is perhaps the highest altitude for any woody plant in the world. It has the scent of eau-de-Cologne, *R. cinnabarinum* (brick-red flowers) and *R. thomsonii* (blood-red flowers) occur in the E Himalaya.

Stony deserts and perpetual snow. These are seen at over 4800 m and have a characteristic flora of cushion-forming herbs and plants with dense woolly hair as an adaptation against strong winds and intense cold. The striking plants of this type are: *Saussurea gossypiphora,* a rare wedge-shaped herb with a rounded head covered in a woolly blanket, distributed in the Great Himalaya. Edelweiss is the national flower of Austria and also occurs in the Himalaya. Botanists call it *Leontopodium* or lion's foot. The star-shaped flower-head, covered

9

with felt, resembles a lion's foot. Cushion plants, *Thylacospermum rupifragrum*, forms hemispheric mounds up to 0·5 m in diameter, which is a growth of centuries. These occur in Ladakh and other high-altitude cold deserts up to 5700 m. *Arenaria musciformis* is also a cushion plant common in the plateau near the source of the Teesta river at 5500 m. Brahma Kamal, *Saussurea obvallata*, is not a plant of the cold desert but occurs in the inner valleys. It is a favourite for offering in the shrines of Badri and Kedarnath. Probably there is no altitudinal limit for plants. One of the Everest expeditions found the Himalayan edelweiss at 6096 m. *Ermania himalayensis* was collected on Kamet at 6400 m and is the highest flowering plant recorded.

K.C.S.

See plate 20 facing p. 257 and plate 27 facing p. 384.

ALTRUISTIC BEHAVIOUR. Any instance of social behaviour involves two or more individuals to whom some benefits or costs may accrue as a consequence of that act. In modern parlance social behaviour is categorized into altruistic, cooperative, selfish, or spiteful depending on how the actor and the recipient share these benefits or costs.

Thus altruistic behaviour is that social act in which the initiator of the act sacrifices something for the benefit of the recipient of the act. Since natural selection is believed to maximize the chances of survival and reproductive success of an individual, altruistic behaviour in which an individual sacrifices its chances of survival of reproduction poses something of a problem for evolutionary theory. The classic instance of such altruistic behaviour is the assumption of sterility by the workers in a honey-bee hive. Such workers even throw away their lives in the defence of their colony. Charles Darwin himself posed this as a problem for his theory of evolution through natural selection and saw that the answer lies in the fact that all members of a honey-bee hive are closely related, and the sterile workers are generally helping their own mother to produce more offspring. Thus the altruistic worker is helping another individual who shares many of his genes. So long as the benefit to the recipient is sufficiently large compared to the cost to the altruist, such altruistic acts towards related individuals can be shown to be favoured by natural selection.

This explanation of altruistic behaviour as involving blood relatives is known as the theory of kin selection.

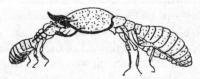

Termite workers feed soldiers who beg them for food.

Most of the instances of altruistic behaviour from the animal kingdom involve such interactions amongst blood relatives. The most primitive organism exhibiting this kind of altruism is the one-celled soil amoeba, *Dictyostelium*. When food is abundant this soil amoeba divides to form many cells. When food is scarce these genetically identical cells come together to form a fruiting body. In such a fruiting body the stalk cells accept sterility to carry aloft other cells which are permitted to fruit.

The highly social wasps, bees, ants and termites are all societies based on altruism towards blood relatives. In these societies many individuals accept sterility and assume duties of foraging for food, defending the nest, caring for the brood and so on. In the wasps, bees, and ants the workers are all females, whereas in the termites they are both males and females.

Instances of altruistic behaviour towards blood relatives in birds include nest-helping amongst birds such as babblers and bee-eaters. In many species of these groups the juveniles from the previous year stay on with the parents and help them in feeding the current year's young. Many social mammals are similarly altruistic. In elephants and lions the core of the herd is a group of related females who protect all the calves and often nurse them communally. In wild dogs all the pack members cooperate in the hunt and will regurgitate meat to the pups at the den (see DHOLE).

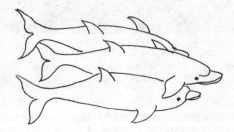

Porpoises lift a wounded comrade to the surface to breathe.

But altruism need not always be based on blood ties. In societies where long-lived animals continue to exchange social interactions with known individuals, it is possible that a network of reciprocal altruistic behaviour comes to evolve, even among unrelated individuals. This has been demonstrated in the olive baboon where males help each other in fights. These coalitions of males tend to be such that if *A* tends to respond to *B*'s call for help, then *B* in turn tends to respond to *A*'s call for help.

A great deal of altruism in human societies is based on such reciprocal exchanges. In fact a whole gamut of human psychological traits including friendship, loyalty, gratitude, moralistic aggression and guilt may be part of such a reciprocal altruistic complex.

See also SELFISH BEHAVIOUR, WARNING SIGNALS.

AMLA, *Emblica officinalis*, a small to medium-sized deciduous tree common throughout the subcontinent including Burma and Sri Lanka up to 1300 m. Its feathery light-green foliage makes it look very graceful. The bark is greyish-brown, peeling in scroll-like patches, exposing the yellowish-buff underlayer. It flowers in the hot season and fruits resembling the gooseberry ripen during winter and probably are the richest known natural source of vitamin C. They are often eaten as thirst-quenchers and are made into sweetmeats.

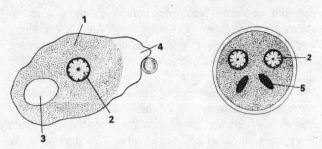

ENTAMOEBA HISTOLYTICA,
ACTIVE AND CYST FORMS
(greatly magnified)

1 Cytoplasm. 2 Nucleus. 3 Vacuole. 4 Pseudopod about to engulf food particle. 5 Nutrient material in cyst.

AMLA
Immature fruits like gooseberries are shown clustered along the branches, and a mature fruit is shown separately and several months later. Both × 1.

AMOEBA. Amoebas are single-cell, animal organisms, belonging to the protozoa. Most species are free-living and can be seen in water from any small freshwater pond, the larger forms being just visible to the naked eye. Parasitic forms occur in many animals, inhabiting the intestinal canal, and are mainly saprophytic, though a few are causative of disease, like *Entamoeba histolytica* which is responsible for amoebic dysentery in man.

Within a single cell, the amoeba contains all the capacities necessary to maintain animal metabolism. The amoeba moves by throwing out cytoplasmic buds, or pseudopodia, into which the rest of the cytoplasm flows in a streaming movement. When a suitable nutrient particle is encountered, the amoeba throws out pseudopodia around it and engulfs it. The particle may remain in a bubble, or food vacuole, into which digestive juices are poured. When the nutrient material has been digested and absorbed from the vacuole, which acts like a primitive intestine, the amoeba moves away, leaving behind any indigestible remnants.

Reproduction is normally by binary fission. From time to time, a type of sexual reproduction, or conjugation between two individual amoebas, with some exchange of genetic material, occurs. The conditions necessary for conjugation to occur are not properly understood. Under unfavourable conditions, drying of the pond, or, for a parasite, increased resistance of the host, the amoeba resorts to encystment for self-preservation. The cyst consists of a chitinous shell with which the amoeba surrounds itself, at the same time dividing into 2, 4 or 8 daughter cells, enclosed with the cyst. When conditions are again favourable, the cyst bursts open to release the contained cells, each of which becomes an individual amoeba.

AMPHIBIANS. The name Amphibia means 'having two lives'; an apt description for a class of animals who live an aquatic fish-like existence when young, and become terrestrial when adult. There are of course exceptions to this rule, but even these had amphibious ancestors.

Thus the name 'amphibia' is applied to a class of vertebrates that fall between the fishes and the reptiles. They differ greatly in form, and range from the legless, worm-like caecilians to the lizard-like salamanders and newts. Some of the major differences that separate amphibians from the other vertebrates are: a body covered with generally moist skin without scales, fur or feathers; soft toes with no claws; a two-chambered heart in the larval stage and a three-chambered heart in adults; external fertilization of eggs; and the process of METAMORPHOSIS.

Amphibians were the first land animals. The first records are from the early Devonian and late Silurian times. These remains are associated with freshwater fish-like animals. The rise of the Amphibia from their fish-like ancestors required radical changes. For instance, the evolution of limbs from fins, a breathing apparatus that could function effectively in air, and a skin that could remain moist. These changes were accomplished very slowly.

The early forms were all four-legged and tailed but were mainly aquatic. The habitat of *Eogyrinus,* one of the earliest known amphibians, seems to have been temporary pools in arid areas, where the drying of the pools would have caused it to make overland journeys to other pools. From these modest beginnings, amphibians have evolved to occupy practically all known habitats except some of the more climatically extreme, such as the polar areas.

At present the Class Amphibia is represented by three Orders: the Apoda, which encompasses the worm-like CAECILIANS, the Caudata or 'tailed amphibians' that are represented by NEWTS & SALAMANDERS, and the Salientia which includes the FROGS & TOADS.

In most amphibians, eggs are fertilized externally. The female may lay the eggs individually or in strings. The eggs absorb water rapidly after they are laid, and after a certain stage cannot be penetrated by the sperm. This requires the sperm to be in contact with the eggs immediately after laying. The familiar clasp or embrace of mating frogs enables the male to release the sperm almost directly the eggs are released.

The eggs vary in size from 2 to 8 mm. Large eggs are common in tailed amphibians, which live in swift water, and in those frogs and toads with direct development (having no tadpole stage).

On hatching, a larva emerges from the egg. In the frogs and toads, this is the tadpole. In the other orders it may be larvae that resemble the adult. Immediately after hatching, the tadpole remains immobile, eating the remnants of the yolk sac. Most amphibian larvae usually start off with external gills but tadpoles generally have internal gills. A few days after hatching the tadpole moves about feeding actively.

The next major changes in the tadpole are usually seen when the adult stage is being reached. The tadpole demonstrates the most striking changes with the appearance of limbs and disappearance of the tail. The major changes in the tailed and legless amphibians at metamorphosis is usually the disappearance of the external gills. Before metamorphosis, the lungs develop and are functional by the time the gills begin to disappear.

In some amphibians, the larval stage is accomplished while in the egg. This allows fully formed young frogs or newts to emerge from the egg. The species exhibiting this adaptation usually breed away from water. In the amphibians in the Indian region this condition is found in a Sri Lankan frog *Rhacophorus microtympanum,* which lays its eggs in depressions on the damp floor of montane forests. Some amphibians even carry their

eggs and tadpoles on them. Though no clear records exist for Indian amphibians carrying eggs or tadpoles, this behaviour has been observed in some South American tree frogs and in the Surinam toad.

Usually, amphibians lay their eggs in a suitable habitat and the eggs are left to hatch and develop by themselves. There are some species that have developed 'nests' to afford the eggs more protection. The tree frogs of the family Rhacophoridae are one such group. They are foam-nest builders. The male uses his hind legs to whip up the albumen extruded with the eggs into a thick froth. The whole frothy mass (with the eggs within it) is suspended from a tree or bush above a suitable body of water, into which the tadpoles will drop when hatched. The caecilian *Icthyophis* is known to excavate a burrow and coil round its eggs to protect them till hatching.

Tadpoles often have modifications for their specialized habitats. Torrent-dwelling tadpoles have their mouths modified into suckers with which they hold on to rocks against the force of rushing water. Certain cliff-dwelling frogs have grooved tails to afford a better hold on the moist rock faces. Amphibians have a wide variety of mouth shapes to suit their habits, ranging from efficient filters for plankton feeders to hooked beaks for carnivores.

During courtship and breeding, male frogs and toads have been noted for their song. It may be that some of the first vocalizations were produced by the voices of amphibians. Fossil evidence suggests the existence of ears in the amphibia as far back as the CARBONIFEROUS. Amphibians possess a vast repertoire of sounds from monosyllabic 'grunts' to complicated bird-like calls.

Of the three living orders of amphibia, only two have been known to produce any vocalization—the frogs and toads, and the tailed amphibians. However, vocalization in the newts and salamanders is rare. All the amphibian calls one hears on the Indian subcontinent come from frogs and toads.

Although calls to attract mates (mating calls) are the commonest uses of voice in amphibians, there are five other conditions which may give rise to sounds. These sounds are called release calls, warning sounds, rain calls, screams, and territoriality calls. Release calls seem to function as a signal by which males can distinguish the sex of the partner. In large breeding concentrations of frogs and toads, when sexual excitement reaches a high pitch, males will attempt to clasp any object that is the size of the female. Often it is another male. The 'clasped' male struggles to escape and emits a croak or chirp: this is the warning vocalization and he is released. A female, on the other hand, remains silent and mating progresses.

Warning sounds serve a similar function as release

calls but are softer and non-vocal sounds. They are probably made by vibrations of the cartilage in the throat and transmitted to the body musculature.

Rain calls. Many species have been known to call sporadically even when they are not engaged in mating activity. These calls are often feeble or partial renditions of the mating call. They usually occur with a rise in humidity before the rains. But often, this call can be elicited by other external stimuli such as the sound of music at a certain pitch and even the drone of airplane engines. Among Indian amphibians, rain calls are often given by *Rhacophorus maculatus* and *Bufo melanostictus*.

Screams. Injured or grasped frogs often give vent to a loud cry that is termed a 'scream'. The probable purpose of the scream is to warn other individuals of danger.

Territorial calls. Frogs that spend most of their time hunting in a small, set area have been known to give voice to short calls which have no relation to the sounds described so far. These sounds are produced almost at random without any apparent stimulus. Some studies have shown that these calls can be associated with a primitive sort of territoriality. The South Indian frog *Rana temporalis* which doesn't migrate or move around much, has been known to produce such calls.

The mating call is of course the epitome of amphibian vocalization. Each species that uses a mating call has its distinctive call. This difference helps to act as a pre-mating barrier. In closely related species sharing the same habitat it is mainly the specific calls which prevent misrecognition at breeding time.

The difference in the mating calls of various species is accentuated by the fact that the male and female function as a single unit, the male being the transmitter (the singer of the first phrase) and the female the receiver (singer of the second). Too much variation from the norm may result in his signal being unrecognizable to the female. The individuality of the call is consistent enough for the call to be used as a taxonomic character in species identification.

When the female has approached the male, the clasping reflex is one of the major means of recognition in frogs and toads. The responses of the clasped member and tactile cues determine if the male will maintain his hold till egg-laying. The tailed and possibly the legless amphibians use other cues for sex recognition. Newts have been known to respond to the colour as well as chemical signals of their partner.

The importance of the clasping reflex is so great to frogs and toads that it takes precedence over the animal's protective reflex during breeding seasons. At other times than the breeding period various defence reactions can be observed. A disturbed Firebellied Toad will bend its body upward to display the brightly coloured under-surfaces. The Common Toad *Bufo me-lanostictus* and the Indian bullfrog *Rana tigerina* will, when threatened by a predator, inflate the lungs and bow their heads. This action increases the apparent size of the animal and has been recorded to deter attack by snakes that were actively hunting till the defence reaction was encountered. It is possible that this inflation and back-arching causes the secretion of toxins from the amphibian's back to repel predators.

In the case of the toad, worms or wormlike creatures under a certain size will elicit a feeding response while similar creatures over that size produce defence reactions. Very large creatures, over a certain threshold size, bring no reaction. Thus it seems that there is a minimum and maximum level of stimulation which produces the defence reaction.

Amphibians are subject to enormous predation. From the egg to adult, they form a part of the diet of innumerable enemies. Very few individuals from a single brood live to reach maturity. The eggs are eaten by insects and even by some salamanders. The tadpole stage falls prey to dragonfly nymphs, water-beetles, giant water-bugs, fishes, birds and crustaceans. When emerging as a froglet there is heavy predation by birds and by larger frogs. The adult frogs are preyed upon mostly by snakes; monitor lizards, birds, and carnivorous mammals are other predators.

The length of life in the amphibia has been recorded to vary from as short as two years in the spadefoot toads to as long as fifty years for some salamanders; the larger animals generally live longer than the smaller ones.

The economic value of amphibians can be assessed in two ways: the visible and the invisible. Of the visible economic value, the contribution to human food is the most important. Frogs' legs are eaten in many countries. In our area the edible green frog and the Indian bullfrog are collected for export. The industry in India wastefully kills tens of millions of frogs. In 1977 1150 tonnes of frog legs were exported to the USA. In Asia, especially in Southeast Asia, dried toads and frogs are used for 'medicinal' purposes. Perhaps one of the most significant contributions made by the amphibia is to science, as laboratory animals.

The invisible economic value is more difficult to assess. Large numbers of insects are eaten every day by frogs and toads, including mosquitoes and cockroaches. Toads especially, being indiscriminate feeders, help to keep down numbers of agricultural pests. Thus amphibians provide an economic benefit which is often not readily seen by the communities they benefit.

Due to their intolerance of salt water, the way amphibians are distributed often provides important clues to past ecological and geological connections between different areas. The fact that hurricanes can move trees and other large objects to islands may sometimes account for the presence of amphibians on islands. However, the major distribution patterns suggest overland dispersal.

See CAECILIANS, NEWTS & SALAMANDERS, FROGS & TOADS.

ANATOMY, the science of the structure of the body, is derived from two Greek words meaning 'to cut apart', indicating that structure is studied through dissection. In the most primitive protozoa and coelenterata structure or shape is maintained only by a firmer layer of cytoplasm. The structure of the body in higher animals consists of a supporting framework or skeleton, to which are attached the muscles, responsible for movement. The internal organs with special functions, like the heart and the digestive organs, are contained within the body cavities. The skeleton is made of stiffer substance (chitin, cartilage, or bone) and is articulated, i.e. its different portions are attached together by joints, ligaments or fibrous bands, so that movement can occur between the different parts. Insects and arthropods have an exoskeleton, a tough chitinous structure which covers the whole body. Vertebrates have a cartilaginous or bony internal skeleton which serves as a rigid support. Animals with exoskeletons can grow only in steps by moulting, i.e. casting off the exoskeleton and then forming a new, larger exoskeleton over the enlarged body, whereas vertebrates are able to grow gradually, the internal skeleton enlarging to keep pace with body growth.

The muscles, joints and ligaments or fibrous bands, together with the skeleton, are called the locomotor system. They are responsible for movement of the animal from one place to another, and also act to maintain the normal posture of the body.

The internal organs, the brain, heart, lungs, digestive and other organs, are contained within spaces or cavities which may be partially or wholly enclosed by the skeletal frame. These internal organs perform special functions for the whole body and are described under the appropriate systems.

The fine structure, or microscopic anatomy, of a tissue or organ can give a better understanding of its function, and of the way this function is performed.

A special branch of anatomy, embryology, is the study of growth and development of the complex animal body from a single cell, the fertilized egg or zygote. The study of embryology, especially in the higher animals, yields insights into the evolutionary patterns which have resulted in the development of the species concerned, since the development of the embryo follows, briefly, some of the rudimentary phases in the evolution of the animal. The embryos of all air-breathing vertebrates, for example, pass through a phase when gills are formed, and then atrophy, indicating that these vertebrates have

all evolved out of aquatic, gill-breathing, forms of life.

See BRAIN, DIGESTION, ENDOCRINE GLANDS, EXCRETION, HEART, RESPIRATION, SENSES.

ANCHOVIES are small fishes of the family Clupeidae, mostly with their abdomen laterally compressed into a sharp edge usually serrated (dented), a comparatively large deeply-cleft jaw, and a single dorsal fin. Their small size of 12 to 20 cm is compensated by huge numbers; about 40,000 tonnes of fish from the east and west coasts of India are landed annually. Several genera, such as *Stolephorus*, *Thryssa*, *Setipinna*, *Coilia* etc., constitute this group. *Stolephorus*, known as Whitebait, is a major component and is more common on the east and south-west coast, while the *Coilia* is more abundant on the northwest coast and is termed Golden Anchovy because of its colour. They are a favourite food, both fresh and dried.

ANGLER FISH. So called, because part of their dorsal fin is modified to form a long angling 'rod' complete with 'bait', called the illicium. When the fish is hungry, it remains motionless on the sea bottom, waving its illicium near its gaping mouth. An unsuspecting fish investigating the worm-like wriggling movement of the bait finds itself sucked into the mouth and devoured. Near the seashore, a typical example is *Antennarius*, found on the west coast.

In the deep sea, we have nearly a hundred species of anglers of the suborder Ceratioidei. Their illicium varies from a stubby lobe to a whip several times longer than the fish. Usually the tip is luminous and emits flashes of light to attract the prey.

In the dark oceans where life is never abundant, a major problem is finding a mate. Four genera of anglers have, however, solved it in a unique manner. The tiny male seeks a female, and when it finds one, its strong pincer-like jaws clamp on to her body. Its jaws and

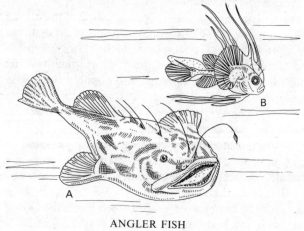

ANGLER FISH
A Adult, B Young stage. × 1/4

tongue grow into her tissues and fuse with them. The male's digestive tract now degenerates, and it gets its nourishment via the female's blood circulation. In fact the male virtually becomes a sperm-producing factory. As many as four males may fuse with a female, anywhere on her body—the belly, head or gill-cover—and fertilize her eggs when laid.

ANIMAL AGGREGATIONS. It is a general rule that under natural conditions animals tend by and large to occur closer together than would be expected on grounds of chance alone. They thus occur in clumps or aggregations of varying degrees of size, density and cohesiveness. Many aggregations are really chance aggregations, of passive animals which have drifted together by the force of some other agency, like leaves collected in a region of stagnation in a flowing stream. Such are the aggregations of the free-floating little sea animals or zooplankters which collect at certain levels, for example, where there is a steep gradient of temperature in the sea.

Yet others aggregate at concentrations of a resource they all need, but have come together without any more specific social tendencies. A few minutes after an elephant deposits its bolus of dung it may be covered by hundreds of dung-flies. The females of these dung-flies come to the dung for it represents a food source necessary for the survival and growth of their larvas, and they come together only incidentally without any attraction for each other. But the male dung-flies come to the bolus because that is where the females are going to come and that is where the males will get a chance to mate.

The bats that come together in their thousands in large caves such as the Robber's Cave at Mahableshwar in the Western Ghats also come together because the cave represents a safe retreat from predators during the daytime. But here they do develop more elaborate interactions, defending their favourite roosting places against others. The females may leave their nursing young in these spots when they go out foraging at night and return accurately back to the very same spot in the enormous caverns.

Some animals swarm together for the purpose of procreation as does the palolo worm of Polynesia. On

Animals like the gaur, the wild buffalo and the elephant, stay together because they can cooperate in actively defending themselves and their calves against predators who are no bigger than themselves.

15

a certain day of the lunar calendar every year millions of these worms come out swimming to lay eggs and fertilize them. Other aggregations owe their origin to the fact that a female had laid a bunch of eggs together and out of these hatched young which form a swarm. Such are the swarms of millepedes or of many caterpillars.

But these are all passive aggregations. Many animals which are otherwise asocial come together at the time of long-distance migratory movements. Thus many birds of prey, such as kestrels, always hunt by themselves but form good-sized flocks at the time of migration.

Thousands upon thousands of migratory waterfowl aggregate at Bharatpur in north India because this jheel represents an island of safe aquatic habitat in a vast dry countryside. But the heronries at which our egrets, storks, ibises, cormorants, spoonbills, darters and herons congregate to breed appear to be of somewhat different type. For the breeding sites—for instance those on islands of Ranganathittu—are by no means the only islands suitable for breeding to be found on the river Kaveri. Only some seven of the myriads of apparently equally suitable islands are chosen by over 2000 birds to build their nests in awfully crowded conditions. Furthermore, the birds that crowd together seem not to cooperate with each other in any way whatsoever. So it appears as if there is value in crowding for its own sake. This value seems to lie in saturating the ability of predators to take a toll of the prey population. For the more than five thousand eggs and chicks produced within an area of a few hectares over just four months represents a very large prey population. It is preyed upon mainly by jungle crows, brahminy kites and bonnet macaques. But the number of predators using such a small area is limited, for the predators have their own dispersion patterns, often including territoriality. So only a small fraction of the eggs and chicks is ultimately preyed upon. If some birds now left and nested on an otherwise suitable island at some distance, they would probably be in the territory of another bonnet macaque troop which could destroy all of their relatively few eggs and chicks. There may therefore be pressure to nest as close to other birds as possible to share in the advantage of saturation of the predator's ability to consume their eggs and chicks.

Herds of smaller antelopes and deer seem to derive from a similar cause. For these animals can neither hide from nor actively resist their predators. A stray animal stands a greater statistical chance of falling a prey than animals in an aggregation. These animals come together in what are aptly termed purely selfish herds: their only response to danger is to bunch together and try to get to the centre of the herd, where they have the best chance of safety.

Other animals, for example pigeons, seem to come together because ten or twenty pairs of eyes and ears are more likely to detect the approach of a predator at a distance from which they stand a good chance of making an escape. Yet others like the gaur, the wild buffalo and the elephant, stay together because they can cooperate in actively defending themselves and their calves against predators who are no bigger than themselves.

Lastly many animals come together because they can feed more efficiently than if they were by themselves. The advantage may merely lie in that this enables them to avoid feeding in areas which have been only recently exploited and hence are likely to be low in food density, as is believed to be the case with certain flocks of grassland and desert finches. The advantage may lie in their flushing insects more effectively, for instance, by flocks of insectivores such as the mixed foraging parties of woodpeckers, drongos, mynas, flycatchers and chloropsises in our tropical forests. In a similar fashion the cormorants fishing in flocks may be rounding up schools of fish more effectively. And finally carnivores, such as wild dogs and hyenas, can bring down prey much larger than their own size by hunting in packs. There is reason to believe that man also owes his social tendencies to the advantage of hunting in a group.

ANIMAL FLIGHT. The first flying animals in evolutionary history were probably insects, which are known to have existed around 300 million years ago: ability to fly may have improved chances of survival in an environment which was marked by the emergence of tall plants, offering new sources of food. Insects were followed by birds (defined as feathered bipeds) about 100 million years ago, and then by bats and other flying mammals (50 million years). (There are birds which cannot fly: the ostrich is an example.)

Flying animals constitute more than three-quarters of all known species—living and fossil. They range in size, among those surviving, from parasitic wasps with a mass of a few micrograms and wing-span of 0·1 mm, to large birds like the black albatross and the Kori bustard (mass about 15 kg, wing-span exceeding 3 m).

Birds perform many impressive feats of flying. They are the fastest animals: the alpine swift (*Apus melba*) has been reported to achieve speeds of 250 kmph. Everest expeditions have observed alpine choughs accompanying them at altitudes of 8·5 km. Some birds travel enormous distances during their annual migration, seeking food in warmer climates when their normal sources disappear during winter: thus, the Arctic tern has been known to fly all the way to the Antarctic and back!

Such extraordinary feats are rendered possible by a

variety of remarkable mechanical developments in the flying apparatus of birds and insects. Hollow bones give birds a structure of great strength and lightness; and the streamlined shape of wings and body provide either high aerodynamic efficiency or excellent control (or both). Feathers possess both aerodynamic and structural advantages; and several subtle aerodynamic devices, whose principles of operation are not always completely understood yet, are used by insects as well as birds. At the other end of the spectrum, however, are some animals capable only of rudimentary flight; flying frogs, fish, lizards etc. barely justify their appellation, as they cannot achieve sustained flight. Their abilities are confined to the performance of what may be called aerodynamically assisted leaps. A flying frog (*Rhacophorus dulitensis*) parachutes, spreading out its large webbed feet; and the gliding lizard (*Draco volans*) has a wing (or sail) that can be furled when not needed. Flying fish (of the family *Exocoetidae*) have fins which are folded under water, and spread out like wings soon after the fish breaks the water surface, following a vigorous swimming effort slantingly upwards. The fish may be airborne for several seconds, and cover a distance of several tens of metres. The analogue of these

leaping animals is the schoolboy's paper glider launched by hand.

Any object in flight experiences two basic aerodynamic forces. One of these, called drag, is a resistance offered by air, and acts to oppose flight. The second, called lift, acts in a direction perpendicular to that of flight, and can be used to sustain the body against gravity. When a roller or pied kingfisher wishes to catch an insect it has spied on the surface, it folds its wings and puts itself in a configuration that produces little lift and minimum drag, and dives down almost vertically ('dropping like a stone'). But, to sustain itself in horizontal flight at a certain height in still air, a bird has to generate enough lift (by forward motion) to counteract its weight, and enough thrust (by spending energy) to overcome the drag it experiences at the speed required to generate the necessary lift. The flapping wings of a bird like a pigeon provide *both* the lift *and* the thrust required for flight: these wings therefore serve the functions that are managed separately by the wings and propeller in aircraft driven by a propeller. In this respect birds are more nearly like helicopters, in which a single rotor generates both lift and thrust.

These aerodynamic forces depend only on the *relative*

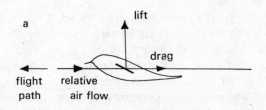

a. The aerodynamic forces on a bird in steady flight as seen by an observer moving with the bird. The air then flows past the stationary bird, and imposes one component of force called drag opposing the flight of the bird, and a second component called lift normal to the direction of the flight.

b. A bird hovering in an up-current. The bird may be stationary with respect to the ground, but is gliding down relative to the air which is moving up. This relative motion generates lift and drag as shown, which together balance the weight of the bird.

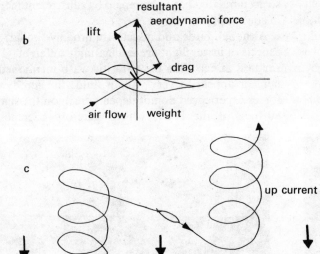

c. When the ground gets warm, as on a hot day, the air near the ground becomes warmer and lighter and so rises, in what are known as 'thermals'. These thermals of rising air often occur next to each other in 'streets'. Many soaring birds (like man-made gliders) utilize the up-currents in these thermals to climb without expenditure of energy, then glide down to the next thermal, climb again and so on, often travelling hundreds of kilometres this way.

velocity between body and fluid. Such birds as eagles or vultures that we often see near hill-tops, suspended in air without so much as a flicker of their wings, are utilizing the strong rising winds that prevail on the windward side of such hill-tops. *Relative* to the wind, the bird is actually gliding down, although relative to ground it is at rest.

Technically, 'gliding' is flight without expenditure of energy by the flyer. Some birds (e.g. woodpeckers) alternate a downward glide holding wings steady, with a short climb beating the wings for a while, producing a characteristically undulatory flight path. When there are up-currents, either because of hills as we have just mentioned or because of 'thermals' (resulting from convection from a heated ground surface), a bird can even 'soar', i.e. climb without expenditure of energy. Kites and vultures can often be seen spiralling up on these thermals: they then glide down, across regions where there is a down-current, to the next thermal where they soar up again. As thermals often occur in 'streets', i.e. next to each other, vultures have been known to travel distances of the order of a hundred kilometres hopping from one thermal to the next. To exploit small up-currents requires aerodynamically efficient wings, with high lift-to-drag ratios. The larger birds often possess such wings, and when gliding or soaring maximize the lift:drag ratio by spreading out the wing to its maximum extent. Albatrosses have wings with an aspect ratio as high as 20, as in some (man-made) gliders. Vultures, although given to soaring, have wings of lower aspect ratio, possibly because of adaptation to short take-off and landing. When the wings are not large, or winds not sufficiently strong, a bird may keep steering itself into wind and quiver its wing-tips to maintain height; kestrels (*Falco tinnunculus*) may be seen doing this while patrolling in search of prey. Mosquitoes hovering above a man are also using the local thermal he sets up by his body heat, but insect flight — as we shall see shortly— is not always governed by the same dynamical principles as bird flight.

The smaller birds cannot generally soar as efficiently as the bigger ones, but some of them (e.g. humming-birds) can 'hover' at a fixed point in still air by standing on their tail and beating their wings vigorously. The energy spent in hovering flight is necessarily much higher than in normal flight, so that hovering is possible only for limited periods. Otherwise motion is necessary for a bird to sustain itself in air—just as for conventional aircraft. Take-off may often be achieved by a leap followed by vigorous wing flapping; many heavy aquatic birds, as well as others such as the blackfooted albatross and the great Indian bustard, need to run on the ground to acquire enough speed for their wings to generate the necessary lift. The effort that the wing of a bird can exert depends on its structure. It has been shown that for the kind of bone and muscle that make up animals, the power output available cannot much exceed 40 watts per kilogram of bird mass. On the other hand the power *required* by the bird for flight increases more than proportionately to the mass. Detailed calculations based on these estimates show that a bird weighing much more than about 12 kilograms cannot fly: the biggest flying birds known, such as the Kori bustard *Ardeotis kori* and white pelican *Pelecanus onocrotalus*, are indeed of about this size.

The power output mentioned above, generated continuously by the burning of carbohydrate or fat in the body through the absorption of oxygen, is about 10 times higher than for human muscle. During hovering or a sprint, however, even more power is required by a bird, and can only be obtained by drawing on reserves and by anaerobic oxidation; the 'oxygen debt' that is so accumulated has to be repaid by subsequent increases in consumption.

Although birds and insects are broadly similar, some aspects of insect flight are fascinatingly different. To the smallest insects, like the parasitic wasp mentioned earlier, air appears a very viscous fluid: the aerodynamic forces experienced do not depend much on the shape or attitude of the body, which therefore becomes less

Brahminy Kite soaring

Caspian Tern

Grasshopper taking off

important than for the bigger flyers. (For the same reasons, the resistance we experience when we stir a pot of honey does not depend very much on the shape of the spoon or its orientation.) Again, compared to most birds, a butterfly which lifts itself off with an apparently effortless flick of its wings may be using an aerodynamic principle not previously known in technology. Similarly wasps are known to obtain, by a characteristic clap-and-fling motion of their wings, a maximum lift of as much as three times what would be expected from the same wings operating in a conventional way. The dynamics of this method of generating lift is only now beginning to be understood. Again, the wing-beat frequency of insects is generally much higher than in birds (being e.g. 1000 times a second in gnats). Such rapid oscillations are possible only because insect wings are not in forced motion (as bird wings are): they operate more like a swing (which has its own natural period of oscillation and uses gravity to store energy) than a fan (which is forced at a definite frequency, by hand or electric motor). This 'resonant' operation is possible because of a material at the wing-joints called resilin, which is made of giant molecules up to a millimetre long and stores elastic energy with a loss rate (of only 4%) that is still unmatched by any man-made material.

Bats utilize as wing a taut fold of skin stretched across elongated fingers; the shape of this wing can be controlled in an intricate way, and the wing can crinkle

to reduce its area by about 25%. These features give bats remarkable manoeuvrability, which they put to good use while pursuing flying insects or in avoiding obstacles when they fly in the dark using their well-known sonar devices. (Incidentally moths pursued by bats go into random manoeuvres and even emit sound waves which tend to confuse or jam the bats' sonar system!)

Among the many interesting aerodynamic devices used by birds, there is room to mention only a few. During certain phases of flight birds spread out their wings in a characteristic way, with feathers at the tip

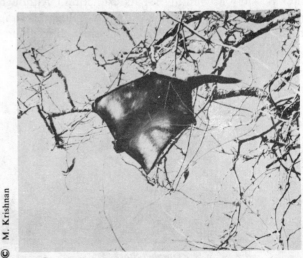

Large Brown Flying Squirrel gliding

Greylag geese in formation

splayed out radially. Tests on wings with what are known as 'sail tips' or 'winglets' show that a part of the drag of the wing can be reduced significantly by such surfaces. Like aircraft, a bird in flight leaves a system of trailing vortices behind (under certain conditions of humidity such vortices can be seen behind aircraft). These vortices create small up-currents in their neighbourhood. Migrating birds (such as geese and cranes), flying in V-formation, exploit these up-currents to reduce the drag they experience and hence the energy they expend in flying. (The leader of the formation however has to spend more energy than the others, and his position is occupied by different birds of the flock in rotation.) Feathers on wings are nearly but not completely airtight, and it is possible that they allow a controlled leakage of air across them enabling the generation of higher lift than would otherwise be possible. (Incidentally, a wing of feathers has enough structural redundancy that it can enable the bird to fly even after it has suffered considerable damage. In contrast, a single tear in the membrane wing of the ancient pterosaur was probably enough to disable it, thus possibly accounting for its extinction.) Although good gliding requires efficient wings with high lift:drag ratios, birds sometimes need to steepen their gliding

Openbill Stork

angle during descent to land. This can be done by increasing drag: the feet of many birds serve this purpose, sometimes (as in pelicans) with webs between toes which, when held into wind, act as effective air brakes.

Many aspects of animal flight, in particular relating to the methods by which control is effected, are still not well understood.

R.N.

ANIMAL LICE. These highly specialized wingless insects of the order Phthiraptera have undoubtedly arisen from psocid ancestors. There are three distinct suborders: Anoplura (sucking lice), Mallophaga (chewing or biting lice) and the peculiar Rhynchophthirina or elephant lice that includes only *Haematomyzus* spp. parasitizing elephants and the wart-hog in Africa. Around 3000 species of these highly modified, parasitic insects (0·5 to 10 mm long) are known. They are dorsoventrally flattened, lack wings and their tarsi are

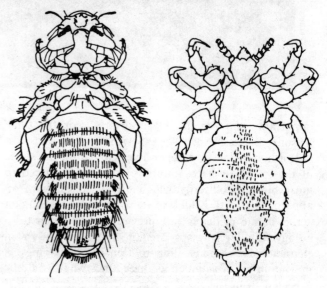

Bird louse × 25   Human body louse × 15

adapted for clinging to the host body. Some are hard-bodied or just pigmented and many have hairy bodies while others are bare.

In general, the Mallophaga feed on feathers of birds while the Anoplura feed exclusively on blood of mammals (except monotremes and bats). They are highly host-specific and it is felt that the relationships of birds may be deduced by studying the affinities of the lice species attacking them. More than 5 or 6 species of lice may parasitize a single bird species, each occupying specific body regions of the host concerned. Mammal lice are less restricted and generally move slower than bird lice, both being negatively phototropic (shunning light). Transfer of lice from one host individual to another is solely by direct contact, some moving from predator to prey or vice versa, and some exhibiting phoresy (transport by use of the body of other animals) by attaching themselves to the hippoboscid or dog flies (see MOSQUITOES AND BITING FLIES). The number of lice on a single host may vary from less than 10 to over 1000 individuals, depending on the host size usually. Adult lice live only for around two weeks, but the human lice may survive for a month.

Eggs are laid cemented to host hairs or feathers and high humidity is not particularly suitable to development. Larvae hatch within a week and live on the host as do the adult lice. Lice have almost no natural enemies except their hosts, which kill lice by grooming, preening and scratching, or even by eating their lice parasites. The dust-bathing and 'anting' habits of birds probably are directed to ridding themselves of excessive lice.

The human louse, *Pediculus humanus*, is found in two forms depending on the host region it inhabits: the head louse and the body louse. The pubic louse, *Phthirus pubis*, is another sucking louse attaching itself to hair in the pubic or perianal regions of man. Human lice are vectors of endemic typhus, french fever and relapsing fever in man. Livestock heavily infested with lice may cause considerable production losses in wool, meat, dairy and poultry industries. *Lipeurus tropicalis*, *Haematopinus tuberculatus*, *Trichodectes canis* and *Columbicola columbae* are the more common species.

See also WOOD-LICE.

ANTS. Perhaps no single acre of land on this earth, save most deserts and the freezing arctic zones, is without one of these most numerous and familiar insects. Comprising one of the most highly evolved social forms of life on our planet, the ant colony consists of winged males and females and the apterous castes of workers. These wingless, terrestrial insects are characteristic in possessing a noticeable 'waist', formed of the first two or three abdominal segments. Another well known feature of ants is the presence of a sting at the tip of the abdomen. Man is by now very familiar with the general appearance of an ant and it would not be incorrect to say that both man and the ants are competitors today for the status of the dominant form of society with a complex social life. It is estimated that from 10,000 to 20,000 species of Formicidae (the ant family) exist all over the world and already about 700 species are known from our subcontinent, which is probably less than half of the actually existent ant fauna here.

The caste system is well developed in ants, as in other social insect groups like the bees and termites. The female castes are the queen and the worker. The queen is the largest individual in the colony, bears wings and a bulky gaster. The worker caste lacks wings and apart from the 'neck' the entire body is one solid piece without any flexible joints besides the wings and legs which are the appendages. There may be major, minor and inter-

mediate forms of the worker caste with functional mandibles and eyes sometimes absent. The soldier phase, if present, is larger than the worker and has a relatively bigger head, sometimes very disproportionate to the rest of the body. They function mainly as defenders of the nest, usually against other ants. In some species, various wingless intermediates occur between the minor worker and the queen castes. These are called the ergatoids, which may live along with the true queen in the nest and even replace her if necessary as the reproductive. The males are mostly winged throughout their life and are usually in between the queen and the worker in size, have large eyes and relatively small heads. One other noticeable feature of all castes of ants is the geniculate or elbow-like antennae.

Like termites, winged males and females emerge from the nest and fly in a mass 'nuptial flight' in particular months of the year, fixed for each species. Mating occurs in flight and then the pair descends to the ground, the male wandering about until he dies while the queen-to-be searches for a suitable place to nest. She constructs a chamber and lives there to create a new colony. Some queens, sometimes with a few workers, may invade the nest of another ant species, and take it over by killing that colony's queen, and living as social parasites. This may only be a temporary take-over or ultimately the queen that took over the nest may replace

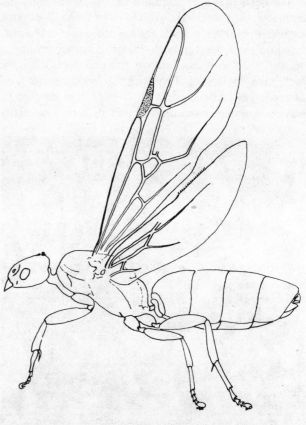

Queen Ant × 10

that nest with her own progeny to make it an independent colony of her own. However, in some cases, the invading ant species may be a permanent parasite and lives in the nest of the other species by making them its own 'slaves'. This interesting association between two ant species is in many cases obligatory, the parasite species not being able to survive and reproduce without its 'slaves'. When the slave species numbers in the nest deplete, workers of the 'master' species raid nearby nests of the slave species and carry back to their nest the pupae of the slave species to rejuvenate the slave colony.

Communication in ants and the way they orient themselves has been studied with great interest by scientists. Known methods of communication are those by smell through the release of a chemical known as a pheromone (being volatile) which is secreted by a gland, and, when released upon a specific stimulus, evokes the desired response (trail-following, alarm-behaviour, etc.). Ants also communicate by the medium of taste, during exchange of food by workers or other castes. Tapping, and stridulation (by rubbing portions of the body against each other) also communicate sounds that mean something to the ants, while other signals may be shared by rubbing antennae together between two ants, and, in species that have well formed eyes, visual communication is another source of contact.

The nests that ants build are usually permanent, and formed mainly under the ground or in wood or existing cavities in rocks, etc. Each ant has a peculiar nest structure and some nests have characteristic ornamentations around their entrances above the soil surface. Some ants, like species of *Polyrhachis*, *Crematogaster* and *Oecophylla*, construct nests on trees with plant fibre or leaves. Another interesting abode of ants (especially

Tetraponerinae) is in the stems and hollows in other parts of some epiphytic plants that act as 'ant houses' and some sort of symbiotic relationship is known to exist between the plant and the ant.

Ants are usually omnivorous scavengers, feeding on plant and animal matter alike. But some species may be specific predators, concentrating on special prey like spiders and their eggs, collembolans, termites, etc. Some genera like the very abundant *Pheidole* are called 'harvester ants' because they depend to a great extent on seeds that are 'harvested' by the colony and stored in the nests for food. Again, ants tend their 'cattle'

© Kumar Ghorpadé

Camponotus sericeus *ants tending* Aphis gossypii *aphids on* Tridax procumbens; *a calliphorid fly on a flower* × 2

(mainly aphids) in the nest or outside and in return for the sweet honeydew that aphids and other bugs produce, they voluntarily protect these insects from their predatory and parasitic enemies. Several plants have their extrafloral nectaries which supply ants with food, and ants are also adept at sipping nectar from flowers of a great variety of plants. Some species of ants have workers that lay special eggs that are used only as food for the youngest larvae and the queen.

The habit of foraging for food is one of the most interesting of all ant activities. The way the ants orient towards the food source and come back with their 'harvest', vegetable or animal, involves orientation to and from the nest. This is done in two main ways: by each individual ant learning locations of restricted feeding zones and finding its way to and fro repeatedly, and by following 'ant-trails' using a chemical substance. The trail is made by workers which find spots with plenty of food and mark their paths to it and back to the nest by a chemical scent. Depending on the particular species, foraging is usually restricted either to daytime or to nights, but it seems that temperature and relative humidity also are important factors.

The size of a colony of an ant is very variable, from as few as a dozen individuals to as many as a million

© Kumar Ghorpadé

Crematogaster *ants attacking newly emerged robber-fly* × 2½

of them. The average size of a colony is around 2000 ants. As many as 200 species of ants may inhabit a square kilometre in tropical rain-forest and a single rotting tree log may harbour 25 or more ant nests. Considering their abundance in sheer numbers and the considerable ecological niches (habitats) that they exploit on earth, ants are certainly very important elements in our environment and deserve more attention than has been bestowed on them by man so far. Though most species are general feeders, many others are undoubtedly important natural enemies as predators of several other insects and other forms of life. Their role in the turnover of soil is little appreciated though they do a lot of good to man in bringing subsoil to the surface and in formation of soil. On the negative side, some harvester species of ants may cause some damage by restricting their area of activity near agricultural areas and raiding the crops for plant matter. However, their habit of 'tending their cattle' like aphids and mealy-bugs (see BUGS) and protecting them from their enemies, is far more injurious to man's interests. But the presence of ants in man's habitations is what is most annoying to him, especially as many of these species possess painful stings.

Ants have a number of their own predatory enemies, most common being the well known 'anteaters', several birds and lizards. Among insect enemies, the 'antlions' that form an ingenious conical pit in the sand devour many a hapless ant. Almost all other insect groups have species that prefer feeding on ants, but ants themselves are in the habit of eating other ants! There are some tiny parasitic wasps that attack ants and ants are host to several other parasites including nematodes and fungi. Undoubtedly the most important enemy of the termites or 'white ants' are the ants, and many species specialize in termites for their major food item. So many other insects, animals and even plants are closely associated with ants that it would be foolhardy to attempt even a broad listing of the ant associates here.

A short summary of the Indian species or genera of ants that are most common would suffice here. The most important 'harvester ants' belong to the genera *Phidologeton*, *Pheidole* and *Holcomyrmex*. Those that tend aphids, coccids and even larvae of lycaenid butterflies mainly comprise *Oecophylla*, *Crematogaster* and *Camponotus*. One type of ant that is extremely ferocious and hunts in huge 'drives' usually commencing in the late evening and continuing through the night, is the genus *Leptogenys*, which can be observed (from a good distance!) to sense the occurrence of a tasty morsel underground, excavate it and then carry parts of it away (root grubs, caterpillars, etc.) after meticulously shredding the poor victim to tiny pieces convenient to carry back home. Species of *Monomorium* and *Solenopsis geminata* are commonly introduced by man into his warehouses to control termites! Other common genera

found in India are *Tetraponera*, *Dorylus*, *Polyrhachis*, *Aenictus*, *Tapinoma* and *Acantholepis*, to mention just a few.

Lastly, ants are mimicked by a peculiar group of spiders, sphecid wasps and several other wasp families, lygaeid and coreid bugs and even beetles! Beetles of the subfamily Paussinae share a special relationship as they are carefully looked after by ants in return for goodies which they give ants!

See SOCIAL SYSTEMS.

APPLE-BLOSSOM SHOWER, *Cassia agnes* (misnamed *C. javanica*), is a medium-sized Malaysian tree (widely planted in the tropics) which after a bare season, puts on fresh foliage in April-May, accompanied by masses of rosy pink flowers resembling apple blossoms. It is like *C. nodosa* but has larger flowers and spines on its trunk. The latter has a few white petals and hence is called Pink-and-White Shower. The former is indescribably beautiful, and brown cylindrical pods 30–61 cm long hang from its feathery drooping branches.

ARACHNIDS are lower animals with a segmented body and jointed appendages. Unlike insects they have no mandibles or true jaws, feelers (antennae), or compound eyes but in their place a pair of false jaws (chelicerae), a pair of pedipalps variable in form and function, and a number of simple eyes. Their head and thorax are in most cases fused to form a cephalothorax which bears below it four pairs of jointed legs ending in claws. The abdomen is usually marked off by a distinct constriction or waist. In a few cases the abdomen is merged with the thorax. They all breathe by air-tubes or book-lungs. The sexes are separate. There is no metamorphosis or change in form during embryonic development, as in insects. Arachnids live on land with the exception of a few species which live in water, sea or fresh.

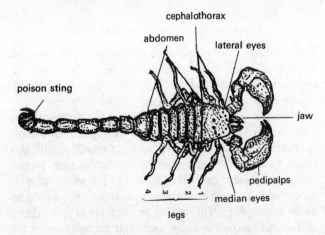

A SCORPION, A TYPICAL ARACHNID, × 1

Arachnids include groups of members of diverse appearance, structure and habits. The common examples are SCORPIONS, WHIP SCORPIONS, SPIDERS, SCORPION SPIDERS, FALSE SPIDERS, SUN SPIDERS, GARDEN HARVESTMEN, mites, TICKS and KING-CRABS.

Arachnids are a much disliked group—since the painful stings of scorpions, poisonous bites of some spiders, the sinister look of whip scorpions, the diseases caused and spread by mites and ticks have all been abhorred by man.

ARCHER-FISH. This fish (*Toxotes jaculator*) is one of the aquatic inhabitants of the northeastern region of India and further east. Being slightly yellowish and with prominent stripes on its body and small in size (about 8 to 12 cm) it makes a good aquarium fish. In natural

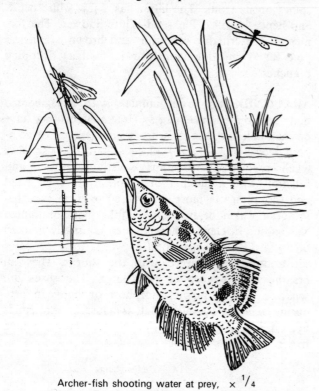

Archer-fish shooting water at prey, $\times \frac{1}{4}$

surroundings it feeds on small aquatic animals near the surface of water but its speciality lies in its hunt for terrestrial insects and flies. When it notices a dragonfly or any other insect sitting on the overhanging weeds out of water, it approaches it quietly and, taking careful aim, shoots a jet of water through its pointed mouth on to its unwary prey and drenches it thoroughly, with the result that it falls into the water, the waiting archer making a quick meal of it. The aim of the water-jet is so quick and accurate that the producer has acquired the name archer-fish. This is also an example of specialized method of feeding in fishes and their capacity to locate accurately an object in air.

ARK SHELLS are widely distributed, thick-shelled bivalves very common along our shores. The shells are readily recognized by their long and straight upper (hinge) margin which bears numerous, tiny plate-shaped teeth. This arrangement is present in arks alone and not found in any other group of bivalves. An ark-shell with closed valves looks like a small box and that may be why it is called *Arca*, which in Latin means a box.

Arks are more primitive than oysters and clams. Their foot is normal and bears a byssus. Their breathing organs are simple gills formed of thread-like filaments arranged parallel to one another. The lower parts of the filaments are folded upwards and connected by hair-like cilia. The mantle consists of two lobes covering the gills as in other bivalves but their lower edges are free, i.e. they are not united at any point to form a siphon as in higher bivalves like clams. The animal keeps the mantle margins gaping to allow fresh water to enter into and the used water to escape from the mantle cavity. This kind of simple gills and open mantles are found in mussels and scallops also.

There are many species of arks varying widely in appearance, size and habits. Their shells are invariably ribbed, generally white, grey or brown in colour. In life most of them are covered by a greenish or dirty black bristly protective coating (periostracum). Their habits vary. Many members live buried where mud is mixed with sand. Some attach themselves to mud-covered rocks and coral reefs, some to masses of pebbles and clusters of broken shells, and some find their way into crevices of stones and hardened mud. Their habitat —the nature of surface on which these animals settle and grow—affects the symmetry and appearance of the shells. Arks are not gregarious and do not form regular beds as oysters or mussels do.

A few species which have something special about them are:

Arca granosa is a handsome, heart-shaped ark 4 to 5 cm long. Its valves are thick and white with knobs along their ribs. Their blood and flesh are reddish in colour instead of light bluish as in other molluscs: hence they are called Blood Arks. The flesh is said to be tough and highly nutritious. This species is found in plenty in knee-deep mud by seashores, in backwaters and estuaries. Some ascend rivers far into the interior.

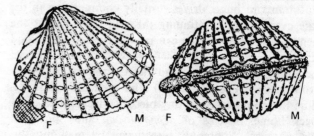

Arca granosa in motion, and seen from the side × 1. F, foot. M, edge of mantle.

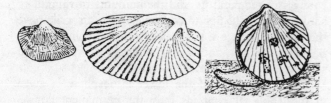

From left to right *Arca symmetrica.* × 1¹/₂ , *A. fusca* × ²/₃ and *Glycimeris taylori* × ²/₃ leaping on the sea floor.

Arca symmetrica. These are the smallest of all common arks, scarcely 1 cm long. Shell valves are symmetrical, dainty, finely ribbed and pure white in colour. Arks are not generally gregarious but this species congregate under stones in the littoral region or attach themselves in assemblages to blocks of dead coral reefs.

Arca fusca, popularly known as the Banded Ark, is marked by deep chestnut colour, three whitish bands radiating from the centre of the upper margin and finely sculptured ribs. In life the shell is mostly covered by a black hairy coating. This species lives 4 to 5 fathoms deep, lying prone or half buried in the muddy sand bottom.

Arca tortuosa is the most curious ark. Its shell is longer than broad, having a twisted appearance: the front portion is twisted to the left and the hind portion to the right. The cause for this abnormal distortion is not satisfactorily explained but it is generally assumed that it is connected with the animal's habit of adapting its form to fit into crevices in rocks and mud banks.

Glycimeris taylori is a species closely allied to arks in its possession of a ribbed shell with numerous plate-like teeth along the hinge. It differs from the arks in many respects: the shell valves are flat and round and not box-like; the hinge margin is arched and not straight. Again it moves freely and is not attached by a byssus to any holdfast. The foot is sickle-shaped and strong like that of a cockle and not tongue-shaped as in an ark. The animal is burrowing in habit. It lives in depths of a few fathoms on sandy bottoms. When disturbed it uses its foot as a lever and leaps like cockles.

See also MOLLUSCS.

ASHOKA, *Saraca asoca,* is a handsome, small evergreen tree, wild along streams in the Eastern Himalaya and Western Ghats. The leaves are dark green and glossy with large, drooping, pointed leaflets. New leaves dangle in tassels of pink or purple for several days before stiffening and straightening to maturity. Orange-red clusters of flowers open about March, leaving large, flat, leathery, purple, 5-seeded pods which split into two halves and coil up. Buddha was born under an Ashoka tree, hence it is sacred to Buddhists. Hindus

Asoka leaves and flowers, × ¹/₂

regard it as sacred, being dedicated to Kama Deva, God of Love. Its fruits are chewed as a substitute for betel nuts.

ASIATIC WILD ASS (*Equus hemionus*). This animal belongs to the family Equidae, and is a close relation of the Horse and of the African Zebra. It differs from the even-toed Ungulates, which include cattle, deer, and sheep, in being single-hoofed and non-ruminant, i.e. non cud-chewing. Within the subcontinent the wild ass is found only in the Little Rann of Kutch and has probably only become extinct in Baluchistan within the past forty years. It has a larger cousin, the Kiang, living on the high plateaus of Ladakh and Tibet. The Indian subspecies, *E. h. khur,* stands about 115 cm (48 inches) at the shoulder, therefore considerably taller than the domestic donkey. It is fawn or pale chestnut in colour with a blackish stripe along the ridge of the back. The male is larger and sturdier than the female. They live in mixed troops of 10 to 30 animals except for 2 or 3 months after the young are born when the mares accompanied by the foals live apart and the stallions keep singly or in scattered twos and threes. The typical habitat of the Indian wild ass is the flat salt desert around Dhrangadhra and Jhinjuwada in the Little Rann of Kutch which gets inundated during the monsoon, leaving exposed little 'islands' or *bets* of slightly raised ground supporting the scanty grasses which comprise the principal food of this animal. One of the largest of such *bets*—the HQ of the wild ass, as it were—is Pung Bet, about 12 km long and 3 km at its widest. They also raid rice and wheat cultivation along the edge of the Rann during night and do considerable damage by browsing on the ripening ears of the crop.

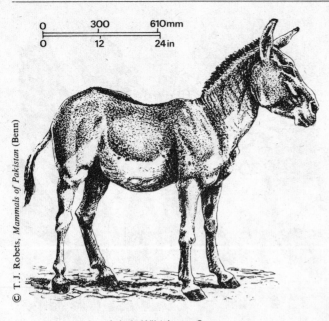

0   300   610mm
0   12   24in

© T. J. Robets, *Mammals of Pakistan* (Benn)

Asiatic Wild Ass or Onager

The wild ass is a hardy animal and fleet of foot, being capable of maintaining a maximum speed of about 50 kmph over considerable distances when chased by a jeep. They have no serious natural enemies and are normally not molested by local villagers, but within recent years epidemics of *surra* and South African Horse Sickness have brought down their population from an estimated 3000–5000 in 1946 to under 1000 in 1962. The census of 1976 put the population at 720. The Indian wild ass is now included in the Red Data Book of IUCN as an Endangered Species; it is afforded total protection under the Indian Wildlife (Protection) Act of 1972, and through effective conservation measures by the Government of Gujarat State, the population is happily showing a steady upward trend. Considering their size and sturdiness and their potential as draft animals, efforts have been made to break them in and also to breed mules from them for use in mountain transport by the Army, but they have not been successful.

S.A.

See UNGULATES
See plate 25 facing p. 368.

ATMOSPHERE. The gaseous envelope that covers the planet Earth, or any other planet or even a star, is called an atmosphere.

Man lives, breathes and flies in the earth's atmosphere. Its ability to sustain life depends upon a very delicate balance of several features, physical and chemical.

Scientific phenomena manifested in the earth's atmosphere were the first to arouse the scientific curiosity of man, which has finally led to the various developments in science and technology. The phenomena occurring at the surface of the earth or near it, such as winds,

changes of temperature and the amounts of rainfall or snowfall, were the first to be studied. The air well above the earth's surface was studied with the help of kites, balloons, aircraft, and is now studied with rockets and satellites.

The composition of the lower atmosphere is given in the Table. It contains only the permanent constituents. Water vapour is a varying constituent; it can vary from 20 gm or more per kilogram of air in the tropics to less than 0·5 gm per kilogram of air in cold, dry, polar continental air.

Composition of Air in the Lower Atmosphere
(Permanent Constituents)

Constituent	Formula	Volume percentage
Nitrogen	N_2	78·084
Oxygen	O_2	20·946
Argon	Ar	0·934
Carbon dioxide	CO_2	0·034
		99·998

The remaining 0·002 percent is composed of very small quantities of gases including neon, helium, krypton, xenon, hydrogen, methane, nitrous oxide, sulphur dioxide (variable), and carbon monoxide (variable).

Another important constituent is ozone; this is also variable. If all the ozone present in a vertical column of the entire atmosphere, extending from sea level, were compressed in a container having the same cross-section as the column, the thickness of the pure ozone sample would vary between 0·2 cm and 0·4 cm, i.e. between 200 and 400 Dobson units. This much of ozone absorbs all the ultra-violet solar radiation harmful to life on earth. The ozone amount is maintained around this value by a series of photochemical processes occurring in the atmosphere, mainly between altitudes of 20 and 40 km.

The earth's atmosphere can be divided into a number of horizontal thick layers with diffuse boundaries; each layer has its own distinct characteristics.

The lowest is the *troposphere*, 15–16 km thick over the tropics, 10–12 km thick over the middle latitudes, and about 7 km thick over the polar regions. In this layer the temperature decreases with height at the rate of about 6°C per kilometre. This is the region where the drama of weather is enacted with its associated clouds, winds, rain or snow.

Above the troposphere is the *stratosphere*, and the transition layer is called the tropopause. In the stratosphere, the temperature generally increases with height; it extends to an altitude of 40–45 km where we encounter another transition zone called the stratopause, where the temperature is around 0°C.

Above the stratopause we have the *mesosphere*, where the temperature decreases with height, up to an altitude

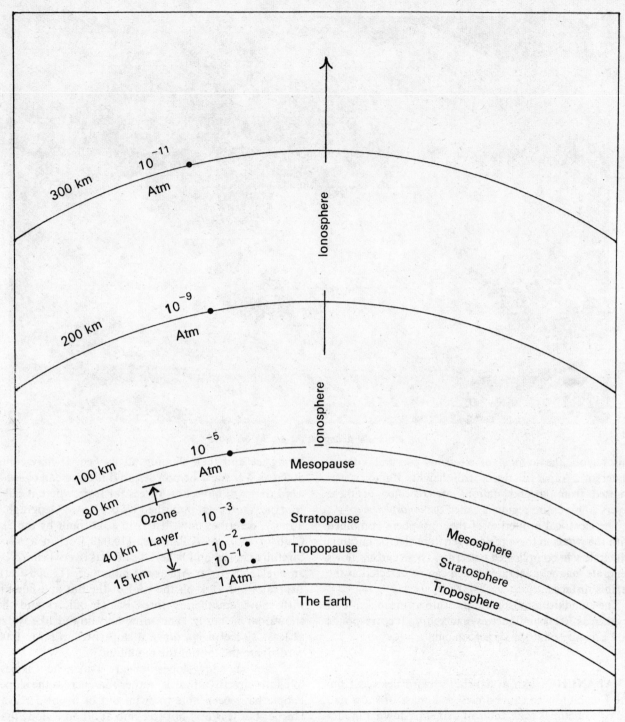

300 km

10^{-11}
Atm

Ionosphere

200 km

10^{-9}
Atm

Ionosphere

100 km

10^{-5}
Atm

Mesopause

80 km

Ozone

10^{-3}

Stratopause

Mesosphere

40 km Layer

10^{-2}

Tropopause

Stratosphere

15 km

10^{-1}

1 Atm

Troposphere

The Earth

TROPOPAUSE Temperature Minimum; about 80°C above the tropics
STRATOPAUSE Temperature Maximum; about 0°C above the tropics
MESOPAUSE Temperature Minimum; about —80°C above the tropics.
Above the Mesopause the temperature steadily increases to more than
1000° C.

of about 80 km. There is another transition layer here called the mesopause, with temperatures around—90˚C. Thereafter we have the *thermosphere*, with temperature continuously increasing with height, to about 1000˚C at an altitude of 1000 km. In the thermosphere, the at-

mospheric composition changes, the molecules are broken down, and electrons and charged lighter atomic particles, called ions, predominate.

The region above 80 km is electrically conducting due to the presence of electrons. At specific layers in

An avalanche falling on the glacier in centre

this region, the number of electrons per unit volume undergo a rapid increase with height. Radio waves emitted from ground stations are reflected at these layers. This entire region is called the *ionosphere*.

The electrical properties of the ionosphere combined with the winds in these regions generate electric currents which flow in complicated patterns. These current flows generate magnetic features which can be detected at the earth's surface.

The standard atmospheric pressure at mean sea level is taken as 760·00 millimetres of mercury. It corresponds to a pressure of 1·013 kg per sq. cm.

AVALANCHES. An avalanche, a word derived from the French for 'to come down', is a mass of snow and ice or sometimes rock and soil that slips down a mountainside, often at great speed. It may be caused in different ways, but is often the result of a thaw or rise in temperature after a heavy snowfall. Snow and ice may collect till their weight carries them down the mountainside taking with them masses of rock and carrying away trees, roads and villages in their path. Such an avalanche may form a dam across a valley, holding up water behind it. When the dam gives way the water rushes down the valley causing great damage. An avalanche familiar to mountaineers is caused by soft snow overlaying ice and then slipping off its bed. It may come down as wet snow or powder, and if large can cause a wind that will blow down trees far from where it falls. *Hints to Travellers*, published by the Royal Geographic Society, describes how: 'In 1936 a hut built by the Ski Club of India at Killanmarg (10,000 feet) in a place carefully chosen and believed safe was overwhelmed by an avalanche from Apharwat mountain (13,600 feet). A great wind took off the roof of the hut and filled it with snow, smothering three British officers and the chowkidar instantly. The snow had flowed like water at least 20 feet deep, down a slope of 1 in 6 for 1000 yards from the foot of the mountain.'

If climbing snow slopes liable to avalanche it is best to climb directly upwards, rather than across the slope. People have been able to swim out of a powder avalanche; but it is best not to climb at all on a slope that might avalanche, or under an overhang of snow. Even shouting may cause an avalanche.

See also Landslides.

AXLEWOOD, *Anogeissus latifolia,* is a tree in tiger areas in Himalayan foothills and in C India. The leaves turn coppery-red in winter and are shed in the hot weather. The fruits are 4 x 5 mm, narrowly 2-winged. The wood is tough and in demand for skis, axles, etc.

BABBLER. Arboreal and terrestrial birds belonging to the large and heterogeneous subfamily Timaliinae, a component of the complex Flycatcher family Muscicapidae which, besides this, includes the true flycatchers (Muscicapinae), the warblers (Sylviinae) and the thrushes (Turdinae). The subfamily Timaliinae includes also the laughing thrushes, scimitar babblers, wren-babblers, parrotbills, and a mixed assortment of other groups of diverse appearance, habits and habitats. They range in size from Myna to less than Sparrow, and in coloration from drab earthy brown (e.g. Jungle babbler) to the resplendent multihued leiothrix, known to fanciers as 'Pekin Robin'. Some live in open savanna country and scrub jungle while some are confined to dense evergreen forest. Some species are largely insectivorous, while the food of others consists principally of vegetable matter. The best known genus of the subfamily is *Turdoides* which contains such familiar species as the Jungle Babbler (*T. striatus*) and the Large Grey (*T. malcolmi*). They are both about Myna size and earthy brown or greyish brown; the former is normally found in wooded country, the latter around semi-desert cultivation. Their characteristic habit of invariably keeping in flocks or 'sisterhoods' of half-a-dozen or more has earned them the name of *Sat bhai* in Hindi ('Seven Sisters' in English). They feed on the ground hopping about in amicable company, searching under fallen leaves and mulch for insects etc. They build open cup-shaped nests in moderate-sized leafy trees and are well known as 'corporate nesters' which means that more birds than the breeding pair assist in all the nesting chores. The eggs, 3 or 4, are turquoise blue, and the nests are commonly brood-parasitized by the Pied Crested and Hawk-Cuckoos. Perhaps equally well

known is the Common Babbler (*T. caudatus*) which is similar but somewhat smaller and slimmer, with a longer and pointed tail. Its earthy brown plumage is dark-streaked. It also lives in 'sisterhoods' but chiefly in semi-desert country, scuttling about under thickets like a rat.

See also BROOD PARASITISM, NIDIFICATION.

BABUL, *Acacia nilotica*, is a moderate-sized tree with dark-brown vertically fissured bark. Characterized by sharp-pointed, white, paired spines, bipinnate leaves with small oval leaflets, yellow, sweet-scented flowers (June-September) and necklace-like downy pods. Indigenous to Sind (where there are the largest tracts) and the Deccan, and naturalized in the rest of India, Sri Lanka and introduced in Burma. A prized fuel and tanning material. Its twigs are used as decoys by fishermen, and the spines serve as fishing hooks and as paper pins.

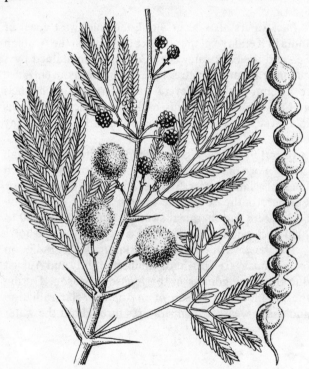

Babul. The sharp spines, bipinnate leaves and globose flowers, with the necklace-like fruiting pod containing 8–12 seeds, × 2/3

BACKWATERS. The term 'backwater' or 'backwaters' in India refers to a system of shallow, brackish-water lagoons and swamps found along the coastline. They occupy an area of hundreds of square kilometres. Although the backwaters have their own special physical characteristics, some of them are part of estuaries. The areas which are close to the permanent connection with the sea are influenced by the regular tidal rhythm and have been referred to as tropical estuaries.

The hydrography of the backwater is mainly influenced by two factors: the short-term changes induced by the tides and the seasonal changes brought about by the monsoon system. The magnitude of variation within the backwater depends chiefly upon the place of observation (nearness to the sea or freshwater source). The backwater remains dominated by sea water for about 6 months, and then, with the commencement of the rains, it becomes freshwater-dominated and continues to remain so with varying degrees for the next 6 months.

The most extensive occurrence of backwaters of different shapes and sizes in our area is to be noted on the west coast as far north as Karwar. These backwaters on the west coast, locally called *kayals*, are almost invariably associated with streams, i.e., within them are the lower reaches of the streams. All the *kayals* are marked by tidal flow, i.e., inflow of sea water at the time of high tide and seaward outflow at the time of low tide.

Backwaters also occur along the southwest coast of India in Kerala; e.g. the connection between the Arabian Sea and the backwater near Cochin is maintained by a channel, about 450 metres wide, which forms an entrance to Cochin harbour. Several rivers, irrigation channels and sewers open into this backwater which terminates in a large lake called Vembanad Lake. There are many others.

In certain tracts, e.g. between the latitudes of Kavali and Nellore in Andhra Pradesh, there are disconnected lines of water-bodies 3 to 8 km from the seashore. Further to the south the lagoon belt is marked by backwaters and marginal swamps ranging from 800 to 2000 m in width and about 1·5 km from the seashore.

The backwaters receive maximum solar radiation from January to April and minimum in July and August due to the cloudiness in the monsoon season. During the pre-monsoon season of maximum solar radiation and warm weather, temperatures throughout the water remain uniform and reach their maximum. With the onset of rains in May, the water temperatures decrease and water at the surface is several degrees warmer than the shallow bottom and a clear thermal gradient develops in the water-column. This difference persists until about September. Salinity remains homogeneous during the pre-monsoon season, indicating that the water is well mixed, but during the monsoon months large quantities of fresh water enter, resulting in low salinity at the surface and high salinity at the bottom.

The halophytic vegetation that occurs in the backwaters is called 'mangrove vegetation'.

See plate 29 facing p. 416.

BAEL, *Aegle marmelos*, occurs throughout the subcontinent up to the Himalayan foothills. It is a small, spinous tree with gland-dotted trifoliate leaves, bearing short panicles of small, greenish, fragrant flowers. The sweet nutritious pulp of the woody fruit is drunk as a sherbet. It is popular with farmers and sacred for the Hindus.

Bael, flowers and fruits, × 1

BAMBOOS or tree grasses comprise a diverse group of plants in the grass family. These giant grasses are separated from the other grasses by certain 'bambusoid' characters which are considered primitive. The prominent rhizome, woodiness of stem, branches and stalked leaves are some of the distinguishable characters, though some floral and anatomical characters are also regarded as features of difference. Because of their

© S. Elamon

Aakkulan lake near Trivandrum

One of the hill bamboos (Chimonobambusa falcata), *locally called
Ringal, which is a mere shrub with culms no thicker than a pencil*

large size these arborescent grasses are also called 'elder brother of grasses'. They have age-old connections with the material needs of man and are fascinating to the artist, the poet, the craftsman and the scientist. Aptly called the 'poor man's timber' bamboos are of great importance to the people of the East where they are found in greatest abundance and variety.

To a layman all bamboos look alike but actually there are more than 1000 different kinds in the world. Although they occur in almost all the continents there are none in Europe. Their largest concentrations are in SE Asia—the continent of bamboo groves—and S America. While 150 species are found in the Indian subcontinent, India alone accounts for more than 100. Spread over an area of 10 million hectares or 13 per cent of the total forest area of the country, perhaps the world's largest reserves of bamboos consisting of over 115 species both wild and cultivated exist in India, areas particularly rich being the northeast region and the Western Ghats. They have a wide range of distribution and are found in all parts of the country except in Kashmir valley. As an understorey they form rich belts of vegetation in well-drained parts of tropical and sub-tropical habitats and grow up to 3700 m in the Himalaya. The distribution of bamboos, however, has been greatly altered by human intervention and natural

stands have at places been more or less cleared off for shifting cultivation. The other intervention comes from the paper industry which cuts or grows bamboos according to its needs.

The structural foundation of the plant is the underground, segmented and condensed rhizome which goes on propagating vegetatively. The aerial part (stem) is called the culm and several culms arising out of the ramifications of the rhizome are collectively called the clump. When they occur in dense impenetrable thickets they are called 'congested clumps', which is the general feature of most bamboos. However, there are others that have spreading rhizomes; here the culms arise singly at intervals and form 'open clumps'.

Bamboo-culms are branched at the nodes. The branches are sometimes spiny as in the case of Spiny bamboo (*Bambusa arundinacea*). Depending on the species they may be mere shrubs with culms no thicker than a pencil as most hill bamboos are, or they may become giants reaching a height of 37 m and a diameter of more than 0·25 m as in the case of the Giant bamboo of Burma (*Dendrocalamus giganteus*), which is cultivated at Dehra Dun and some other places. Whereas most of the bamboos are erect, quite a few are scramblers and even climbers stretching over the crowns of tall forest trees. Nearly all species are green when fresh but some like

Spiny bamboo (Bambusa arundinacea). *The prominent spines at the nodes of this common Indian bamboo make the clumps so congested that it is often difficult to harvest the culms.*

One of the climbing bamboos (Malocalamus compactiflorus) *stretching over the crown of a tree. The fruits are fleshy and full of starch.*

the pantropical Tiger bamboo (*B. vulgaris*) are of a beautiful golden colour with green stripes or otherwise variegated. An occasional species has a near black colour. Most species have hollow culms but some like *D. strictus*—the Male bamboo (so called because of its strength)—have solid culms. The hollow internodes are separated by a transverse septum at the node. The internodes may be condensed, forming bulges as in Pitcher bamboo (*B. ventricosa*), or they may be very long as in *B. polymorpha*. The culms are provided with culm-sheaths which remain attached to the nodes till the culms become hard and then they fall off. A white powdery bloom covers the tender skin of the culm which also disappears as the culms mature. The culm-sheaths are distinctive for each species and therefore they have been effectively used as an aid in bamboo identification (because the flowers are seldom available). The leaves are stalked, flat, lance-like, in two vertical rows and are provided with a sheath.

The growth of a bamboo plant is remarkable. During the monsoons new shoots emerge from the underground rhizome in the form of cone-like buds which again are characteristic for each species and are used as a tool in their field-identification. These are wrapped over by strong, sharp-pointed sheaths covered with minute irritant hair. The tender shoot inside is very

delicate and luscious, but at the same time is poisonous (containing hydrocyanic acid). Perhaps this is how nature protects them from the ravages of grazing cattle or wild animals. Those shoots are a great delicacy and serve as supplementary food. The poisonous effect is destroyed after boiling. Bamboo shoot is then cooked or made into pickles. The popular belief is that new shoots sprout only when there are thunder and lightning. Besides the fact that during the monsoon lightning is common there is a scientific basis too. The mass of shallow bamboo roots makes a compact mat and exhausts the soil of its nutrition all around the clump. The physiological effect of lightning is that (1) it converts the atmospheric nitrogen into nitrates which are washed down into the soil with the rain-water, thus replenishing the impoverished soil and providing much-needed nitrogen to the sprouting buds, and (2) strong intermittent flashes break the dormancy of the bud. Elongation of shoots takes place gradually in the initial stage and then they become so active as to grow in leaps. In some of the tall species like *D. giganteus* they may grow up to about one metre in a day. If one has patience to sit for hours by the side of the growing culms one can even see the actual process of lengthening. The bamboo shoot grows to its full length in three months' time and by then the rains also come to an end.

1 A Moist Deciduous Forest

A rivulet winds its way through Mudumalai's moist deciduous forests, ideal habitat for tiger, leopard, wild dogs, elephants, bison and deer.

Painted by J.P. Irani

(1) Indian Salmon, a Thread Fin *(Polynemus tetradactylus)*, × 1/7 (2) Squirrelfish, a sea perch *(Holocentrus rebra)* × ½ (3) Young stage of the Yellow Grouper, a sea perch *(Epinephelus lanceolatus)*, 1/16 (4) Silver Pomfret *(Pampus argenteus)* × ¼ (5) Rosy Snapper, a sea perch *(Luteanus reseus)* × 1/5 (6) Speckled Grouper *(Epinephelus malabaricus)* × 1/16 (7) Spotted Dory, a sea perch *(Drepane punctata)* × 1/16.

A clump of the commonest Indian bamboo (Dendrocalamus strictus). The culms are solid and very strong, hence the species is called Male bamboo. The arrows A and B and the numbers 1 and 2 show the emerging young cone-like shoots elongating in leaps.

The Pitcher bamboo (Bambusa ventricosa) which has short, constricted internodes and forms bulges or pitchers in a series. This curio is a worthwhile introduction in any botanical garden.

Bamboo clumps go on adding shoots and become impenetrable thickets. Within 3 to 5 years the culm matures and turns brownish.

While grasses flower annually and most commonly reproduce by seeds, bamboos usually reproduce from root-like rhizomes that give rise to new long-lived culms. They generally flower only after many years, at fixed intervals, and not just every year, at which time they produce seeds and then die (compare CONE-HEADS which also exhibit this alarm-clock flowering behaviour). In many bamboos flowering occurs after 20 or 60 years; in several others it may not occur until 120 years after the last flowering. In some species, however, the interval may be shorter, 3 years or so; a few may flower even annually. As a matter of fact flowering in bamboos is erratic and varies between the physiological extremes of constant flowering and constant sterility as evidenced by *B. atra* and *B. vulgaris* respectively. The majority of bamboos fall between these two extremes and represent flowering cycles of several to many years. On the basis of their flowering behaviour bamboos can be classified into those that flower annually or nearly so, those that flower gregariously and periodically, and those that flower sporadically or irregularly. No culm in a plant generally lives through two consecutive flowerings. At the most a

culm reaches 15 years of age and then dies—but the whole bamboo clump is a continuous plant that dates back to the original seed. Each cell must contain a mechanism that dates the age of the genetic stock, for when the right time comes the entire species switches over from vegetative growth to the production of

The Giant bamboo of Burma (Dendrocalamus giganteus). It is the largest of all bamboos. The growth is very rapid, up to 1 m a day. The internodes are large enough to be used as petrol or water containers.

A clump of Mala bamboo (Bambusa nutans) *showing gregarious flowering. When general flowering takes place the spectacle can be breath-taking.*

A close-up of flowering twigs of Bambusa nutans, × ²/₃. *The flowering occurs in this species after every 35 years and the clumps die after seeding.*

flowers. Such a life-history is unique among flowering plants and scientists are still at a loss to explain the reasons for this cyclic flowering or the mechanism by which it occurs. When general flowering takes place the spectacle can be breath-taking; it is all the more exciting when we consider that this flowering takes place only once in its lifetime. No matter where the species grows or whatever be the age of the clump, it happens all so spontaneously. The whole plant, culm,

branches and sometimes even the rhizome, are transformed into a gigantic inflorescence. It looks as if, before its death, the bamboo sacrifices everything to the production of flowers and fruits, and these it produces in abundance. Although the exact cause leading to the general flowering is still a mystery, it is believed that a short rainy season followed by a long spell of hot weather promotes flowering. Perhaps this may explain the popular superstition that bamboo flowering is an

A juvenile shoot of the Tiger bamboo (Bambusa vulgaris) *wrapped over by sharp pointed sheaths covered with minute hair* (× ¹/₃). *The tender* *shoot inside is edible (after boiling) and is considered a great delicacy.*

On the left a large fleshy fruit of the Muli bamboo (*Melocanna baccifera*) which often germinates *in situ*, which is a rare phenomenon. On the right a wheat-like grain of the Male bamboo (*Dendrocalamus strictus*), a general feature in most bamboos. Both × 1.

indication of an approaching large-scale drought, famine or epidemic.

Congenial flowering conditions are indicated by the cessation of all vegetative activities; no fresh leaves or culms are produced for a year prior to flowering, they lose their foliage, and when finally stripped of leaves the culms burst forth in flowers. The flowering and seeding activity in this wind-pollinated group of plants continues for a long period, sometimes as long as 2 years. Such enormous quantities of grain-like seeds are produced that the entire forest floor gets covered with grains. Where famine conditions prevail in the countryside people from the neighbouring villages collect in large numbers and gather the seeds. Some bamboos, like the Muli bamboo of Assam (*Melocanna baccifera*), bear large fleshy fruits, the size of guavas. These too are produced in large quantities and are all starch. People collect them for food and eat them ravenously. Whatever is left on the ground begins to germinate. Some of these large fruits do not get dislodged easily from the parent plant and germinate *in situ*, but this viviparous condition is a rare phenomenon.

All the species do not flower at the same time but it is certain that all the plants of a given species belonging to the same seed-source would flower simultaneously no matter where they are or where seeds or offsets have gone to for plantation. *Thyrsostachys oliveri* flowered in Burma in 1891; seeds were sent to Calcutta, and Dehra Dun, approximately 1500 km apart, and clumps raised at both places where they flowered simultaneously in 1940. More recently in 1961 simultaneous flowering of Muli bamboo was observed in Assam and Dehra Dun approximately 1000 km apart, which aroused a lot of interest in botanical and forestry circles. The Spiny bamboo, which flowered almost throughout India in 1970-71 after a lapse of 45 years, is yet another example of this strange phenomenon where the germ-plasm of every seed is provided with an alarm-clock set to go off at a fixed point of time. In localities where a pure forest of a single species exists over extensive areas the flowering occurs in waves, starting from one end of the forest and ending at the other. The seeds germinate readily and in a short time a fresh crop springs up. For the first few years the seed-grown plants look like grass. Quite a lot of them are destroyed by animals but sufficient numbers survive to replenish the forest stock. Interestingly, since the flowering of different species takes place at different intervals, there is practically no chance of producing hybrids in nature.

Myths surround the bamboo, many of them born in communities used to living with calamities. It has already been mentioned how the prodigal flowering of bamboo is taken to be of great significance because it portends famines and epidemics. In such difficult times the rare occurrence of general seeding of bamboo is a godsend, for it provides supplementary food. One of the popular beliefs is that the rat population increases when the bamboo has flowered and the seed lies on the ground. A rational explanation is that where there is a bountiful supply of seed available there is also a large influx of rodents from the neighbouring areas where severe drought conditions have caused the crops to fail. Reproduction in rats is rapid, and the well-fed offsprings quickly give birth to millions of young ones. The result is that the rodent population reaches an astronomical figure within a short time in a small area. The following year, when there is no bamboo seed to feed upon this army of rats attacks the crops and plays havoc in the fields and villages. The people attribute a direct connection between gregarious flowering of bamboos and plagues of rats. Thus are many of the myths born.

The popular association of bamboo flowering with floods too has some basis. Rats store quantities of foodgrains in long tunnels that they burrow into the mudbanks along the rivers. Millions of such holes drilled in the embankments of rivers are a leading cause of the dykes being weakened, resulting in inundation of the land around. There is also the eventuality of forest fires. The culms which are dry and congested often rub against one another when there is wind. The friction creates enough heat to cause fire, and it does not take much time for a whole forest to be set ablaze.

Cultivation of bamboos is done by seed or offsets, in some cases by cuttings or layers. If the seed is good, which is seldom the case, it germinates easily and the seedlings are reared and transplanted in the field; but as general seeding years are scarce, it is necessary to wait for seed and, therefore, propagation by vegetative methods is quicker. However, the disadvantage of vegetative propagation is that the resulting clumps would flower along with the parent clump, thereby rendering their life short or of uncertain duration. Propagation, therefore, is best done by seed when possible. Offsets consist usually of a portion of old culm with its roots, cut off above a node at 30 to 60 cm above the ground, and the shoots come from dormant buds at the base of the culm. Such offsets are best taken and planted in the season of rest, so that the season of active growth which usually begins with the rains may find them well in position and capable of taking root easily. Offsets taken in the late rainy season after the new growth has started usually fail. Cuttings are normally made by Planting one or more internodes, while layers are made by partly cutting and laying a clum in the ground so that it may take root at the nodes. When the shoots

have appeared the internodes are cut and layers planted separately.

Management of bamboos is based on the physiological development of the clumps. The new culms are produced from the rhizome along the periphery of the clump. Although the culms attain their maximum size in one season they are not ready for utilization for 2 or 3 years because of lack of strength. Further, for proper development of new culms and their support a number of mature culms have to be left during fellings. Thus, although bamboos behave like an annual crop, it is not possible to harvest all the culms annually. A felling cycle of 3 or 4 years is generally prescribed. The centrifugal manner in which the clump expands leaves the younger culms on the periphery, enclosing the older ones in the interior of the clump. It is, therefore, difficult to approach the latter without sacrificing a few younger culms. Each clump is to be worked in the form of a horseshoe, the bamboos inside the 'shoe' are felled, and the clump is allowed to grow outwards.

The number of ways bamboos enter into the diverse phases of human life is too well known. It has been said that these giant grasses are one of those providential developments in nature which, like the horse, the cow, wheat and cotton, have been indirectly respon-

A plantation of the graceful Betua bamboo (Bambusa polymorpha).
*In NE India, where this species is commonly found, the womenfolk use
the internodes for carrying water.*

sible for man's own evolution. Bamboo is a material that is sufficiently cheap and plentiful to meet the vast needs of the human population—from the child's cradle to the dead man's bier. The qualities which make bamboos so versatile are the strength of culms, their straightness, lightness combined with hardness, range in size, abundance, easy propagation and the short period in which they attain maturity. The culms can be easily split with ordinary hand tools. In the humid tropics whole houses are built entirely of bamboo without using a single iron nail, and huge suspension bridges made solely of canes and bamboos are marvels of indigenous engineering skill typical of tribal effort. In fact there is no limit to the varieties of articles that can be made out of this natural plant resource. Edison had used the carbonized filament of bamboo for his early electric lamps and the razor-sharp peel has been, at times, used in place of a surgical knife.

Among the more sophisticated uses are the manufacture of a large variety of writing papers, charcoal for electric batteries, liquid diesel fuel obtained by distillation, enzymes and media for culturing pathogenic bacteria from shoot extracts and the white powder produced on the outer surface of young stems for the isolation of a crystalline compound similar in nature to the female sex hormone. *Tabasheer* or *banslochan*, the fine siliceous matter deposited in the hollow stems of some species, has excellent properties as a catalyst for certain chemical reactions, though in the subcontinent it is prized as a restorative tonic and aphrodisiac. Another aphrodisiac use, though a nefarious one, is attributed to the rhizome of Rhino bamboo (*D. hamiltonii*) which is an exact replica of a rhinoceros horn that fetches a fabulous price; only an expert perhaps can identify the imitation rhino horn from the real. Recently a new use of bamboo, 'Bamboo Reinforced Cement Concrete Construction', has been evolved where bamboos have been used as reinforcing material replacing steel in the construction of roof-slabs, beams, electric posts, etc. The overall economy achieved is 33 per cent as compared to steel reinforced constructions. Bamboo leaves are used for thatching and are also valued as fodder; elephants in particular are fond of it. Dried and mature leaves are also used for deodorizing fish oil. Bamboo sheaths are made use of in lining of hats and sandals. As a popular ornamental, bamboo is used for hedges and in landscape gardening. It is valuable as a wind-break and is particularly useful for preventing soil erosion on account of its interwoven root system.

Although in view of the extraordinary range of uses to which bamboos are put and have assumed world importance, the bamboo industry in India is still in its infancy, especially in the handicraft sphere. But India leads the Asian countries in the utilization of bamboos for paper manufacture, which is the most important industrial use of this raw material. Of an annual production of 9·5 million metric tonnes of bamboos in India more than half (4·9 million metric tonnes) is consumed by the paper industry, leading to a production of 600,000 tonnes of paper per year. Bamboo resources in the subcontinent are plentiful but they are not being utilized to the maximum. Although India alone has nearly 100 native bamboos only 10 of them are being commercially exploited. These are: *B. arundinacea*, *B. balcooa*, *B. nutans*, *B. tulda*, *D. hamiltonii*, *D. strictus*, *M. baccifera*, *Ochlandra ebracteata*, *O. scriptoria* and *O. travancorica*. The bamboo industry in India, therefore, requires to be developed to the fullest extent so that the renewable bamboo resources may be put to the best use and to maximum benefit. The development of bamboo industry may go a long way in solving the problem of rural unemployment and also in earning good foreign exchange. The present revenue derived from bamboo resources, estimated at about 7 crores per annum, can be further increased provided the bamboo industry is developed on right lines. In this connection an important point to note is that bamboos are specific as far as their properties and uses are concerned. Although, taken as a whole, they are a versatile

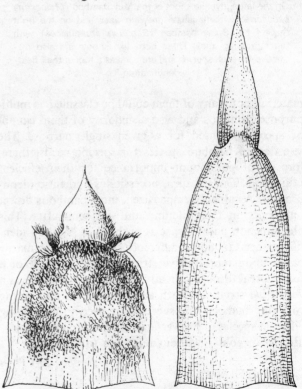

On the left a culm-sheath of the Tiger bamboo (*Bambusa vulgaris*) and on the right that of the Male bamboo (*Dendrocalamus strictus*). The culm-sheaths are distinctive for each species and thus help in their identification. Note the presence of ear-like structures in the former and their absence in the latter.

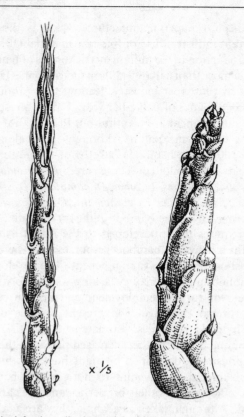

On the left a juvenile shoot of the Muli bamboo (*Melocanna bacciferra*) with flagellate growing apex and on the right that of the Spiny bamboo (*Bambusa arundinacea*) with blunt growing apex. These cone-like shoots are also distinctive for each species and are used as a tool in their field-identification.

material, not many of them could be classified as multi-purpose bamboos and also a majority of them cannot be specifically used for a given single purpose. The selection of bamboo species for specific needs, therefore, is of paramount importance for their efficient utilization. This, in turn, necessitates authentic identification. As already emphasized, most bamboos flower only once in their lifetime and die soon after. This phenomenon poses a special problem in bamboo identification as taxonomic differences in plants are primarily based upon reproductive structures. As a result of a recent breakthrough, already mentioned earlier, certain vegetative structures such as culm-sheaths and young cone-like buds can be successfully used in identifying different species of this versatile economic commodity—'the green gold'—without reference to flowers.

K.N.B.

See also GRASSES.

BANDICOOT RAT. The Bandicoot Rats of Asia are true rodents and should not be confused with the rat-like marsupials of Australia which are called simply Bandicoots. Bandicoot rats are similar to Sewer Rats (*Rattus norvegicus*) but can be distinguished by long guard hairs scattered over the back and by their more protruding incisors. Bandicoot rats are highly aggressive, emitting harsh grunts when disturbed. Strange individuals caged together are likely to fight to the death within a few hours.

There are two species of bandicoot rats: the Lesser Bandicoot (*Bandicota bengalensis*), which at 250 grams body-weight are slightly smaller than a typical sewer rat, and the Indian Bandicoot, which may weigh one kilogram or more and is the largest rat in Asia. Both species are found over most of the Indo-Malayan region, but within this area the Indian Bandicoot Rat is the less widespread, being common in rural areas but rare or absent in cities.

The Lesser Bandicoot Rat is undoubtedly the most serious rodent pest in South Asia. It has become the most abundant rodent species in many of the large cities within its range while other rat species have declined proportionally. It can carry plague as well as typhus, leptospirosis, salmonellosis and rat-bite fever. Since bandicoots live in houses and contaminate food with faeces and urine, they can easily spread these diseases to man. Infestations of this rat in grain godowns in Calcutta are so great that the number of rats feeding in a godown was estimated to exceed two rats per square metre of floor space. In a godown with 244 square metres of floor space, Lesser Bandicoots consumed an estimated 4226 kilograms of grain in a year, or enough to feed about 7000 people for a day. Lesser Bandicoots are also extremely well adapted for exploiting crops in the field. Individuals move great distances to find a field of ripening grain. These rats store large quantities of grain underground. In one field, an estimated 10 percent of the total rice production was hoarded by bandicoots in their burrows.

At harvest, female bandicoots produce litters of 8 to 19 young, double the normal litter size for the species. Bandicoots can build up large fat reserves when food is abundant. They may weigh twice the normal adult weight after several weeks of good feeding. They survive months of drought by remaining underground and feeding on stored food and tuberous roots and utilizing their fat reserves. When the rains return and the next crop is planted, bandicoots will be ready to take their share.

These rats can be controlled, however. On the farm, removal of weeds from the crop and from field edges means less cover for rats and less food before the crop ripens. Ploughing fields immediately after harvest breaks up rat burrows. Use of rat poisons as the crop matures can kill most of the rats that remain. In the city, it is essential to keep food in places that are rat-proof.

G.W.F.

See also RODENTS.

BANDIPUR. The Bandipur Forest in old Mysore was a famous hunting ground of the Maharajas. The first notification in connection with its reserved status was passed as early as 1898. Subsequently its status has been progressively upgraded in conservation terms, and it was constituted as a Tiger Reserve in June 1973.

It has an area of 700 km² and lies between 1000 and 1500 m above sea level. The terrain is undulating and the area is drained by the Kabini, Nugu, and Moyar rivers.

The forest consists of several types of vegetation. In the easternmost portion of the Moyar State Forest there is scrub jungle consisting of stunted trees of sandalwood, neem, myrobalan and many more. The trees are interspersed with lantana bushes and open grasslands. In the southern tropical dry deciduous section the dominant species of trees are teak, *Terminalia* species, bijasal (*Pterocarpus marsupium*), sissoo, gamari, palas, amla, and others. In the southern tropical moist deciduous type there are figs, amla, ber, kanchan and several others.

Because of its varied vegetation, and the many ecological niches which these provide, the Bandipur forest is rich in bird life, and over 150 species have been recorded here during the winter months.

As far as the larger mammals are concerned, since its constitution as a Tiger Reserve the tiger population has apparently been increasing and numbers exceed 30. It is claimed that the bison (*Bos gaurus*) attains its finest development in the Bandipur forest. The other mammalian species include elephant, sambar, spotted deer, barking deer, fourhorned antelope, wild boar, mouse deer, porcupine, blacknaped hare, common langur, bonnet macaque, panther, wild dog, jungle cat, jackal, sloth bear, flying squirrel, mongoose, pangolin, and crocodile.

There is considerable tourism in Bandipur and curiously it is around the Tourist complex that herds of chital congregate to spend the night. It is suggested that the deer are seeking protection from predators.

See plate 21 facing p. 304.

BANYAN, *Ficus benghalensis*, a huge evergreen tree with wide-spreading branches, sending down aerial roots which enter the ground and afterwards become trunks. It is found in the sub-Himalaya and peninsular India, and widely planted as a shade tree. It has smooth bark and large and leathery leaves. The fruits are fig-like, red when ripe. The Great Banyan of Calcutta has one thousand trunks and a walk round the tree is about a quarter of a mile. The specimen is one of the largest in the world in terms of sheer bulk of tissue. About 200 years old, the canopy covers 4 acres of ground. Banyans develop from seeds dropped by birds on walls or on other trees. They sometimes strangle the host tree, which gradually dies, but the two co-exist for years. An allied tree, the Pipal (*F. religiosa*), is also wild in the sub-Himalaya. Sacred to Hindus and Buddhists, it is visited by flocks of starlings in April for its figs in NW India and Pakistan. It is cultivated as a shade tree but is destructive to buildings.

BAOBAB (*Adansonia digitata*). A large fat tree with a bottle-shaped trunk which startles observers by its great girth of 12 m or more. Leafless in the dry season, single white flowers appear in June and oval fruits with edible pulp, 15–20 cm long, ripen by February. It is an African tree, brought to India by the Arabs centuries ago and introduced on the west coast. One of the longest-lived trees in Africa, it is reported to have the property of preserving human corpses.

BARBET. Barbets (family Megalaimidae) are arboreal birds with a stout bill, coloured predominantly green like the foliage of the trees where they often congregate during the fruiting season. Their typical call is an inherent feature of the tropical forests, ranging from the resonant *piao...piao...* of the Great Hill Barbet of the Himalayan foothills to the monotonous *kutru... kutru...* of the Large Green Barbet of the peninsular wooded biotope, and the ventriloquistic duet *kut... kut...* of the Crimsonbreasted or Coppersmith Barbet of the countryside. Some species like the Bluethroated, the Yellowfronted and the Goldenthroated, as the names suggest, are resplendent with gaudy colours.

Barbets are chiefly fruit-eaters, but they also take insects and beetles, often sallying into the air to catch their prey. The breeding season is from April to July. Three or four eggs are laid in a nest-hole excavated in a dead branch of a tree. Both sexes take part in incubation and rearing the young.

BARK. Bark is the non-technical term applied to all components of stems and roots of perennial plants placed external to the vascular cambium including the phloem, cortex and periderm. As such in the older trees, it may be divided into the outer *dead* and inner *living* regions—the latter consisting exclusively of secondary phloem. As expected, the outer layer of bark acts like a shield so that the mature parts are protected from the attack of desiccation, fungal or bacterial infection and/or that of insects. In the more extreme climates, such as those of temperate regions and in areas liable to be swept by fire, it assumes great biological importance owing to the additional mechanical support that it furnishes.

Generally speaking the bark is a more complex tissue system than the wood in both development and structure. It is also technically less accessible for study and often considerably less reliable than that of the wood.

COMPOSITION AND CELLULAR DIFFERENTIATION
IN THE VARIOUS LAYERS OF BARK

All greatly magnified

Cc	Cortical cells	Pe	Periderm	Scc	Suberized cortical cells
Co	Cork	Ph	Phellogen	Sp	Secondary phloem
Crc	Crushed cells	Phe	Phelloderm	Sx	Secondary xylem
Cu	Cuticle	Pi	Pith		
Ep	Epidermis	Sc	Storied Cork		

A, B Position, shape, and extent of the additional periderms formed in a woody stem on which the first-formed periderm develops as an entire cylinder close to the epidermis.

C. D Origin of periderm beneath the epidermis

E Radial section of the stem of *Cordyline australis* showing the position of the layers of storied cork, which enclose patches of suberized undivided cortical cells

F Cross-section of the stem of *Cordyline indivisa* showing a superficial layer of crushed cells and alternating tangential bands of suberized, undivided cortical cells and storied cork.

G Cross-section of the outer part of cortex of *Curcuma longa* to indicate the storied cork.

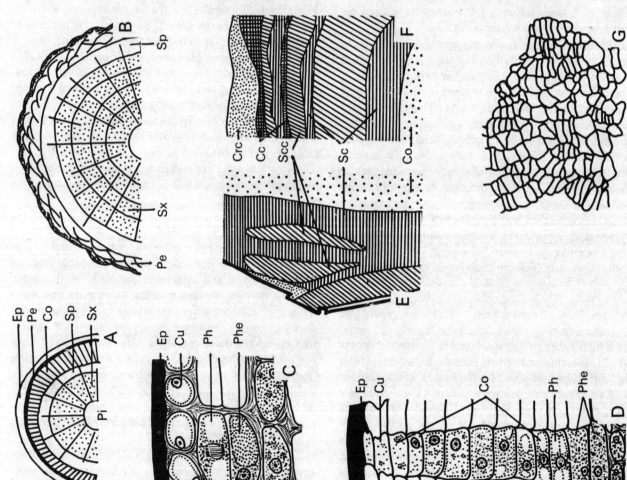

Differences in its appearance, colour, and smell are of a great help in identification, particularly in wet-tropical forests where the foliage and flowers are not easily available because of their immense height. Marked variations also occur in its structure and organization related to the differences in the development of the component tissues. At various times, it includes the primary and the secondary tissues, especially the phloem, cortex, epidermis, and periderm (Fig. 1 A,B).

In a few trees, for example, the cork oak (*Quercus suber* of Europe), the outer region of the bark can be removed in the form of slabs without damaging the inner tissue. The barks of some species of birch (*Betula*; e.g. *B. utilis*) as well as *Eucalyptus* get stripped away in papery layers. More recent studies carried out for several timber-yielding species suggest that the thickness of the bark increases with the stem diameter. When successive layers of the bark are formed in progressively deeper regions of the stems, concentric rings appear to surround the entire stem. In some other plants it is formed as overlapping, scale-like layers so that the outer tissues break up and are sloughed off in patches.

A word may also be mentioned about girdling, which essentially consists of removing strips of bark down the cambium all round the bole. This causes some trees to die easily, others reach this fate slowly, and still others tend to recover gradually from such an effect.

Another important aspect of bark, worthy of mention, emerges from the fact that the majority of the latex-bearing vessels (for example, in the rubber tree, *Hevea brasiliensis*; Scholar tree, *Alstonia scholaris*; and the INDIARUBBER tree) are mostly distributed in it. Usually these are branched and may also show anastomoses. Their specialized, excretory nature is proved by the thick cell walls surrounding them, abundant mitochondria, endoplasmic reticulum, and numerous nuclei distributed all over the lumen. Interestingly enough, these provide one of the best systems for the study of synchronization in the mitotic divisions of the plant cells.

In the dead region of the bark, usually tannins become accumulated in large quantities which impart a brownish-black tinge to it. Some barks, such as those of conifers, possess resins; *Boswellia* has oleo-resins and from the bark of *Cinchona*, the malarial drug-yielding tree, alkaloids are obtained.

Since secondary phloem is the dominant tissue of the bark, the most conspicuous variations in its structures are reflected from differences in the composition and the distribution of the phloic elements. The sclerenchyma cells impart to it a characteristic pattern and these may be distributed irregularly or in tangential bands. When we relate the structure of the bark to that of the phloem, we actually refer to that part of phloem which has ceased to be concerned with longitudinal trans-

location. The conducting phloem constitutes only a small part of the bark and its recognition depends on the knowledge of the developmental history of the sieve elements, particularly that of their sieve areas. In the major survey of the condition of the phloem in the bark of dicotyledonous trees, the sieve elements have been found to be functional for one season only, although instances of their continued longevity for more than five years are also known in the literature.

The changes that occur in the phloem after the elements cease to conduct vary considerably in different species. Their collapse may cause a large reduction in the width. Crystals are very common in the conducting as well as non-conducting phloem components.

Besides originating from phloem the principal components are also contributed by the periderm. With regard to the origin of phellogen, which contributes towards periderm formation, it is necessary to distinguish between the first and subsequent periderms that arise below the previous one and replace it as the axis continues to increase in circumference. In most of the stems, the first phellogen arises in the subepidermal cells (Fig. 1 C,D). In a few plants, the epidermal cells give rise to the phellogen directly and sometimes the phellogen is formed partly from the epidermal derivatives and partly from the subepidermal cells. In still other instances the phellogen arises near the vascular region or directly within the phloem.

In the monocotyledons with pronounced secondary growth, a special type of protective tissue is formed by repeated divisions of parenchymatous cells and subsequent suberization of the products of divisions. Thus the cork cells are produced. Cases are known where several layers of storied cork are formed further towards the centre of the trunk. Between these occur alternating layers of undivided suberized cells and non-suberized elements. In this way a tissue analogous to rhytidome of the dicotyledons is produced (for example, in *Cordyline*; Fig. 1 E,F). In *Curcuma longa* (the turmeric; Fig. 1 G), a special kind of protective tissue is developed by the secondary activity of storied meristem. Here the elements of the outer cortex undergo 3 to 8 periclinal divisions and thus radially arranged layers of suberized cells are formed which cover the stem from all sides.

See also STEM.

BARNACLES are closely allied to water-fleas and copepods and when young are free-swimming with three pairs of appendages. While still free-swimming they develop additional limbs, feed, moult and change their forms. Later they attach themselves by their first pair of antennae to rocks or timber and after a resting stage become shelled adults. The shell is in divisions which overlap and when closed are thoroughly protective.

A barnacle is therefore described as a kind of shrimp standing on its head, attached to some support and with its legs projecting upwards. These legs serve to produce currents in water enabling the animal to respire and capture food.

There are two kinds of barnacles—stalked or Goose barnacles and unstalked or Acorn barnacles. *Lepas* is an example of the former and *Balanus* of the latter. Both the kinds, when adults, measure 3 to 5 cm across. Barnacles are so abundant and so widely distributed that all rocks on the littoral, pilings, shelled animals like molluscs and crabs, the keels and hulls of ships and canoes are all encrusted by these crustaceans. In the Goose barnacles, found in clusters on logs of driftwood or on the hulls of ships after long voyages, their fleshy stalks grow from 2·5 cm to more than 30 cm long!

See also CRUSTACEANS.

BARRACUDA. These are large, rapacious, tropical marine fishes of the genus *Sphyraena* with a slender, streamlined body. The long, powerful jaws are equipped with sharp, dagger-like teeth, those at the front being much longer and forming strong fangs for holding their prey. They grow to three metres and weigh 25 kg. For their size, they are much more ferocious than sharks, and divers are more afraid of the utterly fearless barracuda than of sharks. The dorsal and anal fins are situated far to the rear. Their young congregate in shoals, but the adults are solitary. Eight species of barracuda occur in Indian waters but large individuals are rarely seen.

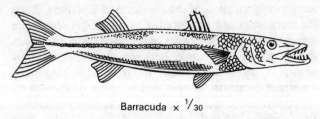

Barracuda × ¹/₃₀

BATS. Bats are classified in the mammalian order Chiroptera which word is derived from Greek and literally means hand (*cheir*), wing (*pteron*). This wing development is the unique characteristic shared by all bats, giving them the power of true flight and also distinguishing them from all other mammalian orders. The wing of a bat actually consists of two layers of very elastic skin stretched between the finger-bones (phalanges) with little or no interconnecting tissue. This membrane usually extends down to the ankle (carpal) joint of the hind leg and in most bats also encloses the tail to form an elastic pouch of skin, known as the interfemoral membrane or uropatagium.

Next to the rodents, bats are the most diverse and abundant order amongst mammals, with over 2000 different species having been described (Miller).

Advances in taxonomic knowledge have led to a simplification of earlier classification and modern works list only some 800 species (Walker). At least 73 species have been described from the Indo-Pak region covered by this encyclopedia. Despite these facts, our general knowledge of bats is probably less than for any other mammalian order and these very diverse and fascinating creatures are associated in most people's minds with feelings of revulsion, as well as numerous superstitions and attributes or traits based upon folklore which are without foundation. This may partly be due to the fact that many species emit a smell which is repugnant to humans and they are only likely to be actively encountered during the hours of darkness, and they can avoid obstacles even while being pursued by an irate householder around a darkened bedroom.

Bats are very difficult to classify because though sharing a similar wing-pattern they vary greatly in the number and development of their teeth as well as in their size, skull shape and external ear and nose appendages. Broadly speaking however they can be divided in two main groups (sub-orders): the fruit bats or flying foxes known as Megachiroptera, and the insect-eating or smaller bats known as Microchiroptera. Some Megachiroptera are actually smaller in size than insect-eating bats but they are all adapted to a diet of soft pulpy fruit or pollen and nectar from flowers and their wing-structure is different, retaining some independence of the second digit, often terminating in a separate claw to aid in clambering over fruiting branches. The Microchiroptera are divided into sixteen families and exhibit great diversity of external form, including fur colouring, wing-shape and dentition. Some Microchiroptera are adapted to feed exclusively on pollen and nectar (i.e. the Phyllostomatidae of South America). They include the notorious blood-sucking family of Desmodontidae which are the true vampires and again confined only to South America. A few species occurring in the New World prey largely on fish which they scoop off the surface of streams with their well-developed hind feet. But the vast majority are adapted to feed largely upon insects. Bats occur throughout the world except in the circumpolar regions and beyond the limit of tree growth. In Palearctic regions many microchiropteran bats exploit the summer abundance of insects in these northern areas, passing the colder weather in prolonged hibernation or migrating southwards to warmer climes. The Fruit Bats or flying foxes are confined to the warmer tropical regions of southeast Asia including Australia and most of the Pacific islands.

Flying Foxes reach their greatest diversity in the Austro-Malaysian region with the Malayan Flying Fox being the biggest known species, having a wing-span of 1·7 meters. In our region seven species occur, from the relatively small Short-nosed Fruit Bat *Cynopterus sphinx*, which is quite a pest of fruit gardens around Bombay, to the large Flying Fox *Pteropus giganteus* so familiar in many Indian villages from its habit of forming large colonial roosts in the exposed branches of tall trees such as the banyan or pipal wild fig trees. Another species of *Pteropus* has been recorded from Assam and the Andaman Islands, while there are two species of intermediate-sized fruit bats belonging to the genus *Rousettus* which roost in comparatively dark caves or similar man-made excavations such as cliff temples or deep open wells. Two other little-known fruit bats just come into our area in northeastern Assam, the Dawn Bat (*Eonycteria spelaea*), specialized to feed upon pollen and nectar, and Blanford's Fruit Bat (*Sphaerias blanfordi*), being unique among fruit bats in having no tail and only a vestigial interfemoral membrane.

All these fruit bats have well-developed vision and are thought to locate ripe fruit trees largely by smell. Their flight is comparatively slow and flapping but they are capable of traversing distances of up to fifteen kilometres between their daytime roost and suitable fruit-laden trees. They are capable of clambering over branches and sucking ripe fruit *in situ* or by clasping a fruit to their breast with one wing and clawed-thumb to bite it off and fly to a more convenient feeding perch. They do not actually chew fruit but only squeeze out and swallow the juice and the soft pulp; their simplified grooved molars and premolars are specially adapted for this mode of feeding. The Short-nosed Fruit Bat is capable of hovering in front of a ripe fruit and of plucking it in mid-air. Fruit Bats are often important agents in assisting in the pollination of some trees such as mangoes, but on balance they are more destructive, as they are voracious feeders consuming quantities of ripe fruits, especially bananas, guavas, mangoes and sapodilla plums. Unable to penetrate the hard rind, they do not feed on citrus fruits.

With very few exceptions, all species of bats confine their feeding and hunting activities to the hours of darkness or dusk. Paradoxically, most insectivorous bats have only weak or restricted vision, but they have an extremely acute sense of hearing and have also developed an extra-sensory method of locating moving or stationary objects, through a system of echo-location, or sonar. This involves emission by the bat, generally while in flight, of very high-frequency short bursts of sound. These sound waves are mostly inaudible to the human ear, but upon striking an object (whether an electric wire cable or a flying moth) will bounce back echoes which the bat is capable of receiving in such a way as to locate the exact direction and even movement of the object. Even fruit bats have been shown to use echo-location but this is comparatively poorly developed in the genus *Pteropus*. The remarkable efficiency of echo-location by bats has been demonstrated by an extraordinary experiment which was carried out on *Myotis* bats artifically blindfolded and released in a darkened room. Vertical piano wires of 0·12 mm diameter were strung from floor to ceiling, at intervals varying from a few metres to as close as 30 cm, this being the wing-spread of the species under experiment. It was found that the bats could easily fly around the room unimpeded and that they manoeuvred successfully between the closely spaced wires. When, however, wires, of 0·12 mm diameter were removed and substituted with much finer wires of 0·07 mm diameter the bats frequently collided with these obstacles, which were apparently too narrow to accurately reflect sound impulses.

Whilst foraging, different bat species employ various hunting techniques. Some bats have a relatively swift and direct flight and concentrate on locating easier-to-find but harder-to-catch large insects such as coleoptera and noctuids. Others have a more controlled but slower zigzag type of flight, able to change direction and speed rapidly in order to seize smaller but weaker flying insect prey. The powerful Sheath-tailed Bats of the genera *Tadarida* and *Taphozous* belong to the former category. Many of the Vespertilionidae bats, such as the Pipistrelles, adopt the latter hunting technique. A large but very homogeneous group comprises the Leafnosed Bats of the family Rhinolophidae. They are characterized by elaborate foliaceous nose appendages, which differ slightly from one species to another. In flight they emit their sonar calls through their nose, keeping their mouths closed. The noseleaf appendages are believed to help in channeling these sound emissions and in locating the direction of returning echoes. The Rhinolophidae are known as the Horseshoe Bats, and several species occur in the subcontinent. They have large ears which can be rotated independently and have a delicately perceptive system of echo-location enabling them to hover over bushes and trees and pick off small crawling insects or spiders.

The family Megadermidae comprises five species of rather large bats which are celebrated for being carnivorous. Two species, *Megaderma lyra* and *M. spasma*, are found in our region. They hunt low over the ground surface or over cliffs and the faces of buildings and can pick off lizards, night-roosting birds and even amphibia on the edge of ponds (see FALSE VAMPIRES).

High-speed photography has enabled some understanding to be achieved of a bat's flight mechanism which is basically very similar to that of a bird, with the downstroke of the arm or wing-beat providing uplift. Bats can glide for short periods but their flight for the most part involves a fairly high expenditure of energy. It has been shown that even their legs are involved in a sort of lateral swimming movement during some phases of flight. When an insect is captured either by the mouth or with a blow from the carpal joint on the forewing, it is often transferred to the inter-femoral membrane where it is held in a cup while the bat further immobilizes it in flight. The prey, if large, may be taken to a convenient roosting place to be consumed, but often it is eaten in flight. Obviously bats must rest a large part of the time and it has been shown that most insectivorous bats can find sufficient food in just one or two hours of foraging in the early part of the night and again just before dawn. The rest of the time they may return to a favourite roost or resting-place, hanging like all bats do, upside down by their hind feet.

During the hours of daylight all bats resort to a particular type of shelter or diurnal roost. In most species this is a darkened and sheltered cavity. Many bats are highly gregarious in their roost and therefore must seek out a suitably sized cave or deserted old building. The Sheath-tailed bats are an example, often sheltering in groups of several hundred strong in some disused darkened cellar or old tomb. Some species are quite tolerant of light in their roost, such as the Fulvous Fruit Bat (*Rousettus leschenaultii*), whereas the False Vampire (*Megaderma*) and Horseshoe Bats (*Rhinolophus*) prefer the furthest and darkest recesses of a cavern or cellar. Some species roost singly and select natural holes in trees, such as *Noctule* and *Barbastella* bats found in the Himalayas. *Scotophilus* bats, such as the Yellow-bellied Desert Bat, need bodily contact within the parameters of their roosting site, i.e. they like to squeeze into crevices. The Molossidae also exhibit this preference. The unique habit of the flying foxes of roosting on the underside of open tree branches has already been described. Very few insectivorous bats roost on open trees, but the Painted Bat (*Kerivoula picta*), with bright orange and black wings does roost among leaves and in south India, where this bat occurs, it roosts among the dried leaves of the *longan* tree whose decaying leaves are not dropped and which turn bright black and orange, perfect camouflage for this uniquely colourful bat. At least one insectivorous bat, *Asellia tridens*, has a preference for subterranean holes or tunnels such as old rodent burrows or natural fissures in the earth.

With many bat species, there is still hardly any information about their breeding biology. It has been observed however that most species generally only bear one young at a time and that there tends to be marked seasonality in the breeding cycle with young all being produced about the same time. This is particularly true of most colonial-roosting species such as the Mouse-tailed Bats, Rhinolophidae and Sheath-tails (*Taphozous* sp.). The breeding season or rut is often marked by an increase in squabbling and activity in the diurnal roosting place. The testes of the male become enlarged and descend into an external scrotal sack. In many species of vespertilionid bats, there is delayed implantation of the fertilized ovum and the female bats form a separate maternal roosting colony away from the males when the young are about to be born in the spring. At the time of parturition it has been observed that the emerging baby is cradled in the cupped interfemoral membrane. After the mother licks its fur dry the baby crawls up her ventral surface until it reaches one of the two teats which are located pectorally, i.e. on the breast. Unlike many other mammals such as insectivores or rodents, no nest is made by the mother so the defenceless newborn young must be constantly carried around by its mother firmly attached to the pectoral teats. In some genera of bats there is a second pair of false teats located in the inguinal region and this offers the advantage of an alternative position for the baby bat to cling to its mother's belly; *Hipposideros*, *Taphozous* and *Megaderma* genera of bats have these false teats. With most bats, after a few days, the young are left behind in the diurnal roost while the mother forages. Young fruit bats, almost the size of their mothers, have however been observed still being carried around during night-time flight and perhaps this is necessary because of their exposed diurnal roost and the fact that speed in foraging for their food is not essential. Despite their small litter size (twins have been recorded in only very few cases in certain insectivorous bats) the long life-span of most bats apparently compensates for this low reproductive potential. In captivity, fruit bats have lived nearly twenty years and insectivorous bats for thirteen years. Furthermore most bats shelter in inaccessible caves or crevices and are relatively safe from natural predators.

Cases have however been regularly recorded of certain birds of prey such as the Hobby (*Falco subbuteo*) and the Red-capped Shaheen (*Falco peregrinus babylonicus*) preying successfully almost exclusively on bats. Snakes also have been recorded preying on tree-roosting bat species. In many parts of Africa the large fruit bats are regularly captured by man and offered for sale as meat in the markets. In India and Pakistan the fat from fruit bats is supposed to possess certain medicinal properties. Man's interaction with bats also includes the exploitation of their accumulated droppings (guano) for fertilizer. The *Tadarida* bats of Mexico roost in underground caverns in numbers totalling millions and the famous Carlsbad caves in New Mexico (U.S.A.) are a classical example, where bat guano has also been harvested as a nitrogen-rich fertilizer.

On the adverse side must be mentioned the disease-carrying role of bats. Most species, especially the colonially roosting ones, are infested with many blood-sucking ecto-parasites and some of these are known to be capable of transmitting rabies as well as trypano-somes (blood parasites). However it is only the blood-sucking bats of the New World which transmit rabies. There is no evidence of Indian bat species acting as carriers of this disease or even of filariasis.

See plate 39 facing p. 544.

BEACH SAND. Deposits of sand that are found on beaches are called beach sand. These sand deposits appear stable under fair-weather conditions but tend to be removed during stormy weather. For example, the beaches of the west coast of India remain without major change during fair-weather months of October to May but undergo extensive alteration during the southwest monsoon months of June to September. The size of the sand grains on the beaches varies from very fine sand (diameter 1/16 to 1/8 mm) to very coarse sand (1–2 mm). Larger size material called granules (2–4 mm), pebbles (4–6 mm) and cobbles (64–256 mm) are found on some beaches. Almost all the sand that is found on the beaches has come from the sea floor. Much of this sand had been previously brought to the ocean by the rivers. Erosion of cliffs also provides sand to the beaches.

Beach sands consist of a wide variety of material. The mineral quartz is very abundant on most beaches. Other minerals such as felspars, mica, magnetite, ilmenite, monazite, and zircon also occur. Beach sands consisting of material not derived from land occur in tropical areas and oceanic islands. On these beaches the sand is composed of calcareous material, usually the skeletons of organisms living in the waters close to the beach. In areas with coral reefs, for example the Lakshadweep Islands, the beach sands consist of pieces of coral and on some beaches in the tropics the sand consists only of shells. Some of the beach sands are of great economic value and the source of many rare minerals. The beaches of the southern parts of Kerala for example have long been the source of valuable deposits of ilmenite, rutile, zircon, monazite and magnetite. On the east coast, the beach sands of Orissa have large ilmenite deposits. The quartz sand found on most beaches is used in the building industry and large quantities are removed for the purpose. The removal of sand from the beaches by artificial means for such purposes in large quantities often upsets the natural stability of the beach.

BEARS (Ursidae). This very distinctive family within the order of Carnivora has three representatives within the Indo-Pakistan subcontinent.

Altogether ten species are now recognized, divided into eight genera and occurring in North and South America, Europe and Asia. No bears exist in Africa, Australasia or the Antarctic continents.

Bears are familiar in appearance to most readers, either from zoos, or even the performing bears of itinerant gypsies or rural circuses. They have relatively huge broad heads for a carnivore with small eyes and a short stumpy tail almost hidden in their body fur. Of a uniform brown or black colour (with the exception of the Polar Bear and the Giant Panda, now classified amongst the Ursidae), bears have sloping hind-quarters, massive thick-set limbs, and they walk on the full sole of their feet with the fore-paws turned markedly inward and all digits armed with long powerful non-retractable claws. Bears are unique amongst carnivores in being largely adapted to an omnivorous diet, subsisting in season upon wild berries and tree-nuts, as well as insects, roots, tubers and any rodents, young birds or other vertebrates which they can catch. Adaptations for feeding include protrusible lips separable from the gums, broad-crowned and tuberculated molars (unlike the shearing carnassial teeth of meat-eaters), and a very developed acute sense of smell. Their eyesight and hearing are relatively poor. Most bears have delayed implantation of the fertilized egg with gestation periods varying from as short as seven months in the non-hibernating Sloth Bear to as long as 250 days in the European and Himalayan Brown Bear. Mating takes place in the summer and the young are born early in the winter, in most species while the mother is hibernating in a cave or den.

Perhaps the most familiar of our region is the Sloth Bear (*Melursus ursinus*). They average smaller and lighter in weight than the other two species but an adult male is nevertheless a powerful animal weighing up to 145 kg and they are greatly respected, if not feared, by all people living in forested areas where they occur.

The Sloth Bear is found in the plains or lowlands of India, being absent from Pakistan, but being one of the notable larger endemic mammals which has reached and colonized Sri Lanka. Avoiding the Himalayas, they are nevertheless forest animals, preferring broken rocky hills clad with dry deciduous jungle, and being absent from open savanna country.

The Sloth Bear has long coarse hair all over its body with a creamy white irregular V pattern on its chest. Its huge claws are ivory white in contrast to the black claws of the Himalayan Black Bear. Their mouth and lips are peculiarly modified to enable the animal to feed on insect larvae by suction. Termites form a very significant part of this bear's diet. Their nostrils can actually be closed at will and their lips are very mobile and protrusible, with a hollowed palate and absence of the inner pair of upper incisors which creates a tunnel effect when the animal sucks or blows. This method is used for cleaning out a shattered termite nest and sucking up the insects from their galleries. As stated above, Sloth Bears do not hibernate in winter, and are believed to form stable or lasting pair-bonds, unlike other bear species.

The Asiatic or Himalayan Black Bear (*Selenarctos thibetanus*) is a more handsome-looking species, whose distribution extends westwards from Iran and across the Himalayan mountain chain up into China, Siberia, Korea and northern Japan. Typically they are forest animals associated with mixed broad-leaved deciduous and coniferous forest, but in the northeast they have extended their range down into the tropical forests of northern Bangladesh and Assam. In Waziristan and Baluchistan a unique subspecies, much smaller in size than the Himalayan population, has adapted to live in precipitous mountain country only with scattered juniper and thorn scrub. Adult males, before hibernation in winter, may attain weights of up to 180 kg and a head-and-body length of 1·6 m. They have a creamy white V pattern on the chest but their fur is much shorter and denser than that of the Sloth Bear, with a noticeable lateral crest of longer hairs down each side of the neck. They are well adapted for tree-climbing with naked soles to their feet and depend heavily for their food on ripe acorns, mulberries, apricots and maize cobs according to season and availability. Occasionally an individual learns to kill domestic stock and becomes largely carnivorous. They hibernate in winter and two cubs are usually born in December. Studies have shown that the 'hibernation' of bears is not the true hibernation of rodent species, as bears maintain their body temperature even during winter sleep and can easily be re-awoken.

In the inner mountain ranges and higher alpine slopes of the Himalaya above the tree-line will be found a smaller paler subspecies of the Holarctic Brown Bear (*Ursus arctos*), which includes the North American Grizzly and the European and Russian Brown bears. The Himalayan subspecies *U. arctos isabellinus* lives largely upon grass and herbs in the short alpine summer, feeding on rowan berries (*Pyrus aucuparia*) and wild currants (*Ribes* sp.) in the autumn, as well as on the starchy roots of wild celery (*Ferula* sp.). Like the Asiatic Black Bear, many individuals learn how to kill sheep, goats and even ponies and become addicted to such a carnivorous diet when the summer grazing flocks provide opportunity. Brown Bears hibernate in winter and the female generally gives birth to two cubs while in her den in December. A captive specimen has been recorded as living up to 47 years, and they are believed to attain sexual maturity at three years of age.

The Malayan Sun Bear *Helarctos malayanus*, widespread in southeastern tropical rain-forest, just comes into our area in the Manipur area and Mizo Hills south of the Brahmaputra river in Assam. More arboreal in living than other bears, it sleeps by day in a rough nest built high up in trees and lives on termites, fruits and

birds. The Polar Bear, *Thalarctos maritimus*, is unique amongst the bears in being the largest species, almost wholly adapted to hunting in an aquatic environment. Confined to the north circumpolar arctic, the males surprisingly do not hibernate in the winter, though the females lie up in a den when producing young. They live largely on seal and fish, and unlike other bears are diurnal not nocturnal in hunting.

See also GIANT PANDA. See plate 33 facing p. 480.

BEE-EATERS are insectivorous birds of the Afro-Asian family Meropidae, with a longish slender, downcurving black bill and pointed wings. Some species are migratory, others resident or local migrants. All the 6 species occurring in the subcontinent are chiefly a brilliant grass-green and vary in size between a sparrow and a dove. The commonest are the little Green Bee-eater (*Merops orientalis*), the Chestnutheaded (*M. leschenaulti*) and the Bluetailed (*M. philippinus*). The largest, Bluebearded Bee-eater (*Nyctyornis athertoni*), is a forest-dwelling species, less common and less social, keeping singly or in pairs. Bees and wasps taken on the wing

© E. Hanumantha Rao

Common green bee-eater × ¹/₂

in swift graceful sallies form their staple diet. Bee-eaters love to perch on telegraph wires and bare branches, and they congregate in large numbers to roost communally in leafy trees. All species nest in self-excavated tunnels in earth-banks and roadside cuttings.

See plate 31 facing p. 432.

BEES. The honey bee is apparently universally known to man through its produce, honey, that the colony produces and stores in substantial quantity. However, apart from the three species of social bees that store honey in their combs (*Apis dorsata*, *Apis cerana* and *Apis florea*), more than 80 to 90 percent of the 20,000 species of bees that exist all over the world are solitary in habit. This vast assemblage of bee diversity is presently classified into some 250 genera arranged in nine families (with about 50 subfamilies). Bees evolved in unison with the flowering plants and several examples of 'co-evolution' (one plant species being intimately associated with a single species of bee which is solely responsible for its cross-pollination) are now facts of nature. Bees are thought to have split off from the sphecoid wasps and turned to a strict nectar-and-pollen diet from their predominantly carnivorous ancestors. This was in the Cretaceous period about 100 million years ago. Bees are sometimes difficult to distinguish from wasps, flies and other insects which mimic them, but the presence of branched hairs on the body and the basal tarsal segment of the hind leg being wider than the succeeding segments, plus the purely flower-feeding diet, enables their identity to be maintained.

Their external features are multifold: some bees may be without any hairs and slender, resembling wasps. Many of them are fuzzy and rotund with a loud buzzing noise while in flight. Most bees are strikingly coloured (this being a warning to their predators), adorned with stripes, spots or patches of red, green, blue, yellow, black, white, orange or brown. In size they may vary from 1·5 mm to over 50 mm in length, and most female bees possess a sting, while all males lack this organ.

As stated above, the great majority of bee species are solitary and each female constructs her nest without cooperation and attends to the functions of worker and queen simultaneously to produce her own brood. Bee nests are very variable in location and form; they may be made in soil, in wood or in pith of stems where a burrow is tunnelled; or it may be a nest made of resin, clay, mud etc., on branches of trees or other protected or exposed situations. The nest may consist of a single tunnel or may have lateral burrows extending from this main tunnel that end in cells. The nests and cells are made of wax secreted from bee bodies or collected from plants, silk produced by the adult bee, dung, mud, resin, pebbles, sand, cut portions of leaves or petals, saliva, plant hair or fibre or secretions from the Dufour's Gland located in the sting. The nests may be positioned or built on trees, rocks, walls, or in existing cavities anywhere, in trees, rocks, holes made by other insects, or in man-made things like gun barrels, keyholes and the like! Most bee species excavate their own nests which may be as deep as 5 m in soil! Nests are constructed solely by females, no males usually helping in nest-building, defence or in provisioning with food. Normally a cell is completed, then provisioned with pollen and honey and sealed before another cell is started on. In some Halictidae however, more than one cell is worked on simultaneously. The mother bee usually does not survive to see her offspring emerge from the cells.

Bees that tend their nests on their own, though there may be more than a thousand nests made by a single bee species in a small area, are called solitary bees. Other species of bees live in small colonies, from two to several females sharing one nest, and a few others live in nests with very largely populated colonies and having queen and worker castes sharing the chores in the nest and outside. Like the cuckoo birds that parasitize other birds' nests to raise their young, the bees too have their 'cuckoo' species, which do not make their own nests but lay eggs in the nests of another bee, whose provisions are fed upon by the cuckoo bee's larvae which also eat the host bee's eggs or larvae. Roughly 10 percent of all bees are inquilines of this sort, whose females have formidably long stings and tough integument as protection against their host bee species.

Bees take their food from flowers, the nectar being their chief source of carbohydrates, and the pollen giving them the proteins they need. Some bees are very specific in the flowers that they visit, depending on only a single species of flowering plant or maybe a genus. These are called monolectic bees. Some others may gather pollen and nectar from a series of related genera (oligolectic) and some more visit flowers of many unrelated plants (polylectic). Though most bees fly by day (diurnal), some venture out to flowers only at night (nocturnal) because the flowers that they depend on open only at night. Those that visit flowers that have their nectaries located deep inside, possess long tongues, and bees also carry pollen in tufts of hair located either on the hind leg or on the underside of their abdomen.

Since bees contain some species that exhibit a highly advanced form of social behaviour, this has attracted a lot of attention. As mentioned earlier, about 80 to 90 percent of all bees are solitary, each female making and provisioning her nests without any assistance. However, some solitary bees have thousands of nests aggregated in an area. The different types of social behaviour can be categorized as follows: *parasocial* bees that form small groups of female adult bees of the same generation that usually share a nest; in these are included communal bees where each female constructs her own cells, provisions them and oviposits in them. The *quasisocial* bees (all female bees lay their own eggs but share the task of cell construction and provisioning) and *semisocial* bees (which have a division of labour—some females laying eggs and others constructing cells and provisioning) are slightly advanced in their social behaviour. Then, the *subsocial* bees have a family group that comprises an adult that feeds her larvae by progressive provisioning. The *eusocial* bee female constructs her own nest and performs all other duties by herself initially; later, when her daughters mature, they remain in their nest and perform the duties of workers while the foundress female becomes the egg-laying queen bee. Lastly, the highest form of social behaviour is exhibited by the *hypersocial* bees (honey bees) where there is a very specialized social arrangement where the queen bee is specialized for egg-laying and cannot, by herself, begin a new nest without the help of her swarm of worker bees.

Now each family of the superfamily Apoidea (bees) will be treated with examples of some Indian species. Firstly, the Colletidae, comprising of the genera *Hylaeus* and *Colletes* in the Indian subcontinent. These construct nests made of silk or transparent, silk-like polymers. *Hylaeus* makes its cells in hollow plant stems or other existing cracks but *Colletes* digs its tunnels in soil. *Hylaeus* carries pollen to the nest in its crop, thus being unique among bees. No Indian species of *Hylaeus* has been studied until now for its biology, but two species of *Colletes* have been studied: *dudgeonii* which tunnels nests into the hard earth of roadside banks, and *nursei*, an important pollinator of cruciferous crops which nests in sloping sand dunes or even in irrigated or dry fields.

The Andrenidae are better known in India and only the genus *Andrena* is found here. This genus of solitary bees makes its nests in soil with uniform, vertical or horizontal secretion-lined cells at the ends of the laterals. *Andrena mollis* is parasitized by the 'cuckoo-bee', *Nomada adusta*, and *A. ilerda* is usually attacked by the sphecid wasp *Philanthus depredator*.

The Halictidae have varied nesting habits and *Halictus ducalis* nests in earthen banks and is parasitized by *Nomada adusta*. Other genera of this family are *Lasioglossum* and *Nomioides* that construct crooked, vertical burrows about 10 inches deep and with as many as ten branch tunnels. *Systropha punjabensis* which is apparently monolectic on *Convolvulus*, makes a vertical burrow in a sandy road embankment with four oval horizontal cells. *Nomia capitata* exhibits quasisocial behaviour and nests in soil.

The biology of the small family, Melittidae, is relatively unknown, but most of these solitary bees nest in soil. *Ctenoplectra chalybea* nests in beetle burrows in wood that have been abandoned, and it collects clay and circularly cuts leaf pieces for its nest.

The next family, the Megachilidae, are the most brightly coloured and abundant family of bees. They are commonly seen around human dwellings where they construct their nests made of mud, resin, leaves or other fibres. They carry pollen beneath their abdomen. The two major genera are *Megachile* (leafcutter bees) that cut leaf pieces with their specially formed mandibles and line their nests with these leaf bits; and the other genus is *Chalicodoma* (mason bees) that make mud or resin nests. *Megachile lanata* is a very common Indian bee and is commonly associated with buildings where the nests made of glutinous clay are located in such odd situations as between paper folds, book spines, handles of teacups, keyholes, gun barrels, fans, rings, etc. This bee is often parasitized by the cuckoo-bee *Coelioxys basalis* and it is an important pollinator of Leguminosae and Cucurbitaceae plant families. Other genera found in India but whose habits are poorly known are *Lithurgus*, *Anthocopa*, *Osmia* and *Anthidium*.

Except for *Xylocopa* (carpenter bees) and *Pithitis*, the biology of species of the family Anthophoridae is little known in India. Most members of this family make their nests in soil; the cells are usually lined with a waxy waterproof secretion. *Amegilla cingulata* is a common bee and burrows into hard earth or in irrigated fields; it is very shy and takes nectar from flowers while hovering in the air. Many genera of the subfamily Nomadinae, especially *Nomada*, are cleptoparasitic (larva feeding on larval food of host). The large black (usually) carpenter bees of the genus *Xylocopa* are a dominant and commonly noticed group of bees and they nest in bamboos, wood, dry branches, tree trunks, etc. Some species (*X. rufa*, *X. tranquebarica*, *X. proximata*) are unusual in being nocturnal or crepuscular. These bees are parasitized by the larvae of a robber-fly of the genus *Hyperechia*, which mimics the bees in adult stage. Many species of *Xylocopa* are no doubt important pollinators of many leguminous crops. The common, small, green bees of the genus *Pithitis* make nests in the pith of stems, and are polylectic. The common species are *P. smaragdula* and *P. binghami*, the former being an effective pollinator of lucerne. Other genera of this family in India are *Ceratina*, *Braunsapis* and *Habropoda*.

© Kumar Ghorpadé

Ceratina *sp. on nigerseed flower* × 2

3 Amphibians and Reptiles

© Romulus Whitaker

© Shekar Dattatri

© Romulus Whitaker

© Shekar Dattatri

Saltwater Crocodile (*Crocodylus porosus*)

Fang of Russell's Viper. The fang is 15 mm long

Painted Frog (*Kaloula pulchra*)

Indian Bullfrog (*Rana tigerina*)

Common Jezebel *(Delias eucharis)*

Blue Tiger *(Danaus limniace)*

Tailed Jay *(Graphium agamemnon)*

Common Tiger *(Danaus genutia)*

Carpenter bee (Xylocopa *sp.*) *boring into bamboo × 1*

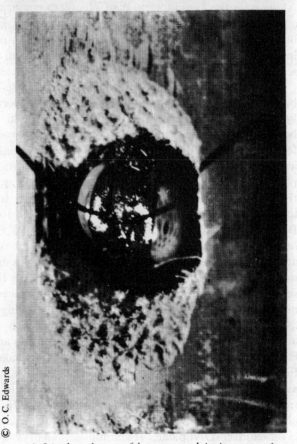

A few days later safely ensconced in its nest × 1

Lastly, the family Apidae contains the social bumble bees, stingless bees and honey bees of the genera *Bombus*, *Trigona* and *Apis* respectively. Bumble bees make their annual cluster of waxy brood cells and honey pots in the litter of abandoned bird or mice nests. Though much is known of their pollinating activities and biology, the Indian species are poorly studied. *Bombus* species are restricted to the higher elevations on the Himalaya where they are insulated from the cold temperatures by their thick hairy bodies. The cuckoo-bee *Psithyrus* parasitizes these bees. The stingless, sweat, or dammar bees (*Trigona*) are mainly tropical in distribution and

their hives are constructed in hollow logs. They are good pollinators and are harmless to man and yield honey, wax and resin and could well be managed as a cottage industry in India.

The honey bees (*Apis*) are by far the dominant and well known bees in our country and a lot of work has been done on them. Our main honey bee is *A. cerana* which replaces the temperate honey bee *A. mellifera* here. The dwarf honey bee, *A. florea*, makes single-combed hives and though honey is meagre, it is supposed to be of medicinal value. *Apis dorsata* or the rock bee is a large species which builds huge combs even in towns, but is yet to be 'domesticated'. These honey bees are the most highly specialized members of the Apoidea with a complex social organization. The queen mates when she ascends on a special flight, accompanied by a swarm of drones. The fertilized queen then either returns to the hive or forms a new colony. She assumes the role of the sole reproductive female. Special queen cells are

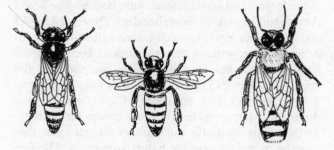

Queen × 1½, worker × 2, and drone × 2 of honey bee

located at the bottom of the hive and larvae fed on the special secretion (royal jelly) become new queens. The larvae in the ordinary cells that are smaller in size are fed on the 'bee bread', and produce the new worker bees. A pheromone secreted by the queen (queen substance) serves to integrate the colony and is eagerly sought by the workers, which when they imbibe it become impotent. Several complex behavioural patterns are exhibited by bees that involve the famous 'bee dances' and also involve several responses to visual stimuli, odours and vibrations. So much is known of honey bees that it would be trite to go into the details here but the interested may find ample reading matter in the many books now available.

Lastly, bees are the major group of insects that are intimately associated with the pollination of crop and wild plants and have evolved in conjunction with several species of flowering plants. However, our bee fauna is so poorly known, that almost any bee family (except for the genus *Apis*) needs a vast amount of study through field observation: we do not even know perhaps 30 to 40 percent of our actual bee fauna so a lot of collection of Indian bees and their description and naming is yet to be done!

See also COMMUNICATION, SOCIAL SYSTEMS.

BEETLES. If insects account for more than three-fourths of all forms of life on earth, just one of the 27 insect orders, the Coleoptera, includes more species than all other animal life put together, and makes up about 40% of all insects. The latest estimates suggest that almost four lakh beetle species have now been named and, undoubtedly, many more still await even to be collected rather than just to be given a name. Whether the hardening of the fore wings into protective case-like structures called elytra is responsible for the success of this insect group is yet open to question. But every amateur insect collector or professional entomologist will agree that beetles are the insects least susceptible to damage, either by hand or when pinned or pointed and preserved in a collection. The earliest fossil record of beetles dates some 250 million years ago, from the Lower Permian, but most Coleoptera probably evolved 50 million years later in the early Mesozoic. Being such a large and diverse group, it is not surprising that some of the largest and smallest insects are beetles. The South American Dynastinae (Scarabaeidae), *Dynastes hercules* and *Megasoma elephas*, which exceed half a foot (16 cm) in length, are perhaps the largest, and beetles of the families Corylophidae and Ptiliidae are some of the smallest; a Mexican species (*Nanosella fungi*) being just 1/50 of an inch (0·4 mm) long! Coleoptera are predominantly terrestrial insects, some groups being secondarily aquatic; with the exception of the open sea, they have colonized all possible habitats on earth. The fore wings hardened into elytra which cover almost all of the abdomen and the membranous hind wings (used in flight), and the mandibulate mouth-parts will separate these insects as adults. Larvae are of varied form, but also strongly mandibulate and with three pairs of thoracic legs (only in a few families are they legless).

Beetles and their larvae occur on all sorts of vegetation, on flowers, moss, grass tussocks, in plant debris, soil, under stones and fallen logs, in plant galls, beneath tree bark, in fungi, in hollow stems and the like. Many species occur also in carrion and in nests of mammals, birds, termites, ants, wasps and bees. Others live in stored products, in dung, in or on the wood of dead, burnt or dying trees and also in caves. The aquatic or semi-aquatic groups live in fresh water, from temporary pools to mountain torrents, brackish water, beneath high-water debris on the seashore, in rock clefts in the intertidal zone or in sand, mud or gravel on margins of streams and rivers. Unlike most other flying insects, beetles are generally rarely seen on wing. However, when necessary they are able to fly strongly (e.g. Cetoniinae) and for long distances (Coccinellidae undertake annual migrations).

A large number of beetles feed on plants, the more common of man's pests being species of the Curculionidae, Chrysomelidae, Elateridae, Scarabaeidae, Bupres-tidae, Dermestidae, Bostrichidae, Nitidulidae, Coccinellidae (Epilachninae), Meloidae, Cerambycidae, Bruchidae, Anthribidae, Brenthidae and Apionidae. Many beetle families are also predacious, the significant ones being Carabidae, Dytiscidae, Gyrinidae, Silphidae, Staphylinidae, Histeridae, Lampyridae, Cantharidae, Cleridae, Meloidae (larvae) and Coccinellidae. The Rhipiphoridae larvae are all parasitic, as are the curious Stylopoidea (treated as a separate order Strepsiptera by some entomologists). The Lampyridae, which are called glow-worms as larvae, feed on snails and earthworms and the larvae of Meloidae attack bee larvae and eggs of grasshoppers. A few beetles of the Leptinidae and Staphylinidae live as ectoparasites of mammals. Along with the Lepidoptera, the Coleoptera comprises a major share of the crop pests of man. However, many beetles also help man as natural enemies of injurious insects and several others are of assistance to him as scavengers.

Many beetles have aggregations of symbiotic yeasts or bacteria in their stomachs each species having a special type of microorganism. These are usually found in beetles that feed on materials like wood, hair, wool, feathers, humus and dry cereals and they help the beetles in their digestion. The 'ambrosia' beetles (scolytids, platypodids) exhibit a highly specialized symbiosis with the ambrosia fungi which they help transmit from tree to tree as they bore into it and feed on the fungus too. Besides some mycetophilid flies, only beetles of the families Lampyridae, Phengodidae and Elateridae are able to produce light. This light produced is remarkable in that no heat is involved, and it helps in bringing the sexes together. Beetles are also great sound-producers, mainly by stridulation which is effected by rubbing a file-like area on one part of the exoskeleton with an adjoining part. Larvae of Lucanidae, Geotrupidae and Passalidae, and adults of Cerambycidae, Scarabaeidae, Endomychidae, Nitidulidae, etc., stridulate. On an average, about 2% of all beetles are adapted to life in fresh water, the chief groups being Amphizoidae, Haliplidae, Noteridae, Dytiscidae, Sphaeriidae, Gyrinidae, Hydraenidae, Dryopidae, Helminthidae, Georyssidae and Hydrophilidae. Subsocial behaviour is exhibited by the dung-rollers (Scarabaeinae) and Geotrupidae.

The chalcidoid wasps parasitize beetle eggs and larvae, the larvae also by Braconidae, Ichneumonidae, Scoliidae, Tachinidae and others. Predators of beetles and their young include vertebrates (especially birds), spiders and aculeate wasps besides many other insect groups. Fish certainly feed on aquatic beetles, and fungi, bacteria, viruses, etc., cause diseases.

The diversity of beetles is now divided into four suborders: Archostemata, Myxophaga, Adephaga and Polyphaga, the last including more than 88% of all species of beetles. More than 150 families are now recognized. The most primitive suborder Archostemata has a single superfamily Cupedoidea with two families. The Cupedidae were apparently common in the Mesozoic, and the species of *Cupes* are the dwindling relics of this ancient family, whose larvae bore into rotting wood. The Myxophaga also have a single superfamily, Sphaerioidea, with four extant families, each of them containing a single genus. One species each of *Hydroscapha* (Hydroscaphidae) and *Sphaerius* (Sphaeriidae) occur in India.

The suborder Adephaga consists of 10 families of mostly predacious beetles. The Rhysodidae are small, elongate, black beetles whose larvae also live in decaying wood; *Rhysodes* is a common genus. The huge family Carabidae, with over 25,000 world species, is popular with amateur collectors and is therefore relatively well known, even in the Indian subcontinent. The species could be broadly separated into three ecological groups: 1) geophiles, which live on the ground surface and are not associated with surface water; 2) hydrophiles, which live at edges of streams or lakes or swamps; and 3) arboricoles, which live on the foliage or trunks of trees. The subfamily Paussinae includes peculiar beetles that live in ant nests and possess beautiful, broad, antennae. The Indian genera include *Protopaussus*, *Pleuropterus*, *Platyrhopalus*, *Euplatyrhopalus* and *Paussus*, among others. The tiger beetles or Cicindelinae are an immensely pretty and admired group. The beetles are often brightly coloured and generally long-legged with elongate antennae. Species of *Cicindela*, *Therates*, *Prothyma*,

Tiger beetle, Cicindela catena × 1½

etc., generally inhabit sandy stream margins, seashores, rivulets and paths in forests and the like. Species of *Collyris*, *Neocollyris* and *Tricondyla* are arboreal and generally metallic blue in colour. The other Carabidae are divided into many more subfamilies and are called ground beetles, many of them helpful to man in keeping injurious insect populations in check. Many carabids are nocturnal and hide under stones, logs, etc., during the day. *Anthia sexguttata* is a very large black, white-spotted carabid that feeds on millipedes; *Pheropsophus*

A carabid, Anthia sexguttata × 1

hilaris is a pretty yellow and black 'bombardier' beetle that keeps scarabaeid root grubs in check and has been used in the Pacific. Other common genera in our subcontinent include *Bembidion*, *Tachys*, *Harpalus*, *Calosoma*, *Chlaenius*, *Amara*, *Scarites*, *Oxylobus* and *Pterostichus*.

The Haliplidae are boat-shaped beetles that occur among aquatic vegetation at the edges of ponds and slow streams; *Haliplus* is a common genus. The water beetles of the families Dytiscidae and Gyrinidae are smooth and boat-shaped, and are a common feature of most inland water expanses. The Dytiscidae have flattened, paddle-like hind legs that they use in swimming, and both adults and larvae occur in both flowing and still water. They feed on a variety of aquatic animals, including molluscs and small fish, and beetles are strongly attracted to artificial light sources at night. *Cybister*, *Laccophilus*, *Dytiscus*, *Hydaticus* and *Copelatus* are some Indian genera. The Gyrinidae are called 'whirligig' beetles and have their fore legs long and modified for grasping. They mostly swim on the water surface in tight circles and some Indian genera include *Gyrinus* and *Dineutes*.

The Polyphaga is the largest suborder of beetles. Five families comprise the Hydrophiloidea, of which the Hydrophilidae is the largest. Adults are mostly aquatic and are oval and smooth, while larvae also live in damp soil, decaying plant matter or in dung. Common genera in the subcontinent are *Hydrobius*, *Sternolophus*, *Berosus*, *Hydrophilus* and *Sphaeridium*. The Histeroidea contain the Histeridae which are flattened, quadrate dull-coloured beetles that are carnivorous as are the larvae. They are found in carrion, dung and decaying vegetable matter where they feed on larvae of other scavenging insects. Some species are myrmecophilous and termitophilous and also inhabit bark of dead trees. *Hister*, *Hololepta*, *Platysoma* and *Saprinus* are some Indian genera. Two other families are known. The Staphylinoidea include ten families, of which the Staphylinidae and Pselaphidae are the most abundant. Most are carnivorous or mould-feeding and myrmecophilous, only a few feed on plants. The rove beetles of the family Staphylinidae are one of the largest groups of beetles, numbering over 27,000 described world species. They are characteristic in having the elytra very short, leaving more than half of the abdomen exposed. They are found in a variety of habitats, e.g., in decaying vegetable matter of all sorts, under stones, at edges of nullahs, in moss, fungi, dung, carrion, seaweed, under bark, in flowers and in nests of termites, Hymenoptera, birds and mammals. *Tachinus*, *Paederus*, *Bledius* and *Quedius* are some common Indian genera. The small reddish or yellowish beetles with short elytra belonging to the Pselaphidae are mostly forest insects, living in leaf litter, under bark of trees, beneath logs, etc., some being found in ant nests; species of *Ctenistes* and *Claviger* occur in India.

The Scarabaeoidea comprise six families: Lucanidae, Passalidae, Trogidae, Acanthoceridae, Geotrupidae and Scarabaeidae, together totalling about 20,000 world species. The larvae are all soft and white to greyish white, and usually of the curl-grub type and feeding on roots or soil organic matter underground. They are also associated with rotting wood, fungi, carcasses or dung. The adults also feed on the same food but also on leaves and flower-nectar; the biology is very variable. Several species assume pest proportions and, like dung-rollers that have been introduced from Africa to Australia to consume dung of introduced cattle and prevent flies from using the dung to breed unchecked, many other species are beneficial to man. The largish Lucanidae beetles come to light at night and their larvae feed on fallen logs in forests, *Lucanus* and *Hemisodorcus* occur in India. Passalidae are dark, flat large beetles that are mainly tropical and both adults and larvae feed on decaying wood. Geotrupidae are stout, strongly convex reddish brown beetles which feed on dung: *Bolbocaras* occurs in India. The Scarabaeidae is a very large and diverse family of usually stout beetles of sombre colours. Larvae live concealed and feed on roots, dung or decaying vegetable matter. Some are found in association with termites. Several subfamilies are recognized. Species of *Aphodius* are dung feeders and belong to Aphodiinae. The dung-rollers belong to the Scarabaeinae (Coprinae) and are a fascinating group, feeding almost exclusively on faeces of various animals, some species being specific to dung of a particular animal. The ancient Egyptians revered what they called a 'Sacred Scarab'. Both male and female together roll a dung ball, after laying an egg inside, and tuck it away in a tunnel in soil for the

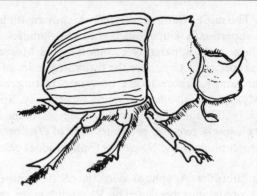

Dung beetle × 3

young to complete development. *Heliocopris bucephalus* is a huge Indian species which specializes on elephant dung. *Copris* and *Onthophagus* are the other common Indian genera. The Melolonthinae, called chafers and root-grubs, is another large group with many economically important species of the genera *Anomala*, *Adoretus*, *Holotrichia* and *Melolonthus* whose adults eat leaves in swarms at night and whose larvae feed on roots. The Dynastinae include some very large brown or black beetles with ornamented thorax and head; *Xylotrupes gideon* and *Oryctes rhinoceros* are injurious to coconuts. The pretty Cetoniinae include some of the most attractive beetles; *Anatona*, *Oxycetonia*, *Protaetia*, *Clinteria* and *Chiloloba* being abundant in India. Adults are sometimes injurious to crops, but most larvae feed

© Kumar Ghorpadé

Cetoniid beetle, Oxycetonia versicolor *on flower* × 3

Cetoniid beetle feeding on cashew peduncle × 1¼

mainly on humus in soil. A species of *Spilophorus* feeds on membracid Hemiptera in India, being a rare case of a carnivorous Scarabaeidae.

The jewel beetles of the family **Buprestidae** are some of the most prized insects for a collector and are abundant in the humid tropics. Many are found on flowering trees in some numbers and their larvae bore and feed on the wood or roots of trees, a few species being of economic importance. Common Indian genera include *Agrilus*, *Sphenoptera*, *Belionota*, *Trachys* and *Spiloptera*. Click beetles of the family Elateridae are another large

A jewel beetle (Buprestidae) *laying eggs on tree trunk* × 1¼

and important group. Larvae that feed on plant roots are called wire-worms and are sometimes pests of crops, but others also feed on larvae of other insects. Adults are fairly large and attractively patterned, and most species have the posterior corners of the pronotum produced into noticeable spines. Species of *Alaus*, *Camposternus*, *Melanotus*, *Drasterius* and *Cardiophorus* are frequently collected in the subcontinent. The widely known 'fireflies' and glow-worms belong to the family Lampyridae which has almost 2000 world species. Adult males and females have a luminescent organ on the lower portion of the abdomen which emits flashes of light at night and some species do this in some numbers with synchronized frequency. The larvae are mostly predacious on gastropod molluscs; common species are *Lampyris* and *Lamprophorus*. The small beetles of the family Dermestidae have larvae that feed on dry animal material and some are pests of stored products, like *Trogoderma granarium*; and species of *Anthrenus* are injurious to household material such as carpets, blankets and the like. Larvae of Bostrichidae are wood-borers and adults are very similar in appearance to the scolytids. *Rhyzopertha dominica* is an almost cosmopolitan stored-grain pest and other genera in the subcontinent include *Dinoderus*, *Heterobostrychus* and *Sinoxylon*. The Cleridae is another abundant family of smallish, elongate beetles which are predacious on other wood-associated insects. *Ommadius*, *Necrobia*, *Tillus*, *Cladiscus*, *Opilo* and *Dasyceroelerus* are some Indian genera.

The superfamily Cucujoidea is the largest and includes more than 50 families with almost 50,000 species. The Nitidulidae are small brown or black beetles that are frequently found on flowers. *Cybocephalus* species are predatory on coccids and other frequent genera are *Meligethes*, *Carpophilus*, *Nitidulus*, *Cametis* and *Brachypeplus*. The Erotylidae are associated with fungi and frequently patterned bright yellow and black. Common genera in India are *Megalodacne*, *Triplatoma*, *Episcapha*, *Anadastus* and *Macromelea*. The Coccinellidae is the second largest family in the Cucujoidea with over 5000 world species. They are usually called ladybirds or lady beetles and are generally brightly coloured. The majority are beneficial to man, feeding on many homopterous insects, many of which are pests. The genus *Stethorus* is predacious on mites and the Epilachninae are phytophagous and mainly tropical in distribution. The more important genera include *Coccinella*, *Pseudaspidimerus*, *Scymnus*, *Epilachna*, *Brumoides*, *Chilocorus*, *Jauravia*, *Menochilus*, *Micraspis*, *Pharoscymnus*, *Hyperaspis*, *Rodolia*, *Sticholotis*, *Nephus*, *Synia*, *Telsimia* and *Calvia*. *Illeis* includes beetles that feed on mildew fungi and *Synonycha* and *Anisolemnia* are some of the largest lady beetles. The Endomychidae are similar beetles, usually found in fungi on trees; *Eumorphus*, *Eucteanus*, *Amphisternus*, *Stenotarsus*, *Ancylopus* and *Trochoides*

Larvae of coccinellid beetle feeding on hard scales × 4

are some Indian genera. The Tenebrionidae, with more than 15,000 world species, is the largest cucujoid family, and one very diverse in extent. Many species are abundant in the arid parts and occur in a wide variety of habitats. Many species are found under stones or logs, in rotten wood or in fungi and also in nests of ants or termites. A few species attack stored products and some Indian genera include *Blaps, Setenis, Cossyphus, Gonocephalum, Opatroides, Seleron, Tribolium* and *Hyprops*. The Mordellidae include humped and smoothly tapered species that are found on flowers as adults and larvae being either predatory or parasitic or found on plants; *Mordella, Mordellistena* and *Glipa* are some of our genera. The Rhipiphoridae are also similar beetles that have unique endoparasitic larvae that attack scolioid and apoid Hymenoptera and Blattodea. These beetles are very similar to the Stylopoidea (Strepsiptera) and some local genera are *Emenadia, Symbius* and *Macrosiagon*. The blister beetles containing the toxic cantharidin belong to the Meloidae and common Indian genera are *Mylabris, Lytta, Psalydolytta* and *Epicauta*. These beetles are very colourful and while adults feed on plants and may be harmful, the larvae, like those of the Rhipiphoridae, exhibit heteromorphosis, and in one stage are predacious on eggs of bees or grasshoppers. The Anthicidae, or ant-beetles, are small ant-like beetles that are found on flowers and on ground in arid area. *Formicomus* and *Anthicus* are some of the Indian genera.

The superfamily Chrysomeloidea is also very large and as large as the Cucujoidea. The longicorn beetles of the family Cerambycidae are generally large and

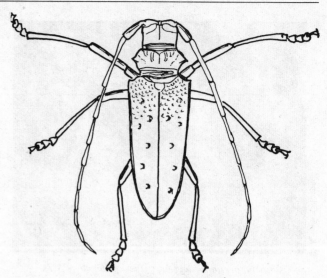

Cerambycid beetle × 1¼

attractively coloured and their larvae are borers in wood. Other larvae live in stems of herbs or even feed on plant roots. Many species attack cultivated and forest plants and some Indian genera include *Batocera, Chloridolum, Chelidonium, Xylotrechus, Sthenias, Apomecyna, Olenecamptus, Monohammus* and *Acanthophorus*. The seed weevils of the family Bruchidae are small brownish insects whose larvae live in seeds of all sorts of plants, mainly of Leguminosae and Palmaceae. *Bruchus, Callosobruchus* and *Caryedon* are some common Indian genera. The large family Chrysomelidae are called leaf

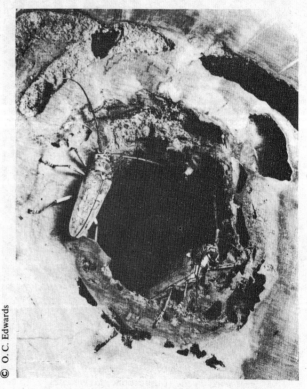

Longicorn beetles, Batocera rufomacula × ½

Larvae, pupae and adult of tortoise beetle, Aspidomorpha miliaris × 3

beetles and a number of them are crop pests. Several subfamilies are recognized and most species are brightly coloured. The majority are foliage feeders though some bore stems, mine leaves or feed on flowers. Some of the Indian genera include *Sagra, Eumolpus, Leptinotarsa, Cryptocephalus, Hispa, Aspidomorpha, Scelodonta, Rhapidopalpa, Monolepta, Lema, Longitarsus, Phyllotreta, Haltica, Chirida* and *Leptispa*.

The Curculionoidea which contain the weevils are the largest superfamily of the Coleoptera and very diverse in form and habit. The antennae are usually elbowed and the beetles are very tough with a rostrum, and all species are plant-feeding. Nine families are known of which the Scolytidae and Platypodidae are now considered related to weevils. These are called bark beetles and many are crop pests or even household pests of furniture. *Xyleborus, Hypothenemus* and *Stephanoderes* are scolytids, and *Platypus* is a platypodid genus found in our subcontinent. The Apionidae are beetles with a long, slender and curved rostrum; *Apion* and *Cylas* are some familiar genera. The Curculionidae have more than 60,000 world species with apodous (legless) larvae that usually feed in a concealed situation and which pupate within the food plant or in the ground. More than 100 subfamilies have been proposed and the classification of this family is in need of thorough study and revision. Many of man's beetle pests of crops belong to this family and, whereas the larvae may feed on roots underground or tunnel into stems, the adults are exposed feeders, attacking almost every part of the plant. A short synopsis of the more frequent genera in the subcontinent would include *Bagous, Myllocerus, Episomus, Baris, Hypolixus, Alcidodes, Cosmopolites, Rhynchophorus, Pempherulus, Ceuthorhynchus, Alcides, Echinocnemus* and *Balaninus*.

The superfamily Stylopoidea includes some 300 species of highly specialized beetles very similar to the Rhipiphoridae. Some entomologists prefer to recognize a separate order, the Strepsiptera, for these peculiar insects whose adults lack a trochanter in their hind legs. Males are small, black or brown, with flabellate antennae and a distinctively shaped head. They are free-living with a pair of broad hind wings, and the fore pair reduced to haltere-like structures, differing from most Diptera (flies) whose hind wings are modified into halteres. Females of the primitive family Mengeidae leave their hosts and move about, but those of the other four families protrude from the integument of their hosts enclosed in their puparia. These insects are endoparasites of Thysanura, Blattodea, Mantodea, Orthoptera, Hemiptera, Diptera and aculeate Hymenoptera. Hosts of these stylops that are parasitized are said to be 'stylopized'. In India, *Pyrilloxenos compactus* on Cicadellidae, *Indoxenos membraciphaga* on Membracidae and *Tridactylophagus mysorensis* on Gryllidae are some of the species reared from hosts.

BENEFICIAL INSECTS. In INSECT PESTS we look at those insects that compete with man for the earthly resources. Here we have a look at those that we term 'beneficial' to our livelihood. It is only when man steps off his self-made 'throne' and takes a real hard, objective look at his fantasy of being the 'ruler of the world' and the most exalted creation of God, that he realizes that all he really is, is a bipedal, naked ape who has evolved, circumstantially, into the commercial human being with the controversial scientific name *Homo sapiens* (=Wise Man)!

Reality in nature (of which man is an integral part) teaches us the fact that while insects do not need man for their survival, we would face certain extinction if insects were to be removed from our ecosystem. We have for so long been under the warped propaganda that insects are all harmful or destructive, that this article on beneficial insects is expected to surprise but correctly inform the layman, especially one who lives in the artificial world of the concrete jungle. I have classified, in ten ways, the uses of insects to man, from the most important to the least essential, for his survival.

Pollination, not only of man's commercial plants, but also of many wild plants that make up the local flora

BENEFICIAL INSECTS

The dwarf bee, Apis florea, *an important pollinator of many wild and cultivated plants* × 2

which are important components of the ecosystem, is perhaps the most important beneficial act that the insects perform in man's favour. Though wind pollination takes care of most of our cereal crops (like wheat and barley), millets (like jowar and bajra), besides rice, which is the most important of our food grains, insects are responsible for many, if not most, of our fruit, vegetable, ornamental and field crops setting fruit after they pollinate the flowers.

In general, flowers that do not depend on insects for pollination are small and inconspicuous, with poorly developed petals, unisexual flowers, no nectar, dry and light pollen, and brush-like stigmas. Those that do need insects to pollinate them have colourful, showy and large corollas, a marked odour, sticky pollen grains and stigmas, and nectaries that secrete a sweet liquid that serves to attract insects. The actual structure of flowers is monitored to attract certain species of insects that are specific pollinators of particular plants. Plants do not possess beautiful flowers to delight man's senses, but to attract the insects for a very important purpose, pollination. In fact, some plant species have 'co-evolved'

with particular insect species to such a degree that both the plant and the insect need one another to ensure each other's survival (see also BEES).

Some notable examples of insect and flower mutualism are cited here which would give readers an idea of the importance of this reciprocal relationship in the environment. A small wasp, *Blastophaga psenes,* is solely responsible for pollination of the edible Smyrna Fig, native of Asia Minor, and for giving the fruits that form the quality and flavour that have made them of commercial use to man. Similarly, species of *Nomia* and *Megachile,* which are wild bees, are important pollinators of the lucerne (or alfalfa) plant. Though honeybees (*Apis* spp.) are used, in conjunction with their ability to provide honey, by man, for pollinating many of his vegetable and orchard crops, like the example of the wild bees on lucerne that are significantly more efficient pollinators for that crop, many other cultivated plants need insects other than honeybees for effective pollination. Unfortunately, in the Indian subcontinent, we have a very limited idea of which of these wild bees and other insects are actual pollinators of our commercial crops. In fact, we are even ignorant of more than 40 percent of our fauna of these insects.

The next most important manner in which insects do us a great amount of good, is by fighting among themselves. The weapons that are used by man to keep in check those insect species that are his enemies, mainly the poisonous chemicals that he is compelled to employ (even if they are hazardous to him and his environment in the long run), are insignificant tools compared to the multitudinous hordes of his insect friends that kill and feed upon his enemies as a daily chore! For our convenience we can group these insect friends of ours into those that are parasites (or parasitoids) and those that are predators. Parasitoids are insects that live in or on bodies of their hosts which they may kill ultimately in the process of at least one stage of their life. Either one

Braconid wasp ovipositing in moth larva inside stem × 10

Parasitized aphid mummies by braconid wasp × 4

© C.I.B.C., Bangalore

A tachinid fly, its puparia and pupa of danaid butterfly host × 3

or many individuals of the parasite may complete development in or on the body of their host, and even these may have other parasite individuals of another species that, as secondary parasites, feed on these primary parasites! Then again, some parasites like animal lice may be permanent parasites, spending all their time and life-stages on the host, or, like the bot-flies, may be transitory parasites, where the adult fly is free-living whereas the larvae live inside the host body. The bedbug is an intermittent parasite, as is the mosquito, since it approaches its host only for feeding on blood and spends other times away from it. Parasites like some animal lice which live only on one species of host all the time, are called obligatory parasites. Those like fleas that can live free of their hosts for part of the time and also shift from one individual host to another, are called facultative parasites. Monophagous parasites, of which there are a majority, are restricted to only one species of host, while polyphagous ones are able to develop on a few closely related host species. Though beetles are the largest insect order today, it is estimated that the order Hymenoptera (Ants, Bees and Wasps)

is actually the one with the largest number of existing species, especially when all the minute parasitic wasps are collected or reared.

The predacious insects belong mostly to the orders Coleoptera (beetles), Diptera (flies), Neuroptera (lacewings & antlions), Odonata (dragonflies & damselflies), Mantodea (mantids), Mecoptera (scorpionflies), some Hymenoptera (ants, bees & wasps) and Hemiptera (bugs). Unlike parasites, insect predators kill and eat part of or the entire prey or suck out its internal fluids, and one predator needs more than one prey individual to complete its life-cycle (if a larva/nymph). There can be no doubt that the greatest single factor that prevents plant-feeding insects from out-competing and overwhelming the rest of the living world is that they are attacked and fed upon by other insects. As a hypothetical example, if just one pair of house-flies (*Musca domestica*) were able to reproduce normally and (like modern man in the developed countries) were able to resist disease and combat their natural enemies, they would, in just five or six months, cover the entire planet Earth 50 feet high with their progeny! Needless

© Kumar Ghorpadé

Larvae of Apanteles *wasp emerging from* Papilio demoleus *caterpillar to pupate under it × 1¼*

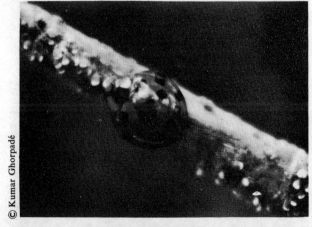

© Kumar Ghorpadé

Giant lady beetle, Synonycha grandis *feeding on woolly bamboo aphid × 1½*

Carabid beetles (Pheropsophus hilaris) *attacking grub of coconut rhinoceros beetle* (Oryctes rhinoceros) × 1¾

to say, the balance that exists in nature (without man's interference) never allows this to happen.

The third way that insects are useful to us, is in their value to humans as food, direct or indirect. Owing to their huge numbers, though of small size, insects probably exceed all other animal matter (biomass) on earth in weight on land. The birds alone probably depend on insects for two-thirds of their food requirements. Many of our commercial fish species subsist largely on aquatic insects. Many game birds, including turkeys and fowl (not the mass-produced broilers or layers that are now being churned out to supply the burgeoning human population, which seems now forced to accept quantity over quality, without any regard for the inevitable grim consequences) feed naturally upon insects and can be raised almost exclusively on such a diet under proper conditions. Many animals, especially those like pigs (meat) and fur animals, eat white grubs and other insects. Man has survived on insect food in his early evolutionary history, and even now some of our primitive and tribal races delight their palate by eating insects such as termites, grasshoppers, crickets, beetles, caterpillars, ants, bee larvae, aquatic bugs and many others.

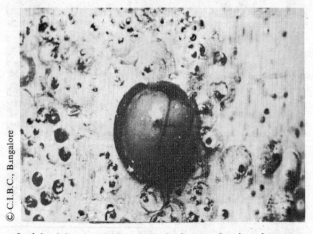

Ladybird beetle (Chilocorus) *feeding on hard scales* × 7

The role of many insects that feed on plants that are harmful to man's interests (weeds) is also something that we must be thankful for. On the other hand, man must be conscious that owing to his shortsighted policy of what he calls 'clean cultivation', resulting in his weeding or destroying by poisonous chemicals what he considers undesirable plants, many insects that were happy and content to feed on these non-commercial, wild plants, have been forced to attack related plants that man crops, and have thus become serious 'pests'. Man's 'evolution' has been from one who was in harmony with nature and concerned with his 'bare necessities' to a greedy and commercially motivated individual who has become completely out of tune with the natural world (that he is a part of and cannot run away from) and confined, miserably, to artificial 'concrete jungles' which emit great amounts of noxious gases and pollute even his drinking water.

Another important manner in which insects help mankind, is by millions of its individuals, that live below ground (as immatures or adults, or both) helping to improve the physical condition of the soil and in promoting its fertility. Insects help to break up rock particles and expose them to the action of water and other weathering influences by bringing them up to the soil surface. The numerous underground tunnels made by insects facilitate the circulation of much-needed air into soil that is essential for the good health of plants. Insects are also of tremendous importance in adding valuable organic matter and humus to soil. Even their dead carcasses accumulating on the soil surface are a great source of natural fertilizers to plants. Their excreta, in chemical content and in mere volume, far exceeds anything that man or any of the larger animals, in unison, can incorporate into soil.

Some of the most interesting and helpful insects are those that dutifully perform their role as scavengers of nature's 'waste' (nothing is wasted in the natural world; it is only man's artificial 'modern' products that are unassimilable by nature's scavengers). Their yeoman service is twofold: first, they remove from the surface of the earth the dead and decomposing bodies of plants and animals, converting them into simpler and more assimilable compounds, removing what otherwise would be a health menace. Second, they convert the dead plants and animals into simpler substances that could then be reused by growing plants as food. Man may find these insects and other such scavenging animals repulsive, but he ought to realize that without them the world would certainly be a cesspool.

Insects and their products have limited use in human medicine as well. In the seventeenth century, almost every insect was supposed to be of medicinal value. Many of these beliefs have now been found to be incorrect and founded on superstition. However, some of their products (like stings of honeybees which have remedial value for rheumatism and arthritis and extracts from their products like the 'royal jelly' of honeybees) are used to some extent as medicine; maggots of certain flies, reared aseptically, have also been used in the treatment of soldiers' wounds. The best-known blister beetle, *Lytta vesicatoria* (the so-called 'Spanish Fly'), occurs in abundance in France and Spain, and has a chemical known as cantharidin in its blood and internal organs. In early days, it was used as an external local irritant or blister, but nowadays it has a place in the internal treatment of certain diseases of the urogenital system and in animal breeding.

Insects have also taught man a great many things and have helped him solve some of the most puzzling problems in natural phenomena. They have also led the way to some of man's remarkable inventions. The ease of handling them, their rapidity of multiplication, great variability, and low cost of maintenance and rearing, have made insects the ideal experimental animals for the study of physiology, biochemistry and ecology. The foundations of modern genetics have been derived from studies of the lesser fruit-fly of the genus *Drosophila*. Studies of variation in populations of a single species, geographical distribution, and the relation of colour and pattern to ecological habitat or other surroundings have been greatly advanced through the study of insects, as has the geological history of the earth (continental drift) and a better picture of the planet's and its living inhabitants' evolution. Principles of polyembryony and parthenogenesis have also been discovered by the study of insects. The behaviour and psychology of higher animals (including man) have been illuminated by a study of the reactions of insects such as the honeybee, and valuable lessons in sociobiology for us have been deduced from a study of the economy of social insects. Insects are also used as an index to stream pollution and such important factors in the conservation of our natural resources. Cockroaches and stored-product insects have been the subjects of many nutritional studies, and insects like the house-fly and mosquito larvae are used frequently for the bioassay of extremely small amounts of insecticide residues on fruits and vegetables.

Many insect products are used by man as his articles of commerce. Perhaps the most important insect of value for its product is the silkworm (*Bombyx mori*). The saliva or spittle of the caterpillar of this moth (truly 'domesticated' by man for over 35 centuries, and now unable to survive in nature without his protection and assistance) is what we use as silk. Sericulture, or the commercial production of silk, is an important industry in China, India, Japan and western Europe. The caterpillar uses its spittle to spin its cocoon within which it pupates. For just one pound of raw silk, over 3000 caterpillars pupating in their cocoons must be sacrificed!

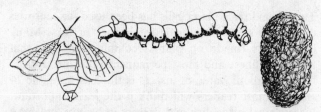

Silkworm (*Bombyx mori*) moth, caterpillar and cocoon × 1½

Several other species of *Bombyx* that feed on mulberry as larvae have also been domesticated. Wild silk is produced by at least 30 species of Asian moths of the family Saturniidae, India possessing several of these, which, however, are of limited value as silk producers. The other insect that man has domesticated, for gain, is the honeybee, which gives him honey and beeswax, besides helping him to pollinate some of his crops too. The bees obtain nectar from flowers, mix it with their saliva, swallow it and store it in their crops (honey sacs) and bring it to their hive. Here the nectar is further masticated with saliva, stored in cells in the comb and these then fanned by their wings to remove a large proportion of water in this honey. When the honey is 'ripened', the cell is capped with wax. Beeswax is used by bees to construct the cells within which they store honey, and it is made by the bee's secretion which is produced in wax glands on the underside of the bee's abdomen.

Some other insects that yield man his necessary resources, are the coccids (see BUGS). Shellac is produced by the lac insect (*Kerria lacca*) that lives on some forest trees in India and Burma. It secretes this substance called lac over its body to protect this sedentary coccid

LEFT: Male lac insect × 5;   RIGHT: Female lac insects with encrustations of lac over them × 4

from adverse weather conditions and its natural enemies. A related insect of the same superfamily produces cochineal, a beautiful carmine-red pigment or dye. This cochineal is the dried, pulverized bodies of the mealybug, *Dactylopius indicus* (in India) and *D. coccus* (native of Mexico). The light-producing secretion of the giant firefly (a beetle of the family Lampyridae) of the tropics is used in minor ways for illumination and may point the way for the synthesis of a substance giving brilliant light with almost no accompanying heat. Many aquatic insects are widely used as fish bait by anglers, and the best artificial 'flies' are modelled after insects.

Tannic acid from insect galls has been used for centuries to tan the skins of animals for leather or furs. Many insect galls (made on plants) contain materials that make the finest and most permanent inks and dyes.

The aesthetic value of insects is the least tangible of all, and most readers will think it the least important. Insects rival birds, fish and flowers in beauty and have been the inspiration of many an artist. Insects are also widely used as ornaments. Their shapes, colours, and patterns serve as models for artists, florists, milliners and decorators. The more highly coloured and striking forms are much used as ornaments in trays, pins, rings, necklaces, and other jewelry. The collector's pride has been enhanced by many beautiful insects, especially beetles, moths and butterflies. Even insect songs have been found extremely interesting by man, and hundreds of poems have been written on them. Gambling by using crickets or mantids (and spiders), trained for fighting, has a great following in China and other oriental countries.

Perhaps the most useful way that insects have helped and will help man is by affording him entertainment and education by watching, collecting and studying them. Throughout their 300 million years of survival and evolution on earth, insects have most certainly traversed many more paths than man (only 2-3 million years old) has managed to do up until now. Learning from insects would give man ideas and methods to counter the forces of nature and its living inhabitants, including man himself.

BHABAR. The word 'bhabar' literally means porous. The name is thus directly derived from the permeable nature of the bhabar lands. It is an elongated belt flanking the Siwalik hills of the Himalaya extending from 77° E to 88° E longitude. The sequential transition from the Siwaliks down to the plains is *bhabar*, *terai* and *bhur*.

Bhabar land is made up of stones and soil that have fallen or been washed down from the mountains above. Due to the porous nature of the surface, the streams running down the mountains are lost in the bhabar and seep into the marshy terai. The river-slope in bhabar region is about 20 to 25 metres per kilometre while it abruptly eases to about 3 mpkm in the terai area.

The general width of this belt varies between 12 and 25 km. It is widest in the west (about 32 km) and in the east it shrinks to about 3 km. In the earliest times perhaps this belt was full of dense forest. It still has a rich flora possessing some of the finest sal forests, particularly in its upper region. With expanding colonization from the south, cultivators are trying to inhabit the area in scattered patches.

See plate 29 facing p. 416.

BHARAL. The bharal or blue sheep is essentially an animal of the Tibetan Plateau. But along the southern limit of its distribution it has penetrated the great Himalaya around such peaks as Dhaulagiri in Nepal and Nanda Devi and Trisul in India. Preferring grassy pastures near cliffs, on which it seeks safety when pursued by snow leopard and wolf, the bharal generally remains above timber-line, from 3500 m upward to the limit of vegetation at 5500 m.

A male bharal with his slate-blue pelage and black chest is a handsome creature, weighing about 60 kg, quite unlike the small, drab females. His body is stout, designed for climbing, and his massive horns sweep out and back in the manner of sheep, as the animal's English name implies. In fact, when discovered in 1833, the bharal was placed in the genus *Ovis*. But it also has certain goat-like traits—a broad, flat tail, no prominent glands between the hooves, to name just two. Because of this mixture of traits, the bharal was then placed into its own genus, *Pseudois*, with one species, *nayaur*, and considered to be an aberrant goat with sheep-like affinities.

Bharal are social, living in herds of about 5 to 20 with as many as 60 or more. Males may be solitary or in bachelor herds when not in rut. For much of the year, bharal are placid, feeding on grasses, herbs, and leaves of low shrubs, but during the rut they become active, males then striving for dominance which in turn gives them priority to estrous females. Bharal, like soldiers, carry their rank symbols with them—impressive horns, large body, conspicuous coat—and these help males to evaluate each other's fighting potential. Smaller animals usually accept their subordinate positions. However, sometimes a fight is necessary to settle a dominance problem. Both males may rear bolt upright on their hind legs and run at each other to bash horns with a loud crack. Goats fight like bharal, whereas sheep rear up only partially or not at all. Behaviour patterns such as this can reveal taxonomic relationships between animals. After a fight, a low-ranking male may use a unique gesture to express friendliness: he rubs his face on the rump of the dominant animal.

The rut begins in late November, at the onset of winter, and lasts several weeks. Most mating is done by dominant males, those 5 years old and older. Such a male follows a female closely, sometimes trailed by other males who try to horn in. Competition may be fierce, the male with the largest horns winning. After a gestation period of about 160 days, one young is born.

G.B.S.

See UNGULATES.

BILLFISHES are large swift, pelagic, circumtropical, oceanic fishes belonging to the suborder Scombridae. They are called billfishes or swordfish because the upper

Swordfish, *Xiphias gladius* × ¹/₅₀

jaw extends forward in the form of a long, pointed snout as a bill. This is a flattened blade in swordfish (*Xiphias gladius*), which may be as long as $1\frac{1}{2}$ m in a $4\frac{1}{2}$ m fish, but is round in cross-section in the sailfish and marlins.

The Sailfish (*Istiophorus gladius*) has a huge dorsal fin, normally retracted into a deep groove on the back when the fish is swimming fast. Off the west coast they are accidentally caught in gill nets along with the shoals of mackerel upon which they prey.

Among marlins, the Blue Marlin (*Makaira nigricans*) is the largest, weighing up to 900 kg. The Black Marlin (*Istiompax indicus*) can be distinguished by its rigid pectoral fins, which cannot be folded without breaking the joints. In Indian seas there is also a third species—the Striped Marlin (*Tetrapturus brevirostris*).

All billfishes are noted for their propensity to attack wooden boats, ramming their snouts into the timber. But the fishermen of Tuticorin have a unique way of handling them. After gaffing a fish, they immediately hold its snout firmly, which action immobilizes the fish. Swordfish are the delight of big-game anglers for their capacity to put up a stiff fight after being hooked. Commercially they are caught by trolling lines or with long lines, which latter may be several km long with 2000 hooks. Their body is well adapted for rapid movement in the water, being firm, slightly laterally compressed and streamlined, offering least resistance.

BIOLOGICAL CONTROL. Biological control of pests (both animal and plant) is that method which relies on the activities of other living organisms to limit the numbers and density of the pests. In the case of animal pests, often insect pests, these controlling agents are usually, though not always, parasitic or predaceous insects or diseases. With plant pests (weeds), either plant-feeding insects or possibly plant diseases can be effective in checking their density and spread.

The most homely example of biological control is, perhaps, that of the cat keeping the numbers of rats and mice down, but other common examples are legion —involving spiders, ladybird beetles on aphids, pre-

datory mammals and birds, and the less obvious examples of some wasps provisioning their nests with paralysed caterpillars, parasitic insects which lay their eggs on or in their host (another insect species) where their progeny develop, ultimately killing their host and thus reducing its population, often in a highly effective manner. Pertinent research can usually pinpoint a species which will effectively limit the numbers of another. Similarly, with plants, there are usually insects which feed on the various plant parts, either destroying it or reducing its competitiveness in relation to other plant species.

This method is particularly applicable to EXOTICS, species which are usually introduced into a new area without the natural enemies which keep them in check in their native homes, and thus acquire greater competitive vigour and become pests. Organisms attacking the pest species are studied in its native region and the most effective ones are introduced into the area where it has become a pest. With weed species, it is essential, prior to introduction, to ensure that the natural enemies to be introduced are strictly host-specific and will not attack other valuable plants in the region of introduction. If these introduced natural enemies become established, they keep the pest species under check thereafter so that their numbers do not increase to damaging proportions as before.

Whatever the degree of control of the pest achieved by this method, and it is really unpredictable *a priori*, it is usually self-propagating without any further effort (and cost), and does not upset the environment except in so far as it reduces the disturbance caused by the exotic that became a pest.

There are some well known examples of biological control in India but with only a few exotic insects having become pests here, these are not too many. The control of the prickly pear cactus (*Opuntia* spp.), of South American origin, by means of an introduced, highly specific mealy-bug, *Dactylopius* sp., was carried out by about 1850 and has continued to suppress the cactus *O. vulgaris* effectively ever since. *O. dillenii* was controlled in southern India by means of the North American *D. opuntiae* introduced from Sri Lanka in 1926, and this subsequently spread northward, controlling the cactus in Delhi, Rajasthan, etc. The cottony-cushion scale, *Icerya purchasi*, a pest, particularly in southern India, of fruit trees and roses, was introduced around 1928 and has been completely controlled biologically by the subsequent introduction of a predacious ladybird beetle, *Rodolia cardinalis*, originating, like *Icerya*, in Australia. With the biological control of weeds, there are several other examples of exotic species which have become pests in India and which have been successfully controlled biologically elsewhere or where such prospects of control are promising. In the former category are *Lantana camara*, against which the bug *Teleonemia scrupulosa* and the BEETLES *Uroplata girardi* and *Octotomma scabripennis* have already been established in India, but there are also several other BENEFICIAL INSECT species available. In the latter category, there are the water weeds, *Eichhornia crassipes* and *Salvinia molesta*, where several natural enemies from South America have been and are being tried in India and elsewhere, but again other tested natural enemies are available. There is also *Chromoleum odoratum*, where to date in India trials with known biotic agents have been unsuccessful.

Perhaps the most pressing need, as was pointed out before 1975, is to investigate fully the possibilities of biological control of *Parthenium hysterophorus*, a plant of South American origin which has become a pernicious weed around some cities. Here the introduction of host-specific natural enemies from South America might well put it under effective control, but meanwhile the weed has continued to spread unchecked.

The prospect for major widespread success with biological control in India is not very promising, but it is a natural method which does not, like other methods of pest control, interfere with the environment.

BIOLOGICAL RHYTHMS. Periodic change with the time of day, month and year is a most spectacular feature of the environment of the earth. The three major periods thus prevalent are those of the solar day (24 hours), lunar month (29 days) and the calendar year (365 days). Most living creatures have adapted themselves in many ways to this temporal order of their environment, giving rise to a variety of biological rhythms. These rhythms enable the plants and animals to carry out their various bodily functions at the most advantageous time of the day, month or year.

Thus cold-blooded reptiles like lizards must raise their body temperature that necessarily falls during the night by sunning themselves in the morning. The most appropriate time for hunting for prey for them is therefore late morning, when they can be warm and active. Crabs on the seashore must adapt their feeding times

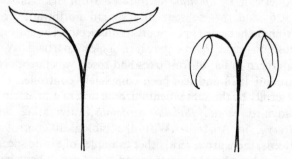

Biological rhythms enable the plants and animals to carry out their various bodily functions at the most advantageous time of the day, month or year.

in accordance with tides which depend on the rotation of the moon. Insectivorous birds must adjust their breeding seasons to correspond with the yearly period of maximum abundance of insects to satisfy the requirements of their fast-growing chicks, and so on.

Animals fall into two broad categories of day-active and night-active, depending on the time during which they actively seek food. On land, insects have large membership in both these categories. Cold-blooded reptiles are constrained to be day-active, and the primarily visual birds are also largely day-active. Amongst the ground-dwelling mammals, both habits are quite common. While the flying mammals, bats, are all active at night, taking advantage of the paucity of night-active birds.

Even within these categories, however, not all animals are equally active throughout the day or night. Their activity patterns vary, depending both on changes in the physical environment, as well as in response to the activity patterns of other animals in their habitat. Dawn and dusk are periods of the most rapid change in light intensities, and most animals use these as cues to initiate or terminate their periods of rest. Thus birds become active and bats go to rest at dawn, and the reverse occurs at dusk. If we look further, the Jungle Crow becomes active at lower light intensities than the Indian Myna, and the pipistrelle bats at higher light intensities than the Flying Foxes. Apart from the fact that dawn and dusk are periods of most rapid rate of change in light intensity and hence most suitable as time cues, they are also the times at which the air is least turbulent, permitting sound to be carried farthest. That is why birds indulge in their most intense vocalization at these times, giving rise to dawn and dusk choruses; as do monkeys like the Hanuman Langur.

Animals also adjust their periods of activity to minimize competition with other species. Thus various species of bees have peaks of flower-visiting activity at different times of the day, and different species of mosquitoes have peaks of blood-sucking at different times of the night.

Such rhythms have naturally greatly fascinated physiologists who have attempted to study them under experimental conditions. It has been shown that the rhythms are not merely imposed from outside, but persist even under totally unchanging conditions. Under these conditions, however, the period of rhythm is not precisely 24-hours, but only nearly so, hence these rhythms are known as circadian (*circa*, about, *dies*, day). They are daily adjusted to the diurnal rhythm of light, temperature etc. through the external cues. It is now known that animals use the social cues provided by other animals as well in adjusting their rhythms. Thus bats confined to a deep part of the cave with no environmental cues of light or temperature can still synchronize their activity with the day–night regime by picking up cues from the vocalization of the other bats in the cave.

Marine animals too exhibit a number of biological rhythms. The zooplankters migrate towards the surface at night and move down deeper during the daytime. The animals on the seashore adjust their periods of activity in relation to the tides. Furthermore, the tides change not only once or twice in a day, but vary in their magnitude with the phase of the moon and time of the year. Certain marine animals such as the famous Palolo worm of Fiji seem to synchronize their breeding with these tides. Thus in Fiji the palolo worm swarms to reproduce every year 7 to 9 days after the full moon in November.

At the other end from the palolo worm, the entire population of which breeds on just one day in the whole year, is an animal such as our Asiatic elephant which seems to breed, and also to come in musth, at any time of the year. The Chital has an extended breeding season, its rutting coinciding with the monsoon and the season of birth of calves peaking from January to March, although some calves are born in every month of the year. This coincidence of birth of calves with the most difficult season of the year in terms of food availability is truly puzzling. In Mysore, the major predator of chital, the Wild Dog, breeds from January to March, presumably because its food is most plentiful at the time of fawning by chital. Amongst our birds, the small insectivores such as Warblers breed during the monsoon, apparently because this is the time of maximum abundance of insects to feed their chicks. The birds of prey, on the other hand, breed mostly during December-March, again apparently because this is the time of maximum abundance of their rodent prey which multiplies following the seeding of grasses and cereal crops towards the end of the monsoon.

The breeding of herons, storks and other colonially breeding water-birds coincides with the monsoon. Thus at Bharatpur in Rajasthan or Ranganathittu near Mysore they breed from July to October, the southwest monsoon bringing most of the rains in those parts of the country. On the other hand, at Vedanthangal near Madras they breed from January to March, this region receiving most of its rainfall during the northeast monsoon. However, this rule is not without exception. Night Herons breed at Ranganathittu from April to August, but near Bangalore, hardly 120 km away, they breed from January to March; similarly Little Cormorants breed at Ranganathittu from July to October, but hardly 80 km away they breed from January to March.

The migratory birds show a remarkable annual rhythm of long-distance movements. Many of our ducks, teals and waders breed in Siberia in summer, from April to September. In autumn they migrate south to India, staying here from October to March, moving north again in the spring. The migratory impulses of these birds is known to be controlled by changing day-length. They migrate southwards after breeding in response to decreasing day-length and north after wintering in response to increasing day-lengh.

Finally, a most spectacular example of biological rhythm is furnished by some species of tropical bamboos. Our commonest species, *Bambusa arundinacea*, flowers and seeds only once in its lifetime at an age of 45–48 years, after which it invariably dies. Moreover, the flowering is synchronized for a whole population so that all the bamboos of a species flower and die over a region of several thousands of hectares within the space of three to four years. The significance of this seems to lie in the fact that when seeds are very occasionally produced in such large quantities, predators on the seeds such as rodents can only devour a small fraction of them. If on the other hand a much smaller seed crop was produced every year, a much greater fraction of the seed could be destroyed by the predators. Hence, it is likely that massive seeding in a few years has been favoured by natural selection.

Animals not only respond to external cues, but as the persistence of their rhythms under constant conditions shows, they also have endogenous rhythms—circadian, as well as circannual, and perhaps of much longer duration as well in the case of some bamboos. The precise nature of these biological clocks is yet unknown. Animals also use these clocks for purposes other than adjusting behavioural rhythms. Thus honey-bees, as also some fish and birds, are known to use the sun for navigation. However, the position of the sun varies with the time of day. These animals make fine adjustments for such movements of the sun by using their biological clock.

BIRD CALLS AND SONGS. Birds, like men, rely on sight and sound. Their unique 'syrinx' allows them a tremendous range of articulation to convey messages to other birds. The baya weaver bird, for instance, uses ten calls in various contexts. Young birds emit *begging calls* when hungry, and send their parents *juvenile location calls* when dispersed. Weavers use *social contact calls* when feeding in flocks, and *flight calls* for co-ordination in flight, especially at take-off and landing. They emit *alarm calls* and *mobbing calls* as appropriate, and *agonistic cries* when fighting. Of the varied calls used by the males in the breeding season, the commonest is the *song* which is sometimes synchronized among several individuals. The males also have a high-pitched *copulatory call* while the females have a sibilant squeaking *solicitation call* for copulation.

Interestingly, the acoustic structure of each call has evolved to suit its particular function. For instance, there are two types of warning calls. When a predator is stationary, the mobbing call is loud, repetitive and of varying frequency. This serves to attract neighbours to join the mob. However if the predator is in the air, and thus an immediate threat, the bird seeks shelter and emits a call on a constant frequency. Such calls are difficult of location for medium-sized birds like hawks or owls. Various species have evolved mobbing and alarm calls with similar acoustic structures, and these are understood and responded to across the species.

The most attractive vocalizations are the males' songs during the breeding season. They advertise the male's possession of a breeding territory, in order to attract females and keep out rival males. Our thrushes have the most melodious songs, for example the ditty of the Shama. There is enough variation within such a song for Shamas to know their neighbours individually.

A remarkable instance of the use of calls for individual recognition and localization is furnished by some scimitar babblers. Pairs keep in touch through duets; the male's flute-like call is answered so promptly by the female that the whole duet sounds as if sung by a single individual.

Finally, we must mention the two famous Indian mimics, the hill myna and the racket-tailed drongo. The hill myna has a fantastic repertoire of whistles, wails, shrieks, gurgles, groans, and squeaks. Each individual uses 3 to 13 such calls, none of which are shared with its mate, but many of which are shared with its neighbours.

However, the hill myna never imitates other species in the wild. That is a speciality of the racket-tailed drongo and the Green Magpie of the Eastern Himalayas. The drongo, apart from its own rich repertoire, mimics the calls and even the complete songs of other species ranging from the serpent eagle, grey junglecock, koel, black woodpecker, grey hornbill, scimitar babbler to the shama. The imitations are sometimes echoed right after the original call, and nobody has the faintest notion why the bird indulges in this fantastic performance.

M.G.

See COMMUNICATION.

BIRD MIGRATION is the periodic movement of birds between their breeding and non-breeding areas or/and any regular movement of birds that forms part of their life-cycle. The term generally denotes periodical extra-limital movement of birds, i.e., across the borders. Movement within the borders is described as Local Migration, and seasonally up and down mountains as Altitudinal Migration.

Of the 2100 species and subspecies of birds in the Indian subcontinent about 300 are migrants from lands beyond the Himalaya—central and northern Asia, and northeastern Europe. There are many others whose palearctic breeding ranges encroach on our northern, northeastern and northwestern borders. These species also behave like the true migrants, and keep the same time-schedule for arrival and departure. Besides, there are local migrants some of which may move a few kilometres and some only a few hundred metres. Others move altitudinally from maybe near the snow line in summer down to the foothills or adjacent plains in winter.

It was believed that the extralimital migrants negotiated the Himalayan chain on the northwest through the Indus Valley and on the northeast through the Brahmaputra Valley. The two migratory streams then advanced southwards and converged at the tip of the peninsula, and finally hopped over into Sri Lanka. However, recent evidence by mountaineers shows that many migrants fly directly across the Himalaya, sometimes even as high as 8500 metres, thus reducing the distance considerably.

The major groups of Indian migrants are: ducks and geese, waders or shore birds, swallows, flycatchers, warblers, thrushes and chats, wagtails and pipits, finches and buntings.

The phenomenon obviously has a survival value, as otherwise such a hazardous journey, covering thousands of kilometres, and requiring such an enormous amount of energy, could not have persisted through so many generations.

In the Palearctic Region, beyond the Himalaya, the winter is severe and day is short. These conditions are uncongenial for finding food for the young that have hatched during the summer, and also for the parents themselves. Hence, by September/October the birds move southward to warmer regions, sometimes more than 5000 kilometres away, to where the winter is moderate and the days longer. By March/April they return to the northern breeding area where temperature and competition for food and nest-sites are comparatively less severe. Thus it can be said that migration is accomplished mainly to exploit the most suitable living conditions throughout the year.

Physiological readiness for the spring migration is linked with the development of the reproductive organs. Colourful plumage, a substantial quantity of subcutaneous fat (which is burnt up while travelling), increased body-weight and enlarged gonads are noticeable prior to migration.

The journey is undertaken singly, in small parties or large flocks, at times of several species together. Migrants are either diurnal, nocturnal or both. Orientation is presumed to be by the position of the sun and topographical features during the day, and by the position of the moon and stars at night. The entire phenomenon may be explained as instinct developed through countless generations over millions of years.

Ducks unlimited!

The fact that inexperienced juveniles often reach the traditional destination without guidance, much ahead of the parents, lends support to this view.

V.S.V.

BIRD-RINGING. Ringing or banding is a technique of marking wild birds for individual recognition. Chiefly two types of rings are used—the split ring and the colour ring.

Split rings are bands of aluminium, or its alloy, shaped like a C. The inside diameter may range from 1·8 mm to over 30 mm to suit the different species. Rings used in India bear the inscription INFORM BOMBAY NAT. HIST. SOCIETY with a serial number prefixed by a letter of the alphabet denoting the size.

The ring is placed on the tarsus and pressed to form a circle. The location, date, name of the species ringed and other relevant data such as age, sex, weight and measurements are recorded in a schedule. Subsequent recoveries of such ringed and documented birds at distant places and dates provide life-history information, otherwise unobtainable.

When a ringed bird is caught or killed *en route* or at its destination, the ring is returned or its serial number intimated to the address on the ring. This enables the ringer, among other things, to chart the migratory route and destination.

Colour rings are mainly of celluloid and sometimes of anodized aluminium or plastic. Permutations of different colours are used on one or both the legs for identifying individuals at a distance. Colour-ringing facilitates identification in the field and is essential in population dynamics and life-history studies of birds.

Ringing in India, though initiated in 1926 by the Bombay Natural History Society, did not make much headway until 1960, since when several lakhs of birds have been ringed, mainly wagtails, waders and ducks. The study confirms the origin of our extralimital migrants.

See MIGRATION.

BIRDS. On the basis of certain distinctive anatomical characters, common to both the groups, it is now generally accepted that birds are derived from primitive reptiles. The two classes together form the division of vertebrate animals known as Sauropsida. The earliest fossil 150 million years old of an undoubted prototype of the modern bird, the Archeopteryx—about the size of a pigeon—clearly shows that feathers, the 'hallmark' of the modern bird, are merely modified reptilian scales. Feathers and the ability to fly are the essential attributes of birds, but there are some birds, like the ostrich and the penguin, that have lost the power of flight through long disuse of their wings. Structural modifications in the skeletal frame of a bird, such as hollow tubular bones for lightness, and their fusion at points of greatest stress for rigidity and for attachment of the powerful flight muscles, enable fast and sustained aerial locomotion. Internal air sacs help to increase buoyancy and regulate body heat (thermo-regulation). These and other adaptations for flight have been copied by man in the designing and construction of modern aircraft.

The birds of the world have been arranged by scientists in 28 Natural Orders which in turn are divided and subdivided into lower categories such as Families, Genera, Species and Subspecies or Geographical Races. It is reckoned that there are altogether some 8600 species of birds living today. Thanks to the great diversity in its physical features and its climate, rainfall, temperature, humidity and vegetation, the Indian Subcontinent is exceptionally rich in bird life. About 1200 species (in 20 Orders) are found in this area which comprises Pakistan, Nepal, Bhutan, Bangladesh and Sri Lanka. It falls within the Oriental Region which is one of the six classical zoogeographical divisions of the world. The Indian avifauna includes some 900 resident species and about 300 that migrate to the subcontinent chiefly from trans-Himalayan lands, mostly in winter, between October and March. Analysis shows that among the resident species 180 or so are endemic, that is, restricted to the Indian Subregion or its immediate environs. Well over half this number are Indochinese in character and have their closest relations in southern China and SE Asia. Seven species are of uncertain origin, while the rest are derived in nearly equal proportion from the Palearctic Region (=Africa, Europe, and Asia north of the Himalaya) and from the Ethiopian Region (=Africa south of the Sahara desert). An interesting point about bird distribution (as well as that of some other animals and plants) within the Indian peninsula itself is the occurrence of several typically east Himalayan genera and even species—e.g. *Garrulax* (laughing thrushes), *Buceros bicornis* (Great Pied Hornbill), *Irena puella* (Fairy Bluebird), *Baza lauphotes* (Lizard Hawk)—in the Sahyadri complex of hills near the extreme southwestern tip. The two disjunct populations are now separated by some 1500 to 2000 km of incompatible biotope, and in explanation it is postulated that the distribution, which must have been continuous at one time, was disrupted by earth-moulding forces such as subsidence or erosion of the connecting high land, wiping out the intermediate population. Hora's Satpura Hypothesis is more fully described under GEOGRAPHICAL DISTRIBUTION.

This is not the place for a systematic list of the avifauna: the interested reader is referred to *A Synopsis of the Birds of India and Pakistan* (1982) by S. D. Ripley, and to the ten volumes of *Handbook of the Birds of India and Pakistan* (1968-74) by Sálim Ali & S. Dillon Ripley, for detailed accounts of the birds. For the average sportsman and field naturalist the following rough-and-ready divisions will suffice: *Water birds* which include Swimming Birds and Shore Birds (or Waders) such as grebes, ducks and geese, pelicans, cormorants, herons, storks, egrets, ibises, spoonbill, flamingos, jaçanas, curlew, sandpipers, plovers, gulls and terns, and others. *Land birds* which include in addition to a large assortment of families and species of terrestrial and arboreal birds, both passerine and non-passerine, such well-marked groups as the raptors or Birds of Prey (eagles, vultures, hawks, falcons, owls) and the so-called Game Birds or Sporting Birds like pheasants, junglefowl, partridges, quails, cranes, bustards, sandgrouse and others.

The fact that, like reptiles, birds also lay eggs lends further support to the thesis of their reptilian ancestry. Birds' eggs differ from most reptiles' in being hard-shelled (calcareous) instead of with a leathery or parchment-like covering. Typically they are pyriform in shape—broad at one end, narrow at the other. Some eggs are white like reptiles' eggs, but most are variously coloured and many are patterned with contrasting spots, blotches or scribbles which serve as protective camouflage when laid in exposed situations. While some birds like partridges and lapwings lay on the bare ground or in a grass-lined saucer-like scrape or depression, most species build nests of twigs and/or grass in trees or bushes, or on cliffs (see NIDIFICATION). The architecture of nests varies from a simple twig platform (crow) or open cup of grass or fibres (bulbul) to the elaborate, intricately woven retort-shaped hanging nest of the Baya or weaver bird. Birds' eggs are incubated by the heat of the parent's body as it squats on them in the nest. The incubation period, i.e the time taken by the embryo to develop till the egg hatches, varies with the size of the bird: for example it is 9 to 10 days in our Small Sunbird (*Nectarinia minima*) weighing 5 grams, and 45 days in the Whitebacked Vulture weighing 4500 g. Some hatchlings (e.g. domestic fowl, duck) emerge fully clothed in down, eyes open and capable of running about and feeding themselves (nidifugous or precocial).

But the majority are born naked and helpless, with eyes closed—little blobs of flesh—and need to be brooded and fed by the parent until they are fledged. Hatchlings that stay in the nest are said to be nidicolous.

Some birds are monogamous in their sexual relations, either pairing for life (e.g. Sarus crane) or maybe for only a single season, or even a single brood (House Sparrow). Some other species are polygamous: either polygynous (several females to one male, e.g. peafowl, pheasants) or polyandrous or promiscuous (several males to one female, e.g. bustard quail, painted snipe, jaçana). Certain other families—in India notably the cuckoos (koel)—have developed the habit of BROOD PARASITISM, which means that they do not build any nest of their own but surreptitiously lay their eggs in the nests of other species and foist upon the hosts the responsibility of incubating them and of nurturing the hatchlings. In the process the imposter normally manages to eject the rightful chicks, thus monopolizing the food brought by the parents. An exception to the general pattern of nidification is the Megapode of the Nicobar Islands (and of the Australian Region) which buries its eggs in a large mound of scraped-up sand and leaves them to be incubated by the heat generated within by fermentation of the humus. The nidifugous chicks are self-reliant on emergence and able to fly and fend for themselves.

Territory, Courtship and Song. On the approach of the breeding season many male birds who have lived sociably up to that time become individualistic and intolerant of others of their own sex and species. The (hormonal) changes within the body—gonads, endocrine glands, pituitary—rekindle Song which has lain dormant. A male selects some exposed or elevated perch whence by loud outpourings of full-throated song he proclaims to all and sundry that he is King of the Castle and in occupation of a 'territory' around the singing-post. The prime function of song is now interpreted as a 'distant threat' to warn a would-be rival to keep off the owner's territory. If the warning is unheeded the interloper is met with bellicose posturings and 'sabre-rattling' to intimidate him into withdrawing before physical chastisement becomes necessary. At the same time song serves the dual purpose of announcing to a prospecting female that a desirable partner and unoccupied nesting site are available. Considering the acute sense of hearing possessed by birds it is reasonable to assume that the song is appreciated by the ready-to-breed female and produces a stimulating effect upon her.

Courtship display which follows after a mate has been acquired takes a variety of forms. It may be on the ground (junglefowl, pheasants), on the water (some ducks), or in the air (birds of prey). It may be an individual performance, mostly by the male as typified by the 'dance' of the peacock, or mutual like the leaping, bowing and prancing indulged in by both sexes of the Sarus crane. The essential purpose of courtship or nuptial display is to awaken or stimulate the breeding impulse in the opposite sex and to keep it keyed up. In cases like the last where both sexes take active part, the display serves to sustain mutual stimulation. Thus even birds that pair for life, and who have no need to attract a new mate each breeding season, indulge in the exercise, continuing it long after copulation is achieved. In species where the male takes little or no part in the domestic chores such as nest-building, incubation, and rearing the brood, and therefore has no special need for protective camouflage at this critical period—he is more brightly and showily coloured in plumage and often has the most bizarre or spectacular and ostentatious nuptial display. In colonial nesting species (e.g. gulls, weaver birds) displays by individual breeding males are taken up by others of the flock and serve as mutual stimulants to the entire colony and ensure successful coordinated breeding among the members.

Rare and Endangered Birds. There is evidence that some species have suffered marked diminution in numbers in the last hundred years or so, or have been, and are, under increasing pressure from the exploding human population. The decline has been steep and rapid since the Second World War, particularly in the last 30 years since the advent of our Independence. Among the principal causes for the decline are: destruction of forests and bird habitats for rehabilitating the hordes of refugees and repatriates from Burma, Sri Lanka, Pakistan, Bangladesh and elsewhere; the mounting cost of living which has exposed edible birds such as junglefowl, partridges and quails to inordinate pressure from poachers and professional trappers; and to similar pressure for lucrative export of ornamental species to foreign countries. The indiscriminate use of toxic chemical pesticides has aggravated the situation though the damage from this source is as yet comparatively restricted. In addition there is of course the natural process of gradual extinction of species that have, for reasons difficult to pinpoint, reached such a low threshold of numbers that their extinction can at best only be delayed by judicious conservation measures, but not averted. The last free-living example of the Pinkheaded Duck—a rare endemic species that was already on the decline (scarce) in the last century— was seen in 1935. Another interesting endemic, the Great Indian Bustard, is presently under increasing threat from poaching and from the deprivation of its natural grassland habitat. Three of our other birds must be considered to have vanished within recent times, since in spite of diligent organized search they have eluded rediscovery. These three are the Mountain Quail of the Kumaon Himalaya last authentically reported in 1876, Blewitt's Owl of the Satpura highlands last

reported in 1914, and Jerdon's Courser of the Deccan plateau, in 1900. The scarce Whitewinged Wood Duck of the swampy jungle pools of Assam is at serious threat chiefly by the deprivation of its specialized habitat through processes of land development.

Migration. Migration is perhaps the most widely known and intriguing attribute of birds in the popular mind. Its regularity and orderliness have excited the wonderment of man from earliest times, so much so that some primitive societies like the Eskimos and Red Indians of the circumpolar north named the calendar months after the arrival times of migrant birds. Many romantic and fanciful notions were entertained even by otherwise knowledgeable bird students up till comparatively recent times concerning the seasonal appearance and disappearance of individual species. Only a more scientific approach to the problem within the last hundred years or so has helped in unravelling some, but only some, of its 'mysteries'. The basic purpose of bird migration is obvious. Birds migrate primarily in order to escape adverse life conditions due to unfavourable climate and shortage of food—thus from the icy conditions of the circumpolar lands in autumn and winter towards the warmer and more hospitable environments of the tropics—and vice versa in spring and summer. The technique of marking individual birds with numbered and self-addressed metal rings or bands round their legs, adopted universally since the beginning of the present century, has been the chief instrument in elucidating many of the riddles of migration. Recoveries of ringed birds in distant places have revealed precise information concerning the routes and destinations of migrants, their speed of flight and distances travelled, the longevity of different species, and many other details of their life-history unobtainable in any other way. Little factual information of this kind existed in India before the Bombay Natural History Society launched a pilot scheme for ringing wildfowl in 1926 and intensified this activity with international aid after 1960. In the last two decades over a quarter million birds have been ringed in India, several thousand of which—especially migratory ducks—have been recovered, some as much as 7000 km away, yielding scientific data of inestimable consequence.

Ecology and Economic Ornithology. As an integral part of the biota, birds play a vital role in maintaining a healthy natural environment. Their impact on human ecology however is double-edged. While many species are economically beneficial in the highest degree—as destroyers of insect and rodent pests of agriculture and animal husbandry, or as flower-pollinating and seed-dispersal agents, or as scavengers and in other ways— some species are harmful to human interests as direct despoilers of monocultured food crops and orchard fruit, or as hosts of external and internal parasites and as carriers of pathogenic viruses often fatal to man and his livestock. Also, in recent years, with the mounting air traffic, birds—chiefly vultures and kites in the case of the Indian subcontinent—have posed an increasing hazard to fast-flying jet aircraft in the shape of mid-air collisions or 'air strikes', resulting in very serious damage necessitating expensive repairs or the complete scrapping of costly engines and aircraft, and sometimes loss of human life.

BIRDS' SENSES AND SENSORY ORGANS. Birds perceive their external environment chiefly by the same five senses (namely vision, hearing, smell, taste and touch) common to most vertebrates. Their receptors or sensory organs are also basically the same, though birds' eyes are the best evolved among the vertebrates.

Taste appears to be poorly developed, there being relatively very few taste buds in a bird's mouth. The fleshy-tongued parrots as a group are probably an exception. Generally, the sense of smell is also poor. Olfactory lobes are however, well developed in the nocturnal Kiwi (*Apteryx*) of New Zealand, which, with its external nares located nearer the bill tip, probably locates its prey of earthworms by smell. Geese, petrels and the New World Turkey Vulture are probably others with a well-developed sense of smell. A few other 'senses' which could be postulated on theoretical grounds (but of which the nature is not clear) are those of navigation and chronological time, which seemingly play a vital role in bird migration, and the kinaesthetic sense which is probably a complex one, by which birds make out their bodily movements (as in flight) and position of head and wings in relation to the body. The receptors co-ordinate with one another to facilitate perception. Eye and ear help in localizing a distant object. Vision aids the semicircular canals in maintaining equilibrium.

BIRD VISION. The visual acuity, or sharpness of vision, in birds is the highest among all vertebrates. Certain features are common to most species of birds, for example:

Their eyes are large in proportion to body size, flattened, with a globular or tubular shape. These features, and a large lens, favour good distant vision and excellent accommodation, especially as compared to the human eye (Figs. 1 and 2). This is useful especially to birds of prey.

The retina of birds has got a special structure called the Pecten, as well as densely-packed and precisely arranged rods and cones (Fig. 1). The latter contain special oil droplets, which improve the quality of the visual image, and increase the capacity for distinguishing contrast and tones. Thus they can easily locate insects, worms, or game nicely camouflaged in their surround-

ings. A notable exception is the Kiwi of New Zealand, which is nocturnal and short-sighted.

Birds have a highly-developed colour sense, which is an obvious advantage in mating, and also for the preferential selection of coloured flowers and fruit to feed on.

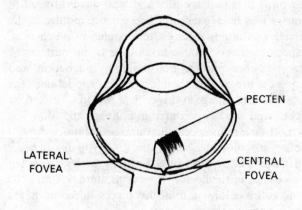

1. THE EYE OF A BIRD

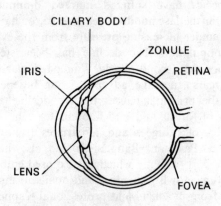

2. THE HUMAN EYE

There are some aspects of vision wherein there is considerable variation between different species of birds, for example: several species, especially birds of prey, have got two foveas in their retina, instead of the usual single central fovea. The second (temporal) fovea is used for binocular straight-ahead vision, while the central fovea permits a panoramic uniocular view on each side of the head.

Similarly, the extent and the overlap of the visual fields shows a corresponding difference depending on the placement of the eyes. Thus most birds which have laterally-placed eyes enjoy a large total field of vision, with a small frontal area of binocular vision, and a small 'blind area' at the back of the head (Fig. 3).

On the other hand, owls and certain hawks have forward-placed eyes, with a smaller total field of vision, but a large frontal area of binocular vision, somewhat similar to the visual fields of the primates, including man (Figs. 4 and 5).

BINOCULAR FIELD

3. VISUAL FIELD OF PIGEON

4. VISUAL FIELD OF OWL

5. VISUAL FIELD OF PRIMATES

Most birds see better in the day or in bright light, whilst others such as owls are chiefly nocturnal, with their eyes structurally adapted for night vision.

H.S.Z.

BISON. The Indian Bison or Gaur (*Bos gaurus*) is one of the most impressive of the world's wild oxen with adult bulls weighing up to 1000 kg and standing 190 cm at the shoulder. The young are a golden fawn colour, whilst cows and young bulls are dark coffee-brown colour with distinctive white stockings on their legs. The bulls when mature are covered with short black glossy hair and have a high bony ridge extending from their shoulders and ending abruptly over the loins. They have a yellow or grizzled area on the forehead and white-stocking feet. Both sexes bear smooth cylindrical upsweeping horns, which are orange yellow in young animals becoming olive in older bulls. The horn-spread may be as much as 85 cm between tips with individual horns up to 80 cm in length. There are two distinct and separate dewlaps, a small one under the chin and the other one on the brisket. There is no white caudal patch as in the Banteng or wild ox of Malaysia.

Gaur are more or less confined to Malaysia, Burma and India and are always associated with extensive forested areas in hilly tracts. In India they are now reduced in numbers, being limited to scattered populations in the Western Ghats, in Madhya Pradesh and Orissa and in the Himalayan foothill regions of Assam and Bhutan. They are shy and retiring beasts despite their massive size, and their numbers have been reduced by successive felling of forest, grass burning and other human disturbances. They graze during the wet season but can subsist upon browse during the dry season and eat a variety of herbs, fallen fruits and bamboo leaves. They feed mostly during the hours of darkness, lying up to chew the cud by day in the shelter of the forest. Generally they keep in small herds of two or three animals ranging up to 40, usually with an old cow being the dominant animal. Mature bulls commonly live a solitary existence or travel in pairs. The rut or breeding season in Karnataka is from November to March whilst in Kanha the rut lasts from December to January. A single calf is born after a nine-month gestation period and cows appear to mate for the first time when two-and-a-half years old and to produce one calf each year.

Like all oxen, gaur have an acute sense of smell but rather poorly developed eyesight. They are always very alert for predators, but both tigers and leopards take a toll of the newly-born calves, with adult animals occasionally being attacked. They are also very susceptible to the diseases which affect domestic cattle, and in

Indian Bison or Gaur × ¹⁄₂₆

the famous Bandipur Sanctuary in Karnataka the magnificent herds of bison were practically wiped out by an epidemic of rinderpest in the late 1960s, from which they have yet to recover.

See UNGULATES. See plate 25 facing p. 368.

BLACKBUCK (*Antilope cervicapra*). Like the Nilgai, this beautiful animal is a true antelope, and is uniquely endemic to the Indian subcontinent and has been placed by taxonomists in a mono-typic genus. They are however much smaller and gazelle-like in build when compared with the Nilgai, adult males rarely exceeding 40 kg in weight with females averaging about 35 kg. Young males and females are a yellowish fawn in colour with white bellies, chests and a large patch around the eye. Very occasionally females bear thin straight spiky horns. The bucks are one of the few examples of sexual polymorphism amongst the antelopes, being strikingly patterned in black and white after they reach maturity at about 3 years of age. However this colour tends to fade to dark brown towards the onset of the summer and individuals from South India are always a dark brown, not black. The adult buck carries a pair of spirally twisting horns which attain a length of 50 to 60 cm measured straight and are ringed throughout with

Part of a buck party, Rajasthan × ¹⁄₁₅

prominent ridges or annulations, and are altogether very handsome. Possessed of keen eyesight and fleetness of foot, they were highly prized quarry for the sportsman in former times when they were much more abundant than today.

Unfortunately, this once abundant animal has declined dramatically in numbers since the early 1950s and is nowadays confined to a few scattered semi-protected herds. One prominent Indian naturalist estimated that their numbers were probably as high as 80,000 in 1947 whereas by 1964, in India, the total population had dropped to as low as 8000. Blackbuck are mainly grazers and do not like browse and they are typically associated with semi-desert or open grassland country, avoiding forest or hill tracts. They have a fondness for feeding in millet, sorghum and other cereal crops, and therefore frequently come into conflict with man; they also prefer a habitat which is cleared of tree cover and brought under cultivation.

Blackbuck are quite gregarious, preferring to live in herds of 10 to 30 individuals, and they are diurnal in feeding activity. Only the young males are driven away from such herds by the master buck, and forced to live singly or to roam in small all-male bands. The rutting season is usually in February and March with fawns being dropped after 180 days' gestation period, at the onset of the monsoon. The bucks guard their harem of does most zealously at the time of rut, and besides driving off any other rival male, have been known even to attack human intruders into their territory. This so-called territory is often marked out by regular dung piles or places where the male deposits his faeces. This behavioural trait is also shown by the Nilgai and the Chinkara Gazelle, and may serve partly to warn off rival males.

See UNGULATES.

BLIND FISHES. Fishes living in a completely dark environment tend to lose their sight, while other senses, especially that of touch and vibrations perceived by the lateral line, become highly developed. Typical cases are the numerous blind fishes found in the dark zones of the ocean, on the continental slope and abysses (see DEEP-SEA FISH).

A similar loss of eye function also occurs in fishes living inside caves. Together with a tendency to lose their body coloration and turn dull grey or white, the fishes are totally blind and even the eyeballs are absent. Most such blind cave fishes are, however, found to have relatives living in the neighbourhood (but outside the caves and hence exposed to sunlight) with perfect eyesight.

The most well-known blind cave fish is *Anoptichthys jordoni* from America. It is popular in aquaria, so is found in many homes in our area. Though blind it can

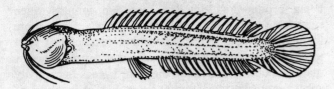

Horaglanis krishnai, a blind fish, × ²/₃

know where food is available and breed like any other fish. The young are born with eyes, but these start sinking into the head after three weeks and get covered over by fat and skin.

In India a blind catfish, *Horaglanis krishnai*, was first discovered in a well at Kottayam in Kerala. It is remotely related to *Clarias*, a catfish, but has not yet entered the aquarium world.

BLINDING TREE, *Excoecaria agallocha*, is a small mangrove tree of the subcontinent and Burma. It is called the Blinding Tree on account of the irritating juice in the stem, which is said by wood-cutters to cause blindness if it enters the eyes. The trunk is often inclined with vertical branches sticking out. Its bark is grey, with red-brown lenticels forming diagonal or vertical stripes. The leaves turn yellow or red before falling in the hot season. It is a good fuel with high calorific value.

BLOOD. In single-celled animals exchange of nutrients and discharge of waste material is a direct process between the cell and its environment, aided by currents of flow within the cytoplasm of the cell. In small multi-cellular aquatic animals, the flow of water through the body also serves as a vehicle from which nutrients are absorbed and into which waste materials are excreted. With increasing size and complexity, a degree of specialization occurs where certain organs are concerned with special activities such as digestion or excretion. To enable the whole body to function effectively, it is necessary to have a transport system which will carry material from one part of the body to another, for example nutritious material from the digestive organs must be transported to the various parts of the body. Similarly waste materials from various parts must be removed and carried to the excretory organs from where they can be discharged. In insects and many worms, this transport is provided by the body fluid, which is propelled by a primitive heart or pump to ensure its circulation to all parts of the body. In higher animals, this is achieved by the blood, which circulates through a complex system of blood vessels. The blood carries water, oxygen, and nutrition, as well as the hormones (secretions of the endocrine glands) to various parts of the body. From the tissues, waste materials and carbon dioxide are taken up by the blood. As the blood cir-culates, waste materials are collected from it by the excretory organs for discharge and the carbon dioxide is removed by the lungs to be thrown out in the breath. Thus each tissue and organ takes up its requirements from the circulating blood, and discharges waste or unwanted material into it, to be further transported to those tissues equipped to deal with them.

The blood has liquid and particulate elements which circulate throughout the body. The liquid element, the plasma, contains nutrients and salts in solution. It also contains proteins concerned with defence, the anti-bodies, which are manufactured by the body against infectious agents. One protein, fibrinogen, has the capacity of forming fibres when exposed to tissue juices from an injured tissue. These fibrous threads entrap the cells of the blood and together with the platelets (see below) form a blood clot, which seals off the injured tissue to prevent bleeding. The clot also serves as a mechanical barrier against entry of infection from outside.

The particulate or cellular elements of the blood are of three main types: (*a*) The oxygen-carrying cells. In higher animals these are reddish in colour because they contain a substance called haemoglobin. Haemo-globin belongs to a group of substances having the capacity to form a loose combination with oxygen in conditions of high oxygen pressure, i.e. in the lungs, and to release this oxygen in conditions of low oxygen pressure, i.e. in the tissues. It also picks up carbon dioxide from the tissues and releases it into the lungs. (*b*) The leucocytes or white blood cells which are concerned mainly with defence against injury or infection. (*c*) Small cell fragments called thrombocytes or platelets which play a part in the mechanism of clotting of the blood.

The blood may be colourless, or coloured. Generally the colour depends on the colour of the oxygen-carrying substance which it contains. In most vertebrates the oxygen-carrying compound is haemoglobin which colours the blood red.

The circulation of blood is maintained by the pumping action of the heart and by the muscular walls of the blood vessels. All parts of the body are supplied with numerous thin-walled channels, the capillaries, which pervade every tissue in the body. Exchange of substances from the blood to the tissues takes place across the thin walls of these capillaries. In normal conditions only a few capillaries are open, the rest being collapsed, potential channels. When infection occurs the collapsed capillaries open up and the blood supply of the tissue is increased immediately, which brings more blood and more defensive cells to the area to fight the infection.

BOAR. Strictly speaking the term Boar is applied to an adult male pig, but the wild pigs which inhabit the sub-continent are commonly referred to as wild boar. There

Indian Wild Boar, Maharashtra × ¹/₁₉

are two species, the widespread Indian Wild Boar *Sus scrofa cristatus* and the very rare and recently rediscovered Pigmy Hog *Sus salvanius* which occurs in northern Assam. The former gets its scientific name (*cristatus* means 'crested' in Latin) from the mid-dorsal crest of longer black hair which runs down from the nape to the loins. Large males have been authentically recorded weighing up to 165 kg and standing 91 cm at the shoulder. Females are smaller and lighter in build but both sexes have elongated snouts, thick muscular necks, and short practically naked tails terminating in a lateral tuft of longer paler bristles.

Wild boars have a number of characteristics which sharply differentiate them from other even-toed and hooved animals with which they are classified in the order Artiodactyla. They are not ruminants with complicated stomachs and they do not chew their cud. They have well developed canines in the upper and lower jaws which are known as tusks and which are used in both sexes as weapons of offence. The molars or grinding teeth, instead of wearing down to the distinctive crescentic pattern of ruminants, have their surfaces capped with rounded hillocks or cusps enabling them to chew and crush a variety of food. They have comparatively well developed outer toes with separate phalanges. In the wild they show a marked preference for succulent roots and tubers for which their tough cartilaginous nose-discs and muscular necks are adapted. They are particularly attracted to carrion and will even cannibalize dead members of their own kind. They are serious pests of sugarcane and maize crops in parts of the northwest and will also wreak havoc in potato and groundnut crops as the tubers or seed pods are developing. Normally they lie up during the daylight hours in thick cover, such as is afforded by natural swamps or canefields. At night they roam long distances in search of food and generally forage in small family parties. Females bear litters of four to eight but there is a high mortality rate and it is rare for more than 3 to 4 young to be successfully reared. Litters can be born throughout the year but in the northwest some seasonality has been observed with the majority of litters being produced during the monsoon months when succulent vegetation and insect larvae are more abundant. Studies have shown that females can start to breed at about 12 months of age but probably few males are able to mate until their third year. In captivity specimens have lived for 20 years but probably 8 to 10 years in the wild would be an exceptional life-span. Wild boars are found in the lower Himalayan valleys, in tropical deciduous forest and thorn scrub habitats, but they avoid extensive desert or evergreen rain-forest.

The Pigmy Hog is so secretive and limited in distribution that it was thought to be extinct until 1971 when an animal was authentically sighted and then specimens were captured. They are diminutive in size, adult males weighing hardly 9 kg and standing 23 to 30 cm at the shoulder, whilst adult females weigh about 6 kg. They are shy and secretive, with family groups spending the day burrowed under a nest which they construct of piled-up chopped sedge and grasses hidden in some thicket. They move with lightning rapidity through the thick vegetation and, like their larger congeners, when confronted are bold and aggressive and can inflict severe lacerations with their razor-sharp incisors. The entire world population is presently believed to survive along a narrow foothill belt in the extreme northeastern border of Assam. Recent studies have shown that females produce only one litter a year and that this is born in April or May during the so-called dry season.

See UNGULATES.

BOMBAY DUCK. The fish, scientifically known as *Harpodon nehereus*, has a halo of mystery right from its name. How the English name 'Bombay duck' has been derived is hard to understand. The regular English name 'bummalo' and the Portugese 'bambulim' are similar to the Gujarati and Marathi names 'bumla' and 'bombil' respectively.

It is a small fish, about 25 cm in average length, with a peculiar soft, fleshy and cylindrical body. When living, the body is translucent. The jaws are red, the pectoral pelvic fins pale brown, and the dorsal, anal and caudal fins dark black, a rather gaudy appearance. Behind its normal dorsal fin it has a small adipose fin without fin-rays. The lateral line scales on the mid-body extend characteristically beyond the tail, making it tri-pointed. It is common from Diu in Saurashtra to Harnai in Maharashtra. Thereafter, it disappears from the southwestern and southeastern coasts of India and appears again on the Andhra, Orissa and Bengal coasts in small numbers. Biologists have not been able to explain satisfactorily this discontinuous distribution in relatively similar climatic regions of the tropics.

Another feature of this fish is the soft noncalcareous bones and large jaw with recurved teeth like those in deep-sea fishes, though the fish in fact thrives

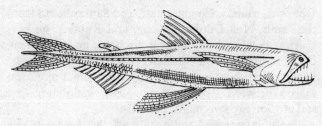

Bombay Duck, × 1/3

in comparatively shallow coastal seas and sometimes enters the mouths of creeks. Its flesh, too, is jelly-like and without any firm muscles, like a deep-sea form. Its body emits a wonderful phosphorescent glow at night, when fresh fish are kept drying.

Bombay ducks are gregarious in habit, moving in large groups and feeding on small shrimps and other fish. They are at times so voracious that they gulp down fish longer than their own body, their stomach getting fully distended. They commence breeding when they are about 21 cm in length and have a long breeding season from November to February. However, exact information about the place of spawning, nature of fertilized eggs and early hatchlings is not yet available.

It is noteworthy that this single species gives an annual yield of about 100,000 tonnes and thus is of good commercial importance. Although fresh Bombay duck is relished, most of the catch is dried on scaffolds, tied in bundles, and sent upcountry where rural people appreciate it to flavour curries.

BOMBAY NATURAL HISTORY SOCIETY. The Society was founded in 1883 by eight individuals among whom were Edward Hamilton Aitken (better known as EHA) and Col. Charles Swinhoe, the entomologist. The objective was: 'To promote the knowledge amongst the public of Natural History in all its branches, including particularly the study of Animal and Plant life of the Oriental Regions and the Zoogeographical Regions adjoining thereunto, both alive and otherwise.' Immediately it was established one of its most important tasks was the publication of a journal which has over the years become a leading publication of its kind the world over. In its first issue, published in January 1886, the Editors said 'in accordance with the character which this Society has assumed from the beginning, the aim of this journal will be, as far as possible, to interest all students of nature, ever remembering that there are many naturalists in the highest sense of the term who have not much technical knowledge of any particular branch of Science as to be able to enter with interest into questions of nomenclature and the discrimination of closely allied species'. Indeed the *Journal* has been a happy mixture of technical papers of the highest order, as well as notes of general interest for the common man.

For nearly a quarter of a century prior to his retirement in 1948, Stanley Henry Prater was the moving spirit behind all the activities of the Society including the exhibition galleries of birds and other animals in the Prince of Wales Museum. His *Book of Indian Animals*, first published in 1948, was largely instrumental in educating the people of India about the wild life of this subcontinent and the need for its preservation.

Similarly *Butterflies of the Indian Region* by M. A. Wynter-Blyth, first published in 1957, made it possible for the non-specialist to identify and take an interest in these beautiful creatures.

But the Society's principal interest has centred around birds, and the *Book of Indian Birds* by Sálim Ali, first published in 1941, and now in its 11th edition, has occupied a unique position in the Natural History literature of this subcontinent. This book has been largely instrumental in creating and sustaining an interest in birds amongst a sizable section of the people of this country. The Society's reference collection of birds is among the best of its kind, and the continuing field trips to study bird migration as well as for collection, ensures that information about Indian bird life is as complete as possible. While the Society has been collaborating for research with many institutions in India and abroad, it has had a special relationship with the Smithsonian Institution in Washington, whose Secretary, Dillon Ripley, has co-authored with Sálim Ali the *Handbook of the Birds of India and Pakistan*.

With the growing interest in the environment it was natural that the BNHS should extend its horizons beyond taxonomic work and collection of specimens in its reference collections, to the equally significant aspects of conservation and preservation of the habitat at large. Through its reports to governments at Central and State levels, it can claim to play a significant role in ensuring that the wilderness and wild life of this country will be preserved for posterity.

The BOTANICAL SURVEY OF INDIA was established in 1890 with headquarters in the Royal Botanic Garden at Howrah. There was great expansion in 1954 and a large staff is now working thoughout the country with the objects of (i) preparing an inventory of the plant wealth in the form of national and regional floras, (ii) assessing the economic potential, and (iii) conserving endemic, rare and endangered plants, particularly those which are the wild relatives of cultivated plants.

The Indian Botanic Gardens, established in 1787, is one of the oldest and best designed Botanical Gardens in the world. These Gardens have been responsible for the introduction, acclimatization and cultivation of many economically and medicinally important, as well as ornamental, plants now seen in India and adjoining countries. The garden provides a rare opportunity to see the tropical flora of the world presented countrywise in 25 sections, and around 26 lakes. The famous Banyan tree with 1430 props is one of the largest trees in the world. The Central National Herbarium, formerly a wing of the Indian Botanic Garden, has more than a million specimens arranged in systematic order.

The Survey has conducted several hundred explorations and has published 7 fascicles of the Flora of India, 21 volumes of Records, 21 volumes of its Bulletin, 8 fascicles of *Icones Roxburghianae* (pictures of species described as new to science by William Roxburgh, Superintendent of the gardens in 1800), and 6 books containing the results of its research activities.

BRAHMAPUTRA. With a course of nearly 2900 km and a drainage area of 580,000 km² traversing southern Tibet in China, the states of Arunachal Pradesh and Assam in India and Bangladesh, the Brahmaputra is one of the largest rivers of the world. Rising from a glacier located about 100 km southeast of Manasarowar Lake in west Tibet (82°10′ E and 30°30′ N) at an altitude of 5150 m it flows eastwards for nearly 1870 km in a shallow valley through the southern part of Tibet as the Tsang-Po, keeping a course roughly parallel to the main Himalayan ranges at an altitude of about 4000 m. Beyond Pe the river turns to the northeast passing through a number of rapids between the high mountains of Gyala Peri and Namcha Barwa (altitude 7756 m), and then turns dramatically south to enter India in the mountains after cutting through the deep gorge of Dihang. It now assumes the name of Dihang or Siong. Near Sadiya it takes another sharp turn towards the west and meets with its principal tributary, the Luhit river, from the east.

Beyond Sadiya it has entered the plain stage and is known as the Brahmaputra. It has now become a mighty river of large expanse. The sprawling river flows down the flat, low valley of Assam Plains from east to west and then southwards for a distance of about 720 km, before entering the Bay of Bengal as an estuary. The Assam Plains extend for about 600 km from Sadiya to Dhubri where the river skirts the western edge of the Garo Hills and flows into Bangladesh, assuming the names of Jamuna and, later on, after its confluence with the Ganga river, that of Padma.

The relief is very low and uniform both in Assam and Bangladesh. The river is carrying a large volume of water and is braided into innumerable interlacing channels forming several lakes and jheels. These channels constantly shift between the sandy shoals. The flood discharge and silt brought by the major tributaries influence the shape of these channels. The plain is perpetually replenished by river-borne alluvium of sands, silts and clays. Between Sadiya and Dhubri its major tributaries from the north are the Subansiri, Dhansiri, Manas, Champamati, and Sankosh and from the south, Noa Dihing, Disang, and Kopili. The northern tributaries bring greater discharge and silt during the flooding season, pushing the river southwards. Floods occur occasionally during the summer monsoon season between April and July. Heavy flooding and soil erosion are two major problems.

The uninterrupted uniformity of the Assam plains and Bangladesh is relieved only by the low-lying hills of hard, gneissic outcrops between Gauhati and Dhubri. Elsewhere, it is a uniform, level plain of river-borne sediments. In Assam the river is bounded on either side by stretches of marshland covered with thick grasses, reeds, bamboo clumps, palms and fruit trees. The view is altogether picturesque! On a clear day one can see the towering, snowy peaks of the Himalaya to the north glistening in the sun, and on the south, lush, verdant hills of Meghalaya with the mighty river in the middle rolling in a narrow, flat valley.

The climates of the hills around the valley are largely affected by altitude. In general, winters are cold, with ample snowfall, and summers mild to hot, with very heavy rainfall. The Tibetan Plateau remains dry due to its leeward location in respect to the Himalaya. In the Brahmaputra Plains and in Bangladesh, rainfall is adequate to heavy, between 2500 mm to 4000 falling mostly between April and July, the amount varying with distance from the flanking hills and the Bay of Bengal. Summers are hot and winters cool. Temperature ranges are moderate in Bangladesh and higher in the Assam Valley. Humidity is always high everywhere but winter fogs are common in Assam.

R.T.

BRAHMINY KITE. Affects well-watered areas such as flooded paddyfields and the sea coast, particularly tidal creeks and fishing villages. Feeds on frogs, fish or crabs, sometimes also raiding poultry yards. Juveniles are brown and resemble the pariah kite, but possess a rounded, not forked tail. Nests from December to April in large trees usually near water. Two or three eggs are laid.

BRAIN. In unicellular organisms such as the amoeba the protoplasm reacts to stimuli in a manner which is called a 'nervous response'. The amoeba's nervous responses include withdrawal from light, vibrations, touch, heat or chemicals. In the jellyfish, though some cells are specially sensitive to light or touch, the remaining cells react to such stimuli and the whole organism moves as a single unit. In more complex invertebrates such as Spiders and Flies, a group of cells cluster into an organ, such as an eye for seeing, or antenna for feeling, and respond in more elaborate ways to different modalities of sensory impressions. At the same time some cells cluster together, to form a nervous complex (ganglion or brain) which can interpret these impressions and control the response made to them by the animal. With increasing complexity, particularly in vertebrates, certain functions are localized in particular areas of the brain.

In primitive vertebrates such as Fish and Amphibia, certain sense organs develop predominantly. In such forms the brain is mainly concerned with vision (optic), or smell (olfactory), and areas concerned with these senses are proportionately enlarged.

A large part of a bird's brain is concerned with vision, while dogs and wolves have highly developed that part of the brain which is responsible for the sense of smell.

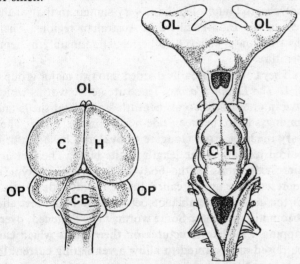

Fig. 1. COMPARISON OF BRAIN OF PIGEON AND DOGFISH
(WHICH RELY ON SIGHT AND SMELL RESPECTIVELY)

Pigeon's brain. Very small olfactory lobes (OL), a cerebrum (CH, standing for Cerebral Hemispheres), enlarged and prominent optic lobes (OP), and a cerebellum (CB).

Dogfish's brain. Large olfactory lobes, small cerebrum and optic lobes.

In the most evolved brains, these functional areas can be demarcated as 'sensory area', 'optic area', 'olfactory area' etc. The concentration of nerve cells gives the brain a grey appearance and areas rich in nerve cells, like the surface or cortex of the brain, are called the grey matter. The function of many areas of the grey matter are known, but others are still not understood. These are known as the silent areas of the brain, since damage or removal of such areas does not appear to cause any obvious disability. Nerve cells give out several processes or fibres which interconnect with those of other nerve cells. Some long fibres, the axons, connect with distant cells in other parts of the brain or the spinal cord. Bundles and sheets of axons (F in fig. 3), passing from one area to another, or to the spinal cord, appear glistening

white and are known as the white matter. Axons also pass from nerve cells in the spinal cord to peripheral nerve cells outside the Central Nervous System and from these cells to peripheral organs of the body.

The brain is that part of the nervous system which is enclosed in the skull and the part that is enclosed in the vertebral column is called the spinal cord. Grey matter is found only in the brain and spinal cord. Groups of nerve cells are also found elsewhere in the body. Groups which are not within the central axis of the brain are called Ganglia. There are sympathetic and parasympathetic ganglia which are concerned with autonomic functions, performed without conscious effort, for example breathing, which continues during deep sleep or even when the animal is unconscious, or the beating of the heart. Voluntary actions like walking, sitting, writing are performed at will.

The mammalian brain (fig. 2) is divided into three

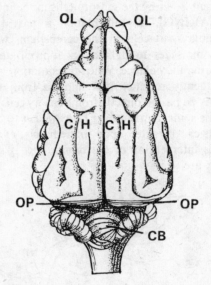

Fig. 2. DOG'S BRAIN —MAMMALIAN

The optic area (OP, shaded) is under the surface of the cerebrum (CH) and the olfactory lobes (OL) are well developed.

parts (1) The Forebrain, called the cerebrum (CH in fig. 2), (2) the Midbrain, and (3) the Hindbrain. The Hindbrain continues down as the Spinal Cord. Superimposed on the Hindbrain is a special organization of the nervous tissue called the cerebellum, or small brain, as against the cerebrum which is the big brain. The cerebrum is the seat of intelligence and is responsible for the smooth elaborate control of all the senses and actions and also of autonomic functions. The cerebellum is concerned with the maintenance of the body balance in various postures. In fish the cerebellum is responsible for maintenance of posture in turbulent water. In reptiles its role is rather subsidiary. A well-developed cerebrum and cerebellum are the characteristics of a highly developed species. In very highly evolved animals the volume of grey and white matter increases with the increase of interconnections between different areas of the brain.

Function.

When a sensory impression is received by a peripheral sense organ (SO in fig. 3) (for example, the eye observes a predator in the vicinity) the information is transmitted along the nerve fibres F, through relays of nerve cells (not shown in the figure), to the nerve cells concerned with sensation in the spinal cord. These transmit the message to nerve cells in the sensory area of the brain, which interpret the sensation and transmit appropriate messages to nerve cells in the motor area of the brain. From here, the message passes to the motor cells in the spinal cord, and thence to the peripheral organs (e.g. muscles) for appropriate action, which, in the example quoted above, would be flight. Similarly when a loss of equilibrium is felt by the sensory cells in the internal ear (see HEARING), the information is passed through various relays, to cells in the cerebellum, which will transmit messages to tighten up appropriate muscles for restoring the balance. When the situation is familiar and frequently repeated, the impulse from the sense organ may be passed directly from sensory cells to motor cells in the spinal cord along F1, (as well as to the brain) so that it can be acted upon immediately, even before conscious intervention of the brain. This is known as the reflex arc.

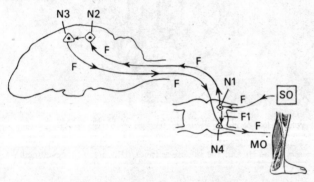

Fig. 3. SENSATION AND RESPONSE

The sensory and motor organs (SO and MO) are on the right, connected by nerve fibres (F, F, F) to nerve cells (N1, N2 sensory N3, N4 motor) in spinal column and brain. The arrows show the direction of transmission and F1 is a reflex arc to permit direct response to a stimulus.

As complexity increases, the number of nerve cells and fibres increase. The surface area to accommodate the increased nerve cells is provided by infoldings of the cortex to form deep fissures and ridges between them, the human brain showing a highly convoluted appearance.

There is evidence to show that increased mental capacity is accompanied by greater complexity of interconnections between nerve cells and groups of nerve cells, and that the complexity of interconnections increases through life. 'Consciousness' appears to depend on the maintenance of such interconnections of the brain as a whole, and cannot be associated with any particular area or structure. The exact relation of consciousness with the anatomical structure of the brain, if any exists, is still an enigma.

BRISTLEWORMS (Polychaetes). These have segmented bodies, and a distinct head often with tentacles, eyes and other sensory structures. Each segment may bear appendages or feet which have many chaetae (bristles)—hence the name 'polychaete' (*poly* many). They are cosmopolitan in distribution. Indeed, the polychaete fauna of India is very similar to that of the rest of the Indo-Pacific and Australian regions. Their distribution is believed to be regulated mainly by temperature.

They could be broadly divided into two major groups:

1. *The Errant Worms.* These are active worms which live in crevices of coral, beneath stones and shells and amongst algae between tide-marks on the shore. They vary in shape from elongate to oval and have uniform paired feet along the length of the body. The feet are mere extensions of the body-wall. *Nereis* is a typical representative and is called the Sea Centipede because of the general resemblance between the body, feet and locomotion of both. Some worms have flattened, overlapping scale-like structures on their backs which can be raised and lowered to allow a ventilating current to pass over the body.

They are vicious and carnivorous and have powerful jaws at the end of an eversible proboscis which can be shot out to capture a luckless sea creature. Besides eyes, their heads also have other sensory structures. The sexes are separate. The eggs are shed into the water where they are fertilized. The embryo develops into a planktonic larva which is drifted about by the ocean currents. Some worms can reproduce asexually by budding off segments from various points.

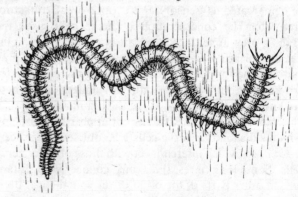

The sea-centipede *Nereis*. It is about 20 cm long when fully grown.

Some Indian species have been found in Chilka Lake and other fresh or partially brackish waters.

2. *The Sedentary Worms*. These worms are highly specialized for a sedentary existence. They live in temporary burrows, interconnecting galleries or permanent tubes made of lime, mucous, sand, pieces of coral and sometimes spicules of sponges. Since they are confined to their tubes, they have developed different ways of exploiting water currents for food. Some worms live in permanently fixed tubes, others carry their tubes about with them.

Serpula, Sabella and many others are called the Feather-duster or Fan worms as only their heads crowned with tentacles protrude out of their tubes like the corolla of a flower. The tentacles sieve out tiny food particles from the water and these are carried by elaborate ciliated tracts towards the mouth, individual particles being sorted out according to size and shape.

Chaetopterus lives in a U–shaped tube and has a bizarre shape. A pair of its feet are winglike and help to suspend a mucous net in which food particles are trapped. It creates its own water current by the pumping movements of special semicircular fan-shaped appendages which abut along the sides of the burrow. The mucous net with the trapped food is periodically swallowed whole. Then the worm goes about the process of secreting another mucous trap.

Terebella has long tentacles which emerge from the top of the tube and crawl and creep over the ocean floor. Detritus adheres to the mucus on the tentacles and is conveyed in a special groove towards the mouth. These tentacles may even be wiped across the mouth!

These ingenious polychaetes have mastered the art of adaptation. Many of them even have giant nerve fibres which can bring about a very rapid withdrawal reflex, ensuring that the worms have retreated securely into their tubes as soon as the faintest shadow falls over special receptor cells on their tentacles.

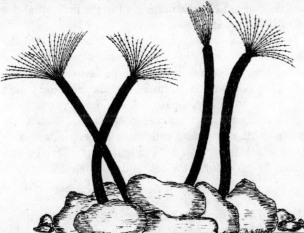

The Fan-worm *Sabella* × 1, spread out at the top of the tube it has made from mud and sand.

Another amazing attribute of the polychaetes is their brilliant coloration. The scale-worm *Aphrodite* has a covering of felt-like hair which displays all the colours of the rainbow in changing light. Many bristleworms are gaily striped and spotted and have brightly coloured gills. Along with sea slugs, they are the equivalent of tropical birds in a rain-forest.

See also WORMS.

BRITTLE STARS are a lesser group of starfish. They have a central disc more sharply marked off from the arms than in starfish. The arms are comparatively very long, tubular and tapering without open grooves. For this reason they are also called Serpent Stars. The arms are muscular, covered by four rows of plates—one row

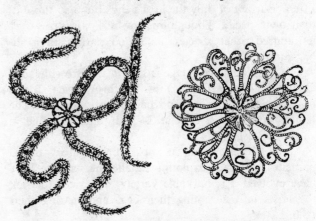

Brittle stars: left, *Ophiocnemis* × ½ ; right, *Astrophyton* × 1/10 (From the Madras Museum Bulletin and Parker & Haswell).

each above, below and on either side. The plates on the sides bear spines. The tube-feet along the arms have no ampullae: they are not used for walking. They function more as feelers. There are a few large tube-feet around the mouth and they scoop out mud and collect prey. Brittle stars move not by tube-feet but by twisting and wriggling the arms which also serve to capture food. When attacked the animal's arms break into several pieces. This enables it to escape from enemies. It can regenerate the lost arms in the course of a few days.

In one family, called the Basketfish, the arms are so profusely branched and curled as to form a sort of network in which fish and crabs are caught. *Ophiocnemis* and *Astrophyton* are common genera found on our sea coast.

See ECHINODERMS and STARFISH.

BROOD PARASITISM: Obligate nest parasitism is practised by about 80 bird species belonging to several unrelated families, e.g. the American Cowbirds, the

African Cuckoo-Weaver, the Old World cuckoos (subfamily Cuculinae), Honeyguides, and the Blackheaded Duck.

The possible evolutionary origin of brood parasitism can be inferred by comparing the parasitic species with their closest non-parasitic relatives. The habits of the Monogamous Baywinged Cowbird (*Molothrus badius*) suggest the beginnings of parasitic tendencies. In exceptional cases they build their own nests, but they usually defer breeding until the nests of other species are available, which they acquire by fighting and then use for laying eggs and rearing their young.

A further stage is illustrated by the Shining Cowbird (*Molothrus bonariensis*). Most females simply lay many eggs on the ground which are variable in colour and shape, while some lay in the nest of host birds. Females often destroy not only the host's eggs but also those of their own species. Though this cowbird parasitizes over a hundred other species, its brood parasitism is ill-organized.

The most advanced stage is seen in the Brownheaded Cowbird (*Molothrus ater*) of North America. Males outnumber females, and no pair-bond is formed. The female lays 4 to 6 eggs at 24-hour intervals, one in each host nest.

About 50 species of cuckoos are obligatory brood parasites, generally exploiting smaller passerines. They have evolved a remarkable variety of devices to trick the hosts into accepting their eggs and rearing their young.

The Indian Brainfever Bird (*Cuculus varius*) and the Large Hawk-Cuckoo (*Cuculus sparverioides*), closely resemble the Shikra (*Accipiter badius*) and Besra Sparrow-Hawk (*Accipiter virgatus*) respectively in plumage. The Brainfever Bird mimics the shikra also in its flight and has been observed to lure its host away from the nest. The adaptive significance of this manoeuvre may lie in the fact that song birds flee from accipitrine hawks passing overhead but are less likely to leave their nest unguarded otherwise.

An elegant ruse is employed by the Indian Koel (*Eudynamys scolopacea*) to outwit its host the crow, *Corvus splendens*. The male approaches the host nest, calls loudly and allows itself to be chased by the owners; meanwhile the female koel slips in quickly and lays her egg.

The female cuckoo has an unusually extrusible cloaca which functions like an ovipositor, permitting her to drop eggs into nests hidden in crevices and holes that some smaller fosterers occupy. The eggs of the Pied Crested Cuckoo (*Clamator jacobinus*) have a thicker chalky shell than its host the babbler's. Some cuckoos' eggs have double shells, separated by a duplicate membrane to reduce the danger of breakage when they are dropped, as opposed to laid, into the nests.

© Loke Wan Tho

Young cuckoo being fed by redstart

The eggs of cuckoos are close in size and coloration to those of the hosts, often small in proportion to the female's body size, which also permits a larger number to be laid. The rejection or weeding out of cuckoo eggs by the foster parents if they are unlike their own has brought about this mimicry.

Another adaptation is the shorter period of embryonic development. The young cuckoo hatches out after only twelve and a half days, and as the egg is deposited in the nest almost when the host starts laying its own, it hatches sooner or at the same time as the nestlings of the foster parent. Upon hatching, the young cuckoo usually eliminates the eggs or nestlings of its host. When one of the host eggs or nestlings presses against the back of a cuckoo-nestling in the first day or two, the little interloper crawls backwards up the side of the nest and heaves it over the edge. Young parasitic honeyguides have sharp hooks on the tips of their mandibles with which they kill the host siblings. No passerine nestlings have evolved such a mechanism for doing away with sibling nest competitors—an adaptation that would be advantageous for the individual but harmful for the species.

In two cuckoo species, *Clamator glandarius* and the koel (*Eudynamys scolopacea*), the young resemble those of the host. The host species of the latter are corvids which are relatively large birds, and the parasitic young are raised in the company of the host young. The significance of this mimicry seems to be that if the host parents recognized a nestling as suspiciously different from their own they might eject the parasite.

See CUCKOO for the Koel, and NATURAL SELECTION.

A.J.T.J

BUGS. The true bugs comprise the dominant insect order of the Exopterygota and are extremely variable in form and habits, but possess mouth-parts that characterize this group of mostly plant-feeding insects. Measuring from 1 to 90 mm, these insects have piercing and sucking mouth-parts that help them to obtain plant fluids, or, as some of their members have evolved to feed on animal matter, to suck animal juices and blood. The Hemiptera, that has been divided into two very distinct suborders, the Heteroptera (many families carnivorous) and the Homoptera (all phytophagous), is a very old insect order, perhaps originating in the Carboniferous, about 300 million years ago.

The bugs feed on all parts of the plant, from underground parts such as roots to exposed parts like leaves, stems, flowers and fruits. Some specialized forms also feed on plant seeds, either on the plant or fallen on ground. The Homoptera are all phytophagous and denizens of the land, while the Heteroptera have many carnivorous groups and also those that are aquatic or semi-aquatic. Most of these 'watery' groups are carnivorous, but if they are not, they usually feed on algae and other small plant particles in water. Some of the larger heteropterous families, like the Miridae, Pentatomidae and Lygaeidae, have both carnivorous and phytophagous members. As man well knows, there are also blood-suckers, like the horrifying bedbugs and some assassin bugs that transmit deadly diseases. Another interesting character of these true bugs is that many of them can produce sound like the grasshoppers and crickets and some other insect orders. All kinds of their body parts are utilized for making sounds, including mouth-parts, legs, wings, etc., as well as tymbals which are well developed in the familiar cicadas that make their presence 'heard' in any forest in the tropics. Bugs also possess special glands that secrete toxic substances which they use to repel predators. Several heteropterous bugs are good mimics, especially of ants, while most other Homoptera and Heteroptera blend with the plant part on which they usually occur, many homopterous bugs being covered with waxy secretions and others producing the 'honeydew' that is sought by many insects, especially ants, which then develop a mutually beneficial partnership with such bugs.

Most Hemiptera lay eggs, though ovoviviparity and viviparity is common in some groups such as the Coccoidea and the Aphididae and Polyctenidae. Some Aleyrodidae and Coccoidea also exhibit parthenogenesis and hermaphroditism. The eggs that are laid are often glued to the plant parts, and are laid in large masses. Other groups of bugs may insert their eggs in plant tissues, embed them between two adjacent plant parts, or lay them in soil, in litter, in crevices or in material in close surroundings of the host animal. The newly hatched nymphs are gregarious in most bugs and these gradually segregate and move independently after the early instars. In bug groups which are restricted to sucking plant sap from the phloem vessels, the nymphs, as well as adults, may continue to live in small to large colonies through-

out their life-cycle. As in many exopterygote insects, the young resemble the adults both in manner of feeding and in general external appearance. In case of the predacious species, the young may eat less prey than the adults and in the phytophagous forms, the young may prefer different parts of the plant than the adults or even different host plant species. Some families of bugs have young living underground and feeding on roots while the adults are free-living above ground and feed on external parts of the plants, like in the familiar cicadas. Many homopterous groups like the aphids, psyllids and coccids cause plants to form galls, within which their young undergo their transformations to the adult stage.

The Hemiptera have a number of natural enemies, both parasites and predators, the latter of the vertebrate and invertebrate groups of animals. Besides birds, reptiles and others, the arthropod predators of bugs include bugs themselves, beetles, larvae of flies, mites, spiders, lacewings, larvae of butterflies and moths, digger wasps, and ants, among other groups. Among their parasites that attack eggs, nymphs and adult bugs, the principal insect order that contains such enemies is the Hymenoptera. Many chalcidoid families, such as the Encyrtidae, Aphelinidae, Eulophidae, Elasmidae, Trichogrammatidae and Mymaridae have developed a 'liking' for bugs as their main host groups. The larger wasp families like the Braconidae and several distinct genera of the fly family Tachinidae also parasitize bugs. The Stylopoidea also attack many auchenorrhynchous Homoptera as well as Pentatomidae. Bugs are also known to be parasitized by nematodes and prone to attack by fungi.

The Hemiptera include some of the major pest species of insects that ravage crops all over the world, and many of them have been introduced to non-endemic regions of the world by man. Though both suborders contain pests, perhaps the Homoptera possesses the larger number, Auchenorrhyncha and Sternorrhyncha having many pest species. The sap-sucking insects are especially harmful when, like the aphids, they aggregate in huge colonies on the young plants and drain them of their well-needed nutrition. The loss of sap tends to make these plants stunted in growth or otherwise abnormal in form. These and other bugs that feed on leaves and stems, besides on phloem, inject toxins into the plant-tissues that make the plant parts die or susceptible to entry by other organisms, like fungi. The phloem-feeders that form large colonies on the plants, especially in the growing regions, like the aphids and the coccids, for example, produce copious honeydew for their own reasons, which, besides being sought after by other insects, also helps the growth of a sooty mould fungus, which being black in colour, completely covers the green leaves and restricts their functioning. Some of these

© Kumar Ghorpadé

Aphids clustered on an inflorescence × ³/₄

bug families also transmit plant viruses which actually do a greater amount of damage to the plant than the bugs do by themselves. Besides roots, leaves and stems, some bug groups also attack seeds, both formed and developing as well as fibrous plant parts. Others, like the cicadas, damage branches and stems of trees by slitting them open for laying eggs. As mentioned above, bedbugs sustain themselves on the blood of man and the triatomine reduviids attack him and transmit a disease called trypanosomiasis.

Among the predacious families or groups of the Hemiptera, there are some species that are useful in controlling populations of pest species of insects and thereby helping man's needs. The more important bug families that possess such predators are the Anthocoridae, Miridae, Reduviidae and the Nabidae. Other groups are of use to man because of their products, like the lac insects (shellac) and the cochineal insects (dye). Others are used as biological control agents against 'man-made' weeds such as *Lantana* which is kept within limits by a tingid, *Teleonemia scrupulosa*.

The order Hemiptera is generally subdivided broadly into two suborders, the Heteroptera and the Homoptera, this latter again divided into the Auchenorrhyncha and Sternorrhyncha. There are probably more than 5000 described species of Indian bugs, but owing to lack of updating of current knowledge and little systematic research being done on several families, the exact number is not available. The known bugs are, again, only a fraction of the actually existing ones, and unless field surveys to collect and identify our wealth of bugs and other insects are made, we may not get to catalogue all our species, many of them being put under threat of extinction by man's reckless slaughter of vegetation and the general environment.

Owing to a large number of families being recognized in the Hemiptera, most of them small and either not being encountered by most of us, or of no importance to us directly, I will only deal with families that are both large and of 'economic' importance to mankind. Of the Homoptera, the section Auchenorrhyncha includes the superfamily Fulgoroidea, which are plant-hoppers of various kinds, that either suck plant sap or feed on fungi. Ten thousand species of this superfamily alone are now known throughout the world. The major families are treated here. What are commonly known as 'plant-hoppers' to the student belong to the family Delphacidae, of which the Brown Plant-hopper (*Nilaparvata lugens*) is now in public notice because of its ravages of the paddy crop in India. Several species of this family transmit diseases caused by viruses in plants and others infest crop plants like sugarcane, sorghum, maize, etc., in large numbers and feed on their sap. The common Indian species are *Sogatella longifurcifera* and *Peregrinus maidis*. *Eurybrachis tomentosa* belongs to the family Eurybrachidae and is a common insect in India noticeable by the frothy white mass that the female carries behind her. The 'lantern flies' (Fulgoridae) are a diverse group and now segregated into many smaller families that once contained familiar bugs like *Pyrops*, *Dictyophara*, *Drona*, *Phenice* and *Ricania fenestrata*. These are of curious form and generally medium sized to large bugs that are found on exposed plant parts, some being very beautifully coloured and flying to light, thus their name. The Cercopidae, Aphrophoridae and Machaerotidae are commonly called spittle bugs or froghoppers, the adults and nymphs living together and the young being confined in spittle-like white frothy masses or calcareous tubes on the stems of plants, for protection from evaporation. They are more abundant in the tropics and many cercopids are very brightly coloured insects. The Machaerotidae often have a spine-like structure on their back and in this resemble the Membracidae of the superfamily Cicadelloidea. Some common Indian species and genera are *Clovia bipunctata*, *Ptyelus nebulosus*, *Cosmoscarta*, *Abidama* and *Callitettix*.

The superfamily Cicadoidea contains some of the largest members of the Homoptera, the cicadas, that are a familiar sound in the forest. They are abundantly represented in the warmer regions of the globe and possess tymbals, by vibrating which each species produces a species-specific sound, that curiously begins and ends abruptly even though hundreds of individuals produce it at the same time! How this timing is coordinated is a mystery to man even now. The female cuts slits in branches of trees and lays eggs in them and the nymphs drop to the ground on hatching, burrow into it and feed on roots. The nymphal stage may take several years and one, called the '17-year Cicada', takes those many years to complete one life-cycle! *Platypleura mackinnon* and *P. octoguttata* are the common Indian species, though many more are known and just as common in different parts of the subcontinent.

5 A Praying Mantis Eating a Dragonfly

The Praying Mantis is one of the insect world's most beneficial predators. It is seen here making a meal of a dragonfly.

6 Decorated Chital

One of Nature's unsolved mysteries is why stags, during the period of rut, 'decorate' their crowns with grasses and shrubs dug out of the ground with their antlers.

The next superfamily is the Cicadelloidea of which the leafhoppers belong to the huge family Cicadellidae that are among the most common and abundant of all insects, using many kinds of plants as their hosts. Almost every individual plant, of many species, when disturbed, will result in many of these leafhoppers leaping out from it! This family is subdivided into many subfamilies and some common Indian genera and species of this group which comprises many pest species, including a few vectors of plant viruses, are: *Amritodus atkinsoni*, *Idioscopus clypealis*, *Ledra mutica*, *Copana spectra*, *C. unimaculata*, *Krisna strigicollis*, *Mukaria splendida*, *Penthimia*, *Hecalus*, *Goniagnathus*, *Nephotettix*, *Deltocephalus*, *Typhlocyba* and *Empoasca*. The Membracidae possess bull-like 'horns' and are generally much stouter plant-hopping bugs, that could be termed the 'bull hoppers'. Many species are intimately associated with ants and live in groups, especially on thorny trees like *Acacia*. *Oxyrhachis tarandus* and *Leptocentrus taurus* are some of the common Indian species.

We now come to the next section of the Homoptera, the Sternorrhyncha, containing four superfamilies that contain the peculiarly 'colonial', gregarious sap suckers, many of whom are important pests of crop plants. Superfamily Psylloidea contains a single family, the Psyllidae, commonly called psyllids, of which more than 150 species are now described from our subcontinent, whereas the world fauna now totals some 1250 species. These psyllids are mostly specific to particular plant hosts that they prefer, and like most other Sternorrhyncha that produce honeydew, have some species that are attended by ants. Psyllidae are notable gall makers and many plant galls are characteristic in shape according to the species that feeds on that particular plant and causes the gall, so that, based on the shape of the gall the species of psyllid that caused it can be known. *Ctenaphalara*, *Trioza*, *Alstonia* and *Psylla* contain species that are common in India.

The Aphidoidea is made up of four families with nearly 4000 described species from the world and probably in the region of 800 Indian species. These insects are those that were called 'plant-lice' in early times, and even perhaps now, and their habit of forming small to huge colonies on the fresh plant growths had made them both well known and studied. A large percentage of aphids (Aphididae) are known to be vectors of plant viruses and many species assume pest proportions on many of man's cultivated plants. Some species have been accidentally transported by man to other regions of the world and a few have therefore become almost cosmopolitan in distribution. Most species of aphids are polymorphic and males are either absent or produced only at certain times during the year. Aphid females are both winged and apterous, may lay eggs or nymphs and may require fertilization by

Aphis craccivora *aphids clustered on inflorescence of glyricidia* (Glyricidia sepium), *tiny wasp on extreme left flower is a proctotrupoid* (Callaspidia rufithorax) *parasitoid of predatory hover-fly larvae* × 1

males or be parthenogenetic. The life-cycle of many aphid species is complicated and has been given a lot of attention by scientists. Some species have alternating host plants and generations that may be winged at one time and apterous at other times. Most aphids have seven or more generations per year, each lasting from 2 weeks to 2 months. A single viviparous female aphid may give birth to as many as 150 or more young in her lifetime. These insects live as colonies on various parts of the plant, like leaves, buds, shoots, stems, and roots. Many aphid species are attended by ants which tend them assiduously so that they have come to be called 'ant-cattle'. Some of the more common Indian aphids, that are also pests, are *Rhopalosiphum maidis* (on maize), *Myzus persicae* (on tobacco), *Macrosiphum rosaeiformis* (on roses), *Aphis gossypii* (on cotton), *A. craccivora* (on groundnut), *Lipaphis erysimi* (on crucifers), *Toxoptera aurantii* (on coffee), *T. citricidus* (on limes, oranges) and the peculiar underground *Tetraneura nigriabdominalis* (on finger millet). The woolly aphid which is an important pest of apples on the Himalaya, *Eriosoma lanigerum*, belongs to the related family Pemphigidae, which also has root-feeding species, *Smynthurodes betae* and *Geoica* spp. These insects alternate between their primary and secondary plant hosts which are specific to each species. The genera *Adelges* and *Pineus* of the family Adelgidae and *Phylloxera* (Phylloxeridae) are important pests of conifers in the temperate regions.

The superfamily Aleyrodoidea again contains a single family, the Aleyrodidae or what are called whiteflies. Some 100 Indian species are now known out of a world total of more than 1000 species. The adults are winged and have a white powdery meal over their bodies, while the young are scale-like and occur in groups on the undersides of leaves of their favourite plant hosts. Since the adults are generally very similar in many different species, the classification of the whiteflies is based principally on the cuticle of the last larval instar! As in aphids, whiteflies also reproduce sexually or parthenogenetically. Again, a single female may lay over 200 eggs on the underside of a leaf and the nymphs pass through four instars on the same plant. The common Indian species are *Trialeurodes ricini* (on castor), *Neomaskellia bergii* (on sugarcane), and the polyphagous virus vector, *Benisia tabaci*.

The mealybugs and scale-insects belong to the other large superfamily Coccoidea that contains more than 6000 species all over the world and probably more than 150 or 200 in India. Many species have been carried about from region to region by man. Several families, the common ones being the Pseudococcidae, Coccidae, Margaroridae, Eriococcidae, Asterolecaniidae and Diaspididae, are recognized but for our purpose I will deal with the superfamily as a unit. These insects are broadly divided into the armoured scales, unarmoured scales and the mealybugs. The former are black or brown and generally hard to the touch, occurring as huge masses on leaves or stems of plants. The genera *Aspidiotus*, *Aonidiella*, *Chrysomphalus*, *Quadraspidiotus* and *Parlatoria* are the common ones in the country. The unarmoured scales that lack a hard exterior and those with a mealy white wax secretion include species of genera such as *Icerya*, *Pseudococcus*, *Coccus*, *Chloropulvinaria*, *Saissetia*, *Ceroplastes*, *Planococcus*, *Dactylopius*, *Eriococcus*, *Orthezia* and *Monophlebus*. The lac insect, *Kerria lacca*, is actually a scale-insect that produces the lac on stems that is used by man to make shellac and that has grown into a valuable product.

The other suborder of Hemiptera is termed the Heteroptera and this can be divided into three subgroups: (1) Hydrocorisae, including the aquatic or littoral bugs of the superfamilies Corixoidea, Notonec-

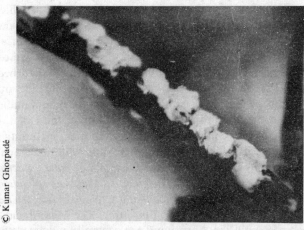

A mealybug (Chloropulvinaria psidii) *on guava stem* × 2

Armoured scale insect (Saissetia *sp.*) *on twig of sandal* (Santalum album) × 1

toidea and Ochteroidea; (2) Amphibicorisae, including only the superfamily Gerroidea, which includes bugs that move about on the water surface; and (3) Geocorisae, that includes all the other superfamilies whose members are mostly terrestrial but some are also littoral or semi-aquatic in habit. We shall commence with the families of this group and move backwards ending with those of the Hydrocorisae.

The Enicocephalidae that solely comprise the superfamily Enicocephaloidea have bugs that look like the assassin bugs, belonging to the family Reduviidae, but are a more primitive group of insects. These bugs live mostly on the forest floor occurring under logs and stones and some adults may be flightless. The ones that can fly often form flying swarms at dusk and these insects generally feed on tiny insects of many families and orders. The genus *Enicocephalus* has some species in the subcontinent.

The superfamily Cimicoidea includes the bedbugs that belong to the family Cimicidae and includes wingless insects that suck the blood of birds and mammals including man. The common bedbug in India is *Cimex lectularius*. The Polyctenidae are similar in that besides lacking wings (the fore wings may be very short), they also lack eyes and are parasites of bats. *Polyctenes* is a common genus of this family. The Anthocoridae are closely related and either feed on other insects or their eggs, or suck blood of vertebrates. Species of *Orius* and *Triphleps tantilus* are found in India. Another small family, the Nabidae, with the prettily marked *Prostemma flavomaculata* being common, contains bugs predacious on other insects. The largest family of this superfamily and one of the largest families of the Heteroptera, the Miridae, includes members that are sometimes very numerous and abundant in favoured areas. Unlike the other families of this superfamily, the Miridae are mainly phytophagous, but quite a few species are predacious. The family includes both pest species and those that are beneficial predators of injurious insects. *Calo-*

coris angustatus, Helopeltis theivora, H. antonii, Disphinctus politus, Cyrtorhinus liridipennis and *Halticus minutus* are some common Indian species.

The lace bugs, as they are called, are very beautifully formed insects and belong to the Tingidae (superfamily Tingoidea), a family that contains plant-feeding bugs, some of which are pests of cultivated plants. *Teleonemia scrupulosa* is a useful species attacking the *Lantana* plant, while *Monanthia globulifera* and *Urentius echinus* feed on crops. The Reduviidae belong to the Reduvioidea and are the familiar 'assassin bugs', that are wholly predacious in habit and some even biting man, though not living on blood. Several groups of arthropods are attacked and fed upon and many reduviids are very beautifully marked, though many others are cryptic in coloration and harmonize with the plant they inhabit. The common Indian assassin bugs are *Acanthaspis quinquespinosa, A. rama, Triatoma rubrofasciata, Holoptilus, Myophanes, Polytoxus, Tribelocephala, Sastrapada, Oncocephalus, Pirates sanctus, Ectrychates, Physorhynchus marginatus, Harpactor costalis, Sycanus versicolor* and *Isyndus heros.*

The superfamily Saldoidea contains one very commonly found family of insects, the Saldidae, whose members frequent the shores of inland waters or estuaries near the sea and are very active. *Salda dixoni* and *Valleriola cicindeloides* are common Indian species, the latter sometimes being placed in a distinct family. The Aradidae (Aradoidea) are peculiar-looking bugs that occur on the forest floor and also in rotting logs or on fungi. *Aradus* and *Kumaressa* are two of the genera. Of the superfamily Coreoidea, the large family Coreidae contains many common and abundant bugs that are plant-feeding, mainly on fruits and some species are notorious pests, for example, *Leptocorisa varicornis, Clavigralla gibbosa, Anoplocnemis phasiana, Riptortus pedestris* and *Corizus bengalensis.* Some of these genera, like *Leptocorisa* and *Riptortus*, are now put in the family Alydidae.

The Lygaeoidea also contains a large number of very common plant bugs and the dominant family is the Lygaeidae with species like *Oxycarenus hyalipennis, Lygaeus militaris, Geocoris tricolor, Nysius minor* and *Elasmolomus sordidus* being frequently observed in fields. The speciality of its members is that most of them feed on seeds of plants, but species of *Geocoris*, for example, are predacious, many feeding on plant mites (Acari). The Pyrrhocoridae include the commonly seen reddish *Dysdercus cingulatus* that is a pest of cotton, and another species, called *Odontopus nigricornis*, is also frequent.

The last, and the most numerous, superfamily is the Pentatomoidea, commonly called shield bugs or stink bugs, and includes the Pentatomidae that includes some very beautifully marked bugs. Again, most are phyto-

© O. C. Edwards

Stink bug (Catacanthus incarnatus) *cleaning its antenna on canna leaf* × 3

phagous but some species are carnivorous too. This large family is currently divided into a number of smaller families and the following species are the commonly noticed ones in the country: *Coptosoma cribraria, Cantao ocellatus, Chrysocoris stolli, Asopus malabaricus, Bagrada cruciferarum, Dolycoris indicus, Cydnus* spp., *Halys dentatus, Nezara viridula, Cantheconidea furcellata, Aspongopus janus, Tessaratoma javanica* and *Cyclopelta siccifolia.*

The second group includes the family Gerridae that are commonly called water-striders and are found on the surface of water on lakes and streams. They feed on small animals, live or dead, as do the other families of this superfamily. *Gerris* and *Halobetes* (Gerridae), *Velia* and *Microvelia* (Veliidae), *Hydrometra* (Hydrometridae) and *Mesovelia* (Mesovelidae) are the common Indian insects of this group.

The last group comprises the Ochteridae (Ochteroidea) which are found commonly on sandy soil near water bodies and are predacious, *Ochterus* being a common genus. The superfamily Notonectoidea also contain all carnivorous, aquatic insects that live for most of the time (all nymphs do) under water. *Notonecta* (Notonectidae), *Nepa, Ranatra, Laccotrephes* (Nepidae), *Cheirochela* (Naucoridae) and the giant bugs which are the largest known, being more than four inches long and feeding on frogs and tadpoles, *Belostoma* (Belostomatidae), are the common Indian genera. The Corixoidea contains only the Corixidae or the 'water boatmen', which are very abundant and often come in huge numbers to lights. They feed on plant matter mostly and *Corixa hieroglyphica* and *Micronecta striata* are common in India.

BULBUL (family Pycnonotidae). Bulbuls are a widely distributed tropical family ranging from Africa to SE Asia from the sea coast to the high hills. There are nearly 30 species in our area. In ecologically suitable areas several species are often found together, while elsewhere each species or group of species is isolated to its own specific habitat. Though soberly dressed, chiefly in greys and browns and olives, many have jaunty crests and bright red or yellow or other markings which give them a perky well-groomed look. Their jaunty temperament and cheerful musical calls make them most attractive creatures. They can be easily tamed and are popular cage birds. They live on fruits, insects and spiders.

Our two most widely distributed and common bulbuls are the Redvented (*Pycnonotus cafer*) and Redwhiskered (*P. jocosus*). The symbolic 'bulbul' of traditional Urdu poetry refers, in fact to the Persian nightingale, not found in our area.

BUSTARDS are three-toed cursorial birds belonging to the family Otididae (order Gruiformes) which has 22 species in the Old World, 16 of which live in the Ethiopian region. One species occurs in Australia and 5 in the Indian subregion, two as migrants and three as residents. The majority are polygamous, the females alone incubating and brooding. The main threats to the bustard populations are the supplanting of their habitats by cultivation, modern farming methods, chemical pesticides and tourism.

The Great Indian Bustard (*Choriotis nigriceps*) is the largest of our bustards, standing over 1 metre. The male is the larger, with a thicker, whiter neck. The population is scattered over India and spills into Paki-

stan. It faces extinction since its natural habitat of savanna and stony scrub is being agriculturized. Breeding is slow, as the birds do not mature till 3 to 5 years, and usually lay a solitary egg. This varies in colour with the terrain, and the incubation period is 24–28 days. While breeding, the cock develops an air-sac in the neck, which when inflated emits a loud boom. During courtship display, the back and tail feathers, wing and tail-coverts, are raised. The prolonged breeding season starts with the monsoon. These omnivorous birds are shy and silent, with cocks, hens and juveniles forming separate groups. The flight is low, with continuous wing-beats.

Our second largest bustard, the Bengal Florican (*Eupodotis bengalensis*), ranges from Kumaon to Arunachal Pradesh. It inhabits mainly Assam and favours tall grass near riverine forest. The male is pied, the female fawn with black markings. The two olive-brown eggs are laid between March and August. Similar but smaller is the endemic Lesser Florican *Sypheotides indica*, whose local movements are influenced by the monsoon. The arid grassland and scrub in which it breeds has largely come under-cultivation, so the birds are now oftener seen in standing crops. The cock has a spectacular courtship display in which he springs into the air, exposing himself above the tall grass. At this time cocks are vulnerable and heavily poached by hunters. Between June and October, the bird lays 4 to 6 olive-brown or green eggs.

The Houbara or Macqueen's Bustard (*Chlamydotis undulata*) migrates from the West Asian region to the semi-desert areas in NW India. In spite of its concealing coloration, heavy hunting with falcons (e.g. by Arab potentates) has reduced the population considerably of late. Distinctive are its black-and-white ruff, and its habit of ejecting a disabling sticky faecal fluid on attacking birds of prey.

The Little Bustard, *Otis tetrax*, is a rare straggler into N India. Its 45 cm length makes it one of the

© Asad Rahmani (BNHS)

Great Indian Bustard, male in partial display × ¹/₁₀

© Asad Rahmani (BNHS)

Bustard habitat, Jaisalmer, Rajasthan

smallest bustards, recognizable in flight by its black neck, black-and-white pectoral collar, white underparts and hissing butterfly-like wing-beats.

BUTTERFLIES. Of all insects, perhaps the butterflies captured the interest of man first. Owing to their comparatively large size, beautiful colour patterns and universal occurrence in most parts of the world where man himself lives, these 'winged beauties' have monopolized popular attention by man and become the most sought-after 'collector's item'. Being but a small part of the large insect order Lepidoptera that includes the more extensive and abundant moths too, their habit of flying by day rather than at night like most moths, is an important distinction between these two lepidopterous groups. The other character that easily distinguishes a butterfly from a moth is the 'clubbed' antenna of butterflies, whereas moths usually have pointed antennae among their several forms, but almost never a clubbed antenna like the butterflies. The skippers (Hesperiidae) are somewhat intermediate between the butterflies and the moths, and though their antennae are thickened at the tips, they also have a hook-like structure on the 'club'.

Tropical forests, with their wealth of plant life and abundant sunshine, are a natural haven for most species of butterflies. But even temperate regions of the earth possess a diversity of butterfly species and it is only in the more arid, uncultivated lands that we find their numbers limited. On mountains, butterfly species are restricted to altitudinal zones just like others are restricted to certain latitudes, from the equator to the poles. The Indian subcontinent, with its rich diversity of habitats, from the ice-capped Himalayan mountains to the parched Thar Desert, and from the lush, steamy tropical forests in the Assam region to the mangrove-lined sea coasts in peninsular India, is the home of around 1000 species of butterflies.

So much has been written about butterflies, even of Indian ones, and since quite a few well illustrated books are available, I will not go into too much detail here. As in most holometabolous insects, butterflies have four life-stages—the egg, the larva, the pupa and the adult butterfly. Unlike most other kinds of insects, the life-histories of butterflies are fairly well known and have been observed and documented for many living species. Like the birdwatcher, the 'butterfly-hunter' is dealing with an animal group that has been studied in great detail. In India, however, we know even the immature stages for probably less than half of our species. The other fascinating aspects of butterfly study, like their duration of life-stages, host plants, enemies, 'courtship' and other behaviour, migration, hibernation/estivation, mimicry, flight periods, distributional range, preferred flowers that each species visits, and their inter-

relationships with other animals and plants (ecology), etc., are a veritable 'open book' for the butterfly enthusiast here. I may mention here that the days of amassing huge collections of 'every' species of butterfly in a given geographical area are gone; collectors must take care to help protect populations of these wonderful creatures, many species of which are now pushed by man into restricted habitats. What is needed is to use the naked eye and the binocular, rather than the net and the poison bottle to needlessly kill these pretty insects for the 'collector's pride'! In many tropical countries 'butterfly farms' have emerged, where the more popular and attractive species are reared on cultivated host plants, and then slaughtered for the use of man's fancy. Perhaps the danger to threatened species of butterflies (and most other forms of life) is not in the actual 'catching and killing' for pleasure or profit, but in man's shortsighted destruction of natural habitats.

The length of the life-cycle of a butterfly species usually takes three to four weeks: the egg and pupal stages occupying about a week each, while the larval (feeding) stage may vary from a week to two weeks. The adult life-span is fairly long and may even exceed a year. Due to unfavourable climatic conditions or food scarcity, many butterflies undergo hibernation in any of their four stages. Since a life-cycle takes a month or so, theoretically there can be 12 generations in a year for most tropical butterflies. But certain optimum breeding requirements of temperature and food prevent most butterflies from producing so many broods; some may have just one brood a year and others may produce four or five. Most butterflies feed on the foliage of plants, especially fresh shoots, while some may bore fruit or pods or even be carnivorous and feed on other insects. The adult butterfly's diet usually consists of flower nectar and free water, and some species are well known for their mass assemblages at their 'salt-licks'! Besides this, some butterflies are also fond of animal dung and urine, fallen rotting fruit, honeydew of Homoptera, tree sap, or even wood ash. Among their enemies are small parasitic wasps and flies that feed on the eggs, larvae or pupae, and the adults and larvae are prey for birds and many other insects and higher animals. For this reason, the larvae of many species of butterflies have protected themselves by warning or protective mimetic coloration or by developing 'hairs' that carry poisonous chemicals.

The Rhopalocera or butterflies (moths are called Heterocera) are divided into many families, and systematic entomologists differ in their opinions as to how many 'natural' groups there actually are! However, for convenience, I will deal with Indian butterflies, sorting them into as many distinct groups as we can make out in the field. Currently, the butterflies are treated as belonging to two superfamilies of the Lepidoptera, the

Papilionoidea and the Hesperioidea, the latter containing the Hesperiidae only. Even in such a well-known group like the butterflies, the classification and names of many taxa remain inadequate and uncertain; I have tried to use the currently operative ones in this and other insect groups but they may actually be incorrect or outdated.

The Papilionidae, or Swallowtails, belong to the subfamily (group) that includes some of the largest and most beautiful of all butterflies. Many possess 'tails' on their rounded or elongate hind wings and their fore wings are generally dark brown or black with a variety of colourful markings on these somewhat broad pointed wings. The legs of swallowtails are all well developed and their claws are large and simple. The pupae have a girdle and are attached at the tail end, with their heads upwards, and the larvae are usually smooth without hairs, though some processes may be present. The Lime Butterfly (*Papilio demoleus*) is one of the common species we find on the plains sometimes causing some damage to citrus orchards. Its young larvae are remarkable mimics of bird droppings, the older ones changing to a vivid green with brown bands. Some other common or beautiful Indian Swallowtails are: *Troides helena* (Common Birdwing), *T. aeacus* (Golden Birdwing), *Parides philoxenus* (Common Windmill), *Pachliopta aristolochiae* (Common Rose), *Papilio clytia* (Common Mime), *P. polytes* (Common Mormon) [the female of this species mimics the female of the Common Rose], *P. iswara* (Great Helen), *P. palinurus* (Banded Peacock), *Graphium sarpedon* (Common Bluebottle), *G. agamemnon* (Tailed Jay), *G. nomius* (Spot Swordtail), and *Lamproptera meges* (Green Dragontail).

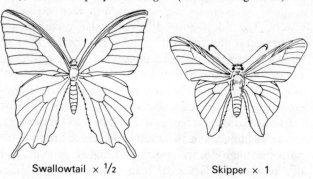

Swallowtail × ½   Skipper × 1

The Pieridae or what are called the Whites and Yellows is the one family of butterflies that contains many species that abound in the colder regions of the world. Most species are either medium-sized or small and as their name suggests are clothed in white, yellow or orange-yellow with dark markings. The hind wing is never adorned with a tail, the legs are fully developed with bifid or toothed claws. The larvae are usually smooth-bodied, some species with very fine hair, and the pupae are girdled and either upright or horizontal

The Common Emigrant, Catopsilia crocale, *with pupa* × 2

in position. Some of our commonest species on the plains belong to the genus *Catopsilia*, whose larvae (some solitary wasps being very fond of them) feed on *Cassia* trees and shrubs. Our common pest of cruciferous vegetables like cabbage are *Pieris brassicae* and *P. canidia*. The other genera and species that are common or noticeable are: *Leptosia nina* (Psyche), *Appias nero* (Orange Albatross), *Delias hyparete* (Painted Jezebel), *Hebomoia glaucippe* (Great Orange Tip), *Catopsilia pyranthe* (Mottled Emigrant), *C. crocale* (Common Emigrant), *C. pomona* (Lemon Emigrant) and *Eurema hecabe* (Common Grass Yellow). An example of the problems taxonomists have in deciding the limits of a species is the case of *Catopsilia crocale* and *C. pomona*, where it is still not certain that these two are really distinct, reproductively isolated species or are varieties of the same species.

The Danaidae have some of the most frequently noticed and abundant of all butterflies in India. The Tigers and Crows are large and either orange or black in colour, with markings of white. They are a distasteful group of butterflies with almost leathery wings and an unpleasant smell and taste making them useful models for other species of butterflies in other families to mimic and share their protection from predators. The males have bright orange or yellow tufts of hair-like extrusions at the end of their abdomen which are used in courtship. These butterflies have weak fore legs that are useless for walking and there is no claw. The larvae usually have spiny processes on each segment or on the head and tail, and the pupa is suspended from the tail end. These butterflies are sometimes included in the large family Nymphalidae. The Common Indian Crow is *Euploea core* and the graceful, large black-and-white Tree Nymphs, which belong to the genus *Idea*, are rare and usually seen flying high above tree-tops in rain-forests.

© Kumar Ghorpadé

A satyrid, Melanitis leda, *feeding on fallen cashew peduncle* × 1

© O. C. Edwards

A bevy of Dark Blue Tigers, Danaus melissa, *resting on* Casuarina × ¼

The Orange Tigers whose caterpillars feed on the milk-weed plant, *Calotropis gigantea* or the twining herbs, *Ceropegia* spp. (Asclepiadaceae), are, respectively *Danaus chrysippus* (Plain Tiger) and *D. genutia* (Common Tiger). The Blue Tiger that is most common is *Danaus limniace.* These butterflies are seen forming large groups on trees in certain seasons of the year and then become an interesting sight.

The Satyridae or what are commonly called the Browns (and Arguses in recent books) were also merged with the huge family Nymphalidae. However, they are distinct enough to be treated as a separate group. These butterflies are usually medium to large sized, with their brown or grey wings bearing ocelli or 'eyes'. They also have degenerate forelegs and their caterpillars and pupae resemble those of the Danaidae and the Nymphalidae. However, unlike these sun-loving butterflies, the satyrids prefer shade and usually keep flying in a slow jerky manner very close to the ground, and have the characteristic habit of suddenly settling on the ground, usually in leaf litter, closing their wings and then leaning down at an angle with the surface, suggesting a leaf being blown by wind! The most common Brown that we see close to our habitations, usually flying about at dusk, is the Evening Brown, *Melanitis leda.* Many species of this family have caterpillars that feed on grasses and some have taken to man's cultivated grain crops (e.g. rice) with glee! Other striking or familiar species are: *Ypthima baldus* (Common Five-ring), *Lethe rohria* (Common Treebrown), *L. verma* (Straight-banded Treebrown), *Mycalesis perseus* (Common Bushbrown) and *Elymnias hypermnestra* (Common Palmfly).

A small but very distinct family, also close to the Nymphalidae and often grouped in it, is the Amathusi-idae, called just Amathusiids. These are very large and brilliantly colourful butterflies, many being a superb metallic blue. They inhabit the lush rain-forests of the Western Ghats, lower Himalaya and Assam and like the satyrids have a jerky flight and weak forelegs. They are crepuscular in habit, flying either early in the morning before the sun becomes too bright and hot, or late in the evening. Some of the more remarkable Indian species are: *Stichophthalma camadeva* (Northern Jungle Queen). *Amathuxidia amythaon* (Kohinoor), *Thaumantis diores* (Jungle Glory) and *Discophora sondaica* (Common Duffer).

The Nymphalidae is one of the largest families of butterflies and includes many colourful species that are stoutly built and fly by day in the sunshine. Unlike the Browns, these butterflies settle after flight with expanded wings, as if to 'show off' their beauty. Their larvae are cylindrical and variously adorned with spiny processes and the pupae hang by their tails. Their forelegs are also weak and many species have dimorphic sexes, the males being coloured differently from the females. The caterpillars feed mostly on dicotyledonous plants and some may assume 'pest' proportions on crops like mango and castor. The Oakleaf butterfly is one of the more popular nymphalids and quite a few species are mimics of other butterflies, especially the danaids. The following are some of the better known and beautiful nymphalids: *Ariadne ariadne* (Angled Castor), *A. merione* (Common Castor), *Phalanta phalantha* (Common Leopard), *Vagrans egista* (Vagrant), *Cethosia biblis* (Red Lacewing), *Precis iphita* (Chocolate Pansy), *P. atlites* (Grey Pansy), *P. almana* (Peacock Pansy), *P. lemonias* (Lemon Pansy),

P. orithya (Blue Pansy), *P. hierta* (Yellow Pansy), *Vanessa indica* (Indian Red Admiral), *V. canace* (Blue Admiral), *Cynthia cardui* (Painted Lady), *Symbrenthia hippoclus* (Common Jester), *Hypolimnas misippus* (Danaid Eggfly), *H. bolina* (Great Eggfly), *Doleschallia bisaltide* (Autumn Leaf), *Kallima inachus* (Orange Oakleaf), *Neptis hylas* (Common Sailor), *N. viraja* (Yellowjack Sailor), *Parathyma selenophora* (Staff Sergeant), *Limenitis procris* (Commander), *Parthenos sylvia* (Clipper), *Euthalia aconthea* (Baron), *E. dirtea* (Archduke), *Apatura parisatis* (Black Prince), *A. ambica* (Indian Purple Emperor), *Prothoe franck* (Blue Begum), *Polyura schreiberi* (Blue Nawab), *P. athamas* (Common Nawab) and *Charaxes polyxena* (Tawny Rajah).

The principally African family, the Acraeidae, has *Acraea vesta* (Yellow Coster) as one Indian representative. The caterpillar is spiny and like the adult butterfly is protected by its oily, bad smelling fluid. Acraeid butterflies are yellow or brownish yellow in colour and their forelegs are degenerate. The Libytheidae (Beaks) is another small family closely related to the Nymphalidae. However, the males have poorly developed non-functional forelegs, while the females have them normally formed. They are called 'Beaks' because their palpi (one of the mouth-parts) are long and beak-like. *Libythea myrrha* (Club Beak) is one of the more common species of the family. The next family, the Nemeobiidae or what are called Punches or Judies, is intermediate between the Nymphalidae and the Lycaenidae, which I shall deal with next. These butterflies are generally small and, like the Libytheidae, are sexually dimorphic with respect to their forelegs. They are mostly dark brown on the upper wings, marked with white or yellow-brown spots, and may be mistaken for the lycaenids.

The familiar Indian species are: *Dodona egeon* (Orange Punch), *Abisara fylla* (Dark Judy) and *Taxila harquinus* (Harlequin).

The Blues, Coppers and Hairstreaks belong to the large family Lycaenidae which are mostly small butterflies with many species having tails on their hind wings. Their forelegs are perfect or slightly reduced and the male has a single claw. Their pupae are usually attached to leaves with a girdle and their larvae are smooth-bodied without long hairs and of a generally flattened shape. Like their names, these butterflies are usually blue or coppery above with brilliant metallic reflections and their wings are cryptically coloured on the lower surfaces. The colours of the upper and lower wings being so radically different, perhaps it helps to confuse the predator when the butterfly settles and closes its wings, and becomes very differently coloured from what it is in flight, with the showy upper surfaces of its wings in full view. These butterflies are very fond of the sun and will not fly on cloudy days. Some peculiar species feed on other insects like mealybugs and many species are tended by ants. A few lycaenids are known to cause damage to legume plants as they bore into the young pods. Another common 'pest' is the species that bores into guava and pomegranate fruits. Of the extensive representatives of this family in India, the following may be mentioned: *Poritia hewitsoni* (Common Gem), *Miletus boisduvali* (Common Brownie), *Spalgis epeus* (Apefly), *Castalius rosimon* (Common Pierrot), *Everes lacturnus* (Indian Cupid), *Celastrina puspa* (Common Hedge Blue), *Zizina otis* (Lesser Grass Blue), *Freyeria trochilus* (Grass Jewel), *Syntaracus plinius* (Zebra Blue), *Euchrysops cnejus* (Gram Blue), *Catochrysops strabo* (Forget-me-not), *Lampides boeticus* (Pea Blue), *Jamides*

Tawny Costers, Acraea violae, *mating* × 1¼

A Hairstreak, Spindasis vulcanus *on flower of* Tridax procumbens × 2

celeno (Common Cerulean), *Nacaduba kurava* (Transparent 6-Line Blue), *Heliophorus epicles* (Purple Sapphire), *Curetis thetis* (Indian Sunbeam), *Amblypodia anita* (Leaf Blue), *Narathura eumolphus* (Green Oakblue), *Surendra quercetorum* (Common Acacia Blue), *Loxura atymnus* (Yamfly), *Spindasis lohita* (Longbanded Silverline), *Tajuria cippus* (Peacock Royal), *Neomyrina hiemalis* (White Imperial), *Cheritra freja* (Common Imperial), *Eooxylides tharis* (Branded Imperial), *Themala marciana* (Cardinal), *Virachola isocrates* (Common Guava Blue), *Rapala dieneces* (Scarlet Flash), *Bindahara phocides* (Plane) and *Liphyra brassolis* (Moth-Butterfly).

The last family of the butterflies and one very different in general structure, is the abundant Hesperiidae, or Skippers. These are brown, thick-bodied butterflies with the antenna hooked beyond the club, forelegs fully developed and short and thick claws. The larvae have a distinct 'neck' and are smooth-bodied and slightly flattened. The pupae are mostly located in a leaf fold. The Skippers are crepuscular insects, though some species may fly by day and some even in the night! Except for a few members, most Skippers feed as larvae on grasses and palms. Owing to their period of activity and confusing species distinctions, there are probably more unknown and undescribed Hesperiidae than any other butterfly family. The more striking or abundant Skippers are: *Hasora chromus* (Common Banded Awl), *Badamia exclamationis* (Brown Awl), *Capila phanaeus* (Fulvous Dawnfly), *Celaenorrhinus aurivittata* (Dark Yellow-banded Flat), *Odina decoratus* (Zigzag Flat), *Mooreana trichoneura* (Yellow Flat), *Tagiades menaka* (Spotted Snow Flat), *Caprona ransonnettii* (Golden

© Kumar Ghorpadé

A skipper × 3

Angle), *Spialia galba* (Indian Skipper), *Ampittia dioscorides* (Bush Hopper), *Ctenoptilum vasava* (Tawny Angle), *Koruthaialos sindu* (Bright Red Velvet Bob), *Notocrypta paralysos* (Common Banded Demon), *Udaspes folus* (Grass Demon), *Suastes gremius* (Indian Palm Bob), *Cupitha purreea* (Wax Dart), *Gangara thyrsis* (Giant Redeye), *Erionota thrax* (Banana Skipper), *Telicota augias* (Pale Palm Dart) and *Iton semamora* (Common Night).

See plate 4 facing p. 49 and plate 34 facing p. 481.

CACTUS. The cactus family Cactaceae is native to tropical America. Some have been introduced in the subcontinent and have become naturalized. *Cereus hexagonus*, with tall columnar stems sharply 6-ridged, and large white solitary flowers, is common. The arborescent cactus-like Euphorbias are wrongly called cactus by laymen. They can at once be separated from cacti by having a milky latex which oozes out when cut. The more common cacti which have run wild are the Prickly Pear (*Opuntia dillenii* and *O. monacantha*) which can be grown from cuttings and is seen as a hedge plant in many areas. A night-flowering cactus, *Epiphyllum*, opens at 9 p.m. and starts closing at about 1 a.m., never to open again. It has a splendid white flower 15 cm across which is a source of much excitement at late-night parties as it is short-lived and spectacular.

CADDISFLIES. The Trichoptera are small to medium sized (1·5 to 50 mm) stream-frequenting insects that look like the smaller, dull, hairy moths, to which group they are closely related. The adults are generally not seen on wing, but usually seen resting on vegetation or stones along running water or lakes and small ponds. Their antennae are thread-like and relatively long and their grey/brown wings folded roof-like over the abdomen at rest. Over 6000 species are known from the world, of which roughly 10 percent should be found in the Indian subcontinent. In India, caddisflies are commoner on hills and regions with temperate climate. The adults mostly do not feed, except some that take flower nectar, if at all. The majority of species fly only at dusk and in the night, though a few species are diurnal. During the day most caddisfly adults seek cool, damp, thickly forested areas, where high humidity prevails, as do some species in the summer when the watercourses dry up.

Eggs are laid in or near water and it is the larvae which are interesting in that each species constructs typical cases within which it lives its 6 to 7 instars and which are characteristically singular of each species. Most case-making larvae cannot swim but some, who make cases of fine plant material, can swim well, using their brush-like hind legs. Larvae usually occur beneath logs, stones or other objects in water, or on foliage of water plants. Some lie exposed on sand or gravel, or even buried in soil. Different species prefer distinct habitats, stream bottoms or types of water. The cases are made of small stones or sand grains, water weeds, or even shells of small molluscs. Pupation takes place in the case itself in case-making larvae and the non-case-making ones construct special pupal cells of sand grains. The adult emerges from the pupal case on the surface of water and is able to fly immediately.

The caddisflies are an important link in the food-chains of stream life. The larvae are generally omnivorous, some feeding on simple plant tissue. Others are carnivorous, capturing and feeding on smaller insects and other animal life, plankton, etc. Larvae of some caddisflies build silken nets or tubes in which they trap their minute animal prey. Some species are destructive in paddyfields where they feed on young shoots of rice plants. Others are important diet for freshwater fish,

A caddisfly × 5

especially trout. Some other larvae are beneficial predators of black-fly larvae in streams. The familiar Indian species are *Hydropsyche asiatica*, *Dinarthrum ferox*, *Notanatolica vivipara* and *Rhyacophila anatina*.

CAECILIANS. Caecilians are limbless, long-bodied and eel-like amphibians. The eyes are lidless, small and usually indistinct; probably a response to their essentially underground, burrowing habit. A distinctive feature of these animals is the protrudable tentacle on the side of the face located between the eye and the nostril.

The body has a series of transverse grooves within which minute scales may be found. Due to their primitive and unique characters it is presumed that caecilians arose from a different fossil group than did the frogs and salamanders.

The most prominent genera in the Indian region are *Ichthyophis* and *Uraeotyphlus*. They are distributed throughout the Indian region, being found from the hills of Sri Lanka to the Himalayas. However, they are not met with in the drier parts of their range. The largest concentrations are in the hilly regions or regions of high rainfall. The preferred habitats of these animals are in moist areas near streams and swamps.

The caecilians lay large yolked eggs, but the aquatic larva stage may or may not be present depending on the species. Some species hatch directly into a young adult. *Ichthyophis* provides a good example of a form with an aquatic larval stage. In *Ichthyophis*, the female makes a small burrow in the soil by a running stream and lays 10 to 24 eggs in it. The eggs are protected by the female who coils herself around them till they hatch. When the eggs hatch the larvae swim into the stream and from then on fend for themselves. There are no gills in the aquatic larva stage but there is a tail fin that is lost when the adult form is reached.

See also AMPHIBIANS. See plate 3 facing p. 48, plate 19 facing p. 256 and plate 38 facing p. 529.

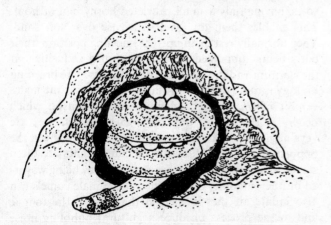

Female coiled protectively around eggs until they hatch, × ¹/₂
(*Ichthyophis*)

CALIMERE. Point Calimere Sanctuary (10°18′N, 79°51′E, the *Calligicum* of Ptolemy, is a low promontory on the Coromandel coast. The only human habitations in the area are in two villages, Kodikkarai on the seashore and Kodiakkadu, further inland. Extensive salt swamps, the winter resort of countless migratory waterfowl, lie to the west of the villages. About three miles east of Kodikkarai and approximately at the head of the promontory, a lighthouse has been in existence since 1902. The sanctuary has an area of about 2500 hectares or 25 km² of dry thorn scrub jungle and open sea coast. The Kodiakkadu Reserve Forest forms the major part of the sanctuary, approximately 1800 hectares of scrub jungle with thorn and other xerophytic vegetation predominating. The forest is not continuous but is intersected by numerous tidal inlets and creeks of varying length and width which are bone-dry in May. Dense thorn-scrub forest of an average 10 to 12 ft height covers the raised land in between the creeks. The area gets

© S. A. Hussain

Flamingos taking off at Point Calimere

most of its rain from the northeast monsoon, but is also within the range of the southwest monsoon. Most of the rainfall is between the months of October and December. The mean annual temperature is in the region of 27°C. Cyclonic storms of high intensity often occur during the northeast monsoon.

The flora consists of evergreen thickets of species with wide geographical distribution and large ecological tolerance. The Sanctuary holds the largest population (about 1000) of Blackbuck (*Antilope cervicapra*) in south India. The forests also hold a few Chital (*Axis axis*) and Bonnet Macaques, both introduced. There is no predator larger than the Jackal and Mongoose; both are common. Pigs occurring in the Sanctuary are presumed to be feral.

Point Calimere Sanctuary is of particular significance in the movements of resident and migratory species of birds in the southern parts of the peninsula and from and to Sri Lanka, 30 miles from the Sanctuary across the Palk Strait. The peak migration month is October when 'waves' of several translimital and Indian species pass through the sanctuary forest. The surrounding salt flats and swamps support large numbers of flamingos, both Large and Lesser, and migrant waders during the migratory season.

CAMEL'S FOOT CLIMBER, *Bauhinia vahlii*, is a gigantic graceful climber, with a stem 80 cm in diameter, which can climb a 30 m sal tree at the speed of 8 m in one season. Its deeply cleft leaves resemble the footprint of a camel, and are auctioned by the Forest Department as wrappers for pan leaf. The pods twist open with a loud pop. The stems are used for making suspension bridges and ropes.

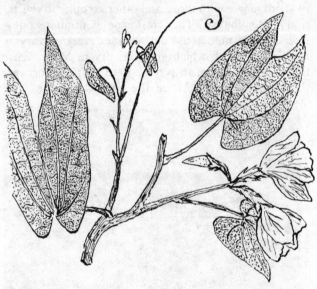

The climber is named after John and Caspar Bauhin, both botanists and brothers. The dual brother idea is evident in the leaf with two lobes, shaped like a camel's foot.

CAMEL. There are no wild camels within the territories covered by this encyclopedia, but a remnant population of wild two-humped, Bactrian camels still survives in southwestern Mongolia (Soviet Union) and in Sinkiang (Chinese Turkestan) in two widely separated regions of the high cold desert plateau of Central Asia. According to the IUCN these wild camels were much more widespread and abundant even in the earlier decades of this century and their decline has been attributed to competition with domestic stock for watering-points as well as continuous hunting.

The domesticated Bactrian Camel is still used as a beast of burden in Tadjikistan in the Soviet Union, Sinkiang in China as well in Tibet and northern Afghanistan. It is a larger animal with more pronounced humps and a thicker, more luxuriant fur than its wild relative.

Throughout the drier desert regions of northwest India and Pakistan, the single-humped dromedary is an important domestic animal still widely used for transporting freight. No fossilized evidence of wild dromedaries has ever been located and available records indicate that this camel has been domesticated by man since the dawn of history. In modern times camels are still used by the armed forces for border patrols and to control smuggling, and though motorized transport is rapidly replacing the camel even in the desert regions, an adult male is still capable of transporting loads of 400 kg on its back for distances of up to 10 or 16 kilometres. Karachi City is perhaps unique in being one of the few large modern cities where camel-drawn carts are still a familiar sight around the docks and railway goods yards.

In recent times many sophisticated techniques have been used to study the physiology of camels and to try to understand the adaptations or processes which enable them to function so efficiently in a desert environment. Contrary to popular belief they do not store water in their humps, which are largely composed of fatty deposits. They are however capable of surviving two to three days without water provided they are not worked during this period. If a camel is used for travelling it does need to replenish its body fluids by drinking at least once a day, but camels can withstand a much higher level of dehydration than other large mammals, and when water is available they are capable of imbibing and assimilating astonishingly large quantities at one draught. Other valuable adaptations include much reduced development of sweat glands (to conserve water loss), and an ability to withstand a daytime rise in body temperature without affecting normal metabolic processes. During the cooler desert night they can lose this stored body heat and regain their normal temperature, a process known as facultative hypothermia.

Camels are ruminants and have large multichambered stomachs and an ability to regurgitate and re-masticate their food during periods of rest by the familiar process of chewing the cud. Their feet are modified by a fusion of tarsal bones into only two toes, but unlike the Bovidae, their feet are broad and cushioned with naked soles, having only a small restricted horny nail or hoof. This enables them to travel with ease over soft sand. They normally rest in an upright kneeling position, their body being supported by well-developed callosities on the sternum and the knees of the hind legs. The breeding biology is quite different from the Bovidae, as the males exhibit *musth* during a definite rutting season which occurs at the end of the cold weather when they are very aggressive and will attack other males if given the opportunity. In the Punjab such camel fights are often staged at country fairs, with sums of money being wagered on the outcome. During the rut the male camels can also inflate air sacs located at the base of the tongue and in the process produce a guttural bubbling noise accompanied by much frothing at the mouth: to the human observer a somewhat revolting spectacle but no doubt attractive to the female camels! The gestation period is eleven months and newly-born camels are capable of standing and following their mothers long distances within two hours of birth.

CAMOUFLAGE. Camouflage, or concealing coloration, has great survival value in a natural world that is largely predatory. Often both predator and prey have need for concealment against each other, as can be noted in the obliteratively coloured tiger, and its protectively coloured prey species, the various deer. It is often difficult to make out a stationary animal in its natural surroundings, and only movement gives its presence away.

Among birds, a general relationship between colour and pattern and favoured habitat can be noticed. Many seabirds are black, grey and white in colour, to match the sea and the sky. Birds of dry grasslands tend to be brownish and often streaked. Many arboreal species have green in their plumage. There are exceptions of course: certain brightly coloured sunbirds live in the drier open forests and are conspicuous. Experiments conducted by Cott (1946) on the edibility of birds indicated that the more conspicuously coloured birds are less edible. Selective pressure could have forced the more edible species into greater camouflage. The case becomes more complicated because colour in birds plays an important role in display. Certain otherwise cryptic species flaunt the more brightly coloured parts of their plumage for advertisement; for instance the red vent of the male Indian robin. It is also rather fortunate for the prey that many predators seem sensitive to only certain colours: the owl for instance is believed to be sensitive to only black, white and grey.

The colour and patterning of certain birds bears a remarkable resemblance to their surroundings. The blotched dead-leaf markings of the woodcock completely obliterate it against its semi-marshy habitat. Many ground-living birds like the various nightjars cannot be seen unless one nearly steps on them. Arboreal birds like the green pigeons, parakeets and leaf birds (*Chloropsis* spp.) are green in colour and blend perfectly into the foliage. The colouring, pattern and stance at rest of the frogmouth make it look exactly like a broken-off snag of a dead branch.

Disruptive coloration is seen in certain plovers, the avocet, the blackwinged stilt, certain lapwings and the pheasant-tailed jaçana. Strong dark markings, usually black against white in unlikely parts of the plumage, break up the bird's outline and make it almost invisible when stationary. Less specific camouflage, known as countershading, is more universally prevalent. A darker shade is used on the upperparts to underplay the effect of light, and a lighter shade beneath to offset the shadow. The bird is thus difficult to make out from above and below.

Certain species affect camouflage only when threat-

ened. The bittern straightens up and freezes on alarm, its streaked brown plumage then merging with the surrounding reeds. Birds of prey like many owls and certain forest-dwelling eagles and hawks are also cryptically coloured, especially when unobtrusiveness makes for a successful hunt.

A classical recent example of the selective value of camouflage is the case of the peppered moth of Britain. In heavily polluted industrial areas the pale-coloured moth was subject to heavy predation due to its conspicuousness against the darkened tree-trunks. A dark mutant appeared and soon spread through the population. The light-coloured moth must have been selectively preyed upon as against the more cryptic mutant. This phenomenon, known as 'industrial melanism', is believed to have affected approximately 10 percent of the larger moth species in areas of heavy industrial activity. A case of natural selection in action before the biologist's eye.

P.D.

CAMPBELL'S MAGNOLIA, *Magnolia campbellii*, a magnificent snow-white magnolia, starring the hillsides of the Eastern Himalaya from Nepal to Manipur at 2100–3000 m. This spectacular tree, leafless when in bloom in April, has white, fragrant, goblet-shaped flowers 15–25 cm across. The yellowish-white wood presents a beautiful colour in panelling.

CAMPHOR, *Cinnamomum camphora*, a medium-sized tree introduced in the subcontinent over a century ago

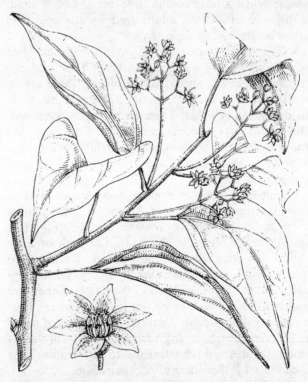

Camphor, leaves and flowers (×1), with enlarged flower.

from China and now common. It has small light green flowers and handsome foliage. The ovoid dark green fruits, 0.5 cm in diameter, turn black on ripening. All parts of the tree contain camphor. Masses of camphor are sometimes found in the trunk and therefore the wood is not attacked by insects.

CANNIBALISM. Cannibalism is the killing and eating by an animal of individuals belonging to its own species. Such cannibalism may serve as a source of nutrition on a regular basis, as an occasional source of nutrition, as a source of nutrition in an emergency or may be a competitive device, serving a nutritive function only incidentally.

There are certain lakes in Scandinavia which support populations of perch in which the sole source of food for the adults is the young of their own species. The young have other suitable sources of food in the form of zooplankton which they consume. The adults lay a large number of eggs and a fraction of the young that grow are converted into adult food. Termites are also notable for routine cannibalism, certain individuals in the colony merely serving as convenient packages of food for the rest of the colony.

When the chicks from a herring gull colony wander away from their nests, other gulls take the opportunity to eat them, although the total amount of food so obtained must be quite negligible. Males of many carnivores including the domestic cat behave in a similar fashion.

Cannibalism is resorted to in emergency by paper wasps. When a wasp colony is disturbed and is about to be abandoned, the adults feed on the eggs and larvae before leaving the colony.

Social animals which hold group territories often fight fiercely to defend such territories. This is as true of many species of ants as of the hyena. There are no holds barred at these territorial fights and many individuals may be killed. These are promptly consumed. This would serve not only as a source of nutrition, but also to weaken the competitive ability of the rival colonies or packs and thereby to secure the integrity of the group territory.

Cannibalism has been noted in many human cultures as well. Among the Incas of America a large number of victims captured during raids on neighbouring tribes used to be sacrificed to the gods. The flesh of these victims was routinely consumed and no doubt served as a regular source of nutrition at least to certain segments of the population such as the priests. Others are known to resort to cannibalism in emergencies. Thus northern Eskimos depend entirely on hunting of mammals during the winter, and occasional years will witness serious failures in this hunting due to bad weather. The Eskimos may then resort to cannibalism.

CARBONIFEROUS AGE. Geological time, spanning over 4500 million years since the birth of our planet Earth, has been divided into smaller units of time by geologists. Among these divisions the Carboniferous age refers to a period of time-span 350 to 270 millions of years before the present, that is, it lasted for about eighty million years.

Land plants existed since 400 million years ago and it took quite some time before varieties of species evolved. Plants on land were originally along river banks, swamps and areas of plenty of rainfall. Subsequently they spread to areas of scanty rainfall also when species evolved that could thrive on a little moisture. But during certain periods of geological time a larger number of varieties of plants spread themselves far and wide over the earth. This was possible because the climate was warm and humid, permitting luxuriant vegetation. But as has always happened in evolution, most of these slowly perished. It can be said that vast forests disappeared because of a large-scale death or extinction of a large number of plants in many parts of the earth. These were covered over by sand, silt and clay and got entombed. A coastal plain covered by forest would sometimes be buried when parts of the sea advanced over this land and the sediments deposited in them covered the entire vegetation. Where this phenomenon is repeated one will come across alternate layers of sediment and plant material. Similarly in inland areas, vast flood-plains may be covered with vegetation and later get buried by sediments brought by the rivers. Or both sediments and plant material may be carried away together and get deposited in inland basins.

We know that charcoal, that is used as fuel in many houses in villages, is obtained by burning logs of wood in a heap covered with bricks, out of contact with air. In almost a similar way, the vast forests buried by later thick sediments have slowly been altered with the passage of time because they have been under enormous pressure from the rocks above, and the temperature of rocks increases with depth. From an original peat, the forests sometimes become lignite, and ultimately coal, in which the main constituent is carbon. It is these that form coal beds mined all over the world. Hence the rocks in which these beds occur in abundance are referred to an age called the Carboniferous. It is to be remembered, however, that there are coal beds in rocks of younger age also.

In the southern continents of the world, just before the period of luxuriant vegetation, there was extensive continental glaciation; ice covered many areas in the same way that during Pleistocene times, much later, ice covered much of the northern hemisphere. There were extensive oceans also containing many invertebrate animals in them.

Among the land animals the most noteworthy were the amphibians that could live on land as well as water, mainly in swamps adjoining rivers and coastal areas. There were reptiles too of pretty big size.

Carboniferous rocks containing coal occur in India along parts of the Damodar, Mahanadi and Godavari valleys, and in a few other places. Coal is an important source of energy.

R.V.

See also CONIFERS.

CARNIVORA. This is a very diverse order of mammals generally considered to be near the topmost branches of the evolutionary tree when the course of animal evolution is pictured as a tree. This is because they exhibit the traits of more highly developed mammals, not only with respect to relative brain size and complexity of cerebral hemispheres, but also because of their basically predatory role upon less developed forms of mammalian and other animal life, including the vast assemblage of herbivores. The word 'carnivora' means flesh-eating, but in actual fact only a minority of this order depend solely upon fresh meat, which they are capable of hunting and killing themselves for their subsistence. Perhaps the best examples of the archetypal carnivore are the cats, as well as some of the wild dogs and smaller weasels. A typical carnivore is therefore armed with strongly developed canine teeth for holding and killing prey and with sharp claws on their forefeet to assist them in striking down their quarry, or in the case of the dog family long legs, coupled with speed and endurance.

Returning to the diversity of the order, it is nowadays grouped by scientists into seven distinct families (if the seals and sea-lions are included), with over 100 genera. Six of these families are represented on this subcontinent. Amongst the Carnivora are some animals which are almost wholly insectivorous, such as the Aardwolf (*Proteles cristatus*) of South America, many of which are wholly piscivorous (otters), the almost wholly carrion-eating and scavenging Striped Hyena, the omnivorous Black Bear, and almost completely herbivorous animals such as the Lesser Panda and Binturong, as well as the vast majority of smaller carnivores which can subsist on ripe berries in season, insect larvae, reptiles and even a limited amount of leaf and stem tissue, when bird or mammal prey is unobtainable. They have also adapted to live in and exploit almost every ecological niche, from the cold dry mountain steppes of Central Asia e.g. Pallas's Cat (*Felis manul*) and the Marbled Polecat (*V. peregusna*), to the hot and steamy tropical evergreen forests of southeast Asia e.g. Temminck's Cat (*Felis temmincki*) and the Binturong (*Arctitis binturong*). The Sea Otter (*Enhydra lutris*) of the northern Pacific lives almost its entire life in the water, whilst the Jackal (*Canis aureus*) and Jungle Cat (*Felis chaus*) have adapted to survive from the great deserts of Rajasthan to the Himalayan broad-leaved deciduous forest. Typically, carnivora produce small highly dependent young which are born blind and helpless, though well covered with hair. The gestation period varies from as short as 49 days in some badgers up to 113 days in the larger cats. Carnivores are mainly cared for by their mothers and fiercely protected until able to hunt for themselves. Because of the nature of their hunting methods and food prey, most carnivores lead solitary lives except during the mating season, but a few species have evolved complex social lives with cooperative care of young having been clearly demonstrated in recent studies of the Hunting Dog or Dhole and even in the Canadian Wolf and African Lion. Delayed implantation of the fertilized egg occurs in some of the mustelinidae (martens) and the bears, whilst nearly all carnivora reach sexual maturity in two years, and many of the larger species have life-spans extending up to thirty years.

All carnivores can swim well, but some are adapted to an arboreal existence and like the Palm Civet or Marbled Cat live almost entirely in the trees, whilst the swimming and diving ability of Polar Bears (*Thalarctos maritimus*) and Minks, which are largely aquatic in hunting, is rivalled only by that of the Otters.

Some carnivores have well developed, elongated legs and capture their prey by extended chases, such as the Indian Wolf (*Canis lupus*) and the Desert Fox (*Vulpes vulpes pusilla*). Others hunt by stealth, a sudden pounce from a tree-branch ambush, or after a short stalk (typical of the smaller cats and some civets). The grace and speed of the Cheetah, adapted to hunt the fleet gazelle, has been well documented and is oddly contrasted with the lumbering almost clumsy gait of such carnivores as the Brown Bear (*Ursus arctos*), or the Hog Badger (*Arctonyx collaris*), both of which are omnivorous in diet and quite fearless and savage if provoked or attacked. Before recounting briefly the main groups in the Indian region, it is interesting to note size variation in this complex order. The Least Weasel (*Mustela rixosa*) of North America is the smallest carnivore though surprisingly fierce and wholly carnivorous in feeding habits. An adult weighs as little as 40 gm with a head and body length of 15 cm. The Alaskan subspecies of the Brown Bear (*Ursus arctos*) is the largest living carnivore with adults measuring up to 280 cm in body-length and weighing up to 780 kg.

Other curious anatomical features shared by most carnivora are the presence of a baculum (bone) in the penis, as well as the development of complicated scent-glands in many genera. Scent marking of territory is characteristic of such species and is thought to be related to the importance of spacing individuals as well as avoiding intra-specific encounters; both hazards to survival in animals which have to range widely to compete for food prey, and which are more than adequately armed in the event of aggressive encounters.

As mentioned earlier, six of the seven families of carnivora are represented in the region covered by this encyclopedia. The family Canidae includes the Indian and Tibetan subspecies of the Wolf (*Canis lupus*), the closely related Jackal (*Canis aureus*), the Red Dog or Dhole (*Cuon alpinus*) as well as the mountain and desert Foxes (*Vulpes vulpes*), Indian Fox (*Vulpes benghalensis*), and King Fox (*Vulpes cana*).

The second family which is almost world-wide in distribution, like the Canidae, comprises the Bears or Ursidae. In our region the dry deciduous forest of the plains and peninsula is inhabited by the Sloth Bear (*Melursus ursinus*), whilst in the Himalayas down to the foothills of Bangladesh the Himalayan Black Bear (*Selenarctos thibetanus*) still survives. Further north and at the edge of the birch forest or alpine zone the Brown Bear (*Ursus arctos*) is found. Most recent opinion in taxonomy has placed the Giant Panda (*Ailuropoda melanoleuca*) in the same family with bears, leaving the Lesser Panda or Bear Cat (*Ailurus fulgens*) as the only representative in the region of the family Procyonidae. In the New World of the western hemisphere this family includes several species of Racoons, Coatimundis and the Kinkajoo. The Lesser Panda eats a large quantity of bamboo shoots and is believed to be largely herbivorous. Like other members of the family they walk on the whole sole of the foot and the Lesser Panda is almost unique amongst carnivores in having all the pads of the sole densely covered by hair. With its needle-sharp claws and well developed digits, it is however well able to climb trees. They occur in Nepal and Sikkim. The next family, Mustelinidae, is almost the largest assemblage within the order, and includes all the Weasels, Stoats, Martens, Polecats, Otters, Ferrets and Hog Badgers and true Badgers. In India and adjoining countries are found the Yellow-throated Marten (*Martes flavigula*), and the Stone or Beech Marten (*Martes foina*). In the innermost northern Himalayan regions the Stoat (*Mustela erminea*), and Altai or Pallid Weasel (*Mustela altaica*), are both found, whereas further east the very rare Stripe-backed Weasel (*Mustela strigidorsa*) and the smaller Yellow-bellied Weasel (*Mustela kathiah*), both occur in Nepal, Sikkim and eastwards to Assam. The larger Himalayan Weasel (*Mustela sibirica*) is an inhabitant of the high cold plateaus of central Asia. The Martens are larger than the Weasels, more omnivorous in diet and diurnal in hunting activity and inclined to be more arboreal. The Weasels have smaller, low-set ears and do most of their hunting on the ground or even inside rodent burrows, as does the beautiful Marbled Polecat (*Vormela*

© M. Krishnan

The Smoothcoated Otter, Lutra perspicillata, *at Delhi Zoo ×* $\frac{1}{9}$

peregusna) which subsists on the Libyan Jird, in Baluchistan. Confined to the northwest Himalayas the Stoat is an alpine denizen living exclusively off Voles of the genus *Alticola* and *Hyperacrius*. Three species of Otters are also found in the region as well as the thick-set Hog Badger (*Arctonyx collaris*) and smaller Ferret Badger (*Melogale moschata*) of Assam and neighbouring Burma. The Ratel or Honey Badger (*Mellivora capensis*) is widespread in the plains of India, especially favouring the dry sub-tropical thorn forest and savanna biotope. Their fondness for bees' honey and larvae is well known and their rather coarse fur, with a very tough leathery skin, seems to make them immune to wasp and bee stings. Two other carnivores are extremely fond of

The graceful unison display of the Sarus crane. These large birds are in India regarded as symbols of marital fidelity.

An effective display of aggression by the Lesser Florican serves to make it look larger than life.

robbing wild bees' nests, the Yellow-throated Marten and the Sloth Bear, and indeed this predilection is shared by most of the Ursidae.

The next big family comprises the Viverridae which includes all the Civet Cats and Mongooses. The Small Indian Civet (*Viverricula indica*) is the most terrestial of the Civets being adapted even to treeless desert areas. The large and handsome Large Indian Civet (*Viverra zibetha*) is more of a forest animal confined to Nepal, Sikkim, Assam and Bangladesh. In the Himalayas the Masked Palm Civet (*Paguma larvata*), takes the place of the Toddy Cat or Common Palm Civet (*Paradoxuros hermaphroditus*). With semi-retractile claws and long thick muscular tails, all the civets are well adapted for tree climbing, but they will hunt and feed upon a wide variety of prey including crabs, snakes, lizards, orthopterous insects and hymenoptera, and they are particularly partial to ripe fruit and nectar. The Himalayan Palm Civet has been observed stealing tomatoes from a household store and the Stone Marten stealing apples from inside a building in Baluchistan. All the Civets have scent glands in the anal region and some species secrete a civet or musk oil from this gland which is used in perfumery and for medicinal purposes.

The Mongooses include the Common or Grey Mongoose (*Herpestes edwardsi*), the Golden or Small Mongoose (*Herpestes auropunctatus*), and the much larger Crab-eating Mongoose (*Herpestes urva*). The first two are widespread in drier parts of north and central India, whereas the Crab-eating Mongoose is confined to Bangladesh and Bengal.

In south India and Sri Lanka the Stripe-necked Mongoose (*Herpestes vitticollis*) and the Brown Mongoose (*Herpestes fuscus*) also occur. The former is a shy forest-dwelling animal and the latter in Sri Lanka is quite commonly associated with villages. In contrast to the Civets, Mongooses have a naked saccular gland surrounding the anus and they are more fossorial in habits, with powerful long claws, well able to excavate their own burrows, and thus able to exploit a rather different ecological niche when compared with the more arboreal civets. One remaining civet has to be mentioned: the large tree-dwelling Binturong (*Arctitis binturong*) of Assam and Sikkim. It is covered with long coarse black hair somewhat grizzled on the back and head, its tail is tapered, but very muscular and bushy at the base and has the unique distinction, amongst carnivores, of being prehensile and adapted as a gripping tool. They are largely omnivorous, like other civets, and appear generally to be rather lethargic and deliberate in movements. Spending the day curled up in some tree hollow in the evergreen tropical rainforest, these animals are rarely seen even where they are comparatively common.

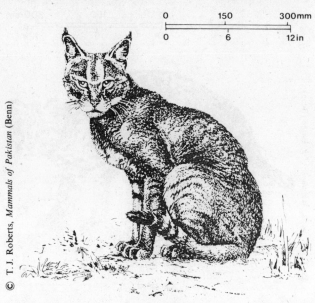

© T. J. Roberts, *Mammals of Pakistan* (Benn)

Jungle Cat

The next family comprises the Hyaenidae of which only the Striped Hyena (*Hyaena hyaena*), occurs in peninsular and central India and Pakistan. They scavenge mostly on bones from carcasses and with massive jaw muscles are even able to chew and ingest cattle hides.

The last large family comprises the Felidae or cats. In many respects remarkably homogeneous in appearance and anatomy, all have foreshortened skulls and wide-domed brain cases, with elongated canines and wide-set ears. They have developed completely retractable claws which remain ensheathed and thus protected from wear when they are travelling, as they can stealthily do, on their cushion-padded toes in a digitigrade manner. The subcontinent is unusually well represented by members of this magnificent and wholly carnivorous family, the Asiatic Lion (*Panthera leo*) and Indian Tiger (*Panthera tigris*) being the best known and

© T. J. Roberts, *Mammals of Pakistan* (Benn)

Pallas's Cat

Leopard Cat

most spectacular representatives. Among the medium-sized cats are the Panther (*Panthera pardus*), Clouded Leopard (*Neofelis nebulosa*), Snow Leopard (*Panthera uncia*), and a possibly surviving Cheetah (*Acinonyx jubatus*) population, in the border region between Iran and southern Baluchistan.

The Caracal (*Felis caracal*), Himalayan Lynx (*Felis lynx*), Golden or Temminck's Cat (*Felis temmincki*) and the Fishing Cat (*Felis viverrina*) are all intermediate in size between the former and the domestic cat.

A host of smaller cats include the Jungle or Swamp Cat (*Felis chaus*), the Desert Cat (*Felis libyca*), Marbled Cat (*Felis marmorata*), Pallas's Cat (*Felis manul*), Sand Cat (*Felis margharita*) and Leopard Cat (*Felis bengalensis*) and these are usually very secretive nocturnal animals, rarely encountered by man in the wild.

See also BEARS, DOG FAMILY, HYENA, LEOPARD, LION, OTTERS, SMALL CATS, SNOW LEOPARD, TIGER. See plate 18 facing p. 241 and plate 33 facing p. 480.

CARPS. These are a group of freshwater fishes belonging to the family Cyprinidae, which is one of the most widely distributed families, covering Europe, Asia and even North America. They are large-bodied fish some of them growing even up to 54 kg (e.g. mahseer). Small-sized members of the family are called minnows. Their body is covered with distinct scales but scales do

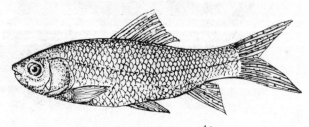

Rohu, *Labeo rohita*, × ¹⁄₁₀

not extend to the head. They have no teeth in the jaws but have tooth-like structures in the throat (pharyngeal teeth) and tactile barbels around the lips. Their air-bladder is connected to the gullet by a small duct and they have only a single fin on the back.

Many freshwater fishes in India belong to this family, well-known examples being the Rohu (*Labeo rohita*), Catla (*Catla catla*), and Mrigal (*Cirrhina mrigala*). These are common in north Indian rivers and are cultivated in tanks and lakes throughout the country for food: their spawn is collected from rivers, grown to fry stage and then supplied to fish-farmers. Carps feed on minute plant and animal life in water and grow fairly rapidly. They do not usually breed in stagnant waters of tanks and lakes but in running waters of flooded rivers. Several other species of carps such as olive carp (*Puntius sarana*), tambir (*Labeo filamentosus*), Calbasu (*Labeo calbasu*) and even the cold-water snow trout (*Schizothorax* sp.) occur in India. Most of the above carps can now be bred artificially in ponds with the help of pituitary hormone injections and the resultant fry used for fish culture.

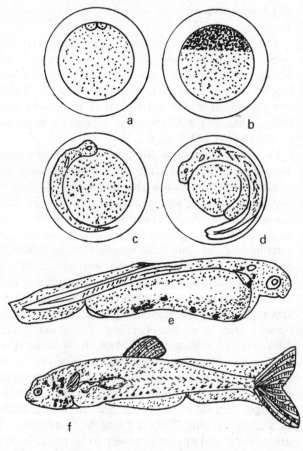

DEVELOPMENTAL STAGES OF A TYPICAL CARP

a, b, c, d developing eggs; e hatchling or larva; f fry, or mini-adult

In addition to these species other exotic carps such as the Common Carp (*Cyprinus carpio*), Silver Carp (*Hypophthalmichthys motitrix*), Grass Carp (*Ctenopharyngodon idella*), goldfish (*Carassius auratus*) and the English Carp (*Carassius carassius*) have been introduced and are now thriving in India. The Common Carp has three varieties, the mirror carp with large scattered scales, the scale carp with normal scales, and the leather carp with very few scales. The mirror carp was introduced into the Nilgiris in 1939 from Sri Lanka where it was brought from Prussia in 1914. From the Nilgiris it was taken to the Kumaon hills and other parts of India where it is thriving as a food fish. It is interesting to note that the Common Carp, an original inhabitant of Central Asia and China, spread to European countries in ancient times and was introduced into north America and to south Asian countries mainly because of its extremely adaptable nature and quick growth and usefulness as a food fish.

The English Carp, known also as the Crucian carp, is a small fish growing only up to 1·5 kg. It was introduced into the Ootacamund lake in 1874 where it is still thriving, but has not been successfully acclimatized to lower altitudes. It feeds on small water-fleas and helps in keeping water clean.

The importation of Grass and Silver Carp only took place in the late fifties of the present century, but because of their rapid growth and ability to feed on aquatic plants and algae they are much in demand by fish-farmers. A grass carp may weigh as much as 32 kg. It matures in its second or third year, when it is about 4 kg, and may gain 5 kg in a single year, if sufficient food is available.

See also FISHES, MAHSEER, MINNOW, GOLDFISH.

CASHEW NUT, *Anacardium occidentale*, was introduced by the Portuguese from Brazil centuries ago and is now common in the coastal areas. The flowers are yellow with pink stripes. The nut is kidney-shaped, seated on the flower stalk enlarged into a fleshy orange-red 'apple'. The enlarged stalk is juicy, rich in vitamin C, and makes a beverage called feni, which is popular in Goa. It is a useful tree for the reclamation of sand dunes.

CASUARINA, HORSETAIL TREE (*Casuarina equisetifolia*), is a large evergreen tree with feathery foliage looking like a conifer. The ends of its branches bear long, slender, jointed, needle-like leaves, with spikes of flowers at the ends of branchlets. It is a littoral tree distributed from Australia to the Andamans and Nicobars and widely planted on the coastline of the subcontinent and in Sri Lanka to reclaim sand dunes and check erosion. It is one of the best fuel-woods of the

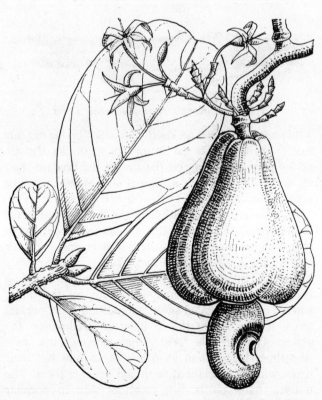

Cashew fruit and nut (×1), with a branch sprouting new leaves behind

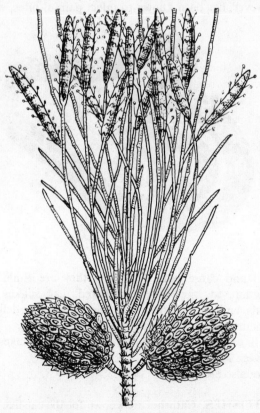

Casuarina shoot with flowers. The egg-shaped fruits have twelve rows of seeds (achenes), × 1.

world and is the principal fuel-wood in the city of Madras. It grows to maturity in 30 years and is able to fix atmospheric nitrogen through bacteria in its root. The stems are used for beams and the bark in tanning.

CAT SNAKES. These snakes, of the genus *Boiga*, form an interesting group. The large round eyes and long feathery tongue indicate nocturnal habits and cat snakes are often seen at night, hunting for mice, small birds and lizards, their main prey. The Common Cat Snake is found throughout the subcontinent and often chooses to pass the day in the cool of a palm-leaf branch or hut; in fact their Tamil name means 'palm-leaf snake'. They may also spend daylight hours under tree bark or in a shady bush. Cat Snakes are often mistaken for sawscaled vipers because of the similar colour and markings, but they are longer, thinner and harmless. Also, cat snakes have symmetrical shields on the head while sawscaled vipers have small rough scales.

There are eleven species of cat snakes in India, with colour variations from green to bright orange and brown. Their distributions vary; one, the Andaman Cat Snake, is found only in the Andaman and Nicobar Islands. The Himalayan Cat Snake is found up to 3000 m above sea level. They have fangs at the back of their mouth with a very mild venom for dealing with their small prey. The Forsten's Cat Snake, which grows up to 2 m however, has been known to inflict painful bites, which

© Rom and Zai Whitaker

Cat snake × ¹/₂

swell and cause pain. Generally they are timid and reluctant to bite. When provoked, cat snakes coil tightly and vibrate their tails, rearing back in defence. Sometimes they will turn over and play dead!

All cat snakes lay eggs; common cat snakes lay between 3 and 8.

See also SNAKES.

CATFISHES. Catfishes are known for the cat-like long barbels around their mouth. They belong to the super-family Siluroidea which includes families such as Ari

idae, Siluridae and Bagridae and others. Most catfishes are bottom-dwellers and use their barbels for finding food. Their body is scaleless, but some have protective bony plates or shields on the head or near the dorsal fin. Some of them like the Shingi (*Heteropneustes fossilis*) and Magur (*Clarias batracus*) have strong pointed spines on their dorsal and pectoral fins. Some have, in addition to the dorsal fin, a small fleshy fin without fin-supports (rays or spines) called the adipose fin. They inhabit both marine and fresh water, a large variety being found in fresh water. Usually they are dull grey coloured but some are small and colourful and are used in aquaria. The largest catfish in India is the goonch (*Bagarius bagarius*), which grows to about 250 kg.

The commonest riverine catfish is the Wallago. It grows to a length of 2 m and weighs as much as 20 kg. It has a very large mouth, studded with thin pointed teeth and is so voracious and destructive of other small fish that it is often called the 'freshwater shark'. It breeds in rivers and is common thoughout our region. Other freshwater catfishes such as *Mystus aor*, *M. seenghala*, *Onipok* spp. are also good food fish.

Some marine catfish are known for their parental care. The marine Shingala (*Tachysurus sagor* or *T. sona*) male hatches the eggs in his mouth; and the parents of the freshwater shingala (*Mystus seenghala*) make saucer-like depressions on the margins of rivers and lay their eggs therein for protection.

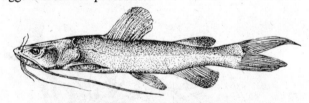

Freshwater shingala, *Mystus seenghala* × ¹/₁₀

In nature, catfishes are detritus-feeders, taking organic matter available near the bottom, or small fish, the size of the prey depending on the size of the predator and its mouth. Fishes such as Shingi and Magur will take artificial feeds also and are useful in fish culture.

Another interesting catfish of the family Schilbeidae is *Neotropius khavalchor*. It is a small fish about 15 cm long with prominent eyes, scaleless, silver-grey in colour and having long whisker-like barbels similar to other catfishes. The mouth is ventral but the outstanding feature is the exposed rasp-like teeth on the under surface of its elongated snout, in addition to its normal teeth. It has a remarkable habit of feeding on the scales of other fish. Swiftly approaching other fish, sometimes bigger than its own size, it pulls out their scales with its specialized teeth and devours them. The unusual habit of the fish coincides with its peculiar adaptations, namely the wide ventrally situated mouth, the specialized teeth on the lower exposed portion of

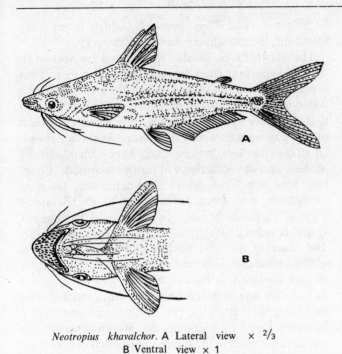

Neotropius khavalchor. A Lateral view × ²/₃
B Ventral view × 1

the snout, and strong hooked pectoral spines; the latter serving to grip the body of its victim. The British Museum has also recorded a similar lepidophagous (scale-eating) habit in two species of *Corematodus* of Lake Nyasa in Africa, stating: 'In each case stomach and intestine contained nothing but hundreds of minute scales exactly like those which cover the caudal fins of so many cichlid fishes. It seems that *Corematodus* specialises in this curious diet.' This indicates that although the aforesaid habit is unusual, it is by no means unique.

N. khavalchor (the specific name meaning scale-stealing in Marathi, and the local name of the fish) lives in the Krishna river in Satara District. It is interesting to note that a somewhat similar fish, *Pseudeutropius acutirostris*, occurs in the river Irrawaddy of Burma. This lends support to the theory of the Malayan origin of the freshwater fish fauna of peninsular India and its dispersal along the Satpura range of hills, suggested by Dr Hora of the Zoological Survey of India in 1944.

CAVES. Caves are natural cavities in the ground including all subterranean voids, except mines and tunnels which are man-made. A cave may be composed of a number of chambers, in which case it is described as a series of caverns. A cave system is an assemblage of caverns interconnected by smaller passageways through which at least water or air can be interchanged.

In India natural caves are found in the Simla Hills, in Mahabaleswar, S. India and Uttar Pradesh, as also in limestone country, for example in parts of Assam. Generally, caves of the oriental region do not possess

the interest of those of some parts of Europe and America. As a rule they are not of vast size or impressive exterior. In Europe are found highly ornate solution caves, many of which have yielded priceless remains of early man and anthropoids. Oriental caves have not yielded paleontological deposits of importance, and neither have they been the home of races whose civilization is extinct.

However subsequent findings, especially the discovery of the Singhanpur rock shelter, put India on to the map of paleolithic art. Many rock paintings are found in India possibly 1000 years old, whose exact dates are unknown. Rock paintings, many of which have been recorded from the Kaimur range and the Mahadeo hills around Pachmarhi in central India, are comparable with those found in Europe dating back to the end of the ice age.

Few caves in India, Burma and the Malay Peninsula have streams or lakes. One exception is the Siju cave in the Garo Hills, Meghalaya. This is probably the largest limestone cave in India, having many caverns and a stream running through it. The largest cavern is at a distance of 1400 ft from the opening. Its height is 180 ft above stream level and it is 200 ft long and approximately 100 ft broad. In 1919, a long-legged Reduviid bug not hitherto known to science was discovered in it.

The best known cave and cave systems of the world principally result from chemical dissolution of a soluble, weakened host rock. Soluble rocks include limestone, marble, dolomite, gypsum and salt. Most caves in India are of limestone. The principle underlying the solution process is that all rocks dissolve in certain natural acids such as carbonic and sulphuric acids. The rocks dissolve not only on their exposed outer surfaces, but also on their interiors, where fractures and partings permit solvents to circulate. Rocks of limestone are particularly soluble in natural solvents, and caves formed in it, although preconditioned by fracturing and abetted by weathering and mechanical erosion, enlarge mainly by chemical solution.

Limestone cave formation occurs in three stages. The first stage is due to solution by water *under pressure*, i.e. below the local water-table. Often it will have commenced beneath a peneplain towards the close of a previous cycle of erosion. Following uplift of land introducing the present cycle of erosion, the system of cavities begins to drain out. In the air-filled cavities, stalactites and stalagmites and other calcite deposits start forming. In the third stage, the water in the cavities is not confined under pressure but has a free surface. This stage commences soon after the joints or fissures are sufficiently open to admit rain-water freely into the air-filled cavities below. If a stream discovers the cave and pours into it, this stage proceeds at an accelerated pace. The surface streams and flood waters wash in land and water animals, some of which may successfully adapt to cave life.

Life in caves. The deep interiors of caves provide an environment of total darkness and relatively constant temperature and high humidity, to which many forms of animal and plant life, and micro-organisms, have adapted. Species lacking eyes and pigment in adulthood have evolved, since these are properties not needed for survival underground. The European cave salamander *Proteus unguinus*, and the cave fish *Amblyopsis spelacea*, found in Mammoth cave, Kentucky are the best known troglodytes (cave-dwellers).

Some of the animals who live successfully in caves, although common in outside habitats, include swiftlets, bats, flies, moths, and reptiles. Bat droppings and the moulds growing thereon make up the food for innumerable parasites, for example, mites, myriapods, insects, molluscs, and isopods, which in turn are preyed upon by beetles and their larvae, spiders etc.

A 'living' cave is one where stalactites and stalagmites are in active formation. The percentage of humidity is over 90, and the air is clean and fresh. These conditions are ideal for troglodytes. The complete darkness, uniform temperature and uniformly high humidity, are responsible for special adaptations like the increase in the size of sensory organs, for example, enlargement of special setae; of antennae, palpi etc., of legs where they are used for tactile purposes. Other changes are the loss of colour and pattern, degeneration, reduction or elimination of eyes, and marked intolerance of reduced humidity.

In the Batu caves, Kuala Lumpur, it is the presence of bats that provides food for animal life in the cave. At Niah, in the great cave in Borneo, are found some half a million bats of seven resident species. Here also roost some 4,500,000 swiftlets (Apodidae) of three species. It is interesting to note that the swiftlets fly out for food at dawn, which is when the bats return, and that in spite of the total darkness there are no collisions.

The fauna of caves in India does not appear to be highly specialized. True troglodytes have not been reported. In the Siju cave, Assam, some 102 species have been recorded, out of which only 23 species were found beyond 1500 ft. The only aquatic animal there which shows adaptations to cavernicolous conditions is the prawn *Palaemon cavernicola*. A chamber 500 ft from the opening is full of bats. Guano has been found deep inside the cave, but bats no longer inhabit these chambers. Most of the species found here are found outside also. The only species which are known from other caves are *Cubaris cavernosus* from Cherrapunji; *Chelisoches morio* commonly found in other Assamese caves and in those of the Malay Peninsula; *Alrichopagon cavernarum*, which is also known from the Batu caves

in Selangor; and *Tinea antricola*, which is probably found in the Moulmein caves in Burma.

The life in caves, as also the ossified remains of it, give us important clues for unravelling the past. Thus a primitive insect, *Campodea*, has been found in two widely separated caves in the Simla Hills. It is believed that these are survivals from a time when *Campodea* of this size were as widely distributed as surface forms. Of all cave dwellers, bats especially have been intensively studied, and the occurrence of rarities recorded. 'Long-lost' bats like Wroughton's Free-tailed bat, *Otomops wroughtoni*, have been rediscovered in the Barapede Cave at Talewadi, Mysore. In the caves at Krishnapur, some six miles down the valley to the west of Talewadi and close to the Goa border, a few male specimens of *Taphozous theobaldi* have been observed. This species, though common in Burma, Malaya and Java, is rare in India; it was recorded only once from India, during the Mammal Survey in 1911. From such instances it has been noted that animals that would otherwise die out have survived in the stable and isolated environment of caves.

A.S.B.

CHAMPAK, Yellow or Golden Champ (*Michelia champaca*). This large evergreen tree, with large leaves with a wavy margin, is esteemed for its profuse yellow and

Leaves and flowers of the champak (× ⅓). At the bottom, from left to right, a flower with perianth removed showing finger-like stamens girdling the carpels; a pod, split and showing eight seeds; and a cluster of unopened fruit-pods.

scented flowers. It grows wild in Nepal, Sikkim and other parts of E Himalaya, Burma and the Western Ghats and is widely cultivated in gardens and temples. The flowers yield champa oil, used in perfumery. The heartwood is strong and durable, capable of taking a high polish, and is valued as a furniture wood.

CHAMELEON. Chameleons are a unique group of arboreal lizards characterized by a long extendible tongue, watch-spring-like tail, protruding, independently moving eyes and toes opposed as in parakeets'. Africa is the home of these lizards. Only one species, *Chameleo zeylanicus*, occurs in India and Sri Lanka.

Chameleons have a remarkable ability to change colour. A chameleon can change its colour from green and brown to different shades of yellow, white or black (but not red) in the space of a few minutes. Most chameleons are green and live among leaves, but they do not change colour to match their background. Their colour changes with the temperature and often becomes much brighter when chameleon meets chameleon.

The chameleon's eyes are set in protruding turrets and can be rotated quite independently, so that the animal can see forward with one eye whilst looking sideways or backwards with the other. The eyes rotate ceaselessly in all directions till the prey is sighted, and when the chameleon makes up its mind to catch the prey the eyes stop rotation and focus on the object, towards which the lizard advances with great deliberation. About 25 cm away from its prey the chameleon stops, aims with its head, rocks sideways the better to judge the distance, and suddenly the insect vanishes, having been snatched up by a lightning lick of the tongue.

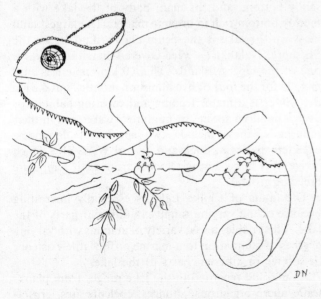

Chameleon × ½

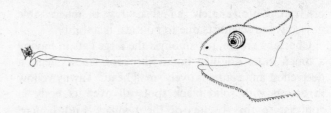

Chameleon with extended tongue catching prey × ⅓

The fore and hind feet of a chameleon are modified to form clasping organs suited for its life in bushes and trees. These grip branches so firmly that even a heavy wind cannot dislodge the animal. The tail is prehensile and helps the chameleon to negotiate gaps between branches.

With all its unique features, a chameleon is a slow-moving creature. When picked up it hisses and can inflict painful but harmless bites.

The chameleon lives mainly upon insects which it eats in large quantities and great variety including agricultural pests such as grasshoppers. The female lays 10–30 eggs (in winter months in South India). Eggs are laid in a hole in the ground dug by the female, and hatch about 120 days later. A fully grown chameleon measures 375 mm including its tail.

See also LIZARDS.

See plate 9 facing p. 128.

CHAPLASH, *Artocarpus chaplasha*, is a large deciduous tree of eastern India, the Andaman and Nicobar Is. and Burma. The leaves from young trees are totally different from leaves of mature trees. In the former they are up to a metre long with lobed and serrated margins, in the latter they are one-third this length with a rough texture and stiff hair. Elephants relish the leaves. The timber is used in tea boxes.

CHEETAH or HUNTING LEOPARD (*Acinonyx jubatus*). Sadly this animal became extinct in the Indian subcontinent in the 1950s. The last authentic sighting was actually recorded with a photograph, in the pages of the Bombay Natural History Society's *Journal*, when the Maharajah of Korwai State (wrongly spelled Korea in the *Journal*) shot three cheetah—presumably juveniles, in one night as they stood transfixed in the headlights of his car. This was in 1948. There is a good deal of circumstantial evidence that a few still survive in the border region between extreme southwestern Baluchistan (Pakistan) and Fars Province of Iran. A skin said to have been collected near Turbat, in this region of Pakistan, was deposited with the Royal Scottish Museum in 1972 and there have been much more recent sightings of cheetah in bordering regions of Iran. Unfortunately all this area is inaccessible nowadays to scientists due to political instability.

Cheetahs are unique amongst the large cats in having a much more dog-like build with long sinewy legs, a deep chest and comparatively small head. Tawny yellow in colour with solid black spots all over its body in contrast to the rosettes of the Leopard and Jaguar, it has a very distinctive black tear-stripe extending from the inner corner of its eye to the corner of its lips. The neck and shoulders have a coarse rough mane of upstanding hair and the claws are only semi-retractile, lacking any sheaths.

In temperament the cheetah is playful and even affectionate and for centuries they have been tamed and used by Indian princes for hunting gazelle. Today the cheetah is rare throughout its range and is confined to savanna country in the more remote regions of south and east Africa, with remnant populations precariously surviving in Iran, Arabia and North Africa. Cheetah prey mainly on antelopes or gazelle and after a careful stalk, run down their victim in a short sprint over open country, when their speed and grace rivals that of any other terrestrial mammal. High-speed photography and other equipment has shown them to be capable of attaining speeds of 110 kmph for short distances.

Unfortunately this beautiful cat still comes into conflict with man and is hunted for its skin, though more and more countries are banning trade in cheetah skins as well as those of other wild cat species.

See plate 18 facing p. 241.

CHEVROTAIN. This mouse-deer (*Tragulus meminna*) is limited more or less to the peninsula in India, and is a true forest animal, seldom being found away from tree growth. It is found both in mixed deciduous and in mixed evergreen habitats, including sholas, and prefers well-watered hill forests. In the undergrowth and litter of the forest floor its olive-brown coat, broken up by pale harness-markings, is remarkably obliterative, and being diminutive and low-to-ground, it often escapes notice. It is solitary, and active both by day and night, and feeds on low herbage and also on many forest fruits, fallen to the ground. It is a thirsty animal and when it is hot it pants openmouthed, but without its tongue hanging out like a dog's. It can climb well, and sometimes lies up inside a hollow bole. It is remarkable not only taxonomically, but also for its lack of vocalizations. In the wild it comes out with no call by which it can be known—a captive individual, when frightened, emitted a low, breathy grunt.

M.K.

See also Deer, Ungulates.

CHILKA LAKE, on the eastern seaboard in Orissa, is a shallow, brackish-water inshore lake, the largest on the subcontinent. Geological and fossil evidence indicate that the lake was originally a part of the sea. Due to the silting action of the Mahanadi river system, which drains into the northern end of the lake, and the northerly currents in the Bay of Bengal which have formed a sand-bar along the eastern shore, Chilka has gradually become a shallow lagoon. Silting and tidal actions have been significantly altering the waterspread area, by changing the relative positions and dimensions of the land masses. The area varies in dry and flood seasons between about 560 and 800 km².

The lake is bounded on its western and southern shores by a range of rocky hills, and merges into cultivated land at its northern end. Roughly pear-shaped, the lake is 75 km long in a NE-SW axis, and 32 km wide at its broadest. Chilka is divided into an outer channel with a narrow 'mouth' leading into the sea, and with a sandy bottom, and the main body of the lake with a muddy bottom rich in organic matter. The largest land mass on the lake is the peninsula Parikud Is. on the SE shore. Nalaban (=weed-covered marsh in Oriya) is one of the biggest islands. This 10 km² marsh is submerged for the four or five monsoon months, but during the winter is a major feeding and roosting habitat for over a hundred species of migratory waterfowl in their thousands; these include two species of flamingos, over ten species of duck and several WADERS which arrive on Chilka during October from their temperate breeding grounds.

The fauna of Chilka Lake is especially interesting because of the varying salinity in different parts of the lake. This enables a vast variety of animals with varying degrees of adaptation to a marine or fluviatile existence to survive in different parts of the lake.

Animal life recorded in the lake ranges from planktonic micro-organisms, sponges, coelenterates, crustaceans, leeches and several other invertebrate forms, to

Mouse Deer, Western Ghats, × ⅓

© M. Krishnan

a vast variety of fish which together sustain the enormous wintering bird populations. A few estuarine turtles and three species of snakes are found in the lake. Mammalian fauna includes otters and a species of dolphin in the lake, and several rodents, bats and sloth bears on the shores and surrounding hills. A flourishing fishery, exploiting crustacea, molluscs and fish is the mainstay of the local population.

CHINAR, *Platanus orientalis,* a majestic tree up to 30 m high and 12 m girth, was introduced into Kashmir by the Moguls and is now a dominant feature of the landscape of the valley. It is characterized by greyish bark peeling off in long flakes, and 5–7 lobed leaves which are deciduous and turn a golden yellow in autumn. The fruits consist of numerous achenes, small, dry, single-seeded fruits that do not split open. Wood white fringed with yellow or red, used in making boxes and trays which are lacquered and painted. Chinars are likely to become more widespread, being tolerant of pollution.

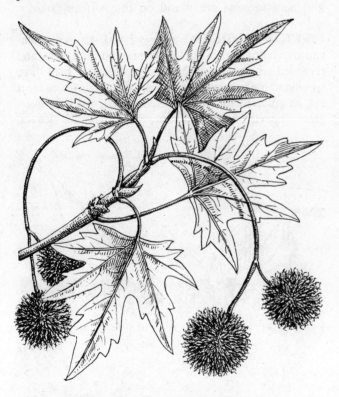

Chinar leaves and pendulous, globose fruits, × ¹/₃

CHLOROPHYLL. Plants owe their characteristic green colour to a pigment called chlorophyll which absorbs red and blue lights, leaving green light to be reflected. This pigment was the first substance recognized as responsible for light-absorption during photosynthesis in plants.

Chemical nature. Chemically, chlorophyll has a

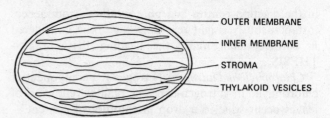

OUTER MEMBRANE
INNER MEMBRANE
STROMA
THYLAKOID VESICLES

Schematic drawing magnified about 10,000 times. Chlorophyll is in the thylakoid vesicles, which have probably developed from the inner membrane.

'head', consisting of four organic rings with a magnesium ion held at the centre, and a long hydrocarbon chain or 'tail'. Interestingly, its structure has a marked resemblance to blood haemoglobin, which at the centre contains an iron atom instead of magnesium. (An ion is electrically charged but an atom is electrically neutral.)

In fact chlorophyll is not a simple substance, but a group of related pigments. About ten chemical types of chlorophylls having small variations in their chemical structure have been isolated. These are called chlorophylls *a, b, c, d* and so on. Chlorophylls *a* and *b* are the most frequent. The others may only be found in a particular group of plants.

Localization. Chlorophyll and other pigments that perform photosynthesis are found in well-defined structures in the plant cells called chloroplasts. Each chloroplast has an outer envelope consisting of two membranes, a liquid (sometimes called stroma) enclosed in the envelope, and a set of flat membrane-line sacs called thylakoids that are embedded in the stroma. The pigments are firmly attached to the membranes of thylakoids.

Photosynthetic bacteria and blue-green ALGAE, which are primitive cells without a nucleus (called prokaryotes), are also present either in cell membranes or in special vascular structures called chromatophores.

Chlorophyll and photosynthesis. The wave-lengths of light absorbed maximally by chlorophylls (the absorption spectrum) coincide closely with the wave-lengths at which photosynthesis proceeds at the maximum pace (the action spectrum). This type of evidence proved that chlorophyll is the predominant light-trapping molecule during photosynthesis.

The ionizable chlorophyll *a* molecules occur at sites in the thylakoid membrane called 'reaction centres'. Each centre consists of many chlorophyll molecules and each thylakoid may have several hundreds of these reaction centres.

Two light reactions of photosynthesis (Hill's reaction and photosynthetic phosphorylation) are operated by energy primarily captured by chlorophylls. A quantum of light falling on a chlorophyll molecule gets absorbed, thus making it excited. The full quantum of excitation energy is passed from one pigment molecule to another in the reaction centres and photosynthesis commences instantaneously. Light reactions have been considered in further detail under the heading PHOTOSYNTHESIS.

Chlorophyll and nitrogen. Nitrogen is a component of chlorophyll. In nitrogen-deficient plants, yellowing of leaves occurs due to a drop in chlorophyll content.

K.K.N.

CICHLID FISHES. The family Cichlidae (pronounced sick-lid-ee) comprises perch-like freshwater and estuarine fishes mainly from Africa, and Central and South America. Only two species come from South Asia, both of which are found in Indian waters.

They are intelligent fishes, noted for the way in which they assiduously protect their eggs and care for their young. The family is well-known since many species are popular as pets in home aquariums, such as angel fish and firemouths, which are extensively bred in India though originally coming from South America.

The Orange Chromide (*Etroplus maculatus*) is found in estuaries in South India and Sri Lanka. It grows to 8 cm and has an earthy brown body with three large, round, dark spots with a bluish or yellowish border. In the breeding season, rows of tiny red spots appear on the body, which assumes a yellowish orange coloration. The newly-hatched young feed on a secretion on the parents' skin.

The Pearl-spot, also called Green Chromide (*Etroplus suratensis*) is a brackish-water fish and grows to a much larger size (40 cm) and can thrive in purely fresh water. It has a bluish or greyish green body with six to eight dark vertical stripes; moreover, the iridescent white spots on the scales look as if rows of pearls have been affixed to the body and hence the name pearl-spot. The eggs are laid in small cup-like hollows and, when hatched, the parents shift the young from hollow to hollow by sucking them into the mouth and spewing them out again. In case of danger they are again sucked in the mouth and shifted to safety. It is a popular food fish and occurs on the southwest coast up to Malvan.

The African Mouth-breeder (*Sarotherodon mossambicus*, formerly known as *Tilapia mossambica*) was originally from Mozambique but has been introduced all over south and southeast Asia as a food fish. Growing to over 35 cm, it can live and breed in brackish or fresh water and has spread into most of the coastal tracts in India. It is omnivorous, feeding on aquatic vegetation as well as animal life, and lays over 300 eggs at a time which hatch in 6 to 8 days. The fish is a prolific breeder, repeating the act three times a year and becoming sexually mature when four months old. The female takes the fertilized eggs into her mouth and keeps them there until they hatch. Even after birth the young leave her mouth only to forage for food, but at the

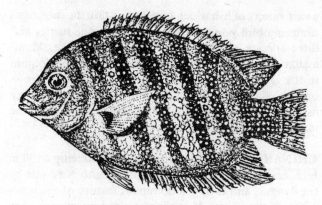

Pearl-spot, *Etroplus suratensis* × ¼

slightest sign of danger they swim back into their mother's mouth where they are safe from enemies.

This Mouth-breeder has so successfully established itself on the coastal areas that it is considered unwise to introduce it elsewhere, as it is feared that it might compete for food with cultivable native fish because of its prolific breeding. The remaining five species of *Tilapia* and *Sarotherodon* are found on the African coast.

CIVET FRUIT, DURIAN, *Durio zibethinus*, is a Malayan tree cultivated in India, Burma and Sri Lanka and regarded as the most famous tree of the East. The specific name is derived from the Italian *zibetto* or civet cat, an animal with an offensive smell. It is a lofty evergreen, heavily buttressed tree, with drooping leaves,

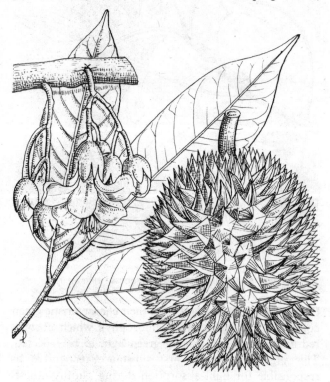

Flowers and fruit of the Civet tree. Both grow from the stem. × ⅓

silvery brownish on the underside. The tufts of cream-yellow flowers open between 2 and 3 p.m. (summer) and fall off at 2 a.m. apparently self-pollinated or pollinated by bats. The fruits are round or egg-shaped, like a football covered with thick spikes. The seeds are embedded in custard-like pulp which, mixed with cream and sugar, has a delicious taste but a putrid smell. A taste for durian is quickly acquired, and some trees have been planted in the Nilgiris. The smell of the fruit attracts wild animals: elephants have the first pick, the tiger, deer, rhinoceros, and monkey enjoying what is left. A red-flowered variety *roseiflorus* has the best taste. It is one of the most beautiful flowering trees.

CLAMS are an immense group of edible bivalves widely distributed in all coastal waters. There are many kinds of them belonging to different families. There are the purely sea forms and also the backwater forms. The former includes Venerid clams, Surf-clams, Giant clams, Paper clams, Wedge-shells and Soft-shelled clams. The backwater forms are Venerid clams belonging to the genus *Meretrix*. The flesh of all clams is valued as food and their shells are of intrinsic beauty, variety and colour.

Clams are far more developed bivalves than the primitive ARKS. Their gill filaments, instead of being free, are united at regular intervals by cross-bars containing blood vessels. Again the mantle edges are not open all round as in arks but have openings only for the siphons and the foot. In the deep-burrowing clams such as soft-shelled clams the two siphons are fused to form a long tube.

India has a wealth of different genera of clams both along the seashores and in the backwaters.

Venerid clams form the largest bulk of sea clams. Their foot is comparatively short and compressed and their siphons are free or partly united. They live below half-tide mark, buried a few inches deep in sand beds and mud-flats. They are more or less localized and do not make any long excursions in search of food. They feed on minute plant and animal organisms washed back and forth by the waves. Seashore birds feed on them at low tide, and those living near rocks are attacked by carnivorous WHELKS.

Nothing much is known about the development of the

various clams. In the case of one genus *Sunetta* it has been observed that the very young ones appear in December, and by the following October have reached their full size of about 33 mm. The clam population shows a definite increase during the dry season February to May when they are collected in basketfuls from sand below shallow waters.

There are several genera of Venerid clams: *Circe*, *Gafrarium*, *Venus*, *Paphia*, *Katelysia*, *Tapes*, *Dosinia*, *Pitar*, *Sunetta*, etc. All these genera are alike in their internal structure, habits and habitat. They are however remarkable for their shells which are well-known for their varied shape, concentric grooves, radial bands, elegant sculpture, delicate colours and chevron designs. It is because of the exquisite beauty of these shells that their entire family is named *Veneridae* after Venus the Goddess of Beauty. They are also called Tapestry shells because of the elaborate network and chevron designs on many of the shells. Of all the species, special mention may be made of *Pitar erycina* which occurs in large numbers near Adam's Bridge and the north coast of Sri Lanka. Shells of this species fetch a high price in the booths in front of Rameshwaram Temple. The shell is very large, 8 cm long and 7 cm high, thick, with its apex inclined forwards. There are concentric grooves over the shell and the surface colour is orange brown. Several broad radial bands of chestnut tinge diverge from the apex marginwards.

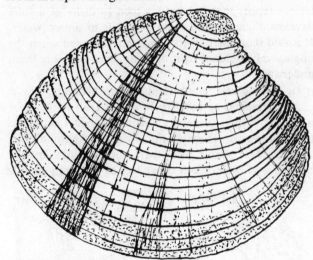

Pitar erycina shell × 1.

Surf-clams (Mactrids) are shallow burrowers in sandy shores, among the surf. The common Indian species number but a few all belonging to genus *Mactra*. They vary in size from 2 to 6 cm. The shells have their two valves similar; in shape triangularly oval. A shell with closed valves, when viewed from above, looks bulging and roomy. Hence the name *Mactra* which means in Latin a kneading-trough. In habits and structural details Mactrids agree with Venerids but their

Clams moving; left, *Sunetta*; right, *Mactra violacea*, both × 2/3

shells are very thin and fragile. Their siphons are united and the foot tongue-shaped, both adapted to live buried in sand with the tips of their siphons alone exposed above the surface. They feed on minute organisms contained in the surf, i.e. the thick white foam of crashing waves. Soon after storms one can see these clams dislodged from their burrows and scattered along the shore. These molluscs are considered more tasty than Venerid clams.

The two large Mactras of our shores are *Mactra turgida* and *M. violacea*, both 6 to 7 cm long and 5 cm high. The shell of the former is yellowish cream in colour and finely concentrically grooved; the latter is smooth-surfaced and violet, tinted with white blotches. There is a third, *Mactra mera*, of smaller size, straw-coloured and with a number of white radial bands.

Giant Clams. Among the sea clams, the Giant clams (*Tridacna gigas*) are the largest and heaviest of living bivalves. The shell surface is ornamented with thick, broad, crested, radiating ridges and the margin deeply indented. The interior is thickly enamelled and polished. The animal grows to about a metre in length and weighs 100 kg or more. The mantle of this animal has wide openings. Its foot is short but it attaches itself with strong byssus threads to dead coral or rocks. It has only one adductor muscle connecting the two valves but its pulling power is so great that anything caught between the valves can scarcely be pulled out. Collectors report that it takes six men to dislodge a fully developed giant clam from reefs 20 m under water. It is said that pearl divers have been drowned when the mighty valves closed tight on their arms, trapping them underwater.

In Indian seas giant clams occur here and there, chiefly attached to coral reefs in the Gulf of Mannar and in larger numbers around the Laccadives (Lakshadweep). These clams, living a stationary life among coral reefs, rival their surroundings by the brilliance of their colour: the edges of the fleshy mantle appearing between the gaping valves resembles a serpent brilliantly coloured in shades of red, blue and green.

The two valves of *Tridacna gigas* exhibited in the British Museum weigh 70 and 71 kilograms and are almost a metre long and the animal inside would weigh about 9 kg. In the Madras Government Museum also two huge valves of *Tridacna cumingii* obtained from the Laccadive Islands are exhibited.

Sometimes pearls of inferior quality and size are produced in the interior of giant clam shells.

Paper shells (Tellins), Wedge shells (Donacids) and *Soft-shelled clams (Myids)* are close allies of clams. On all our sandy shores they are as abundant as Venerid clams. The one remarkable feature of these bivalves is they have very long siphons which can extend up to twice the length of the shell. Their foot is wedge-shaped and strong. They live in shallow waters buried a few inches deep in sand with just the tips of their siphons in level with the surface. At low tide the mouths of their burrows get exposed and the shore birds peck into these holes to catch them. The birds succeed in seizing a few, but most manage to escape by withdrawing their siphons into the shells and burrowing deeper down into the mud with their nimble foot.

Tellins include many species. A score or more are found along our coasts. Their shells are diverse in size ranging from 0·5 to 8 cm in length but all are relatively thin and flat and delicately tinted white, light pink, yellow or purple. Over a dark background of mud, a number of empty Tellin shells lying scattered appear like bits of paper: hence the popular name Paper Shells.

Donacids are of a stouter build—the shells are thick

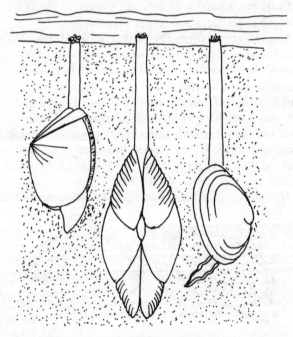

Clams buried in sand. Left to right, *Tellina ala*, *Donax scortum* and *Mya* sp. All × 1.

A giant clam, × ¹⁄₂₀, shown exposed on a coral reef.

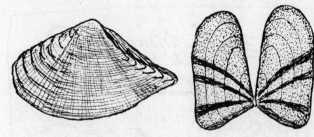

Two *Donax* shells, *D. scortum* and *D. cuneatus*, both × 1.

and wedge-shaped (hence the name). Different species are coloured differently and even one and the same species may have different colours—yellow, chestnut, grey or blue. Many of them carry radial rays. The valves washed ashore often remain attached like the spread-out wings of a butterfly and therefore Donacids are also called Butterfly shells. Donacids play where the waves wash up and recede, riding up and then suddenly digging into the soft wet sand. They dig so fast that they disappear in a few seconds. They are collected in large numbers along with other clams for food. The largest Indian Donacid is *Donax scortum*, 6 to 7 cm long. It has a curious shell with a corrugated surface, brownish grey in colour outside and purple inside; its hind end is beaked.

Soft-shelled clams form an interesting group. They burrow more deeply. Their foot is almost cyclindrical and their siphons very long and united. They live in shallow muddy bottoms. At low tide fishermen locate them as they squirt like RAZOR-FISH. Their flesh is edible, tasty and popular.

Backwater Clams belong to the genus *Meretrix*. All the backwaters and saline creeks along both coasts of India abound in this clam. In Cochin backwaters and Vembanad lake members of the species *Meretrix ovum* grow so abundantly that they form regular beds or layers over the bottom. These shells are dredged in boatloads for the manufacture of quicklime and cement. The flesh is widely eaten. The shell can easily be distinguished by its heart-like shape, rounded umbo, smooth grey surface and two broad rays running from the umbo to the lower margin. An adult clam measures about 4 cm long and 3 cm high.

Along the east coast the common clam that abounds in the muddy sands of estuaries is *Meretrix casta*. This is a thicker and larger shell than the west-coast *M. ovum*. The shell is dull brown outside and purplish inside. Their flesh is also eaten, but not with so much relish as the west-coast species.

There is another species of backwater clam, *Meretrix meretrix*, which is very large, 8 cm × 5 cm, called the Great clam. The shell exhibits considerable variation in colouring and at least five well-defined colour varieties are known. It is found at the mouths of rivers and creeks opening into the sea, irregularly along both the east and west coasts. These Great clams do not form regular beds. Their flesh is a delicacy.

See also MOLLUSCS.

CLASSIFICATION. The purpose of biological classification is twofold. By arranging groups of different categories into a hierarchical system, the lower categories contained in the higher, one can build up a framework in descending order beginning from the highest (Kingdom) to the lowest (Taxon), which can facilitate identification at different levels of groups and help in memorizing each by the similarities and differences, and by assigning distinct names to each category. A more scientific and serious purpose is to assess the natural affinities between different organisms and groups of organisms and to build up a picture of both the amazing diversity and the fundamental unity of life on earth.

Classification solely intended for convenience in identification of different categories is called 'artificial classification'. In such an arrangement the constituent categories at different levels of hierarchy do not necessarily reflect biological affinities. On the other hand, a system of classification that is based on the natural affinities between different categories, and which seems to fit in some sort of evolutionary pattern, is called a 'natural classification', and is valid as long as the theory on which it is based is valid. In other words, artificial classification is more subjective and the natural one more objective.

Though Aristotle gave birth to the science of animal classification three centuries before Christ, the present system of biological classification, now in universal usage, was applied by Linnaeus in 1758. The Linnaeus system has two great advantages: it gives a uniform and scientific procedure for naming biological groups at different levels of hierarchy, which ensures uniformity in naming animals and plants irrespective of linguistic or national barriers. Secondly, the Linnaean hierarchy also projects natural affinities of groups at different levels and can be used for assessment of phylogenetic relationships of constituent groups in a category and of different categories.

Linnaeus, in naming species (the basic unit of his hierarchy), used a binary nomenclature. For example, the biological name of man, *Homo sapiens*, has two elements, the first—a noun—is the name of the genus, and the second—an adjective—is the trivial or specific name which, in combination with the generic name, denotes the species to which the organism belongs. The system of naming species, called binomial nomenclature, is now universally followed; and whatever be the local name of an animal or plant, in biological literature it is given only one name with two components.

The various constituents in the framework of animal classification are called taxonomic categories. Linnaeus used only five in classifying the animal kingdom, namely, *classis, ordo, genus, species,* and *varietas.* Soon, however, two additional categories were adopted, namely, *family* (between *ordo* and *genus*) and *phylum* (above *classis*). The category *varietas* (variety) was eventually discarded and is now replaced by *subspecies,* making the actual nomenclature trinomial (generic + specific + subspecific names) in groups like birds where the basic unit is often recognized as subspecies. With the phenomenal increase in our knowledge of the animal kingdom and the number of animal species, it became necessary to indicate precisely the position of a species in the hierarchy. This necessitated the adoption of additional categories either by combining the original category names with prefixes like 'super' or 'sub', or by using new category names like *tribe* and *cohort.* Consequently, now as many as 18 categories or taxa are generally used in classification: Kingdom—Phylum—Subphylum—Superclass—Class—Subclass—Cohort—Superorder—Order—Suborder—Superfamily—Family—Subfamily—Tribe—Genus—Subgenus—Species—Subspecies. According to the rules of zoological nomenclature, the name of a superfamily must end in 'oidea', that of a family in 'idae', subfamily 'inae' and tribe in 'ini'.

K.K.T & B.B.

CLASSIFICATION AND NAMING.

The Latin names for plants and animals seem very clumsy but they are full of meaning for biologists and are used all over the world. The animal called *tendua, cheeta, wagh, yuz* or *chinna puli* in various parts of our area has only one international name, *Panthera pardus. pardus* with a small p is its specific name and *Panthera* its generic name and by the use of two-word Latin names all the millions of living plants and animals can be identified, and such binomial names have even been given to many microscopic organisms such as bacteria and viruses.

Classification, arrangement into groups, is necessary before a generic name can be given and the system of classification used for plants and animals was devised by a Swedish naturalist, Carl Linnaeus, more than two hundred years ago. Linnaeus was a tireless collector of specimens, journeying himself to Lapland, France and England and inspiring others to collect in many other parts of the world, including Japan and Ceylon. At the time of his death in 1778 he had about 19,000 sheets of pressed plants, 3200 insects, 1500 shells, over 700 pieces of coral, 2500 books and many letters and notes. His genius lay in recognizing similarities and grouping them logically. Before his time whales had been considered to be fishes because they lived in water and bats were thought to be birds because they could fly. Linnaeus recognized that they were like many other animals in so far as their young were fed from the mother's breast. He invented the term Mammalia to include all such

A SKETCH FROM ONE OF LINNAEUS'S DIARIES

In Lapland he saw a beautiful red flower growing far out in the middle of a pond. It reminded him of the Greek story of Andromeda chained to a rock in the sea, to be eaten by a dragon, so he named a genus of water-plant *Andromeda.*

animals, *mamma* being the Latin word meaning breast.

Animals are divided into two main divisions according to whether they have backbones or not, are vertebrate or invertebrate. Some of the invertebrates are described under PROTOZOA, SPONGES, COELENTERATES, ECHINODERMS, MOLLUSCS, CRUSTACEANS, INSECTS and ARACHNIDS. Vertebrates are described under FISH, AMPHIBIANS, REPTILES, BIRDS and MAMMALS.

Plants too are divided into two main divisions, those with flowers and those like ALGAE, FUNGI, MOSSES and FERNS with no true flowers: they are labelled 'secret marriers' (cryptogams). Flowering plants reproduce by seeds while the lower plants (cryptogams) reproduce by spores. The FLOWERING PLANTS (phanerogams, or 'open marriages') are further divided into those with naked or protected seeds (gymnosperms, e.g. CONIFERS, and angiosperms), and the angiosperms further divided according to whether the seed sprouts one leaf (monocotyledenous or monocot, e.g. GRASSES) or two leaves (dicotyledenous or dicot). There are about 250,000 species of flowering plants so without a well-organized system of classification and rules of naming plants (nomenclature) the study of the plant kingdom is difficult or impossible.

The main divisions are called phyla (plural of phylum) and they are subdivided into smaller groups, so that the whole arrangement is hierarchical. The main groupings are

Kingdom (Regnum). Plant or Animal
Division (Phylum)
Class
Order
Family
Genus (plural genera; adjective generic)
Species
Subspecies

The species is the unit and its name is never changed. As knowledge of the course of EVOLUTION increases, the position of a plant or animal in the hierarchy may change, in which case the generic name must be changed too. Generally groupings are made from the most primitive, such as the single-celled amoeba, to the most highly developed, man (*Homo sapiens*), and organisms are referred to as 'higher' or 'lower' accordingly.

Returning to the leopard, *Panthera pardus*. Its two-word, binomial Latin name tells us it belongs to the genus *Panthera*. There are four other members of the same genus, the tiger, the lion, the jaguar, and the snow leopard; and the genus is one of the twenty in the family Felidae, cats great and small. Its full description according to the Zoological Code, distinguishing the leopard from 4500 other mammals, would be:

Kingdom	Animalia	Animals
Phylum	Chordata	Vertebrates
Class	Mammalia	Mammals
Order	Carnivora	Meat-eaters
Family	Felidae	Cats
Genus	*Panthera*	Leopard-like
Species	*pardus*	

Populations of leopards living in different places may develop different characteristics, though they would still be able to breed with other leopards. They are known as subspecies, or races, and have three-word, trinomial names. The subspecies of leopard are called *Panthera pardus saxicolor* (stone-coloured), *P. p. sindica* (living in Sind), and *P. p. millardi* (Millard was the name of a former secretary of the Bombay Natural History Society). Botanical names may be even more complicated, with five components.

Scientific names have to be chosen in accordance with the rules of the Zoological, Botanical or Bacteriological Code and any change of name, or the name of a new species, has to be approved at an international congress. The scientific study of the variation of living organisms is called Systematics or Taxonomy. Taxon usually means a group, such as the species Leopard, but it can also be applied to an individual. The plural of taxon is taxa. Specific names are sometimes called trivial names.

R.E.H.

See also CLASSIFICATION, SCIENTIFIC NAMES.

CLASSIFICATION OF FLOWERING PLANTS. Attempts at classification of the Flowering Plants have been made since ancient times. In the earlier period, plants were classified depending upon their utilitarian value, whether edible, poisonous or medicinal. Aristotle and Theophrastus, who had knowledge of about 500 plants, classified them into herbs, shrubs and trees. Pliny and Dioscorides, during the first century A.D., described a number of medicinal plants and recognized some plant groups among them. It is only about a thousand years later that Albertus Magnus gave descriptions of a large number of plants and he is credited with the recognition of differences between plants which begin with a single leaf (monocotyledons) and those whose seeds sprout two leaves (dicotyledons). The Renaissance in Europe witnessed increased interest in plants, their form and structure, and the science of Botany came to be firmly established. Adventurous explorers of the 15th to 17th centuries brought knowledge of exotic plants and large collections of plants to Europe from far-off lands. Sexuality in plants was recognized and Carolus Linnaeus in the 18th century published his Sexual System of Classification of plants based on the reproductive structures in the flower. He also standardized the naming of the plants in Latin form for each species (e.g. *Mangifera indica* is the Latin name for the common mango, where *Mangifera* is the name of the genus to which the species, *indica*, belongs and the combined, binomial form represents the name of the biological unit, the species). His classification was, however, arbitrary and artificial and botanists tried to look for more natural relationships among the plants. George Bentham and Joseph Hooker devised a Natural System of Classification in the 19th century and this has been widely used in our area and elsewhere. In Europe, the German botanist Engler's classification was more popular. Darwin's theory of Evolution and Mendel's enunciation of the Laws of Inheritance in the latter half of the 19th century brought about a radical change in the thinking among botanists and during the present century Hutchinson in England, Bessey in the United States of America, and more recently, Takhtajan in Leningrad and Arthur Cronquist of the New York Botanical Garden, have published their systems of classification based on evolutionary principles.

The Angiosperms are divided into two Classes, Dicotyledons and Monocotyledons, based primarily on the fundamental difference in the number of cotyledons ('seed-leaves') in the two Classes. The Classes are further divided into Sub-Classes, Orders and Families. Some common floral characters form the criteria for division at each level. About 400 families of Flowering Plants are now recognized among the Angiosperms (the family names end in -aceae). The Families comprise the Genera and each genus is made up of Species. The Species is the unit in the hierarchical system of classification.

The Indian flora is estimated to include about 15,000 species of flowering plants. Selected families from this flora, including some familiar and useful plants, are briefly described here. Generic names or binomials are in parentheses and the sequence of the families is as in Bentham and Hooker's system of classification.

CLASSIFICATION OF FLOWERING PLANTS

SELECTED FAMILIES OF INDIAN FLOWERING PLANTS

DICOTYLEDONS: Leaves usually net-veined; floral parts usually in 4's or 5's; cotyledons 2.

RANUNCULACEAE: Mostly herbs with numerous spirally arranged stamens and carpels. Many colourful herbs like the buttercups (*Ranunculus*), columbines (*Aquilegia*), larkspurs (*Delphinium*), monkshoods (*Aconitum*), virgin's bowers (*Clematis*), wind-flowers (*Anemones*); mostly in the Himalaya and on the hills.

MAGNOLIACEAE: Trees and shrubs with showy flowers with indefinite number of spirally arranged and free carpels on a raised torus. Magnolias of eastern Himalaya and Michelias (for flowers and timber).

ANNONACEAE: Floral parts usually in 3's; stamens and carpels many in number, spirally arranged. Custard-apple (*Annona*), Mast tree (*Polyalthia*), extensively planted; ylang-ylang (*Cananga odorata*), source of perfume.

NYMPHAEACEAE: Aquatic herbs, leaves round, with stalk attached to the back (peltate), flowers attractive, floral parts many. Sacred lotus (*Nelumbium*), water-lilies (*Nymphaea*).

PAPAVERACEAE: Herbs. Flowers conspicuous, many attractive to look at like blue-poppies (*Meconopsis*) of Himalaya, Californian poppy (*Eschscholtzia*) in gardens, opium poppy (*Papaver somniferum*), source of the drug morphine and narcotic heroin.

BRASSICACEAE (CRUCIFERAE): Flowers have 4 petals placed crosswise (cruciform), stamens 6, 2 short, 4 long; fruit pod-like. Well known vegetables, cabbage, cauliflower, knol-khol, mustard (also seeds for oil), turnip (all species of *Brassica*), radish (*Raphanus sativus*). Many occur at high altitudes in Himalaya. *Ermania himalayensis* collected at 6400 m, highest altitude recorded for occurrence of a flowering plant.

THEACEAE: *Camellia sinensis*, the Tea plant.

CLUSIACEAE (GUTTIFERAE): Many useful plants. Poon tree (*Calophyllum*), flowers fragrant; ironwood, nagkesar (*Mesua ferrea*), mangosteen (*Garcina*), Balgi (*Poeciloneuron*), well known timber tree found only in peninsular forests.

DIPTEROCARPACEAE: Trees with winged fruits. Wood of many valuable. Gurjan (*Dipterocarpus*), sal (*Shorea robusta*), dammars (*Vateria, Vatica, Hopea*).

MALVACEAE: Stamens fused below into a tube, anthers single-lobed. Cotton (*Gossypium*), Shoe-flowers (*Hibiscus*), hollyhock (*Althaea*), Indian Tulip tree (*Thespesia*); closely related family,

BOMBACACEAE has silk-cotton tree (*Ceiba*), Baobab (*Adansonia*) and Civet Fruit or Durian. The cocoa family (STERCULIACEAE) with species of *Sterculia*, valuable for wood and gum (*Sterculia urens*) and the jute family (TILIACEAE) are grouped with Malvaceae and Bombacaceae in a common Order.

RUTACEAE: Oil glands present in leaves; flowers regular, parts in 5's, prominent disk below ovary. Orange, lime, lemon, pomelo and other fruits, all species of *Citrus*, elephant apple (*Feronia*), sacred bael (*Aegle marmelos*), curry-leaf (*Murraya koenigii*), east Indian satinwood (*Chloroxylon*).

MELIACEAE: Several useful plants. Neem (*Azadirachta indica*) for wood and oil; mahogany (*Swietenia*), rohit (*Aphanamixis*), toon (*Toona*), all of timber value.

ANACARDIACEAE: Fruits one-seeded, surrounded by fleshy part of fruit wall with resin canals. Astringent juice common and many are toxic (e.g., poison ivy, *Rhus toxicodendron* of North America). Mango (*Mangifera indica*), cashew-nut (*Anacardium occidentale*), widely cultivated; pistachio nuts (*Pistacia*), *Rhus* species in tanning and dyeing, hog-plum (*Spondias pinnata*), marking-nut tree (*Semecarpus anacardium*). The well known cultivated litchi (*Litchi chinensis*) of North India belongs to the allied family, Sapindaceae.

FABACEAE (LEGUMINOSAE): One of the largest families of Flowering Plants. Gynoecium characteristically of one carpel, developing into a Legume. Divided into three subfamilies:

PAPILIONOIDEAE: Corolla characteristic with large standard, two lateral wings and boat-shaped keel of remaining two petals; stamens 10 bundled in one or two groups. Several useful Pea (*Pisum*), Sweet peas (*Lathyrus*), Green and Black Grams, french beans, lima beans (all species of *Phaseolus*), cowpea and horse gram (*Vigna*), chickpea (*Cicer*), soya bean (*Glycine*), Agati (*Sesbania*), groundnut (*Arachis hypogea*); fodder-yielding lucerne, alfalfa, clovers, trefoils (*Medicago, Melilotus, Trifolium*), fenugreek, methi (*Trigonella foenum-graecum*), sunn-hemp (*Crotalaria*) for fibre; for timber, Indian rosewood and sissoo (*Dalbergia*), Red Sanders, kino and padauks (all species of *Pterocarpus*) and tragacanth gum (*Astragalus*) among several other useful plants.

CAESALPINIOIDEAE: Corolla not distinguished into standard, wings and keel, stamens free. Cassias for attractive inflorescences, some yield senna; Bauhinias, *Saraca*, the ashoka tree and Flame Amherstia, copper pod (*Peltophorum*) are all planted widely for flowers, as also Gold Mohur (*Delonix*); Tamarind (*Tamarindus indica*).

MIMOSOIDEAE: Flowers regular, stamens often many. Acacias for wood, gums, barks for tanning, soapnut, some fragrant (*Acacia farnesiana*, cassie perfume), Albizias for wood; *Mimosa*, Elephant Creeper (*Entada*).

ROSACEAE: Stamens many and carpels free in many. Roses (*Rosa*), fruit trees, apple (*Malus*), pear (*Pyrus*), almond, apricot, cherry, peach, plum (all species of *Prunus*), strawberry (*Fragaria*), raspberry (*Rubus*), quince (*Cydonia*).

COMBRETACEAE: Trees, shrubs, climbers. Fruits winged in some. Terminalias for timber, fruits medicinal (myrobalans), axlewood (*Anogeissus*) for firewood.

MYRTACEAE: Trees, shrubs, many with oil glands. Stamens indefinite, ovary inferior. Eucalypts, planted widely; guava (*Psidium*), bottle-brushes (*Callistemon*) in gardens, jaman (*Syzygium cumini*), cloves (*Syzygium aromaticum*).

LYTHRACEAE: Many ornamentals (crape myrtles), some of timber value (all species of *Lagerstroemia*)) henna (*Lawsonia*).

CUCURBITACEAE: Mostly herbaceous climbers. Flowers unisexual, ovary inferior, fruit fleshy with coalesced receptacle (pepo). Gourds, melons, pumpkins, squashes as edible fruits and vegetables. Colocynth (*Citrullus colocynthis*) medicinal.

APIACEAE (UMBELLIFERAE): Inflorescence of umbels, ovary inferior, walls of fruit with oil canals. Many aromatic, used for flavouring, in medicine; some poisonous. Anise (*Pimpinella*), caraway (*Carum*), coriander (*Coriandrum*), cumin (*Cuminum*), dill (*Peucedanum*), fennel, *saunf* (*Foeniculum*), parsley (*Petroselinum*), parsnip (*Pastinaca*) etc. Roots of *Ferula* yield asafoetida. Carrot (*Daucus carota*).

RUBIACEAE: Large tropical family. Flowers regular, tubular, ovary inferior. Many with alkaloids in roots, bark (*Cinchona*, *Cephaelis* (ipecacuanha), both introduced from Brazil); coffee (*Coffea*), madder (*Rubia*), kadam (*Anthocephalus*), yellow teak (*Adina cordifolia*). Some with fragrant flowers (*Pavetta*, *Ixora*).

ASTERACEAE (COMPOSITAE): Largest and most specialized family of Dicotyledons. Inflorescence, head or capitulum, ovary inferior, calyx in most modified to scales or tuft of hairs (pappus), which aid in the dispersal of fruits (seeds) by wind (e.g. Dandelion). Asters, calendulas, chrysanthemums, cosmos, daisies, dandelion (*Taraxacum*), everlastings (*Anaphalis*, *Helichrysum*), inulas, sawworts (*Saussurea*), senecios, sunflower (*Helianthus*), tansy (*Tanacetum*) have attractive heads, many are garden favourites. Chicory roots (*Cichorium intybus*); safflower (*Carthamus*), niger-seed (*Guizotia*), sunflower yield edible oils. Lettuce (*Lactuca sativa*), leafy vegetable, and wormwoods (*Artemisia*), aromatic and medicinal.

ERICACEAE: Rhododendrons and PRIMULACEAE (Primulas) of the Himalaya have attracted world-wide attention for their charming flowers.

EBENACEAE: Trees, shrubs, some with dioecious flowers. Ebony wood (*Diospyros* species), some with edible fruits (persimmon).

SAPOTACEAE: Trees, some with sticky latex. Sapodilla (*Achras*), for fruits and chicle-gum; mahua (*Madhuca*), flowers edible, seeds yield oil; several with valuable wood, Indian gutta-percha tree (*Palaquium*).

OLEACEAE: Trees, shrubs. Flowers regular, tubular, stamens 2. Jasmines, olive (*Olea*), ash (*Fraxinus*), coral jasmine (*Nyctanthes*), Himalyan lilac (*Syringa emodi*) belong here.

APOCYNACEAE: Trees, shrubs, climbers with latex. Many with showy flowers. *Rauvolfia*, *Holarrhena*, *Catharanthus* (*Vinca*) are medicinal (alkaloids), many poisonous.

ASCLEPIADACEAE: Milkweeds with latex. Many twiners. Anthers united in a ring and joined to stigma, pollen massed in pollinia (collections of pollen grains in waxy masses and carried as such by insects). Indian sarsaparilla (*Hemidesmus*), *Tylophora*, are medicinal. Some with succulent stems (*Caralluma*, *Ceropegia*).

CONVOLVULACEAE: Many twiners with attractive tubular flowers. Sweet potato (*Ipomoea tuberosa*).

SOLANACEAE: Economically very important. Potato (*Solanum tuberosum*), tomato (*Lycopersicon*), tobacco (*Nicotiana*), brinjal (*Solanum melongena*). *Hyoscyamus* (henbane), *Datura*, *Atropa* are medicinal; some are poisonous. *Capsicum* (chillies) for fruits. *Petunia*, *Cestrum* ornamentals.

BORAGINACEAE (Heliotropes, forget-me-nots),

SCROPHULARIACEAE (calceolarias, snapdragons (*Antirrhinum*), Linarias etc.,) and

ACANTHACEAE (Thunbergias, Coneheads, Barlerias, Crossandra, Beloperone) are all popular in gardens. Many wild ones in the Himalaya.

VERBENACEAE: Most valuable timber tree, teak (*Tectona grandis*); Verbenas, *Petrea*, Clerodendrums are seen in gardens.

LAMIACEAE (LABIATAE): Herbs, shrubs, many aromatic; flowers 2-lipped, stamens 4, 2 short, 2 long; fruit of 4 nutlets. *Ocimum* species (basil, *tulsi*), marjoram (*Origanum*), patchouli (*Pogostemon*), mint (*Mentha*), lavender (*Lavandula*), thyme (*Thymus*) all yield oil, medicinal. Salvias are garden favourites.

PIPERACEAE (Black pepper, betel vine), and MYRISTICACEAE (nutmeg, mace) have incomplete, unisexual flowers and yield the species named.

LAURACEAE: Trees and shrubs; oil cavities in leaves, floral parts in 3's. Cinnamon (bark of *Cinnamomum*), camphor (by distillation of wood chips of *Cinnamomum camphora*), avocado (*Persea*), several of value for wood.

EUPHORBIACEAE: Flowers unisexual, cyathium in *Euphorbia*, ovary 3-carpelled. Many with milky latex Castor (*Ricinus communis*), rubber (*Hevea*, *Manihot*), amla (*Emblica*), ornamentals like *Poinsettia* (coloured leaves near cyathia), crotons (*Codiaeum*), *Acalypha* and others.

MORACEAE: Flowers unisexual arranged in fleshy, hollowed out receptacles (syconia) along with sterile flowers and associated with ovipositing insects, figs (*Ficus*). Several species of *Ficus* planted along avenues and near villages. *Ficus benghalensis* (banyan) and *F. religiosa* (pipal) are regarded as sacred. Mulberry (*Morus*) in feeding silk-worms. Jack and Bread fruits (*Artocarpus*). *Antiaris toxicaria* is the upas tree (arrow poison). Allied family, CANNABIACEAE, has only the hemp plant (*Cannabis sativa*).

BETULACEAE (Birch, *bhojapattra*), JUGLANDACEAE (walnut) and SALICACEAE (willows) are all catkin-bearing with reduced unisexual flowers. Valuable woods. Mostly Himalayan in distribution. FAGACEAE is represented by the oaks (*Quercus*) in Himalaya.

MONOCOTYLEDONS: Leaves parallel-veined; floral parts usually in 3's, with distinction into sepals and petals lacking; cotyledon 1.

ORCHIDACEAE: See ORCHIDS.

MUSACEAE: (banana, Traveller's palm (*Ravenala*), Bird-of-Paradise flower (*Strelitzia*).

CANNACEAE (*Canna*).

ZINGIBERACEAE: ginger (*Zingiber*), turmeric (*Curcuma*), cardamom (*Elettaria*), *Amomum*, *Costus*.

MARANTACEAE (*Maranta* for arrowroot). These families are all economically important or useful as ornamentals.

LILIACEAE: Perianth conspicuous, stamens 6, ovary 3-carpelled. Lilies, fritillaries, *Gloriosa*, *Hemerocallis*, tulips (*Tulipa*), Trilliums and *Yucca* are all of ornamental value. Alliums (onion and garlic), *Colchicum* (for colchicine), *Asparagus*, Aloes are other useful plants.

ARECACEAE (PALMAE): See PALMS.

ARACEAE: The aroids (Anthuriums, Calla, Arum, Monstera, Rhaphidophora, Philodendron) are all popular garden plants. *Amorphophallus* and *Colocasia* edible. *Acorus* is medicinal. Some poisonous. Inflorescence, a spadix often with attractively coloured spathes.

POACEAE (GRAMINEAE): Perhaps the most important family of flowering plants from the economic point of view: cereals, sugarcane, fodder grasses, bamboos find universal use. See GRASSES.

There are many curious plants among the Angiosperms, like the parasites, saprophytes and insectivorous plants. The families OROBANCHACEAE, BALANOPHORACEAE and CUSCUTACEAE (dodders) are exclusively parasitic, the former two on the roots of hosts and the third twines around the host's stem. *Cassytha* of the Lauraceae is also a total parasite, twining around the host stem. *Sapria himalayana* (RAFFLESIACEAE) is a rare parasite of the Mishmi hills in eastern India; its flowers have the same structural features of the Monster Flower, *Rafflesia arnoldii*, but are smaller in size.

The mistletoes (*Viscum*) and several genera of the Loranthaceae are semi-parasites on branches of trees. The attack of some of these parasites causes extensive damage to the host, as in the case of Mango. A member of this family, the minute *Arceuthobium minutissimum*, causes severe damage to the blue pine (*Pinus wallichiana*) in the Himalaya. *Santalum album* (SANTALACEAE), famed for its sandalwood and oil, is a semi-parasite on the root of the host.

Saprophytes are chlorophyll-less plants thriving on decaying leaves and humus in the forests. The Indian-Pipe (*Monotropa*) and some orchids lead a saprophytic existence. Plants which depend on insects for their source of nitrogen have developed curious adaptations for the capture of insects. The pitcher-plant, *Nepenthes* (NEPENTHACEAE), the sundew, *Drosera* and *Aldrovanda* of the DROSERACEAE, bladderworts (*Utricularia*), butterworts (*Pinguicula*) of the LENTIBULARIACEAE, are the INSECTIVOROUS PLANTS seen in India.

M.A.R.

CLIMATE. The climate of any meteorological station, or area, or region is an integrated summary or a synthesis of the features of the day-to-day weather at the station, area, or region. The integration refers to features such as temperature, wind, pressure, cloud, rain, fog, thunderstorms, etc., etc. The integration or synthesis means more than determining the mean (or the median or the mode). It depends upon the so-called distribution of the variable under consideration, whether it be temperature or rainfall. If the variables are distributed according to the 'normal' or 'gaussian' distribution, the 'mean' and the 'standard deviation' together give a complete picture of the variable. In many of the climatological tables, we usually find the mean as well as the extremes. It is reasonable to assume that the standard deviation is roughly a third of the difference between the mean and the extreme maximum in the case of all the variables, except rainfall, whose distribution is not normal.

Our area presents great contrasts in its climatic elements. Within its boundary are found places with annual rainfall as low as 15 cm and as high as 1100 cm. Similar contrasts are found in temperatures also. On the Kerala coast the average temperature is about 27°C and the average relative humidity 60 to 80 per cent. The annual and daily ranges of temperature and humidity in this area are also small. In the northwestern plains summer temperatures soar up to as much as 47°C, while the winter temperatures drop down to 0°C. Owing to the diversity of climatic conditions in different areas it is not possible to treat our area as a single unit for climatological description.

There is however one major climatic feature which is common and this is the alteration of seasons known as the monsoons.

The Southwest Monsoon season, June to September, is the chief season of rainfall over most of the land. The southwest monsoon sets in over Kerala and Bengal early in June or late May and extends over the rest of the area by the end of June. It continues till the end of September or early October, its withdrawal commencing from the northwest. The activity of the monsoon is stimulated by a series of depressions which travel westwards from the north Bay of Bengal. With the withdrawal of the monsoon from northern India, clear autumn weather prevails there in October, November and the first half of December. In the southeast of the peninsula, however, the chief rains of the year occur in this season, accompanied by winds blowing from the northeast, and are associated with the retreat of the monsoon over the Bay of Bengal. They are often called Northeast Monsoon rains. In winter the weather in northern India is affected by a series of low-pressure systems which enter Pakistan from Iraq and Iran, and travel in an easterly direction. These weather systems, which are generally preceded by clouding and a rise of air temperature, cause snowfall over the hills and light to moderate rain in the plains. They are sometimes followed by cold waves in the rear. The winter season which ends by March is soon followed by hot weather with its rapidly increasing temperatures and dust haze in the atmosphere. This hot season is also characterized by the appearance of occasional duststorms and thunderstorms, which continue till the setting in of the monsoon in June.

See also MONSOONS, SEASONS.

CLIMBING PERCH. This fish (*Anabas testudineus*) is known for its accessory respiratory organ which enables it to remain out of water for a long time. Some old records had claimed that the fish was seen climbing a palmyra and hence it received the name 'climbing perch'. However, closer observations indicate that it is capable of leaving a sheet of water and travelling on grassy lands over long distances, swaying its head sideways and forward, with the help of its opercular spines. If a palmyra palm comes in the way the fish may cover a part of the climb for a short distance but this accidental travel cannot be called climbing the palm.

Climbing perch are found all along the coastal belt of southern India and are very common in northeastern India, where they are known as *koi* and are very popular as food. They grow to about 15 cm and feed on small fish and other aquatic animals. Having the accessory respiratory organ, they can thrive in very stagnant and polluted waters and breed in ponds and tanks by making a small nest of aquatic weeds and laying eggs therein.

See AIR-BREATHING FISHES.

CLIMBING PLANTS. Plants require sunlight for the manufacture of food by PHOTOSYNTHESIS. They cannot survive if their leaves are not exposed to the sun in such a way that they can absorb maximum light. To achieve this most plants develop stout herbaceous or woody stems which support the leaves and carry them towards the light. However, in nature when plants grow together there is a struggle for more light, and to solve this problem plants develop several ingenious devices. One such device is the development of a climbing habit, which enables them to reach the light without growing a stem strong enough to support the leaves. They depend on the strength of some other objects for their support and by developing long slender stems shoot up through the dense vegetation and reach the light more quickly than the plants which have to develop stems thick enough to support them. To know how successful such plants are one has to see a Bougainvillea climbing the tallest trees or a CAMEL'S FOOT CLIMBER covering the top of several trees in a tropical forest.

In order to climb, these plants use several means to attach themselves to the support. In many cases the plants simply twine around a support in a clockwise direction to the right as does a Yam (*Dioscorea alata*), or in an anticlockwise direction to the left as in Mussel-shell creeper (*Clitoria ternatea*). Many twiners growing in tropical forests such as Madhvi lata (*Hiptage benghalensis*), Malha bel (*Butea parviflora*), etc. have long

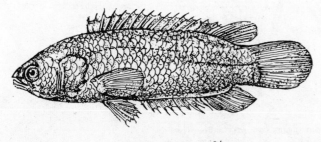

Climbing Perch × 2/3

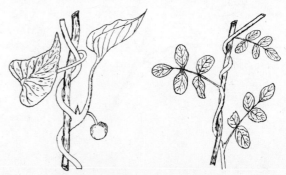

Yam twining clockwise round its support and Mussel-shell Creeper twining anticlockwise

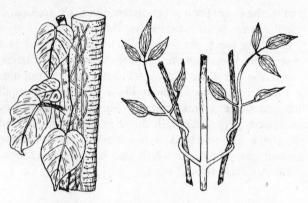

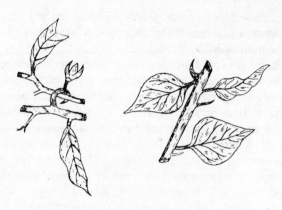

The Money Plant climbs by putting out roots as it grows and Travellers Joy twists its petioles round the support.

Bougainvillea sticks to its support by woody prickles and Kantili Champa has hooked thorns

and woody stems which climb up big trees, often extending from one tree to another forming loops and wreaths. These woody climbers are called lianas and often twine around the trees with a very tight grip, bending and distorting the stems, so causing considerable damage to timber trees.

Other plants have developed special climbing organs. The Nepal Ivy (*Hedera nepalensis*), Money plant or Devil's Ivy (*Epipremnum aureum*) climb by means of adventitious ROOTS which develop along the side of the stem in contact with the support, often forming an intricate network of roots. Bengal Traveller's Joy (*Clematis gouriana*) and the Pelican flower (*Aristolochia grandiflora*) climb by twisting the petiole round the support. Kantili champa (*Artabotrys odoratissima*), so commonly planted in gardens for its very fragrant flowers, climbs by means of curved hooks which develop from the flower stalk. The climbing roses, Himalayan raspberries, Bougainvilleas, Rattans and many other plants stick to the support by means of woody prickles and spines.

Probably the most important of the climbing organs however are tendrils. They are specialized parts of stems or leaves or parts of leaves and are very sensitive

to contact, coiling round the support to help the plant in climbing. In rare cases the tendrils bear sticky discs at their tips which stick to the support. In the Gourd family, and Passion flower, the twining tendrils are modified branches. In the coral creeper (*Antigonon leptopus*), the tendrils are modified flower shoots, while in peas some of the upper leaflets of the compound leaves become climbing tendrils. In Yellow Vetch (*Lathyrus aphaca*), the whole leaf is modified into a tendril, and to make up for the loss certain outgrowths of the leaves called stipules become greatly enlarged to take over the normal food-manufacturing function of green leaf. In Smilax on the other hand these stipules turn into tendrils and help the plant to climb. In Glory Lily (*Gloriosa superba*) the tip of the leaf is a tendril and coils round the support.

S.D.

CLIMAX VEGETATION. One of the most important consequences of biological regulation in a community (a group of one or more populations of plants and animals) as a whole is the phenomenon of ecological succession. When a cultivated field is permitted to lie fallow for some time, it produces an annual crop of

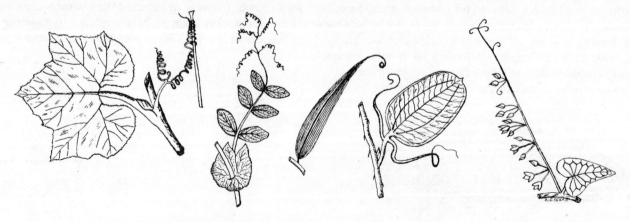

(From left to right) Tendrils modified from shoots (gourd), leaflets (pea), leaf-tip (Glory Lily), stipules (Smilax) and flower shoots (Coral Creeper).

weeds in the first year with some grasses; numerous perennials in the second year and a community of perennials thereafter. In forest areas, however, the perennial herbs are soon replaced by woody plants, which become dominant. These successive changes of vegetation may be either gradual or fast depending on many other factors like geographic location, climate, soil, seed source, etc., and will continue until the vegetation is made up of such species which are in complete equilibrium (steady state) both with the general environment and the micro-environment that determines the success and failure of reproduction of these species. This orderly series of vegetational changes on a single site through time is called plant or vegetational *succession*. The most striking characteristic of succession is the progressive development of vegetation on the same site resulting from the successive replacement of one plant community by another of different **growth** form until it finally terminates in the highest **type** of vegetation possible under the given environment. The successive stages of vegetation are entirely different in structure and function from the vegetation that eventually develops on the site.

It was not until the seventeenth century that any systematic studies of the changes in vegetation were made which were continued into the eighteenth century when for the first time the term 'succession' was applied to such vegetational changes. A detailed consideration of the relationship of organisms to their environment suggests that vegetational changes on a given site follow changes in the environment. Two general types of habitat changes may result in the modification of community structure or composition: (i) development of the community causes parallel developmental changes of the environment, and (ii) physiographic changes can likewise modify the environment substantially. Succession may, therefore, be defined in terms of three parameters, namely, (i) that it is an orderly process of community changes—which are directional and, therefore, predictable, (ii) these community changes result from the modification of the physical environment by the community, and (iii) they culminate in the establishment of as stable an ecosystem as is biologically possible on a particular site. It is important to note that succession is *community-controlled*, each set of organisms changes the physical environment thereby making conditions favourable for another set of organisms to follow. When the site has been modified as much as it can be by the biological processes, a steady state develops. The species involved, time required, and the degree of stability achieved depend on geography, climate, soil and other physical factors, but the process of succession itself is biological and not physical, i.e. the physical environment determines the pattern of succession but does not cause it.

The successive developmental stages of succession are known as *seral stages* and the entire series of communities in the process is called a *sere*. Two kinds of succession are recognized, namely, (i) primary, and (ii) secondary. Succession that begins on bare sites where no vegetation has grown before is termed as *primary succession*. It may also be observed on newly exposed sites (new alluvial deposits, new estuarine deposits, sand dunes, land-slips) available for colonization by vegetation, water and bare rock representing the two extreme types of habitats upon which primary succession is initiated. Evidently such habitats are unsuitable for the growth of most plants, and consequently, the pioneering species that do establish themselves must have adaptations permitting survival under extreme conditions. Moisture conditions usually control their ability to invade the new area. If the habitat is extremely dry, it is described as *xeric*; if wet, *hydric* and if intermediate, *mesic*. The successional trends are accordingly referred to as being *xerarch*, *hydrarch*, or *mesarch* succession.

Whatever the conditions of the initial habitat may be, reaction of vegetation tends to make it favourable to more plants by reducing the extremes, which is reflected in improved moisture conditions. Thus drier habitats become moister and wetter ones drier as succession progresses. Because of the diversity of habitats upon which succession may begin, there are an almost equal number of possible pioneer communities. Within a climatic area, however, the number of communities decreases as succession progresses because the trend is towards mesophytism from both drier and wetter habitats. Thus unrelated habitats may eventually support similar vegetation and may even undergo identical late stages of succession.

The successive stages leading up to a steady sal (*Shorea robusta*) forest of the Gangetic alluvium in Uttar Pradesh are fairly well known and provide a typical example of primary succession. These are broadly as follows, and also indicated in Fig. 1.

(1) *Acacia catechu—Dalbergia sissoo*
(2) *Acacia catechu—Holoptelea—Adina—Albizia procera*
(3) *Holoptelea—Adina—Lagerstroemia parviflora—Bombax—Terminalia bellirica.*
(4) *Adina—Lagerstroemia parviflora—Terminalia—Shorea.*
(5) *Shorea—Lagerstroemia—Terminalia—Adina.*

If the soil is rather damper, *Trewia nudiflora* and *Toona ciliata* appear at stage (3) and *Terminalia tomentosa* in stages (4) and (5), with *Syzygium cumini* mainly below the top canopy. In the temperate mixed coniferous forests, a typical succession at about 2400 m is (1) shrub associations, (2) blue pine, (3) mixed coniferous forest

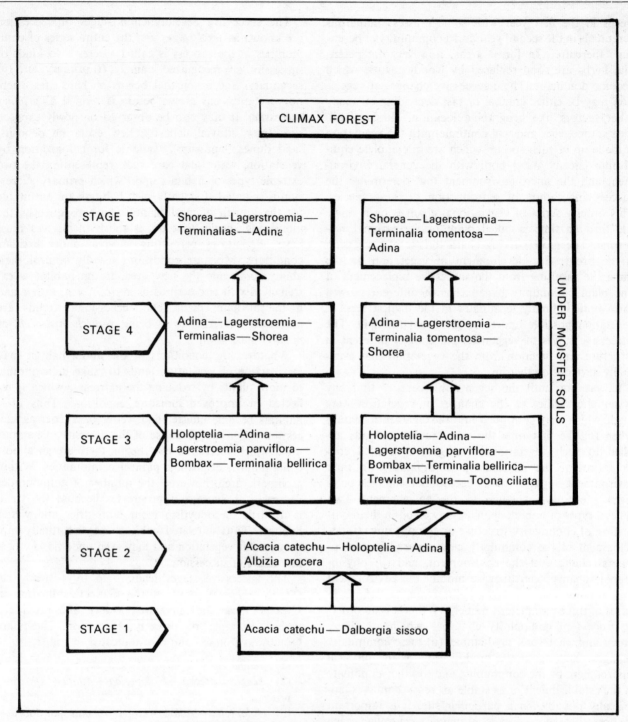

CLIMAX FOREST

STAGE 5	Shorea—Lagerstroemia—Terminalias—Adina	Shorea—Lagerstroemia—Terminalia tomentosa—Adina
STAGE 4	Adina—Lagerstroemia—Terminalias—Shorea	Adina—Lagerstroemia—Terminalia tomentosa—Shorea
STAGE 3	Holoptelia—Adina—Lagerstroemia parviflora—Bombax—Terminalia bellirica	Holoptelia—Adina—Lagerstroemia parviflora—Bombax—Terminalia bellirica—Trewia nudiflora—Toona ciliata
STAGE 2	Acacia catechu—Holoptelia—Adina Albizia procera	
STAGE 1	Acacia catechu—Dalbergia sissoo	

UNDER MOISTER SOILS

Fig. 1. TYPICAL SUCCESSION ON GANGETIC ALLUVIUM IN UTTAR PRADESH

of deodar (*Cedrus deodara*), spruce and blue pine, (4) mixed coniferous forest of spruce, fir and deodar.

When an existing type of vegetation at any stage in the primary succession or an established community of vegetation which is in a steady state, is destroyed by adverse influences like fire, grazing or felling of trees, the successive progression that has hitherto been taking place is altered to a more or less marked retrogression. When these adverse influences, which have brought about this retrogression, are removed or when the area on which the vegetation has been destroyed is left to itself, revegetation of the area, termed as *secondary succession*, starts which is similar to the primary succession but different from it in many ways. To what extent the development of vegetation in the secondary succession resembles the primary succession depends on the degree of disturbance. Although the first communities of vegetation that develop in secondary succession may not be typical of primary succession, the later stages may be similar. Barring exceptions, it is, therefore, not always possible to distinguish between corresponding stages of primary and secondary succession and retrogression stages.

As mentioned earlier, eventually, succession terminates in communities whose complex of species is so adjusted to each other, in the environment that has developed, that they are capable of reproducing within the community, especially the potential dominants, and of excluding new species. Both as to composition and structure, the community is now stabilized. All community processes continue, possibly at a greater rate than in earlier phases of succession, but they result in no major modifications to the organization. This vegetation is in equilibrium with the climate, the soil and the herbivorous fauna. So long as these environmental parameters do not change, vegetation will continue through time with essentially the same species pattern. The vegetation is thus in a state of dynamic equilibrium, in a steady state with its environment, and entrance of other species is almost impossible. Barring disaster, severe external influences or a change of climate, the community will continue indefinitely, because individuals that are lost for any reason are replaced by their own progeny. Such a community of vegetation is recognized as *climax*.

Climax is, therefore, the termination of successional process or culmination of development on a site and is determined by the whole environment—climate, soils, animals, fire, man, etc. acting upon the available regional flora. It is possible for an entire climatic region to have a single type of climax vegetation, but because of variations in local environments (e.g. soil) such a region is likely to have several or many climax vegetation types. In fact, such a region will be a mosaic of both successional and climax types.

In practice, however, it is convenient to recognize various climax types. While the general climax over an area is known as the *climatic climax*, if in any succession a stage immediately preceding the climax is long persisting for any reason (such as fire) but which is more or less stable under the prevailing conditions, it is called a *subclimax*. When the disturbance is such that the true climax becomes modified or largely replaced by new species, the result is an apparent climax, called *disclimax*. In a general climax type exceptionally unfavourable sites (low moisture availability) carry a vegetation type which is more drier than corresponding to the general climate of the area, e.g. chir pine on ridges in subtropical broad-leaved forest. Conversely a more favourable climate (locally cooler or moister sites) carry a more mesophytic vegetation (sheltered damp sites in a sal forest carrying semi-evergreen or evergreen forests). Therefore, for any particular climax the

© F. R. I.

A climax moist teak forest, South Chanda, Maharashtra

contiguous climax produced by less favourable conditions is termed as *preclimax* and the one produced by more favourable climate is known as *postclimax*.

Edaphic climax is the result of soil conditions usually differing from those of the area as a whole and demonstrably dependent on soil properties. For example, in the sal climatic climax on the Gangetic alluvium, patches of heavy clay carry edaphic climax forest in which bael (*Aegle marmelos*) or *Terminalia* predominates. *Physiographic climax* is a product of local environment associated with peculiarities of topography while *Pyric climax* is obtained under recurrent fires. These are some of the various climaxes that have been distinguished.

O.N.K.

CLOUD. A visible aggregation of minute water droplets in the atmosphere, above ground, is called a cloud. When the tops of clouds extend much above the freezing altitude (where the temperature is zero degrees Centigrade), the droplets will be frozen and will consist of ice particles.

Clouds can be broadly classified as low clouds (about 0·5 km to about 3 km altitude), medium clouds (about

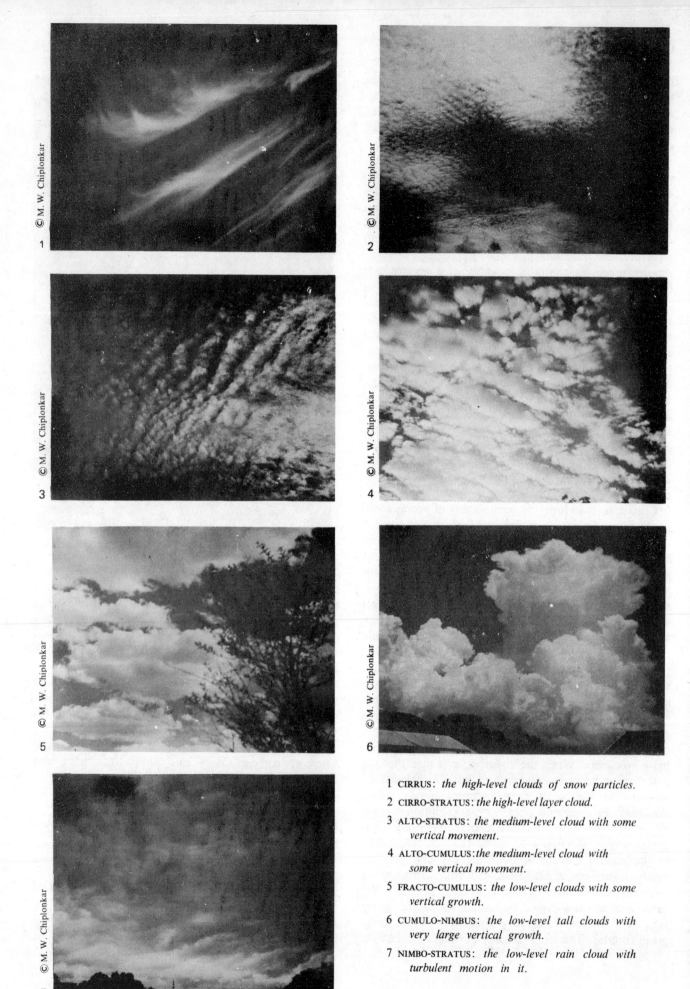

© M. W. Chiplonkar

1

© M. W. Chiplonkar

2

© M. W. Chiplonkar

3

© M. W. Chiplonkar

4

© M. W. Chiplonkar

5

© M. W. Chiplonkar

6

© M. W. Chiplonkar

7

1 CIRRUS: *the high-level clouds of snow particles.*

2 CIRRO-STRATUS: *the high-level layer cloud.*

3 ALTO-STRATUS: *the medium-level cloud with some vertical movement.*

4 ALTO-CUMULUS: *the medium-level cloud with some vertical movement.*

5 FRACTO-CUMULUS: *the low-level clouds with some vertical growth.*

6 CUMULO-NIMBUS: *the low-level tall clouds with very large vertical growth.*

7 NIMBO-STRATUS: *the low-level rain cloud with turbulent motion in it.*

© **M. W. Chiplonkar**

3 km to 5 or 6 km altitude), and high clouds (about 6 km to 10 or 15 km altitude). High clouds mostly consist of ice particles. These can be further classified as layer clouds, without much vertical extent, and without any appreciable vertical motion in them; and as cumulus or heap clouds, with a large vertical extent, sometimes from 2 km to 10 km or even more, and with large vertical currents active in them.

Clouds are formed when moist air ascends in the atmosphere and cools to a temperature when the moisture available in the ascending air is sufficient to saturate it. The now visible cloud continues to rise, becoming cooler as it does so, until it reaches its ceiling. Such an ascent can be brought about by several circumstances. The most easily understood cloud-formation occurs when moist air blows across a mountain range and is obliged to ascend up the slope. These are called *orographic clouds*. There are occasions, particularly summer afternoons, when the solar radiation heats up the air near the ground. The hot air ascends in vertical columns as smoke does. If the air is moist it condenses as the air ascends to a sufficiently great height. The condensation imparts the latent heat of condensation to the air, and enables that air to rise further and further. This is called convection, and the associated clouds are called *convective clouds*. There are occasions, when the winds at the first one or two kilometres over a region above the earth's surface blow towards each other, come together or converge: then air is forced up. The upward motion is encouraged, when the winds in the upper atmosphere, at 6 to 10 km altitude, blow away from each other or diverge. This usually happens when the atmospheric pressure near the ground is low over a wide area, hundreds of kilometres across. Under these conditions an extensive cloud formation occurs. Most of the clouds during the monsoon form this way.

It is a long step from the formation of clouds to the formation of rain. The cloud droplets are very small— a few thousandths of a millimetre across. The smallest rain-drop which can fall down should be about two-tenths of a millimetre across.

See RAIN.

Spectacled cobra × ¹/₁₆

Monocled cobra × ¹/₁₆

Cobra on eggs × ¹/₅

COBRAS. The cobra is one of the most talked-about snakes in the world, a fact that is exploited by snake-charmers, who display sick, defanged cobras, making them 'dance' to the music of their pipes. (Actually the snake sways to the movement of the pipe, carefully following it as a potential enemy.) When frightened and on the defensive, cobras have one of the most spectacular displays of any reptile in the world. Spreading elongated ribs they are able to expand their necks to show the dramatic hood, in order to startle enemies. There are three cobras found in our area; the Spectacled Cobra, found throughout most of the country, the Monocled Cobra (a subspecies) in the north and northeast, and the Black Cobra of the northwest. Cobras feed mainly on rats and frogs, small birds and toads, though the Monocled Cobra often takes prey such as fish and other snakes. Cobras lay eggs, usually between 12 and 30, in a rat hole, dead tree or in a termite mound. The female stays with the eggs until they hatch. The young feed on insects, frogs, lizards and small snakes. The hatchling

Cobra hatching × 1

snakes are in turn preyed on by monitor lizards, birds of prey and even bullfrogs! Cobras favour ricefields, with their abundance of rats, and rat holes for shelter. They are wary of humans and quick to escape if given a chance. When provoked they will raise the hood and flash the markings to intimidate the enemy, biting only as a last resort.

In spite of bans that have been imposed on the export of snake-skins, the trade continues to flourish and cobra skins are very popular. Thousands of cobras are killed every year. Agriculturally this is a disaster, as cobras are efficient controllers of rodents which destroy large quantities of grain.

See also SNAKES.

COCKLES are edible bivalves easily recognized by their elegant, heart-shaped shells with strong, radiating ribs. The animals live on the seashore in sand and mud and also in creeks and brackish water. In view of the heart shape characteristic of cockles, the genus and the family to which they belong are called *Cardium* and Cardiidae respectively. There are several handsome species distri-

A cockle on the surface of the sand (left)
and leaping (right) × ½

buted along both the east and west coasts of India. The animals do not congregate in beds like backwater clams or mussels. Cockles possess long siphons. Their gills are as well developed as those of clams. Their foot is large, elongated and bent in the middle at right angles, the latter portion shaped like a sickle. They can burrow in sandy mud. The pointed free end of the foot is pushed into the sand and dragged like a ploughshare. Using the foot as a lever they make long leaps. When the high-tide waves enter the littoral sand-banks one can see cockles leaping. They are not sedentary like ARKS but move about freely. If the food supply is inadequate at one place they can move to another place where it is available. They are therefore more adventurous than clams. From pure sea water, cockles sometimes find their way into brackish creeks and thence into fresh water. The influence of changed waters and the adaptability of these bivalves to the altered environment result in small variations and in the long course of time in the formation of new species. Zoologists have a good example in *Cardium* to prove how migration from sea to brackish water and the incidental change in environment can give rise to new species or even new genera.

In our area the commonest cockle is *Cardium asiaticum*, distinguished by its more or less round, inflated shell, smooth radial ribs, creamish colour, tinged pink at the umbo and along the toothed, interlocking lower margin. In some specimens the radial ribs appear crested. *Lunulicardia retusa* is a small white cockle with radial ribs and a very prominent keel extending from the umbo to the middle of the lower margin. This species is very common along the Coromandel coast.

Apart from the true cockles, there are the false cockles represented by the very common *Cardita bicolor*. Its shell is thick and heart-shaped, deeply ribbed, white with red or orange dots or spots. It prefers to live on sandy and gravelly bottoms under shallow waters. As in true cockles the foot is powerful and sickle-shaped, with the aid of which the animal can plough and leap.

In some countries cockle shells are as broad as 10 or 15 cm and the animals multiply in fishable quantities; but in India cockles are rarely found in large numbers and their size does not exceed 5 cm.

See also MOLLUSCS.

COCKROACHES. These familiar ground-frequenting insects are the 'scavengers' of the insect world, feeding on any kind of dead animal or plant matter. More than 3500 species are known all over the world, but it is estimated that this represents only 50 percent of actually existing species! *Periplaneta americana, P. australasiae*,

Blatta orientalis, *Blatella germanica* and *Supella longi-palpa* are the common household cockroaches which have now become cosmopolitan (or at least tropico-politan) with the help of man. They occupy man's houses, stores, markets, and almost any other structure with abundance of food, shelter and the right range of temperature and humidity. Besides eating whatever food is left accessible to them by man, they also spoil stored products with their excreta as well as eat book bindings, container labels and the like. Some species even are known to harbour pathogens like *Salmonella* and poliomyelitis virus harmful to man. Cockroaches, like rats, are so universally omnipresent around the habitations of man and ecologically so similar, that they are the most important economic and health hazards to man.

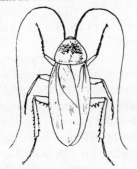 

American cockroach, *Peri-*   German cockroach, *Blatella*
planeta americana, × 1   *germanica*, × 2

The majority of cockroaches are nocturnal, hiding in leaf litter, under bark or stones and in thick grass. The diurnal species, however, are found on foliage of trees and shrubs. Most cockroach species are found uncon-nected with man and some of them, like *Theria peti-veriana* and *Panesthia regalis*, are strikingly beautiful species with spots of white or bands of orange on a dark coloured body. The characteristic oothecae or egg-cases are laid by most species, but in some others live young ones are 'laid' by the female or even parthenogenesis (asexual reproduction) is recorded. The life-cycle is long, certainly more than a year in many species. The adult longevity is also protracted, species living for one to four years. *Neostylopyga rhombifolia* is a common 'rural' species, inhabiting the thatched roofs of huts and breeding in them. The most common 'urban' species, *Periplaneta americana*, was stated by H. Maxwell-Lefroy (whose *Indian Insect Life*, published in 1909, is the only fairly comprehensive, profusely illustrated popular insect book written on our insect wealth even today!) to be considered a 'reliable weather prophet'; flying about in houses just before rain fell!

Some species are attracted to light in the nights but most of them are rarely visible, keeping close to soil, seldom venturing out by day. Besides their frog, reptile, bird or mammal predators, cockroaches have many insect, protozoan and helminth parasites that help to keep their populations in check. The WASP family Evaniidae are specialized parasites of cockroach egg-cases, and several other chalcidoid families attack them. The members of the sphecid wasp subfamily, Ampuli-cinae, are also specially fond of adult cockroaches, which these beautifully metallic wasps hunt, sting and drag to their nest-holes. The female wasp lays an egg on the benumbed cockroach, the larvae hatching out com-pleting its life-cycle feeding on it.

COELACANTH, a group of ancient fishes, was believed to have inhabited the rivers and lakes of the Old World some 30 million years ago and was known to scientists only through the study of their fossils. But strangely enough, a living fish of that group was accidentally caught in a trawl net in December 1938 off the coast of east Africa at a depth of 80 m. It was a bright blue, oily-looking creature 1·6 m long and weighed 58 kg. Its identification created great interest among ichthyologists and after careful examination it was concluded it was a 'living fossil'. Another specimen was caught in the same area in 1952 and a few more thereafter. So far the fish has not been recorded from seas around the Indian subcontinent.

The peculiarity of the coelacanth lies in its fins, known as 'lobe fins', wherein each fin has a short scale-covered lobe as base with finrays attached to it like a fan. The fin rays are hollow. In place of a bony vertebral column it has a cartilagenous tube-like notochord, giving the fish its name coelacanth, meaning 'hollow spine'. Some of the internal organs also are of primitive nature. Its scales are large and considerably overlapping (almost three-layered), thus giving greater protection to the body. These are the characteristics of ancient fossilized fishes.

The structure of the fins and the basal lobe indicate that the fish is a slow-moving bottom form. It appears to have migrated from fresh water to the sea and adapted to the new circumstances without much structural change. This is contrary to what normally happens to any species during its long evolutionary history, and hence the fish is considered to be an astonishingly queer species.

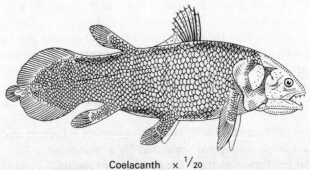

Coelacanth × ¹/₂₀

COELENTERATE. The word comes from two Greek words, *koilos* meaning hollow and *enteron* meaning intestine, and in English is pronounced see'lenterate. Coelom, the large body cavity in ECHINODERMS is also pronounced see'lome, and is not to be confused with colon, part of the large intestine, which is pronounced with an initial k-sound.

Coelenterates are many-celled, mostly radially symmetrical, aquatic animals, of a low type of organization. HYDRAS, CORALS, JELLY-FISHES, SEA-ANEMONES, COMB-JELLIES and OBELIAS are examples. They are widely distributed in all seas between tidemarks and at varying depths. Jellyfishes float in open sea; others remain attached to some substrata. *Hydra* is the typical freshwater genus of sedentary habit.

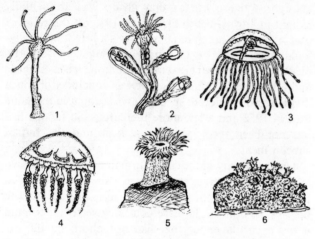

EXAMPLES OF COELENTERATES

1 *Hydra*, ×10. 2 *Obelia*, × 1. 3 Medusa jellyfish, × 10.
4 Jellyfish, × $^1/_{10}$ 5 Sea-anemone, × 1. 6 Coral, × 1.

There are a few other freshwater genera which are free-swimming medusas: *Limnocnida indica* lives in rock pools of flowing rivers; *Mansariella lacustris* is reported from the sub-Siwalik Mansar lake, 400 m above sea level; and *Craspedacusta sowerbyii*, which has a wide distribution in tanks and lily ponds elsewhere in Asia, has also been reported from an aquarium tank in Poona!

Coelenterates exist in two forms (1) tubular polyps, single or colonial, and (2) saucer-shaped, free-swimming medusas or jellyfishes.

Despite diversity in form, their body essentially consists of a single sac—the gastrovascular cavity—opening to the outside by an aperture which is both the mouth and anus. In jellyfishes this cavity has numerous pouches and branching canals. The mouth is ringed by a number of stinging tentacles which stun and capture small prey such as worms and minute larvae and carry them to the mouth.

The body wall is of two layers—outer ectoderm and an inner endoderm separated by a non-cellular layer of gelatinous matter. The outer layer consists of conical cells and small interstitial cells, the latter giving rise to stinging cells, muscles and nerve cells. The inner layer, lining the cavity, consists of cells with whip-like threads (flagella) and glandular cells which waft and digest the prey. Waste matter and undissolved food are squirted out through the mouth. The sense organs are in the form of nerve cells—few and scattered in polyps but numerous and concentrated in jellyfishes.

In colonial coelenterates different polyps are modified to carry out different functions: some polyps are for feeding alone, some for reproducing, some for feeling, some for offence and defence, some for locomotion and some for floating. This division of labour is an example of polymorphism.

Coelenterates are capable of replacing lost parts.

Many colonial polyps are protected and supported externally by a thick cuticle, and jellyfishes by thick jelly. The coral anemones secrete within an axial skeleton of horny or limy matter.

Reproduction is both asexual, i.e. by formation of buds or by splitting (fission), and sexual by means of ova and sperms which are produced either by one and the same or by different individuals. In jellyfishes and sea-anemones sexual reproduction is the general rule where the ova and sperms are produced by different individuals. The ova and sperms, when mature, are discharged into the water where they unite and become long ovoid, ciliated larvas (or larvae). The larvas swim for some time and settle on suitable supports and develop into fresh individuals.

In the lower coelenterates like *Hydra* and *Obelia* both asexual and sexual reproduction occur but the former is dominant. A process (or bud) is given off from some part of the parent animal. This bud develops and forms a complete animal. The bud may get detached from the parent either before or after development as in the case of *Hydra* or may remain in permanent vital connection with the parent form as in *Obelia*.

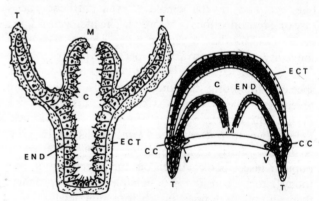

LONGITUDINAL SECTIONS OF A POLYP
AND A MEDUSA

M mouth, C gastrovascular cavity. T tentacle. ECT ectoderm.
END endoderm. V velum. CC circular canal.

It is this latter mode of bud formation and the buds remaining adhered to the parent that account for the formation of colonies. In some species of *Hydra* one and the same individual produces both sperms and ova at two different spots along its trunk and these spots look like knobs or bulges. In other species there are separate male and female *Hydras*. The sperms when mature are shed into the water in which they swim and

some manage to reach knobs containing mature ovaries. A few sperms penetrate the ovaries. Fertilization takes place, followed by the formation of free-swimming larvas which eventually settle down and develop into fresh hydras.

In the colonial forms like *Obelia*, apart from the asexual buds which remain attached to the parental body and enlarge the colony, certain cylindrical structures are also formed along the stem which asexually produce tiny saucer-shaped bodies called medusas. These, when ripe, escape into the water where they float and swim freely with the aid of their bell-like body and fringe of cilia. These medusas are sexually active. There are male and female medusas producing either sperms or ova. The germ cells when mature are shed into water where they unite to form tiny free-swimming larvas which in due course settle down and develop into a fixed, asexually active *Obelia*. Thus in the life history of *Obelia* and its allies and also in a few jellyfishes like *Aurelia*, a stage in which reproduction takes place by a process of budding or fission alternates with a stage in which there occurs a true sexual mode of reproduction. This phenomenon is called 'alternation of generation' or 'metagenesis'.

COMB-JELLIES or SEA-WALNUTS are peculiar jellyfishes more organized than other coelenterates. In view of their special organs of locomotion—eight 'combs' or rows of ciliated plates—they are grouped under a special Class *Ctenophora* (from Greek *ktenos,* comb; *phoros,* bearing). They are all beautiful, free-swimming animals, floating on the open sea, emitting iridescent colours; many of them are phosphorescent. They are widely distributed and are abundant in all warm seas. Comb-jellies are generally pear- or egg-shaped—hence the name sea-walnuts—but some are long and like ribbons.

The animal has an oral end where the mouth is situated and an opposite (aboral) end bearing sensory organs. Eight comb-like ciliary bands are arranged over the body longitudinally, extending from near the oral end to near the aboral end. By a rhythmic movement of the cilia along the bands the animal is propelled through the water. With a few exceptions comb-jellies have a pair of long tentacles, each one arising from a pouch on either side of the aboral end. These tentacles are sensory. They have no stinging cells but a series of adhesive glue cells which capture food—minute organisms floating on the surface—and carry it to the mouth. The mouth opens into a narrow gullet where most of the food is digested. The gullet opens into a funnel which branches radially and distributes nutriment to all parts. Part of the undigested food is ejected through the mouth and part through the anus.

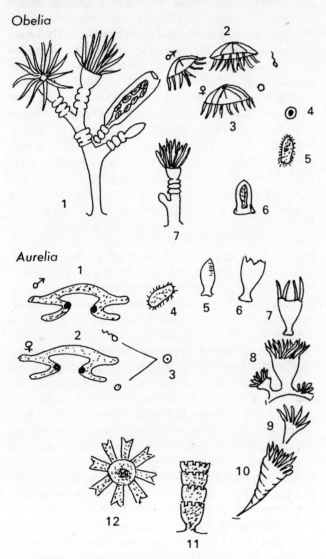

ALTERNATION OF GENERATION IN *OBELIA* AND *AURELIA.*

Obelia. 1 Polyp producing and releasing a medusa. 2 Male medusa with sperm. 3 Female medusa with ovum. 4 Fertilized ovum. 5 Free-swimming ciliated larva. 6 Larva settling as an embryo. 7 Embryo growing into adult polyp.
Aurelia 1 Male jellyfish producing sperm. 2 Female jellyfish producing ovum. 3 Fertilized ovum. 4 Free-swimming ciliated larva. 5 Larva settling as an embryo. 6 Embryo developing. 7 Young fixed polyp. 8 Fully grown polyp producing bud asexually. 9 Released bud settling as a polyp. 10 Polyp growing vertically. 11 Polyp splitting transversely into saucer-like jellyfish. 12 Released young jellyfish.

Each comb-jelly produces both male and female germ cells. These, when mature, are first discharged into the gullet from where they escape through the mouth to outside water. Here some of the eggs get fertilized and develop directly into adult comb-jellies.

Some of the common and interesting comb-jellies found along the Indian coast are:

Pleurobrachia, with an ovoid body somewhat compressed laterally with eight prominent ciliated bands of equal length. The body is 2·5 cm long and the tentacles nearly 14 cm. The body is transparent and therefore as it moves in water only the tentacles are seen waving.

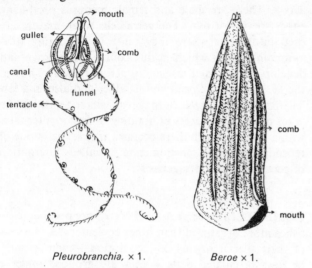

Pleurobranchia, × 1.   *Beroe* × 1.

Beroe is a thimble-shaped comb-jelly without tentacles. The body is long and conical, laterally compressed and measures 10–11 cm long.

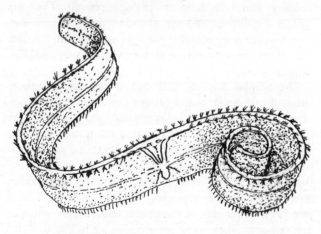

Venus's Girdle (*Cestus*) × ½.

Cestus (Venus's Girdle), is a long ribbon-like comb-jelly more than 2 cm broad and 30 to 35 cm long. It is a very attractive animal, exhibiting green, blue and violet colours when swimming.

See COELENTERATE.

COMMUNICATION. For animals, a most important component of their environment is other animals, particularly of their own species. Their behaviour is continually influenced by what the other animals are doing, and in turn they affect the behaviour of their fellow animals. Sometimes this effect may be direct, as when monkeys physically push each other to gain a favourite perch. But often one animal affects another in a marked way, without physically expending energy commensurate with the effect. Thus, a threatening face, directed by a dominant male monkey at its subordinate, may send it whimpering away, without the dominant male actually pushing or chasing it away. When an animal thus influences another without physically forcing it into that action, we say that an act of communication has taken place. Such communication may take varied forms: a female House Sparrow solicits copulation by squatting and depressing her tail, a male Shama attracts females to his territory and repels other males by his melodious song, an ant recruits other ants to the defence of the colony by releasing a chemical, a rhesus monkey offers to groom another by nudging it, and so on.

Animals use all the various sensory modalities at their disposal in the act of communication; communication in each sensory modality having its own special features. Thus, the visual signals have a limited range in space, cannot go round obstacles and fade away instantly, so that they cannot be used in the absence of the signaller. The vocal signals can have a much longer range, can go round obstacles, but also fade away instantly. The chemical signals, on the contrary will not fade away instantly and can continue to convey a message long after the signaller has left the particular spot at which the message has been deposited. The visual signals are then particularly appropriate for short-range communication in cases where the message conveyed has to change rather rapidly, for instance in co-ordinating the orientation of swimming movements of fish in a large school such as that of sardines or mackerel. The chemical signals on the other hand are more appropriate when a single message has to be continually broadcast for a long time, for example, in marking the boundaries of territory through the use of special compounds in the urine in many carnivorous mammals such as the tiger.

Apart from the fact that particular sensory modalities are suited for particular kinds of messages, any given signal is also adapted to its message. Thus ants use an 'alarm' pheromone to alert the colony on defence, and a 'trail' pheromone to recruit other workers to food sources. Now, if the alarm were to last too long, the colony would be thrown into panicky activity, and the alarm pheromone is a substance which breaks down rather rapidly, within seconds. The trail pheromone

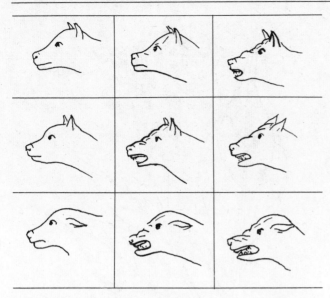

Facial expressions of a dog that result from a superposition of various intensities of fighting and fight intentions.
(From K. Lorenz, 1953)

In the upper figure a dog approaches another animal in a fully aggressive posture. The lower figure shows the same dog in a conciliatory stance, in which virtually all of the signals of the aggressive display have been reversed.

on the other hand lasts somewhat longer, for tens of seconds. Finally the sex pheromone by which a female silkworm moth attracts her mates, lasts much longer, for several hours.

Animals deal always in here and now, and their messages tend to be evocative, urging another animal to follow a particular course of action. Monkeys urge other monkeys to keep away from a food source they want for themselves, move in a particular direction, groom or be groomed, mount or be mounted and so on. There is however one remarkable exception, that of the famous bee dance. For the dance of the honey-bee communicates coded information about the direction, distance and value of a distant source of nectar. It is therefore the single notable example of a symbolic message. In brief, a honey-bee worker, on return from a successful foraging trip, performs a waggle dance in the dark interior of the hive. The dance is in the form of a figure of 8. The angle that the axis of this figure of 8 makes to the vertical corresponds to the angle that the source of nectar makes to the direction of the sun. The tempo of the waggle dance codes information about the effort involved in the flight to the nectar source. This effort, of course, depends on the distance to the nectar source, but also varies with whether the wind is in favour of or against the forager. At the same time, the chemical cues released by the dancing bee convey information about the species of the flower yielding nectar. While the dancer is performing, other bees follow her dance and receive information so that they can successfully locate the nectar source without having to actually follow the dancer who also often returns to the same source.

While the bee dance has thus evolved to symbolically code very detailed information, its origins lie in the simple messages like 'let us move together' that are common throughout the animal kingdom. Social animals, such as flocking birds, co-ordinate their movements through such messages, for example, through flight-intention signals before taking off. A bird about to fly must necessarily perform some preparatory movements such as crouching, flapping the wings a couple of times to warm up and so on. Such intention-movements naturally convey the information to other birds that one of their flock is about to fly off. This message can be enhanced by these intention-movements becoming more striking and constant in form. This has happened in many animals, where the elaborated intention-movements have become signals co-ordinating movements. In pigeons, for instance, flight is preceded by wing-flapping and the number of wing-flaps is proportional to the distance flown. In a similar fashion, the flight-intention movements of primitive bees are a much simpler form of dance, followed by all the bees including the dancer flying off to the nectar source. This process of elaboration, of a simple locomotory pattern incidental to a particular behaviour into a communicatory signal, is known as ritualization. The elaborate courtship dance of the peafowl is such a ritualization of food-pecking movements.

A fascinating debate has arisen in recent times on the extent of cheating during the course of animal communication. In threat displays, for instance, it would always be advantageous for an animal to try and appear much more powerful than it actually is. There would however be an equal pressure for the opponent to see through the pretence. On balance, the signals that evolve would tend to be true indicators of the capabilities of the threatening animal. Thus, in bullfrogs the pitch of the croak depends on the size of the animal and is thus a good indicator of its capability as an opponent. Experiments have shown that in a croaking match the frog with a deeper croak almost invariably wins.

There are however many known instances of cheating as well. For example, in some fish, certain males take on the coloration of the female, and having thereby managed to enter the territory of a male, mate with his female. Deception in communication is however far commoner in interspecific encounters. Many edible butterflies mimic the coloration of other poisonous species and thereby deter predation. Angler-fish lure prey to their mouths by dangling a fleshy bait which looks like a worm to their unsuspecting prey.

The repertoire of communicatory signals is rather limited in all animal species, being in the range of 20 to 60. It is only in man with his symbolic language that this limitation has been completely overcome, and an infinite number of communicatory messages have become available. It is indeed his communication system that sets man completely apart from all other animals. However, even in man non-verbal communication in form of facial expressions, gestures and so on is now known to play an important role in social intercourse.

CONEHEADS, the species of *Strobilanthes*, are ornamental herbs or shrubs of Malagasy and tropical Asia. Several species are grown for their attractive foliage and flowers. Most of the species are aromatic, some are employed in medicine, while others are poisonous and cause several diseases and even death. Goats and sheep browse mature leaves which are generally safe. The genus has been split into many genera, e.g. *Strobilanthes* proper, *Carvia*, *Nilgirianthus*, *Perilepta*, and *Phlebopyllum*. *Strobilanthes* in the strict sense does not occur in our area and about 85 species previously described under this genus from the subcontinent have now been transferred to the other newly erected genera. The most outstanding feature of *Strobilanthes* is gregarious flowering, with a definite periodicity which is independent of weather conditions, only sporadic flowering, or no flowering, taking place in the intervening years. They have a sharply defined flowering cycle of 2, 3, 6, 7, 12, 17 or more years, at the end of which time

Nilgirianthus ciliatus, synonym *Strobilanthes ciliatus*, × 1

all plants of a given generation flower gregariously and then die. BAMBOOS also manifest this alarm-clock flowering behaviour. Conehead flowers vary from purple to white, often being blue or yellow, and present a glorious sight with scented floral abundance covering entire hillsides. The commonest Indian species are: *Perilepta auriculata* in the north in the Himalayan moist temperate forests and *Nilgirianthus ciliatus* in the south in the southern montane scrub forests.

K.N.B.

CONES, SLIT-LIPS and AUGERS. These three groups of gastropods belong to one tribe, 'Poison Teeth' or Toxoglossa. They are characterized by the possession of a large poison gland in their throat which is connected by a duct with a barbed tooth on the radula. The animals are carnivorous and feed on small crabs, fish and bivalves. They paralyse the prey, secreting the poison while biting. They live among rocks near low-tide mark. Their shells vary very greatly in shape and architecture.

The sexes are separate. Eggs are fertilized inside the female body. The female lays the fertilized eggs in capsules which look like flattened grains. These capsules are attached to a sheet and fixed to some rock. Each capsule contains besides the eggs a quantity of an albuminous fluid which serves to nourish the developing

9 Snakes and a Chameleon

© Romulus Whitaker

Indian Chameleon (*Chameleo zeylanicus*)

© Romulus Whitaker

Flying Snake (*Chrysopelea ornata*)

© Romulus Whitaker

Sawscaled Viper (*Echis carinatus*)

© Shekar Dattatri

Russell's Viper (*Vipera russellii*)

10 Kingfisher, Flowerpecker, Magpie-Robin, Lapwing

The Small Blue Kingfisher (*Alcedo atthis*)

Tickell's Flowerpecker (*Dicaeum erythrorhychos*)

Magpie-Robin (*Copsychus saularis*)

Redwattled Lapwing (*Vanellus indicus*)

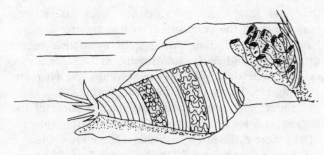

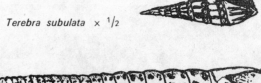

Terebra subulata × ¹/₂

T. tenera × 6

TWO AUGUR SHELLS

A Cone (*Conus textile*) moving with siphons projected × 1. Her egg capsules also × 1 are on the rock behind.

embryos. In many cases only a few—sometimes only one—of the embryos contained in the capsule develop, the rest serving as nutriment for the survivors. The developed embryos look like little adults with a thin shell. When the nourishment within the capsule is exhausted the embryos come out of the capsule to lead their independent lives.

Cones have conical shells with a short suppressed spire, the coils of which are almost concealed by the overgrowth of the body whorl. The mouth of the shell is a long and narrow slit through which the elongated flat foot emerges when the animal starts moving. The snout and the two pointed tentacles are protruded through the tubular anterior end of the shell. While moving, the position of the shell is parallel to the ground. Cones are commonly found half-buried in mud in the vicinity of isolated rocks and coral reefs at extreme low-tide mark. The chief variations among cones are not in their shape or habits but in their colour markings. Several beautiful species with variegated colour designs are common amidst rocks and corals of the Indian seashore and around Lakshadweep. Living cones should be handled with care, for although a bite from these animals is rarely fatal, it causes severe pain and illness.

Slit-lips (Turrids) possess whelk-like shells—fusiform or pear-shaped with a prominent conical spire and a long anterior canal. The outer lip of the shell invariably carries a notch or slit: hence the name slit-lip. They vary in size, colour and sculpture. The animal crawls

A Slit-lip (*Surcula javana*) crawling × 1.

by the side of mud-covered rocks exposed at low tide. There are over 50 species of Turrids distributed along our shores. Most of them are small. The largest and the most beautiful Indian slit-lip is the brown Tulip-shell, *Surcula javana*.

Auger shells are at once recognized by their long, slender tapering shells with numerous twists ornamented with spots or patches of colour. In appearance they look like the boring tool with a screw point: hence the name auger shell. They live partly buried in sand in moderate depths around the mainland coasts. They are much more common around Lakshadweep. The tip of their long shells is so slender that it sometimes gets broken off. This does not affect the animal in any way as the adult animal lives only in the anteriormost three or four large whorls. Zoologists say that some species purposely break off the shell-tips to lighten the burden: but it is not known how they do it. The twists or whorls of augers (as in all spiral-shelled molluscs) do not bear any relationship to their age.

Nearly 30 common species of augers exist along the Indian shores. They vary widely in size, ranging from 0·5 (*Terebra tenera*) to 16 cm (*T. subulata*).

See also MOLLUSCS.

CONIFERS (Gymnosperms). This group of trees includes well-known trees such as Deodar, Christmas trees (Spruces and Firs), Pine, Larch, Cypress, Juniper, Yew, Hemlock, etc. These are well represented in the Himalaya. Only one conifer is native to South India viz. a podocarp (*Podocarpus wallichianus*) in the Nilgiris. Conifers or 'Cone bearers' belong to the division Gymnosperm where the seeds are not protected by a fruit wall, but remain naked. The gymnosperm line is one of great antiquity, reaching far back in geological history, at least two or three hundred million years. As a group they are interesting because of their antiquity, and their excellent fossil record, which is a nearly unbroken record of the past vegetation with a fascinating variety of genera which are in existence today. Although several of their members have become extinct, and some like the Giant Redwoods of California are vulnerable to extinction, they are still the dominant

© K.C. Sahni

Himalyan Larch, Yumthang, Sikkim at 3000 m. At leaf-fall it is attractive with foliage turning a rich gold brown. In spring the new flush of leaves are fresh larch-green and the young female cones are coppery red and look orna-mental.

their old leaves and growing new ones not at one season but all the year round, so that the tree is never bare of leaves. Indian Conifers, or even their bigger division the Indian Gymnosperms, are all evergreen, except Larch which sheds its leaves completely in winter, growing new leaves in the following spring.

Conifers are usually large and include some of the biggest and longest-lived of all trees. In the museum of the Forest Research Institute, Dehra Dun, there is a 2·5 m wide section of Deodar more than 750 years old, dating back to the 13th century when Qutb Minar was built. There are some immense stands of DEODAR in Manali in Himachal Pradesh. Redwoods of California, for example, are so huge that a tunnel wide enough for a motor-car to drive through has been cut through the trunk of one of them. The diameter of the trunk at ground level may reach 10 m. Considered to be the tallest tree in the world, trees up to 120 m (400 ft) high occur in the fog belt of the Pacific coast of America. Their age is estimated to be 3000 years. The Qutb Minar is 73 m high and the tallest deodar of 76 m would overtower it.

Some specimens of Bristle-cone Pine or Foxtail Pine (*Pinus aristata*) of N America are reckoned to be the oldest living things on earth today, having germinated even before the Indus civilization, some 6000 years ago. The Sun tree (*Chaemaecyparis taiwanensis*) from Taiwan is claimed by scientists there to be more than 6000 years and measures 25 m (88 ft) in girth. Giant Sequoia (*Sequoiadendron giganteum*), a near ally of *Sequoia sempervirens* both classed as Redwoods, is estimated to weigh 2000 tonnes and is regarded as the most massive tree of the world. The Great BANYAN of Calcutta is a rival in its sheer bulk of tissue. Dawn Redwood (*Metasequoia glyptostroboides*), discovered in 1941 in China, was considered to have become extinct about 20,000,000 years ago. Its discovery aroused worldwide interest, and to save it from extinction, seeds were distributed to several Botanical Gardens, and trees are now flourishing in Darjeeling and Almora. This is regarded as the ancestor of the present Redwood (*Sequoia*). It has a potential in forestry because of its fast growth of 1 m a year. These, as well as the Maidenhair Tree (*Ginkgo biloba*) introduced in the Himalaya, are designated as 'living fossils'. Ginkgo is a unique tree in many respects and is in fact the only surviving member of a large order of plants that was once widespread but now exists as fossilized remains.

Certain other genera and species as the Sequoias are probably making their last stand along the Pacific coast, and some sudden environmental or climatic alteration would probably bring about their total elimination. Excessive demands for the timber of redwoods has denuded many of the original forests but some have been preserved by conservation efforts and magnificent

forestmakers of the world, and are an outstanding feature of the environment of the Himalaya. The extinct gymnosperms seen as fossils today were contemporaneous with the extinct gigantic reptiles, the dinosaurs which were ferocious-looking giant-sized lizards of the past.

Conifers do not bear flowers, but produce pollen and seeds in cones and hence the name 'conifer'. They are the simplest seed-bearing plants. Seeds and pollen are usually winged. A second distinguishing feature is that their leaves are narrow, often needle-like (except *Podocarpus* where they are broad or lance-shaped). This reduces the surface area through which transpiration takes place and enables them to thrive under cold or dry conditions. Economically they are the most valuable group of the gymnosperms. Cycads with palm-like habit are quite distinct from other gymnosperms (see CYCADS).

Almost all are evergreen trees, rarely shrubs, shedding

Fir forest (Abies pindrow) *in Sonamarg, Kashmir, at 2000 m. Note columnar crowns of firs. Coniferous forest, given time, regenerates itself crowding out broad-leaved* intruders. *This is well illustrated in the NW Himalaya as in this picture showing gregarious habit.*

stands exist in the Sequoia National Park. For the survival of spectacular, rare and endangered flora and fauna, National Parks and Sanctuaries are in evidence in many countries. However, pines, spruces, firs, etc. are well established in Europe, N America and Asia and apparently are not facing extinction from natural causes.

The native conifers of our area are Pine, Hemlock, Spruce, Fir, Deodar, Larch, Chinese Coffin Tree (in Burma), Juniper, Cypress, Incense Cedar (in Burma), Yew, Podocarp and Plum-Yew (rare); Japanese Cedar (*Cryptomeria japonica*), exotic tropical pines and other conifers have been introduced for industrial and afforestation purposes. Before these are described, their reproduction as exemplified in chir pine, the most typical conifer of the subcontinent, is described. In conifers, the least specialized of the seed-bearing plants, sperms and eggs are produced in cones, and the seeds are not enclosed inside an ovary but are naked. The male and female cones are separate but on the same tree. The male cones are much smaller, barrel-shaped and catkin-like. The female cones bear spirally arranged bract scales, in the axils of which arise the cone-scales, bearing two ovules. The male inflorescence is visible in January. As it matures, the scales open, showering in abundance sulphur dust-like pollen grains. Each pollen grain has two balloon-like wings which aid dispersal by wind and alight on the female cones. At least a year elapses between pollination in the spring and fertilization. At the time of pollination the scales

of the female cones open to receive the pollen (sperm). The female cones grow fast to adult size after fertilization, the scales thickening, bending back and closing tightly. They mature in about 25 months and open about May in the dry weather to disgorge the winged seeds which do not fall at once but take 2 or 3 weeks to escape completely. The fertilized winged seeds are carried far and wide by the wind. Seeds falling on the ground germinate when the monsoon breaks. Some are eaten while still on the tree by Flying Squirrels. Chir trees growing in the open produce male and female cones in 10 to 15 years and much later in closely growing trees. Flowering continues till after maturity, i.e. for 60 to 90 years. Early flowering is a symptom of unhealthiness or unsuitable site. For purposes of timber other than pulp most conifers like chir, deodar, etc. mature in 60–90 years, i.e. the period of active height growth. Thereafter, the height remains constant till death, a definite leading shoot is lost and the crown gradually becomes umbrella-shaped. The tree continues to grow in diameter and volume for a further period which may be offset by decay.

A brief account of conifers of our area follows (also see FOREST TYPES):

Pine: There are five members known as Chir pine, Blue pine, Chilgoza pine, Khasi pine and Merkus pine. A sixth, new to science, called *Pinus bhutanica*, was recently discovered from Arunachal and Bhutan (see PINE).

Himalayan Hemlock (*Tsuga dumosa*), is a large tree

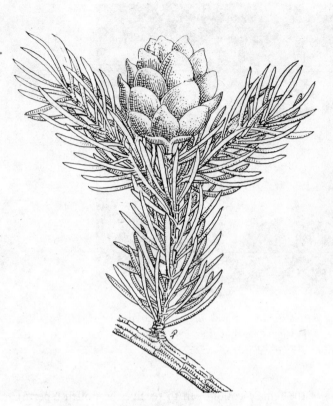

Himalyan Hemlock, shoot with foliage and female cone × 1

West Himalyan Spruce, cone with winged seed × 1

from E Kumaon to Arunachal and Burma. It resembles Fir in foliage but is at once recognized by its much smaller female cones which are about 2·5 cm long. The timber is used for making shingles.

Spruces are tall trees, with sharply pointed needle-like leaves and pendant brownish cones. There are two species, *Picea smithiana* (West Himalayan Spruce) and *P. spinulosa*, the East Himalayan Spruce. In the former the needles are quadrangular in coss-section, in the latter they are flat. The former up to 60 m tall is found from Afghanistan to Kumaon and the latter even taller is found from Sikkim to Arunachal.

Fir. There are four in the Himalaya, Low-level W Himalayan Fir (*Abies pindrow*), High-level Fir (*A. spectabilis*), E Himalayan Fir (*A. densa*) and *A. delavayii*, a rare Chinese tree in Arunachal. *A. pindrow* is a tall tree up to 60 m, with a cylindrical crown with dark purple, erect, barrel-shaped cones which break up on the tree itself on ripening. Found from NW Himalaya to Nepal at 2300–2600 m. The High-level Fir occurs at 2800–3300 m: it is a smaller tree with a broader crown, with upturned branches and shorter leaves. The East Himalayan Fir is characterized by a pagoda-shaped crown forming pure forests above 3000 m.

DEODAR is classed as one of the most valuable timbers of the subcontinent.

Himalayan Larch (*Larix griffithiana*) resembles Deodar in branch and leaf arrangement. In the former the leaves are deciduous and the pendulous cones hang for a long time on the branches. In the latter, the cones are much bigger, erect, and break up on the tree. Larch is the only deciduous conifer of our area, not valuable as timber but a graceful tree found all through the eastern Himalaya at 2400–3600 m up to Burma.

Chinese Coffin Tree, *Taiwania cryptomerioides* of Taiwan, also occurs in N Burma. Logs are exported from N Burma to China. Female cones are 13 mm across. In foliage it is like Japanese Cedar, which is common in Darjeeling.

Common Juniper, *Juniperus communis*, is the most widely distributed shrub of the N Hemisphere and forms the upper limit of woody vegetation in the Himalaya up to 4300 m. Cones are bluish black and the leaves are small and needle-like. The wood and leaves are used as incense in monasteries and the cones in making gin. Himalayan Pencil Juniper (*J. polycarpos*) forms large forests in Kagan valley, Baluchistan and in Lahul. Trees up to 10 m girth, and estimated to be 1000 years old, are recorded from Lahul. The cones are brownish-purple.

Himalayan Cypress (*Cupressus torulosa*), the only native cypress of our area. Tall tree up to 45 m high

West Himalyan Fir, cone with foliage, × ³/₄

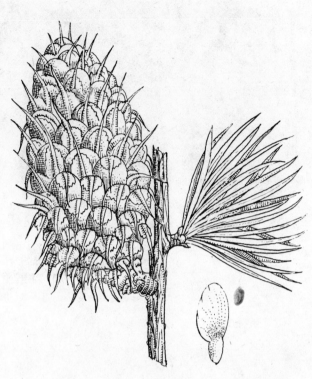

Himalyan larch, shoot with foliage, female cone and winged seed × 1

with drooping branches. Cones globose 12 to 17 mm across occurring in clusters. Leaves, small, closely set on the twigs. Timber is used in making pencils.

Incense cedar, *Calocedrus macrolepis*, a Chinese conifer reported from Burma. In appearance like the Chinese Arbor-vitae (*Thuja orientalis*) called 'Mor Pankhi' or Peacock feathers, which has been in the Taj Mahal gardens for a long time. The former has linear-oblong cones while in *Thuja* they are oblong.

Yew, *Taxus baccata*, a small, slow-growing, long-lived tree attaining an age of 2000 years. Characterized by linear leaves and solitary seeds enclosed in a scarlet pulpy covering. The seeds are poisonous and children should be cautioned from eating the tempting seeds. The bark is used as tea in Ladakh. Distribution: Himalaya, Khasi, Naga and Manipur Hills.

Podocarp (*Podocarpus wallichianus*) is the sole native conifer of S India in the Nilgiris. Unlike other conifers its leaves are ovate. It was recently recorded from Great Nicobar Island, and is also found in Burma. The Oleander Podocarp *P. neriifolius* is a small tree of Sikkim, Arunachal and the Andamans. It has oleander-like leaves and globose seeds.

Plum-Yew (*Cephalotaxus*), a rare Yew-like tree of Arunachal, Naga, Khasi, Manipur Hills and Burma.

Many exotic conifers have been introduced mostly

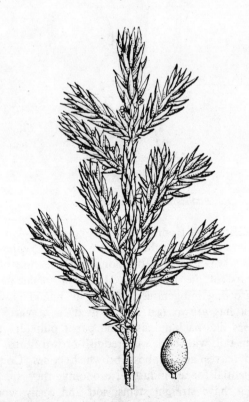

Common Juniper, branchlet with foliage and berry-like cone × 2

133

Fossil gymnospermous leaf of Glossopteris browniana (×·2), *from Tattitola, Pachwara coalfield. Horizon, Raniganj formation. Age, Permian (230–280 million years ago).*

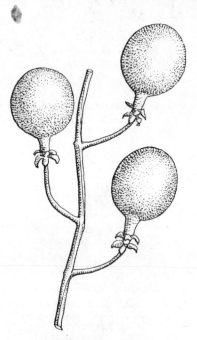

Podocarpus wallichianus, seeds × 1

in the Himalaya from Japan, Mexico and N America, the commonest being the Japanese Cedar (*Cryptomeria japonica*) which is now the dominating feature of the landscape of the Darjeeling Himalaya, replacing the slow-growing indigenous trees. In its humid environment it has grown fast and is used for a variety of purposes like packing cases and paper pulp. It grows to 45 m tall, with bark with reddish-brown fibres, awl-shaped leaves and globose brownish cones. Conifers are essential for an industrial economy; they are fast-growing, have straight stems, soft and easily worked wood, useful in construction, and produce long fibres needed in the paper industry. For industrial and af-

forestation purposes many fast-growing exotic tropical pines, particularly from Mexico, are under trial in the Himalaya and in the hills of peninsular India. The most promising so far is *Pinus patula*. Conifers are important in forestry as they cover vast areas, and are of great significance from the standpoint of the production of timber for railways, buildings and resin, which together are a source of immense revenue.

Gymnosperms had a million years' start over the angiosperms and are a major constituent of many coal deposits. Most of the workable coal seams of the Peninsular Lower Gondwanas of India also contain a significant proportion of gymnospermous plant remains.

K.C.S.

CONSERVATION. To conserve means to protect or to save, and though conservation can apply to anything, it is now associated in the public mind with the protection of nature and all the natural forms of life which exist on our planet. In the past India had a strong tradition about respecting all forms of life and there has also been evidence of the protection of trees with religious fervour in Sacred Groves. But by and large there is very little respect left for trees anywhere in India and in the rest of the subcontinent today. This has been largely due to the fact that wood is such an important source of fuel for the common man. 68 percent of the rural population, and 45 percent of the urban population, still depend on firewood for cooking and heating.

There have been several simultaneous influences which have led to the destruction of our forests and of our natural areas. The per capita area of forests in India is now as low as 0·13 hectares, compared to 20 in Canada, 3·60 in the USSR and 1·04 in the world as a whole. The increase in the human population has been the principal factor and with more and more land required for human settlements, for industrial activities and for agriculture, the natural forests, grasslands, swamps, and marshes are all tending to disappear. This process continues today and poses one of the most important challenges to government officials as well as to public-spirited conservationists.

We are of course not unique in this regard. In almost every country of the world the destruction of natural resources has resulted from the pressures referred to above. In the highly developed countries of the West where population has ceased to grow to any significant extent, damage to natural resources has diminished. But they have in the last few centuries succeeded in destroying most of their natural assets anyway, and have seriously polluted large areas of their countries. In the developing world, and in countries such as India and Pakistan, there is still the opportunity to retain a great deal of the natural wealth provided that we act effectively to reverse the trend of the recent past, and to organize our economic life to operate in harmony with nature. It has now become well known that conservation and development can go hand in hand and that it is in the long-term interests of a country to conserve its natural resources for they are the foundation of material welfare. This is the message which is being put forward by the International Union for Conservation of Nature and Natural Resources and by the United Nations Environment Programme. More specifically, the World Conservation Strategy which was launched by several countries of the world, including India, in March 1980, provides the guidelines for the conservation of the planet's natural resources. Its three main principles are: Preservation of ecosystems; maintenance of genetic diversity; and sustainable use of natural resources. It is important to note that conservation does not mean the sort of protection which prevents the utilization of natural resources. Conservation implies the wise use of natural resources on a sustained-yield basis. After the second world war, in 1948 a group of ecologists and scientists formed the International Union for the Protection of Nature. After some time they realized that 'protection' did not adequately convey the purpose of the organization and consequently the name was changed to the International Union for Conservation of Nature and Natural Resources.

Since nature is a living and complicated entity its protection is not as simple as protecting a monument. For example, the Taj Mahal and the temples at Khajuraho can remain in good shape if they are safe from physical violence and from the corroding pollution of human activities in the neighbourhood. As far as nature is concerned, because of the myriad influences operating on it, and because of the fact that human activities in the past centuries have already upset the balance so severely, it is now necessary to take remedial action. This entails a careful understanding of the factors which have upset the balance in the past and restoring it by scientific measures. Sometimes of course the damage is irreversible and there is no way of undoing the mistakes of the past. The clearest example of this is that any species of animal or plant which has become extinct cannot be recreated. But even in the case of habitats, it is not always possible to restore the pristine conditions of the past.

The discipline associated with conservation is ecology, and ecologists tell us that if any individual animal or plant is to be saved we must save the entire complex of life around it as well as the habitat in which it exists. This is known as the ecosystem, and the principal concern of conservationists and ecologists is to identify the various ecosystems of the country and to ensure that they are protected against damage of any kind. That any form of life cannot be protected in isolation was well demonstrated by the case of the tiger. It was found that if the tiger population in this country had to be saved from extinction then it was essential that the deer and wild boar on which the tigers feed should also be protected and be available in adequate numbers. For the deer to survive there had to be adequate grazing land for food as well as forest cover for their shelter. Project Tiger therefore involved the establishment of 11 large reserves where natural forests and their associated vegetation and animals would all be protected, for only on this foundation could the tiger be saved.

The extinction of the hunting leopard (the cheetah, *Acinonyx jubatus*) is a warning for our country. The cheetah died because of over-hunting and simultaneously through habitat destruction. The open grasslands which were its favoured habitat have been brought under the plough and the blackbuck on which it fed have become very depleted in numbers. Without territory and food no animal can survive. The same race of cheetah as existed in India continues to exist in Iran in small numbers, and one can only hope that in course of time it will be reintroduced into the subcontinent. Such operations have been carried out in the case of the Nene duck and the Arabian Oryx. The duck, a native of Hawaii, was bred in captivity at the Wild Fowl Trust at Slimbridge in the UK and later released in Hawaii. Similarly the Arabian Oryx were bred in captivity at Phoenix and San Diego Zoos in the USA and they are now being reintroduced into Arabia.

For conserving the animals of India it was thought necessary to establish a central advisory body which would recommend to the State Governments what action needed to be taken. Consequently the Indian Board for Wild Life was established in 1952. It meets regularly and advises the various State Governments about its decisions. The 150 or more sanctuaries which have been created in India are largely the result of the recommendations made by the IBWL. The Wildlife Protection Act of 1972, now prevalent in all the States of the Indian Union, takes into account the legal measures required for saving wild animals both inside and outside sanctuary areas. Much wildlife, including many of the smaller forms and birds, exists outside sanctuaries, and the Wildlife Act is designed to save them too from needless exploitation.

The conservationist is concerned with saving all the natural resources. These consist of renewable natural resources like air, water, soil, fauna, and flora as well as non-renewable resources like petroleum, coal, manganese, and iron ore. At the present rate of exploitation our coal reserves, which are considered to be substantial, will only last for about 75 years. It is evident therefore that these resources must be used with great circumspection.

On the other hand with regard to the renewable resources, conservationists have been emphasizing the need to utilize them on a sustained-yield basis. So long as we exploit only the annual growth of trees and other forms of vegetation, or harvest animals not beyond the level of their annual growth, we can look forward to having these resources with us on a permanent basis. The oxygen in the air which we breathe is being constantly produced by natural forces. Vegetation absorbs carbon dioxide which emanates from animals and produces oxygen which is released into the atmosphere. As long as there is an adequate amount of vegetation on our planet, and so long as we do not pollute the atmosphere with unnatural chemicals, pure air which is vital for all forms of life will continue to be available. The same is the case with water. The total quantity of water on our planet is fixed, but there is a continuous natural cycle of which it is a part. Water falls on the land in the form of rain, is absorbed into the soil, runs down into the streams, rivers, and wells. Finally, it finds its way back to the ocean. There is a lot of transpiration of water from the leaves of the trees which also finds its way into the atmosphere. With the arrival of the monsoon the water comes back to the land. The quantity of water which India receives in the form of rain has been calculated to be 3000 million acre feet. This is adequate for our needs but unfortunately because of the cutting down of the forests, and the destruction of ground vegetation the water, instead of percolating into the ground, runs into the oceans too rapidly. When this happens human beings are unable to make the best use of the water resources. Our attempt has to be to delay the water cycle, and to keep the water on the land as long as possible, and the only natural way to achieve this, is to see that the land is well covered with vegetation. Dams and engineering works do help, but are not as effective as the natural remedies.

The lack of forest and vegetation cover not merely affects the water regime but it also results in a calamitous loss of the top soil, which is by any reckoning our most precious asset. It has been estimated that 6000 million tons of soil is lost every year through erosion by wind and water in India alone. It is equally pertinent to remember that one inch of top soil may take as long as 1000 years to build up, and therefore for all practical purposes it is a non-renewable resource.

Conservationists have also a significant role to play with regard to agriculture. To get the maximum yield of crops from the soil we have to use chemicals, fertilizers and pesticides. But in many cases this use is overdone, with the result that we unnecessarily kill a host of living species which are beneficial to agriculture. Quite often in just one square yard of land there are hundreds of insects and worms which help to aerate the soil and which play their part in keeping the soil productive and help to renew it in course of time. While pesticides and fertilizers have become necessary in our largely artificial world we have to use them with caution, so that the natural fabric of life is not ruined or disturbed.

Every continent has gone through millions of years of evolution and the species of life which are extant in various regions are therefore attuned to the climate and the soil and have withstood competition from others in the surroundings. These populations rise and fall according to natural laws, and within our time-scale, no species becomes extinct through natural forces. When man introduces EXOTIC species, however, which are not part of the environment, he sets in motion a chain of disturbances which often lead to calamitous results.

All in all, the role of the conservationist is to recognize the ecological needs of all species of animal and plant life, and to create those conditions which will ensure their permanent survival.

See also ECOSYSTEM, IUCN, INDIAN BOARD FOR WILD LIFE, TIGER.

CONSERVATION IN BHUTAN. Bhutan's rich wildlife has been well protected in the past because of the Buddhist beliefs of its population. The Bhutanese also have a fear of using firearms, believing that this will offend the deities of the woods and valleys, and will cause rainfall.

The first sanctuary to be created in Bhutan was Manas Wildlife Sanctuary in 1966. Located on Bhutan's most important river, it adjoins India's Manas Wildlife Sanctuary in Assam state and adds 400 km² to its 2900 km². Its hilly terrain is covered with dense forest, and it harbours among other species elephant, tiger, deer species, the golden langur, the golden cat, black bear, and wild buffalo. This sanctuary is one of the greatest tourist attractions of Bhutan.

In 1974, eight more Sanctuaries/Parks/Forest Reserves were created. Of these, the Goley Game Reserve adjoins the Manas Sanctuary, has similar flora and fauna and adds another 200 km^2. The Pachu Reserved Forest is situated between the Rydak and Sankosh rivers adjacent to the border of southern Bhutan, and covers 140 km^2. This continues into the Mochu Reserved Forest which is situated between Sankosh river and Single village and has an area of 280 km^2. The Doga National Park covers about 20 km^2 and is the home of the goral around Dobji Dzong. The Khaling Reserved Forest, situated at the southeast corner of Bhutan, has an area of 235 km^2, and contains dense vegetation and rich wildlife.

In 1979, three sanctuaries were combined to form the extensive and important Jigme Dorji Wildlife Sanctuary, which covers the entire northern part of Bhutan from East to West. It has an area of nearly 8000 km^2. The terrain is hilly to precipitous, and the northern reaches abound in glaciers. A large number of alpine pasture lands are scattered over the region. The Jigme Dorji Sanctuary is the home of the Musk Deer. Other species include blue sheep, tahr, goral, Tibetan antelope, Himalayan bear, wolf, leopard, snow leopard, takin and marmot. Among birds, the tragopan, the monal pheasant, the blood pheasant and the snowcock are of special interest.

CONSERVATION IN PAKISTAN. From time to time during the 1950s and 1960s, individual sportsmen, hunters and wildlife lovers expressed public concern about the dwindling wildlife resources of the country and the almost total disappearance of the larger game animals. But it was not until 1967 and 1968, when a team of foreign wildlife specialists carried out surveys in Pakistan, that attention was properly focused on the need for the Government to take positive action. These original surveys were sponsored by the Pakistan Tourist Development Board and their leader Mr Guy Mountford, an International Trustee of the World Wildlife Fund, submitted a report in November 1969 to the President of Pakistan. Resulting from this a Wildlife Enquiry Committee was set up in 1971, and more detailed information was collected about the status of various birds and animals and the possible areas for establishment of wildlife sanctuaries. The National Council for Conservation of Wildlife was also established in July 1974 in Islamabad, to devise and implement a national conservation strategy. During this time a good deal was also achieved by the efforts of private individuals, and in 1971 concerned sportsmen and conservationists banded together to start a national appeal

of WWF, which has subsequently initiated and sponsored the re-introduction of recently extinct indigenous species such as Blackbuck and Marbled Teal to Lal Sohanran Park and Chir pheasants (*Catreus wallichii*) into the Murree Hills, as well as the first reliable censuses of Wild Goats and Sheep in areas of their former range.

The first national park to be established was in 1972 at Lal Sohanran in the Punjab, an area of 31,354 hectares located some 25 miles northeast of Bahawalpur city, and comprising a representative sample of the Cholistan Desert, with a large seasonal lake and an irrigated forest plantation in one corner which offered additional shelter and attraction for winter migrant wildfowl on the lake, and Nilgai (*Boselaphus tragocamelus*) and Hog Deer (*Axis porcinus*) in the plantation. The Dutch WWF section donated fencing which enabled the re-introduction of 10 Blackbuck donated by a game ranch, from of all places, Texas, USA. These animals introduced in 1971 have since increased to some 26 head.

Shortly after this the second national park was established in Sind Province at Kirthar. Located in the foothill region west of the Indus, this park of 308,733 hectares area was created in 1974. It contains several very large herds of the Wild Goat (*Capra hircus*), estimated in 1979 to number over 1100 animals. Besides the Sind Ibex there is a viable population of the Baluchistan Urial (*Ovis orientalis blanfordi*), the Chinkara Gazelle, the Hyena and an occasional Leopard. The Sind Government has built access roads, a tourist complex for visitors and the area is patrolled by game wardens.

Around the same time on the initiative of WWF, two private Hunting Reserves were given Government recognition and official game reserve status. In both cases private individuals had effectively preserved endangered game species and Dr G. Schaller, the wildlife specialist, published his detailed behavioural studies of wild sheep and goats in these reserves which focused world attention on their importance. The one at Kalabagh in the extreme southwest corner of the Salt Range in the Punjab contains about 500 Punjab Urial (*Ovis orientalis punjabiensis*) as well as a sizeable Gazelle population. The other, called Chitral Gol, was the former hunting preserve of the Mehtar of Chitral. A natural amphitheatre in the hills to the west of Chitral town, it contains about 200 head of the Pir Panjal subspecies of the Markhor *Capra falconeri cashmiriensis*. This area encircled by peaks rising to 16,000 ft and clothed over its lower slopes by scrub-oak forest of *Quercus baloot* also contains Black Bear and Snow Leopards.

Two other national parks have been created. The Margalla National Park, actually inaugurated by Prince Bernhard of the Netherlands during his visit to Pakistan in 1973, covers some 11,635 hectares including the Rawal

Lake and the 3000-foot-high range of hills which are clothed with the last remnant of tropical dry deciduous forest in Pakistan. Ecologically unique and very interesting because of its comparatively high annual rainfall (82 cm), the Margalla Park contains a small population of Grey Goral and Barking Deer as well as Leopard, Wild pig, Rhesus macaque and an extremely varied population of birds. The latest checklist of bird species sighted within the Margalla Hills now totals over 300 species including such rare ducks as the Smew (*Mergus albellus*) and Goosanders (*Mergus merganser*) on Rawal Lake, and in winter many Himalayan passerine migrants such as Blueheaded Redstarts, Redmantled Rosefinches and thrushes.

Perhaps the most significant and dramatic park to be created is the one called Khunjerab National Park, located in the far northern area of Pakistan, covering 22,675 hectares. It was established in 1975 and extends over most of the extreme north-western part of Hunza State, right up to the border of Chinese Turkestan. The upper reaches of the Khunjerab Valley itself provide the wintering grounds for small herds of the fabulous Marco Polo Sheep (*Ovis ammon polii*). Over 60 animals were reliably sighted in 1979, the last available census. The Park contains besides some of the world's loftiest peaks in the Karakoram Range and spectacular scenery, also the Himalayan Ibex (*Capra ibex sibirica*), the Red Bear (*Ursus arctos*), the Himalayan Lynx (*Felis lynx*), the Snow Leopard, and a few Blue Sheep. Also Snow Partridges, Monal Pheasants and many little-known smaller passerine bird species such as Stoliczka's Tit-warbler (*Leptopocile sophiae*), Pleschanka's Chat (*Oenanthe pleschanka*), Güldenstädt's Redstart (*Phoenicurus erythrogaster*), and the Redbreasted Rosefinch (*Carpodacus puniceus*).

In addition to this national park the Naltar Valley, just north of Gilgit main valley, has been given special reserve status. It has during the past 30 years been maintained by the Pakistan Air Force as a hunting reserve and as a result of the exclusion of domestic grazing flocks and control of poaching it contains a much better population of Himalayan Ibex than any of the surrounding valleys.

In Baluchistan the Government has two game reserves, one covering the Chiltan range where the Chiltan Wild Goat survives and another at Mashlak to the north of Quetta where gazelle and a few wild sheep survive and the Houbara Bustard occasionally breeds.

Besides these national parks, the provinces have also enacted Wildlife Protection and Preservation Ordinances carrying lists of scheduled bird, mammal and reptile species which are totally protected from any form of hunting or exploitation. Punjab and Sind both promulgated these ordinances in 1972. A Wildlife Management Board has been set up in Sind Province which has been most active and energetic in developing a number of conservation projects. Largely as a result of Sind Province's initiative, a country-wide Houbara Bustard (*Chlamydotis undulata*) Conservation Project was established in September 1980. Like all the Bustard species in the world, this winter migrant to Pakistan has suffered a drastic decline in numbers, in this instance accelerated by an intense effort by various dignitaries from Arab Gulf States who visit Pakistan to hunt the Houbara each winter. In each of Sind, Punjab and Baluchistan provinces now two separate and extensive areas have been officially designated as Bustard Sanctuaries and the IUCN has given generous financial support for the Project. Much still needs to be done in this field if indeed it is not too late for the Bustard, but after years of frustration, an active conservation project is being supported by the President himself. Sind Province has taken the initiative also in creating a sanctuary for the Indus Blind Dolphin (*Platanista indi*) along the river north of Sukkur and reaching to Kandkhot below the Guddu Barrage. Reduced to an estimated 170 dolphins during the early 1970s the total ban on catching dolphins has now resulted in a gradual increase of this unique freshwater mammal to an estimated 280 individuals (February 1981 census).

The Karachi sea coast forms one of the world's eleven known nesting sites of the huge marine turtle *Chelonia mydas* known as the Green Turtle. Throughout the monsoon and winter months females haul out on to the suitable sandy beaches around Hawkes' Bay and Sandspit to lay their eggs. The Sind Wildlife Management Board, after receiving reports of heavy nest predation by feral dogs, seagulls and even by picnickers, initiated a conservation project. Clutches of eggs are now dug up and hatched in a protected enclosure. The newly hatched turtles are released into the sea after nightfall so that neither dogs nor seabirds can kill them. So far over 59,000 eggs have been transplanted and over 33,000 hatchlings returned to the sea. The Olive Ridley Turtle, less common along Karachi beaches, is also being studied and included in the next transplanting scheme.

It would only be fair to conclude that wildlife conservation in Pakistan still has a long way to go before the continuous depletion of both wildlife resources and destruction of the last remnant of natural ecosystems is going to be slowed down if not arrested. Like neighbouring countries, all the national parks are subject to much illicit cutting of fodder grass and fuelwood, illegal grazing by domestic stock which is an especially serious problem in the mountain parks, and even from occasional poachers. Funds for adequate development of access roads, for game wardens and anti-poacher control are still totally inadequate and detailed studies of these last wilderness areas and their fauna need to be conducted before proper park management plans can be devised. But there is a growing public awareness in Pakistan for the need to preserve these wilderness areas inviolate, and in recent years the Government has given increasing attention and financial support for establishment of these parks and for conservation programmes aimed at saving particular species.

T.J.R.

CONTINENTAL SHELF. The continental shelves are the shallow part of the sea floor immediately adjacent to the land. They are generally smooth platforms that extend seaward. At a distance from the coast, which varies from region to region, this underwater platform undergoes a sudden change in slope called the shelf break. The shelf break therefore marks the seaward edge of the continental shelf. Beyond the shelf break, the floor of the sea descends to deeper regions called the continental slope (Fig.1). The total area of the continental shelf forms 8 percent of the ocean floor. Defining the boundaries of the continental shelf is important in determining the free passage of ships, mineral and fishing rights and matters of defence. The average width of the continental shelf is 40 nautical miles and the average depth to the shelf break is 72 fathoms. There is, however, considerable variation in the width of the shelf and the depth to the shelf break. For example, the continental shelf off the west coast of the Indian peninsula is as much as 280 km off Bombay and becomes as little as 60 km off Cochin (Fig. 2). The depth of the shelf break on the Western shelf is about 90 metres (50 fathoms) in the northern regions and becomes deeper in the south. The total area of the Indian Continental Shelf is about 450,000 km².

Of this total the Western Shelf is about 300,000 sq km, the balance being on the east coast. As can be seen, Pakistan and Sri Lanka have narrow continental shelves, whereas Bangladesh and Burma have relatively wide shelves.

The nature of the sea floor (topography) of the continental shelf is highly variable. Numerous depressions and ridges are present. In addition, there are many examples of valleys, apparently extensions of river valleys on the land which cut across the continental shelf. These underwater valleys are called submarine canyons. These are found on many shelves of the world including the eastern continental shelf of India and on the Pakistan shelf. The sediments of the seabed are derived from a variety of sources. The two main sources are the land areas of the earth from which boulders, cobbles, pebbles, sand and mud are derived, and from the sea itself in the form of shells of dead organisms. The agents which transport sediments from land to the continental shelf are rivers, wind and glaciers. Minor quantities come from outer space in the form of meteoritic dust.

The continental shelf is a zone of which bottom sediments are shifted by the waves and currents present in the ocean. The distribution of sediments on the shelf is based on a sea level lowered during the recent Ice Age. During this Ice Age large quantities of water were removed from the ocean by natural processes and stored in the form of ice on land. This removal of water from the ocean resulted in lowering of the sea level to as much as 100 m below the present sea level and converted the continental shelf into dry land. Near regions where glaciers occur, as in the northern latitudes, large quantities of boulders and clay are

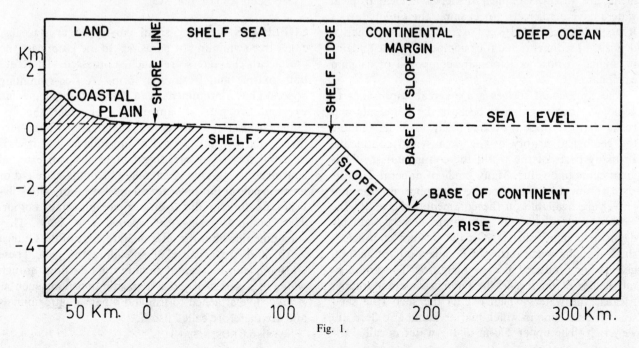

Fig. 1.

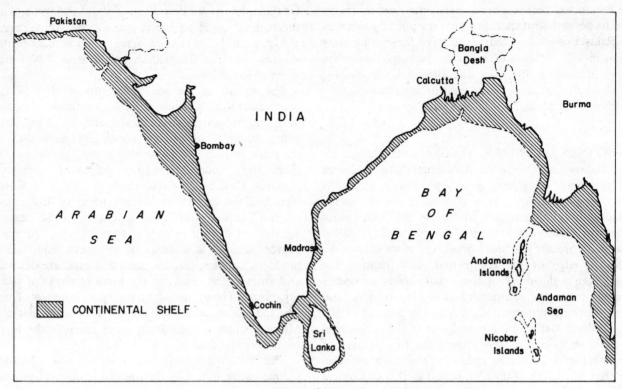

Fig. 2.

Continental shelves around India, Burma, Pakistan, Bangladesh and
Sri Lanka. Continental shelves around Lakhshadweep and the Andaman
and Nicobar islands are present but are extremely narrow.

deposited on the shelf. Off large rivers, for example off the Gulf of Cambay into which flow the rivers Narmada and Tapti, the continental shelf is mud-covered over a large area. Where strong ocean currents are present the shelf is narrow, and fine-grained sediments are absent. Continental shelves in clear tropical waters have coral reefs, for example the Great Barrier Reef of the Queensland Coast of Australia, and extensive deposits of sediments and rock composed of the remains of animals whose skeletons are composed of calcium carbonate.

The continental shelves are of great importance to geological oceanographers because the sediments and rocks that are present on them help in understanding the geological history of the oceans and continents. In many parts of the world the continental shelf has great economic value. Many kinds of mineral deposits such as tin, diamond, ilmenite, phosphorite, sand and gravel are available on the continental shelves of the world. Oil and natural gas forms a very valuable deposit on many shelves. On the western continental shelf of India, particularly off Bombay, oil and natural gas is being produced in large quantities. The continental shelf is of interest to the biological oceanographer because the waters which overlie the shelf (the shelf sea) forms a zone in which marine fauna and flora are very rich. The upper 200 m of the water is called the euphotic zone where all forms of life are found in great abundance. More than 70 percent of the total fish catch of the oceans comes from the waters of continental shelves.

R.R.N.

See PLEISTOCENE ICE AGE.

COPEPODS are very small oar-footed crustaceans, either free-swimming or parasitic. In the latter case the crustacean characters are much suppressed. There are both marine and freshwater forms. A free-swimming copepod has a prominent head bearing a single median eye and five pairs of appendages of which the first antennae are large and serve as swimmerets. The abdomen is narrow and devoid of appendages but it ends in a tail fork. The breeding female has two external conspicuous egg-sacs. Generally the animals are red or green in colour but those found in association with jelly-fishes floating in the open sea are bluish in colour. The common freshwater copepod is *Cyclops*. They sometimes harbour the intermediate stage of the Guinea worm and are a cause of spreading the pest. These copepods can be seen as minute dots darting through water in a series of jerks. Parasitic copepods such as *Argulus* live attached to fish and other aquatic animals and are therefore called fish-lice.

See also CRUSTACEANS.

140

CORAL JASMINE, *Nyctanthes arbor-tristis*, a small deciduous tree with drooping quadrangular branches, and rough opposite leaves distributed in the sub-Himalayan tracts and forests of Madhya Pradesh. Flowers fragrant white, with a salver-shaped corolla and orange coloured tube, opening towards the evening and dropping the next morning. The flowers are gathered for Hindu ceremonies, also for dyeing silk and cotton. Its bark is used for tanning and the rough leaves for polishing wood.

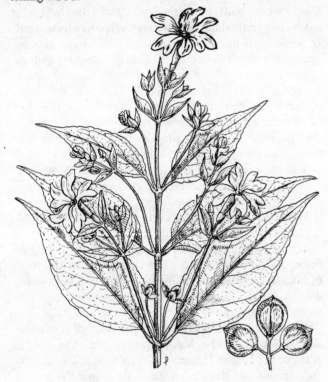

Coral Jasmine, flowers and seeds, × ½

CORAL REEFS, ISLANDS and ATOLLS. Corals are calcareous skeletons of minute marine organisms or polyps. The principal builders of coral reefs and islands are the true or stony corals also known as Scleraltinia, which secrete the skeleton externally at the base and lower part of the body. As the colony grows, the polyps on the outer edges multiply, smothering the lower ones and growing upon their skeletons, continuously withdrawing calcareous material from the water. Thus coral reefs are masses of carbonate built up from the sea floor by the accumulation of the skeletons of a profusion of animals and algae.

The role of organisms in the construction of a reef is threefold: whole skeletons may form a loose or tight framework; broken-up and overturned skeletons form the vast quantities of lime, sand, and mud that fills the interstices of the framework; and organisms such as encrusting algae, particularly red algae and blue-green algae, cement the framework and the interstitial material together.

Reef-forming corals are dependent on warmth and sunlight, and grow best in shallow sunlit water between the low-water mark and six fathoms (36 ft or 11 m), though they may have a sparse existence between 22 and 30 fathoms (132 and 180 ft). It must however be stated that corals are not able to survive even a short exposure in the air to the sun's rays, so that their upward limit of growth is determined by that of lowest water at spring tides. They prefer water of normal salinity and with an annual maximum temperature above 22°C but below 28°C, though they may remain alive in waters whose minimum temperature in winter is as low as 15°C.

Coral polyps usually remain inactive in daytime, feeding only at night on plankton which rise to the surface. Because they require warm shallow waters, coral polyps are limited to continental and inland shores in tropical and subtropical zones. They are almost absent from the western coasts of South America and Africa, because of the cold currents from the Antarctic. They are abundant on the eastern coasts and inhabit two regions—the Atlantic and Indo-Pacific regions.

The several reef types found in these regions include Fringing Reefs, Barrier Reefs, Patch or Platform Reefs, and Atolls. Fringing reefs are formations which grow along the shores of a continent or island. In the Indian ocean such reefs are found around southern India, Ceylon, Andamans, Mergui and the Malay Archipelago. Barrier reefs are separated from the land by a deep channel—10 miles or more in width. The Great Barrier Reef off the northeast coast of Australia is over 1200 miles long, and an estimated 500 million years old. Its dimensions are such that it can be seen from the moon. Patch or platform reefs occur as isolated patches on a continental shelf, roughly parallel with the coastline. These are found around Antarctica, Patagonia, and the Falkland Islands in waters 2° to 6°C in temperature. Some isolated platform reefs are found in the Indian ocean. Atolls are ring-shaped reefs that in their grandest oceanic form mark the rims of sunken, truncated volcanic cones. They are typified by the Maldive and Lakshadweep atolls, which encircle shallow lagoons or bays, ranging from 1 to 50 miles across.

Origins

The first acceptable theory of reef origin and development was put forth in 1842 by Charles Darwin, and his concept is valid even today. According to Darwin, a reef forming an atoll, and a reef forming a barrier are *not* different in 'general size, outline, grouping and minute structure'. Barrier reefs bega as reefs fringing the land around which they now form a barrier and oceanic atolls began as reefs fringing a volcanic island.

The subsidence of the land fringed, allowed the reef to grow upward and outward. Making use of the 'subsidence theory' Darwin explained both the formation of atolls, and also why encircling barrier reefs stand so far from the shores they front. He cited as evidence the observations of the inhabitants of the Maldive archipelago and of Keeling island, who had observed water levels rising over the years.

Because maximum growth of corals occurs at the seaward edge, lagoons develop between the ascending barrier, or atoll, reef and the land or volcanic cone. With the complete submergence of the volcanic cone, the atoll lagoon contains only coral islands. The oceanic atoll reefs of the Pacific rise from volcanic cones that have subsided probably intermittently, in areas of oceanic deeps. According to the Darwinian Subsidence theory, the annular atoll reefs extend and grade downward into barrier reefs, which originated as reefs that fringed the volcanic cone. On the other hand, the compound atoll reefs, such as the Maldive islands and Lakshadweep, are believed to have grown above foundered continental crustal segments. While Darwin's concept is still valid fundamentally, many consider submergence by the rise of sea level, following melting of Pleistocene ice sheets, to be a better explanation of the latest upward growth of many reefs, particularly on continental shelves.

A.S.B.

CORALS are fixed coelenterates very closely related to sea-anemones with which they agree in most of the structural and developmental details. Corals, with the exception of a few genera, form extensive colonies and they all secrete a horny or calcareous skeleton, the colour and consistency of which vary with different families.

A typical coral polyp looks like a sea-anemone with a cylindrical column, a crown of tentacles and a gastric cavity with a grooved gullet. In addition to these it has also a cup of calcareous matter at its base. This cup is the skeleton or 'coral' which increases in size as the animal grows.

Corals feed on small organisms such as algae, proto-

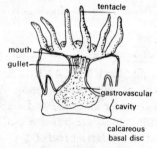

Longitudinal section of a
coral anemone × 1½

zoa, hydroids, worms, crustaceans and molluscs. In ingesting these organisms into the mouth the tentacles play the same role as in sea-anemones.

There are only a few genera of solitary polyps. A majority of them are colonial. The colonies show variation in external form brought about by the different modes of budding in the different genera. In the simple type of colonies the original polyp sends out a horizontal branch (stolon) from which a limited number of fresh polyps bud out. Another common type is like *Obelia* where buds are formed as lateral growths from a somewhat vertical stem resulting in a tree-like formation. In some cases many buds grow parallel to one another producing dense colonies of closely adhering polyps.

Corals occur along the coast from Gujerat southwards but are concentrated in the waters near Tuticorin, Rameshwaram, Krusadi Islands, the coast of the Gulf of Mannar and around Lakshadweep and the Andaman and Nicobar groups of islands. In the Maldives also there exist many reefs and atolls.

The more common and interesting local corals are:
With 8-branched tentacles: *Alcyonium, Corallium, Gorgonia* and *Pennatula*
With many simple tentacles: *Flabellum, Madrepora, Cavernularia* and *Astrangia.*

Alcyonium, or Deadman's Finger, forms a colony of thick, soft leathery lobes. The skeleton is in the form of calcareous spicules of different shapes and sizes which do not form a solid support but allow the fingers to move and sway in the water.

Corallium, which produces the red coral of commerce and is found in the Mediterranean sea, has several allies in the Indian Ocean. It forms a plant-like colony and consists of a thick branching axis. From the surface of this axis, which is red, pure white polyps emerge. The axial rod is thick and consists entirely of lime spicules embedded in a cement-like deposit of lime.

Several species of *Gorgonia* (Sea Fans) are common where reefs abound. *Gorgonia* looks like a plant with long delicate branches springing from its short main stem. The polyps are small with eight short tentacles. The entire animal is flexible and brightly coloured. Dried specimens are beautiful objects esteemed as curios. Young bivalves of the genus *Pteria* live attached to the Gorgonian branches for protection. Both the coral and shell-fish are orange-coloured and the ears of the shell resemble the branches of the host. Similarly yellow and red ovulas (see COWRIES) lurk in the branches of Gorgonians of corresponding colours.

Pennatula (Sea Pens) form bilaterally symmetrical colonies, the polyps arranged on either side of a long axis supported by a horny or calcareous skeletal rod. The colonies live partly buried in mud.

Cavernularia is very common along the east coast, and is closely allied to Sea Pens. It is neither feather-

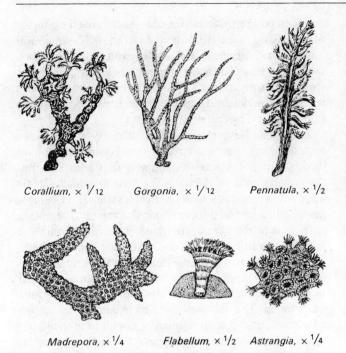

Corallium, × ¹/₁₂   Gorgonia, × ¹/₁₂   Pennatula, × ¹/₂

Madrepora, × ¹/₄   Flabellum, × ¹/₂   Astrangia, × ¹/₄

like nor has it an axial skeleton. When thrown ashore by waves it is a yellowish lump with a hard, narrow furrowed end and a broad bulging end with small pits marking the position of the contracted polyps. In life the furrowed end is fixed in the mud and the colony expands vertically. Through the pits the beautiful polyps spread out.

Flabellum is the simplest form of stony coral. It is a solitary polyp resembling a small conical cup, the tapering part being the stalk by which the young living polyp is attached. The broad side of the cup is somewhat flattened or oval in cross-section. It is of calcareous composition with radial septa, rough and brown outside and white and smooth within. Such a solitary coral is called a corallite.

In the more complex genera like *Astrea*, *Astrangia*, *Madrepora* and *Dendraphylla* the corallites produced side by side by thousands of polyps get cemented together to form the common skeleton of a coral colony.

Millions of polyps living attached to one another and secreting coral for millions of years have resulted in fringing reefs, barrier reefs and atolls. These harbour within them several kinds of organisms which provide food for many sea animals and for fishes reared in fisheries.

Coral reefs check sea erosion and act as bolsters against cyclones and storms.

See COELENTERATE & CORAL REEFS.

CORBETT. The Corbett National Park is situated in the folds of the Garhwal and Kumaon Siwalik ranges an area made famous by the writings of Jim Corbett. It covers an area of 500 km² and the forests consist of both dry and moist Siwalik Sal together with dry mixed deciduous forest. It has spectacularly large meadows which provide excellent grazing and viewing of wildlife. The Ramganga river cuts across the Park and a large artificial lake has been formed by damming the river at its northern extremity.

Corbett National Park was the first National Park of India, created in 1935 at the behest of Sir Malcolm Hailey the then Governor of the U.P. Prior to its creation as a National Park it was a hunter's paradise. The herbivores include elephant, chital, sambar, barking deer and hog deer and on the slopes of the Himalayan foothills goral can be seen. There is a fair number of tigers and panthers, and panther kills are often seen on tree branches where they are hidden for safety as is the custom with this predator. There are also wild boar, leopard cat, Indian civet and jackal. The Ramganga river abounds with mahseer and is hence very popular with anglers. Both the marsh crocodile as well as the gharial are found in the Ramganga river and in the newly created reservoir. Other reptiles include monitor lizard, python, and a large number of snakes. The Zoological Survey of India has listed more than 400 species of birds in the Park.

CORMORANT (family Phalacrocoracidae). A gregarious black-plumaged water-bird inhabiting inland waters, brackish lagoons and tidal creeks. It has a hook-tipped bill and elastic gular pouch. It catches its prey (fish) by underwater pursuit, brings it to the surface, tosses it in the air and swallows it. When not hunting it is usually seen perched on a rock with wings spread open to dry. Nests communally with egrets and other water-birds. Three species occur within the subcontinent.

COROMANDEL EBONY, *Diospyros melanoxylon,* medium sized tree of the peninsula with a greyish-black bark peeling off in rectangular scales. Its leaves are egg-shaped to oblong, leathery, hairy beneath. Flowers appear in 3–12 flowered velvety drooping cymes. The fruits are globose, and yellow. Its leaves are in demand for wrapping bidis and are collected in government forests. The heartwood is black, streaked with purple or brown, and the outer wood is also valuable and takes a good polish: it is used in cabinet and inlay work, for brush-backs, etc. *D. ebenum* is the EBONY with jet-black heartwood, not growing in sufficient quantity in the Deccan but more frequent in Sri Lanka. *D. marmorata* is the ANDAMAN MARBLE WOOD, an ornamental, rare and endemic timber of the Andamans characterized by dark grey heartwood with black streaks.

See drawings on next page.

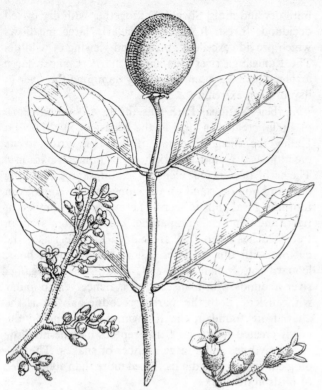

Leaves, flowers and fruit of Coromandel Ebony, × 1

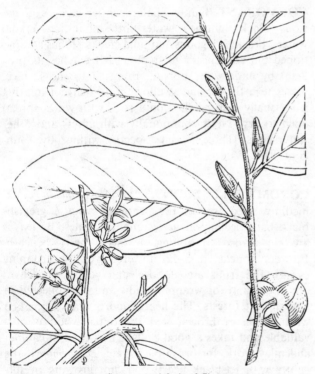

Leaves, flowers and fruit of Ebony × 1

COURTSHIP. Baya weaver bird males fluttering their wings and singing vigorously at the females from their half-completed nests—street dogs fighting with each other while pursuing a receptive female who keeps yelping all the time—greyling butterfly males chasing

and then bowing before a female—these are all examples of courtship. Courtship is the series of behavioural interactions amongst the males and females which eventually leads to their copulation. Courtship is widespread among animals, particularly in those species where the males and females form a pair-bond at the time of sexual receptivity.

Zoologists have postulated a diversity of functions for courtship. But what really stands out as common throughout the animal kingdom is that courtship has the consequence of delaying the time at which a male achieves a copulation with the female. It is almost invariably a period during which the male is attempting to persuade the female to mate with him, while the female is waiting before she gives the final consent.

This correlates with the fact that there is little of courtship in species such as the Hanuman Langur in which one male may hold absolute sway over a group of females. The females then have almost no choice of a mate, and receptive females copulate immediately after indicating their willingness through appropriate posture. Moreover, it has been noted in some cases that females who are not in heat may solicit copulation to please the aggressive tendencies of the male.

Courtship, on the other hand, is much more elaborate in species in which there is a choice open to the females from amongst a group of males. It has been particularly well studied in the European fish, the ten-spined stickleback. In this fish, during the breeding season, the males establish territories in shallow streams. In his territory the male constructs a nest in the sand in which the

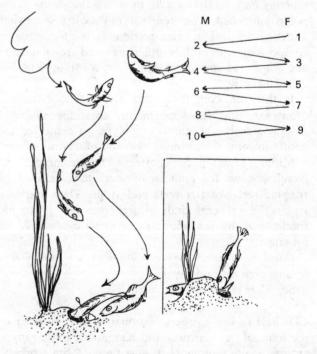

In the stickleback the courtship is a sequence of mutually releasing actions.

11 Delta and Terai Grassland

The Ganga delta, known as the Sundarbans, supports tigers, wild boar, deer and a multitude of other life forms including crocodile, lizards, snakes, and crustaceans such as lobsters and crabs. The habitat also attracts a wide spectrum of water birds.

The grasslands and forests of the terai in Manas, Assam, shelter elephants, tigers, leopards, swamp deer, capped langur and the very rare golden langur. Also seen in Manas is the merganser, a fish-eating duck.

Grasshoppers are the most well represented of the Orthoptera. Their highly developed hind legs allow them to leap great distances. Grasshoppers are herbivorous and, when they congregate in large swarms, can be quite harmful to standing crops.

eggs will be laid. The territorial males have a red belly, while the females ready to lay lack the red colour on the belly, which is swollen. When a red-bellied stranger approaches the territory of a male he is chased off. But when a female with a swollen belly appears she is courted. The courtship involves the male swimming away from the female. If she follows, the male leads her to the entrance of the nest in a zigzag approach. At the nest the male indicates the nest-entrance and withdraws. The female may then enter the nest and lay. On her leaving the nest, the male enters himself and fertilizes the eggs. The female then leaves the territory and the male stays on, guarding and aerating the eggs and taking care of the young after they are hatched.

This courtship of the stickleback illustrates many interesting points. It involves a long chain of alternating behaviour patterns by the male and the female. The next pattern in the series is released only when the partner responds appropriately. Otherwise the chain is broken off. Thus the chain cannot be completed even if one of the responses of one of the partners is inappropriate. Since different species differ in the details of the behavioural sequences during courtship, this serves to prevent hybridization and consequent loss in fitness, since interspecific hybrids are almost always less viable and less fertile.

Secondly, many of the behaviours during courtship are released by a few simple stimuli. Thus a territorial male will attack a dummy with a red belly, even though the dummy may lack any other features of a stickleback, but will fail to attack a faithful replica of a male stickleback without a red belly. Similarly, his courtship is released by the single stimulus of a swollen belly.

The fact that very simple stimuli will invoke a whole behaviour sequence has led to the development of some very interesting cases of deceit during courtship. A three-spined stickleback male may occasionally notice another fish at the periphery of his territory while a female is laying eggs in his nest. This fish, lacking a red belly, may stimulate the male to court it in a hurry without fertilizing the first clutch of eggs just laid. The second fish may respond appropriately and enter the nest after the first female has left it, and even take the attitude of a female laying eggs. In certain cases however it turns out to be a male without a red belly who is merely fertilizing the eggs laid by the first female!

What function does courtship perform? The older theories stressed prevention of hybridization amongst species and reduction of initially aggressive tendencies of the partners as the major functions. The modern interpretation however stresses the role it plays in enabling a female to choose a male of high quality as a mate. The process of courtship is viewed as a process which permits a female to judge the physiological vigour as well as a male's ability to hold a high position in the

Courtship displays of junglefowl, pheasants and peacock have originated from food calls

social hierarchy of males. It would be of advantage for a female to select such a mate as he will pass on to her offspring, particularly her sons, qualities which in turn will ensure their high reproductive success.

The persistence and vigour of male courtship are then supposed to serve as indicators of a male's physiological vigour. The female may judge his social status by watching the outcome of contests amongst males. Thus in the elephant seals the females merely watch the males fighting for them, and then mate with the winner of the fight. Alternatively, males may compete for some other reward such as the more central territories in a lek of grouse or in a colony of kittiwakes and the females may merely prefer whichever males happen to be occupying the central territories.

In many birds of prey a male feeds the female during courtship.

COWRIES. In India cowries are perhaps the best known group of shell-fish. They are world-famous for their handsome, highly polished, oval or globular shells. The young cowry has an elongated shell with a distinct conical spire and a long wide mouth aperture bounded by a thin outer lip. As the animal becomes adult, the mantle lobe on each side expands and grows over the shell, the edges of the two lobes meeting along a line. In this position the inner surface of the mantle secretes a thick layer of highly polished enamel of different colours over the shell. The last whorl of the shell grows somewhat excessively and rolls round the other whorls. This growth and the enamel deposit over the spire by the mantle completely conceal the actual spiral nature of the shell. Side by side with this transformation, the outer lip of the shell, which was originally thin and sharp, gets curled inwards, thickened and toothed. The inner lip also thickens and its edge assumes a furrowed appearance. As a result of the thickenings of the two lips the mouth aperture of the shell, previously wide, becomes narrowed considerably. Thus the adult cowry shell is formed into a highly polished, oval shell with all trace of the young spiral masked by the deposit of enamel and with a long narrow mouth aperture on the ventral side.

A cowry moving over a rock, underwater × 1.

Not only the shells but also the animals within them are beautiful. Few objects are more attractive than a large cowry crawling in a coral-reef pool, the mantle bright with scarlet and yellow, beset with gracefully branched filaments. Cowries feed on tiny animals and plants. Coral reefs, the underside of boulders, mud-covered rocks where seaweeds abound, are all favourite haunts of these animals. Most cowries prefer moderately deep water. There are several species occurring in India out of which at least a dozen are exceptionally attractive. These are commonly found along the coast of Bombay, in the Pamban area and around the Laccadive islands (Lakshadweep).

A few of the more outstanding Indian cowries are: 1. The Giant Cowry (*Cyprea testidunaria*) measuring nearly 12 cm long, surface and sides pale brownish violet, decorated with chestnut spots mingled with numerous white dots; 2. the Tiger Cowry (*C. tigrina*) measuring nearly 10 cm, pear-shaped, heavy, pale yellowish-white, decked with large brownish-black spots; 3. Mole Cowry (*C. talpa*) 8 cm long, long oval, yellowish-cream with brown transverse bands; 4. Serpent's head Cowry (*C. caput-serpentis*) about 4 cm long, broadly oval, border brownish-black, central portion pale brown with numerous whitish dots; 5. Eyed Cowry (*C. ocellata*) 3 to 4 cm long, oval, back dotted with greenish spots ringed round with yellow; 6. Money Cowry (*C. moneta*) the most popular and abundant cowry, recognized by its uniform cream or shining yellow colour, medium size (3 to 4 cm long), thickened border and slightly conical front. Several other curious species are also common.

There are many uses of cowries. In ancient India money cowries had exchange value and were used as coins as they are now being used in some villages in southern Africa.

Closely allied to cowries are some interesting forms called Ovulas which look like thin, small, oval cowries. The main difference is that in ovulas the inner lip of the aperture is not toothed. Further the shell is actually colourless and translucent but appears coloured in harmony with the surrounding in which the animal lives. The colour perceived is that of the animal's mantle refracted through the translucent shell. Ovulas live among coralline growths of different kinds. Those that live amidst red and yellow corals are red and yellow and those amidst pink and orange are tinted correspondingly. The colours of the ovulas, blending harmoniously with the background, effectively protect them from their enemies. Another feature of ovulas is that they can crawl up and down the coral branches with the aid of their foot which is modified suitably. Ovulas of different kinds are common among the coral reefs in the Gulf of Mannar, Gulf of Kutch and around Lakshadweep. *Primovula pudica* is a common ovula, looking more or less like a thin white cowry. *Volva sowerbyana* is a spindle-shaped ovula with both ends pointed. *Ovulum volva* is a fairly large shell 8 cm long, the body elongated, ovate and inflated with the anterior and posterior canals greatly protruded. In view of its shape the shell is popularly called the Weaver's Shuttle.

See also MOLLUSCS.

CRABS are the most dominant and highly organized decapod crustaceans, very widely distributed along sea-shores, sandy bottoms, sides of rocks, in brackish and fresh waters, in estuaries and mangrove mudflats, water-logged rice-fields and on moist land.

In a typical crab the fore-part of the body is depressed

and covered by a broad shield expanded laterally, under which its water-breathing organs or gills are well protected. In land crabs the gills function using the moisture in air. In some forms the upper part of the gill chamber gets modified into a sort of air-breathing lung.

Of the five pairs of legs the first pair is curved forwards and is invariably stouter and larger with the extremities developed into formidable pinching claws or chelae. The other four pairs are disposed laterally or backwards and used for crawling and digging. In the swimming forms the ends of the hind one or more pairs are flattened into oars or paddles. In masked crabs the tips of the last two pairs have hooks developed to hold a foreign object such as a bivalve shell.

Along the front portion of the shield (carapace) there are two pairs of pits or sockets through which one pair of stalked movable eyes and a pair of antennules emerge. Both these organs can be retracted into the sockets.

In all true crabs the abdomen is greatly reduced—insignificantly small and short compared to the forepart of the body. It cannot be seen from above as it is inturned—tucked beneath the shield. In male crabs the abdomen consists only of one or two segments whereas in the female it is larger and has four segments. The larger size of the female abdomen enables her to accommodate the developing eggs. In Mole crabs and Hermit crabs the abdomen is large and pronounced.

There are many families of crabs. They vary very widely in size, shape and architecture, colour, habits and habitat. In size they range from one centimetre to nearly a metre. In colour some are dull brown or black and some green, blue, red, yellow, orange and scarlet. Some have chevron patterns, streaks or dots on the shield. The shape and architecture of the shield differ in the different families. Certain crabs plant on their shields seaweeds, sponges, coelenterates etc. to conceal themselves from their enemies and at the same time to capture their prey by surprise. Several crabs live inside corals, sea-cucumbers etc. Such an association for sharing common food is a good example of commensalism. The enormous growth and bright colour of the chelae of the male 'calling crab' serve to attract the female and to fight the sexual battle.

In crabs there are separate males and females. The fertilized eggs, when mature, first escape as curious larvae—zoea—looking very much different from the adult. A zoea has long spines on the carapace and a slender but well-developed abdomen. This stage is soon followed by the second or *Megalopea* stage, which has two stalked eyes, a rounded carapace, five pairs of legs and a long abdomen. With the next moult the adult form is reached with a reduced and inturned abdomen.

Like most other crustaceans crabs are omnivorous. The food consists of small animals such as snails, tadpoles, insect larvae, worms, decaying matter, tender

Birgus, the Coconut Crab × ¼

plants etc. In all cases the pinching claws are used to catch and crush the prey and convey the broken pieces into the mouth. One genus of giant crab, *Birgus*, common along the Pacific shores, climb up coconut trees and with their stout pincers bore holes in coconuts, extract the kernel and feed on it. Mole crabs have no developed chelae. They live in sand, moving in and out with the tide. They have however plume-like antennae which catch organic matter on which they feed.

Crabs have many enemies. Most of the larger sea animals feed on them. Seashore birds—cranes, gulls, eagles, crows etc.—swallow them. In paddy fields dogs, jackals, wild cats, ducks and waterfowl eat them up. Man is also destructive to the medium and large sized crabs; crabs being considered by many to be tastier than prawns or fish.

India has several interesting genera of crabs. There are many species of curious crabs along both shores, and around Krusadi and Lakshadweep many large-sized varieties exist in abundance.

Some of the interesting genera are:

Emerita is a mole crab having an oval body about 3 cm across. Its curved carapace is barrel-shaped and pink in colour. It resides buried in sand in the littoral region exposing its plume-like antennae above the sand. Its

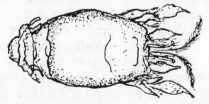

Emerita, ×1

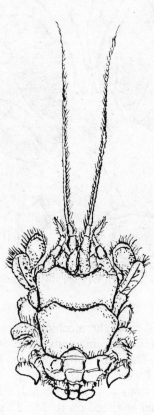

Albunea, × 1

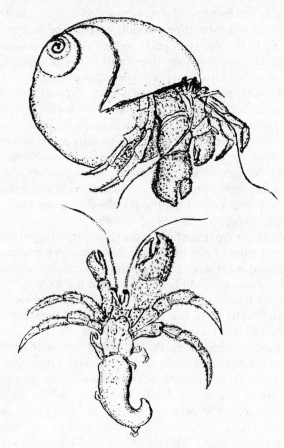

Pagurus, × 1
Inside the shell and out of the shell.

legs are adapted for digging. When the tide moves over them, the antennae collect organic matter on which the animal feeds. All its operations, viz. digging, crawling and swimming, are in the backward direction. The receding wave often dislodges the animal from its hole but it manages to dig another hole and get out of sight. Not infrequently many members get entangled in weeds and debris thrown ashore by advancing waves.

Albunea is another mole crab of stronger build and its body is rectangular. The antennae and other appendages are very prominent and fringed. It lives below low-tide level and often gets entangled in fishermen's nets.

Pagurus is the common hermit crab. Unlike other crabs its body is soft and abdomen long without any shelly armour. The animal however protects itself by inserting the hinder part of its body into any empty gastropod shell. Consequent upon the adaptation of the shell and mode of life the abdomen gets twisted and the lower appendages get atrophied. One of the chelae assumes abnormally large size. Hermit crabs are very abundant on all rocky and muddy shores where most of the empty shells of whelks, horn and top shells and periwinkles are occupied by them. When two hermit crabs wish to obtain possession of the same shell there is a regular tug-of-war between them. When the crab

grows larger it only has to change the old shell for a larger one. At this stage it is amusing to see a naked hermit crab slipping first into one shell and then into another until it obtains a good fit.

Uca is popularly known as a 'calling crab', 'dhobi

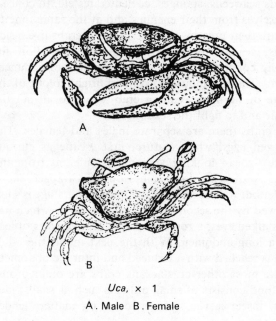

Uca, × 1
A . Male   B . Female

crab' or fiddler crab'. It is a stout crab burrowing in mud or sand between tide marks. Its eyes are set on long pedicles. The male crab has one of his chelae enormously developed and brightly coloured in metallic scarlet. He sits at the mouth of his hole and waves his red chela like a washerman beating clothes on a stone or someone beckoning another. Hence the names 'dhobi crab' or 'calling crab'. These crabs occur in large numbers in all mangrove mudflats and estuaries.

Ocypoda, sometimes called 'ghost crab' or 'sand crab', though the name means 'swift-footed', is seen in large numbers burrowing in sand near or above high-tide mark. They are generally of uniform sand colour but some of them are red or yellowish. They feed on organic matter in the sand. They can be seen on the beach running rapidly sideways or bringing up pellets of sand from the depths of their burrows and making attractive patterns on the beach.

Ocypoda, × 1

Dorippe is the common 'masked crab'. It is light-coloured and slender and has a peculiar habit of covering itself with a bivalve shell. Its last two pairs of legs are specially modified into hooks for holding the shell. The carapace is somewhat oblong and beautifully sculptured.

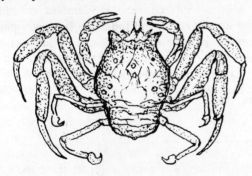

Dorippe, × 1

Calappa is an example of a box crab found plentifully in backwaters. When at rest with all its limbs tucked in, it looks like a small rugged cube of wood. The chelae are strong and crested and the other four pairs of legs adapted for swimming and crawling.

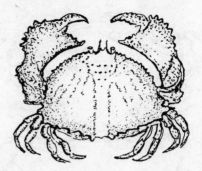

Calappa, × 1

Matuta is another box crab living in the same habitat as *Calappa* but differs in having its body and legs more flattened, well adapted for swimming. The carapace is shield-like with a triangular spine projecting on either side.

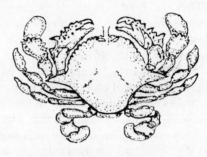

Matuta, × 1

Scylla is a large-sized edible crab with prominent, massive chelae. The last pair of legs is flattened for swimming and the others are normal for walking. This is available on the east and west coasts of India in large number and forms a major fishery.

Scylla, × ½

Neptunus is another edible crab growing to a size of nearly 20 cm across. The carapace is beautifully coloured with three bright red spots at the hinder part. The last pair of legs is well modified for swimming.

Paratelphusa. The genus includes fresh-water or paddy-field crabs. They live largely in rice-fields or on moist land near sheets of fresh water such as ponds, streams, etc. or in fresh water itself. Those living on land

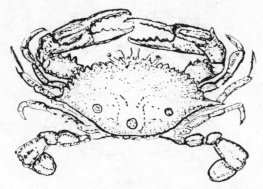

Neptunus, × ½

have their breathing chamber modified to hold the maximum quantity of moisture, in which the gills function. The moisture is occasionally replenished by dips in nearby water-pits or ponds, thus these crabs lead an amphibious life. They are about 8 cm across the carapace when fully grown and are creamish yellow in colour with sometimes a dash of black in the middle. Those found in hilly tracts are dark brown and larger in size. In summer, they hibernate in their intricate burrows which are usually near some source of water. Small mounds caused by these burrows are a common sight in wild low-lying lands in the Western Ghats around Poona. During early monsoon, they lay eggs which are attached to the abdominal segments and the latter directly hatch out into tiny young ones, without any larval stage. The young remain clinging to the abdomen of the mother till they get scattered into different pools or pit-holes in the neighbourhood and live an independent life.

These crabs usually feed on insects, worms, fish spawn, young shoots of vegetation, organic debris, moss, slimy algae, etc. They are thus partially scavengers but by the burrowing habit prove destructive to rice-field bunds, pond bunds, internal lining of wells and even irrigation canals. They are considered a delicacy by the local hill people. They occur in the Western and Eastern Ghats, in Bihar, in the Rajasthan canal area and probably elsewhere, but have been little studied.

Common Field (freshwater) crab × ½

CRAB'S-EYE CREEPER, *Abrus precatorius,* common in the tropics of the subcontinent and Burma. The seeds are scarlet with a black eye, and are exceedingly poisonous. Jewellers use the seeds as weights, each seed weighing 1·75 gm. They also make necklaces and bracelets.

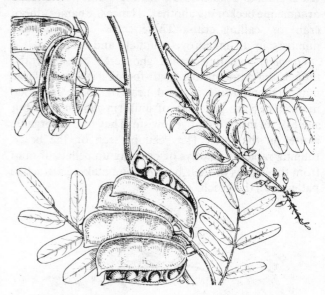

Crab's-eye Creeper. Leaves, flowers and seed-pods, × 1

CRANE. Like Bustards, Cranes belong to the Order Gruiformes. They are long-legged terrestrial birds with regular habits, often seen feeding in marshes or cropland, on vegetable or animal food. They are believed to mate for life. Apart from their resonant trumpet-like calls they may hiss or stamp like bustards, in alarm or aggressiveness. Of the total 14 or 15 world species, six occur in the Indian subcontinent of which two are resident. Of the latter, the endemic Sarus Crane (*Grus antigone*) is the largest in the world. This greyish crane with a bald red head is common around the jheels of north and central India. Sarus cranes pair for life, their courtship displays of leaping wildly in the air, bowing and duetting loudly, are spectacular, and they are protected by local sentiment.

The Common Crane (*Grus grus*) and the small handsome bluish-grey Demoiselle (*Anthropoides virgo*) are both abundant winter migrants which particularly favour the Kathiawar peninsula and the Deccan. The Siberian Crane (*Grus leucogeranus*), a seriously endangered species, is a winter migrant from the USSR, now regularly seen only at Keoladeo Ghana in Bharatpur. It is pure white with black flight-feathers, with a red 'face' and pinkish legs. The likewise endangered Blacknecked Crane (*Grus nigricollis*), breeding in Ladakh and Tibet, has a dark red crown, black neck and legs. The Hooded *Grus monacha* only infrequently visits the northeast. Its white head and neck are conspicuous on the grey body, the red forecrown being covered with stiff black feathers.

See plate 7 facing p. 96.

CROCODILIANS. Of the 22 crocodilian species in the world, covering three families, three occur on the Indian subcontinent. Of these, *Crocodylus palustris*, commonly known as the mugger or marsh crocodile, and *Crocodylus porosus*, the saltwater or estuarine crocodile, belong to the family Crocodylidae. *Gavialis gangeticus*, the gharial or gavial, is the only surviving member of the family Gavialidae.

Crocodiles (i.e. crocodilians) have been on the earth for over a hundred million years, evolving from 60ft giants like the *Gavialsuchus* of the Mesozoic period, a long-snouted forerunner of the present-day gharial. Like all reptiles they are 'cold-blooded' and derive energy by basking in the sun. The basking time required to attain an optimum temperature (one suitable for feeding, swimming and breeding) varies with the size of the animal. Open-mouth basking perhaps aids temperature control. Additionally, gaping dries out parasites such as leeches which are often present in the mouth.

Crocodiles move to catch prey, change position, chase a rival, or find a mate, in one of the following ways. They swim with lateral strokes of the powerful tail, with the fore and hind feet held close to the trunk. The 'high walk' is used to cover short distances on land: the trunk is raised clear off the ground and the crocodile proceeds with a slow, waddling gait. The more usual land movement is a belly shuffle, with the legs used as propellers. These are the main modes of locomotion; but the Nile crocodile has been seen galloping like a horse. The New Guinea crocodile and Johnson's crocodile regularly break into a gallop when surprised away from water, their natural retreat.

High walk of mugger × $^1/_{25}$

From the egg to about the one-year stage, crocodiles are eaten by a variety of predators which include birds, bullfrogs, lizards, snakes, jackals and wild pig. Young crocodiles feed mainly on insects, frogs, fish and snakes, and later eat little except fish and mammals. Food caught on land is usually taken to the water, and swallowed whole; when too large, pieces are broken off by twisting and shaking. The theory that kills are buried underground until putrid has no proof. Cooler temperatures slow down the digestive process and crocodiles take less food in cool weather. In India crocodiles are unpopular with commercial fisheries, as they are alleged to consume large quantities of fish; but they also aid fisheries, since they feed on slow-swimming diseased fish, and on predators such as catfish which eat the fry of commercially valuable fish. Crocodiles also eat other predators such as cormorants, otters, turtles and crabs.

Crocodiles have excellent senses of smell, hearing and vision. They can detect prey by tactile and possibly chemical receptors in the form of skin pores located in the scales of the jaws. Crocodiles live for over 100 years. They are vertebrates with red blood and a four-chambered heart. They are more active by night, spending most of the day basking and sleeping in the sun or submerged in the water. They can stay under water for hours without breathing, but normally come up for a breath several times an hour. Their rough scales are a protective armour against predators and their own kind. In the wild the sex ratio is generally equal. Crocodiles don't need to eat every day, and in captivity it is demonstrated that even fairly large animals (over 3 m) need only five to ten kg of food per month. Crocodiles actively hunt for their prey, usually at night. They usually stalk and make a fast grab with their jaws or leap at their prey but have been observed trapping fish with body and tail at the edges of ponds or rivers.

The breeding season of every crocodile species differs, to coincide with local factors such as suitable water levels, temperatures and food availability for hatchlings. Mating takes place in water and involves a complicated ritual of bubble-blowing, circling, and head movements. The female excavates a hole or constructs a mound nest (depending on species) roughly $1\frac{1}{2}$ months after mating and guards the eggs, perhaps even dampening the nest by releasing water from the cloaca. The eggs, which are white, brittle and nearly twice the size of duck eggs, hatch in 60 to 90 days and the young stay grouped with one or both parents for up to several weeks before dispersing. They grow rapidly in the first two years, sometimes reaching two metres. Subsequently growth slows down.

Crocodile skin is much sought after in Europe, Japan and America for making leather goods such as bags, belts and shoes. The result is that throughout the world crocodiles have been hunted ruthlessly during the last fifty years or so and now their very existence is endangered. International trade in skins and live specimens of all three Indian species is banned under the Convention for International Trade in Endangered Species and the killing of crocodiles in India is banned under the Wildlife (Protection) Act of 1972.

The broad-snouted, stocky **mugger**, which today rarely exceeds 3·5 m in length, is distinct from the saltwater crocodile in having four prominent post-occipital scales on the neck. These are absent in the saltwater crocodile. Mugger were formerly found from

Mugger at Madras Crocodile Bank × ¹/₂₅

Mugger courtship × ¹/₁₀

Pakistan to Assam, and over most of peninsular India and Sri Lanka. Today the mugger is rare in Pakistan, Nepal and Bangladesh and probably extinct in Burma. India and Sri lanka hold the last viable populations. Mugger have been comparatively successful in recent years because they can live wherever there is water and food: they are found in streams, ponds, lakes, reservoirs, rivers and marshes. In Sri Lanka mugger even inhabit tidal lagoons.

With the occasional exception of the nesting female or rare man-eater, mugger do not attack man. They live in groups. Undisturbed populations can consist of 20–30 animals with a strict 'pecking order'. Territorial fighting between males, specially during the breeding season, sometimes results in serious or fatal injury. A distinct social hierarchy is followed, with large males dominating sub-adult males and females. Subordination is signalled by raising the head, exposing the throat. They travel great distances in search of water during drought, and dig deep tunnels in the river banks to cool themselves.

In the breeding season (December in S. India) the female digs a 50 cm hole and lays her 15 to 50 eggs, usually at night. When, about 60 days later, the young are ready to hatch, the *umph, umph* sounds from the egg are a cue for the mother crocodile to dig up the nest. She will gently carry them to the water in her jaws. Hatchlings are 26 cm long at birth.

The **saltwater crocodile** is sleeker than the mugger with a longer and narrower head. The bright yellow-and-black colouring of the juvenile may fade in adults but is much brighter than that of mugger. 'Salties' grow bigger than any other species—specimens over 8 m have been shot—and large individuals have been known to take men as prey. A widely distributed animal, *C. porosus* has been reported from India and Sri Lanka through southeast Asia to Australia. It is more solitary in nature than is the mugger, and also strongly territorial.

Estuarine creeks and mangrove are typical saltwater crocodile habitats. In India they were found in suitable areas along the east and west coasts, but hunting and human pressure have confined the present small populations to the Sunderbans in West Bengal, the Bhitar Kanika islands in Orissa and the Andaman and Nicobar Islands in the Bay of Bengal.

Like the American alligator (*Alligator mississippiensis*) and the Malayan gharial (*Tomistoma schlegeli*), salt-

Mugger hatchling × ²/₃

Mother carrying young × ¹/₁₂

Saltwater crocodile and nest × ¹/₄₀

water crocodiles construct mound nests, gathering and piling vegetation and earth into a dome-like structure. A nest in North Andaman was constructed of bamboo leaves and grass, 1 m high, 2 m wide at the base and contained 51 eggs in a chamber at the top. It was located about 10 m from a small stream in the grassy fringe just inland from the mangroves. In the Andamans egg-laying begins in early April, in Orissa in May. After the 20–90 eggs are laid the female rests on top of or close to the mound in a wallow, dug during the nest-making period, for the 75–80 days until hatching. She digs up the young when they are ready to hatch and carries them to the water in her jaws. Hatchlings stay with the mother for a few weeks before dispersing.

Feeding habits are similar to those of mugger. Where abundant, fish, frogs, crabs, water monitor lizards, deer and pig are regular items of prey. Due to its conspicuous size, reputation as a man-eater and dwindling habitat, *C. porosus* is in danger of extinction throughout its range. The International Union for the Conservation of Nature has recommended that this species be totally protected in every country it lives in, with the exception of Papua New Guinea where it is hoped to save the animal from extinction through controlled exploitation.

The long, slender jaws, bulging eyes and sleek olive-

Saltwater crocodile eating bandicoot × ¹/₆

green trunk distinguish the riverine **gharial** from the other two Indian species. Specimens over 7 m long existed but no very large specimens remain due to heavy hunting. In fact the gharial is one of the most seriously endangered Indian animals, with an estimated total adult wild population of about a hundred to two hundred individuals.

Formerly abundant in the major river systems of Nepal, India and Pakistan (its past and present existence in Burma is not confirmed) the present remnant groups are mostly confined to a few isolated areas on the Ganga, Mahanadi and Brahmaputra rivers. The southernmost point of the range is Orissa and the largest single population (34 animals) is in Katerniaghat, U.P., on the Girwa river. One reason for the failure of the gharial to survive is its specialized habitat requirements. Gharials inhabit deep pools (where fish are plentiful) in big rivers, using sandbanks for basking and nesting.

Gharials are almost exclusively fish-eaters and the long, narrow jaws are efficient for catching and swallowing fish. The sword-like jaws are jerked sideways to snap up a passing fish, which is then swallowed by several backward jerks of the head. Their fish-eating habits and timid nature render them harmless to man though it is said that breeding females will charge and bite intruders near the nest.

Almost nothing is known of the natural history of the gharial. They are social and live in groups usually made up of a single dominant male, several females and several sub-adults. As with other crocodilians, the young live apart after a one- or two-month nursery period.

40–80 eggs are usually laid, in nest-holes on river banks. Gharials may fail to breed where there is excessive disturbance. On the Karnali river in Nepal seven adult females failed to nest in 1976, possibly due to disturbance by men investigating the site for a dam. Normally, egg-laying begins in late March and most nests hatch in June.

The Government of India has, with aid from FAO/UNDP, initiated a long-term programme to save the three Indian crocodilians from the imminent threat of extinction. Artificial incubation of wild eggs (which raises normally low survival chances) and captive rearing of juveniles has been the basis of this programme. Eight states are involved and sanctuaries and national parks have been constituted, the most extensive being the 2500 km² Chambal River National Gharial Sanctuary which involves U.P., M.P. and Rajasthan. Hundreds of captive-reared gharial, mugger and salt-water crocodiles have been released.

The Madras Crocodile Bank is a private trust run by the Madras Snake Park and World Wildlife Fund—India. The aim of the Trust is to breed crocodiles for

Male (with knob) and female gharials × ¹/₁₆

release in wild, protected areas. So far it has bred over 1300 mugger and supplied 500 for release. It is also rearing gharial and breeding the saltwater crocodile.

See also REPTILES.

See plate 3 facing p. 48 and plate 16 facing p. 193.

Z.W.

CROW. A member of the passerine family Corvidae which includes in addition choughs, magpies, jays and nutcrackers, many of which are superficially very different in size, shape and coloration. But they all possess certain anatomical and behavioural characteristics in common which determine close relationship. The typical crows of the subcontinent all belong to the genus *Corvus*. They are predominantly glossy black in colour, ranging in size from a kite's (Raven) to a pigeon's (House Crow). The ubiquitous House Crow (*Corvus splendens*) may be taken as the archetype of the genus, and the other related genera though less closely associated with man, share many of its typical habits and behaviour. The House Crow is a highly intelligent bird inasmuch as it has managed to adapt itself successfully to a life of unsolicited commensalism with man and to survive under conditions that call for great sagacity and cunning. It is impudent, inquisitive and cheeky, yet exceedingly wary and alert at all times, and with an uncanny 'hunch' for distinguishing a harmless man from one not to be trusted. Its omnivorous diet and ability to thrive on a city's garbage makes it a useful scavenger and thus a welcome ally to inept municipal administrations. But it is a bully and a menace to defenceless and aesthetically desirable garden and countryside birds and a ruthless plunderer of their eggs and young. In suburban areas and around outlying villages this knavery is taken over by the larger all-black Jungle Crow (*C. macrorhynchos*), often wrongly referred to as a 'Raven'. The true Raven is the largest and most powerful of the crow tribe. It is a dweller of the High Himalaya, but descends regularly into the northern plains in winter.

CRUSTACEANS are lower animals (i.e. non-vertebrates) such as shrimps, prawns, lobsters, crabs and barnacles. They form one of the principal classes of the great phylum Arthropoda (animals with jointed limbs). Crustaceans are mostly aquatic in habitat whereas the other arthropods mostly live on land. Crustaceans are so called because their body is protected by a chitinous or subcalcareous 'crust' or cover which is cast off periodically. Many of the small crustaceans such as *Daphnia*, *Cyclops* and even small prawns serve as important food of fishes and other larger aquatic animals.

In a typical crustacean such as a crayfish or Indian Spiny Lobster the body is elongate and is formed of 19 segments (somites) plus an anteriormost portion and a terminal *telson*. The body may be divided into three regions—a head, a thorax and an abdomen. The head consists of the presegmented region and five segments, and the thorax eight segments. The segments of both the head and thorax are fused to form a rigid 'cephalothorax'. Behind the cephalothorax is the flexible abdomen composed of six movable segments and the hindermost telson.

Appendages

Each segment bears a pair of appendages. During embryological stages each appendage is typically *biramous*, i.e. with a basal part and two branches arising therefrom. In the adult, however, the appendages get greatly modified to perform different functions.

The thirteen segments of the cephalothorax bear the following appendages.

First – a pair of many-jointed, filamentous antennules for feeling and keeping the equilibrium of the body.

Second – a pair of long many-jointed antennae for feeling.

Third – a pair of jaws or mandibles for crushing the food.

Fourth & Fifth – each a pair of maxillae for creating currents of water in the gill-chamber.

Sixth, Seventh & Eighth – each a pair of foot-jaws or maxillipeds for feeling and holding the food.

Ninth to Thirteenth – each a pair of walking legs for walking and feeling. In the spiny lobster all these legs are more or less of equal length and size and end in simple claws. In true lobsters and crayfishes, the first three pairs are chelate and especially the first pair is enormously enlarged and has powerful pincers. Each of the abdominal segments (14th to 19th) carries a pair of appendages modified to function as swimming organs or swimmerets.

The pair of appendages on the 14th segment are reduced in the female but in the male they function as organs for transferring sperms. The appendages on the 15th in the male are used for transferring sperms and in the female for attachment of the eggs. The appendages on the 16th, 17th and 18th segments are in

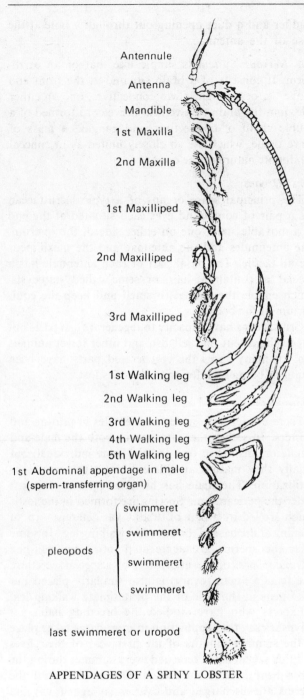

Antennule

Antenna

Mandible

1st Maxilla

2nd Maxilla

1st Maxilliped

2nd Maxilliped

3rd Maxilliped

1st Walking leg

2nd Walking leg

3rd Walking leg
4th Walking leg
5th Walking leg

1st Abdominal appendage in male
(sperm-transferring organ)

pleopods {
swimmeret

swimmeret

swimmeret

swimmeret
}

last swimmeret or uropod

APPENDAGES OF A SPINY LOBSTER

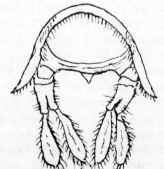

An abdominal segment with a pair of swimmerets.

the females used both for swimming and for the attachment of eggs. The last pair of swimmerets (uropod) are wholly for swimming.

External Skeleton

The entire body of a typical crustacean is protected by a chitinous coating impregnated with lime. Strictly speaking every body segment or *somite* is covered by a separate chitinous ring or *sclerite*. These rings, however, get fused to form a shield or *carapace* over the rigid cephalothorax. The anterior part of the carapace is produced forwards to form a pointed beak or *rostrum* bearing a row of spines. Over the carapace a groove can be seen which is the mark of separation between the head and the thorax.

Over the abdominal segments which require movement, the chitinous rings remain separate but they are united to one another by soft chitinous membrane. By this arrangement the animal is enabled to bend and straighten its abdomen in a vertical plane.

The circular rings cover not only the dorsal side but also extend laterally and ventrally and over the appendages.

The number and structure of the body segments and appendages and the extent of the exoskeleton show great variation in the different orders.

Members of the smaller crustacean groups have as many as 41 to 63 thoracic segments and appendages. Parasitic forms have neither segmentation nor appendages in the adult. Between these two extremes all gradations are common. Sedentary forms like barnacles have reduced appendages and a hard calcareous exoskeleton resembling that of a shell-fish. In all the highly organized crustaceans like lobster and crabs the carapace completely covers the thorax and they have their characteristic well-developed walking legs, the first pair of which have pinching claws (except in spiny lobsters).

Physiological System

Crustaceans possess all the important physiological systems of higher animals, viz. digestive, circulatory, respiratory, excretory, nervous, reproductive and sensory.

The digestive system consists of a mouth opening at the front end on the ventral surface between the jaws. A short gullet connects the mouth with the stomach which is sac-like but with a constriction which separates it into a larger anterior chamber and a smaller posterior chamber. The interior of the stomach is lined by a series of chitinous ossicles which grind the food and hence are collectively called a gastric mill. Opening into the posterior chamber on either side is the duct of the digestive glands or liver. From the hinder end of the posterior chamber starts the intestine which is a small tube passing dorsally and backwards, finally opening to the outside through the anus at the end of the abdomen.

The principal food of crustaceans includes small animals like snails, insect larvae, eggs and tadpoles, plankton organisms and decaying organic matter. They even prey upon each other. A few of them are wholly plant-eaters. It is during the early morning and at dusk that these animals are most active at catching and eating their prey.

The maxillae and the maxillipeds hold the prey which is crushed by the jaws or mandibles. The finer crushed pieces pass through the gullet to the stomach whereas the coarser particles are ejected through the mouth. In the stomach the gastric mill further pulverizes the food particles which get mixed with the digestive fluid or a secretion from the liver. The food now passes on to the intestine, the walls of which absorb the nutritive part. The undigested matter escapes through the anus. The digestive organs including the intestine are enclosed in a cavity which contains blood and communicates with blood vessels.

Respiration

Respiration is performed chiefly by a series of gills contained in a special chamber on either side of the thorax. The gills are protected by the wall of the thorax on the inner side and the gill cover externally. Each gill is plume-like, having a tubular stem bearing numerous hollow filaments. The filaments are in contact with the blood vessels contained in the stem. From the water current circulating in the gill chamber oxygen is taken to purify the blood which passes into the cavity covering the heart—the *pericardial sinus*.

Circulation

The circulatory system consists of a heart and several principal arteries and a number of spaces called sinuses into which the blood passes from the arteries. The heart is sac-like and muscular and lies in the space in the upper part of the thorax—the pericardial sinus from which the aerated blood enters into the heart through special valvular pores. The heart distributes the blood throughout the body by various arteries which end in capillaries. The capillaries open into a series of venous sinuses from which the blood enters the ventral sinus. From here the impure blood gets purified by the gills and returns to the pericardial sinus completing the circulation. The blood is colourless and contains slightly bluish coloured *haemocynin* in place of the *haemoglobin* of higher vertebrates.

Excretion

The excretory organs are a pair of glands placed laterally on either side of the gullet. Each gland has a bladder and a duct opening out through a hole at the base of the antennae.

The Nervous System is simple like that of an earthworm. It consists of a brain situated at the front end above the oesophagus. Two connectives, one on either side, join the brain to a ventral nerve cord formed of a double chain of connectives and ganglia, a mass of nerve tissue, which are so closely united as to conceal the double nature of the cord.

Sense Organs

The principal sense organs of a typical crustacean are a pair of compound eyes each situated at the end of a movable stalk, one on either side of the rostrum. The antennules and the antennae and the maxillipeds are all tactile. The basal part of each antennule has a special sac containing hairs or 'setae' called 'statocysts' which enable the animal to smell and keep the equilibrium of the body.

Crustaceans have capacity to regenerate lost parts but not to the extent that jellyfish and other lower animals can. In many cases the regenerated parts have been found to be much inferior to the ones lost.

Reproduction

There are separate males and females in prawns and lobsters. In some other crustaceans both the male and female organs are in one and the same individual, but usually they mature at different times to avoid self-fertilization. The male has his sex organ (testis) just under the pericardium. Sperms are formed in the testis which are led by a pair of tubes—vas deferens—to an opening at the base of the last (5th) walking leg. In some cases the sperms are aggregated into vermicelli-like *spermatophores* or sperm capsules by a special secretion. The female organ (ovary) is also similarly placed but this opens at the base of the penultimate walking leg.

Experts who have studied the breeding habits of crayfishes say that copulation in most cases takes place in the summer months of the first year of their lives and this sexual act is repeated every summer during the subsequent three or more years. During copulation the sperms are discharged and ovaries lay eggs (ova) and the eggs get fertilized. The fertilized eggs are bound by a sticky glandular secretion and the whole glutinous mass is protected by the mother underneath her belly. It is recorded that 100 to 600 greenish eggs are laid by a female at a time. The eggs develop only slowly. In temperate waters it takes several months for the eggs to hatch. In the case of marine prawns, the eggs are laid directly into the water. The life history of the several orders of crustaceans varies widely. In most cases several larval types and stages make their appearance before they become adults. The embryo generally passes through a free-swimming 'nauplius' stage marked by

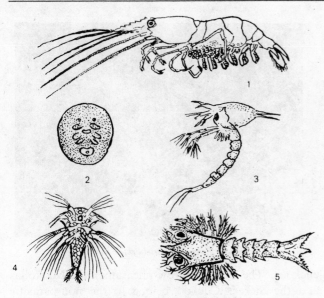

1 *Palaemon*, a common fresh-water prawn × ²/₃ showing eggs under her belly

2 Nauplius stage of higher crustaceans passed within the egg membrane, × 10

3 Zoea or Metanauplius stage of higher crustaceans; free-swimming, × 10

4 Free-swimming nauplius of lower crustaceans, × 10

5 First larval stage of a lobster × 6

the possession of three pairs of appendages which later become the antennules, antennae and mandibles. In the higher crustaceans the nauplius stage is rare in the sense that it is passed through within the egg membrane or in the brood pouch—the young one hatching as a metanauplius or 'zoea'. According to observations made by experts a young crustacean in temperate waters sheds its skin four times in a year in summer for two or three years before it attains full size; in tropical seas these stages are gone through in much shorter time. The life of an individual has been calculated to extend for three or four years and even five years in the case of lobsters.

Crustaceans form an immense class including some 25,000 species grouped under 5 sub-classes and 20 orders. They are very widely distributed—mostly in the sea at varying depths down to 3000 fathoms as abyssal, pelagic and littoral forms, in inland salt waters, in fresh water and on moist land. They are active swimmers, crawlers and walkers with requisite appendages suitably modified.

There are many instances of protective and aggressive adaptations among crustaceans which enable them to conceal themselves from enemies or to attack them under disguise. Certain crabs plant seaweeds, sponges etc. over their carapace for concealment. There are also instances of commensalism—living together and sharing the same food. Males of some develop special colour or extraordinary girth in their appendages for attracting the opposite sex. Some can produce sound with the

movable fingers or their chelae. Some are sedentary like barnacles with a protective shelly coating like a mollusc. Many lead a parasitic life and they have reduced limbs. *Sacculina* is a strange parasitic crustacean, sac-like in appearance, about 2 cm across, with no trace of limbs, mouth or food canal. It lives attached under the tail of a crab by means of a system of branching, thread-like roots which penetrate into the legs of the host. Through these roots the vital fluid of the host is sucked by the parasite.

In size crustaceans vary from microscopic forms to those measuring with their legs extended 1·5 m long. There are many curious forms among them. *Apus* is a primitive genus found in freshwater ponds and pools. It measures 5 to 10 cm long. The fore part of its body is covered by a broad shield-like carapace behind which is a tail of many segments ending in a pair of long filaments. It has more than 60 pairs of leaf-like limbs!

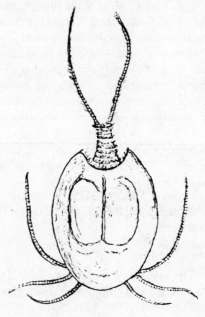

Apus × 1

Brine shrimps (*Artemia*) and **Fairy shrimps (*Branchipus*) are not really shrimps but curious** lower crustaceans resembling tiny shrimps: the former live in salt-water lakes and the latter in fresh water. They have no carapace and their appendages are much flattened. They row themselves about on their backs by means of their flat limbs. Brine shrimps can tolerate very high salinity and can breed parthenogenetically (without fertilization by a male). Their eggs are capable of remaining viable for years if they are kept in dry condition. The eggs hatch out in fresh or saltish water in 24 hours and the larvae are used as food for other animals in laboratory cultures or fish-breeding.

The Opossum shrimp (*Mysis*) looks like a small transparent shrimp less than 6 mm long. It lives in

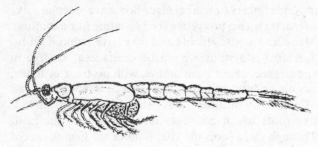

Mysis × 1

coastal waters. The carapace covers the major part of the thorax. Thoracic appendages, although leg-like, are all biramous. It is called the opossum shrimp because it carries its eggs in its brood pouch till the young hatch out, as mother opossums carry their young in a pouch.

Lobsters and crayfishes are solitary; others like shrimps and most of the lower crustaceans are gregarious, occurring in shoals. Most of the smaller crustaceans —copepods, amphipods etc., form the chief food of larger marine animals. The larger ones—shrimps, prawns, lobsters, crabs etc.—are the food of man and are therefore of great economic significance.

Since all crustaceans feed on decaying matter along with other food they are scavengers of the sea.

A few of the more common and interesting examples are:

Lower crustaceans WATER-FLEAS, COPEPODS, WOOD-LICE, BARNACLES, and SAND-HOPPERS.

Higher crustaceans DECAPODS including prawns, shrimps, lobsters and CRABS, and STOMATOPODS —mantis shrimps.

CUCKOO. The cuckoos, of which we have 21 species in 10 genera, are divided into two distinct groups: the parasitic (see BROOD PARASITISM) and the non-parasitic. Well-known among the former are the Koel, the hawk-cuckoos, the Indian, and the Pied Crested

Male and female Koel, *Eudynamis scolopacea* × ¼

Female Koel at palm-berries × ⅟₇

cuckoos. The Coucals (crow-pheasants), the Malkohas and the Sirkeer Cuckoo belong to the non-parasitic group.

All members of the family have zygodactyle feet with the fourth (outer) toe reversible. Our cuckoos range in size from the sparrow-sized Violet Cuckoo (16 cm) to the Large Greenbilled Malkoha (60 cm). Some (the Clamator cuckoos) have conspicuous pointed crests; a few, like the Emerald and the Violet cuckoos, have resplendent plumage; but most species are rather drab and better identified by their loud and monotonous calls than by their appearance.

The parasitic cuckoos are all arboreal. Their diet consists of insects, predominantly caterpillars. Many have a preference for hairy ones. The Koel, however, is primarily a fruit-eater although it will also eat insects and caterpillars. Most of them eat the eggs of other birds as well.

Compared with the fast-flying, arboreal true cuckoos, the non-parasitic members of the family are rather clumsy and sluggish. The familiar crow-pheasant feeds mostly on the ground, taking lizards, frogs and insects. All the non-parasitic cuckoos build their own nests and are devoted parents. One of them, the Redfaced Malkoha, has become extremely rare and may be on the verge of extinction.

K.K.N.

CULTURE. Human culture involves traditions handed down from generation to generation by imitation of behaviour and through spoken and written language. The culture of animals is similarly that part of their behavioural repertoire that is not coded genetically but is transmitted through a process of learning. The cultural component is therefore expected to be negligible in most of the lower animals whose behaviour is largely instinctive, that is programmed genetically. Much of the behaviour in the more highly evolved birds and mammals is however learned and therefore subject to cultural transmission and change.

A notable example of a culturally transmitted component of behaviour is song in certain bird species. In some birds song is fully hereditary and expressed in the full adult form without the bird ever hearing it. In other birds, however, it is expressed only in a very rudimentary, primitive form unless the bird has been exposed to it. In such species of birds the song develops regional variants, or 'dialects', which are culturally transmitted in a given region. The Hill Myna has a fantastic repertoire of calls which is developed by imitation of calls of neighbouring hill mynas of the same sex. The racket-tailed drongo, which imitates the calls not just of its own species, but also those of other species, must possess a most interesting vocal culture which remains to be explored.

In many bird species the food habits are rather flexible and involve a great deal of learning and hence cultural variation. The herring gulls on the northeastern coast of North America originally fed mostly on animals in the intertidal zone and had a restricted population. Early in the century however a population of gulls learnt to feed on city garbage, which opened up a tremendous new food resource to this species. Gradually this habit spread through imitation and permitted an explosion of the population of this species.

In a similar fashion, tits in Europe at one point discovered that milk bottles left on doorsteps represented a good source of food if only they puncture the foil cap and eat the cream collected near the top. The rapid spread of this behaviour through the European populations gave clear evidence of cultural transmission.

But the best evidence for cultural transmission comes from a coastal population of Japanese macaque monkeys. This population was artificially provided with wheat and potatoes by the scientists studying it. To begin with the monkeys used perforce to eat the sand with the wheat grains strewed on it and the mud on the skin of the potatoes. But it turned out that amongst these monkeys was an inventive genius—a female named Imo. Imo discovered at two different times several months apart that sand could be removed from wheat by floating it on sea-water and mud from potatoes by washing them in sea-water. Other monkeys picked up these habits by imitating Imo. Notably enough, monkeys of different age and sex showed different rates of learning. The young were quickest to learn, and old males were most resistant to learning.

CUSCUS. Cuscus or khas (*Vetiveria zizanioides*) occurs usually in moist places such as the beds of rivers in the plains and lower hills and is well known for the

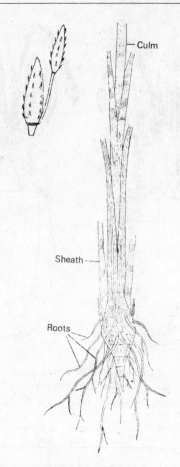

Basal portion of cuscus plant (× 1) and spikelet showing strong curved spines (× 10)

oil of vetiver extracted from its spongy roots. The roots are largely used for making screens (cuscus tatties) which, when soaked with water, cool the rooms and provide a pleasant aroma. Such screens are also used in electrical room-coolers. Oil of vetiver is useful in medicine and perfumery; it is a favourite perfume in many scents, soaps and syrups. The oils obtained from wild plants and cultivated plants differ in aroma and in physical and chemical properties. Khas grass is also cultivated in certain parts of peninsular India; rich well-drained sandy loam soil is said to be more suitable.

The grass is easily recognized by its large flowering branches and spikelets covered with strong curved spines.

CUTTLEFISHES are two-gilled cephalopods living between stones and in rock fissures in the sea usually at a depth of a few fathoms. They often come into shallow waters also. They belong to the genus *Sepia*. The animal has a distinct head, bearing eight stumpy sucker-clad arms, two long arms with suckers at their ends, and a pair of prominent and highly developed

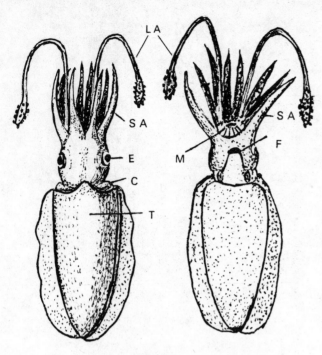

DORSAL AND VENTRAL VIEW OF
A CUTTLEFISH, $\times \frac{1}{2}$

LA long arm. SA stumpy arms. E eye. M mouth. F funnel.
T shield-like trunk. C cavity into which head and arms can
be withdrawn. The anal opening is hidden under the skin
below the funnel opening.

eyes, resembling in size and appearance those of a fish.
The free end of the head bears the mouth. The head
is connected with the body by a neck. The trunk or
body is shaped like a small elongated shield, the base
of the shield being towards the head. The body of the
animal is supported internally by a long calcareous
piece which is laminated, porous and brittle and popularly known as cuttlebone. The position of this 'bone'
is just beneath the dorsal side of the body which is
darker in colour and harder to touch than the ventral
side. The thick skin of the mantle covers the entire
body. The mantle fold assumes the form of a concave
ridge or pouch round the base of the neck into which
the head and the arms can be withdrawn. Fringing
each side of the body is a thin, frill-like, muscular fold—
the fin—which is used as a swimming organ. The funnel,
which is the outlet of the mantle-cavity, is situated at
the middle of the ventral side near the anterior end.
By ejecting water through this funnel the animal darts
rapidly through the water.

All cuttlefishes have an ink-sac placed near and
connected to the anal opening. They protect themselves
by emitting a dark fluid from the sac and hide from
their enemies in this smoke-screen. The name *Sepia*
given to these molluscs refers to the brown pigment
called sepia made from the dried and pulverized contents of the ink-sac of cuttlefishes and their allies.

The living cuttlefish exhibits frequent changes of colours.
These changes are due to the contraction and expansion
of certain pigment cells contained under the skin of
the animal. In some cases 'iridocytes', i.e. cells which
display rainbow-like colours through the skin, are
also present.

In cuttlefishes the sexes are separate. Mating takes
place during the breeding season. Fertilization is internal. The female has certain special glands that secrete
a viscid material by means of which the eggs when
deposited adhere together in masses.

The eggs of cuttlefishes are curious objects. They
contain yolk. A capsule which is large, black and pear-
shaped holds many cuttlefish eggs. Several such capsules
are laid. Each one is tied to some place of attachment
by a kind of ribbon at the upper end of the capsule,
the whole forming a large group like a bunch of grapes.
The bunch is usually attached to some foreign body
such as the stem of a seaweed. The eggs develop within
the capsule nourished by the yolk. As the embryo
increases in size, the yolk is gradually absorbed and
when the embryo leaves the egg it has all parts developed and looks like a miniature adult.

The cuttlefish eggs have no free-swimming larval
stage and in this respect all cephalopods differ from
other molluscs.

The soft calcareous cuttlebones have commercial
value. They are gathered in the months of February
and March, i.e. the pre-southwest monsoon months,
when large numbers of them are cast ashore by waves.
These soft bones are used as fine sandpaper for rubbing
down paint, in cleaning glass and polishing bronze
and silver.

See also MOLLUSCS.

CYCADS are trees or shrubs, distinct from other gymnosperms by their palm-like habit, with large feathery
leaves forming a crown above an unbranched trunk.
Male and female plants are separate. The male flower
is an erect cone borne at the apex of the stem. It has an
axis bearing spirally arranged scales (stamens), on the
undersurface of which are numerous pollen-sacs. The
female flower consists of hairy sporophylls or carpels
bearing the ovules. The wind-borne pollen alights on
the ovule where it is caught in a secretion and germinates
to form two sperms that swim to the egg, propelled
by cilia. On fertilization the ovule develops into drupe-
like coloured seeds which are the size of duck's eggs.
Cycads are among the most primitive of seed plants
and have remained virtually unchanged for some 200
million years. It was once believed that cycads were
the link connecting ferns with seed plants. Current
opinion based on fossil evidence is that they are true
gymnosperms. The only genus native to our area is
Cycas with five species of which three are interesting:

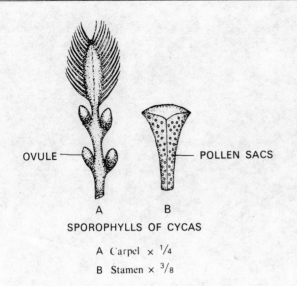

OVULE — POLLEN SACS

A   B

SPOROPHYLLS OF CYCAS

A Carpel × ¼

B Stamen × ³⁄₈

C. circinnalis is a palm-like tree of S India and Sri Lanka. *C. pectinata* is found in the E Himalaya, Bangladesh and Burma; its orange fruits and tender shoots are relished by tribals. *C. rumphii* is a coastal tree of the Andamans, Nicobars and Burma. The fruit is poisonous when raw and is edible after cooking.

See also CONIFERS.

CYCLONES. The term 'cyclone', derived from a Greek word meaning the coil of a snake, was first used by Henry Piddington about the middle of the last century for describing tropical revolving storms encountered in the Bay of Bengal and the Arabian Sea. The word was chosen as appropriate to denote the weather systems in which the winds blow anticlockwise, along an inward-coiling spiral, spiralling about a centre. The heavily raining cloud-bands associated with such a system are also in the form of anticlockwise inward-coiling spirals. (The movements in the southern hemisphere are clockwise round the centre.) The diameter of the coil can be anything from a hundred kilometres to six or seven hundred kilometres. The angle that the spiral makes with circles concentric with the centre of the cyclone is about 20°. However, towards the centre this angle gradually decreases and the motion becomes circular at a distance of 20 to 50 kilometres from the centre of a severe cyclone, depending on the severity and size of the cyclone. This picture of a cyclone with its heavily raining cloud-bands was originally inferred from scattered observations from ships within the field of several cyclones. It has now been verified from pictures of cyclonic clouds taken by weather satellites.

Fig. 1 is the plan of such a cyclone as seen from above, and Fig. 2 is a vertical section across the centre of the cyclone. Fig. 3 is a satellite photograph of the same.

Such tropical revolving storms are known as Hurricanes in the Atlantic and eastern Pacific, as Typhoons in the Western Pacific and as Willy-willies in Australian waters.

Generation and Movement of Cyclones. Tropical cyclones do not form within about five hundred kilometres on either side of the geographical equator. They move from east to west with a significant poleward movement. They form over oceanic areas where the surface temperature of the sea is 27°C or more. Such a zone is called the thermal equator and it shifts north and south with the seasons. The cyclones weaken as they move over land areas where there is no energy-feed from a water surface, and where increased friction contributes to the weakening of the winds. Usually the winds aloft at 8–12 kilometres altitude determine the direction of movement of the cyclone. During the winter months in the Northern hemisphere, the tropical cyclones recurve towards the north or northeast on arriving at latitudes varying from about 12° to 20°N.

During the formative phase of a cyclone there is only an area of weak winds and heavy showers and squally weather. A large part of the atmosphere, about a thousand kilometres across, and about 6 kilometres high, gets warmed up and the atmospheric pressure over this area falls by about half-a-percent of its normal value. This is sufficient for the winds to develop a feeble circulatory motion in an anticlockwise sense, if the area is 500 kilometres or more north of the equator. This circular component of the wind motion is a consequence of the rotation of the earth's surface around the local vertical. This rotation is zero at the equator and gradually increases towards the poles. Within about 500 km of the equator the rotation is too feeble to generate a circular component in the wind motion in a low-pressure area. Once the circular motion starts, there is an indraft of moist air into the area; this leads to an updraft within the area, an increase in the rainfall activity within the area, an increase in the warming up of the atmosphere, an increase in the pressure fall, an increase in the circular motion, an increase in the indraft, and so on. The system goes on intensifying. Not always. Only on 20 percent of the occasions do the systems intensify, in the rest of the cases the indraft serves to fill up the low-pressure area. The reasons for the continuous intensification process are not known precisely. There are several hypotheses. One is that there should be a mechanism in the upper air at the 6–10 kilometres levels, over the low-pressure area, adequate to remove the air that comes up in the updrafts.

Main Features of a Cyclonic Storm. A tropical cyclone when fully developed is a vast violent whirl 150 to 800 km (100–500 miles) across, 6 to 10 km (4–6 miles) high, spiralling round a centre and progressing along the surface of the sea at a rate of 300 to 500 kilometres (200–300 miles) a day. The speed of the wind in a mature storm can be occasionally 160 kmph (100 mph),

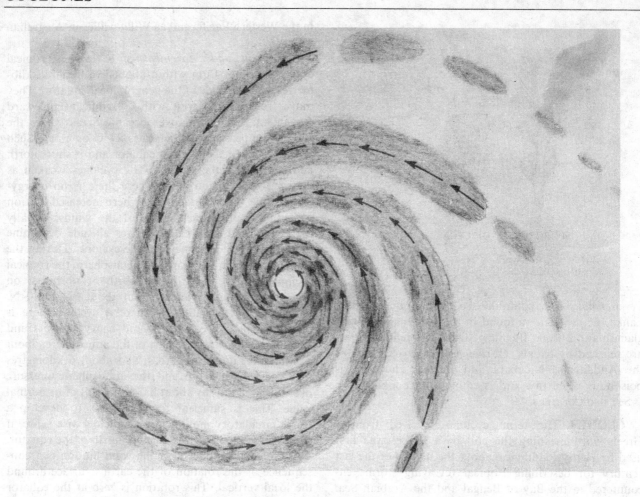

1. Schematic diagram of a cyclone showing the inward spiralling cloud-bands (shaded). The eye at the centre is cloud-free. The arrows indicate the wind direction.

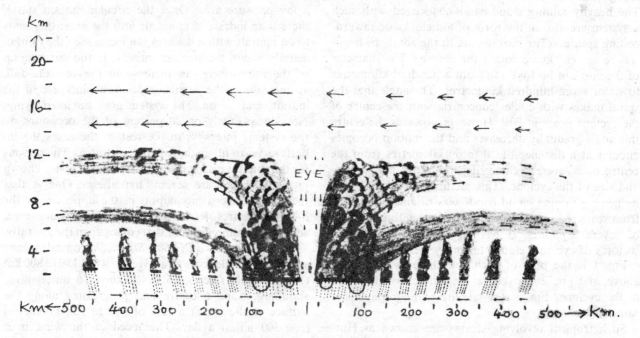

2. A VERTICAL CROSS-SECTION (SCHEMATIC) ACROSS THE CENTRE OF A CYCLONE

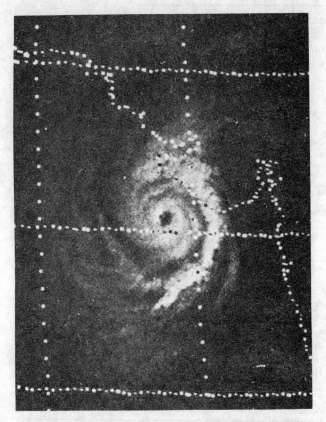

3. *Satellite photograph of a severe cyclone that affected Porbander in November 1976.*

and very rarely even more, about 240 kmph (140 mph). The winds encountered in the cyclones of the Pacific and Atlantic are usually stronger than those encountered in the cyclones of the Bay of Bengal and the Arabian Sea. There are exceptions like the one which struck the Andhra coast in November 1977.

According to an accepted international convention, a tropical revolving weather system is called a cyclonic storm when the wind speed reported or expected in it increases to 60 kmph. When the wind speed is less it is called a depression. It is called a severe cyclone when the wind speeds reported or expected increase to 85 kmph or more. In all cases there are occasional squalls, brief spells of stronger wind, when the speed may be as much as 50 percent more. Unfortunately there is no expression other than 'severe cyclone', even when the wind speed is as high as 200 or 250 kmph (120 or 140 mph) with squalls where the wind speeds can reach still higher velocities around 300 kmph.

Mature Phase of a Severe Cyclone. In the mature phase, a severe cyclone is found to consist of four parts:

(i) A calm central area varying between 10 and 30 km in diameter, where absolutely calm air or light winds with partly clouded or clear skies prevail. This is called the 'eye' of the storm.

(ii) An inner ring of very strong winds (90 kmph or more) 50 to 100 km in width, within which fierce squalls and torrential rains occur.

(iii) An outer storm area within which the winds are strong (50–60 kmph), and where rainstorms alternate with non-rainy periods. This annular area can be 100 to 300 kilometres in width.

(iv) The outermost area of weak cyclonic wind circulation. The winds are weak. There are widely separated patches of rain.

Detection of a Cyclone. When the cyclone is out in the sea, the coastal areas experience the weather prevailing in the outer periphery. This, along with data from a few ships nearer to the centre of the cyclone, enabled meteorologists to detect the existence of a cyclone. However, with the advent of weather satellites, clouds with a spiral band structure reveal the existence of a cyclone. The 'eye' of the cyclone appears as a dark spot when the pictures are developed so as to show the clouds as bright.

There are radar stations emitting radio waves of 7 or 10 centimetre wavelength, which can detect the spiral bands of a tropical cyclone within about 300 kilometres of the radar. Such radars have been installed at coastal stations at Karachi, Bombay, Goa, Madras, Vishakapatnam, Paradeep (Orissa), Calcutta, Chittagong and at Dacca.

Sequential pictures show the speed and direction of movement of a cyclone; forecasting the future movement is generally an extrapolation of the movement in the immediate past. The error in the location of the coastal point where a cyclone strikes, when forecasted 24 hours ahead, is still of the order of about 120 kilometres.

Effect of Strong Winds and Heavy Rains. The thrust per unit area exerted by a strong wind on any structure is proportional to the square of the wind speed. It is given by the equation:

$$T = 0.056 \, V^2$$

where T is the thrust in kilograms per square metre and V is the wind-speed in kilometres per hour. When a wind blows parallel to a surface like that of a roof, it exerts a lifting effect similar to the force which provides a lift to an aircraft speeding through air. It is this force, nearly equal to T in magnitude, which rips the roofs off of buildings. There are special 'building codes' which give specifications for the construction of buildings capable of withstanding the thrusts and lifts of maximum wind speeds of 80 kmph, 100 kmph, 120 kmph, and so on.

The heavy rains, occasionally torrential, which fall over the land areas in association with cyclones cause severe floods. Yet on the whole the rains by themselves are beneficial.

Storm Surge. As is well-known, a strong wind acting on a sea surface generates waves, whose periods,

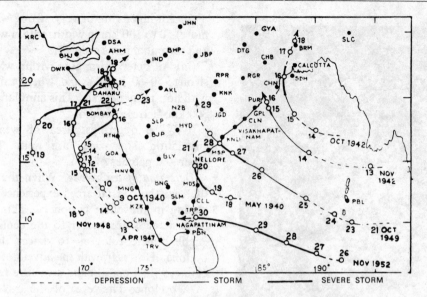

4. A few typical tracks of the cyclones in the Bay of Bengal and the Arabian Sea. During the period 1891–1970 408 cyclones formed in the Bay of Bengal of which 96 intensified into severe cyclones, and 83 formed in the Arabian Sea of which 40 became severe cyclones.

amplitudes and energy depend upon the strength and duration of the wind. Besides these undulatory waves, the wind produces a drift in the body of the water extending to a depth of one or two hundred metres, determined by the wind-speed and latitude of the place. The magnitude of the drift is proportional to the square of the speed of the wind.

When the storm is far out at sea, such a drift produces no visible effect. However, as the storm approaches land, the depth of water decreases, and the drift piles up the water on the shelving coast and the sea level rises. The gentler the slope of the sea bottom, the greater is the rise. The rise is also large when the strong wind blows parallel to the coast, that is when the track of the cyclone is perpendicular to the coast. The rise in the sea level becomes a maximum as the ring of hurricane winds hits the coast, and continues to be so till the centre of the cyclone moves inland, and winds begin to blow in the opposite direction. This duration can be about four to six hours.

This rise in sea level, and the drift of water inland perpendicular to the wind is called the 'storm surge', or occasionally the 'storm tide'. The rise in sea level caused by a cyclone can be as much as seven or eight metres, but more often it is two or three metres. If the cyclone crosses the coast around the time of the usual luni-solar high tide, the rise in the sea level is enhanced, although it is not a case of simple numerical addition.

The rise in sea level and the drift of sea water inundates a narrow belt of the sea coast on both sides of the track. The heavy damage and suffering caused by a severe cyclone is through such a storm surge. If the winds in a cyclone are measured or correctly estimated, it is possible to compute the likely rise in sea level along any given coast whose underwater topography is known.

Storm Tracks. Some of the typical tracks followed by cyclones in the Indian Seas are given in Fig. 4. Cyclonic disturbances varying in intensity and magnitude originate in the Bay of Bengal and the Arabian Sea, mainly during the period April to December.

TABLE

Statistics of cyclones and severe cyclones which occurred in the Bay of Bengal and the Arabian Sea during the 70-year period, 1891–1960.

| | Bay of Bengal | | Arabian Sea | |
	All Cyclones	Severe Cyclones	All Cyclones	Severe Cyclones
January	5	1	1	–
February	1	–	–	–
March	4	1	–	–
April	18	7	5	3
May	32	17	13	8
June	43	5	14	7
July	58	8	2	–
August	41	–	1	–
September	47	8	5	–
October	70	19	17	5
November	61	20	22	16
December	28	10	3	1
Total	408	96	83	40

The cyclones which form during the pre-monsoon months April and May, and the post-monsoon months October, November and December are generally of great intensity and often have an inner belt of hurricane winds and a calm centre. They usually form between latitudes 10° and 14°N. After moving in a westerly to northwesterly direction in the earlier stages, they sometimes turn towards the north or northeast.

The cyclonic disturbances, which form during the monsoon period June to September, are usually of small intensity. They form in the north Bay of Bengal and move in a west-northwesterly direction.

Handbooks of Precautionary Measures. The India Meteorological Department issues timely warnings against cyclones. The warnings specify the winds likely to be encountered and the possible heights of storm surges if any. However, they can be of service in minimizing loss of life and property only if these warnings are disseminated quickly to all concerned and adequate arrangements exist for taking appropriate precautionary measures immediately on receipt of the warnings.

This will require a well planned and co-ordinated programme, written down in the form of detailed *Handbooks* separately for each Coastal District. Such Handbooks should contain, besides other information, the topography of the area, indicating the areas which will be *above water*, when storm surges of 3 metres, 5 metres, and 7 metres occur. They should also contain extracts from building codes, indicating the buildings, which will *not* be blown away when winds of 80 kmph, 100 kmph, and 120 kmph occur. District Collectors and village officials should read the Handbooks and conduct rehearsals in the first weeks of May and of September every year. Through proper returns, the State secretariats should check that the rehearsals have been conducted. Once this is done the loss of life and property can be reduced.

DACHIGAM, only 21 km NE of Srinagar, is an area long preserved for the maharaja's hunting and was declared a sanctuary in 1951. Though only 140 km² in area it includes an unusually wide range of habitats with elevations from 2000 to 5000 m. A great variety of deciduous trees and shrubs including many fruit species in the lower valleys are replaced by pine and birch forests and alpine meadows higher up. Dachigam is the last refuge of the hangul (Kashmir Deer) whose very survival as a species was threatened by habitat destruction, indiscriminate grazing and mass poaching after 1940. The Dachigam population of about 3000 hangul had been reduced to less than 200 by 1970 but has since then more than doubled. The grazing of sheep and other livestock is now restricted and other provisions of the Jammu & Kashmir Wildlife (Protection) Act 1978 observed. The stags spend the summer near the snows, regrowing their antlers. Around October they rejoin the hinds and fight for the possession of a harem. Fawns are born in April, when antlers are shed. At Dachigam, besides hangul, Musk Deer, Serow, Leopard Cats, Langur, Otters, Brown Bear and Long-tailed Marmots are to be seen.

See DEER, MARMOT and Plate 15 facing p. 192.

DAMSEL or ANEMONE FISHES (family Pomacentridae) are another group of colourful fishes as the names suggest. The Damsel fish *Amphiprion* is the most exquisitely coloured fish, a bright chocolate body with dazzling orange stripes. Another peculiarity is that it takes shelter in a giant sea anemone, moving about flirtingly among the tentacles and sometimes entering and popping out of its mouth, as if to kiss. The interesting part of the behaviour is that the giant anemone, which catches other unwary fish with its tentacles, sucks their juice and ultimately kills them, is quite friendly with the *Amphiprion*, which in return cleans the tentacles of their mucus and helps himself to it, an excellent example of symbiosis. The fish is therefore also called Anemone fish.

DARTER (family Phalacrocoracidae), also known as 'Snake Bird'. Differs from the closely related Cormorant in being of solitary habit and having a long slender brown snake-like neck and straight sharp-pointed bill. Swims half-submerged, with only the sinuous neck showing. Fish are speared by underwater pursuit, brought to the surface, tossed in the air and swallowed. Usually nests in mixed heronries.

© Sálim Ali

Darter × ⅛

DECAPODS. Prawns, shrimps, lobsters and crabs are all decapods—highly organized crustaceans characterized by a well-developed carapace covering all the thoracic segments. There are 5 pairs (10) walking legs and hence the name of the order. The first pair of walking legs are usually well developed and provided with pinching claws except in the Indian Spiny Lobster. Most of them are type forms and the general description of crustaceans applies to them.

Shrimps and prawns have their body more or less compressed laterally and their legs slender, used more for swimming. The eye-stalks are long. The exoskeleton is not calcified and therefore fairly flexible. Both are gregarious and occur in large assemblages. Shrimps have

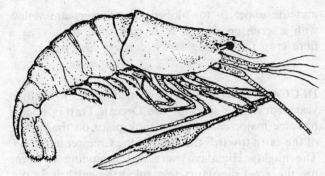

Typical Indian fresh-water prawn × ⅓

a depressed carapace and a small rostrum. They are sand-loving forms living in coastal waters. They are rapid burrowers. Their body colour agrees with that of the sand and mud in which they live. Genus *Crangon* is found in most tropical waters but in India, *Acates indicus* is most common on the west coast. Its length ranges between 6 and 15 cm. Boiled shrimps are brown in colour, a feature which distinguishes them from prawns which turn red when cooked.

Prawns occur both in salt water and in fresh water and there are several species of them. *Penaeus indicus* is a salt-water species common all round the coast and is known as White Prawn. *Macrobrachium* is a fresh-water genus. In life, prawns are white, yellow, brown or greyish in colour with scarlet markings. Some forms attain a length of 40 cm. There are two types of prawns, those which lay eggs in water and others which carry eggs attached to their abdominal appendages. Most of the sea prawns in Indian waters which belong to the genera *Penaeus, Metapenaeus, Parapeneopsis* etc. are of the first category. They spawn in the coastal waters and after the eggs hatch out and pass through their larval

stages, the tiny prawns (post-larvae) travel shorewards. They then enter estuarine backwaters and feed on the organic matter abundantly available in such environment. After about four or five months they return to the sea and thrive there to breed and repeat the cycle. The other egg-carrying group is largely freshwater, in India represented by the genus *Macrobrachium*. They breed in the estuarine waters and after passing through several larval stages, the young ones migrate into fresh water for further growth.

Prawns are always popular as food and are commercially important both within our area and abroad. As they are good foreign-exchange-earners, prawn farming is being taken up at several places on the east and west coast to augment the supplies available from the sea.

About 52 species of prawns and shrimps are found in India out of which a dozen species have been reported from the coastal waters of Madras and its neighbourhood.

In true lobsters (*Homarus*) the body is not laterally compressed and their limbs are adapted for creeping. The first pair of legs are very greatly developed. The Indian spiny lobster resembles a true lobster but the first pair of legs have no pincers. Its antennae are whip-like and armed with spines at the base and these appendages are both tactile and prehensile. Both kinds of lobsters are solitary in habit. Some of them grow to more than 60 cm, weigh 1·5 kg and live for 5 years.

At least three species of Indian spiny lobster occur along the Indian coast—*Palinurus fasciatus*, *P. ornatus* and *P. dasypus*. Their difference lies in the number and arrangement of spines over their carapace. *P. ornatus* is 50 cm long excluding the antennae and 15 cm across the shield.

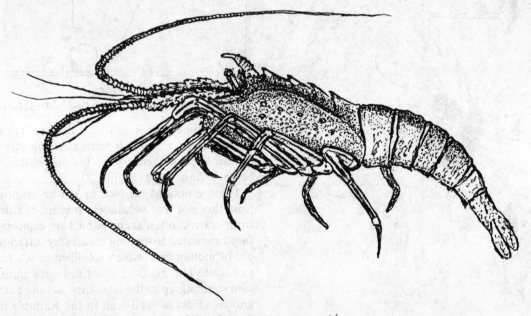

Spiny lobster (*Palinurus*) × ¼

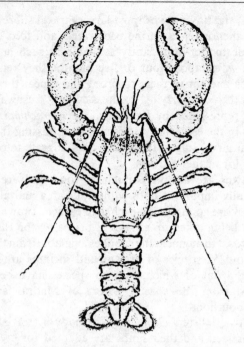

Lobster *Homarus* (decapod) × ¹⁄₆

CRABS are highly organized decapods with many families of curious and interesting forms.
See also CRUSTACEANS.

DECCAN HEMP, *Hibiscus cannabinus*, is a tall herb with a prickly stem. The lower leaves are heart-shaped

Deccan Hemp, × 1

and the upper 5- to 7-lobed. The flowers are yellow with a crimson centre. It is cultivated mainly as a fibre crop for making sacks, cordage and paper.

DECCAN TRAP. It all happened some 70 million years back. The story of the Deccan Trap is related to some major changes that took place on the surface of the earth towards the close of the Cretaceous period. The mighty Himalaya were just beginning to come up, the great dinosaurs that ruled the earth were disappearing, and the crust of the earth in western India was fractured. This was a remarkable event because large quantities of molten rock (lava) were poured out from these fractures filling up the valleys and covering the hills of the existing topography. The eruptions continued intermittently, with periods of quiescence, building up a huge volcanic plateau covering at least 500,000 km² with a thickness of at least 2500 m (Fig. 1).

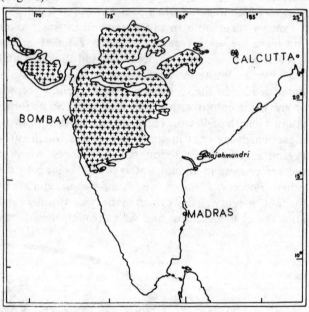

Fig.1.
The distribution of the Deccan Trap lavas.

These rocks are known as Deccan Trap, a name derived from a Swedish term meaning steps, because of the step-like scenery of the formation, each step being a flow of lava (Fig. 2).

These eruptions, known as fissure eruptions, differ from those of true volcanoes. Instead of being erupted from craters, often accompanied by explosive activity, they originated from long fissures by a tranquil welling out of molten lava, which solidified into a rock known as basalt. The fissures, now filled with solid rock, are seen standing up in the landscape as long narrow ridges known as dykes, well seen in the Rajpipla hills in the Broach district.

Fig. 2.

DECCAN TRAP SCENERY

As the lava cooled and contracted it weathered into characteristic rounded masses by the exfoliation of concentric shells. At places it developed columnar jointing, giving rise to symmetrical prismatic columns, so strikingly developed at Gilbert hill, Bombay (Fig. 3).

Fig. 3.

Columnar prismatic columns in the Deccan Trap.

Lavas such as formed the Deccan Trap are commonly full of liquids and gases. These are the last part to solidify and are now seen as cavities known as vesicles. They may be lined with chalcedony and agate, or with crystals of quartz and of minerals known as zeolites (Fig. 4). They are abundant in the Poona district, and beautiful examples were found when the railway tunnels were being constructed, and so they were called 'railway diamonds'.

During the eruptions there were intervals of quiescence during which small lakes and swamps could form on the surface of the cooled lava. In these depressions thin beds of sediment were deposited, and freshwater plants and animals were able to live in the shallow waters. These were buried when the next flow came, and are now found as 'Inter-trappean' beds with fossils of plants and animals. These include freshwater mollusca such as *Physa* and *Lymnaea*, and vegetable remains such as the palm *Nipa, Chara,* water-ferns and various fruits and seeds are fairly common in certain localities. At Bombay the small frog *Oxyglossus* and the freshwater tortoise *Hydraspis* have been found in St. Mary Islands, off Mangalore. (Fig. 5).

Fig. 4.

9″ × 7″ geode from the Deccan Trap, filled with quartz crystals.

Fig. 5.

Fossil gastropod (*Physa*) and fossil frog (*Oxyglossus*) from the Inter-trappean beds.

The Deccan Trap basalts, though sombre in colour, are used as a building stone. Agates, cornelians and onyx, obtained from the cavities in the rock, are polished to provide ornamental stones. A deposit of fluorite, the largest in India, occurs in the Deccan Trap at Ambadongar in the Baroda district. The subaerial alteration of the higher flows has led to the formation of laterite, and its alumina-rich variety known as bauxite is a valuable source of aluminium.

Geologists are of the view that the entire volcanic activity responsible for the Deccan Traps lasted many millions of years, though the exact period of its duration is still uncertain. There can be no doubt, however, that it was one of the major episodes in the geological history of India.

W.D.W.

DEEP-SEA FISH. As sunlight does not penetrate beyond 500 m from the surface, conditions in the deep sea are not conducive to life—utter darkness, immense cold, tremendous pressure, and scarcity of food. Yet living beings, including fishes, have been found even in the deepest part of the ocean, at depths exceeding 10,000 m.

Deep-sea fishes may be defined as those which live beyond the continental shelf, that is, at more than 200 m depth. Characteristic features are: some kind of bio-luminescence; as identification marks, a weak, almost brittle skeleton and soft, almost liquid flesh; small size; a wide mouth with large jaws studded with sharp teeth, having the capacity to eat a large meal whenever prey is available.

Bioluminescence, the emission of light by the deep-sea fishes themselves, is of crucial importance. Lights glow in rows on the sides of the body or on special spots, and are specific in each species and sex. They are made up of glands in the skin consisting of lenses, light-producing cells and behind these reflector cells. Two kinds of chemical enzymes are involved in the production of light, luciferin and luciferase. When these interact in the presence of oxygen in the blood, light is produced. It is without heat, unlike our common electric lights. In some fishes when the enemy is near, it is frightened away by the light glands and luminous slime producing a haze of light. In some fishes colonies of special light-producing bacteria are found in patches on the skin and they produce a special glow in the darkness.

In the upper layers of the deep sea live the Lantern fishes (family Myctophidae). Up to 15 cm long, they have two rows of pearly light organs running along the belly. Although they live at 150 to 500 m depth in the day, at night they ascend almost up to the surface.

Stomiatoid fishes resemble lantern fishes in having a double row of light organs. They are particularly noted

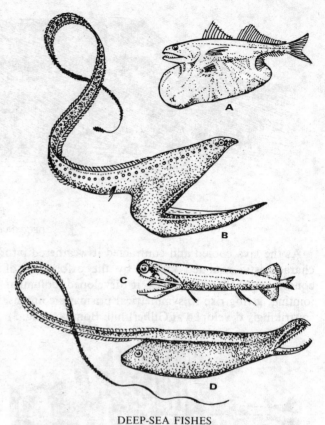

DEEP-SEA FISHES

A × ⅙. The stomach is enormously distended with food. B × ¼. The jaw is abnormally large. C. Normal appearance. D × ⅒. A swallowed fish can be seen through the stomach wall of the gulper, the skin of which is also extremely thin and distensible.

for their watery gelatinous flesh and brittle bones with weak calcification. Most of the stomiatoids carry a barbel hanging down from their chin or throat; this may end in a simple tip or have elaborate branches like a bouquet of flowers, a bunch of grapes or a christmas tree. The young of one of these fishes (*Idiacanthus*) has grows, the eyestalks become shorter and are tightly coiled into a knot inside a bony capsule in the skull. One 25-cm-long fish, *Chanliodus*, has curved, sharp, spine-like teeth so long that the fish cannot close its mouth; it is aptly called the Sabre-toothed Viper fish.

The Hatchet fishes are aptly named, as their compressed body with sharp-edged belly resembles an axe with the tail forming the handle. They are tiny fishes, about the size of a rupee coin, the largest being only 8 cm long. Some of the hatchet fishes have elongated telescopic eyes.

In the Gulper, or Pelican, (family Saccopharyngidae), living at 800 to 2500 m depth, the enormous head is joined to a ridiculously thin body tapering to a thread-like tail. It can swallow fishes much larger than itself.

The deep-sea ANGLER FISHES are described separately.

DEER. 'Deer' is the name applied to members of the ruminant families Cervidae, Moschidae and Tragulidae, though especially to the first of these. The other ruminant families are the Bovidae (cattle, antelopes, sheep and goats) and Giraffidae (giraffes, of Africa). Unlike Bovidae, the animals grouped as Deer do not have hollow horns; and they generally (not always in the Cervidae) have canine teeth in the upper jaw, which Bovidae never do. There are also differences in the feet. However the three families of Deer are not closely related to one another.

The Tragulidae include only the Mouse-deer or Chevrotains. These are among the smallest ruminants, only 25–30 cm high; they have a less complex stomach than other ruminants, large lateral hooves, big tusks in the upper jaw in males, and a wedge-shaped body suitable for scampering quickly through dense undergrowth. Two species of the genus *Tragulus* live in Southeast Asia; South Asia has just one species, *Tragulus meminna* (sometimes placed in a separate genus, *Moschiola*); and a different genus lives in West Africa. The Indian Chevrotain, olive-brown speckled with yellow and with white spot-rows along its flanks, lives in moist forest country in Sri Lanka and in India as far north as about 24°N.

The Moschidae include only the Musk-deer; there is one living genus, with three species, two of which are found in India. The Forest Musk-deer (*Moschus chrysogaster*), from the wooded slopes of the Himalayas in NW India, Nepal, and Bhutan, is smaller and darker; the Alpine Musk-deer (*M. sifanicus*), found above the tree-line, is bigger (50 cm or more in height), longer-faced, lighter coloured (grey-brown) and has ears rimmed with yellow. All musk-deer have thick quilly fur which stands out from the skin; a tiny tail, not visible externally, with a gland on it; very large lateral hooves; and—distinguishing them from Cervidae—a gall-bladder. The males have long dagger-like canines in the upper jaw, slightly movable when the jaw opens and closes, and a musk gland in the groin which is the source of most of the musk used in perfumes.

The Cervidae, though very diverse, are all more highly evolved than Tragulidae and Moschidae in many respects: they have a more convoluted brain, a different type of placenta, a penis without the long thread-like process seen in primitive ruminants, no gall-bladder, and glands on the face (in all species, even if only rudimentary) and on the metatarsals and between the hooves (in most species). With the exception of the rather distinctive water-deer (*Hydropotes*) of China, all deer also have antlers in the male (one species, the Reindeer of the far northern tundra and coniferous forests, has them in females as well). Antlers are like horns, and often wrongly called 'horns', but they differ in two important ways: first, they are usually branched (not

invariably), and secondly, much more importantly, they are solid and are shed and regrown each year. In a medium-sized deer the growing buck will sprout simple spikes at a year of age; larger ones with a small branch as well as the main spikes at two years; and gradually add on more branches, or tines, with age until at about seven years he has a long back-sweeping beam with three or four major tines and perhaps a number of small twigs at the branching-points. The antler is like bone in its consistence; when it grows it is covered with thin, finely furred, highly vascularized skin called Velvet. When the antler has reached its full growth the velvet is shed, the blood-supply having been cut off by growth of a ring, the Burr, at the base of the antler; so the velvet dries up and falls off, or is rubbed off, in patches which during the shedding hangs off the antler like Spanish moss. The sole function of the antler appears to be in sexual display and ritual male-male combat; after the rutting season is over, the antlers themselves are shed, and the male deer has none until they begin to grow again somewhat later in the year.

Why, it has often been asked, do antlers shed and regrow each year? It has been pointed out that they grow very rapidly—they are perhaps the most rapidly growing tissue to be found anywhere in the animal kingdom—and this takes an enormous amount of energy, which must be supplied by extra food. In addition, while growing the very intricate network of blood-vessels so near the surface of the skin must cause the animal to lose a great deal of body heat: it has even been suggested that their purpose is as 'radiators', though their periodic shedding makes this a rather dubious explanation. Again, though they are used in fighting other males, would they not, even outside the rut, be useful in self-defence against predators? The answer, it has been suggested, is that they must be shed in order to make the males less conspicuous; during the rut they exhaust themselves by their fights and continuous vigilance, and would afterwards be easy prey—and predators would learn this. So by shedding their antlers they come to resemble females and so escape being the special targets of predators.

The branching of antlers seems to be, at least in part, an adaptation to non-lethal combat. The antlers of the two interlock, and they push and wrestle until one is exhausted, disengages and runs away.

In the largest deer, the Moose and Reindeer of Northern Eurasia and North America, the males are called bulls, the females cows. In the medium-sized deer the males are stags, the females hinds; in the smaller ones, the males are bucks, the females does. Young deer are known as fawns.

The family Cervidae can be divided into three subfamilies: Hydropotinae for the Water-deer alone;

Odocoileinae for the American deer and the Reindeer; and Cervinae for the rest. All South Asian deer belong to the Cervinae, which range from the small Muntjac to the enormous Moose. Only two genera live in our area: *Muntiacus* and *Cervus*.

The Muntjac or Barking Deer (*Muntiacus muntjak*) is the smallest of the true deer of the subcontinent. 'Muntjac' (properly 'Muncak') is an Indonesian name; in Hindi it is called *kakar*. It is some 50–75 cm high. The most noticeable characteristic is that the antlers are short, with only a single branch, but the pedicels on which they stand are up to 10 cm long (nearly as long as the antler), and continue down on to the face as ridges which meet on the nose in a V. The Indian species is remarkable in having the lowest chromosome number in mammals; the female has 6, the male 7; and this is quite different from the closely similar Chinese species which has 46 in both sexes! Male muntjac have long canines, which are used in self-defence; the female lacks them. Muntjac live in pairs with a small home range; they appear to have no special breeding seasons but bear young all year round. The fawns are spotted until they are 6 months old, but adults are uniform rich reddish above with white belly, inside of thigh, underside of tail, and chin. They inhabit thickly wooded hilly country extending westwards as far as Pakistan. Their dog-like alarm-bark resounds loudly; this bark may also function as a spacing call between the pairs.

They are quite aggressive, very courageous animals which will fight even leopards if attacked.

The genus *Cervus* contains several species which are placed, for convenience, in various subgenera. All have a short tail, relatively short lateral hooves, a hairless rhinarium and a large face-gland; the antlers are distinguished by having a large branch, the brow tine, pointing forward and originating immediately above the base.

Deer of the subgenus *Axis* are the most primitive in the genus. Except for the antlers of the buck, there is little sexual dimorphism; and the antlers are simple, with just a brow tine and one other fork, slender, smooth and white. In our region there are two species; one of them, the Chital or Spotted Deer (*Cervus (Axis) axis*), is the most familiar of all the deer in this area. About 90 cm high, weighing 85 kg, the chital is bright reddish-fawn in colour with numerous white spots at all ages and seasons. The antlers are long—up to 1 metre—though slender, and may have extra 'twigs' apart from the three main points. They are found in light forest and plains country, requiring some open land for grazing; being mainly diurnal they are easy to see. The herds number usually 5 to 10, but they may gather in grazing aggregations of 200 or more. There is no special rutting season, though in different places there are often peaks of mating; males grow and shed antlers at different times, and when their antlers

© M. Krishnan

Axis Deer × ¹/₁₉

are fully developed and have lost their velvet they roam widely, entering and leaving many herds in search of females in estrus. A herd therefore contains both sexes. When the buck has found an estrous doe he defends her against rivals and 'tends' her, then departs when mating is completed. In more seasonal areas, such as the Kanha National Park, mating tends to be concentrated in the hot season so that fawns are dropped mostly in winter (December to February).

© M. Krishnan

Barking Deer or Indian Muntjac × ¹/₆

Quite closely related to the Chital is the Hog-deer or Para (*Cervus (Axis) porcinus*), which is smaller (60 cm high) with antlers only about 35 mm long but resembling the Chital's except that they stand on elongated pedicels rather like a Muntjac's. Hog-deer are also stouter in build and shorter-legged; the coat is unspotted, brown and minutely speckled with white on each hair, but young have spots. The species lives in the alluvial grass plains of northern India, Pakistan and Sri Lanka, where it is found in pairs; during the mating season (September/October) the pairs seem to congregate into herds, each male trying to monopolize an estrous female. It is interesting that two other species, closely related to the Hog-deer, occupy a relict distribution on small islands in Indonesia and the Philippines.

The subgenus *Rusa* contains a number of species which are somewhat less primitive than *Axis*; the males are considerably larger than the females and have throat-manes, and the antlers are much stouter, with a shortened beam, often rather complex, and rough-textured. The commonest species in South Asia is the Sambar (*Cervus (Rusa) unicolor*). This is up to 150 cm high, short-bodied and high-legged; the stags weigh up to 320 kg. Antlers are relatively short, and have usually only three points though very commonly extra little snags in the forks; longest antlers—over 1 metre in length—are seen in Central India. Like Chital, Sambar occur all over India, Nepal, Sri Lanka, and Bangladesh, but not in Pakistan. They inhabit deep forest, especially in hilly country, and being more nocturnal are less often seen. The colour, corresponding to the forest habitat, is dark brown, unspotted; the coat is coarse and shaggy. No permanent herds are formed; random associations, very unstable in composition, are all that are seen. There is a more marked mating peak than among chital, but even so mating is not restricted to this peak; most mating occurs from November to February, so that births are mostly in the late summer. The stags fight for territory, rather than directly for hinds, and those which acquire the best territories (with the most browse) attract more hinds and so mate more often. It is possible that two subspecies are found in our area: the Indian Sambar (*C. u. unicolor*) is found over most of the area, but the Southeast Asian Sambar (*C. u. equinus*) may occur in Assam and other north-eastern states, and in Bangladesh: more information is needed to establish this. It has more heavily built, less spreading antlers in which the two terminal tines are not subequal as in the Indian race, but the posterior tine is much shorter and at more of an angle to the beam.

The SWAMP DEER or Barasingha also belongs to the subgenus *Rusa*; at one time it was put in a special subgenus, *Rucervus*, but the work of the Soviet zoologist Flerov has shown how basically similar it is

© M. Krishman

Swamp Deer × $^1/_{16}$

to *Rusa*. There is a separate headword for this species.

The Brow-antlered Deer (*Cervus (Rusa) eldi*), called Thamin in Burma and Sangai in Manipur, has also sometimes been placed in 'Rucervus' or in its own subgenus *Panolia*, but is clearly in *Rusa*. Smaller than either Sambar or Barasingha—stags are about 120 cm high—it has a coarse brown coat in stags which is light fawn in hinds; fawns are spotted. The antlers are very unusual: circular when seen from the side, with the brow tine curving forward in direct continuation of the beam, and there are many small snags at the forks. The species is now very rare over its whole range, which extends from Vietnam in the east to Manipur in the west; and the Indian subspecies, the Manipur Brow-antlered Deer (*Cervus eldi eldi*) is the most endangered of all. A survey by Ranjitsinh in 1975 found that only 14 remained in the wild; they live on Keibul Lamjao, a small marsh of 35 km² at the southeast corner of Logtak Lake in the Manipur Valley. Here they live on big islands of floating decaying vegetation on which grasses and sedges grow; human activity has reduced the area of the available habitat to 14·4 km², and there is an urgent need to prevent incursions. Fortunately in 1962, when the numbers were higher, a pair was sent to Delhi zoo and another to Calcutta zoo; since then there have been more than 40 births, and breeding groups have been established in seven Indian zoos. It is planned to return some to the wild to broaden the gene-pool of those remaining there.

The Manipur Brow-antlered Deer rut in March and April, whether in the wild or in captivity; births occur 8 months later. They differ from all other races of the species in that the hind surface of the pasterns are horny, not hairy, and are applied to the ground when walking: an adaption for their marshy habitat.

Finally the subgenus *Cervus* contains the most highly specialized of the genus. Males have neck manes, and are much larger than females; the antlers have at least four tines; there are differences in the skull. Of the two species, the Sika (*Cervus (Cervus) nippon*) is found in Japan, Taiwan, the Soviet Far East, eastern China, and northernmost Vietnam; while the much larger Red Deer (*Cervus (Cervus) elaphus*) extends over the whole of northern Eurasia and North America, and reaches the very northern edge of our area. The big northeast Siberian and North American races are called Wapiti. The three South Asian races are closely related to one another and, together with a fourth race from Szechwan and Kansu, form a group: they are all mountain-living deer, fairly large in size with many-branched antlers which, however, have no 'crown' at the top unlike wapiti; they are grey-brown in winter, lighter in summer, with lighter underparts and a creamy-white rump-patch; fawns are spotted. All three are in grave danger of extinction; unlike the Manipur Brow-antlered Deer, they have not been bred in captivity.

The best-known of the three South Asian races is the Hangul or Kashmir Stag (*C. e. hanglu*), often called Barasingha but not to be confused with the Swamp Deer. This is some 120–125 cm high in stags, with a small rump-patch tending to orange-white in tone, bordered by a dark brown stripe at the sides and with a dark stripe going through it from the back to the base, or even to the tip, of the tail: the lips and chin are white. At one time this deer was found over the Vale of Kashmir and southeast across the border of Himachal Pradesh; its range is now fragmented—there are 20–25 in the Gamgul Siya–Behi Sanctuary in Himachal Pradesh, about 20 in the Desu Sanctuary some 115 km southeast of Srinagar, and (perhaps the only viable population) 150–200 in the Dachigam Sanctuary. In this latter region they winter at low levels and go higher, sometimes above the tree-line, in summer; the stags, after shedding their antlers, have even been seen as high as 3650 m. They rut at the end of September and beginning of October, and fawns are dropped the following May.

Somewhat larger, about 140 cm high, are the two more easterly races, both of them known locally as Shou. One, Wallich's Deer (*C. e. wallichi*) distinguished by its very large rump-patch with little or no dark line through it or bordering it, lives in the upper Matsang (Tibetan Brahmaputra) and Sutlej valleys and extended formerly into the Mustang salient of Nepal where it has not been recorded for about half a century. The other, Hodgson's Deer (*C. e. affinis*), has a smaller rump-patch—larger than the hangul's but not extending on to the backs of the thighs—but, again, little trace of the dark lines; it still occurs, reputedly, in the Tsangpo valley and its tributaries south of Lhasa in Tibet, extending south to the Chumbi Valley, the Raidak valley of Bhutan, and Sikkim.

South Asia thus has a good selection of the living Old World deer. Three—Muntjac, Sambar and Chital —are widespread and in no danger of extinction. Others—Mouse-deer, Musk-deer, Swamp Deer and Hog Deer—are more restricted in range and their status needs watching although at present they are not endangered; while the Brow-antlered Deer and the three races of Red Deer are nearly extinct and only intensive action can save them; in the case of the former, captive breeding may have turned the scales at the last moment. The Hangul was formerly a favourite target of hunters, for its magnificent antlers; musk-deer are economically useful and could perhaps be farmed; chital and sambar are essential components of their ecosystems and, especially, are useful as prey sources for endangered carnivores, especially tiger and dhole.

C.P.G.

See also CHEVROTAIN, MUSK-DEER, SWAMP DEER, UNGULATES. See plate 6 facing p. 81 and plate 25 facing p. 368.

DEODAR, *Cedrus deodara*, is a large evergreen tree distributed from Afghanistan to Garhwal, and in the Kurnauli Valley in W Nepal. It is common between 1700 and 2400 m. Trees up to 76 m tall and 13 m girth have been recorded. The largest and most magnificent are found near temples, where they are venerated. The spreading, table-like branches are distinctive, and

© F.R.I.

Grove of large deodars near a temple

the bluish green, pointed leaves grow in tufts along them. The female cones are barrel-shaped and grow erect on the branches. The timber is valuable for railway sleepers, buildings and furniture.

See also CONIFERS.

DESERT PLANTS. The word 'desert' suggests to many people a dry, flat, sandy area, empty, hot and desolate. There are such places, but deserts may be cold, hilly, gravelly or rocky too. They are always dry and desolate.

Plants have evolved in many different ways in order to survive desertic conditions. The roots of trees often go down three times the height of the visible stem in their search for underground water. And to catch the moisture from light showers some plants have developed 'rain roots', a network of fine roots that spread out just below the surface of the ground. To reduce the loss of moisture through transpiration the leaves of desert plants are usually small and often coated with resin or hairs. But sometimes, as in cacti, the stems and branches develop into storage bags for water, leaves disappearing completely or being modified into strong spines or prickles (see PLANT DEFENCES).

While the Thar is the most populated hot desert of the world, cold deserts like those of Ladakh, Lahul and Spiti are sparsely populated. The livestock population in both areas exceeds their carrying capacity, that is to say, that the soil is not capable of producing enough to feed the goats, yaks, camels and cattle. Though the formation of deserts is a natural process, human activities such as extensive cultivation during the rainy season, overgrazing, cutting of trees and shrubs for fuel have intensified desertification. Man and his agents have contributed to desertic conditions. The vegetation is scanty and very much sought by grazing animals and human beings. Desert plants have special protective arrangements to survive extreme climatic conditions but some of the adaptations—such as the development of water-storage tissue—make the leaves and stem of desert plants tempting for browsing animals.

In a hot desert, such as the Thar, the growth of vegetation depends on the soil and moisture characteristic of different land forms. The plants associated with hilly, rocky areas are gum arabic (*Acacia senegal*) and *Anogeissus pendula*. Where the hills join the plains, *Acacia leucophloea* and *Salvadora oleoides*, with *Ziziphus* and *Capparis* shrubs are the characteristic vegetation. On the plains themselves are *Prosopis cineraria* and *Salvadora oleoides*, in old river-beds *Acacia jacquemonti* and *Tamarix*, on the gravelly areas called 'bajdas' *Calligonum polygonoides*, on sand dunes *Panicum turgidum* and in the saline depressions between sand dunes, which are shallow pools in time of flood, *Cap-*

paris decidua, Haloxylon salicornicum, Prosopis cineraria and *Capparis decidua* and on the flood plains themselves *Tamarix dioica, Salvadora persica* and succulent shrubs such as *Haloxylon salicornicum* form the salient vegetation.

Plant adaptations to desert environment are morphological, physiological and anatomical. Plant growth forms change from tree to straggling scrub type due to continuous grazing, browsing and lopping, though where the tree and shrub vegetation is protected oasis-like conditions prevail. From these areas grazing and removal of dead and dry vegetation only is permitted.

Short-lived herbs called ephemerals constitute 50–60 percent of the desert flora. The ephemerals start with the onset of the monsoon (mid-July) season and complete their life-cycle by September-October when the monsoon tapers off. Density and growth of tree and shrub vegetation depend primarily on the moisture supply from the underground aquifer which is reached by means of a long tap root. The most common tree, called 'khejri', possesses a 30–40 metre deep tap root. Root systems develop both vertically and laterally. Desert plants such as *Aerva pseudotomentosa* and *Haloxylon salicornicum* have even lateral roots 4 or 5 metres long. A common plant of the extreme arid environment, *Calligonum polygonoides* (phog), possesses fine surface roots, 'rain roots', on the woody root parts just below the ground surface and extracts surface moisture from light showers. A low shoot–root ratio, of 1:3 to 1:6, is common for desert plants.

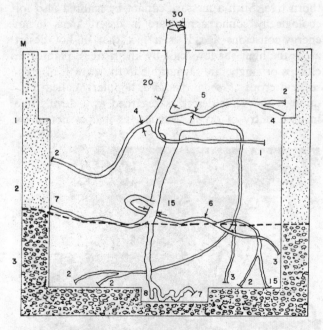

The root system of a 6 m high khejri tree. The figures along the roots indicate their diameter in cm. The upper soil section is sandy, and the lower is a kankar zone. Though khejri is only 6 m high its roots may go down 30 m to find water.

To reduce transpiration desert plants are distinguished by small leaves, no leaves, shedding of foliage, shedding part or whole of shoot branches, and rolling of leaves. Morphological plasticity is also exhibited in desert plants. Small spiny species produce broad leaves and reduced spines under less arid conditions, e.g. in *Fagonia cretica* and *Alhagi pseudalhagi*. The roots of many desert plants produce additional buds, sprouting into new branches when partly exposed by erosion.

Heavy covering of the surface of leaves by a wax-like substance or resinous varnish-like coating; richness of woody elements, high ratio of volume to surface (compactness); abundant hairs and special arrangement of stomata in recess of grooves are common features of desert plants. An impermeable seed coat, germination with alternation of temperature from 10° to 30°C, in preference to constant temperature, inhibitors affected by light and water, high transpiration rates, water retention to both structural and physiological entities, special process of metabolism in succulent plants are some physiological adaptations found in desert plants.

The commonest of desert plants, khejri (*Prosopis cineraria*), an evergreen tree, is revered and protected in farms and other places. Its leaves are small and when offered to Lord Ganesha are called *sona* (gold)—an intriguing symbolism.

One cannot but be impressed by the age-old and deep-felt regard held by the people for this sacred thorn tree. Such a message cannot be ignored also for ecologically sound agriculture in desert areas in an energy-conscious age. It is a fact that in hot desert areas the light shadows cast by khejri trees, planted in clumps or singly, are favourable to the growth of grasses and crops. *Prosopis juliflora*, popularly called mesquite in the Thar desert, is regarded as a devil's tree in its country of origin, Mexico, as it does not allow

Khejri being lopped for fodder and branches for fencing

any other vegetation to grow under it; but it provides much needed fuel to the local population in the Thar desert.

Wind erosion due to lack of plant cover in hot deserts is also active in cold deserts. A desertic formation predominantly with dwarf scrub generally adjoins a dry temperate forest. The stocking of the forests is poor and most of the areas support only scattered or stray trees. Pockets of deodar, kail (blue pine) and birch forests are seen. In the past these have been over-exploited. Good canopy forests occur only in few locations. Biota and extreme climate too have played their adverse role to further lessen their occurrence. There are frequent blanks along the nullahs due to snow slips. Even shrubs that remain are being rapidly exterminated by digging up for fuel. Drier conditions give way to thorny species. Plants (*Arenaria musciformis*—a cushion plant, *Acantholemon* sp., *Caragana pygmaea, Saussurea gossypifora*) show desertic adaptation with low spreading habit and hairy surface. Some of these occur in great masses with their flowers lending a brick red and yellow hue to the landscape that catches the eye. Along the streams a characteristic vegetation, 1–3 m high, is common. The scrub gives place to stony deserts with small moss-like herbs and spreading bushes (juniper, willows, Berberis, rhododendron, Caragana, etc.). Crops such as barley, wheat and buckwheat are sown on fertile areas. *Kuth* and several medicinal

Exposed roots of Prosopis cineraria *(khejri), a common tree in the Thar desert*

13 Seagulls and Shikras

Sunrise at Point Calimere silhouettes a flock of seagulls and a herd of domestic cattle.

The shikra *(Accipiter badius)* and its fledgling young photographed in the Rajpipla forest, Gujarat. Habitat destruction and pollution adversely affect birds of prey almost everywhere on the Indian subcontinent.

14 The Atlas Moth

The order Lepidoptera is made up of several thousand species, with moths constituting 90 per cent of this large insect group. The Atlas Moth (*Attacus atlas*) depicted above, is the largest known moth.

Caragana pygmaea, *a plant characteristic of the cold deserts of Ladakh*

herbs are collected and brought down to the plains in exchange for oil, jaggery, and utensils and transported on pack animals such as sheep and goats along traditional migratory routes. Afforestation with suitable species is the urgent need both to support life and for environmental conservation.

R.K.G.

DEW. When a surface is cooled by night radiation to a temperature lower than the dew-point of the moist air with which the surface is in contact, water vapour is condensed on the surface as minute water drops. This is called dew. The dew-point is the temperature of a moist air sample, to which the air must be cooled in order to make it saturated, or attain a relative humidity of 100 percent, without any addition of water vapour.

Dew is formed on grass or exposed surfaces on a morning when the previous night has been clear and the wind movement very small, almost calm, and the air in the previous evening was not very dry. The necessary cooling is effected by the surface warmth radiating away during the night. If there is appreciable air movement, the ground does not cool much; for the radiation removes heat from a greater thickness of the atmospheric layer near the ground.

Dew is common during the cold weather over most parts of our area.

In desert areas, the cooling at night is large, and there is formation of dew on many mornings. This is a valuable source of moisture especially when the dew forms over a roof and can be collected via appropriate channels.

DHOLE or THE INDIAN WILD DOG. The Indian Wild dog or dhole (*Cuon alpinus*) is a pack-hunting canid. An adult normally weighs 18 kg, stands around 50 cm at the shoulder and the total length, including the 40–45 cm long tail, is approximately 130 cm. Females are slightly lighter in build. The fur is rusty sandy in colour with a somewhat paler ventrum. The bushy tail is black-tipped and may have a mixture of grey and white hairs.

Ten subspecies have been reported ranging from Amurland and the Altai mountains in the U.S.S.R. and over the whole of Continental Asia and occurring in the islands of Sumatra and Java but not in Japan, Sri Lanka or Borneo. In India there are three subspecies; *Cuon alpinus dukhunensis* found south of the Ganga, *C. a. primaevus* in Kumaon, Nepal, Sikkim, and Bhutan and *C. a. laniger* which formerly occurred in Ladakh and northern Kashmir. A questionnaire survey conducted in 1978–9 showed that only the subspecies *C. a. dukhunensis* is fairly common, the others being on the point of extinction.

As in other pack-hunting canids, the Dhole pack is an extended family unit and the number in the pack rarely exceeds 20 animals, whilst groups of 5 to 12 are most common. Dholes are largely social and true loners are rare. Packs are thought to be territorial and the area covered by the pack in its hunting is correlated with the abundance of prey. Dholes maintain an almost constant number in a given area and this is because breeding is restricted to one female in the pack, and emigration or mortality limits the numbers of both adults and pups. Not much is known about hierarchy in dhole society. Pups are highly quarrelsome, subadults occasionally fight, whilst adults are amicable. One of the fascinating aspects of the dhole society is the play indulged in by the adults and pups as well.

Dholes feed on a wide variety of prey ranging from smaller mammals such as rodents to large deer. Common prey animals in India are chital, sambar, wild pig, hare and calves of domestic cattle. Dholes are predominantly diurnal and hunt mainly in the morning and evening. As a prelude to the hunt the dholes assemble at a defecation site, rub their bodies against others, nuzzle each other and after this mood-synchronizing ceremony, set off to hunt.

Dholes commonly adopt two strategies while hunting. One is to kill prey as the whole pack moves through the scrub in an extended line and the other is for some members of the pack to remain on the periphery of the jungle so as to intercept the fleeing prey flushed out by other members of the pack. Small mammals like hare and chital fawn are seized on any part of the body and are killed with a single head-shake. Larger animals are attacked from behind and the usual points of attack are the rump and the flank. This leads to severe shock, evisceration and loss of blood which eventually kills the prey.

© M. Krishman

Dhole or Indian Wild Dog × ¹/₁₁

Soon after killing, dholes begin feeding excitedly. Fawn kills are torn to pieces within seconds after the kill is made and each dhole runs away with its share to eat it away from others. Even when the kill is as large as an adult chital stag, subtle hierarchical disputes do not allow all the members of the pack to lie around the kill and eat together. Each dhole runs away to eat the meat dismembered from the carcass away from the other members. Heart, liver, rump, foetus and eyeball are the first parts eaten. Dholes eat fast and within 60 or 90 minutes each consumes nearly 4 or 5 kg of meat. Feeding dholes frequently drink water, if water is nearby. If water is some distance away, they head for the nearest water-hole soon after eating. Dholes do not cache food but come back for the remains. Whenever available, they scavenge on tiger and leopard kills also.

Communication involves vocalizations, olfactory and body language. Vocalizations include whine, growl, growl-bark, scream, whistle, and squeaks by the pups. Of all these the whistle, which is a contact call, is the most prominent. It is still not understood how the dhole produces this whistle call which is mainly used to reassemble after the pack members get dispersed after an unsuccessful hunt. Olfactory communication includes interdigital gland secretions, urination, and defecation of the group at the intersection of trails. Such 'latrine sites' are the best clues for locating dhole packs in the jungle. The repertoire of their body language is by contrast rather limited.

Dholes become sexually mature in about one year. Mating involves a copulatory tie. Gestation lasts for 60 or 62 days and in India, whelping occurs between November and March and the litter size is around 8. Prior to parturition the bitch prepares a den, usually an existing one, dug by porcupines on the banks of stream-beds or under or between rocks. The entire pack takes part in feeding the lactating bitch and the pups,

which are fed with regurgitated meat even when they are three weeks old. Sometimes a second dhole, other than the mother, is seen near the den-site when others are away hunting and it is popularly known as the guard dhole. After they are three or four weeks old, pups spend considerable time playing in the morning and in the evening near the den-site. The hunting range of the pack, when the pups are in the den, is considerably smaller than their normal range.

Pups at the age of 70 to 80 days leave the den permanently, and if the den-site is disturbed earlier than this, the pack will move the pups to another den. During the early days of the pups' wandering life the pack continues to care for them by regurgitating meat and by leaving escorts to accompany them and allowing them to dominate at the kills. Also they hunt either very early in the morning or late in the evening to avoid the heat of the day or man or both. While hunting the pups are left either alone or in the company of one or two adults in a secluded place and soon after the kill is made the pups are brought to the kill. When five months old pups actively follow the pack and even at the age of 6 months very often beg for food. At around 8 months of age, the juveniles take part in killing even large prey like sambar.

In their natural habitat dholes frequently come in contact with other animals including man. Tigers and leopards mostly avoid them. Elephants chase them and many birds such as Mynas, Redwattled Lapwings, Peafowl and Grey Junglefowl sound an alarm whenever they see dholes. Man's relationship with the dhole, which is shy of man, is always one of negative interaction. The residents of jungle hamlets exploit dholes by appropriating their kills and until the recent past man tried to eliminate them by poisoning their kills and by paying bounties.

Now the dhole is a protected animal under the 1972 Indian Wildlife Act, and has become the main predator in many of the Indian sanctuaries.

A.J.T.J.

See plate 40 facing p. 545.

DIGESTION. The arrangement of the digestive system is such that it facilitates the intake of food, breaks down this food into simple (molecular) forms which the body can easily utilize, and throws out the unutilized parts. This breakdown is achieved both by mechanical action of the teeth, jaws, etc. and by chemical action of various enzymes which are secreted by the different parts of the digestive tract.

Food is captured before it enters the digestive tract. The amoeba takes in food by osmotic process. Birds use their beaks as one would use a pair of forceps. Chameleons, toads, ant-eaters and cattle use their tongues, swans and giraffes have long necks to reach their food, and elephants use their trunk to pick up food.

The digestive cavity of sponges is lined by cilia, the rhythmic beating of which carries food-laden water along the digestive cavity. Coelenterates have a digestive sac open at one end which serves as both mouth and anus. Echinoderms have two ends—but the anal end is nominal as very little waste is left for expulsion. The starfish everts its stomach from the mouth and wraps it around the food. After digestion, the stomach is withdrawn and indigestible parts are left behind.

As the bulk of an animal increases, a proportionate increase of the digestive surface is achieved by increase in diameter and length of the system and by having internal folds and diverticula. Cheek pouches are well developed in animals like duckbills and squirrels who need temporary storage of food in order to grab a maximal amount of food in minimal time.

In lower, sedentary forms of life feeding is a continuous process while in higher animals, feeding is intermittent.

The digestive process starts in the mouth by the action of saliva. Salivary glands are absent in fishes.

The oesophagus in birds has a crop which acts as a temporary storage for food hastily secured. This is because the stomach or gizzard is very small. Food from the crop can be passed down to the gizzard in small quantities, or can be regurgitated for feeding the young. Pigeons produce a cheesy nutritious substance from the lining of the crop, called Pigeon's Milk, which is fed to the young ones in the nest. Seed-eaters have a muscular grindmill, the gizzard (see Fig. 1).

The next step in digestion is achieved by the digestive juices of the stomach, and the intestines. Extreme sub-

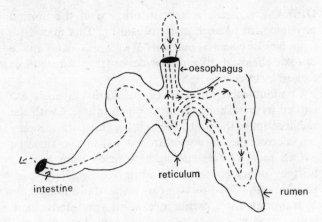

Fig. 2. A RUMINANT'S STOMACH

division of stomach is seen in ruminants (Fig. 2) for regurgitation of swallowed food. The vampire bat's stomach has a spacious reservoir adapted to its blood-sucking habit. In camels the rumen and reticulum are lined with water cells which enable the animal to store water and endure prolonged periods of dryness. Further digestion is carried out in the intestines which are long and coiled up in the abdomen. The undigested residue is finally thrown out through the anus.

Speed of digestion varies according to the diet and the size of the animal. In some frugivorous species the fleshless stone of a berry may be voided less than fifteen minutes after it is swallowed. Small carnivorous birds are said to digest their prey within a few hours.

DIPPER. Starling-sized, stumpy-tailed and thrush-like, the dipper (family Cinclidae) is unique among the passerines in being aquatic. It lives on rushing Himalayan streams, and feeds by diving in icy water and walking along the bottom, on insects, snails, minnows and trout-fry. Its call is a piercing *djiit* that carries far over the roaring torrent. Both sexes are alike. A dome-shaped nest of grass and leaves is built among rocks bordering torrents. Eggs 3 to 6, white. Dippers are found in both the Old and New Worlds.

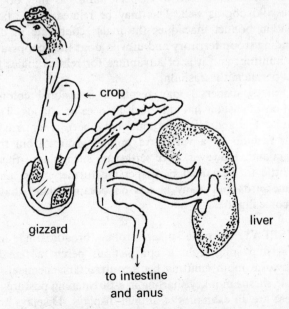

Fig. 1. A BIRD'S DIGESTIVE SYSTEM

© Loke Wan Tho

Dipper × 1/3

DISEASE. Disease is any disturbance of the normal physiological balance of an organism. This may arise from lack of essential nutrients or vitamins, from intake of toxic chemicals like insecticides or from disturbances caused by infection.

Communicable diseases or infectious diseases are caused by parasitic organisms which interfere with the metabolism of the host. Such pathogenic parasites may be microscopic or visible to the naked eye (insects, worms etc). The disease agent spreads from host to host by direct contact, by indirect contact (e.g. contamination of food or water by infected droppings), or is transmitted by carrier organisms, or intermediate hosts, usually insects, which are called vectors for transmission of the disease. Communicable diseases capable of being spread very rapidly, affecting many individuals in a community at one time, are said to be *epidemic* diseases. Epidemics arise from increased virulence of the disease agent, from contamination of a common source (food or water), or, in insect-borne diseases, from rapid increase of the vector, after its breeding season.

Various epidemic diseases are known to affect animals in the wild, often following an epidemic in domestic animals. Epidemics of foot-and-mouth disease, of rinderpest and of African horse sickness have been reported in recent times. Large herds of wild animals are said to have been wiped out during an epidemic of rinderpest in the African National Parks during the late 1960s. No accurate information is available for Indian wild life.

Parasitic disease agents have evolved a variety of mechanisms to ensure survival. Some have complicated life-cycles, passing through a variety of hosts, with a different stage of development in each. Others infect reservoir species. A reservoir is a host species which suffers little or no disability but allows the parasite to multiply in its body so that large numbers of organisms are available for further spread.

DISPERSAL. Dispersal refers to movements of animals, or mobile stages in the life of plants such as seeds, away from where they were born. These are to be distinguished both from routine food-searching movements within a restricted area that may constitute an animal's home range, as well as from the periodic migratory movements such as those of birds from the temperate zones to the tropics and back. Dispersal serves the purpose of relieving pressure of overcrowding in a given area and gives the animal a chance to find a better habitat. It also reduces the risk of inbreeding.

Dispersal is generally the function of particular stages in the life history of animals. Thus, barnacles or oysters are marine invertebrates which are fixed to a single spot on the substratum throughout their adult life.

Their larval stages however are free-swimming and disperse widely. In birds, the chicks tend to disperse in the first few months after fledging and then are relatively sedentary for the rest of their adult life. In many social mammals, the young may disperse on approaching sexual maturity. In the insect group of aphids, there are two forms, winged or wingless. In the wingless stage the aphids remain sedentary, multiplying, for instance, on a single leaf. Several generations of wingless aphids may thus be produced, increasing their level of crowding. At high levels of crowding then the winged forms are produced which can disperse away to fresh habitats, thus relieving crowding on habitats on which several wingless generations had been passed.

Apart from reducing the pressure on crowded habitats and making possible the colonization of fresh ones, dispersal also tends to reduce the chances of mating amongst close relatives. This may be favoured but inbreeding leads to the production of inferior progeny. Avoidance of inbreeding can be achieved by the dispersal of either of the two sexes, and in many animals either males or females disperse leaving the individuals of the other sex in the vicinity of the place of birth. In birds the sex that disperses more tends to be the female sex. This seems to be due to the fact that males are at an advantage in staying near the place of birth because they often inherit their father's territory, while the females settle in their mate's territory.

In mammals the sex that disperses is generally the male sex. Thus in elephants the female relatives tend to stay together for life, constituting the basic unit of elephant society. The males are driven off as they approach maturity. Lions and the Hanuman langur also behave in a similar fashion. In the bonnet macaque however the females tend to disperse more, as they do in the wild dog as well. This may be related to the fact that in bonnet macaques the males cooperate in defending troop territory and in wild dogs they cooperate in hunting; and it is of advantage for related males to cooperate in this fashion.

Finally dispersal may result in occasional colonization of distant habitats such as oceanic islands. The Andamans and Nicobars must have been colonized in this fashion by a variety of animal groups from the mainland. However, once settled on such islands there is little to be gained by dispersing further away, and some birds and many insects on oceanic islands have become flightless.

DISPLAY. A male magpie-robin broadcasting his courtship song from a conspicuous perch, a female silkworm moth emitting her sex-attractant chemical, a male rhesus monkey snarling in a threatening posture— these are all examples of animal displays. Displays are highly stereotyped sequences of movements (or highly

Rhesus macaque males communicate their status through the postures displayed while walking. *Top*—dominant; *bottom*—low-ranking

specific emissions of sounds etc.) that serve as social signals to other animals. The male magpie-robin advertises his possession of a territory and his readiness to mate, the female silkworm moth her sexual receptivity, and the monkey his intention to attack physically through these displays. Such displays or signals are devices that the animals use to manipulate the behaviour of other animals around them. For instance, the magpie-robin uses his song to persuade other males to leave him alone and females to approach him. The other males use the vigour of the song to judge whether an effort to dislodge this male from his territory is likely to succeed or not, while the females use it to judge how capable a defender of territory and provider of food this male is likely to be. It is to the advantage of the singer to pretend that he is more vigorous than he in fact is, and to the advantage of other males and females to see through any such deception. At the same time predators such as hawks will make use of the signal to locate their potential prey. This information the magpie-robin would rather not convey, but then the goals of conveying his possession of territory to other magpie-robins and of disclosing as little of his location to predators as possible may be incompatible, and the male magpie-robin has to accept a compromise.

Displays have thus evolved because it is of advantage for any animal to acquire information about the true potentialities and intentions of its neighbours, and because it is of advantage for any animal to convey to its neighbours either true or sometimes false information of its own potentialities and intentions. The origins of these displays can be best illustrated by displays based on intention movements. In a flock of pigeons, for example, the birds stick together because they can thereby acquire an early warning of the approach of a predator and hence can flee with a better chance of escaping. Such a flock derives its safety from all of its members sticking together. It is therefore of advantage for every bird to acquire information on the intended movements of other birds and to adjust its own movements accordingly. Now a bird must perform certain preliminary movements, for instance stretching of legs and some flapping of wings before it takes off. These are the so-called flight-intention movements. Now it is of advantage for any pigeon to detect such flight-intention movements in other members of the flock and thereby prepare himself for flying with them. At the same time, it is of advantage for a pigeon to inform other pigeons of its intention to fly, so that they will also join it. It would therefore be of advantage for a pigeon to exaggerate its flight-intention movements to permit its easy detection by fellow pigeons. This process can gradually lead to the use of very stereotyped and exaggerated intention-movements as displays conveying reliable and detailed information on the intention of an animal to move, and even the direction and distance to which it plans to move. In some species of pigeons, for example, the number of flappings of the wing prior to take-off provides a reliable indication of the distance to which the pigeons intend to fly.

The most interesting example of the development of an intention-movement into a display is the dance of the honey-bees. Honey-bee workers returning from feeding at a rich source of food perform characteristic dances in a figure of 8. The inclination of the axis of this figure to the vertical direction inside the hive indicates the horizontal angle that the food source makes to the direction of the sun from the hive. The tempo of the dance and the rate of abdomen-wagging provide clues to the distance of the food source from the hive. It is as if the returning foraging bee is announcing to other workers: 'There is a rich source of food in such and such a direction and at such and such a distance from the hive, let us all fly in that direction.' A number of other bee species show intermediate stages of development of this dance display.

These displays involve a situation where it is of advantage to both the signaller and the receiver to ensure transmission of correct information. There are however many other situations in which the signaller would find it of advantage to convey false information, while the receiver would wish to surmise the true state of affairs from the signal conveyed. Thus when two males are competing for a scarce resource, such as a female in heat, each would like to avoid a fight if the chances of success in a fight are low. Each male would therefore like to assess the true strength of the rival, while conveying an exaggerated impression of its own strength. This is why two males threatening each other take a posture which exaggerates the impression of their size, raise their hair or feathers to achieve the same effect and then display their weapons such as canines to the best effect. But since the other males will be selected for their ability to see through such exaggerations, they will have little ultimate effect on the outcome of the contests. On the other hand, displays which are a reliable index of the relative strength of males will be used to decide the outcome of the contest. Thus in certain species of bullfrogs, the pitch of the croak depends on the size of the male. When two rivals encounter each other, they have a croaking match, and the male with a less deep croak accepts defeat.

Displays have thus developed in a variety of contexts ranging from courtship, threat, co-ordination of movements, alarm and so on and are based on every sensory modality at the disposal of the animal. A number of non-verbal displays, akin to animal displays, have been identified in human communication as well. They range from overt ones, such as shaking a fist in a threat-display, to erect or stooping postures to convey more subtle information on the social status of an individual.

DOAB. A doab is the alluvial tract of land between two rivers, the best known being between Ganga and Yamuna. There are many Doabs in the Punjab Plain. The depositional processes operating for a long time have united the doabs of its five rivers (*panj ab*) into a homogeneous geomorphological entity. These doabs have maintained their clear identity ever since the first settlement of the Aryan tribes in what was known as the Sapta Sindhu, the land of the seven rivers. From east to west these doabs are Bist doab, lying between the Beas and the Sutlej; Bari doab, between the Beas and the Ravi, Rechna doab, between the Ravi and the Chenab; Chaj doab, between the Chenab and the Jhelum; and the Sind Sagar doab, between the Jhelum-Chenab and the Indus.

DOCTOR FISHES are tiny creatures of about 8 to 15 cm in size, slim and elongated in body and coloured blue and black with horizontal stripes. They are found on the west coast of India in the offshore region. One of them, *Fissilabrus dimidiatus*, is the most interesting individual, cleaning the teeth and infected soft tissues of other fishes and removing their parasites. This is done to obtain a meal for itself and in this effort it enters the jaws of large groupers, the vicious moray eels and poisonous scorpion fish. The most intriguing part of this commensalism is that the aforesaid fishes which are carnivorous and devour a large number of other small fishes, open their mouth quietly when the doctor fish arrives and the latter without any fear of the ferocious teeth, enters the mouth cavity and completes its cleaning duties with its pointed jaws and supple action. It attends to several such patients in an hour's time and provides a wonderful example of mutual understanding and symbiotic behaviour.

A blue-green Surgeon fish has a lancet-like sharp bony knife at the base of its tail, depressed in a groove, with which it can inflict injuries on enemies by erecting it and lashing the tail sideways.

DOG FAMILY (Canidae). Because dogs have been domesticated by man and trained to assist in many different functions, the whole family is of special interest. It comprises 14 genera and about 35 species occurring almost thoughout the world except New Zealand, Australia (the Dingo was introduced by aboriginal immigrants) and Oceania.

Intelligent and bright-eyed, restless and cunning, all members bear close similarities to domestic dogs, with well-shaped heads, long pointed muzzles, upstanding ears, bushy tails and deep-chested bodies. They are especially adapted for hunting and catching live food prey through an extended chase, and the speed and endurance of some species is phenomenal. A steady loping trot is the method by which most canids eventually exhaust their fleeting quarry. They have an acutely developed sense of smell and can follow the tracks of their prey even over hard sun-scorched ground, with ease. Representatives of the family include the well-known Wolf (*Canis lupus*) with the closely similar Jackal (*Canis aureus*) and Coyote (*Canis latrans*), and many species of foxes. Less typical canids are the Racoon Dog (*Nyctereutes procyonides*) of Japan which is the only winter-hibernating canid, which can swim and catch fish quite well. The tiny Fennec Fox (*Fennecus zerda*) of North Africa is wholly adapted to hot sand-dune desert, where it subsists largely on lizards and locusts. At the other extreme is the Arctic Fox (*Alopex lagopus*), found in north polar regions where it has been recorded in Greenland surviving 450 km from the nearest ice-free land. They will eat any vegetable or animal matter but subsist mainly upon lemmings and they do not hibernate even in mid-winter. Most of the Canidae form stable pair-bonds (unlike the domestic dog) with the male assisting in bringing food to the cubs. Litter sizes can be large, up to 13 offspring, and they are normally fed by both parents by regurgitation until the pups are old enough to swallow raw meat. The gestation period is 49 to 56 days in foxes and up to 60 to 65 days in Wolves, Coyotes, and Jackals. A peculiar feature observed in the breeding biology of

Chiloscyllium indicum × ⅛

many canids is the copulatory 'tie' whereby the male is not released by the cervical muscles of the female until some minutes after mating.

In our region the Indian Wolf is largely confined to desert or steppic mountain regions and is now quite rare. The Jackal is adapted to both forest, desert and the outskirts of human settlements and has penetrated the Himalayas up to elevations of 2500 m. Both the Jackal and the Wolf will mate with the domestic dog and produce fertile offspring, indicating their very close genetic relationship. Jackals tend to be sociable creatures and in rural areas the dusk-time chorus of wails from jackals as they emerge from their day-time burrows is, to the writer, one of the most evocative and appealing sounds of India. Jackals are very versatile and adaptable and will hunt for frogs in quite deep water, clamber into bushes to reach ripening fruit and raid the town refuse-dump for kitchen scraps. They will eat beetles and grasshoppers and are adept at stalking and catching small rodents. Litters of two to seven pups are normally born in the winter or spring months and they are suckled for up to ten weeks and hunt with their parents until they are eight months old. The Red Dog or Dhole (*Cuon alpinus*) is the only example, from our region, of a canid which lives and hunts cooperatively in a pack. This method of hunting has the obvious advantage of enabling larger and more dangerous prey species to be overcome and is typical of the hunting methods of the European and North American populations of wolf, but not of the Indian Wolf. Cape Hunting Dogs (*Lycaon pictus*) in Africa also hunt cooperatively. Four species of fox are found in our region, including the Tibetan race of the Red Fox (*Vulpes vulpes montana*) with a thick woolly pelage, and the large-eared desert-adapted Rupell's Fox. (*Vulpes rupelli*) in southern Makran region of Baluchistan. Rupell's Fox has adapted to hunt over soft sand-dunes by developing tufts of hair which cover and protect the pads of its feet.

As a last tribute to the intelligence of this family, a partial catalogue of the skills for which man has trained domestic dogs includes pulling carts in Belgium, collecting and driving sheep flocks over mountainous terrain, guarding mountain flocks against theft or animal predators, protecting property including factory premises and ammunition dumps, retrieving game and finding game for sports gun-shooters, detecting drug smugglers and apprehending criminals, and last but not least as companions and faithful friends for many otherwise lonely humans.

See also DHOLE, FOXES, WOLF.

DOGFISHES are cartilagenous fishes of the Shark family, occurring in shallow waters of both coasts of India and have two genera, *Chiloscyllium* and *Scyliorhinus*. They are also called cat sharks. Another small shark, *Scoliodon*, is also sometimes mistakenly styled as dogfish. *Chiloscyllium* is a bottom form, has flattened teeth, a blunt rounded snout, a short trunk and a slender shark-type tail. It feeds on crabs, prawns and molluscs inhabiting the bottom or shallow shores. Being a cartilaginous fish, the male has claspers for mating and holds the pectoral fin of the female in his mouth as if biting it, to steady her. After fertilization she lays her eggs in a mermaid's purse and leaves it in the sea for hatching.

See also FISHES.

DOLPHIN. What is the difference between a dolphin and a porpoise? Strictly speaking neither word is used in a purely scientific sense, and in some cases the two terms are applied to the same species. Generally porpoises have blunt or rounded snouts and heads whilst dolphins have a beak-like snout which is technically known as the rostrum.

All toothless or baleen WHALES, as well as dolphins and porpoises, belong to the order Cetacea but the dolphin and porpoise, together with many other smaller forms of whale such as the Beluga and Killer whales of the Arctic Sea, belong to the sub-order Odontoceti, meaning 'toothed whales'. They are all characterized by having one single orifice in the top of the head for breathing air. The Baleen whales have a pair of external nostrils or blow-holes.

There are two species of purely freshwater dolphin unique to the subcontinent: *Platanista gangetica* inhabiting the Ganga and Brahmaputra rivers and the Irrawaddy in Burma, whilst the Dolphin of the Indus river has recently been recognized as a distinct species *Platanista indi*, having developed slightly different bones in its skull. These river dolphins have narrow and quite long snouts, with a large number of needle-sharp teeth. They live exclusively in the turbid silt-laden waters of great river systems, catching fish and crayfish by a system of echo-location or sonar, as they have lost the ability to see in their muddy water environment.

Like all mammals they give birth to a fully developed young which must come to the surface to breathe and which is suckled by the mother. They feed mostly at night-time and do so by swimming on their sides, their spade-shaped square-cut flippers aiding them to pass through quite shallow water. From the coastal waters

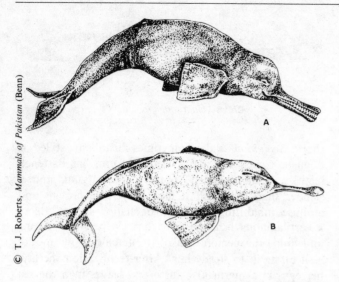

© T. J. Roberts, *Mammals of Pakistan* (Benn)

Indus Dolphin, A *Lateral view,* B *Dorsal view showing spiracle open to exhale.*

of Sri Lanka up to the Sunderbans in the east and the Arabian sea coast in the west there are two fairly common species which are considered neritic in ecological terms—preferring to haunt shallow estuarine and coastal waters. One is the Plumbeous Dolphin *Sousa plumbea,* and the other the Finless Black Porpoise *Neomeris phocaenoides.* The former is a purplish-grey colour with a pronounced narrow rostrum and a peculiar hump in its mid-dorsal region, from which the small dorsal fin protrudes. It is about 2·75 m in body-length. They travel in parties of 2 to 6 and feed largely upon squid and sole. The Finless Black Porpoise, as its name implies, has no dorsal fin and is a purplish-black colour and rarely exceeds 1·5 m in total body-length. Its teeth are peg-like, not conical as are most dolphins', and its head is bluntly rounded. There are several other species of dolphins which may often be encountered in huge schools in the coastal as well as deeper waters of this region. These include the Cape Dolphin *Delphinus capensis,* the Common Dolphin *Delphinus delphis,* the Bottlenosed Dolphin *Tursiops aduncus,* and the Electra Dolphin *Lagenorhynchos electra.* All these species have fairly prominent sickle-shaped dorsal fins and are very swift swimmers capable of leaping bodily out of the water when excited or displaying. Unlike their congeners in the river systems, these pelagic dolphins have well-developed eyes and can see their prey. All dolphins however employ a hunting system based upon the emission of high-frequency sound waves which bounce back variable echoes indicating whether the object is moving and what size it is, and by this means food prey can be located with unerring accuracy, a method of sensory perception hard for us visually-oriented humans to comprehend.

See also WHALES & DOLPHINS.

DOMESTICATION. Man has brought under his full control a number of plant and animal species since the beginning of the New Stone Age ten thousand years ago. These plants and animals not only live near his habitation, but breed under his control and make available a number of products that man utilizes at his will. But prior to such domestication, the hunter-gatherers must have routinely brought to their dwellings the young of many animals and brought them up so that these animals became habituated to man. Even today, tribals of the Nilgiris regularly keep animals such as Giant Squirrels as pets near their huts. And Asiatic Elephants are rarely bred under captivity, more often they are captured wild and then trained to obey man's commands.

Animals thus captured wild and tamed would retain their natural physical and behavioural attributes. The animals bred generation after generation under captivity would however entirely change under the influence of selective breeding by man. These are of course the truly domestic animals ranging from silk-worms and ducks to dogs and camels. A number of attributes may make an animal particularly suitable for domestication. Firstly, most domesticated animals are social in their wild state. This makes them tolerant of living at high densities in domestic herds; it also means that they have plastic behaviour and often high levels of learning abilities. The cat is perhaps the only asocial animal that is extensively domesticated, and of course it is the least tame of all domestic animals. The domestic animals are mostly herbivorous; this makes it easier to feed them. The domestic animals also derive from stocks that share the ground-dwelling habits of man; thus gallinaceous birds such as chicken, quails and turkey figure prominently amongst domesticated birds, and the domesticated Mallard duck is the most ground-dwelling of all ducks.

Man selectively breeds animals which are easy to tame and control, which serve best the use for which they are domesticated, and which breed most prolifically under conditions of domestication. Such selective breeding has markedly changed the attributes of animals under domestication. They have come to tolerate the presence of, and manipulation by, man, considerably lose their aggressive tendencies, tolerate high levels of crowding and lose inhibitions against ready breeding. Thus domestic ducks have altogether lost the elaborate courtship which precedes copulation in wild Mallard ducks. Man has also evolved a variety of breeds of domesticated animals suited for specific purposes. This is particularly the case with dogs whose breeds range in size from lapdogs smaller than a hare to huge mastiffs bigger than a wolf.

At least seventeen truly domesticated animals are maintained today on the Indian subcontinent. The

Characteristics of domestication. In each case the wild form is on the left, the domesticated forms derived from it on the right. (From Lorenz, 1965a)

honey-bee is native to India and wild hives of this species were traditionally tapped everywhere. Modern bee-keeping is however rather restricted to a few forested hill tracts. The wilder silkworm moth is maintained to produce tasar silk based on forest trees in central and eastern India, and the fully domesticated silkworm based on Mulberry, introduced from China, is maintained particularly in Mysore. The lac insect is commonly maintained in central and eastern India.

Of the domestic birds, we have chicken and duck. The chicken is closely related to the Red Junglefowl and may have originated in India. It is widely kept throughout the country. The duck is derived from the Mallard which is a winter migrant to India. It is kept in large flocks in Kerala where it supports a large group of pastoral people. It feeds largely on grain strewn in paddy fields after harvest, and the duck-keepers migrate with their flocks in search of harvested fields.

The dog is perhaps the oldest domesticated animal and is still widely used in India by tribal groups in hunting and by shepherds in protecting their sheep. Twenty-eight breeds of dogs have been described from the Indian subcontinent, for example, Bhotia and Dhangari sheep dogs, and the Mudhol and Banjara hounds. Cats are kept as domestic pets and also because they hunt rats.

Large flocks of sheep are maintained by professional shepherd communities in the northwestern Himalaya, in Rajasthan and in the semi-arid tracts of peninsular India. These shepherds migrate over long distances with their flocks. In the Himalayas they move to the alpine pastures in summer and to lower altitudes in winter. In peninsular India they move to low-rainfall tracts in the monsoon and to adjacent higher-rainfall or irrigated tracts in the dry season. Goats are maintained in smaller flocks by agriculturists or landless labourers throughout the rural areas of the country.

Cattle with a characteristic hump is the principal draught animal for Indian agriculture and rural transport and is maintained in small numbers by cultivators throughout India. Larger herds of cattle are maintained by pastorals in Rajasthan and in some parts of the peninsula. The buffalo is our principal milch animal and is maintained in small numbers by cultivators. Professional pastorals such as Rabaris, Gavlis and Gujars maintain larger herds of buffalo in forested hill tracts. The Indian buffalo is closely related to our wild Buffalo and freely interbreeds with it where the two come together as in Kaziranga National Park in Assam. The Yak is maintained as a domestic animal in some Himalayan tracts. The Mithun is a species of *Bos* maintained in large semi-wild herds in northeastern India.

The pig is traditionally maintained as a scavenger and as a source of meat by some tribes of northeastern India, and is also maintained sporadically in other parts of the country. The donkey is maintained sporadically as a beast of burden for carrying clothes, stones and mud. The horse used to be maintained in large numbers for military purposes and for transport, but its use is declining rapidly. The camel is maintained in large numbers for transport in the arid tracts of Rajasthan and Gujarat.

Domesticated animals have gone feral in some localities. For example, there are feral cats all over the country, feral cattle on the island of Sriharikota off the east coast of India, and feral Toda buffaloes in the Nilgiris. The Sri Lanka wild buffalo is also believed to be feral.

DOMINANCE BEHAVIOUR. Many animals have the habit of living in groups. We are all familiar with the terms *schools* of fish, *flocks* of birds, *troops* of monkeys, *packs* of dogs and *prides* of lions. While living in groups undoubtedly confers certain advantages such as protection from predators and the sheer advantage of numbers in tracking down prey or finding food, it must also have its associated disadvantages (see SOCIAL SYSTEMS). One of the obvious disadvantages of being part of a group of animals of the same species is that there would be keen competition for everything—all the animals in the group would need the same kind of food, shelter, mates, etc. This competition may result in fighting and it often does. However, settling all disputes by actual bitter quarrels irrespective of the disparity in the fighting abilities of the contestants would be very uneconomical for both parties. As if in conformity with this logic we see that animals often do not have to indulge in fights to settle disputes. Many animal groups organize themselves into dominance hierarchies (see SOCIAL HIERARCHY) where each animal in the group has a rank and a higher-ranking individual always has an advantage

Submissive behaviour in wolves. *Left*—Appeasing by food begging; *right*—passive submission by rolling on the back (After R. Schenkel, 1967)

over lower-ranking individuals in the case of a dispute. High-ranking individuals keep communicating their higher status to the lower-ranking ones by means of what we might call dominance behaviour.

It is easy to recognize the dominant male in a pack of wild dogs by the way in which he holds his head, ears and tail and by the way in which he approaches his subordinates and the females in the group. Look at a troop of rhesus monkeys and you will immediately see that there is always one unmistakably dominant male who holds his head and tail high and does not hesitate to look into the eyes of other males in the troop. Going back by millions of years in evolutionary time, watch carefully a colony of paper wasps that are busy building a little honeycomb-like structure under your window and you will immediately recognize the dominance behaviour of one or a few wasps over others. Most often you will see that the dominant individual unhesitatingly approaches the subordinate while the subordinate hesitatingly draws back especially if it does not have any food in its mouth to offer. More rarely you will see that the dominant individual climbs right on top of its opponent and tries to bite it while the subordinate one keeps as still and compact and crouched as low as possible.

Dominance behaviour is not necessarily visual. The honey-bee queen prevents the subordinate workers from developing their ovaries and laying eggs by secreting a certain chemical substance. If the queen dies or is removed the concentration of this chemical falls and the workers begin to lay eggs. Now, expose the queenless worker to extracts made from the queen or to a few workers who have been in contact with the queen and they will no longer lay any eggs.

Dominance behaviour can also be acoustic. A dominant blackbuck not only approaches its subordinates with raised neck and nose, folded ears and curled tail, but also gives a series of harsh calls.

What makes an animal dominant over another? Obviously there must be a host of factors. Many years of research with a variety of animals have revealed that size, age, hormones, fighting experience, strength, familiarity with the habitat, influence dominance behaviour but the exact contribution of each factor still remains obscure. An intelligent Chimpanzee once discovered that empty cans make a great deal of noise and from then on he used this knowledge to scare all the others in his group. Guess what happened? His rank immediately rose in the dominance hierarchy.

R.G.

DRAGONFLIES and DAMSELFLIES. The Odonata is an archaic and unique insect Order comprising some of the most commonly noticed and spectacular flying insects. The elongate abdomen, the two equal or subequal pairs of membranous (usually hyaline but sometimes beautifully marked in colour), reticulately patterned wings, and the large multifaceted eyes characterize them. The more robust dragonflies (suborder Anisoptera) have the eyes usually contiguous with each other, the fore and hind wings dissimilar in venation and usually in shape, and held horizontally in repose. The other major suborder (Zygoptera) consists of more slender damselflies with eyes well separated, fore and hind wings petiolate and similar in size and venation, and held together vertically when at rest. A third peculiar, almost extinct suborder, the Anisozygoptera, involves only two known living species which possess characters intermediate to the above two orders. The wings are like those of damselflies but the general body shape and larvae are like dragonflies. *Epiophlebia superstes* is probably the only living representative, found in Japan. The other species, *E. laidlawi*, is currently known only from a larva, found in a stream below Ghoom in Darjeeling District and described in 1921. Over 5000 species of dragonflies and damselflies are now known from the world, of which about 500 exist in the Indian subcontinent.

These structurally specialized insects possess complex behavioural traits, making them an excellent group for study by naturalists. The adults being large (mostly 30 to 90 mm long, a few less than 20 mm or more than 150 mm) and flying by day, the methods of a bird-watcher with a pair of binoculars and lots of patience

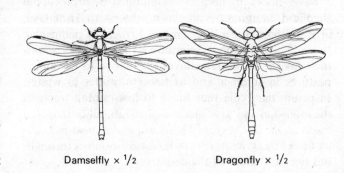

Damselfly × ½   Dragonfly × ½

© O. C. Edwards

Damselflies resting on toadstool mushroom × ½

© Ajai Ghorpadé

Dragonfly resting with characteristic wing posture × 1

can reveal very interesting insights into the life and habits of these birdwatchers' insects. As with birds, they have complicated territorial, sexual and other behaviour, with visual stimuli playing a large part. Some species, such as *Pseudagrion decorum* and *Pantala flavescens*, exhibit local migratory flights each year. In spite of their powerful flight, dragonflies and damselflies are peculiarly local, some rare species like *Chloroneura apicalis* being confined to a small stretch of river in the Western Ghats. Though found all over the Indian region, Odonata are particularly diverse and abundant in the tropical forested areas in Sri Lanka, Western Ghats, southern slopes of the Himalaya and in Assam and Burma. Best months for observing or collecting them are those either preceding or following the South-west Monsoon. Some species are very seasonal, flying only for a few weeks in the entire year and being single-brooded and rare. Dragonflies are typically diurnal but a few are crepuscular or even nocturnal, and some species form large aggregations, many others occurring in large swarms.

Dragonflies and damselflies breed in fresh water (or sometimes even brackish water). Any river, stream, rivulet, lake, tank, pond, marsh or even a swimming pool or man-made cemented reservoir or well, has its own fauna. Usually most damselflies of both sexes, when mature, haunt areas in and around their breeding water stretch. Most dragonflies, however, wander far from the source of water they emerged from, males usually returning soon and rarely departing, females on the other hand mostly staying away and reappearing only to mate and lay eggs. In *Anaciaeshna martini* mating is in forest away from water where males pre-ponderate, the females conversely staying near water. Dragonfly males usually occupy 'territories' that they defend, like birds, and wait for a favourably inclined female to copulate with. Members of the families

Gomphidae and Libellulidae guard circular areas by short darting flights from a medial, chosen perch: those of Aeshnidae and Corduliidae patrol linear territories over a flowing watercourse, flying up and down over the 'beat'.

Visual courtship behaviour may precede mating but tactile and chemical stimuli are also involved. Sustained flight in a 'tandem position' usually precedes copulation in flight or at rest, the male often escorting the female during oviposition, even under water! In one group of Odonata, the eggs are inserted into tissues of submerged, emergent or terrestrial plants, or even into soil slightly above water level. In the other group, eggs are extruded singly or in sticky masses or strings by the female and are washed off by flicking her abdominal tip into water while in flight. Some species are commonly seen to do this on macadamized roads, mistaking them for water! The entire life-cycle, from egg to adult, is usually from 2 to 18 months, some taking several years. Diapause (a period of suspended development) is recorded in egg, larval and adult stages for many species. The mature larva moves to the surface vegetation or edge of water and metamorphosis occurs usually in the night or about dawn, the soft emerged adult flying away from the breeding haunt to dry out and harden and assume typical adult colour. Many damselflies have prettily coloured bodies and the living 'jewels', *Rhinocypha* spp., have brilliantly iridescent wings too. Most dragonflies are clear winged and soft-coloured, but some, like *Rhyothemis variegata*, *Megalogomphus superbus* and others (males usually prettier than females!) are very attractive. The only other insects dragonflies and damselflies can be mistaken for are antlions or possibly some lacewings, but dragonflies and damselflies can be readily distinguished by their almost inconspicuous antennae.

Dragonflies and damselflies are voracious predators,

the adults capturing by their prehensile legs almost any kind of flying insect, some species occasionally taking a heavy toll of honey-bees and some, like *Orthetrum sabina*, feeding exclusively on other smaller species of Odonata. Dragonflies evidently consume large numbers of mosquitoes and house-flies and are important agents in the BIOLOGICAL CONTROL of those noxious insects. The aquatic, dull-coloured larvae patiently wait for or actively stalk small animals (other insects, even of their own kind, fish, tadpoles, etc.) which they grasp by rapidly extending the prehensile labial 'mask'. Rarely may some species acquire 'pest' status by feeding on trout fry or other young fish in cultivated fishponds. However, the dragonflies and damselflies are an overwhelmingly interesting and important faunal element on earth and deserving of more scientific and aesthetic attention than has been bestowed upon them so far in India.

See LACEWINGS & ANTLIONS. See plate 26 facing p. 369.

DRONGO. This family (Dicruridae) comprises some 20 species spread over Africa, SE Asia, the Philippines and south Pacific islands east to the Solomons. The Indian subcontinent possesses 9 species, of sizes ranging from bulbul to myna. The predominant coloration is glossy black, with sexes alike. Our commonest species are the Black, Grey, Racket-tailed, Haircrested and White-bellied drongos. The Black Drongo is also called the King Crow.

Found in all types of habitat from wooded savannas and gardens to moist forest, their primary food is insects—termites, dragonflies, cicadas, grasshoppers—and spiders etc. Drongos attack flying insects with great ferocity, usually catching them in the air but sometimes striking them on the ground. Known for their courage, they brook no intrusion into their nesting territory and audaciously attack and put to flight marauders such as crows, hawks and even eagles from the vicinity.

DRUMSTICK TREE (*Moringa oleifera*). This small tree of the sub-Himalayan tracts of the northwest is widely cultivated in the subcontinent. The bark is soft and corky. It is very popular now in agro-forestry as it provides food, fodder, fibre, and is fast-growing. The leaves are fed to cattle and are rich in vitamins.

DUCK (family Anatidae, which includes also geese and swans). Aquatic birds characterized by webbed feet, flat bills with fringed comb-like edges, tubby but streamlined bodies, and pointed wings suited for power-sustained flight. Divided into two groups: Dabbling, or Surface-feeding, Ducks and Diving Ducks.

Dabbling Ducks can usually be recognized by an iridescent patch of colour on the wing (the 'speculum'). They have a narrow unlobed hind-toe *contra* a broadly lobed one in the Divers. Their diet is largely vegetarian; they frequent and feed in shallow water on emergent plants or by 'up-ending' to reach the bottom mud for seeds, molluscs, worms, etc. Of the fifteen odd species of this group in the subcontinent five are resident, the rest winter visitors from their palearctic breeding grounds. Among the residents the commoner species are the Spotbill (*Anas poecilorhyncha*)—a large mottled brown duck easily recognized by its yellow bill-tip and two red spots at the base of the upper mandible; the Nakta or Comb Duck (*Sarkidiornis melanotos*), slightly larger, glistening blue-black and white, with a distinctive knob or 'comb' on the bill of the male; the chestnut-coloured Whistling Teal (*Dendrocygna javanica*), commonly seen on large lotus-covered village tanks; and the Cotton Teal (*Nettapus coromandelianus*),—also known as Goose-teal from the resemblance of its bill to a goose's—which is our smallest duck, being only slightly larger than a pigeon. All the last three usually nest in hollows in ancient tree-trunks standing in or near water. The Whitewinged Wood Duck (*Cairina scutulata*), found only in the dense swampy jungles of the NE areas, is facing extinction largely due to increasing human pressure on its specialized habitat. In size and coloration it is rather like the Nakta. Of the

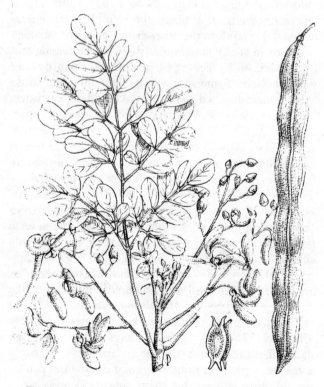

The thrice-pinnate leaves and white flowers of the drumstick tree (×1/2). Below is a winged seed (× 1/3) and to the right one of the greenish, pendulous pods (×1/3).

Male Pintail × ⅛

migratory Surface-feeders the commonest species are the Pintail (*Anas acuta*), the Common and Garganey teals (*Anas crecca* and *A. querquedula*), the Gadwall (*A. strepera*) and the Wigeon (*A. penelope*). The Mallard (*A. platyrhynchos*), which enjoys almost global distribution and is the ancestor of almost all domestic breeds of duck, is restricted chiefly to the NW parts of the subcontinent.

Diving Ducks have dumpy rotund bodies, short backwardly placed legs, and large feet with a broadly lobed hind-toe adapted for diving and underwater propulsion when feeding in deep lakes. Typical Diving Ducks are the Pochards, of which several species are winter visitors to the subcontinent. They are mostly sober-coloured birds and lack a metallic speculum. The drake Red-crested Pochard, however, has a brilliant golden-orange head on a brown and black body and is often misreported as the endemic Pinkheaded Duck (*Rhodonessa caryophyllacea*) which is now probably extinct. Pochards are awkward on land but they fly well, with rapid wing-strokes, and are regarded by hunters as good sporting birds. Mergansers or Sawbills are a small segment of the diving ducks distinguished by a very narrow serrated bill adapted for holding slippery fish which is their main diet. They live on torrential mountain streams in the Himalaya. Ducks constitute a valuable food resource and are hunted on a large scale for meat or sport in every country through which they pass on their bi-annual migrations. As a result of over-exploitation their world population is seriously impaired.

S.Q.A.

DUDHWA. The Dudhwa National Park was constituted in February 1977. It has an area of 480 km² and is almost flat. The Park is situated in the northern part of Uttar Pradesh along the Indo-Nepal border. It consists of a vast alluvial plain with perennial streams and has an average rainfall of 1600 mm. The vegetation consists of tropical semi-evergreen, tropical moist deciduous, swamp forests, and dry forests. The dominant trees are sal, laurel, haldu, sissoo, and semul. Dudhwa is famous for the large numbers of Swamp Deer, perhaps the largest congregation of this species anywhere in India. The other species of mammals found here include tiger, panther, sloth bear, jackal, mongoose, ratel, hare, elephant, antelope, sambar, spotted deer, barking deer, hyena, otter, langur, rhesus monkey.

As in most other protected areas in India grazing by cattle and poaching are serious causes of disturbance.

DUGONGS. The mammalian order Sirenia, popularly known as the sea-cows, includes two surviving genera—the monospecific dugong, *Dugong dugon* and the manatee, of which three species exist. The African continent separates the habitats of the two genera.

Dugongs occur along parts of the Indian Ocean coastline and, more commonly, in the western Pacific. Together with manatees, they form the only group of herbivorous marine mammals in existence.

In Indian waters, the largest dugong population exists probably between India and Sri Lanka, in the Gulf of Mannar and in Palk Bay, where abundant pastures of seagrass meadows grow in the shallows, providing food. Dugongs are also known from the Gulf of Kutch and from the Andamans. They have been recorded in lengths ranging from roughly one to three metres and in weights reaching an estimated 400 kg.

Not infrequently, dugongs swim in family groups consisting of a single young and its parents. Their attachment to each other often results in the capture of the entire family if any one of the members is netted or harpooned. The young suckle from prominent nipples situated high up on the mother's chest. The position of these (in the armpits under the flippers) is probably responsible for the old belief that sea-cows were mermaids. Sea-cows, in fact, are most closely related phylogenetically to elephants, sharing with them such traits as heavy bones, enlarged upper incisors (these miniature tusks occur only in the male) and anteriorly-placed mammae.

Dugongs do not stray far from coastal waters, and surface to breathe at intervals ranging from 30 seconds to 8½ minutes. These habits render them easy prey for man.

Dugongs have usually been hunted for meat. The fat underlying their skin, the skin itself and the tushes and bones have also been put to a variety of uses. Their keen sense of hearing has probably been instrumental in the continued survival of dugongs in areas where they were once abundant. Substantial populations survive today only in north Australian waters. Twenty years has been suggested as the dugong's longevity.

The present-day awareness of the possibility of man-engendered extinction of the dugong, for which the species has been listed in the Red Data Book by the IUCN and, in India, in Schedule I of the Wildlife (Protection) Act may give it a better chance to survive than that accorded to its larger cousin, Steller's sea-cow, which was hunted to extinction less than a century after its discovery in 1742.

S.B.

DUN. Though Professor D. N. Wadia said 'The Kashmir valley itself may be taken as an exaggerated instance of a dun in the middle Himalaya', the name is usually given to those shallow valleys between the Himalayan foothills and the Siwaliks. The Siwaliks are formed of very deep beds of sands, gravels and conglomerates eroded during and after Pleistocene times from the Himalaya and pushed up to about 1000 m by later earth movements. There are a number of duns westwards from the Punjab, the best known being the Dehra dun with its main town Dehra Dun (660 m) about half-way between the Yamuna and the Ganga and more or less on the water-parting between them. Into these rivers drain the mountain torrents from the north after rain. Those that flow into the Song join the Ganga, and those into the Asan the Yamuna. Because of the soil the water-table is low and the dun is irrigated from small canals.

EARLY MAN. We depend upon written records to arrive at a picture of the history of a country or a place or people. When we go back to earlier evidences like coins, inscriptions on rocks, excavated materials, they help in reconstructing ancient history. But man was there even before all these. That period since the advent of man till he deliberately left records of his activities is called prehistory. At that time there were many animals on land, mostly wild. Among the well-known animals of prehistoric times are the woolly mammoth (elephant), rhinoceros, reindeer, stag, bison, bull and wild boar. That these existed is known partly from a study of the fossils of these animals and partly from sketches of them made by early man in caves and on rocky surfaces.

It is generally accepted that man is also an animal evolved out of certain earlier animals, and the indications are that the apes were our immediate ancestors. As a matter of fact the mammalian order PRIMATES includes prosimians, monkeys, apes and man. There was always a controversy as to when man did appear first. The generally agreed human characteristics are speech, vertical posture, functional differentiation of hands and feet and great intellectual qualities. By a detailed study of the bones of primates in different states of preservation in different parts of the world and the implements man has used, it has been possible to trace his history. From what he has left behind one can arrive at a knowledge of his companions, his habits, his environment and also the climatic changes during his period of stay.

The earliest implements used by him were made of stone (flint, chert, quartzites, slaty shale). He had shaped them primarily to throw at and kill game to satisfy his hunger. Then he shaped small pieces into sharp arrow-heads, and hand-axes were also common. Some of these are called paleoliths, neoliths and microliths. These were found mixed with sediments in the terraces of flood plains of rivers, in lake beds, cave deposits, indicating his habitat. Some of them were called chipped stones (thunder-stones as some mistakenly named them) and polished stones. By a process of dating it was noted that early man migrated from place to place. This may be due partly to change of climate and partly due to lack of enough game.

The sites where the implements are found may be the actual habitats, or basins into which the implements were washed, or an area close to a hill where these were made because one would find implements in different stages of make. Jacques Boucher de Crevecoeur of France (1788–1868) is recognized as the father of prehistoric studies and he initiated many excavations of sites of these implements.

In India the most important areas are in parts of Kashmir (Pahalgam), in Punjab Himalaya (Beas valley), in Central India (Narmada and Wainganga valleys), Bagor in Rajasthan, Lekhania in Bihar, Betamcherla in Andhra Pradesh, Attirampakkam in Tamil Nadu etc. In Pakistan the Indus/Soan valley in the Northwest Frontier Province is well known besides Sanghao, Jamalgarhi and Khanpur. In Sri Lanka implements have been found in Ratnapura, Bandarawela and Balagonda, and in Yenangyaung in Burma.

By and large all these belong to middle to late Pleistocene in age, that is about sixty to twenty thousand years ago.

R.V.

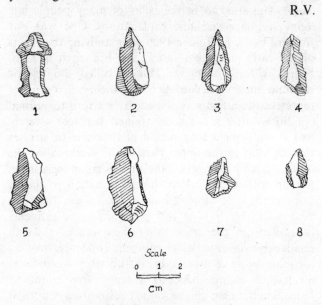

MICROLITHS FROM POONA

No. 1 is a unique tanged arrowhead. (From Damodar Dharmanand Kosambi, *Introduction to the Study of Indian History*.)

EARTHQUAKES. You know that the earth shakes and shivers at times, as we do in the cold winter months? These shocks are called earthquakes. Such tremors may pass off peacefully if they are very mild, but can destroy everything around if they happen to be very severe—pull down buildings, twist the rails, crack open hillsides, displace roads (Fig.1), produce huge waves on the surface of the sea (and also land) and usually lead to flooding and fires. A few thousand people lose their lives. There is no way of stopping earthquakes but today we have learnt to build better houses that will not collapse when the earth 'quakes' and are trying hard to find a method of predicting them.

Fig.1. *Earthquake damage after the Quetta quake of 1935 (31 May).*

Why does the earth tremble like this occasionally and at certain places? Is it because the Almighty God gets angry with a part of erring mankind and so punishes them? This used to be the belief of many people but today we have scientific explanations. One was put forward by H. F. Reid in 1906 after studying the effects of the earthquake that shook San Francisco in California, U.S.A., severely. This is called the 'elastic rebound' theory. According to this theory, earthquakes represent the attempt of the earth to return to normal conditions after it has been strained at places severely and slowly over a long period of time due to stresses that develop in the upper part of the earth called the 'crust'. Such places, where the stresses go on increasing, put up with them, undergoing deformation in stages, but snap up, developing huge cracks, when they can bear no more. It is at this point that readjustments take place in the crust using these cracks through relative movements, most of them suddenly, just as the mainspring of a clock breaks up when wound too much and springs back to a loose, free condition immediately. This, almost sudden, restoration of normal conditions brings about an earthquake at that place, called the 'focus', within the crust; and the point on the surface directly above the focus is called the 'epicentre' (Fig. 2).

2. DIAGRAM SHOWING THE FOCUS AND EPICENTRE OF AN EARTHQUAKE

The point on the surface immediately above the focus is called the epicentre. The intensity of tremors decreases with increasing distance from the focus. Lines drawn between places where tremors are of the same intensity (e.g. AA', BB', CC') are called isoseismal lines.

Minor adjustments take place for some time before and after the major shock; these are the series of 'foreshocks' and 'aftershocks' which may last a few hours, days or even months.

Nearly all earthquakes have their focus in the shallow crust at depths of less than 10 kilometres. However, there are earthquakes with their focus lying at depths of a few hundred kilometres.

What happens during an earthquake?

During an earthquake, essentially two types of waves or vibrations are produced at the 'focus' and travel through the earth. They are: (i) the P waves, also called the 'primary' or 'push-pull' waves, which are longitudinal or compressional in nature, with a to and fro movement like an earthworm; these travel 8 km a second on the surface of the earth and (ii) the S waves, also called the 'secondary' waves, which are transverse or shear waves with a transverse movement like the waves that are produced on water in a pond when a stone is dropped or like those sent down a rope by shaking it; these travel 4 km a second (Fig.3). The velocities of both these waves increase with depth in the earth's crust.

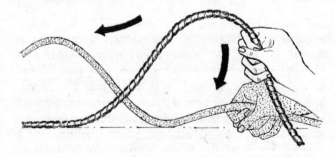

Fig. 3. S waves in a rope

Recording an earthquake

The recording of earthquakes is achieved by using the principle of inertia. Inertia is the tendency of a heavy body, such as a pendulum, to remain stationary when its support moves. The instrument used for this purpose is called a 'seismograph': it consists of

15 A Variety of Habitats

Mahabaleshwar, Western Ghats, once rich in animal life, now largely degraded.

Bhimashankar, Western Ghats, a hilly area teeming with wild boar and barking deer.

The Radhanagri Bison Sanctuary, near Kolhapur, also supports leopard, wild dogs and wild cats.

Habitats in the Western Ghats are suffering from the pressure of an ever-increasing human population.

The Moyar river in Mudumalai, Tamil Nadu, home to large herds of wild elephants and gaur.

Mountain stream in Dachigam, Kashmir. Black bear and the extremely endangered hangul are found here.

16 Crocodiles and a Star Tortoise

Marsh Crocodile (*Crocodylus palustris*) laying eggs

Marsh Crocodiles hatching

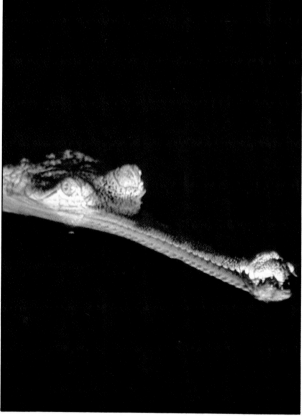

Gharial (*Gavialis gangeticus*)

Star Tortoise (*Geochelone elegans*)

the 'seismometer' that is sensitive to the ground movements and the recorder which registers them. Different types of seismometers and different methods of recording are available. In the simplest type (Fig. 4), a pendulum is suspended from a support and a rotating drum carries waxed or smoked paper; levers are used to enlarge the pendulum movement and a pin or stylus is fixed to the last lever which scratches the paper as it moves over the drum. Such scratched portions can be made more prominent by using a suitable chemical. Thus 'seismograms' are obtained.

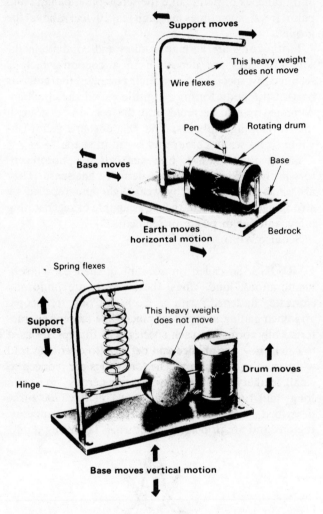

Fig.4. The principle of the seismometer. Horizontal motions are detected by the mechanical arrangement shown at the top, and vertical motions by the arrangement shown at the bottom.

© 1971. The Open University, Bletchley, England

There are other types with horizontal and inverted pendulums and with photographic recording either directly or indirectly. In these a mirror attached to the pendulum or a galvanometer reflects a beam of light on to a photographic paper moving over the rotating drum.

Reading the records

The seismologist is a scientist who studies earthquakes. He is able to recognize the many phases of the waves that are recorded in the seismogram during an earthquake—not only the P and S waves but the many other phases that are produced due to many changes which they undergo in the earth's interior, such as reflections and refractions. Since these waves travel with different velocities, he uses the time difference between their arrivals in the recording station to find out the distances they have travelled. This not only

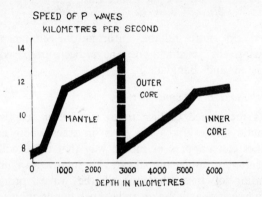

Fig.5. Wave speeds vary with depth due to variation in composition and rigidity.

© *Investigating the Earth* (Houghton Mifflin Co.)

helps to locate the epicentres of the earthquakes, but more significantly, the behaviour of the earthquake waves provides the most detailed X-ray picture of the earth's interior structure (Fig.5). If we are able to say today that there is an upper crust a few tens of kilometres thick, followed in depth by a mantle down to 2900 km, and an inner liquid core to the centre of the earth, all formed of different materials, this is almost entirely due to the study of earthquake waves.

Where do they happen?

There are two well-defined belts over the world where earthquakes happen repeatedly—the belt around the Pacific Ocean and the zone running from the Mediterranean through the Alps and Himalayan ranges to the Far East (Fig.6). India falls in the second zone and so is visited by earthquakes occasionally. It was proposed by Prof. W. D. West in 1936 that the Indian subcontinent could be divided into three earthquake zones, almost coinciding with the physiographic units, namely, the Great Ranges, the Great Plains and the Peninsular Plateau, with decreasing possibility of earthquakes. P. C. Hazra and D. K. Ray proposed in 1962 four zones in India alone composed of rocks of different ages, the youngest (Himalayan) zone having the greatest chance of earthquakes. Recently, the University of Roorkee has proposed five earthquake zones in India

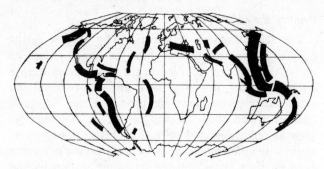

Fig.6. Most earthquakes occur along the mobile belts of the earth's crust.
© *Investigating the Earth* (Houghton Mifflin Co.)

based on more reliable information. Similar maps showing the seismic zones have been prepared for Pakistan and Bangladesh.

Earthquake-proof construction

As mentioned earlier, today we are in a position to build houses and dams, even in areas where earthquakes can happen, in such a way that these will not be damaged by earthquakes. This involves an understanding of the behaviour of the waves produced during severe earthquakes and the behaviour of buildings when set in motion, and consists essentially of tying all parts of the building together into a single body during construction so that they will not break even when subjected to earthquake vibrations. This is done in houses by placing thin steel plates, about 10 cm wide, on the walls and roofs connecting the corners. The Bhakra Dam across the Sutlej river in Punjab is a very high dam (225 m) that has been constructed using such principles.

See also FAULTS.

V.S.

EARTHWORMS (Oligochaetes). These are familiar inhabitants of every garden and can be seen in great abundance during the rains when they become very active. They are cylindrical, segmented worms without any appendages. However, each segment has a few hook-like chaetae embedded in the skin with which the worms gain purchase on the substratum. Hence the name Oligochaete (*oligo*, few *chaetae*, bristles). They are sometimes iridescent due to the optical effect of fine striations on the cuticle.

Their cartridge-shaped ends are ideally suited for burrowing. They tunnel extensively, literally eating their way into the soil, and as they extract organic material from the soil that passes through their alimentary tract, they eject the rest at the top of their burrows in the form of misshapen lumps or pellet-like castings. In this process, they turn the soil over, break up decaying vegetation, aerate the soil allowing plant roots to penetrate freely, and help in the formation of

humus—rich top-soil. Farmers, consequently, are eternally grateful to them. There is a record of a mound of castings three and a half feet high in the Botanical Gardens at Calcutta.

They cannot tolerate dryness and their burrows may be as deep as 10 feet to reach moisture and are often lined with mucus, or sometimes stuffed with leaves and paper which the worms drag in. They are mostly nocturnal and emerge from their burrows at nightfall, though they retreat rapidly when disturbed. Some are arboreal and inhabit accumulated detritus in the axils and branches of plants like the screw-pine, banana and palm trees, sometimes as high as 20 feet above the ground.

Earthworms are hermaphrodites and copulate in the soil. The eggs are deposited in a cocoon which is secreted by special tissue which encircles the anterior part of the body forming a girdle called the clitellum. Albumen is also secreted into the cocoon to nourish the developing embryos. The cocoons are yellowish-white to brown in colour and ovoid in shape.

Earthworms lack eyes, however scattered photoreceptors in their skin give them a dermal light sense. They are generally attracted by weak light and repelled by strong light. If injured, they are capable of regenerating a few anterior or posterior segments.

See also WORMS.

EARWIGS. So called on account of their ear-shaped, membranous, hind wings, these insects are uniformly elongate, flattened, with a freely moving, telescopic abdomen ending in a pair of pincer-like forceps characteristically formed in each species. The forceps are used to catch prey, for offence and defence, to open and fold wings, and in copulation. The fore wings are reduced to small, leathery cases and earwigs are generally 7 to 50 mm long, buff to black in colour. Most of the 1200 or so described species occur in tropical or warm temperate regions and are principally nocturnal, favouring damp,

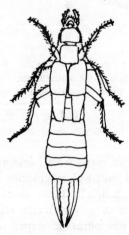

An earwig × 5

confined spaces (in soil, under bark and stones, in fallen logs), and are commonly attracted to lights. Their food consists of a wide range of living and dead plant and animal matter: usually animal food is preferred. The female earwig exhibits parental care of her eggs and young as long as necessary, and her behaviour can be likened to that of a brooding hen. Species of *Labidura* are commonly found near habitations, especially frequenting spaces under flowerpots. Two distinctive suborders of Dermaptera consist of wingless, viviparous forms, *Arixenina* parasitic on bats and *Hemimerus* on rats.

ECHINODERMS are spiny-skinned, radially symmetrical lower animals without backbones. They have highly developed locomotory, alimentary, vascular and nervous systems. They exhibit a starlike pattern, some with radiating arms. In general form they vary greatly: some distinctly starlike, some merely pentagonal, some with feather-like arms, some globular and some worm-like. They all possess a leathery coating over a skeleton of lime plates often with spines. The plates may be connected at movable joints. The plates may be closely united to form a continuous covering or scattered as particles or spicules. Other characteristics of echinoderms are that they all possess an extensive body cavity or coelom, a water-circulating system, and numerous organs called tube-feet, along their arms. These tube-feet help the animals to grip the surface, move about and obtain food. They are connected to a system of internal tubes through which water circulates.

Echinoderms include STARFISHES, SEA LILIES

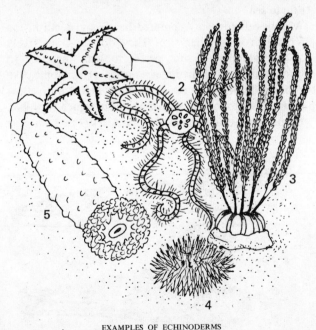

EXAMPLES OF ECHINODERMS

1 Starfish. 2 Brittle star. 3 Sea lily. 4 Sea-urchin. 5 Sea cucumber.

or Feather Stars, BRITTLE STARS, SEA-URCHINS and SEA CUCUMBERS. They are all widely distributed in all seas from the shore between tide-marks to varying depths.

Most of the echinoderms can throw off parts of their body on provocation, and as long as the animal is alive new parts will grow in replacement.

All echinoderms develop from a fertilized egg, which passes through a tiny, ciliated larval stage adapted for life in the open sea, in contrast to the adult which almost always lives at the sea bottom. Although adults are radially symmetrical the larvae are only bilaterally (two-sidedly) symmetrical. This shows that the flower-like symmetry of the adult is not primary, but secondary, and shows that echinoderms are related to higher animals with backbones.

ECOSYSTEM. The living community of plants and animals in any area together with the non-living components of the environment such as the soil, the air and the water constitute the ecosystem. The living and the non-living matter interact with each other, and the resulting flow of energy makes for the productivity of our world. In a particle of surface loam there are 60 million bacteria; in an acre of agricultural land there can be 30 million invertebrates, and in an acre of land there are over 50 thousand earthworms. Then there are the larger insects, reptiles, mammals, birds and plants. All these together form the biotic community, and as Raymond Dassman points out: 'These relationships are never static; growth and death, change and replacement, go on continuously. Energy pours down from sunlight and is captured by green plants and transformed . . . water moves through the ecosystem in an intricate cycle starting from and returning to the atmosphere.'

The functioning of an ecosystem depends on the energy which is poured into it and the prime source of energy is the sunlight. Green plants are the only effective converters of sunlight energy into forms useful to animals and to man.

No energy can be produced without the consumption of matter, and this results in the pyramid of life having a wide base of vegetation, a smaller quantity of herbivores, which feed on the plants, and a considerably smaller number of carnivores which live on hooved species. Because of immutable natural laws one kilo of chital meat cannot produce one kilo of tiger, for a great deal is lost in the process of converting the meat of the deer into the metabolism of the carnivore.

The area or extent of an ecosystem depends on the objectives of the investigator. The entire stretch of the Ganga could be considered as one ecosystem, ranging from the melting snows in the Himalaya to the Hooghly where the waters ultimately reach the Bay of Bengal. Of course this mighty river consists of many smaller ecosystems, each one of which has got to be preserved in the interest of the river. The removal of vegetation on the banks, and particularly on neighbouring hillsides, leads to siltation which, apart from reducing the capacity of the river, alters the quality of the water and subsequently the insect and fish life. Even a village pond is an ecosystem by itself and many of these are getting silted up, polluted, or bereft of life because of the destruction of the natural complex of vegetation around the pond.

It is noteworthy that one of the most productive ecosystems is at the point where sea water meets fresh water. Therefore our estuaries need special protection because it is here that the spawning of fish and of other aquatic life often takes place.

Conservationists have now realized that to save the natural world, the focus should not be on saving individual species of life, but on saving the ecosystem of which that species forms a part. Unless the entire ecosystem is preserved the individual species will die in course of time.

EELS. Though eels are long, cylindrical and serpent-like, they are true fishes. When present, their paired fins are very much reduced in size, the dorsal and anal fins being considerably elongated and sometimes continuous with a reduced caudal fin. Their scales are minute and embedded in the skin, with the result that superficially they look scaleless. Their movement in water or on wet grass is also typically serpent-like. Some eels are marine, while others inhabit rivers and lakes and belong to widely separated families. However, a large number belong to the suborder Anguilloidea, and of them *Anguilla bengalensis* and *A. bicolour* are found in our rivers. There are many gaps in our knowledge of the life history of eels in the Indian region but the life account of their European cousin *A. anguilla* is very interesting. For a long time no one knew when and where the fish were breeding. It was imagined that when eels became old, they rubbed against rocks and that new eels were born from the small pieces rubbed off. However, scientists found that when they are about 12 years old eels migrate from European waters straight to the Sargasso sea, about 7000 km across the north Atlantic ocean, and breed in that area. After the eggs hatch, the larvae, which are thin small leaflike creatures called leptocephali, quite different from their cylindrical parents, feed there for some time and then take a homeward journey on the warm waters of the Gulf Stream. This long journey takes about three years, during which time they slowly change into mini eels, and enter European rivers to grow into adult eels. After about 12 years the same process is repeated.

It is interesting to note that eels from American rivers, which belong to another species, also migrate to

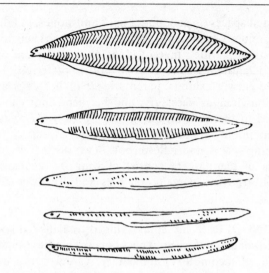

Developmental stages of *Leptocephalus* larva (top) into mini-adult eel (bottom) × 4/5

the same breeding area but the leptocephalus larvae of those eels always move towards American rivers and do not get mixed up with the European eel larvae.

Another freshwater eel of our rivers, *Mastocembalus armatus*, belongs to a different family (Mastacembalidae) and is of different nature. Though snake-like, the dorsal fin has short saw-like spines and hence it is called Sawback eel. It hides in crevices and gulps unwary prey.

The marine eel, *Muraenesox talabon*, a large yellow eel which sometimes weighs as much as 10 kg and measures 3 m, has a long pointed snout and ferocious canine-like teeth. It has a long tubular air-bladder and it forms an important fishery in the northeastern Arabian sea.

Moray eels too, have pointed heads and teeth in the jaw though of smaller size. They do not hunt freely like the talabon eels but prefer to wait for prey to approach them in their underwater caverns. Some, like the painted moray, have striped or black-and-yellow patterns resembling the surroundings in which they live.

Some small eels such as the green eel (*Pisoodonophis*

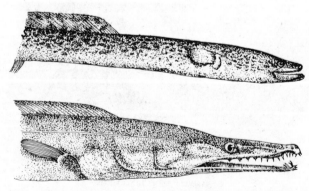

Above, head of *Anguilla bengalensis* × 1/12
Below, head of *Muraenesox talabon* × 1/20

boro), which are common on the west coast, are known to be burrowing eels. They are small (25 cm) and greenish in colour and have a tough muscular tail with the help of which they are said to burrow holes in the bunds of the estuarine partially submerged areas known as 'khajan' or 'gazni' lands.

See also DEEP-SEA FISHES

EGRETS (family Ardeidae) are white marsh birds with long legs and necks. Three common species are the Large Egret (*Ardea alba*), the Median (*Egretta intermedia*) and the Little (*E. garzetta*). During the breeding season egrets develop ornamental lacy plumes on their back, the commercial demand for which by fashionable ladies in Europe had once brought the birds near to extinction. The Cattle Egret (*Bubulcus ibis*) is similar to the Little, but with a yellow instead of black bill. In breeding plumage it acquires an orange head, neck and back. It is usually found in dry localities attending grazing cattle for the insects etc. they disturb. Egrets are sociable by nature and commonly seen feeding in small parties, roosting gregariously often with crows and mynas, and nesting communally. Their food is chiefly insects, frogs and reptiles.

Egret × ¹/₁₀

ELECTRIC FISHES are a few specialized species which are capable of producing electricity in their bodies and are thus unique in the animal kingdom. For

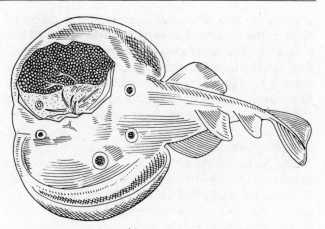

Torpedo marmorata × ¹/₃. Part of the head has been cut open to expose one of the two electric organs.

this power-generation they possess electric organs which consist of a group of hexagonal muscular tissues separated by fibrous ones which are connected by sensitive nerve endings serving as positive and negative poles at opposite ends. These muscles on stimulation by nerves create electric charges which can be measured and vary from 60 to 220 volts, or even more in some cases. They are used for stunning the prey or driving away enemies. Man also sometimes becomes a victim if he happens to touch the fish.

In another type of electric generation, low-frequency fields are created around the body for recognizing the mate in muddy waters as well as for sensing the approach of enemies or obstructions in the neighbourhood. The electric eel (*Electrophorus*) and catfish (*Malapterurus*) of the Amazon river are capable of lighting small electric bulbs. In Indian waters the small ray fishes *Torpedo* and *Narcine* are known for their capacity to administer electric shocks of low intensity. They have their electric organs situated on both sides of their backbone, behind the gill slits. In the specialized Stargazers, which are also capable of giving electric shocks, these muscles are situated near the eyes in the head region. All the three species inhabit the muddy coastal waters of both coasts and sometimes startle fishermen by their shocks.

See STARGAZER.

ELEPHANT. The elephant of Asia and India, known by the scientific name *Elephas maximus*, is the second largest land mammal living on earth today. Only its relative the African Elephant (*Loxodonta africana*), is bigger. These two species are the sole survivors of a once rich and varied group, and both are now threatened with extinction. This is particularly true of the Asian Elephant, of which less than 40,000 remain in the whole world.

Most people are familiar with the general outward appearance of an elephant, but may not know how to distinguish the two surviving species. The African is the

© R. A. Krishnaswamy

larger of the two, a big bull standing 3·35 m (11 ft) or more at the shoulder and carrying tusks about 2·13 m (7 ft) long. The Asian species seldom stands more than 3·05 m (10 ft) at the shoulder and its tusks average about 1·37 m (4·5 ft) in length, with the longest known pair measuring 2·59 m (8·5 ft). Whereas the female of the African species carries quite long tusks, the Asian female is either tuskless or has very small tusks, known as 'tushes', which often do not project beyond the jaw. Many Asian males are tuskless and are known as 'mucknas'. A large African bull may weigh over 7112 kg (7 tons), an Asian bull about 6096 kg (6 tons); the females of both species weigh between 1000 and 2000 kg (1 and 2 tons) less than the males.

The silhouette of the two species, looked at from the side, is very different. The back of the African elephant shows a marked dip between the fore and hind quarters. whereas that of the Asian forms an unbroken convex curve. The African has a comparatively elongated face with a flat forehead; the Asian has a more rounded face, with a twin-domed forehead. The ears are most distinctive, being so large in the African species that they cover the whole of the neck and shoulders and reach as low as the breast; the ears of the Asian species are comparatively small. The African's trunk is marked by repeated horizontal ridges, and ends in two fleshy processes or 'fingers'; the Asian's trunk is smooth, and has only one such finger.

The differences mentioned so far will be sufficient to distinguish the two animals, but there are other less obvious ones. For example, secretion from the *musth* glands, situated high on each cheek between eye and ear, is more or less continuous in both sexes of the African elephant. On the other hand the female Asian elephant rarely, if ever, shows *musth*, while in the Asian male *musth* is a discontinuous, temporary condition of varying duration, and is associated with unpredictable and sometimes dangerous behaviour. The function of *musth* is poorly understood in both species.

The elephant is entirely vegetarian, and subsists on a wide variety of woody plants, palms, bamboos and other grasses. An adult requires about 270 kg (600 lb) of such food daily. This is gathered and prepared for ingestion by the combined use of the trunk, tusks and feet. Elephants also drink by drawing water into their trunks, up to 9 litres (2 gallons) at a time, and then releasing it into their mouths. Large elephants have been known to drink 227 litres (50 gallons) of water in one day.

After mating the pregnant cow carries the foetus for nearly 2 years before giving birth, which is one of the longest gestation periods for any mammal. The new-born calf is only 90 cm (3 ft) high but weighs 90 kg (200 lb). Usually one calf is born, but twins and even triplets have been recorded. In the wild, baby

© M. Krishnan

Adult bull with long tusks, Western Ghats × $^1/_{38}$

elephants are born into relatively stable family groups of about 6 animals. The members are all related to each other. The groups consist of adult females with their immature calves of both sexes. Because of their slow growth-rate, there is a long period of attachment between mother and calf, and it is not unusual to see one or more calves of different ages and sizes at heel behind their mother at the same time. As they mature the bond of attachment weakens and calves stray farther from their mothers. The females will never leave their orginal family group, but the teenager bulls eventually separate completely.

Bulls are commonly found on their own, and tend to occupy ranges different to those of the families. Sometimes they may form temporary groups with 2 or 3 other bulls. Bulls will join the different family groups they happen to encounter for short periods. During such temporary spells with the females a bull may mate with any cow happening to be in estrus at the time.

There are other unique features to elephant dentition apart from their tusks. Each jaw carries two molars, each of which is replaced six times during the animal's life by a progressively more massive and complex tooth. As the food they eat is coarse, elephants' teeth are worn away quite fast, and the replacements are therefore essential. However when the sixth and last set has been worn down, elephants cannot eat properly and will eventually die. Usually however an elephant will die from other causes before reaching this stage.

Apart from the tiger, which may take the occasional calf, the Asian elephant's only serious enemy is man, thanks to his greed for ivory and space. Man's development of elephant habitat is the biggest threat to the species in Asia. The Asian elephant was once found from Iraq in the west, throughout Asia south of the Himalaya, and east as far as China. Today, Pakistan and the countries to its west have lost the elephant, which now has its last strongholds in India, Sri Lanka, Burma and other parts of southeast Asia. The elephant undoubtedly roamed over all but the most arid tracts of the Indian subcontinent a few thousand years ago, having been recorded even from the dry tracts of Punjab and Saurashtra in the fourth century A.D. Today its range has shrunk to parts of forested hilly tracts of the Western and Eastern Ghats in South India, the eastern fringes of forest tracts in Central India, the foothills of the Himalaya and the hills of the north-eastern states. The total population in India is around 15,000 elephants.

Throughout India the elephant and man live in constant conflict with each other; the elephant regularly raiding the cultivated crops of man, and killing several human beings every year, and the cultivator harassing and killing an equal number of elephants. Indians have captured and domesticated elephants for at least the last 3000 years; a large number are still captured and used for hauling timber, for riding and as a prestige symbol for temples. A large community of artisans and traders (estimated to be 7200 in 1980) depends on ivory for their livelihood and a number of tuskers are regularly poached for this purpose. Man is also continually encroaching into the remaining habitat of elephants which is ever shrinking and fragmenting. There is clearly a real threat to the continued existence of the Asian Elephant in India.

R.C.D.O.

See plate 25 facing p. 368.

ELEPHANT CREEPER, *Entada pursaetha,* a gigantic climber with disjunct distribution, found in the Eastern Himalaya, Western Ghats, Burma and Sri Lanka. Reported to attain 1·5 km in length, it is perhaps the largest climber in the world, with enormous bean-shaped pods over 1·5 m long and 10 cm wide, containing chocolate-coloured seeds. Small, yellowish flowers appear in summer in long spikes. The leathery pods are like elephant's skin and these pachyderms relish its foliage. The stems contain drinkable water which is obtained by cutting it into sections. The bark is used for killing fish. Plants like this are becoming rare as vast areas are being cleared to plant industrial woods like teak, eucalyptus and pines. To save such fantastic plants from extinction, they are protected in National Parks.

ENDEMISM. An evolutionary phenomenon in which isolation leads to the development of unique species confined to a particular area. Isolation can be *geographical,* as in the case of oceanic islands or in landmasses separated by insurmountable barriers like high mountains or big rivers; *ecological* when specialized ecological niches prevent intermixing of populations between different habitat zones; or *physiological* when incompatible breeding seasons separate populations reproductively. In all cases the result is lack of exchange of gene pools, and consequent diversity of evolutionary trends leading to the formation of endemic species. Endemism is most pronounced in oceanic islands; on landmasses it is less marked, and spread over larger zoogeographical zones or their ecological subdivisions.

In spite of their long-distance flying ability, birds also exhibit endemism. While elusive of explanation, it is clear that some recent endemism is the result of climatic and geographical changes in an originally wide range. Of the subcontinent's 1200 species, 176 are endemic. Two percent of these have untraceable affinities; 18% were clearly related to Palearctic birds till the Pleistocene and later Ice ages; and another 18% to the Ethiopian region via the now extinct Gondwanaland. But 62% of the birds, like other Indian animal groups, have Indo-Chinese affinities, bearing out the theory that the India-China-Burma trijunction was the Indo-Malaysian evolutionary cradle. Some of these endemic species have become extinct or nearly so through hunting and habitat destruction. Such are the Pinkheaded Duck (*Rhodonessa caryophyllacea*), the Mountain Quail (*Ophrysia superciliosa*), and Jerdon's Courser (*Cursorius bitorquatus*).

B.B. & K.K.T

ENDOCRINE GLANDS (HORMONES). Glands are spaces, or potential spaces, lined by secretory cells. These cells are able to prepare special extracts or secretions which are collected in the gland space and then passed out, through special ducts, or flow into the blood stream. Each gland or group of glands prepares a particular substance, e.g. the mammary glands prepare milk, the sweat glands prepare perspiration etc. Glands whose secretion passes out through ducts are known as exocrine glands. Those whose secretions are carried away by the blood, and not passed out of the body, are the endocrine glands.

The endocrine glands are important because their secretions, known as hormones, act on special cells or organs in other parts of the body, whereby they control and regulate the metabolic functions, growth and maturity of the body. Metabolism refers to the absorption, utilization and excretion, i.e. the turnover in the body, of the various substances ingested, as well as to the role these substances play in growth and energy production.

The important endocrine glands are: (1) the Pituitary; situated at the base of the brain, which produces several hormones concerned with growth, sexual maturity, stimulation of milk secretion, etc. It also exercises a regulatory function over the activity of the other endocrine glands; (2) the Thyroid, situated in the neck, which controls metabolic activity and also growth; (3) the Adrenal or suprarenal, situated near the kidneys, which is concerned with sexual maturity and also with metabolism especially in regard to salt and water balance. There are other endocrine glands, concerned with sugar and glycogen metabolism (the islets of Langerhans), with calcium metabolism and bone formation (the parathyroids), and with sexual activity and procreation (the ovaries and testes). These sex glands produce internal secretions in addition to production of the ova and the sperms, which are necessary for reproduction.

The endocrine secretions also act on the other endocrine glands, to stimulate or inhibit their activity according to the requirements of the body. Various abnormalities of development can be traced to deficiency or malfunction of certain endocrine glands.

The thyroid hormone which is responsible for metabolic activity and maturity is a compound containing iodine, which cannot be produced if iodine intake is deficient. Usually the iodine dissolved in water is sufficient, but in areas where dissolved iodine is absent or inadequate, various abnormalities are seen in the population. A striking example is the axolotl of Mexico. This is an amphibian which never develops beyond the larval or eel-like form, but matures, mates, and dies in the larval form. However, on addition of iodine to the water in which they live these eel-like animals develop into mature amphibians. Human populations in iodine-deficient areas are known to suffer from goitre, an abnormal enlargement of the thyroid, or with more severe deficiency, from cretinism, a failure to develop, both mentally and physically.

The secondary sexual characters, distribution of hair, development of breasts in mammals etc., are produced by endocrine activity, as also the brilliant colours and display found during courtship in many animals and birds. In most species, the courtship display occurs before actual sexual activity, and the production of sperms to fertilize the ovum occurs after the courtship display is over. Thus the 'breeding plumage' in birds, is really a 'courtship plumage', which may fade by the time actual breeding occurs.

ENVIRONMENT. Literally, the word means surrounding objects or circumstances; but the term has come into prominence in recent decades as a result of the fact that the conditions in which human beings now live have been deteriorating over the years. In olden times it was assumed that natural gifts like clean air and fresh water would be available for the asking. But with industrialization, the creation of an infrastructure for industry which proliferates even in the countryside, and the large-scale use of synthetic chemicals for public health and agriculture, natural resources such as air, water, soil, flora and fauna have been seriously affected in every part of the world. As a result of this, our en-

vironment has also deteriorated and has affected the quality of human life.

To redress the imbalance to the extent possible, the first UN conference on the Human Environment was held in Stockholm in June 1972, and since then, with the forming of the United Nations Environment Programme, almost all countries have undertaken to monitor the quality of air, water and other components of the natural world. In the United Kingdom, for example, the Thames was but a few years ago a highly polluted river, and over long stretches there were no fish or birds. Today the waters are so clear that there are more birds and fish than there have ever been. Similarly, smog, the London combination of smoke and fog, has disappeared. In the early 1950s there were several deaths due to smog in London, but today the environment has been cleared of this menace.

The environment consists of living forms of life as well as non-living substances like the soil, rocks, water and air. It is the interaction of the living and non-living which consitutes our environment. It is important to remember that a seemingly healthy environment can rapidly be destroyed by the infusion of some unhealthy element. Many of the components of the environment have the capacity to travel far and wide through the food-chain unless their progress is carefully monitored and arrested. For example, a chemical like DDT used for the killing of mosquitoes finds its way into the water, is eaten by fish, then by birds, and then travels on their wings to faraway places where, eaten by other predators, it may again cause insidious damage.

In our area, though industrialization is only patchy, there has already been very considerable deterioration of the natural environment. Forests have been cut down or submerged for the creation of multipurpose projects in mountain areas. The rising population demands more and more firewood for cooking the evening meal, and already less than a fifth of our area is under forest cover, while at least 30 percent of the land ought to be forested. The grasslands have been overused over many decades due to increase in the cattle population, and the absence of any rotational grazing. The wetlands have become eroded and polluted, and in short the environment needs to be intelligently rehabilitated in the shortest possible time. A Department of Environment was established by the Government of India in 1980 and it has been authorized to oversee the activities of all the other sectors of the government, to ensure that avoidable environmental mistakes are not made.

ESTUARINE SNAKES. Five species of watersnakes inhabit our island and coastal estuarine marshes. Dog-faced watersnakes (*Cerberus rhynchops*) are the best known and are found on the shores of the Indian

Ocean and the Bay of Bengal. They grow to 1¹/₂ m and are the largest of the five. All have fangs at the back of the mouth with a mild venom and all have eyes and nostrils positioned on top of their heads so as to allow seeing and breathing with the rest of the body submerged. Estuarine snakes use their sensitive tongues under water to feed on fish and eels.

See also SNAKES.

EUCALYPTUS. The eucalyptus or gum trees are the dominant trees of the Australian continent and were observed by naturalists on Cook's expedition when samples were collected. The name Eucalyptus is derived from the Greek *eu* meaning 'well' and *kalyptos* meaning 'covered', referring to the cap which covers the flowers before they open. Eucalyptus are recognized by their giant size, mostly smooth white, talcumpowder-like bark and evergreen, alternate, gland-dotted aromatic leaves. The sepals and petals are united to form a cap which splits transversely. Foresters are constantly in search of fast-growing trees, that are adaptable to different climates, and are useful for a variety of purposes. Eucalyptus are such trees and therefore planted all over the world. *Eucalyptus* has 600 species growing in a range of habitats from the snow-clad Australian Alps to the semi-arid deserts. In size they range from shrubs to the giant Mountain Ash (*E. regnans*), 90 m in height and the tallest hardwood tree of the world. Over a hundred species have been introduced into the subcontinent. Along the railroad track of the 'toy trains' from Kalka to Simla and Coimbatore to Ootacamund groves of different eucalyptus are seen, some planted over a century ago. The more important or interesting eucalypts planted in our area are: Blue Gum, Lemon-scented Eucalypt, Mysore Gum, Rainbow Eucalypt and the Red Gum.

Blue Gum (*E. globulus*) is widely planted in the Nilgiris or Blue Mountains. The bluish foliage of Blue Gum and the undergrowth of the blue CONEHEADS gives a bluish tinge to these mountains. The Blue Gums were planted in the mid-nineteenth century to meet the fuel-wood requirements of the Nilgiris. They are tall soaring trees 45–65 m in height, with smooth silvery to steel-blue bark. The bluish-white leaves turn dark green when mature and the flowers are whitish. The leaves are a source of eucalyptus oil.

Lemon-scented Eucalypt (*E. citriodora*) is so named because of the pleasant lemon aroma of its leaves. The trees are 20–40 m high, with smooth, white shiny bark. The leaves are lance-shaped. Fruits pitcher-shaped to ovoid or top-shaped.

Mysore Gum is a hybrid, widely planted because of its fast growth: 13 m in 3 years has been recorded.

Rainbow Eucalypt (*E. deglupta*), is so called because

Blue Gum shoot with two flowers and the four-ribbed fruit, × ²/₃

of its attractive multicoloured bark. It is endemic in Timor and in cultivation at the Forest Research Institute, Dehra Dun.

Red Gum (*E. camaldulensis*) gets its name from the red colour of the wood. The trees are 20–50 m tall; with silvery fresh bark contrasting with the dull brown of the previous years. It is easily recognized by its beaked fruits.

In Rajasthan and the Punjab eucalypts have been used in millions to afforest treeless wastelands and semi-arid areas.

K.C.S.

EVOLUTION. The origin of life on earth is a matter for speculation. Recent researches in biological chemistry and, in particular, the synthesis of nucleic acids and portions of genes in the laboratory, have given some indications of how it might have occurred. It can be postulated that as the earth cooled and its crust solidified there were periods of storm, stress, earthquake and lightning with drastic changes of temperature and pressure and electrical discharges, conditions favourable for the synthesis of various elements to form complex compounds, including the long-chain structures of linked carbon atoms described as organic compounds. Among these would be the ribose and deoxyribose nucleic acids which are necessary for ordering the structure and functions of a living organism to make it a specific entity (see HEREDITY : GENETICS). Most living organisms reproduce and multiply very rapidly. The progeny of a single bacterial cell could cover the entire surface of the earth in a period of weeks if allowed to multiply unchecked. Such massive overproduction is necessary to ensure that some members of the species will find conditions suitable for survival. Of the millions of progeny only a few find themselves in favourable circumstances. The rest perish. Further, as the numbers of organisms increase, there is competition for the available sources of nutrition, both between different species and between members of the same species.

All members of a species are marked by identifiable common characteristics, but they are not all identical and a considerable range of variation may be seen among members of a species. Those variations which fit the individual for life in its environment will naturally predominate since more individuals having this type of variation will flourish, whereas individuals who lack this favourable variation will die out. When large-scale climatic changes, such as an ice age, occur, variants which were previously favourable may become unfavourable in the changed environment. A species which has greater adaptability and can produce a wider range of variations, may still produce variations better fitted to the altered environment and survive. Species which are more fixed, with less capacity for variation, may die. The fact that they once existed can be discovered from fossils—the remains of dead animals and plants which get mineralized and buried between successive geological strata where they remain preserved until accidentally discovered. Fossil records are obviously incomplete, since relatively few organisms die in circumstances in which their remains are mineralized and preserved. Many thousands of fossils have, however, been uncovered and studied. From these it is possible to understand something about forms of life which previously existed on earth, and in some cases, to trace connections between these and the species existing today. For example, a line of descent can be traced from a fossil animal with toes instead of hoof, which was probably the precursor, through intermediate fossil forms, of both the horse and the ass (Fig. 1). The dog-sized fossil horse *Eohippus* would be able to move silently through the dense undergrowth in a primeval forest. It was not strong enough or fast enough to escape from predators in open country. A change in habitat gradually brought about dominance of a form which could run fast over grassland and rough, sometimes rocky, terrain and also had the strength to escape from large predators. Such conditions favoured the evolution of a large horse with hard horny hooves.

Up to the 19th century most scientists accepted the

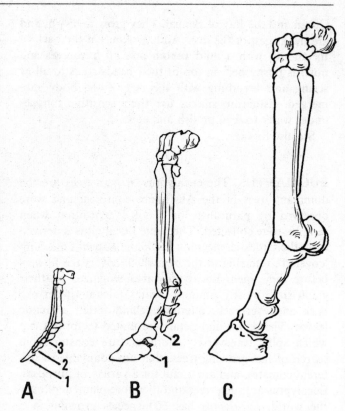

FIG. 1. STAGES IN EVOLUTION OF THE
HORSE'S HOOF
(A, B, C TO SAME SCALE)

A Foot of *Eohippus* (fossil) showing three of its four toes

B Foot of *Merychippus*, (fossil) with enlarged central digit (1) and rudimentary fifth digit (2) lifted off the ground.

C Foot of modern horse *Equus*, with big central hoof. The fifth digit has disappeared.

belief that all species of life had been created individually and continued to exist in their present form. As biologists discovered the tremendous diversity of living organisms and attempted to name and classify them, they saw that these living organisms fell into natural groups, and that some members of a group were more closely allied than others. Certain closely allied forms shared all but one or two characteristic features. These few differences could often be related to the environment in which these species lived, in being better fitted to live in that environment. This led people to speculate whether one species could modify and change into another. Experiments aimed at producing such changes were unsuccessful. These were necessarily short-term experiments, whereas advances in the geological sciences now indicated that life on earth could be measured in millions of years. Darwin suggested that changes of the type necessary may appear as cumulative changes over thousands of generations, even if not apparent during the few generations of a laboratory study.

FIG. 2. VARIETIES OF PIGEONS PRODUCED
BY SELECTIVE BREEDING

A White fantail, B Brunswick, C Nun.

Charles Darwin was appointed the biologist on the naval survey ship H.M.S. *Beagle* in 1831. During the voyage he collected thousands of specimens of plants and animals from various parts of the world. He saw that there were species found on an isolated island resembling, in all aspects except one or two, similar species found on the nearest mainland. The features by which the island species differed were such that it was better fitted for life in the island environment, and he presumed that they may have been produced by variations over centuries and millennia of geological time, the cumulative effect of such established difference leading to formation of a new species. Darwin spent many years in the study of his collections and found more evidence which favoured his theory. In 1859 he published his book *The Origin of Species* which expounded his theory of evolution.

Darwin expressed the view that change in a living species was selected, in the sense that a variant form which was more efficient, i.e. better able to survive in the environment in which it was placed, would naturally do better and thrive and outnumber other variants. From the range of normal variations among its progeny, circumstances would again favour survival of those with favourable characters while others would perish. Over the generations, as more and more of the total surviving population were those with the favourable variations, the progeny would show larger proportions with the favoured character and that would become the norm for the species. In short, an automatic process of *Natural Selection*, through *Survival of the Fittest* (best adapted to its environment), was continually at work and was responsible for the variety of forms and diversity that we observe. He also pointed out that artificial selection, as practised by plant and animal breeders,

had in fact resulted in forms so varied that they could be considered as different species if their origins were unknown (Fig. 2). It follows that a species which is capable of producing a greater number and/or variety of forms is usually better able to produce a favourable variant when the environment changes. Thus over-production of progeny may not be as purposeless and wasteful, as at first glance it appears. For the same reason, sexual reproduction where each generation carries a greater potential for variation, also offers greater possibility of producing a favourable variant.

Darwin's work aroused a storm of controversy and opposition, but gradually it began to be accepted as sound, and to a large extent, to fit the observed facts. However, there were many questions still to be answered.

Variant forms are still variants of the *same* species. How could they change their genetic structure to form a new species? One possible mechanism of such change is mutation. A mutation is a change or rearrangement of the genetic structure. Mutations occur spontaneously in most species, and a small proportion of the progeny in any generation may be mutants. Most mutations are harmful, and the animal or plant with mutated genes cannot develop from the zygote, or if it develops cannot survive. Occasionally a mutant may arise which is more hardy or better adapted to the environment and better fitted to survive. Its progeny will then grow to replace the unmutated form. The incidence of mutations increases when the organism is exposed to certain chemical or physical agents, and in times of stress (storms, lightning and pressure changes), the possibility of mutations occurring is increased. It is also known that viral infections, which are not uncommon, can induce changes in the gene structure. The frequency with which such

203

changes occur in nature cannot yet be estimated, but the fact that they occur, and can be artificially induced in the laboratory, is significant.

From time to time, people have suggested that evolution cannot be a random process, that it must be motivated by some force or agency, or Life Force. This Life Force motivates organisms to evolve in a particular way, according to a pattern already destined. The force is inherent in every living being and all evolutionary change proceeds according to this unconscious motivation. For instance, we cannot conceive of any random process of natural selection which would favour development of the complicated mechanism of sexual reproduction, since the results, in terms of increased variety, could only act to improve the efficiency of the species some time after the change has taken place. The Darwinian might believe that this change occurred, among many others, and persisted because it proved to be favourable. The Creative Evolutionist would say it was a change motivated by the life force. Motivation, or a guiding force, is a concept which can neither be proved nor disproved. To explain a phenomenon which we cannot yet understand by another incomprehensible factor, like a Life Force, does not appear to make matters clearer.

We may thus postulate that development of a complex molecular structure, with a capacity to reproduce itself, may have been the earliest form of living organism. Such an organism may develop various modifications in the process of synthesis and breakdown. Where such modifications are favourable for survival, they become established forms. Further development proceeded along certain patterns, each line evolving further, or if not well adapted to its environment, becoming extinct. Darwin's theory of Natural Selection represents the most coherent effort to explain the main outlines of the evolutionary process, though many details still require elucidation. Since every species is continually varying, the process of evolution is continuous and will continue as long as life exists on earth.

See also HEREDITY: GENETICS.

EXCRETION (URINARY SYSTEM). Animals in the living process collect useless and harmful end-products of metabolism. These end-products can be gases, solids or liquids. The animal's body gets rid of these by various ways. Waste gases are thrown out by the respiratory system, solids by the digestive system and the waste liquids by the urinary system and skin.

The simplest form of urinary system is seen in protozoans which have a contractile vacuole which throws out waste products. More complex animals have special cells which extract the waste products from the body fluids, and these are connected to ducts through which the waste products are thrown out. In vertebrates, the special excretory cells are grouped into a separate organ, the kidney, which passes the excreted waste matter through ducts to an excretory chamber, the cloaca, or, in more developed forms, the urinary bladder.

The composition of urine varies in different animals. Carnivores generally have acid urine, while that of herbivores is alkaline. The smell of the urine is characteristic, and in many species is used as an identifying feature of the individual, to mark its territory by voiding urine at the borders. In species which need to conserve water, the urine is concentrated and may be semi-solid, as in birds, and in animals like turtles which drink sparingly.

The sweat glands also play a role in excretory activity regulating the quality and concentration of perspiration.

EXOTICS. In considering the fauna and flora it is interesting to note the elements that are not of Indian origin and which have arrived here either accidentally and become well established, or have been brought here for some specific purpose in comparatively recent times—although this may encompass several centuries (see ENDEMISM). Amongst exotics, there are of course a very large number of cultivated plants, both crops and ornamentals, which are usually so familiar that their origins are never contemplated. For example, maize, millets, wheat, oats, potatoes, papaya, chikoo, avocado, tomato, tea (at least in southern India), tobacco and coffee have all been introduced at some time. Similar introductions have been made all over the world. Many of our common flowering and timber trees, including some fast-growing pines, all species of *Eucalyptus*, gulmuhr, jacaranda, silver oak, etc. are exotics. Naturally, a large number of ornamental shrubs and garden flowers are exotics, both in the tropics and the temperate hill tracts of this country.

Among the exotic animals, of which there are comparatively few, there are also those brought in purposely, e.g. breeds of domestic animals which are protected by man from competition, and those brought in accidentally, a few of which have become pests, e.g. the Norwegian Rat, and the San José Scale insect which attacks apple trees in Kashmir. In this respect, India has been most fortunate when compared to other areas of the world.

Of particular interest in relation to the environment are those plants (and the few animals) that have found our conditions so favourable that they have been able to multiply tremendously and assume the status of pernicious weeds or pests. Some of the plants were brought into the country purposely as ornamentals! One might cite the ubiquitous lantana (origin: Central America) which was originally introduced as a hedge plant for its pretty flowers and foliage. Another scourge is the water hyacinth, *Eichhornia crassipes*, brought to

India in the latter half of the 19th century because of its beautiful hyacinth-like flowers. Its ability to cover, very quickly and effectively, large areas of inland water was not realized until it was too late. Another land weed now of some consequence here is the South American eupatorium weed (*Chromoleum odoratum*) which covers large hill tracts of Kerala and Coorg, to the detriment of other indigenous vegetation. The prickly pears, *Opuntia* spp., originating from South America, were introduced as ornamentals in the 18th century and spread to become unpleasant weeds in many parts of the country.

Recent intruders into the Indian flora are the water fern (*Salvinia molesta*) which certainly came from Sri Lanka (though of South American origin) around 1960 and now covers many of the inland waterways in both Sri Lanka and Kerala, including many of the backwaters of the Cochin Harbour complex. The most recent weed to gain well-earned notoriety is undoubtedly *Parthenium hysterophorus* (the so-called Congress weed), which was first recorded near Poona around 1950, and which is now rampant, being the dominant plant in ever-increasing areas around Delhi, Bombay, Poona and Bangalore. It is, moreover, toxic to animals and produces a severe allergic reaction in a proportion of humans handling it.

Another category of exotics is that of stored product pests where, with the commercial transport of dried vegetable material of all kinds to and from all parts of the world, pests (mostly insects) attacking such produce, have become cosmopolitan. India has both received and given its share of such pests.

Thus it is seen that the subcontinent has been really fortunate in that it has received many valuable crops and domestic animals from abroad, and that very few exotics have become weeds or pests.

Exotics are invariably introduced without the natural enemies which attack them in their native land. Hence, a possible method of controlling those which become pernicious weeds or injurious pests is to establish some of the most effective of these natural enemies against them by introduction. This forms the main object of BIOLOGICAL CONTROL.

EXTINCT and VANISHING BIRDS. Between A.D. 1600 and the present day 94 species of birds have become extinct on earth and more than 190 species face imminent extinction. Of these less than 25 per cent declined or died out from natural causes. Most of them were or are inhabitants of oceanic islands. But even on the continents some birds have vanished or are vanishing, mainly due to man's activities.

On the Indian subcontinent at least four birds have presumably become extinct since 1870: the Mountain Quail, Jerdon's or the Doublebanded Courser, Blewitt's or the Forest Spotted Owlet, and the Pinkheaded Duck. The last reliable records of the above birds appear to have been in 1876, 1900, 1914 and 1935 respectively. But the Pinkheaded Duck, which in 1864 Jerdon described as quite common in Bengal, was extremely rare even at the time of its discovery in 1790. In every case the causes of their extinction are obscure.

Jerdon's or Doublebanded Courser × ⅓

Pinkheaded Duck, male × ¼

Since 1900 the Indian landscape has changed so much that the habitats of many birds have shrunk or been drastically altered. Pollution and pesticides have been at work insidiously, reducing animal populations still further.

The state of our ignorance precludes a detailed survey of our endangered species. But it is generally agreed that the numbers of all the Himalayan pheasants, the larger birds of prey, waterfowl such as the ducks, and those forest birds that require large old trees to nest in, have dwindled alarmingly. Some of these are already on the Red List along with the Great Indian Bustard. The major cause for the decline is the rapid alteration or loss of habitats due to human activity.

A simple instance is the practice of removing as many over-mature trees as possible from managed forests. The hornbills and the larger woodpeckers, which need such trees to nest in, inevitably suffer. Now that, in addition to opening up even the most inaccessible areas, there is also large-scale deforestation for monoculture of exotics, how long can the Great and Malabar Pied hornbills, the Great Black Woodpecker and some of the larger owls continue to find suitable nesting sites?

Many of the larger natural reedy tanks or jheels that were not only a cornucopia to the heron and egret tribes, but afforded refuge to hordes of migrant ducks, have been filled; some have been heavily polluted, and almost all have become subject to constant disturbance. This has deprived enormous numbers of birds of food and shelter. Even the best of our bird sanctuaries, the Bharatpur Ghana, is not immune to the threat of reclamation, diversion of its water supplies and continual disturbance.

Among the larger birds threatened with extinction, the Tibetan or Blacknecked Crane and the Great Indian Bustard have now won wide publicity. But the plight of another large bird, the Grey or Spottedbilled Pelican practically confined to India, has received little attention. What has happened to the 'millions' that till about 1940 used to nest near Shwe-gyen in Burma till the spread of agriculture deprived them of their breeding place is anyone's guess. However, in the 1940s a few thousands were found nesting near Tadepalligudem in Andhra Pradesh. Owing to constant human disturbance, tampering with their feeding grounds and inaction of the authorities concerned, this pelicanry was wiped out of existence by 1970. The few other known breeding places of the Grey Pelican account for perhaps less than a thousand pairs.

K.K.N.

FAIRY BLUEBIRD (*Irena puella*). A gorgeous thrush-like bird occurring in disjunct populations in the heavier forests of the Western Ghats, northeastern India and the Andamans. The male is brilliant ultra-marine blue and deep black, the female a duller blue-green. In the non-breeding season they forage in small flocks, keeping to the upper forest stratum, and communicating by rich and mellow call-notes. Their food is chiefly arthropods, fruits and nectar.

FALCONRY. Falconry is the art of training birds of prey for sport. In Asia and Africa it is known to have flourished thousands of years ago, and was carried to medieval Europe by returning crusaders. There it became a status symbol with kings and the nobility. It demands great skill and patience.

Earlier, falconry was used for getting food; today it survives as a field sport. Since the invention of firearms, only a handful of enthusiasts have kept the art alive, and enabled it to stage a comeback.

History tells us that Chengiz Khan, Alexander the Great, and the Mogul emperors were all keen falconers, as were religious leaders like Guru Govind Singh and various Muslim saints.

During the World Wars, the British Intelligence Service used falcons to intercept carrier pigeons loaded with information by enemy agents. Today, trained falcons assist in ridding airfields of birds and reducing the very costly bird-strike menace.

Falconry can hardly be charged with cruelty since it repeats a natural phenomenon between evenly matched participants. Prey is either left unscathed or killed outright; and falconers are more concerned with the style of a single kill than with a large bag.

The birds of prey used in falconry fall into three broad groups on the basis of the colour of their eyes, the shape of their beaks, the structure and shape of their wing and tail, and their scaly or feathered legs. Foremost amongst them are the Accipiters. In falconers' parlance these are known as short-wings. They have, in common, yellow eyes, round-tipped broad wings, and long tails. The fourth primary feather in each wing is the longest. Their beaks are slender. Unlike falcons they have long tarsi and, generally, short powerful toes. This group comprises the *Baz* (goshawk), the *Basha* (sparrow-hawk), the Shikra and the Besra (sparrow-hawks). All accipiters have similar habits, and because of their body and wing structure are intended by nature to hunt in wooded country. Their long tails and rounded wings give them the added advantage of extreme manoeuvrability. With age, their yellow eyes progressively turn orange, then deeper orange and finally a copper-tan shade.

Goshawks, after training, can be flown at partridges, crows, pheasant, junglefowl, grouse and hares. The male goshawk is almost a third the size of the female, but nevertheless a proficient hunter.

Next in the Accipiter group of birds is the *Basha* sparrow-hawk (*Accipiter nisus*). When properly trained, these brave birds are capable of tackling prey almost twice their own weight. The beginner is, however, cautioned to handle this species with great care because all *basha* sparrow-hawks have unusually delicate constitutions. A Persian saying goes: 'With the Goshawk for king, and the Gos-tiercel for prime minister, the Basha sparrow-hawk is a prince in the accipiter hierarchy.' This ranking is accepted by all falconers.

The Besra sparrow-hawk is a rare hybrid accipiter. In the order of prominence it stands second to the *basha*. It is an exceptionally rare bird and very little is known about it from a falconer's standpoint. On the other hand the Shikra (*Accipiter badius*) is the most common and abundant accipiter in the Indian subcontinent. It is really the bird for the aspiring falconer as it can stand inexperienced handling, without any ill effects.

The second group consists of long-wings or falcons, which are very popular with all falconers. These belong to the subfamily Falconinae and amongst them are some of the fastest birds known. In the genus *Falco*, the second primary feather of the wing is the longest, the first and the third being of equal lengths. These hawks have dark brown eyes, and because of this they are called *siah chashm* in Iran. They all have very powerful beaks that have a barb or notch on the upper mandible. They also have long toes which help them to grasp and carry prey long distances.

Broadly, there are two kinds of falcons. One group comprises the Peregrine Falcon and its several subspecies, and the other belongs to the Saker Falcon

(*Falco cherrug*) and the Gyr Falcon (*Falco rusticolus*) tribe. To these may also be added the tiny Redheaded Merlin or Turumti (*Falco chicquera*). True peregrine falcons breed in the Arctic and migrate to the tropics in winter.

The black Shaheen (*Falco p. peregrinator*) and the red-naped Shaheen (*F. p. peregrinoides*) are the migrant peregrine's smaller brothers. They are smaller than the Peregrine, and migrate only locally. They can be trained to hunt mallard, partridge, grouse, plover and other game of similar size. Another falcon in this group is the Laggar (*Falco jugger*), which is extremely cooperative and possesses the capacity to 'wait on' (keep soaring high above the falconer's head) for long periods. It is very common in India and easily available with bird dealers.

The Saker falcon (*Falco cherrug*), a close relative of the Gyr falcon, is also abundant in India. It is found in two colour phases, a dark phase and a light one. Surprisingly both phases are found living side by side. During the heyday of falconry Saker falcons were trained to catch hares, gazelles, bustards, crows, horned owls, kites, etc. Lastly we come to the Redheaded Merlin. These stout-hearted falcons can be trained to hunt almost all game that is brought to the bag by a Sparrowhawk. They can also hunt plovers and rollers.

The third group is the very powerful eagle subfamily Aquilinae. All the eagles that are used by falconers are of two categories; (*a*) the true eagles or *Aquila* group and (*b*) the hawk-eagles or *Spizaetus* group. As in the case of Accipiters they also have rounded wings; however the legs of all eagles used for falconry are always feathered down to the toes.

In training, a raptor is first blindfolded with what is known as the Rufter's hood, and jesses and bells are secured to its feet. It is then very gradually introduced to its new surroundings by being hooded and unhooded at short intervals over a period of some days. It is regularly fed to the accompaniment of a whistle, so that in the trainee's mind food and whistle get closely associated. Getting the predator used to the presence of man becomes the trainer's primary objective, and when the bird begins to feed unhooded from the falconer's fist, it is encouraged to jump from its perch to the extended fist on to an offered piece of meat. This jumping distance is daily increased till the hawk is seen flying to the fist whenever called from afar. This training applies equally to all hawks and eagles. Soon afterwards the predators are entered for game, and hunting starts. The common method is to flush game from cover and to cast the hawk after it.

With all falcons, once they start jumping to the fist from a little distance, they are encouraged to fly to the lure, which is either a dead pigeon or a dummy concealed in feathers. The lure is tied to a long string held by the falconer at one end. Accompanied by the whistle, the lure is swung round the falconer's head, and thrown out on to the ground. Whenever the falcon flies to it, it is rewarded with a morsel. Ultimately the lure is concealed as the falcon flies to strike it. The falcon will then start to soar above the trainer's head, looking out for the hidden lure. At such times if game is flushed from cover, it will be immediately attacked by the falcon which, in falconer's terms, has been 'waiting on'. It is important that the hawk or falcon be in peak condition. Starving the bird in the hope of it hunting better that way should never be attempted.

In America and Europe, there are falconers' clubs, which meet annually to exchange experience and information and fly their hawks. The problem of obtaining hawks and falcons for future falconers is receiving attention from specialists in many countries, including India. Successful attempts have been made to breed hawks in captivity.

Accessories such as hoods and bells may be had from dealers through falconry clubs, although most falconers make their own.

M.O.

FALSE SPIDERS are minute arachnids, in length not exceeding 7 mm. Outdoors they live in debris, dry leaves, under stones and ant mounds. Within houses they are found among old files, bundles of papers and leaves of unused books. Hence they are also called Book Scorpions. They feed on small insects and mites.

False spider (*Chelifer*), × 3

Under a lens they look like scorpions with prominent pedipalps, four pairs of legs and an oval segmented body without a tail. Most of them are sluggish but some are active and move like crabs. They have silk glands in their cephalothorax opening near the jaws. These are however used only for making cocoons and a nest to live in when they cast off their skin (moult). One peculiar habit of false spiders is that they cling with their claws to the bodies of flies and get transported to distant places—a factor accounting for their wide distribution.

Regarding their mating and other habits little is known.

See also ARACHNIDS.

FALSE VAMPIRES. False Vampires are characterized by their peculiarly-shaped elongated nose-leaf and very large rounded ears joined across the forecrown. The genus *Megaderma* is not related to the blood-sucking true vampires of the family Desmodontidae which is confined to South America, but they are nevertheless celebrated for being the only bats in Asia which feed upon warm-blooded prey. Two species are found in the Indian subcontinent. The commoner, *M. lyra*, is distributed in the subcontinent, China, Burma and Malaya, and *M. spasma* occurs in peninsular India, Sri Lanka and Burma, extending into Malaysia, Indonesia and the Philippines.

During the day, false vampires hide in caves and old buildings, and they prefer very dark places for their diurnal roost.. Their colonies are smaller as compared to those of other micro-chiropterans and though exclusive colonies occur, they frequently roost with other bats such as the Mouse-tailed Bats (*Rhinopoma* species) and the Sheath-tailed Bats (*Taphozous* species). Invariably however in such roosting places, they are found in the darkest part of the cave. They are the last to leave the hide, well after sunset. They do not hibernate during winter but migrate to warmer localities.

In addition to feeding upon insects, *M. lyra* captures and consumes fishes, lizards, birds and smaller bats. It hunts flying low over surfaces searching the ground, pools and walls. It has a characteristic habit of carrying its prey to perching sites where it can be devoured at leisure.

M. lyra females produce their young from February to May and usually only one is born at a time. There is only one report on the occurrence of twins in the species. The gestation period is about 150 days. The young suckles for about three months. Sexual maturity is reached in about two years.

I.P.

FAULTS. Faults are fractures in rocks along which a considerable amount of movement has taken place. The movement causes the features of the rocks to become discontinuous and end abruptly at the fault (Fig. 1).

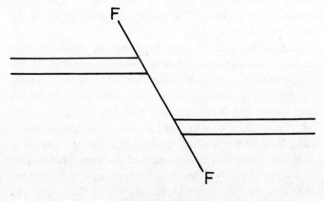

Fig. 1. F–F Fault

If a stick is bent by applying pressure by both hands the stick bends and if the pressure is released it regains its original shape. It is said that the stick has undergone elastic deformation. On the other hand if the pressure is increased the stick finally yields by breaking. In this case the elastic limit of the stick is exceeded and therefore it has yielded by breaking. Similarly the rocks below the surface which are under great pressure due to the overlying weight of rocks also undergo elastic deformation because of the movements in the upper layers of the earth. If the forces involved exceed the elastic limit of the rocks they break resulting in the formation of a fault. The energy thus released is propagated in the form of an earthquake wave. Most earthquakes are associated with active faults. Movements along faults and the resulting earthquakes are responsible for many disasters and loss of life and property. Large dams, bridges, tunnels, and mines have collapsed as a result of movement along the faults. Therefore knowledge of the presence or absence of the faults is needed in many human activities such as mining, exploration for mineral resources, and in civil engineering constructions.

Faults may vary in size from as small as those extending to a few centimetres to those that may extend to hundreds of kilometres. The largest fault and the best known in the world is the San Andreas fault of California in U.S.A. which extends to nearly 1000 km.

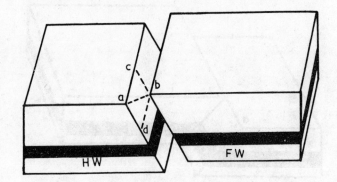

Fig. 2. FW Foot Wall. HW Hanging Wall. ac strike slip. ad dip slip. ab net slip

The movement of the rocks on either side of the fault is essentially parallel to the plane of the fault (Fig. 2). Granulation and polishing of the rocks in the fault plane are common features formed due to rubbing of blocks of rock along it. In Fig. 2 a fault separates two blocks. The block above the fault plane is called the Hanging Wall whereas the one below the fault plane is known as the Foot Wall. The horizontal displacement is known as the Strike Slip, the displacement down the fault plane is called the Dip Slip and the maximum displacement is termed Net Slip.

A fault in which the hanging wall has moved down relative to the foot wall is a normal fault (Fig. 3 A). If

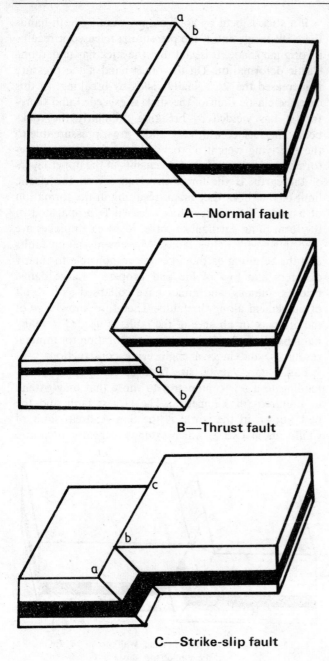

A—Normal fault

B—Thrust fault

C—Strike-slip fault

Fig. 3. FAULTS

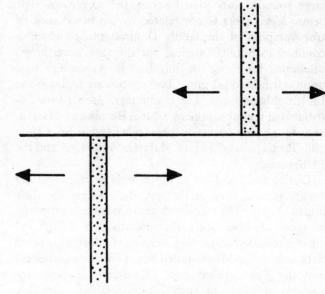

Fig. 4. Transform Fault

layas and Alps and involve compressional forces for their formation. Normal faults are usually formed under the influence of gravity and are commonly associated with Rift Valleys as those of East Africa.

See also EARTHQUAKES.

FEATHERS. Birds are the only animals with feathers. They have evolved from cold-blooded ancestors (see FOSSIL BIRDS) which were covered with scales and a few filament-like feathers or *cryptoptiles*. When birds became warm-blooded the filaments evolved into proper feathers or *teleoptiles* to prevent the loss of body heat. Feathers made flight easier by increasing the surface area of the wings and by streamlining the contour of the birds' body. The feathers of nestlings are called *neossoptiles*.

Structure

The long tapering central *shaft* which supports the web or vane of a feather has a short and hollow basal portion, the *calamus*, and a larger and solid distal *rachis*. The calamus has two holes (*upper* and *lower umbilicus*) at its ends. The rachis supports two rows of small branches or *barbs*.

The barbs bear smaller branches or *barbules*. The barbules which point toward the tip (anterior) of the feathers have a series of projections or *barbicels*. In the middle of the underside of the feathers the barbicels bear hooks called *hamuli*. The barbules of the posterior side have ridges on which the hooks of the hamuli interlock, making the feathers flexible and strong. Except in ostriches, penguins and screamers feathers grow along distinct tracts or *pterylae* which are separated by regions of sparse feather growth (apteria) where they cover the entire body.

the relative movement is such that the hanging wall has moved up relative to the foot wall it is called a thrust fault (Fig. 3 B). The inclination of the fault plane in the case of thrust faults will usually be low. A fault with a lateral movement but with practically no vertical movement is called a strike-slip fault (Fig. 3 C). Faults associated with the oceanic ridges are known as Transform faults (Fig. 4). These faults differ from those described earlier in that they are due to differential spreading of the ocean floor.

Thrust faults and strike-slip faults are usually associated with the folded mountain belts such as the Hima-

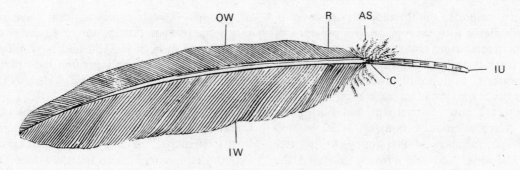

A TYPICAL FEATHER (QUILL)

R rhachis. IW inner web of vane. OW outer web of vane. AS aftershaft.
C calamus or quill. IU inferior umbilicus where pulp enters.

Types of feathers

The *contour feathers* which cover the body have large, firm vanes and fluffy bases. The large contour feathers of the wing (*remiges*) and tail (*rectrices*) are the *flight-feathers*. Quills of the hand are the *primaries*. Flying birds have 9 to 12 primaries; the flightless cassowary has 3, and ostrich 16. The *secondaries* are the quills of the forearm. They vary from 10 (Bobwhite Quail) to 32 (Wandering Albatross). In some primitive groups the 5th secondary is absent (e.g. Charadriiformes), making their wings *diastataxic*. Wings with all secondaries intact (e.g. Passeriformes) are *eutaxic*.

The rectrices are counted in pairs from the centre of the tail outward. Most species have 12, humming-birds, swifts and most cuckoos 10, the peacock 20, and ostriches 50 to 60. The hairlike *bristles* and loose-webbed

semi-plumes are also contour feathers.

The small hairlike feathers with or without minute vanes underlying other feathers are the *filoplumes*. Beneath the contour feathers lie the soft *down feathers* which have very small or vestigial rachis; their barbs fan out in tufts at the tip of the calamus. Down covers the nestlings of fowls and hawks, and is abundant in water-birds.

On their breast and lower back some birds like herons have *powder downs* which grow continuously and crumble at their tips. The powder so formed is used for cleaning and drying of their feathers (preening).

All feathers are renewed (moulted) at least once a year.

D.N.M.

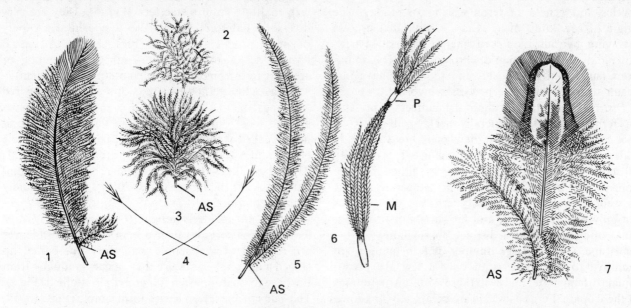

DIFFERENT KINDS OF FEATHER

1 A feather with free barbs and aftershaft (AS). 2 A nestling down-feather. 3 A permanent down-feather with aftershaft (AS). 4 Filoplumes. 5 An emu's feather with very long aftershaft. 6 A mesop- tile (M) pushing out its predecessor, a protoptile (P). 7 A back feather of a 'game bird' (pheasant) with aftershaft (AS).

FERNS. Ferns belong to the Pteridophytes, a group of plants which shares with seed plants the presence of well-developed conducting tissues (xylem and phloem) but differs from them in producing no seeds. Today, the living members of this group nowhere on the earth's surface constitute the dominant vegetation: seed-producing plants are now predominant. But during the Coal Age (Carboniferous Period), 230–280 million (10^6) years ago, members of Pteridophytes—the tree Lycopods, Calamites and the ferns—constituted the major forests on the earth's surface and the present coal reserves are believed by geologists to have been formed by stratification of that vegetation during the course of evolution on this planet involving millions of years. Impressions of many ferns and fern-allies are found in most coalfields. During evolution, when the first plants appeared on the earth, some developed complex leaves and came to be known as ferns while others, which possessed simple needle- or scale-like single-veined leaves, are the plants commonly called 'fern-allies' (horsetails, club-mosses, spike-mosses, etc.).

Fossil History and Evolution: The fern line is one of great antiquity reaching far back in geological history to c. 240 million years, about 100 million years before the flowering plants appeared on the earth's surface. Interestingly, there are no fossils at all of most of the modern ferns. How did they evolve? Most people agree that there was a common ancestral stock which during the course of evolution has undergone many transformations, both on the 'convergent' and 'divergent' patterns. This has led to the formation of the different species and genera of ferns which exist today. The origin of new kinds of ferns is a complicated process involving hybridization (crossing of different types), polyploidy (multiplication of chromosomal sets), mutations (sudden changes), etc. and today we have just begun to understand the processes involved.

Distribution: Estimates about the numbers of present-day ferns vary greatly. Generally speaking, about 10,000 species are recognized but new ones are still being added from unexplored tropical regions. The majority of ferns are met within the tropics although these flourish well in the warm and cold temperate regions as well. In wet tropical forests ferns are conspicuous not only by their number and the variety of forms but quite often they lend a distinct physiognomy to the landscape. Quite a large number of fern families are exclusive to the tropics and only a few are predominantly temperate: and even those occur on mountains in the tropics. Ferns as weeds are completely unknown except for a water fern, *Salvinia molesta*, which is the most aggressive weed of water channels and paddy-fields in south India. It multiplies very quickly, spreads fast and is a real threat to agriculture in the region.

Because of their capacity to grow well from spores or through vegetative means, ferns have generally a very long-distance distribution, even inter-continental, being found in both the Old and New Worlds. From the ecological viewpoint ferns find their best homes in areas of high humidity and rainfall such as the Eastern Himalaya or the Western Ghats.

In the subcontinent about 600 species of ferns are found. They are abundant between 600 and 2100 m. In the Himalaya, they may grow up to about 4200 m altitude, even extending to the snow-line. Occasionally, ferns are found in the plains even in extremely dry conditions. Indian ferns grow near the timber-line in the Himalaya (Figs. 1 and 2), in the forests (Figs. 5, 9, 10), in the crevices of rocks and stone walls (Fig. 3), as epiphytes (Figs. 4 and 6), or as climbers (Fig. 7). But most ferns grow on the ground or on stones, although epiphytic ones are very conspicuous in the warm temperate forests of the Himalaya, particularly in the eastern region.

Fern-allies number about 150 and amongst them spike-mosses (*Selaginella*), club-mosses (*Lycopodium*) and horsetails (*Equisetum*) are equally common in the warm temperate as well as tropical regions.

Water ferns, as the deeply rooted *Marsilea minuta*, commonly known as the 'water-clover' (Fig. 11) or free-floating *Salvinia natans* (in Dal lake, Srinagar) and *S. molesta* in south India (Fig. 8) or *Azolla pinnata* with bright orange-red plants in north Indian plains, are conspicuous.

Plant Body: Within the ferns the range of plant size is enormous. It varies from a few millimetres in a filmy fern to 10–15 m in a tree fern (Fig. 5). They are very diverse in habit and form. Also they exhibit a variety of spore-producing organs (Figs. 13–18). A typical fern possesses roots, a stem—upright, ascending or creeping (a rhizome, the most common form of stem)—and leaves but no flowers, fruits or seeds. As compared with most plants ferns' growth is simpler. The leaves are the most obvious part of the fern and are commonly called fronds which usually consist of a petiole (stipe) and a leaf (lamina). These are simple (Figs. 21, 23) or lace-like with divided laminae, representing pinnate (Fig. 22), bipinnate (Figs. 20,24), or compoundly pinnate conditions.

Except for Adder's tongue (*Ophioglossum*), Grape fern (*Botrychium*) and their allies (Fig. 21), fern leaves are rolled over the petiole with their tips curled up (Fig. 1) and as they mature they uncurl gradually from the base to the top, when full growth is attained (Fig.2). The veins of most fern leaves form simple (Figs. 13, 17, 26) or reticulate patterns (Fig. 19) usually without cross veins. The rhizomes are generally small, thick, thin or even thread-like in filmy ferns. Their growth is usually initiated by a single or several large apical cells which are well protected by hairs, scales or paleae (Figs. 27a,b,

HABITATS OF INDIAN FERNS

1, 2 *Osmunda claytoniana*, Valley of Flowers, 4200 m, Garhwal Himalayas, × ¹/₂₀ . 3 *Asplenium dalhousiae*, Kulu valley, 1500 m, × ¹/₅ . 4 . 7 *Asplenium nidus* and *Stenochlaena palustris* respectively in low hill tropical forest, Teesta, 150 m, Darjeeling × ¹/₂₀ and ¹/₅₀ . 5 *Cyathea* *spinulosa*, a tree fern in Lloyd Botanic Garden, Darjeeling, × ¹/₅₀   6 *Drynaria mollis*, a common polyploidy of Naini Tal, 2100 m, × ¹/₁₀. 8 Free-floating *Salvinia molesta* in Botanic Gardens, Punjabi University,

213

FERN HABITATS AND MORPHOLOGY

9 *Pteridium aquilinum*, a bracken fern, Kodaikanal, 2100 m, × ¹/₂₀. 10 *Diplazium esculentum*, edible fern, Teesta, Bengal plains, 150 m, × ¹/₂₀. 11 *Marsilea minuta*, 'water clover', in Botanic Gardens, Punjabi University, × ¹/₁₀. 12 Reduction division in a spore mother cell from sporangium of *Diplazium muricatum* showing 41 bivalent chromosomes (n-41), × 2600. 13-18 Fronds showing soral conditions in *Lygodium*, *Polystichum*, *Athyrium*, *Pteris*, *Marattia* and *Adiantum* respectively, × 5.

28a). Likewise, the rhizome apex, stipes and often the lamina may be clothed to various degrees in hairs and scales. The vascular cylinders (steles) of ferns present an array of different structures, unparalleled in any other plant group. The most common form in advanced ferns is the 'dictyostele' representing a number of inter-connected vascular strands ultimately connected with leaves and roots (Fig. 25).

The reproductive structures are called the sporangia, the spore cases. These are borne along the margin (Fig. 16) or on the back of the leaf blade (Figs. 13–15, 17,18) and are usually aggregated into small structures called sori (singular, sorus). Each sorus may contain from a few to several hundred sporangia and is quite often covered by a protective organ called the indusium (Fig. 26). Hairs of paleae often protect the sporangia if there is no indusium (Fig. 28b,c,d). Generic distinctions are quite often based on the characteristic ar-rangement, shape and size of sori (Figs. 13–18) as well as the structure of the sporangia which are very variable amongst different fern groups. At maturity the sori are brownish and may appear as small pustules (Figs. 19,23). The indusia, if present, either curl back or are shed away. In highly developed ferns (*Asplenium, Dryopteris, Polypodium,* etc.) within each sporangium usually 64 spores are produced, whereas in the primitive ferns (*Ophioglossum, Angiopteris, Gleichenia,* etc.) the number is much higher, even in hundreds or thousands. The spore production per fern individual is fantastically high. In a simple-leaved *Phymatopteris griffithiana* the number of spores per plant is 1·7 million and in com-pound-leaved *Dryopteris chrysocoma* nearly 60 million spores are produced. These are either bean-shaped (bilateral or monolete) or tetrahedral (trilete). Each spore is a minute single-celled structure covered by protective walls, the outer wall being extremely resistant

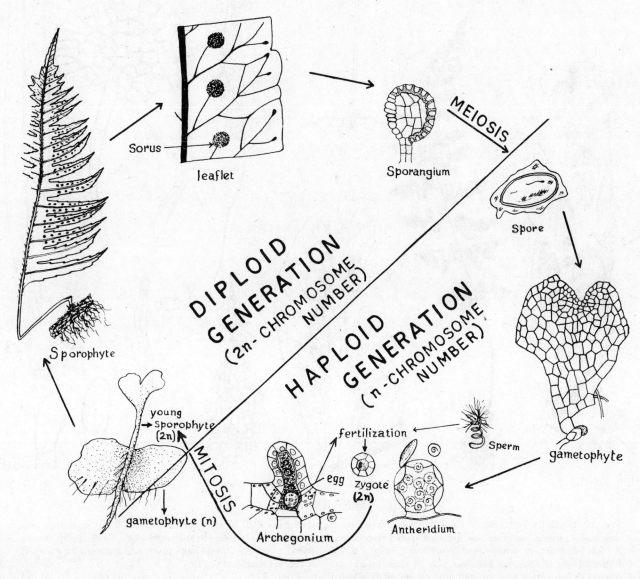

Fig. 19 FERN LIFE CYCLE

FERN LIFE CYCLE

Figs. **20, 21, 23, 24** Fern habits: *Osmunda regalis, Ophisglossum vulgatum, Lepisorus excavatus* and *Athyrium* species respectively, × ¹/₁₀. **22** *Asplenium normale* with proliferating apex. × 5. **25** Extracted stele of *Diplazium latifolium*, × 2. **26** Pinna lobe of *Asplenium dalhousiae* showing sori partially covered by flap-shaded indusium, × 10. **27** (a) Rhizome scale, × 30. (b) Scale apex with glandular cell, × 100. **28** (a) Dermal hairs, × 100. (b) Sporangial palea, × 100. (c) Stellate hairs protecting sporangia in *Pyrrosia*, × 100. (d) Soral hairs, × 100.

to decomposition. In rocks millions of years old spores have been found with their walls intact and almost unchanged.

The majority of ferns contain only one kind of spore but water-ferns produce spores of different sizes. Spores are very characteristic structures in many groups of ferns, and are of importance in identification and classification.

Life Cycle: Each cell of a typical fern sporophyte (the dominant generation) possesses two sets of chromosomes (2n), one set from the egg (n) and one from the sperm (n). A cell with two sets of chromosomes is called diploid and a cell with only one set is called haploid. Normally, sexual production of a fern plant is accomplished through the production of spores. During spore production, the number of chromosomes is divided in half (Fig. 12), i.e. each spore has only one half (n) as many chromosomes as the spore mother-cell or any other cell of the mature plant (2n). When spores are mature, the sporangium opens slowly but closes suddenly, thus throwing out the spores. On release, the spores are carried by wind currents. Most of them do not germinate and only a small number fall at moist shady places and germinate to give rise to the gametophytes or prothalli (prothallus). These haploid structures (n) are green and manufacture their own food. At maturity gametophytes produce haploid sex organs—antheridia and archegonia, the former at the base and on the margins of the wings and the latter near the notch. When the antheridium is mature the antherozoids or sperms are released which swim their way to the neck of the flask-shaped archegonium and down to the bottom to fertilize the egg and form the zygote. The sperm has one set of chromosomes and the egg has also one set, thus the resulting fertilized egg (zygote) has two sets. Repeated divisions of the zygote result in the production of a young sporophyte with 2n chromosome number. The sequence of events is diagramatically represented in Figure 19 wherein the prominence of diploid and haploid generations is demarcated. There is alternation of generations: the spore-producing plant body (sporophyte) is followed by a free-living gamete-producing prothallus (gametophyte) which gives rise again to the plant body.

The young coiled leaves of some ferns are eaten as green vegetables in parts of the Himalaya. Ferns are also popular as outdoor and indoor decorations. In the dim past they became the constituent of coal, and today ferns are extremely useful in biological research, as their two independent sporophytic and gametophytic generations can be manipulated in the laboratory with great ease.

See also CONIFERS.

S.S.B.

See plate 37 facing p. 528.

FIRST-AID and HEALTH IN THE FIELD. People working under field conditions are sometimes faced with medical problems in places where no medical aid is available. It is not possible to deal with all the emergencies which could arise. However some guidelines to help in avoiding problems and in dealing with some of the common ones, should be understood by field workers.

1. *Preparation.* All persons working in the field must be immunized against common infections, and must carry first-aid kits, water disinfectants etc. as indicated below.

1.1. Preventive immunizations against typhoid, cholera and tetanus are essential precautions which must be completed at least one week before departure. Vaccination against smallpox is also advisable.

1.2. A general medical check-up and completion of any treatment required, especially for teeth, ear, nose, throat and eye complaints, is of the greatest value in avoiding trouble in the field.

1.3. When working in areas of intense heat or at high altitudes, salt tablets should be carried and taken in proper dosage, to avoid heat stroke.

2. *Common problems*

2.1. Gastroenteritis. A condition which can be caused by a variety of factors, and very common during field work in India.

The symptoms are diarrhoea, vomiting, nausea etc., often together with severe griping or colic.

Gastroenteritis is usually due to infection acquired through contaminated food or water, and can be largely avoided by consuming only hot, freshly cooked food, and by disinfecting the water before drinking. Water disinfectants, in tablet or liquid form, are available at most chemists. They impart an unpleasant taste to the water which can be minimized by letting the water stand for a couple of hours after adding the disinfectant, to allow it to settle.

Treatment. Weakness and prostration is mainly due to loss of fluids. This can be avoided by drinking plenty of water (previously disinfected). Small drinks should be given at frequent intervals even when vomiting persists. In severe cases, common salt, enough to give a definitely saline taste, should be added to the water. Kaolin powder is a simple remedy to control diarrhoea, and, to some extent also the nausea and vomiting. 1 or 2 tablespoons of the powder are made into a smooth paste with a little water, enough water then mixed in to give a drinkable consistency, and the mixture drunk before the powder settles down. Kaolin is not absorbed and the dose can be repeated 3 to 4 times a day. Mild cases can be controlled in a few hours to a day. Many drugs are available to control vomiting and diarrhoea. Such preparations, safe for self-medication, can be kept for emergencies under advice of and with detailed instructions from a doctor. Antispasmodic tablets can be used to relieve griping if it is very severe. If there is no relief of symptoms after a day of such treatment arrangements must be made to carry the patient to the nearest hospital or doctor, since further delay may be dangerous.

2.2. *Accidental injuries*. Treatment of injuries is adequately covered by several compact manuals on first aid to the injured, one of which should be carried and consulted when required.

In general all cuts, wounds, bruises and burns should be thoroughly cleaned and washed with soap and water. Burnt areas should be kept moist after cleaning, with a sterile dressing soaked in slightly salt water. Large deep wounds which are difficult to scrub can be flushed out with large quantities of a mild antiseptic solution. A dilute solution of Potassium permanganate ($KMnO^4$), pink, not red in colour, can easily be prepared in luke-warm water and is relatively painless. A liquid antiseptic should then be applied with a sterile swab. The wound should be left open as far as possible. Large, deep or lacerated wounds or wounds which are bleeding should be covered with a dressing of sterile gauze and a light bandage applied. In bleeding wounds the bandage should be firm enough to prevent excessive blood loss. The patient should then be taken to the nearest clinic for expert attention. Tetanus may complicate any deep or contaminated wound and medical advice must be sought on anti-tetanus precautions. Note: Antiseptic creams, ointments and pastes should be avoided for open wounds as they tend to seal in any residual infection. These are useful for wounds with unbroken skin or after surgical cleaning of the wound.

Where a fracture or head injury is suspected it may be dangerous to lift the patient or make him sit up. He should be transferred with minimum of movement, to a stretcher, wooden board or any other rigid support on which he can be lifted and carried to the nearest clinic. Excellent pain-relieving drugs are available in tablet form and one of these may be used to relieve pain until medical attention is available. Immobilization of a fractured limb helps to reduce the pain. In the absence of some form of splint, the limb can be simply bandaged to the nearest sound part, e.g. the lower limb can be tied above and below the fracture to the opposite limb which becomes a sort of splint; the upper limb, with bent elbow, can be tied across the chest.

2.3. *Wounds from animal bites and claws*. Since most animals scrabble in the earth these injuries carry the same risk of tetanus and other infections and should be cleaned and disinfected as described for other wounds. In addition, animal bites may carry a risk of rabies infection, against which precautions must be taken (See RABIES).

2.4. *Insect bites and stings*. These usually cause irritation by injection of an organic acid, and can be neutralized by application of an alkaline solution. Strong Ammonia, a solution of NH_4OH, is safe to use and can give remarkable relief if applied immediately. Local application of ice, and pain-relieving drugs can be used, according to severity of the pain. Scorpion stings may be painful and severe enough to incapacitate a person for several hours; bites and stings of other species are rarely serious.

2.4.1. *Malaria* is spreading in areas which had become relatively malaria-free. Precautions against mosquitoes, which are also responsible for transmission of diseases like filaria (elephantiasis) and encephalitis, are therefore necessary, even if not always entirely successful. Mosquito-nets and insect repellants are both useful but have their limitations. In areas where malaria is endemic, a prophylactic maintenance dose of an antimalarial drug may be taken. Your doctor or the local health department would advise you on this point.

2.4.2. *Leeches* are generally regarded with a dread which is disproportionate to their importance. Leeches drop off if common salt is applied at the point of contact, or with application of heat like a lighted cigarette-end. The leech injects a chemical which prevents clotting of the blood, and bleeding may continue for some time. Washing the wound followed by application of dry sterile gauze or cotton usually stops bleeding in a short while. Styptic pencils or burnt cotton wool can also be used.

2.5. *Snake bites*. Prevention: The use of strong boots, keeping to well-trodden paths and general vigilance, especially in wet weather, will generally avoid confrontation with snakes. Most snakes are non-poisonous. Poisonous snakes have two large poison fangs which leave puncture marks like those of a large injection needle. Such paired puncture wounds at the site indicate a poisonous snake bite. If the snake is captured or killed, more definite identification is possible by reference to a snake chart.

Injection of Antivenene is effective against all poisonous snakes found in India. The injection is not free from risks and should not ordinarily be attempted except under medical supervision. Since village dispensaries may not always have Antivenene in stock, it is advisable to keep one dose of two ampoules in the first-aid box. It may be noted that a delay of a few hours is generally tolerated without danger if general measures of prevention are carried out. In bites of the King Cobra, however, the venom is very potent and may take effect in half-an-hour. In such cases, if no medical aid is immediately available, the risk of injection of Antivenene is justifiable. Detailed instructions are provided with each ampoule, which must be carefully followed. A strong dose of an antihistaminic preparation should be given to reduce the danger of side effects of the injection.

General preventive measures are aimed at preventing the venom, which is carried in the blood stream, from spreading through the body and attacking the central nervous system. A tourniquet, viz. a constricting bandage or rubber or elastic band is tied tightly round the upper part of the affected limb (thigh or upper arm) where the soft tissues and blood vessels can be constricted against the single central bone. The tourniquet should be tightened until the pulse is almost imperceptible in the limb below. Prolonged interruption of the blood supply can lead to gangrene of the limb. To prevent this the tourniquet must be released for a few seconds every two or three minutes to allow some fresh blood to reach the part. The tourniquet must be applied immediately after snake bite and this procedure continued until medical aid is available.

(See also SNAKES.)

3. *First-aid kit*. A first-aid kit must be kept by every team working in the field and the contents inspected and replaced as they are used or become old. It should contain the following items:

3.1. Instruments. Scissors and forceps (tweezers)
A sharp knife or blade
Syringe and injection needles.

3.2. Dressings. Bandages—2·5, 5 and 12 cm width, 2–5 m length
Gauze
Cotton wool
Swab sticks.

(Most of these items are available in small, presterilized sealed plastic packs which remain sterile until the seal is broken.)

3.3. Spirit lamp and small pan for boiling water.

3.4. Medications. Liquid antiseptic (e.g. Savlon) 100 — 250 ml.
Potassium permanganate crystals in securely stoppered bottle.
Pain-relievers. Aspirin and some stronger preparation for emergencies.
Kaolin powder.
Anti-diarrhoeal tablets.
Anti-spasmodic tablets to relieve severe griping.
Strong ammonia solution.
Any other medication required by individual members of the team.

3.5. Antivenene (I.P.). 2 ampoules.

3.6. Tablets or solution for disinfection of water.

(Medical advice must be taken for selection of medica-

tions and detailed instructions on their dosage and use must be written and kept in the first-aid box, with each medication.)

The suggestions for contents of the first-aid box must be adjusted and modified according to requirements like number of persons, duration of camps, distance from nearest medical centre etc.

A.N.D.N.

FISHES are aquatic, cold-blooded vertebrates which normally breathe (i.e. take oxygen dissolved in water) through their gills and have fins on their bodies. They constitute one of the most ancient groups among vertebrates and are presumed to have evolved from small primitive forms, known as Protoohordates, from which all the higher vertebrates including the man have evolved. Paleontologically, fishes are known to have originated as early as about 500 million years ago. Several have evolved into modern fishes while others have become extinct during the course of evolution. However, at present about 20,000 to 30,000 species are reported to be existing. In waters round the Indian subcontinent about 2000 species occur. There are even more insect species, but the number of fish species far exceeds the number of all other vertebrate species—mammals, birds, reptiles and amphibians—all put together.

Fishes are found almost wherever there is water, whether fresh water, mixed estuarine water or pure salt water, except in some of the crater lakes of Africa which have no communication routes to any other outside watercourses. Consequently, they are extremely varied in size, shape, length and in every other way. The variations range from the smallest (11 mm) and lightest (0·08 g) *Pandaka pygmea* of the Philippines to the largest (16 m) and heaviest 25-tonne Basking Shark of tropical seas such as ours. For fresh waters, the records for weight are comparatively lower: 1000 kg for the sturgeon (*Acipenser*) of the Volga river and 256 kg for catfish, *Silurus glanis*.

Several ancient groups of fishes have now become extinct and are only found as fossils. Existing fishes are divided into three main groups: (1) Elasmobranchii, cartilaginous fishes like sharks, skates and rays, (2) Holocephali or Chimaerae, cartilaginous but with single-pointed, rat-like tail, and (3) Teleostomi, the bony fishes. Cyclostomes such as lampreys and hag fishes, though dealt with along with fishes (Pisces), are not considered by some authors to be true fishes. The last group (bony fishes) is further subdivided into (*a*) lung-fishes (*Dipnoi*), air-breathing forms, (*b*) lobe-fin fishes like the Coelacanths, and (*c*) the true bony fishes of modern times (Teleosts) which form a very large majority of the existing fishes, such as the carps, catfishes, eels, sardine and mackerel.

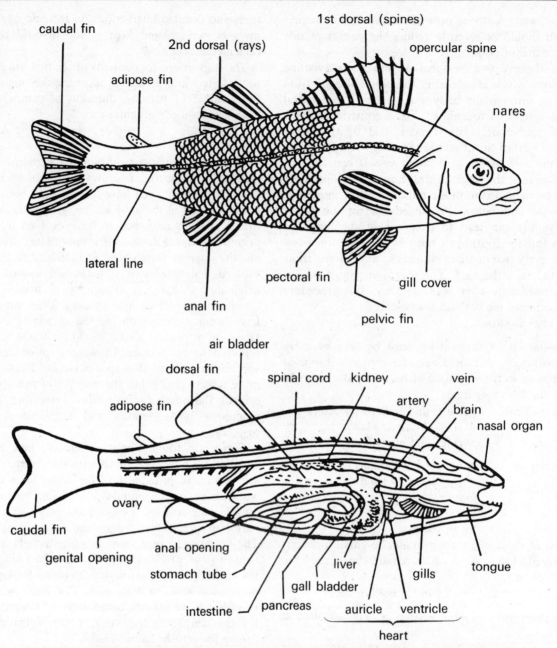

Fig. 1. A TYPICAL BONY FISH SHOWING EXTERNAL AND INTERNAL CHARACTERS

External characters. The body of a fish comprises a head, a trunk or body, and a tail (see fig. 1). The foremost part of the head constitutes the jaws which are of different shapes and sizes in different fishes. In some forms like the CARPS (rohu, catla etc.) the jaws are toothless while in sharks, barracuda, perches, murrels etc. teeth of various sizes and shapes are found. Above the jaws are two pairs of nasal openings called nares, equivalent to the nasal apertures of higher vertebrates, having taste-buds for understanding the smell and taste of the water.

In some fishes tiny projections or tendril-like structures known as barbels occur in conjunction with both the jaws as in carps. In others these barbels are long, whisker-like, as in CATFISHES which have derived their name from these long cat-like whiskers. The barbels are tactile (sensitive to touch) and help in finding food, obstacles and enemies.

In addition to jaws, nares and barbels, other important characters such as well-developed eyes, gill-covers, paired and unpaired fins, gill-silts etc. occur as shown in the illustration. Fins are of various sizes and shapes in different fishes. Bony fishes have a common gill opening for all the gills, while sharks and lampreys have independent gill-slits, one for each gill. Fishes have no external ear but they have an internal ear which

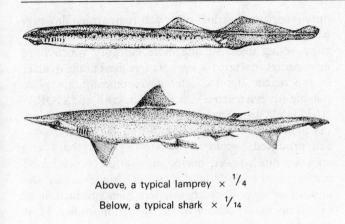

Above, a typical lamprey × ¹/₄

Below, a typical shark × ¹/₁₄

can receive sound-waves conducted through water and which helps them to maintain their balance. On the sides (mid-line) of the body of most fishes, and also on the head of some, rows of minute pores are perceptible. These constitute the specialized lateral line sense organs for detecting obstructions in water and coordinating movements.

The bodies of most fishes are covered with thin, oblong and overlapping scales like tiles on a house. These are embedded in the skin and are of many different types. Catfishes have no scales but in some cases have bony flattened protective shields on the upper part of the body while eels have minute scales embedded in the skin.

Most fishes have a torpedo-shaped, streamlined body, tapering at both ends so that it offers least resistance while moving in the water. A fantastic variety of other

shapes are found, some of which are described under EELS, DEEP-SEA FISHES, and SUNFISH.

The capacity of fishes to adapt to different ecological conditions and to tolerate different hydrostatic pressures, water temperatures, salinities etc. is also very wonderful. Some of them are capable of thriving in the low pressures prevailing at an altitude of 4500 m in lake Titicaca in South America; while others stand the tremendous pressures occurring at ocean depths of nearly 11,000 m. Similarly some fish subsist at the freezing temperature of − 2°C while others can live in hot-spring waters of 40°C. As regards salinity, their capacity ranges from 0·1 to 50% salinity.

Internal Structure. The shape of any fish is usually formed and supported by its skeletal structures, the head by its skull and allied bones, the trunk by its central vertebral column and other dermal bones and the tail portion by its bony pattern (fig. 2). The skull is associated with the jaw bones and gill apparatus. The skull bones protect the vital central brain, and the vertebral column harbours the delicate nerve cord and remnants of the notochord. The muscles are attached to the bony structures of head, body and tail and their contractions are responsible for the various movements of the different parts of the body. The different fins also have supporting bones and are attached to them for facilitating movement.

The aforesaid skeletal structures are made of hard bony material in bony fishes (teleosts) while they are of cartilage, flexible material like that of a human external ear, in elasmobranches and cyclostomes.

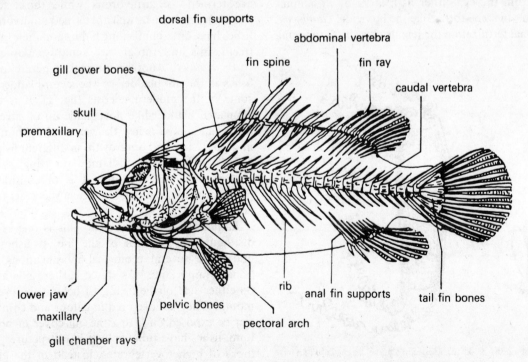

Fig. 2. SKELETAL STRUCTURE OF A TYPICAL BONY FISH

Functional account. Although fishes are genealogically the lowest and earliest of the vertebrates, they possess most of the anatomical and physiological systems that are found in higher groups (fig. 1). A well-developed brain is present with its different lobes, along with other cranial and body nerves spreading right into the tail region. Similarly, the digestive system, starting with the mouth, opens by the anus situated at the end of the body cavity and before the tail region. The respiratory system is concerned with circulating blood through comb-like gills, effecting its oxygenation by absorbing oxygen dissolved in water and releasing carbon dioxide. The blood has colourless plasma with red blood corpsucles floating in it. The heart is two-chambered (auricle and ventricle) and is connected with arteries and veins which spread over the entire body and take part in the blood circulation.

The excretory system is composed of a pair of kidneys attached to the dorsal wall of the body-cavity: they drain into the urine tubes (ureters) which join the common urino-reproductive duct and open outside by a vent. The reproductory system is also well developed. In normal cases, the reproductory products, i.e. eggs and sperm of female and male respectively, are evacuated into ducts and after joining the common urino-genital canal, pass outside by a vent situated behind the anal opening. Considerable variations occur in the size, shape and nature of eggs and larvae, depending on the evolutionary, genetic and ecological conditions. After fertilization of the egg by the sperm, whether in the body or outside in the water, an embryo develops and emerges from the egg either as a larva or in a mini-adult form. In some forms like the larvicidal *Gambusia*, after internal fertilization the female gives birth to young

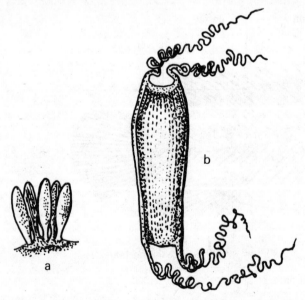

(a) Eggs of a goby × 4. (b) Mermaid's purse, an egg-case of a dogfish × 1/3.

ones (viviparous), while other fish (oviparous), lay eggs which later develop into larvae and then into free-swimming young ones. In one more category which is of intermediate nature, the eggs are fertilized in the oviduct of the female but the later development takes place outside (ovoviviparous), as in the dogfish and HORAICHTHYS. In oviparous fishes large numbers of eggs (as many as 28 million were found in *Molva molva*) are produced; while in viviparous types the young ones are much fewer, maybe only 40 or 50 per brood (as in *Gambusia*). Where parental care prevails, the number of eggs is much smaller: high production of eggs is invariably accompanied by high mortality. Most of the Indian bony fishes have no external sex organs but the sexes are separate. In some, secondary sexual characters appear only in the breeding season.

Food. As a general pattern, fish feed on small aquatic plant and animal life including fish; larger ones preying on smaller individuals, and some are purely vegetable feeders. Minute plant and animal life, collectively called plankton, is taken whole along with the water in which they live, the water being strained out with the help of hard extensions of the gills, gill-rakers, rather like a garden rake. Carps have no teeth in the jaws but have teeth-like bony projections (called pharyngeal teeth) in the throat. Pointed teeth of fish like barracuda and perches are used for grasping and tearing, while flat teeth in the jaws of skates and rays are useful in crushing bivalves and other shells to suck juicy material inside. Mullets and grass carp feed on leafy aquatic vegetation. Fishes feeding on other fishes, insects and other animals are termed as carnivorous while those feeding on aquatic vegetation as well are termed omnivorous. Thus, fishes have different feeding habits and their mouth parts (teeth, lips and throat) are suitably adapted.

Air bladder. Another internal structure peculiar to fishes is the air bladder or the swim-bladder found in some of the modern teleosts (fig. 1). It is usually an elongated bladder-like float made up of silvery fibrous material situated below the vertebral column. It varies in shapes, sizes and outgrowths in different fishes. When dried it is known as 'fish maw' or isinglass and is of commercial value. For fish it is of considerable importance in maintaining balance in the water at different depths.

Respiration. Respiration by gills is one of the main distinguishing features of the life of fishes and has undergone several ecological modifications of evolutionary importance. As stated earlier, gills are located on either side of the head just behind the eyes and the mouth. They are deep red in colour and comb-like and can be exposed on lifting the gill cover of bony fishes. Lung-fishes have tubular lung-like structures similar to those of higher vertebrates, instead of the gills. Some specialized bony fishes such as climbing perch and mud-skippers and some catfishes are capable of living out of water for several hours. This is because they possess accessory respiratory organs of different types (see AIR-BREATHING FISHES).

Locomotion. Fishes are known for their smooth and speedy movements in the water. This is largely due to the action of the various bundles of muscles on both sides of the body which by their alternate contraction and expansion (i.e. lateral flexion), generates the typical locomotion seen in fusiform fishes like the mackerel. The caudal fin works more or less like a rudder for direction and balance. Great speeds, ranging even up to 100 kmph, are attained in this manner in fishes like the Bonito, a large mackerel. In the Cow or Trunk fishes, where the body is encased in a hard inflexible tunic and lateral flexion is not possible, the locomotion is achieved by the sideway flopping action of the fins, especially the caudal fin, the resultant movement being very limited and slow, and such fish do not cover long distances. In the vertically swimming fish such as the seahorse, it is largely the dorsal fin which propels it through the water. In the bottom-dwelling flattened-bodied skates and sting rays (*Trygon*), locomotion is achieved by the wave-like action of the modified pectoral fins which surround their bodies. Propulsion through air is also a special mode of transport.

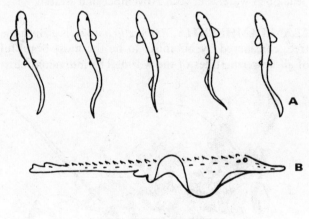

LOCOMOTION IN FISHES
A Movements of a fusiform fish
B Movements of a flat-fish

The long-distance swimmers and shoaling fish like sardines or mackerel rarely restrict themselves to any particular locality but those inhabiting tidal pools or submerged rock caves have their own localities or territorial boundaries. Slow-moving fishes like the cow fish and seahorse necessarily select and stay in their protected habitat in coastal areas.

As regards the zonal movements of fishes, it can be stated that the specialized deep-ocean forms always remain at the bottom as they are adapted to live in the intense water pressure. They would die automatically if brought to the surface. The surface or pelagic forms (such as sardines, mackerel, tuna, barracuda), monopolize the upper layers of the sea. The middle-layer inhabitants (such as sharks and pomfrets), constitute a very large group which traverse vast tracts both vertically and sideways. In the coastal shallow seas, the bottom-dwelling forms like the sole, skates and rays, dominate the zone.

Mass movements of fish, known as migrations, occur either for food or for propagating the race, i.e. breeding (see SALMON, HILSA, EEL) and on rare occasions on account of lack of oxygen or adverse temperature changes. Surprisingly enough, these migrations are not always in the same quality of water, being sometimes from salt water to fresh water (anadromous) or vice versa (catadromous). Similarly from deeper seas to tidal areas, a small sea-fish, the grunion, actually comes ashore to spawn in California.

Movements in riverine fishes are not restricted to any particular area, though mahseer and some large catfishes have preferences for certain deep pools in the course of the rivers where they wait for their favourite prey. Similarly, some mahseers are known to visit certain fish sanctuaries near temples in certain seasons, for example Hardwar and Rishikesh in north India and Dehu and Alandi villages on the Indrayani river near Poona, where they feed voraciously on puffed rice, peanuts and pelleted ata given by the pilgrims, although it is not their natural food.

Longevity. The longevity of fishes depends largely on their natural size, food, protection and other ecological conditions, large-bodied fish living longer than the smaller ones. The common carp is reported to live for about 50 years in sheltered ponds. Small fish such as the *Pandaka pygmea*, *Horaichthys* and *Gambusia* live hardly for a year.

Coloration. The fascinating colours of fishes have long attracted attention, especially since the aquarium hobby came into vogue. The varied and dazzling colour patterns are due to the presence of pigment-containing cells (called chromatophores) and reflecting tissue cells (known as iridocytes) in the dermal layer of the skin above or below the scales. Three main pigments, red, yellow and black, are deposited in these cells and by the combination of the colours in depth and intensity different shades and patterns are produced. These shades of colour remain permanent or change according to the sensitivity of the fish to fear, love, anger or other internal impulses, endocrinal secretions or external stimuli of heat, light or colours of surroundings. Camouflaging habits of fishes simulating the colour of the surroundings in which they live are also examples of such colour changes.

Special adaptations. Some fishes like the electric ray (*Torpedo*), and certain catfishes and eels of S. American waters can produce their own electricity through specialized tissues and give shocks to their enemies both for offence and defence. Some deep-ocean fishes carry their own lighting arrangement, i.e. have bioluminescent lights. These organs consist of light-producing cells, lenses, and reflector surfaces situated on different parts of the body. On the ocean floors where perpetual darkness prevails, the light organs serve in finding food, attracting prey or a mate, and also warn enemies.

Some fishes are known to be poisonous (see POISONOUS FISHES).

Conflicts. Fishes rarely fight as a group for food or shelter but they do take shelter as a group when approached by any large predator and also run away in groups. In this group behaviour, they seem to seek safety in numbers. Some individuals, like the male Siamese fighter, an aquarium favourite, fight with another contending male quite aggressively. The GIANT GOURAMI, an introduction into India from Indonesia, while guarding its nest fights and drives away intruders pugnaciously.

Conditioned response. Some fish such as carps respond to habitual signals and are capable of being trained to a limited extent, mainly for a reward of food. On creation of certain sounds they gather at a particular place for their usual meal. A shark is reported to have learnt to strike a target and swim away for a promised morsel of food. Even a light signal is understood with reference to food supply. Some fish are reported to recognize a baited hook and avoid it.

Special behaviour. Some fish are capable of creating gurgling or grunting sounds which are magnified in association with the air bladder and the connecting bony parts. These fishes are known as grunters or croakers and are common on the Indian coast where fishermen listen for these sounds, dive to the bottom and catch the fish.

Sleep. A question often asked is whether fish sleep, like higher animals. Because of absence of eyelids in most fishes, they do not seem to be sleeping; but they go into a state of suspended animation which can be termed as sleep. In this condition they remain motionless and sometimes lie on their sides. This has been observed especially in the aquarium.

See plate 2 facing p. 33 and plate 22 facing p. 305.

FISHING EAGLE. Pallas's Fishing Eagle occurs at inland waters in Pakistan and northern India. This majestic eagle is brown with paler head and a broad white band across the tail. It hunts its own prey—fish, snakes, small mammals and crabs—or pirates prey from other raptors, forcing them to surrender their lawful prize. Also takes carrion. Three white eggs are laid on a massive stick-nest on an isolated tree growing near or in water, between November and March.

FLAME AMHERSTIA, *Amherstia nobilis,* is a Burmese tree, considered by botanists to be the most beautiful of all flowering trees of the world. The individual blos-

© M. Krishnan

Pallas's Fishing Eagle taking off × ¹/₄

A cluster of Flame of the Forest flowers, × ²/₃

soms look like humming-birds, each mounted on slender, intensely red stalks and arranged into long, drooping, candelabra-like sprays, so that the bird-like flowers stick out elegantly in all directions. The flowers have 5 red petals, 3 large with blobs of gold, reversedly heart-shaped, and the other two minute. They come into bloom from January to March. The evergreen leaves resemble those of Ashoka, which taper to a fine tip called a 'drip tip', an adaptation for draining rain-water off the leaves quickly. This is seen in many trees from high-rainfall areas, e.g. Indiarubber. The tree is so rare that it seems to have been seen only about twice in a wild state, but is in cultivation in Bengal and Sri Lanka.

FLAME OF THE FOREST (*Butea monosperma*) is so named as the massed crowns of bright orange flowers suggest a forest in flames. It is a small deciduous tree with hard leathery leaves, widespread in the subcontinent, including Burma and Sri Lanka. In spring, when leafless, begins a vermilion riot of spectacularly beautiful blooms, great clusters of pea flowers burst open contrasting vividly with the jet-black velvety calyces and stalks. The flowers are fertilized by babblers, sunbirds and other birds. The blazed trunk exudes a red juice which hardens into a gum.

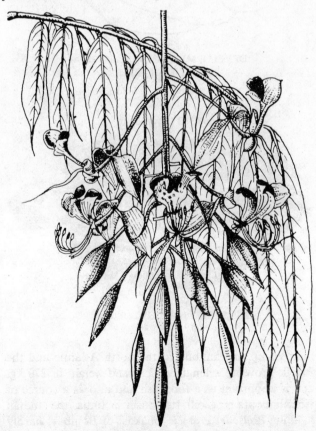

Flame Amherstia, leaves with pointed tips and candelabra-like flowering sprays, × 1/3

FLAMINGO. The two species of flamingo—the Greater *Phoenicopterus roseus* and the Lesser *Phoeniconaias minor*—found in the subcontinent have a wide distribution outside our region. The former ranges across Iran and the Middle East north into western Siberia, west to Spain and S France and south into East Africa. The Lesser Flamingo occurs along the seaboard of the Arabian Sea from India, the Persian Gulf and the Red Sea south into East Africa and the island of Madagascar.

With their long legs, fully webbed toes and long necks, flamingos are adapted to feed in shallow water by reaching down to the muddy bottom. The bill is a specially adapted organ for filtering out organic matter in the bottom ooze and also surface water. While visiting freshwater lakes and large rivers, flamingos of both species have been recorded as far inland as Delhi. Their ideal habitat is brine lakes, and tidal estuaries. In India, consequently, the best locations for watching both species are the Gujarat coast, Sambhar Lake in Rajasthan, Point Calimere in South India and Chilka Lake in Orissa. To all these locations, the birds are foraging, non-breeding visitors.

Both species are highly gregarious, and particularly so when nesting. Huge concourses assemble in favourable years in the Great Rann of Kutch where exists the largest nesting site of the Greater Flamingo in Asia. The Lesser Flamingo, long suspected of nesting in the same area, was conclusively found doing so with the larger species in 1974. The most important nesting locations, mainly of the Lesser Flamingo, are on the soda lakes of East Africa. Smaller nesting colonies of the Greater exist in Iran, western Siberia and Spain.

Flamingo nesting colonies are a remarkable sight with thousands of birds packed together. Nests, usually constructed in shallow water, are mounds of mud from a few centimetres to half a metre in height, with a depression at the top. A single chalky white egg is laid, or occasionally 2, which both parents take turns to incubate, settling on the egg in the conventional manner

© Sálim Ali

Flamingo colony × 1/16

with legs folded under the body. Newly hatched chicks are covered with fluffy white down, which later becomes blackish. The chicks are fed by a secretion from the crop of the parents. Growing chicks are herded into tightly packed 'crèches' cared for by a few adult birds, while the rest of the parents are away foraging. In favourable years, nesting activity extends over several months while in drought years no nesting occurs: this adaptation is necessitated by the need for such large congregations of birds to be ensured an optimum food supply. Flamingo chicks have straight bills which gradually develop into the characteristic downcurved bill of the adults. Juveniles have brownish plumage and black legs and bills, and the adult white plumage washed with pink is developed after the first moult.

S.K.

FLAT-FISHES. As the name suggests, these fishes have flattened, asymmetric bodies spreading out in the horizontal plane on the bottom of the sea where they usually live, with both eyes on the upper side of the head. The dorsal and anal fins are modified to skirt the body and the caudal fin is flattened in the same plane as the body, unlike the normal erect position. The side on which flat-fishes lie is white and is called the blind side; the other, upper, side is exposed to light, and is dark-coloured and sometimes patterned with markings.

Despite the aforesaid unsymmetry in the adult condition, the young fish are born with a symmetric body like any normal fish; but after a month, when they are about 10 mm long, the head starts twisting during its growth and during this development one of the eyes starts moving until it reaches the opposite side (upper) of the head (see figure). The jaws also twist to some extent. In some flat-fishes the twist is on the right side, while in others it is on the left. In the dextral flat-fishes such as the plaice (*Pleuronectis*) or sole (*Solea*) the eyes are on the right side of the head. The Tongue soles (*Cynoglossus*) are sinistral, having their eyes on the left. At the time of migration of the eye the dorsal fin also extends and grows along the edge of the head above the eye, the anal fin covering the other edge of the body. In the primitive *Psettodes* one eye remains on the top of the head instead of migrating completely to the dark side and its dorsal fin starts at the nape.

Flat-fishes are masters of camouflage, cleverly imitating the colour and pattern of the background (of sand, mud or gravel) on which they lie. Their movements are feeble and limited, being accomplished by undulations of the body and its fringing dorsal and anal fins. In the Tongue soles both these fins are united with the caudal to form a continuous fringe around the body.

As a group flat-fishes are not of much commercial importance in Indian waters but in the temperate zone plaice and sand dab are popular. The largest member of

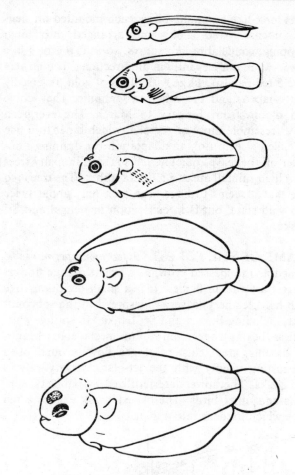

DEVELOPMENTAL STAGES OF A
FLAT-FISH (SOLE)

At the top the second eye is hidden (lateral view)

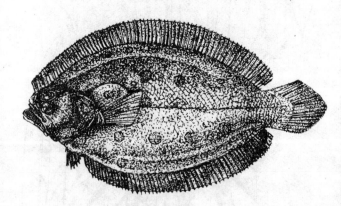

Full-grown flat-fish, the Tongue sole,
Pseudorhombus arsius, × $\frac{1}{3}$

their group, the halibut, in the north Atlantic and the Pacific grows to a length of 3 m and weight of 270 kg, and is important as a food fish and also as a source of vitamin-containing oil. Its cousin in India, the 'Indian halibut' *Psettodes erumei*, 'bhakas' of Bombay, hardly reaches 65 cm. Other flat-fishes in Indian waters are much smaller.

FLEAS. These small, strongly sclerotized, laterally flattened, yellowish brown to blackish, spectacularly jumping insects (1 to 6 mm long) are strongly specialized and endowed with backwardly directed hairs and spines, enabling them to progress with ease through the hairs or feathers of the host: More than 2000 species are so far described from the world, about 80 species known from India, mainly in cooler areas. Their strong legs with spines endow them with characteristic leaping propensities, some individuals even covering 30 cm in a jump, many times their own size! Adults are parasitic mainly on warm-blooded mammals (also a few on birds) and striking biological adaptations enable them to withstand unfavourable environmental conditions for long periods.

Most fleas are not as closely associated with their hosts as are ANIMAL LICE for example, and are more nest-specific than being host-specific. The blood-sucking adults are only intermittently parasitic, taking blood mostly to mature their eggs and maintain fertility. Fleas also leave the body of a host that dies and seek a new one, not necessarily of the same species, thus gaining significance as potential vectors of diseases. Fleas can survive more than a year without food, waiting patiently until another host individual comes to the nest, or spending considerable time searching for another warm-blooded host mammal or bird, no matter what species. Flea species associated with migratory birds overwinter in the nesting places as adult or immature stages. Adults of many species mate on the host or in the nest almost immediately after emergence and a meal of warm blood. Eggs are laid in the nest and those laid on the host body soon fall to the ground. The composition and microclimate of the nest is of critical importance for development of young stages and the larvae mostly require fairly high temperatures and humidity for development. Domestic species, however, living in house dust and similar habitats, are more resistant. After hatching from eggs, larvae attain adulthood in two weeks to eight months.

Fleas are of considerable medical importance, transmitting several diseases besides being domestic pests causing annoyance and irritation through their movements on the body and their bites. Some fleas are intermediate hosts of tapeworms and the like. The most dreaded disease, bubonic plague (which killed over 10 million people in India between 1898 and 1918), is transmitted by fleas from rats to man. Other bacterial infections transmitted by fleas are brucellosis and salmonella. Rickettsial infections and viral infections like tick-borne encephalitis are other diseases for which fleas are vectors. The common species of importance in India are *Xenopsylla cheopis*, *X. astia*, *Ctenocephalides canis*, *C. felis*, *Pulex irritans* and *Echidnophaga gallinacea*.

FLIES. 'Flies are among the most familiar of insects, and yet among the least understood. We talk about "the fly", as we talk about "the ant" and "the frog", and then we have in mind one species, the common house-fly, *Musca domestica*. This highly adaptive insect happens to like the sort of litter that man produces wherever he lives, and while man has been domesticating the pig, the goat, the horse and the dog he has also unwillingly domesticated "the fly".' Thus begins one of the best written books on flies, *The Natural History of Flies* (1964), by one who was, undoubtedly, a master fly-expert, the Englishman Harold Oldroyd.

When we see, however, that there are a little less than a lakh of species of flies, classified under some 120 families, and that they are so diverse as to include some large, brightly coloured forms as well as dull, tiny midges, we realize that we really do not know much about this interesting group of insects. Oldroyd goes on to state: 'It is true that most of them breed in decaying organic matter of some kind, but we must remember that disgust is purely a human reaction. All animals must have protein food. The fact that many flies have learned to get it from the products of organic decay is a sign of their biological efficiency. Decay there must be. New organisms cannot be built unless old ones disintegrate. In exploiting sources of food that will always exist, flies have made their future secure.'

The order Diptera includes what are called the true flies, gnats, midges and mosquitoes. Insects of several other orders are also commonly called 'flies'—dragonflies and damselflies (Odonata), mayflies (Ephemeroptera), stoneflies (Plecoptera), scorpionflies (Mecoptera), butterflies (Lepidoptera), caddisflies (Trichoptera), sawflies (Hymenoptera) and fireflies (Coleoptera). From all of these 'false' flies, the Diptera can be distinguished by the presence of only one pair of functional wings and the reduction of the hind pair into halteres that are used as gyroscopic stabilizers in flight. Flightless (and therefore wingless) Diptera can usually be identified by

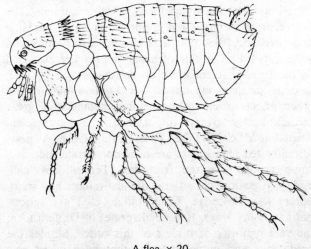

A flea × 20

the well-developed middle segment of the thorax, which accommodates the flight muscles. Flies range remarkably in their size and appearance. The largest may be some 3 inches long, with an identical wing-span and the smallest, hardly visible to the naked eye, may be 50 times shorter, and about 125,000 times smaller in bulk. The ancestral flies (like other life forms) branched out into various evolutionary lines, many of which have now become extinct. The living flies are members of some branches that did survive, some already declining, whereas others are still advancing. However, it is an evolutionary fact that, like the fate of the dinosaurs, all large animals (including man?) and plants seem to be unable to survive for long, and the largest flies are slowly, but surely, on the way out. The future is with the small acalyptrate flies, like *Drosophila*, which have not yet committed themselves too far in any direction, and obviously still retain great evolutionary vitality.

Fly larvae are all without true legs and are all very different in form from the adult flies they grow up into. The larval and the adult stages, therefore, are able to live completely independent modes of life and are thus able to utilize natural resources maximally. The larva is the principal feeding-stage in the Diptera, although in some groups the adult fly is more competitive. The fly imago (adult) usually feeds only to make up the deficiencies of the larva, besides for obtaining energy for flight and daily activity, or to help maturation of eggs, by a necessary blood meal, as in the female mosquito. The adults also perform the important functions of reproduction and dispersal with their well-developed powers of flight, sight and smell. A paper by D. Keilin, published in the journal *Parasitology* in 1944, contains an excellent description of the habitats that fly larvae have adapted to, and I can do no better that to reproduce this in full:

> There is hardly any type of medium capable of supporting life from which dipterous larvae have not been recorded. In fact they occur in every kind of watery medium, such as streams, rivers, waterfalls, ponds, lakes, brackish water, hot mineral springs, and the water-reservoir of terrestrial and epiphytic plants: they are found in a variety of dry and semi-fluid media such as sand, earth, mud, animal, and vegetable substances in different degrees of decomposition, excrements of animals, and in festered wounds of animals and plants. They are met with as scavengers in the nests of mammals, birds, wasps, bees, ants, and termites. As parasites they are found in different portions of plants, i.e. roots, stems, leaves, and flowers, boring galleries or forming galls. They are known also to live as true parasites of a great variety of animals such as Oligochaetes, Molluscs, Crustacea, Arachnids, Myriapods, practically every large Order of Insects, Amphibia, Reptiles, Birds and Mammals, including Man. Finally, in a certain number of species the larvae grow to different stages of development within the uterus of the mother, leaving it in some cases only just before pupation.

Most fly larvae live concealed, usually in a medium that is more or less humid. Very few larvae live openly on vegetation exposed to light and air.

The pupae of most of the more primitive groups of flies are curious mummy-like objects with external projections and swellings that contain the head and locomotory organs inside. Pupae are motionless except for a few like those of mosquitoes that swim about and those of a black-fly, *Simulium*, that feed and spin cocoons as well! The pupae also have well-developed adaptive structures like spines and extensions of prothoracic spiracles to suit their needs, in the environment they live in, before emerging as adults. The more advanced fly families possess larvae that are called maggots and the pupal stage process is simplified. The last larval skin becomes hard and dark and forms a tough pupal case, called puparium. The adult fly emerges from the puparium and then to the surface of soil, or other matter it lies buried in in the pupal stage, through the help of a large inflatable sac on the head called the ptilinum.

Adult Diptera are ubiquitous and predominantly free-living. They inhabit practically all kinds of terrestrial habitats and occur in aerial plankton. One remarkable chironomid midge of the genus *Pontomyia* even inhabits the sea. Some families at least form major elements of the fauna of rain-forests, deserts, intertidal zones and snowfields. Generally, most species of flies are associated either with forest or water, and their considerable powers of flight enable them to be wide-ranging within each habitat. However, some ectoparasitic flies are more restricted, living on the skin of their hosts. Except for a few groups of flies with non-functional mouth-parts (e.g., Oestridae), most adult flies are polyphagous, taking up only liquid food for which their mouth-parts are specially adapted. Such foods include free water; a variety of animal and plant secretions; products of decomposition of organic matter; soluble solids, which are first liquified by salivary secretions; and the tissue fluids of other animals. The last refers to the large and important group of parasitic and predatory flies that attack other arthropods and vertebrates. Body fluids of hosts or prey are extracted by means of different kinds of mouth-parts that pierce, lacerate or even masticate. The Diptera outrank all other insect orders in possessing members of great medical and veterinary importance. These groups are treated in MOSQUITOES & BITING FLIES. As pests of cultivated plants, flies are of minor significance. But some groups like the fruit-flies (Tephritidae), gall-midges (Cecidomyiidae) and the leaf-miners and stem-borers (Agromyzidae, Chloropidae, etc.) are serious pests in many areas. Beneficial species of Diptera form the vast majority of insects of this order. Besides the predacious and parasitic forms that keep insect pest species in check, flies are almost as important as many hymenopterous insects (especially bees) as pollinators. The potential of some plant-feeding flies as weed killers is as yet little exploited by man.

The natural enemies of flies include birds, bats, reptiles, amphibians, fish, spiders, dragonflies, beetles, bugs, wasps and other flies that predate on larvae and adults besides the other immature stages. The ecto-parasites of flies are mainly mites, algae, fungi and proto-zoa. Endoparasites comprise viruses, bacteria, fungi, protozoa, helminths and insects like several chalcidoid and other microhymenopterous families, ichneumonid and braconid wasps, as well as tachinid and bombyliid flies.

The diversity of flies was hitherto subdivided broadly into three suborders: Nematocera, Brachycera, and Cyclorrhapha. This grouping was considered not to be 'natural' by the late Russian entomologist, Boris Rohdendorf, who separated the curious family Nympho-myiidae into the suborder Archidiptera and put all the rest into the Eudiptera which he again divided into 13 infraorders, like the Bibionomorpha, under which the Nymphomyiidae are now included. A stable phylo-genetic classification of the Diptera has not yet been achieved. This is due to many factors: (1) the long evolutionary history of the flies, earliest records being from the Permian, more than 225 million years ago; (2) some groups like the muscoid and acalyptrate flies being in the process of radiation even now; (3) several evolutionary lines having become extinct and the poor fossil record of these at hand; and (4) even though Diptera are the fourth largest insect order known today, they have been relatively poorly collected and studied, especially with respect to immature stages and their natural history. Harold Oldroyd recently suggested re-grouping of the Diptera into three suborders, based also on their ecology and habits, rather than purely on struc-tural characters. He proposed that the ancestral flies had larvae living in moist media, being neither truly aquatic nor terrestrial. The crane-flies and related families he called Superstata, whose larvae again were not pre-dominantly aquatic or terrestrial. This left two large groups of more advanced flies which he thought con-veniently separated themselves into an 'earthy' group and a 'watery' group. Most of the remaining Nemato-cera and the more primitive Brachycera related to the tabanids and stratiomyids, he called Madescata. These are the water midges, gnats, mosquitoes and horse-flies whose larvae are predominantly aquatic and whose females suck blood by specially developed mandibles; these include all the 'watery' families. The 'earthy' families, most of which are highly advanced, he called Arescata. These include the higher Brachycera and all of the Cyclorrhapha, and are what are the land midges and 'flies'. Their larvae live in drier media though some have become secondarily aquatic; those which have also

re-acquired the bloodsucking habit use structures other than the mandibles. I will adopt this classification here as it appears distinctly more 'natural' than earlier groupings.

The Superstata contains the large family Tipulidae, the Pachyneuridae and some related families. The 'daddy-long-legs' or crane-flies of the family Tipulidae are immensely versatile in their choice of larval habitats. It is also one of the largest families of flies (13,500 + species) and cosmopolitan in distribution. More than 1500 Indian species have been described and the majority occur in the Himalaya and the moist forests of Assam, Burma and the peninsula. They are usually found resting on foliage in damp, shady places on their extremely long legs. Some are restricted to special habitats, others being crepuscular, and many come to man's artificial light sources. Among the diverse larval habitats, most occur in damp soil or decomposing vegetable matter; some are injurious to lawns in more temperate climes. *Eriocera*, *Limnobia* and *Conosia* are some genera found in the subcontinent.

Of the Madescata, the mosquitoes and other biting flies have been dealt with elsewhere. These include the superfamilies Psychodoidea, Culicoidea, Chironomo-idea and the Tabanoidea. The other superfamilies that are included in this suborder are Blephariceroidea, Dixoidea, Deuterophleboidea, Orphnephiloidea, Rhae-tomyoidea and the Stratiomyoidea. The first five com-prise a single family each: Blephariceridae, Dixidae, Deuterophlebiidae, Orphnephilidae and Perissommati-dae. The last is a peculiarly southern temperate group with five known species restricted to southeastern Australia and Chile. The adults have eyes separated into dorsal and ventral sections and occur in wet sclerophyll and rainforests, while the larvae breed in decomposing fungi. Less than 40 species are known from the Indian subcontinent of the other four families, which contain generally small flies that are found on vegetation or rocks near flowing streams in wet forest.

© Kumar Ghorpadé

A soldier-fly (Stratiomyidae) on flower of nigerseed (Guizotia abyssinica) × 2½

The Stratiomyoidea comprise mainly the Stratiomyoidae, or soldier-flies, which are a cosmopolitan group with a wide range of forms, recognizable by their characteristic wing venation. Many are strikingly marked in colour and are generally scarcely observed and collected. The larvae are either aquatic, or found in damp soil and rotting vegetation. Along with the related Xylomyiidae (Solvidae) and Rachiceridae, some 150 Indian species are known; *Odontomyia*, *Sargus*, *Ptecticus* and *Clitellaria* being familiar genera. Of the Chironomoidea, the water midges of the large family Chironomidae are common and often mistaken for mosquitoes. Almost 200 Indian species are known and the crepuscular or nocturnal flies usually abound in the vicinity of water in enormous numbers as 'mating swarms'. The larvae are aquatic, lying buried in the bottom debris or free on vegetation. One of the rare marine insects belongs to this family — *Pontomyia*. Some of the widespread genera are *Chironomus*, *Tanytarsus* and *Orthocladius*.

All the rest of the Diptera, that are essentially terrestrial though some have returned to an aquatic habitat secondarily in the larval stage, are grouped under the suborder Arescata, and make up more than 75% of all fly species described. Around 30 superfamilies in the following nine infraorders are recognized: Bibionomorpha, Asilomorpha, Musidoromorpha, Phoromorpha, Termitoxeniomorpha, Myiomorpha, Braulomorpha, Streblomorpha, and Nycteribiomorpha. Families now under the Bibionomorpha were placed earlier in the Nematocera. A little more than 50 species of Bibionidae, often called march-flies, occur in the subcontinent, especially in temperate or high-elevation areas. The flies are sluggish and fly weakly, mostly in the late evenings and night. Larvae occur in damp soil or rotting vegetation, some feeding on plant roots, probably on dead tissue. Species of *Bibio*, *Plecia* and *Dilophus* are the more common bibionids in the Himalaya. The Cecidomyiidae (Itonididae) or gall midges is a large family of considerable economic importance, many species being plant pests and a small group feeding on injurious Homoptera. *Zeuxidiplosis giardi* is an example of a beneficial weed-killing gall midge. The flies are all small or even minute, with reduced wing venation and hairy wings. Larvae of most species have a 'sternal spatula' on the prosternum, and many live in plant galls or are scavengers in decomposing organic matter. Some larvae are paedogenetic: reproduction occurring inside them! Other gall midges are endoparasites or live as inquilines in other insect galls. *Contarinia sorghicola* (sorghum), *Orsculia oryzae* (paddy), *Asphondylia sesami* (sesame), *A. ricini* (castor) and *Dasyneura lini* (linseed) are the more serious crop pests in India. Over 250 species are currently known from our subcontinent. Another large and widely distributed family of these flies is Mycetophilidae

(Fungivoridae) which contains the fungus gnats; some 120 species are known from India and adjacent countries. The larvae are associated with fungi, living inside the fruiting bodies or outside, in webs or mucilaginous tubes. The adults usually have the abdominal base constricted and are more numerous in wet temperate forest; *Mycetophila* is a large genus.

The Asilomorpha represent nearly all of the higher Brachycera, especially the ones with non-aquatic larvae. Some of the largest, robust, striking coloured flies belong to this infraorder. The Nemestrinidae, with about a dozen Indian species, are called tanglewing-flies because of the crowded wing venation. Larvae of *Hirmoneura* are parasitic on longicorn and lamellicorn beetles and those of *Trichopsidea* attack grasshoppers. The Acroceridae (18 Indian species) include peculiar flies with a small head in relation to the enormously large thorax. Their larvae are endoparasites of spiders; *Ogcodes* and *Pterodontia* being some genera. The Therevidae or stiletto-flies (16 Indian species), with *Phycus*, *Psilocephala* and *Thereva* being some of the genera, contain many wasp-mimics, and adults are usually found in dry habitats like sand dunes and beaches. Larvae are predacious and occur in soil. A family of these flies with adults of distinct 'personality' is the Asilidae. It is a large family of aggressive, predacious Diptera, called robber-flies or assassin-flies; some 225

© Ajai Ghorpadé

A dashing robber-fly poised on her perch × 2

species have been named from our subcontinent. Most flies are bristly with a very conspicuous pair of well-separated eyes and a sharp proboscis. The prey (usually other flies and Hymenoptera) is captured in flight and held by the powerful legs on alighting when its body juices are sucked up. Some species are beautifully metallic in colour and others are certain mimics of sphecoid and vespoid wasps. *Hyperechia xylocopiformis* is a remarkable mimic of the carpenter bee, *Xylocopa*, its larvae developing on those of the bee in its nest in wood. *Laphria*, *Clephydroneura*, *Microstylum*, *Ommatius*, *Xenomyza*, *Machimus*, *Maira*, *Promachus* and *Philodicus* are some of the more numerous genera in India. Larvae occur in soil or rotting wood. *Leptogaster* (and related genera) is a distinctive asilid resembling the damselflies (Odonata). These slender flies with an elongate abdomen hunt their prey in long grass and similar habitats.

The Bombyliidae are called bee-flies owing to their remarkable mimetic resemblance to bees. They inhabit arid regions with abundant sunshine though some of them prefer cooler, forested areas. Many species exhibit a strong hovering flight and larvae are all either parasitic or predatory on immatures of grasshoppers, beetles, bees, wasps, moths, etc., and form an important natural control component. The 125 species so far known from the Indian subcontinent can be divided broadly into two groups: (1) furry-bodied flies with a long proboscis for sucking flower nectar; and (2) scaly-bodied flies with a short proboscis and feeding possibly on flower nectar, honeydew of aphids and coccids, etc.; also perhaps on organic liquids from wet and rotting vegetation and stagnant pools as well as on perspiration from animals. **Some of the better known Indian genera are** *Bombylius*,

A bee-fly, Exoprosopa bengalensis, *nesting* × 1¹/₂

Systoechus, *Systropus*, *Toxophora*, *Exoprosopa*, *Villa*, *Petrorossia*, *Exhyalanthrax* and *Geron*. The family Empididae consists of a large number of species that look like miniature robber-flies. Adults are predatory as are most larvae, which also make do with carrion or even vegetable matter; some are aquatic. They are termed dance-flies because of the habit of *Hilara*, essentially, of dancing in swarms over water. A dominantly temperate group, we know of a little less than 200 Indian species. They exhibit complex mating activity in swarms and some species have a very elaborate courtship behaviour: males offering captured prey or bubbles of frothy secretions to attract the female! A related family of long-

legged, metallic blue, green or bronze flies is the Dolichopodidae, which also have predacious adults found running about on foliage, tree-trunks, sandy borders of watersources or even on water; some have very prettily marked wings. Larvae are very versatile in their choice of habitat, which may be terrestrial, freshwater or even marine; many larvae are also predacious. A similar number of species are known from India as the empids: *Chrysosoma*, *Sciapus*, *Diaphorus*, *Hercostomus* and *Sympycnus* (Dolichopodidae) and *Empis*, *Drapetis*, *Syneches*, *Hybos* and *Platypalpus* (Empididae) being common genera.

The Phoromorpha includes one large family, the Phoridae, which are called coffin-flies because many larvae are scavengers living in carrion and decomposing organic matter. Adult flies have a hunchbacked appearance and the small flies have a characteristic manner of moving in a quick jerky manner on leaves. Species of *Megaselia* are pests of cultivated mushrooms, and some others will even attack a collection of dry insect specimens! Some 135 Indian species have been described, but many more await discovery. Two other families are placed in this infraorder, and one each in the Musidoromorpha and Termitoxeniomorpha, that are small and uncommon.

This takes us to the largest infraorder, Myiomorpha, inclusive of about 20 superfamilies and over 50 families. These correspond to the Cyclorrhapha proper, which were divided into the Aschiza and Schizophora, and the latter again into the Acalyptratae and Calyptratae. The Syrphidae and Pipunculidae possess flies that are among the most superb hoverers in the insect world. It is the males that hover either in swarms or individual-

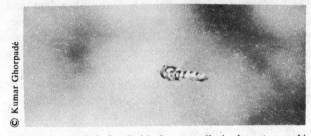

A male syrphid fly, Ischiodon scutellaris *hovering* × 1¹/₂

ly, in a particular spot, where females find them, preparatory to mating. The hover-flies, as the syrphids are called, are also one of the most diverse families, in larval and adult habits, as the Tipulidae. Many species are mimics of a variety of bees and wasps and the majority of adult flies visit flowers for sustenance. Larvae are predacious on various Homoptera, and thus are important natural control agents, while many others are either injurious to bulbous plants, saprophytic in liquid media or in rotting vegetation, or even living in nests of ants and wasps. The more common predacious species in India (over 300 species known) belong to the genera *Ischiodon*, *Allograpta*, *Episyrphus*, *Allobaccha*, *Metasyr-*

© Kumar Ghorpadé

A hover-fly, Phytomia crassa, *on flower of nigerseed,*
Guizotia abyssinica × 2

© C.I.B.C., Bangalore

A tephritid fruit-fly × 10

phus, Meliscaeva and *Paragus*; saprophytes to *Eristalinus, Ceriana, Rhingia, Phytomia* and *Milesia*; some *Eumerus* and *Merodon* occurring as larvae in bulbs of onion and lilies; *Microdon* living as larvae in ant nests and *Volucella* and *Graptomyza* in nests of wasps. The Pipunculidae are small, dark, big-headed flies that are endoparasites of fulgoroid, cercopoid and cicadelloid Hemiptera as larvae. Only around 70 Indian species have been described, but the fauna is certainly much larger. *Pipunculus, Tomosvaryella* and *Verrallia* are the more familiar and wide-ranging genera.

The Schizophora include flies with the calypteres at the base of the wings either poorly developed (Acalyptratae) or well formed and prominent (Calyptratae). The acalyptrates are numerically more abundant and also diverse. The Conopidae are wasp-mimicking flies that are endoparasites of adult wasps, flies and grasshoppers in the larval stages; the 55 Indian species belong mostly to *Conops* and *Physocephala*, with *Stylogaster* being a peculiar fly. The fruit-flies belonging to the Tephritidae are some of the more economically important groups of flies to man. It is an essentially tropical group and one subfamily, Dacinae, possesses some of the most serious pests of fruits and vegetables known, the genus *Dacus* being very large. Other fruit-fly larvae breed in flowers and feed on seeds, such as *Dioxyna* and *Tephritis*; more than 300 Indian species are known. The snail-killing flies of the family Sciomyzidae are important natural enemies of aquatic and terrestrial molluscs, their larvae being specific parasites or predators of ground or freshwater snails, many of which are intermediate hosts of liver-flukes and vectors of schistosomiasis, leishmaniasis, etc. Only a dozen Indian species are known, most of them belonging to *Sepedon*. Another large acalyptrate family, widely distributed in the tropics, is the Lauxaniidae (over 50 Indian

species), abundant in moist forests, grassland and seashore vegetation. The Sepsidae (17 Indian species) are the small ant-like flies found on animal dung rapidly waving their wings. Chamaemyiidae are small silvery flies whose larvae are important predators of aphids, coccids and mites; less than 10 species have been described from our subcontinent. Drosophilidae are another numerous group of flies, more than 140 Indian species have been described. Larvae are mostly fungivorous, some eating yeasts growing in decaying fruit. Many species of *Drosophila* are known and have been used as experimental animals by geneticists.

The Agromyzidae (190 Indian species) is another economically important family, larvae of many species being leaf- or stem-miners and gall-makers on crop plants. Species of *Melanagromyza* and *Phytomyza* attack redgram, cowpea, beans and pea. Each species makes a distinctively patterned leaf-mine. A little over 100 species of Chloropidae are known from India, the most familiar members being the irritating small, black eye-flies of the genus *Siphunculina*. *Oscinella* and *Thressa* attack oats and cardamom, while species of *Siphonella* are parasitic in spider egg-masses. The shore-flies or Ephydridae have many aquatic larvae and adults are found near or on bodies of salt and fresh water. *Brachydeutera, Hydrellia* and *Dryxo* are some Indian genera; some 50 species are recorded from the subcontinent. Before ending this section on acalyptrates, mention may be made of the Celyphidae or beetle-flies. Sometimes treated as a subfamily of Lauxaniidae, the dozen or so Indian species are often mistaken for beetles, with their enormously enlarged scutellum that covers the entire abdomen.

Finally, we will deal with the calyptrates or what are the muscoid (house-fly like) flies. Six species of Gasterophilidae are known from India; these are the horse-bots, somewhat bee-like, with larvae that are endoparasites of horses, donkeys, mules, rhinos and elephants. The Anthomyiidae (40+ Indian species) are mainly north temperate and with phytophagous larvae; *Hylemyia* is a genus with cosmopolitan pest species damaging vegetable seedlings. The Muscidae is a very large (almost 400 Indian species) and varied family containing the house-fly and its many relatives. Medically important members are dealt with in MOSQUITOES & BITING FLIES. *Atherigona* includes many species that are important stem-borers of millets. Some species of *Fannia* occur in houses feeding on organic refuse, as do some other species of *Musca* like *vicina* and *nebulo* (treated as subspecies of *domestica*, generally). The bazaar-fly is *Musca sorbens* and other species that breed in dung and fly to animal sores are *pattoni*, *ventrosa*, *fasciata*, *conducens*, etc., while *crassirostris* sucks blood of animals. *Morellia*, *Orthellia*, *Phaonia*, *Dichaetomyia*, *Xenosina* and *Limnophora* are some other

© Kumar Ghorpadé

Fly (Musca *sp.*) *on flower of nigerseed* (Guizotia abyssinica) × 2¹/₂

muscid genera. The metallic Calliphoridae (125+ Indian species) are called blue-bottles (*Calliphora*, parasites of earthworms) and blowflies (*Lucilia*, endoparasites of sheep) and many other testaceous species also occur. Species of *Bengalia* oversee ants carrying immatures from one nest to another, pounce on them and feed on the young ants: *Hemipyrellia*, *Pollenia*, *Chrysomyia* and *Stomorhina* are other common genera, those of the last hovering in swarms under the shade of flowering trees like mango; many adult calliphorids feed on flower nectar. The Sarcophagidae are the grey and black flies with a checkerboard pattern on their abdomen which are seen near carrion, in which their larvae breed. *Sarcophaga* contains most of the 60-odd Indian species described. A specialized subfamily, Miltogramminae, contains small flies that live as food-parasites in wasp nests in soil.

The Tachinidae is an immense family with almost all of its species (about 325 in India) being parasitoids of several other insect orders, mainly Lepidoptera, Coleoptera, Hemiptera and Orthoptera, many of which are injurious to man's crops. Tachinids are almost as important as potential biological control agents as are the ichneumonoid and chalcidoid Hymenoptera. Adult flies are very abundant on flowers, especially in the Himalaya, and present a great variety of structural forms, being generously bristled all over. Their taxonomy is difficult and perhaps not even 50% of the existing species in the subcontinent have been collected, let alone reared. Some species lay eggs directly on or in their hosts; others lay them in places frequented by the host which is tracked by the larva that emerges; still others lay eggs on the host's food-plant (microtype eggs) which are ingested by the host during feeding. Some species that are of importance in India are—*Blepharipa zebina*, *Palexorista solennis*, *Chaetexorista javana*, *Carcelia corvinoides*, *Peribaea orbata*, *Sturmiopsis inferens*, *Compsilura concinnata* and *Bessa remota* (on Lepidoptera); *Prosena siberita* and *Medinodexia morgani* (on Coleoptera); *Alophora indica* and *Euthera mannii* (on Hemiptera); *Koralliomyia portentosa* (on Hymenoptera); and *Acemyia indica* (on Orthoptera).

The Oestridae (7 Indian species) have vestigial mouthparts and are parasites of sheep. The last three infraorders, Braulomorpha (Braulidae, parasites on bees), Streblomorpha (Streblidae, ectoparasites of bats, 12 Indian species), and Nycteribiomorpha (Nycteribiidae, ectoparasites of bats, 40 Indian species) are specialized offshoots of the muscoids.

Finally, two verses on flies, one by Ogden Nash and the other by Christopher Mendola, which appeared as a response to TIME magazine's cover story on 'The Bugs are Coming' (July 12, 1976), are reproduced here:

God in His wisdom
Made the fly
And then forgot
To tell us why.

—OGDEN NASH

There was once a fly named the tsetse
Who thought that life was just peachy.
She looked at the earth
At her moment of birth
And said, 'Veni, vidi, vici.'

—CHRISTOPHER MENDOLA

FLOCKS. Flocks are aggregations of birds which may come together for a variety of reasons. It may be merely by chance at a concentration of a resource they all want, it may be for large scale movements, it may be to better defend themselves against predators, or to improve their ability to get at a scarce resource, or, again, it may be to defend a common group territory against conspecifics.

Many species of bats find safety during the day in caves and may come together in millions in some of the larger caves. A large number of cattle egrets may aggregate around a herd of grazing buffaloes to feast on the insects flushed by the beasts, and hundreds upon hundreds of swifts, swallows, bee-eaters and drongos may hover over the progressing front of a forest fire to feast on the insects flushed by the fire. Otherwise solitary birds, like kestrels, may band together for long-distance migrations.

Many bird flocks may help the birds to better defend themselves against predators. This they mostly accomplish through a more efficient early-warning system. Thus the distance at which a sparrowhawk is detected increases with the size of the pigeon flock, and any individual pigeon in a flock needs to spend much less time looking around for a predator than does a solitary pigeon. But in a flock the pigeons lower down in the hierarchy are harassed while feeding by pigeons higher up in the hierarchy and have to keep moving all the time. In a very large flock therefore the birds lowest in hierarchy have to pay considerably in terms of reduced feeding efficiency although they gain in efficiency in avoiding predation. At some point these disadvantages will outweigh the advantages and the pigeons at the bottom of the hierarchy will be better off if they leave the flock. Such processes must be setting an upper limit to the size of bird flocks.

Other flocking birds defend themselves more actively against predators, whom they may mob. Our crows are very fond of this and seem to spend a great deal of their time and energy in this effort.

Many bird flocks may enhance the feeding efficiency of the birds. This may be passive through exchange of information about patchy and temporary food sources. Thus at a communal roost of parakeets, some birds would have located good fruiting trees on the previous day, while others may have just exhausted a food source and be in need of finding a new one. These latter may be able to detect the confident flight of parakeets who know of a good fruiting tree and could follow them.

Then again birds in flocks may together flush many more insects than by themselves. This may be why many insectivorous birds such as white-eyes and minivets feed in flocks and why large mixed hunting parties form in our forests. The latter may include a whole array of species from nuthatches, woodpeckers, drongos, mynas, minivets, tits, chloropsises and so on and are a great treat to watch.

Thirdly, flocking birds may actively cooperate in

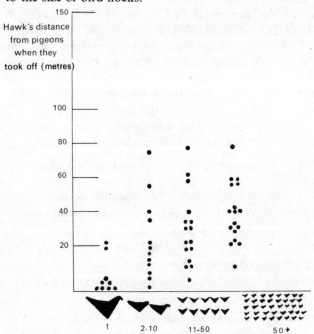

The distance at which pigeons become aware of the approach of a predator increases with the size of the flock.

Response of a flock of rosy pastors to a sparrow-hawk flying overhead. The sparrow-hawk finds it difficult to attack the tight bunch; for a high-speed collision with such a bunch may injure it.

herding their prey together like beaters on a shikar, and cormorants fishing in large flocks are known to do this. Lastly flocking may merely help birds avoid foraging in a region which has been recently exploited and is therefore poor in food, and this is believed to be the case with flocks of desert finches.

Some species of babblers (such as the Common, Jungle, Large Grey and Whiteheaded) occur in small cohesive flocks that are really family parties of closely related individuals. This is of course why they keep babbling away to each other and make such a chatter. These flocks seem to defend group territories against other babblers, and have helpers who feed chicks at nests along with the parent birds. From such small cohesive parties of babblers to loose aggregations of thousands of Rosy Pastors at a locust swarm, the variety of bird flocks is indeed fantastic.

FLOWERING PLANTS. The Angiosperms or the Flowering Plants are seed-bearing plants in which the seed-forming ovules are enclosed in an ovary that later develops into the fruit. They are thus distinguished from the other group of seed plants, the cone-bearing Gymnosperms, in which the ovules are exposed at the time of pollination. The CONIFERS, cycads, and ginkgos are all gymnosperms. They were the dominant components of the world's flora during the Jurassic period of the Mesozoic era. The Angiosperms appeared on the scene some time during the Cretaceous period, about a hundred million years ago, though the exact time and place of the origin of this group are uncertain. Fossil evidence indicates that the Angiosperms were widespread at the end of the Cretaceous period, about 65 million years ago and, during Tertiary times, they attained dominance in the world's flora.

In the present-day flora, it is estimated that there are about 250,000 species of flowering plants. They occur in every climatic zone and known habitat. They present a wide variety of form and structure and range in size from the minute duckweeds to the giant redwoods and eucalypts, some of which may reach a height of nearly 100 m. The Angiosperms are the most conspicuous elements in the flora of tropical rain-forests, of deserts, of high mountain ranges, of lakes, ponds and marshes, of coastal beaches and extend even to arctic wastes and marine habitats. In their life-span, they range between the extremely short-lived ephemerals of the desert (*Boerhavia repens*, in the Mali Republic of Saharan Africa, is reported to complete its life cycle in 8 to 10 days) to the long-lived ones which exist for hundreds of years. Trees which are dated to be more than 3000 years old are known.

Flowering Plants are generally distinguished as Herbs, Shrubs or Trees. Soft-stemmed plants with very little or no growth of wood are herbs. Plants living for only one season, as for example, the pulses, cereals, seasonal flowers and others, where the stem remains soft throughout and which flower, fruit and die at the end of the season, are Annual Herbs. Longer-lived herbs often possess roots which develop woody tissues and produce annual shoots every season; these are Perennial Herbs. Most of the larger plants that grow around us live for many years and develop woody tissues. Among them are the trees and shrubs. Those plants which develop a prominent columnar stem with branches appearing on them at some distance above the ground are Trees. Woody plants in which several branches arise from above the soil line and give the plants a bushy appearance are Shrubs. Many plants climb around some support and these may have soft stems (Herbaceous Climbers) or woody stems (Woody Climbers). Plants are often seen perched on the branches or trunks of trees. These are the Epiphytes.

In the case of trees, there is a seasonal increment in the growth of the trunk involving the addition of woody, fluid-conveying (xylem) tissue in the form of rings. Since ordinarily one ring of xylem is formed each year, the number of rings in a tree indicates its approximate age.

All flowering plants produce flowers at an appropriate phase in their development, there being considerable variation in the duration, periodicity and extent of flowering. Since the vegetative characters of a species vary, the flower, whose structure remains constant for any given species, is relied upon for classificatory purposes in the Angiosperms.

M.A.R.

See CLASSIFICATION OF FLOWERING PLANTS.

FLOWERPECKER. Flowerpeckers (family Dicaeidae) share many characters with sunbirds, especially the serrated tips of the mandibles. Generally drab-coloured e.g. Tickell's Flowerpecker, the males of some, like the Firebreasted and Scarletbacked, have patches of brilliant red in their plumage.

Flowerpeckers are arboreal; they feed on insects, and many with tubular tongues also largely on flower-nectar. But their main food is fruit, especially the berries of the mistletoes *Loranthus* and *Viscum*. Throughout their range they are the most important disseminators of *Loranthus* seeds. These plant parasites pose a serious economic problem to foresters and horticulturists. The birds are also instrumental in pollinating many *Loranthus* as well as other ornithophilous flowers. They build a small, hanging pouch-like nest of felted vegetable down and fine fibres with a side entrance, in trees, 3 to 10 m from the ground. Their eggs, 1 to 3 in a clutch, are usually plain white.

See plate 10 facing p. 129.

FLOWERS. Flowers are the characteristic reproductive organs in the Angiosperms or Flowering Plants. They are concerned with the production of fruits and seeds. Structurally, a flower is regarded as a condensed shoot with highly modified leaves. A typical flower consists of four parts arranged in whorls. The outermost whorl is the calyx, comprising green, leafy structures called the sepals. Inner to the calyx is the corolla, consisting of petals. The petals are usually brightly coloured and are often associated with honey glands or nectaries. The two inner parts, which are actually concerned with production (essential organs) are the stamens, collectively termed the androecium, and the pistil(s) or the gynoecium. The pistil has the ovary at its base which matures into the fruit. Within the ovary are the ovules which later develop into seeds. All the above parts of the flower arise on the floral receptacle (sometimes called the thalamus or torus) which is the condensed axial part at the summit of the flower stalk, the pedicel (Figs.1–4). There are many variations from the typical condition described above. Floral parts vary in their number, in the number of whorls and often exhibit fusion among themselves or with members of the adjacent whorls. Floral parts also vary in their relative position on the receptacle; three forms are generally recognized, the hypogynous, perigynous, and epigynous conditions (see Figs.5–7). In the first two the ovary is said to be in a superior position and in the third, the ovary is inferior. Flowers are regular when the floral parts within a whorl are all similar in size and shape and exhibit radial symmetry (actinomorphic); if irregular, the flowers may exhibit bilateral symmetry and are then known as zygomorphic. A flower which has all the four parts, sepals, petals, stamens and pistil, is known as a Complete flower; when it lacks one or more parts, the flower is Incomplete. In some flowers the distinction into sepals and petals is wanting and there may be one, two or more whorls of a uniform kind of floral lobes which are collectively termed the perianth (e.g. Lily flower).

Most flowers have both the sexes in them and are termed bisexual flowers; sometimes one of these organs may be absent and the flowers are then unisexual, male or female as the case may be. The unisexual flowers, both male and female, may be borne on the same individual plant (monoecious), as in the case of castor plant, or on different individuals (dioecious) as in the date palm.

The flowers may appear singly in the axils of leaves or, as is commonly seen, they may be produced in clusters or on common axes or grouped in various ways. These clusters or groups of flowers appear at the ends of branches or in the axils of leaves and are known as Inflorescences. In the inflorescence the individual flowers are borne in the axils of reduced leafy structures known as bracts. Some of the inflorescences charac-

teristic of certain families are: the *umbel* (long-stalked flowers arising at the same point on similarly branched axis) seen in the coriander family (Umbelliferae), the *head* or *capitulum* in the family Compositae, where a large number of sessile flowers are massed on a common disk, *cyathium* in the Euphorbiaceae, where numerous male flowers represented by single stamens are arranged around a single female represented by only the pistil, the whole group surrounded by an involucre with a honey cup (as in *Poinsettia*), the *catkin* seen in birches, willows and others where unisexual flowers appear crowded on pendulous, elongated spikes and the *spadix* of Aroids and Palms where the unisexual flowers are arranged on a fleshy or fibrous axis which is protected by a large sheath, the spathe (coconut palm).

Flowers range in size from the minute ones of duckweeds (about 1 mm wide) to the largest flower known to science (nearly a metre wide) produced by a parasitic plant of the Sumatran jungles, named *Rafflesia arnoldii*.

The distribution of flowers on the plant, as also the time of their appearance, duration and periodicity vary greatly. Flowers may be produced only seasonally or may appear throughout the year. There are plants which flower only once during their lifetime as in the case of most annuals. Larger and longer-lived plants which flower only once during their lifetime and immediately die thereafter are exemplified by the Talipot Palm and a Yucca (*Yucca whipplei*, the Spanish Dagger). Some of the tropical bamboos exhibit gregarious flowering at periodical intervals, once in several years. A very unique bamboo, *Melocanna baccifera*, the *muli* bamboo of Assam, has a flowering cycle of 30 to 35 years and once it flowers, the flowers continue to be produced for about two years. The last time it flowered was in 1958–60. This is the only member of the grass family to produce fleshy fruits. Another striking example of gregarious flowering is seen in the dicotyledonous Strobilanthes or Coneheads (Acanthaceae), whose species flower at intervals of 4, 7 or 12 years and in those years, whole hillsides in the peninsular hill ranges are covered by the enchanting blue flowers of these shrubs. Some plants flower only at night, the most noteworthy examples being Civet or Durian of Malaya and some of the Cacti of the Mexican and North American deserts. The fragrant night-queen (*Cestrum nocturnum*) also blooms late in the evenings or at night, a device to attract insects. The fragrance of flowers is one of the most delightful of all experiences. Roses, Jasmine, Gardenia, Hyacinth, Lavender, Cassie (*Acacia farnesiana*), Mignonette (*Reseda odorata*), Rosemary, Tuberose (*Polyanthes tuberosa*). Violets, Ylang-ylang (*Cananga odorata*) and several others figure in the perfume trade. The fragrance is due to the presence of minute quantities of altered essential oils in the petals. Interestingly, there are also flowers which emit

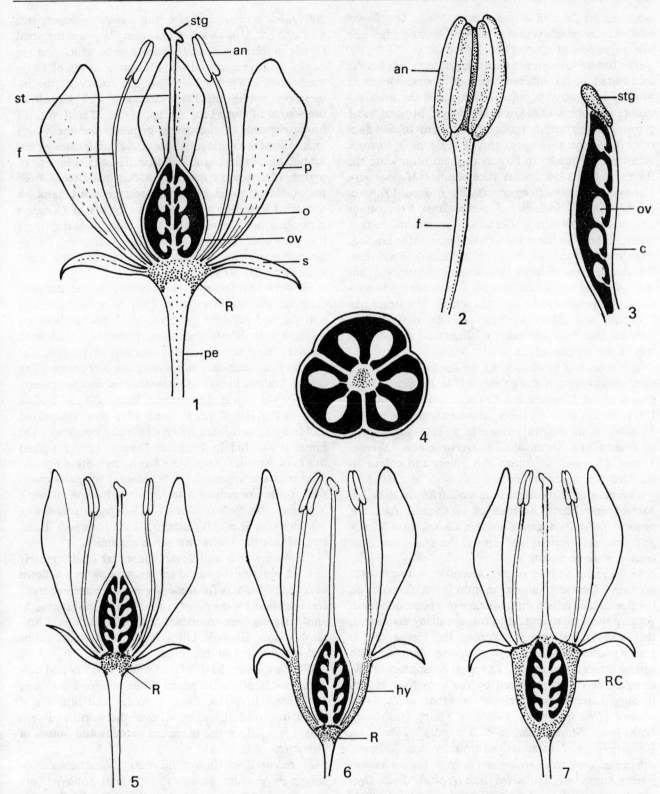

1 Sectional view of a representative flower showing the parts. pe pedicel; R receptacle; s sepal; p petal; f filament; an anther; o ovary; ov ovule; st style; stg stigma.

2 A stamen (front view) with filament and two lobes of anther.

3 A carpel (c), one of the many free carpels in a buttercup flower (sectional view).

4 Cross-section of an ovary of 3 fused carpels (syncarpous ovary).

5 Hypogynous flower: ovary is placed in line with or above the level of origin of the other parts on the receptacle (ovary superior).

6 Perigynous flower: bases of sepals, petals and stamens fused to form a cup-like structure, the hypanthium (hy); ovary wall free from hypanthium (ovary superior). 7 Epigynous flower: ovary wall fused with the receptacular cup (RC) and the other floral parts appear to arise from above the ovary (ovary inferior).

All figures diagrammatic

offensive smells and attract carrion flies. The flower of *Rafflesia arnoldii* is said to smell of decaying fish and that of Sapira emits mephitic odours.

The Indian subcontinent has many fascinating flowers distributed in its different climatic zones. Mountain flowers are always popular and some of the most exquisite ones are on the Great Himalaya. Most of those growing at high altitudes are reminiscent of the flora of north temperate regions and of the Alps in Europe. Among the Himalayan flowers of great beauty are the Asters, Balsams (*Impatiens*), Blue-poppies (*Meconopsis*), Canterbury Bells (*Campanula*), Dandelions (*Taraxacum*), *Dianthus*, Edelweiss (*Leontopodium*), Everlastings (*Anaphalis*), Fritillaries, Gentians, Geraniums, Inulas, Irises, Lilies, Louseworts (*Pedicularis*), Monkshoods (*Aconitum*), Morina, Peony, Potentillas, Primroses, Rhododendrons, Saussureas, Sedums, Senecios, Tansies (*Tanacetum*), Wind-flowers (Anemones), Wormwoods (Artemisias) and a host of others. The Primroses (*Primula*) and Rhododendrons are the most sought after by collectors and many of them have found their way to the gardens of the West.

The monsoon flowers of the peninsular forests have their own charm. Among them are the Balsams, Begonias, Cassias, Commelinas, Crinums, Curculigo, Glory Lily (*Gloriosa superba*), Lilies, Mussaenda and numerous orchids, aroids, zingibers and others. Trees and shrubs in forests like Terminalias, Lagerstroemias, *Shorea*, *Mesua*, *Ougeinia*, Sterculias and others add colour to the landscape when in bloom.

The deserts here are not so colourful as those of Mexico and North America or of Central Asia but species of *Aerva*, *Alhagi*, *Capparis*, *Crotalaria*, *Calatropis*, *Pluchea*, *Prosopis* and several Acacias attract attention when in bloom.

In the plains and dry forests, several attractive flowers are seen. The early summer months bring blossoms on the trees which enrich the landscape in a blaze of colour. Among those contributing to this are many Bauhinias, the Indian coral tree (*Erythrina*), the Flame of the Forest (*Butea*), *Cochlospermum*, a spectacular plant with yellow flowers, and others. The most conspicuous flowering is, however, produced by the numerous exotics in our country, particularly in the urban areas. Gold Mohur (*Delonix regia*), Colville's Glory (*Colvillea*), both from Madagascar, the Bignonias, Tabebuias, Bougainvilleas, Jacarandas, all from tropical America, the Indian cork tree (*Millingtonia*), from Burma as also Flame Amherstia; *Spathodea* from tropical Africa, *Grevillea*, the bottle-brushes (*Callistemon*) and Eucalypts from Australia are among the many which adorn our avenues, parks and buildings.

Besides the importance of flowers in the perfume industry already mentioned, flowers also provide other useful products. The dried, unopened flower buds of *Syzygium aromaticum* give the cloves of commerce and also oil. Hops used in flavouring beer are the dried female inflorescences of *Humulus lupulus* which contain complex bitter essential oil. Flowering heads of *Chrysanthemum cinerariaefolium* are the source of the insecticide, pyrethrum. Plant nectar is well known as the source of sugars for the honey-bee. The flavour of honey depends on the species of plants the honey-bees visit. Flowers containing substantial quantities of essential oils impart characteristic flavours. Flowers of mahua (*Madhuca*) containing nectar are directly used for edible purposes or fermented. Resinous hairs on the dried flowering tops of female plants of *Cannabis sativa* are used in medicine and more importantly as the source of the narcotics going under the names bhang, marijuana, hashish, etc. Saffron is the dried stigmas of *Crocus sativus*.

Flowers have figured prominently in the cultural, religious and social spheres. They have been used as national and religious symbols. The lotus has been associated with Hindu goddesses, particularly Lakshmi. Flowers have been used for worship in temples and are essential requisites in all religious ceremonies. The famed 'brahma kamal' (*Saussurea obvallata*) is regularly used for worship at the shrines of Badrinath and Kedarnath in the Himalaya. Flowers have been designated as National and State flowers in many countries. The Lotus is the Indian National Flower. In the United States of America, each State has its own State Flower. Flowers have appeared on the stamps of many countries. Girls are named after flowers: Pushpa (flower), Kamala and Padma (lotus), Champaka (*Michelia*), Mallika and Shephali (jasmine), Lata (creeper), Rose, Iris, Hyacinth, Violet are some examples.

Flower festivals and flower shows are held regularly in most parts of the world and exotic flowers are flown over long distances for these shows. An extensive trade has been built for the supply of cut flowers, floral wreaths and garlands even to distant lands. Carnations, Chrysanthemums, Gladioli, Lilies, Jasmines, Orchids, Tulips and others figure in this trade.

Plant explorers have braved many hazards and collected seeds of exotic plants from remote mountains of China, Himalaya, Pamirs, Andes and others and also from tropical jungles all over the world and such plants now adorn the botanical gardens and homes of all nations of the world.

In recent times flower decoration has become increasingly popular as an art form and hobbyists are engaged in creating enchanting designs not only using natural flowers but also creating artificial flowers in paper, silk, cocoons, corals, sandalwood, shells and other objects. Perhaps, the most unique achievement in this creative art is that of the father-son team, Leopold and Rudolph Blaschka of Dresden, whose Glass

Flowers are now housed in the Botanical Museum of Harvard University.

See also POLLINATION, FLOWERING PLANTS.

See plate 20 facing p. 257, plate 27 facing p. 384, plate 28 facing p. 385 and plate 35 facing p. 496.

M.A.R.

FLYCATCHERS are a diverse group of small insectivorous passerine birds with compressed bill belonging to the family Muscicapidae. All Old World flycatchers are found in Europe, Asia, Africa and Australia up to New Zealand. The ecologically equivalent New World flycatchers of the family Tyrannidae have no taxonomical relationship with them. Forty species are found in the Indian subcontinent, including eight endemics and a few winter migrants from the Himalaya and beyond. The Pygmy Flycatcher (8 cm) is the smallest in size whereas the Paradise Flycatcher is the largest (20 cm). They live in all habitats, though the majority are forest-dwellers. The Redbreasted Flycatcher (*Muscicapa parva*) is the commonest migratory species and in winter may frequently be found in gardens even in congested cities like Bombay and Calcutta.

Most species gather food by aerial sallies or 'flycatching', some by flitting amongst the foliage or from the ground. A few take small soft berries in addition to insects which are their staple diet. They usually live in pairs; some occasionally join the mixed-species hunting flocks of insectivorous birds. Many maintain a territory all the year round, some during the breeding season only. Some like the Black-and-Orange Flycatcher (*M. nigrorufa*) are highly parochial.

Breeding activities start around March when the males sing and perform courtship displays, spectacular in some species like the ribbon-tailed Paradise Flycatcher (*Terpsiphone paradisi*). In many species the females alone build the nest; in others both sexes share all the breeding activities. The nests are normally placed on shrubby plants and trees; some in crevices or in banks of streams or under jutting rocks, etc. An open cup-shaped nest is the rule though some prefer holes or depressions in tree-trunks, rocks or walls of buildings. A few make globular nests with a side entrance. Much cobweb is usually employed in the open nests. The clutch-size varies between two and four eggs, exceptionally five or only one. Incubation and nest-feeding are either by the female alone or by both sexes equally.

See plate 31 facing p. 432.

FLYING FISHES. These are so called because during their normal swimming movements in water, they occasionally take to aerial flight by jumping into the air and gliding horizontally. The distance thus covered is at times as long as 400 m. These fishes occur on both east and west coasts, but more so on the southwest where

Flying fishes × ⅙

they form a fishery in winter months. They belong to the family Exocoetidae. Ten species are recorded from Indian seas, of which three, *Cypselurus altipennis*, *Exocoetus volitans*, and *Paraexocoetus brachypterus* are common.

These are small 25-cm-long fish with elongated roundish body and long broad pectoral fins extending up to the tail fin, the lower lobe of which also is elongated and pronounced. In some species the pelvic fin too is enlarged. Being pelagic, they swim usually in the upper strata of water. When they are about to fly, they accelerate their speed for a short distance, then suddenly jerk their head out of water and with powerful vibrations of the caudal portion of their body and the tail fin kick the water surface and become air-borne like a plane, spreading out their large pectoral fins for gliding. The pectoral fins are not flapped like birds' or butterflies' wings but kept distended.

The aerial flights are undertaken usually to escape enemies or when alarmed by approaching ships. But sometimes they may be purposeful, probably to cover distances quickly while catching small pelagic fish for food.

Fishermen off the southwest coast tie a broad piece of cloth to two poles on a catamaran or a small boat while moving in the fishing area. The flying fish in their flight strike the cloth and fall into the boat, to be merely collected by the fishermen.

The leaping action of fishes such as chanos, mullets and rohu across nets are merely to escape the nets and although they cover some distance in air, the action is different from the aforesaid flying or gliding.

The eggs of flying fish have long tendril-like processes by which they are entangled with each other or attached to weeds. They remain floating till the young ones hatch out.

FLYING SNAKES. Two species of 'flying snakes' are found in our area. The better known species is *Chrysopelea ornata* found in the hilly forests of the southwest and northeast. The other, *Chrysopelea paradisi*, is found only on Narcondam Island of the Andamans group in the Bay of Bengal. Flying snakes are so called because by extending the ribs and thus flattening the

Flying snake × ⅕

body, they can glide from high trees to reach the next tree or to escape enemies.

The colour and markings of the Flying Snake are beautiful. The black back has white or yellow bands and blotches, and red rosettes. The head is barred with yellow. It is a thin and vibrant snake, active by day. Not much is known about its life-cycle except that the female lays 6 to 12 eggs. Adults feed on frogs, lizards, birds and perhaps birds' eggs. Although they prefer dense forest habitat, in Sri Lanka and Thailand flying snakes are often seen in open places, around houses and gardens. The average length of a Flying Snake is 1m; but specimens of up to 1·75 m have been recorded.

See also SNAKES.   See plate 9 facing p. 128.

FOG. The deposition of dew on grass and other objects on some mornings, as well as the sweating of containers of cold drinks on humid days, show that atmospheric moisture condenses upon any surface whose temperature is below the dew-point. This is known as volume condensation of atmospheric water vapour.

Similarly, such condensation occurs when a large volume of humid air cools down till the air becomes saturated or very nearly saturated: when the relative humidity increases to about 100 percent. And that volume condensation of water vapour is Fog. It is cold air in which there are minute droplets of water, each less than a twentieth of a millimetre in diameter, and a few thousands of them in each cubic centimetre of air. The meteorological services declare a fog when the transparency of the atmosphere decreases to such an extent that large objects at 1000 metres and beyond cannot be observed with the unaided eye. A strict distinction between a fog and a cloud is difficult. In the case of clouds the droplets are bigger in size—ten to a hundred times more in diameter; the number of droplets in the clouds, however, is less, only about 100 or 200 per cubic centimetre of air.

The cooling necessary for the formation of a fog over the plains of India and Pakistan is often the night-cooling caused by the outgoing heat radiation emitted by the land surface under a clear sky and light winds. This is called radiation fog. Fog occurs in the early morning, following a night when the sky clears rapidly on a wintry evening after some light rain. On a few of these occasions, advection of cold air to the rear of the rain-producing weather system adds to the local cooling effect. Over hilly terrain fog occurs as moist air is forced up along a sloping land surface. In such cases it is usually a cloud sitting on the hill which is taken as a fog by those stationed on the hill. It appears as a cloud to the persons below. There are rare occasions when a fog is produced by moist air streaming over land already cold. This occurs along the coast and is called advection fog.

The occurrence of fog over India is much less common than over the middle-latitude countries in Europe. The frequency is maximum in the northeast, where fog may occur on about ten mornings in January and about half that number during the months of December and February. Over the plains of Punjab and of Uttar Pradesh, the frequencies are about a half of what they are in northeast India. Elsewhere fog is even rarer.

FOLDS. Most rocks that we find on the Earth's surface are deformed from their original shapes and positions. Only very rarely do we find undeformed and undisturbed rock exposures. The deformations generally found are in the form of bends, flexures, and crinkles. Such structural forms in rocks are known as folds. The best display of folds is exhibited by the rocks which have some planar structure in them, such as bedding in sedimentary rocks or the contacts in a sequence of lava flows. The occurrence of single and simple folds is generally rare. Folds usually occur as a series, in which each fold need not be an identical repetition of others. There may be differences in bending and curvature of the folds in all the three spatial directions. Moreover a region may be subjected to several periods of deformation during which earlier folds may be folded again giving rise to highly complex twisted and contorted shapes of the rocks.

Folds may be observed directly or inferred from various types of data. Determination of the dips (inclinations) of the strata is a most useful method for inferring folds. On the mountain belts such as those of the Himalayas very large folds which extend to hundreds or even thousands of metres may be found or inferred. Along with these, folds of small scale which are measurable in centimetres may also be found. Where rock exposures are limited to small hillocks the dimensions of the exposed folds are also limited.

Upward bends in rocks are known as Anticlines whereas the downward bends are called Synclines.

© E. Bharucha

The Kestrel (*Falco tinnunculus*) will hover in the air before dropping down at blurring speed to snatch up its prey—a field mouse, lizard or small ground bird—in its powerful talons.

© E. Hanumantha Rao

The Peacock (*Pavo cristatus*), graces wild places throughout India and has, in fact, learned to live in close proximity to man as well.

© B.N.H.S.

India's coastline provides winter feeding grounds to seagulls and a host of other migratory birds.

18 The Great Cats

The Cheetah (*Acinonyx jubatus*), extinct in India.

The Tiger (*Panthera tigris*), returned from the brink of extinction.

The Lion (*Panthera leo*), less than 200 left alive in India.

The Panther (*Panthera pardus*), a most adaptable cat, relatively secure.

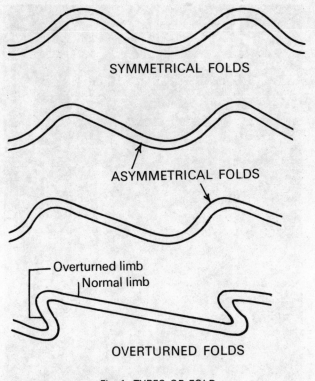

Fig. 1. TYPES OF FOLD

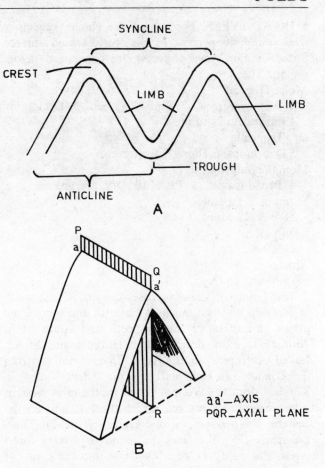

Fig. 2. Different stages in the development of overturned folds and showing progressive shortening of the beds.

The different parts of folds are shown in Fig. 1. Symmetrical folds are those in which the two limbs of a fold dip with the same angle but in opposing directions. Asymmetrical folds are those in which one limb dips at a greater angle than the other. A fold with one of its limbs overturned is known as an overturned fold. Different stages in the development of an overturned fold are shown in Fig. 1. In this figure progressive shortening of the bed due to folding can be seen. In strongly folded regions the normal limb of an overturned fold may break from the overturned limb by a low-angle thrust fault and may be displaced over a great distance from its original place of origin. Such structures are known as nappes and are common in the Himalayas and Alps. Very old rocks exposed in peninsular India exhibit a very intricate type of folding. Folds and faults are usually associated with each other.

The great mountain chains of the world, the Himalayas, the Alps, the Andes, the North American Cordillera and the Appalachians are all folded mountain chains with nappes and thrust faults. In addition to folding and thrusting they are associated with metamorphism and igneous plutons (bodies of rock exposed after solidification at great depth). These mountain chains are linear features generally formed near the edge of a continent. Though the Himalayas are not found along the edge of a continent it is believed by geologists that a great sea called Tethys existed between the Indian and the Asian land masses very long ago

in the geological history of the Earth. It is further believed that the Himalayas have formed in response to the northward movement of the Indian landmass and its thrusting under the Asian continent. As a consequence of this the sediments in the Tethys were folded and crumpled and thrust upwards. Into the cores of these folds the granitic plutons have intruded. It is thus the great Himalayan mountain belt was formed.

In the folded regions ore deposits are usually found concentrated in the parts of folds. Commonly the crests and troughs of the limbs are important places. Ore deposition may be earlier or later than folding and in each of these cases the location of the deposit in the folds may be different. Therefore in an exploration for mineral deposits a detailed investigation of the folds is necessary. Such investigations are also necessary in all types of large-scale civil engineering constructions such as dams, bridges, tunnels, and highways since the design of the foundations and abutments has to be suitably modified depending on the presence or absence of folds and other structures in the area.

C.V.R.K.P.

241

FOREST TYPES. Forests are the climax vegetation over about 40 percent of the world's land surface. Sixteen major types of forest are recognized in our region. They are:

Moist Tropical
    1 Wet evergreen, 2 Semi-evergreen, 3 Deciduous, 4 Littoral and swamp.
Dry Tropical
    5 Deciduous, 6 Thorn, 7 Evergreen.
Montane Subtropical
    8 Broad-leaved, 9 Pine, 10 Dry evergreen.
Montane Temperate
    11 Wet, 12 Moist, 13 Dry.
Subalpine
    14 Subalpine.
Alpine
    15 Moist, 16 Dry.

The Tropic of Cancer crosses the region just south of Karachi and just north of Calcutta and there is no major change in vegetation north and south of it. Altitude is more decisive than latitude, and broad-leaved subtropical forests (type 8) occur not far from the Equator, an example being the 'Silent Valley' in Kerala. Along the Western Ghats, on the rainy western side are typical evergreen forests (type 1) and on the eastern, drier slopes, Semi-evergreen (type 2). There are many trees that may be found in several forest types, and on a trek of a few days from the plains of Assam to the High Himalayas one would pass through many different types.

1 *Wet evergreen*

The Southern and Northern forests constitute two sub-groups which are virtually identical in forms.

It is characterized by tall, straight evergreen forest of Dipterocarps (see GURJAN) and *Hopea wightiana*. Dipterocarps (meaning 2-winged fruits, although many are 5-winged) have a buttressed trunk at the base like a tripod to help hold the tree during storms, and then, rise, often to a great height, smooth and unbranched, before reaching the open, cauliflower-shaped crown. *Hopea* is also a straight-boled tree in this type. In this upper layer other associates are Jaman, Jack tree, *Elaeocarpus*, mango and Poonspar. In the middle storey the Fish-tailed Palm (*Caryota urens*) and Betel-Nut Palm (*Areca triandra*) are met with; rattan palms or canes (*Calamus*), which at first are ordinary palm rosettes, develop into lianas of extraordinary length, climbing over trees in their struggle for light. *Strobilanthes* (see CONEHEADS) with blue flowers and *Ixora* form the ground floor. In the south this type is met with along the Western Ghats, Sri Lanka, Andamans and Nicobar Islands. In the Nicobars, Dipterocarps are completely lacking. In the Western Ghats of Kerala and Sri Lanka

© Forest Research Institute

WEST COAST TROPICAL EVERGREEN FOREST

A group of Dipterocarpus bourdilloni *(various sizes)*, Kingiodendron pinnatum *(large tree)*, Polyalthia *sp.*, Vitex altissima *with evergreen undergrowth and canes.* Shencottah, Kerala.

karani (*Cullenia rosayroana*) also occurs in the top storey and is the habitat of the rare Lion-tailed monkey.

In the north this type occurs in the NE region including Bangladesh and Burma. The forests generally present a many-tiered appearance of which the top storey is constituted by Dipterocarp, Chaplash, *Tetrameles nudiflora*—the last-named characterized by enormous plank buttresses, a favourite nesting tree of hornbills. These trees, not all of which are evergreen, tower above the rest of the canopy. This storey is followed by large trees of Hollock, most striking when in fruit with a coppery red tinge, Red Silk-Cotton, Queen's Crape Myrtle (ornamental), *Bischofia javanica* (bark favoured by tigers for cleaning their claws), *Sloanea assamica* (with jet black spiny fruits), etc. The middle storey is formed of gregarious species like the white-flowered *Mesua ferrea* (ironwood), *Phoebe goalparensis*, etc., all valuable species. This storey normally determines the economic value of the forest. The lowest storey comprises small trees, viz. *Syzygium formosum*, *Dillenia indica* (with large handsome white flowers and large globose fruits dispersed by elephants), *Talauma*

ASSAM VALLEY TROPICAL WET EVERGREEN FOREST
Sal-Dipterocarp association, Lakhimpur division, Assam.

ANDAMANS SEMI-EVERGREEN
A large buttressed Parishia insignis *carrying a tangle of woody climbers and ferns, Middle Andamans.*

hodgsonii (with large magnolia-like flowers) and canes. Tree FERNS are also seen associated with SCREW-PINE. The wild banana is a conspicuous feature and occurs gregariously towards the upper limit in evergreen forests at 1000 m. The most conspicuous epiphytic elements are the gorgeously coloured ORCHIDS, ferns and fern allies which cover tree-trunks and branches. A strikingly handsome 'bamboo orchid' stands out majestically amidst wild plantain. Cauliflory (i.e. flowers developing on the main trunks and branches) is common.

2 Semi-evergreen

This type is variable, being intermediate between tropical evergreen and moist deciduous, and includes both evergreen and deciduous trees—the former predominating. There is a tendency to gregarious habit.

In the upper storey, IRUL and LAUREL are typical with frequent gurjan. Evergreen *Myrtaceae* and *Lauraceae* constitute the middle layer, with buttressed stem. Bamboos with ground cover of evergreen shrubs, ferns and orchids are typical. The type occurs in the eastern slopes of the Western Ghats, Andamans, Nicobars (without dipterocarps) and East Himalayan foothills (without irul).

ARUNACHAL SEMI-EVERGREEN
Wild mango, extreme left, Jaman (behind) and Albizia. Note the closed dense forest across the river which has nesting colonies of hornbills. Kameng District.

3 Deciduous

The trees are of large girth, about 25 m in height, heavily buttressed with dominant trees turning leafless in the dry season. There is a definite second storey of deciduous trees and shrubby evergreen undergrowth;

243

ANDAMANS MOIST DECIDUOUS

A group of Terminalia bialata. *The winged fruits are butterfly-shaped.*

bamboos, climbers and epiphytes are restricted to wet situations. This type is seen throughout our regions except Western and NW India and constitutes the valuable sal and teak forests.

A typical forest of this type will have:

Top storey: *Terminalia paniculata*, Red Silk-Cotton, Mango, Rosewood, Irul and sporadic Teak.

Middle storey: Amla, Elephant apple (*Limonia elephantum*), *Careya arborea*, Spiny bamboo (*Bambusa arundinacea*).

Bottom storey: Indian Screw Tree.

Ground flora: Coneheads, Ixora.

4 *Littoral and swamp*

Littoral forests consist of Whistling Pine (CASUARINA) forming fringes on sandy beaches along the sea shore. In the Andamans and Nicobars, Bulletwood (*Manilkara*) occurs in bays where *Casuarina* is absent.

Mangroves are seen in the deltas of Ganga, Brahmaputra and other rivers. Their breathing roots, i.e. stilt roots and knee roots, consist of soft spongy tissue and are adaptations to meet inadequate aeration due to waterlogged or submerged soil. Fruits of mangrove germinate on the tree itself, fall down on the mud and strike roots.

LITTORAL FOREST

Bulletwood and Poon trees in background with cylindrical trunks, Andamans.

MANGROVE FOREST

Rhizophora mucronata (*stilt roots*) *and* Bruguiera gymnorrhiza (*knee roots*). *Andamans.*

The Mangrove Date and Nipa PALMS are also seen in tidal forests. As the ground level rises the Blinding Tree (*Excoelaria agallocha*) takes over. Its leaves turn red or yellow before falling in the hot season.

5 *Deciduous*

This occurs throughout the northern area of our region except the eastern. Deciduous in the dry season, the canopy rarely exceeds 25 m. The common trees are sal, axlewood, *Terminalia alata*, khair (*Acacia catechu*) and riverine sissoo.

In the southern peninsula, presence or absence of teak divides the forest into two classes. Teak is associated with axlewood and *Terminalia*. In non-teak forests, the characteristic tree is axlewood associated with *Terminalia*. *Diospyros* is common with salai (*Boswellia serrata*) widespread and some karaya. *Shorea*

SOUTHERN TROPICAL DRY DECIDUOUS FORESTS
Nearly pure Red Sanders (Pterocarpus santalinus) *in foreground with some Axlewood* (Anogeissus latifolia) *in the background with dense undergrowth of grass. Cuddapah Division, Andhra Pradesh.*

talura occurs locally. The chief bamboo is *Dendrocalamus strictus.* Excluding the Western Ghats, this type is seen in Madhya Pradesh, Gujarat, Andhra, Karnataka and Tamil Nadu.

6 *Thorn*

Spiny acacias under 10 m with low crowns, spurge (*Euphorbia*), caper (*Capparis*), and naturalized cactus are typical. The hottest month is June (45°C) and the coldest is January (4°C). This type is seen on black cotton soils and in Pakistan, North, West, Central and South India.

CARNATIC UMBRELLA THORN FOREST
Pure Acacia planifrons *association. Kanyakumari, Tamil Nadu.*

7 *Evergreen*

Hard-leaved evergreen trees like *Manilkara hexandra,* and Maulsari (*Mimusops elengi*) with fragrant cream flowers, *Memecylon edule,* with a sprinkling of deciduous *Diospyros,* jaman and *Albizia amara.* Restricted to Karnataka coast, Andhra and Sri Lanka. The rainfall is maximum in October and November.

8 *Broad-leaved*

There is considerable difference between the southern and northern forms. The floristics of the Silent Valley include the following interesting species, POONSPAR, *Cinnamomum sulpharatum, Rhododendron arboreum* ssp. *nilagiricum* and FRAGRANT WINTERGREEN with disjunct distribution also present in the E Himalaya. There is a wide range of rainfall and dry season from one to seven months, which is offset by high humidity. The type is restricted between 1000 and 1700 m in the South Indian hills.

The northern form on the lower slopes of the E Himalaya from 1000 to 2000 m is subject to shifting cultivation and fires, which adversely affect the flora. The wet hill forests consist mostly of evergreen with a sprinkling of deciduous trees about 30 m tall. It is recognized by alder (*Alnus nepalensis*), oaks, chestnuts, cherry

NORTHERN MONTANE SUBTROPICAL BROAD-LEAVED
Jaman tree in flower along stream, wild banana, tree ferns and canes. Alder trees higher up in the picture with white stems. Kameng District, Arunachal.

and birch. Spectacular ORCHIDS festoon the tree-trunks, as many as 600 colourful species being known in the eastern region. Bamboos and numerous climbers are also present. The rainfall is over 1000 mm. At Cherrapunji with 10,800 mm it is the highest recorded in the world.

9 Pine

Characterized by predominance of chir. In moist conditions, oak is the typical broad-leaved tree. There is a sprinkling of wild ornamental trees such as *Rhododendron arboreum*, *Pyrus pashia*, *Lyonia ovalifolia*, Indian laburnum, amla, sandan and sal in the lower reaches. The shrubby growths are the wig plant (*Rhus*), berberis and raspberry (*Rubus*) and the musk rose (*Rosa moschata*). This type occurs on steep dry slopes on Siwalik conglomerates and sandstones in the Western and Central Himalaya. In the Khasi, Naga and Manipur Hills chir is replaced by Khasi PINE. The spectacular associates are *Quercus griffithii*, *Schima wallichi* and *Rhododendron arboreum*.

SUBTROPICAL PINE FORESTS IN ARUNACHAL

Chir, the ornamental Albizia *in the foreground is a new find,* A. arunachalensis. *Kameng District.*

10. Dry evergreen

The dry evergreen associates generally are *Dodonea viscosa*, with shining as if varnished leaves, olive, wild pomegranate, and oleander. The type is characterized by a prolonged hot and dry season and a cold winter.

11 Wet

In this, two subgroups are distinguished, Southern and Northern.

Over the century, this type has shrunk due to the replacement of slow-growing indigenous trees with fast-growing eucalypts to meet the fuelwood require-

SUBTROPICAL AND DRY EVERGREEN
Acacia modesta *scrub, Jammu Division.*

ments of a rising population. The original flora, as seen in small pockets called sholas in the Nilgiris, consists of *Rhododendron arboreum* ssp. *nilagiricum, Syzygium arnottianum, Symplocos,* etc. The ground flora is formed by *Nilgirianthus* (see CONEHEADS) with blue flowers which come into gregarious flowering at well-set intervals of 3 to 12 years or more, like an alarm-clock set to go off at a certain time. The young leaves of trees give a range of colours. Mosses, ferns, lianas and epiphytes clothe the tree-trunks. The forest is generally in patches on sheltered sites on rolling grasslands in the Nilgiris, the high ranges of Kerala and Sri Lanka from 1500 m upwards. The rainfall is from 1500 to 6250 mm.

The Northern Montane Wet Temperate Forest is distinguished from S Indian hill forests by the annual range of monthly mean temperature, and the distribution of rainfall (maximum in October in the south, instead of July or August in the north). The northern forest has three stages, two of them mainly coniferous, associated with either evergreen, oaks or mixed deciduous species, and the third is predominantly evergreen. The oaks form the large tree canopy, Lauraceae being relegated to the middle storey though numerically predominant. The profile is below:

Top storey: Buk Oak (*Quercus lamellosa*), *Q. lineata*.
Middle storey: *Machilus, Litsea* (both Lauraceae), *Engelhardtia spicata, Acer campbellii, Schima wallichii.*
Ground storey: Champ (*Michelia doltsopa*), *Alcimandra cathcartii*, CAMPBELL'S MAGNOLIA (three ornamental Magnoliaceae) starring the hillsides, *Rhododendron* spp., cherry.

The trees are of medium height with branches covered with mosses, ferns and orchids. A dwarf bamboo undergrowth with woody climbers occurs at higher

NORTHERN MONTANE OAK FOREST

Oak (Quercus lamellosa, Q. lineata), *Maple* (Acer campbellii), *Champ* (Michelia doltsopa) *and* Castanopsis tribuloides, *over* Magnolia, Symplocos, Rhododendron, Machilus, *underwood including hill bamboo* (Arundinaria maling). *Darjeeling.*

elevations. Under less wet conditions and at still higher elevations consociations of silver fir are met with. This type occurs from E Nepal eastwards to Arunachal from 1800 to 3000 m with a minimum monsoon rainfall of 2000 mm. The winter months November to March are relatively dry. Dense mist is characteristic of these forests during the monsoon. This type has been affected by shifting cultivation, grazing and fires which have resulted in the destruction of the forests and the extinction or near extinction of rare flora and fauna. Such areas of spectacular rare, primitive and endemic flora are being demarcated as biosphere reserves, sanctuaries and national parks.

12 *Moist*

The formation extends along the Himalaya with a rainfall between 1000 and 2500 mm, also in the form of snow in the west.

W Himalaya: Broad-leaved ban oak, brown oak, *Rhododendron arboreum, Lyonia ovalifolia, Cornus capitata,* walnut, etc.

CONIFERS: Low-level fir, blue pine, deodar, spruce, cypress, yew. The first three are generally extensive and gregarious. Shrubs include *Indigofera, Viburnum, Skimmia* and *Daphne.*

E Himalaya: Broad-leaved: *Quercus pachyphylla* (very large), *Q. lineata, Acer campbellii, Betula utilis, Engelhardtia spicata,* Campbell's Magnolia, *Rhododendron arboreum, R. grande,* etc. Conifers: hemlocks on drier ridges, red fir (*Abies densa*). Bamboo: *Arundinaria maling, Thamnocalamus aristatus.* Shrubs: *Daphne, Rubus, Berberis, Piptanthus nepalensis.* Ferns: Abundant.

In the altitudinal range 1500 to 3300 m the limits of this group vary with aspect and configuration. In the E Himalaya, heavy rainfall is responsible for the much less gregarious conifer flora though there is a greater variety of species. Coniferous trees unknown in the western Himalaya are larch, *Cephalotaxus, Podocarpus,* East Himalayan spruce, *Abies delavayii, A. densa,* etc. The mixed deciduous forests are generally found in depressions and ravines. At lower elevations alders border the watercourses.

13 *Dry.* Conifers predominate with broad-leaved trees like oak, maple, and ash. They are of low height, chilgoza, deodar and juniper being characteristic, with fir

DRY BROAD-LEAVED AND CONIFEROUS FOREST

Chilgoza with Holly (Quercus ilex) *understorey. Bashahr, Himachal Pradesh.*

and blue pine at higher elevations. *Artemisia* (worm-wood) and Ephedra form the ground vegetation.

In the W Himalaya, there are five climax types (see CLIMAX VEGETATION) containing (1) Chilgoza and holly (as in picture above), (2) Chilgoza and dry deodar, (3) Himalayan dry temperate deciduous, (4) Himalayan high level dry blue pine forest, (5) Himalayan dry juniper forest. In the E Himalaya, there are two climax types, (1) dry temperate coniferous and (2) dry juniper/birch forest.

In the inner ranges, the monsoon is feeble and precipitation is usually under 1000 mm and is mainly in the form of snow in winter. This type is found in Kashmir, Lahul, Kinnaur, Sikkim, etc.

14 *Subalpines.* It extends from Kashmir to Arunachal from about 2900 to 3500 m.

The shrubs in the W Himalayan subalpine are *Sorbus*, juniper, *Rhododendron lepidotum*, *R. hypenanthum* and *R. campanulatum*, blackcurrant and willows. There is a gorgeous flush of ornamental herbs including primulas, potentillas, and blue-poppies. which also occur in the E Himalayan type.

In the E Himalayan subalpine Red Fir (*Abies densa*), Black Juniper (*Juniperus wallichiana*), a shrub in the W Himalaya which turns into a tree in the

E Himalaya, Birch and Larch are the common trees. Due to high humidity and heavy rainfall the timber-line in the E Himalaya is about 600 m higher than in the W Himalaya. Rhododendrons completely dominate the landscape in the Eastern Himalaya, particularly in the subalpine and alpine zones, and display a much greater richness of species. About 80 species are known to occur in the E Himalaya which make their first appearance in type 8 and spread over in ascending order to other types on the basis of altitude: the Western Himalaya has only about 5 species of rhododendron. The spectacular rhododendrons in type 14 (E Himalaya) are *R. dalhousiae* (fragrant white flowers), *R. thomsonii* (blood-red flowers), *R. cinnabarinum* (cinnamon-red to brick red), etc.

BIRCH/RHODODENDRON SCRUB FOREST

Birch (Betula utilis), Rhododendron campanulatum, *the former much damaged by snow. Near tree limit 3500 m. Upper Bashahr, Himachal.*

WEST HIMALAYAN SUBALPINE FOREST

Birch (Betula utilis) (*at bottom*), *with high-level Fir and Prunus and Maple on slopes. Sonamarg, Kashmir.*

DECIDUOUS ALPINE SCRUB

Rhododendron campanulatum *and deciduous shrubs bent by snow; near tree limit 3500 m. Upper Bashahr.*

15 *Alpine (moist)*. This forms a low dense, scrub evergreen forest, almost entirely of rhododendron with some birch, the stems all curved owing to snow-slides.

The western form is similar to the eastern. Mosses and ferns cover the ground. Gorgeous shrubs, mostly rhododendrons, and herbs are abundant (see ALPINE PLANTS). Snowfall is ample, a thick layer of black humus is present and the soil generally wet. This type is distributed throughout the Himalaya and on the highest hills near the Burma border.

16 *Alpine (dry)*. It is a xerophilous formation in which dwarf shrubs predominate. Characteristic plants are common and black juniper, *Caragana* sp., *Artemisia maritima*, honeysuckle, *Potentilla* spp., along streams, willow, *Myricaria* and *Hippophae* are typical.

In the E Himalaya, drooping juniper (*Juniperus recurva*) at 3000 to 4600 m succeeds black juniper at 4300 to 4900 m and associations are pure at higher elevations.

DRY ALPINE SCRUB
Mat-like Juniper on northern slopes at 3100 m. Lahul, Himachal Pradesh.

Mixed composition of forests. The most usual condition is a mixture in the top canopy of varying number of species. Simple mixtures are presented by conifers, e.g. deodar and spruce with silver fir, or deodar with blue pine or chilgoza (Type 13), but more commonly in tropical forests there are several to many species contributing to the top canopy (Type 1). In tropical deciduous forests which cover so large a part of our area the greater part of the top canopy is formed by 6 to 10 species, including *Terminalia* spp., RED SILK-COTTON and JHINGAN etc. A mixture of species is beneficial for the humification of the soil and favours the breaking down of leaves and needles into humus. Forests of a single species have the risk of epidemics by damage by insects and other pests, e.g. the two Himalayan Dwarf Mistletoes attacking dry juniper and

blue pine forests in Lahul (Type 13), see PARASITIC PLANTS.

Three-storey composition. In addition to these top-storey mixtures, we frequently come across vertical mixtures in which one species or set of species occupies the top storey, and others form an underwood. Fir forests often have a complete underwood of oak or rhododendron. Many-tiered forests of three or more storeys are described under Type 1. Here generally there is a top layer of tall emergent trees with buttressed stems followed by a main canopy layer of trees with dense crowns. The buttressed stems and dense crowns of the main canopy (middle storey), protect the tall emergent trees against severe wind and rain storms. The bottom storey consists of small trees.

Spectacular Flowering Trees in various Forest Types
Forest
Type
1. Poonspar, Hollock, Red Silk-Cotton, Queen's Crape Myrtle, *Mesua ferrea*, *Syzygium formosum*, *Dillenia indica*, Flame Amherstia.
2. Red Silk-Cotton, *Terminalia paniculata*, Kadam, Hollock, *Elaeocarpus* sp.
3. *Terminalia paniculata*, Red Silk-Cotton, *Careya arborea*, Indian Screw Tree, Ashoka.
4. Poon Tree.
5. *Shorea talura*, Indian Laburnum, Flame of the Forest, Coral Jasmine. Indian Screw Tree, *Holarrhena antidysenterica* (also fragrant), Kanchan.
6. *Ixora arborea*, Indian Laburnum, *Tecomella undulata*.
7. *Mimusops elengi* (also fragrant), *Memecylon edule*.
8. Poonspar *Rhododendron arboreum* ssp. *nilagiricum*, Wild cherry, *Schima wallichi*, *Bauhinia purpurea*, Champak, *Talauma hodgsonii*.
9. *Rhododendron arboreum*, Sandan, Indian Laburnum, *Lyonia ovalifolia*, Campbell's magnolia.
10. *Punica granatum*, *Erythrina*.
11. *Rhododendron arboreum* ssp. *nilagiricum* (southern forests). *Michelia doltsopa*, Campbell's magnolia, *Schima wallichi*, *Engelhardtia spicata*, *Rhododendron* spp. (northern forests).
12. *Rhododendron arboreum*, *R. grande*, *R. hodgsonii*, *R. falconerii*, Campbell's magnolia.

Rainfall and forests. The amount of rainfall is the primary factor determining the luxuriance and type of vegetation which can exist. The distribution of rainfall is characterized by two dry zones along the 30° parallel of latitude both north and south, and an increase of rainfall outwards from the zones towards the poles and the equator. With the exception of the extreme west and northwest, the rainfall is almost entirely brought by the SW Monsoon over most of the subcontinent, and to a less extent in the southeast by the retreating or NE monsoon.

Buttressed roots. In many trees of the wet tropical forest (padauk, hollock, mangrove, etc.) and some in the drier tropics, e.g. Silk-Cotton, the swelling at the base is pronounced and takes on a buttress form which may extend 15 ft or more up the stem. In extreme cases these are in the form of wings, thin enough for planks to be hewn directly out of them, e.g. *Elaeocarpus* sp. and sometimes Red Silk-Cotton. A further development is seen in *Rhizophora* (Type 4) comparable with the flying buttresses of architecture, the base of the stem being supported by stilt roots.

Associated fauna. Most animals are able to, and do adapt themselves to a wide range of habitats, and there are few exceptions when either climate or vegetation restricts an animal to a particular type of forest.

V.R. & K.C.S.

FOSSIL BIRDS. In relation to the remains of other vertebrates, the global record of fossil birds is rather poor. The fossil avifauna from the Indian subcontinent is even more inadequately known. The reasons for this are manifold: Firstly, birds have a delicate, light-weight skeletal structure which easily decomposes before the process of fossilization sets in. Secondly, except for some primitive types such as *Archaeopteryx* and *Hesperornis*, birds lack teeth, a feature which generally represents the best method for deciphering taxonomy and evolutionary lineages. Taxonomic differentiation of fossil birds is consequently based mainly on osteological differences. Thirdly, birds (as indeed other animals) tend to be best preserved at sites closest to their natural habitats, thus water-dwelling birds are best represented in the fossil record in depositional basins occupying lakes, rivers and coastal areas.

As yet, there is no substantiated record of a fossil bird from the Indian subcontinent prior to about 12 million years. The most comprehensive review of the fossil avifauna was made nearly a century ago by Lydekker (1884) from the Siwalik Group of rocks outcropping all along the Himalayan foothills from the Potwar Plateau in Pakistan across India and extending up to Burma. The fossil birds recovered from the Siwaliks are mostly water-dwellers, except for the remains of some rather large terrestrial flightless forms. Most of the genera of fossil birds so far known from the Indian subcontinent still exist today, though the distribution of *Struthio* and *Dromaius* has shrunk considerably. The fossil avifauna includes pelicans, cormorants, storks, mergansers, ostriches and emus.

The remains of struthious birds are relatively more common because of the robust nature of their bones.

Siwalik ostriches are specifically indistinguishable from the Recent *S. camelus.* A collection of cervical vertebrae with significantly larger dimensions than *S. camelus* indicate that the Siwalik ostrich had a stouter neck. The presence of an emu, *Dromaius sivalensis*, considerably larger than *D. novae-hollandiae*, is also indicated. A third ostrich-like genus, distinct from *Struthio* and *Dromaius*, and represented by the second phalangeal of the third digit of the foot of a tridactyl struthioid, was also present. The remains of fragmentary ostrich egg shells found at some archaeological sites in western India were probably brought over by traders and do not represent an indigenous form.

Both the Siwalik pelicans, *Pelecanus cautleyi* and *P. sivalensis*, represented by fragmentary ulnas, are somewhat smaller than the existing *P. mitratus* ($=$ *roseus*) common to India and Africa. The cormorant, *Phalacrocorax* is similar to *P. carbo*. Siwalik storks were of various dimensions ranging from considerably large forms to those similar in size to *Leptoptilos crumeniferus*. The Anatidae family is represented by a form closely allied to the present-day smew, *Mergus albellus*.

The limited number of taxa known from the Siwaliks does not give a true indication of the wealth of taxonomic diversity that must formerly have existed.

A.S.

FOSSILS. All living species, animals or plants, die after a period of time. Some of these get completely decayed beyond recognition and get washed off or dispersed in water or earthy material. But some, whether in the sea or in the land, get covered by sand or clay or any other inorganic material. It can be called a burial of the organism by natural processes. As time elapses and when this buried material is brought to light at the surface, either by natural erosion or by artificial digging up, it is found that an imprint or a cast of the original living species, whether plant or animal, is seen. Hence it can be said that a fossil is any remains, trace or impression of life on earth. The word fossil is derived from the Latin word *fossilis* which means something dug out of the ground.

The branch of science that deals with the study of ancient life on earth is called Paleontology, though in later usage this is strictly confined to the study of the ancient animals only, and the science of Paleobotany deals with ancient plants.

To the early scientist the fossils indicated different things. To some they were discarded pieces thrown out by the Creator in his attempt to evolve perfect living organisms. Perhaps it was the Greek historian Herodotus in 450 B.C. who first correctly inferred their significance. He not only said they were evidences of ancient life but also that most of them have lived in the sea, then got buried in sand and then lifted up to the position where they are now. In some cases the sea might have sunk to lower levels, leaving the shells embedded in sands at higher levels.

There are different types of preservation of fossils. Rarely the whole animal is preserved. There are however a few cases of woolly mammoths (elephants) that lived in Siberia about a million years ago getting entombed in ice. When dug out, a majority of the parts were seen preserved exactly as they would have been in the original animal. Similarly some tiny insects, creeping over the bark of a tree, get covered by the resins oozing out of the tree and are completely preserved. But the most common type is the case where the hard part of the animals (bones, shells) or plants (stems) get covered and in many cases the replacement is so gradual and perfect that the original structure of the animals or plants can be clearly seen in the fossils. Occasionally the tracks (dinosaurs), trails (worms) and imprints (feet of animals, leaves of plants) are also seen in rocks giving clear evidence of their earlier existence. The fossil collector usually hunts for fossils in sedimentary rocks because they are preserved best in them.

Why do we study fossils? A fossil is to the geologist what ancient coins and inscriptions are to the historian. Both attempt to reconstruct past history. The former goes back millions of years, but the latter, a couple of thousand years only. A detailed study of the fossils in different areas of the world has established that there is change in the nature of species with passage of time, not always uniform or regular. This is called evolution. A biologist is interested in this. Since a geologist records the history of rocks, the fossils aid him as time indicators. Occasionally mass-scale extinction of a great number of species may denote past climatic changes or tectonic movements in the earth. It should however be remembered that though the planet earth was born about 4500 million years ago, and life is known to have existed since about 2500 million years ago, identifiable fossils which could be studied and used extensively by an earth scientist have only existed since about 600 million years.

In India numerous plant fossils are found in some of the strata of the Gondwana System in Central and Eastern India, vertebrate fossils in rocks of the Siwalik System along the foothills of the Himalaya and the Gondwana System, and marine fossils within the Himalayan strata and in some of the coastal deposits.

R.V.

See also DECCAN TRAP

FOXES. Nine species of foxes belong to the genus *Vulpes* which is distributed through nearly all the major zoogeographical regions of the world from the arctic to South America. Five species occur within the region covered by this encyclopedia, of which only two are relatively widespread and common.

These are the Red Fox *Vulpes vulpes*, and the Indian Fox *Vulpes bengalensis*. The former occurs almost all over the Northern hemisphere. It is very variable in pelage—so much so that more than forty different subspecies have been described. Three of these, all quite distinctive in appearance, occur within our region. They are *Vulpes vulpes pusilla*, the Desert Fox, *V. vulpes griffithi* the Hill Fox, and *V. vulpes montana* the Tibetan Fox. Only the last has a wholly rufous coat with grey-black chest and legs and the backs of the legs velvety black. The Hill Fox has a luxuriant coat much prized by the fur trade, with a mixture of creamy buff hairs giving a general silver-and-chestnut appearance. The Desert Fox generally has the backs of its ears dark brown or greyish sooty with white patches on the feet, and very little rufescence in its greyish buff fur. All have a white tip to their bushy tails.

© M. Krishnan

Desert Fox, Rajasthan × $^1/_9$

The Indian Fox has a black tip to its tail and is slimmer in build with rufous legs and the backs of its ears not contrasting darker. Its body fur is greyish buff. It occurs throughout India from Cape Comorin to the Himalayan foothills. The Sand Fox or Desert Fox *Vulpes rüppelli* occurs in the extreme southwestern regions of Pakistan, bordering the Makran coast. This fox has a white tip to its tail, enormous ears which are not dark on their dorsal surface, and the soles of its feet are covered with long hairs which conceal the naked pads.

In the mountainous regions of the western borders of Pakistan, the very rare Blanford's Fox occurs. This also spreads into Iran, Afghanistan and the U.S.S.R. It is the smallest of the foxes in this region, with black-tipped guard hairs and a very thick luxuriant fur, much prized by furriers. The tail sometimes has black and sometimes white hairs on its tip. Finally, in Tibet and the northern borders of Nepal, the Tibetan Sand Fox,

Vulpes ferrilata, occurs. Few specimens have been collected and it may prove to be no more than a distinctive subspecies of the Red Fox.

An adult Red Fox weighs 5 kg with head-and-body length 60 to 70 cm, whereas the Indian Fox weighs 2 to 3 kg and measures 45 to 60 cm in head-and-body length. Blanford's Fox weighs barely 2 kg and measures only 40 cm in body length.

Foxes usually hunt singly and are not gregarious, but they do form stable pair-bonds and most species are thought to be monogamous. The Red Fox vixen produces her litter of 4 to 6 pups during the late winter or early spring in our region. These are born blind and relatively helpless in a nest-chamber excavated in a burrow, often adapted from an existing porcupine burrow or a natural cave. The cubs are fed upon regurgitated food brought by both parents for the first few weeks of their lives. They emerge at 5 weeks and then spend a good deal of time playing outside their den in the early morning and evening. They are independent in hunting by about 5 months of age.

Foxes will subsist upon a variety of food, depending upon wild fruits such as *Zizyphus* in season, as well as lizards, orthopterous insects, and any small birds or rodents which they can catch. In coastal areas, both fish and crabs are hunted, and in Rajasthan scorpions have been found in their stomachs. When not disturbed or in cloudy weather most foxes will hunt partly by day,

but close to human habitation they are strictly nocturnal in activity.

They have a variety of vocalizations, from a staccato bark given by the male as an advertising call, to a chattering cry when excited, and various whines and growls not unlike their domestic congeners. Males mark their territory by urinating on conspicuous vegetation clumps or stones. The Indian Fox generally breeds during the monsoon period, and bears 3 to 4 cubs after a gestation period of 52 days. As in domestic dogs, there is a copulatory tie. Foxes have great endurance and can hunt or run for many kilometres at speeds of over 30 km per hour.

FRAGRANT WINTERGREEN (*Gaultheria fragrantissima*). This evergreen, fragrant shrub, with orange-brown bark, grows in Nepal, Arunachal Pradesh, the Khasi Hills, the Nilgiris and the Western Ghats. The flowers are greenish-white and the hemispherical fruits are enclosed in a fleshy deep blue calyx. The oil from the plant is used as a pain-killer in aspirin and is a flavouring agent. The fruits are edible.

FRANGIPANI (*Plumeria acuminata*) is planted near places of worship and therefore is also known as the Temple or Pagoda flower. It grows to 4–6 m and has a soft bark and non-tapering branches which on cutting exude a milky latex. It is leafless in winter. Broadly

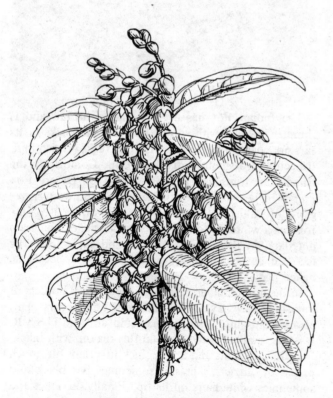

Fragrant Wintergreen. Shoot with flowers, × 1

The parallel-veined leaves and funnel-shaped flowers of the frangipani, × ¹/₂

lance-shaped leaves with parallel veins sprout in crowded spirals at the tips of branches. The flowers grow in upright clusters, creamish with a golden centre, funnel-shaped with five spreading petals and are profuse from February to October. The tree is an exotic from Mexico, naturalized in our area for two centuries.

FROGS and TOADS. The Salientia comprise the amphibians that one is more familiar with. This order encompasses the frogs and toads. They all possess short, tail-less bodies and long legs. The hind legs have four segments which function as efficient levers to enable these animals to progress by leaps. This order has a large representation in the Indian Region, usually separated into three subgroups, the hill toads, true toads and true frogs.

This division is done according to natural groupings that occur in skeletal and muscle characteristics but is usually perceived in the physical appearance too.

The hill toads are a widespread group that stretches from North America to Southeast Asia. In the Indian Region they are restricted to the northern hills and the Himalayas. They are represented by four genera in this region: *Ophryophryne*, *Aelurophryne*, *Scutiger* and *Megophrys*. These animals vary in appearance from the small, toad-like *Scutiger* to the large, ornate *Megophrys*. Some hill toads demonstrate an extraordinary ability to withstand low temperatures. The genus *Scutiger* has been collected at an altitude of 5000 m in Tibet. Ag-

gressiveness towards predators as a part of the defence reaction is unusual among amphibians, but one genus of the hill toads, *Leptobrachium*, are known to open their mouths wide and bite when cornered or threatened.

The true toads and tree frogs are a familiar group. The suborder is well represented in the Indian Region by the genus *Bufo* which contains the common toad *Bufo melanostictus*. In the field the commonest character

Common toad (Bufo melanostictus) × 1

is their relatively dry warty skin. Most of the toads possess parotid glands on the sides of their head or body. These glands produce an acrid, toxic secretion when stimulated. The active component of this, bufotoxin, has been demonstrated to be toxic to vertebrates.

All the Indian toads lay their eggs in long strings. The tadpoles are usually dark-coloured. The call consists generally of a trill though some species such as *Bufo kelartii* have a piping song. Bufonids are generally nocturnal but may be met with in the day during monsoon showers. They are indiscriminate feeders and will attempt to eat any small animal under a given size. The size threshold varies in relation to the size of the toad.

The true frogs comprise three well-represented families in the Indian Region. The family Ranidae or the true frogs, the family Rhacophoridae or the tree frogs, and the family Microhylidae or the narrow-mouthed frogs.

Ranidae or the true frogs are generally characterized by a fairly smooth, moist skin. They are differentiated from the other two families by spinal and digital structure.

The Ranids usually lay their eggs in loose clumps but utilize all types of water bodies for their breeding. Thus they range from the typically pond-dwelling Indian Water Skipper. *Rana cyanophyulyctis* to the Cliff Frog *Nannophrys ceylonensis* that breeds on moist, vertical cliff faces. The tadpoles of the mountain stream breeders, such as of the genus *Staurois*, have an adhesive disc on the ventral surface behind the mouth, for holding on to rocks in the fast streams. Another unusual feature for a tadpole is seen in *Rana alticola* of Sikkim. This tadpole has well-developed parotid glands on both sides of the head.

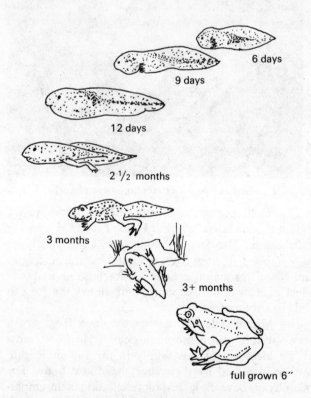

6 days

9 days

12 days

2 1/2 months

3 months

3+ months

full grown 6″

THE DEVELOPMENT OF A FROG

Bullfrog (Rana tigerina) × ¹/₂

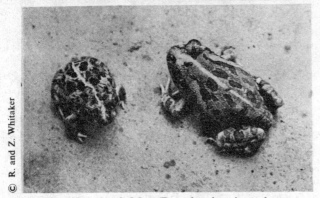

Short-headed frog (Rana breviceps) × 1

Golden frog (Rana temporaria) × 1

Ranids are the frogs most extensively used by humans as food. In the Indian Region the principal species exported to be eaten are *Rana hexadactyla, R. crassa* and *R. tigerina.*

This family, the most primitive of the suborder, has adapted to a wide range of environmental conditions. The small, toad-like, Short-headed Frog *Rana breviceps* lives in arid regions, estivates during the drought and emerges with the rains. The Golden Wood Frog *Rana aurantiaca* has small dilated tips on its toes and ascends bushes much in the manner of a tree frog. The Corrugated Frog, *Rana corrugata*, is completely aquatic, spending all its life in water.

The tree frogs (Polypedatidae) are readily distinguishable by the tips of the digits being expanded into discs. Though the majority of species in the family are arboreal by habit, some species such as the Small-eared Tree Frog *Rhacophorus microtympanum* have become ground-dwellers. This species has also evolved the interesting habit of undergoing the tadpole stage in the egg so that a fully formed froglet hatches out of the egg. There are a few other exceptions to the arboreal habit such as the waterfall-dwelling *Rhacophorus nanus*. An unusual feature of this family is the habit of laying eggs out of water. This is accomplished by making a foam 'nest', the construction of which is described under AMPHIBIANS. The foam nest dries, forming a crust on the outside that is resistant to water loss. Thus the eggs inside are kept moist. The eggs hatch within the nest; the nest liquefies from within as the tadpoles develop and the crust is broken by the liquid mass at about the time the tadpoles are ready to take to the water.

Tree frogs are widely distributed in the Indian Region. However, a high diversity of species is found only in the wet or hilly areas, *Rhacophorus maculatus* and *R. leucomystax* are the species most often met with in the drier areas. Often they are the only members of the family to penetrate dry or arid habitat.

Common tree frog (Rhacophorus maculatus) × 1

The most spectacular member of this family in the Indian Region is the Malabar Flying Frog, *Rhacophorus malabaricus*. This frog has long toes, all of which are fully webbed. It leaps from trees and extends its fingers and toes; this action creates a parachute or umbrella effect at each limb extremity, and allows the frog to glide to another tree.

The Microhylidae or narrow-mouthed frogs are a very variable group morphologically. However, most of the members of this family in the Indian Region can be identified by their short head and body. The microhylids have the largest representation in the amphibian fauna of dry, arid areas. They are excellent bur-

Flying frog (*Rhacophorus malabaricus*) × 1

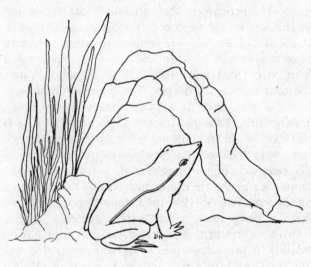

Ornate narrow-mouthed frog (*Microphyla ornata*) × 1

Painted frog (*Kaloula pulchra*) × 1

rowers and when the rains cease all members of this family that inhabit dry areas burrow into the ground. Here they curl up into a ball and remain in a torpid state till activated by the next rains. The Painted Frog *Kaloula pulchra* is one of the more striking members of the family, being coloured grey with patches of orange or crimson on the back. The Balloon frogs, *Uperodon* sp., are unusual in the fact that when breeding they blow their bodies out and float on the water. They use their bloated bodies as resonating chambers in making their mating call.

The rare Indian Black Toad *Melanobatrachus*, a small coal-black frog with a squarish head and rough skin, also belongs to the family Microhylidae. The genus *Microhyla* is essentially a lowland genus in India while the genus *Ramanella* seems to represent the hill or montane forms.

See also AMPHIBIANS. See plate 3 facing p. 48, plate 19 facing p. 256, and plate 38 facing p. 529.

FROST. Frost is the formation of an icy deposit on any outdoor surface exposed to a night sky. It occurs when the temperature of the air in contact with the surface, often a vegetation surface, falls below the freezing point of water, 0°C. Vegetation provides enough moisture. When there is abundant moisture, and little wind, then frost can be severe.

Frost when severe can cause much damage to fruit trees and grape vines in orchards. One method of preventing a night frost is to churn the air layers. Then the night cooling is distributed over a great depth of the air and the air temperature close to the vegetation surface does not fall below 0°C. Another method is to provide heating through a number of oil burners, distributed in the orchard during the night a few hours before the expected formation of frost. Pakistan and northern India are liable to frost during the cold weather.

FRUITS. FLOWERING PLANTS produce fruits and seeds for their propagation. Popularly, different parts of the plant are termed as fruit. But a fruit is strictly an outcome of a FLOWER or its accessory parts such as the receptacle, the sepals or the whole inflorescence. A fruit is the result of stimulus by fertilization, which immensely activates secondary growth of the ovary or of other parts of a flower. A fruit is a storehouse of seeds and a means of transport for their dispersal. In a mango or banana, for example, the ovary is transformed into a fruit but in a jackfruit or pineapple the whole inflorescence proliferates into a bulky fruit. The ovules which play the most important role in reproduction of the plant are nourished by the fruit till they become mature seeds. During evolution plants have adopted elaborate protective measures for the ovule within the carpel or fruit, but in non-flowering gymnosperms the ovules are naked and seeds are borne in cones.

About 250,000 species of flowering plants bear fruits but only a negligible number are useful to man and have been brought under cultivation. Some cultivated fruits such as oranges, bananas or grapes are seedless and develop without fertilization of the ovule, this process being known as parthenocarpy. Such fruits have been developed by man and are rarely found wild. Most of our commmon fruits, cereals and pulses have undergone prolonged and repeated trials for their improvement over their wild ancestors till a required quality had been achieved. Nowadays many hormone and hybridization techniques are practised for such improvement. Some important fruits, including the apple, guava, litchi, papaw and pineapple, are widely cultivated. They are not indigenous but were introduced in the distant past.

The variety of shapes and sizes, armaments and sculptures, in fruits is endless. Their colour, smell and taste have helped their dispersal by animals. A fruit has three distinct concentric layers; the outer thin skin is called the pericarp, the central fleshy portion is the mesocarp and the inner harder layer, where the seed is enclosed, is the endocarp. These layers greatly help in spreading the species.

The number of seeds in a fruit varies from one in the small, dry fruits that do not split open (called achenes), to many thousands in the capsule of an orchid. Seeds are either released by the fruit or are carried away along with the fruits. Fruits and seeds develop simultaneously till maturity. This may take only a few days, or nearly ten years for a Double Coconut seed. A fruit may develop from a flower having a solitary carpel, or more than one carpel may fuse together to give rise to fruits like melons and oranges.

Flowering and fruiting in plants usually take place once a year, regulated by seasonal changes. Some flower and fruit in the hot season, and others (e.g. cereals, pulses and oilseeds) are winter-fruiting, known as rabi crops. A few plants (e.g. coconut) set fruit throughout the year; and bamboos and some palms produce fruit only once in their life-cycle, after several decades. Fleshy fruits generally grow in summer and ripen during the rainy months when seeds have ideal conditions for germination. Due to variation of temperature, fruiting in plants is much earlier in peninsular India than in the northern plains.

Fruits have been variously classified but are broadly divided into dry fruit or fleshy fruit. They are divided into three categories, namely simple, aggregate or composite. A composite fruit develops from an inflorescence but other fruits develop only from a single flower. Simple fruits are grouped into two kinds according to their condition on maturity, dry or fleshy, and dry fruits may be either dehiscent or indehiscent. Dehiscent

fruits generally open to shed their seeds. In indehiscent fruits the seeds are only released following decay, or germinate directly from the fruit.

Dry Fruits

(a) 'Achenes' are dry indehiscent one-seeded fruits with a hard wall. In Traveller's Joy or a Buttercup (Fig.1) the whole fruit is an aggregate of achenes developed from a single carpel. Until the hard pericarp wall is ruptured the seeds inside will not germinate.

(b) 'Samaras' are dry, winged, one- or two-seeded fruits. These are also achenial fruits and the wings are formed by proliferation of the wall of the ovary: examples, Maple, Ash, Indian Elm (*Holoptelea*) (Fig.2).

(c) A 'cypsela' is a dry one-seeded fruit developing from a bicarpellary ovary. The fruits of Sunflower family are of this type (Fig.3). Here the fruits are generally light, small, air-borne on a flat receptacle. Most of them are crowned with a calyx modified into a hairy or feathery process, to facilitate floating in the air.

(d) 'Nuts' are also dry, one-chambered, single-seeded fruits. The pericarp of these fruits is hard, woody and indehiscent arising from one or more carpels. Though Coconut and Betelnut are commonly called nuts, they possess a fibrous pericarp and the endocarp forms a hard coat. Other 'nuts' like Walnut, Almond, Waterchestnut are actually drupes. The common Groundnut is not a true nut but a two-seeded pod developing underground. Examples of true nuts are Oak (Fig.4) and Cashew. In Oaks the nut is formed from a tricarpellary ovary but only one seed develops. A large cuplike receptacle supporting the acorn is formed from the minute leaves and the nut is shed, leaving the cup intact on the tree. In CASHEW the kidney-shaped nut grows upon a fleshy stalk called cashew apple.

(e) A 'schizocarp' is a dry indehiscent fruit having more than two carpels and splits into one-seeded segments. The schizocarp of hollyhock or mallow (*Malva*) (Fig.5) are common examples where the carpels are radially arranged in a compact ring: they split into wedge-shaped parts when mature. The schizocarp of *Geranium* splits suddenly into one-seeded fruitlets, which may be ejected or may remain hanging on the style. In the Carrot family the bicarpellary schizocarp ruptures into single-seeded fruitlets called mericarps. The commercial spices, Coriander, Fennel and Cumin, are examples of schizocarp fruit.

(f) A 'caryopsis' is a special type of dry fruit, characteristic of grasses and cereals including paddy (Fig.6), wheat and barley. Here the fruit is indehiscent and the pericarp is entirely fused within the seed-coat, so that the seed is inseparable from the fruit. In maize the fruit, called grain, is bare but in paddy, wheat and oat the fruits are enclosed by a husk. Their falling off with

19 Snakes and a Lizard

Common Krait (*Bungarus caeruleus*)

Andaman Watersnake with eggs. The snake is 60 cm long.

Common Garden Lizard (*Calotes versicolor*)

Spectacled Cobra (*Naja naja*)

20 Flowering Trees and a Rhododendron

Variegated Bauhinia (*Bauhinia variegata*)
Found wild in the sub-Himalayan tract, cultivated elsewhere.

Indian Laburnum (*Cassia fistula*)
Common in deciduous forests.

Silk-cotton tree (*Salmalia malabarica*)
Flourishes throughout the subcontinent except in the most arid tracts.

Rhododendron (*Rhododendron kendrickii*)
Shrub with red flowers in temperate rain-forests around Tawang (Arunachal Pradesh) at 2600 m. Rhododendrons completely dominate the landscape in the eastern Himalaya.

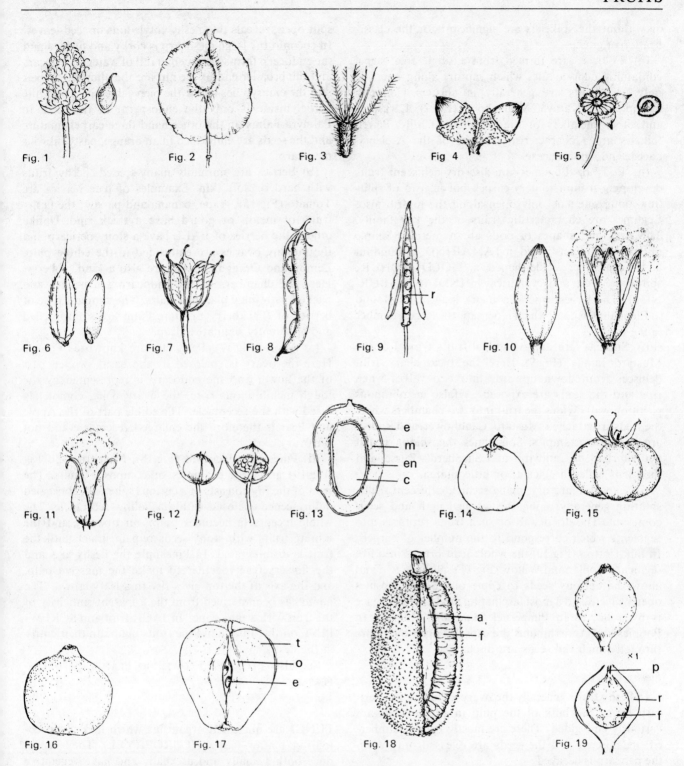

1 Buttercup, an aggregate of achenes. 2 Indian elm (*Holoptelea*). Single-seeded winged fruit. 3 *Tridax*, a cypsela. 4 Oak, a typical nut called an acorn. 5 Schizocarp fruit of *Malva*. On the right a single fruitlet.

6 Caryopsis of rice. On the right a grain. 7 An aggregate of follicles in Monkshood 8 Pod of Pea split open. 9 A siliqua of Mustard showing seeds on the replum. 10 Un-

opened and opened capsule of Silk-cotton. 11 An opened bivalved fruit of Willow. 12 Un-opened and opened fruit of Pimpernel: opening by a lid. 13 Cross-section of Mango showing e epicarp or skin, m mesocarp, the fleshy edible part, en stony endocarp, c fleshy cotyledon. 14 A drupe of Indian Jujube. 15 A berry of Tomato.

16 A hesperidium of Orange, which is a type

of berry. 17 Apple cut open; t thalamus, o ovary, e endocarp. 18 A sorosis of Jackfruit, the fruit cut open, showing the a core or axis and f individual fruits. 19 Synconus of Fig, and cut open fruit showing p aperture at the tip through which insects enter for pollination, r receptacle. f fruits inside.

or without the spikelets are significant for the classification of grasses.

(g) 'Follicles' are formed from a single free carpel containing many ovules which rupture along one side only. The fruits are generally in clusters. Common examples are Karaya, Monkshood (Fig.7), Larkspur, and other members of the Buttercup family. Paired follicles are a characteristic feature of the Asclepiadaceae and Apocynaceae.

(h) 'Pods' or 'Legumes' are also dry dehiscent fruits developing from a solitary carpel, but instead of splitting along one side only open along the midrib also. Legumes are characteristic fruits of the Pea family. The sizes and shapes of pods are numerous. Simple in Pea (Fig.8), constricted in TAMARIND, Groundnut and Coral tree; sickle-shaped in GOLD MOHUR; spiral in Medic and cylindrical in INDIAN LABURNUM. The largest pod is a metre long and is found in Elephant Creeper, which is perhaps the largest climber in the world.

(i) 'Siliquas' are dry cylindrical fruits typical of the Mustard family (Fig.9). Here the bicarpellary fruit dehisces from down upwards into two valves when ripe and the seeds are exposed. A false membranous partition wall divides the fruit into two chambers where the seeds are attached. Mustard, Cauliflower, and Radish are common examples. Sometimes the siliqua is flat and short as in Candytuft and Shepherd's Purse, and then it is called a silicula, or little siliqua.

(j) 'Capsules' are dry, many-seeded, dehiscent fruits, splitting generally along the whole length and seeds come out. The dried wall of such fruits ruptures into segments which correspond to the number of carpels. In Silk-Cotton (Fig.10) the whole fruit breaks into five segments. Poplar and Willow (Fig.11) split into two and allow the cottony seeds to come out. Some capsules open by teeth and a most fascinating mode of dehiscence is by a lid, as in Pimpernel (*Anagallis*) (Fig.12). In Poppies and Antirrhinums the capsules open by a pore through which the seeds are discharged.

Fleshy Fruits

In fleshy fruits generally the ovary including the pericarp forms the bulk of the pulp in which seeds are variously embedded. These are mostly many-chambered, indehiscent and the seeds are released only when the pericarp is decayed.

(a) 'Drupes' are fleshy fruits developing from a single or multicarpellary ovary where the pericarp is differentiated into epicarp (the skin), mesocarp (the fleshy part) and endocarp (hard and stony). These are also known as 'Stone Fruits'. Mango (Fig.13), Coconut, Indian Jujube (Fig.14), Peach and Palmyra Palm are common examples of drupes. The eatable pulp of the mango is actually mesocarp and the rejected stony seed, when

split open, reveals two fleshy cotyledons or seed-leaves. In coconut the large mesocarp is corky and fibrous and the endocarp forms a stony ball full of water. Two scars and the broader end of the nut are the abortive carpels and the embryo lies below the pierceable pit. The white kernel inside is only the endosperm of the seed. In Palmyra Palms on the other hand three carpels mature and the seeds are embedded in an orange, pasty, fibrous mesocarp.

(b) Berries are normally many-seeded, fleshy fruits with hard or soft skin. Examples of true berries are Tomato (Fig.15), grape, banana and papaw; the fleshy fruits of melon or gourd have a thick rind. Unlike others, the berries of BAEL have a stony pericarp and the gummy ovaries are embedded in the edible pulp. Lemon and Orange (Fig.16) are also a kind of berry. Here the chambers of the endocarp are papery and packed with juicy placental hairs. The simplest type of berries of Blackberry or Date Palm are single-seeded and apparently simulate a drupe.

(c) Apple (Fig.17), Pear, etc. are known as Pomes. Here the ovary is enclosed by the basal swollen part of the flower and the endocarp is represented by the tough membranous core, the ovary being completely fused with the receptacle. The edible part of the Apple and Pear is therefore the enlarged receptacle and not the ovary.

(d) Pineapple, Jackfruit (Fig.18), Mulberry and Fig (Fig.19) are ideal examples of composite fruits. The fruit of the Fig consists of a curious spherical, concealed inflorescence enclosed within a hollow receptacle. The whole receptacle becomes pulpy on ripening and the minute fruits with stony seeds remain intact until the fruit is disintegrated. In Pineapple the fleshy axis and the flowers fuse together to make the massive pulp, and the axis at the top gives rise to a leafy crown. The name has been derived from the apparent similarity of the fruit with a pine cone. In JACK fruit and SCREW-PINE on the other hand the fruits maintain their entity in the fleshy spadix.

See also FLOWERS, FLOWERING PLANTS, SEED DISPERSAL.

U.C.B.

FUNGI are nucleated organisms which lack green colouring matter (see CHLOROPHYLL). They reproduce both sexually and asexually and have vegetative structures usually consisting of branched filaments surrounded by cell walls containing cellulose or chitin or both and obtain their food by infecting living organisms as parasites or by attacking dead organic matter as saprophytes. In the absence of chlorophyll, they require already elaborated food in order to live. But given carbohydrate, most fungi can synthesize their own proteins by utilizing nitrogen and various mineral elements.

Carbon, oxygen, hydrogen, nitrogen, phosphorus, potassium, magnesium, sulphur, boron, manganese, copper, molybdenum, iron and zinc are required by almost all fungi. Many fungi are capable of synthesizing the vitamins required for their growth but some have to get thiamine or biotin or both from outside. They store their excess food in the form of glycogen or oil. Fungus consists of microscopic filaments called hyphae, and a mass of hyphae is called mycelium. Fungi have well developed nuclei, each with a nuclear membrane and nucleolus. There are 45,000 species of fungi belonging to 5100 genera.

Asexual reproduction occurs by fragmentation of hyphae into component cells, which may remain thin-walled or become thick-walled before separating from each other; by simple splitting of a cell into two daughter cells by constriction and formation of a cell wall; by budding as small outgrowths from parent cells; and by spores. Sexual reproduction may take place through distinguishable male and female sex organs or may be delegated to somatic hyphae.

Fungi can be seen with the naked eye as cottony growth on bread; blue, green or black moulds on pickles, citrus, leather goods and textiles; leathery, corky, or hard fruit-bodies or crust-like growths on standing trees and logs; the well-known mushrooms, toadstools, bird's nest-like fruit-bodies on soil and organic matter and coral-like, club-shaped, ear-shaped and charcoal-like fruit-bodies growing on decaying stumps of trees in the forests. Dispersal of fungi takes place mainly as spores, which are microscopic structures formed in well-defined fruit-bodies or directly on mycelia. Fungi thus occur in soil and in the air everywhere from the equator to the poles up to the limit of vegetation. One gramme of soil may contain up to 50,000 fungi living in dynamic biological equilibrium with other soil microflora and microfauna.

They grow best at temperatures between 20 and 30°C but some thermophilic fungi can grow at temperatures up to 60°C. Still others such as *Sporotrichum carnis*, *Penicillium expansum* and *Torula botryoides*, causing spotting of meat, grow readily at 0°C. *Cladosporium herbarum*, causing black discoloration of meat, can grow at −10°C. The minimum relative humidity under which they can grow is directly related to the suction pressure exerted by hyphae and this suction pressure in some species of *Penicillium* and *Aspergillus* may be as high as 217 atmospheres, enabling these fungi to grow at very low humidities.

Every living being on earth, animals (including humans) and plants, are harmed or benefited directly or indirectly by fungi. Crops worth billions of rupees are destroyed annually all the world over by the ravages of fungal parasites. Fungi are the chief agents of destruction of wood in usage. Wood decayed by fungi such as *Armillariella mellea* turns fluorescent. Forest dwellers get frightened on seeing glowing wood at night as they consider this as the fire of evil spirits.

Fungi also destroy annually plant and animal products worth millions of rupees. The moulds destroy fruits and seeds in storage. *Chaetomium* species are responsible for destruction of fabrics containing cellulose. Various *Aspergillus* species impart a mouldy appearance to leather goods and cloth fabrics in humid weather. *Aspergillus niger*, the common black mould, attacks exposed food and causes decay. *Neurospora* species, the pink bread mould, can cause much trouble in bakeries. *Geotrichum candidum* causes moulding of food containing lactic acid. The fungus also attacks pickles (forming white skin-like growth), ripe fruits of tomato, peach, water melon and musk melon.

Fungi also cause a variety of diseases of humans and animals. Various species cause diseases, the symptoms of which resemble tuberculosis and which also attack birds, cattle, sheep and horses. Some cause diseases of mucus membranes, skin, nails, lungs, beard, moustache, scalp of children and the gastrointestinal tract. Some species cause diseases of fish and fish eggs and may do significant damage to fish hatcheries.

Fungi need not be considered as agents of destruction only. They are useful to plants and human beings in many ways, Leaves, branches, logs, roots of trees, carcasses of animals and birds, waste foods, animal and human wastes, droppings of birds are all converted to simple organic and inorganic salts, adding to the fertility of the soil by the action of hundreds of saprophytic fungi in close association with bacteria and other organisms. Some fungi enter into symbiotic association with the roots of higher plants. Such association is called mycorrhiza. It protects plant roots from diseases, provides resistance against frost and drought and helps plants to establish themselves in infertile soils. Subterranean termites are known to cultivate certain fungi in their nests in the form of fungus gardens. A lichen is

Fruit-bodies of a fungus causing decay and disintegration of a tree-stump, × ⅓

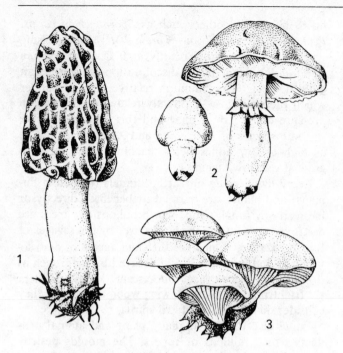

1 The true morel commonly called guchhi, *Morchella esculentua*, × 2/5 2 The common field mushroom, *Agaricus campestris*, × 2/5 3 The Oyster mushroom, *Pleurotus ostreatus*, × 2/5 These mushrooms are considered delicious by humans.

A poisonous mushroom, Fly agaric, *Amanita muscaria*

basically a self-supporting symbiotic association of a fungus and an alga.

Some fungi are good to eat. *Morchella*, the true morel and *guchhi* of commerce, is collected in Jammu and Kashmir and Himachal Pradesh and sold at fabulous prices. Truffles of commerce, which occur in France and Italy, grow underground. These are hunted with the help of dogs and pigs, who can smell them and dig them out. The 'mushrooms' commonly available in the market are the spore-bearing fruit-bodies of agarics, morels, truffles and puff-balls which develop from an underground network of hyphae which flourish as long as there is moisture and decaying matter to nourish them. They are fast-growing, appearing overnight in the rainy season, often in the same place year after year. They are gathered and offered for sale in the markets. Cultivation of mushrooms as food has developed into an industry in USA. and Japan and is becoming popular in India. Mushrooms as soup, stew, in pulao, as dry and curried dishes are popular. As they are rich in proteins containing essential amino-acids in good proportions, they are a good tonic. Some are highly poisonous. *Amanita verna*, the destroying angel, *A. phalloides* the deadly agaric and *A. brunnescens* are deadly poisonous. *A. muscaria* is poisonous if eaten in quantity. Individuals differ in their reaction to mushrooms—what is one man's meat may be another man's poison. *Amanita muscaria* and species of *Psilocybe* are hallucinogenic and produce ecstatic effects when eaten. *Psilocybe mexicana* is a mushroom used by Mexican Indians in religious rites.

Fungi are also used to make various antibiotics and other commercial products. *Penicillium notatum* and *P. chrysogenum* are sources of penicillin, the wonder drug. *Rhizopus nigricens* is used in the manufacture of cortisone. *Claviceps purpurea*, the fungus causing ergot of rye, is used for the preparation of a powerful abortifacient which is also utilized in controlling hemorrhage during childbirth. Yeasts, a group of fungi, are employed in breweries for the manufacture of alcohol with carbon dioxide as a waste product, and in bakeries the carbon dioxide is the important product and alcohol is the waste. The compressed and dry yeast cakes of commerce are made by pressing into cubes a great number of yeast cells together with inert matter such as a starch. Citric and gluconic acid are commercially manufactured by the use of black mould, *Aspergillus niger*. *Rhizopus nigricens*, common bread mould, is used commercially for the manufacture of fumaric acid. In Japan, *Aspergillus oryzae* is used to make sake, an alcoholic beverage concocted from rice. It is fungi which give flavour to cheese.

Fungi are thus important constituents of our microcosm without which life on earth would come to an end. While the losses due to activities of fungi are enormous, the benefits accruing from their activities to both plant and animal kingdoms cannot be underestimated.

S.S.

GAMARI (*Gmelina arborea*) is a medium-sized deciduous tree found scattered throughout the subcontinent up to 1300 m. Its leaves are mostly heart-shaped, 8–15 cm long. The yellow flowers are produced in profusion when the tree is leafless in spring, and are ornamental. The fruits are egg-shaped. The wood does not warp and therefore is in demand for tom-toms, musical instruments and dugouts. It is a fast-growing tree, and therefore popular in tropics of the world for its valuable timber.

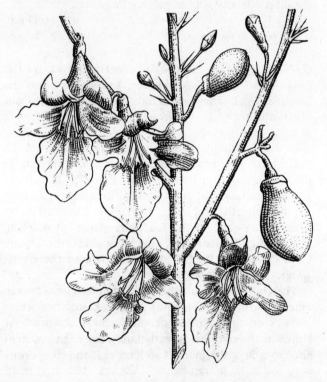

Gamari. Flowers and fruits, × 1

GAME BIRDS. Over 150 species from 8 widely differing avian families are included in the term Game Birds. They cover the Phasianidae (pheasants, partridge, quail and junglefowl), the Otididae (bustards and floricans), the Pteroclidae (sandgrouse), Anatidae (ducks and geese), the Charadriidae (snipes and godwits) and others.

The numerous species, both migrant and resident, vary widely in their habitats, ranging from arid plains to Himalayan snowy ranges, and from dense forest to open expanses of water and marshland; yet they share a marked set of characteristics. Primarily, the flesh of game birds is valuable as food. This characteristic puts them high on the preference list of their predators including man, who has hunted them from time immemorial.

As a result, extreme wariness is a distinguishing mark of all game birds, with an uncanny ability to sense and evade danger. So near-perfect is the camouflage of most species that when the bird freezes it is virtually impossible to spot, even on open ground. Only when it suddenly takes off with a whirr of wings, almost as if conjured out of bare ground, is it seen — and then, with its quick acceleration, it is already speeding away, from the shikari or stooping raptor, at a velocity unexpected in a bird so dumpy and short-winged — the characteristic build of many game bird species.

Since hunting has traditionally been popular among the upper classes it has been customary, from early times, to set aside reserves and to guard the game therein for the benefit of the privileged few. This has undeniably played a vital role in the conservation of game. The old tradition has now changed to suit prevailing ideas, and in most civilized countries this has resulted in a powerful conservation movement supported by a set of game laws applicable to all.

In India the enforcement of realistic game laws has become a matter of urgency. Due to continuous pressure from a fast-growing rural population, areas of scrub, pasture and marshland, the favourite haunts of the game bird, are shrinking rapidly. Another threat to game survival is the expertise of professional trappers catering to the demands from urban kitchens and restaurants. Prospects are bleak indeed unless authority can enforce the kind of management practices which have enabled other industrialized countries to build up and maintain a flourishing population of game birds.

Because of the many problems in the enforcement of our game laws, the ideal solution would be the establishment of a well co-ordinated countrywide agency with the single object of wildlife conservation and management.

The varied and widespread habitats of the game birds does, however, have a happier aspect. Unlike game animals which require extensive areas of undisturbed wilderness for their preservation, game birds can subsist

Bustard quail approaching eggs × ⅕

© O. C. Edwards

Painted sandgrouse at water × ⅓

© O. C. Edwards

Tibetan sandgrouse and chicks × ¼

© Sálim Ali

on small reservations, in areas of mixed cultivation and scrub and even in hedgerows between cultivated fields. The numerous species are each in their own way adapted to utilize specific sections of the food available, and thus make use of the complete gamut of resources of the habitat, ranging from seeds gleaned from stubble fields, through wild vegetation to insects and other small animal life. Moreover, the reproductive cycle of birds being short, usually annual re-establishment of game to optimum limits, even in badly run-down areas, is possible within a reasonably short time. The vast areas of the country capable of supporting healthy populations of game birds without interfering with agricultural activities holds great promise of increasing their numbers.

The fundamental task in game management is to ascertain the breeding periods and ensure undisturbed conditions by a strongly enforced closed season. A more difficult, and equally important, task is to understand the population dynamics of game species and aim at maintaining numbers to match the food resources of the habitats, which fluctuate markedly from season to season. An area can hold a much larger population during the lush months than during the lean season, and the optimum population for both periods has to be

estimated. The difference in the numbers for the peak and lean seasons is the harvestable surplus that may be taken during the open season by regulated hunting.

Obviously, over-hunting would deplete the stock. But under-utilization, apart from being a waste of a valuable resource, also has a deleterious effect. Insufficiently spaced nesting territories, paucity of food and cover and the resulting proneness to disease, are factors particularly damaging during the vulnerable breeding season. This is a pointer to the danger of imposing blanket or haphazard restrictions on all hunting in the facile belief that, regardless of other conditions, the less a game species is hunted the more it flourishes, and that a total ban on hunting is therefore the most effective way of preserving game.

D.J.P.

See plate 8 facing p. 97.

GANGA, the Ganges, rises in the Himalayas, on the Indian side of the Tibet border, and empties into the Bay of Bengal. The general direction of its flow is from north-north-west to southeast. At its delta, the flow is generally southward.

Ganga is usually mapped under this name from the junction of two headstreams at sacred Devaprayag in the mountainous part of Uttar Pradesh. The Alaknanda, the larger of the two headstreams, rises about 30 miles north of Nanda Devi; the Bhagirati flows from an ice-cave in the Gangotri glacier at 4000 m. Three other headstreams, the Mandakini, the Dhauli Ganga, and the Pindar also contribute to the river's water.

The Ganga then cuts through the outer Himalayas to emerge from the mountains at Rishikesh, after which it flows on to the plains from Hardwar onwards. At Hardwar the river has already fallen more than 3000 m and from here the plain is so level that the river drops only 360 m in its remaining 2200 km. Although there is seasonal variation in the river's flow, its volume increases markedly as it receives more tributaries and enters the region of heavier rainfall. From April to June the melting Himalayan snows feed the river, while in the rainy season from July to September the monsoon causes floods.

In Uttar Pradesh, the principal right-bank tributary is the Yamuna which flows past Delhi to join the Ganga near Allahabad. The junction of the two rivers is most sacred to Hindus and the festival of Kumbh Mela is held once every ten years, when over a million people bathe in the river. Tons, which descends from the Vindhya Range in Madhya Pradesh, joins the Ganga soon after the Yamuna. The left-bank tributaries in Uttar Pradesh are the Ramganga, the Gomati and Ghaghra.

In Bihar the tributaries on the north bank are the Gandak, and the Kosi, and its most important southern tributary is the Sone. After skirting the Rajmahal Hills to the south, the Ganga flows southeast to Farakka at the apex of the delta. Its westernmost distributary is the Hooghly, on the east bank of which stands Calcutta. The Hooghly itself is joined by the Damodar from the west, and by the Brahmaputra from the north near Goalundo Ghat. The combined stream, now called the Padma, joins with the Meghna river above Chandpur, after which the waters flow through various channels into the Bay of Bengal.

In the delta region, the Ganga and its tributaries and distributaries, constantly change course. The seaward prolongation of silt deposits from the river valleys covers an area of about 50,000 km² and is composed of repeated alterations of clays, sands, and marls with recurring layers of peat, lignite and beds of what were once forests. The seaward side of the delta has a stretch of swamp lands and tidal forests, called the Sundarbans.

The natural history of the Ganga is chronicled from very early times. Its fauna is mentioned in the edicts of the emperor Asoka and in writings of the Greeks and the Romans. Though somewhat fanciful these mention such animals as the dolphin, peculiar to the Ganga. The fauna of the Ganga shows three distinct zonal divisions—the cold-water fauna specialized for the fast-flowing cold waters before it emerges from the Himalayas, the fauna of the plains section, and the fauna of the estuaries. Each is adapted to its own peculiar situation; for instance, the fish of the hill section are provided with suckers to prevent them from being washed downstream with the flood waters, the animals of the turbid waters of the plains section of the river, such as the porpoise and the goonch or freshwater shark, depend mostly on their sense of touch for finding food. The estuarine fauna has in many instances a high tolerance of salinity variation. Particular mention needs to be made of some of the animals which though not exclusive to the Ganga are mainly known in relation to

it. Among these are the Gangetic Dolphin, a unique freshwater relative of marine dolphins, and the Gharial, or long-snouted fish-eating crocodile. Both are believed to be relict fauna of the mighty Indobrahm of the Pleistocene era, which broke up into the Indus and the Ganga. The dolphin and the gharial now occur in both river systems. The Ganga also has several fishes with adaptation for breathing air, necessary for fishes likely to be stranded by flood waters. Another adaptation is for aerial vision, as seen in the freshwater MULLET, a useful adaptation for a fish which spends most of its time on the surface and is likely to be stranded. Another group which is numerous in species and numbers is that of freshwater turtles.

GARDEN HARVESTMEN (Phalangids) are arachnids with a small flattish, oval body 6 to 9 mm long and four pairs of thin long legs measuring 60 to 90 mm long. Because of their long legs they are called Daddy-Long-Legs. They are often mistaken for spiders from which they sharply differ in the absence of silk glands and spinnerets and in having their abdomen segmented and fused with the cephalothorax.

They appear in huge assemblages during the harvest season, whence their name 'garden harvestmen.' It is a sight to see hundreds of them clinging to the herbage with their legs and rocking their bodies up and down in unison. At the slightest interruption they break the assemblage and scurry helter-skelter. They are good hunters. With their long legs they speedily span over leaves and grass and outmanoeuvre their prey—mainly insects. They are generally darkish grey but in many species the colour matches with the surroundings.

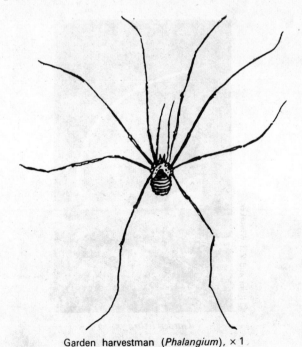

Garden harvestman (*Phalangium*), × 1

They also feed on insect larvae, spiderlings, mites and their own young ones. They not only suck the juice of their prey but also chew up and swallow solid particles. Juices of fruits and vegetables are also sucked by them.

During the breeding season males fight among themselves to win the female. The pregnant female spins no cocoon for her eggs but lays them in crevices on the ground, under stones or fissures of bark. In cold and temperate countries this takes place towards the end of autumn, after which the adults die. The eggs survive during winter, hatch out in spring and become adults in summer. In this country hot summer and heavy rains reduce their population. They are however found in plenty during the months August to March. Common genera are *Phalangium* and *Liobunum*.

GARDEN LIZARDS. The lizards of the *Calotes* group are characterized by the presence of crests on the neck and back, a dewlap in males and an extremely long, slender tail. The commonest, the Indian garden lizard *Calotes versicolor*, is found on bushes and shrubs all over the country. This harmless lizard is called 'blood-sucker', for its trait of displaying scarlet breeding colours. Though they may station themselves on tree-tops to heights of 8 m and more, they are quite at home on the ground. They feed mostly on insects, larvae and spiders. The female lays about 12 eggs which hatch two months later. The average length of a *Calotes* is 370 mm, of which 210 mm makes up the tail. Of the 25 species of *Calotes* occurring in southeast Asia, nearly 15 species are found in India.

Male Sita's lizard with dewlap displayed × ²/₃

Green garden lizard × ¹/₃

Garden lizard × ¹/₃

Garden lizard laying eggs × ¹/₃

Sita's lizard *Sitana ponticeriana* found in the drier parts and coastal plains of India and Sri Lanka derives its specific name from Pondicherry where it abounds. It looks like a small *Calotes*. However, it has no dorsal crest and has a very long, slender tail. The male of *Sitana* has a bright blue dewlap which is repeatedly fanned out when the animal is excited, such as during the breeding display. It has five toes on its front feet and four on the hind feet. This little lizard runs so fast that its front feet often leave the ground.

Sita's lizard preys mostly on ground insects. The female lays six to eight eggs in a small hole at the base of a bush. The eggs hatch in 50 to 60 days.

Two pretty garden lizards commonly seen in the South Indian hills are Horsfield's lizard *Salea horsfieldi* and the Anamalai lizard *S. anamallayana*. They are characterized by a dorsally flattened body and tail and large dorsal scales. Besides having dewlaps the males have pointed crests on the neck and back. They attain a length of 300–325 mm. A third species of *Salea* occurs in Assam.

See also LIZARDS.

GARFISH are represented in the seas of the Indian subcontinent by two families, Belonidae and Hemirhamphidae. The first family (suborder Belonoidei) comprises full-beak gars, i.e. fish having both jaws equally prolonged to form a beak armed with sharp teeth. Hence they are also known as 'needle fish' or 'long toms'. Their body is roundish and elongated like a rod, coloured sea-green or bluish green on the back, turning gradually to silvery white beneath. Active and gregarious in habit, they often thrust themselves out of water to skim the surface at great speed, with only the tail remaining submerged, and pounce on shoals of small fish. Generally they grow up to about 45 cm but larger species are reported to grow even up to 150 cm. Although eight species occur in our seas, *Strongylura crocodila* is most common. One species, *Xenantodon concila*, is a freshwater fish and is found in most of the rivers of India. Marine forms are normally residents of the open sea but sometimes stray into estuaries also for breeding, as is evidenced by the availability of their characteristic spawn in these waters, as a globular mass of jelly, light pink in colour with eggs each measuring 3 mm in diameter held together by sticky tendril-like outgrowths on the eggs and adhering to seaweeds. Gas bubbles entrapped in the mass sometimes keep it floating.

The other family, Hemirhamphidae, includes half-beak fishes, in which only the lower jaw is drawn out, the upper remaining short. Taxonomically they belong to a different suborder (Exocoetoidei) and are closer to the flying fish. Three genera occur in the Indian seas but *Hemirhampus* is the commonest. In habit they are similar to Belonids and are equally prized as food but the former are algae and plankton feeders while the latter are mainly carnivorous.

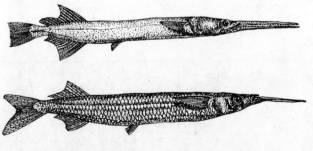

Upper, *Xenentodon concila* × ¹/₄
Lower, Half Beak, *Hemirhamphus* × ¹/₄

GECKOS. Found everywhere in India up to about 4000 m, most geckos are characterized by a soft skin, large unblinking eyes with a cat-like pupil, and expanded adhesive tips of the toes which allow them to walk on smooth, vertical surfaces. They are mostly nocturnal. The large genus *Hemidactylus* comprises 65 species and is widely distributed throughout the warmer parts of the globe. About 20 species occur in India, Burma, Bangladesh, Pakistan and Sri Lanka. Of these, the commonest species are the Brook's house gecko, *Hemidactylus brooki*, the common house gecko *H. frenatus* and the yellow-bellied *H. flaviviridis* of north India and Pakistan. The tree gecko *H. leschenaulti* of South India, Rajasthan and Pakistan is common in human dwellings.

The geckos of the genus *Cyrtodactylus* found in India and Pakistan have clawed toes and mainly dorsoventrally compressed bodies. The fat-tailed gecko, *Eublepharis macularius*, inhabits the arid areas of northwestern India and is strictly nocturnal.

Several geckos are restricted to hilly and forested tracts. Of these the commonest are tiny geckos of the genus *Cnemapsis* of South India and Sri Lanka. The Andaman day gecko *Phelsuma andamanense* found throughout the Andamans often enters houses. Another interesting forest gecko is the so-called 'flying gecko', *Ptychozoon kuhli* of the Nicobars, which can parachute small distances by means of the membranous expansions on the sides of its head, body, limbs and tail.

The tokay, *Gekko gecko* of northeastern India and Bangladesh, derives its name from its cry *toukday*. This red- or orange-spotted lizard is the largest of all Indian geckos. Its usual prey is cockroaches and other large insects, and other geckos. Occasionally it will take a small bird or rat.

See also LIZARDS.

GEOGRAPHICAL DISTRIBUTION. Animals and plants are not erratically distributed. Their occurrence in different parts of the globe shows an understandable pattern which has a general correlation with physiographical features of the earth. The branch of zoology dealing with the study of distribution of animals in space and the factors governing it, is known as zoogeography or the geographical distribution of animals.

Animals are mobile by nature and disperse from one area to another in search of suitable living conditions. Different animals have distinct ecological preferences; and since these conditions vary from area to area, the movements, dispersal and consequent colonization of new areas by animals are governed by a series of complex factors that are historical (distribution in time), geographical, climatological and ecological. The present day distribution of animals in space is the product of geological and evolutionary processes, and can be better understood in the context of the history of this planet and the origin, evolution and diversification of life on earth through aeons.

In studying the geographical distribution of animals, the general range of a taxon is considered and not its local distribution. When a large number of taxa of different groups of animals are investigated, the cumulative result fits in a general pattern defining the limits of faunal distribution in a region depending upon the faunal features.

The criteria used for delimiting zoogeographical regions are different for land and marine animals, because the set of factors that govern their respective dispersal are different. Distribution of marine animals is influenced by salinity, temperature, currents, availability of food and such other factors that may affect their mobility. The land animals, however, have to contend with different situations, but temperature and precipitation are important in their distribution because these two influence vegetation types—the primary source of food and shelter to all land animals.

The zoogeographical regions mentioned here are based on the distribution of land vertebrates, particularly mammals and birds. The limits separating these regions are not necessarily geographical, and are often determined by topographical features like high mountain ranges, rivers, climatological characters, vegetational diversity, etc. The factors that prevent dispersal of fauna from one region to another are called barriers, and these restrict animals to particular regions and isolate them from each other. The nature of barriers that affect the distribution of marine animals is naturally different from that of continental and insular landmasses. For the spread of the insular terrestrial fauna, particularly of oceanic islands, the surrounding seas act as effective barriers and consequently these faunas display a high degree of endemicity (see ENDEMISM).

For zoogeographical purposes the entire land surface

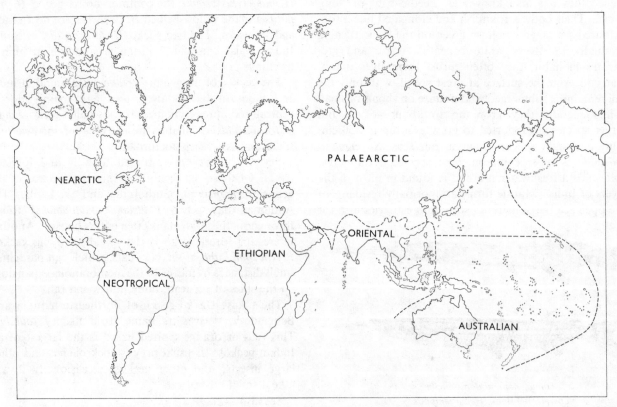

Fig. 1 Zoogeographical Regions of the World

of the earth has been divided into a number of Regions, chiefly on the basis of distribution of warm-blooded vertebrates. These are (Fig. 1), in the Old World: The Palearctic Region comprising Europe, North Africa, and northern Asia; the Ethiopian Region consisting of Africa south of the Sahara; the Oriental Region comprising tropical Asia; and the Australian Region that includes mostly Australia, New Zealand and Polynesia. In the New World the zoogeographical regions are the Nearctic comprising North America above the tropics, and the Neotropical that embraces tropical North America and the whole of South America. The Palearctic and Nearctic Regions are sometimes treated as a single Holarctic Region.

Birds are to a large extent adapted to certain climatic conditions, particularly in respect of food supply. In the northern temperate and arctic zones the winter season creates inhospitable conditions for birds, necessitating departure of practically the entire bird populations temporarily to warmer areas with favourable food supply. This seems to be at the root of the phenomenon popularly known as bird migration, the annual double-journey between two areas, the breeding grounds during the summer and the winter quarters (see BIRD MIGRATION).

The migratory as well as the routine bird flights are generally affected by certain barriers that limit their range of distribution. Sea, for example, serves as an effective barrier in the dispersal of many land birds. Yet there are some land birds which can circumvent the water barrier to establish themselves on far-off islands. On the island of Tristan da Cunha, separated from the nearest mainland by some 3200 km, there are five species of land-birds independently descended from mainland ancestors. Conversely, several American (mainland) birds are absent from the not-too-far-off West Indies.

High mountain ranges also act as barriers, both physical and zoogeographical, since such mountain ranges (e.g. the Himalaya) also serve as boundary between two regions.

The relation between temperature and distribution of birds becomes apparent as one sees the bird life in the tropics and the higher latitudes. The tropics have the most abundant and diversified bird life. The northern temperate zone has, however, lesser bird life, both numerically and in diversity, and in the subarctic and arctic areas true land-birds are not many, but there is an interesting concentration of many migratory waders and waterfowl. In the Southern Hemisphere there is a gradual infiltration of tropical forms into the South Temperate Zone, much less in magnitude than in corresponding northern latitudes. Antarctica has only sea-bird populations. In Australia and New Zealand there is a localized bird fauna.

The avifauna of the Indian subcontinent comprises mostly birds of the Oriental Region, with a sizeable proportion of Palearctic and Ethiopian elements. The formidable Himalaya mountain broadly forms the boundary separating the Oriental fauna from the Palearctic. However, the temperate climate in the Inner Himalaya has allowed incursion of Palearctic elements, and a good chunk of the avifauna there is more Palearctic than Oriental. The transition zone between Oriental and Palearctic Regions in the Himalayas varies from east to west corresponding with the latitudes and height. It lies approximately between 2400 m and 2700 m altitudes in the east, but somewhat lower in the west. In the Indian Subregion the areas to the west of the Indus valley in Pakistan and the northern part of the Himalayas are in the Palearctic Region, the rest being in the Oriental.

On the basis of evolutionary divergence among warm-blooded vertebrates, the Oriental Region has been subdivided into three subregions (Fig. 2).

1. *Indian Subregion*: covering the whole area south of the Himalayan foothills and Sri Lanka, west to the Indus valley and east to the Ganga-Brahmaputra river deltas.

This subregion has been further subdivided into

a) *Southwestern Province*: comprising the humid, tropical, heavy-rainfall area of the Western Ghats from its northern end in Khandesh south to Kerala and southwestern Sri Lanka. Though small, zoogeographically this is a very interesting area for it exhibits striking similarities of its fauna with those of the Indochinese and Indomalayan Subregions located far away.

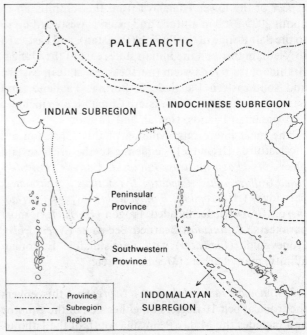

Fig. 2 Zoogeographical Subregions of the Oriental Region

On the basis of the similarities in the hill-stream fishes between this Province and the above-mentioned Subregions, S. L. Hora postulated the 'Satpura Hypothesis' to explain that dispersal of these fishes took place along a once-continuous mountain chain from the eastern Himalaya and Garo Hills through the Rajmahal Hills, Satpura Range and the Western Ghats to Kerala and Sri Lanka. This might explain the discontinuous distribution of such birds as the Large Goldenbacked Woodpecker, Great Hornbill, laughing thrushes, Spinetailed Swifts, frogmouth, some woodpeckers (*Hemicircus, Dryocopus* and *Picumnus*), Baza, etc. occurring in this Province and the Indochinese Subregion. Sri Lanka, which had been alternately joined to and separated from Peninsular India several times during the Pleistocene era, received faunal dispersal waves from the mainland, and due to prolonged isolation many endemic forms have also developed there. Some of the birds listed above, such as the Great Hornbill and the laughing thrushes do not occur in Sri Lanka, and strangely enough, the vultures (*Gyps* spp.), so common on the Indian mainland, are completely absent.

b) *Peninsular Province*: covering the rest of the Indian Subregion. Eastern Oriental avifauna is almost absent in this area, but there is a preponderance of Palearctic and Ethiopian elements. Among them many examples may be mentioned: the Rain Quail, Wood Snipe, Rufoustailed Flycatcher, Large Pied Wagtail, as Palearctic in origin, and the Painted Partridge, Bustards (*Choriotis, Sypheotides*), Indian Courser, the probably-extinct Jerdon's Courser, Grey Hornbills, several babblers (*Turdoides* spp.), etc. are with Ethiopian affinities.

2. *Indochinese Subregion*: this includes the southern aspect of the lower Himalaya from the foothills up to about 2000-2700 m altitude and extends westward up to to the Salt Range in the Punjab (Pakistan), and eastward to Vietnam through the Indian states east of the Ganga-Brahmaputra river system (northeastern states), eastern and southeastern Bangladesh, Burma, Thailand, and parts of southwestern and eastern China. It also includes the Andaman Islands, Hainan and Taiwan.

The most interesting avifauna of this subregion are the leafbirds (Irenidae)—endemic to the area, several pheasants (*Pavo muticus, Pucrasia, Catreus, Polyplectron, Gallus gallus, Lophura, Syrmaticus, Argusianus*), partridges (*Arborophila, Bambusicola*), mynas (*Acridotheres, Gracula*), Redheaded Trogon (*Harpactes*), most parakeets (*Psittacula*), Bearded Bee-eater (*Nyctyornis*), Honeyguide (*Indicator*) with unmistakable Ethiopian affinity, many barbets (*Megalaima*), etc.

3. *Indomalayan (Malaysian) Subregion*: this covers the tropical belt 10 degrees on either side of the equator and includes the Malay Peninsula south of the Isthmus of Kra, all the islands west of Weber's line, and the Nicobar Islands.

This Subregion is characterized by the complete absence of birds with Ethiopian affinities and of the Palearctic bird element of the temperate zone of the Himalayas.

B.B. & K.K.T

GEOLOGY. For many thousands of years the mind of thinking man has tried to understand how the planet Earth began and how its various features were formed, and even today the Earth poses a variety of problems both large and small, of the present and of the past.

Some 5000 million years ago there was only a cloud of matter in the form of dust and gas. Later, Earth gradually evolved, and still later oceans, continents, and a primitive atmosphere was formed.

The study of Earth involves not only events and processes which we observe now, but also those that have acted through long periods of time. We cannot separate the past from the present, and the present becomes understandable only in relation to the past.

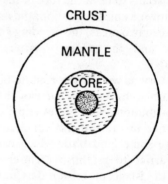

THE MAIN DIVISIONS OF THE EARTH, AS INFERRED FROM SEISMIC DATA

The crust is only a few kilometres thick below the oceans, though 30 or more kilometres thick below the continents. The mantle extends about half-way to the dense core.

The deep interior of Earth is inaccessible and no rays of light penetrate to enable us to see what lies below the surface. The deepest borehole is only about 8 km; just a pinprick on the Earth. But rays of another kind can penetrate and carry with them their messages from the interior. The Earth has been found to have elastic properties that allow movement set up at the source (focus) of an earthquake to radiate into the interior and to spread over the surface. At some distance from the focus, depending on the strength of the shock, the movement is no longer perceptible, but sensitive seismographs can record the waves that emerge at the surface. From such seismic studies we have learned that the Earth consists of a *core* surrounded by a *mantle* on which there is a thin *crust*. It is on this crust that man has his abode.

Every minute the face of Earth is changing. Old

landscapes are destroyed and new landforms are created. The panorama of Earth history is not something which is dead and static, but something which is very much alive and pulsating with throbbing vitality.

Destruction and construction are unceasing in their operation. Rocks crumble into dust and from this dust new rocks are formed. Hills were once considered everlasting and Earth features to be permanent, but we know now that they are worn down by the action mainly of water and wind.

Geology has brought new conceptions of time just as astronomy has revolutionized ideas of space. Hampered by human standards of time, it is difficult to visualize the vastness of geological time. A day of 24 hours is to man an important period of time and a person living for 100 years makes history, but during the whole of this time may be formed a sedimentary deposit only about 2 to 10 cm thick according to the coarseness of the material. The total sedimentary strata are nearly 150 km thick and one can imagine the time necessary for forming such a deposit. If we can imagine a geological clock in which the whole of geological time is represented by 24 hours, the total period of man's existence on Earth covers only about 3 seconds!

Geological time is the measurement of Earth's physical history. Geological processes are very slow, especially when compared to human history. Processes that appear to have no effect when observed from day to day often produce tremendous results when continued over periods of geological time.

Geology is the science which deals with the composition, structure, and character of Earth's crust, and with the processes that are continuously altering it. It tries to find out how the earth's surface features have developed, and the way in which each continent was built up. It is one of the characteristics of geology that it relies also on other sciences and interprets the earth in terms of knowledge of chemistry, physics, biology, astronomy, etc. But while drawing freely from other sciences, geology claims as its peculiar territory the rocky framework of the globe. It is mainly by the remains of plants and animals embedded in the rocks that the geologist is guided in unravelling the chronological succession of geological changes. He has found that a certain order of appearance characterizes these organic remains, that each great group of rocks is marked by its own special types of life, and that these types can be recognized, and the rocks in which they occur can be correlated even in distant countries, and where no other means of comparison would be possible.

One of the working principles of geology is that the present is the key to the past; that only as we understand the present, where everything is open on all sides to the fullest investigation, can we expect to decipher the past, where so much is obscure, imperfectly preserved, or not preserved at all.

While, however, the present condition of things is thus employed to understand the past, we must obviously be on our guard against the danger of unconsciously assuming that the phase of nature's operations which we now witness has been the same in all past time, that geological changes have always or generally taken place in former ages in the manner and on the scale which we behold today, and that at the present time all the great geological processes, which have produced changes in the past eras of Earth's history, are still existent and active. As a working hypothesis we may suppose that the nature of geological processes has remained constant from the beginning; but we cannot postulate that the action of these processes has never varied in energy. The few centuries wherein man has been observing nature obviously form much too brief an interval by which to measure the intensity of geological action in all past time.

In dealing with the geological record, as the accessible solid part of the globe is called, we have to realize that at the best it forms but an imperfect chronicle. Geological history cannot be compiled from a full and continuous series of documents. From the very nature of its origin the record is necessarily fragmentary, and it has been further mutilated and obscured by the revolutions of successive ages. From such an imperfect record, the geologist has to piece together, arrange in an orderly sequence, and reconstruct the history of the planet Earth.

Geology includes the study of the atmosphere, its movements, and its reactions with the mineral make-up of rocks; it includes the study of rocks and their relationship to each other; and the effect of water upon rocks.

Besides providing us with more scientific knowledge of the world we live in, geology has also helped to make available many of the raw materials that are basic to modern industry.

C.S.P.

GIANT GOURAMI (*Osphronemus goramy*), belonging to the family Anabantidae, is so called in contrast with a small fish known as Dwarf Gourami found in the northeastern region of India. The Giant Gourami was one of the food-fishes imported into Madras as early as 1865 from Indonesia, its country of origin, an earlier importation into the Botanical Gardens, Calcutta having failed in 1841. More consignments were imported into Madras in 1916 and bred successfully in fresh water. The fish was brought to Bombay in 1937 and after a few years, fingerlings were sent to Lahore also.

The Giant Gourami is marked by the possession of a labyrinthic accessory respiratory organ for aerial respiration in addition to its normal gills. Consequently, it is tenacious of life and can survive in foul water. It

Giant gourami with its nest slightly upturned × 1/16

Dwarf gourami with its air-bubble nest × 1

has a laterally compressed elevated body, teeth in its jaws and a pair of much elongated pelvic fins (figure). Growing to a length of about 60 cm, it weighs about 5·5 kg, though a maximum weight of 9 kg has been reported. Normally a fish of maritime climate, it thrives best up to about 400 m above the mean sea level in temperatures varying from 22 to 32°C, though records of its thriving at an altitude of 800 m are also available.

The fish is essentially a herbivorous creature feeding mainly on aquatic plant life; nevertheless, it is known to devour insects, frogs, flesh and small fish too. It is fond of lotus (*Nymphea*) leaves and avidly takes skinned bananas also. Besides these, it can also take cabbage leaves, turnip, tomato, beet and other vegetable refuse from garden and kitchen waste such as boiled rice. Feeding on such a variety of items, it can serve as a good biological control in weed-infested ponds and in turn produce good human food.

The Giant Gourami becomes sexually mature in the third year, the males being distinguished by the presence of a small hump on the nape. The fish is renowned for its habit of building a large nest out of aquatic weeds and other fibrous material. Both parents take part in the construction of the nest, the building material being carried in the mouth and fixed in the nest, just as crows and sparrows do. The egg, about 2·4 mm in diameter, is lemon-yellow in colour, glistering remarkably because of a quantity of oil at the upper pole. When the eggs hatch out after a couple of days, the larvae remain attached to the weeds in an inverted position (figure) and slowly develop into fry in about ten days. During this period, when the eggs hatch out and larvae develop, the parents keep careful guard near the nest and drive away any intruders.

Being comparatively a slow grower, the Giant Gourami takes about eight to ten years to reach maximum size. Its longevity is not clearly known but may be about 20 years. Being a natural breeder in ponds, omnivorous, and good as food, it is a welcome addition to our fish fauna.

The Dwarf Gourami, *Colisa lalia* of the same family,

is a beautiful aquarium fish, especially the male, which is hardly 5 cm long and smaller than the female. It is more colourful, its deep body striped with oblique bands of vivid red and deep blue-green. The fins, which develop brighter colours during the breeding season, are red and blue. Both sexes are credited with the habit of building nests of bubbles blown through the mouth and covered with saliva (figure). Eggs are laid underneath the bubble nest and taken care of by the male, who prevents the female from approaching the eggs and larvae.

In nature, the fish inhabits small puddles and ponds covered with weeds and subsists on small crustaceans such as *Daphnia* and *Cyclops* and insect larvae including those of the mosquito. It is a partial air-breather like its greater cousin.

GIANT PANDA (*Ailuropoda melanoleuca*). The Giant Panda does not occur within the zoogeographic regions specifically covered by this encyclopedia but through the publicity of the World Wildlife Fund it has become a well-known and recognized animal both as a symbol of the need to preserve many of the larger and more spectacular of the earth's vanishing creatures as well as for its attractive appearance and behavioural characteristics.

The Giant Panda is confined to a very limited region of montane bamboo-clad forest in western Sichuan (formerly Szechwan) Province of China, in the region around Lake Hzinlu and the Yunwa Mountains, at elevations between 2500 and 3500 m.

In appearance the Panda is a thick-set bear-like animal, adults weighing up to 250 pounds and being snow-white in colour except for black limbs, two spectacle-like black eye-patches and upstanding ears and a broad band over the shoulders. The Chinese call it *Pei-hsiung* or white bear. It was first discovered by the outer world when the famous French naturalist Père Armand David brought a live specimen back to France in 1869. Its affinity with other mammals has puzzled scientists and at one time it was thought that its nearest relative was the Lesser Panda which inhabits the montane forests of Nepal and Assam. More recent researches indicate this animal's close relationship to the true bears, though unlike the latter, which are basically carnivorous, the Panda is almost wholly herbivorous and has adapted itself to subsist almost entirely upon the young shoots of the bamboo *Sinarundinaria*. Such fibrous and tough food has led to the development of different cranial characteristics from that of the true bears (to enable the attachment of stronger chewing muscles), as well as huge tuberculated grinding molars and a sixth 'thumb' in the shape of a sensitively developed pad opposing the first digit which enable it to grasp the thinnest bamboo stems in a human hand-like manner.

When the international WWF was created in 1961, its founders chose the Giant Panda as their symbol because it was already a much loved and popular zoo exhibit besides being extremely rare and in danger of extinction. Latest estimates place the total world population between 400 and 1000 individuals. There is no doubt that this animal in the twenty years of WWF's existence has played a more than symbolic role in helping to enlist the support of the current one million regular known donors, as well as the $55 million which have so far been collected by WWF and subscribed towards nearly 3000 different conservation projects. It is very fitting therefore to be able to record that the Chinese Government have launched a long-term project in collaboration with the WWF to try and study the environmental needs of the Giant Panda so as to ensure its future and ultimate survival; that a one billion dollar project has already been launched in China for building a suitable research centre, and that Dr George Schaller (a contributor to this encyclopedia) has been put in charge of the research project.

Like all the bears the Giant Panda gives birth to undeveloped tiny offspring. The Peking Zoo captive bear produced a baby which weighed only five ounces, but which has since grown up in a normal and healthy way. They are believed not to hibernate in winter but to be able to subsist even during snowfall upon bamboo twigs. They can climb trees well and the latest information gained by Dr Schaller through radio-tracking of two free wild individuals shows them to be remarkably active also during the hours of darkness.

GIR. The Gir Sanctuary for the Asiatic Lion was established in September 1965. It is in the District of Junagadh in Gujarat 90 km east of Keshod airport. It is well connected by road and rail. Its total area is approximately 1400 km² and consists of undulating low hilly country with elevations ranging from 152 to 530 m above sea level. The highest peak is Mandivala.

The main species of trees are teak, salai (*Boswellia serrata*), palas, fig, and thorn forests of acacia and ziziphus.

Besides lions, other mammals to be seen are panther, sambar, spotted deer, nilgai, fourhorned antelope, chinkara, boar, hyena, jackal, langur, porcupine and the blacknaped hare. There are over 100 bird species including the Painted Sandgrouse and several species of raptors. Being on the main north-south migratory route birdlife is particularly rich in the winter months.

Steve Berwick and the herbivores in the Gir

Because it is the only area where the Asiatic Lion still exists the sanctuary is of particular significance. The Asiatic Lion was once widely distributed and its range included the Punjab, Bengal, and a large part of the Kathiawar peninsula. The last lion killed outside Kathiawar was in 1884. Since then this species has been restricted to the neighbourhood of the Gir forest. Around the year 1900 there were possibly no more than 100 lions and it is largely due to the foresight of the Nawab of Junagadh that the species has been saved. A census taken in 1936 indicated a population of 289 lions. The estimate for 1968 was about 177 while by 1980 the population had gone up to about 210. The improvement in the status of the lion is largely due to the effective protective measures taken by the Government of Gujarat. The core area of the Sanctuary has been converted into a National Park and the illegal grazing by cattle within sanctuary limits has been greatly reduced and this has enabled the prey species of the lion to multiply. The Maldharis, the cattle graziers, and their buffaloes, have been moved out of the sanctuary limits and settled elsewhere. This operation has been recognized as one of the most important moves by any State Government in the interests of wild life.

Because the Asiatic Lion is now located in only one small area within the Gir Sanctuary it is obviously in a very vulnerable position and attempts are being made to translocate a few of these lions into another area so that a second home for this endangered species is established. A pair of lions was also taken in 1972 to the Jersey Wildlife Trust in the hope that they would breed in captivity.

See also LION.

271

GLACIERS. Glaciers, so named from the French for ice, are often called 'rivers of ice' because many are found in valleys and slip or flow down these valleys. In colder lands than India they may end at the sea—the great glaciers of Antarctica, where they cover most of the continent, end as cliffs of ice from which icebergs break off and float away. In our area glaciers start where snow collects at heights where it does not melt. This snow consolidates into ice whose weight and gravity cause it to fall in massive avalanches from cliffs or to move slowly down a valley at speeds ranging from one or two centimetres to more than a metre a day. Such valley glaciers will stretch to well below the heights where snow melts and end in what is called the glacier snout where the supply of ice from above is not enough to make further progress against the heat that melts it.

On its way off a plateau, down a valley, or round a corner, moving over uneven ground at speeds which are not the same in the middle and edges or the base and surface of the ice, openings in the ice known as crevasses are caused. If these are covered by new snow they can be difficult and dangerous to cross, and anyone walking over or climbing a snow-covered glacier should be roped to companions.

Walking on the terminal moraine of Khatling Bamak, in Garhwal. Another glacier is seen on the left

© H. Kapadia

The longest glaciers in our area are in the Karakorams where the Siachen stretches for 72 km. In India the longest is the Gangotri of 44 km ending in the famous Gaumukh. From an ice cavern hung with icicles issues the Bhagirathi which becomes the Ganga. Many people are familiar with the crevasses in the Khumbu Glacier in Nepal from pictures taken by parties climbing on Mt Everest. One of the easiest glaciers to visit is the Pindari beyond Almora.

Glaciers leave behind stories that tell us something of the history of our earth. They carry down rocks and other debris that has fallen on to them, and at their sides and where they melt this is left as heaps called moraines; they scrape the rocks over which they move, leaving scratches on them; they carve their valleys in a U-shape steeper sided than the V of a river; and they leave hanging tributary valleys joining the main valley above its floor. Where these are found we can tell that the area was once covered by glaciers in one of the ice ages.

There is plenty of such evidence well below the present ends of Himalayan glaciers. Peninsular India was too near the equator to be covered by snow and ice, but there are interesting plants and animals in the high hills of the southwest similar to those of colder latitudes which show that there too the climate was altered during the ice ages. The sudden extermination of the great animals whose fossils are found in the Siwalik Hills is thought to be further evidence of the cold of glacial ages.

GOBIES (Gobiidae) form a large group of small fishes mainly occurring in the shallow coastal waters, coral reefs, rock pools as well as in estuarine waters. Some live in purely fresh water also. Most of them are small, generally within 20 cm, and one of them, living in the Philippines, has the distinction of being the smallest fish in the world, hardly 11 mm (see FISHES).

Gobies are marked by their prominent eyes placed close together almost on the top of the head; two dorsal fins, large pectorals, and pelvics joined together to form a cup-shaped disc by which they can adhere to rocks or other substrata. Some of them can live out of water (see MUD-SKIPPERS) and possess strong muscular pectoral fins with the help of which they hop about. Many are brilliantly coloured to match the colourful environment in which they live.

Female gobies are usually smaller than the males and less colourful. Unlike most other fish, their eggs are oblong or in some cases even more elongated, and adhesive. Many gobies carefully select and clean with their teeth rocky surfaces or undersides of shells or tiles by removing the mud etc. for better attachment of the eggs by one end. A few small gobies lead a symbiotic life by living in the cavities of live sponges while others inhabit the gill-chambers of larger fishes. A blind goby is known to harbour permanently in the tunnels of burrowing shrimps. A freshwater goby (*Glossogobius guiris*), occurring in most rivers, lakes and ponds of peninsular India, grows to a maximum length of 30 cm. Pale yellow in general colour with black dots over the entire body and fins, it is attractive and is also considered good eating. It has a large mouth studded with sharp teeth and feeds on small fish, prawns and insect larvae. This goby becomes sexually mature when 12 cm long and breeds in ponds, and margins of lakes and rivers, during the early part of the monsoon. Many other gobies are found on both east and west coasts of India and other parts of the subcontinent including the Andaman and Nicobar Islands.

GOLDEN EAGLE (family Falconidae). One of our largest and most powerful raptors with a huge wing-span and fully feathered legs; dark brown with a white band across base of tail. Old birds have the nape and neck-hackles golden. Ranges over temperate Europe, Asia and N America, breeding in high mountains including Himalaya. Its natural food is pheasants, snow-cock, hares etc., but it can be trained to hunt large animals like gazelle and is esteemed 'the Falconers' Dream'. That it carries off human children is a myth.

GOLDFISH. Although we are familiar with the common goldfish kept as a pet in garden ponds or home aquaria, it is not realized by many that they have been selectively bred from dull-coloured wild ancestors. If left on their own in a pond without weeding out the undesirable dull-coloured throwbacks, in a few generations they will revert to their earthy-brown wild colour.

In China, where it originally belonged, fish-breeders selected only the coloured individuals out of the wild stock and crossed them over and over again for generations to obtain the desired colour patterns. Simultaneously, they also selected for breeding individuals having different forms of tails, body patterns, shapes of eyes and jaws and so on, ultimately to create as many as 120 new varieties which are being perpetuated by connoisseurs. Consequently goldfish can now be bought in many colours and shapes; apart from the typical golden form, there are black (called moor), white, silvery and multicoloured (called calico) colour strains. Depending on the development of fins, there are the *shubunkin* (ordinary tail), comet (with a flowing tail longer than the body), fantail (double, short tail), and veil-tail (long, flowing, double lobes).

Selective development of the eyes gives rise to telescopic-eyed goldfish, in which the eyes protrude from the sides of the head, and the bubble-eye, which has a swelling below the eye looking as if water has accumulated below the skin. In the celestial goldfish, the eyes are turned upward, so that the fish appears to be perpetually staring at the sky.

Then there is the *oranda* or lion-head, in which the skin around the head develops wrinkled, swollen nodules appearing like a stylized lion's mane. In the pearl-scale, the white body has its scales protruding like roof tiles, while the pompom has two rattle-like protuberances hanging below the nostrils.

Young goldfish are usually dull brown in colour, and the other colours develop only after it is six months old.

In the breeding season, the gill-covers of the male are covered with prickly tubercles, while the body of the female becomes greatly swollen with eggs. The fish are capable of adapting themselves to different climates and are now thriving all over the world, endearing themselves because of their fantastic shapes, sizes and colours. In very cold weather even if a fish is frozen solid in a garden pond, it can be brought to life after warming to normal water temperature.

GOLD MOHUR (*Delonix regia*). The tree grows up to 15 m and has an umbrella-shaped crown with feathery foliage. When in bloom in early summer it is a cloud of gorgeous crimson flowers. The individual flowers have four scarlet spoon-shaped petals and the fifth (called 'standard' by botanists) is variegated with gold and vermilion. It produces sword-like woody pods, 60 cm long. It is a fast-growing tree of Madagascar, where it was discovered over a century ago. It has spread to all frost-free areas of the globe, but has disappeared from the wild in its native Madagascar and is on the endangered list as a wild plant.

Gold Mohar. Shoot with flowers and leaves, × 2/3

GOOSE (family Anatidae). Geese are aquatic birds larger than ducks but smaller than swans, with webbed feet for paddling and a comb-like fringe alongside their flat bills for straining food particles from water. Their wings are adapted for swift and long-ranging flight. There are 14 species of worldwide distribution, many of them migratory, nesting around the Arctic Circle and migrating to Asia and Africa in winter. Their spectacular migratory flights in wedge formation are a delight to watch.

Of the seven species recorded in our area, the Greylag (*Anser anser*) and the Barheaded (*A. indicus*) are common in the north, wintering on the larger jheels and rivers, but rare in the peninsula. The Greylag, ancestor of most domestic varieties, breeds from Asia Minor to Kamchatka; the Barhead, believed to be endangered, breeds in Ladakh and Central Asia around high-altitude lakes. They nest and feed gregariously. Their food is mainly vegetarian: young shoots of winter crops etc., obtained by grazing in fields and marshes; also some aquatic vegetable matter. Both species are very wild and alert in their winter quarters, where they are eagerly sought after by hunters.

GORAL. These are a group of goat-like animals with close affinities to the true antelopes, which have been placed in one sub-order, Rupicaprinae. They are all inhabitants of mountainous areas often at lower elevations and in areas of higher rainfall, but always in terrain characterized by rugged cliffs and steep precipitous slopes. The Takin and Serow of Southeast Asia, the Chamois of Europe and the Rocky Mountain Goat of North America all belong to the same group.

Two subspecies of Goral (*Nemorhaedus goral*) occur within our limits. The Grey Goral (*N. g. goral*) is found in the northwestern outer hills from Swat and the

Margalla range in Pakistan across to Kashmir and Himachal Pradesh. The Brown Goral (*N. g. hodgsoni*) occurs in Nepal, Sikkim and Assam.

Gorals have small conical backward-curving horns (10–12 cm only in length) of almost equal length in both sexes. They have large deer-like ears and a naked moist nose-pad, but their short legs and powerful hindquarters are more characteristic of the true goats. Grey Goral have brindled greyish fur with a conspicuous white throat-patch and a white patch on the lower jaw. The Brown Goral has a rufous brown coat and a black spinal stripe and an indistinct black stripe up the hind portion of the thigh. Adults stand about 66 cm at the shoulder and weigh up to 30 kg.

Goral often live in regions close to human settlements because they favour lower elevations and areas with some vegetative cover. They survive because of their secretive habits, their ability to lie concealed in some crevice between boulders and to confine their feeding periods to the hours of darkness. Where not disturbed however they have been observed grazing during the day and if by chance a person actually stumbles upon their daytime retreat, they will bound away in zigzag leaps, soon disappearing behind some bushes or rocks. When disturbed or sighting danger they emit a loud staccato sort of sneeze and continue to repeat this warning call long after they have retreated to inaccessible safety.

Females reach sexual maturity at two years of age and produce one young, rarely twins, after a gestation period of 240 days. In the northwestern Himalayas the young are born from mid-April to early May. They are not very gregarious and usually graze singly but the young accompany their mother until the next offspring is born, at which time she drives them away.

See UNGULATES, MOUNTAIN SHEEP & GOATS for account of Serow.

GRASSES. Grasses are by far the most important single source of wealth in the world. They provide the bulk of the cereals which form food for man and cattle as well as herbage and dried fodder for wild and domestic animals. Grasses play an important role in conservation of soil and moisture, particularly in mountainous regions. Sugarcane and bamboos are also large grasses. Many grasses provide fibres, paper, thatching and building materials, essential and edible oils, alcohol, gums, medicines and many other useful products. They also support the production of dairy products, such as wool, meat and leather.

Economically, the most important grasses in India are the cereals such as rice or *dhan* (*Oryza sativa*), wheat or *gehun* (*Triticum aestivum*), millets [such as pearl millet or bajra (*Penisetum typhoides*), maize or *makai* (*Zea mays*), giant millet or jowar (*Sorghum*), barley

0   15   30 cm
0   6   12 in

Grey Goral sub-adult in summer coat

© T.J. Roberts, *Mammals of Pakistan* (Benn)

or *jaun* (*Hordeum vulgare*) and common millet (*Panicum miliaceum*)], sugarcane, *ikh* (*Saccharum officinarum*) and bamboos.

In rural areas, many lesser important millets such as Finger Millet or *ragi* (*Eleusine coracana*) and *kodo* (*Paspalum scrobiculatum*) are also grown. Grasses of comparatively restricted cultivation are finger grass (*Digitaria cruciata* var. *esculenta*) in the Khasi Hills and the oat or *jai* (*Avena sativa*) in northwestern Himalaya.

Among the wild fodder grasses, thatch grass *siru* (*Imperata cylindrica*), *kans* (*Saccharum spontaneum*) and *dub* (*Cynodon dactylon*) are the commonest in the plains. In moist regions in the plains, species of *sandhor* (*Bothriochloa*), *anjan* (*Cenchrus*), *bamna* (*Chloris*), *dub*, *makra* (*Dactyloctenium*), *khel* (*Dichanthium*), *sawan* (*Echinochloa*), *Eragrostis*, panic (*Panicum*), foxtail (*Setaria*) and *ula* (*Themeda*) and in drier regions and on the seashore, species of *Aeluropus*, *lappa* (*Aristida*), chloris, lemon grass (*Cymbopogon*), *dheb* (*Desmostachya*), *dhanua* (*Ischaemum*) and *Sporobolus*, etc. are more common.

In moist hilly areas, species of white bent grass (*Agrostis*), brome (*Bromus*), rye grass (*Lolium*), napier (*Pennisetum*), canary grass (*Phalaris*), *Poa* and *Trisetum* occur commonly. In dry regions in the hills, species of brome, *sandhor*, fescue (*Festuca*), *Pennisetum* and *Stipa* are common.

Though several exotic species are being cultivated for fodder and have given good results, many indigenous species are also suitable for the purpose. Kikuyu (*Pennisetum*) has been introduced in many areas in the hills, particularly Sikkim.

Some species yield useful essential oils. The genera *Cymbopogon*, *Vetiveria* (see CUSCUS) and *Bothriochloa* are particularly important. The aromatic oils in these species are obtained by steam-distillation.

BAMBOOS are the main source of paper; in addition, species of *Themeda*, *Saccharum*, *sabai* (*Eulaliopsis*), *nal* (*Arundo*) and reed (*Phragmites*) have been shown to be suitable for the purpose.

Grasses are the main material forming lawns for houses, gardens or for sports. *Dub* is the commonest grass used for lawns. In areas of high rainfall, thatch grass and *kush* (*Chrysopogon aciculatus*) provide good lawns. *Axonopus compressus* is also suitable. But these grasses need to be regularly mown to avoid growth of tall culms and inflorescences.

Grasses have been widely employed for soil conservation. In drier areas, *Panicum*, *Lasiurus* and *Cenchrus* have been recommended for stabilizing sand-dunes.

It is not commonly known that grasses not only bind the soil, but even build it up. Certain fast-growing species like reeds build up soil by growing at a fast rate and by filling marsh lands.

On the coasts the SEASHORE PLANTS *Spinifex*

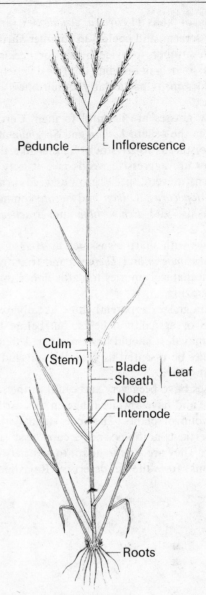

A grass plant, × 2/3

and *Thauarea*, help in stabilization of sand. *Brachiaria* and *Urochloa* are also good sand-binders.

Some grasses are medicinal. These are put to a number of other miscellaneous uses such as for making brooms and ropes, for matting and thatching purposes and for stuffing pillows. In our area, being predominantly agricultural and the population mainly rural, materials for thatching and walls of huts, as well as ropes and mats, are in daily use. The common grasses used for thatching are *Imperata*, *Andropogon*, *Saccharum spontaneum* and *Desmostachya*. Bamboo strips are the most commonly used material for the framework of walls. Culms of *nal* and reed, *kans* and species of *Themeda* are also largely used as reeds for the framework of huts.

Bamboos provide the bulk of the material for matting but reeds and *nal* are also used. *Sabai*, *kans* and *Desmostachya* are largely used for making ropes.

Reeds of khas (*Vetiveria zizanioides*) are used for making screens and coolers in summer. Fruits of *Coix* and a few others provide beads for necklaces. Many wild grasses help in breeding improved varieties of cereal and fodder crops, particularly for hardiness and disease-resistance.

A few grasses are harmful to man. Certain species such as in the genera *Lolium* and *Sorghum* are reported to be poisonous to livestock. Species like thatch grass and *kans* are aggressive weeds in a variety of habitat conditions. Several species grow as weeds in agricultural fields. *Alopecurus*, *Lolium* and *Secale* commonly infest wheat-fields, and *Echinochloa* and *Isachne*, the rice-fields.

Grasses with sharp awns such as *Heteropogon*, *Aristida*, *Chrysopogon* and *Themeda* and fruits of *Cenchrus* cause mechanical injuries to cattle either through contact or grazing.

Certain grasses apparently look like obnoxious weeds in lawns or agricultural fields; but before their large-scale removal, it should be remembered that their presence may be essential to keep the soil and that piece of land together.

Grasses have been very successful in the struggle for existence and have a wider range on this earth than any other group of plants. They can stand both cold and hot desertic conditions and occupy vast stretches of country. They are able to adapt to a variety of climatic conditions; from the hot deserts of Rajasthan (see DE-

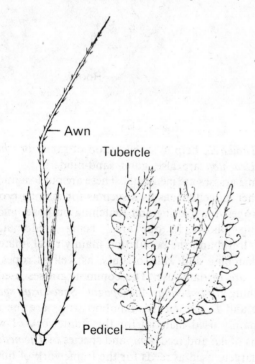

Spikelets showing awn (*left*) and teeth called tubercles on margins of glumes (right) which help in dispersal (× 8)

SERT PLANTS) to the cold deserts of Ladakh and alpine pastures in the Himalaya; they grow on sea shores, in water (see WATER PLANTS), in dry sands and in dense humid forests. They are abundant in open sunny places and form vast grasslands on mountain slopes as well as in the plains.

The very wide distribution of grasses is due to the numerous ways of dispersal of seed and vigorous vegetative reproduction. The dispersal of seeds by wind is the most common method. The light weight of seeds and presence of such structures as wings, silky tufts or feathery awns in many species enables the grass seeds to fly to great heights, and grass-seeds have been located in the atmosphere up to several thousand metres. Some grasses have awns, bristles or teeth on margins of glumes and lemmas or paleae to help in dispersal. In the spikelets of some grasses, small plantlets are produced in place of florets, i.e. proliferation takes place. The most common example is *Poa bulbosa*.

The earliest fossil records of grasses are from the Upper Cretaceous period. Certain fossil records earlier than this period and suspected to be grasses were based on impressions of leaves or small bits of inflorescences but they could not be identified with certainty. Grasses can be easily distinguished from all other groups of plants by their characteristic habit, leaves and inflorescence. Some grasses superficially resemble sedges, but differ in having jointed, round and hollow stems and leaves placed alternately on opposite sides of the stem.

Grasses belong to the monocotyledonous group of plants, i.e. they are characterized by an embryo having a single seed-leaf (cotyledon) and by their stems having irregularly scattered woody fibres; these stems do not increase in thickness with age. These characters are in contrast to the dicotyledons which have an embryo with two cotyledons, stems with woody fibres in a zone between pith and bark, and increase in thickness with age.

The family of grasses, namely the Gramineae (alternative name: Poaceae), is one of the largest families of flowering plants, having about 650 genera and 10,000 species in the world, and in number of individuals grasses are by far the largest group of plants. About 1100 species of grasses are found in the subcontinent.

The majority of grasses are herbaceous; only a few are shrubby or arborescent with woody stems such as bamboos. They are annual, biennial or even perennial in duration. Some bamboos have a life-cycle of over 100 years.

The erect ascending stems bearing an inflorescence are called culms. They are simple or branched. Some branches of the stem may become horizontal and develop below the ground (rhizomes) or trail on the surface of the ground (stolons).

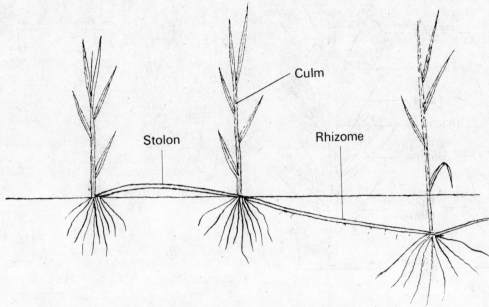

Stolon and rhizome connecting three grass plants, × ¹/₂ .

In annual grasses, all or most of the shoots bear inflorescences and rhizomes or stolons are not formed. In perennial grasses, a few or many leafy sterile shoots are present at the base of the plant and rhizomes or stolons are formed. Rhizomes are distinguished from roots by the presence of joints, and leaves which are bladeless, reduced and scale-like. Grasses having rhizomes are called rhizomatous, and those having stolons, stoloniferous. These qualities are useful in soil conservation.

Leaves are borne alternately in two rows on opposite sides of the culm. A well-developed leaf comprises leaf-sheath and leaf-blade. On the inner side of the junction of the leaf-sheath and leaf-blade, develops a small, usually membranous or hairy, structure called the ligule. The outer side of the junction is called the collar. The leaf-blade is mostly linear, much longer than broad and with parallel nerves.

The inflorescence in grasses shows more variation and complexity than in any other group of plants. It is made up of small units known as spikelets borne on one or more axes. The spikelets vary in arrangement, structure, size, shape, sex and in hairiness. The spikelets may be sessile on an axis forming a spike. Such spikes are sometimes solitary (as in wheat), paired (as in *kodo*), or digitate (as in Finger Millet or *dub*). Sometimes the spikelets are pedicelled and borne on an axis.

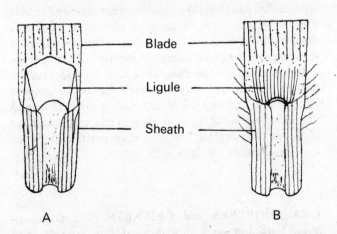

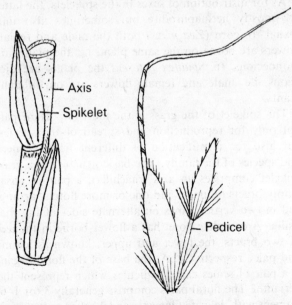

Grass ligules. A membranous, B hairy, × 10

Two sessile spikelets on an axis and a pair of spikelets: (*left*) one sessile awned spikelet, the other (*right*) a pedicelled awnless spikelet, × 8

277

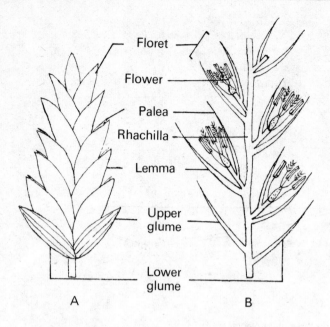

A SPIKELET WITH SEVERAL FLORETS

A A closed spikelet, × 10.
B diagrammatic interpretation of the same spikelet.

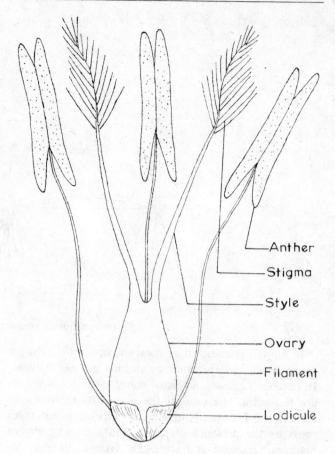

The grass flower showing lodicules, the male and female reproductive parts. The anthers are male and pollen-bearing and the female part consists of the stigma, style and ovary.
(× 100)

When pedicelled spikelets are borne on a branched axis, a panicle is formed (as in khas or in Lemon Grass). Various modifications and mixtures of the above conditions are met with. One interesting type is the false panicle where a group of spikelets is seated on a peduncle arising from a spathe-like sheath and terminating a branch of the panicle.

As for distribution of sexes in the spikelets, the latter are mostly hermaphrodite but sometimes also unisexual. In corn (*Zea mays*) both the male and female flowers are borne on the same plant, i.e. the plants are monoecious. In *Spinifex littoreus* the plants are dioecious, i.e. male and female flowers are on different plants.

The spikelet of the grass is the most important part not only for reproduction and spread of the species, but also for identification of different tribes, genera and species of the family. The basic structure of a grass spikelet comprises an axis (rhachilla), a pair of basal empty bracts (glumes) and one or more florets, arranged in two vertical rows on alternate sides of the rhachilla. A complete floret has a flower borne in the axil of two bracts, the lower and upper, known as lemma and palea respectively. At the base of the floral organs is a pair of scales called lodicules, which represent the perianth. The floral parts comprise generally 3 (or 1–6) stamens with delicate filaments and 2-celled anthers and a unilocular ovary topped by 1–3 styles and generally feathery stigmas.

The fruit is mostly 1-seeded. From the simple basic structure described above, the various parts of the spikelets show great variations, reductions and occasionally total absence.

In the majority of grasses, the lodicules swell, pressing the bracts apart to cause the flower to open. Cross-fertilization can then take place between male and female flowers on the same or separate plants. In extremely dry conditions even self-fertilization within an unopened flower may take place to ensure survival of the species.

Grasses are so common, so useful and so much a part of our life that they are taken almost for granted like sunlight and air. But in order to ensure healthy and continued regeneration of grasses on this earth, we have to avoid over-exploitation and take positive steps for the conservation of their rich variety.

S.K.J.

GRASSHOPPERS and CRICKETS. The grasshoppers, both long-and short-horned, the crickets, and their allies belong to the Order Orthoptera, of which more than 20,000 species are known from the world.

These ground-frequenting insects are commonly found all over the earth, except in the coldest regions, and are numerous in the tropics. They are both vegetarians or carnivores, active by day or night and occur on vegetation, on or beneath soil, in caves or in nests of ants. A grasshopper is well known for its prodigious power of leaping and the tremendous damage done by locusts needs no introduction.

Most Orthoptera are solitary insects, sometimes gregarious, and indulge in jumping either to escape or to launch themselves in flight. In some groups, the ability to fly has been lost. Most long-horned grasshoppers and crickets are nocturnal, well known for their habit of 'singing' at night, whereas the short-horned grasshoppers are diurnal. Many Orthoptera are very wary insects, possessing a keen sense of sight or hearing, and, as we have experienced as schoolboys, very difficult to catch by hand. The ones that are sluggish, unable to fly or inactive, are those that rely on their cryptic colouring to escape detection or hide in inaccessible places.

The habit of stridulation, carried out by long-horned grasshoppers and crickets by rubbing the specialized veins in the anterior parts of the forewings, and by the short-horned grasshoppers by this method and by friction between rows of teeth or ridges on the hind femora, is characteristic of this order of insects. Other kinds of these insects stridulate using other parts of the body like the tibiae, abdominal ridges, hind wings, or even the mouth-parts. As this habit is associated with sexual excitement and courtship, only the male stridulates or does it better than the female. The 'songs' produced by one species are different from those produced by any other species, and though two closely related species may be indistinguishable from each other externally, they 'play different tunes'.

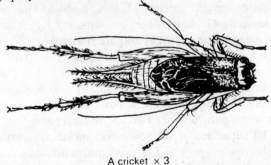

A cricket × 3

The most abundant kinds of grasshoppers and crickets are those that live exposed on the ground or on vegetation. To protect themselves from predators, many of them are cryptic in form or colour, resembling leaves, twigs, stones or tree bark. Some are semi-aquatic, living on plants growing in water, in damp soil along watercourses, lakes or tanks, and even sometimes swimming in water. The second group of these insects spend most of their time in burrows in soil or in rotten wood, or conceal themselves beneath stones, loose bark, fallen leaves, or other such debris. Thus, they are very drab in coloration, with their forelegs usually modified for digging. Another group that lives in caves has even less colour and small eyes. Lastly, subterranean forms are pale, sluggish creatures with vestigial eyes.

Most gryllacrids, tettigoniids and gryllids (long-horned grasshoppers) insert their eggs in leaves or in stems of plants and some cement them in rows to twigs. The acridids (short-horned grasshoppers) lay eggs in soil, but a few may deposit them in soft, dead wood, in stem pith or in tussocks of grass. The semi-aquatic species attach their eggs to water plants below the surface of water and the burrowing types lay them in special chambers within their burrows. The long-horned grasshoppers and crickets lay single eggs but the short-horned grasshoppers may lay them in batches of 10 to 200, and often in what are called 'egg pods'. The nymph that hatches out, through the earlier instars and later ones, resembles the adult except in some features, like possessing wing-stubs only. Some young nymphs of a few species may mimic insects like ants or tiger beetles, but in later stages (instars) look almost like the adult, but without fully formed wings.

These insects are usually of economic importance, in the case of locusts being highly injurious. Most pests belong to the superfamily Acridoidea, but a few long-horned grasshoppers, mole crickets and field crickets may assume pest status in certain conditions and geographical areas. Some species of short-horned grasshoppers, under favourable conditions of temperature and general environment, may assume the state of a 'locust'. In areas with special features like vegetation, soil, etc., which are highly conducive to the rapid multiplication of the grasshopper species, it multiplies immensely and is restricted to special favourable feeding and breeding areas like these. This forced 'concentration' then leads the 'solitary' phase of the grasshopper to assume the 'gregarious' or 'migratory' phase, when it is termed a 'locust'. These huge aggregations of this species then migrate to other areas before the food supply in their original habitat becomes scarce. This habit of flying away to a distant area may also help the species to disperse and to locate other environments suitable for its survival and growth.

There are three species of 'locusts' in India: the Desert Locust (*Schistocerca gregaria*), the Bombay Locust (*Patanga succincta*) and the Migratory Locust (*Locusta migratoria*). Most of these are usually found living in the areas surrounded by and including the Thar Desert, in the Sind province of Pakistan and in the Indian states of Rajasthan and Gujarat. The 'plague' periods are usually followed by periods of 'recession' when the population of these grasshoppers is small.

The migrations usually proceed outward from this area either across the country to Assam or southward and westward. The locust is a hungry insect and consumes its own weight of plant matter every day. It is estimated that a square mile occupied by a locust swarm that has settled includes 300 tons by weight of these insects! An average swarm may occupy 10 square miles, but in some cases the swarm may be as large as 300 square miles in extent!

The natural enemies of grasshoppers, tettigoniids and crickets are many, especially birds. Our Common Myna (*Acridotheres tristis*) is an efficient grasshopper eater, like its scientific name, and has even been released by man on Mauritius island in the Indian Ocean to keep down the destructive numbers of local grasshopper species there. This was one of the very early attempts at 'biological control' by man. Other insect enemies of these grasshoppers are digger wasps, robber-flies, ants, mites, and several different kinds of tiny wasp and fly parasites.

Coming to the classification, and hence to the diversity of these insects, the order Orthoptera is divided into two suborders: the Ensifera, containing the super-families Gryllacridoidea, Tettigonoidea and Grylloidea; and the Caelifera, comprising the Acridoidea, Tetrigoidea and Tridactyloidea. The Ensifera are the long-horned grasshoppers (each antenna with more than 30 segments), and crickets. The primitive Gryllacridoidea or cave crickets are usually pale to dark brown creatures, hiding by day in tree holes, under bark, beneath debris, or in hollow logs or burrows in ground, and come out only at night. More than 1000 species are known from the world and these insects are at home in both dry and humid regions of the earth. Some species live only in caves and a few are wholly subterranean, or live in soil. Many live on plants where they may construct shelters by rolling leaves around and securing them in suitable shapes with the aid of their own secretions. The common families are the Stenopelmatidae, Gryllacrididae and Rhaphidophoridae. In India species of *Oryctopus* and the curious *Schizodactylus monstruosus*, with its tegmina rolled up and living in deep burrows, are the familiar gryllacridids.

The largest superfamily of the Ensifera is the Tettigonioidea, or what are commonly called the long-horned grasshoppers, of which more than 5000 species have been described. Their antennae are usually longer than their body and the fore tibiae have the tympana. Many species are green and blend well with the colour of the vegetation on which they occur, but others are brown or speckled. Some are found on aquatic vegetation or even on soil. Most species feed on plant foliage but others are carnivorous and these grasshoppers are well hidden among their places of occurrence by their mimetic coloration. Besides the primitive family Pro-

phalangopsidae, with only 3 living species, there is only one other family, the Tettigoniidae, which includes 19 subfamilies. The Conocephalinae includes many species which are active in the day and frequent grasslands. Species of *Conocephalus* are common in India. The largest subfamily is the Phaneropterinae with the genera *Caecidia* and *Alectoria* being well represented in India.

The superfamily Grylloidea includes the true crickets that are familiar to every one of us by their incessant chirping in the night, especially in the monsoon season near forests. More than 2000 species are known which have the tympana in their fore tibiae that are used in stridulation. Many species have short wings or either further reduced or no wings. They are usually black or brown insects, sometimes being pale brown or yellowish in colour. They come out at night, spending the daytime in protected situations mentioned earlier. They seem to prefer living in humid environments like wet forests, and their chirping is a familiar sound during warm nights especially after a shower. As for food, most species are omnivorous, feeding on an assortment of things. The Gryllidae is the largest family, with the genus *Metioche* being abundant. Most species are found in cultivated areas also and some may be injurious to crops. Some of the common genera of Indian Gryllidae are *Brachytrypes* (which are often prey of species of the sphecid wasp genus *Sphex*), *Gryllus*, *Loxoblemmus*, *Liogryllus*, *Nemobius*, *Scapsipedus*, *Trigonidium*, *Cyrtoxipha*, *Oecanthus*, *Arachnomimus*, *Meloimorpha*, *Madasumma* and *Corixogryllus*. The Myrmecophilidae is a small family with peculiar crickets adapted to sharing ants' nests. Their eyes are reduced, they lack wings and live on secretions produced by the ants which mistake the mimetic insects for their own kind, and may even feed these imposters directly! *Myrmecophila acervorum*, *Ornebius guerini* and *Pteroplistus platycleis* are the common Indian species of this family. The mole crickets, or Gryllotalpidae, are familiar large insects that enter houses at night and are noticeable with their large fore legs that are modified for digging. They live underground in burrows which they dig. *Gryllotalpa* is a cosmopolitan genus of this family and one species, *africans*, is known to be a pest, accumulating seeds and storing them in its burrow underground.

The superfamily Acridoidea comprises the majority of all grasshoppers and crickets and more than 10,000 species are now known. These are the short-horned grasshoppers and locusts, and are mainly confined to the ground, especially in arid or cultivated areas, though some species may inhabit short herbs or shrubs as well. Most of these grasshoppers are active only during the day, feeding on vegetation. These Acridoidea, along with the Tetrigoidea and Tridactyloidea, comprise the suborder Caelifera. The most primitive family of the Acridoidea is the Eumastacidae, their members living

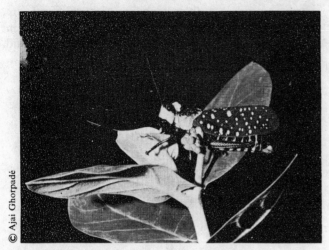

© Ajai Ghorpadé

Aularches miliaris *grasshopper on* Calotropis × ³/₄

© O. C. Edwards

Hooded grasshopper, Teratodes monticollis × 2

grasshoppers that lack the specializations of other families that have set them apart. Some species, like those of the subfamily Oxyinae, are associated with semi-aquatic vegetation, the species *Oxya velox* being common in paddyfields. The other subfamily, Cyrtacanthacridinae, possess the large species that fly powerfully. Most of the known 'locusts' belong to this subfamily. The common grasshopper genera belonging to this group are *Catantops, Hieroglyphus, Chrotogonus, Cyrtacanthacris, Teratodes, Epacromia* and *Acridum.* One grasshopper that is very widely known in India is the brilliantly coloured *Poekilocerus pictus* which is curiously attached to its host plant, *Calotropis,* which incidentally has other insects also that are specific to it.

The Tetrigoidea is a superfamily with only one family, the Tetrigidae, whose members are often termed 'grouse-locusts'. They are usually found on damp soil, in sandy or grassy terrain and are semi-aquatic, being able to swim well. They are believed to feed on algae and other cryptogams, or on wet mud from which they probably obtain dead or living vegetable matter. The common Indian genera are *Paratettix, Scelimena, Criotettix* and *Acantholobus.*

The last superfamily of the Acridoidea is the Tridactyloidea, which contains less than 100 species all over the world. There are two families, the Tridactylidae and the Cylindrachetidae. The tridactylids frequent the shores of water-sources and other bodies of water where they make galleries in sandy soil. Like the previous group, these insects feed on algae and other vegetable matter found in or on the soil. In India we have a few species of the genus *Tridactylus,* a couple of them being *variegatus* and *thoracicus.* The Cylindrachetidae are a peculiar group not found in our subcontinent, but the less than ten described species occur in Australia, New Guinea and Patagonia.

See plate 12 facing p. 145 and plate 26 facing p. 369.

GREBE (family Podicipetidae). Diving birds of lakes and ponds with dense silky plumage, vestigial tail, small wings and pointed bill. Some have crests. Their legs are placed far back, and the toes lobed. They feed on aquatic insects, frogs, crustaceans, fish etc. The sexes are alike, and they build floating nests of decomposing vegetable matter. Great Crested and Blacknecked grebes are winter visitors to the subcontinent; the Little Grebe or Dabchick is resident.

GRIFFON VULTURES are the typical Old World vultures of the genus *Gyps* feeding on large mammal carcasses. Their powerful hooked bill, long bare neck and bare head are well adapted for the purpose. Of the 7 species, four, namely the Fulvous griffon, the Himalayan griffon, the Longbilled vulture and the Whitebacked, occur in the subcontinent, the first only as a

on shrubs and grasses, common in dense forests. *Choroetypus fenestratus* is a common Indian species. The Pyrgomorphidae is a large group, with most included grasshoppers having a conical head. Some species like *Colemania sphenarioides,* called the Deccan Wingless Grasshopper, a pest of sorghum, have reduced wings. The other common species are *Aularches miliaris* and *Atractomorpha crenulata.* The largest family of the Caelifera is the Acrididae, which includes a variety of

winter visitor and the rest resident and also endemic.

They are gregarious out of necessity, since their food and other requirements are specialized and limited. Individuals locate carcasses from the air by direct vision by following the movements of their fellows or of other scavengers. A gathering of feasting vultures can dispose of a bullock carcass in an astonishingly short time. The Whitebacked is the commonest griffon in the plains; it is an efficient natural scavenger around villages and in forests. The huge Himalayan griffon nests on cliffs, whereas the Whitebacked prefers trees, the Longbilled using trees in some places and cliffs in others. Normally a single egg is laid and the young is looked after by both parents.

GULL (family Laridae). A grey-and-white gregarious aquatic bird seen flying about in shipping harbours, or riding the surf, or following off-shore fishing boats. It is largely a scavenger, feeding on dead fish and miscellaneous jetsam. Several species occur on our seaboard, nearly all as winter visitors.

See plate 13 facing p. 176.

GURJAN, *Dipterocarpus turbinatus*, is a lofty tree in the evergreen forests of Assam, Bangladesh, Andamans and Burma. Its bark is light grey and the leaves leathery and glossy, egg-shaped to lance-shaped, their stalk covered with a waxy bloom. The attractive flowers with pinkish-white petals are over 7 cm across. The spindle-shaped fruits are propelled by two papery wings. The wings help in dispersal of fruits which is important as Gurjan seeds lose their viability early. Gurjans are some of the largest trees of our forests, up to 60 m in height, and yield valuable reddish-brown timber and gurjan oil.

A gurjun tree growing near Chittagong, about 1850. The botanist J. D. Hooker thought it was the most superb tree he had seen in Indian forests. The sketch is based on one in Hooker's *Himalayan Journals*.

HABITAT SELECTION. Animals dwell in a very complex and heterogeneous world, with each animal species adapted to living in a rather restricted segment of it. Thus the Racket-tailed Drongo is adapted to live in denser, more moist forest in comparison with the Whitebellied Drongo. Moreover, where they occur together, the Racket-tailed occupies lower levels of the tree canopy in comparison with the Whitebellied Drongo. In South India the Bonnet Macaque prefers open countryside, the Hanuman Langur deciduous forest, the Nilgiri Langur thicker, wetter forest and the Lion-tailed Macaque evergreen forest rich in trees of the genus *Cullenia*. On the rocky sea coast oysters, mussels and barnacles occupy specific levels in relation to the fluctuations of the tide.

It is thus very important for any animal species to find and occupy a habitat suited to its particular ecological requirements and animals employ a variety of behaviour patterns to this end. For a moth with a colour pattern blending with lichens it is important to land on a patch of lichen when settling on a tree-trunk, and for a stick insect to arrange itself to look like a naturally sticking-out stick on a plant. In fact a number of experiments have demonstrated that animals do possess the appropriate instinctual tendencies for habitat selection.

Animals use a variety of clues to select appropriate habitats. For instance, a cockroach would move around in an experimental enclosure till its back comes into contact with a solid substratum. This enables it to get into crevices where it is safer from predators. Similarly wood-lice move around more and turn less frequently when they are in a dry region; they slow down and turn more frequently in a region of high humidity. This mechanism enables them to congregate in regions of high humidity. Larvae of mussels initally move towards light and hence open water, but at the time of settling they swim away from light towards solid substrata. Furthermore their settlement on the substrate is governed by its chemical nature.

A very interesting experiment on habitat preference was carried out when a number of Bonnet Macaque troops were trapped from the Bangalore city and released on a hill with scrub savanna about 40 km away. The city monkeys evidently preferred the urban to a scrub habitat and gradually moved back to the city!

But of course this habitat could have become available to the bonnet macaques only in the past few thousand years. Before that they must have been essentially forest animals, as many populations still are. Animals however have an excellent ability to adjust to new habitats. Thus as the terai forest has been brought under plough, the tiger has learnt to live in the sugarcane field. Sugarcane with its dense cover indeed simulates the old swampy habitat, and this is presumably why the switch has been made.

HAIL. Solid lumps of ice or balls of ice, when they fall along with rain, are called hail or hailstones. Their size varies from a few millimetres diameter to that of an orange or cricket ball, occasionally even more. They occur during severe thunderstorms, which are then called hailstorms. The clouds from which they fall have a great vertical extent, ten kilometres or more. Hidden in these clouds are large upward currents or updraughts with velocities of ten to twenty metres per second, powerful enough to tear the wings off a glider or even a large aeroplane. Birds are not victims. They are small compared to the size of the up and down draughts and are simply carried up or down. Also thunder and lightning frighten them and keep them away.

In most cases the hailstones consist of alternate layers of clear ice and opaque ice. The opaque layers have air bubbles trapped with the frozen particles of ice. In the upper half of the hail-forming cloud, a large amount of liquid water-drops exist at temperatures well below the freezing temperature, that is well below zero degree centigrade. These are called supercooled water drops. Near the top of the cloud there are supercooled droplets as well as solid ice particles. The initial ice particles grow into larger pieces of ice by the freezing of supercooled water droplets on them. A growth into large sizes is permitted by the strong updraughts of air within the cloud, which cause the ice particles to encounter more and more supercooled water droplets. Whether a clear ice layer or an opaque ice layer is formed depends upon the relative abundance of supercooled water drops and ice particles around the piece of hail. The hail eventually falls out of the cloud, when it has become so heavy that the updraught within the cloud can no longer hold it up.

Hailstones inflict considerable damage to standing crops and to fruits and fruit-bearing flowers on apple and mango trees. Cattle in the open may be battered to death by hailstones of large size. The glass in windows and doors may be smashed by hail.

Soviet scientists claim that they have an operational system of preventing the formation of hail by shooting silver iodide particles into threatening clouds.

Many places in the north, particularly near the hills, experience one or two days of hailstorms each year. They occur mostly towards the end of the cold weather, in February–March. Hailstorms are much rarer in southern India, where the warmer air temperatures below the cloud level serve to melt the hailstones unless they are sufficiently large in size when they leave the cloud.

HARES. What we call in Hindi a *khargosh* is not a rabbit but a hare. True rabbits (*Oryctolagus*) do not occur in the subcontinent. The Indian Hare, *Lepus nigricollis*, is widely distributed, from Pakistan to peninsular India, Sri Lanka and in Nepal, Sikkim, Bhutan and parts of Assam. Another widely spread species, *Lepus capensis*, the Cape Hare, has its eastern limits (*L. c. tibetanus*) in Baluchistan, Pakistan, and Kashmir. The Arabian Hare, *L. arabicus*, considered by some authorities to be conspecific with *Lepus capensis*, occurs in Baluchistan; the Woolly Hare, *L. oiostolus*, at higher elevations in Ladakh, Sikkim and Nepal and the Hispid Hare, *Caprolagus hispidus*, is found along the foot of the Himalayas from Uttar Pradesh to Assam. Very little information is available about this latter species, which is rare and local in distribution.

L. nigricollis is a non-social animal of open country but also inhabits low hills, taking refuge in *Euphorbia* bushes or *Saccharum* grass thickets near cultivated fields. It inflicts appreciable damage to irrigated crops such as wheat, gram and mustard. It does not live in burrows like the true rabbits, but hides under shrubs or tall grass. Sometimes it rests in the open in a self-cleared shallow scrape called a 'form'. Its nocturnal activity is bimodal, soon after dusk and before dawn. It grazes on grasses, or browses young leaves of shrubs and trees. In the desert, during summer hares congregate to feed on dub grass, *Cynodon dactylon*, in dried *nadis* (rain-water tanks). *Zizyphus* leaves and *Indigofera* spp. are preferred by them. The hare also debarks young trees during summer, probably to maintain the water balance in its body.

L. nigricollis dayanus breeds all the year round with peaks during monsoon and winter. The litter size varies from 1 to 4, the average being 1·84. The availability of green vegetation appears to have significant influence on its breeding activity, as larger litters of up to 3 or 4 young are encountered only during the monsoon season. Young born during the monsoon are more likely to survive than those born in the first half of the year.

Hares are timid animals and rely upon instant flight to escape predators. Foxes, cats, jackals and wolves are all formidable predators, besides man. Owls and eagles also sometimes prey on young hares.

I.P.

HAZARIBAGH. This national park situated in the Hazaribagh District of Bihar has an area of about 200 km², and is situated about 6 km north of the national highway connecting Ranchi and Patna. It has an altitude ranging from 300 to 600 m above sea level, and a few hills rise slightly higher. The annual temperature varies from 25 to 27·5 C and the rainfall is between 1000 and 2000 mm.

The forest is of the tropical dry deciduous type consisting mostly of sal with fairly dense bamboo growth in the western part. Nearly half of the geographical area is under forest cover.

There is a great deal of movement of wildlife between this national park and the adjoining forested areas, and the animals which occur here include tiger, hyena, wolf, leopard, chital, sambar and barking deer.

There are several forest rest houses in and around the national park, and the Divisional Forest Officer, Hazaribagh, West Division, can be contacted for accommodation.

In the winter months particularly, when the migrants arrive, bird life is plentiful in this national park.

HEARING and BALANCE. Sound is appreciated by a special organ of hearing, the Ear, though to a varying extent sound can be sensed as vibration by the skin through superficial sense receptors, which are also concerned with the sense of equilibrium.

Hydra, corals and other stationary animals have no organ of equilibrium, but possess receptors, like gravity receptors which enable them to assume a position with reference to the centre of the earth. In fishes, the ear-like Weber's organs are connected by a transverse canal and also connected to a series of receptors placed along each side of the body, the lateral line system, which is associated with vibration sense and equilibrium.

The sensitivity of snakes to airborne sounds is limited or absent, and the ear cavity in such legless animals is very small. Direct contact with the ground enables them to receive vibrations. The rattlesnake cannot hear its own rattle though the rattle sounds when the snake is excited. A snake's response to the snake-charmer's pipe appears to be due to the swaying movements rather

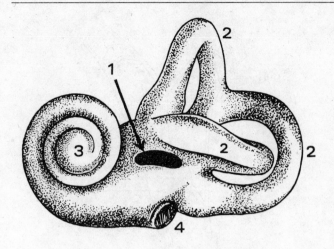

THE INNER EAR (enlarged)

1 Oval opening which engages foot of 3rd ossicle. 2,2,2 Semicircular canals in three planes at right angles to each other, for sense of equilibrium. 3 Cochlea, a spiral tube which registers sound vibrations. 4 Opening for auditory nerve to brain.

than to the sound or the gay colours of beads on the pipe.

Vibrations are appreciated by all animals irrespective of their mode of hearing. Amphibians receive airborne sounds and have well-developed organs of hearing. In vertebrates, the auditory membrane or ear-drum transmits sounds through a system of bony levers, the middle ear, to the internal ear in which an arrangement of

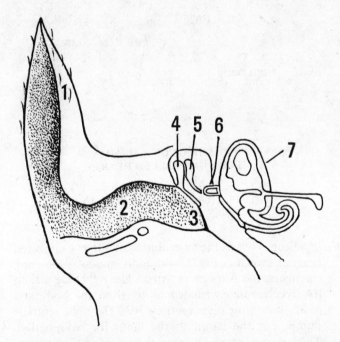

THE EXTERNAL EAR

1 Pinna. 2 External auditory canal. 3 Ear-drum or membrane 4, 5, 6. Ossicles which amplify and transmit vibrations to the inner ear (7).

sensitive receptor cells are connected to the auditory nerve. The inner ear also contains a system of three semicircular canals set in planes at right angles to each other, which contain a number of small granules floating in the canals. Any movement or change of position causes movement of these granules, which is perceived by sensitive cells within the semicircular canals. These sensations provide a sense of balance and of orientation with respect to gravitational force. Thus the inner ear is an organ both of hearing and of equilibrium. Hearing is further refined in mammals by the addition of a pinna or external ear (the ear in common parlance) which can be used as an ear-trumpet, and, in most species, can be turned in various directions to better appreciate the direction from which sound is heard.

The pinna or external ear is well developed in fast-running animals of open spaces (deer, horse etc.) and is rudimentary in aquatic or burrowing mammals.

In birds, hearing and equilibrium sense are highly developed. The air sacs of birds also serve as ballast organs for maintenance of equilibrium.

The range of hearing varies considerably in different species. A dog can hear sounds of considerably higher frequency than humans, and bats can appreciate supersonic waves.

HEART. The distribution of nutrients and oxygen to provide energy and other requirements of various parts of the body has to be maintained by continuous circulation of these substances. The nutrients and oxygen in primitive animals are contained within the body fluid which is kept in motion by pulsatile movements of the body as a whole. In somewhat larger animals these movements are supplemented by a pump or heart, a hollow muscular organ which pumps the fluid in one direction, to maintain a constant current of movement. The insect heart is a serie of cones situated on the dorsal aspect of the body, open at both ends. Its pumping action maintains a continuous unidirectional movement of the body fluid, which pervades the tissue spaces of the body cavity (Fig. 1). The worn has one dorsal and one or

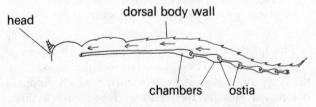

Fig. 1. DIAGRAM OF AN INSECT HEART

two ventral tubes (vessels), connected by a series of tubes. Five of these near the head end are contractile and function as primitive hearts (Fig. 2). The dorsal and ventral vessels have openings and branches along their course, from which the blood passes to the tissues.

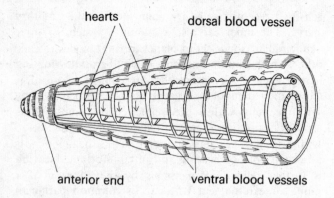

Fig. 2. DIAGRAM OF EARTHWORM'S HEART AND
CIRCULATORY SYSTEM

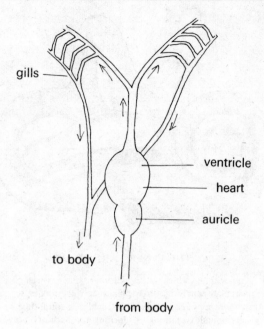

Fig. 3. DIAGRAM OF TWO-CHAMBERED
FISH HEART

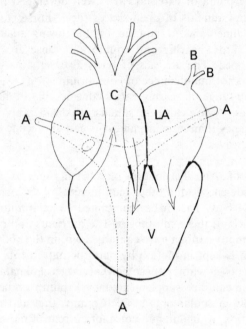

Fig. 4. DIAGRAM OF AMPHIBIAN HEART

A Blood vessels from body which empty from behind into
   right auricle (RA).
B Blood vessels from lungs which empty into left auricle (LA).
   Both streams of blood are mixed in the ventricle (V).
C Aorta carries blood from the ventricle to body and lungs.

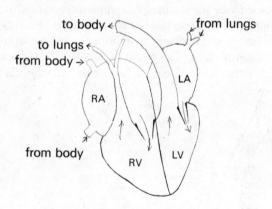

Fig. 5. DIAGRAM OF A MAMMAL'S
FOUR-CHAMBERED HEART

Right atrium (RA).   Right ventricle (RV).
Left atrium (LA).   Left ventricle (LV).

The heart of primitive fishes has two chambers as
shown in Fig. 3. The auricle receives the blood from
the body and passes it to the ventricle which pumps it
out to the gills for oxygenation. Frogs have a three-
chambered heart as shown in Fig. 4. It has two receiving
chambers called auricles. The right auricle RA receives
impure deoxygenated blood from the body while the left
auricle LA receives pure oxygenated blood from the
lungs. These two separate blood-streams pour their
contents into one ventricle V which now pumps out
mixed blood to the body. In birds and mammals such a
handicap is overcome by evolution of a four-chambered
heart as shown in Fig. 5, with more muscular receiving
chambers, the Auricles or Atria. The right side atrium
RA receives deoxygenated blood from the body and
passes it into the right ventricle RV. The right ventricle
pumps out the blood to the lungs for oxygenation.
The oxygenated blood from the lungs now returns to
the left atrium LA and is passed to the left ventricle
LV, which pumps it out to the body. There is no
mixing of the blood, the two streams remain separate.
Thus mammalian tissues get pure oxygenated blood.

HEDGEHOGS. These relatively primitive mammals belong to the order of Insectivora, and are mainly dependent upon an insect diet which they hunt largely by smell and during the hours of darkness. Hedgehogs have adapted to live in a comparatively arid environment and there are four species found in the subcontinent, of which one occurs in the plains of South India (the Pale Hedgehog *Paraechinus micropus*) extending up to the Rann of Kutch and southern Sind in Pakistan. Two species are associated with the dry rocky foothill country immediately west of the Indus valley—these are *Hemiechinus megalotis*, the Afghan Hedgehog, and *Paraechinus hypomelas*, the Migratory or Blanford's Hedgehog. In the Indus valley and extending into Rajasthan the Long-eared Hedgehog *Hemiechinus auritus* is widespread and it overlaps with the Pale Hedgehog in lower Sind Province, and the Rann of Kutch.

Hedgehogs are active largely during the monsoon season and warmer months of spring and summer. In the northwestern hilly regions they survive the cold dry winters by hibernation and cope with the extreme heat of their desert environment by spending the daytime in burrows or crevices which they are capable of excavating themselves. Even in periods of food shortage or excessively high temperature during the summer months they can relapse into a period of torpidity with varying degrees of metabolic deceleration. This phenomenon, described as estivation, requires further study. Hedgehogs are voracious feeders and have been recorded as attacking and overcoming venomous snakes, scorpions, lizards, arachnidae, coleoptera as well as subsisting upon ripe fruit and even, in Baluchistan, ripening melon crops. Any chance encounter with a ground-nesting bird affords them the opportunity to devour the eggs or nestlings. As is well known, the animal when approached immediately rolls itself into a tight ball which entirely conceals the head, limbs and vulnerable belly. Its dorsal region is covered with a protective armoury of spines. They produce 4 to 6 young in a litter, which at birth are blind and are largely naked, having only a few soft and scattered spines. These spines rapidly harden, and new ones sprout within two or three days.

If encountered undisturbed, during their night-time foraging, it will be noted that they travel with a surprisingly rapid trotting gait and that they snuffle audibly as they poke their nose into likely crevices in search of any hiding lizard or mole-cricket. Foxes and jackals have been recorded as being quick enough to kill hedgehogs before they can roll themselves into a ball. Hedgehogs are usually heavily parasitized by ticks.

See INSECTIVORA.

HENNA, *Lawsonia inermis*, is a thorny or unarmed shrub cultivated in gardens for its fragrant flowers, wild in Baluchistan and on the Coromandel coast. The

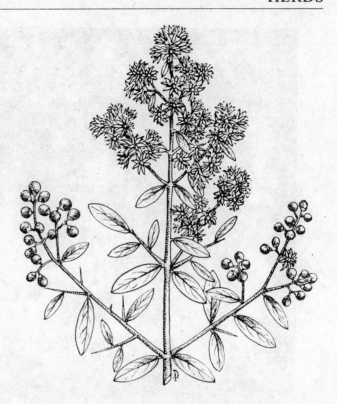

Flowers and fruits of Henna, × ³/₄

flowers are white or rose-coloured. The leaves are the source of a dye (*mehndi*) used for colouring finger-nails and hands. The nails of Egyptian mummies have been found to be dyed with henna.

HERDS. Herds is a term generally applied to the social groups of larger herbivorous mammals such as deer, buffaloes and elephants. In addition, we talk of *troops* of monkeys, *sounders* of pigs, *packs* of wolves, *prides* of lions and so on. But here I shall generalize the term 'herd' to apply to all social groups of mammals.

A variety of mammals occur in social groups. And they occur in social groups for five kinds of reasons: (i) by chance at a resource they all wish to make use of; (ii) because they find it more convenient to move long distances in groups; (iii) because they find safety in numbers; (iv) because they can exploit some resource more effectively as a group; and (v) because they can defend some resources better as a group against their competitors.

A good number of elephants move into the Bandipur Tiger Reserve on the Mysore plateau during the rainy season. During the day they may be seen grazing in herds of ten to twenty. Often at noon, but invariably at dusk, they visit the tanks to drink and wallow in water. Of an evening over a hundred elephants may congregate to do so at a single tank. These are merely chance aggregations of the first kind as are the aggregations of bats in large caves.

287

© M. Krishnan

Part of a Chital herd in a forest glade

Again this same animal, the elephant, although we know more about this from Africa than from India, forms large herds to move from one area to another in its seasonal migrations. There are many such well-known elephant migration routes in India too, though most are now disrupted, and animals used to congregate to undertake movement along these routes. This was perhaps because the presence of a number of experienced old animals was a great advantage while moving along such relatively unknown areas.

The elephants certainly find safety in numbers, not so much for the adults as for their calves. The bulls often live and wander by themselves, but the females always live in groups of relatives—grandmothers, mothers, aunts, daughters, nieces and their calves. Such a group is always vigilant, for although the adults may be safe, the calves can fall victim to tigers. At the slightest alarm the cows rush together to form a protective front for the calves. The cows may also nurse each other's calves.

A very different kind of social group that finds safety in numbers is that of smaller deer and antelopes such as chital and blackbuck. Herbivores of this size cannot defend themselves against predators and species that live in forested habitats—such as barking deer and fourhorned antelope—are in fact largely asocial. But species that live in more open habitats cannot hide themselves behind shrubs and trees, so they hide themselves behind each other! A stray chital is most susceptible to predation, so they come together. When confronted by a predator, their response is to bunch together. The animals on the periphery of such a bunch run the greatest risk of being killed, so there is a scramble to get to the centre of the herd and naturally enough the big stags, who have no chivalry, force their way into the very centre.

Fourthly, mammals come together to feed themselves more effectively. This is the case with chimpanzees whose fluid groups have no definite territory, but who call each other on discovering good sources of food. The pack-hunting carnivores such as wild dogs co-operate even more actively in bringing down victims larger than themselves.

Finally, mammals may band to defend a territory over which they feed against other groups of the same species. The spotted hyena in Africa is known to do this, as apparently are monkeys such as our Hanuman langur. At Mount Abu, Hanuman langur females fight with females of neighbouring troops at territorial boundaries and find definite strength in numbers.

HEREDITY: GENETICS. A living organism reproduces itself. While the progeny, generation after generation, consists of individuals recognizably of the same species, they are not all identical. Certain character differences appear at random while others are constantly repeated within groups or families.

Gregor Mendel, an Austrian monk whose interest was plant-breeding, set out to investigate these characters in peas. He selected two varieties of peas, one tall and the other short. On cross-breeding these two types he obtained in the first (F1) generation, all tall plants.

When these were further inbred, the second (F2) generation showed 3 tall to 1 short plants. However the tall plants were dissimilar, because, on further inbreeding, the F3 generation showed that (*a*) 1 of 3 tall plants bred only tall plants; (*b*) 2 of 3 tall plants bred mixed progeny similar to the F2 generation, i.e. 3 tall and 1 short plants, The F2 short plant continued to breed only short plants (Fig. 1).

Mendel repeated these studies for other characters like colour of flowers, etc., and was able to deduce from these studies certain general rules of inheritance. His

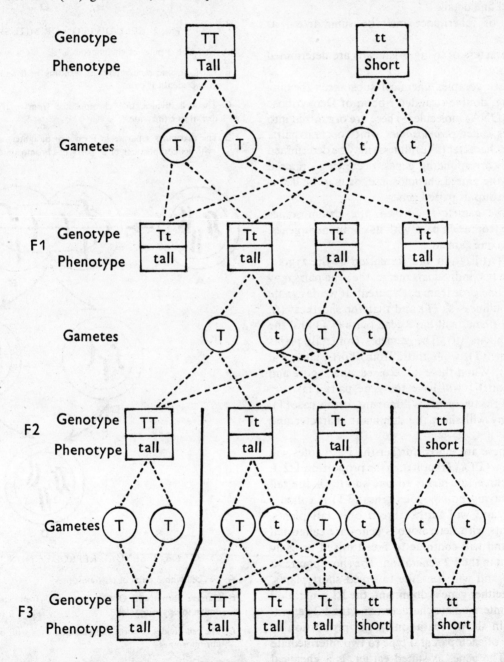

FIG. 1. MENDELIAN INHERITANCE

Crossing of tall and short pea plants.

work remained unnoticed until, about 80 years later, his records were found in the papers of the monastery where he had worked. They aroused great interest and several workers repeated his studies on a variety of plant and animal species. These workers established that the rules of inheritance in sexual reproduction, established by Mendel, were substantially correct. Mendel used only his eyes. Later workers were able to use more advanced techniques. Using an electron microscope, with a magnification of several thousand times, one can visualize not only the nucleus, but also individual chromosomes and fine details.

The rules of inheritance may be summarized as follows:

1. The characters of living organisms are determined by their *genes*.

2. Genes are complex chemical substances forming part of a long, double-stranded ribbon of Deoxyribose nucleic acid (DNA) molecules. These are organized into rodlike bodies, called *chromosomes*, which occur in pairs.

3. A single character (like tallness) may be determined by a pair of corresponding genes, occupying opposed positions on the paired chromosome, or may be determined by a group of paired genes.

4. When the gamete is formed, the chromosomes separate, one from each pair, with its component genes, going to form one gamete.

5. During fertilization and formation of the zygote, gametes from two individuals merge, the gene pairs now consisting of one gene from each parent. If we designate the gene for tallness as (T) and that for shortness as (t), the adult tall plants will have gene pairs (TT) and the adult short plants (tt). The gametes from tall plants will carry gene (T), while those from short plants will carry gene (t). When these are crossed, the zygote and first generation F1 will have the gene pairs (Tt).

6. If one of the two genes is dominant, in this case (T), all F1 progeny will show the dominant character and appear tall.

7. When these are inbred further the F2 zygotes will have gene pairs (TT), (Tt) or (tt), in the proportion 1:2:1, resulting in three tall plants to one short. Of the tall plants, one having homologous genes (TT) is called a tall *genotype* and will further breed only tall plants. Two tall plants with heterologous genes are called tall *phenotypes*, and will continue to breed mixed progeny like those seen in the F2 generation. The short plant is a genotype (tt) and will continue to breed short plants.

8. Where neither gene is dominant, the F1 generation will show an intermediate character, and the F2 generation will again show separation in the proportion of one genotype of each parental type to two intermediate phenotypes. The gene, as stated earlier, is a chemical complex forming part of a long double strand of deoxyribose nucleic acid. Each strand may carry hundreds

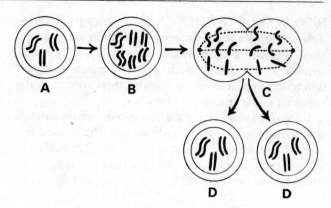

FIG. 2. CELL-DIVISION OR MITOSIS

A Cell with 3 pairs of chromosomes.

B Each chromosome replicates resulting in 6 pairs or 3 sets of two identical pairs.

C The cell divides, one chromosome from each pair being drawn to either side.

D The two parts separate to form two daughter cells, each with its complement of 3 pairs of chromosomes.

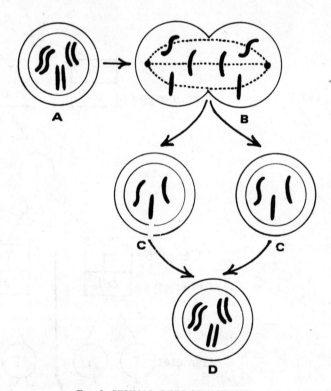

FIG. 3. SEXUAL REPRODUCTION

A Sex cell with 3 pairs of chromosomes.

B Gamete formation. One chromosome from each pair is drawn to opposite sides.

C The cell divides to form two gametes, each with one set of 3 chromosomes.

D Two gametes, of different origins, combine to form a zygote with 3 pairs of chromosomes. This further divides by mitosis to develop into a new individual.

of genes, and every gene has its own fixed position or locus on the strand.

The deoxyribose nucleic acids possess a special property. The nucleic acid forms combinations with elements in the environment and also induces formation of links between these new elements, so as to organize them into a chain complementary to itself, a mirror image. This mirror image gets detached and induces, in turn, a complementary chain, an exact replica of the original nucleic acid. The process is similar to preparing a block from a picture and using the block to reproduce a copy of the original picture. The nucleic acid thus exactly replicates the genetic structure of the organism.

During cell-division (mitosis), the paired chromosomes replicate themselves, resulting in four chromosomes. One pair goes into each daughter cell, ensuring that the genetic characters of the species are transmitted to the progeny (Fig. 2).

In the formation of the gametes for sexual reproduction, the chromosomes separate, but do not replicate. Each gamete receives one chromosome from each pair, half the genetic constitution of the parent. When this merges with another gamete, usually from a different individual, (and, when sexes are separate, from an individual of the other sex) they form a zygote, having a full complement of genes, which then develops into a new individual with characters from both parents (Fig. 3).

During separation and replication of the genetic strands, accidental changes or mutations can occur. 'Crossing over' where portions of the complementary chains are interchanged, or 'detachment', where a portion of one chain remains adherent to the other and is carried away with it, resulting in two unequal gametes, are not infrequent (Fig. 4). Mutations are also caused by viral infection, by exposure to physical agents like ultra-violet rays, X-rays, etc., and by certain toxic chemicals. Most mutations are harmful and individuals having them do not survive.

In organisms which reproduce asexually, a similar transfer of genetic material from parent to progeny must take place to ensure that the species characters are continued. However, the genes are not organized into identifiable physico-chemical configurations, and have not been studied to the same extent. Recent interest in genetic engineering, however, has led to intensive study of the genetics of, and genetic transmission in, unicellular organisms like *Escherischia coli*, the bacteria commonly found in the human intestines.

See also REPRODUCTION.

HERONS (family Ardeidae) are largish marsh-inhabiting birds with long bare legs and unwebbed toes. The neck is long and flexible, shaped like an S and tightly 'telescoped' in flight. They stand at the water's edge, head sunk between the shoulders, alert for prey which is impaled or forcepped with a lightning jab of the bill. Their food is frogs, fish and crabs. The only sound they make is a harsh croak. The Grey Heron (*Ardea cinerea*) has a long black occipital crest in the male. Other common members of this group are the Purple Heron (*A. purpurea*) and the Night Heron (*Nycticorax nycticorax*). The best known is the Pond Heron or Paddy Bird (*Ardeola grayii*)—buff-streaked earthy brown with snow-white wings and tail which flash into prominence only in flight.

HIGH RANGE. The High Range Game Preservation Association was formed in 1928 with the object of regulating shooting in the High Range. The area under the jurisdiction of the Association was approximately 556 km^2, the concession area of the Kanan Devan Hills Produce Co. Ltd. The Company was responsible for providing the funds and the membership came mainly from the many tea estates in the area.

The Association concentrated its activities on the Eravikulam/Rajamallay, and Karunkulam areas and Rajamallay was maintained as a Sanctuary. Muduvan tribal watchers were engaged to assist shikaris and naturalists and a hut was maintained in Eravikulam for the benefit of members. A sister association, the High Range Angling Association, undertook responsibility for stocking the rivers and dams with trout and now maintains a flourishing trout hatchery.

In 1969–70 when the Government of Kerala initiated steps under the Resumption of Surplus Lands Act the

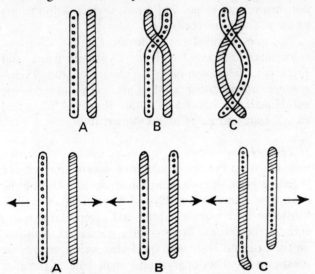

FIG. 4. CROSSING OVER AND DETACHMENT

A, B, C are three pairs of chromosomes. When drawn apart during gamete formation, they usually separate completely, as in A. Rarely, portions of chromosomes adhere to each other resulting in exchange of the overlapping parts, 'crossing over', as in B, or in part of one chromosome being carried away with its pair, 'detachment', as in C.

fate of the wild area hitherto looked after by the Association was jeopardized. However in 1975 the Eravikulam/Rajamallay area spreading over 88 km² was declared a wildlife sanctuary by the Government of Kerala.

The plateau has a mean elevation of 2133 m and contains many peaks including Anaimudi (2744 m), the highest south of the Himalaya. The Eravikulam area is unique in many ways for it is almost untouched by human activity and is perhaps the only place in India which has been totally free from cattle intrusion. The clear streams and the splendid ground cover are good indicators of such untouched areas. In 1977–8 Eravikulam was upgraded into a National Park. Though the overall control is now with the Government of Kerala the Association continues to manage the Park and has been accorded representation on the Indian Board for Wildlife and the Kerala Wildlife Advisory Board.

The most important species of this region is the Nilgiri Tahr, for the sanctuary has nearly half the known world population, the rest being distributed in the adjoining hill ranges. Other animals include tiger, sambar, wild boar, and elephant. There is rich birdlife in the shola forests.

HILL-STREAM FISHES.

The streams on the lower reaches of the Himalayas, Nilgiris and Garo hills are so torrential that they move even large boulders and wash away silt and soil, leaving rounded stones and pebbles lying in cold, clear, well-oxygenated water. Waterplants which we are accustomed to see in stagnant pools cannot take root in these streams, and the only plant life is the algal slime covering stones and rocks. Yet there are many species of fish which inhabit such streams, although their body-shape and behaviour undergo modification to ensure their survival.

Hill-stream fishes live on rocks at the bottom and on slopes. Their head and body are greatly flattened across the belly and only slightly arched along the back. Scales are greatly reduced on the under-surface, and in some cases disappear altogether.

The paired fins—especially their outer rays—are modified as organs of adhesion and the number of inner rays consequently increased. The outer rays become greatly thickened and flattened. The fins, which in other fishes are situated on the under-surface, are pushed outward and placed horizontally on the sides of the body, so that the chest and belly can be firmly applied to rocks. The fin on each side of the body is semicircular and shaped in such a manner that the two together form an adhesive disc.

The two lobes of the tail fin are unequal, the lower being longer and stronger than the upper. The root of the tail, where it joins the body, becomes long and narrow.

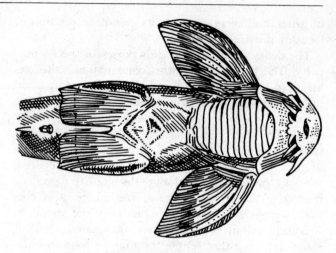

HILL-STREAM FISH × 1

Ventral aspect showing pleats of the adhesive organ and large paired fins

The mouth becomes adapted for browsing on the algae covering stones and also for adhesion. Instead of being a transverse cleft at the tip of the snout, it is displaced ventrally and backward. The jaws are strong, with sharp cutting edges, and are often covered with a horny layer. Barbels are reduced. The lips, together with the mouth, form a rounded sucker.

The eyes are displaced from the side of the head and are situated close together on the top. The gill openings are restricted to the sides. The air bladder is reduced in size and wholly or partly encapsulated in a bony case. Finally, the skin on the under-surface is not smooth but thrown into ridges and grooves which help the fish to cling closely to rocks.

Examples of hill-stream fishes are *Balitora, Bhavania, Garra* and *Psilorhynchus* in the Cyprinoid fishes, and *Erethistes, Glyptosternum, Glyptothorax* and *Pseudecheneis* among catfishes. Most of these are found in the sub-Himalayan streams of Uttar Pradesh, Bihar and Bangladesh and parts of the Nilgiri hills.

HILSA (*Hilsa ilisha*) is the only Indian fish which migrates from the sea or estuarine waters up rivers for breeding: such fish are called anadromous. During the southwest monsoon, i.e. July and August, shoals of these fish leave their marine habitat and ascend large rivers such as the Ganga, Brahmaputra, Godavari, Krishna, on the east and Narmada and Indus on the west coast for spawning. They migrate on a small scale into smaller rivers also. At one time it was believed that after breeding most of the adults return to the sea but observations during the late fifties indicate that at least one group, as well as the juvenile fish, remain in the vast estuarine waters of the Sundarbans (Ganga and Brahmaputra). These observations have not been confirmed in respect of other rivers.

The fish is known as Indian Shad, being very similar to the American Shad. *Alosa sapidissima*. Hilsa, which is a common name in the northeastern region of India, is called Palla on the northwest coast and is one of the large Clupeids (family Clupeidae) growing to a length of 60 cm and weighing 2·5 kg. The males are slightly smaller. They spawn at the age of 2 or 3 years. In the Ganga the fish used to ascend beyond Kanpur and beyond Agra in the Yamuna but although a channel has been made to by-pass the recent Farakka barrage, it remains to be seen whether the hilsa will use it in full strength. In the Godavari and Krishna rivers their migration is obstructed by the anicuts except during heavy floods. In the Indus the Sukkur barrage stops them. In all these rivers the number of migrants is seriously declining year after year. Production of fish which was 18,000 mt in 1955 dwindled to 5000 mt in 1981. In the Narmada they ascend up to about 160 km from the sea where their migration is halted by the forceful downstream flow of the river over the rock boulders, though a few strong swimmers do cross this limit too. By and large they are observed to spawn in this river beyond about 70 km from the sea. In the Ganga such spawning spots are several hundred kilometres away from the sea and spread over wide areas, depending probably on age, size of fish and strength of flow of water.

A small second migration upstream in April and May was also reported in the Ganga when there were very few barrages on the river and its tributaries. A small-scale migration was also observed in the Narmada in April–May but this was not concerned with breeding as the females were not gravid. This aspect of fish behaviour needs further investigation.

During the spawning migration the hilsa feeds only sparingly on diatoms and other green algal matter. If the downstream current is strong it travels near the bottom, otherwise through the middle column of water. Actual spawning has not been observed so far but fertilized eggs have been collected and grown in nursery ponds. A female hilsa produces 280,000 to 1,800,000 eggs each season but tremendous mortality prevails in nature at different stages. The eggs are about 1·8 to 2 mm in diameter and float in water due to an oil globule in them. The hatching period is about 26 hours and the newly hatched larvae (2·6 mm) after a week-long larval stage become free-swimming.

Another fish very similar to *Hilsa ilisha* and mistaken for it on many occasions is *Hilsa toli* known as 'giant herring' and on the Balasore coast as 'Chandana Hilsa'. The latter has large scales of slightly bluish hue as compared to smaller scales of yellowish tinge in the case of *H. ilisha*. The caudal lobes of *H. ilisha* are short and stubby while those of *H. toli* are elongated and drawn out. An important ecological difference is that the latter is always a coastal marine fish rarely entering fresh waters or even the estuaries. It is larger than the true hilsa but its life history has not been studied so far. It is less appreciated as food compared to *Hilsa ilisha* which is most popular in the north-eastern region and fetches the highest price in the markets.

HIMALAYA. The Himalaya, the 'Abode of Snow', stretches from the northwest, where the Indus divides it from the Hindu Kush and Karakoram ranges, for 2400 kilometres eastwards to where the Brahmaputra cuts through the mountains of Tibet. With their foothills and minor ranges the Himalaya are in width from about 160 to 400 kilometres, and the passes across them are nearly all over 5000 metres. The snow lies permanently at heights that vary from below 5000 metres up to the summits, the highest, Mount Everest, being 8848 m and the next, Kangchenjunga, 8598 m. The highest mountain wholly in India is Nanda Devi, a little under 8000, and in the Karakoram K2 is 8610 m. In the west 'the snows' are mostly hidden from the plains by foothills, but on a clear day there is a wonderful view of them up the Ganga Canal from Roorkee. In the east, where they rise rapidly from the plains, the views can be splendid. The slope from Tibet is more gentle than that from India, and it is therefore more difficult to defend India than to attack it across this mountain barrier which has for so many centuries cut it off, except for restricted contacts with Tibet.

The Himalaya are not only a wall difficult to cross by man, but they contain the summer monsoon winds, deflecting them so that they shed much of their rainfall on the plains south of the range. The snow that falls on the high mountains and the rain at lower altitudes give rise to the many great rivers that feed the Indus and Ganga and so make possible the irrigation of the Indo-Gangetic plain, and the generation of hydro-electricity.

The Himalayas are geologically 'young'—millions rather than hundreds of millions years old—and they are still apparently rising. This uplift is now thought to be caused by movements of the earth's crust known as Continental Drift. At one time peninsular India was separated from Asia to the north by a great sea known as the Tethys. Then the 'plate' on which peninsular India floats over the semi-fluid rock beneath the earth's crust began to move northwards, and the bed of the Tethys was pushed upwards in great folds and the Himalaya emerged. Fossil sea animals are found in their rocks at above 5000 m.

Because these mountains are young they are still so high, but very gradually, as they rise, they are also 'weathered' or worn away by the action of ice, water, avalanche, and even wind; and earthquakes may cause great rock and land slips. Their rivers have also carved great gorges in them and carry away silt that is dropped in the plains, thus making them so fertile.

The changes in temperature that take place with changes in height are much more sudden than those which result from changes of latitude, so the differences in a vertical climb of four or five miles are as great as in a journey of 200 miles north or south from the equator. This makes the Himalaya a wonderful area for naturalists. Up to 1000 m are found plants and animals of the tropics; between there and roughly 3000 m those of temperate lands; and thereabouts and higher the lovely Alpines: blue poppies, primulas, gentians and a profusion of others. With height the trees and birds and animals change from the amaltas to rhododendrons and azaleas and all the pine and fir trees, from the tiger to snow leopard, from the peafowl to the snowcock. The people, too, at different heights and in secluded areas, change and live differently, and some communities have until lately been so cut off as still to attract anthropologists and sociologists.

The infinitely varied rocks and land forms of these mountains attract geologists and searchers for valuable minerals as well as zoologists, ornithologists and botanists. More disturbing, their forest wealth has tempted timber merchants to fell more trees than they should. Water power brings in the dam-builders, and the needs of defence the engineers with their road-building. Above all they are the happy hunting grounds of growing numbers of mountaineers, trekkers, pilgrims and tourists. They are so vast that one might think that all this mattered little; but changes have been greater during the last thirty years than in centuries before except at the highest levels, and even at the South Col of Everest there is a collection of junk. Thirty years ago you had to walk for at least a week to reach areas now accessible in a day by bus or jeep and this makes a holiday in the mountains possible for growing crowds. They cut wood for their fires and too often leave their tins and rubbish unburied. Increasing numbers of mountain villagers need more land to till and more grass for their animals, and they encroach on the forests. The demands for more and more wood and consequent over-felling means that rainfall, instead of being absorbed in the soil held by the roots of shrubs and trees, now courses down the hillsides increasing floods in the plains and filling the lakes behind dams with silt. Explosives have been used in the rivers to kill fish, and automatic weapons on the slopes to kill animals. If man cannot learn to control his greed and to care for his surroundings he may spoil one of the world's great assets: the approaches to the Abode of Snow. The Snows themselves are likely to outlast man.

J.T.M.G.

HIMALAYAN POPLAR, *Populus ciliata*, is a lofty tree found all along the Himalaya at 1200–3500 m. It has dark grey bark, vertically fissured. The leaves are heart-shaped, 7–18 cm long, with hairy and serrated margins. The flowers are catkins, both male and female catkins being 10–15 cm long. It is a fast-growing tree and its wood is used for water troughs and as fuel in the cold deserts of Ladakh.

HOLLOCK, *Terminalia myriocarpa*, is a gigantic deciduous tree of the foothills of Nepal, Darjeeling, Sikkim, extending eastward to Arunachal and Burma up to 1400 m. The greyish brown bark peels off in long flakes. This handsome tree with long drooping branches is spectacular when in flower or fruit. The whole tree is covered with masses of tiny creamish flowers in autumn which turn coppery red in November as the small winged seeds appear. It yields a valuable plywood timber, much in demand for making tea-chests.

HOME RANGE. The Hanuman langur illustrates very well the concept of home range. Seven troops consisting of both males and females have been extensively studied in Mount Abu in Rajasthan. Over a four-year period, members of each troop were always found within their respective, approximately ninety-acre, plots and each such plot is called the home range of a particular troop. Each of these home ranges contains several suitable sites such as banyan and mango trees where the langurs roost and where they get a certain amount of protection from the leopard, their most dangerous predator. Each home range also contains several shrubs and trees such as lantana, coral tree (*Erythrina*), kachnar (*Bauhinia*) and jacaranda, whose leaves form the staple diet of the langurs. In the case of langurs, territory is not different from home range (see TERRITORIALITY) and the home range itself is defended against intrusion by members of other troops. The home ranges of different troops overlap to some extent and a great deal of fighting occurs when two troops meet at the borders of their home ranges. This fighting is however very ritualized and often leads to the victory of the stronger troop without any injury to any animal in either of the troops.

Many animals, especially mammals, restrict their movements to definite home ranges in a manner similar to that described for the Hanuman langur. The bonnet and rhesus monkeys among primates and the chital and blackbuck among ungulates are other examples from India where the home ranges have been mapped.

What is the optimum size of a home range? This of course depends on the number of animals that it has to sustain, and the quality of the habitat. The home range should be large enough to provide sufficient food

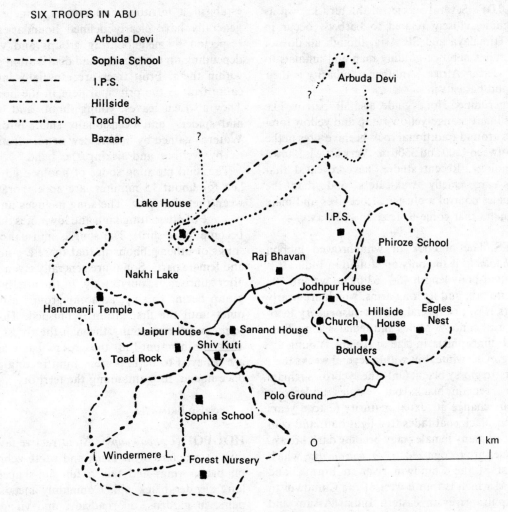

SIX TROOPS IN ABU

· · · · · · Arbuda Devi

- - - - Sophia School

- - - - I.P.S.

- · - · - Hillside

- · · - · · Toad Rock

——— Bazaar

Arbuda Devi

Lake House

I.P.S.

Phiroze School

Raj Bhavan

Nakhi Lake

Jodhpur House

Hanumanji Temple

Hillside House

Eagles Nest

Jaipur House

Sanand House

Church

Shiv Kuti

Boulders

Toad Rock

Polo Ground

Sophia School

0 ——————— 1 km

Windermere L.

Forest Nursery

Home ranges of six troops of the Hanuman langur in the vicinity of Mount Abu in Rajasthan.

for all the animals dependent on it and yet not too large because the animals would be unnecessarily exposing themselves to the dangers of predation if they traversed a very large area. Studies on some species of squirrels have shown that their home ranges are just large enough to provide sustenance to the animals occupying it. Thus the same animals would need a larger home range if the habitat is poor than they would need if they occupied a rich habitat. One would also expect that the size of the home range would increase with the number of animals. It does if the quality of the habitat is constant. Sometimes, however, the size of the home range is not correlated with the number of animals due to a curious set of circumstances. Some parts of the habitat are often much better than the others. A large and therefore often stronger troop succeeds in occupying the best part of the habitat. Here such a troop would defend enough area to sustain itself. The weaker, though smaller, troop is thus forced

to settle for the poorer portion of the habitat and this troop will now need a much larger home range to sustain even the small number of animals in it.

Why is it that some animals maintain a home range and defend it against intruders while others don't seem to? The answer to this question lies in the difference in the nature of their food supply. If an animal's food source is rather fixed in space and time, such as a large banyan or mango tree, it pays for the animal to establish itself or its troop around this resource and defend it against other animals because these large trees will continue to supply food for a long time to come. On the contrary, consider an elephant that has to move over a great distance before it can get enough to eat. The food supply is so sparsely distributed and so unpredictable that nothing is gained by spending energy in driving out other animals from a region that does not promise any definite and long-term returns.

R.G.

HONEYGUIDE. Several species of the bird known as the Honeyguide, closely related to barbets, occur in Africa, the Himalaya and SE Asia, though all do not have the peculiar habit of guiding man and animals to beehives as seen in Africa. Another peculiarity is their reported brood-parasitism.

The Orangerumped Honeyguide, a small sparrow-like bird with brilliant orange-yellow rump and yellow forehead, occurs around traditional rock-beehive sites in the Himalaya between 1500 and 3500 m. Very little is known about its biology. Recent studies have revealed that honeyguides are chiefly wax-eaters and that the dominant males control a cluster of beehives and mate with the females that come to feed on the wax.

HOOLOCKS. The hoolock or whitebrowed gibbon (*Hylobates hoolock*) is the only ape found in India. It is small (head-and-body length 45·7–63·0 cm, weight 7·0–7·9 kg), has no tail, and its long arms, which are nearly as long as its legs, are specialized for suspensory locomotion in forest canopy. Males and females show little or no sexual dimorphism in size or weight. Young are born a pale greyish white, but within several weeks their coat darkens to glossy black. Only the eyebrows remain white. Males remain black, but females undergo a second colour change at sexual maturity (7 to 8 years of age). Their black coat fades to a light buff, and with age the chest of many females may become dark brown. Two subspecies are recognized: *H. h. leuconedys*, which is found east of the Chindwin river in Burma, and *H. h. hoolock*, which is found west of the Chindwin to the Brahmaputra river in eastern India (Assam and Nagaland).

Hoolock gibbons once occurred throughout much of Assam, Burma, Bangladesh and parts of southern China, but their present distribution has been reduced considerably. This is attributed to increasing human populations and expanding *jhuming* (slash-and-burn agriculture) that has eliminated much of the evergreen and mixed evergreen rain-forest upon which these strictly arboreal apes are dependent. Extant populations of hoolocks are restricted to where forests still persist in hilly terrain of North Cachar, Garo, United Khasi-Jaintia and parts of Lakhimpur and Tirap provinces of Assam. They also occur in the Chittagong Hill Tracts of Bangladesh bordering Arakan in Burma.

Hoolock gibbons live in small groups comprised of a monogamously mated adult pair and up to four offspring. Adults mate for life, which may be 30 years or longer. Young are born singly, every 2 or 3 years, remaining with the group until they are sexually mature. Births appear to be seasonal, generally occurring in the winter months. At maturity offspring disperse, partly in response to the gradual increase in aggression by the adults. Males must then find sufficient space to establish a territory and attract a mate. Territories generally have clearly defined boundaries which are respected by neighbouring groups and vary in size, depending on the number and distribution of food trees within them. Fruit trees are especially important because fruit is the principal item in the hoolock's diet. They also eat leaves, young shoots and buds, insects and spiders, and occasionally small birds and eggs. Water is gained by licking dew or rain from vegetation or by wetting and licking the hand.

The loud morning songs of hoolock gibbons, which last for about 15 minutes, are an elaborate duet between the adult pair. The song includes an accelerated passage of alternating high and low notes that gradually become more shrill. The song is unusual compared to songs of other gibbons in that there are no clear male and female parts. Songs are generally given 2 or 3 hours after sunrise, or, only rarely, in the afternoon. Once a group begins to sing, adjacent groups usually remain quiet until the first group is finished. Thus, hoolock songs pass sequentially through the forest, one group after another, until all the groups have called. Songs are believed to be important in attracting a mate, and once mated, in maintaining the territory.

R.L.T.

See also PRIMATES.

HOOPOE (*Upupa epops*). An attractive fawn-coloured myna-sized bird with black-and-white zebra markings on back, wings and tail, a full fan-shaped crest and long slender black bill. Commonly met, singly or in pairs, in gardens, on roadsides and village outskirts, walking about with a waddling gait, probing in the soil for insects and grubs. Nests in holes in walls or tree-hollows lined with straw and rubbish. Call: *oop-oop-oop*.

HORAICHTHYS. The full scientific name is *Horaichthys setnai* Kulkarni, commemorating Dr Sundar Lal Hora, the eminent ichthyologist, Dr S. B. Setna, then a fisheries officer of Bombay, and C. V. Kulkarni, the

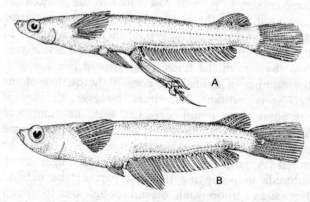

Horaichthys setnai. A Male, B Female × 4

M.Sc. student who identified the fish in 1937. At first it appeared like a young goby or minnow, but after careful study it proved to be an adult of a kind so peculiar that a new family of the top minnows (Cyprinodonts), Horaichthyidae, had to be created for it. This thin, translucent, wisp-like inch of fish is packed with special features. The male and female forms of most bony fishes are alike, but in *Horaichthys* the male has a structure called a gonopodium which is utilized to transfer spermatophores (bundles of sperms) from male to female. The spermatophores are barbed, get attached near the genital opening of the female, there break open to allow the sperms to fertilize the ova, and thereafter development into miniature adults proceeds in the water. *Horaichthys* breeds throughout the year and is a useful fish as it feeds on the early larvae of mosquitoes. It is found all down the west coast of the peninsula and in Bombay area is called *motake*.

HORN SHELLS are a group of adventurous and gregarious univalves plentiful along the seashores and in brackish and fresh water. The sea forms are common on mud-banks and pools in the intertidal area and among the seaweeds on which they feed. The brackish-water forms abound in the oozy slime flats of estuaries; and the freshwater forms on the sand and muddy banks of rivers and lakes. Horn shells have a tendency to migrate from the sea to the land and adjust themselves to the new environment. They are often seen moving on semi-fluid slime, slowly creeping over low-lying land and browsing on the tiny vegetable growths on the bases of the trunks of coconut trees and mangrove roots. They return to the slimy beds after feeding. These univalves are capable of remaining out of water for several days because their breathing system is specialized—they have a well-developed siphon and their gill filaments are broken up into a close network of blood tubes. Zoologists say that some species can live out of water for three months.

These univalves possess shells which vary in size and sculpture, but all are elongated, tapering towards the apex and broad and bell-like at the mouth end. They resemble small 'horns' or 'trumpets' and hence the name Horn shells.

The more common and abundant genera found in India are: *Cerithium, Potamides, Telescopium* and *Tiara*.

Cerithium is a purely marine form. It is found in plenty in tide-pools and muddy rocks at half-tide. The shells are about 3 cm long with many whorls, their mouth semi-circular and channelled at both ends. The whorls are slightly rounded, bearing a series of smooth nodules on the outside, black or red in colour. Sometimes thin lengthwise thickenings (varices) are also present.

Potamides is a brackish-water genus often found in multitudes on the mud flats of estuaries. Especially in the months of February and March the weed-grown shallows with slimy bottom will be found to be teeming with *Potamides*. They closely resemble *Cerithium* in size and sculpture but the outer lip is larger and turned out. The whorls are regular and rounded. The general colour is brown or blackish red, with a few whitish spiral lines.

Telescopium is a large brackish-water genus, measuring nearly 12 cm. It has no tubercles or ridges. The aperture is strongly channelled. The surface colour is greyish brown but the interior is smoothed over with a thick layer of plum-coloured enamel. The animal thrives in soft exposed mud of backwaters.

Tiara looks like a small *Potamides* with rounded whorls, the surface of which is ornamented by thin transpiral ribs. The shell mouth is not channelled. It is a freshwater genus. Living specimens are found in plenty on the muddy and sandy banks of rivers and canals and in the shallow ditches dug around coconut palms for watering.

Economically Horn shells are very important and useful. In many parts of India—Konkan districts of

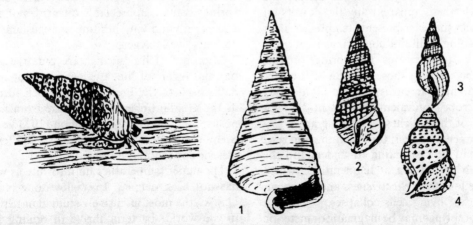

Potamides crawling over mud, and the shells of **1** *Telescopium*, **2** *Potamides*, **3** *Tiara* and **4** *Cerithium*, all × ½.

the west, Godavari and Krishna in the east and Tinnevelly and Tuticorin in the south—the small shells of *Cerithium* and *Potamides* are collected in basketfuls, burnt in kilns and converted into quicklime. This lime slaked with water is ideal for whitewashing. The flesh of the large *Telescopium* is wholesome food.

See also MOLLUSCS.

HORNBILL. Hornbills (family Bucerotidae) are easily recognized by the shape of their bill which, though it looks heavy, is actually hollow and spongy inside. Ten species occur within the Indian subcontinent. They prefer heavily forested areas, except the smallest, the Indian Grey Hornbill, which is also found in opener country. The largest, the Great Pied Hornbill, is primarily a dweller of tropical rain-forests of NE and SW India. The unique endemic Narcondam Hornbill is confined to a 15 km^2 volcanic island of the Andaman group in the Bay of Bengal.

The plumage varies from dull greyish brown to vivid rufous, yellowish, white and black. Four species are sexually dimorphic. Their flight is noisy and laboured. The Great and Lesser Pied Hornbills have prominent casques while the Rufousnecked and Narcondam Hornbills have ridges and furrows on the upper mandible.

Hornbills are chiefly fruit-eaters but they also take insects, reptiles and small mammals. The breeding season is between February and April, sometimes up to June. A natural hollow in a tree is used as the nest site. The female enters the hole and plasters up the entrance with her own excreta mixed with mud and wood bark, leaving only a slit-opening through which the male feeds her throughout her 'confinement'. One to 4 eggs are laid and during the incubation period the female moults her flight-feathers. The male continues to feed her till she is ready to come out, after which both feed the young till they break the plaster and emerge.

HOT SPRINGS. Hot springs in India attract the average man with their apparent mystic powers of healing and hence the ever-present temple in their vicinity. To the biologist, life in hot springs is of great interest for more esoteric reasons. Hot springs provide conditions probably akin to those existing at the very dawn of life, and thus the earliest forms of life may resemble the microbes prevalent in present-day hot springs. Analyses of these niches will enable an assessment of upper temperature limit for life in the universe. Moreover, it would be fascinating to understand the mechanisms which permit life at temperatures where normally enzymes lose their effectiveness and the membrane architecture of living cells collapses.

The origin of hot springs may be magmatic or meteoric or a combination of both. Below the earth's crust the molten substance called magma exists at temperatures as high as 1200°C and at pressures reaching several thousand atmospheres. Hot springs may originate from the magma and spout through crevices or channels in the crust. Alternatively, ground (meteoric) water or rain water may meet the magmatic surface and the mixed liquid would then emanate as a hot spring.

Hot springs are not as common in India as in some other countries such as Japan and New Zealand. But India has a fair number of hot springs, including the hottest spring in the world at Kulu in Himachal Pradesh. There are over 350 separate localities in the Indian subcontinent where hot springs occur in four broad belts. The first is in Bihar, more or less parallel to the boundaries of the coalfields, and in Rajgir and Monghyr areas, the second along the western coast in the Ratnagiri, Thana, Kolaba and Surat districts and the third and fourth in the Sind-Baluchistan or Himalayan regions respectively. In addition, there are other smaller areas, e.g., the Mahanadi valley of Orissa; Birbhum and Darjeeling districts of West Bengal, parts of Assam, Sikkim and certain portions of southern India where the disposition of the springs follows the general tectonic trend of the country.

Hot springs are mostly alkaline to near neutral in pH, but there are several which are acidic (pH 4·0). The hot springs in Western India are all alkaline, the pH ranging from 7·4 to 8·8. They are generally rich in minerals varying from 1 gram per litre to 196 g/l. The spring at Lasundra (Kaira district, Gujarat) has the highest mineral content among those in western India with total dissolved solids of 7·5 g/l. Some springs have salinity higher than that of sea water. The chloride and sodium contents of the Lasundra spring are 3500 milligrams per litre and 1340 mg/l respectively compared to about 1500 mg/l and 250 mg/l respectively in the Ganeshpuri springs near Bombay. Most hot springs are also enriched in sulphur (reduced), silica and calcium. Fluoride is present in mg/l quantities, the values for western India hot springs being 1.2 to 5.7 mg/l. Hot springs contain appreciable quantities of arsenic, manganese, aluminium, lithium, magnesium, bicarbonate and dissolved oxygen.

Microbes of hot springs are putative relics of primordial forms of life. Evidence for hot-spring activity dates back to the Precambrian. Rock formations such as the Gunflint chert which arose from hot-spring deposits, and which are two billion (10^9) years old, teem with fossil microbes resembling the Flexibacteria of present-day thermal springs.

The upper temperature limit for life as we know now has not been defined. In Yellowstone National Park, U.S.A., the most intensively studied hot spring location in the world, bacteria thrive in boiling spring water (boiling temperature 92°C, because of the high mineral content). Living non-photosynthetic filamentous and unicellular bacteria have been recorded up to a temperature of 95°C. The upper limit for O_2 evolving photosynthetic cyanobacteria (blue-green algae) is 73–75°C, the most prevalent form being the unicellular rod-shaped *Synechococcus lividus*. Filamentous cyanobacteria abound at temperatures of 55–65° C, and the cosmopolitan *Mastigocladus laminosus* has been found to occur in more hot springs around the world than any other cyanobacterium. Most hot springs in Western India harbour this microbe, which grows into large sheets at 61°C in Tooral hot springs about 320 km from Bombay on the Bombay-Konkan-Goa road. Thirteen other cyanobacteria are found at 60–62°C in the hot springs of Western India including filamentous and unicellular genera. Sulphur-tolerant bacilli are found at 60°C in acidic hot springs.

Cyanobacteria are also the major group of organisms in hot springs of lower temperatures. But true-celled organisms such as certain fungi and the acidophilic algae of uncertain taxonomic status, *Cyanidium caldarium*, are found in hot springs at temperatures between 56° and 60°C. Photosynthetic purple sulphur bacteria also occur in this temperature range. Green algae (Chlophyceae) and diatoms (Bacillariophyceae) are common up to about 50°C. Many microscopic protozoans and specialized crustaceans, flies and beetles also live in hot springs at temperatures between 45° and 50°C.

J.T.

See also ALGAE.

HUNTING SPIDERS.

Hersilia savignyi is a hunting spider common on old walls and trunks of large trees. The abdomen is oval and flat with two long posterior spinnerets. The legs are long and spiny, the third pair very short. Their colour harmonizes with the surface on which they move. These spiders adhere to the surface with their flat body so closely that a casual look can never detect them. Again at the slightest disturbance they scurry so fast in a spiral or zigzag direction as to escape notice. They deposit their long oval cocoons in fissures of bark or crannies of walls.

Oxyopids are hunters found in the ears of grass and inflorescences of herbage. They have a conical abdomen, raised cephalothorax and strongly spined legs. The light green or yellow colour of their body and black spines blend harmoniously with the ears and corns of grass and adequately hide their presence.

Thomisids (Crab Spiders) are small spiders living concealed inside flowers. The abdomen is round or pentangular and the legs short and strong. They move sideways like crabs. They are generally coloured pale green, yellow, white, or pink to suit the colour of the flower in which they reside. Some species can change

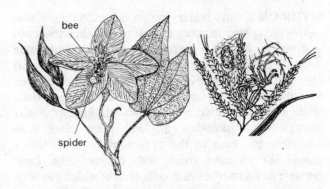

Left, a Thomisid spider catching a bee in a Bauhinia flower. The visiting bee is pounced upon and caught in the spider's jaws, × 1

Right, an Oxyopid with her cocoon among ears of grass corn, × 1

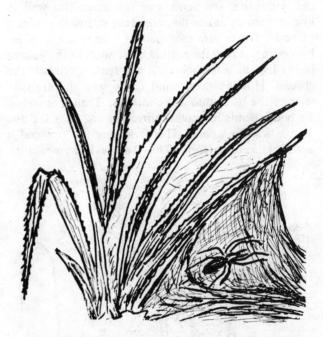

Hippasa pantherina, ×1 by the side of an aloe plant.

their colour.

Heteropoda venatoria (the Huntsman Spider) is common in houses, especially in bathrooms and unfrequented corners. It is a large brown spider with strong, radially arranged legs. It is nocturnal in habits and feeds on crickets, cockroaches and mosquitoes. Its cocoon is like a flat disc which the female carries underneath her belly.

Hippasa pantherina belongs to the family of outdoor hunting spiders (Lycosids). The female builds a complicated funnel-shaped web with a platform by the side of trees, old walls, pineapple suckers, etc. She sits at the mouth of the web but if disturbed retreats deep into the funnel. She carries her oval cocoon beneath her belly, attached to the spinnerets.

See also SPIDERS.

HYDRA is a tiny, solitary polyp not exceeding 6 mm in length. It lives in ponds and rivers attached to stones and weeds. Under a lens it looks like a short tube fixed at one end by a sticky disc and the other end free and conical. The cone is called the hypostome or manubrium, at the summit of which is the mouth leading into the cavity of the tube—the gastrovascular cavity—where digestion of food takes place. Surrounding the base of the hypostome is a circlet of thread-like tentacles armed with stinging cells. Each tentacle has numerous such cells each of which contains a coiled whip bathed in a poisonous fluid. At the slightest pressure the cell bursts and out shoots the whip, piercing the prey. The tentacles can extend, capture prey and take the food into the mouth.

The body wall of hydra is of two layers—the outer one protective and inner one digestive. The wall is firm and muscular and the animal can stretch or contract its body. It can also move slowly from place to place by looping. From the normal erect position the animal bends like a horseshoe until the tentacles touch the ground. Holding the ground firmly with the arms, it releases the basal disc and contracts. Then it stretches its body, bends in another direction and fixes the disc at a convenient point. Then releasing the free end it assumes its normal form. Thus it moves like a measuring

A HYDRA LOOPING. × 2.

worm. Some species move by gliding the basal disc slowly.

From the attached position a hydra can hold or adjust its body in such a way as to secure maximum oxygen and food supply. A hydra living at the bottom of a tank stands erect; attached to the side of a piling it grows horizontally; and if on floating weeds it hangs directly downwards.

Some hydras are symbiotic: green algae live in their cells and feed on the waste of the hydras which in their turn receive a copious supply of oxygen from the photosynthetic activity of the algae. Hydras have great power of regeneration. Even if cut into pieces each part develops into a whole. Biologists have produced hydras with many heads by grafting pieces of a cut animal to the trunk of a living hydra.

See COELENTERATE.

HYENA. The family Hyaenidae is represented by a single species in our area, the striped hyena, *Hyaena hyaena*. About five fossil species have been recognized in the Siwalik beds. Remains of the African spotted *Crocuta* have been found in caves near Karnool, Tamil Nadu. The species has a vast Saharo-Indian range, from northern Africa to India. However, it is absent from Assam, Burma and Sri Lanka, and is uncommon in the sandy desert. Its preferred habitat is foothills bordering the plains, nullahs and ravines where it shelters in caves, among boulders, and in burrows.

The hyena is a nocturnal animal and can wander long distances in search of food. It mainly thrives on carrion, carcasses of animals and bones of prey left over

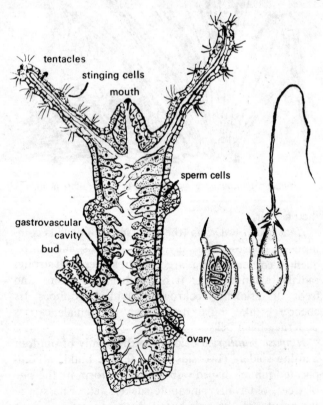

tentacles
stinging cells
mouth
sperm cells
gastrovascular cavity
bud
ovary

Longitudinal section of a hydra, × 20 with a stinging cell before and after use, × 100. Used whips are discarded and new ones grow.

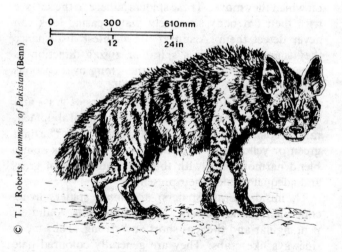

© T. J. Roberts, *Mammals of Pakistan* (Benn)

Striped Hyena

by larger carnivores. Thus it is a useful scavenger. In spite of its reputation of cowardice, it can deprive panthers of their kill. Sheep, goats, calves and more often dogs are preyed upon. In northern Uttar Pradesh hyenas are dreaded as 'baby-lifters'.

When attacked by wild dogs and held beyond hope of escape a hyena 'shams dead', an instinct very rare in animals. Its voice has been described as a chattering laugh or weird kind of howling.

The young, 2 to 6, are born with their eyes and ears closed, usually during spring and summer.

I.P.

IBEX. Several forms of wild goat are called Ibex (the Walia and Nubian Ibexes for example), and are found in the more arid mountain ranges from North Africa, Ethiopia, Arabia and the Caucasus stretching across Iran and Afghanistan to the Himalaya and Tien Shan mountains of the Soviet Union. Within the subcontinent two distinct species occur. In the high inner Himalayan ranges, the Siberian Ibex *Capra ibex sibirica*, and in the southern mountain ranges across to the Sind Kohistan, the wild goat or Sind Ibex *Capra hircus*. Like all the goat tribe they are sturdy beasts with relatively strong legs and are adapted to subsist upon the sparsest of xerophytic vegetation, depending for survival against predators on their uncanny ability to traverse the most precipitous cliffs and to climb into inaccessible crags.

The Sind Ibex is the same species as the Cretan wild goat and Persian Pasang and is believed to be the progenitor of domesticated goats. They are adapted to live in hot climates at altitudes as low as 450 m in cliffs along the Makran coast, and they can subsist on browsing such thorny bushes as *Acacia senegal* and *Zizyphus mummularia* (ber) when grass is scarce in years of drought. The Siberian Ibex is confined to altitudes above 3500 m, preferring to remain near the limit of the permanent snow-line. There they can subsist on grass and *Artemesia maritima*, for which they can dig in the snow with their forefeet.

Both species are gregarious, young males and females of the Sind Ibex forming herds of about 70 or 80 individuals, though the Siberian Ibex is rarely encountered in herds of more than 20 to 30. Old males tend to keep in small groups on the periphery of the main herd or to live a solitary existence until the rutting season. The gestation period is 5 to $5\frac{1}{2}$ months in both species and in years of good food-supply most females produce twins. In the Sind Ibex the rut is at the end of the monsoon in September and young are born in February. In the Siberian Ibex the rut is in late December/early January and young are born in May.

To see one of the older bucks silhouetted against the skyline is an unforgettable sight. The Sind Ibex has a silver-grey body with distinctive black belly, chest and a black line along its spine and in a collar around its shoulders. The fur is short and harsh with a long black beard and sweeping scimitar-shaped horns—sharply keeled in front, and in older bucks measuring 107 cm (the world record is 133·2 cm). Females lack any beard, have short smooth spiky horns of about 15 cm in length and their body coloration is yellowish brown. Newly-born kids are a silvery tan and most attractive.

The Siberian Ibex is an altogether more massive beast, adult males having a dense woolly coat which is dark coffee-brown with whitish cream saddle and rump patches. The females and young males are more reddish brown and have short horns measuring 15 cm which bear corrugations or annulations. A trophy head may have horns of 127 cm or more with regularly spaced ridges or bosses on the flattened anterior surface. The fleece of the Siberian Ibex is so fine and soft that it rivals the Vicuna of the high Andes, and legend has it that shawls woven from *pashm* would pass through a signet ring.

See UNGULATES, MOUNTAIN SHEEP & GOATS.

INDIAN CASSIA LIGNEA, *Cinnamomum tamala*, is a medium-sized evergreen tree confined to the wetter parts of northeastern India and Bangladesh. The bark

Triple-nerved leaves of Indian Cassia lignea × ½

is dark brown, wrinkled, thin; blaze reddish brown, very aromatic. The leaves are highly scented, used in flavouring food. It produces panicles of white flowers and fleshy, ellipsoid fruits, 1·3 cm across, black when ripe.

INDIAN CORAL TREE, *Erythrina variegata*, is a moderate-sized tree of Eastern India, its sub-Himalayan tracts, Andamans, Nicobars and Burma. Branchlets are armed with black prickles; leaves trifoliate. Clusters of dazzling scarlet flowers with projecting stamens are gorgeous when in bloom from February to April when the tree is leafless. The fruiting pods are 15–30 cm long, black, constricted between the 3–12 reddish seeds. The tree is used as a support for pepper and betel vines and the wood for making dugouts.

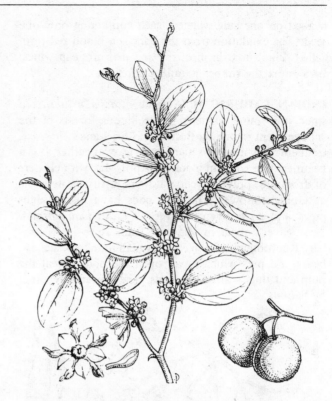

Indian Jujube. The sweet yellow or brown drupes (× ¹/₂) are relished by birds and humans. The tree is useful to foresters, for its prickles protect seedlings planted beneath it.

Indian Coral Tree. Flowers and leaves, × ¹/₂ . Mynas, babblers, bulbuls and sunbirds are attracted to the nectar of the spectacular flowers and effect pollination.

INDIAN JUJUBE, *Ziziphus mauritiana*, a small tree with drooping branches armed with curved spines, is distributed in the scrub forests of the subcontinent where it tolerates a high degree of drought. The flowers have small hood-shaped petals alternating with larger sepals. The stamens opposite to the petals and embraced by them are features helpful in identification. The leaves are fed to silkworms.

INDIAN KINO TREE, *Pterocarpus marsupium*, is found scattered in deciduous forests mostly in the peninsula. It has large leaves, with 5 to 7 leaflets. The pale yellow flowers grow in panicles. The pods are nearly circular,

Indian Kino Tree. Flowers, leaflets and pods, × 1. The pod shaped like a kangaroo's pouch explains the specific name.

beaked on one side, winged, each containing only one seed. The exudation from the bark is a blood-red resin called 'kino' used in medicine and in European wines. In S India the timber is ranked after teak.

INDIAN LABURNUM, Amaltas, *Cassia fistula*, is a small tree found scattered in deciduous forests of the subcontinent including Burma and Sri Lanka; in areas frequented by monkeys many trees grow together. It is a beautiful tree when in bloom in April-May with clusters of drooping sprays of yellow flowers justifying its name of Golden Shower. The long pods hang like straight pipes and have given the tree its Latin name *fistula*. New leaves appear in May and fruits ripen in December. Reproduction is effected through monkeys, jackals, bears and pigs which break open the pods to eat the pulp and thus scatter the seeds. The flowers are eaten by Santals

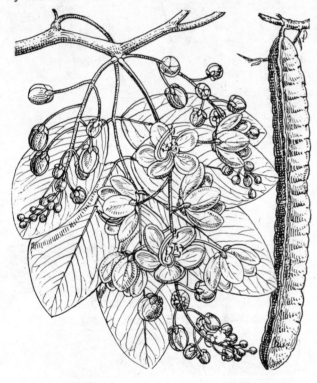

Indian Laburnum. Leaves and flowers × ½. Fruit × ⅓. The long pods are relished by monkeys and therefore called monkey sticks.

INDIAN LOTUS. The sacred lotus of India (*Nelumbo nucifera*) is common throughout our area in ponds, tanks and lakes. Unlike the water-lilies (*Nymphaea, Euryale* or Foxnut), the lotus is unarmed, without spines or stinging hairs, and its leaves are carried by the petioles well above the water level. The decorative red to white flowers are used for religious purposes. The abconical ovary contains a number of seeds embedded in a swollen receptacle. It is supposed to have been introduced

Indian Lotus (Nelumbo). *Note that the leaves are raised above water level.*

into Egypt around 50 B.C. Because of excessive exploitation of its edible seeds, the lotus is showing a gradual decline.

See WATER-LILIES.

INDIAN OLIVE, *Olea ferruginea*, is a small to medium-sized evergreen tree characteristic of the dry hills of the western Himalaya up to an altitude of 2400 m. The bark peels off in irregular strips. Leaves are leathery and pointed at the tip; flowers whitish; drupes, black when ripe, edible but not particularly pleasant to eat. The wood is used locally.

Flower × 5, leaves and fruit × 1. Olives are long-lived trees, aided by a durable wood that resists attacks by fungi and insects. The flowers require about eight months to develop into the dark olive fruit.

21 More Habitats

A rock cleft by nature's awesome forces. Today the Deccan Trap's dry scrubland habitat supports many raptors, but not many larger animals.

In Bandipur, Karnataka, as in so many Indian jungles, chital (*Axis axis*) are an important prey base for predators.

Kaziranga, ideal habitat for rhinos, elephants, tigers, swamp deer and buffaloes. One of India's largest pelicanries is also located here.

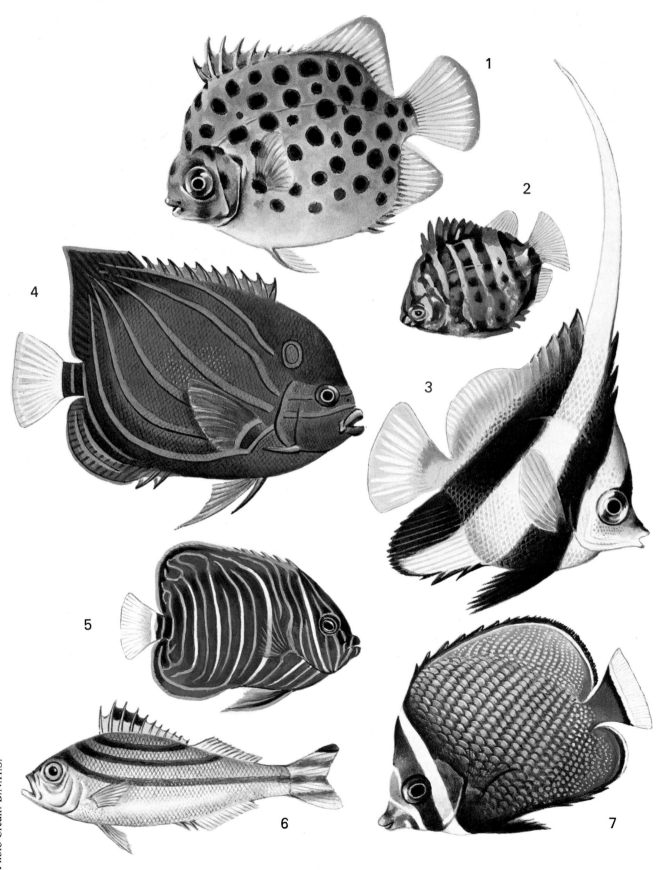

(1) Adult Scat, a perch *(Scatophagus argus)* x ¼ (2) Young Scat x 1 (3) Banner Fish, a sea perch *(Heniochus acuminatus)* x ½ (4) Adult Blue-ringed Angel Fish, which will attack and eat any other fish in the tank *(Pomacanthus annularis)* x ⅓ (5) Young Blue-ringed Angel Fish x 1 (6) Target Perch *(Therapon jarbua)* x ½ (7) Butterfly-fish, a sea perch *(Chaetedon collaris)* x ¼

Himalayan Poplar, in December 1980.

Flowers (female above, male below) and leaves of the Indian Poplar, × 1. The straight, linear-shaped leaf on the right develops into the toothed, rhomboid-shaped leaves.

INDIAN POPLAR, *Populus euphratica*, a medium-sized tree found in the plains of the Punjab and Sind along river banks and ascending up to 4000 m in Ladakh. It has an extraordinary range of latitude, being found from the Altai Mountains, 45°N, to the Equator. Flowers in catkins. Its outstanding feature is its ability to withstand very high and low temperatures and scanty rainfall. In the cold desert of Ladakh its bark helps to keep wildlife from starving. It also occurs along the river Euphrates and is known there as the Euphrates Poplar. The wood is suitable for shoe heels and cricket bats and the twigs are chewed for cleaning teeth.

INDIAN SCREW TREE, *Helicteres isora*, a shrub or a small tree of the Siwaliks, Western Ghats and the Andamans, with slantingly heart-shaped leaves, rough-surfaced above, hairy beneath, and with a toothed margin. Flowers red with bent petals. The fruit is like a bunch of cords twisted and tapered to the apex. The flowers appear in March–September and fruits ripen from December–January. The fibre from the bark is used in canvas. Charcoal from the bark is used as gunpowder.

INDIAN TARPON, *Megalops*, belongs to the same family as the MILK FISH *Chanos* (Chanosidae) and has almost the same habits as regards breeding and

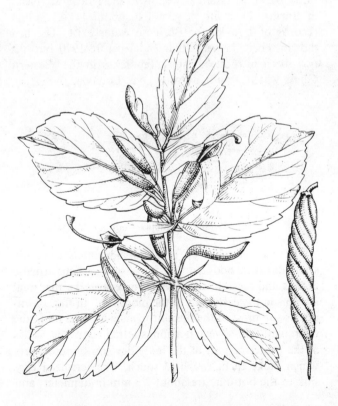

Indian Screw Tree. Leaves, flowers and fruit, × ²/₃

movement into estuarine waters except that it is carnivorous. It is a somewhat laterally compressed fish, with large silvery shining scales, and an upturned lower jaw. A prolonged last fin-ray of the dorsal fin is its characteristic. It grows to about 60 cm and weighs 3 kg. Its outstandingly large eyes have earned for it the name 'big eye' or 'ox eye'. As a fingerling, it feeds exclusively on *Cyclops* and *Daphnia* and as such can be used as a biological control for Guinea worm. The fish breeds during the early part of the monsoon, and innumerable ribbon-like larvae about 2 cm long enter creeks and freshwater puddles where they get their favourite food. *Megalops cyprinoides* is a smaller cousin of the famous tarpon of American waters and hence is called the Indian Tarpon. It has a similar fighting capacity when hooked. Being a partial air-breather it can sustain life even in foul waters.

See AIR-BREATHING FISHES.

INDIAN TROUT, of the family Cyprinidae, has no taxonomic relationship with the real trout of cold waters which belongs to a different family Salmonidae, except that the former is honoured with the suffix 'trout' given to it by anglers because of similarity of shape and the habit of taking a fly or fly spoon on a rod and line like a real trout. Indian trout, *Barilius bola*, is a warm-water fish of the rivers of the Indo-Gangetic plain. Its distribution extends to the Brahmaputra basin in Assam as well as to the Irrawaddy basin in Burma. Normally it grows to about 1 kg, though records of 1·36 kg for Burmese water exist. The fish did not occur in peninsular India before 1970 but has now been introduced into a few lakes in the Western Ghats which do not overflow, e.g. Lonavla.

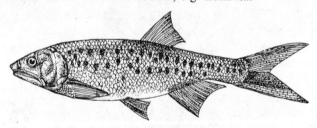

Barilius bola, the Indian Trout × ¼

The fish, growing up to about 35 cm, is elegantly streamlined in body and feeds on smaller fish, aquatic insects and flies. In the lakes, it breeds in the marginal streams in the early part of the monsoon. In the spawning season most of the fins of the male become pink, the opercular (gill-cover) portion spotted, and the scales of the posterior part of the body develop minute warts which make them rough to touch. The fertilized eggs sink to the bottom, are about 2·3 mm in diameter, and creamy or straw-coloured. The hatchling period is about 65 to 70 hours in water of 26°C temperature. After a post-hatching semi-quiescent period of four days, the hatchling becomes free-swimming and starts fending for itself on minute animal life in the water. The juvenile fish becomes sexually mature at the end of the second year and if in captivity, where growth is stunted, even when it weighs hardly 250 gm.

Another similarly nicknamed trout is the 'snow trout' of the family Cyprinidae. Species of three different genera, *Schizothorax*, *Schizothoraichthys* and *Diptychus*, are given this epithet, the reason for which is not very clear as no form or habit except habitat is similar. They are residents of mountain streams of Kashmir and Himachal Pradesh and breed in various months from October to June when the water temperature is between 18° and 21°C. Eggs are laid in shallow pools of clear-water streams amidst stones and gravel, a habit almost similar to the natural breeding of real trout. They are all small fish growing up to about 30 cm, feeding largely on aquatic insects and other organisms and algae attached to submerged rocks.

See also TROUT.

INDIARUBBER TREE, *Ficus elastica*, is a large buttressed tree from Assam where it was also cultivated as a source of rubber. It is much inferior to the rubber-tree of Brazil, and when the latter was introduced in Kerala and Malaysia the tapping of *F. elastica* ceased. With its large shapely leaves it is now a popular foliage plant.

INDIVIDUAL RECOGNITION. It is important for all animals to distinguish an individual of their species from that of another, for conspecifics play an important role in their life, for instance as mates or competitors. But for a vast majority of animals the recognition of a conspecific stops at identifying it as a member of their own species, of a particular sex and age, and perhaps belonging to a particular social group. Thus an ant worker does distinguish workers belonging to her own colony from workers belonging to another colony or soldiers belonging to her colony, but does not discriminate between individual workers.

There are however a number of species of more highly social animals in which individuals recognize each other as specific individuals. The commonest case of this is in species with some form of parental care. In such species it is important for parents to discriminate and dispense parental care preferentially to their own young. Indeed such animals show remarkable powers of individual recognition of their own young even in a whole crowd of young such as that in a breeding colony of sea birds. Hume's account of his visit to the Laccadives is very interesting in this regard. He recounts how he put a dozen Sooty Terns in a bag and kept them on board his vessel for 36 hours, but as they would not eat anything, he brought them back to Cherbaniani island where they had been captured and released them. He writes: 'We took the rest back and let them loose again on their native soil. In less than three minutes, amidst the pattering uproar of thousands of pairs of wings and through the haze of the dense flock of birds with which we were encompassed, the parents of every one of the "lost children" had found them out, and were busy feeding and caressing them.'

But individual recognition in birds, now known to be dependent largely on fine differences between calls and songs, is not restricted to parents and their young. Many male birds establish breeding territories, which they defend against rivals of the same species. After some initial testing out of each other's strength, the territorial boundaries tend to become stabilized in each season. It is now known that a bird knows the individual songs of all its territorial neighbours and pays little attention to them. If, however, the song of a stranger is broadcast near its territorial boundary, or of a neighbour from a direction different from that of the neighbour's territory, it becomes highly excited.

The Hill Mynas have a very rich repertoire of calls. It turns out that every individual has its own specific repertoire overlapping with that of neighbours of the same sex to varying degrees. In certain birds like the Scimitar Babblers, members of a pair keep in touch, through each singing part of a duet.

Individual recognition is most highly developed in species that live in moderate-sized social groups with a marked social hierarchy such as in troops of monkeys or packs of wolves. Monkeys are visual animals and their facial features show much variation that renders individual recognition easy. Wolves, on the other hand, employ smell to a much greater degree. In such social groups not only appearance, but behaviour also characterizes specific individuals, so that a dominant rhesus macaque has a different upright gait contrasting with the stooping gait of a subordinate individual.

INDIVIDUAL VARIATION. In all sexually reproducing species every individual is genetically different from every other except in rare cases such as identical twins. Such hereditary variation expresses itself in many ways, for example in the number of bristles on the legs of fruit-flies or the choice of species of foster parent in parasitic cuckoos. To such genetic variation is added a further component of variation due to differences in the environment in which the animals grow. Thus in many species of predatory birds such as owls,

the young hatch one after the other, so that at any time the chicks in a nest differ in size from each other. When the parent birds come to feed the chicks they preferentially feed the bigger chicks that beg more vigorously. As a consequence, when there is a food shortage the bigger chicks hatched earlier get all the food, while the younger chicks starve, die and may be actually eaten by their older siblings.

Variation in size amongst individuals is thus quite common and often plays an important role in social animals in which the bigger animals generally tend to occupy higher positions in the social hierarchy. Thus in chital, the rut calls are given exclusively by the biggest males who also monopolize all sexual mountings. In elephant herds, the nucleus of the society is formed by the females. Amongst these the biggest females tend to dominate the other animals. Variation in size, and in other attributes, can thus differentiate the behaviour of animals and the role they play in their social group. In rhesus macaque, for instance, the bigger dominant male has a special gait, first access to food and mates, mounts subordinate males, but is seldom mounted by them in turn, and is groomed by subordinate males and females, but seldom grooms them in turn.

The most marked individual variation is however developed in insect societies such as those of ants and termites. In an ant or termite colony, there often are a number of specialized castes, each performing its own special role in the colony. These may be foragers who collect food, soldiers who guard the colony, sometimes leading to bizarre forms such as honeypot castes which do nothing but store the nutrient liquid tapped from aphids in their enormously distended bellies.

INDUS. The Indus is the principal river of Pakistan. In length, volume and drainage area it is one of the greatest rivers in the world. Over one-third of its drainage area lies in the mountains in western Tibet, and in Jammu and Kashmir state of India, and the remaining in the semi-arid plains of Punjab in Pakistan. The river's annual flow is about 275 million cubic yards, nearly twice that of the river Nile. The Sanskrit word *sindhu* means a river, and from it the country names Hindustan and India arose.

The river rises in the Kailash range in southwestern Tibet at an altitude of 5100 m north of the major Himalayan ranges. For nearly 320 km it flows in a northwesterly direction before crossing into Kashmir at an altitude of about 4600 m; then it continues another 240 km or so in the same direction. The major rivers which join it up to this point are Zaskar, Shyok, Hunza and Gilgit. At Bunji it turns decisively southward in an almost perpendicular shift to skirt Hunza Parbat in a spectacular gorge as deep as 4600 m and 19 to 25 km wide. This most dramatic shift has remained a geologic mystery. From now on the river flows in a swift mountain stream in Swat and Hunza region of Pakistan. It enters its plain stage near Attock after receiving the tributary of Kabul from Afghanistan on its right side. It has become a broad, meandering silt-laden stream by now, and has fallen from the altitude of 5100 m in Tibet to 600 m. After Attock it flows south. At Mithankot it receives the accumulated waters of the Punjab rivers. Near Karachi it takes a southwest direction making a small delta of 3000 km² which merges with the great salt waters of the Rann of Kutch. The delta coast is fringed with dead creeks, mangrove flats, shady beaches and ridges.

The major Indus tributaries are all snow-fed. Their flow varies seasonally. In winter (December to February) the discharge is at a minimum and floods occur during the rainy season between July and September. Often at this time there are flash floods. The flow is at a maximum as the river emerges into the plain. After that it decreases in volume as it follows a course of declining precipitation, increasing evaporation and seepage.

Considerable historical and physiographic evidence attests to the shifting nature of the river in its lower course below Attock, particularly in its deltaic section. The river began to flow into the Rann of Kutch about 200 years ago. It has been moving to the west about 16 to 32 km during the last 700 years.

Climatically, the river lies in a semi-arid to arid land. From its source to the mouth rainfall varies between 125 and 250 cm annually. Except for its mountain section, the Indus basin is the driest part of the Indo-Pakistan subcontinent. Northwest cyclones bring some winter rains (about 12 cm). Most precipitation is received from the southwest summer monsoons between July and September. Climate varies from the dry semi-desert in Sind and Punjab to the severe high-altitude climate in Hunza, Gilgit, Ladakh and West Tibet. In the plains, winter temperatures touch the freezing mark, while the July temperatures may even reach a maximum of 42° C with occasional readings of 45° C. Jacobabad, one of the hottest spots in the world, located west of the Indus river, has often recorded temperatures of over 49° C (120° F).

R.T.

INFANTICIDE. Infanticide is the killing by an animal of young individuals of the same species. This may occur either when the young has little chance of successfully surviving and is then consumed to recover what might otherwise be a complete loss; or it may occur when an infant which is quite likely to survive is killed, probably as a competitive device.

In the case of many birds of prey the clutch may be of three eggs while in most years the parents can successfully feed only two chicks. The third chick is successfully raised in occasional years, when the rodent population is unusually large. But the three chicks hatch at different times and the last one to hatch is at a disadvantage as the parents preferentially feed the two older and bigger chicks which beg more vigorously, with larger gapes. Eventually the third chick begins to starve and is then killed and eaten by its bigger siblings.

In a similar fashion many carnivore females produce large litters. The less healthy runts or injured young are often killed and eaten by the mother.

But the infanticide which has raised much greater controversy is the killing by males of infants quite capable of surviving, apparently as a competitive device. This has been well documented in our Hanuman LANGUR and is now reported from several other primates as well as the lion in Africa. In all these species, one or a few males control a group of adult females. As young males approach maturity, they are driven out of the group. These ousted males then try to take over another group of adult females by driving out the male or males resident with such a group. The tenure of resident adult males in such female groups is in fact very short and they are driven away in a few months or years after taking over a group of adult females.

It is in animals with this kind of social structure that we find that males which have taken over a group of adult females tend to kill the infants being nursed by the females. The killing of such infants is believed to have the effect of hastening the next conception by these females. The male or males who have newly taken over thus bring forward in time the opportunity for themselves to impregnate these females and thereby propagate their own genes. Since the average tenure of the males in any group of females is rather short such infanticide could help them in significantly enhancing the number of offspring that they would sire during their lifetime.

This interpretation thus considers infanticide by male langurs and lions as a reproductive strategy. Since natural selection favours the maximization of the reproductive success by any individual, such a strategy is expected to be favoured by natural selection.

An alternative interpretation treats infanticide not as a normal behaviour favoured by natural selection but as a manifestation of social ill-health due to crowding. This interpretation, for example, points to the fact that infanticide among Hanuman langurs has been recorded only from the more crowded populations. However, the timing of infanticide and the reported persistence with which the male stalks the infant as a specific target, appears to favour the first interpretation.

INSECT EVOLUTION and BIOLOGY. Insects evolved from their wormlike ancestors some 350 million years ago, the earliest known hexapod fossil being that of a probable collembolan, *Rhyniella praecursor*, from the Middle Devonian Geological Period, found in Scotland. The next fossil remains of an insect is from the beginning of the Upper Carboniferous, the age of the great coal forests, by the end of which insects had radiated extensively and ten orders, Blattodea (cockroaches) dominating, having become established. A primitive dragonfly ancestor, with wings 2½ feet across, was another of these creatures ruling the air. Even though the endopterygote orders and the hemipteroid group (Psocoptera, Phthiraptera, Hemiptera, Thysanoptera) are not known as fossils from the Carboniferous, they must have been present then as both groups had radiated widely by the Lower Permian. The reptiles emerged in the Permian to become a dominant living group on earth, but insects had achieved mastery of air before any other living creature did and for at least another 50 million years were the sole inhabitants in air. All orders of insects present in the Carboniferous were still present in the Permian, and several other orders had also appeared, including many that have survived to the present day. A great burst of adaptive radiation of insects occurred about this time, when land plants were also developing rapidly. The Palaeoptera (only Odonata and Ephemeroptera surviving now) reached their climax, and, of the Neoptera, the blattoid-orthopteroid orders (Blattodea, Isoptera, Mantodea, Zoraptera, Dermaptera, Grylloblattodea, Plecoptera, Orthoptera, Phasmatodea, Embioptera) attained their greatest diversity. But these groups were no longer dominating the world's fauna, as the Psocoptera and homopterous Hemiptera, besides the Mecoptera and Coleoptera, were abundant and well established. This was the end of the Palaeozoic Era and the beginning of the Mesozoic, around 225 million years ago. Nine of the 12 insect orders that became extinct did not survive beyond the Palaeozoic. But all existing orders except three (Meganisoptera and Paraplecoptera died out in the Triassic; Glosselytrodea in the Jurassic) were present at this time, which is an example of staying power unparalleled among the higher animals. There was diverse evolution among these surviving orders, especially in the Neuroptera, Mecoptera and Diptera. The Thysanura, Dermaptera, Orthoptera, Phasmatodea and Hymenoptera appear in the fossil record for the first time in the Triassic, but undoubtedly many are much older. In the Cretaceous, towards the end of the Mesozoic, some dinosaurs still dominated the earth and reptiles had begun to fly. The first flowering plants like *Magnolia* had evolved and, except for nine orders, most of the others were well established. The Caenozoic Era began some 70 million years ago and first fossil records of the Isoptera, Mantodea, Zoraptera, Grylloblattodea, Phthiraptera, Strepsiptera, Siphonaptera and Lepidoptera are from the Tertiary or even the Quarternary, the latter being less than a million years ago. However, except for the white-ants (termites), the Strepsiptera (treated here as Coleoptera) and the fleas, most of these orders were probably more ancient. Thus, for at least 300 million years, insects have been on earth and adapting to the changing environment. They certainly evolved from a *Peripatus*-like ancestor on land, and when the amphibians came ashore, developed the ability to fly and perfected complete metamorphosis which enabled them to escape predators, and exploit new resources and habitats more efficiently. They then took advantage of the rise of flowering plants (especially bees, wasps, flies, moths and butterflies), and the ants, termites and bees developed a complex social life.

The overwhelming success of insects is due to at least six major assets that they developed in the endless quest for survival: an external skeleton, small size, flight, metamorphosis, specialized system of reproduction, and adaptability. By wearing the skeleton outside of their body, insects have used the best way of protecting themselves by the use of a limited (though specialized) amount of material. Insects are a living example of the validity of what man now appears to have grasped as a truism—'small is beautiful'. Unlike ourselves, the demands of insects from our environment (with mostly non-renewable resources) are meagre. Like the huge dinosaurs that once ruled the earth, for a relatively brief span of time, most of our larger mammals (elephant especially) are now in real danger of extinction. Science fiction, which nowadays is becoming more and more ridiculous, depicts insects of sizes as large as us humans or larger, which is an impossibility. The fact that insects were the first animals to develop wings for flight, and that most have still retained, if not perfected them, is a great asset to them in their overwhelming success. Flight has enabled them to escape from enemies in a jiffy, to traverse large distances to find food and to search efficiently for their mates, besides other obvious advantages. The development of metamorphosis (especially by the endopterygotes) has enabled insects to divide their life stages into four distinct phases and structural adaptations. This kind of pattern has allowed insects to adopt at least two completely different life-styles—a sort of 'Dr Jekyll and Mr Hyde' character, so to speak; the larval and adult stages being able to exploit entirely different food sources and life-styles, to distinct advantage. Unlike us humans, who spend only a fifth of our life-span as 'immatures', insects spend almost all their time as inconspicuous and admirably adapted larvae or nymphs; the adult period, efficiently, being used only for males and females to find each other, mate and reproduce, ensuring the next generation, which

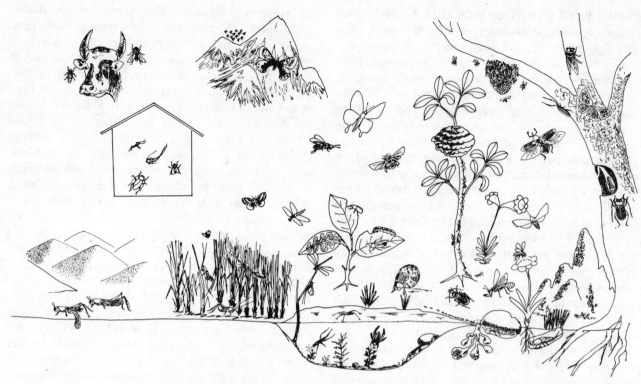

DIVERSITY OF INSECT HABITATS

is all that life is really about. On this critical require-ment for a generally bisexual living entity, insects have achieved wonders that man still is fumbling for. Winged adults are able to delay fertilization of the egg, even after mating has occurred (by storing the male's sperm cells in a little sac called a spermatheca until the female is able to find proper environmental conditions and food for her young). As we shall see in the complex social insects (termites, bees, ants, wasps), they even have developed ways and means to produce either 'boys or girls' and even different 'castes' to suit, not their fancy, but their genuine requirements! Finally, no other form of animal life has been able to adapt to such extremes of living conditions. Just the things insects have disciplined themselves to feed upon provide a panorama of their marvellous adaptability. As a necessity, they have modified their mouth-parts to feed on these varied substances, whether plant or animal, and some, like termites, have even employed living bacteria to inhabit their alimentary canal and help them digest even cellulose!

Plants provide food for a great host of insect groups. Depending upon the number of plant species a single insect species can develop upon, it is either called mono-phagous (only one species of food plant), oligophagous (a few related plant species), or polyphagous (a wide range of plant species). Leaves are a common plant part that insects consume and some species are re-markable defoliators of whole trees. Some other phyto-phagous insect species consume only part of the leaf, browsing on only one side, leaving the other side intact, some 'skeletonizing' the leaf, leaving the veins and veinlets intact. Other specialized insect larvae feed in a 'mine' that they construct between the intact upper and lower epidermis (or 'skin') of leaves, and each species can be identified based on the form of this mine itself! The piercing-sucking insects, of which the true bugs comprise a majority, do negligible mechanical damage to the plant by their stylet-like mouth-parts, but some may cause injury by the toxic action of their salivary secretions. In many others, their ability to

© Kumar Ghorpadé

Adult and larva of phytophagous lady beetle (Epilachna *sp.) skeletonizing leaf of watermelon* × 2¹/₂

310

transmit plant pathogens, like viruses, is much more injurious to the plant than by their direct feeding and sucking of plant sap. An intermediate type of feeding is exhibited by the thrips which rasp the plant part and suck the oozing juices. Like thrips (several species), many other insect young induce plants to produce abnormal growths called galls, within which they live and feed. Galls may occur on stems, leaves, flowers or roots and are characteristic of the species of insect that causes them. Many other insects live on or inside the bark or timber of trees and their characteristic pattern of bores generally gives away their identity. Many species of insects specialize in being 'undertakers', responsible for feeding on dead plant matter, from whole forest trees to fallen leaves and fruit pulp and skin. The abundance of insects has made plants also use them for their own benefit. Most plants that flower have come to depend on (and use) special kinds of insects to help them in pollination and hence in their regeneration. A few kinds of plants have attempted to provide 'shelters' for particular insects (ants and species of *Acacia*, for example) in return for these insects' symbiotic activity. On the other hand, insectivorous plants (e.g. pitcher plants) have developed which attract and feed on insects that happen to be enticed.

Insects also have associated themselves with vertebrate animals, either as their food or as their hosts. They may also be intermediate hosts of nematode worms that later attack vertebrate animals and even man. Some insects have developed into blood-feeders (see MOS-QUITOES & BITING FLIES, INSECT PESTS) and these cause irritation by their bites in addition to loss of blood. More importantly, insects also assume the role of dangerous vectors of a variety of animal and human diseases. From the internal feeding bot-flies to the external parasites like animal lice and fleas, many different groups of insects have independently acquired the ability to associate with vertebrate animals, their homes, products and even excreta.

Carnivorous insects are termed predators or parasites (correctly, parasitoids) based on whether the insects feed on more than one prey during development (predators) or just one host or part of one host (that usually survives) that a parasite attacks. Predation is widespread among insects and it takes several forms according to the insect group in which it occurs and the prey they attack. Both larvae/nymphs and adults may be predacious, or either one only. Like ladybirds or hover-fly larvae feeding on colonies of aphids, the act may be open and very evident. Mantids, however, wait inconspicuously and motionless (like spiders) for their prey to come within reach of their prehensile forelegs. Dragonflies are master predators of the air, consuming their prey while in flight, whereas other winged predators like robber-flies catch them in the air but return

© Kumar Ghorpadé

Larva of hover-fly (Dideopsis aegrota) *feeding on bamboo leaf aphid* × 1¹/₂

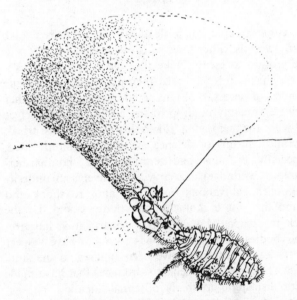

Antlion larva feeding on an ant in its conical pit in sand × 6

to their perch to devour the hapless animal. It seems curious that many insects which spin webs have not learnt to utilize them like spiders to catch their prey, though some caddisflies and mycetophilid flies have. One of the best known traps that insects have devised are the conical pits made by larvae of antlions in sand. The aculeate wasps have learnt to use their prey to feed their young after paralyzing them with their toxic secretions and leave them fresh enough for the time that their young require to feed on them and complete development. As prey to other animals, insects also perform a significant part in the balance of nature. Many freshwater and terrestrial vertebrates (ant-eaters, fish, frogs, birds, bats, rodents, etc.) depend to a large degree on insect food. Invertebrates (besides insects themselves), especially spiders, other arachnids and myriapods are also extensive predators of the abundant insect life on earth. Insects have developed ingenious methods

© Kumar Ghorpadé

A procryptic crab spider on identically coloured yellow flower of Guizotia abyssinica *catching a bee,* Apis florea × 2 ½

for either escaping or misleading their predators. Like many weevils, other insects also feign death or drop to the ground to escape. Like the common house gecko, some phasmatids also shed their appendages to escape enemies, if necessary. Many lepidopterous larvae either wriggle violently or drop to the ground by themselves or with the help of a silken 'rope' when disturbed. Chemical means of defence are also commonly used, especially by ants and termites. The 'bombardier' beetles (Carabidae) discharge, explosively, an undesirable chemical vapour from their anus to shock and immobilize for a moment their hunters, and use the trick to escape. Many moths have urticating hairs on their bodies that even man has learnt to have respect for! And we certainly have been exposed to the stink bugs which secrete repellent substances that have made them distasteful to many potential enemies. The assumption of colours or their patterns by certain insects to either warn predators or to conceal themselves from them, and the ingenious ways by which insects have developed into mimics, is a fascinating study. These are

© Kumar Ghorpadé

Hyperechia xylocopiformis, *a robber-fly mimicking the carpenter bees,* Xylocopa spp. × 1 ½

adaptive responses by these insects to mostly bird predators that hunt by sight rather than by smell or other senses. A type of coloration, structure or behaviour that helps it to escape the attention of a predator, is termed procryptic. Some insects that are genuinely distasteful or are formidable in possessing stings, etc., are flamboyantly and suggestively coloured with aposematic patterns (like danaid caterpillars or vespid wasps) which predators like birds learn to recognize and then avoid. Sometimes, groups of such distasteful or injurious species exhibit a common colour pattern to 'educate' predators, which is called Müllerian mimicry. This warning method adopted by such disagreeable insects has led some more or less uncommon and unarmed insects (like some nymphalid butterflies or hover-flies) that may not be distasteful, to model themselves on these aposematically coloured insects and evade predation. This is termed Batesian mimicry.

Many insects have become parasitic, especially on other insects which they help to keep in tolerable population limits. Much of this parasitism is of a special type, which results in the host being completely consumed (except for the skin) and in its death. This is essentially different from what we normally term parasitism, where the host is allowed to survive by the parasite, which is in its favour. Thus, insect parasites are usually given a different term, and are called parasitoids. These parasitoids are mainly larvae and unlike other insect parasites of vertebrates (animal lice, bedbugs, fleas, biting flies, etc.) attack other insects which they kill. Besides insect hosts of parasitoids, they also attack spiders, ticks, chilopods, terrestrial isopods, earthworms and snails. Most insect parasitoids belong to the Hymenoptera and Diptera (see BENEFICIAL INSECTS, INSECT PESTS, WASPS, FLIES) and these may attack any stage of the host insect (usually beetles, moths and butterflies) from egg to adult. These parasitoids themselves may be parasitized by other insect parasitoids which are called hyperparasites. Parasites of insects may also be arachnids, helminths, protozoa, fungi, bacteria, rickettsias and viruses. The arachnid parasites of insects are mainly mites (Acari), but some mites (and insects) use insects only for dispersal (phoresis).

There are only a few terrestrial habitats and niches that have not been occupied by some group of insects, and few climatic conditions to which none have become adapted. The arctic zones do have a sizeable insect fauna (especially flies) even though activity and reproduction in these extremes is limited to a few months only. The Antarctic continent is the least inhabited, though in early geological periods, it was an important dispersal route for insects of Gondwanaland before the continents drifted. Deserts, islands, cold mountain tops, grasslands, freshwater lakes, rivers, rivulets ('nullahs'), tropical forests, coniferous belts, and even caves have their special insect fauna, specially adapted to these habitats in the course of their and the planet's evolution. Insect diapause (in egg, larval, pupal or adult stages) has been an important factor in the ability of these creatures to survive fluctuating and inimical environmental conditions. Insects are known to hibernate during adverse cold periods in temperate climes and to estivate through the extremely hot summers in the arid tropics. Most groups of insects have also adapted to aquatic or subaquatic life. Larvae/nymphs of all mayflies, dragonflies, stoneflies, most caddisflies, many bugs, beetles and flies, and a few lacewings, scorpionflies, moths and wasps are aquatic. All stages of some bugs and beetles and a few members of other orders are also aquatic, though adults of these can leave water to disperse by flight. The majority of these aquatic insects live in relatively still waters, though a few are found in places where either wave action is strong or the water runs rapidly. Only the sea is almost free of insects, though many insects fly over it to disperse and a specialized chironomid fly skates over its surface. Estuaries and mangroves are inhabited by special kinds of insects that have adapted to this transitional area between fresh and salt water.

Like individuals in human beings that are so distinctly different in face and form from one another, although of the same interbreeding, biological species, different individuals of the same insect species exhibit variation and polymorphism. Physiological changes may be for short periods as the temperature rises or falls or it may be due to a diurnal rhythm. Among morphological changes, the best example is of the solitary and gregarious phases of some grasshoppers which become locusts, mainly because of changes in population density. Aphids exhibit a cyclical polymorphism, where distinct forms appear in successive generation as the annual cycle proceeds. These forms, of which there may be as many as 20 in some species, may be parthenogenetic and viviparous or sexually reproducing and oviparous and may be wingless or winged. Genetic polymorphism, wherein the genetic heterogeneity of a population is maintained by natural selection, favours the establishment of locally adapted subpopulations, which may be of considerable importance in speciation and long-term evolution. Sexual dimorphism is another common characteristic of insects and generally females of such species are also very variable in colour patterns, leading to mimicry and procryptic coloration.

Insect architecture has to be seen and understood to be appreciated. From the silk that we use as clothing, to the silk used to spin webs, this substance is secreted by a wide variety of insects and extruded from glands from the abdominal opening, mouth or even fore-feet. Silk may be used to construct shelters, to weave silken tunnels or traps. Caddisflies and caseworms (psychid moths) construct portable cases that are specific to each species, and may be made of twigs, leaves, sand or pebbles. Spittle bugs (cercopids) live in a spittle-like foam on plants and many coccids have ingenious protective covers. Termites construct some of the largest insect 'houses' which are inclusive of 'air-conditioning' inside them. Wasps, bees and ants use clay and mud to make, as our potters do, exquisite homes for themselves or their young. Paper wasps first taught man the use of paper and the social insects (see TERMITES, WASPS, BEES, ANTS) could teach us how we could improve our own 'primitive' ways to develop a more useful and naturally compatible social organization, that each one of them have been following for the last 50 million years or so. 'Space-age Man' may still do well to reconsider the advice (the 'wisdom of the ancients' that we neglect at our peril) of King Solomon: 'Go to the ant, thou sluggard — consider her ways and be wise. . . .'

See also INSECT IMMATURES, METAMORPHOSIS

INSECT IMMATURES. What is a species? The most reasonable answer to this 'million rupee question' would be: A species is a *kind* of living organism, animal or plant, that is reproductively isolated, in nature, from other such kinds of living organisms. But what is a 'kind'? The best answer to this question that I know of is: A kind (species) is a *life-cycle*. This leads one to the proverbial question — 'What came first, the chicken or the egg?' — which takes us all the way back to the compound DNA.

That the species category (see CLASSIFICATION, CLASSIFICATION AND NAMING) is the most (or only?) objective, the most real of all categories used by taxonomists is an undebatable fact. All other categories, be they supraspecific (Kingdom, Order, Family, Genus, etc.) or infraspecific (subspecies) are subjective, artificial and essentially man-made, to help him classify the diversity of life. Living things are dynamic, very diverse and immensely variable; so is the study of Natural History.

Identifying insects, especially, is very similar to solving a mystery. The clues (diagnostic characters) have to be searched for, found, and the mystery solved through inductive reasoning. For a vast majority of the presently named insect species, we know only the adult stage of the life-cycle; in some cases only one sex is known, and some other species names may be based on just a single specimen or even a fragment of it. Though a few new species have been proposed based on an immature stage of the insect kind, most species have an adult specimen as the primary type. However, in some groups like the whiteflies (Aleyrodidae: see BUGS), an immature stage, rather than the adult, is used in its classification and naming. Since every species is a life-cycle, and we know only a small percentage of the immature stages of insects, it becomes apparent that our present classification of most insect groups is based on very few 'clues' and is therefore highly tentative.

It is not only for the basic requirement of a stable classification that we need to study all the immature stages of each insect species discovered so far; keys to the principal stage(s) responsible for damaging our crops or products are also essential. The life-cycle or life-history of an insect species is a record of all the insect's habits and changes of form from the egg stage to its death, including the habitat, and seasonal occurrence (of each life stage, and, the time spent in each stage). The study of a life-cycle is a very important and fascinating pursuit, most necessary for us to understand a species completely and to gather data to deduce relationships. Each species must be studied principally by *observation* of all its stages, both independently and in relation to its environment, in its natural habitat. Experimentation, in the field or in the laboratory, is only secondary and is liable to be overvalued or misinterpreted by students.

Brief notes on the immature stages of each insect order have been provided in each of the headwords treating them; METAMORPHOSIS has also been individually treated. Here I will deal succinctly with insect development and mention the diversity of each immature stage of both hemimetabolous and holometabolous insects. Except for a few parthenogenetic insect species, where fertilization of the egg by a male sperm is not necessary, the majority of insect species reproduce their 'kind' by fertilization of the female's egg by the male's sperm through copulation. The mating behaviour of insects is another aspect of their natural history that is, to say the least, absorbing!

Though eggs of related species of insects are distinctly different, they are not as variable as the adults that lay them. Since they are very small and difficult to find in nature, most entomologists have not given much time and effort to the study of insect eggs. Recently, the development of the scanning electron microscope and the need to distinguish eggs of many insect groups for both basic and applied studies has led to the improvement of our knowledge and understanding of insect eggs. The ability to find eggs of important pest insects in the field and to recognize them is being used both to predict future damage by them and in their control (see INSECT PESTS). Besides the sculpturing of the eggshell (visible under very high magnification), its size, shape, colour, place of occurrence, mode of laying and attachment, kind of scar made if inserted in plant or animal tissue,

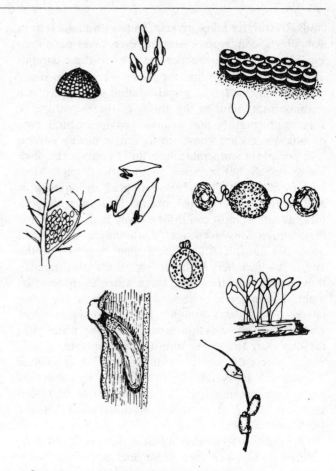

Several types of insect eggs (after *Destructive and Useful Insects* by Metcalf & Flint

whether laid singly or in groups, the precise arrangement with respect to each other, etc., are useful pointers to the specific identity of an insect egg.

The life-cycle of an insect begins with fertilization, when the sperm and egg fuse to form a single cell. When death comes, the insect body is a highly organized, complex living machine made up of millions of such cells. All that happens from fertilization to maturation of the adult insect we call development and growth (or metamorphosis). This period can be neatly divided into two parts: the part before hatching from the egg, called embryology, and that after hatching termed postembryonic development. Embryonic development may occur wholly after oviposition (egg-laying), or partly while the eggs are still inside the mother's body, or entirely inside the mother's body in case of viviparous species. Almost every alteration between these conditions occurs in different insect groups, notably in the more advanced true flies. After egg-laying the egg may hatch in less than a minute (as in some Sarcophagidae flies), take about half a day (as in house-flies), nine months (some Lepidoptera), or even as long as a couple of years (as in some Phasmatodea). In *oviparous* reproduction, the development of the egg needs food that is provided in the form of egg yolk stored inside the egg by the mother insect. No more nourishment or even incubation (as in birds) is necessary until the young larva or nymph hatches out (eclosion).

In some insects, like aphids, the mother does not always lay eggs but retains them inside her body until they hatch and then she 'lays' active young. This is termed *ovo-viviparous* reproduction which is not analogous to *viviparous* reproduction (like in man and other mammals). The young insect here receives all its nourishment from the yolk of the egg and not from the parent's circulation (through the placenta in mammals). In the insect it is simply delayed oviposition or premature hatching. Thus, the two common forms of reproduction that occur in insects are oviparous and ovoviviparous. The condition of viviparity, analogous to that in man, is known only in the case of a few flies like the tsetse (Glossinidae) and the sheep tick (Hippoboscidae).

Insect eggs are usually laid in a situation where the young that hatches out expends as little energy as possible for its immediate food and shelter. After laying the egg(s) the mother insect pays no further attention to them, usually dying shortly afterwards. In the majority of insects there is very little parental care or family life, and the young are left to lead an independent and self-supporting existence right from birth! In some moths whose larvae feed on grasses or their roots, the female may just drop her eggs randomly while flying low over grass. In most instances eggs are laid singly or in batches on the food plants (some beetles, heteropteran bugs, moths and butterflies). In homopteran bugs and long-horned grasshoppers, eggs may be inserted into plant tissues; this insertion, if deep (as in sawflies and cynipid wasps) produces plant galls. In green lacewings, eggs are laid borne on stiff pedicels attached to the plant surface. In other insects they may be laid glued to the plant surface or beneath a web or cottony covering. The cockroaches and mantids lay them enclosed in a firm capsule or ootheca. Many aquatic insects like chironomid flies and caddisflies surround their eggs with a gelatinous secretion that swells and forms a jelly-like spawn in water. Cicadas lay eggs in twigs of trees and many insects insert their eggs in soil (crickets, many beetles). The parasitoids oviposit on or within the host body (tachinid flies, parasitic Hymenoptera); those that attack vertebrates fasten eggs to feathers or hair of hosts (animal lice, some flies).

As mentioned earlier, insect eggs are generally very tiny and are only conspicuous to the naked eye when they are laid in groups or in obvious masses. About a dozen of the smallest eggs would occupy the head of an ordinary pin, and the largest eggs are not more than 3 or 4 mm in diameter. A single female may lay just one egg (like in true females of certain aphids), or as many as a million of them, at least. On an average, most insects lay just a little over 100 eggs. The honey bee queen is known to lay over 2000 eggs every day, while the queen termite lays one every second until millions have been produced, days on end! Eggs may be laid one at a time, or a few of them laid every day for many days. There may also be a number of successive 'batches' of eggs produced at intervals as by the house-fly: it lays from two to seven lots of eggs at intervals of two to five days, each lot consisting of about 125 eggs.

On emergence (eclosion, hatching) from the egg, some insect young are very similar to the adult while others are distinctly different. These are, respectively, what are generally called *nymphs* (Hemimetabola) and *larvae* (Holometabola), which undergo simple or complete METAMORPHOSIS (see also INSECTS). Though young of hemimetabolous insects are commonly termed nymphs, the young of all insects (except the pupal stage of Holometabola) could be spoken of as larvae. Other terms used by the common man for larvae are: caterpillars (larvae of moths, butterflies, sawflies), grubs (of beetles), maggots (of flies), worms (especially of moths and some flies), etc.

Insect larvae assume an immense variety of forms, many of which are clearly adaptive. Some of these have been dealt with in other headwords on insect orders (see also BENEFICIAL INSECTS, INSECT EVOLUTION AND BIOLOGY, INSECT PESTS). Here, the diversity of larval form will be treated under four main groups: protopod larvae, polypod larvae, oligopod larvae and apodous larvae. The *protopod* larva is a little more than a prematurely hatched embryo, without any abdominal segmentation and with rudimentary

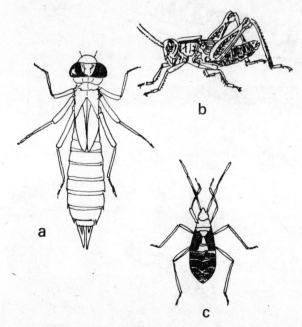

Nymphs—(a) Odonata (b) Orthoptera, (c) Hemiptera

315

appendages on the head and thorax. This type occurs in the early instar of certain parasitic Hymenoptera, especially the families Platygastridae, Figitidae, Dryinidae and Scelionidae. The typical examples of the *polypod* type are the caterpillars of most Lepidoptera, sawflies and scorpionflies which are termed *eruciform* larvae. They have a well defined body segmentation, presence of abdominal limbs or prolegs and poorly developed thoracic legs and antennae. These larvae are relatively inactive, living in close proximity to their food. This type of larva is also present in some parasitic Hymenoptera like the Figitidae and the Proctotrupoidea. The *oligopod* larvae are characterized by the presence of more or less well developed thoracic legs and the absence of abdominal appendages. There is considerable variation in the general appearance of these larvae but two common types can be distinguished. The *campodeiform* larvae possess an elongate, more or less spindle-shaped, somewhat depressed body which is often well sclerotized, long thoracic legs and usually a pair of terminal abdominal processes. They are mostly active

predators with well developed sense organs. Campodeiform larvae occur in Neuroptera, some Coleoptera (especially Adephaga and Stylopoidea) and in the Trichoptera. The second type are the *scarabaeiform* larvae which are stout, cylindrical and C-shaped with shorter thoracic legs, a soft, fleshy body and no processes at the tip of the abdomen. They lead a less active life, occur in the vicinity of abundant food supply, mainly in soil. The Scarabaeoidea possess the majority of this type which also occur in the Anobiidae, Ptinidae, etc. The *apodous* larvae are mainly derived from the oligopod condition and here the trunk appendages are totally suppressed. They may be divided into three types depending on the degree of development of the head. The *eucephalous* larvae (most Nematoceran Diptera; families Bruchidae, Curculionidae, Buprestidae, Cerambycidae, etc., of the Coleoptera; and Aculeate Hymenoptera) have a more or less well sclerotized head whose appendages are relatively little reduced. In the *hemicephalous* larvae (Tipulidae and most Brachycera of Diptera) there is much reduction of the head and its appendages with a marked retraction of the head into the thorax. The *acephalous* larvae (Cyclorrhaphan Diptera) have no obvious head capsule or appendages.

The skin of an insect does not expand like that of a mammal and growth in this inexpansible shell of an exoskeleton cannot be regular and continuous like in a human child. To achieve any increase in its size the insect's shell must be split open. Thus every insect during its growth has to shed its skin one or more times, this process being known as a moult or ecdysis, the cast skins being called the exuviae. The intervals between the moults are known as stages or stadia and the form assumed by the insect larva in each stadium is termed an instar. On hatching from the egg, the larva is in its first instar, at the end of which the first moult occurs and the larva assumes its second instar, and so on. The final instar is the fully mature adult or imago. This is preceded by the pupal or pharate adult stage in the endopterygotes, which we shall deal with a little later. There may be an appreciable interval between the separation of the old cuticle from the hypodermis (outer skin) and its subsequent rupture and casting off during each moult. During this time, the instar within the old cuticle is known as the pharate instar. In some insects the old cuticle is not ruptured but retained as a protective covering for the new instar, as in the puparium of Cyclorrhaphan Diptera, the larval sac of dryinid wasps, etc. The successive shrivelled exuviae may also be retained at the hind end of the body as in some tortoise beetles and parasitic Hymenoptera, or, may be added on to the 'scale' covering of some coccids.

The number of moults that different insects undergo vary from an average of 3 to 20 in most insects. In extreme cases, as many as 60 moults have been recorded

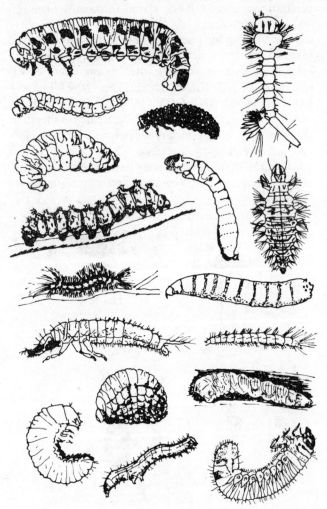

Diverse kinds of insect larvae (after *Destructive and Useful Insects* by Metcalf & Flint)

for a collembolan. Though little importance is given to the phylogenetic significance of the number of moults in each group of insects, it is generally considered that specialized taxa have fewer instars than do more primitive ones. There is also considerable variation in the number of moults in a single species of insect, and this may be hereditary, influenced by environmental temperature, inadequate nutrition or other adverse conditions.

The imago or adult is the fully developed mature stage of an insect with functional reproductive organs and associated mating or egg-laying structures, besides fully formed and usable wings in alate groups. The only known exception to the winged adult stage is the mayfly order Ephemeroptera, in which group the stage before the winged reproductive adult also has wings which it uses. This curious preadult instar with functional wings is called a *subimago*. Other peculiar immatures are found in the thrips, whiteflies, male coccids and winged female chermesids and phylloxerids which are sometimes regarded as aberrant Hemimetabola or Holometabola. Their early immature stages are without rudiments of wings and are very distinct from their adult form, while one or more inactive or semiquiescent stages (suggestive of the holometabolan pupa) precede the adult instar.

In most insects the larval instars of each species are similar in general appearance and in their feeding habits, differing only in size. However, in some groups, two or more of the instars are very different in form, being specialized for different modes of life in their respective instars. This phenomenon is called larval *heteromorphosis* or hypermetamorphosis. It is common in parasites that lay their eggs far from their host individuals. The first instar larva (called planidium or triungulin) is a mobile form with bristles, tails or other processes and is able to stand erect and is often capable of jumping on to its host or another insect that will carry it to its ultimate host! There is remarkable convergence (similarity in form) in planidia of several different insect groups that undergo heteromorphosis. From the second instar onwards, when the host is found, the larvae are sedentary with none of the modifications of the first instar and become stout and grub-like. Heteromorphosis of this type occurs in Neuroptera (Mantispidae), Coleoptera (all Meloidae, Rhipiphoridae, Drilidae and Stylopoidea; some Carabidae, Colydiidae and Staphylinidae), Diptera (Acroceridae, Bombyliidae, Nemestrinidae, some Sarcophagidae and Tachinidae), Hymenoptera (many Parasitica, especially subfamilies Eucharitinae and Perilampinae of Pteromalidae), Lepidoptera (ectoparasitic Epipyropidae), and Coccoidea (*Margarodes* and allies), besides one specialized Siphonaptera. In most parasitic Hymenoptera the eggs are laid on or in the host, so in these the first instar is not a plani-

dium. Some agromyzid flies, bruchid and cerambycid beetles and leaf-mining moth larvae of the families Gracillariidae, Phyllocnistidae, etc., also exhibit major differences in at least two of the larval instars. In any case, when all groups of insects are considered as a whole, there is a complete series of intermediates between the heteromorphic form and what is normal.

In all holometabolous insects there is a resting, inactive instar that precedes the adult stage which is termed the pupa. In most endopterygote insects the change from larva to adult is so profound that the insect cannot possibly attend to any other functions besides that which is going on inside it in the pupal stage. In this period of transformation the insect cannot eat or move about, its breathing is reduced and it is for all purposes quiescent externally, but internally it is as active as at any period after embryonic development. Since the pupal stage neither feeds nor moves about it is neither beneficial nor injurious to any other organism. In this transitional period the body of the larva and its internal organs and systems are being remodelled to the extent necessary to adapt them to the requirements of the future imago. Though the term pupa usually denotes the entire pre-imaginal stage, it should be noted that for a varying period before the adult emerges from it, the 'pupa' actually represents a pharate adult. The limited locomotion that pupae are observed to achieve is usually the result of movements by the pharate adult which lies inside the old pupal cuticle separately. The prepupa is nothing other than the pharate pupa, but this term is used in more than one sense.

Except for some specialized Coleoptera, Diptera, Hymenoptera and Lepidoptera, most pupal or pharate

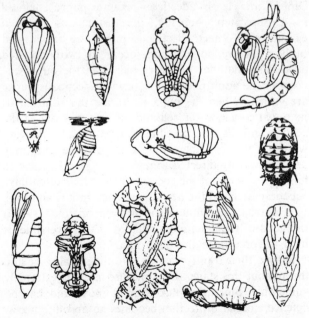

Different kinds of insect pupae (after *Destructive and Useful Insects* by Metcalf & Flint).

adult stages of endopterygotes are passed in a cocoon or cell of some kind. In the cyclorrhaphan flies the last larval cuticle (puparium) serves as the cocoon. The manner in which the adult effects its escape from the cocoon is used to differentiate the major types of pupae. In primitive endopterygotes the pupae have powerful, sclerotized, articulated mandibles by the use of which the adult escapes from the cocoon after which it sheds its pupal cuticle. This occurs in the Neuroptera, Mecoptera, Trichoptera and some primitive Lepidoptera and is termed a *decticous* pupa. The pharate adult of groups that have decticous pupae also uses its feet besides the pupal mandibles to wander far from the site of the cocoon through soil or water before it sheds its pupal cuticle. The other major type of pupa is called an *adecticous* pupa and here the pharate adult does not have mandibles to use to escape from the cocoon or cell. Two main forms of adecticous pupae are known but intermediates are also recorded. An *exarate* pupa is one in which the appendages are free and not attached to the body wall. They are found in Siphonaptera, most Coleoptera (including Stylopoidea) and Hymenoptera and in the cyclorrhaphan and most brachyceran Diptera. An *obtect* pupa is one in which the appendages are strongly attached to the body, the exposed surfaces of the pupal appendages being much more strongly sclerotized. Such pupae are prominent among the higher Lepidoptera, some Coleoptera, chalcidoid Hymenoptera and nematoceran Diptera. Decticous pupae are never obtect and may evolve directly into the adecticous exarate pupa (e.g., some Trichoptera, probably Siphonaptera, most Coleoptera and Hymenoptera). The so-called *coarctate* pupae of cyclorrhaphan Diptera are clearly adecticous exarate pupae enclosed in a puparium. Actually, the distinctions between exarate and obtect adecticous pupae are trivial and both types are polyphyletic (occur in two unrelated taxa). The methods of emergence from the pupal cocoon of the adult pharate instar with adecticous pupae are varied. When the pupa is exarate the adult sheds its pupal cuticle in the cell and bites its way out of the cocoon with its own mandibles, as in most beetles and wasps. When the pupa is obtect, the pharate adult may use backwardly directed spines on the pupal cuticle to force its way out of the cocoon. Such pupae often have cocoon-cutters on the head as in many moths and flies. In cyclorrhaphan Diptera, the ptilinum or frontal sac on the head of the emerging adult fly, that is eversible, is used to force open the puparium.

During their transformation into the pupal stage and throughout this stage, the immature insect is extremely vulnerable. Since the pupa cannot move and has no defence, special protection becomes an absolute necessity. Concealment from their enemies and from adverse weather conditions, shock and other mechanical disturbances is carried out by the last larval instar. Many moth and beetle larvae burrow into ground and there construct earthen cells in which to pupate. The majority of insects construct cocoons of silk that may be used to bind over extraneous material like wood chips, body hair, pebbles, fragments of plants, skins of prey, etc. Others may be procryptic or enclosed into a protected situation under bark, inside logs, under stones, grass or leaf mould, etc. Like moths of the family Saturniidae whose pupae are enclosed in a thick cocoon, many other insect pupae are similarly covered. A few specialized species of Coleoptera, Diptera, Hymenoptera or Lepidoptera may not even have a cell, cocoon or puparium, but be strongly obtect and sclerotized or protectively coloured, or, be enclosed in the host integument as in chalcidoid parasitic wasps.

When the time comes for the eclosion (emergence) of the imago, the pupa either darkens noticeably in colour or the colours of the adult become apparent. By means of movements of the legs and body the pharate adult succeeds in rupturing the cuticle of the pupa and emerges fully formed except for the wings. Crawling up the nearest available support, the teneral adult rests in a position to allow the wings to expand by forcing blood into them from its body. It also excretes liquid drops (meconium) from the anus which represent the waste products of the pupal metabolism. The first trial flight is then undertaken when the wings are completely expanded and the insect dried and hardened. The time of emergence is also peculiar for different species and may be early in the day or towards evening. In certain aquatic insects, the adult is able to fly immediately after emergence. Adults are seldom sexually mature immediately after emergence and in most cases males require a few days to mature and the females take longer. In short-lived insects like mayflies for example, mating occurs within a matter of hours after emergence and eggs are laid shortly afterwards. The life of insect adults varies considerably. On an average the ones with a short life-span live only for a couple of weeks or more while long-lived forms may survive for a year or more.

Two peculiar phenomena in insect life-cycles may also be mentioned. In some gall-midges (Cecidomyiidae) there occurs precocious reproductive maturity in larvae or pupae which results in their producing eggs or living young. This is called *paedogenesis*, which is always combined with a complex or irregular cycle of generations. The other phenomenon of *parthenogenesis* is the ability to reproduce without fertilization in certain adult insects. It may occur only irregularly in some species, and in Hymenoptera, females that lay eggs without being fertilized (mated) produce only males while eggs of inseminated females produce either males or females. In some other parthenogenetic insects no males are normally produced and their females lay unfertilized eggs that produce only females.

When conditions are favourable, the development of many insects proceeds without any interruption. Even if it is stopped temporarily because of adverse conditions, it is immediately resumed when they improve. On the other hand, many insects can, under certain conditions, pass into a state of *diapause* (dormancy) when development is inhibited even though external conditions appear suitable for growth. Diapause may consist of retardation of growth for a few weeks or complete cessation of development for several years, but it is ultimately 'broken' and the life-cycle then continues normally. Diapause also occurs at a definite stage of life-history, which may be the egg, larva, pupa or adult, depending on the species. Some species enter diapause every generation and others do it every other generation regularly. In some moths like the silkworm (*Bombyx mori*) different strains behave differently. The reasons for diapause are many, mainly external factors, which also interact. Variations in light intensity and temperature, desiccation and unfavourable nutrition are known to be responsible. Exposure to low temperatures usually results in breaking diapause but mechanical injury, etc., is also effective. Insect metamorphosis is controlled by hormones, of which the 'juvenile hormone' produced by the corpora allata in the prothoracic glands is responsible for larval development. In the last larval instar, production of this hormone ceases and development is directed towards formation of the adult, under the influence of the 'growth and differentiation hormone'.

The enormous increase in size of many insects from birth to maturity is astonishing, especially when compared to that in man, where the increase is around 20-25 times only. Some of the following ratios for increases in weight of the newly hatched larva to the full-grown stage will indicate the existing range in insects: aphid, 16; confused flour beetle, 85; locust, 126; honey bee, 1576; silkworm moth, 8417; and cossid moth, 72,000.

When we realize that almost the entire life of an insect is passed in its immature stage, when only a quarter or a fifth of ours is, not only the fact of the insects' efficient use of energy but also of the enormous labour ahead of us in first finding and then observing and learning about insect young strikes the entomologist. From the time of Aristotle (384–322 B.C.) up until now, many humans have spent whole lives studying insects, but what is known and correctly recorded is but an insignificant part of the whole. In my teens I was struck by the words of a song in an American film, which, I think, summarized the meaning of life:

> Multiplication, that's the name of the game,
> And each generation plays the same . . .
>
> —'Come September'

INSECT PESTS. Most courses in Entomology, or Insect Science, deal with insects as enemies of man. Besides the all-too-rare hobby of man of collecting and preserving beautiful insects (mainly butterflies and beetles), we have made an attempt to study insects in the field, classroom and laboratory mainly with the objective of finding ways and means of dealing with the pestiferous species that have hounded us from time immemorial. I could do no better than quote the American entomologist, S. A. Forbes:

The struggle between man and insects began long before the dawn of civilization, has continued without cessation to the present time, and will continue, without doubt, as long as the human race endures. It is due to the fact that both men and certain insect species constantly want the same things at the same time. Its intensity is owing to the vital importance to both, of the things they struggle for, and its long continuance is due to the fact that the contestants are so equally matched. We commonly think of ourselves as the lords and conquerors of nature, but insects had thoroughly mastered the world and taken full possession of it long before man began the attempt. They had, consequently, all the advantage of a possession of the field when the contest began, and they have disputed every step of our invasion of their original domain so persistently and so successfully that we can even yet scarcely flatter ourselves that we have gained any very important advantage over them. Here and there a truce has been declared, a treaty made, and even a partnership established, advantageous to both parties of the contract—as with the bees and silkworms, for example; but wherever their interests and ours are diametrically opposed, the war still goes on and neither side can claim a final victory. If they want our crops, they still help themselves to them. If they wish the blood of our domestic animals, they pump it out of the veins of our cattle and our horses at their leisure and under our very eyes. If they choose to take up their abode with us, we cannot wholly keep them out of the houses we live in. We cannot even protect our very persons from their annoying and pestiferous attacks, and since the world began, we have never yet exterminated—we probably never shall exterminate—so much as a single insect species. They have, in fact, inflicted upon us for ages the most serious evils without our even knowing it.

However, our existence itself is due to the fact that only an insignificant percentage (less than 0·1%) of living insects have bothered to compete directly with us; the vast majority are either beneficial to us or are in no way concerned with man (see BENEFICIAL INSECTS). From our early attempts at 'extermination' or 'eradication' (some urban pest control agencies still style themselves as 'Pest Eradicators') of insect competitors, we later tried to just 'control' pests, now we hope to only 'manage' these organisms, but, ultimately, we shall have to learn to 'coexist' with them—that is, if we succeed in controlling our own numbers first and then stop fighting with each other!

So much is written about insect pests, even in our subcontinent, that I consider it unnecessary to delve in any detail into the 'major' and 'minor' insect pest species in India and neighbouring countries. Instead, I will consider the many ways insects are harmful to us, our animals and crops, with examples. Later, the ways and means that we have found to combat our insect enemies will be covered, ending with views of our future against our most serious competitors.

Insects that had selected plants either before man began tilling the soil after his emergence from the apes, or that have subsequently become habituated to his crop plants as an easy means of achieving their daily food requirements, are perhaps the most important of our insect enemies. Those that compete with us for the plants that we have found economical or useful could be differentiated into several groups according to the plant part they attack, the way they consume its parts, or by their mode of making the plant suffer or die.

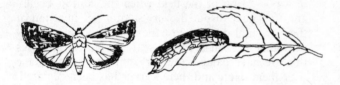

Biting and chewing pest: moth and caterpillar × 1

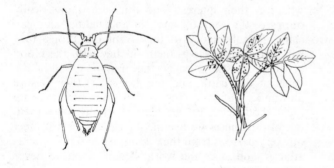

Piercing and sucking pest: aphid × 10 and its colony on plant

According to the mouth-parts of insects, they could be divided into those that ingest plant matter by biting and chewing, and those that pierce and suck. Very much like a grazing cow or horse, many insects chew off external plant parts, grind them with their mandibles and swallow both the solid and liquid parts together. Almost all parts of a plant are chewed up by several groups, which eat leaves, buds, stems, flowers, fruit or bark, besides roots and other plant parts. Either or both the larval and adult stage of the chewing pest

insect may feed on any part of the host plant. The Deccan Grasshopper (*Colemania sphenarioides*), which is unable to fly, and restricted to the Indian peninsula, feeds mainly on the leaves of jowar (*Sorghum bicolor*), but also on other millets such as bajra, navane and ragi. The Pink Bollworm (*Platyedra gossypiella*) that attacks and feeds on flower buds and later, the bolls, of cotton is, on occasion, a serious pest of this fibre crop. The larvae of the Pea Stem Fly (*Ophiomyia phaseoli*) burrow into the stems of cow pea and some other pulses, causing them to wither and dry after they feed on the soft parts of the stems. The brilliant black and red blister beetle (*Mylabris pustulata*) feeds on flowers of many plants in large numbers, in the appropriate season every year. The Common Guava Blue butterfly (*Virachola isocrates*) has caterpillars that bore into and feed on fruits of pomegranate and litchi, besides guava. Caterpillars of the Bark Borer Moth (*Indarbela tetraonis*) are commonly seen (rather, their frass) on trunks and branches of trees where they feed on the bark at night. The white grub of a common chafer beetle (*Holotrichia serrata*) feeds on the roots of several crop plants.

The other group of plant-feeding insects pierce the epidermis and suck the cell sap. Only the liquid portions of the plant are ingested through a sharp point of one of the mouth-parts which pierces the plant and the liquid is then sucked through the tubular body of the piercing-sucking mouth-part. The Bugs (order Hemiptera) comprise the majority of this type of plant-feeding insects, although insects of some other orders also have piercing-sucking mouth-parts, but feed on animal blood (see MOSQUITOES & BITING FLIES, FLEAS, ANIMAL LICE). As with the biting-chewing insect pests, these piercing-sucking ones also feed on leaves, buds, stems, fruits, etc. One group of moths, that are fairly large and attractively patterned, pierce ripening fruits of *Citrus*, guava, etc., with their sharp proboscis and suck the fluids. They belong to the subfamily Ophiderinae (family Noctuidae) and are mainly tropical in distribution, with *Othreis fullonia* and *O. materna* being the common 'fruit-sucking moths' in India. The aphids or plant-lice are the most common and cosmopolitan insect pests that form their colonies of mostly females and nymphs on leaves, stems, buds, flowers and fruits of our crop plants and pump out the plant's juices. There is hardly any wild or cultivated plant that does not have at least one aphid species addicted to its sap. The Cotton Aphid, *Aphis gossypii*, is one of the most widely distributed, common, and polyphagous species in our subcontinent, with many coloured forms. Some other aphids, such as *Tetraneura nigriabdominalis*, live underground and suck sap from roots of ragi and other millets. Several other insects of the order Hemiptera (see BUGS) assume pest proportions on man's cultivated

23 Two Moths

The Tassar Silk Moth has been bred for centuries for the value of the fibre produced from its cocoons x ¼

As can be seen from this photograph of *Spiramea rotata*, moths are normally cryptically coloured. On account of their nocturnal habits moths are not as well known as butterflies.

© E. Hanumantha Rao

The Blackwinged Stilt (*Himantopus himantopus*) can be seen wading in the waters of marshes, jheels, village tanks, salt pans and mudflats all over the subcontinent.

© E. Bharucha

The Purplerumped Sunbird (*Nectarinia zeylonica*) is seen flapping her wings in defence of her nest.

© E. Bharucha

The Blacknaped Blue Flycatcher (*Monarcha azurea*), a sparrow-sized bird, holds on to a slender stalk while keeping a watchful eye out for some likely insect prey.

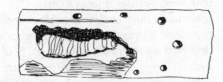

Stem borer: cerambycid adult × 2 and larva × 1½ in bore

© Kumar Ghorpadé

Longicorn beetle. Many cerambycids are injurious to forest trees × 1¼

© C.I.B.C., Bangalore

Fruit-flies ovipositing × 4

which do more damage than the insect's feeding itself. The piercing-sucking pests are also more difficult for man to control.

Many insects damage crop plants by boring and tunnelling in the bark, stem or twigs ('borers'); in fruits, nuts or seeds ('worms' or 'weevils'); or between the surfaces of the leaves ('leaf-miners'). Several longicorn beetles bore into trunks (*Anoplophora versteegi* on orange), stems (*Batocera rufomaculata* on mango), twigs (*Chelidonium cinctum* on citrus), and roots (*Dorysthenes hugelii* on apple). Several caterpillars of moths (*Euzophera, Leucinodes, Nephopteryx,* etc.), larvae of fruitflies (*Dacus* spp.), and weevils (*Sternochaetus mangiferae* on mango nuts), bore into and feed on fruits and nuts. Many weevils (*Calandra, Sitophilus,* etc.) attack grain and seeds of various crops. Among leaf-miners, several agromyzids (*Melanagromyza* spp.) and caterpillars of microlepidoptera (e.g., *Phyllocnistis citrella*) are commonly found in India. Some insects cause cancerous growths on plants, within which they live and feed. These are called gall insects, and the more common are thrips and flies among others. Many cicadas, tree

Gall wasp × 14 and its plant gall

hoppers, tree crickets and weevils damage cultivated plants by laying their eggs in some part of the plant. Leaf-cutter bees and leaf-cutting and other ants carry leaves and other parts for constructing shelters or lining their nests. Some ants carry coccids and aphids to the plant and establish them there, usually close to their nests for their own use (feeding on honeydew), and thus enable these noxious insects to infest crop plants. Many scolytid beetles, flies, bees, grasshoppers, other beetles, aphids, thrips, whiteflies, etc., disseminate organisms that cause plant diseases (fungi, bacteria, protozoa, mycoplasma and viruses) by injecting them into the tissues of the plant as they feed, or carry them into their bores or tunnels, or make wounds through which such disease organisms gain entrance. Another very interesting kind of association that insects are responsible for is their role in bringing about cross-fertilization of certain rusts that cause diseases of crop plants. Black stem rust of wheat, *Puccinia graminis*, growing on barberry, and the related *P. helianthi*, on sunflower, is assisted to produce spores only by visiting insects.

plants and suck sap from several plant parts. Some of the more common pests are the coccids, whiteflies, psyllids, mirids, leafhoppers, stink bugs, lace bugs, chinch bugs, squash bugs and others. Among the biting-chewing and piercing-sucking kinds of insect pests, the latter are evidently the more dangerous, especially as quite a few of them also help transmit plant viruses

Besides the damage insects do to man's cultivated plants, they also annoy or injure him and his livestock or pets (and wild animals). The many ways by which insects cause annoyance to man is by their presence in places where we object to them (several kinds of insects); by the sound of their flying about or 'buzzing' (e.g., bees, wasps); by the foul odour of their secretions or decomposing bodies (cockroaches); by the offensive taste of their secretions and excretions left upon fruits, foods, dishes, tableware (flies, cockroaches, store pests); by irritations as they crawl over the skin (ants, small beetles, flies); by chewing, pinching, or nibbling the skin (ants, small beetles, wasps, bees); by accidentally entering eyes, ears, nostrils, alimentary canal, causing myiasis (flies); or by laying their eggs on skin, hairs, feathers (bot flies, animal lice).

The general unpopularity of insects stems mainly from an irrational fear and hate of the 'creepy, crawly' things. However, there are some insects that, in defence, bite or sting painfully. Venoms of insects can cause bodily pain and illness in the following ways: (1) by biting with their mouth-parts (flies, bugs, animal lice, fleas); (2) by stinging, with a sting located at the abdominal tip (ants, bees, wasps); (3) by nettling with hollow poison hairs located on the bodies of some caterpillars (families Limacodidae, Lymantriidae); (4) by leaving caustic or corrosive fluids on the skin, when they are crushed or handled (blister beetles); and (5) by poisoning animals when they are accidentally swallowed (blister beetles, rose chafer beetle).

By making their homes on or in our bodies and those of our animals, as external or internal parasites, insects do a lot of damage. Insect parasites, those that are either beneficial to us or harmful, are of several kinds (see BENEFICIAL INSECTS). The animal lice and fleas are all parasitic on animals and man, and several groups of true flies and bugs are also inimical to their health. The majority of these insects live on the skin and their constant movement on it causes nervousness, restlessness, loss of sleep, failure to feed, and thus a general poor physical condition and increased susceptibility to diseases. With their excreta, they also contaminate animal hair, fur and feathers, leading to interference with the excretory function of the skin. Animal lice (except Mallophaga, bird lice), fleas, mosquitoes, other biting flies, bedbugs, etc., either bite or pierce the skin and feed on blood. The bird lice that feed on all sorts of birds including poultry, and some mammals like horses and cattle, are incapable of drawing blood and feed on solid particles like dry skin, parts of feathers or hairs, clots of blood, and the like, and their chief injury is perhaps due to nervous irritation through their nibbling of the skin and running about over it. The internal insect parasites that tunnel into muscles, nasal, ocular, auricular, or urogenital passages, cause mechanical injury and promote infections. They are more dangerous than the external parasites and belong to the true flies, besides mites (Acari). The cattle grubs, horse and sheep bots and ox warbles (family Oestridae) and the screw-worm fly (family Calliphoridae) are fly larvae that feed on the internal body parts of animals and even man (*Dermatobia hominis*, in South America). A few other kinds of insects, habitually, and some others, accidentally, live in the alimentary canal of animals. The horse bots belong to the former group, and many other flies (bluebottle flies, flesh-flies, house-flies and rat-tailed maggots) of the families Calliphoridae, Sarcophagidae, Muscidae and Syrphidae, occur accidentally, causing myiasis, being swallowed with impure drinking water, milk or infested food. Their presence in the stomach generally causes symptoms of nausea, vomiting and fever.

Insects also disseminate diseases (bacteria, protozoa, fungi, viruses, parasitic worms) from sick to healthy animals, from some wild animal (reservoir) to man and domestic animals, or from a diseased parent or earlier life-stage, in one of the following ways: (1) by accidentally conveying pathogens from filth to food; (2) by transporting pathogens from filth or diseased animals to the lips, eyes or wounds of healthy animals; (3) by the insect host of a pathogen being swallowed by the larger animal, in which the pathogen causes disease; (4) by inoculating the pathogen hypodermically as the insect bites an animal; or (5) by depositing the pathogen upon the skin, in faeces, or through its proboscis, or in its crushed body, the pathogen entering through the insect's bite, or a scratch, or the unbroken skin. Some of the more common or important diseases prevalent in the Indian subcontinent are dealt with briefly here. Malaria, a protozoan disease transmitted by the bites of around 100 species of anopheline mosquitoes, is, undoubtedly, the most important disease of man. Species of *Aedes* mosquitoes transmit dengue in man and several other mosquitoes are responsible for encephalitides in man, horses and birds. Filariasis is also transmitted through bites of several genera of mosquitoes, while Papataci fever is due to sand-flies of the genus *Phlebotomus*, which also transmit leishmaniasis and kala-azar. Bubonic plague, or the 'black death' of the Middle Ages, is a bacterial plague that is transmitted by bites of the oriental rat flea (*Xenopsylla cheopis*), and other fleas. In India, between 1896 and 1917, almost 100 lakhs of people died from this disease.

Finally, insects destroy or depreciate the value of stored products and other possessions of man, including food, clothing, drugs, animal and plant collections, paper, books, furniture, bridges, buildings, timbers of mines, telephone poles, telegraph lines, railroad ties, trestles, and the like, (1) by devouring these things as their food; (2) by contaminating them with their secretions, excretions, eggs or their own bodies, even though the actual product itself may not be eaten;

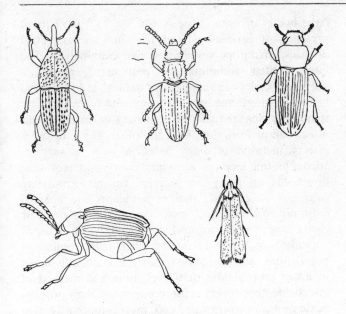

Store pests: (clockwise from top left) — rice weevil, *Sitophilus oryzae* × 10; saw-toothed grain beetle, *Oryzaephilus surinamensis* × 7; confused flour beetle, *Tribolium confusum* × 7; pulse beetle, *Callosobruchus* sp. × 7; rice moth, *Corcyra ciphalonica* × 5

(3) by seeking protection or building tunnels within or on these substances; and (4) by increasing the labour and expense of sorting, packing and preserving foods. The more frequent of these household or stored product pests are ants, termites, powder post beetles, silverfish, cockroaches, several moths and other beetles, including weevils.

We shall now look at ways that man has found to keep populations of his insect enemies within tolerable limits, either by directly killing them or by making life miserable for them and preventing their increase. Control measures that depend on man for their application or success are termed applied control, and those that do not need man for their continuance or success are called natural control measures. When certain insect species that are harmful to man have not been held in check by natural factors, or when the natural balance has been interfered with or upset by man himself, applied control measures become necessary. These can be differentiated into (1) chemical control by the use of insecticides, repellents, attractants, and auxiliary substances; (2) physical and mechanical control by specially designed machines or other manual devices, and the special manipulation of physical factors of the environment; (3) cultural control by variations in the usual farm operations; (4) biological control by the introduction and establishment of natural enemies of insect pests; and (5) legal control, by regulating commerce, farming, and other human activities that affect the prevalence and distribution of dangerously destructive insects, the success of insect control operations, or the health of man.

Applied control is, as a rule, expensive, and the amount that one can reasonably expect to save by using these methods needs to be weighed against the expense involved. Generally, only crops or articles of high value to man (horticultural, stored product, cash crop or household) could be economically protected by applied methods of control that have been reasonably well developed. However, practical applied methods of control have not been satisfactorily worked out for many insect pests attacking field and forest crops, and those attacking livestock and man. The five groups of applied insect control measures are dealt with very briefly here, as more detailed literature on these aspects for the interested reader are legion. Chemical control involves the use of insecticides and accessory agents formulated as dusts, granules, pellets, sorptive dusts, water suspensions, solvents, emulsifiers, wetting or spreading agents, stickers, deodorants, masking agents, stabilizing agents, etc., and applied by dusting, spraying, fumigating and even by use of aircraft. Insecticides may be grouped into three general classes: (1) stomach poisons, (2) contact poisons, and (3) fumigants. Another way of classifying them is by their chemical nature and source of supply, such as (1) inorganic compounds, generally effective only as stomach poisons; (2) synthetic organic compounds which may act as contact, stomach or fumigant poisons; and (3) organic compounds of plant origin, largely employed as contact poisons. Stomach poisons are generally used against biting and chewing insects, but in some cases may also be employed to control piercing-sucking pests. The main compounds that are used as stomach poisons include arsenicals, fluorine compounds, organic stomach poisons, minor stomach poisons, systemic poisons and poison baits. Contact poisons are more efficient in killing insect pests with piercing-sucking mouth-parts, and include nicotine alkaloids, pyrethroids, synergists (activators), rotenoids, synthetic organic insecticides, dinitrophenols, organic thiocyanates, DDT and derivatives, BHC and lindane, chlorinated terpenes, cyclodiene insecticides, organophosphates and carbamates. Systemic insecticides (e.g., Dimethoate, Phosphamidon) when applied to seeds, leaves, stems or roots of plants are absorbed and translocated within the plant to all its parts and have the advantages of minimizing the inequalities of spray coverage to some extent, protecting new plant growth subsequent to application, increasing the length of residual control, and being less damaging to beneficial predatory and pollinating insects. Systemic insecticides like Phorate (Thimet) are useful as seed or soil treatments to protect young plants from insect pests like aphids, leaf-hoppers, thrips or mites. Similar systemic compounds are useful against animal parasites, when fed to them or topically applied. Acaricides like Kelthane are specially effective against mite pests. Other insecticides and accessory agents used include miscellaneous organic insecticides, soil poisons, wood preservatives, sulphur and lime-sulphur compounds, oils, emulsions and soaps. Fumigants are gaseous poisons used to kill insect enemies and are useful against these in enclosed spaces, regardless of the type of mouth-parts. Methyl bromide, carbon tetrachloride and ethylene dibromide are some of the fumigants used among many others. Repellents like Bordeaux mixture (mainly a fungicide but useful against some insect pests) are used on plants, as repellents to bloodsucking insects, and mothproofing materials are employed indoors to protect clothes, books, etc., against insect pests. Among attractants that are used mostly as food or sex attractants, those such as methyl eugenol employed to attract fruit-flies (Tephritidae) are very useful in reducing populations of males. Chemical control through the use of insect poisons does have limitations like residual problems and development of resistance by the target organisms. However as a rapid and convenient tool for control of most of his insect enemies, chemical control has been a 'boon' to man.

Physical and mechanical control measures are special operations conducted by man, either manually (destruction of eggs on plants by hand, digging of trenches as physical barriers) or through use of mechanical gadgets (e.g., use of low temperatures, radiation, superheating), and generally give immediate results. Cultural control measures involve use of ordinary farm practices and are usually preventive and indirect methods differing from physical/mechanical control measures. Some of the cultural measures used are crop rotations, tilling of soil, destruction of crop residues and weeds, variation in planting/harvesting time, and use of resistant crop varieties. BIOLOGICAL CONTROL involves the introduction and encouragement of natural enemies of generally non-endemic and accidentally introduced insect pests. The natural enemies used may be other predacious and parasitic insects, vertebrates, nematodes, or diseases and this may be integrated with insecticide treatments, the latter being selectively applied to protect the natural enemies. Legislative control measures that include quarantine and inspection laws, compulsory clean-up measures, insecticide laws, etc., are used to prevent introduction of new pests from foreign countries, to prevent spread of pests within the country or state, to enforce control measures to deal with established pests, to prevent adulteration and misbranding of insecticides, to determine their permissible residue tolerances in foodstuffs of man and domestic animals, and to regulate the activities of pest-control operators and the application of hazardous insecticides.

Natural control measures, however, include control by (1) climate factors like sunshine, rainfall, heat, cold and wind movement; (2) physical characteristics of the country and its topography such as mountain ranges, large bodies of water, type of soil, etc.; (3) the *natural* presence of predacious and parasitic insects, insect diseases, predatory vertebrates, and cannibalism. The latest control technique of 'pest management' involves an understanding of such natural and ecological factors in the immediate environment. We now have, thankfully, realized that present data on many of our insect competitors (let alone the other 99·9% of other insects, inclusive of many beneficial ones) is very inadequate; that levels of 'economic injury' and 'economic thresholds' of pests are required before formulating strategies of control; that some plants are able to naturally tolerate insect 'pests' without effect on their yield; that 'weeds' around cultivated fields/orchards actually help maintain natural enemy and pollinator populations, and that their removal may actually result in lower crop yields; that level, time and method of insecticide application is more important than prophylactic pesticide treatments; and, most importantly, that short-term benefits, monetarily, could irreversibly upset man's long-term requirements. An insect species is, *per se*, never a pest; insect pests are usually man-made.

What of our future against our insect competitors? As I started, I could do no better than to end with one of the quotations (by the noted early naturalist, E. B. White) used by the history-making, forthright American lady naturalist, Rachel Carson, in her devastating book, *Silent Spring* (1962):

> I am pessimistic about the human race because it is too ingenious for its own good. Our approach to nature is to beat it into submission. We would stand a better chance of survival if we accommodated ourselves to this planet and viewed it appreciatively instead of skeptically and dictatorially.

The chemical poisons that we are now supposedly 'dependent' upon are termed 'ecological narcotics' with justification, and, like the ominous (and unnecessary!) nuclear weapons that we are irresponsibly playing with, these could spell the doom of mankind and most of the earth's living inhabitants, proving Albert Schweitzer's forecast: 'Man has lost the capacity to foresee and to forestall. He will end by destroying the earth.'

INSECT STRUCTURE. Insects belong to the Phylum Arthropoda, the word meaning 'joined legs'. Insects differ from other arthropods (which include spiders, centipedes, millipedes, woodlice, crabs, lobsters, etc.) in possessing only three pairs of legs. Almost all other arthropods have more—spiders having four pairs and millipedes as many as 300 pairs! The segments of legs follow the pattern of our own limbs and are called coxa, trochanter, femur, tibia and tarsus, from the body outward. The insect body is also characteristically divided into three parts—head, thorax and abdomen. Besides

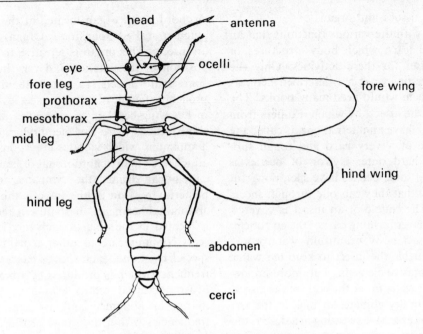

EXTERNAL PARTS OF AN INSECT

this, insects are the only arthropods that have wings (no other invertebrate group possesses them) but some insects may have them reduced or almost modified into stubs.

The head bears a pair of antennae (or 'feelers'), which carry a variety of sense organs. Also present on the head are a pair of simple and/or compound eyes and a set of laterally symmetrical mouth-parts, which include a pair of mandibles, a pair of maxillae and a labium (or lower lip) and labrum (or upper lip). These are the more important, basic parts of an insect's body. The other parts that will be used are shown in the accompanying illustrations. Both external structure and internal anatomy of the insects are made easier to understand through the figures supplied, with the main organs or structures marked.

The internal anatomy of the insect has the same major organs (digestive, circulatory, respiratory, reproductive, excretory, etc.) that we have. The nervous system lies in the ventral region of the insect's body and the circulatory system composed of the long, tubular heart is dorsal in position. The digestive system from the mouth to the stomach and gut to the anus is located in the centre. Many insects have what are called 'crops' and 'gizzards' in the fore part of the digestive tract, meant for crushing and grinding and even storing of food. The excretory system of the insect consists of a bundle of thin tubes called 'Malpighian tubules' attached between the mid gut and to the hind gut. The reproductive organs, testes (male) and ovaries (female), are located in the abdomen. The circulatory system is different from ours in that there are no blood-carrying vessels, but the heart merely pumps it through large vacant spaces within the body cavity so that the blood bathes the various organs, takes up nutriment from the intestine and enables the Malpighian tubules to absorb the excretory products from blood. As in mammals and other vertebrates, the major function of blood is not to

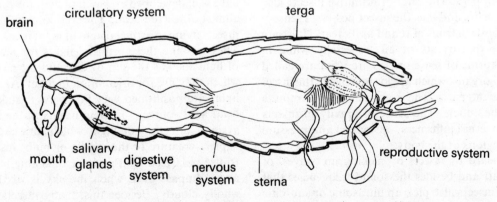

INTERNAL SYSTEMS OF AN INSECT

carry oxygen to the tissues and organs.

Now we will deal with the various functions that an insect performs and learn which body structures or internal organs it uses for these activities. Only the major or most interesting ones are described here, as dealing with all of them would need many books! The external skeleton of the insect, by which it differs from most vertebrates, who have an internal one, is called the cuticle, and is made of a very hard and tough substance, chitin. The hard outer elytron of beetles is heavily slerotized and some beetles have such powerful mandibles (or 'jaws') that they can bite through sheets of copper or zinc! The cuticle of an insect is covered by a thin waxy outer layer forming part of the epicuticle, which makes the insect body beautifully waterproof! Not only does this enable the insect to keep *out* water and other toxic or abrasive elements, but, more importantly, it conserves the water *inside* the body of the insect, even when they live in dry climates or high in the air. What we could lose by excessive sweating in a desert, the insect's exoskeleton conserves, to great advantage. One disadvantage is that the hard outer skeleton prevents growth—therefore the process of 'moulting' in insects wherein they undergo several 'instars', or larval stages, which is preceded and followed by 'shedding' of their skin, and, in the time that the outer skeleton is soft and is hardening, they 'grow'.

The insect sense organs are very different from the ones we have and though they perhaps react to the same sort of stimuli as we do, their senses of touch, smell and sight are received by organs peculiar to them. These sense organs occur in the insect cuticle and each sense cell (or a group of such cells) is connected to the central nervous system by nerves. Since the cuticle is discarded at each larval moult and during pupation to the adult stage, the entire nervous system of the insect changes at each step, the structures being more complex in the adult than in the immature stages. The hairs on the insect body are the simplest kind of sense organs that respond to touch. The antennae have a concentration of such sensory hairs as well as other organs resembling small pegs. The cerci of primitive insects also carry a lot of hairs and thus the insect possesses sensory appendages both in front of it and at its rear. Hairs are also sensitive to currents of air, especially while in flight. Other forms of sense organs are what are called 'chordotonal' organs, which are located inside the insect body in different parts. These are sensitive to minute stresses on the cuticle. The sense of gravity in insects depends on external influences, on contact and pressure at different regions of the body.

Sound is heard by insects by picking up 'waves' of compressed air and besides the sound frequencies that we can hear, insects also pick up ultrasonic or subsonic waves, that are too high or too low for our ears to detect.

A chordotonal organ located at the base of the insect antenna, called Johnston's Organ, occurs in almost all insects. This organ is sensitive to any change in the position of the antenna and very highly developed in some male insects, to enable them to detect an approaching female of their own species from the frequency of her wing-beat! In grasshoppers and crickets and in some cicadas there is a membrane in the cuticle called 'tympanum' which vibrates in response to sound, and is attached to many chordotonal organs beneath it. Often called 'ear drums' the tympana are located on the posterior portion of the insect's thorax or in the first abdominal segment. In some crickets, the tympanum is located in the tibia of each fore-leg. Since these insects communicate to other individuals of their own species by sound, each species has a very critical sound frequency which is produced by vibrating these organs. Most nocturnal moths also have tympana on their thorax or abdomen and are sensitive to ultrasonic frequencies within the range of those emitted by bats, and quite inaudible to us.

Organs that are receptive to light stimuli are located on the head and are of two forms, eyes (compound, multifaceted) and ocelli (with a single lens). In some insect larvae and pupae, ocelli are their only 'eyes' and are located on the sides of the head. In adult insects, like moths, flies, etc., that have a pair of compound eyes, there are also present three ocelli on top of the head. Their exact function is not well understood but if their use is lost, the insect's vision reduces even though the compound eyes are functional. There may be more than 20,000 facets or ommatidia in each compound eye, and in some dragonflies and, in flies and moths, there may be many more than this number. However, in other insect groups, like the ant workers, there may be only 10 or so and it is known that ants are almost 'blind'. The images that these ommatidia produce can be likened to a black-and-white photograph when seen through a strong hand lens. Different kinds of insects see only different shades of colour and not all colours are seen by all insects. Insects like most flies and dragonflies have well developed eyesight, in flies, the eyes occupying almost all of the head. The well known habit of many insects flying to artificial light at night is easily explained. They mistake these artificial lights for natural sources of light such as the sun by day and the moon by night, which they use for orientation in flight. As these natural lights are positioned very far away, mistaking the artificial light sources for them results in the insect taking a spiral direction of flight which ultimately leads the hapless creature to the candle or bulb. Insects are able to use polarized light and are thus able to use the sun as a compass even when the sky is partly or almost wholly cloudy. Besides this, many insects also have a 'biological clock' that works all 24 hours, called 'circadian rhythm'. This sort of an 'internal clock' enables the insect to know the 'time' of day or night for its usual chores. Though we also possess such a physiological 'clock', our routine use of mechanical time suppresses this natural 'clock'—the 'jet-lag' of human beings that travel a lot by air across the world is proof of this natural 'time-keeper' in us.

The sense of taste and smell that man separates cannot really be distinguished in him or the insect. A chemical substance floating in air is 'smelt' while another dissolved in liquid or present in food is 'tasted' by us. Insects also have strong senses of taste/smell and will reject unsuitably smelling or tasting food substances. These chemoreceptors are located in the mouth-parts of many insects and in some may even be located in the tarsi of the fore leg! These senses are in general much more developed in insects than in man. These strong chemoreceptors enable insects to find food, to determine, in the case of parasitic wasps, if the host egg has already been oviposited in or not, to smell flowers that are acceptable by the particular species, to locate sex pheromones released by females of their own species, and other things. The antennae are also structures that enable the insects to smell and taste.

The insects breathe very differently from the way we do. The oxygen is carried by internal tubes called tracheae to the tissues. These tracheae open out of the body through spiracles in terrestrial insects, located on the sides of the body segments in larvae and on the thorax and abdomen in adult insects. These spiracles are capable of being opened and closed by valves. Since this means of diffusing oxygen into the body is very inefficient, only if the body size were small would such a respiratory system work suitably. It is perhaps by devising this means of respiration that insects have succeeded in limiting their overall body size, which is an important factor in their tremendous success as an animal group.

Locomotion in insects is carried out by walking or flying, generally. The three pairs of legs most insects have are used either for walking or for more progressive methods of terrestrial movement, like jumping. Larvae of insects like the familiar caterpillars of butterflies, have the 'true' legs in the fore part of the body and the hind part has several unjointed clasper-like feet on the abdominal segments. With these the larva can crawl about. In some moth larvae, 'feet' are missing from the major central portion of the body so that these larvae are called 'loopers' from the way they progress with their limited 'feet'. Adult insects like fleas and grasshoppers have peculiarly developed legs (usually hind pair) that enable them to do prodigious leaping. Many aquatic insects and their larvae are able to swim, using modified structures on their legs which they use as oars or paddles. Many insects live in soil and can be

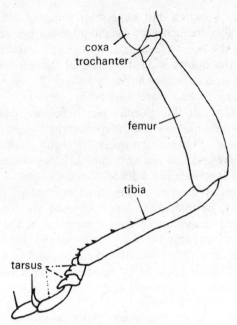

PARTS OF AN INSECT'S LEG

called 'burrowers'. Like the mole cricket with powerful 'fore arms' that enable it to dig into soil, and the digger wasps, there are many other insects that possess structures that help them get in and out of soil that is their abode.

Of all arthropods, only insects have attained the power of flight and were perhaps the first animals to fly, as far back as 300 million years ago. At that time, dragonflies had wing-spans of almost two feet! Typically, insects have two pairs of wings which arose from lateral expansions of the thorax. Wings of insects are extremely diverse and the two pairs function in different insect groups in different ways. In some, like many moths, the fore and hind pair are locked together by structures at their bases so that they can be beaten together in flight. The wings are formed of strong rods or veins and a reticulate pattern of cross-veins that strengthen the sclerotin that makes up the wing membrane. In butterflies and moths and some other insects, the wing may be covered with scales of iridescent colours or, like in the dragonflies, they may be transparent (hyaline) or bare. In many flies for example, the wing membrane bears hundreds of small hairs called microtrichia that almost uniformly cover the wing expanse. In beetles, the fore wing is hardened and used as a protective cover for the hind wings and the abdomen below it. In true flies, like the house-fly and the horse-fly, the hind wings have been reduced to what are called halteres or knob-like structures that serve as gyroscopes or balancing organs. In these flies, the thorax is well developed to house the strong muscles that are needed to operate the single pair of functional wings. Large insects like the dragonflies and locusts beat their wings about

20 times a second, but some small flies can achieve a remarkable frequency of over 1000 times per second! Besides the sailing flight that a dragonfly exhibits, recalling the double-winged aircraft of the First World War, some flies like the hover-fly and bee-fly can remain stationary in one place and hover, somewhat like a kestrel bird can, though much more efficiently. Naturally, we would like to know the speeds which these insects can attain. The fastest insects are again dragonflies, horse-flies and hawk-moths, and such larger creatures that fly at speeds of 15 to 50 kilometres per hour or even more. With the amazing structures that are carried by insects or that insects are composed of, they have succeeded in colonizing almost all parts of the earth in soil, water (only the sea remains almost totally free from insect activity), air, above ground and in vegetation, and have become the most overwhelmingly dominant form of life on our earth, making up more than three-fourths of all forms of life.

See also INSECT IMMATURES, METAMORPHOSIS

INSECTIVORA, Insect-eating mammals. Though geographically widespread (except in Australia and Antarctica) this is one of the less diverse orders of mammals, generally considered to include the more primitive and less specialized forms. Nowadays they are classified into eight families and about 63 genera, with a total of less than 800 species compared with over 2000 described species of bats and about 5000 described species of rodents.

Insectivora, as their name implies, have a mainly insect diet but they are a confusing group taxonomically and can only be described by a combination of characters, as they do exhibit considerable anatomical variation as well as ecological adaptability. Most insectivores are relatively small mammals, all have five toes on fore and hind feet. Most have weak, poorly developed eyesight but an acutely developed sense of smell and elongated rather mobile pointed snouts. While the water shrews and some tenrecs have webbed feet and are adapted to an aquatic existence, the moles are almost sightless and live and feed subterraneously. The vast majority of insectivores, however, live and forage on the ground, scurrying rather clumsily on the heel and forepart of their feet (plantigrade), having relatively short legs. Most are omnivorous in feeding, often tackling prey as large as themselves, and they are characteristically voracious in appetite and aggressively hostile when they encounter other members of the same species. Their teeth are relatively unspecialized and undifferentiated so that it is difficult to distinguish between canine, pre-molar or molar teeth, but they are rooted and usually sharply cuspidate, enabling them to crush the hard exo-skeletons of insects and crustacea.

In the region covered by this encyclopedia three families are represented. The Erinaceidae includes Moon Rats and Gymnures found outside our area in Malaysia and southeast Asia, as well as all the HEDGE-HOGS. Through their ability to hibernate or estivate, hedgehogs have adapted to withstand periods of food shortage or adverse weather and they have therefore been able to colonize many of the more inhospitable arid and mountainous regions of northwest India and Pakistan. Five species are known, divided into two genera; *Hemiechinus*, with relatively long ears and no parting of the spines on the forecrown; and a second genus *Paraechinus*, characterized by a peculiar deep furrow of naked skin on the forecrown which produces a parting of the spines. The Pale Hedgehog (*Paraechinus micropus*) is the only species which extends its range down into peninsular India.

All hedgehogs, as is well known, have coarse hair on their face, flanks and belly but their crown and backs are covered with short stiff pointed spines. Newly-born offspring of the Longeared Hedgehog (*Hemiechinus auritus*) have a sparse scattering of whitish 2 mm-long soft spines in their body at birth. Within five hours these spines have hardened and darkened in colour and grown to 8 mm in length, and fresh spines sprout within a day of birth. Litter sizes vary from 2 to 7 and the young are weaned between 3 or 4 weeks of age. The gestation period is about 35 to 45 days and only the female takes part in rearing the family. In captivity Hedgehogs have lived for up to 10 years. Many species are fond of ripe fruit and they will fearlessly attack and eat scorpions as well as small snakes.

The second family found in our region comprise the Shrews or Soricidae. With 20 genera and nearly 200 described species these relatively tiny mammals form the major group within the order. The smallest living mammal, which has been found in widely scattered parts of India as well as Pakistan, happens to belong to this family and is the Etruscan or Pigmy Shrew (*Suncus etruscus*). Adults weigh as little as 7 grams and newly-born young, which are blind and naked, are hardly bigger than house-flies. They live mostly in leaf litter and are thought to feed on small soil-born crustacea and insects. The well known House Shrew (*Suncus murinus*) also belongs to this genus. About the size and shape of a small rat, they are widespread in southeast Asia and occur often in association with man and his dwellings in nearly all regions of the plains of India. Males have a scent gland on the flank from which a powerful musky odour is emitted during certain seasons, hence they are also called Musk Shrews. Clothed like all the shrews in soft velvety fur, they are largely nocturnal in hunting, scurrying along the base of walls or inside drains, seizing whatever edible creature comes in their path. Litter sizes vary from 1 to 5, with 3 being the average number of young born, and breeding can be at any time of the year. The gestation period in most shrews is very short, 17 to 28 days, and in the House Shrew the young are weaned and fully covered with fur at about 3 weeks of age and young females are able to breed at about 2 months of age. The young accompany their mother on her nocturnal forays, trailing behind her in a train, each holding to the tail of the one preceding it and the first in line holding on to the mother's tail.

In the Himalayas the Lesser Shrew (*Sorex minutus*) occurs in the extreme northwest of Pakistan and Kashmir. These tiny animals survive in the alpine pastures and are distinguished by the cusps or crowns of their teeth being red in colour. Another genus of so-called white-toothed shrews is *Crocidura*, represented by 3 or 4 species in India and Pakistan. *C. russula* is found in the Himalayas, and *C. pergrisea* is confined to the dry hilly regions in Baluchistan and the North West Frontier Province. Another genus, *Soriculus*, is represented by several species occurring in the north-eastern Himalaya from Kumaon to Sikkim.

Another unique shrew (*Feroculus feroculus*), the Long-clawed Shrew, is found only in the mountainous regions of Sri Lanka, whilst a Water Shrew (*Chimmarogale platycephala*) has been collected from Kashmir and the northern Punjab hill regions. In Sikkim, Bhutan and adjoining Tibet the Webfooted Water Shrew (*Nectogale elegans*) also occurs. Because of the difficulty of attracting these small animals to conventional rodent traps and baits, very few specimens have been collected of these various localized species of shrew, and even less is known about their habits and life history.

The third family represented in our region is the Talpidae which includes Moles, Desmans and Shrew-moles. In the extreme northeast of the Himalaya and the Assam foothills the Indian Short-tailed Mole (*Talpa micrura*) occurs. This animal lives its entire life in subterranean burrows and, like the European mole, lives mostly upon earthworms. They have short, very velvety, black fur and almost vestigial eyes, with very broad powerful spade-shaped forefeet adapted for digging and permanently turned outwards at right angles to the body. They are believed only to be able to distinguish between light and darkness, having lost the need for sight, and they are active in foraging both by day and by night, able to move as rapidly backwards as forwards, along their narrow earth tunnels. Another subspecies with a slightly longer club-shaped tail is known as the White-tailed Mole (*Talpa micrura leucura*) and this has been collected from the Khasia and Naga Hills.

Recent researches have indicated that the Tree Shrews (of the family Tupaiidae) have closer affinities with primitive primates than with Insectivores despite their common name.

See also HEDGEHOGS, TREE SHREWS.

INSECTIVOROUS PLANTS, also known as Carnivorous Plants, are interesting because of the wonderful methods they have adopted for catching their small animal prey. Although they trap insects and consume them, the prey never forms more than a very small fraction of the food of these rare plants, all of which possess green leaves and obtain their main food supply in the normal way by photosynthesis. Why then do these plants trap insects? They are mostly found in swamps and bogs where there are few plants and animals. This causes nitrogen deficiency in the soil. Insects and small aquatic animals have a high protein content and so the insectivorous plants make up for deficiency of nitrogen, especially of protein, by digesting their prey. It is, however, quite possible for them to live healthily without these animal-supplements to their food. Out of over 450 species of carnivorous plants in the world, about 40 species are found in the subcontinent. These are classified as Sundew, Butterwort, Venus' Fly-trap, Water Fly-trap, Pitcher Plant and Bladderwort.

Sundew, the species of *Drosera*, of which three viz. *D. peltata*, *D. burmannii* and *D. indica* are found in the subcontinent including Burma, are small herbs whose leaves are covered with numerous glandular tentacles. The reddish glands secrete a viscous fluid which glitters

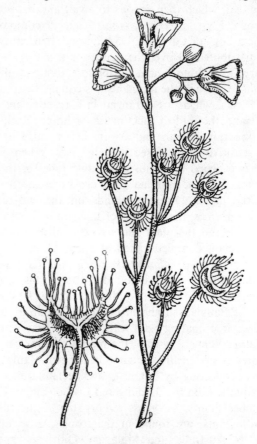

Sundew × 1

in the sun like dewdrops, hence the name 'Sundew'. The glands are sensitive and react only to chemical stimuli. When any insect mistaking the glistening substance for honey alights on the leaf, it gets entangled in the sticky fluid and the tentacles stimulated by the digestible compounds present in the insect-body bend down on it from all sides and cover it. When the insect is suffocated to death digestion begins by means of a protein-digesting enzyme. After the process is completed, the leaves resume their normal shape and the indigestible parts of the insect are blown away by the wind. The tentacles are then ready to capture another insect.

Butterwort is represented by only one species in India, *Pinguicula alpina*. It is called 'Butterwort' because the upper surface of the leaves is covered with a pale-yellow sticky substance which looks like a thin layer of butter. This herb with feebly developed roots grows in the alpine Himalaya. An unwary insect, alighting on the sticky leaf, gets caught like a fly on a fly-paper. The margins of the leaf are incurved and when an insect is caught they roll more inwards so as to hold it. The upper surface of the leaf is covered with numerous glands which give off a juice containing enzymes that help in splitting up the proteins of the insect-body to make them digestible. The digested proteins are absorbed by the surface of the leaf, which then unrolls to await another victim.

Venus' Fly-trap, *Dionaea muscipula*, is common in the peat-bogs of Carolina, USA and is often cultivated and distributed all over the world for teaching natural history. Like Sundew and Butterwort it catches insects with its leaves and the mechanism is equally fascinating. The leaves, which are arranged in a rosette, are two-lobed and the midrib functions as a hinge. Each lobe bears marginal teeth and on its upper surface three long sensitive hairs jointed at their bases. When one of these is touched by an insect the two lobes of the leaf snap together, the marginal teeth cross one another and trap the insect. Digestive glands on the leaf-surface secrete enzymes, the proteins of the captured insect are absorbed, and the plant gets its extra nitrogen. When the process is completed the leaf opens again.

Water Fly-trap, *Aldrovanda vesciculosa*, is a widely distributed plant which is found in the salt-lakes of the Sunderbans salt-marshes south of Calcutta, freshwater jheels of Bangladesh and in ponds in Manipur. It is a rootless, free-floating plant with whorls of leaves. It is a miniature Venus' Fly-trap; the mechanism for catching the prey is practically the same in the two plants. Due to over-collecting by botanists and reclamation works this aquatic is on the 'Endangered List' of Indian Flora.

Pitcher Plants, the most remarkable of the insectivorous plants, are found in the tropics, especially in Asia, N Australia and Malagasy. Only one species, *Nepenthes khasiana*, is found in India (Khasi, Jaintia and Garo Hills). It is a climbing undershrub which climbs by means of the tendrillar stalk of the leaf. The pitcher, stalk of the pitcher and laminated structure are the modifications of the leaf-blade, the petiole and the leaf-base, respectively. The pitchers vary from 10 to 20 cm or even more in height and are brightly coloured. When young the mouth of the pitcher remains closed by a lid which afterwards opens and stands more or less erect, thus leaving the mouth open. The rim of the mouth has an incurved margin with a firm shining surface. The plant exudes a sweet substance which acts as a bait for the insects. The insect, attracted by the bright colour of the pitcher and the sweet exudation, crawls up to the mouth of the pitcher and tries to get the sweet food material just inside. In doing so it slips on the shining rim and tumbles down through smooth and sharp hairs pointing downwards. The struggling insect is thus trapped, becomes tired out after a while, falls down into the liquid and is finally drowned. The liquid contains protein-digesting enzymes secreted by the digestive glands on the inner surface of the pitcher. The insect is digested and ultimately the nitrogenous food absorbed (see PLANT DEFENCES).

Bladderworts are common insectivorous plants with about 30 species in the subcontinent, of which *Utricularia aurea* is the commonest. They are mostly floating or slightly submerged rootless aquatic herbs with finely dissected leaves looking like roots but for their green colour. Some of the leaf-segments get modified into bladders.

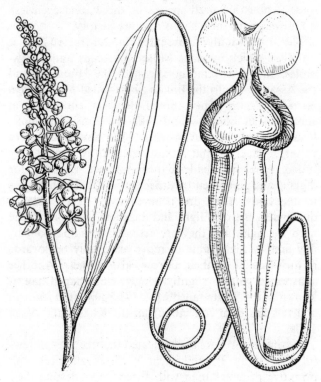

The Pitcher Plant (× 1/3) with its leaf modified to capture and digest insects.

Each bladder is 3 to 5 mm in diameter and is provided with a trap-door entrance which could be pushed open only inwards but never outwards. Small aquatic animals enter by bending the valve which easily gives way. After their entrance, the valve shuts itself automatically, thus preventing their escape. Numerous digestive glands present on the inner surface of the bladder secrete enzymes and absorb the digested food.

K.N.B.

INSECTS. The Class Insecta (or Hexapoda) is the largest of the Phylum Arthropoda, containing invertebrate animals. At least 85 percent of all known animal species (in essence, of all forms of life on earth) are insects. Estimates of the actually existing number of insect species have risen meteorically (as a result of more areas of the world, and habitats, being sampled more thoroughly) from a modest 625,000 species in the 1940s to a figure of 20 to 30 million species today.

Insects are, obviously, the most successful and abundant animals on earth and have evolved and adapted to colonize almost every single land and fresh water habitat. They are man's major competitors for the planet's resources, and, whatever we may think, are as highly evolved (certainly better adapted for survival) as ourselves. They are the only invertebrates that have acquired the ability to fly and their small size and hardened exoskeleton, besides many other ingenious structural, physiological and behavioural adaptations, have enabled them to rival us humans as the dominant form of life on our green planet.

Adult insects are different from the other members of the Phylum Arthropoda in having the body divided into three regions (head, thorax and abdomen); in the presence of three pairs of legs and one or two pairs of wings on the thorax; in the absence of any appendages

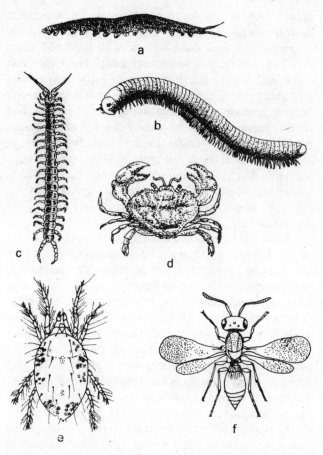

Members of major classes of phylum Arthropoda — (a) Peripatus, (b) millipede, (c) centipede, (d) crab, (e) mite, (f) insect

on the abdomen used for walking; in having only a single pair of antennae on the head; and by breathing through tracheae or branching air-tubes that open through spiracles along the sides of the body. The other

© Kumar Ghorpadé

A few of the larger members of major Insect orders

main classes of the Arthropoda include the Arachnida (mites, scorpions, spiders, ticks, etc.), which have no antennae and four pairs of legs, with the body being divisible in only two regions at most. The Chilopoda (centipedes) have the body not divisible into thorax and abdomen, one pair of antennae, each body segment bearing a single pair of legs and spiracles, and the first pair of legs modified into poison claws. The Crustacea (crabs, shrimps, woodlice, etc.) have two pairs of antennae and at least five pairs of legs, and the body segments divided into two regions only. The Diplopoda (millipedes) resemble the Chilopoda, except that the apparent body segments each carry two pairs of legs and spiracles.

I have used each headword to deal with one Order of the Insecta, and have tried to summarize essential information about each one of these. The headwords Metamorphosis, Insect Structure, Insect Evolution & Biology, Insect Immatures, Insect Pests and Beneficial Insects deal briefly with these aspects and should serve to inform readers and provide needed perspective. With the word space provided for this section in the encyclopedia, it has not been possible to deal as efficiently with insect diversity as perhaps was possible in other sections. The insects section of the encyclopedia is therefore only an introduction to these fascinating creatures and for the more inquisitive and interested reader, I suggest the following books which I myself have used liberally in the preparation of many headwords:

1) *The Insects of Australia*—Edited by D. F. Waterhouse, Melbourne University Press, Carlton (1970).
2) *Indian Insect Life*—By H. Maxwell-Lefroy, reprinted by Today & Tomorrow's Printers & Publisher, Delhi, (n.d., *c.* 1970).
3) *Destructive and Useful Insects*—By C. L. Metcalf and W. P. Flint, revised by R. L. Metcalf, McGraw-Hill Book Company, Inc., New York (1928, international student edition 1962).

The following table gives the general classification of the Insects into Subclasses, Divisions and Orders, with the common names used for each headword and, in parentheses, the percentage of each Order among the Insecta (= 100%) in number of known species. Some Orders have been treated in more than one headword, e.g., the Hymenoptera covered by the headwords Ants, Bees, Wasps.

Class INSECTA

Subclass APTERYGOTA: (primitive, wingless insects with metamorphosis slight or absent)

Order   PROTURA: Primitive Insects (0·06%)
  DIPLURA: Primitive Insects (0·04%)
  COLLEMBOLA: Primitive Insects (0·25%)
  THYSANURA: Silverfish (0.09%)

Subclass PTERYGOTA (winged or secondarily wingless insects with metamorphosis varied, rarely slight or wanting)

Division EXOPTERYGOTA (or Hemimetabola; metamorphosis simple, sometimes slight, with pupal instar rarely present, wings developing externally, and immatures similar to adults in structure and habits, generally called nymphs)

Order   EPHEMEROPTERA: Mayflies (0·25%)
  ODONATA: Dragonflies & Damselflies (0·6%)
  PLECOPTERA: Stoneflies (0·2%)
  ORTHOPTERA: Grasshoppers & Crickets (2·5%)
  MANTODEA: Mantids (0·2%)
  PHASMATODEA: Stick & Leaf Insects (0·3%)
  DERMAPTERA: Earwigs (0·12%)
  EMBIOPTERA: Webspinners (0·3%)
  BLATTODEA: Cockroaches (0·4%)
  ISOPTERA: Termites (0·25%)
  ZORAPTERA: Zorapterans (0·003%)
  PSOCOPTERA: Psocids (0·3%)
  PHTHIRAPTERA: Animal Lice (0·307%)
  THYSANOPTERA: Thrips (0·4%)
  HEMIPTERA: Bugs (9·3%)

Division ENDOPTERYGOTA (or Holometabola; metamorphosis complete, with pupal instar, wings developing internally, immatures differing from adults in structure and habits, called larvae)

Order   NEUROPTERA: Lacewings & Ant lions (0·6%)
  MECOPTERA: Scorpionflies (0·05%)
  LEPIDOPTERA: Butterflies, Moths (15·3%)
  TRICHOPTERA: Caddisflies (0·75%)
  DIPTERA: Flies, Mosquitoes & Biting Flies (12·2%)
  SIPHONAPTERA: Fleas (0·25%).
  HYMENOPTERA: Ants, Bees, Wasps (18·25%)
  COLEOPTERA: Beetles (37·0%)

INSTINCTIVE BEHAVIOUR. An oyster or barnacle shutting up when a shadow suddenly crosses over it, a panther stalking a chital, a peacock displaying before a peahen and a chameleon changing its colour as it moves: all of this wide variety of activities will be considered behaviour. Behaviour is essentially the activity of an animal's effectors—primarily muscles, but also others such as colour-cells. Behaviour then is essentially any activity of muscles, and such activity can generally be traced as a response to some environmental stimulus. For an animal's receptors or sense-organs are continually furnishing it with information about its surroundings and an animal uses this information to appropriately adjust the activity of its muscles. If you clap loudly near another person's face he will invariably wink. This is because a loud noise near one's face has a high chance of being associated with injury to the eyes, so that we have evolved to instantaneously protect the eyes by closing them. This action is released from the youngest age and invariably follows whether the person is hungry or well fed, angry or calm, healthy or sick. Charles Darwin went on to try another experiment in front of a glass cage containing a cobra. When Darwin stuck his nose to the glass, the cobra was provoked to strike. Although Darwin was perfectly safe with the glass in between them, he invariably jerked his head back as the cobra struck. And in spite of special efforts he could not keep himself from drawing back as the cobra struck. Such stereotyped behaviour, which is expressed the very first time that a situation presents itself, and which is not subject to change with repeated experience, is the most typical instance of what is termed instinctive behaviour.

Such instinctive behaviour involves a stereotyped response to a given environmental stimulus irrespective of an animal's internal state. But much other instinctive behaviour depends on the animal's internal state. Thus a gazelle walking close by does not release hunting behaviour in a lion which has just had a big meal, but does release it in a lion which has not eaten for a long time; or a Hanuman langur female does not solicit copulation when she is not in heat, but does so near the peak of her estrus. Much instinctive behaviour thus involves stereotyped responses to particular environmental stimuli, but whether the response is made or not depends on the animal's internal state. In the appropriate physiological state the response is invariable: a hungry lion always hunts a gazelle, and a female Hanuman langur in heat always solicits copulation.

Much animal behaviour however changes in response to the animal's experiences, and changes in an adaptive fashion. Thus a barnacle shuts when a shadow suddenly passes over it. But if this keeps on happening repeatedly, it stops shutting after a while. Birds learn to utilize new sources of food. Thus a milk bottle on the doorstep did not initially release a feeding response in European tits. But then some tit discovered that puncturing the foil and sucking the cream collected at the top did indeed provide a good food source. The habit rapidly spread over Europe, with tits watching other tits feeding from milk bottles adopting the habit themselves. Such adaptive changes in response to stimuli as a result of experience are known as learning; and such learned behaviour is distinguished from innate or instinctive behaviour.

The relative significance of instinctive as opposed to learned behaviour used to be a much debated and highly controversial issue. It was a part of the general Nature versus Nurture controversy. It is now recognized that such a dichotomy is artificial, and that all behaviour is a result of an interaction between the genes and the environment. There is then a spectrum of behaviours, from those which are essentially genetically programmed such as the winking reflex, to those which involve a great deal of learning, such as feeding preferences.

A striking property of instinctive behaviour is that it is often released by very simple stimuli which constitute only a part of the overall stimulus situation. The European robin males set up territories in the breeding season and attack other males who enter their territory. Experiments with dummies have shown that it is the red feathers on the breast that release this attack. Thus a stuffed male with the red feathers painted brown evokes no aggression, while just a tuft of red feathers on a piece of wire is attacked vigorously. Similarly, in the ten-spined stickleback, aggressive behaviour of a territorial male is released by the red belly characteristic of territorial males, and courtship behaviour by the swollen belly characteristic of a female ready to spawn.

Why is it that only a few stimuli of the stimulus object are employed to release a particular kind of behaviour? At any time a stimulus object can present a plethora of stimuli. Not all these stimuli would however be available all the time because of a variety of infering factors from the environment. While choosing some combination of stimuli to identify an object, the animal could make two kinds of mistakes. It could, for instance, conclude that a rival male is present when it is in fact absent, or it could conclude that a rival male is absent when it is present. There would be a certain cost attached to each kind of mistake. For instance, it is known that there are certain non-territorial males who lack a red belly and are responded to as females and who take advantage of the situation to fertilize the eggs of a female who has spawned in the nest of the territorial male. The animal would then evolve to use those particular stimuli which minimize

Shore-birds instinctively prefer a giant artificial egg to their own normal-sized egg.

the overall cost attached to such mistakes. By and large, this would lead to the most easily and regularly available stimuli assuming the role of the sign stimuli which release a given behaviour.

These sign stimuli would be particularly simple and stereotyped in those cases where the behaviour is a response to other members of their own species such as mates or rivals. For the attributes of the conspecifics will be very constant as opposed, for instance, to the attributes of potential prey animals. The development of simple sign stimuli is therefore expected to be most pronounced in behaviour patterns involving conspecifics. Such sign stimuli are then known as social releasers. For example, the red breast-feathers of the European robin serve as a social releaser, releasing territorial behaviour. Very often animals develop specific behaviour patterns in conjunction with the sign stimuli to render these stimuli more conspicuous. Thus the Pekin Duck male has some particularly striking feathers in his tail. These feathers serve as social releasers during the courtship behaviour of the duck. During courtship the male performs special preening movements which draw attention to and render these feathers all the more conspicuous. Thus in this case, preening which normally serves the function of maintaining the barbs of the feathers in proper alignment serves to heighten the effectiveness of a social releaser. This phenomenon whereby a structure or behaviour normally functional in some other context assumes the function of a social signal is known as ritualization.

This process of ritualization is particularly well documented in the case of intention movements which precede any movement in animals. Thus birds stretch their legs and flap their wings before flying off. Such movements have been exaggerated and have become stereotyped to serve as a social signal to co-ordinate flight in a flock of birds. Thus in the case of some species of pigeons the wing-flaps not only coordinate movement, but also serve to communicate the distance to which the pigeons will fly; this being proportional to the number of times that wings are flapped during the signal.

Besides intention-movements, another category of behaviour which has evolved to serve a signalling function is known as displacement behaviour. Displacement behaviours are those behaviours expressed when the animal simultaneously experiences two equally powerful opposite tendencies. Thus during the territorial fights of herring gulls, a herring gull may be equally motivated by an urge to attack and an urge to flee. Instead of doing either, it may then pluck grass. It has been suggested that some social signals (such as courtship preening in the Pekin Duck, mentioned above), may in fact have evolved by ritualization of displacement behaviour.

Apart from the external stimuli, an animal's be-haviour may change with internal factors. Thus a vulture which has gorged itself well on a carcass during the morning will stay perched on a tree, whereas a vulture which has not had any food in the morning will soar up and search for carcasses. Such variation in behaviour due to internal factors is said to be due to variation in the motivational state of the animal. An animal is said to be driven by a number of different drives such as hunger and sex. The behaviour expressed at any moment is supposed to depend on the relative strength of the different drives. The behaviour prompted by any drive is classified into two components: appetitive and consummatory. Appetitive behaviour is the behaviour involved in searching for the stimulus object which is the goal of the particular drive activating the animal. Thus if a vulture is driven by hunger, it will soar in the sky and look for carcasses on the ground. It will also watch the other vultures soaring in its neighbourhood. This behaviour is relatively variable. When the vulture locates a carcass on the ground, or another vulture missing from its patrol in the sky, the appetitive behaviour comes to an end. For either of these stimuli imply that food has been located. If the vulture has itself located a carcass, it will directly swoop down to it. If on the other hand it finds one of the neighbouring vultures missing from its patrol, it will move to where it should have been and look for a carcass on the ground. It is likely to locate the carcass on the ground with the missing vulture descended on it. It will then swoop down to it.

This will end the appetitive component of behaviour propelled by the hunger drive. This behaviour of searching for the stimulus object ends when the stimulus object is located and releases another set of behaviour termed as consummatory behaviour. This involves the vulture settling near the carcass, approaching the carcass, perhaps fighting with other vultures and dogs for access to meat, tearing away and feeding on pieces of meat, etc. Compared to the appetitive behaviour this consummatory behaviour is relatively stereotyped. With the accomplishment of the consummatory behaviour, the hunger drive will be discharged and some other drive will perhaps propel the vulture's behaviour.

Interestingly enough, if the normal sequence of behaviours is thwarted, the animal may nevertheless go through some of the normal behaviours in the absence of appropriate stimuli. Thus if a flycatcher is maintained in a cage and fed artificially its hunger will be satisfied, and will not go through the normal hunting behaviour. Yet occasionally such flycatcher will sally forth from its perch, snap at a nonexistent fly, act as though beating it and then swallowing it. Such behaviour is known as vacuum behaviour.

One may think of instincts as being organized in a hierarchical order. Consider the baya weaver bird. During the winter the males and females feed in flocks on grass seeds, insects and other prey. On approach of the breeding season, hormonal changes in their bodies arouse the reproductive instinct. This may be thought of as driving in the female a whole series of behaviours, from location of trees where males are building nests, choice of a mate, completion of the nest, laying of eggs, incubation and feeding of the chicks. The nest-building instinct may then be thought of as one of the subsidiary instincts at a lower level of hierarchy than the reproductive instinct. The nest-building instinct of the male will itself be composed of components such as locating suitable grass fronds or leaves, tearing them and bringing them to the nest, weaving them in the nest. The nest-weaving instinct will further be made up of components such as pushing the grass frond through, tying a knot, pulling the knot taut and so on. The generalized reproductive instinct has to be activated before the female will look for a half-completed nest by a male. The nest-building instinct will be released only when she has accepted one such male as her mate and is in possession of a half-complete nest. The instinct to tie a knot with a grass frond will not be released until she has come to the nest with a collected frond in her beak and so on.

How far are we sure that such exceedingly complex patterns of behaviour as the building of a perfectly shaped baya weaver nest is an instinctive, that is, genetically programmed activity? There are many lines of evidence supporting this, in particular the fact that animals raised totally in isolation from birth exhibit many of these behaviours. For example, in a number of migratory species of birds, the parents leave the summer breeding homes in Europe before the young. The young then make the journey on their own, navigating over hundreds of kilometres of terrain they have never seen, following the clues from stars and from the landform features without fail, and finally arriving at their Asian and African wintering grounds.

The evidence for the inheritance of complex behaviour patterns has been further strengthened by breeding experiments. In honey-bees a dead pupa is removed from its cell by first uncapping the cell and then throwing the corpse away. It is now known that the trait for uncapping and for throwing away the corpse is controlled by one gene each. By breeding experiments one can produce honey-bees which will open a cap but will throw the corpse away only if the experimenter removes the cap, as well as honey-bees which will do neither or both of this set of activities.

The relative importance of instinctive versus learned behaviour in the animal kingdom has always been a matter of great interest. The relative importance of learning is particularly high in the higher vertebrates, including primates and of course man. Even in these animals however learning is constrained by genetic bounds. Thus rats can learn to associate the size of

food offered, but not its taste, with an electric shock. A number of instinctive patterns of behaviour persist even in the higher animals including man. Thus, as was mentioned at the beginning, Charles Darwin could not overcome the instinct of withdrawing his head from a striking cobra, even though the cobra was on the other side of a glass pane. But it is highly likely that instincts underlie much of other more complex behaviour as well. The protective and nursing behaviour of mothers towards their infants seems to have a large instinctive component.

At an even higher level of complexity, it has been argued that human males will have the genetic predisposition to dispense their paternal care to children with the highest degree of blood relationship with them. In monogamous or polygynous societies, there are of course the children of a man's wife or wives, although, as the saying goes, maternity is a fact while paternity is always a conjecture. In a polyandrous society however the situation may be different. If the many husbands of a woman are all brothers, then this woman's children are still quite closely related to him. However, in case the husbands are unrelated, many children of a man's wife will be fathered by unrelated males and may bear no blood relationship with him. On the other hand, his sister's children are always guaranteed to bear a certain blood relationship to him. Hence in polyandrous societies with several unrelated husbands, a man's closest blood relations may be his sister's children. We therefore expect that in such societies maternal-uncle care may be the prevalent form of care of young by males. In fact an examination of the anthropological data does reveal such a correlation and suggests that even rather complex culturally conditioned behaviour in humans may have an instinctive basis.

IORA. The Iora (*Aegithina tiphia*) is a sparrow-sized arboreal bird commonly found in low and deciduous forests and semi-urban areas. It is easily identified by its distinctive musical whistling calls. The male is bright

© E.H.N. Lowther

Iora sheltering young from the sun × ½

black and yellow with two white wing-bars, the female chiefly yellowish green. Its food consists chiefly of insects and spiders. Marshall's Iora, with much white in its tail, is confined to northwestern India and Pakistan.

IRUL, *Xylia xylocarpa*, a lofty deciduous tree in the Western Ghats, extending into Madhya Pradesh, Bihar and Orissa. Its feathery leaves, pale yellow flowers in globe-like heads and flat woody pods that open suddenly are characteristic. The timber is hard, heavy and durable and therefore valued in bridge construction and as railway sleepers.

I.U.C.N. The International Union for the Conservation of Nature and Natural Resources is an international organization consisting of states, government departments, international entities such as UNESCO and UNEP (United Nations Environmental Programme) as well as private conservation organizations. 'Its broad purpose is to foster the maintenance of the biosphere and its diversity by rational management of the earth's resources.' It was established in 1948 at an international conference at Fontainebleau which was sponsored by UNESCO and the Government of France. Originally established in Morges in Switzerland its headquarters was moved to Gland near Geneva in 1980 where it shares common premises with its sister concern, the World Wildlife Fund (W.W.F.).

The I.U.C.N. operates through several permanent missions which include Species Survival, National Parks and Protected Areas, the Ecology Commission, Education Commission, Environmental Planning and others.

Its most important work has been done through the Species Survival Commission of which Sir Peter Scott is the Chairman. SSC has been responsible for the production of Red Data Books on endangered species of animals, birds and plants. Information about each species is included on the sheets, and this information is being continually updated. Only species which are considered to be seriously threatened are included in these data books, and for ensuring that these endangered species survive, projects are formulated for protecting their habitat. In extreme cases where there is too much destruction of, or in, the natural habitat, captive breeding is recommended. Some examples of species which have recovered after coming to the brink of extinction are the Arabian Oryx, the Nene Goose, and the Hardground Barasingha. It was in the General Assembly of the I.U.C.N. in 1969 held in New Delhi that Project Tiger was set in motion. Since its inception the I.U.C.N. has taken great interest in this project, and has arranged for scientists to monitor its progress.

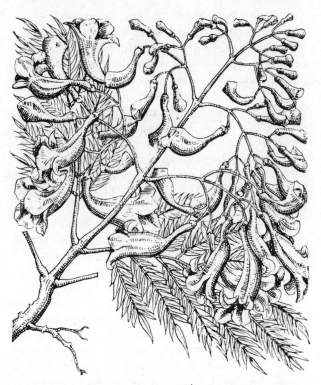

Jacaranda. Flowers and leaves × $\frac{1}{2}$. This exotic tree with its masses of blue flowers is now widespread.

JAÇANAS or LILY-TROTTERS are tropical, water-hen-like birds with exaggerated spidery toes and long hind-claw, living and nesting on floating vegetation in ponds and jheels. They horny spurs on the shoulders; some species also have a horny plate on the forehead. Two of the seven living species occur in the Indian subcontinent: the blackish Bronzewinged with a prominent white eyestripe, and the chocolate-brown Pheasant-tailed with white wings and arching sickle-shaped black tail. The larger polyandrous female lays several clutches of eggs each season, and leaves her several male partners to incubate and tend the young.

JACK TREE, *Artocarpus heterophyllus*, wild in the Western Ghats and grown throughout the warmer

Pheasant-tailed jaçana × $\frac{1}{5}$

JACARANDA, *Jacaranda acutifolia*, is a medium-sized tree of Brazil widely introduced in the subcontinent. The foliage is as finely cut as a fern, symmetrical and elegant. The blooming of the bluish-mauve flowers in May is spectacular. Bluish flowers are extremely rare in Indian trees and jacarandas are therefore an unforgettable spectacle. The fruit is a flat and circular woody capsule ripening in January. Jacaranda thrives from sea-level to 1500 m. Its trunk is a good host for growing orchids, notably the Fox-Tailed Orchid.

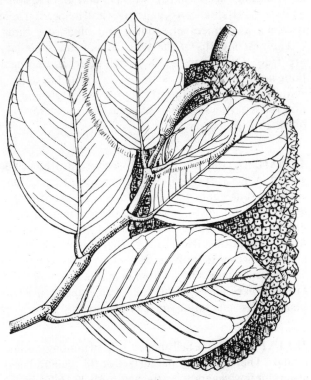

Leaves × $\frac{1}{2}$, jack-fruit × $\frac{1}{4}$. Flowers sprout from the trunk and large branches and develop into enormous hanging fruits.

337

parts of the subcontinent. The fruits borne on the trunk are of enormous size, weighing up to 40 kg, and are much esteemed. The timber is used in furniture, house-building and a dye obtained from the wood is used in colouring the robes of Buddhist priests.

JALDAPARA. Jaldapara sanctuary is situated in Jalpaiguri District of West Bengal and consists of a narrow strip straddling the Torsa river on the west, and extending up to the Malangi river in the east. It is just north of Hasimara on the National Highway from Siliguri to Dispur. The total area of the sanctuary is about 90 km².

The sanctuary is less than 150 m above sea level and has an average rainfall of 2000–4000 mm. The average annual temperature ranges from 22·5 to 25°C.

The forest is of the tropical moist deciduous type and it covers approximately a third of the geographical area. A recent problem with this sanctuary is the proliferation of *Mikenia*, an exotic climber which is smothering the vegetation. Physical removal of this plant seems to be an impossibility in view of the extensive area which it has already covered; and so perhaps biological control methods may have to be tried.

It is not always easy to see wildlife in this sanctuary because of the thick undergrowth. But nevertheless a large variety of animals exist here, including swamp deer, barking deer, sambar, wild pig and an occasional wild elephant, rhinoceros, gaur, tiger and leopard. There is a rich assortment of resident and migratory birds.

The best months of the year to visit this sanctuary are from December to May and there is a tourist bungalow with catering facilities. The Hasimara railway station is only 2 km from the tourist bungalow and the Forest Department hires out jeeps and elephants for touring the sanctuary.

JAMAN, *Syzygium cumini,* an evergreen tree widespread in our area up to 1500 m, is mostly found along streams. It is characterized by opposite, gland-dotted leaves, with a vein running along the leaf margin. Fragrant, greenish white flowers bloom from March to May. The purple fruits ripen from June to August, and are dispersed by birds and flying-foxes. They are also eaten by men and are used in making wine in Goa. Jaman makes an excellent avenue tree and is used to shade coffee plantations in Coorg and the Nilgiris.

JELLYFISHES are bell-shaped free-swimming coelenterates. They float near the surface of the sea and backwaters; a few genera live attached to rocks, mud banks and weeds. They differ from the *Obelian* medusas in having a number of notches along the umbrellar margin

Jaman × ²/₃. Its fragrant flowers are a source of honey. The purple fruits are favoured by flying-foxes and birds.

and in the absence of a velum. They are transparent and glassy, with their tentacles and internal organs brilliantly coloured. Some are also phosphorescent. The shape of the bell differs in different orders—some are goblet-shaped, some like four-sided cups, some conical, some saucer-like and some flat discs. They vary in size from 2·5 cm to 1·5 m. They are a wonderful sight: the large jellyfishes cover the sea for miles and miles and the tiny ones float like glistening specks by millions and break into streaks of light when dashed ashore by waves.

All jellyfishes are carnivorous and feed on crustaceans, fishes and other small sea animals.

Hundreds of genera occur in Indian waters. A few of the more interesting ones are *Aurelia, Charybdea, Cassiopea, Acromitus* and *Pelagica.* COMB-JELLIES are not true jellyfishes, but very beautiful.

Aurelia. These are large, semi-transparent, saucer-shaped forms about 12–30 cm in diameter. The upper surface of the dome is granulated. The mouth is surrounded by four thick, leaf-shaped oral arms somewhat curved. The umbrella margin is comparatively thin and hangs down. The oral arms, radial canals, gonads and tentacles are of beautiful violet colour. They congregate in large shoals on the ocean surface.

Charybdea. The bell is a four-sided cup about 10 cm in height and 5·5 cm diameter. The upper half is pyramidal with a concave top, the lower half expanding towards the base. There are four prominent contractile

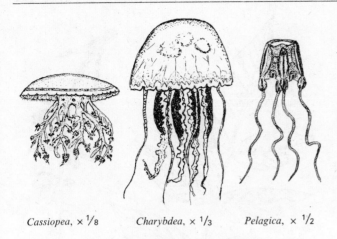

Cassiopea, × ⅛   Charybdea, × ⅓   Pelagica, × ½

THREE VARIETIES OF JELLYFISH

tentacles each 20 cm long. Stinging cells are scattered over the top and sides of the bell. In life the animal is transparent and the tentacles pinkish. It lives in shallow water and voraciously feeds on fishes.

Cassiopea. These are common near Madras sea coast in shallow waters. The bell is flat and disc-like 12·5 cm broad and light brown in colour. Marginal notches are not prominent. Oral arms are eight in number, pinnately divided and darkish in colour.

Acromitus is a small jellyfish about 4 cm in diameter. The bell is almost hemispherical, pale white, covered by yellow spots. There are eight prominent oral arms, each arm ending in a long tapering filament ringed with stinging cells. It occurs in backwaters and local fishermen call it *rabanachatra,* Ravana's Umbrella.

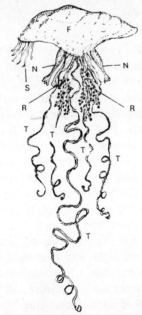

THE PORTUGUESE MAN-OF-WAR
(*PHYSALIA*) × ½

F float; T tentacles; S, N, R sensory, nutritive, and reproductive polyps.

Pelagica is another small jellyfish with a transparent umbrella measuring 5 cm in diameter when expanded. In shape it looks like a small *Aurelia.* It is colourless in life but its tentacles and gonads are violet. Some species are phosphorescent.

Physalia or the Portuguese man-of-war is a very abundant jellyfish. Each colony consists of a small inflated float 5 to 10 cm long, transparent and brightly tinted purple or blue. From the underside of this float growths of different shapes and sizes and performing different functions hang down: nutritive polyps, sensory polyps, reproductive polyps of deep blue colour and tendril-like tentacles with a battery of stinging cells. One of the tentacles is exceptionally long—nearly two metres. The sting of this jellyfish is poisonous and painful. It can paralyse a large fish and is a menace to human swimmers in warm seas. *Physalia* moves gregariously in shoals on the surface of the ocean in association with other blue pelagic jellyfish and violet snails. They feed on fish fry, copepods and other sea organisms.

See COELENTERATE, SNAILS 3.

JERBOA. Jerboas are highly specialized rodents which are adapted to survive in remote desert areas often in a habitat characterized by shifting sand dunes and extremes of temperature. There are three genera which occur within the subcontinent and all of them are confined to the western border regions of Baluchistan contiguous with southern Iran and Afghanistan.

The most striking physical features characteristic of this group are the greatly elongated kangaroo-like hind legs, the diminutive fore-limbs and the shortened neck with fused cervical vertebrae. They progress by hops, using only their hind limbs, and their tails are used as balancing organs. Their tail has a prominent terminal tuft of longer black-and-white hairs which provide a valuable contact signal. Their tiny fore-limbs are only used to assist in collection of food and in burrowing in the sand. Their hind feet are often highly modified with a reduced number of digits and fused metatarsals and their toes are often fringed by thick tufts of bristles which enable them to traverse more efficiently over soft sand.

Extending from North Africa up into the cold deserts of southern Russia and Mongolia, jerboas are classified into a number of distinctive genera of which three occur in Pakistan. These are the large, Threetoed Jerboa *Jaculus blanfordi*, two species of the long-eared, Fivetoed Jerboas, *Allactaga elater* and *Allactaga hotsoni*, and the diminutive Pigmy Jerboa *Salpingotus michaelis*. Only the Fivetoed Jerboa *A. elater* is relatively widespread and occurs in the broader valleys and plateaus of north and central Baluchistan. It is adapted to live and burrow in quite hard stony soil and even occurs in the outskirts of Quetta City. All jerboas have large dark eyes, soft sandy yellow dorsal fur with white throat and belly and extremely long whiskers. They are purely nocturnal in activity, spending the day in underground burrows. With a head-and-body length of 106 mm the Fivetoed Jerboa is slightly smaller than the Threetoed or Persian Jerboa *J. blanfordi* which average 128 mm in body-length, and both species have tails equal in length to the head and body and, when pursued, are capable of covering more than two metres between hops and travelling at estimated speeds of up to 48 km per hour.

Their young are born naked and blind in underground nest-chambers and litter sizes of 2 to 4 have been recorded in all four species. The gestation period is believed to be comparatively lengthy for a rodent (up to 40 days) and the development of the young till they reach independence is equally prolonged, as almost adult-sized and fully furred young of *Jaculus* have been excavated from their nest-burrow, when their eyes had still not opened. They subsist mainly upon seeds and the sparse desert vegetation, obtaining all their water requirements by metabolic processes (oxidation of carbohydrates). During periods of severe cold or summer drought they are also capable of lapsing into periods of torpidity when they remain underground in their burrows for prolonged periods.

See also RODENTS.

Flowers, fruit and foliage × ²/₃. During timber-hauling operations in forests the bark of jhingan is used to protect the backs of elephants from the friction of the chains used for dragging.

JHINGAN, *Lannea coromandelica*, a deciduous tree of wide distribution. Foresters recognize it when a kukricut on the trunk reveals a crimson blaze marked with pink and white. Small pale-yellowish flowers cover the tree when it is leafless in April. The fruits are red when ripe. In Nepal its gum is used in sizing paper and elsewhere in confectionery.

JUMPING SPIDERS (Attids). *Plexippus paykulli* is the commonest jumping spider—strong-legged, stout-bodied, medium-sized, blackish and decorated with white streaks. All jumping spiders have two prominent anterior median eyes like motor-car lamps. The males and females are almost of equal size. Jumping spiders are found everywhere—on the tops and trunks of trees, along walls and fences on open ground, on flowers and in hundreds of other places. They like sunshine and rain and thrive in all seasons. There are several ant-mimics among Attids.

The males are generally more attractive and ornamented than the females. It is a sight to see the male dancing before his partner before mating.

In colouring Attids outshine all the other families of spiders. Some are tar-black with white streaks, some metallic green; some ochre-brown and some iridescent blue. Some can change the colour of their eyes.

See also SPIDERS.

JUNGLEFOWL. Two species of junglefowl are found in India, the Red and the Grey. Both are similar to the domestic fowl in size and build, and the former also in coloration. They are hardy versatile birds of the pheasant family thriving in almost all forest habitats, hill and plain. The Red Junglefowl—the progenitor of all the world's diverse domestic breeds—affects the northern forests of the country from about 2000 m in the Himalaya and down the eastern side of the peninsula. The Grey dominates the western half and all down the W Ghats. Its range is largely coincident with that of the teak tree (*Tectona grandis*) as that of the Red is with sal (*Shorea robusta*). Junglefowl have exceptionally keen sight and hearing and are usually very shy and alert, but where undisturbed they are often seen during the early morning and late afternoon scratching and feeding in forest clearings either singly or in parties of a cock and 3 or 4 hens.

The challenging crow of the Red cock, given chiefly before daybreak and at dusk, is very similar to that of the domestic Bantam breed; the Grey's is more partridge-like but harsher. The breeding season extends over the summer months when the hen raises a brood of 4 or 5 chicks. They are covered with down and fully active at hatching (nidifugous: precocial) capable of running about and feeding themselves under the mother's tutelage. If suddenly come upon while leading chicks too young to keep pace with her in an escape bid, the mother sometimes gives a curious distraction display, rushing helter-skelter as if demented while the chicks freeze among the ground litter and magically melt into their surroundings.

D.J.P.

KADAM, *Anthocephalus chinensis,* a deciduous tree of the evergreen and moist deciduous forests of eastern India, Nepal, Bangladesh, Western Ghats, Andamans and Nicobars. Its horizontal branches with shining leaves (15 to 30 cm long) with prominent parallel veins

Kadam. Fruiting branch × ³/₄. Fruit a golf-ball-like or laddu-like orange mass of closely packed capsules. Note parallel lateral nerves of leaves.

are characteristic. This beautiful tree is admired for its golden balls of flowers. The small scented flowers are combined in rounded heads a little smaller than a golf ball which appear in June and are sacred to the Hindus. The fruits are large and orange-coloured. They ripen in August, and are relished by flying-foxes and monkeys. It is a remarkably fast-growing tree, growing up to 3 m a year, and is valued for matchwood and plywood.

KANCHAN, *Bauhinia variegata,* is an ornamental tree, wild all over our area in the plains and hills up to 1500 m. It has a stocky trunk and smooth dark green leaves which are cleft in the middle. Kanchan is at its showiest in February to April when large fragrant white or purple flowers clothe the leafless branches. Bees visit the flowers and effect pollination. The leaves are used as wrappers for bidis and flower buds are pickled and eaten. A cultivated white-flowering form, *B. candida,* is also common.

KANHA. Kanha was a reserved forest in 1879 and was established as the Banjar Valley Sanctuary in 1935 and as the Kanha National Park in June 1955. In 1972 it became one of the Tiger Reserves under Project Tiger and the present area is 1500 km². Its altitude varies from 500 to 1000 m above sea level.

Kanha forms a part of the central Indian highlands which stretch across Madhya Pradesh from west to east. It is a large amphitheatre of meadow-like grasslands surrounded by flat-topped hills. The wet season begins in late June and ends in October, producing about 1800 mm of rainfall, and this is followed by a cool season from November to February and a hot season from then until the monsoon arrives. The temperatures range from 5°C to 43°C.

Sal forest in Kanha

The tree cover consists mainly of the sal (*Shorea robusta*) interspersed with various species of Terminalias. There are occasional patches of bamboos and thickets. About half the park is covered by dry deciduous woodland of axlewood, kanchan, and amaltas.

The wildlife includes tiger, panther, dhole, hyena, sloth bear, jackal, civets, boar, chital, swamp deer, sambar, mouse deer, blackbuck, fourhorned antelope, nilgai, gaur and Hanuman monkey.

Kanha is possibly the best place in India for the viewing of tigers. Some of the animals get so used to tourists on elephant-back that they allow an approach to within a few feet in the certainty that they will not be harmed. In the census taken in May 1979 the tiger population was estimated at 71.

KARAYA, *Sterculia urens,* a deciduous tree of western and central India, has smooth, white bark which peels off in large, papery plates. The leaves are heart-shaped and 5-lobed, growing at the ends of branches. Yellow flowers appear in crowded erect bunches. The fruits are red and boat-shaped: they grow in bunches of five and burst open when ripe. The bark yields a fibre for making coarse cloth and ropes, and exudes a white gum used in thickening ice-creams and cosmetics.

KAVERI. Famed for its traditional sanctity, its picturesque scenery, and its utility for irrigation, the Kaveri is a principal river of Southern India. Rising on the Brahmagiri hill at Talakaveri (12°25′N, 75°34′E) at an altitude of over 1310 m in the Western Ghats it flows for 760 km through Karnataka and Tamil Nadu, and makes a large delta before entering the Bay of Bengal.

The upper course, lying in the hills of Coorg, is tortuous with high banks and rocky beds. From its source it leaps down the east-facing scarps of the Western Ghats through a series of falls, losing nearly 500 m in altitude within a short distance of 10 km. The gradient then gradually becomes gentler and the stream matures as it enters the Mysore Plateau.

Most of the middle course passes over a moderately undulating topography with average elevations of 750 m above sea level in the Mysore Plateau. The river now meanders in broad loops over the gentle, rounded spurs of hard, gneissic rocks. However, the topography lends itself to the storage of water by anicuts for canal irrigation, some of which were built over a thousand years ago. The two sacred islands of Srirangapatanam (Seringapatam) and Sivasamudram, 78 km apart, lie in this section. The former is close to the city of Mysore and the nearby Krishnasagar Dam, built for the generation of hydro-electricity and water conservation. At Sivasamudram the river descends over a precipice forming the famous Falls in two branches, Gagana Chukki (the 'Cloud Spray') and Bhar Chukki (the 'Heavy Spray'). These falls are also utilized for the production of hydro-electricity. The river gallops over the hard spurs of the plateau in a succession of waterfalls, flowing through a meandering valley.

The lower section of the river begins below its confluence with river Bhawani, a little north of Erode. At Tiruchirapalli, it branches into two major courses, the northern or upper branch, called Coleroon, and the lower one, the Kaveri proper, the two meeting at Srirangam where a grand anicut was constructed in the 11th century. Tiruchirapalli, 90 km from the sea, lies at the head of a delta covering an area of more than 10,000 km². The drainage is highly interesting. It consists of a number of closely-packed parallel-flowing streams. Later on, the two branches become wider apart with several distributaries taking a northeasterly direction towards the coast, resulting from the southwest monsoons which have deflected the courses northward, particularly north of Karikal. South of Karikal the deflection is inhibited by the southern bulge of the delta, west of Point Calimere.

The delta shoreline is backed by a mass of mudflats and salt swamps of 8 to 10 km wide. It is remarkably straight south of Coleroon to Point Calimere. The coast makes a right-angle bend due west of Point Calimere for nearly 30 km, enclosing a large swamp to its north.

The Western Ghats receive as much as 3000 mm annual rainfall, most of which falls between June and August. Mysore Plateau is semi-dry (800-1100 mm annual rainfall). Towards the eastern coast rainfall again increases, but has a distinct winter maximum largely resulting from the retreating northeast monsoons from October to February. Along the coast it is nearly 1440 mm: within 50 km inland it falls to 900 mm.

KAZIRANGA. Kaziranga was first established as a game reserve in 1908, and in 1974 was designated as a National Park. It has been included in Project Tiger since 1972. It has an area of 430 km² and is about 75 m above sea level. It is located in a low-lying area between the Brahmaputra river and the Mikir hills. Except along the southern fringe, much of the park is marshland liable to periodic flooding.

The vegetation is typical of swampy land in north India in which the water hyacinth has become established. There are large pools fringed with reeds, especially of *Phragmites* karka. There are also patches of tall grassland dominated by such species as *Arundo donax, Saccarum spontaneum* and *Alpinia* species. Trees and thickets consist mainly of white siris (*Albizia procera*), hickory (*Carya arborea*), lendi (*Lagerstroemia parviflora*) as well as jarul (*L. speciosa*).

Kaziranga has been for decades one of our most impressive wildlife preserves and it is largely because of Kaziranga that the Great Indian Rhinoceros has survived. Viewing the rhinoceros from elephant-back is one of the highlights of visiting the national park. The large herds of Swamp Deer make a most impressive picture against the background of the Himalayan ranges. The larger animals include the tiger, leopard, elephant, pig, hog deer, swamp deer and wild buffalo.

Kaziranga is rich in birdlife which includes grey pelicans, blacknecked storks, lesser adjutant stork and numerous species of waterfowl. The hill slopes have many woodland species of birds including red junglefowl.

See plate 21 facing p. 304 and plate 29 facing p. 416.

KEOLADEO. The Keoladeo Ghana Bird Sanctuary, now a National Park, rated to be one of the finest waterfowl reserves in the world, was a shooting preserve of the Maharaja of Bharatpur. It was notified as a Sanctuary in 1956, and now enjoys total legal protection.

Its area is only 29 km² and it is less than 200 m above sea level. It is bisected by a metalled road which connects it to Bharatpur city and the entire area is divided into small compartments by boundary roads which are jeepable. Through a system of gates the depth of the water is maintained at about 1·5 m in the interests of the various types of waterfowl.

The vegetation belongs to the tropical deciduous type and the sanctuary is covered with shrubs and medium-sized trees consisting of babul, khejri, ber, jamun and others. Several of the trees, such as babul, are well adapted to surviving under extreme conditions. They do well when the jheel is at high-water mark and the trees under several feet of water; equally well in drought periods when the ground around them is totally dry. It is on these trees that egrets, herons, cormorants, darters, spoonbills, and other birds nest during the monsoon period.

More than 300 species of birds are found within the sanctuary, the largest congregation being of migratory ducks. Some indication of the numbers can be gauged from the fact that in 1938 the Viceroy and his party shot 4273 birds in a single day.

A particularly noteworthy visitor is the Siberian Crane, whose only known wintering ground in India is this National Park. Though Keoladeo is famous as a bird sanctuary it has a wide variety of mammals, including the rhesus macaque, Hanuman monkey, chital, sambar, hog deer, nilgai, porcupine, boar, fox, jackal, civets, fishing cat, leopard cat, jungle cat and otter.

This Sanctuary lies amidst surroundings which have been totally denuded of all vegetation, and consequently there is great pressure on its ecosystem as a result of grazing by cattle and buffaloes and the illicit cutting of trees.

KING COBRA. King Cobras are the most intelligent of all snakes. Growing to 5 m in our area (6 m in Thailand) and living in dense evergreen forests of the Western Ghats, Orissa, Bihar, Bengal and the Andamans, and the northern hill forests, King Cobras are fascinating animals that remain largely unstudied. The name is misleading; King Cobras, *Ophiophagus hannah*, are a different genus entirely from cobras. Although King Cobra venom is not rated very toxic, this is balanced by the quantity. An adult snake can yield up to 7 cc's of venom—enough to kill a full-grown elephant! King Cobras eat other snakes, mainly rat snakes and water-snakes, which are envenomated and paralyzed before they are swallowed. The only enemy of King Cobras is man, who generally shoots when he meets them in the forest or on tea and coffee plantations. When

© Rom and Zai Whitaker

King cobra × ¹/₁₀

threatened, King Cobras put up a splendid defence, by raising the hood (which is longer than the cobra's, and has a V marking), growling and charging the intruder with open mouth if caught or injured. In the Andamans, King Cobras nest in May and June. They are the only snakes in the world that build a nest. The female scrapes leaves and humus together with her tail and body and constructs a conical nest about 30 cm high. Having laid eggs at the bottom of her nest, she guards them for the sixty or more days it takes them to hatch. During this period she usually does not feed: it is only when hatching time nears that the female, thin and emaciated, hunts again. The hatchlings disperse soon after they hatch, climbing up bamboo stands and looking for young snakes and probably skinks for food.

Although not directly exploited in India, King Cobras are threatened by human encroachment. Large tracts of their pristine forest habitat is being cleared and destroyed. Although they are becoming increasingly rare, they are killed whenever possible.

In Hong Kong, King Cobras are eaten, and in Thailand used for venom extraction in order to produce antivenom serum for their bite.

A better name for them is hamadryad.

See also SNAKES.

KING-CRABS (genus *Limulus*) are the only marine arachnids. They came into existence in this world some 200 million years ago. It is a wonder that they have survived so long without any change—a living fossil.

Their body consists of a cephalothorax and a segmented abdomen, the former covered by a horseshoe-like carapace and the latter by an abdominal shield bearing six jointed spines on either side. To the posterior end of this shield is fitted a stiff tapering tail or telson. Thus the animal is completely armoured on the upper surface exposing the simple eyes alone to the view. The underside of the animal reveals a pair of small pincer-like jaws, a pair of pedipalps, the mouth between them, three pairs of walking legs with clawed extremities and a pair of legs with flattened, oar-like toes. In the abdominal region are the genital opening, the breathing organs (or gills) and the anus. Judged by its horny shield, awkward movements, marine habitat and breathing gills, earlier naturalists called it a king-crab or horseshoe crab although it is really an arachnid.

King-crabs are found along the sea coasts of many countries. They live in moderately deep waters on a sandy bottom, 2 to 8 fathoms down. In temperate countries their breeding season is spring when they appear in shallow waters to spawn. They are active during the night. During the day they live partly buried in mud or sand. They move about by 'swimming hops', propelled by gill plates and the last pair of paddle-like legs. They feed on sea worms and molluscs which they secure by their jaws while burrowing. They burrow by lowering the front part of the carapace and then, thrusting their telson vertically down into the sand, they raise the body. In this position they scoop out the sand below by their walking legs and in a short time disappear beneath the sand. The worms or shell-fish secured during this operation are broken into

particles by the ends of the pedipalps and walking legs and conveyed to the mouth.

The females are larger than the males and measure about 50 to 55 cm. Although dreadful to look at, they are all harmless to man.

Two species of *Limulus* are common on the coast of Bengal—*Limulus moluccanus* and *L. rotundicauda*. The former is purely marine, living on sandy bottoms from tide-line down to 20 fathoms. The females come to shallow waters during the breeding season which is March, i.e. the end of the cold season. Their eggs, green in colour and each 3 mm in diameter, are laid and safely carried by the abdominal appendages of the mother. *Limulus rotundicauda* is estuarine. From the sea it travels up into the interior through the river for a distance of nearly 90 miles and is quite at home in fresh water.

The freshly hatched young ones look like small adults but without the tail.

See also ARACHNIDS.

KINGFISHER. A family of predominantly fish-eating birds, Alcedinidae, belonging to the order Coraciiformes which, in addition, includes bee-eaters, rollers, hoopoes and hornbills. Five genera and twelve species are found in the Indian subcontinent, ranging in size from less than a sparrow (Threetoed Forest Kingfisher), to almost as large as a pigeon (Storkbilled) and varying in colour from pied black-and-white to brilliant blue and rufous or resplendent purple and amethyst. They possess short tails, stout straight dagger-shaped bills, short legs and weak zygodactyl feet. Most species are of solitary habits and sedentary but some, like the diminutive Forest Kingfisher (*Ceyx tridactyla*), make seasonal local migrations which are as yet little understood. Kingfishers live near freshwater streams, lakes and ponds or brackish lagoons and mangrove swamps, and capture their fishy prey by plunging headlong from an overhanging branch or while in scouting flight above the water. A specially spectacular method is practised by the Pied Kingfisher (*Ceryle rudis*): during its scanning flight, 8 or 10 metres above a placid stream, the bird suddenly stops dead in mid-air with body tilted upright and wings flapping rapidly, and plummets into the water below, presently to emerge with a silvery minnow in its beak. However, the Whitebreasted Kingfisher and some of its close relations (genus *Halcyon*) have taken to an overwhelmingly insect diet and are oftener met with in open forest than near water. Besides insects and fish they take crabs, frogs and lizards etc., stooping to the ground like a shrike to seize the quarry. Kingfishers nest in self-excavated burrows in roadside earth cuttings or in the steep sides of kutcha wells and borrow-pits. The tunnel may be half a metre or more in length and ends in an unlined bulbous egg-chamber. Termite

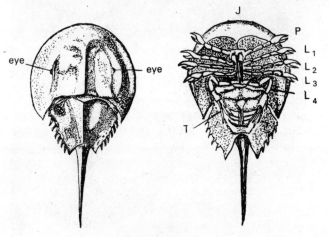

KING-CRAB. × ¹/₁₀
Left, view from above. Right, view from below.
J jaws. P pedipalp. L$_1$, L$_2$, L$_3$, L$_4$, legs. T toes.

mounds and carton nests of tree ants are sometimes also used for the purpose. Four to seven roundish white eggs (according to species) are laid, which both sexes incubate. The hatchlings are naked and blind (nidicolous) at birth, and are tended by both parents.

See plate 10 facing p.129.

KOKAM BUTTER TREE, *Garcinia indica,* is an evergreen tree with drooping branches, found in the tropical rain-forests of the Western Ghats. The leaves are egg-shaped or oblong, dark green above and the fruits spherical, *c.* 3 cm in diameter, dark purple, with 5 to 8 seeds. It flowers from November to February and the fruits ripen in April-May. The fruit is used to flavour curries and also for preparing cooling syrups. Kokam butter and edible fat is obtained from the seeds and is suitable for use as confectionery butter.

KRAITS. Krait venoms are among the most toxic venoms of any in the world. The potent neurotoxin can, in a very short time, cause respiratory failure, and kraits therefore have the reputation of sucking away a person's breath. Several hundred deaths a year are caused by the Common Krait which is found throughout the subcontinent up to 1700 m. It is less common in parts of the northeast which is the home of the beautiful Banded Krait, which has bright yellow-and-black bands

Common krait × ⅙

on its body. Although potentially dangerous, banded kraits are unwilling to bite, especially during the day, being nocturnal snakes. They hunt after dark for their prey. Both common and banded kraits are snake-eaters. Banded kraits will kill and eat common kraits, which is perhaps why the common krait is scarce in the northeast.

Kraits are nocturnal and hide in rat and mouse holes during the day. They seem to be highly territorial snakes and during the breeding season two or three males can be seen performing the vertical territorial 'dance', a ritual wrestling bout that is often

mistaken for mating (all snakes mate on the ground, in a horizontal position). Kraits are cannibalistic, a trait often noticed in captivity. All kraits lay eggs. Common and banded kraits lay 8 to 15, and the female appears to stay with the eggs until they hatch. Very little is known about the other five kraits found in the region. A newly described species, the Andaman Krait, is found in the Andaman Islands and the other four are distributed over Uttar Pradesh, Bengal, Bihar, Orissa, Assam, the eastern Himalaya, and Bangladesh.

See also SNAKES. See plate 19 facing p. 256.

KUKRI SNAKES. Kukri snakes are small, nocturnal, harmless snakes found throughout our area. They grow to about 50 cm. There are 17 species in our area, only three of which are commonly seen: the Banded or Common Kukri (*Oligodon arnensis*) and the Russell's Kukri (*O. taeniolatus*) are both found throughout the subcontinent and the Travancore Kukri (*O. travancoricus*) found in the southwest. These are shy snakes, glossy-scaled and brightly marked. Many of the species have chevron or inverted V-shaped marking on the head. They are good at burrowing into soft soil and spend the daylight hours asleep in a crevice of bark, a dead tree or under leaves.

Kukris feed on geckos, skinks and small mice. They forage at night, frequently near human dwellings in

Kukri snake

which numerous geckos live. Kukris lay 3 to 6 eggs and the young are extremely small, probably feeding on insects and spiders. Kukris look slightly krait-like and the Banded Kukri in particular gets killed by mistake.

Kukris are gentle snakes and will not bite when handled. Secretive habits and mimicry are their de-fences. When picked up, an excited snake may defecate on the catcher's hand or press the end of its tail into the hand, a harmless snake's attempt to convince you that it is dangerous!

See also SNAKES.

LACEWINGS and ANTLIONS. The orders Neuroptera (Lacewings, Antlions, etc.) and Megaloptera (Alderflies, Snakeflies) are the most primitive groups of insects undergoing complete metamorphosis (Holometabola), with a distinct pupal stage. In general, but more particularly Indian species, they are relatively poorly known and studied, so much so that a very small percentage of existing species have even been collected!

The Alderflies are a small group of less than 300 described species mainly occurring in temperate regions and associated with streams, in India on the Himalaya. The larvae are aquatic and their mouth-parts are not modified into a sucking beak, like the Neuroptera, and they also have lateral abdominal gills. Of the two known families, only the Sialidae are reported from India *in extenso*, species of *Sialis* coming to light. Of the Corydalidae, *Acanthocorydalis asiatica* is common in northeast India, males having the mandibles greatly enlarged into formidable 'elephant-like' tusks. Adults are probably predacious, as are the larvae, on stream animals; they in turn are fed upon by fish.

Over 4000 species of Neuroptera are known, this order comprising a variety of forms with wing-spans of 5 to 120 mm. Most of them may be recognized by the end-forking of the main veins in the wings which are held roof-like at rest. These insects are more abundant in the tropics, usually frequenting the warmer, drier regions. Around fifteen families of this order are currently recognized and only the more common and numerous ones are dealt with here. This is one of the insect orders that is almost wholly beneficial to man, feeding on and thus keeping in check many pest species of insects.

The Coniopterygidae are very small insects (100 known spp.) with a powdery meal all over the body and wings, resembling male whiteflies. The larvae have a small head and straight, needle-like jaws, and feed on mites, psyllids and scale insects. The mature larva spins a silken cocoon within which it pupates on a leaf or in crevices of tree bark. *Coniopteryx pusana* and *Piloconis guttata* are two of the well known Indian species.

The Mantispidae are small to medium sized (wingspan 10 to 50 mm) mantid-like species with raptorial fore legs and long, narrow, subequal, reticulate wings with a distinct pterostigma. Some species have wings attractively patterned. The larvae hatch from clusters of eggs laid on tree bark and feed on the egg sacs of spiders, while the adults frequent foliage of trees and grab smaller insect prey. A common Indian species is *Mantispa rugicollis*.

The Hemerobiidae or Brown Lacewings are small (wing-span 10 to 20 mm), delicate insects with bead-like, long antennae and brown, often marked wings. Adults are generally crepuscular or nocturnal and feign death if disturbed during the day. Both larvae and adults prey on aphids and are useful natural enemies of many pest species, especially in temperate climes. The eggs are laid on leaves or bark, attached to them sideways, and the elongate, brownish larvae with strong jaws are very active, pupating in a cocoon, which is a very delicate web. *Eumicromus australis* is common in India.

The Chrysopidae or Green Lacewings are an important family of neuropterous insects, the adult and larvae preying on a large variety of aphids, mites, thrips, leafhoppers, mealybugs, psyllids and other groups of insects with many pest species. The adults are usually larger than brown lacewings and are typically green or yellow, with clear wings, dotted with black in some species. The long, thread-like antennae are also distinctive. Larvae hatch out from whitish eggs borne on a long stalk and are usually found to carry skins of sucked prey or other debris on their bodies. The circular, white pupal cocoon is also covered with the same detritus as the larva. *Chrysopa scelestes* is a familiar Indian species.

Adult green lacewing × 4½

The Myrmeleontidae or Antlions is the dominant family of the order Neuroptera, comprising mostly large species with long, narrow, often pointed wings, beautifully patterned. Adult antlions have comparatively short antennae but not as short as those of the dragonflies which they resemble. Generally crepuscular and

Adult antlion × 1

nocturnal, antlions also avoid high-rainfall areas, the drier regions with sandy soil being especially favourable. Eggs are laid in sand and the larvae, which are the antlions, are endowed with attractive, curved mandibles which they use to capture and feed on prey. The larva constructs an ingeniously shaped conical burrow at the bottom of which it lives and waits for an unwary ANT or other small insects to drop in. *Indoclystus singulare* and *Palpares infirmus* are the 'demons of the dust' in India, while *Dendroleon contractus* has larvae living on mud-covered tree-trunks.

The Ascalaphidae are rather rare, antlion-like Neuroptera with long clubbed antennae and transparent (hyaline) wings, besides the eyes which are divided into upper and lower sections. They also fly mostly by day, hunting prey like the dragonflies they resemble. The larvae live openly on tree-trunks or on the ground, under stones or debris. *Ogcogaster tesellatus* and *Ascalaphodes canifrons* are commonly found in India.

The Nemopteridae is a small family of beautiful insects with elongated, ribbon-like hind wings resembling mayflies in flight. *Croce filipennis* and *Halter halteratus* are common Indian species found around habitation, their long-necked larvae inhabiting dust and feeding on psocids and other insects. The pupa is in a cocoon made of sand and debris and the life-cycle occupies almost a year.

See also BUGS, INSECT STRUCTURE.

LAGGAR (family Falconidae). A widely distributed falcon ranging from Iraq to India. Adult dark brown above, whitish below with elongated brown spots and moustachial stripes; immature largely brown. A resident parochial species, pairs often inhabiting cities and using tall buildings as hunting bases. Food mainly rodents, birds, lizards, locusts etc. Female larger than male, weighing up to 1 kg, and sometimes trained to hunt, but deemed by falconers as inferior to the Peregrine or the Shaheen.

See FALCONRY.

LÄMMERGEIER OR BEARDED VULTURE. Occurs in the Himalaya and other northern mountains, soaring up to 7500 m. Its wing-span is 2·75 to 2·85 metres, the largest for Old World vultures. Distinguished in flight by pointed wings and wedge-shaped tail. Feeds upon carrion and bones. Large bones are carried aloft and dropped on rocks and shattered. Maintains regular ossuaries for the purpose. Nests from December to March on cliffs, laying one or two eggs.

LANDSLIDES. A landslide is the fall of a mass of rock or soil, or occasionally of waste material taken from a mine. It is most likely to happen where rock and soil holding water lies over clay or other rock that is impermeable. The water collects over the underlying impermeable rock making it slippery, and the overlying rock and soil slides downhill. Most landslides occur after heavy rain. Other causes can be earthquakes, or rivers or the sea undercutting cliffs or the sides of valleys.

LANGURS. Langurs, of the subfamily Colobinae, are found throughout India, ranging from altitudes near 4600 m in the Himalayas to near sea level, and living in habitats that vary from moist montane forest to semi-desert. The name 'langur' is derived from the Sanskrit word *langulin*, 'having a long tail'. Langurs and their near relatives (African colobus monkeys and Asian odd-nosed monkeys), collectively called colobines, are distinguished from other Old World monkeys (macaques, baboons and guenons) in three ways: 1) they lack cheek pouches for the temporary storage of food, 2) females do not show sexual swelling in the perineal region during estrus, and 3) they have large complex stomachs. Their specialized stomachs permit ruminant-like bacterial fermentation by anaerobic bacteria, which allows them to consume vast quantities of mature leaves. This is why langurs are also called 'leaf-eaters' or 'leaf monkeys', even though they eat other foods as well, including fruits, berries, flowers, buds, seeds and, where they are in contact with man, cultivated crops and vegetables.

There are five species of langurs (= *Presbytis*) in the Indian subcontinent. The most widely spread and most numerous is the Hanuman langur (*P. entellus*), named after Hanuman the monkey-god and loyal servant to King Rama, whose exploits were chronicled in the *Ramayana* some time around the fifth century B.C. The most recently discovered and least common langur is the Golden langur (*P. geei*) of northwestern Assam. Else-

Common Langur, Sariska, Rajasthan × ¹/₉

Horned Lark, ♀, × ¹/₂

where in Assam and countries to the west the Capped langur (*P. pileatus*) is found. The Purple-faced langur (*P. senex*) is confined to Sri Lanka, while the Nilgiri langur (*P. johnii*) occurs in selected areas of southern India. Because of the ever-increasing deterioration of the Nilgiri langur's preferred habitat, it is the most threatened of the Indian langurs.

Although some differences in behaviour and socio-biology among langurs have been reported in various geographical regions of India, it is not yet possible to correlate these differences with any precision to environmental or genetical gradients. These include differences in the composition of diets, home range sizes, intra- and inter-group relations and, especially, group size and composition. In general, langurs live in social groups, ranging from about 10 to 35 individuals, and, in exceptional cases, there can be over 100 individuals. Two types of groups are formed: 1) bisexual groups, which are comprised of one to several adult males, a number of adult females and their immature offspring; 2) all-male groups, composed solely of adult males of various ages. Bisexual groups are the reproductive units; all-male groups represent the non-breeding population. But every several years the bisexual group may be attacked by the all-male group; then the resident male is driven off, and one of the invading males takes his place. This male then attempts, and often succeeds, in killing as many nursing infants as possible. Such infanticidal behaviour brings the mothers of the infants back into sexual cycling. The new male can thus mate and father several infants before he too is deposed. This behaviour is apparently confined only to colobines.

R.L.T.

See plate 37 facing p. 528.

LARK. Sparrow-like ground birds (family Alaudidae) chiefly streaked sand-coloured, well known especially for aerial song, e.g. the Skylark. Sometimes kept caged as good song-birds. Seen singly or in pairs; sometimes gregariously. They build their nests on the ground, under clumps of grass, etc.

LAUREL, *Terminalia crenulata*, is a large tree of the Western Ghats, with dark-grey, furrowed bark peeling in rectangular pieces. The leaves have a round-toothed margin which gives the tree its specific name. The small, pale yellow flowers appear in April and the 5-winged reddish fruits, 3 × 3·5 cm, ripen in August. The wood is dark brown streaked with black, and provides the figured laurel of commerce. The timber from the related species *T. alata* and *T. coriacea* is also called laurel but does not have an attractive grain.

LEAFBIRDS or CHLOROPSIS. Leafbirds (family frenidae) occur in woods of S Asia including our area. Predominantly green in plumage, they feed on insects, fruits and flower-nectar. They possess a tubular nectar-eating tongue and are important flower pollinators. Leafbirds are accomplished mimics of other birds' calls and make amusing pets.

LEAVES. Green leaves, containing chlorophyll, are the chief assimilating and transpiring organs of plants. Small, brown and membranous leaves without chlorophyll are developed on the underground stems of many plants such as ginger, turmeric and onions, and also around buds, and are called 'scale-leaves'. The first green leaf, or pair of leaves, to come up is called the cotyledon, and 'cotyledenous leaves' are much simpler in form than the 'foliage leaves' which develop later.

The stem exhibits the leaves in such a way that there is minimum overlap and maximum sunlight reaches the surface. The arrangement of the leaves on the stem form clearcut patterns (Fig. 1). However, depending upon the species, trees have dense canopy or open canopy. Closed canopy cuts the sunlight from reaching the

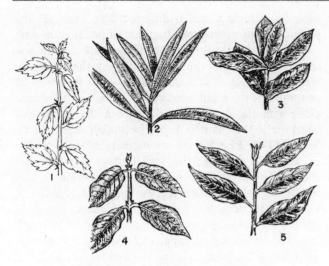

Fig. 1. ARRANGEMENT OF LEAVES
1, 5 alternate; 2 whorled; 3, 4 opposite

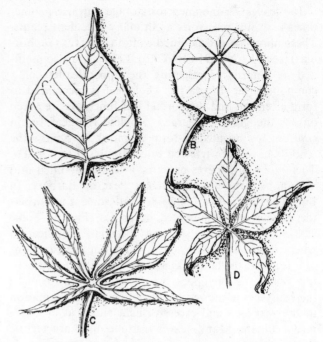

Fig. 2. Leaves of Pipal (A), garden Nasturtium (B), Tapioca (C) and Semul (D).

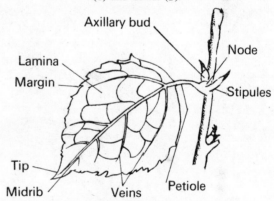

Fig. 3. PARTS OF A LEAF

Lamina. The green expanded portion.
Axillary bud. An undeveloped shoot bud in the axil of leaf and stem.
Node. The point of insertion of leaf on stem.
Stipule. A lateral outgrowth near the base of a leaf.

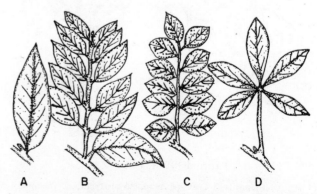

Fig. 4. A simple leaf (A), a branch (B), pinnately compound leaves (C), and a palmately compound leaf (D).

ground and hence adversely affects the undergrowth in a forest, creating dense shade.

Although different types of leaves exist, every plant has its own leaf shape (Fig. 2). A leaf has several parts, a leaf blade, midrib and veins (Fig. 3). This can distinguish stem-bearing simple leaves and compound leaves. Banyan and Jaman have simple leaves with a leaf blade and petiole, while Banana and many grasses have long leaves. The veins in these leaves are parallel, while in Banyan they are in the form of a network. The appearance presented by the lamina depends chiefly on the extent to which its membrane is developed between the branches of the vascular system. Sometimes it is completely developed and the margin of the blade is entire. Usually, however, it is not completely developed. The extent to which it is incomplete varies considerably. Sometimes there are only small irregularities or cuttings of the margin. Frequently larger incisions are produced between the leaf veins or branches. Many terms are in use to describe the forms of outline presented by simple leaves or the leaflets of compound leaves (Fig. 4). Leaflets in a compound leaf could be like a feather or a palm or in pairs along the midrib. There are some leaves which have toothed or serrated edges as in Oak. Similarly, the apex and incision of the lamina may vary. Therefore, leaf shape helps in plant identification (Fig. 2). Shade- and moisture-loving plants usually have thin leaves with poorly developed cuticle. Many plants bear different kinds of leaves on the same plant. In water-loving plants floating and submerged leaves are of different kinds! The former are generally broad while the latter are narrow and ribbon-shaped. Plants exposed to intense sunlight usually have thicker and more resistant leaves (see DESERT PLANTS).

The leaves get modified for storage purposes, sometimes as water reservoirs as in many succulent plants. These may also get modified as food reserves (euphorbias) (see PLANT DEFENCES). Leaves are also modified as tendrils representing the upper leaflets of a compound leaf. In pea the whole leaf is specialized to form a tendril, and the normal leaf functions are taken on by the greatly enlarged stipules. The leaf tip in *Gloriosa* is elongated to form a tendril. In *Smilax* the tendrils have been referred to as stipules (Fig. 5 A-D). There are many leaves or parts of leaves modified into a spiny structure as in Prickly Pear or Barbary. In Babul (*Acacia*), Ber (*Ziziphus*) and some Euphorbias the stipules are represented by spines (Fig. 6). Some acacias from Australia show peculiar modification where the stalk is flattened to carry out the functions of a leaf and the lamina is absent. These flattened petioles, known as phyllodes, are developed so as to place their surfaces in the vertical plane. This is an adaptation to over-exposed and excessive light and heat in a tropical climate. Many leaves assume a characteristic position at night which is different from their daytime position. This night-time alteration of position of sleep, results generally in folding up of the leaflets as in Sorrels. In some plants the leaflets fold together when the leaf is darkened as in some acacias while in others assumption of the night position of leaves normally coincides with the onset of darkness but the movements continue if the plant is kept continually dark. In the SENSITIVE PLANT (touch-me-not, *Mimosa pudica*) in sleeping position the leaflets fold together and the whole leaf hangs down and this position is commonly assumed during the evening (Fig. 7). It can, however, be brought on at any time during the day if the plant is subjected to the stimulus of shock by touching. The movement is rapid and within a few seconds after touch the plant assumes the night position. Many of these movements are important for protecting the stomata at night since blocking by dew is prevented and transpiration commences early in the morning when the leaves unfold.

In various insectivorous plants the leaves develop

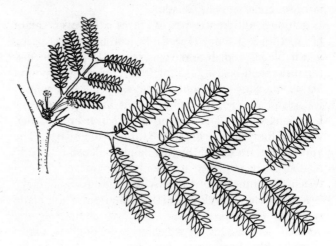

Fig. 6. Spines of Babul × ¼

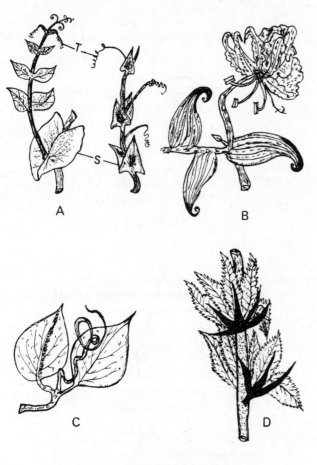

Fig. 5. MODIFIED LEAVES

A Leaf-tendrils. B Leaf-apex modified into tendril. C Terminal leaflet modified into tendril. D Primary leaves modified into spines. T = tendrils S = stipules

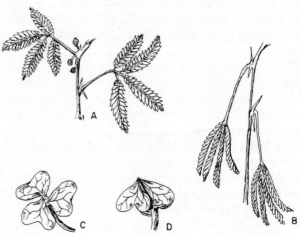

Fig. 7. *Mimosa pudica* leaf: A normal; B sleep position. Sorrel: C normal; D sleep position. (All × ⅔)

into pitcher-like structures and show other peculiarities of form connected with their ability to trap and digest insects.

Plants like all other organisms solve their problem for adequate supply of water. Leaf fall is a periodic phenomenon which follows the formation of an absciss layer at the base of the petiole. Normally plants absorb water from the soil through their roots. Deciduous trees lose their leaves due to seasonal climatic conditions such as on the approach of winter or the onset of a dry season. Evergreens may also shed some of their leaves at other seasons, as in the spring. Tropical trees shed their leaves periodically even in absence of pronounced changes in temperature. The causes leading to leaf fall are not clearly understood. Leaf shedding can also be induced sometimes by exposure to dry conditions or to toxic gases. During extreme temperatures plant life comes almost to a standstill carrying on only essential processes. Slowing down of plant activity is also due not so much to the cold itself but to the plant's inability to absorb water from cold, winter soil. Therefore, plants must conserve water as much as possible during winter. Since plants give off water through their leaves it is therefore natural to get rid of some leaves or all as in deciduous trees. Leaves get thickened and modified to prevent loss of water in some evergreen trees. In herbaceous plants the problem hardly arises, since the shoot itself dies down before winter.

Leaves in autumn begin losing their green colour, turning into shades of yellow, brown and red. This is due to the chemical breakdown of chlorophyll. Since food manufacture can no longer go on in the leaf some cells at the base of the leaf start getting separated from one another and form a point of weakness leaving the leafstalk attached to the twig only by a vein. By the weight of the leaf, by wind action or frost the vein is broken and the leaf falls. Leaf fall is a vital active process which follows the special metabolism of cells at the base of the petiole and is carried out by deciduous trees to get rid of structures which might otherwise cause them to lose much water. It is possible that waste products are passed into the leaf before it falls. The plant used it as a means of getting rid of waste matter formed during its life's process. The leaves on dead branches do not fall, they wither or curl up, no absciss layer is formed and hence they are not shed, thus showing that leaf fall is a living process. Casting of branches, commonly observed in forest trees especially in the dry season, of flowers, inflorescences and fruits is a similar process. Leaf fall is of practical significance for raising crops under different agro-forestry practices without adversely affecting crop yields.

R.K.G.

LEECHES (Hirudinidae). These well-known creatures differ from other annelids in being devoid of chaetae and 'feet'. They usually have a sucker at both ends of the body. The external annulations on the body are far more numerous than the internal segments and the number of annuli corresponding to one internal segment varies from species to species. Indian Leeches are of three main types:

1. The *Rhynchobdellids*—freshwater and marine leeches with a small aperture in the centre of the anterior sucker through which the pharynx can be pushed out in the form of a proboscis to suck blood from the prey.

2. The *Gnathobdellids*—aquatic or terrestrial with a mouth as the centre of the anterior sucker provided with jaws which make a Y-shaped incision on their prey when sucking out blood.

3. The *Pharyngobdellids*—primarily aquatic leeches which are not truly blood-sucking, preferring to swallow worms, insect larvae, snails and other aquatic creatures.

India has many habitats ideally suited for leeches which has led to their diversification and abundance in certain areas. They seem to appear magically during the rains and disappear during the hot, dry and very cold seasons. They have been found hibernating in valleys in the extreme north and moving uphill during the rains. Large numbers of the terrestrial species have also been seen climbing up trees during heavy rains, probably to escape from inundation.

Haemadipsa is the bane of the inhabitants of the moist forests because it hangs down from the vegetation by its posterior sucker and latches on to anybody brushing past and soon begins its bloody feast. *Hirundinaria granulosa* was the so-called Medicine Leech used in the infamous practice of phlebotomy (blood-letting) which was prescribed as a cure for all evils ranging from delirium to gout. This leech can take in up to ten times its weight of blood at a stretch. This remarkable feat is accomplished by a capacious stomach with many pouches. Leeches thus engorged need not feed for several months. The aquatic leeches attach on to frogs, fish, snakes, turtles and even wading birds and water-beetles. *Limnatis* is a troublesome pest when it enters

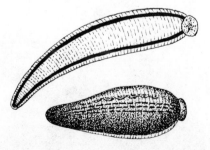

The leech *Hirudinaria granulosa* × 1, ventral aspect above, dorsal below. This is the leech formerly used so much in medicine that doctors were called leeches.

through the nostrils of domestic cattle and even man while drinking water and causes severe inflammation of the nasal passages and epiglottis. Leeches secrete an anticoagulant—hirudin—into the wound and a vaso-dilator which dilates surrounding capillaries, bringing more blood into the area.

Leeches are hermaphrodites and yet cross-fertilization takes place through an elaborate mating sequence where-in sperm packets are transferred either by insertion of the penis into the genital opening of the female or by hypodermic impregnation in which the sperm packet is injected into the body-wall of the female at any point, but usually near the clitellum. The sperms then find their way through body tissues to the eggs. The cocoon is secreted by the clitellum in the form of a girdle into which the fertilized eggs are placed. The eggs are laid a few days or several months after copulation. The leech then slips its front end through the girdle, the ends seal up and the cocoon hardens with a colour ranging from cream to mud-brown. Some cocoons are membranous, are attached to the parent's body and the young after emergence adhere to their parents. Most terrestrial species deposit their cocoons beneath stones and other objects or in the soil. The threadlike young emerge a few weeks later but need to survive for four or five years before they are fully grown, and may be active for another fifteen years thereafter. Some aquatic species ventilate their cocoons, which are attached to the bottom of the river or lake bed, by undulating movements of their body.

Some leeches have a few pairs of eyes at the front end. Most have sensory papillae on one annulus of each segment. Most leeches are smooth, though a few, like *Pontobdella*, are warty and tubercled. Most respire through their general body surface but *Branchelion* and some others have projecting arborescent gills ar-ranged in a row on each lateral side. Their mode of locomotion is either a looping motion, by alternate attachment and release of anterior and posterior suckers, or a swimming movement through water by vertical and not lateral undulations. The colours of leeches

The marine warty leech *Pontobdella* × 1, on a rock

vary from rust brown to olive green, often with striking patterns of patches, dots and streaks.

R.B.

See also WORMS.

LEOPARD (*Panthera pardus*). Possibly the most adapt-able of the higher mammals, the ubiquitous leopard has, over the ages, colonized two continents. According to some scientists there are fourteen subspecies in Africa while there is some difference of opinion as to how many survive in Asia. India has three of them, and a fourth exists in Sri Lanka where they appear to have reached the Island before it separated from the mainland.

From snow-line to sea level, from rain-forest to arid desert, they have adapted their living habits to survive in diverse ecosystems, and unlike the tiger, who has been more hesitant in his advance, they have successfully adapted their existence with co-predator and super-predator alike. The catholicity of their demand for prey species, ranging from birds, and rodents to the lesser mammals, has enabled them to exist in unbelievably harsh conditions. But under acute pressure from the fur trade they are now endangered throughout their range, more so in many areas than the symbolic and more demanding tiger. During the heyday of the skin trade, a matched Somali Leopard Skin coat could sell for 20,000 to 25,000 dollars, and it was estimated that the lifetime breeding effort of a pair of leopards went into the manufacture of such a coat. But now, even with the banning of trade by international signatories to the Convention on the International Trade in Endan-gered Species, and an agreement by the International Fur Traders Federation, an illegal traffic still flourishes, particularly through Far Eastern ports.

The leopard has been maligned, chiefly by sport killers who have suffered from, and resented, his almost uncanny capacity for effective retaliation when wounded. Thus it is often claimed that they are unpredictable and treacherous. Also some zoo keepers maintain that they have on occasions turned on their benefactors. However such behaviour or incidents where man is attacked only reflect the leopard's exceptional capacity for survival and the unnatural stresses of incarceration and crowding. If treated with dignity and under-standing, hand-reared specimens of this solitary and withdrawn cat are entirely tractable and even affectionate.

The largest leopard recorded is 9 ft 1 in. in length shot by the Maharajah of Nepal. However they are con-siderably slimmer than the tiger, with an average weight of 120–150 lb, and a 250-lb animal could be considered the maximum. Their entire body is one of integrated suppleness, and weight for weight no animal can hold a candle to their abilities. Their arboreal capacity is phenomenal and they are as much at home in trees, which they use for caching prey out of the reach of other predators and scavengers, as on the ground, and they have the capacity to haul almost double their body-weight up a perpendicular trunk. They will readily take to water, and though not as dependent on it as the tiger, will cool off in the hot weather if given the necessary shelter and cover as they lack the self-confidence to sit in exposed places when by themselves. They swim expertly with their tail moving in the manner of crocodilians, as against the tiger who generally swims with a rigid tail.

Their breeding dynamics are somewhat similar to the tiger's and after a period of courtship they are static during the actual mating period, normally lasting three days and nights, and it appears that the male initially breaks contact. The gestation lasts ninety-three days and though the leopardess seeks seclusion, the male may visit his family occasionally, for it is wrong to suppose that in free-living conditions the father will pose a threat to the cubs. The period up to weaning is three months, but the cubs usually do not accompany their mother and assist her at kills till the age of 8 or 9 months. Though there has been no visual confirmation in natural conditions, it seems likely, according to strong circumstantial evidence, that the mother may regurgitate food for her offspring. Contact is maintained by a limited vocal vocabulary. Due to their ability to subsist on comparatively smaller prey species there is limited evidence that cubs start hunting early, and could be

© M. Krishnan

Adult male leopard on the prowl × ¹⁄₁₇

abandoned by their mother at the age of a year.

Due to its extraordinary resilience to harsh living conditions, and wide adaptability, there is a widespread belief that the leopard can never be exterminated and public opinion is still not convinced that this magnificent cat needs any special protection, but there is growing evidence that its future survival is as precarious as that of the tiger and that throughout its range it is becoming increasingly rare.

A.S.

See plate 18 facing p. 241.

LILIES. In the strict sense lilies are plants which belong to the genus *Lilum* of the family Liliaceae. However, popularly many other similar plants are called lilies.

1 A stylized lily painted about 1550 B.C.
2 Nilgiri lily, found wild in the Nilgiris and also grown in gardens. 3 *Lilium polyphyllum*, wild in Himalaya and cultivated in gardens.

The lilies are bulbous herbs, with unbranched stems having typically trumpet or turk-cap shaped flowers, white or coloured. The flower parts are three, or groups of three, and the seed-bearing ovary is borne above all other parts.

True lilies are ornamental garden plants and many species are grown. The most famous is the Madonna or sacred lily (*Lilium candidum*), which is a native of South Europe and Central Asia. It was known from very early times and formed a part of a mural painted about 1500 B.C. in Crete. Another common garden favourite is the Easter or Trumpet lily (*Lilium longiflorum*), a native of China and Japan. The large fragrant white flowers are used as offerings particularly during Easter. More than half a dozen species of these true lilies are found wild in Himalaya and in the mountains of South India. The Himalayan giant lily (*Lilium giganteum*) can be up to 4 m tall, which is almost double the height of a normal man, and bears at the top 6 to 12 large white fragrant flowers. Boys make trumpets from the hollow stems of the plants, which give a deep bass note. The Nilgiri lily (*Lilium neilgherrense*) from Nilgiri hills in South India and many other species from Himalaya (*Lilium nepalense*; *L. polyphyllum*; *L. (Notholirion) thomsonianum*; *L. wallichianum*) have been introduced in gardens and are being cultivated all over the world in temperate countries.

One more popular plant which belongs to the lily family but is not a true lily is the Glory Lily (*Gloriosa superba*). It is a climbing plant and climbs with the help of its leaves, whose tips coil around the support. It bears numerous orange-coloured flowers, and is commonly grown in gardens and is also found wild all over the region.

S.D.

See CLIMBING PLANTS.

LIMPETS are primitive gastropods living attached to rocks exposed at low tide. The twisting tendency characteristic of all gastropods has only slightly affected the symmetry of these molluscs. Externally the shell of a limpet is conical and tent-like. The animal has two auricles, two sets of breathing organs, two kidneys. Its radula is well adapted for grinding tiny plants that grow on rocks. It has a broad sole-like foot. In the early stage of development as larva, the limpet shell is spiral but the animal soon settles down on rock. The twisting stops and the shell becomes a flattened cone. Some species of limpets have a hole or perforation at the apex of the shell and some a slit or notch along the front margin of the shell. These holes or slits serve as outlets of used water containing waste matter. Pure water is drawn up by the animal into the gills through the space between the edge of the shell and the rock surface, circulated through the tubular fold of the mantle and finally thrown out along with excreta through the apical or marginal hole. Limpets with apical holes are called Keyhole limpets and those having marginal slits are Slit limpets. Shells of the various species of limpets are differently ornamented by radial ribs, noduled or plain, simple or elaborate sculpture, concentric striae and coloured rays.

During low tide one can find limpets in large numbers adhering to exposed, weed-covered flat rocks. It is however a difficult task to detach them. The foot of a limpet is broad and pad-like. It can work like a sucker, creating a vacuum beneath enabling the animal to cling to the rock tenaciously. It has been estimated that a limpet weighing 5 grams exerts a resistance of more than 17 kilos. This clinging power and the flat conical shell protect them from the dashing waves and other enemies.

Limpets are remarkable for their 'homing instinct'. This means the limpet lives stuck to one particular spot or depression on the rock. Even if it crawls out to feed on the tiny seaweeds and algae, it returns to the original place with great regularity. To enable the animal to feed on sea plants it has a long radula armed with numerous rows of beam-like teeth.

Limpets are fairly common on the rocks, pilings and dead coral along both coasts of India. But they are not as abundant or large as in the cooler waters of the Pacific and Atlantic where they may be 10 cm long. Indian species are comparatively small and seldom collected for food.

Closely allied to limpets are the Ear shells. They differ from limpets in having a shell slightly spiral at the apex and with a much broader and flattened bodywhorl. The left margin of the shell has a row of holes instead of one hole as in Keyhole limpets. The mantle margin inside the shell is also perforated corresponding to the holes in the shell. Of these holes the posteriormost alone is used for throwing out waste water whereas the others serve as inhalent apertures to allow fresh water into the mantle cavity.

Ear shells are poorly represented in our area. Only two or three species of the genus *Haliotis* measuring 2 to 3 cm are sometimes found attached to low-tide rocks. In other countries they grow large—12 to 30 cm—are called abalones, and are a much valued food.

See also MOLLUSCS.

LINE-WEB WEAVERS. *Theridion* sp. and other species belonging to the family of Theridiidae are small spiders with a rounded abdomen and slender legs. Their webs are of irregular strands. They are common within houses in nooks and corners and also outdoors among foliage. Some curl up dead leaves or construct an inverted wine-glass-like silken tube for the protection of their eggs. Generally they are brown but some, such as *Argyrodes*, are metallic silver in colour. Many outdoor Theridiids construct their webs in the vicinity of orb-weavers.

Pholcids are weavers of untidy webs within houses along rafters, angles of ceilings, behind picture frames, etc. Their legs are very long and thin and the abdomen rounded. When their webs are touched these spiders oscillate their body up and down, feign death and drop to the ground. They attach their round cocoon to their mouth.

See also SPIDERS.

LION. The Asiatic lion (*Panthera leo*), which once roamed over a large part of north and central India, belongs to the subspecies *P. leo persica*. It is almost identical to its African cousin in size and appearance, males weighing up to 200 kg and measuring up to 2·4 m in length, and having a thick neck ruff or mane of longer shaggy hair. This is usually of a light tawny-yellow colour like the rest of its body and only very rarely is it black; but there are two deep longitudinal folds of skin along the belly usually well fringed with long hairs, which feature is not well developed in African lions.

Today the Indian lion is confined to a small relict and isolated population in the stunted teak forests of Gir in Kathiawar. The total population, recently estimated to be hardly more than 200, is now totally protected but is still in danger because of the gradual encroachment by human settlements both along the boundaries as well as inside the Gir Forest. Because this magnificent cat is in danger of extinction it has been well studied by wildlife biologists who have shown its almost total dependence upon kills of domestic cattle grazing inside the forest, which brings it into conflict with the local herdsmen. Furthermore such kills are often expropriated from the lion by these herders, who can salvage the hide and other parts of the carcass, thus depriving the lion of its food. Natural mortality amongst the cubs is consequently often quite high, though the male assists in hunting and killing food for the cubs. Usually only 2 or 3 cubs are born in a litter and most are produced during the cold-weather months, though breeding can occur at any time of the year.

By day lions sleep, often sprawled on their backs, and they only become alert in the late afternoon, when they wander off in search of prey, at dusk or just at daybreak. The males often emit a series of resonant deep-toned roars during the night and this is thought to be partly advertising their presence to rival males as well as to potential mates.

Lions normally kill by bounding up to their prey after having stalked to within close range. The hapless animal is usually seized by the throat and pulled to the ground until it suffocates. Two individuals will often cooperate in stalking and killing their prey and will

Male Lion × ¹/₂₄

Lioness

also share the food afterwards, but the lionesses and younger animals have to wait until the adult male, if present, has eaten his fill, before they can approach close to the carcass and start feeding.

See plate 18 facing p. 241.

LITTLE COD, *Brigmaceros atripinnis*, or *B. mcclellandi*, is important only because it is the only representative of the well-known Cod family (Gadidae) of the Atlantic and Pacific Oceans in the Indian seas. In contrast with the Atlantic cods, it is a small slender fish, growing only to about 7 cm. Its tiny size is compensated by the huge numbers in which it is caught off the northwest coast of India, its catch averaging 3000 tonnes per annum. Being small and slender it is easily dried in the sun and sent into the hinterland as non-seasonal food.

LIZARDS. Lizards, classified as Suborder Sauria of the large reptilian Order Squamata, dominated the earth nearly 150 million years ago. Present-day lizards vary in size from a skink 5 cm long to the Komodo dragon of Indonesia, 3 m in length. About 3000 species of lizards from 20 families are known from all over the world. Of these, about 150 species belonging to eight families inhabit India and the adjoining countries.

Most lizards are characterized by a slender, scaly

357

body with four well-developed limbs, a short, flat tongue and an external ear opening. Some limbless lizards superficially resemble snakes. They can, however, be distinguished by their ear openings and movable eyelids.

The scalation among lizards is diverse. Some have coarse, overlapping scales, others have smooth scales. Yet others, like the burrowing worm lizards, have lost their scales and instead have rings of skin encircling the body.

The tongue of a lizard may be thick and fixed, or extensile, or even prehensile. The paired Jacobson's organs situated beneath the nose of the lizard serve as organs of smell and taste for particles transmitted by the tongue.

Most lizards are insectivores. Several exceptions include the spiny-tailed lizard, chameleon, the Himalayan rock lizard and the common house gecko which have been observed eating vegetable matter. Monitor lizards appear to be wholly carnivorous, preying on a wide variety of small animals such as frogs, rats and crabs.

GECKOS, legless lizards and SKINKS have the ability to shed the tail and regenerate a new one in its place. This is a protective mechanism by which the lizard can escape, leaving a wriggling tail behind to divert its pursuer. Lizards generally shed their skin in

Painted gecko

fragments, periodically, though some of the legless lizards shed it in one piece, as a snake does. All lizards change colour depending on light, temperature, and the mood of the animal. Males go through very drastic and abrupt changes of colour during the breeding season.

Most lizards lay eggs; a few bear living young. Eggs, which are oval and white, are deposited in the open, in a sheltered place, against or under rocks, or in tree-trunks. The number of eggs varies considerably. The incubation period is from 40 to 60 days. Some skinks stay with the eggs.

Like fishes, amphibians and other reptiles, lizards are 'cold-blooded' which means that they lack an effective internal mechanism for regulating their body temperature in response to the changes of their surroundings.

House lizard × ¹/₂

Because of their minimal requirements of water, lizards and other reptiles can thrive in arid desert areas.

Except two species, the gila monster of USA and the beaded lizard of Mexico, all lizards are non-venomous. Despite the fact that geckos are the most inoffensive lizards, the unreasoning fear that they are poisonous is still common. On the other hand, to hear the bark of the giant tokay gecko soon after a child's birth is considered very auspicious in a Malayan house.

The lizard's most effective defense mechanisms are flight and camouflage. They intimidate enemies by bluff, and sometimes by colourful display. When excited, the agamid Sita's lizard distends and contracts its dewlap (throat-fan) rapidly, like a flickering light. The hump-nosed lizard, *Lyriocephalus scutatus* of Sri Lanka, deters enemies by merely opening its wide, red mouth.

The families of Indian lizards

Agamids (family Agamidae). The agamids are a large group of ground-, rock- and tree-dwelling lizards. Of the 280 species recognized so far in the family, as many as 38 species belonging to 12 genera occur in the Indian region. Agamids are easily recognizable by well-developed movable eyelids, mobile heads, five-toed limbs and a slender long tail. Their skin is dull, rough or spiny. The tongue is broad and flat. Agamids are mostly insectivorous.

Of all the agamas of India, two rock lizards, namely the Himalayan rock lizard *Agama tuberculata* and the South Indian rock lizard *Psammophilus dorsalis*, are well known for their striking coloration and abundance. The male Himalayan rock lizard measures 140 mm and is gorgeously coloured. Its range covers Uttar Pradesh, Punjab, Kashmir, and extends to Nepal and Pakistan. It is a common garden lizard in and around Simla, Garhwal and Kumaon. It feeds mainly on insects.

The South Indian rock lizard frequents the low hilly areas where it is seen basking on rocks. It is very agile and quick to escape into a rock crevice if it feels threatened. The male attains a length of 140 mm and becomes a brilliant orange-and-black during the breeding season.

The GARDEN LIZARDS are also agamids.

The genus *Draco* comprising the flying lizards is represented in India by two species, namely, *Draco dussumieri* of South India and *Draco norvilli* of Assam and Indo-China. This genus affords the best example of discontinuous distribution among animals. Between the South Indian species and the Indo-Chinese forms, which do not range west of Assam, there is a gap of 1600 km, in which no other species is found.

Uromastix hardwicki which is found in the desert tracts of Uttar Pradesh, Rajasthan, Punjab and Pakistan. It avoids the sun during the hottest part of the day by hiding in a burrow excavated by its powerful claws. Its diet is varied and includes several kinds of insects, flowers, blades of grass and fruit. It defends itself by flicking its spiny tail.

Five agamids are endemic to the mountainous regions of Sri Lanka. Unlike the other agamids, which lay eggs, the tree-dwelling agamid of Sri Lanka, *Cophotis ceylanica*, brings forth its young alive. Another equally interesting agamid commonly seen in the hilly districts of Sri Lanka is the hump-nosed lizard *Lyriocephalus scutatus*, which is recognized by the hump on its nose and bizarre body shape. Local people regard it as evil and venomous. Though it is terrestrial in its habits it sometimes ascends trees.

Snake Lizards (family Anguidae). The snake lizard, or legless lizard, *Ophisaurus gracilis*, the sole representative of the family Anguidae in India, is found in Darjeeling and the Khasi Hills, often at high altitudes from

Flying lizard × ²/₃

© Rom and Zai Whitaker

Legless lizard × ¹/₂

These slender lizards are called 'flying lizards' because they are capable of gliding between trees by means of 'wings' or 'patagia' which are folds of skin supported by six pairs of elongated ribs. The wings of the flying lizards are orange or yellow with dark spots. The male South Indian flying lizard has a long beard-like dewlap. They are all tree-dwellers, seldom descending to the ground, and feed largely on ants. The female digs a shallow hole in the ground and lays 2 to 5 eggs.

Another interesting agamid is the spiny-tailed lizard,

820 to 1500 m. So called because of its brittle tail and lack of limbs, it glides like a serpent by lateral undulations. It probes its way, employing its black tongue in a truly snake-like fashion. However it can be recognized as a lizard by its movable eyelids and an ear-opening. Its body is covered with ring-like overlapping scales. It is a terrestrial species, hiding under logs and stones for most of the day. It is sluggish by nature and becomes active only at night when it wanders in search of snails and small crustaceans. It is quite harmless and does

not bite when handled. When first picked up, it feigns death. The female lays 4 to 6 eggs.

CHAMELEONS are probably specialized descendants of agamid lizards.

Worm lizards (Dibamidae). The glassy-scaled Indian worm lizard, *Dibamus novae-guineae*, occurs in the Nicobars and all the way to New Guinea. It is apparently derived from the skink family. It measures 225 mm and looks like a slender worm. In keeping with its subterranean existence, the worm lizard is blind and devoid of fore-limbs, which are modified into scaly flaps in the male. The tail is short and cannot be detached. The egg has a calcareous shell.

Lacertids (Lacertidae). Lacertids are an abundant group of lizards considered as typical forms because of their slender bodies, well-developed legs and long pointed tail. They are met with in the sandy, grassy and rocky areas of the whole of Europe, Asia and Africa, where they are specially abundant. The fragile tail is easily shed and the broken part is regrown. The majority are insectivorous. Barring the viviparous *Lacerta* of Europe, all lacertids are egg-layers. Of the 150 species contained in the family, nine from five genera (*Acanthodactylus*, *Cabrita*, *Eremias*, *Ophisops* and *Takydromus*) occur in India.

The Indian fringe-toed lizard, *Acanthodactylus cantoris*, has scales on its digits that project along the sides to form a comb-like fringe facilitating the locomotion of the animal on sand. This lacertid of the sandy tracts of North India is brightly marked while young. It is a very agile lizard, disappearing quickly into bushes on the slightest alarm. Another Indian genus, *Cabrita*, is represented by two small lizards, *Cabrita leschenaulti* and *C. jerdoni* found in the forests of South India. They measure 15 cm in length.

Species of the race-runner lizards, namely *Ophisops jerdoni* and *O. beddomei*, are found in North and South India respectively. These 10–18 cm long lizards and the species of *Cabrita* have transparent discs in the centre of the lower eyelid. The astonishing long-tailed grass lacertid, *Takydromus sexlineatus khasiensis* occurs in the Khasi Hills. This six-lined lacertid is 5 cm long and its tail measuring 35 cm supports the weight of the animal in its jumps from stem to stem in grassy areas.

T.S.N.M.

See also REPTILES, CHAMELEONS, GECKOS, SKINKS, GARDEN LIZARDS and MONITOR LIZARDS.

See plate 9 facing p. 128 and plate 19 facing p. 256.

LOACHES are all small freshwater fish belonging to the family Cobitidae, growing at most about 8 cm long. They are thin and cylindrical in structure, with a number of barbels around the mouth. Many are colourful, with black and yellow stripes or pale yellow with black dots. They occur in shallow streams and puddles all over India, usually half buried in the bottom sand. Some occur in ricefields in western India during the monsoon and are therefore called 'ricefield loach' (*Lepidocephalichthys thermalis*). It has a comparatively large tooth-like bony projection below the eye by which it remains attached to the substratum. A loach *Botia striata* from Kolhapur and *Botia lohachata* from U.P. are popular aquarium fishes.

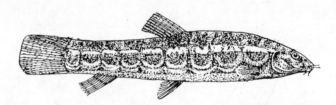

Lepidocephalichthys × ³/₄

Loaches normally feed on organic matter available at the bottom. Their breeding habits are not clearly known but the young fry, which are very tiny and colourless, quickly distribute themselves to new waters and shallow fields where they are a surprise to the novice.

See AIR-BREATHING FISHES.

Adult female Bonnet Monkey grooming young × ¹/₇

MACAQUE MONKEYS. This is a group of rather stocky thickset monkeys belonging to the genus *Macaca* which are distributed throughout the Indian subcontinent. Predominantly olive-brown in coloration with comparatively short tails, these monkeys are adapted to a widely varied diet most of which they glean from the ground so that they are not as dependent upon tall, tree forest as are the purely leaf-eating langurs.

The best-known species is the common Rhesus Monkey (*Macaca mulatta*) which inhabits most of the northern part of India and extends into the Himalayas of Pakistan. Its naked face is livid pink and it is the largest of the five species, adult males reaching up to 10 kg in live weight. Easily recognized by the rusty orange fur on its hindquarters, it is replaced in south India by the Bonnet Macaque (*Macaca radiata*) which lacks the orange fur on its lower quarters but is distinguished instead by a whorl of longer darker hairs radiating from its fore-crown. They also have a comparatively longer tail than the Rhesus, indicative of more arboreal habits. Adult males rarely weigh more than 8 kg. In Assam and Chittagong region of Bangladesh the Assamese Macaque (*Macaca assamensis*) is also a larger and heavier monkey, equally at home in the tropical rain-forest of Sylhet and Assam or the mangrove forests in the Sundarbans, in which region it is largely dependent on a diet of crabs. In the Nilgiri Hills and Western Ghats a remnant population of the much rarer **Liontailed Macaque** is now in danger of extinction due to the gradual destruction of the evergreen rain-forest

Rhesus Monkey, male adult × ¹/₇

Liontailed Macaque × ¹/₇

upon which it depends. Unlike all other macaques this striking monkey has a naked black face framed by a long lion-like ruff of white hairs, whilst the rest of its body-fur is black. *M. nemestrina*, the Pig-tailed Macaque, occurs in Nagaland. Finally in the northeastern extremity of Assam, in the hill forests, the Stump-tailed Macaque (*Macaca speciosa*) is found, which has a tail barely 25 cm long. It is said to be more insectivorous in diet than the other macaques.

Macaque monkeys are important to man in that they have been extensively used for testing of newly developed drugs and in particular for the development of the Salk vaccine against poliomyelitis. The export of Rhesus monkeys for this purpose is now restricted, due to an increasing realization that wild stocks are being rapidly depleted by this trade. All these species are markedly social, living in fairly cohesive bands of from 20 to 70 individuals in which one male is the dominant leader. Studies have shown that an equally large female is often the second ranking individual in the tribe. Diurnal in feeding, the whole troop usually retires to one group of trees or sheer cliffs for the night and where persecuted, as in Pakistan, they are shy and wary. In parts of India they are adapted to live commensally with man in the towns and villages and for reasons of sentiment and religion they are not molested despite their marauding and destructive habits.

See PRIMATES. See plate 32 facing p. 433.

MACKEREL belongs to the family Scrombridae, which includes some of the most important food fishes of the world, second only to Herrings and Sardines in quantity and quality. In Indian waters two species of mackerel, *Rastrelliger kanagurta* and *R. brachysoma*, are found, the latter being restricted to Andaman waters. Recently a third species has also been recorded. The first species represents the Indian mackerel for all practical purposes.

The Indian mackerel, being generally from 20 to 23 cm long (maximum 32 cm), is much smaller than the Spanish or the Atlantic mackerel which grows to more than 60 cm. The Indian species has an attractive stream-lined body, terminal mouth with minute teeth, adipose eyelids, two dorsal fins and a forked caudal fin. Behind the dorsal and anal fins 5 or 6 detached finlets are present as also the normal pectorals. The iridescent

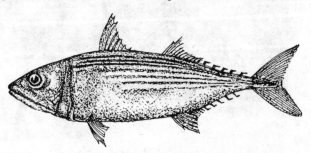

Rastrelliger kanagurta × ⅓

greenish blue colour of the upper part of the body, with greyish oblique stripes, silvery belly and yellow fins make the fish a beautiful denizen of the sea.

Being of oceanic habitat, mackerel migrate to coastal waters in large numbers from September to March on the west coast from Cape Comorin to Ratnagiri when the inshore water is cooler and contains more food material. The shoals are in thick masses of millions of individuals swimming close to each other and so near the surface that they can be easily spotted from a distance. The shoals are usually arrow-shaped. The period of migration varies from region to region, being earlier in the south and later in the north.

Their food consists of minute zooplankton and phytoplanktonic organisms such as copepods, crustacean larvae, diatoms and algae. The period of their first maturity is about two years, at a length of 20 to 23 cm. Of the commercial catch the average size ranges from 18 to 22 cm, indicating that only a few might get a chance to breed. Normally the spawning season is estimated to be April to September on the west coast and October to March on the east coast though supplementary spawnings in other months have also been recorded. Longevity is estimated to be five years. Ovarian eggs are about 1 mm in diameter but the sites of actual spawning and fertilized eggs have not been studied. In early fry stages the individuals have thick spines or spikes on the head, and look quite different from the adult.

About 80 to 90 percent of the total landings of mackerel are from the west coast and the fluctuation in the numbers is so great that the total landings per annum vary from 20,000 to 200,000 tonnes, the average being about 60,000 tonnes.

MAHOGANY, *Swietenia mahagoni*, is the true Mahogany of Central America. It was introduced in India in 1795 and has done well in Kerala. It is a large tree, deciduous here, with a buttressed trunk. The fern-like leaves turn coppery brown before shedding in February. Young leaves are a delicate shade of brilliant emerald, when they look beautiful. The fruits are egg-shaped capsules with winged seeds. It is one of the best cabinet woods of the world.

MAHSEERS belong to the carp family, Cyprinidae. Five species of genus *Tor* and one of *Acrossocheilus* are called mahseers and are found in different parts of the Indian subcontinent. Being marked by large shining scales and four barbels they are also known as 'large-scaled barbels'. One species *Tor putitora* or the common Himalayan mahseer, grows to a length of 2·75 m while another (probably *T. musullah*) from the Karnataka rivers is known to have weighed 54·3 kg.

Mahseers are riverine fish favouring clear waters and deep pools. **They can adapt themselves to conditions** in large reservoirs also by breeding in the feeder streams. They are omnivorous, feeding on a variety of items available in the waters in which they live. One angler-author (H. Thomas, 1897) enumerated their food items as follows: 'Aquatic weeds of all sorts, some taken intentionally, some when grabbing at the insects that live on them; seeds of the *Vateria indica* or Dhup of West coast, which are about the size of a pigeon's egg; the seeds of many other trees also which hang over the river when it is forest-clad; bamboo seeds; rice thrown in by man; and unhusked rice, or paddy as it is washed from fields; crabs .... small fish, earthworms, water beetles, grasshoppers, small flies of sorts, water or stone crickets, shrimps and molluscs or fresh-water snails, the latter shell and all, and smashed to pieces like the 'crabs'. Recent observations on *Tor tor* from Narbada river have shown that the species feeds mainly on vegetation and insects, though other mahseers do take small fish occasionally.

For breeding, mahseers prefer small slow-flowing rivers where water is clear, well oxygenated and the bed is sandy or covered with pebbles. They become sexually mature when about 30 cm in length. In northern India, particularly in the Punjab, the fish is said to have three spawning seasons, (1) January–February; (2) May and June and (3) July to September. Different reports exist about their breeding seasons, but the July–August period seems to be common for most of them and that is considered as the peak period. Spawning in rivers has not yet been observed, but in the waters of Bhim Tal in Kumaon males and females are reported to gather together in shallow marginal areas, where amid the splashing of water eggs are laid and fertilized. The eggs are beautiful glistening lemon-yellow or brownish in colour, 2·5 to 3·2 mm in diameter, and being heavy with yolk sink to the bottom. They take about 80 hours to hatch in water with a temperature of 26°C. The hatchlings or the larvae have a large yolk-sac, yellow in colour and mostly quiescent but they suddenly move with jerky action. They appear to be negatively phototropic, congregating and hiding in crevices till they develop into free-swimming fry. This development takes about six days in the case of the peninsular species *Tor khudree*. This quiescent condition is a critical stage,

when the larvae, huddled up into groups in corners and crevices, are likely to be preyed upon by predatory fish and other organisms. After they become free-swimming they fend for themselves.

Mahseers are great favorites of anglers because of their large size and fighting capacity when hooked. A considerable decline in their numbers has, of late, been reported in the sub-Himalayan rivers as well as in the south, due to illegal dynamiting. Efforts are, therefore, being made to rehabilitate these fish by breeding them artificially and introducing fry and fingerlings into different waters.

MAHUA, *Madhuca longifolia* var. *latifolia*, is a deciduous tree indigenous in the Gangetic plain between the Ravi and Great Gandak in the peninsula. The leaves occur in clusters at the ends of branches. The flowers are scented, sweetish, fleshy, cream-coloured and have an irresistible attraction for bears, deer and fruit- and nectar-eating birds such as mynas, bulbuls, parakeets, flowerpeckers and white-eyes. The ovary has a prominent long style. A spirit distilled from the flowers is a powerful stimulant and appetizer. In Maharashtra they are the main source of spirits of wine. The oil from the seed is used to adulterate ghee and for the manufacture of soap and candles.

Mahua. A shoot with leaves, and flowers with long styles, × 1/2. The petals fall off early, as shown in the sketch.

MALARIA. This disease is caused by a protozoan parasite of the genus *Plasmodium* which lives in the host's blood-stream. It infects, multiplies in and destroys the host's red blood cells, liberating several daughter parasites, each of which infects a fresh red cell (see 4, 5, 6 in figure). The process occurs in regular cycles, of 24,

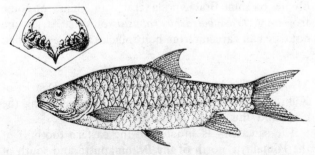

The Deccan mahseer, *Tor khudree* × 1/8
Inset, bones and teeth of the throat × 1/2

48 or 72 hours, depending on the species of *Plasmodium* involved. Each cycle is accompanied by shivering and rise of temperature, a symptom classically associated with malaria. After a time, as the host develops resistance, the cycles gradually cease, and the parasite, unable to multiply freely, forms gametocytes (7 8) which cannot develop further in this host. The gametocytes are ingested with the blood, by a mosquito which bites the host. In the mosquito the parasite passes the sexual phase of its life-cycle (9, 10, 11), resulting in the liberation of hundreds of parasites, *sporozoites*, which lodge in the mosquito's salivary glands. The mosquito, during its next blood feed, injects hundreds of sporozoites into the victim's blood-stream. When the victim belongs to a susceptible species, the sporozoites are taken up by scavenger or phagocytic cells, mainly in the liver, where they undergo several cycles of multiplication, the exo-erythrocytic cycle (1, 2, 3). After some time the para-

sites become capable of infecting the red blood cells and start the cycle of malarial fever in the new host. Some species of parasite directly infect the blood, without passing through an exo-erythrocytic phase.

Protection of the individual may be achieved by prevention of bites (mosquito nets, repellent creams, etc.), but protection of the community can only be through control of the vector, the mosquito. For a time, the use of DDT and other insecticides appeared to be very effective, but resistant mosquitoes have developed and proliferated rapidly, so that the disease is spreading back into areas from which it was thought to have disappeared. Mosquito control is now directed more towards BIOLOGICAL CONTROL such as the use of larvicidal fish which feed on mosquito larvas in ponds and wells where mosquitoes breed, the release of males (rendered sterile by chemical or X-ray treatment) in large numbers to prevent reproduction, etc. Attempts are also made to trap large numbers of mosquitoes, attracted by ultra-violet light, by the use of pheromones (insect hormones) and other means. At present there appears to be no clear solution and, for some time to come, it seems that we may keep the disease comparatively restricted by a judicious combination of various methods of control.

Malaria also occurs in other animals, including poultry and monkeys, the species of *Plasmodium* and of the vector mosquito being different for each host.

MAMMAL. Most people confuse the words 'mammal' and 'animal'. Correctly speaking, all living things are either plants or animals. Insects, birds and fishes are all animals, as well as cows and humans. What makes Mammals unique is their ability to nurse their young with milk. Mammals are also animals with hair. Humans have hair and nurse their young. People are therefore mammals and because of our close affinity with this great Class Mammalia in the animal kingdom, mammals have long since attracted man as a fascinating field for study as well as a rich source of clues as to our own bodily functioning and physiology. Even wholly marine-adapted mammals like the whales and dolphins have a few hair follicles and vestigial bristles around their mouth area at some stage in their development. The bristles on some Brachypoda (crabs) and 'hairs' on some frogs, e.g. *Trichobatrachus robustus* of West Africa, are not true hairs arising from hair follicles but are modified cells.

MANAS. The Manas Tiger Reserve is in the District of Kukrajhar in Assam and the total area including the core and the outlying forests is 2900 km².

The sanctuary is situated in the eastern foothills of the Himalaya, north of the Brahmaputra and south of Bhutan. It includes a variety of landscapes such as

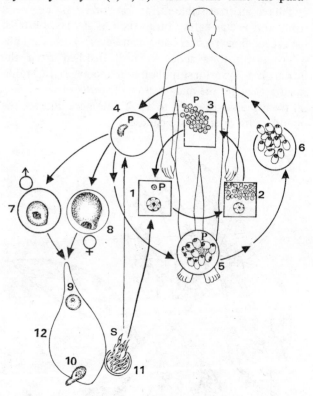

LIFE-CYCLE OF MALARIA PARASITE (diagrammatic)

P Parasite S Sporozoite

Tissue or exo-erythrocytic cycle in man
1 Parasite enters tissue cell. 2 Multiplication of parasites.
3 Cell bursts and releases parasites to infect fresh cells.

Erythrocytic cycle in red blood cells in man
4, 5, 6 As in 1, 2, 3 & 7 (male) and 8 (female) gametocytes.

Sexual cycle in mosquito
9 Zygote in mosquito stomach (12). 10 Zygote penetrates stomach wall and forms a sporocyst (11) which releases sporozoites (S). Sporozoites lodge in the mosquito's salivary glands, to be injected at the next bite.

riverain forests, grasslands, patches of evergreen and savannah and magnificent stands of sal (*Shorea robusta*). Scenically it is one of the most impressive sanctuaries of India and if co-operation is established with Bhutan there will be an extensive natural biotope for a wide range of mammals and birds. Two exotics which pose a problem are *Mikania cordata* from tropical America and the water-hyacinth. *Mikania* interferes with the forage potential and the water-hyacinth smothers the rivers and the lakes.

During 1972 a census indicated that there were 12 tigers in the sanctuary and another 30 in adjoining areas; in 1979 the population had increased to 69. The other animals include boar, barking deer, hog and swamp deer, elephants, rhinoceros, buffalo, black bear and gaur.

See plate 11 facing p. 144.

MANGO, *Mangifera indica*, a large evergreen tree with a dome-shaped crown, wild in the evergreen forests of the eastern Himalaya and the Western Ghats. It has been cultivated in India for 4000 years and originated in the Assam-Arunachal-Burma region. It has run wild throughout the subcontinent. The inflorescence panicle contains 3000 small flowers. It is easy to recognize from other members of the mango family in evergreen forests by the presence of only one stamen. A tree in Chandigarh is 9·6 m in girth with a crown-spread of 2250 m² and yields 450 maunds annually. The mango is regarded as one of the best fruits of the world. In May the branches and trunks of mango trees are covered with the spectacular Fox-tailed Orchid (see ORCHIDS).

Mango. Leaves and inflorescence × 1/2, fruit × 1/4. The Thais eat flowers and the Javanese the tender purple leaves.

MANGOSTEEN, *Garcinia mangostana*, an evergreen tree of Malaya, cultivated in Sri Lanka and in the Nilgiris. Fruit with a thick red rind and white pulp which is delicious. It is a prized edible fruit of Asia.

MANTIDS. These solitary, strictly predacious, generally large insects with raptorial fore legs, usually held in a peculiar supplicatory posture, are popularly known as 'praying mantids'. They are restricted to the warmer parts of the world where more than 1800 species exist. The Indian fauna possibly numbers around 200 species, many of them being very attractive in build. The adults and nymphs are mostly cryptically coloured, green or brown and similar to the type of shrub, herb or tree-trunk they inhabit. They usually stalk prey or lie in wait for them, lashing out with their fore legs armed with spines to capture and hold the smaller insects while feeding on them. Adult mantids are not capable of strong flight, even though they do possess well developed wings (in some species the hind wings being strikingly coloured). The females of most species are flightless and usually devour the hapless, smaller male after he copulates with her. In China 'mantis fights' are popular, with a lot of betting involved.

The female deposits several eggs in a typically shaped ootheca very similar to, though more elegant than, the egg-cases of the related cockroaches. They are attached to tree-trunks, stems, to rocks or walls or in soil or grass. Each such ootheca may contain from 10 to 500 eggs and in some species the female may guard it until the young hatch. The oothecae are subject to attack by small proctotrupoid or chalcidoid wasps and mantid densities are generally limited by such parasites and by the cannibalistic habits of newly hatched nymphs

© Kumar Ghorpadé

Praying mantis on tree-trunk × 1½

365

Praying mantis capturing Longhorned Grasshopper × 2

(as of mating adults). There appears to be some sort of territorial behaviour in many mantids and detailed studies of these fascinating and highly interesting, aggressive creatures would be very rewarding. Being general predators on any suitable insect or spider prey, they cannot be classed as 'beneficial' predators in the agricultural or forest ecosystem, but they are, undoubtedly, a more beneficial than harmful insect order.

Species of *Empusa* are a common type of mantid found in the Old World tropics. The very striking *Gongylus gongyloides* and species of some other genera like *Heterochaetula*, *Phyllothelys* and *Sphendale*, are typically Indian in distribution. *Hierodula westwoodi* and *Schizocephala bicornis* are the commonly noticeable species in cultivation on the plains.

See plate 5 facing p. 80 and plate 26 facing p. 369.

MARKHOR. This wild goat (*Capra falconeri*) is distinguished by the flattened keel of its horns which in males twist in a corkscrew shape. There is much variation in the shape of their horns which has resulted in the describing of a number of subspecies. Adult males stand less than a metre at the shoulder, and the northern races have horns measuring about 150 cm in length. They have a blackish muzzle and ears, and a ruff of longer white hair down their chests and along their spines. Females are reddish-brown with short twisted horns and, like the males, white-stockinged legs.

They inhabit the most precipitous cliffs and crags in the higher arid mountain ranges of Baluchistan and the northwestern Himalayas. A small population survives precariously in the Pir Panjal range in India. Markhor are much prized as a trophy by the local hill people

and their numbers have declined throughout their range due to this hunting pressure. Small populations also occur in Afghanistan and Soviet central Asia.

Markhor graze grass and browse on shrubs and are active mostly in the early morning and late afternoon. Young males and females keep together in small bands, the older males remaining solitary and staying at higher altitudes except during the rut which is in late November. At this time the males emit a pungent goaty odour and fight each other for possession of estrous females, rising high on their hind legs and crashing their horns sideways down on to their opponent.

See also MOUNTAIN SHEEP & GOATS.

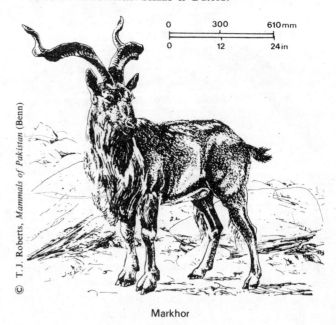

0	300	610 mm
0	12	24 in

Markhor

MARKING-NUT TREE, *Semecarpus anacardium*, is common in the subcontinent including Burma. It is characterized by large leathery leaves, clustered near the ends of branches, and purplish black fruits 2·5 cm long which are full of acrid juice. These can cause blisters on contact with the skin. The fruit is seated on a fleshy orange-coloured receptacle which is edible. The black resin from the fruit is used by Indian washermen as marking-ink for clothes.

MARMOT. Marmots (*Marmota* species) are large heavy-set rodents related to squirrels. There are three species in North America, and six in the Soviet Union but only two occur within our limits: the Bobak (*Marmota bobak*) and the Long-tailed Marmot (*Marmota caudata*). They inhabit alpine pastures or the rock-strewn slopes of upland valleys beyond the tree-line. These fascinating animals live in colonies and are familiar to the mountain trekker by their loud ringing alarm-calls, *bree-bree* pitched almost in the tone of a whistle. They actually hibernate about eight months of the year and emerge only in May when the last of the winter snows have still only exposed a few bare patches of ground.

Adults weigh up to 5 kg and have a head-and-body length of about 45 to 55 cm. The Bobak has a short black-tipped tail less than 15 cm in length but the Long-tailed, as its name implies, has a tail nearly 50 cm long. The Long-tailed Marmot has golden buff or tawny fur with the upper part of the face and mid-dorsal region predominantly black. The Bobak is a more uniform ochraceous colour. They have small low-set ears and short legs but despite their appearance are very agile at clambering over steep rocks. One litter a year is borne by the females, numbering 4 to 6, and this is produced during June after about thirty days' gestation.

See also RODENTS.

MARTENS. Belonging to the family Mustelinidae, Martens are related to otters, weasels and pole-cats but are generally intermediate in size between otters and weasels, with more prominent ears and a body adapted for climbing and hunting in trees.

There are three martens found in the subcontinent, two of which are closely related and comparatively large animals with exceptionally long tails. These are the beautiful Yellow-throated Marten (*Martes flavigula*) of the Himalayas extending across to Assam, Burma, and Indo-Malaysia, and the Nilgiri Marten (*M. gwatkinsi*) confined to southern India, which is more rufescent in coloration where the former is black. The third species, the Beech or Stone Marten (*Martes fiona*), occurs across the inner drier Himalayan ranges from Baluchistan to Sikkim in the east. The Stone Marten also occurs westwards in the mountains of Iran, Turkey and the Mediterranean countries.

Martens have broad flattened skulls with pointed inquisitive faces and short powerful limbs enabling them to run swiftly up vertical tree-trunks. They are extremely agile while hunting in trees but look somewhat clumsy when travelling on the ground, as they arch their spine when running. They are essentially carnivorous, living off birds and small mammals, but are known to have a distinctly sweet tooth, subsisting largely upon ripe fruit and berries in the season. The Yellow-throated Marten is passionately fond of robbing the combs of the wild hill bee.

They usually hunt by day and all three species are adept at running through tree-tops and bounding from branch to branch. The Stone Marten is however often found in more open mountainous country and is not so dependent upon forest cover as the other two species. They usually live and hunt singly. Only the female cares for the young, which are born in the spring, usually

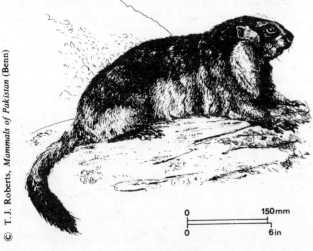

Long-tailed Marmot

Yellow-throated Marten

in a tree cavity or cranny between rocks. Beech Martens have been recorded with litters of 4 or 5 and Yellow-throated Martens with litters of 2 or 3. The young accompany their mother on hunting expeditions until they are 4 or 5 months old.

The Beech Marten has two irregular creamy-white stripes down its throat and the rest of its body is a dark glossy chestnut colour. They measure about 30 to 45 cm in body-length with a tail of 20–25 cm, and weigh about 1·75 kg.

The Yellow-throated Marten has a black face and hindneck with a creamy-white throat darkening to yellow where it meets the back of the neck. The upper part of its body is grizzled silver blond and the long slender tail is equal in length to the body and may measure up to 60 cm. They weigh up to $3\frac{1}{2}$ kg. All martens have quite a repertoire of vocalizations from abrupt *tock tock* contact calls, to growling and hissing threat calls. They are relatively bold and fearless even when encountering man, but generally avoid the proximity of human habitations and in areas which are more heavily populated they hunt by night. All martens possess anal scent glands and mark their territory by rubbing these glands on prominent stones or clumps of vegetation.

MAST TREE, *Polyalthia longifolia*, is a columnar evergreen tree, indigenous to Sri Lanka and S India. It looks like a flagstaff, with its branches and foliage adhering close to the main stem. The drooping, wavy-margined leaves are lance-shaped, tapering to a fine point. The flowers are greenish with petals similar in shape to the leaves. Oval-shaped fruits collect at the end of a common stalk. On ripening they are visited by bats at night which feed on them. The next morning the ground is scattered with the seeds—the remains of the banquet! It is a popular tree for landscaping.

MATING SYSTEMS. Mating systems or, to use an anthropomorphic equivalent, marriage, can be of many types. In most modern post-industrialization societies, one man marries one woman and this is called monogamy. When a member of one sex marries more than one member of the opposite sex the system is called polygamy. In many primitive and tribal societies, for example, a man has several wives, the number often being proportional to his wealth or social status. This is known as polygyny. The Nayars of Kerala on the other hand practised polyandry, a system in which a woman has several husbands. While these are the various systems of mating or marriage that are accepted by a given society it is not uncommon to find allegations of certain individuals of either sex breaking these rules and mating with individuals outside a given system of marriage—something that we may call promiscuous mating.

Each of these mating systems, monogamy, polygamy, polygyny, polyandry and promiscuous mating are all encountered in the animal kingdom. If you are one of those naturalists whose passion it is to study the sexual life of animals you will follow animals during the breeding season and see, for example, that a large number of insects such as *Drosophila*, butterflies or grasshoppers follow no rigid rules about whom to mate with and are rather promiscuous in their mating patterns. Leave these lowly forms for a while and look at birds such as crows, sparrows, egrets or mynas. You will notice that the same male and female will be found together in what might be called a very faithful monogamous marriage. In many instances this pairing lasts for the entire life span of the individuals. Monogamy is also found in a few mammals and the wild dogs are a good example. Now look at the Pheasant-tailed Jaçanas or a pride of lions and you will see clear evidence of polyandry. On the other hand look at the Hanuman langur male who owns a whole harem of females or the troops of bonnet macaques with their male dominance hierarchies and you will see polygyny, i.e. a male mates with a large number of females while any given female ends up mating with only one or at best a very small number of males.

Although one can find examples of each of these marriage systems, it is true that polygyny is far more common than either monogamy or polyandry in the animal kingdom and 70 percent of pre-industrialization human societies were polygynous. In a situation like this it is useful to ask 'why?'. Basically the reason is quite simple. Males, whether human or animal, contribute cheap and superabundant sperm while females make precious few nutrient-filled eggs. How many children can a woman bear after all? The legendary Queen of Troy, Hecuba, is supposed to have had more than twenty children. And how many children can a man father? Ismail, the king of Morocco, is reported to have fathered 1056 offspring!

Polygyny has many important consequences, one of which is that it results in intense competition among males for females. This leads to the development of a variety of structures such as the antlers of chital and the tail of the peacock or to a variety of behaviours such as the songs of the koel—all meant to outcompete other males and/or impress and be chosen by females.

Occasionally however, other factors come into play to offset these ideal conditions for polygyny. In many bird species the survival of the chicks becomes very difficult unless both parents stay together, bring food and defend the nest, creating a necessity for monogamy.

R.G.

MAYFLIES. A 'dancing swarm' of very delicate, light coloured insects with long 'tails' over or near water

Gaur (*Bos gavrus*)

Spotted Deer (*Axis axis*)

Elephant (*Elephas maximus*)

Wild Ass (*Equus hemionus*)

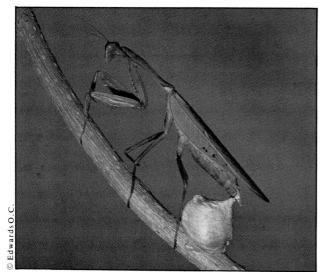

© Edwards O.C.

Preying Mantis at conclusion of egg-laying.

Wasp putting a live caterpillar in her nest.

Dragonfly

© O.C. Edwards

Yellowspotted Grasshopper

Caterpillar

© O.C. Edwards

Red Velvet Mite, an arachnid

A mayfly × 5

will usually be of these frail, short-lived mayflies, green or brown in colour. They have fairly large net-veined wings and after emergence from their watery life as immatures, they collect in large swarms to fly up and down and thus become noticeable. The adults live only a few hours, some a few days, and do not feed. They frequent all sorts of aquatic habitats but are more abundant in flowing streams and in colder regions in India on the Himalaya or the higher hills in the peninsula. Over 2000 species have been described from the world and perhaps a little under 200 Indian species are so far known.

Copulation takes place in flight and the female lays eggs in water. Most mayflies have a long developmental period, some of the larger species taking two years to mature. In some species more than 27 nymphal instars have been recorded! Particular species inhabit definite types of habitat, from large rivers to smaller streams, or even temporary nullahs or tanks. Some occur in marshy habitats or water-choked backwaters or swift-flowing, clear, cold mountain torrents. The burrowing or other nymphs are usually scavengers or plant-feeders, living on vegetable detritus or microscopic organisms in water, especially diatoms. A few are predacious and have well developed mouth-parts. Ephemeroptera usually emerge as winged subimagos, which moult to produce the true adults.

This is a very ancient order of insects and an important element of the aquatic community. The large numbers of mayflies inhabiting different types of water, especially in the colder regions of the world, form an important food of aquatic life, especially of freshwater fish. The nymphs convert the vegetarian aquatic resources into animal form and mayflies are also fed upon in large numbers by other water insects like beetles and dragonflies, besides stoneflies and the like. Some nymphs act as intermediate hosts for certain fish flukes and some nymphs and adults are infested by thread-worms. Mayflies are strongly attracted by artificial lights and the

large numbers that fly to lights in houses near large bodies of water sometimes become a nuisance. The common Indian species are *Cloeon dipterum*, *Ephemera immaculata*, *Palingenia orientalis*, and many species of *Baetis* on the Himalaya.

MELGHAT. The Melghat Tiger Reserve has an area of nearly 1600 km² and is situated in the Amravati District of Maharashtra. It forms a part of the southern section of the Satpura Range known as the Gavilgarh hills. The former Dhakna-kolkaz Wildlife Sanctuary is included in the Tiger Reserve. It is 114 km from Amravati and a convenient rail-head is Badnera on the Central Railway.

Melghat means 'the meeting-place of ghats' and as the name signifies it consists of steep hills and valleys. It has some of the finest Dry Deciduous Teak Forests of our area and other prominent species are haldu, axlewood, laurel and kadam. Lantana was apparently introduced in this area in the last century and since then it has become a menace to the natural regeneration of many local species. The highest point of the sanctuary is 992 m above sea level and the lowest 381 m.

The temperature varies significantly between the day

Forest cleared in Melghat Tiger Reserve for helipad

and the night and it is not uncommon to get a chilly night after scorching heat in the afternoon. In May the temperature goes up to 43°C while in January the minimum is about 12°C. Within the Reserve itself there are wide variations in rainfall ranging from a 1000 mm to 2250 mm.

The geological formation consists of the Deccan Trap, and bouldery soil is the most common type in the greater portion of the Reserve. As a result of this the rainwater quickly drains away and the five rivers—Khandu, Khapra, Sipna, Garga, and Dolar—are dry during the summer months. The few pools which retain rainwater throughout the year are the only source of water during the summer months.

Melghat was a favourite tiger-shooting area until it became a part of Project Tiger in 1972. In 1979 the tiger population was thought to be 63. Apart from tigers other animals to be seen are the rhesus macaque, langur, gaur, fourhorned antelope, nilgai, sambar, chital, boar, porcupine, hare, panther, dhole, sloth bear, jungle cat, ratel, civets, fox, hyena and jackal. There is rich bird life including forest-dwelling species such as the Great Horned Owl and birds found near water such as the Storkbilled Kingfisher.

METAMORPHOSIS. The change in form from the larval to the adult stage is termed 'metamorphosis'. Insect metamorphosis is the most spectacular and widely occurring, and may range from (1) *ametabolous*, where there is little or no change from larva to adult (Subclass Apterygota: Primitive Insects, Silverfish); (2) *hemimetabolous*, where the form of the larval instars (stages) gradually approaches that of the adult (Division Exopterygota: Mayflies to Bugs); and (3) *holometabolous*, where there is a striking change in the external form from larva to adult, and a pupal stage is intermediary to these (Division Endopterygota: Lacewings to Beetles). Because the Insecta exhibit metamorphosis (none, partial or complete) so extensively, metamorphosis has become almost synonymous with them. However, in the frogs (Class Amphibia) there is a distinct change from the tailed and limbless tadpole to the tailless adult frog with two pairs of legs. The other animal group that shows definite metamorphosis is that which contains the lampreys (Class Agnatha), and which is the most primitive vertebrate group.

Like the chicken or duck chicks and ducklings that emerge on hatching from the egg and are so similar to their full grown chickens or ducks (except for size), some immature insects are also so like their parents on hatching that anyone would know that they belonged to the same kind of insect. The Subclass Apterygota, with the primitive insects, especially the springtails (Order Collembola) and the silverfish (Order Thysanura) undergo very little or no metamorphosis, and are col-

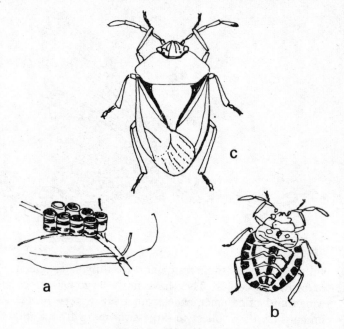

Simple metamorphosis (Hemimetabola) — (*a*) eggs, (*b*) nymph, (*c*) adult of stink bug (Hemiptera)

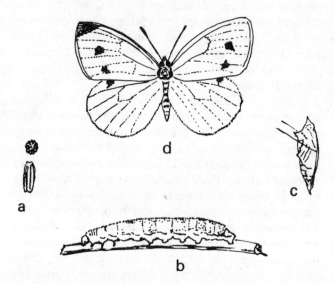

Complete metamorphis (Holometabola) — (*a*) eggs, (*b*) larva, (*c*) pupa, (*d*) adult of butterfly (Lepidoptera)

lectively termed the *Ametabola*. The growth of these insects, from the newly emerged state (from the egg) to the adult, is hardly accompanied by greater changes in appearance from those that take place in us humans from infancy to manhood. These adult insects have no wings.

No immature insect, when it emerges from the eggshell, has visible wings and all winged insects undergo metamorphosis during their development. In many insect groups, the young are very similar to the adult except for the absence of wings and genitalia (organs of reproduction). After a period of growth, however, small wing pads appear on the external surface of the insect larva. The more developed the young insect becomes, the more it resembles its parents. Such a development is called a gradual, partial or simple metamorphosis and the young of such insects are termed nymphs. They usually have the same habits as their parents and all of them are commonly seen living and feeding together, in the same manner, and on the same food source, not unlike a hen and her little chicks. Grasshopper nymphs and adults both eat the same grass or other plant leaf, while hopping about together; many bugs and their nymphs suck the sap of the same plant; and the ghastly bedbug nymphs and adults both live together in the crevices of your bed and suck your blood at night. This group, as a whole, is known as the *Hemimetabola*. It includes insects of the more primitive Orders which are those commonly called Mayflies, Dragonflies, Stoneflies, Grasshoppers, Mantids, Stick Insects, Earwigs, Webspinners, Cockroaches, Termites, Zorapterans, Psocids, Animal Lice, Thrips, and Bugs.

The largest groups of insects and the most advanced have larvae that have very different habits and are distinct in form from the adults. Like, for example, the mosquitoes, whose young live in water and feed on microscopic plant matter, while the adults fly in air and have females that suck blood. Again, like the root grubs (larvae) that live underneath soil and feed on roots of grasses, and have adult cockchafer beetles that fly about and feed on leaves of trees. So also, like caterpillars of butterflies and moths which live and feed on vegetation, inside leaves, stems or trunks, etc., rest as pupae on plants or inside them or in soil, and emerge to fly away and use their proboscis to suck nectar from flowers. Since the young and adults of these insects live in completely different environments or habitats, they evidently require to be of completely different structure. They therefore are very different in form, and except for the informed layman or student, no one else would suspect these larvae and their adults to belong to even the same group of closely related animals! The young of these insects also show no traces of wings or even wing buds during any period of growth. Such young insects are called larvae. When the larva is full grown, a striking change takes place, and after shedding its skin, the insect appears with its distinct large wing pads, long legs and antennae externally visible. This stage is called the pupa. In this stage the internal organs are transformed to the adult condition. On completion, the pupal skin is shed and the adult takes form by rapid expansion of its wings to full size, hardening of the body wall, development of adult colour pattern, and several other changes where the following structures are involved: epidermis, oenocytes, those of the nervous and circulatory systems, muscles, fat body, alimentary canal, Malpighian tubules, etc. Such development is called complete or complex metamorphosis. The largest and most advanced orders of insects including the Lacewings, Scorpionflies, Butterflies, Caddisflies, Flies, Mosquitoes, Fleas, Moths, Ants, Bees, Wasps, and Beetles have such a complete metamorphosis, and the group is called *Holometabola*.

MICROBES. Microbes are very small organisms which cannot be seen by the naked eye, some not seen even with a powerful microscope, widespread in nature, and are found in water, soil, and in living or decaying animal or plant tissue. Only a few hundreds of species have been identified and studied, mainly those which cause disease or are important in agriculture or industry. Microbes play an important role in many natural processes, such as the breakdown and dispersal of dead animal and vegetable matter. They are also important in the process of digestion and absorption of food in intestines of larger animals.

The majority of free-living microbes are aquatic, and form part of the food of sea anemones, fish and other larger animals which filter them out of the water by various means.

Microbes fall into several large groups of which one, the protozoa, are animal, while others belong to the vegetable kingdom, i.e. they are capable of synthesizing their nutrients from simple substances. The main groups are:

1. *Protozoa*. Single-celled animals, free-living or parasitic. Many parasitic forms are pathogenic (agents of disease).

See PARASITES, AMOEBA, MALARIA.

2. *Bacteria*. Single-celled organisms, possessing a rigid cell wall which gives a characteristic size and shape to members of each genus or group. Most bacteria are free-living and there are forms which thrive in conditions generally considered inimical to life, e.g. at high temperatures or in the presence of strong acids or alkalis. Bacteria may be aerobic or anaerobic, that is, they thrive in conditions where little or no free

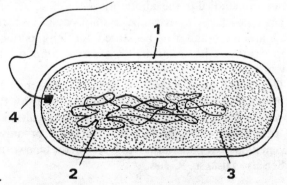

DIAGRAMMATIC REPRESENTATION
OF A BACTERIAL CELL

1 Cell wall. 2 Nuclear material. 3 Cytoplasm. 4 Flagellum.

oxygen is available. Some bacteria, for example those which cause tetanus or anthrax, can produce, under adverse conditions a shell or covering, a spore, which can protect it in a dry state for long periods of time. When favourable conditions recur, the spore germinates to give rise to a single vegetative organism. Spores are resistant to many chemical and physical agents, and material containing such spores must be disinfected by steam under pressure, which can penetrate the spore and destroy the organism (cf. cyst of amoeba, fungal spores, saprophytic or symbiotic sporing in plants). Parasitic bacteria may be pathogenic. (See PARASITES.)

3. *Mycoplasma.* Single-celled organisms, mainly parasitic, lacking some of the characters of bacteria, notably a cell wall. They are possibly degenerate forms of bacteria. Most forms which have been studied are disease agents.

4. *Viruses.* These are widely distributed in nature. It is said that any living tissue, if properly examined, will yield one or more viruses. They are submicroscopic (too small to be visible under a microscope) and completely parasitic. The virus consists only of the specific genetic material, supported by a protein coat and is incapable of performing, independently, any functions of a living cell. When it enters a suitable living cell, it utilizes the metabolic apparatus of the cell for reproducing (replicating) itself. In some viruses which infect bacteria, the virus may become intimately associated with the genetic apparatus of the host, and is responsible for transmission of certain characters of the host species.

Viruses have a specific tropism. A virus will infect, ordinarily, only one species, and, in the larger animals, only one tissue or cell type of that species. This makes them very difficult to study, unless a laboratory animal can be infected and used as a model for study of the infection. Considerable progress in study of viruses was made after the introduction of tissue culture, where cells are grown, in glass or plastic containers. Thus susceptible cells can be cultivated, infected with the virus, and studied in the laboratory.

Many purified viruses can be crystallized; in this form they are inert and can be stored for long periods. The virus can be reactivated by introduction of the crystals into a susceptible cell.

Depending on the point of view, a virus may be considered as the most primitive form of life, on the border between living and non-living, or as the ultimate in parasitism, retaining no part of a living organism except its genetic material as is also seen in the tapeworm.

See also DISEASE, RABIES.

MIGRATION. The march of seasons over the earth implies that different regions are more or less favourable for any animal species at different times of the year. Thus, the flush of insects in the early summer renders Siberia a good habitat for insect-eating birds; it is however totally inhospitable for them during the winters. On the other hand the northern Indian plains are a much better habitat for these birds after the late summer rains and in winter, but are quite inhospitable in late spring and early summer. Many animal species take advantage of such a situation by spending different times of the year in different habitats. Such regular seasonal movements constitute animal migrations, the most spectacular amongst these being the migrations of birds, described under BIRD MIGRATION.

The ELEPHANT is another well-known migratory animal. It has a wide range of habitat tolerance, from thorn scrub to rain-forest. Where possible however it tends to move into dry tracts during the monsoon and wet tracts in the dry season. Thus it used to move on to the Mysore plateau in the monsoon and on to the Wynaad plateau during the monsoon. In recent years, however, it has lost much of its original habitat, and suffered disruption of its migratory routes. It is therefore increasingly restricted to small pockets of forest where it used to concentrate only in certain seasons. This has led to very many problems of habitat degradation and increasing conflict with man.

Many aquatic animals also show periodic migrations. Thus shoals of the marine fish MACKEREL move inshore during the winter and off the coast during the summer, and HILSA fish move up the river Ganga for breeding. Some species of butterflies are also known to perform extensive migrations. However, we have scanty information on all these animal groups.

MILK FISH, *Chanos chanos* (suborder Percoidei), being a typical estuarine form is capable of withstanding considerable dilution with fresh water. It is an elegant, silvery streamlined fish with terminal mouth and widely bifurcated tail-fin lobes. In the family Chanidae, adults grow to about a metre in length and inhabit coastal waters. They breed in the same locality but the larval forms and young ones swim shorewards and enter creeks where they feed on green algae and other organic matter in the estuaries; and when about 20 to 30 cm (10 months old) return to coastal waters to breed after two years. Being a quick-growing fish feeding on algae, and of non-predatory tendency, it is extensively used in coastal aquaculture in Indonesia and S. India. Two hundred years ago Haider Ali of Mysore grew this fish in a tank in Kundapur and since then it has been grown in tanks and acclaimed as an excellent food fish.

MIRAGE. A mirage is an optical illusion of a water surface on the horizon, observed over a hot desert landscape.

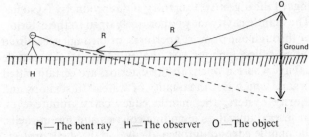

R—The bent ray H—The observer O—The object
I—The inverted image

On a hot clear day the air over a sandy/rocky surface gets heated through contact with the surface of the earth. The heated air becomes less dense than the less hot air above it. Rays of light when passing from a denser medium, the air above, to a lighter medium, the air near the ground, are bent away from the normal to the surface and reach the eye of the distant observer. The rays of light from the earth's surface sand or rock, on entering the denser medium are bent away from the normal, so that they pass above the head of the observer. Consequently, what an observer sees on looking towards the horizon is not the earth's surface—sand or rocks—but the shimmering layers of the less heated air above. This gives the illusion of water near the horizon away from the observer. A Sanskrit word for mirage is *mriga-trishna*,—deer's-thirst.

In extreme cases of intense heating, if there is an elevated distant object like a hill, a faint inverted image of it may be seen, adding credence to the illusion of an intervening water surface. Etymologically the word mirage is derived from the French *mirer*, to reflect.

MINNOWS are a group of small fishes belonging to different genera of the family Cyprinidae which inhabit fresh waters. The minnows are thus found in most of the rivers, lakes and ponds throughout the Indian sub-continent. Most prominent among these are the Chilwas (*Chela* spp. and *Oxygaster* spp.) which are usually found close to the surface of water and the Rasbora slightly below. Both these groups hardly grow to 10 or 12 cm. Small berils (*Barilius* spp.), *Daneo* spp. and *Ambassis* spp. are still smaller. All these minnows feed largely on small crustaceans like *Daphnia*, *Cyclops* etc. as well as mosquito larvae. Though they are of lesser value as food, they serve humanity quite substantially indirectly by destroying its enemies such as the mosquito larvae and *Cyclops*, which latter are the intermediate hosts in the life-cycle of the dreaded Guinea worm parasite. Minnows breed during the early part of the monsoon on the margins of ponds, lakes and rivers and the young ones distribute themselves to different sheets of water both temporary and perennial. Small barbels (*Puntius* spp.) are also included under minnows but for their food they prefer algal matter both floating as well as attached to rocks on the bottom. Most of the minnows have a quiescent larval stage and become free-swimming and start feeding after three days.

Top minnows are also small (less than 8 cm) and belong to a different order, namely Cyprinodontiformes. They are also called Killifishes or Toothed Carps, in contrast with other cyprinoid minnows which have no teeth on their jaws. They include the genera *Aplocheilus*, *Oryzias* and *Horaichthys* and are always skimming the surface of water; the first genus inhabits purely fresh water while the latter two prefer estuarine conditions. Feeding largely on small animal life they are excellent larvicidal fish. Their eggs are round with thin tendril-like outgrowths all over them by which they are attached to weeds. There being no larval stages young ones are fully developed at birth and start feeding as soon as they hatch out. The *Aplocheilus* male is quite colourful and makes an attractive aquarium fish.

MOLLUSCS. Molluscs are one of the oldest and largest groups of soft-bodied animals, generally protected by a calcareous covering or shell of their own manufacture. Mussels, oysters, clams, snails, slugs, squids and cuttle-fishes are all examples. Because they have a shell, molluscs are also called shell-fish. Mussels, clams and oysters have a shell formed of two folding pieces (or valves) and therefore they are called Bivalves. Snails and whelks have a single shell of a conical or spiral shape and so are called Univalves. Some forms such as slugs and squids have no shells. There is also a primitive and rare group of molluscs called Chitons which possess a shell of eight tiny horny plates arranged on their back. But these chitons are not familiar molluscs.

Molluscs have been in existence in this world for more than 500 million years. They are a very large group comprising more than 100,000 species varying widely in habits and habitats, size and shape.

General Distribution. Most molluscs live in the sea; some in fresh water; and some on land. The sea molluscs live along the shores at different depths up to 300 fathoms, on the sea bottom, in shallow waters and on rocks and reefs near low-tide mark, in intertidal (littoral) regions, on scattered rocks, under stones, buried in sand-banks and mud-flats, in tide-pools and also along and above high-tide rocks and walls.

The freshwater molluscs live in rivers, lakes, ponds, pools and waterlogged paddyfields.

The land molluscs are common on wet walls, on stems and leaves of plants and trees generally during the rainy seasons.

Body Structure. The body of a typical mollusc consists of four regions—the upper (dorsal), lower (ventral), front (anterior) and hind (posterior). The anterior portion bears the mouth and organs of touch, sight and smell and is therefore reckoned as the head region.

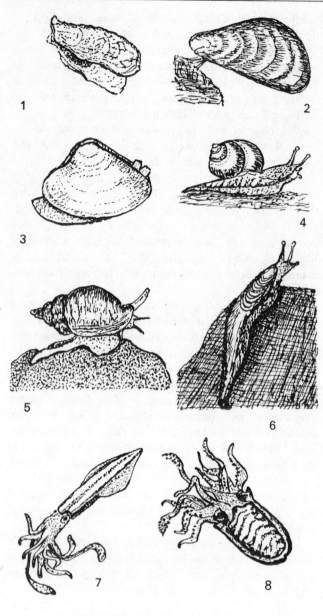

REPRESENTATIVE MOLLUSCS

1 Oyster × $\frac{1}{3}$. 2 Sea mussel × $\frac{1}{4}$. 3 Clam × $\frac{1}{2}$. 4 Land snail × $\frac{1}{2}$. 5 Sea snail × $\frac{1}{2}$. 6 Land slug × $\frac{1}{2}$. 7 Squid × $\frac{1}{6}$. 8 Cuttlefish × $\frac{1}{5}$.

ings of the digestive, excretory and reproductive systems. The mantle cavity may either freely open to the exterior, or through one or more apertures, or through specialized tubes called siphons. At the entrance of the mantle cavity, where it opens to the exterior, are certain pitted organs which test the quality of water. In scallops and thorny oysters, the mantle edges carry minute eyes. In several univalves including cowries and moon-shells, the mantle edges outgrow and envelop the shell, protecting it from abrasion and at the same time accounting for the high polish of the shell surface. In some slugs like *Limax* their small shells become internal being covered up by the mantle. In cuttlefishes and squids, where shells are absent, the thickened mantle protects the animal. Thus the mantle is a very important possession of all molluscs and is modified differently in different forms.

Foot. The next important characteristic organ of all molluscs is the foot. It is a prominent thickening on the ventral side of the animal's body and is modified to serve different forms of motion. In fixed molluscs like oysters, the foot is much reduced; in crawling forms like snails, it is flat; in cockles, crescent-shaped and in octopuses and squids it is modified into powerful arms.

In the general classification of molluscs, the foot is the main feature taken into account.

Molluscs are highly organized animals having all the important physiological systems well developed.

The Digestive system consists of a mouth, a throat leading into a stomach, followed by an intestine, a rectum and an anus opening into the mantle cavity. In symmetrical molluscs like mussels and clams, the mouth is at the anterior end and the anus at the posterior. But in asymmetrical univalves, like snails, the anal end gets twisted, brought forward and placed either to the left or right side of the mouth. Digestion is aided by the secretion of salivary glands into the throat and of the liver gland into the stomach. The mouth may be a circular or semicircular aperture as in snails, with labial palps as in mussels or modified into a snout or proboscis as in carnivorous whelks. Molluscs, other than bivalves, also possess jaws or beaks of different types for tearing and grinding the food. They also possess a flexible, ribbon-shaped tongue (or radula) set with transverse rows of minute horny teeth of varying shapes and sizes in different genera.

PART OF A RADULA, × 100,
showing lateral and median teeth.

Behind this there is a swollen mass—the visceral sac—containing the vital parts—the heart, stomach, liver, kidney and the generative organs.

Mantle. Overhanging from the dorsal region and enveloping the body is a tough muscular skin called the mantle. It is this mantle that secretes the shell. The outer surface of the whole mantle is capable to some degree of secreting shelly matter but the most active part is the margin. The mantle sides enclose between themselves and the visceral mass a chamber called the mantle cavity. This contains the respiratory organs in the form of gills or lung and also the open-

The number of teeth on the radula of a common whelk (*Buccinum*) is 220 to 250 whereas in another univalve (*Umbrella*) it is 750,000! As a rule the plant-eating molluscs have a long radula with many small uniform teeth whereas in carnivorous types the radula is shorter with fewer, large teeth. Molluscs having a radula and jaws possess a distinct head with tentacles, eyes and other sensory organs which are essential for them to find their food and grind it.

Bivalves do not search for food. They feed on organisms such as diatoms, protozoa and broken-down fragments of plant life which float in water and on particles of organic matter derived from dead animal and plant life that settle at the bottom of the sea, rivers or ponds. These minute particles are wafted to their mouth by hair-like filaments called cilia on their gills. A complex head region with snout, eyes, tentacles or radula is not required for such feeding, and bivalves have only a rudimentary head carrying a mouth with lips and two palps covered with cilia (vibratile hairs).

Respiratory, Circulatory and Excretory systems. Oxygen in solution is taken by the gills from water in the mantle cavity (in lung-breathing forms from the atmosphere) and carbon dioxide given off. The gills or lung filaments contain blood vessels. Blood from these passes to the heart and into the ventricle. Thence the blood is pumped both anteriorly and posteriorly through the aortas to the various parts of the body. It is collected in the vena cava, carried through the kidneys; from there to the gills or lung, and back to the heart.

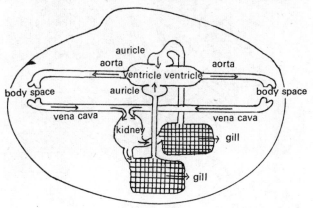

Diagram showing the circulation of the blood in molluscs

The kidneys or excretory organs consist typically of two symmetrical glands placed on the dorsal side of the body in close connexion with the cavity which encloses the heart. Each kidney opens on the one hand into the mantle cavity close to the anus and the other into the space covering the heart. The venous blood returning from the body passes through the vascular valves of the kidneys which are largely formed of cells containing uric acid. The blood thus parts with its impurities before it reaches the breathing organs.

Nervous and Sensory systems. The nervous system in all molluscs generally consists of paired nerve centres or concentrations situated in the regions of the head, foot and visceral mass. They are connected by longitudinal fibres. The anterior ones are also connected transversely.

All molluscs invariably possess two kinds of sensory organs: one for sensing sound and equilibrium (auditory) and the other for testing the water (olfactory). The former is controlled by the nerve centre on the foot and the latter by the anterior nerve centre. The other important sensory organ is the eye.

In bivalves eyes are less developed than in other molluscs. Oysters have eye-spots along their mantle margin which can respond to changing degrees of light and shade. Eyes of scallops and spondyles are a little more developed.

In univalves which have a distinct head, two pairs of tentacles are present and these are tactile (sensory to touch). The longer pair carries well-developed eyes at their tips or base.

In cuttlefishes, squids etc. all the sensory organs are very much developed. They have two large and prominent eyes one on either side of the head. The arms and tentacles of these animals are highly tactile.

Reproduction. The sexes are united in some molluscs and separate in others. Reproduction invariably takes place by eggs which after being developed in the ovary of the female are fertilized by the sperms of the male. The eggs are then laid and undergo their subsequent development outside the parental pouches. In some univalves like periwinkles and most bivalves, the eggs are hatched within the female pouch before they are expelled. The molluscan larvas are minute top-like bodies with a collar of hair-like cilia with whose aid they make spinning movements in water. These larvas are very much like the larvas of marine tube-worms.

Molluscs are all prolific and many of them lay millions of eggs during their breeding season. If all these eggs were to develop and become adults there would be no space in the sea for other animals. But the majority of these eggs become food for fishes and other sea animals.

Enemies. Many fishes and sea animals feed on the molluscs which survive. Even the great squids are attacked by whales. Gulls, kites, vultures, terns and crows feed on clams, mussels and limpets. Rats eat land-snails and freshwater molluscs. Thrushes, chaffinches, frogs, toads and certain beetles feed on slugs and snails. And some molluscs (e.g. whelks) feed on other molluscs (e.g. oysters, mussels and arks).

Protection. Nature has also provided molluscs with many protective measures—colours matching with their surroundings, lids or opercula (in univalves) to close the mouth of the shell tightly, growths such as spines, knobs and ridges on shells, slimy or poisonous secretions, froth of bitter taste, ink-glands to raise 'smoke-screens' in the water. Cockles can make long leaps and escape. Certain molluscs possess the power of amputating some parts of their body when attacked by enemies and afterwards reproducing them. This is noticed in land snails and slugs, in harp-shells and razor-fish.

Uses of Molluscs. Molluscs are useful to man in many ways. Oysters and mussels, clams, snails and squids—all yield valuable food to man. Pearl-oysters, mussels, turbans etc. produce precious pearls and mother-of-pearl. Many shells such as chanks, cowries, cones and volutes are objects of beauty and fetch good prices. Shells are a rich source of raw material for manufacture of quicklime and cement. Certain molluscs have medicinal value.

Oysters and other sedentary molluscs which form regular belts protect the shore from erosion. Molluscs which feed on putrefying and decomposed animal and vegetable matter such as dog-whelks are scavengers which keep the coastal waters clean.

Classification. Molluscs are divided into five Classes. Leaving alone the rare Chitons and the very small group of tusk-shells (*Dentalium*), all the other molluscs fall under the following three important Classes:

1. GASTROPODA (stomach-foot) Molluscs with a disc-shaped foot under their stomach
2. PELECYPODA (hatchet-foot) Foot laterally compressed
3. CEPHALOPODA (head-foot) Foot split into sucker-clad arms and arranged in a circle round the head region

Gastropods are univalve molluscs characterized by the possession of a conical or spiral shell, an asymmetrical body, a flat foot and a radula in the mouth region. No other group of molluscs presents so much diversity in size, shape, habits and habitat and in the shell form. They are the most adventurous—there are sea, freshwater, land and amphibious forms among them. There are gill-breathers and lung-breathers. Some are oviparous (laying eggs) and others viviparous (keeping the eggs in their body cavity until the eggs are hatched and the young ones come out as miniature adults). There are some forms among them without shells, such as slugs.

Gastropods are asymmetrical molluscs—the opposite sides of the body being not similar. As larvas they are symmetrical, but owing to the twisting or coiling growth of the shell (usually in the clockwise direction) the symmetry is affected. The hinder end of the digestive canal, originally posterior to the mouth, is swung down and forward and brought to lie above the mouth. Organs placed on the left get shifted to the right where

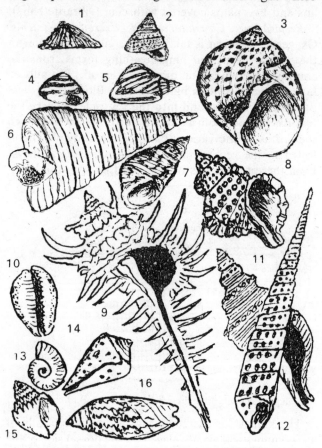

DIFFERENT FORMS OF GASTROPOD SHELL
(APPROX. LIFE SIZE)

1 Limpet. 2 Top (see SNAILS 3). 3 Natica. 4 Button shell (see SNAILS 3). 5 Small Dogwhelk. 6 Horn shell. 7 Periwinkle (see SNAILS 3). 8 Purse shell (see WHELKS). 9 Rock whelk (*Murex*). 10 Cowry. 11 Slit-lip (see CONES). 12 Auger (see CONES). 13 *Planorbis* (see SNAILS 2). 14 Cone. 15 Small Buccinid whelk. 16 Olive (see VOLUTES).

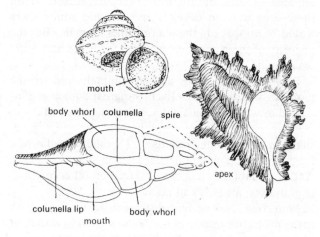

Upper left, snail shell. Lower left, parts of a gastropod shell. Right, exterior of a gastropod shell.

their growth is greatly reduced. Organs on the right side move towards the left. Where the twisting is extreme even the originally paired organs like auricles, ventricles and kidneys become single.

Attempts have been made to explain why snails get coiled in growth and are subject to the disturbance called 'torsion'. Originally at any rate the animal is not curved. It is the shell that curves the animal and the spiralling tendency is inherent in it. Dynamic spiral growth is found in nature not in shells alone but in the florets of a sunflower, scales of a pine-cone, horns of a mountain goat, etc. The original form of univalve shell was only conical—a small tent-like covering to protect the vital parts. As these parts commenced to increase in size and weight with the growth of the animal, they naturally swung down pulling the shell with them. The result of the two forces combined— the increasing size of the visceral hump and the tendency to pull the shell over with it—probably resulted in the conversion of the conical into a spiral shell which gradually came to envelop the whole animal. Where the visceral mass became flattened instead of growing, the conical shell might have been reduced into a simple plate as in slugs.

The spiral form of the shell varies very greatly in gastropods, from the almost flat limpets to the steep and elongated spires of auger shells, with 20 or more oblique rounds. There are also discoid shells coiling round in the same plane. The opening of the shell in most cases can be closed by the animal with a horny or calcareous lid called the operculum.

The common gastropods are: CONES, COWRIES, HORNS, LIMPETS, NATICAS, SLUGS, VOLUTES and WHELKS.

Pelecypods (Lamellibranchiata) are bilaterally symmetrical molluscs. They live in sea and fresh water. There are no land forms. All possess shells which are of two pieces or valves one on either side of the body. These valves are joined along the upper margin by a hinge-like arrangement which can open or close like the covers of a book. The valves when closed completely shut up the body. The closing is done by a pair of muscles attached internally connecting the two valves— one at the front end and the other at the hind end. At about the middle of the hinge margin is an elastic ligament which keeps the valves open.

Bivalves have no distinct head or tentacles. The mantle is bilobed—one lobe covering each side. In the less developed bivalves such as mussels, arks and scallops the mantle lobes are not united along the ventral margin. They open freely to the exterior and no siphon is developed. In the higher forms like clams and razors the mantle lobes are united in such a way as to form two tubes or siphons, one inhalent to admit water inside and the other exhalent to throw out used water.

The siphons may be free or partly united as in clams or completely fused into a long, elastic siphonal tube as in burrowing forms like wedge-shells and razor-fish. The breathing organs (gills) are simple thread-like filaments folded upwards in the primitive arks and sea mussels but well developed with vascular connections in the higher forms like clams and razor-fish.

The foot in pelecypods is generally wedge-shaped. In attached forms like oysters the foot is much reduced but in burrowing forms it is strong, long and cylindrical. In the jumping cockles it is sickle-shaped. In sedentary arks and mussels the foot secretes tenacious threads called a byssus for attachment to rocks and mud-banks.

Common examples of bivalves are: ARKS, CLAMS, COCKLES, MUSSELS, OYSTERS, PIDDOCKS, RAZOR-FISH, and SCALLOPS.

Cephalopods are all sea forms without an external shell, except for the Paper Nautilus and the Pearly Nautilus. The foot is divided into a number of arms at its anterior end and would appear to be attached to the head of the animal. The head bears a pair of prominent eyes. The mouth is armed with jaws or beaks to tear the prey. The head is followed by the body— a muscular sac containing the essential organs and gills. The posterior portion of the foot is modified into a funnel which communicates with the internal mantle cavity. By ejecting water forcibly through the funnel the animal shoots through water like a rocket.

The arms of cephalopods—eight or ten in number— are long, tapering and flexible, furnished with stalked or sessile suckers with or without horny rims.

With the exception of the Pearly Nautilus all cephalopods possess an ink-sac with the secretion of which they can raise a smoky screen in water to confuse their enemies and at the same time to escape unnoticed. All cephalopods are carnivorous.

The common cephalopods are: CUTTLEFISHES, OCTOPUSES, PEARLY NAUTILUS and SQUIDS.

MONGOOSE. Mongooses are terrestrial, active, bold and predaceous animals, well known for their fights with snakes. A single genus, *Herpestes,* occurs in the subcontinent. Two species, *H. auropunctatus* and *H. edwardsi*, are widely distributed in the region whereas the Ruddy Mongoose (*H. smithi*), which is distinguished by the black tip to its tail, is found from Rajasthan to Sri Lanka in the south and up to Bengal in the east. The Brown (*H. fuscus*) and Striped-necked (*H. vitticollis*) Mongoose inhabit southern India and Sri Lanka. The large Crab-eating Mongoose, *H. urva*, is distributed throughout Bangladesh, Nepal and the Assam region, its range extending to southern China, Thailand and Taiwan.

The mongooses are diurnal except when they reside

very close to villages or in the outskirts of towns. They are very adaptable animals and occur in a variety of habitats from desert to cultivated cropland and light scrub forest. They live in burrows, in man-made drains, culverts, and in ruined buildings. They can freely climb trees and swim easily. Most of the mongoose species mark their territory with the secretion from their anal gland. When attacked or cornered they erect their long hair and fight rather fiercely. They consume a variety of food, principally insects, scorpions, centipedes, toads, frogs, lizards and snakes, birds and their eggs, and rodents. They feed on carrion also and are frequently observed feeding on fruits and other parts of plants. In spite of the fact that the mongooses are fairly common, very little is known about their breeding habits. Most of the species breed all the year round and the litter size varies from 1 to 3, the gestation period being 40 to 60 days. A captive female *H. edwardsi* produced five litters during 18 months. Sexual maturity is attained at the age of 8 or 9 months. The mother becomes very fierce before and after parturition. Their life-span varies from 7 up to 13 years for the larger species.

I.P.

Above — Yellow monitor lizard × $^1/_4$
Below — Bengal monitor lizard × $^1/_4$

MONITOR LIZARDS. Monitor lizards are of a striking appearance with elongated head, unusually long neck and tail, and snake-like tongue. All the thirty existing species are so similar that they are included in a single genus *Varanus*, in a range from Africa, throughout southeast Asia to Australia. Monitors are the world's heaviest and largest lizards. The Komodo dragon of Indonesia grows to a length of over 3 m and may weigh 200 kg.

Despite their massive bodies monitors are surprisingly good runners and swimmers. The smaller ones are excellent climbers. Their long, forked tongue is employed as a sensory organ. They hiss loudly in defence and lash with their powerful tails. They are all carnivorous and feed on a wide variety of prey including insects, crabs, molluscs, rodents, birds and carrion. They are particularly fond of eggs, in search of which they may raid bird, turtle and crocodile nests. Food is torn and crushed with their powerful teeth and jaws and swallowed whole or in large chunks. The female lays about 15 to 30 eggs and deposits them in termite mounds, holes in the ground, hollow logs or piles of brush. Though the eggs of the common monitor may hatch in 2 months, water monitor eggs may incubate for 9 months.

Of the four species occurring in our area, only the common monitor *Varanus bengalensis* is widely distributed. It is met with in forest areas as well as the outskirts of villages and suburbs. The yellow monitor, *V. flavescens* of North India and Bangladesh, is a stocky

Monitor lizard at termite mound × $^1/_4$

species common in Bihar. The desert monitor *V. griseus* lives in the arid and rocky areas of Africa, Pakistan, and northwest India. It is greyish yellow in keeping with the sandy soil. The water monitor *V. salvator*, occurring in Sri Lanka and the coastal mangrove regions of Orissa, Sunderbans, and the Andaman-Nicobar islands, is India's largest lizard, reaching a length of 2·5 m. It is blackish above with yellow rings, and white below.

See also LIZARDS.

MONSOONS. The word 'monsoon' is derived from an Arabic word meaning 'season'. It was used by mariners several centuries ago to describe the system of winds which blow from the southwest during six months of the year, and from the opposite direction during the other six months. These winds are obviously associated with the greater solar heating of the Asiatic continent during summer and its greater cooling, during winter, compared to the heating of the oceanic areas: the Indian Ocean in the south and the Pacific Ocean in the east. The summer monsoon (the southwest monsoon of India) may be treated as a gigantic sea-breeze and the winter monsoon (the northeast monsoon of India), as a gigantic land breeze.

The rotation of the earth plays a large part in the monsoon winds as they undergo a large latitudinal displacement. The Indian monsoon winds during the months June to September come mostly from the southern hemisphere; as they cross the equator and move towards the Indo-Gangetic plains, they move northwards from the equator. This results in their acquiring a component from the west, as the earth's surface velocity is fastest at the equator and is from west to east. These winds with southerly and westerly components constitute the southwest monsoon. These winds obviously are moist, as they travel long distances over oceanic areas, both south and north of the equator, over the Arabian Sea and the Bay of Bengal. The winds which come across Burma and South China, from the Pacific Ocean, are also moist.

On the other hand, the winds which blow over the subcontinent during the winter months, November to March, are mostly of Asiatic land origin. As they move from the northern latitudes towards the equator, they acquire a component from the east. These winds blow from the northeast—the so-called Northeast Monsoon. They are comparatively dry, as they are of land origin. It so happens that this winter cold air descends into the North Indian plains as it crosses the Afghan, Himalayan and Chinese Mountains; consequently the Northeast monsoon is dry over North India. It is also not as cold as over the elevated land areas of Tibet and Kashmir. Winds from further east move across the Bay of Bengal and pick up some moisture; they also get mixed with the more moist air from the Pacific Ocean. Consequently the northeast monsoon winds arriving over the Andhra coast, Tamil Nadu, Kerala and Sri Lanka are quite moist.

There are large seasonal variations of winds, associated with the northward and southward movement of the sun across the equator and the changed patterns of heating of the earth's surface. However, over the Atlantic and the Pacific there is a rough symmetry of sea areas, north and south of the equator. Over the Americas also there is a rough symmetry of land areas with the equator passing right across Brazil. (The situation is in great contrast with that prevailing over the section of the equator extending from the East African coast to Malaysia. There is no land south of the equator, and there is considerable land north of the equator.) Hence the monsoons are only feebly developed over the Pacific, the Atlantic, and Americas. The winds of the opposite hemisphere hardly extend beyond five to ten degrees of latitude north or south of the equator. On the other hand, during the Indian summer monsoon, the air of the southern hemisphere extends to latitudes as far north as 25°N, and occasionally even more.

The zone, which runs more or less east–west, over which winds of equatorial origin (northern hemispheric or southern hemispheric) encounter winds of higher latitudes is known as the Tropical Convergence Zone (TCZ). Usually the atmospheric pressure is lower along this zone than on either side of it. The air equatorward of this zone, is the moist monsoon air. This zone moves almost up to 25° N over the subcontinent during the months June to September while over the rest of the world it hardly moves up to 10° N. Usually this zone is the seat of extensive clouds of great vertical extent and also of copious rainfall.

In the case of the Indian monsoon, the orientation of the Himalayan and Burman mountains and the shape of the northern end of the Bay of Bengal are such that some of the air to the north of the Tropical Convergence Zone also comes from the southern side round Bangladesh. The northern Bay of Bengal is warmest during May to September, and serves to make the monsoon air current move northwards across Bangladesh, whereafter it is deflected towards the west and the periphery of the Himalayan massif which is too big a barrier for the air current to cross. However, the moist air-stream ascends over part of the slope and provides abundant rainfall on the southern slopes of the Himalaya, up to an elevation of three to five kilometres. Higher elevations are mostly cloud-free.

The Western Ghats provide a barrier right across the monsoon stream. There is abundant energy in the stream in the form of latent heat of water vapour. The mountains are not too high, less than 2 kilometres on the average and less than 3 kilometres even at the highest peak of Anaimudi in Kerala. The rising moist air provides abundant rainfall along the Western Ghats and along the coastal strip to the west of them. The situation is similar along the coastal regions lying to the west of the Arakan Yomas of Burma and Bangladesh. The Khasi-Jaintia Hills in Assam lie east–west across the northward flowing monsoon current. The world-famous Cherrapunji and the somewhat less famous village Mowsynram both lie on the crest of these hills, and consequently receive an annual rainfall of around 1100 cm, the world's highest in spite of a few rainless months. (A station on the Hawaii mountains has a slightly higher total rainfall, spread throughout the twelve months of the year.)

Onset and Withdrawal of the Southwest Monsoon

Over seventy per cent of the annual rainfall over India as a whole, and around ninety per cent of the annual rainfall over several parts of India, occur during the summer monsoon.

The commencement of the monsoon, the so-called onset, is an important event. On many occasions the moisture-laden air current arrives spectacularly, accompanied by strong winds and a spell of heavy rains. On these occasions it is easy to specify the date of onset. However, on other occasions the monsoon current creeps in. Then it is difficult to specify the exact date of onset. We know when it was not there, and we know when it is there, the transition period being three to four days. Based on a discontinuous but sustained increase in the mean five-day rainfalls over individual observatories, it is possible to specify the mean date of onset

of the monsoon over an observatory. Fig. 1 shows the dates of onset as prepared by the India Meteorological Department. According to this chart, the southwest monsoon arrives over Kerala and Bangladesh around 1 June. It is somewhat earlier over Sri Lanka and Burma, and still earlier over the Andaman Islands. Thereafter the monsoon gradually spreads northwards and westwards into the country. It reaches Bombay around 10 June, Calcutta a couple of days earlier. It covers the whole of India, except the extreme northwest, by 1 July. Peninsular India, Gujarat and Rajasthan are pervaded by the monsoon current from the Arabian Sea, the so-called Arabian-Sea Branch; the belt from Burma to the Punjab and Kashmir is served by the Bay of Bengal Branch of the monsoon.

Fig. 2 represents the dates of withdrawal of the monsoon rains from the country. The withdrawal is less abrupt than the onset. The summer monsoon begins to withdraw from the extreme northwest around 1 September; the process of withdrawal is completed around 15 October. Thereafter, rainfall continues over the south peninsula, but it is due to a moist current arriving from the northeast. It is sometimes called the withdrawing monsoon, and more often the northeast monsoon. The northeast monsoon rains set in around 15 October, and finally withdraw from the entire peninsula by 15 December and from Sri Lanka by about 5 January.

Monsoon Rainfall. The mean annual rainfall over the plains of India is about 110 cm, about 30 cm higher

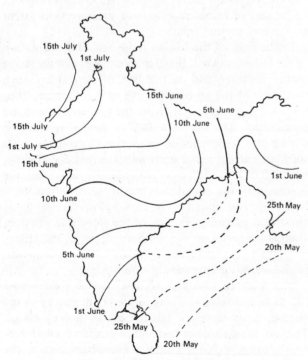

1. NORMAL DATES OF ONSET OF
SOUTHWEST MONSOON

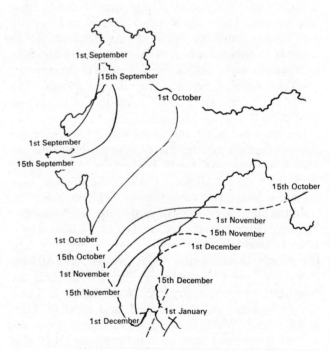

2. NORMAL DATES OF WITHDRAWAL OF
SOUTHWEST MONSOON

than the global average. The total volume of water that falls on the plains of India during the monsoon is about 2500 cubic kilometres. Even in extreme cases of floods or drought, the variation in the total rainfall from this mean is of the order of only ten percent. It is worth mentioning that over the plains of India the average rainfall *per rainy day* is roughly the same, whether it be at Trivandrum, Calcutta, Nagpur, Bombay or Bikaner—it is 2–3 cm per rainy day, about ten times the average rainfall per rainy day in the European and North American countries. Hence the greater proneness of India to floods. This feature is shared by other tropical countries also.

The distribution of rainfall in any one year varies in space as well as in time. There are also variations from one year to another. In years of poor rainfall it can be only thirty or forty percent of normal. The variability is large over the poor-rainfall regions: Rajasthan and North Gujarat. The distribution in time, even when the total monsoon rainfall over an area is not very deficient, can be so bad that the crops suffer and the area experiences a drought. Thus floods and droughts occur, in one part or the other; occasionally the same area experiences floods during a brief spell, and drought during the rest of the same year. Water management needs considerable skill.

The Northeast monsoon, October to November, is the principal rainy season for Tamil Nadu in peninsular India. Averaged over the plains of Tamil Nadu, the total Northeast monsoon rainfall is around 20 cm and represents about 50 percent of the annual rainfall; over the coastal districts of Tamil Nadu the average rainfall is around 25 cm and represents about 60 per cent of the annual.

On 22 June, the sun is at its northernmost position and the subsequent months July and August are the hottest over the northern hemisphere. Over India and Burma, this is the period of monsoon with much cloudiness and rain, and therefore cooler. In the middle latitudes 40°–60°, winter is their principal rainy period. Our rainy period is during this relatively warmer season, and therefore, our crops can grow faster than in continental areas where winter is normally the time of rain; if managed well, India has a potentiality of being one of the world's greatest producers of food.

MOSQUITO FISH, GAMBUSIA. This is not an indigenous fish in India but has been introduced during the early part of the present century, as a biological control for the eradication of malaria. Scientifically known as *Gambusia affinis holbrooki* (family Poecilidae), it is a surface fish and has a special liking for mosquito larvae

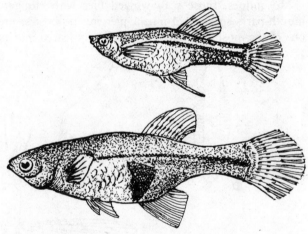

Gambusia affinis, slightly enlarged; male above

on which it feeds voraciously. It is now serving its useful purpose in most of the large cities of India. Mosquito Fish are also called Killifishes or Top minnows.

The fish is a native of Central America and has been introduced into many tropical countries for its larvicidal propensity. Though it hails from natural streams, it is most adaptable and can thrive in stagnant and even partially polluted water.

Being viviparous it gives birth to young ones which start to feed immediately. The male is small-sized, hardly about 2·5 cm, while females grow up to 8 cm, are dominant and more numerous. The male is more colourful and can be distinguished by its elongated intromittent organ which is developed from a few anterior fin-rays of the anal fin.

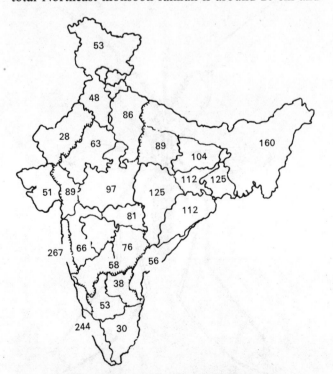

3. NORMAL DISTRIBUTION OF MONSOON RAINFALL IN INDIA, IN CENTIMETRES (after K. N. Rao)

Another exotic fish which is also sometimes called 'Mosquito fish' is the guppy, *Lebistes reticulatus*, belonging to the same family as the Gambusia. Also styled 'million fish' because of its prolific breeding, it is smaller than Gambusia and is bred for aquariums. Males are extremely colourful and give rise to numerous varieties. Originally it is supposed to have come from the West Indies but it is now thriving all over the world.

MOSQUITOES and BITING FLIES. Of all the insect orders, the Diptera or True FLIES are primary in members of medical or veterinary importance. The blood-sucking mosquitoes, black-flies, sand-flies, biting midges, horse-flies and the African tsetse-flies transmit malaria, filariasis (elephantiasis), leishmaniasis, trypanosomiasis (sleeping sickness) and a range of arboviruses including dengue, encephalitis and yellow fever. Various other helminths, protozoa and viruses causing disease in domestic animals, e.g., surra in horses, avian malaria, etc., are transmitted by house-flies and biting flies, including the stable-flies annoying to domestic animals.

Mosquitoes, those scaly-winged flies with elongate mouth-parts specialized into a piercing proboscis, are one of the most common and universally feared insects

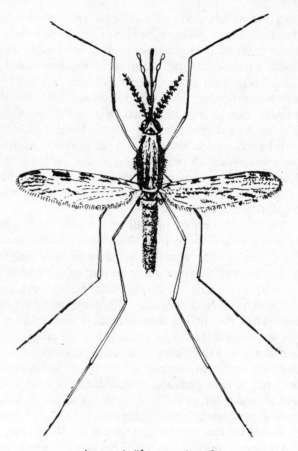

An anopheline mosquito × 8

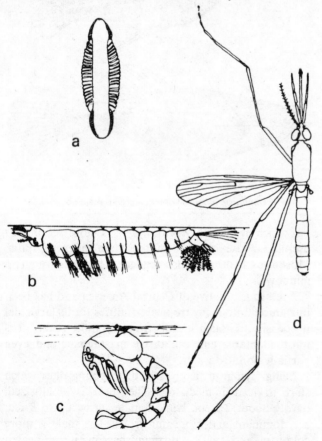

Life-cycle of mosquito — (*a*) egg, (*b*) larva, (*c*) pupa (*d*) adult

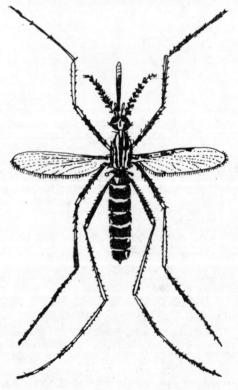

A culicine mosquito × 10

known to man. However, only a fraction of the existing species (over 300 occur in the Indian subcontinent) will attack man, being specially adapted to feed on the blood of other domestic and wild mammals, birds, reptiles or frogs. Only female mosquitoes suck blood, which is required so that their eggs can mature. The males, usually slender insects, with bushy antennae, live only on flower nectar, plant juices, etc., as also do the females largely! Eggs are laid and larvae develop in running or still water habitats; the latter may be any form of water in a natural or artificial container, capable of holding water even for a short period. Both larvae and pupae are active swimmers and dive into water if disturbed. Adults emerging from pupae are mainly nocturnal, hiding in shady, moist habitats during day time though some species may be active during the day in poorly lighted situations. The most important Indian

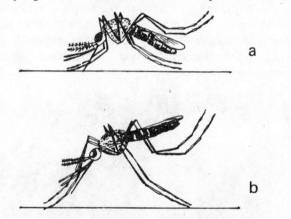

Resting positions of (a) *Culex*, (b) *Anopheles* mosquitoes

species are *Aedes aegypti, Anopheles culicifacies, A. stephensi, Culex pipiens quinquefasciatus, C. bitaeniorhynchus* and *C. tritaeniorhynchus* (the last two transmitting the recently troublesome encephalitis). Species of *Toxorhynchites* are very large, beautifully coloured mosquitoes, whose larvae are promising predators of harmful mosquito larvae and whose adults, both male and female, are non-bloodsucking.

The Simuliidae or black-flies are small (2·5 to 3·5 mm), stout flies found almost everywhere near flowing water in which the larvae occur. Females of many species suck blood and cause intolerable annoyance to man as well as deaths of his livestock. Onchocerciasis, a filarial disease causing human blindness, is transmitted by some species and diseases of birds by others. Around 50 Indian species are known, of which *Simulium aureohirtum* and *S. grisescens* are common.

The Phlebotomidae or sand-flies are another important group of biting flies in India. *Phlebotomus argentipes* and *P. chinensis* are vectors of visceral leishmaniasis (kala-azar) in Assam and Kashmir. Dermal leishmaniasis (oriental sore) is transmitted by *P. papa-*

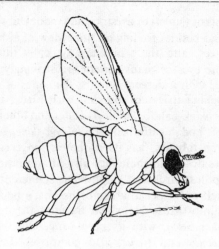

A black-fly × 17

tasi and *P. sergenti* in northwest India. The former species transmits sand-fly fever, a predominently Middle East virus disease. Besides man, species of the dominantly temperate *Phlebotomus* also attack other mammals, while species of the largely tropical *Sergentomyia* feed on blood of reptiles, mainly lizards. Early stages of sand-flies live in soil or leaf litter and only females suck blood of vertebrates.

The tiny Ceratopogonidae or biting-midges (1 to 4 mm) feed partly on plant juices but many females also take animal food and blood. Some non-medically important species predate on chironomid flies, while other species suck blood of larger insects including engorged female mosquitoes. Some others of the genus *Forcipomyia* are important pollinators of rubber and cocoa trees. Several *Culicoides* spp. are known vectors of diseases like onchocerciasis of horses and cattle and several human filariases. The immature stages develop in wet decomposing substrates.

The large and handsome Tabanidae include the horse-flies and deer-flies, that are vectors of loaiasis and try-

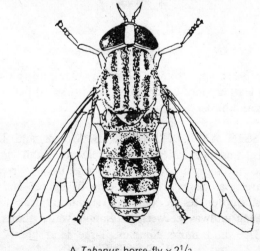

A *Tabanus* horse-fly × 2^1/$_2$

panosomiasis (surra) of livestock and man. Only females suck blood besides feeding on honeydew of aphids and plant juices, and these flies are typically diurnal in habit. The eyes of many genera are strikingly banded or spotted with iridescent colours, the wings of *Haematopota* being intricately speckled with black, those of *Chrysops* black-banded, while *Tabanus* and other genera have the body patterned. *Philoliche longirostris*, a Himalayan tabanid, has its proboscis longer than the body. Besides man, these flies also attack other mammals, reptiles, birds and amphibians. Different species will attack only special areas of its host's body!

The family Muscidae which includes the house-fly, *Musca domestica*, with its perplexing, closely related forms *nebulo* and *vicina*, also comprises the blood-sucking *Stomoxys* and *Haematobia* flies. The house-fly is found wherever man has settled anywhere on earth and can transmit pathogens biologically and mechanically to human food and his body. The young stages develop in dung, any rubbish mixed with excrement, decaying vegetable matter or decomposing food, meat or carcass. *Musca sorbens* or the bazaar-fly is complementary to the house-fly, inhabiting open areas around man's habitations and attracted to sores and wounds on his body. *Stomoxys calcitrans*, the stable-fly, transmits trypanosomes to man and animals and *Haematobia* spp. transmit diseases of cattle and horses, both being very irritating and injurious to livestock.

A Hippoboscid fly × 6

Hippobosca spp. (family Hippoboscidae) are the common leathery flies biting dogs, also horses, other mammals and birds.

MOSSES AND LIVERWORTS. Mosses and liverworts, together called bryophytes, are small, green, shade- and moisture-loving land plants, usually growing in large patches and reproducing by spores and not seeds. Bryophytes are considered significant in evolution as being midway between the simple water plants like ALGAE and more complex, large land plants like FERNS and FLOWERING PLANTS.

The life-cycle of these plants is represented by two generations, i.e. sexual (gametophytic) and asexual (sporophytic), alternating with each other.

There are about 14,000 kinds of mosses and 10,000 kinds of liverworts in the world. In our area about 2000 mosses and 750 liverworts are so far known, and their actual number may be much more, as explorations add more species.

They are abundant in the Himalaya and other temperate regions of the world, on a variety of habitats, such as soil, rocks, boulders, tree-trunks and fallen logs, forming significant and eye-catching carpets. In the Himalaya they are common up to a height of 4500 m. Higher up they are sparse. *Aongstroemia julacea*, a moss, has been collected during one of the Everest expeditions from a height of 6000 m. This is the highest elevation at which a bryophyte has been seen.

It is the tropics, particularly the tropical rain-forests, which abound in a multitude of species. A moss genus *Macromitrium*, with nearly 450 species, is largely confined to such forests, forming large cushions on tree-trunks and branches.

© J. N. Vohra

Moss cushions of Macromitrium *sp. on a tree-trunk.*

Blue Poppy *(Meconopsis aculeata)* grows at altitudes above 3000 m in the western Himalaya.

Epilobium sp. Seasonal monsoon plant found in the Valley of Flowers.

Wagatea spicata, a woody climber growing in the Western Ghats.

Mussaenda glabra, a shrub found in moist evergreen forests.

28 Dendrobium Orchids

Dendrobium nobile, found at about 2000 m in the Himalaya from Nepal eastwards to China.

Dendrobium primulinum, found in Nepal, Sikkim and Manipur. Flowers in March-April. It is a decorative plant, grown in gardens on fern-blocks.

Dendrobium ochreatum, found in the Patkoi Hills, Nagaland, below 2000 m.

Dendrobium infundibulum. Perfect flowers last for about eight weeks.

© E. P. Gee

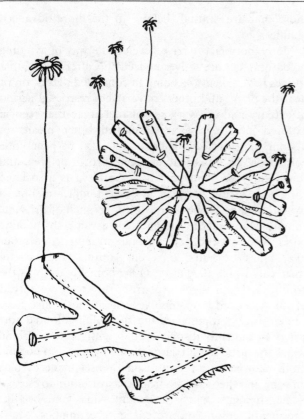

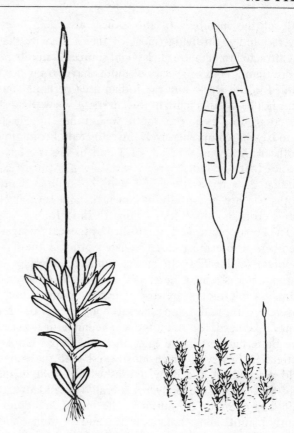

Liverwort growing (× 1), showing female plants. Stalked capsules are borne upside down, on the under surface of the umbrella between the lobes, and are not visible in the drawing. *Below*, part of thallus × 2.

Moss plant, *Hypophila comosa*, × 7. *Right*, capsule × 35, showing spore sacs enclosing columnella. *Below*, Habit (mode of growth) × 1½

Liverworts are so called, because certain forms resemble the liver in shape. They were also believed, for a long time, to cure liver ailments. They may either have a flattened and lobed body, which is several layers of cells thick, at least in the middle, or may be differentiated into stem and one-cell thick leaves, looking like mosses, but unlike the latter they always grow prostrate on the substratum. Both types grow close to the damp surface, whether on the ground or on a tree-trunk and anchored by thread-like structures called rhizoids. They multiply by spores which may be enclosed in capsules within the plants or borne at the end of short stalks, which occur on raised umbrella-like structures in certain forms. When the capsules are ripe, they burst and the spores are blown far and wide, to take their chance of germinating.

Mosses are more developed. The plant body is always differentiated into stem and leaves. (There are no roots. Rhizoids perform the function of roots, as in liverworts.) Some remain prostrate on the ground, forming mats or carpets, bearing capsules laterally on stalks, but others grow as in the figure. In these the spores are right at the top.

Mosses and liverworts also multiply vegetatively by various ways. The most common way is by repeated branching, followed by the death and decay of the intervening older parts, as in *Sphagnum* or peat moss. In certain forms such as *Pohlia nutans* and *Campylopus flexuosus* shoot tips and branches break and fall off and develop into new plants. Specialized bodies known as gemmae (as in *Marchantia* and *Cyathophorella*) and underground tubers (as in *Bryum erythrocarpum*) are means of vegetative propagation in several other forms.

Bryophytes in general are endowed with an exceptional power of regeneration. Under suitable conditions any small part can regenerate into a new plant, enabling them to colonize bare soil, bringing stability to freshly cut slopes, and preparing conditions suitable for the colonizing of ferns and flowering plants.

In recent years, it has been found that they also have great potential in absorbing heavy metals, such as copper, lead, and iron, directly from the atmosphere, and hence can act as indicators of air pollution.

J.N.V.

MOTHS. The Lepidoptera, comprising what we commonly call moths and butterflies, is one of the larger insect orders and one of the better known ones. While butterflies fly by day and are generally noticed feeding on flower nectar, the moths, which make up more than

385

90% of the members of this order, are their more diverse and night-flying (most of them) cousins that get attracted to man's artificial light sources in numbers. More than a lakh of species of moths have so far been named in the world and the Indian area perhaps contains not less than a tenth of this diversity. As with most insect groups, even our moth wealth has not been seriously or comprehensively investigated. From tiny moths with a wing spread of 3 mm or less we have species that are much larger, with a wing spread exceeding 25 cm or more. As with butterflies, many moths too are beautifully ornamented, and several of man's crop and store pests belong to this group.

The Lepidoptera are distinguished by the scaly-winged members whose most closely related order is the Trichoptera (caddisflies). In moths—unlike butterflies—the antenna is almost never clubbed, but is of several forms. Adult moths are also generally more densely covered with scales, on several body parts besides wings, like head, legs and thorax. The immature stages, like butterflies, comprise egg, larva and pupa. Unlike butterflies, however, which mostly rest with the wings held erect above the body, moths usually fold them, roof-like, above the body or laterally expanded and held appressed to the surface. The moths are dominantly plant-feeding, both as larvae and as adults, and they have evolved in unison with the flowering plants. Some moths also perform annual migrations as butterflies do, many of them being frequent visitors to flowers to sip nectar. Some moths exhibit seasonal variation in colour like many butterflies, besides sexual dimorphism and geographical variation. Moths reproduce sexually, except in rare cases where parthenogenesis occurs. Males are attracted to females either by visual stimuli or by chemicals called pheromones that the female releases. Egg-laying is generally induced by special smells that the host plant produces through its essential oils. From a dozen to a few thousand eggs may be laid by a single female, being deposited singly or in batches, near or on the food-plant, externally or internally.

Though most moth larvae are phytophagous, some also feed on animal matter. A few newly hatched larvae make a meal of the chorion of the egg. Others feed on the egg masses of other Lepidoptera or of spiders, on ant larvae or on scale insects. Cannibalism is also prevalent in some moth species and others feed on material of animal origin, like wool, etc., whereas some are ectoparasites of leaf-hoppers also. Few plant habitats have been left unexploited by Lepidoptera, almost every plant part providing food for the larvae: leaves, fruit, seeds (in situ or fallen), flowers, galls, buds, twigs, branches, bark, trunk or roots. Moths that have larvae feeding in concealed situations (borers, leaf-miners, bark-miners, leaf-tiers, case-bearers, etc.) are usually of primitive stock, whereas the exposed feeders, especially those that are diurnal, belong to the more advanced families.

Many moth larvae have developed a way of avoiding predators: they are active and feed only after dusk. Other species have taken to feeding in concealed places, on or near the plant, and others have a very restricted period of activity and are very seasonal. Larvae that feed in exposed situations have developed other means of discouraging or fooling predators. Use of procryptic coloration, becoming either green, brown or grey, and resembling the plant part that the larva is associated with, aids escape. Larvae of some families assume a twig-like shape, or lie flush with the branch of the same colour. Many larvae, if detected, assume threatening poses, with or without special body structures or armour like spines or hairs, or even painful or distasteful chemicals which they eject. Others quickly drop to the ground or lower themselves from the tree on a thread of silk produced by themselves. Adult moths also have similar tricks: rapidly raising their dull-coloured fore wings to suddenly display their brightly coloured hind wings to startle the predator. Some pyralid and noctuid moths can pick up the ultrasonic noises made by bats on wing and then take vigorous evasive action to escape! Some sphingids with transparent wings and bee-like aposematic coloration are efficient bee-mimics. Many moth larvae are adorned with hairy and spiny structures which are modified setae that are capable of giving a stinging sensation. Moths and their immature stages are so abundant and diverse both biologically and morphologically, that it would be presumptuous to attempt even a general summary here.

The parasitic enemies of moths are mostly tiny wasps that belong to the Chalcidoidea. The moth larvae constitute food of spiders, mites, aculeate wasps and a large variety of vertebrates, especially birds. Other insect parasites belong to the Braconidae, Ichneumonidae and Tachinidae, besides nematodes, bacteria and viruses. Since moth larvae are predominantly phytophagous, they cause crop losses to man to a large degree. They are generally called borers (e.g., Cossidae), leaf-miners (Gracillariidae), leaf-rollers (Tortricidae), leaf-tiers and webworms (Pyralidae), forest defoliators (Limacodidae, Geometridae, Eupterotidae), or cutworms and armyworms (Noctuidae). Most of these are indigenous pests, but many other introduced pests through man's commerce are cosmopolitan, injurious to stored products (Gelechiidae, Oecophoridae, Tineidae), orchards (Tortricidae), and vegetables (Gelechiidae, Yponomeutidae). Most Lepidoptera are unconcerned with man and only a few are obviously beneficial. Some feed on coccids (e.g., Eublemma, Stathmopoda) and others (Epipyropidae) are parasitic on leaf-hoppers. Attempts at using certain species of moths in the biological control of noxious weed plants are in progress.

The order Lepidoptera is now divided into four sub-orders: Zeugloptera, Dachnonypha, Monotrysia and Ditrysia; over 75 families are generally recognized. Only the more important or abundant ones will be treated here very briefly. The Zeugloptera are very small, diurnal, metallic moths that are accepted as archaic, and are represented by a single family, the Micropterigidae. The larva has a slug-like appearance and the pupa has pointed mandibles used to open the cocoon. Moths feed on pollen and the minute larvae feed probably on detritus and pupate in a parchment-like oval cocoon. They usually occur in rain-forests. The Dachnonypha are also a primitive group, the moths being very small and usually diurnal. Of the three known families, only one, Neopseustidae, is recorded from India.

The Monotrysia is probably not a homogenous group, but the dozen or so families that it is made up of are associated together owing to the number of primitive features they exhibit in common. The superfamily Hepialoidea consists of three families of small to very large moths. The Palaeosetidae are small moths with no ocelli and reduced mouth-parts, inhabiting rain-forests; the genus *Genustes* occurs in Assam. The Hepialidae is the dominant family with some very large moths whose larvae tunnel stems and roots; *Phassus malabaricus* is common in tea plantations. The Nepticuloidea contain a couple of families and the moths are among the smallest Lepidoptera. The larvae mine leaves and *Nepticula* is a common genus. The last superfamily, the Incurvarioidea, also includes small to very small moths whose larvae are leaf-miners. Four families are known of which the Incurvariidae are fairly common, the moths being either sombre in colour or metallic; *Incurvaria* and *Adela* are common genera.

The Suborder Ditrysia contains the bulk (97+%) of the Lepidoptera, with a great diversity of forms but similar complex female genitalia. The Lepidoptera have been variously divided into (1) the Rhopalocera (butterflies) and Heterocera (moths) or (2) the Macrolepidoptera (Cossoidea, Castnioidea, Zygaenoidea, Hesperioidea, Papilionoidea, Geometroidea, Calliduloidea, Bombycoidea, Sphingoidea, Notodontoidea, Noctuoidea) and the Microlepidoptera (suborders Zeugloptera, Dachnonypha and Monotrysia; Tortricoidea, Tineoidea, Yponomeutoidea, Gelechioidea, Copromorphoidea, Pterophoroidea); or (3) the Jugatae (wings interlocked with a jugum and fore and hind wings similarly veined) and the Frenatae (wings interlocked by a frenulum and fore and hind wings with dissimilar venation), in the past, and one of these being followed even now by some entomologists. The superfamily Cossoidea includes two families, which are the most primitive of the Ditrysia and generally comprise largish, grey or brown, fast-flying moths whose larvae bore into the heartwood or larger roots of trees and shrubs. *Zeuzera*

coffeae is a common pest of coffee in India, and belongs to the family Cossidae. The superfamily Tortricoidea also is archaic and has two families, with Tortricidae the more numerous. The more primitive Tortricinae includes the genus *Cacoecia* with some Indian species that roll and web leaves and are minor pests. The subfamily Olethreutinae has the codling moth, *Cydia pomonella*, which is a worldwide pest of apple. The Tineoidea embraces seven families of which the Psychidae, Tineidae, Gracillariidae and Lyonetiidae are the more abundant. *Phyllocnistis citrella*, the citrus leaf-miner, belongs to the Phyllocnistidae which contain very small, usually shiny white moths. The Psychidae (bagworm moths) are an extremely interesting family whose larvae are case-bearers and whose females are among the most degenerate Lepidoptera. Males are fully winged but females may be brachypterous or even apterous. *Clania* and *Acanthopsyche* are some familiar Indian genera and some species may assume pest proportions in some years. The Tineidae are small moths with larvae feeding on animal fur and plant fibre, a species of the genus *Tinea* being a household 'clothes moth'. Species of *Acrocercops* mine leaves of some crop plants, producing blister-like mines in leaves; these belong to the family Gracillariidae. Six families make up the Yponomeutoidea, the Glyphipterigidae and Yponomeutidae being the largest. The Glyphipterigidae contain mostly tiny diurnal moths with bright metallic or orange markings; *Glyphipterix* is a large genus. The 'diamond-back moth', *Plutella xylostella*, is a serious pest of many cruciferous vegetables; it belongs to the Yponomeutidae.

The Gelechioidea is one of the largest superfamilies of the Lepidoptera and made up mostly of small to very small moths. Larvae are concealed feeders, some bearing cases, leaf-mining, tunnelling into stems or fruits, or feeding beneath leafy or silken shelters, and predatory on spider egg-sacs. About 18 families are now recognized, the Gelechiidae and Oecophoridae containing maximum species. The Angoumois grain moth (*Sitotroga cerealella*), the pink bollworm of cotton (*Platyedra gossypiella*) and the potato tuber moth (*Phthorimaea operculella*) are some common Gelechiidae. Besides species of *Cosmopterix* that mine leaves of plants, the Cosmopterigidae also contain the scale insect predators (*Batrachedra*) that feed beneath a silk webbing and pupate in a flat, elliptical cocoon. The Oecophoridae, which are extremely abundant in Australia, have larvae that tunnel in wood or in flowers or galls, join leaves or live among detritus in soil. The Stathmopodidae are very small, smooth-scaled, shiny moths that include the genus *Stathmopoda*; larvae of some species feed on scale insects or spider's eggs. *Opisina arenasella* (=*Nephantis serinopa*), a Xyloricti-dae, is a serious pest of coconut.

Of the seven families in the Zygaenoidea, the Limacodidae (slug caterpillars) is the most interesting. The moths are mostly colourful but it is the larvae, with no prolegs, stinging hairs and bright aposematic coloration that are intriguing. The pupae are enclosed in an oval or pyriform cocoon with a circular, lid-like opening at one end. The bee-fly subfamily Systropinae contains specific parasites of this moth family. *Latoia lepida* is the mango defoliating insect with bright green slug-like larvae with stinging hairs. Other genera are *Natada*, *Belippa*, *Thosea* and *Altha*, among others. The very tiny, ectoparasitic larvae of Epipyropidae attack the homopterous Fulgoridae, Cicadellidae and Cicadidae. In India, *Epipyrops fuliginosa* parasitizes *Idiocerus* and *E. eurybrachydis* feeds on *Eurybrachis*.

The Pyraloidea consist of five families, of which the Pyralidae is by far the largest, and also one of the commoner families of moths, adapted to diverse terrestrial and aquatic habitats, many being pests of cultivated plants and stored products. Several subfamilies are recognized and the larvae are found in shelters of webbed leaves or shoots, tunnels in shoots, stems, seed heads, galls or fruits, or in silken galleries among mosses, herbaceous plants or fallen leaves, or in shelters and cases amongst aquatic plants in fresh water or in stored products or in nests of Hymenoptera, and rarely predacious on coccids. The pupa also is in a silken cocoon, usually in the larval shelter. Serious crop plant pests include *Scirpophaga incertulas*, *Cnaphalocrosis medinalis*, *Nymphula depunctalis* (rice), *Pyrausta machoeralis* (teak), *Chilo zonellus*, *Stenachroia elongella*, *Marasmia trapezalis* (sorghum), *Chilotraea infuscatella*, *Proceras indicus* (sugarcane), *Crocidolomia binotalis*, *Hellula undalis* (crucifers), *Etiella zinckenella*, *Maruca testulalis* (pulses), *Dichocrocis punctiferalis* (castor, turmeric, ginger), *Leucinodes orbonalis*, *Euzophera perticella* (brinjal), and others of the genera *Hymenia*, *Ephestia*, *Margaronia*, *Sylepta*, *Noorda*, *Hellula*, *Antigastra* and *Nephopteryx*. *Corcyra cephalonica* is a store pest and laboratory host for some parasitic Hymenoptera. *Galleria mellonella* is the wax moth which is injurious to bee hives. *Hyblaea puera* belongs to the Hyblaeidae and is a serious pest of teak.

The Pterophoridae is the sole family of the Pterophoroidea and consists of small moths with usually leaf-mining larvae that later feed exposed or in the plant stem. Species of *Sphenarches*, *Exelastes*, *Trichoptilus* and *Oxyptilus* occur in India. The Geometroidea is noted for procryptic wing patterns of its moths and the remarkable larval adaptations that aid in concealment. Seven families are known, the Geometridae, with the genera *Biston*, *Semiothisa*, *Scopula* and *Hyposidra* being the most numerous and best known. The larvae are called loopers or semiloopers and are usually twig-like, occasionally living in a loose shelter of leaves, feeding on foliage, and pupating usually in a flimsy cocoon in debris or in soil. The Bombycoidea has 13 families which contain medium-sized to large moths with broad wings and usually strongly pectinate antennae, and larvae spinning strong silken cocoons. *Bombyx mori* is the silk-moth of the family Bombycidae. Saturniidae has some of the largest moths existing, and the genera *Philosamia*, *Actias*, *Attacus* and *Antheraea* contain wild silk moths that are very beautiful. The Eupterotidae has species of the genus *Eupterote* that attack cardamom, drumstick and other forest trees. Lasiocampidae is a large family with larvae possessing dense secondary setae and being the dreaded 'woolly bears' of school

© Kumar Ghorpadé

Caterpillar of a lasiocampid moth on ragi (Eleusine coracana) *earhead* × 2

children, called *kamblipoochi* in Tamil. *Metanastria hyrtaca* and *Taragama siva* are common and their larvae feed on *Acacia* leaves.

The Sphingoidea contain the hawk moths belonging to the family Sphingidae. These are generally colourful, large moths that are fast fliers, generally seen sipping nectar from flowers with their long proboscis while in flight. The larvae are also large and brightly coloured, smooth-bodied, with a spine-like horn on the eighth

A pyralid × 3   A pterophorid × 4

A prettily marked moth caterpillar (× 2)

© Kumar Ghorpadé

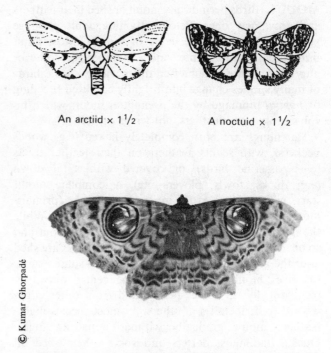

An arctiid × 1½   A noctuid × 1½

A noctuid with prominent eye-spots on fore wings × ¾

© Kumar Ghorpadé

A clearwing hawk moth × 2½

© O. C. Edwards

abdominal segment; they are diurnal feeders. *Cephonodes hylas* and *Macroglossum walkeri* are bee-mimicking species; other common Indian species and genera are *Acherontia styx*, *Deilephila nerii*, *Hippotion celerio*, *Herse convolvuli*, *Theretra*, *Psilogramma* and *Nephele*. The Notodontoidea have three families, the Notodontidae being the most abundant and common, with the crab caterpillar, *Stauropus alternatus* and the genera *Turnaca*, *Dinara*, *Norraea* and *Spatalia* representing familiar Indian taxa.

The last superfamily, Noctuoidea, contains seven of the most advanced and economically important families of moths. The Lymantriidae is a tropical family, the larvae having dense tufts dorsally and species like *Euproctis lunata*, *Perina nuda*, *Dasychira mendosa* and *Orgyia*

postica being pests of *Acacia*, castor, *Ziziphus*, *Ficus*, etc. The Arctiidae is another common family with a lot of economically important species, the larvae having dense secondary setae. The moths are usually pale coloured with colourful spots or bands, common Indian species being *Spilosoma obliqua*, *Amsacta albistriga*, *A. lactinea*, *Pericallia ricini* and *Utetheisa pulchella*. The Noctuidae is one of the most specialized families of Lepidoptera, with larvae generally without secondary setae, mainly phytophagous, but sometimes stem-boring or even predacious on coccids. The pupa is usually in a cell in soil, or in a silken cocoon not incorporating larval hairs. The moths are very diverse in colour pattern and size and most species are nocturnal. Many subfamilies are recognized but this family is notable for the number of crop pests it contains, these being fruit- and stem-borers, leaf-feeders, fruit-suckers, cutworms, etc. *Eublemma amabilis* feeds on coccids and *Heliothis armiger* is a polyphagous and almost cosmopolitan pest. *Spodoptera mauritia* (paddy), *Mythimna separata* (sorghum), *Earias vitella*, *Anomis flava* (cotton), *Agrotis ipsilon*, *Euxoa segetum* (potato), *Sesamia inferens*, *Spodoptera exigua* (finger millet), *Plusia* (*S. lat.*) (indigo), *Othreis fullonia* (citrus), *Perigaea capensis* (safflower, nigerseed), *Achaea janata* (castor), *Spodoptera litura* (tobacco), *Eublemma olivacea* (brinjal), *Polytela gloriosae* (lily), and others like *Adisura atkinsoni* (hyacinth bean), *Orthaga exvinacea* (mango), etc., are major crop pests in the Indian subcontinent.

See plate 14 facing p. 177, plate 23 facing p. 320 and plate 36 facing p. 497.

MOULT. Birds periodically moult or shed their feathers and grow new ones. Before attaining adult plumage young birds undergo a series of moults. Most adult birds moult into a winter and a summer (breeding or nuptial) plumage which often differ widely. Male birds of many species change into brightly coloured breeding or *nuptial* plumage by the prenuptial moult which involves only body-feathers, not wings or tail.

Hatchlings are born completely naked (e.g. woodpeckers), with scanty feathers on their feather tracts (e.g. passerine birds), or covered with natal down (e.g. ducks, fowls, plovers etc). A complete moult starts with the loss of the innermost primary (primary moult) and proceeds systematically outwards. While the first quill is growing the second one drops out and so on. Body-feathers, secondaries and rectrices are shed after the primary moult has progressed to some extent. Rectrices moult in pairs from the centre outwards (centrifugally) in some species, and from outer edges inward (centripetally) in others. In most species flight-feathers moult gradually without impairing flight. Ducks, flamingos, darters and some others shed all their wing-feathers simultaneously and remain flightless for a few weeks. Many female hornbills moult their flight-feathers all at once while incubating, confined in their nest-holes. Hormones influence the growth of feathers by controlling the blood supply to the feather follicles or 'buds'. Breeding usually takes place after completion of the prenuptial moult but some species moult and breed simultaneously.

D.N.M.

See plate 10 facing p. 129.

MOUNTAIN. Any natural surface of the earth that rises to noteworthy height above its surroundings can be called a *mountain*. It is taller than a *hill*. Mountains may occur as peaks usually conical in shape, and rising above their surroundings; they may be irregular groups such as the small mountainous areas in the Eastern Ghats; or they may be ranges or parallel ridges, such as the Himalaya.

Mountains are formed in different ways, and according to their mode of origin may be classified as: residual or erosional mountains, volcanic mountains, folded mountains, fault-block mountains, and domed mountains.

Residual or *erosional* mountains have generally been carved out of a plateau, which is a large flat area of the earth's surface that has been uplifted. Earth movements raise the plateau, while streams wash along and undercut the sides, causing deep canyons to be formed.

Volcanic mountains are created by the eruption of volcanic material. They are usually conical in shape.

Some of the world's highest mountains are of volcanic origin. Among these are Mount Rainier in the United States (4391 m), Mount Aconcagua in the South American Andes (7034 m), Mount Kilimanjaro in Africa (6007 m), and many others.

Folded mountains are formed in places where thick sedimentation has taken place in long narrow depressions in the earth's crust called geosynclines. As sediments are deposited layer by layer, year after year, they pile up thicker and thicker. The weight of these sediments presses down on the lower crustal layers, and causes them to bend downward. Finally the lowest layers formed of sediments may be buried so deep that they may melt, perhaps even coming to the surface as volcanoes. Certain forces, lateral as well as vertical, then cause the sedimentary layers to be compressed and folded, raising them up to form mountains. Thus were created the highest and lengthiest mountain systems of the world such as the Himalaya, the Andes (South America), and the Rocky Mountains (U.S.A.).

Fault-block mountains are created when faulting occurs along one side of a fracture, thus leaving the other side in its original place. Movement of enormous amounts of rocks along a fault is the origin of this mountain type.

Domed mountains are found in areas where intrusive rock is surrounded by sedimentary rocks. They were originally formed when magma was forced into the earth's crust in such a concentrated way that it could not spread out, but instead pushed up the surface of the earth into irregular dome-like mountains.

The Himalaya (Sanskrit *hima*, snow + *alaya*, abode) constitute the highest mountain system of the world. They extend for a distance of 2500 km, are 150 to 400 km broad, and cover about 500,000 km². They are typically tectonic in origin, having been uplifted from the bed of the great mediterranean sea, the Tethys. The Himalaya took probably several million years to attain their present height. These uplift movements have not yet ceased, for this region is still unstable and susceptible to earthquakes.

The Himalayan mountains can be divided into three parallel or longitudinal zones—the Great Himalaya (Himadri) in the north, the Lesser Himalaya (Himachal) in the middle, and the Outer Himalaya (Siwalik) in the south.

The Great Himalaya or the Himadri is a majestic range of mountains which rises above the limit of perpetual snow. Some of the highest peaks are situated in this range—Mt Everest (8848 m), K² (8611 m), Kangchenjunga (8598 m), Nanga Parbat (8126 m), Nanda Devi (7817 m), and others.

In contrast with the Himalayan ranges, the mountains found on the Peninsula are not fold mountains. They

The Western Ghats rising 1000 m beyond the Koyna lake, near Mahabaleswar

represent the erosional remnants of former surfaces uplifted and subjected to more than one cycle of erosion.

The oldest mountain range in India is the Aravalli. It is a typical tectonic mountain which was uplifted and folded during the late Precambrian period. Erosion during the ages has reduced what was once a lofty range into low-lying hillocks. The Aravalli range extends from Delhi southwestwards to near Ahmedabad for a distance of about 800 km. The highest point in the range is Mt Abu (1158 m).

The Vindhya range traverses nearly the whole width of Peninsular India, a distance of about 1050 km in an E–W direction. This range forms one of India's main watersheds, and along with the Satpura range, is often considered as the dividing line between North India and the Deccan.

South of the Vindhya range, and more or less parallel to it, rises the Satpura range, separating the Narmada and Tapti rivers. Several of its peaks rise above 1000 m.

The Western Ghats (Sahyadri) run for about 1600 km along the western border of India from near the mouth of the river Tapti to Cape Comorin. They form the faulted edge of an upraised plateau. The northern 650 km of the Western Ghats are formed of horizontal flows of lava, and exhibit the characteristic Deccan Trap landscape. Some of the prominent peaks are Mahabaleswar (1438 m), Kalsubai (1646 m) near

Igatpuri, and Salher (1567 m), 90 km north of Nasik.

South of the Palghat Gap, the Ghats are steep and rugged. Anaimudi (2695 m) is the highest peak in the Peninsula.

The Nilgiris or Blue Mountains constitute one of the most picturesque spots in south India. This is the meeting-point of both the Western and Eastern Ghats, and forms an extensive plateau that is abruptly cut off on all sides by faults. Two of the highest peaks are Doddabetta (2637 m) and Mukurti (2554 m).

C.S.P.

See also FAULTS, FOLDS, UNCONFORMITIES.

MOUNTAIN SHEEP and GOATS. Broadly speaking, the family Bovidae—the hollow-horned ruminants—can be divided into a Cow-like group and a Goat-like group. The Cow-like group includes the cattle and buffaloes as well as some of the species popularly lumped as 'antelopes', such as the Nilgai and Chowsingha. The Goat-like group includes most of the antelopes, such as Blackbuck and Gazelles, and a group of species specialized for high mountain life, which we can call the Mountain Sheep and Goats. They are allotted a special subfamily, the Caprinae, and as well as the animals we think of as 'true' sheep and goats they include Tahr, Bharal, Takin, Serow, Goral and Chiru.

Special articles deal with some of these species; so here we will give a general survey, concentrating on the general appearance and relationships of the animals.

The goats (genus *Capra*) at present inhabit the mountains and dry country from Central Asia to the Sudan with outlying species in the Alps (W. Europe) and Spain. There are a number of species; exactly how many is difficult to decide, but they can be broadly divided into Goats, Ibex and Markhor: the Goats have narrow horns with a keel up the front edge, the Ibex have broad horns with a broad front surface on which regular transverse knots are placed, and the markhor have spirally twisted horns with a keel on the hind surface, not on the front. The ibex always have scimitar-shaped horns; the goats generally scimitar-shaped, but sometimes slightly twisted (though never as much as markhor).

The Wild Goat (*Capra hircus*) lives in arid desert ranges, from the Indus valley in Pakistan as far west as Turkey and Crete. It is a yellow-brown colour with darker — often blackish — face-blaze, throat, and shoulder—and flank-stripes; the forelegs are also black on the front surfaces, interrupted at the 'knees' and fetlocks by white marks. Although they range right down to sea-level, they specialize in habitats with steep cliffs and precipitous areas, on which they are very agile and sure-footed. They browse as well as graze, rearing up against tree-trunks and even walking up those that are sufficiently sloping. A stomach concretion often developed in this species used to be attributed mystical or curative powers, and it was known as the Bezoar Stone: hence the Wild Goat was sometimes called the Bezoar Goat.

Domestic goats are derived from the Wild Goat; whether domestication first took place in southwestern Asia, Iran or Pakistan is uncertain,—in fact, it most probably occurred independently in different places. While most Wild Goats have scimitar-shaped horns, a spiral twist (which is 'homonymous'—twisting inwards—in some breeds but 'heteronymous'—twisting outwards—in others) is usual in domestic goats' horns; at one time it was supposed that the Markhor must also have been domesticated to provide the spiral, but it is now known that in at least three places on the margins of the range of Wild Goats occur populations with a high frequency of at least partly twisted horns: on the Chiltan and nearby ranges of Pakistan (where they are heteronymous), and on the Caucasus and on Eremomilos island in the Aegean (where they are homonymous).

The Markhor (*Capra falconeri*) is a larger animal than the Wild Goat, standing about 1 metre at the shoulder as compared to 85 cm for the Wild Goat; it is grey-white to beige-red in colour, with the strong dark markings of the wild goat; it has a big throat-ruff, especially marked in winter, and there is a crest along the back which hangs down (the Wild Goat has a very short, upstanding one). The horns are very large, twisted, and diverge in a V when seen from the front; in the northern parts of the range the spiral is loose and open, in the southern parts it is more tightly twisted. Markhor, like Wild Goats, favour low, arid cliff country. In the north, they live along the Indus nearly as far as Skardu, and along its tributaries the Hunza, Gilgit and Astor; in the south, they live as far as Quetta; and west they extend through Afghanistan into the southernmost U.S.S.R. There is a lot of variation both within and between the numerous rather isolated populations of Markhor, and about ten subspecies have been described; but Schaller and Khan recently cut these down to two: the flare-horned northern race (*C. f. falconeri*) and the straight-horned southern race (*C. f. megaceros*), which is also smaller and less long-haired in winter. A third race may be valid in the Soviet Union. The southern race, living on the more accessible desert ranges of Baluchistan, is gravely endangered.

The Asiatic Ibex (*Capra i. sibirica*) is found in the high ranges of Central Asia, from the Altai and Tien Shan and their outliers to the Karakoram; in Ladakh it occurs as far east as Leh, but further south it extends along the Himalayas as far as the Sutlej — but east of here, there are none. About the size of a Markhar, the Asiatic Ibex is prominently bearded, medium brown in colour with a light 'saddle' on the back in adult males, more prominent in summer. It favours much higher altitudes than Wild Goat or Markhor, but the same steep country where great agility is needed.

All goats have anal glands, knee calluses, beards and a strong body-odour in the male; and the tail is flat, and naked underneath. Sheep can look very like goats, but always they lack these goat characteristics and instead have glands on the face, and in the feet and the groin, and the tail is round and hairy. There are some differences in behaviour too; for example, when fighting in rutting season male goats rear up and lunge downward at their rivals, whereas male sheep run at each other with their heads down.

Wild Sheep are found in a great arc from Turkey and Iran through Central and Northeastern Asia through western North America. Over most of their range they inhabit high mountains—though preferring flat or rolling plateau country, not precipitous country like goats—but at either end of it they approach sea level. Of the five species, two occur in South Asia: the Urial (*Ovis orientalis*), about 80–90 cm high, and the heavily-built Argali (*Ovis ammon*), 110–130 cm high. The Argali has a broad ruff on the neck; the Urial, only a narrow one with a 'bib' under the throat.

Argali in this region are of two races. The larger

Tibetan Argali (*Ovis ammon hodgsoni*), with its very heavy horns which curve forward along the sides of the head, is found on the Outer Plateau of Tibet from Nepal and Bhutan west to the Karakoram Pass in Ladakh; here it lives in the high ranges above the Ladakh Urial, which ranges east of Leh—it is said that they sometimes visit the same valleys, but do not interbreed. The Marco Polo Sheep (*O. a. polii*) comes in from the Pamirs to the Tagdumbash Pass in Ladakh, but its range is separated from that of the Tibetan Argali by a 300 km gap; its horns are thinner, but they spiral outwards dramatically.

Certain species of this group have been placed in separate genera as being supposedly intermediate between Sheep and Goats. Notable among these is the Bharal or Blue Sheep (*Pseudois nayaur*), which spills over from Tibet into northern Nepal, and along the ranges as far west as the eastern edge of the Karakorams. This has thick, laterally flaring sheep-like horns, but is strongly marked with black stripes on the flanks and legs like a goat; and its main features, both physical and behavioural, reveal it as a rather divergent goat, not a sheep. It lives at 3500 to 5500 m on flat plateau land more like sheep—so it is perhaps convergent with them in evolution.

Rather more like goats are the Tahrs (*Hemitragus*), which differ mainly in their very short, backcurved horns. In the north, along the ranges from the Pir Panjal east to Bhutan, lives the Himalayan Tahr (*H. jemlahicus*), nearly 1 metre tall, with a thick silky coat lengthened on the foreparts into a heavy cape; it is brown in colour. In the south lives the Nilgiri Tahr (*H. hylocrius*), of the treeless meadows at 1200 to 2600 m, in the Nilgiri and Anaimalai Hills and part of the Western Ghats; now much reduced in number and strictly protected. A third species lives in Arabia.

But apart from these closely related animals, there are some more divergent forms which are often classed as 'Mountain Goats'. In South Asia, they are represented by the Serow and Goral. That they are not Goats, nor yet sheep, is shown by the muzzle, which has a large bare, moist area at the tip unlike sheep and goats; by the simple, unkeeled horns; and by the shape of the skull.

The Serow (*Capricornis*) is a heavily built animal of about 1 metre in height, with very long ears and short, backcurved horns. It is black or red, with a partially white mane along the neck. Though no long-term study has been made, all observers agree that serows live in moist gorges at high altitudes, very precipitous but where there is thick vegetation; they live mainly solitary, and move with amazing speed through the dense vegetation and among the cliffs. The species found in our region is the Common Serow, *C. sumatraensis*, whose total range extends from Sumatra in the south to central China in the north, and to Kashmir in the west; two other species live in Japan and in Taiwan.

Goral (*Nemorhaedus*) are much smaller than serow, but otherwise very similar; in fact, it is doubtful whether they should be put in different genera. Whereas serow have glands in the groin and on the face, the goral has none in the groin and only very small face ones. The colour is less strong, though like serow the legs tend to be lighter than the body; the tail is longer; but the horns and general shape are the same. Goral live in the same kind of country as serow, sometimes in the same valleys, but in small herds at least at some seasons; but in some places—as has been recorded for both Szechwan and the Soviet Far East—they inhabit much drier, more open rocky cliffs at low altitudes. Goral populations seem to live semi-isolated from each other, in nuclei of some two dozen or so. The three species all border on our region. The Common Goral (*N. goral*) is 70 cm high, grey or brown in colour with a whitish or orange patch on the throat, and paler on the limbs; it is found along the Himalayas into Kashmir, southeast, into central Burma, northeast to the Soviet Far East. Hunters used to refer to 'grey' and 'brown' goral as if they were two different species, the grey one living in Kashmir and U.P., the brown in Nepal and Bhutan; but in fact these are all intermediates between the two colours, and all one can say is that goral tend to be greyer to the west, browner to the east.

Two other, poorly known species of goral must be mentioned here. Bailey's Goral (*N. baileyi*) is still known only from one specimen, collected in the early years of the century at Dre, southeast Tibet, north of the great bend of the Brahmaputra, at 2800 metres; it may therefore range into India, and this must be investigated. It is a strong brown in colour, the legs not paler than the body, and with only a small white patch on the throat; the skull is oddly serow-like. The other, slightly better known species is the Red Goral (*N. cranbrooki*), only 58 cm high, a beautiful foxy red in colour with no white on the throat, and with the legs not paler than the body. This species, described only in 1961, is known from the Adung Valley in northernmost Burma, and from the Mishmi Hills, where it lives from 2500 up to 3800 metres—at higher altitudes, that is, than Common Goral.

Difficult to classify, but perhaps related to serow and goral, is the strange Takin (*Budorcas taxicolor*). The hunched withers reach 120 cm high, the back sloping down behind to the haunches; the snout is swollen and ram-faced; the horns enlarged at their bases, turning out then up. The legs are short and stout. Unlike serow and goral there are no glands in the feet or on the face; there are no inguinal glands (in the groin). The colour pattern is basically dark brown with a lighter zone on the back; on the muzzle, underparts, limbs and rump the dark brown may become black, and the extension of the light zone, which tends to greyer in females and golden in males, is also variable—partly geographically. Takin live in very steep areas, above 2000 m; in winter they come down to the sheltered valleys in the bamboo and rhododendron forest, where the small herds move in single file making narrow trails through the thick vegetation; they emerge to graze in nearby meadows in morning and evening. In summer they move out above the tree-line; and here, at the time of the rut, huge herds are said to gather.

The last species we will describe here is the Chiru (*Pantholops hodgsoni*), which is very different from other Caprines. It stands 80 cm high; the head is held high, and the male alone has horns, which are long, thin and pointing up. The coat is dense and crisp, almost like a woolly sheep's; buffy in colour, with a dark brown zone on the face. There are extremely large glands in the groin, but none on the face or in the feet. The hoofs are long and narrow, the lateral hoofs unequal in size— the outer ones very small, the inner ones enlarged and broad. The tooth-row is reduced in the upper jaw, having only 2 (instead of the usual 3) pairs of premolars. The muzzle is swollen, the nostrils very large. Chiru live on the high plateau of Tibet; in the very highest areas, where the Wild Yak live, at 4500 to 4700 metres. In summer, herds enter the Changchenmo valley in Ladakh, crossing from nearby areas of Tibet, to graze on the river flats as the snow melts; at this time, the sexes are usually keeping apart, in separate herds. In winter comes the rut, when males fight, very viciously it is said and sometimes fatally; and the females gather together in herds of up to 200. When resting, a Chiru scrapes out a shallow pit in the ground, giving some shelter from the biting winds; when alarmed, the herd packs together, running very fast with heads stretched low so that the dust rises and does not enter the nostrils, but does not run far.

All of these fine and majestic animals are still to be seen in the high country of the wilder parts of Pakistan, India, Nepal and Bhutan; in some places, alas, they have been pushed out by domestic stock or by being hunted too much. Their complete protection is a matter of urgency.

See UNGULATES, TAHR, BHARAL, IBEX, URIAL.

C.P.G.

MOUSE. Mice are perhaps the most successful and rapidly evolving of all mammals. The term Mouse usually is reserved for members of the genus *Mus* which includes some 15 species in addition to the ubiquitous House Mouse (*Mus musculus*). Mice are closely related to rats. Mice can be distinguished by their enlarged first molar which is bigger than the other two molars combined, and usually, but not always, by their smaller

size. In much of the world, the House Mouse is the only species of mouse and it may live either with man or in the wild. In Southern Asia, including India, however, there are many other species of mice all of which live in the wild. Here the House Mouse lives almost exclusively in or near human dwellings.

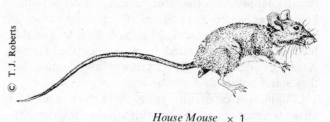

House Mouse × 1

Among the wild species are those that have spiny fur and often live in dry places, those that resemble shrews with long noses and small eyes and live on the floor of montane forests, and those that resemble House mice and live in cultivated fields and grasslands. The native mice of Australia belong to a separate genus.

House Mice usually have 4 to 7 young per litter and can breed when only 35 days old. Besides human food they may eat glue, paste, soap, and other household articles. Mice are difficult to control and some populations have become resistant to certain rodenticides. House Mice have been domesticated and strains of high genetic purity are widely used in medical research.

See also RODENTS.

G.W.F.

MUD-SKIPPERS. Persons who visit inter-tidal mud-flats in the estuarine areas of the eastern or western coasts might have observed some goggle-eyed animals hopping from place to place and sometimes disappearing into their holes. These are the mud-skippers which, though moving about in the air and on the mud are all true fishes belonging to the family Gobiidae. Two important genera are the *Boleopthalmus* and *Periopthalmodon*. The common mud-skippers are known around Bombay as *nivta*. They are small, growing to about 20 cm in length, elongate with rounded bodies and eyes almost on the top of the head with partially closing eyelids. Their bodies are bluish grey, with slightly oblique black stripes interspersed by iridescent light blue spots; the head is darker but its sides are covered with black and blue spots. The second dorsal and pectoral fins are similarly coloured but tipped with orange. The first dorsal fin is tipped bright blue. Another species is dull olive grey with white irregular streaks on the body and a pink anal fin. These colours deepen and become brighter during the breeding season. Mud-skippers possess gills which have become so functionless that if they are compelled to remain under water for a long time they get asphyxiated. They depend on their accessory respiratory organs. It is reported that in some cases the tail also is vascular and aids in respiration.

When high tide submerges the mud-flats, the mud-skippers swim on the surface and are usually seen resting on the stems and branches of mangrove plants

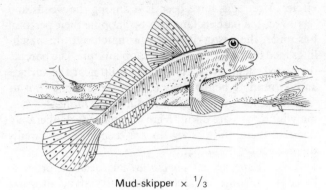

Mud-skipper × $\frac{1}{3}$

(*Rhizophora, Avicinnia* etc.). When the tide recedes they distribute themselves on the mud-flats and make temporary burrows in which they live during the inter-tidal period, and skip about with the help of their large pectoral fins which have an unusually strong muscular base. These hopping movements are mostly in search of food—the copepods, small fish, prawns, crabs and even algal matter which may be stranded in tiny puddles.

Mud-skippers are known for their territorial instinct. A male does not allow other males to enter the area around its hole and fights to drive out the intruders. During such fights, the colourful dorsal fins are raised and the gill-chambers inflated to frighten the opponent. With binoculars an observer can see how the dominant male knocks down his opponent.

Observations on *Boleopthalmus dussumierei* near Bombay indicate that the fish has preference for vegetative matter, especially the diatoms which they obtain on the muddy bottom. They make Y shaped burrows, become sexually mature when about 110 mm long and breed from July to September. Eggs (1000 to 5000) are attached to the walls of the burrows and guarded by the male.

See AIR-BREATHING FISHES.

MUDUMALAI. The Mudumalai Sanctuary in the Nil-giri District of Tamil Nadu was established in 1940 with an area of only 62 km². In 1956 it was extended to 295 km², and at present it covers an area of 321 km². The sanctuary is situated at the trijunction of the three southern States of Tamil Nadu, Karnataka and Kerala, and has been included in the proposed Biosphere Reserve which the Government of India intends to set up.

From the natural history point of view its situation at the base of the Nilgiri range and with an altitude varying from 850 to 1250 m is significant. It contains many forms of life peculiar to the plains as well as to the hills. Moyar Betta is the highest peak in the sanctuary and the lowest altitude is at the picturesque Moyar waterfalls. The perennial Moyar river runs along the centre of the Sanctuary from south to north.

Mudumalai has an impressive range of landscapes and vegetation types, open grasslands, swampy land, well-clothed hill ranges and undulating park land. The western part has rich tree growth while in the eastern part there is scrub jungle. There are thorn forests on the eastern side of the plateau at the foot of the Nilgiris.

Moist bamboo brakes occur along the stream banks and lantana has run wild over most of the area. Unfortunately *Eupatorium*, the undesirable exotic from South America, is gaining ground. The main tree species include teak, rosewood, laurel, bijasal (*Pterocarpus marsupium*), ben teak (*Lagerstroemia lanceolata*) and many more.

The wildlife includes tiger, leopard, jungle cat, toddy cat, leopard cat, jackal, stripednecked and common mongoose, sloth bear, dhole, ratel, fishing cat, pangolin, Malabar civet, fourhorned antelope, hyena, wild boar, Hanuman and bonnet monkeys, Malabar and flying squirrels. There are well over 120 species of resident and migratory birds.

See plate 1 facing p. 32 and plate 15 facing p. 192.

MULLETS. This is a large group of estuarine fishes of the family Mugillidae, two important genera being *Mugil* and *Rhinomugil*. They have a small mouth, a rounded snout and a flattened head. They have a soft second dorsal fin, and their lateral line is indistinct or nearly absent. They occur on both the western and eastern coasts of India and the connected estuarine creeks and backwaters. One of the major species, *Mugil cephalus*, grows to a large size of about a metre, the rest being smaller ones. The adults usually inhabit coastal waters. The exact location of their spawning areas is not yet identified but is expected to be within the coastal belt. On the west coast they breed in the months of March, April and May and after a couple of months the young fry enter the estuarine areas, which are then inundated with rain-water in the early part of the monsoon season. They feed on green algae and other small aquatic plants which at that season flourish at the bottom. In clear waters they are seen browsing on this vegetation with tail upwards. They return to coastal water after four or five months, but those that remain after the monsoon water recedes form an important fishery, as mullet is relished by all classes of people. Being vegetative feeders, they constitute an important species for aquaculture on both coasts of India. Chilka lake on the east coast is well-known for its mullet fishery.

Freshwater mullet *Corsula* chasing caddis flies, × ⅛

One species, *Rhinomugil corsula*, lives mainly in fresh water and breeds there, only occasionally going to estuarine waters. It lives in the rivers of the northeastern Gangetic plain. Other mullets are bottom feeders, but *corsula* is a surface feeder. In swampy areas where caddis flies and other insects fly over the surface of water groups of *corsula* chase them and swallow them in flight. Normally *corsula* feeds on underwater algal matter. In tidal areas it is seen crawling on mud-flats in search of filamentous algae. When introduced in large numbers into reservoirs it is reported to breed in them.

Red Mullets belong to a different family, Mullidae, which is entirely marine, living near the sea-bottom and found offshore, in the open sea. They are slightly elongated and reddish yellow in colour. They have two fairly long processes or barbel-like structures below the chin which are used for exploring the sea-bottom for food. These outgrowths can be folded when not required. Because of these, red mullets are also known as 'goat fish'. In Indian waters, three genera, *Mulloidichthys*, *Parupaneus* and *Upeneus*, are common. They are small fishes about 15 to 20 cm long and are not much valued as food.

MURRELS. These belong to the family Chanidae. They are also known as 'snake heads' because of their depressed snake-like head and elongate body. There are six species of murrels in Indian waters: *Channa marulius* is the largest, growing to a length of about 1·5 m and weighing 12 kg; two other smaller species are *C. striata* and *C. punctata*.

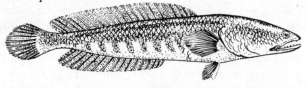

Channa striata × ⅙

Murrels are a popular food fish. They live in rivers, lakes and ponds. Biologically they are interesting because of their capacity to live out of water with the help of their accessory respiratory organs. On the inner upper part of the gill-chamber a small cavity is formed in which a bunch of vascular (blood-circulating) tissue is developed which helps in aerial respiration in addition to the normal respiration by gills. Because of this adaptation murrels can survive even in foul waters

and can migrate overland from pond to pond at night and can remain alive after capture for hours together if kept moist.

Large murrels (*C. marulius*) breed in February and March. They build a simple nest by collecting pieces of aquatic weeds and arranging them together in a small circle. After eggs have been laid among the weeds and fertilized, the parents fan them by flapping their pectoral fins under the nest. After the eggs hatch out, the hatchlings feed for some time on the zooplankton. The parents then herd them into a close-knit group near the surface and, swimming below, they lead the young ones out to feed in richer waters. After about three months, when the young ones are about 8 cm, the parents deliberately disperse them and if they do not disperse, the parents themselves set upon them.

A small species, *C. punctatus*, has been reported to breed from June to October and its larvae measure 1·5 cm. They feed on planktonic organisms such as copepods and other minute aquatic animals.

Commercially, murrels are popular and fetch high prices. They are carnivorous in habit, feeding largely on small fish and insect larvae. Due to this habit they are unwanted in piscicultural ponds but it has been found that exclusive murrel culture based on artificial feeds is practicable.

See FISHES, AIR-BREATHING FISHES.

MUSK DEER. Musk deer (*Moschus moschiferus*), though belonging to the great family of Cervidae, are quite different in many respects from typical deer. In the first place they lack any antlers but males have instead developed backward-curving tusk-like incisors in their upper jaw. Small in size, with the hindquarters distinctly higher than their shoulders, they also have a gall bladder which other deer do not possess and a uniquely developed scent gland in the abdominal region which produces the valuable musk. The females also have only two teats whilst all other Cervidae have four.

Musk deer are confined to the sub-alpine scrub zone in that borderline area where birch forest gives way to dwarf juniper and *Lonicera*, at altitudes ranging from 3000 up to 4000 m. They are always associated with broken rocky slopes having some bushes or thin tree cover and extend from Chitral in the west to Nepal and Bhutan in the northeast.

They are not gregarious, in contrast to most other Cervidae, and the musk gland is probably an important factor in maintaining social contact. Unfortunately the secretion from this gland is a valuable fixative in the perfume trade and this shy little deer has been driven to the verge of extinction in most parts of its range by incessant poaching for the income yielded to local hunters from sale of its musk pod. In China a scheme

```
0       15      30 cm
0       6       12 in
```

© T. J. Roberts, *Mammals of Pakistan* (Benn)

Musk Deer

has been in operation for keeping musk deer in captivity and extracting the musk secretion from the male deer without having to kill them. This would certainly seem to offer the best compromise for conserving wild stocks and meeting the commercial requirements for musk. It is not known whether the Chinese have succeeded in breeding this animal in captivity, which would be essential if large-scale musk farming was to be developed.

Musk deer males will fight savagely during the rut, slashing at each other with their canines, and this takes place during November and December. The gestation period is believed to be 160 days and females are sexually mature at 18 months of age during their second winter. Individuals frequent the same territory for prolonged periods and have the habit of regularly depositing their faeces in prominent places to mark the boundaries of their range. They will eat a variety of young leaves and twigs, and in the winter can subsist partly upon moss and lichen which they scrape off the rocks. Adults have a peculiarly coarse brittle fur with practically no underwool, in colour silvery grey varying to dark sepia-brown, with indistinct horizontal lines of spots on the flanks in younger specimens. They stand 53 cm at the shoulder and weigh about 11 kg.

See also DEER.

MUSSELS are the most familiar edible bivalves, as important and popular as oysters. There are sea mussels and also freshwater mussels. The sea mussels form a large group consisting of several genera including *Mytilus* (Blue and Green mussels), *Modiolus* (Horse mussels) and *Lithophaga* (Date mussels). Of these, *Mytilus* and *Modiolus* are characterized by elongate, ovate shells, narrow at the anterior end and broad posteriorly. The hinge and umbo are close to the narrow end. The foot is somewhat long and narrow and bears strong byssus threads for attachment. Of the two internal muscles that close the valves one alone is well developed; the other one (the anterior) is very small. The mantle cavity opens freely—the siphons are rudimentary or even absent: if present the exhalent one alone is developed. The animal attaches itself to rocks, pebbles, oysterbeds and other substrata along the seashores at low-tide mark below water level. *Lithophaga* shells are long, oval and almost cylindrical. The animal bores holes in rocks and corals and lives in them. These holes are made not by the shell, which is fragile, but by an acid secretion from certain glands of the animal's mantle. The acid corrodes limestone and rocks. The shell however is not affected by the acid as it is well protected by a horny outer layer (periostracum).

A GREEN MUSSEL WITH FOOT EXTENDED BEHIND ITS BYSSAL THREADS, × 1/2 AND DATE MUSSELS IN THEIR ROCK BURROWS × 1.

The Green mussels (*Mytilus viridis*) are so called because of the brilliant bluish green colour of the coating investing the shell surface. This species is widely distributed along both the shores of the subcontinent. It grows to a large size, 6 to 13 cm long and 5 to 8 cm broad. It grows in clusters attached by byssus threads to oyster-belts, rocks and pilings, about 5 metres below water level. It flourishes along the west coast especially from Goa to Quilon and Trivandrum. This prolific species in this region forms into fishable layers for a period of 4 or 5 months preceding the outbreak of the southwest monsoon (January to May). Its flesh is relished by the local people and they call this mussel *kallinmel kaya* meaning 'fruit on stone'. During the peak season, gathering of this mussel is a minor industry for the fishermen of Kerala. *Mytilus* is sensitive to cold and rains and that may be the reason why its population shows a reduction during the monsoon period—June to December.

For the maximum growth of any organism, adequate space and food supply are primary requisites. Under favourable conditions, *Mytilus* grows very rapidly especially if the food supply is abundant and if there is no inconvenient crowding of individuals. In 1862 a buoy newly cleaned and painted was placed in the harbour at Arcachon, on the Atlantic coast of France. In less than a year after, it was found to be covered with thousands of very large *Mytilus edulis*, 100×48 mm, the ordinary size on the adjoining banks being about 50 to 60 mm × 30 mm.

Horse mussels (*Modiolus* spp.), also popularly known as Weaving mussels, are as large as Green mussels. The umbos of the shells are not anteriormost but a little distance away and the periostracum covering the surface is brown and coarse. The byssal threads are dense, bushy and long, substantiating the term 'weaving mussels' given to them. These mussels live deeper down than *Mytilus*. It is along the southern shores of India that these mussels are found in plenty. In Palk Bay and the Pearl Bank region in the Gulf of Mannar, *Modiolus* has been observed to be so abundant that several square miles of sea bottom are covered continuously with a carpet of these brown shells, felted together in a tangle of byssal threads. These brown mussels are also collected regularly for food. The smaller ones are dried, ground and made into poultry meal and fertilizer.

India is rich in freshwater mussels, which are widely distributed in almost all water-clogged fields, shallow rivers and lakes, ponds and pools. The common species is *Lamellidens marginalis*. Its shell is longer than broad, inequilateral with prominent umbos which are often corroded. The outer surface of the shell is invariably covered with black or dark-brown periostracum. The inside is pearly. This mussel grows to 9 cm long and 6 cm broad. Unlike in *Mytilus*, the siphons are more developed in *Lamellidens*, both the valve-muscles are equally developed and the foot is large and wedge-shaped. When it moves on mud, its foot projects half an inch or more beyond the valves. They feed on minute algae and other microscopic organic matter existing in the water in which they live. The flesh of these mussels is considered insipid, but poor people collect them for food. The shells, being nacreous inside, are used for button manufacture, fruit-peelers and for making lime. Tolerably handsome pearls are sometimes found inside the shells.

The sexes are separate in freshwater mussels. Their reproduction and development of larva are particularly interesting. The female is ovo-viviparous. About 200,000 eggs are produced. In the breeding season the eggs expelled from the ovary pass into a chamber in the mantle cavity near the cloacal opening. Here they are fertilized by the male sperms introduced with the respiratory current. The posterior portion of the female shell is enlarged to contain the eggs, and has a special pouch for them. The fertilized eggs are incubated in this pouch where they develop into embryos. From this incubatory chamber the young escape as larvas of a special form known as glochidia, characterized by the possession of a shell armed with hooks at the edge of the valves. Some of the glochidia escaping from the maternal pouch sink to the bottom but many succeed in attaching themselves to a fish by means of their hooks. The stimulus of the hook causes the flesh of the host to grow round the glochidium as a cyst. During the period of encystation the larvas are parasitic and absorb nourishment from the host. This period lasts for three weeks. Then the young ones come out of the host to lead independent lives. Maturity is not attained until several years after birth. One genus, *Anodonta*, is understood to become fully grown only in 12 to 14 years!

Both sea and freshwater mussels have the first enemy in man who consumes them as food. Crows, kites, gulls and other aquatic birds also feed on them. Carnivorous molluscs like whelks, murices and naticas feed on sea mussels either by opening the valves forcibly with the armed lips of their shells or by boring holes and sucking the flesh with their radulas. Rats catch the moving freshwater mussels by biting their exposed foot.

Instances of poisoning caused by eating mussels are many. The mussels themselves are not poisonous but when they grow in stagnant and polluted waters, they are contaminated.

In this context the following instance will be of special interest. Paddy is grown in water-clogged fields. To protect the crop powerful insecticides are sprayed. In the same field many mussels live. After harvesting the crop, ducks are led into the fields to feed on the molluscs. In 1975 hundreds of ducks thus fed in a field near Trichur (Kerala) died in a few hours. Analysis revealed that it was due to the consumption of mussels surcharged with the insecticide. The cattle which ate the straw gathered from these fields also developed symptoms of poisoning.

A general warning is therefore given that before eating mussels, oysters or other shell-fish they must be gathered only from unpolluted waters and before use they must be kept alive in clean salt or fresh water for at least 12 hours. By this time the pure water will displace any contaminated water contained in the mantle cavity.

See also MOLLUSCS.

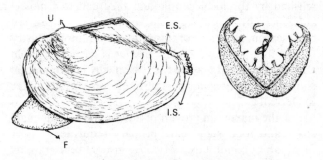

A FRESHWATER MUSSEL

U umbo, F foot, E.S. and I.S. exhalent and inhalent siphons, ×1. On the right is a greatly enlarged glochidium larva, showing the hooks on the shell and, in the centre, the byssal thread by means of which the larvae keep together in the maternal pouch.

MYROBALAN, *Terminalia chebula,* is a medium-sized tree of wide distribution in deciduous forests of the subcontinent and is plentiful in the dwarf, twisted, moss-covered forests of the hill resort of Mahabaleshwar. It is also easy to recognize from its leaves which have a pair of large glands at the top of the leaf stalk. The fruits are egg-shaped, faintly angled, an inch long and are known in trade as myrobalan. Thanks to a durable bark the tree withstands forest fires well, which is helpful as it is an associate of the inflammable Chir pine.

Myrobalan. Leaves, inflorescence and fruit × ¹⁄₁₀. The fruits are one of the most valuable tanning materials for the leather industry.

MYRIAPODA. The group Myriapoda includes certain curious lower animals—their body elongate and worm-like consisting of numerous segments and protected by a series of tough or rigid horny rings. They have a distinct head bearing a pair of jointed feelers (antennae), a pair of eyes and two or three pairs of jaws. Each body segment bears one or two short jointed legs on either side. They breathe by a system of air-tubes (trachea), like insects. The two imporant groups of Myriapoda are Centipedes and Millipedes.

Centipedes (*Chilopoda*) are myriapods, long and worm-like, with their body segmented and flattened. The number of segments vary from 15 to 173 according to species and each segment bears one pair of jointed legs. Each leg ends in a claw. The first pair of legs of the trunk are modified into poison jaws. Bites of all centipedes are painful to man.

Centipedes are found under damp conditions in crevices of old walls, under stones, bark of trees and thick layers of dry leaves. They are very active in the dark, hunting and preying upon insects such as locusts, cockroaches, earthworms and snails. Some species have a special liking for Acridiids and Gryllids (crickets). They crush the prey with their jaws.

Several species of centipedes are common in India. The chief genera are *Scolopendra, Geophilus, Ethmostigmas* and *Scutigera.* Of these the first three have long bodies with many segments and short legs whereas *Scutigera* has a short body with only 15 segments and long spider-like legs. Many species of *Scolopendra* are found ranging in length from 6 to 20 cm. In some the segmental coverings are squarish with a green centre and light blue border and the legs bright orange red. Some are black, some bluish purple and some brown.

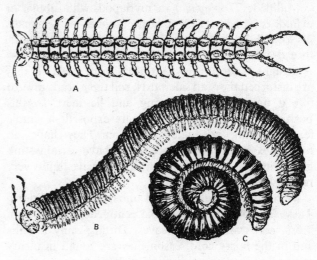

A. A centipede (*Scolopendra*), × 1
B. The giant millipede *Julus,* extended and
C. rolled like a spring, × ¹⁄₂

Some are thin and long and as they crawl over walls during the night they leave a phosphorescent trail behind which is considered as a device for attracting partners or for keeping their enemies off.

Scutigera is strictly cavernicolous. It never exposes itself but lives concealed in empty flower-pots, piled-up tiles, or in crevices of walls. Any slight disturbance makes it scurry away rapidly and disappear from sight. It is able to produce with its legs a kind of squeaking

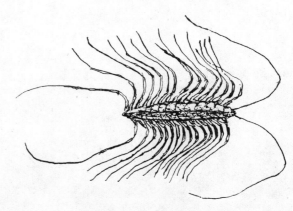

A centipede (*Scutigera*), × 1

sound which is probably intended to attract the attention of a partner or to frighten enemies.

Eggs are laid and hidden by the females, safely covered by earth in burrows. The young ones when hatched have all segments in full.

Centipedes cannot bear extreme heat. In hot summer they prefer moist places to live in, like loose earth around watered plants, inside flower-pots, crevices of stones, along drains and in corners of bathrooms and cisterns.

Millipedes (*Diplopoda*) are myriapods with tubular or cylindrical bodies with a large number of segments, 25 to 100, each segment bearing two pairs of short, thread-like jointed legs. While moving, their progress is straight, smooth, slow and of uniform speed. If they are disturbed they fall sidewards, roll their heads inward like a tightened watch-spring and lie inert. In this position their hard rings alone are exposed: a simple feat affording adequate protection. They have no poisonous jaws like centipedes but have certain stink glands instead, which can emit an offensive liquid very repulsive to their enemies.

Millipedes thrive and breed under moist conditions. They are not carnivorous like centipedes but feed on moss, herbs and rotting leaves. During rainy months and in the misty cold season they are found in plenty over moss-covered compound walls, herbage and decaying leaves. They occur throughout the subcontinent. There are many species among them. They vary in size from 5 to 25 cm in length and 5 to 20 mm in girth. In coastal Kerala there are certain giant species (20 to 23 cm) in which the middle of each segment is girdled with extra thick chitin. They are shining brownish black in colour, the head portion and legs reddish and the thickenings on the rings black. Their secretion is exceedingly pungent and loathsome, staining and cor-

rosive. A uniformly black species with red legs measuring 10 to 13 cm long is found in herds among heaps of rotting coconut-palm leaves. Smaller millipedes with semicircular rings, brownish in colour and about 6 cm long, are abundant among dead leaves. Where the soil is reddish and lateritic small tubular, brick-red coloured millipedes are found both on the ground and climbing over herbs. They are nearly 7 cm long and some of them are ornamented by a scarlet mid-longitudinal streak on the upper side.

Pill millipede, (*Arthrosphaera*)
rolled into a ball, × ½

In the old Cochin State—Parambikulam Forest area—there exists a curious millipede which is short but stout and can roll itself not like a spiring but like a ball. This is a pill millipede of the genus *Arthrosphaera*. It is also capable of producing a squeaking sound probably by the rhythmic vibration of its legs.

Despite the horny armour and the repugnant fluid, millipedes often fall a prey to reduviid bugs. The pill millipede however is not attacked by this bug.

From 60 to 100 eggs are laid in specially dug holes in damp corners filled with pellets of earth. The eggs hatch in two weeks. The young ones resemble adults but have fewer segments. As they grow, new segments appear in front of the anal segment.

See also ARACHNIDS.

NARMADA. Unlike most of the southern rivers of India which flow across the peninsula from West to East, the Narmada and its immediate neighbour to the south, the Tapti, flow East-West into the Arabian sea.

Historically considered as the boundary between Hindustan proper and the Deccan, the Narmada rises near the hill shrine of Amarkantak (1000 m) in the Maikal Range. In its upper reaches the young river frequently changes direction and descends steeply over several rapids and falls. At Jabalpur, taking an abrupt right-angled turn, it flows over the Dhuan Dhar falls ('fall of mist') and cuts a gorge through the famous Marble Rocks.

From here on it enters what is thought to be a rift valley and courses almost due west through alluvial upland valleys hemmed in between the Vindhya range to the north and the Satpuras to the south. These valleys nowhere exceed a width of 50 km but in places are constricted and force the river to run through gorges.

At the western extremity of the valley, at the edge of the Deccan Plateau, the smooth course of the stream is once again obstructed by the Rajpipla Hills. The now full-grown river however twists through this massive barrier and in a series of boisterous rapids descends from its rugged Deccan homeland to the flat fertile coastal plain of Gujarat, through which the now almost mile-wide river meanders placidly for the last 100 km of its total 1300 km length. After washing the southern walls of the ancient, but now unimportant, port of Broach the river widens into an estuary opening into the Gulf of Cambay.

Throughout the Narmada's upper course in the Deccan the area which it drains, the highlands of Central India and slopes of the Vindhyas and Satpuras were, up to 50 years ago, among the wildest regions of the country with extensive jungle cover, mostly deciduous and abounding in almost all Indian species of game.

This wild region of the Narmada basin has not escaped the floral and faunal degradation of the last few decades. In its course through the Deccan the economic contribution of the river to the basin is unimportant. The river floor sunk in between high banks consisting of hard rock, the water level is well below the level of the adjoining countryside. There are no important tributaries and the river's utility either for irrigation or water transport is negligible.

In contrast the flat deep-soiled Gujarat coastal plain through which the wide low-banked river winds its final course is greatly enriched by the alluvial deposits the river brings down from its upper reaches. At times the flood waters can be devastating and submerge vast areas, but normally the moderate annual floods contribute to the fertility of the islets and the marginal lands in the immediate vicinity of the river.

The fishery of the lower Narmada is well known. The sea fish (*Hilsa ilisha*) swim up during the monsoon to spawn and are netted in large numbers, forming an important supply both for the local and Bombay markets.

NATICAS or Moon shells are very common univalves found in large numbers on the littoral sand-beds. They are active sand-burrowing molluscs. In appearance they look like apple-snails (see SNAILS 2). Some species have much flattened shells. They all feed voraciously on bivalves. Their foot is very large, produced before and behind, and it contains a system of water canals that enables the animal to burrow swiftly. The front lobe of the foot is turned back on the head. Eyes are absent or buried. The lobes of the foot are so extensive and developed that in one genus, *Sinum*, it measures nearly 6 to 7 times the size of the shell; and this animal when it moves looks like an elongated flattened leaf. This overdeveloped foot serves the animal in more than one way: firstly, it is singularly well adapted for working its way through a sandy bottom; secondly, the enfoldment protects the animal from abrasion; and thirdly, it is used to engulf and smother the bivalves buried in sand upon which it preys. All naticas possess a long and powerful radula with sharp, file-like teeth. Having engulfed a clam or mussel in the folds of the foot the natica, using its radula, bores a neat hole through one of the valves and eats the flesh.

Natica deposits her eggs in what looks like a long strip of sandpaper curled like a shirt collar. The sand is agglutinated by a copious mucus into a sort of sheet and thousands of eggs are deposited inside this collar, each egg enclosed in a coating of viscid fluid. During

Moon shell with spread-out foot, × 1 and egg case, × 2/3

the breeding season (April and May) such egg-laden collars are found in large numbers firmly implanted on sand flats and in shallows at the mouths of rivers where the high-tide waves are not violent. These egg-cases therefore are not ordinarily swept away. The eggs develop into embryos within the capsule and come out of it when they have developed all parts and are able to move about and feed independently.

See also MOLLUSCS.

NATIONAL PARKS IN SRI LANKA. There are six wildlife reserves and national parks in Sri Lanka varying in size from the 500 square miles (1300 km²) Wilpattu national park to the 14 square miles (36 km²) Lahugala reserve, which is a haven for elephants.

The total area of the parks is about 1200 square miles and the two most important ones are Wilpattu and Yala.

Wilpattu National Park. Wilpattu is situated in the north-western part of the country and stretches inland from the coast to a distance of about 40 miles. Its eastern border is within 15 miles of the ancient city of Anuradhapura. The park is bounded north and south by two perennial rivers, which act as natural barriers. The coastal strip lies in the arid zone and receives only about 30 inches rain annually but the central and the eastern regions have almost 55″ (1400 mm). However, the entire park is subject to severe drought and to surmount this problem water-holes are being restored in many areas.

The park consists of various ecological habitats like the thorn scrub close to the shore, open grassland, high dry-zone forest, riverine forest, rock outcrops and so

on. But the real enchantment of Wilpattu lies in the serene blue waters of its *willus*, which are oval basins of water formed by natural seepage augmented by the monsoon rains.

Although there are perhaps 300 elephants, they are almost nocturnal in habit and visitors see them only during the dry season when thirst forces them out in the heat of the day to drink at a *willu*. Some of these herds are migratory, seeking the sanctuary of the park after raiding the rain-fed crops in the neighbouring villages. On the other hand, there are elephants which leave the park in search of perennial sources of water and evergreens, along traditional routes of migration.

Besides elephants, Wilpattu has excellent representatives of many other mammalian fauna. Large herds of spotted deer frequent the plains, while sambar and barking deer keep more to the denser forest. There is plenty of food for the four species of primates including purple-faced and gray langurs, and the loris. The king of beasts at Wilpattu is the leopard, which in the absence of any predator larger than itself, stalks through his kingdom both day and night. An adult male will annex a territory of about 3 square miles and protect it by regular patrols and scent-marking. During the mating season leopards will be seen in groups of up to five, that is, one female being courted by four males. Many sightings have been recorded of a mother with cubs at dusk on the open plains.

The forest offers a fascinating variety of birds, dominated by the raucous Malabar pied hornbills, barbets, blackheaded orioles and woodpeckers. There are many colourful species for the serious birdwatcher and on a morning's outing one can see scores of the Ceylon junglefowl, the male of which has a remarkable combination of reds, blues, greens, oranges and black.

Yala National Park. Diagonally across the country nestling in the southeastern corner and bounded by sea and river is the Yala National Park. The landscape is a curious mixture of rock outcrops, scrub jungle and plains, laced with freshwater lakes. Its eastern border is scalloped with high sand-dunes bowing gently along the shores and the sea.

This is again the home of the elephant. For in a day's wanderings it can reach river or sea, thorn bushes or lush foliage, a variety of grasses to be scraped up with its toe-nails or huge rocks which it climbs on padded feet. Approximately 300 elephants live in this 375 square-mile strict natural reserve.

At Yala the elephants are easy to observe as they feed on the plains or browse in the thorn scrub or as they cross the lagoons on their way to the sea. A visitor who does not see elephants here is indeed unfortunate.

However, the ease with which elephants were seen nearly ruined this park, as hundreds of vehicles carrying foreign tourists converged on it. Elephants took exception to the invasion and challenged vehicles which came too close; while leopard and bear took to their heels. The timely introduction of restrictions helped to avoid damage to the status of wild beasts.

Yala like Wilpattu is rich in birdlife, more especially the aquatic species and waders. There are painted storks, pelicans, herons and egrets, whistling teal and the little grebe. The niche occupied by the junglefowl in Wilpattu is taken by the peacock here. In November and December the males dance for everybody's benefit, rending the air with their piercing calls.

Separated by a large river is the Yala East National Park of about 70 square miles. While it resembles Yala in many respects it has its own character, with a quieter atmosphere through a lack of access during some months of the year, and the fact that it is 300 miles away from Colombo.

It harbours a concentration of migrant birds like sandpipers, godwits, avocets, turnstone, plovers and even flamingos. It is also the habitat of the resident blacknecked storks, though the number is limited to about nine. There are plenty of bear, leopard and spotted deer, and on a 6-square-mile plain known as Bagura, there are congregations of up to 2000 deer when the grass is good.

NATIONAL PARKS and NATURE CONSERVATION IN NEPAL.

The kingdom of Nepal is situated along the southern slopes of the Himalayas, bordering India in the southeast and west, and the autonomous region of Tibet in China in the north. The total area is about 140,000 km² and extends for about 800 km from east to west and 150–250 km north to south. More than 75 percent of the total land mass is occupied by mountains.

Nepal can be divided roughly into four zones. The first is the Terai region, close to the Indian border. In the old days this belt was known as *char kos jhari*, which means a stretch of eight miles of forest. Most of this region has now been converted into agricultural land. This was unfortunate, because, apart from its role in conserving groundwater and preventing soil erosion, the Terai belt had a spectacular wildlife including tiger, leopard, bear, gaur, wild buffalo, rhinoceros, deer and antelope. The aquatic wildlife included the Gangetic dolphin, gharial, mugger crocodiles, and many species of fish. Nearly 800 species of birds could be found in this jungle.

The Inner Terai zone consists of a series of basins and valleys, the hills ranging from 500 to 2000 metres. The third zone, 2000 to 5000 m, is a meeting place of subtropical and mountain forests. In the lower hills there are broadleaf species of the sal type extending to coniferous trees like maple and rhododendron in the higher ranges. Hence, at the lower levels tropical species

of wildlife such as tiger, rhinoceros, gaur, wild buffalo, and others, are replaced in the higher areas by the black bear, serow, Himalayan tahr, red panda and others.

The fourth, or the extreme northern, region consists of some of the highest mountain peaks in the world including Mount Everest. The wildlife here consists of Himalayan tahr, blue sheep, Himalayan marmot, mouse hare, and several varieties of pheasants. Ecologically, and perhaps even culturally, this zone is the only one in the country which has been able to maintain its heritage because of its remoteness and inaccessibility.

With its magnificent natural features and scenery, it was expected that tourism would become a great force in Nepal. It has now developed into a gold-mine and many trekking agencies have been set up to cater for it. Unfortunately, even ecologically fragile areas have been opened up for the industry, and conservationists are now demanding that a curb should be imposed on unchecked visitation.

NATIONAL PARKS AND WILDLIFE RESERVES

Royal Chitawan National Park

This was the first National Park to be gazetted in 1973. The area is 932 km², and it is situated in the Rapti Valley, south of Kathmandu. Kathmandu is 220 km away by road.

The vegetation consists of both tropical and subtropical types and it is the last remaining habitat of some species, and harbours 36 species of mammals including tiger, leopard, sloth bear, gaur, deer and antelope. More than 250 species of birds have been spotted here, and in the rivers and lakes, there are such rare species as the Gangetic dolphin and gharial.

Sagarmatha National Park

This National Park, situated in Solokhumbhu District in the northeastern region, has an area of 1243 km². Its major attraction is Mount Everest, besides several other major peaks. The vegetation consists of birch, rhododendron, and oak in the higher region, whereas lower down pine forests dominate the landscape. The mammals include Himalayan tahr, serow, marmots, red pandas, and among the birds several species of pheasants can be found. The Sherpa community, which has played such an important part in mountain trekking in Nepal, lives in this region.

Langtang National Park

This was gazetted in 1976, and has an area of more than 1700 km². This park is a paradise for botanists and comprises some of the most beautiful lakes and snow-capped mountains, including Langtang Lirung, and Dorji Lakhpa.

Rara National Park

This is the smallest National Park, with an area of

about 106 km². It is in Mugu District, and its main attraction is Lake Rara which covers 10 km² and is situated at an altitude of 2900 metres.

Phuksundo National Park

This is the largest National Park covering 4144 km², and is situated in one of the most inaccessible areas of the country, at an average altitude of 3000 m. This is the only park beyond the great Himalayan barrier, and consequently has very scanty rainfall, and the vegetation is very sparse. Because of its inaccessibility, the fragile ecology of this region has been preserved. The park abounds in blue sheep, and probably argali (nayan) (*Ovis ammon*), Tibetan antelope, Tibetan gazelle, and wild yak. A notable inhabitant is the snow leopard. A feature of this National Park is the Phuksundo Lake which has the nation's highest waterfall draining out to form the Suligad river.

Suklaphanta Wildlife Reserve. This area of about 307 km² in the far western corner of Kanchanpur District was established for the protection of swamp deer. The vegetation is dominated by grassy maidans and besides swamp deer, there are many species of wildlife, including the tiger.

Koshi Tappu Wildlife Reserve. This reserve of 65 km² is in the far eastern terai region. It is important because it has the last surviving population of wild buffaloes in the country. It has a rich bird-life including many residential birds such as storks, cranes, darters, and cormorants. It is also a significant transit and breeding ground for migratory birds such as ducks, coots, and geese.

Royal Bardia Wildlife Reserve. This reserve is located in the mid-west region adjoining the eastern bank of Karnali river in Bardia District. It is now extended to 950 km². It is an angler's paradise and an important habitat for the tiger and many other varieties of wild animals.

K.S.

NATURE CONSERVATION IN BANGLADESH.

Bangladesh is a country of alluvial flat plains and low hills, mainly in the north and eastern parts — 12,000 km² are legally Reserved Forests, and another 10,000 km² are State Forests. The forest cover is approximately 15 percent of the total land area which is about 144,000 km².

The country is rich in wildlife, containing some 125 species of mammals, 600 species of birds, two species of crocodiles, and a gharial; 80 species of snakes, 20 species of lizards and skinks, 25 species of turtles and tortoises, and approximately 20 amphibians. There are many species of beautiful butterflies and moths, and about 1000 species of fish.

Extinct and Vanishing Species

During the last century, Bangladesh has lost all the three species of Asiatic rhinos, gaur, banteng, wild buffalo, hog deer, swamp deer and Malayan mouse deer, possibly nilgai; pinkheaded duck, common peafowl, black partridge and possibly the Bengal florican, and marsh crocodile.

The species of wildlife which are facing extinction include Burmese brown tortoise, common batagur, ring lizard, gharial, estuarine crocodile; Burmese peafowl, most partridges, masked finfoot, whitenecked, blacknecked and greater adjutant storks, whitewinged wood duck, nakta or comb duck, goshawk, ringtailed or Pallas's fishing eagle, king vulture, Egyptian vulture, laggar falcon, redheaded merlin, sarus crane and greater pied hornbill.

Among mammals, the tiger, leopard, golden cat, clouded leopard, hoolock gibbon, stumptailed macaque, slow loris, giant and Hodgson's flying squirrels, Himalayan and sloth bears, large Indian civet, binturong, hog-badger, pangolin, wild dog, serow, hispid hare and most of the aquatic mammals may be considered as vanishing or threatened with extinction.

There is literally no status report of the fishes, except those which are used commercially. As far as my information goes, the mahseer, goonch (*Bagarius bagarius*) and *Channa barka* are facing extinction.

Wildlife Act 1973

To protect the depleted population of wildlife, the Government promulgated a Wildlife Preservation Act in 1973. Unfortunately, its implementation is not satisfactory, and there is need for establishing a separate Wildlife Wing under the Forest Department.

National Parks and Sanctuaries

Between 1977 and 1982, the Government created a bird sanctuary in Barisal District, three national parks, and eight wildlife sanctuaries. Two of the national parks are within 100 km of Dhaka City. The third one is near Cox's Bazaar, a beautiful sea resort, some 500 km away from Dhaka, on the southeastern side of the country.

The Madhupur and Bhawal national parks are in the coppice and degraded sal forest belts, and the third is in the hilly terrain, devoid of any vegetation cover due to clear-felling operations of the evergreen and mixed evergreen forests during the past few years. All the three national parks comprise only about 200 km².

The Madhupur national park still retains a fair population of wildlife. It supports a dense population of the Capped langur (*Presbytis pileatus*) and some Rhesus macaque (*Macaca mulatta*). Formerly, there were elephants, tigers, leopards, jungle cats, Asiatic one-horned rhinoceros, wild buffalo, spotted and barking deer. At the moment, all that is left are a few sounders of wild boar. The vegetation of the park is being depleted by encroachment.

The eight sanctuaries together cover an area of roughly 1000 km². Three of these are located in the Sundarbans mangrove forest, and the rest are in the Sylhet, Chittagong and Chittagong Hill Tracts Districts. The lone bird sanctuary at Haila Haor in Sylhet District has an area of 20 km².

Public Awareness

As everywhere in the world, public awareness is vital for conservation. In this respect, a team of wildlife biologists in Dhaka University have played an important part in motivating Government policies and in the introduction of conservation courses in primary, secondary and higher education institutions. Because of a public outcry, the number of rhesus macaques exported to the United States for purposes of research has been brought under control. The Government has also recently banned the export of all wildlife products, excepting frog legs and a few species of turtles and crabs.

R.K.

NATURE CONSERVATION IN BURMA. *General*: Burma with a land area of 676,575 km² has unusual ecological diversity ranging from the moist evergreen forest (rain-forest) of Tenasserim in the extreme south to the temperate forests and alpine vegetation of the far north (Kachin State), where there is permanent snow and mountains of over 5800 metres. Mean annual rainfall varies between as little as 600 mm in the central Dry Zone to 6300 mm in parts of Tenasserim and the Arakan coast.

Though a national forest inventory is now being carried out (1983), no accurate data are yet available on the extent of the various forest types. But based on 1973/79 satellite imagery the Tropical Forest Resources Assessment Project (GEMS Programme) estimated total forest cover in 1980 at 47·7%, compared with 57% based on 1953/68 air photographs. However, these figures include all land not used for settled agriculture including logged-over forest, scrub land and vast areas of secondary bamboo in the Arakan and elsewhere. 'Undisturbed forest' was estimated at 21%, but of this a large proportion has suffered some human disturbance. Deforestation through shifting cultivation and other causes was conservatively estimated at about 25,000 ha per year.

Protected Areas. About 13·5% of the land area is Reserved Forest but in Burma Forest Reserves are in effect production forest, and are therefore mostly subject to logging or other forms of exploitation including hunting. There are no Forest Reserves in the northern mountains which comprise the upper catchment of the Irrawaddy system, and where the preservation of forest cover is so vital to control of water-flow and prevention of flooding in the densely populated Irrawaddy basin.

The only other protected areas are the 14 Wildlife Sanctuaries, most of which date from before World War II. Most of these Sanctuaries are too small to be effective and their total area is only 4608 km² or 0·7% of the land area, which compares very unfavourably with the percentage of total land area set aside for nature conservation in neighbouring countries. Under the present law wildlife in these Sanctuaries is legally protected but the habitat is not, with the result that many have been logged or otherwise seriously degraded. Burma as yet (1983) has no National Parks.

Fauna. Most of Burma lies within the Indochinese zoogeographic Subregion of the Oriental Region, with the Arakan and Chin Hills in the Indian Subregion, and the high mountains of the extreme north with their typically Himalayan species in the Palearctic Region.

Elephant, gaur, banteng, sambar, barking deer, tiger and leopard are widely distributed in the less disturbed forested regions of Burma apart from the far north. But in the absence of factual data their status is uncertain. Two species of rhinoceros formerly occurred, of which the Javan rhinoceros is already extinct and the Sumatran rhinoceros probably so.

Among other large mammals with more localized distribution are hog deer, thamin (found in the drier areas of central Burma), musk deer and tufted deer from the northeastern border with Yunnan, and in Tenasserim two species of mouse deer. There are also four species of goat-antelope; takin (which occurs only in the northwest of Kachin State), serow and two gorals. Tapir are found in mainland Tenasserim. They previously occurred up to latitude 18°N but whether their present range extends so far north is uncertain.

Carnivores include Himalayan and Malayan bear, clouded leopard, wild dog, Asiatic jackal. and, in northern Kachin State, red panda and possibly wolf.

Among primates several species of *Macaca* and *Presbytis* are fairly widely distributed, while there are two gibbons: the hoolock of Upper Burma and the white-handed or lar gibbon of Tenasserim.

Marine mammals and reptiles occurring in coastal waters and riverine estuaries include dugong (now very rare) and Irrawaddy dolphin, the saltwater crocodile and possibly five species of marine turtle, of which the most common are the green turtle and probably Olive Ridley, although the latter has in the past been confused in statistical data with the loggerhead and the relative status of the two species is therefore unclear.

About one thousand bird species have been recorded from Burma (Smythies 1953), this relatively high diversity being due to the fact that the country extends into two zoogeographic regions each with different bird faunas, the forests of Tenasserim containing many Malaysian species whereas in the central and northern parts of the country the bird fauna has Indian and

Chinese affinities. A large number of Himalayan species occur in the montane forests of north and west Burma. There are relatively few endemic species.

Species Conservation Priorities. The most urgent priority is the conservation of large mammals, particularly elephant; also marine turtles and the saltwater crocodile. In the almost total absence of reliable data on the present status of wildlife populations in Burma it is impossible to give anything other than a very subjective impression of the degree to which individual species may or may not be endangered.

a) *Elephant.* The elephant is of major economic importance to Burma for extraction of teak and other hardwoods which are one of the country's main sources of foreign exchange. There are approximately 5400 captive elephants, most of which are employed in the timber industry. However, the reproductive rate among captive elephants is insufficient to maintain numbers and it is therefore necessary to continue to capture an average of about 120 wild elephants per year. Estimates of the wild elephant population range from 3000 to 6000. But indications are that the lower figure is probably the more realistic. Mortality in capture operations is officially admitted to be about 20% and may be even higher. There is also a significant amount of illegal capture, and also poaching for ivory.

In nearly all recently surveyed areas the elephant population has been found to be appreciably lower than previous estimates, and it can therefore be safely assumed that the overall population is considerably lower than the official figure of about 6000, and that with the continuing offtake, known, numbers are steadily declining.

b) *Rhinoceros (Dicerorhinus sumatraensis).* This species formerly occurred in Kachin State, Upper Chindwin, Arakan, Mongmit/Sagaing Division, Kayah State and Tenasserim, but there have been no recent confirmed reports of its survival in any of these areas and it may be already extinct.

The only areas where it has been reported to occur in the past 20 years are the Tamanthi Wildlife Sanctuary in Upper Chindwin and Shwe-u-daung Sanctuary on the border between Sagaing Division and the Northern Shan States. But both these areas have been subject to extensive insurgent activity, and that any rhino still survive is doubtdul.

c) *Thamin (Cervus eldi thamin).* The Burmese subspecies of this deer is confined to the drier areas of central Burma. Although fully, protected by law, thamin are extensively hunted, but fortunately appear able to withstand hunting pressure moderately well and also to adapt to habitat changes. However, their range has been considerably reduced and the only population which can still be regarded as truly viable is in the Kyatthin Wildlife Sanctuary about 150 km north of Shwebo where there are believed to be about 2000, while there are also a few hundred in the somewhat larger but much degraded Shwezettaw Wildlife Sanctuary to the west of the Irrawaddy in Magwe Division. Thamin are vulnerable but not yet endangered, through positive conservation measures are needed if they are to survive in the long-term.

d) *Wild Cattle.* Gaur (*Bos gaurus*) and banteng (*Bos javanicus*) occur sporadically throughout much of Burma in areas where there is still good forest cover and little human disturbance, gaur generally preferring more hilly country than the banteng. Although theoretically protected, both species are heavily hunted and are becoming increasingly scarce. Both are vulnerable, if not endangered.

e) *Tiger (Panthera tigris).* Burma is the only country where tiger occur in which the species is not protected by law; a legacy from former times when they were so plentiful as to be considered vermin. But the situation today is unfortunately very different. There are a few isolated areas where they are still relatively plentiful, but elsewhere they are now increasingly rare. This is probably due both to scarcity of prey species such as the heavily hunted sambar, and also to hunting, trapping and poisoning of the tiger themselves for their skins which fetch high prices in Rangoon and Bangkok. Tiger in Burma are vulnerable but probably not yet seriously endangered.

f) *Saltwater crocodile (Crocodylus porosus).* Formerly widely distributed in estuaries and tidal swamps of Arakan, the Irrawaddy Delta and Tenasserim, crocodile have been heavily hunted for skins and are now very seldom seen. Another major factor has been loss of habitat due to extensive clearing of mangroves for rice cultivation. There are, however, apparently still viable populations in the Irrawaddy Delta where about 500 hatchlings a year are collected for the government crocodile farm in Rangoon; possibly also in less disturbed coastal area of Arakan and Tenasserim where there are still extensive areas of suitable habitat among the tidal creeks and mangrove swamps. Crocodile in Burma are endangered and urgent conservation measures are needed.

g) *Marine Turtles.* The following five species occur or are reported to occur in Burmese waters:

—Green Trutle (*Chelonia mydas*). Commonest species but numbers greatly reduced.

—Olive Ridley (*Lepidochelys olivacea*). Fairly common off the Irrawaddy Delta.

—Loggerhead (*Caretta caretta*). Status uncertain but reported to be fairly common in Delta region.

—Hawksbill (*Eretmochelys imbricata*). Rare.

— Leatherback (*Dermochelys coriacea*). Very rare.

There are turtle nesting beaches along the coast and on certain offshore islands in Arakan, the Irrawaddy

Delta and Tenasserim, of which the most important appear to be Thamihla Kyun (Diamond I.) off the mouth of the Bassein river, Kadonlay and Gayedgyi islands off the mouth of the Bogale river, and Aung Bok in the South Moscos Islands (Tenasserim).

Both Thamihla Kyun and the Moscos Islands are legally established Wildlife Sanctuaries but nearly all turtle eggs are taken; from the former by the government-owned Fisheries Corporation and from the latter by a local contractor under licence from the Forest Department. Eggs are also taken from all the other known nesting sites. Past records show that at the beginning of this century 1.5 to 2 million eggs a year were harvested from Thamihla Kyun. The average annual offtake today is only about 150,000: a 90% reduction. Many former nesting beaches are not now visited by any turtles.

It is clear that turtle populations have declined markedly and that two species (Leatherback and Hawksbill) are endangered while the other three species must be considered as seriously threatened.

Administrative Responsibility and Legislation. The Forest Department is legally responsible for wildlife conservation under the Forest Act (1902) but lacks sufficient staff for either effective management of protected areas or for law-enforcement elsewhere. The Wildlife Protection Act 1936, with minor amendments, still remains the basic legislation governing the conservation of wildlife. It is, however, inadequate for present-day needs, lacking any provision for establishment and management of National Parks or other important provisions which should be covered, and with totally inadequate penalties for offenders. The issue of hunting licences has been suspended since 1958.

Future Prospects. In 1981 the Government of Burma, with assistance from UNDP/FAO, initiated a new Nature Conservation and National Parks Project. Several outstanding areas very suitable as National Parks and Nature Reserves have since been identified, and it is hoped that steps will be taken to establish them in the fairly near future.

Proposals for much needed new legislation and creation of a Department of Nature Conservation and National Parks have also been submitted to Government (1983).

Burma is at present less advanced in the field of environmental conservation than most other countries in the region. Nature, as elsewhere, is under increasing pressure and action is urgent if wild areas with their rich flora and fauna are to be safeguarded for the future. The most pressing need is for the establishment of an effective system of National Parks and Nature Reserves. Provided that steps are taken soon, Burma could have National Parks (including Marine Parks) which are among the finest in Asia, and which apart from safe-guarding threatened or endangered ecosystems and species could also eventually play a major role in development of the economically important tourist industry.

J.B.

NEEM, *Azadirachta indica*, is a fast-growing tree, wild in the dry forests of the Deccan and in the dry forests of Burma. It has feathery leaves, toothed leaflets, curved like a sickle. The small fragrant, cream-coloured flowers, which are five-petalled, appear in March–April. The fruits are egg-shaped, and turn yellow on ripening in June–July. It is an excellent avenue tree giving shade. Its leaves repel insects and its mere presence is believed to keep an area free from malaria. Neem oil is efficacious against pyorrhea and is used in tooth-paste.

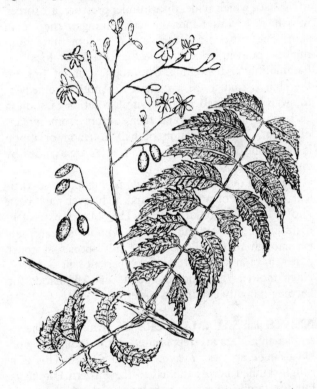

Neem. A shoot, showing leaves, flowers and fruit × 3/4

NEST-WEAVERS. These belong to the family Eresidae represented by the common species *Stegodyphus sarasinorum*. They are medium-sized spiders, greyish white in colour with an oval abdomen and short legs. The males are slightly smaller and darker and account for only 5 to 10 percent of the population. They are social in habits. Many live together in one large, thickly woven bag-like nest. The nest, resembling a squarish waste-paper basket, has several holes which are the entrances. An extensive sheet of white sticky strands is attached to the main nest. Eresidian webs are common features along hedges, over bushes and at the ends of branches of trees. Hundreds of spiders can be seen on the outskirts of a nest, especially towards dusk, industriously moving about, some engaged in repairing the snares, others dragging the entangled insects into the nest for consumption later. An insect which comes into contact with the sticky sheet can seldom escape.

See also SPIDERS.

NETTLES. Protection to plants is afforded by hairs (see PLANT DEFENCES) on stinging nettles (*Urtica dioica*), Nilgiri nettle (*Girardinia heterophylla*) and *Laportea* (*L. crenulata*, *L. pterostigma*). Nettles consist of an elongated tapering hair whose broader round base is embedded in a columnar mass of epidermal cells. When a nettle is touched, the brittle silica-like point of hair breaks and a puncture is made in the skin. The contents of the hair, which evidently are under pressure, are forced into the skin and act upon the nerve endings in the human skin. The hair contains formic acid, or sometimes albuminous poison, and the injection produces local inflammation and swelling. The application of lime to the skin is said to be a palliative. Another remedy is to swab the affected portion with a solution of sodium carbonate or ammonia and then to apply some greasy ointment. The application of half a lemon or onion is also said to remove irritation. Nettles are avoided by deer and other wild grazers.

The sting of *Laportea crenulata* is sometimes so virulent that cases of prostration and high fever have been known to follow contact with it. The pollen is extremely irritating to eyes and nose. Those who have been foolish enough to use the leaves of this species as toilet paper have not forgotten the experience in a hurry. Naga mothers are said to find it very efficacious for keeping naughty children in order.

NEWTS and SALAMANDERS. The Caudata or newts and salamanders are represented in the Indian Region by a single species, *Tylototriton verrucosus*. This is a rough-skinned newt confined to the Eastern Himalaya but the family to which it belongs, Salamandridae, is widespread over Europe and the Holarctic region. *T. verrucosus*, commonly referred to as the Himalayan Newt, is the most primitive member of its family. It is normal with these primitive salamanders for the tips of the ribs to be pointed and protrude through the skin.

The Himalayan newt breeds in cold, clear mountain ponds but is rather terrestrial in habit. Other members of the genus may be completely aquatic. The larvae are aquatic and have long gills, a dorsal fin and a broad tail characteristic of an aquatic mode of life. On reaching adulthood these characters are lost in favour of internal lungs, strong legs, rough skin, etc. The warts that appear on the skin are granular glands that produce an acrid fluid on stimulation. This fluid may serve as a deterrent to predators.

In mating the male uses his front legs to grasp the front legs of the female from behind. It is in this position that the emission of the spermatophore is induced. The female takes up the spermatophore into her cloaca and egg-laying follows a few days later.

See also AMPHIBIANS.

NICOBAR BREADFRUIT, *Pandanus leram*, is a tree up to 15 m high unmistakable with its stilt roots, supporting a repeatedly forked stem. The leaves are up to 4 m long, strap-shaped and armed with fearsome prickles. The fruits weigh up to 18 kg and are pineapple-like in appearance. In the Nicobars they are used for making bread and the leaves are woven into mats.

NIDIFICATION (from the Latin *nidus* meaning nest). All receptacles for eggs laid by birds are called nests. Birds are warm-blooded and their embryos require steady warmth for development. A nesting bird *incubates* or applies its body-heat to the eggs for long periods of time. The nest protects the eggs and the incubating bird from predators and the rigours of climate. Nests are of many types:

Ground nests. Birds like lapwings, most ducks, game birds such as junglefowl and partridges, shore birds (e.g. sandpipers), sea birds (gulls and terns), flamingos, nightjars, larks and others nest on the ground, some with a pad of grass etc., others without. The eggs of many are protectively coloured and marked to match their surroundings.

Himalayan newt (Tylototriton verrucosus) × 1

© R. and Z. Whitaker

© Sálim Ali

Flamingo nests in the Great Rann of Kutch

Cavity nests are situated in rock caves and crevices, burrows in earth banks, holes in tree-trunks etc. Parakeets, rollers, bee-eaters, hornbills, kingfishers and woodpeckers are some of our typical hole- or cavity-nesters.

Platform nests of twigs placed on shrubs and trees are the normal nests of herons, storks, ibises, cormorants etc., also of birds of prey, pigeons, crows and many others. Some waterbirds such as jaçanas nest on floating leaves of water-plants.

Cup-shaped nests are built of grass stems and other pliable plant materials, often plastered outwardly with cobweb, and fixed in the forks and crotches of branches. Ioras, flycatchers, bulbuls, minivets, babblers and drongos build nests of this type.

Domed or *ball-shaped nests* are loosely built of plant material by crow-pheasants and munias and have lateral entrances.

Pendant nests are usually untidy-looking pouches made of materials such as dead leaves, fibres, bark, cobweb, hair, and draped with caterpillars' droppings etc. suspended from shrubs at moderate heights. They are the typical nests of sunbirds. Wren-warblers have rather similar but loosely woven oblong purses attached to tall grass. Weaver birds' nests, usually suspended from date palms or babul trees, are compactly woven retort-shaped structures of grass or palm leaf. The Tailor Bird and some related wren-warblers place their nest within a funnel formed by cleverly stitching two or more large leaves of a shrub along their margins.

Mud nests of Swallows and Swifts are made of mud or straw cemented with the birds' own saliva and stuck to cliff walls and such surfaces. Edible-nest Swiftlets' nests are built often entirely of the bird's inspissated saliva in rock caves; they are commercially exploited as a food delicacy.

Mound nests. Megapodes lay their eggs within a huge mound of rotting vegetable matter collected and piled up by the birds themselves. The eggs are incubated by the heat of the fermenting compost.

The Whitebellied and some other eagles use the same site for a nest year after year, adding more materials each year till it attains an enormous size. Munias often appropriate the old nests of Baya weavers and other species. Crested Tree Swifts lay a single egg in a diminutive saucer glued to the upper surface of a horizontal branch.

Roles of male and female. Both sexes share nest-building in birds like woodpeckers and swallows and most other species. In some, like the weaver birds and jaçanas, the males alone do the building. Nest construction is usually completed within a week, but some birds take longer. Females alone build in many species, typical examples of such being the sunbirds of the Old World and humming-birds of the New. In the pigeons and doves the male usually provides the material with which the female builds.

Copulation. The actual process of introducing sperms into the female is copulation; it may take place on the ground (junglefowl), on a perch (arboreal birds), or in the air (Alpine Swift).

Number of mates. The majority of birds are *monogamous*, i.e. copulation takes place only between the members of a pair. Pheasants and weaver birds are *polygynous*, i.e. each male fertilizes the eggs of several females. The painted snipe, bustard-quail and jaçanas are *polyandrous*, i.e. their females mate with more than one male during a breeding season. Many of the cuckoo family are believed to be promiscuous in their mating.

Nesting Associates. Herons, ibises and darters, and terns of different species breed in mixed colonies. Some small birds build in the proximity of nests of larger or bolder birds, e.g. the oriole near a drongo's nest, for protection against marauders. Other birds nest close to the nests of ants, wasps and bees for the same reason. Some parrots and kingfishers build inside termite nests.

Eggs. The colour, shape, size and number of eggs laid in a clutch vary widely among different species, and sometimes within the same species. The majority of eggs are coloured, especially in the case of open nesters, affording effective camouflage. Eggs usually come in shades of blue, green and brown, variously streaked, blotched or spotted. Cavity-nesters generally lay lightly coloured or white eggs. Egg shells are usually smooth with a matt surface in most birds, but may be glossy or pitted in some families.

Clutch size. The eggs laid by one bird in one nesting attempt is its clutch. A clutch may vary in size from 1 to 28 eggs.

Most species of birds are *determinate* layers, i.e. they lay a definite number of eggs per clutch, contra *indeterminate*, which will continue to lay an indefinite

© Salim Ali

Nests of Cliff Swallow, Hirundo fluvicola

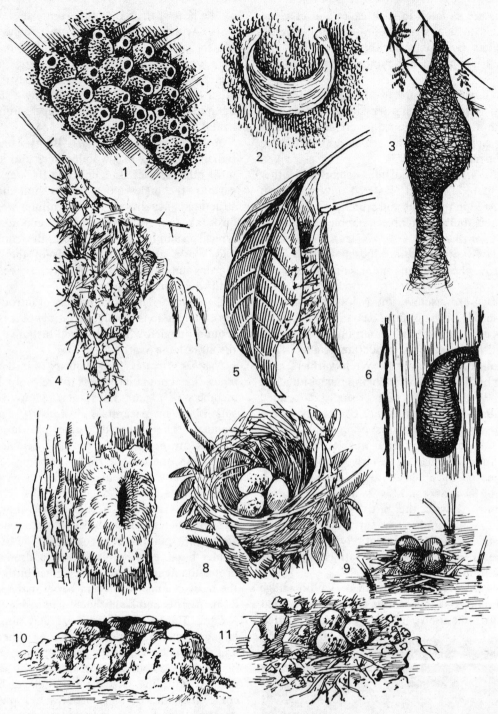

SOME TYPES OF NESTS

1 Cliff Swallow $\times \frac{1}{8}$.	5 Tailor Bird $\times \frac{1}{2}$.	9 Jaçana $\times \frac{1}{4}$
2 Edible-nest Swiftlet $\times \frac{2}{3}$.	6 Woodpecker $\times \frac{1}{8}$.	10 Flamingo $\times \frac{1}{16}$.
3 Baya Weaver Bird $\times \frac{1}{4}$.	7 Hornbill $\times \frac{1}{12}$.	11 Lapwing $\times \frac{1}{2}$
4 Sunbird $\times \frac{1}{2}$.	8 Bulbul $\times \frac{1}{2}$.	

number if the egg is removed each day.

Incubation may start from the first or the penultimate egg. The incubation period is variable; in Indian birds 10 or 11 days (sunbird) to about 45 (vulture). Either or both sexes incubate. While thus engaged most species develop a brood-patch: an area of naked skin on the abdomen which comes in contact with the eggs and which receives increased blood supply.

Hatching. The embryo breaks out of the shell by cutting it with its *egg tooth*, a temporary protuberance on its upper mandible. The parents remove the broken egg shells or eat them. To keep the naked nestling warm they have to cover or *brood* it. Both parents may feed it but one sex is usually more attentive. For nest sanitation the parents in some species remove the faeces voided by the nestlings in the form of pellets or *faecal sacs*. These sacs may be dropped at a distance or swallowed.

D.N.M.

NIGHTJAR. Known also as goatsuckers because of the superstition that they suck milk from goats, night-jars are nocturnal birds. Like the owls, they have very soft plumage. Their flight is swift, silent and distinctive. Their heads, eyes, legs and feet resemble those of the swifts. The beak is short and soft. The enormous mouths, equipped with stiff rictal bristles, enable them to secure their food (mostly beetles and moths) on the wing. Nightjars habitually perch along and not across branches as their feet are short and weak. They roost and lay their eggs on the ground without building any kind of nest. Yet their young are nidicolous.

All nightjars are cryptically coloured. The seven species occurring in our region are more readily identified in the field by their distinctive call-notes than by their appearance. Both nightjars and the closely related frog-mouths are renowned for their camouflage.

Common Indian Nightjar, *Caprimulgus asiaticus* × ¼

NILGAI, BLUE BULL. The Nilgai (*Boselaphus trago-camelus*) is an antelope, believed to be quite closely related to the African Eland, and it is endemic and unique to the peninsular and northern plains areas of the subcontinent, where it avoids true forest and favours tropical thorn scrub or savanna country broken up by rocky ravines. It is absent from Kerala or Sri Lanka as well as northeast India. In Pakistan it survives precariously only in the borders adjacent to India.

In appearance nilgais are quite stocky with rather sloping shoulders and high withers, giving a superficially horse-like outline. Females and young males are a light tan or sandy colour with a white throat-patch and similar spots around the angle of the jaw. As is typical of the Bovidae the male is about 25 per cent larger and heavier and is clothed in steely blue-grey with a prominent white throat-patch and fetlocks, a crested mane of long black hairs down the back of the neck

© M. Krishnan

Nilgai, Blue Bull, Rajasthan × ¹/28

and a curious pendulant tuft of black bristles on the brisket. They bear short, slightly keeled black horns averaging about 20 cm in length. Females lack horns and average about 125 kg and males often exceed 200 kg in weight.

In habits they often associate in small groups of young males and females, but never in large herds, and the adult bulls prefer to remain solitary. They are usually active in grazing in the morning and late afternoon and are quite diurnal. They will graze grass and browse from bushes, being particularly fond of the wild fruits of *Zizyphus* trees and the leaves of *Prosopis spicigera*. Breeding can occur at any time of the year but most calves are born at the beginning of the monsoon season. Post-partum estrus is common so that females become pregnant while still suckling their previous calf. The gestation period is just over eight months. Nilgai come into conflict with man because of their crop-raiding proclivities, and they are especially fond of sugarcane. Though they have often been free from molestation in the past because of their apparent similarity to sacred cows, they are now quite rare and confined to scattered pockets, largely as a result of shrinking of suitable habitat and the increased area of such land brought under cultivation.

See UNGULATES.

NILGIRI WILDLIFE ASSOCIATION. The Nilgiri Wildlife Association was established in 1877. It played a useful role right from the beginning, for from the time of the colonization of the Nilgiri plateau by the British in the 1820s, hunting in the hills was indulged in on a massive scale. It was to check this unprincipled slaughter that the Association was formed.

As a result of a representation made to the Governor of Madras, the Duke of Buckingham, the first Con-servation Law in the country, namely the Nilgiri Game and Fish Preservation Act, was passed. This prohibited the hunting of immature males and females of deer and gaur and introduced closed seasons for many species.

Apart from its interest in the larger animals the Association has made pioneering efforts to introduce trout into the streams and lakes in the Nilgiris.

Since the coming into being of the Wildlife Protection Act of 1972 the Association has lost some of its executive functions, but it continues to act in an advisory capacity, and has been given representation on the Indian Board for Wildlife as well as on the Tamil Nadu Wildlife Board. It receives an annual grant from the government.

The Association runs three lodges on the Nilgiri Plateau and maintains a Register of Wildlife Guides who accompany tourists to see wildlife.

NITROGEN IN PLANTS. Nitrogen is of central importance in the metabolism of all organisms because it is a component of biomolecules such as proteins and nucleic acids. The Earth's atmosphere is 79 percent nitrogen, and different forms of it cycle through the living organisms of the biosphere. This section will consider the fixation of molecular nitrogen for biologically useful purposes and the nitrogen cycle in the atmosphere.

Nitrogen fixation. Nitrogen is a relatively inert element and cannot be utilized by higher plants in its molecular forms. However, reduced forms of nitrogen such as nitrate and ammonia are utilized by living organisms. The reduction of molecular nitrogen of the air, by industrial or natural means, to biologically useful forms is called 'nitrogen fixation'.

The largest part of organic nitrogen in the world comes from nitrogen fixation. A total of about 100 million tons of nitrogen are fixed annually throughout the world by natural processes and another 25 million tons are fixed by industrial means. Nearly 10 percent of the natural fixation occurs in thunderstorms, when the lightning releases nitrogen from the atmosphere, and the remaining 90 percent through the agency of certain micro-organisms.

The micro-organisms that fix nitrogen are of two principal types: free-living (asymbiotic) bacteria that are capable of independent existence, and symbiotic micro-organisms living in the roots of certain plants, mainly legumes.

Asymbiotic fixation. The micro-organisms that fix nitrogen asymbiotically are some heterotrophic bacteria, photosynthetic bacteria and photosynthetic blue-green algae.

The heterotrophic bacteria can fix nitrogen in the dark. These include aerobic bacteria and anaerobic bacteria. Although both types are found in many soils throughout the world, they contribute substantially to the nitrogen content of the soil only under special conditions which include a copious supply of decayed plant tissue and high water content.

A number of green and purple sulphur bacteria as well as non-sulphur photosynthetic bacteria can fix nitrogen asymbiotically using the energy of light.

Certain types of ALGAE, such as photosynthetic blue-green algae, can fix nitrogen. These generally consist of chains of cells in long filaments. Some cells

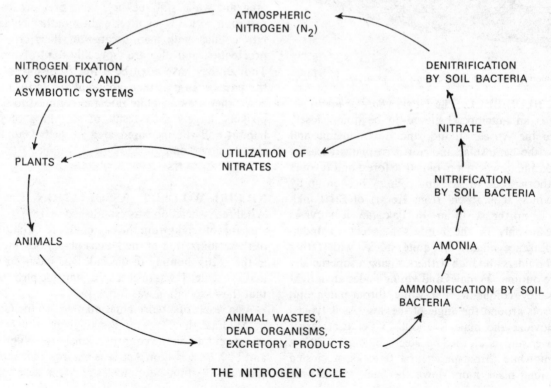

THE NITROGEN CYCLE

in chains known as heterocysts are larger than others, have thick cell walls and are colourless. These are capable of nitrogen fixation. The nitrogen fixers among blue-green algae include about 40 species and they appear to be important mostly in wet tropical soils (e.g. ricefields).

Symbiotic fixation. Symbiotic nitrogen-fixing systems occur in many vascular plants. In fact, over 10,000 species of higher plants are hosts to nitrogen-fixing organisms. Of these, nearly 200 are legumes cultivated as crop or horticultural plants. As commercial nitrogenous fertilizers are expensive for poor farmers legumes are of special significance in our area. The hosts in several other symbiotic systems include non-leguminous plants such as alder and even small water-ferns like *Azolla.*

Nitrogen-fixing micro-organisms which infect higher plants are largely members of the genus *Rhizobium.* These usually live in small, knob-like protuberances known as nodules. The nodules are present mostly on roots of host plants except in a few cases where they occur on leaves. Symbiotic nitrogen-fixation occurs only in the nodules and the host plants themselves do not have the ability to fix nitrogen. In a symbiotic association, the bacterium obtains carbon-containing substances from the host and the host gets the fixed nitrogen from the bacterium. The association is thus mutually beneficial to both the organisms.

Initial infection occurs in root hairs which undergo characteristic curling. Then threads of mucilaginous substance in which rhizobial cells are embedded grow from root hairs to the root cortical region. A mass of host cells, some infected, some not, develop into a young nodule. Nodules are pinkish in colour when cut because of a pigment that develops in the infected host cells.

Symbiotic nitrogen-fixing systems contribute far more to the nitrogen economy of natural communities than do the asymbiotic systems, largely because of the ability of nodules to continue to fix nitrogen for long periods. During this time, fixed nitrogen (in the form of amino-acids) is translocated continually from root nodules to other tissues where it is utilized. Asymbiotic systems, on the other hand, generally fix no more nitrogen than can be used by the micro-organisms themselves. The fixed nitrogen becomes available to the plant only upon the death and decay of the free-living micro-organism that fixed it.

Mechanism of nitrogen fixation. Ammonia (NH_3) appears to be the final end-product of nitrogen fixation. Nitrogen (N_2) is an extremely stable molecule and it is difficult to imagine how biological systems could easily reduce it to NH_3. Nevertheless plant physiologists are on the way to solving this challenging problem. A mechanism involving an enzyme termed as 'nitrogenase' has been proposed for nitrogen fixation which has a good amount of experimental support.

Nitrogen cycle. The amount of nitrogen removed from the atmosphere by biological nitrogen fixation is closely balanced by the amount returned to the atmosphere by micro-organisms that convert organic nitrogen to gaseous nitrogen. Several steps are involved. First, ammonia produced by biological nitrogen-fixation is assimilated by plants and micro-organisms into amino-acids, proteins, and other nitrogenous products. Some plants may be consumed by animals. When animal wastes, dead plants, animals and micro-organisms are subjected to decay, complex organic nitrogenous compounds undergo decomposition into a number of simpler compounds (e.g. amino-acids). Then, by a process called *ammonification,* a group of soil bacteria together with certain fungi convert amino-nitrogen to ammonia. The ammonia thus formed reacts with other chemicals present in the soil (e.g. carbon dioxide and water) to form ammonium salts (e.g. ammonium carbonate). The next step in the sequence is called *nitrification.* Nitrification is carried out by two kinds of soil bacteria: bacteria of the *nitrosomonas* group oxidize the ammonia of the ammonium salts to nitrite (NO_2), bacteria of the *nitrobacter* group oxidize the nitrite to nitrate (NO_3), the most readily utilized of all inorganic nitrogen compounds by higher plants. Finally, nitrates in the soil are reduced by certain soil bacteria to molecular nitrogen, which escapes into the atmosphere. This process is called *denitrification.* Denitrifying bacteria are active only under anaerobic conditions in wet soils containing much organic matter.

The cycle of fixation of atmospheric nitrogen, assimilation of fixed nitrogen by plants and then animals, and return of nitrogen to the atmosphere by denitrifying bacteria constitutes the biological nitrogen cycle.

Nitrogen and photosynthesis. Nitrogen is a component of CHLOROPHYLL and also of the enzymes essential for PHOTOSYNTHESIS. In nitrogen-deficient plants, yellowing of leaves occurs due to a drop in chlorophyll content, thus affecting photosynthesis.

K.K.N.

NUTHATCH (family Sittidae). Sparrow-sized birds with a straight chisel-shaped bill and short tail, remarkable for their agility in clambering woodpecker-like up, down and around tree-trunks and branches (and in certain species on rocks), often upside down. They pick arthropods off the bark and crevices, and are also noted for their habit of cracking nuts to eat the kernel. They frequently associate with the mixed bird parties feeding in forest. Seven species occur in the subcontinent, the commonest being the Chestnutbellied (*Sitta castanea*) and the Velvetfronted (*S. frontalis*).

OAKS (*Quercus*) generally occur in the temperate areas of the Himalaya and strangely enough none are found in the Nilgiris or in the highlands of Sri Lanka. The main use of the timber is as fuel and the leaves are a good fodder and hence they are ruthlessly lopped for this purpose. An indirect value of oaks is the protection which they give to the more important coniferous trees, during the youth of the latter. About 32 species are found in the subcontinent including Burma. The more important of these are the Ban oak, Brown oak, Buk oak, and the Moru oak.

Ban oak, *Quercus leucotrichophora* (*Q. incana*), is a moderate to large evergreen tree with leathery dull green leaves, grey-felted underneath and sharp-toothed on the margin. Young foliage has a lilac or a purple tinge. It is distributed all along the outer Himalaya except the Kashmir valley proper where the full force of the monsoon is not felt. It is capable of growing on the hottest and driest hillsides and in such situations it is stunted and gnarled. In moist valleys it is a tall straight tree. The fruits (acorns) are solitary or in pairs. The nut is egg-shaped, half enclosed in the woody cup. A sweet exudation, known as Oak Manna, is used in confectionery. The fallen seeds are eaten by birds before they ripen. Their survival is therefore threatened unless the ground is worked so that falling acorns get hidden.

Brown oak, Kharsu oak (*Q. semecarpifolia*), is a high-altitude oak of the Himalaya, extending eastwards into Burma and ascending to 3600 m. The foliage has a coppery or brownish tinge in autumn and winter and kharsu forests look spectacular under clear blue skies at that time. The edge of the leaf may be smooth or spiny-toothed; the acorns solitary, the cup, small, flat and thin, covering only the base of the round smooth nut 2·5 cm across. The timber is good building material and the leaves are suitable for feeding caterpillars of the silk-moth *Antheraea pernyi* introduced from China.

Buk oak, *Quercus lamellosa*, is a very large umbrella-shaped tree of the eastern Himalaya, Manipur and Burma where it occurs gregariously between 1800 and 2600 m along with maples and other trees of the Magnolia family. It is recognized at a distance by its large leaves, up to 30 cm long, with a white under-surface. The leaf margin is sharply toothed like a saw. The leaf veins (up to 25 pairs) are prominent beneath. The large flat nuts are characteristic. They are lodged in shallow, saucer-shaped cups 7 cm across. The flowers appear in April-May and the fruits in November. It is an excellent fuel.

Moru oak (*Q. dilatata*) is a large evergreen tree of the western Himalaya and Nepal, growing on cool moist aspects between 2000 and 2750 m. The leaves are spinous and holly-like. The flowers appear at the same time as or before the leaves, as in all oaks. Acorns grow singly, and have an ovoid nut, pointed at the tip and seated in a hemispherical cup.

OBELIA is a typical coelenterate forming small colonies of 3 to 15 cm in height and very common along all sea coasts. It grows on rocks, piles and seaweeds. The main stem looks like a bit of twine bearing short branches arranged alternately. The branches towards the free end carry a conical structure—the hypostome—surrounded at the base by a ring of tentacles. These are the nutritive polyps of the *Obelia* colony. They resemble and feed like hydras and transmit nutriment to the other parts also. The branches towards the lower part of the stem bear cylindrical or long oval bodies called blastostyles which are reproductive in function and asexually give rise to small cup-like buds—medusas.

In *Obelia* the entire stem and branches are protected by a horny tubular coating. This coating extends around the nutritive polyps in the shape of a wine glass and

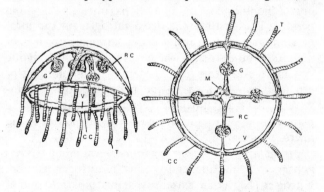

VERTICAL AND TRANSVERSE SECTIONS
OF A MEDUSA

R C radial canal. C C circular canal. M mouth. V velum. T tentacles. G Gondas producing sperms or ova.

414

completely encloses the reproductive polyps (blasto-styles) as a hollow transparent case.

The medusas when mature rupture the apex of the covering and swim away. A medusa is for all intents and purposes a miniature jellyfish. It is free-swimming. Its body resembles a tiny bell with an upper (ex-umbrel-lar) and an under (sub-umbrellar) side. The edge of the umbrella folds inwards and is called the velum. At the centre of the sub-umbrella is the mouth leading into the internal cavity which branches into four radial canals the free ends of which join with a circular canal run-ning close and parallel to the rim of the umbrella. This mechanism serves to distribute the food digested in the central cavity to all parts of the medusa. Along the rim of the umbrella are the tentacles, 16 in the young but many in the adult. At the base of the ten-tacles are minute globular sacs containing sense organs which guide locomotion. The bell-like body enables the medusa to float with ease, and through rhythmic movement of the tentacles it swims freely.
See COELENTERATE.

OBSERVATORIES. Stations equipped with scientific equipment and appropriate staff for making outdoor observations of astronomical, seismological, magnetic, or weather elements are called Observatories.

Atmospheric pressure, air temperature, humidity of the air, wind direction, wind speed, past rainfall or snowfall, kind of clouds, cloud base, cloud amount, transparency of the atmosphere (visibility), duration of sunshine, intensity of solar radiation, and the type of weather (rain, thunder, fog, haze, etc.) occurring at the moment, are some of the observations made at a Meteorological Observatory. Some of them are mea-sured quantitatively, others by nature are qualitative. Such observations are made at fixed times, once or several times a day, depending upon the importance of the station. For example, at most airports, such ob-servations are made 24 times a day. Some instruments at the important stations automatically record the data so that measurements are available for every minute of the day. All these are called Surface Observatories.

At some of the more important observatories, upper winds at several altitudes, extending to 20 or 25 kilo-metres above sea level, are measured with the help of balloons filled with hydrogen or helium. These balloons ascend in the atmosphere and drift with the winds at predetermined levels. They are called Pilot Balloon Observatories. During cloudy days, the balloons cannot be seen by the eye above the cloud levels. In order to be able to make measurements on cloudy days, a few stations send metallic targets up with the balloons, and have radars on the ground for tracking the targets which move with the drifting balloons. These are called Radio Wind Stations.

At a small number of stations arrangements exist for measuring air temperature and humidity at various levels as the balloon ascends into higher and higher altitudes up to 25 or 30 kilometres. Here also radio-wave techniques are employed. The ascending balloon carries a special wireless transmitter, which emits radio waves carrying information about temperatures and humidities and the corresponding heights. These are called Radio Sonde Observatories. In India there are about 500 surface observatories, about 100 Pilot Balloon and Radio Wind Observatories, and about 30 Radio Sonde Observatories. In Pakistan there are about 50 surface observatories. The corresponding numbers for the globe as a whole are about 15,000, 5000, and 1500.

Besides these there are about 5000 rainfall reporting observatories in India which record the rainfall for the past 24 hours, once each day.

In addition to the various kinds of observatories located over land and a few islands, there are about 8000 ships on the high seas, which voluntarily take some meteorological observations, at fixed times, four times a day. The data are transmitted by the ships as coded wireless messages to the nearest coastal radio station.

In recent years space technology has provided satel-lites and special instruments on them for remotely measuring useful meteorological elements over the vast uninhabited areas and the vast oceanic areas. These are promptly transmitted by the satellites, and contribute substantially to the conventional meteorological obser-vations over the inhabited areas, which are in fact limited to about 15 percent of the global surface.

All these observations are standardized and co-ordi-nated by the World Meteorological Organization with the active co-operation of the national meteorological services of the world, about 150 in number. Most of the data are transmitted in standard codes via land-line or wireless communication systems as telegrams or telex to central meteorological offices. Of late, communica-tion satellites also have been pressed into service. In times of peace all the National Weather Services of the world have free access to the data; all of them can make the necessary arrangements to be in the distribution chain.

Each National Service collects its national data at one or two central offices and, after scrutiny, processes them into various kinds of climatological tables. Countries such as the USA, USSR and UK, collect and process data for the whole world, as their commercial and political interests are global.

OCEAN CURRENTS. A general movement of oceanic surface water from one geographical position to another is called an ocean current. The water which moves is constantly replaced, so that the movements are along closed curves of large dimensions, sometimes of the order of a thousand kilometres or more. Wind is one of the principal causes for the motion of sea water. There is a large correspondence between the oceanic surface wind and the ocean surface currents. The wind gets a grip over the surface because of the small ripples on it.

Another cause of ocean currents is the difference in densities in the different parts of the ocean due to differences in temperature and salt contents; the large-scale motions are significantly affected by the rotation of the earth round its axis. This type of circulation is called thermo-haline circulation.

These types of ocean currents are essentially horizontal. They have very small vertical components also. Whenever a current system on the oceanic surface is associated with divergence—more water going away from an area than coming in, cooler and more nutrient water from deeper layers moves up. This is called up-welling. On the contrary, wherever ocean surface circulation is convergent—more water coming into the area than moving out of it, there is down-welling. Warm less nutrient waters accumulate in regions of down-welling.

The sea-areas off the west coast of India and the sea-areas off the Somalia and Makran coasts are regions of up-welling, where there is an abundance of nutrient water and of marine life.

The astronomical tides caused primarily by the gravitational attraction of the moon, produce a significant rise in the sea level. These rises in sea level in the open sea coupled with the rotation of the earth generate ocean currents which move in and out of narrow water-bodies such as gulfs, creeks and river mouths, and are called Tidal Currents.

OCEANOGRAPHY. Oceanography is the application of all sciences to the study of the ocean. It is also sometimes called Oceanology. Modern oceanography is divided into four main parts: (1) Biological Oceanography, (2) Chemical Oceanography, (3) Geological Oceanography, (4) Physical Oceanography. The biological oceanographer studies the occurrence of living organisms in the ocean. The chemical oceanographer studies the chemical properties of sea water. The physical movements of sea water is dealt by the physical oceanographer and the geological oceanographer studies the sediments and rocks present on the bottom of the ocean. These divisions of oceanography are not rigid and the different aspects are closely related.

Oceanography is studied for many reasons. The ocean covers about 72 percent of the earth's surface, therefore, simple curiosity about such a vast region has led to its study. The ocean is necessary for trade, travel and national defence. Much of the trade between countries is carried out by ships and on the ocean

floor are laid electrical cables which connect many countries making it possible to talk to each other across the ocean. Many wars have been fought at sea by surface ships, and submarine and ocean research therefore is helpful to national defence. We also obtain many chemicals, minerals and food from the sea. The chemical resources are iodine, bromine, potassium, magnesium and other elements. The sea water is a source of fresh water for the desert regions of the world. The minerals obtained from the seas are phosphorite, manganese, sand, gravel, tin, and diamond. Important deposits of oil and natural gas have been found beneath the ocean. Biological wealth of the oceans are the fishes of various kinds, pearls, and the accumulations of shells of various organisms which are used in the making of cement. Some of the organisms in the seas are likely to yield chemicals which find use in medicine. The oceans also play an important role in controlling the climate and weather on the earth.

The most important oceanographic instrument is the research vessel. It is the oceanographer's working platform. It carries the oceanographer and many kinds of instruments used in the study of the ocean. Most early oceanographic ships were converted from other kinds of ships for research purposes. Modern oceanographic research requires ships that can perform numerous different tasks and are therefore specially constructed. In addition to surface ships the oceanographer also has underwater vessels called submersibles.

Modern oceanographic research is generally considered to have started with the Challenger Expedition (1872–6). This expedition under the direction of Sir C. Wyville Thomson went around the world. The *Challenger* was a 260-foot 2300-ton steamship. She covered 111,000 km and carried out many different kinds of biological, chemical and geological studies. Previous to this many travellers, explorers and scientists have crossed the various seas but their primary purpose was not to study the oceans. In 1893, the Norwegian Expedition under Fridtjof Nansen allowed the wooden ship *Fram* to be frozen in the ice of the Arctic ocean and drifted across the North Pole. Nansen thus collected valuable meteorological, astronomic, and oceanographic data. From 1925 to 1927, the German ship *Meteor* collected temperatures, water samples and deep-sea sediments from the South Atlantic. Other expeditions on famous ships such as *Discovery, Discovery II, Dana, Willebrord Snellius, Albatross, Galathea* and many others carried out oceanographic studies in various parts of the oceans.

The period after World War II saw a great increase in oceanographic activity. Many new research ships were constructed, new and more accurate methods of oceanographic data collection were developed. Despite the large number of ships, many nations realized that

29 Forests and Backwaters

Kaziranga, one of the richest faunal regions of the Himalayas.

© E. Bharucha

© Shekar Dattatri

The diversity of plant and animal life in the rain-forests of South India needs study. The forests are under siege by a burgeoning human population.

© Suresh Elamon

The tranquil backwaters of Kerala are threatened by over-exploitation and pollution.

30 The Eagle Owl—threat display

The classic threat display of the Eagle Owl is designed to scare away would-be enemies by making the owl look larger than it really is.

© Vivek Paranjpe

because of the large area of the ocean no single country, however large, could hope to carry out the work alone. Cooperative research programmes for oceanographic surveys thus came into existence. One such programme was the International Indian Ocean Expedition (1962–5). This multinational expedition involved the ships of fourteen countries and its purpose was to study in as much detail as possible the Indian Ocean about which very little information was then available. Systematic oceanographic surveys in India began with this expedition. Many voyages to distant parts of the Arabian Sea and the Bay of Bengal were carried out by Indian oceanographers on the ship *Kistna*. When the I.I.O.E. came to an end the Government of India established the National Institute of Oceanography to continue the studies of the ocean around India. The Institute's research vessel is the *Gaveshani*, a 65-metre 1600-ton converted ship. She can accommodate 19 scientists, has a large variety of instruments, and can remain at sea continuously for about three weeks.

Oceanographic studies of a limited nature are carried out in some universities in India but only three of them, Andhra University at Waltair, and Cochin and Mangalore Universities, award degrees in oceanography. Advanced training and specialization is usually obtained abroad.

TABLE I

Area, Volume and Mean Depth of the Oceans

Ocean	Area (million km²)	Volume (million km³)	Mean depth (metres)
PACIFIC	181.344	714.410	3940
ATLANTIC	94.314	337.210	3575
INDIAN	74.118	284.608	3840
ARCTIC	12.257	13.702	1117

R.R.N.

OCTOPUSES are cephalopod molluscs without an external or internal shell. They have a girdle at the forward end of their head bearing eight long, sucker-clad, tapering arms. The body of an octopus is comparatively small, oval or rounded, bearing a pair of prominent eyes at the head region. Giant octopuses are deep-sea forms and are not common in Indian waters. The species found around India are all small, the body and head measuring 5 cm and arms 16 to 18 cm. The largest species, *Octopus herdmani*, has a body less than the size of a closed fist with arms over 70 cm long when fully extended. The small forms live in shallow waters in the littoral region at low-tide mark, concealed

Small octopus near her rock shelter, with eggs on the rock above, × ½.

under rocks or within large empty oyster shells. Medium and small crabs are their favourite food but they also feed on bivalves and fishes. Both in Beyt Channel and Palk Bay, where the shallow sea bottom is rich in seaweeds, jellyfish, corals and oysters and where there are plenty of loose rocks with a rich population of crabs, octopuses exist in large numbers. Unlike squids, which are gregarious and move in company, octopuses are solitary by habit. They often construct a shelter of stones from which they go out in search of prey, always alone. They can crawl over the sea bottom with their arms and can also swim by jet propulsion. During daytime they remain within their shelters but at night they go out to feed. They can be seen swimming actively in shallow waters at low tide after dusk. While moving about, octopuses can change their colour from brown to yellow or dull red. After their night hunt they return to their shelter before dawn. Octopuses are said to be the most intelligent and rapacious of all molluscs. Fishermen who trap octopuses credit them with a great deal of sagacity: when an octopus enters an empty shell it is careful to block the entrance with a shell or piece of stone as a screen against enemies. The octopus is also called 'devil fish' in view of its rapacity when attacking its prey. Its beak is very formidable and it can tear its prey into pieces. Animals caught in its sucker-beset arms can seldom escape.

There are many unproved stories of octopuses having seized men from fishing boats and devoured them. Even discarding such accounts as myths, the fact remains that once caught by the suckers, unless the animal's arm is cut with a knife any attempt to detach the suckers will be futile.

Like most other cephalopods, the octopus has an ink-sac with the secretion of which it can make a screen and escape from its enemies.

Octopuses have some capacity to reproduce a lost

limb, for finger-like projections with suckers grow in place of an arm which has been amputated.

The sexes are separate. The female deposits thousands of small berry-shaped ova attached to a strong, flexible thread which runs along the centre of the mass. The mass is fixed to some stone close to the shelter of the mother. Great is the care with which the mother protects her eggs. She will fight with all her might any predaceous creature which approaches her eggs.

Octopuses are sometimes eaten, and are used as a bait for catching fishes. For first catching the octopus, a number of heavy empty shells are tied to a strong line. Several such lines are suspended during the night: the octopuses get into them and are pulled up and collected by the fishermen.

The development of the eggs is similar to CUTTLE-FISHES'. The young ones emerging from the eggs measure 6·5 mm. They look like a tiny octopus with the arms just appearing as buds around the head region.

Belonging to the octopus family (i.e. cephalopods with eight arms), is the Paper Nautilus or Argonaut. The female argonaut possesses a beautiful shell, discoidal in shape, fragile, corrugated and of papery consistency. All the young argonauts are born naked. The males continue to be naked throughout their life, but the females, 10 to 12 days after their birth, secrete a shell with the enlarged extremities of the dorsal pair of their arms, which contain certain special glands like those in the mantle margin in other molluscs. The shell has no muscular attachment to the body. The shell-secreting dorsal pair of arms are curiously modified to look like sails which hold the shell in position and also protect it. The shell also serves as a receptacle for developing the eggs. The young argonauts remain in the shell until they are capable of leaving the mother. In the living animal the shell is flexible and semi-transparent but when removed it becomes fragile and brittle.

The male argonauts are much smaller than the females. All argonauts swim backwards near the surface of the ocean waters. Over the sea bottom they crawl with the help of their arms.

The Paper Nautilus is very common around the Andaman and Nicobar islands.

See also MOLLUSCS and PEARLY NAUTILUS

OIL AND GAS. Oil and gas is also collectively called petroleum (rock oil, from the Latin words *petra*, rock or stone, and *oleum*, oil). It occurs in many places on the earth as gas, liquid, semi-solid, or solid, or in more than one of these forms at a single place. Chemically petroleum is a very complex mixture of hydrocarbon (hydrogen and carbon) compound with small amounts of nitrogen, oxygen, and sulphur. Liquid petroleum is called crude oil to separate it from refined oil. It has an oily appearance, does not mix with water and floats on it. Petroleum gas, called natural gas, is mostly composed of methane. The semi-solid and solid forms of petroleum are called asphalt, tar, and pitch. Because of its wide occurrence, unusual appearance and character, petroleum has been used by man from very early times for medicinal, religious and economic purposes. In the middle of the nineteenth century petroleum was discovered in very large quantities under the ground. Since then it has become commercially and industrially very important. The first crude oil well was drilled in 1859 in U.S.A. and now it is being produced in North and South America, Africa, Russia, the Middle East and the Far East. The Middle East countries, though small in area, are one of the largest producers in the world. In India oil and natural gas is produced in the states of Assam and Gujarat. In Burma there are many oil wells in the Irrawaddy valley and in Pakistan natural gas found deep under the Bugti Hills is piped south to Karachi and north-eastwards to beyond Islamabad. During the last few years oil and gas has been discovered beneath the sea floor. About 18 percent of the oil and 12 percent of the gas of the total world production now comes from the oceans. In India, oil and natural gas is now being produced from the sea off the West Coast.

The main use of oil and natural gas is to produce energy for power and heat. Oil is used in the crude state for fuel oil or road oil. Its refined products such as gasoline (petrol) and kerosene are used in motor cars, for home cooking and lighting. Petroleum products are used as lubricants and for making chemicals, medicines, paints, plastics, dyes, explosives, and rubber.

OLEANDER, *Nerium indicum*, is an evergreen shrub

Oleander. A shoot with leaves and flowers, × ½. Note the transverse veins in the lance-shaped leaves.

with milky juice, rod-like branches, narrow tapering leaves and fragrant deep rose or white flowers. It is found in the west Himalayan foothills along stream-beds. The plant is poisonous; a small amount if browsed is sufficient to cause death and therefore it is avoided by animals. Rat poison is made from the milky latex.

OOZES. Sediments that are found in the deep parts of the ocean far from land are called pelagic sediments. These deposits are the skeletal material of plants or animals, or fine-grained clays. The term 'ooze' is applied to those deep-sea sediments which contain more than 30 percent of skeletal material. The oozes are named after the organism that is most common in the deposits; common types of oozes are foraminiferal, coccolith, pteropod, radiolarian and diatomaceous oozes. Diatoms and coccoliths are plants and the others are animals. Radiolaria and diatoms have shells made of siliceous material while the others have calcareous shells. Most of the organisms live in the surface waters of the ocean and when they die their shells settle to the ocean floor and form a sedimentary deposit. Calcareous oozes are rarely found below 5000 metres because below this depth the sea water dissolves all calcareous (calcium carbonate) shells. Calcareous oozes cover an area of 54 percent in the Indian Ocean, 36 percent in the Pacific Ocean and 67 percent in the Atlantic Ocean. Siliceous oozes cover areas of 20,15 and 7 percent of the respective oceans. The oozes are valuable to oceanographers because they can be used to learn much about the environment in which the organism lived.

ORB-WEAVERS. All Argyopids and Tetragnathids weave fantastic geometrical webs.

Orb-web construction is fascinating. P1 and P2 are two plants on either side of a narrow pool, about 80 cm apart. The female orb-worker starts emitting a drop of silk at point A and pays out a thread windwards in the direction of P2. The silken threads are glutinous and aided by the breeze are caught at point B. This delicate line is enough for the spider to move along AB and strengthen it by adding one or more fresh lines. At point A she fixes another thread and carrying its free end carefully with one of her hind legs descends down along P1 and at point C stops and fastens the thread. AC is then strengthened. From C another bridge line CD is made in the same way as AB. From D the spider climbs up with a loose thread and connects it to B forming the line DB. Thus the boundary lines are made.

Now the spider goes to E, mid-point of AB, attaches a thread and drops down on CD at point F where the thread is hauled taut and tied (Fig. 2). After strengthening EF, she climbs up to point G and fixes a fresh thread there. Carrying the loose end of the thread and moving to

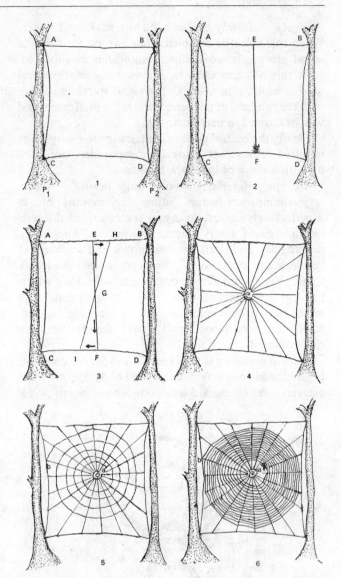

point H via GE, she pulls and fixes the thread at H, resulting in the spoke GH. Attaching a fresh line at G and holding its free end, the spider travels along GF and FC up to point I where she pulls the line taut and fixes it forming the spoke GI (Fig.3). In this way more lines are made on both sides alternately (to keep balance) until all spokes, about 20 in number, are made. Completing the radii a hub is made in the centre by laying a small spiral (Fig.4).

Next the spider, commencing from point 'a' on one of the radii, circumambulates the hub anticlockwise fixing a thread from spoke to spoke and making a spiral ending at 'b' (Fig.5). This spiral is only temporary—a foothold with non-viscid threads for the spider to build the more elaborate viscid spiral.

She now starts at 'b' making a fresh spiral more uniformly and closely interspaced, moving in the clockwise direction. In working this spiral (Fig.6) the spider emits a drop of viscous fluid on every line joining two consecutive radii and carefully twangs it so that each line gets uniformly coated with the sticky substance. When this spiral is completed one can notice that the spiral previously woven for foothold has disappeared. How this has happened is, she used the coarse spiral only as a platform to make the viscid spiral and along with the progress of the perfect spiral the platform spiral was swallowed to make new silk.

Finally the central portion containing too many knots is bitten off and left blank by some spiders but filled up with a mesh of lacework by others.

Argyope pulchella is a fairly large beautiful spider, very common on hedges during rainy months. She is a perfect orb-weaver, sitting in the centre of the web like a vigilant sentry, facing downwards. She has a pentangular abdomen banded brown and yellow and her legs are stretched in the form of St. Andrew's Cross. Very often she decorates the radii along which her legs are stretched with flossy zigzag threads. The bright tints of this spider have a warning significance. Males are very much smaller than females and are dull-coloured.

Nephila maculata is the Giant Wood Spider common in jungles during rainy months. The body of the female measures more than 5 cm long and 1·5 cm wide.

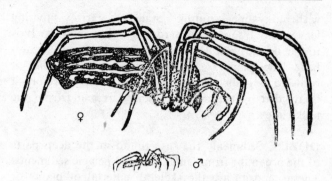

The Giant Wood Spider, *Nephila maculata*, × 1

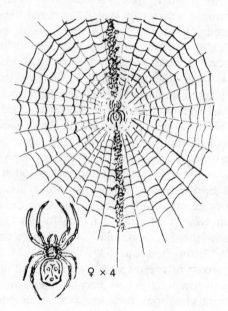

Cyclosa confraga, × ²/₃

Argyope pulchella with a cocoon, × ²/₃

When the legs are stretched fore and aft they cover a length of nearly 16 cm. The abdomen is an elongate, truncated cone, black with longitudinal yellow stripes above and orange dots below. The legs are black with yellow dots at the joints. Males are insignificantly small and dull-coloured. The web of the female *Nephila* is a giant wheel.

Cyclosa confraga is a small brown orb-weaver. Along the vertical diameter of the web she deposits small particles of debris and herself sits in the centre perfectly camouflaged: a wonderful protective adaptation.

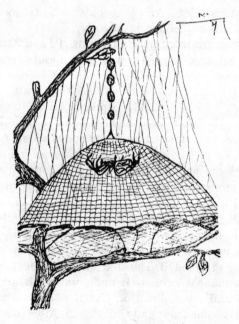

Cyrtophora cicatrosa, × ²/₃

Cyrtophora cicatrosa is another garden spider, medium-sized, greenish black in colour speckled with silver white and with four tubercles on her back. She first weaves a horizontal orb-web with many radii and spirals and then raises its centre to form a dome. Under this dome she builds a flat mesh. She sits inside the top of the dome in an inverted position. Her cocoons are globular, green in colour and are suspended above the dome like a beaded chain. Webs are often built on thorny shrubs. These spiders are of gregarious habits—several of them building their webs in the same shrub, each member with its individual web remaining a separate entity. This is the commonest and most abundant species found in our gardens.

See also SPIDERS.

ORCHIDS. Orchids (botanically Orchidaceae) are one of the largest groups of flowering plants, exhibiting an incredible range of diversity in size, shape, structure, number, density and colour of the flowers. About 17,000 kinds of wild orchids are known, spread practically all over the world, but particularly abundant in the warm humid tropics. In India, there are about 1300 mostly in the Himalaya, in the Western Ghats, and in the northeastern region where nearly 600 species are known, with many of them occupying pride of place in the orchid collections of many famous botanical gardens of the world.

Orchids mostly grow perched on trees, though there are several ground ones too and one climber. Some grow on decaying material (saprophytes), surviving only for a short while. The epiphytic orchids (i.e. those that are on trees) have usually long clinging roots often forming an entangled mass, collecting decaying leaves and other debris which nourish the orchids. These roots can also absorb atmospheric moisture. Further, they lack root-hairs but are often infected by thread-like fungi called mycorrhiza which substitute for the absent root-hairs. In the sub-tropical jungles, it is common to find several kinds of blooming orchids spread over different vantage points on the trunk and limbs of a single tree, each tree a hanging garden in itself. Ground orchids prefer moist, cool, humus-covered soil and have underground tubers and/or stout fleshy roots.

Orchids are perennial herbs with simple leaves, solitary, paired, clustered or scattered all along the stem. They usually have fleshy swollen or bulbous parts called pseudo-bulbs. Some orchids have beautifully marbled, variegated-veined leaves, for which they are valued instead of for their flowers.

The orchid flower, like the lily, has its parts in threes: there are three sepals; of the three petals, the two side ones are similar to each other and sub-similar with the sepals, the third petal is different. This is the labellum or lip. It is this unique lip that is distinctive and lends to each kind of orchid its particular personality. The

Two eiphytic orchids: *Pholidota articulata* with cylindrical jointed pseudo-bulbs and pendant long chain of small flowers, and *Coelogyne cristata* with ovoid pseudobulbs and pendant bunches of large white flowers.

Two land orchids: *Liparis rostrata* with small flowers and *Phaius tankervilliae* with large flowers.

sepals, petals and the lip are atop the flowerstalk with which the ovary is also combined. Within the flower the stamen and the stigma together form a fused complex body—the column, which is another special feature of the orchid flower. Also, instead of the usual pollen dust of individual grains, there is an adhering mass called the pollinium. The varied colour, shape and size of sepals and petals with the more varied lip, and the column, as also the numbers of pollinia, lend to each kind of orchid a uniqueness that enables botanists to name them and has attracted the admiration of botanists and orchid enthusiasts alike.

Orchid flowers show incredible imitations of insects, birds and other animals, giving rise to popular orchid names such as Frog orchid, Dove orchid and Spider orchid. Some orchids are fragrant and a few are malodorous. All these unusual features are to ensure cross-pollination by insects. Most orchid flowers twist on their stalk through 180° to bring the lip into the proper position for the insect pollinator.

Orchids reproduce vegetatively by breaking apart, each piece growing to a new plant, or by seeds. Seeds are minute, millions of them per fruit, and scattered by wind. The seeds are extremely slow to grow and in nature there is a great deal of waste. In the orchid trade, seed is grown in special nutritive media in test-tubes and flasks in aseptic conditions. Recently a new method called 'meristem-culture' has been developed wherein tiny pieces of the growing point of an orchid are incised and grown in test-tubes.

Orchids need aeration, good drainage, much humus and moisture, apart from warmth. Ground orchids are best grown in well perforated pots filled with one-third broken brickbats and the remainder with a mixture of leaf mould, river sand and charcoal. Epiphytic orchids can be grown on blocks of wood with bark, the roots spread on the bark and tied with jute or coir twine. It is advisable to have some moss also packing the roots to help in retaining moisture. Generally, rough-barked

Rhynchostylis retusa. *This is popularly called the Foxtail orchid, possibly in allusion to the dense flowered spike. The long thick flower bunch, called 'Kapau phool', is a bridal decoration in Assamese marriages.*

Pecteilis gigantea, *more familiar under its other name* Platanthera susannae. *This is a large ground orchid with finely scented big blooms, often called the Butterfly orchid because of the spreading lateral sepals.*

trees are good. The hollow of a large split bamboo or cages or baskets of wooden batons also make good receptacles for orchid-growing. A garden with an old mango or jack-fruit tree can form an excellent focus for bringing together many kinds of orchids. Also, drystone walls or loose rock piles can be used for cultivating orchids. They are best watered indirectly by spraying or sprinkling from a distance and also by exposure during rain. Ground orchids can be watered like other plants.

Many wild orchids have almost disappeared or are on the verge of extinction due to ruthless collection from the forests. The Blue Vanda or *Vanda coerulea* is one of these rare ones in northeast India, as also the Lady-slipper orchids or Paphiopedilums.

The International Union for Conservation of Nature (IUCN) has a special survival commission on orchids and many countries have taken measures to protect their orchids. They are also banned in export trade. To conserve our rare and disappearing orchids and to bring together in an 'Orchid Bank' the different kinds of rare native orchids, the Botanical Survey of India has three National Orchidaria, at Shillong, Howrah and Yercaud in the Shevaroy hills near Salem.

An infinite range of new kinds or man-made creations have been produced by breeding in various combinations many pairs of wild orchids. These hybrids can be indefinitely replicated. 35,000 such hybrids have been registered in Sander's *Hybrid Register*! The South American *Cattleya*, the Mexican *Laelia*, the Asian *Dendrobium*, *Paphiopedilum* and *Vanda* have all been extensively used for new creations. Also, the natural variations noticed amongst wild orchids have been carefully nurtured and new 'cultivars' created.

The enjoyment of the subtle beauty of orchids is often marred for laymen by the seemingly difficult botanical names. Orchids, like other plants (and animals), are named in Latin according to a system called binomial nomenclature devised by the Swedish botanist, Linnaeus, more than 200 years ago. The names are in two parts, the first called the generic name, and the second, the specific name. The first name indicates the shared character of one or more species (which together form a genus) and the specific name particularizes the kind of species. This may be in honour of an individual like *Paphiopedilum spicerianum* after Lady Spicer, or after the locality from which it was first collected, like *Malaxis khasiana*, or after some characteristic of the plant itself like *Dendrobium densiflorum*, *Vanda coerulea* and *Bulbophyllum leopardinum*. There are now elaborate directions and rules in naming plants, including orchids, which are reviewed and revised every five years at an International Botanical Congress. There is a separate set of rules governing the naming of garden plants, including hybrids and cultivars. Hybrid names are marked with an X in front of the name and are not to be in Latin but can be from any fancy source. See also CLASSIFICATION AND NAMING.

Although orchids are well-known as ornamentals, some of them are of medicinal value also. The dried pseudo-bulbs of 30 different kinds of orchids are known in the drug trade as salep. The fragrant fruit of the cultivated climbing orchid *Vanilla planifolia* is the source of the essence vanillin.

A.S.R.

See plate 28 facing p. 385 and plate 35 facing p. 496.

ORIOLE. The family Oriolidae consists of 34 species of brightly coloured arboreal song birds spread over most of the Old World. Four myna-sized species occur in the subcontinent, the commonest being the Golden and Blackheaded—chiefly brilliant yellow and black—and the Maroon Oriole of NE India.

Their melodious, flutelike calls are a joy to hear during the breeding season—April to August.

ORNITHOPHILY. Ornithophily, or flower pollination by birds, is common in the tropics but virtually absent in the temperate zone. While feeding on flower-nectar birds contact the flower's essential organs and transfer pollen on their head-feathers to other flowers, thus effecting cross-pollination. As listed by the Austrian naturalist Otto Porsch (1924), the important pollinating families are brushtongued lories, humming-birds, honey creepers, honey-eaters, spiderhunters and sunbirds, flowerpeckers, and white-eyes. The last three are familiar in our area. The leafbird (*Chloropsis* spp.; family Irenidae), peculiar to SE Asia, is also a specialized nectar-feeder along with several others less narrowly adapted for this diet.

The mutually beneficial relationship between ornithophilous flowers and nectar-feeding birds has in some cases evolved highly specialized adaptations in both flower and bird.

Adaptations of flower-birds include a long slender, often decurved bill (to facilitate probing flowers with a long tubular corolla), the absence of rictal and nasal bristles, a tongue made tubular by the curling in of the margins, frayed at the end to make the tip brush-like, and increased protrusibility of the tongue. These adaptations are found in flower-birds in different combinations, but are best developed in sunbirds and humming-birds.

Reciprocal adaptations in flowers include conspicuous colours—red, orange, pink—and abundance of nectar. Their essential organs are oriented so as to ensure contact with the birds' head-feathers. The corolla of most ornithophilous flowers is narrow and tubular, with the nectar near the base of the tube, accessible only to specialized nectar-feeding birds. In such flowers, the

Jungle Myna sipping nectar from coral blossoms × ¹/₃

© Sálim Ali

Jungle Babbler × ¹/₆

© Sálim Ali

regularity in the method of approach of pollinators to the nectar has enabled the flowers to orient their essential organs in the best possible manner to ensure cross-pollination. The tubular corolla of highly evolved ornithophilous flowers has, more than other adaptations, promoted the reciprocal association between specialized nectar-feeding birds and ornithophilous flowers.

An extraordinary ornithophilous adaptation exhibited by many tubular flowers of the mistletoe family of plant-parasites (Loranthaceae) is that their mature flower buds fail to open unless a bird, almost exclusively one of the specialized nectar-feeders, squeezes the tip of the bud with its bill to spring it open. Hence it is clear that these flowers, known as 'explosive' flowers, are exclusively ornithophilous.

Some ornithophilous flowers which secrete nectar on a more or less flat surface (e.g. *Bombax ceiba*, *Erythrina* spp., *Spathodea campanulata*) are largely visited by both insectivorous and frugivorous birds with no elaborate adaptation for nectar-feeding. These birds lick in the nectar, and in doing so may effect cross-pollination. Thus many non-specialized Indian birds, such as drongos, mynas and babblers, also play a useful role.

For feeding on the nectar of flowers, normally pol-linated by insects, sunbirds and other specialized nectar-feeders at times prick a small hole at the base of the corolla to 'rob' the nectar, without conferring on the flowers the benefit of cross-pollination.

P.K.

See PARASITIC PLANTS

OTTERS. Otters occur in all the major continents of the world except Antarctica and Australasia. Related to the weasels, skunks and polecats, they belong to the family Mustelinidae. Otters are distinct in having developed special characteristics enabling them to hunt their food prey in water. Experts recognize about nineteen species throughout the world, and excepting the larger Sea Otter, all are very similar in shape and appearance. Adapted for swimming under water they have dense short velvety fur which traps insulating air, and prevents them from becoming chilled from constant wetting. Their heads are broad and flattened, with small low-set ears which, with their nostrils, close (by muscular valves) when under water. Their large feet are webbed between the toes and their tails are broad and muscular at the base with flattened tips, enabling them to twist and turn with great agility under water. They hunt largely by night, especially in areas subject to human disturbance.

Three species inhabit the subcontinent. The typical otter of the plains, from the Indus river to the Deccan, is the Smoothcoated Indian Otter (*Lutra perspicillata*). Males are slightly bigger than females but not otherwise distinguishable. They weigh up to 10 kg with a total body length of 102 cm, of which the tail is 40 cm. Their fur is silvery brown, being paler on the belly and creamy white around the throat. The Common Otter (*Lutra lutra*) occurs in the Himalayan region from Kashmir eastwards to Assam, and another small population inhabits the higher mountain streams of the Ghats in South India. About the same size as the Smoothcoated Otter, their fur is often slightly darker in colour and males are generally lighter in weight. The third species, known as the Clawless Otter, is much smaller. Of similar coloration, they average 80 cm in total length with the tail 30 cm and males weigh up to 4·5 kg. They do have claws on their digits but these are reduced to rudimentary spicules. They are found in the lower reaches of rivers in the eastern foothills of the Himalayas as well as in a small relict population in the Ghats of south India.

The Smoothcoated Otter often hunts in parties or groups and is quite a sociable animal, whilst the Common Otter is generally more solitary outside the mating season. The former inhabits the larger rivers as well as lakes, and even wanders into canals and swamps. The Common Otter often ascends in summer up to the snow-fed alpine torrents to feed on snow trout. Both feed largely on fish but will take any waterfowl which they can surprise and overcome. The Clawless Otter is believed to subsist much more upon freshwater crayfish and molluscs found in the stony beds of the hill streams which it frequents.

Otters produce blind helpless young in a natural cave or partially excavated hole in a river bank. Usually three to five young are produced which do not venture outside their burrows until 6 or 7 weeks old. Adults are quite vocal, calling to each other with piercing screams when excited. They are also amongst the most playful of carnivores even in the natural wild state. Over land they move with arched back and a clumsy gait but nevertheless are quite at home and can travel for long distances on dry land.

See plate 33 facing p. 480.

OWL. There are 29 species of owls in the Indian region. They range in size from the Eagle-Owls (larger than the kite) to the Pygmy Owlet (hardly larger than the sparrow). They are mostly nocturnal, but all can see quite well in daylight and some even hunt by day. Most owls depend as much on their ears as on their eyes to find their prey. Still their eyes are remarkable for their size, shape and position. Like man's, the eyes of an owl are forward-looking and set in a flat face. Their heads are large since they have to accommodate the huge eyes. Even more remarkable are the ears which in many species are asymmetrical in shape and position. The facial disc of feathers radiating from the eyes is believed to help in locating the sources of sounds with great precision. The soft plumage and some special modifications of the flight-feathers enable them to fly noiselessly and to take their prey by surprise.

Owls have very powerful feet and strong needle-sharp claws with which to capture their prey. As most owls swallow their food whole, their beaks are smaller and less powerful than those of eagles and falcons. The birds are well-known for their habit of casting up the indigestible parts of their food (fur, feathers, bones etc.) in the form of pellets.

Being nocturnal, owls depend upon their call-notes for communication with others of their species. Moreover they need good camouflage to hide during the day from their enemies and persecutors. Therefore most owls are dressed in shades of brown and grey, generously marked with darker streaks, splotches and spots. Some species have curious tufts of erectile feathers on the head which have been misnamed 'ears' or 'horns'. Their functions are obscure, but they have nothing to do with hearing or defence.

Nocturnal, ghostly, mysteriously noiseless in their movements, but endowed with eerie call-notes, owls have always been objects of superstitious awe in our country and greatly persecuted. Nevertheless they per-

Great Horned Owl, *Bubo bubo* and Pygmy Owlet, *Glaucidium bradiei* on about the same scale × $^1/_{10}$

© O. C. Edwards

Mottled wood owl with lizard for young × $^1/_5$

haps have the highest claim for strict protection among all our birds, because of their inestimable services as destroyers of rodents which are amongst our most serious agricultural pests.

K.K.N.

See plate 30 facing p. 417 and plate 31 facing p. 432.

OYSTERS are very widely distributed and are found in shallow waters of all warm and temperate seas and backwaters. Apart from their use as a table delicacy and their value as an easily digestible food for sick persons, there are certain groups of oysters that produce precious pearls and mother-of-pearl which are of great economic value. Also the thick oyster beds and belts protect the coastal areas from being eroded by sea waves. Oysters settle down and grow together to form beds or clusters attached to roots of mangroves by the edge of backwaters. This peculiar habit of living crowded, cementing themselves to the substrata by one of their valves (the left), affects both their external shape and internal structures. Externally, the two shells are irregular in shape and different from each other. Internally the foot is greatly reduced and only one shell muscle is developed to keep the valves closed; the attachment being by the valves no byssal threads are developed; the mantle lobes are united posteriorly to form two siphons, exhalent and inhalent.

It has been estimated that as many as 300,000 to 60,000,000 eggs are laid by a single oyster!

The reproductive activity of an oyster is supposed to commence from the third year. The sexes are united. This means each individual functions both as a male, producing sperms, and as a female, producing ova. Some zoologists believe that an oyster does not fertilize its own eggs. Others are of opinion that the individual oyster does fecundate its own eggs, the ova being first produced and passed on to the mantle chamber. This stage is popularly termed the 'white sick' stage. After an interval the male sperms are let out which fertilize the ova and the 'black sick' stage results.

DEVELOPMENT OF AN OYSTER

1 Eggs, 2 Fry, 3 Young (over six months), 4 Over one year, 5 About 2 years.

Embryos at this stage emerge. They are called 'black spat' and are furnished with vibratile hairs (cilia) for free-swimming life.

Within two days' time these larvas must settle down on suitable substrata or they perish. And in fact the majority of the larvas are washed away by waves or consumed by other molluscs, crustaceans and fish; yet many succeed in safely fixing themselves to suitable holdfasts. The young oysters that are suitably fixed, grow to 9 mm (about ⅓″) in six months but to one inch in one year. It is calculated that an oyster grows by one inch every year and its age is known by its size, i.e., as many years as it measures inches across. Experts say that an oyster's natural life extends up to 10 years and it is at its prime at the age of 5.

There are several species of edible oysters distributed along India's shores. Of these, the one that is most succulent and of good size is *Ostrea virginiana*. It is described as a very hardy species which can withstand changes in salinity, lending itself most readily to cultivation. It thrives in every estuary and backwater. The Indus creeks are so full of this oyster that most of the northwestern districts of this peninsula are supplied with oysters collected from these creeks. On the Kerala coast also, this oyster flourishes and grows to large size. The Madras Fisheries have a small oyster park at Pulicat where this oyster is cultivated and marketed during the non-monsoon seasons. Oysters are affected by heavy monsoon rains which reduce the salinity of coastal waters and cover the beds with rain-water, dirt and debris. On the western coast during the southwest monsoon period (June to September) and on the eastern coast during the north-east monsoon (November to January) oysters are not collected.

Next to *Ostrea virginiana*, there is another purely marine oyster, *Ostrea crenulifera*—the rock oyster, which occurs in such large numbers as to form a regular belt on the rocky shores right down from Sind to Kerala. On new-moon days they are mostly exposed at low tide. The belts on the east coast are less dense. Rock oysters are small, about 8 cm in diameter, and this reduced growth has been ascribed to overcrowding.

Window-pane oysters are at once recognized by their large, flattened, disc-shaped, thin translucent shells, pearly within, and by the bifid tooth ∧-shaped at the hinge margin. The oyster lies more or less prone on the muddy bottom of creeks and bays, and occurs in very large numbers in all important harbours— Karachi, Okha, Bombay, Goa, Cochin, Tuticorin, Madras, Visakhapatnam and Chittagong—as also in most bays and creeks. The oyster possesses a trumpet-shaped foot which is used not so much for locomotion as to sweep out the mud that may enter the gills. The transparent shells are used in panelling windows. It is said that the idea of glazing windows with these shells

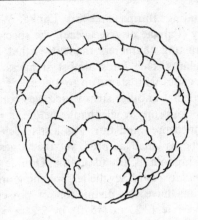

Placuna placenta (Window-pane oyster) × 1

was borrowed by the Portuguese from China and introduced by them into India. In Goa there are still bungalows with windows made of these shells.

Pearl oysters. The Indian genus which produces pearls is *Pinctada*. Its shell valves are roundish, with wing-like projections at the umbo, a rough corrugated exterior and bud or finger-like projections along the lower margin. The interior is highly pearly. It lives under water, attached to rocks and other shells by means of byssal threads. It feeds on microscopic plants and animals and other organic matter contained in the water. If the place of attachment is not congenial, or inadequate in nutriment, this mollusc has the capacity to cast off the threads, crawl away a short distance, and refix itself to some more convenient substratum by producing fresh byssal threads. The thread is said to be formed by a glutinous secretion poured into a groove on the underside of the tongue-shaped foot. The secretion in contact with water coagulates and forms into an elastic thread. The animal attains a maximum size of 9 cm or 3½″ in 3½ years. Therefore just as in the case of true oysters the age of a Pearl Oyster can be known by its size in inches.

The sexes are separate in pearl oysters but according to some authorities each individual functions as a male and as a female, producing sperms and ova alternately. Fertilization takes place outside the mollusc, in water. Development is rapid and at the end of a week the larva settles upon the bottom and makes its first attachment. In this condition it is called 'spat'. The formation of fishable beds of pearl oysters depends upon a rich spat-fall and the spats getting suitably fixed to the bottom. If at the time of spat-fall the waves or undercurrents are strong, most of the spats get washed away and fishable beds are not formed.

Pearl oysters occur sporadically everywhere on the coasts but only in three places, viz., the Gulf of Man-nar, Palk Bay and the Gulf of Kutch, do they flourish in fishable numbers. Pearls are formed between the shell and mantle, caused through various kinds of irritation. Irregular bubble-like pearls are sometimes formed which render the appearance of the shell grotesque. These pearls are called blister pearls and are of no economic value. They have however a protective significance in that blister-pearl-bearing oysters are seldom eaten by predaceous fishes.

Although the pearl oysters of this country are small they occur in large numbers, and pearl fishing from the banks off the Indian mainland at Tuticorin has been going on for over 2000 years.

Enemies of oysters. Murices and their kin (see WHELKS) are the most powerful enemies of all oysters. With their sharp radula they file holes on the upper valves of oysters and using their sucking tube (proboscis) they eat the flesh within. Some birds with strong bills feed on young oysters by breaking the thin upper valves.

Seastar (starfish) eating an oyster. × 1.

STARFISH are destroyers of oysters. Wrapping its arms over an oyster the starfish uses sustained suction of its tube-feet and forces the valves open. Holding the lid open with the arms it applies its mouth to the soft body of the oyster and eats it. Mussels and their allies are not strictly speaking enemies of oysters but they are dangerous to oyster-beds because their presence impedes the free growth of oysters.

Although eyes similar to those of scallops are not present in oysters, their mantle edges contain certain cells with which they can sense the approach of enemies. Fishermen say that whenever their canoe sails by the side of an oyster bank the oysters simultaneously close their valves tight with a clapping sound.

See also MOLLUSCS.

PADAUK, *Pterocarpus dalbergioides*, is a large, but-tressed tree of the Andamans. When blazed with a kukri it exudes a blood-red juice and therefore is also called Andaman Redwood. The compound leaves consist of 5 to 9 egg-shaped leaflets. It has golden-yellow flowers in terminal panicles. The fruit is a circular winged pod with 1 or 2 shining seeds. The wood is valued for panelling. The buttress produces large burrs which when sliced with power-saws turn out beautifully figured veneers used as table tops after polishing.

PALAMAU. The Palamau Tiger Reserve in Bihar, about 175 km NW of Ranchi, has an area of 1000 km². It is a dry deciduous forest, in which the trees shed their leaves in summer. The climate can be most trying, with temperatures going up to 47°C. The reserve is bordered by the Koel river and sal, terminalias and bamboo dominate many areas and provide suitable conditions for a wide range of animals and birds.

From April 1972 when Project Tiger was initiated a core area of almost 300 km² was established where no exploitation of any kind was permitted. As a result of this protection the core area has now a lush grass carpet, the streams are clear and the check dams in various localities retain water throughout the year.

Palamau is visited by a large number of tourists and spotted deer allow a very close approach. The animals have begun to trust human beings and recognize that it is possible to co-exist with man.

The wildlife includes tiger (in 1979, 37), elephant, leopard, hyena, wolf, gaur, sambar, chital, nilgai, barking deer and wild boar.

PALMS are the largest group of monocotyledonous trees and reach their greatest diversity in the humid tropics. Upwards of 100 species occur in the subcon-tinent including Burma and Sri Lanka. Within the palm family (Palmae or Arecaceae), the species exhibit more contrasting characters than any other family of flowering plants. One finds almost stemless palms in contrast to canes with stems up to 169 m (556 ft) high. It is claimed that another stem which was considerably longer was torn up by elephants before it could be accurately measured and thus was lost to science. The seeds may be as small as peas, yet the largest seed in the plant kingdom is found in the Double Coconut (*Lodoicea maldivica*) of Seychelles, weighing up to 23 kg (50 lb). Palms have the distinction of possessing the world's largest leaf, 23 m (75 ft) by 2·2 m–5 m wide. This is seen in *Raphia* sp. from Africa. Before the Red-woods and Eucalypts were discovered, wax palm (*Ceroxylon*) of the Andes was known as the tallest tree, growing to a height of 60 m. The palms are the dino-saurs of the plant kingdom: some are viciously spiny, armed with formidable prickles.

The flowers are white, pale yellow or greenish and are produced in such great profusion as to make the inflorescence a very striking object. The inflorescence is protected in a sheath-like bract or spathe, an organ of importance for botanists in diagnosis and identification as many palms look alike. The flowers are small and variable, bisexual, or male and female flowers may occur on the same plant or on different plants. The floral parts occur in each whorl in threes. The outermost of three sepals is followed by three petals. The stamens are usually six. The ovary consists of 3 carpels. The fruit is a berry or drupe which has often a fibrous covering. The inflorescence produces enormous quantities of pollen. A mature coconut tree is estimated to produce about 400,000,000 pollen grains every year. The soil beneath coconut and date palms is often yellow with pollen shed by flowers. The great majority of species flower and fruit every year, but some e.g. *Corypha*, *Plectocomia*, *Plectocomiopsis*, etc., spend their whole life in preparation for the supreme act of flowering and fruiting. After many years' growth the Talipot Palm (*Corypha umbraculifera*) develops a gigantic terminal bud over a metre high. This in due course bursts with a loud pop and releases a majestic and gigantic panicle 6 m (20 ft) tall by 9 to 12 m (30 to 40 ft) across. The inflorescence containing up to 60 million flowers is the largest known among flowering plants and after flowering the tree dies. In contrast, the inflorescence in *Pinanga* spp. consists of tiny spikes. Pollination is by insects or, in a few cases (Coconut, Date), by wind.

Palms generally have a tall woody stem, bearing a crown of leaves. Some are stemless. In others the stems are thin, very long and carried over other vegetation. The trunk of some species is smooth, with circular scars where the leaf-bases have dropped away. Some are beset with branched spines which are as sharp as needles. Branching of the stem is a rare feature but is a normal feature in the Branching Palm (*Hyphaene*). The palms including the monocotyledons are without secondary thickening (see STEM) and the plant body consists of primary tissue only. Their leaves are very striking. These are of two types—the palmate (fan-shaped) and the pinnate (or feather-shaped)—each with a petiole. In some of the climbing palms called rattans the leaf is produced into a whip-like appendage armed with formidable prickles. These aid the plant in climbing.

The fruit is either a berry or a drupe, and in the latter the endocarp is united to the seed. When the fruit ripens two of the carpels become abortive—as, for example, in the coconut. The fruit of one section, which includes the canes, is covered with closely fitting scales. As already mentioned fruits range in size from very small to very large as in the Double Coconut, which has been introduced in Calcutta and in Sri Lanka. The seed measures up to 45 cm. Their nuts (used as bowls by sadhus) are scattered by the currents of the Indian Ocean on the shores of South India. The endosperm of the fruit may be soft containing oil and proteid as in coconut, or it may be hard as in date.

The coconut may be taken as a general example in germination. The nut is covered with a fibrous husk. The interior is lined with albumen with a watery fluid in the cavity. As it ripens, the fluid is reduced and becomes white and the albumen turns hard. In germination a special absorbing organ is developed on the cotyledon, on the inner end of the embryo (the plumule, or rudimentary stem), and roots push through one of the embryo's pores. The cotyledon, or rather the specialized portion, attacks the white meat and continues to grow until it fills the shell. The roots push out and enter the soil. The young plant continues to grow and finally the connection between it and the nut is severed. The initial leaves are simple and the characteristic foliage is developed after several years.

The use of palms by Man stretches far back into antiquity and is closely woven into the folklore of the countries where they occur. They provide many of the necessities of life (starch, wood, shelter), as well as the refinements of leisure (perfumes, wine, beads). We wipe our boots on coir-mat (husk of coconut). Walking-sticks, baskets are made from cane, brooms, rope, cordage and even cloth can be obtained from palms. Of foods, the dates, oils, fats, sugar, sago and starch are well known. Palm wine or toddy is obtained from palms. *Phoenix sylvestris* is tapped for its sap. After the rains the lateral leaves are removed. A V-shaped cut is made and the dripping sap is collected. One tree yields over 35 kg of sugar in a season, which if fermented gives many litres of toddy. The Date Palm (*P. dactylifera*) also yields toddy. The sap is fermented and yields arrack, the most intoxicating liquor in Asia. The palms have a potential for commerce and are likely to become economic as the price of energy from fossil-fuel is increasing. The oil palm has been introduced into the subcontinent from Africa, and in Malaya it vies with rubber in value as an export commodity. Turning to the canes, they have lost importance in schools, as corporal punishment is no longer in vogue in modern education. However, high-quality cane furniture is back in fashion and is an item of export particularly from Malaysia to N America.

The maximum number of species of palms are found in Burma, northeastern India, Bangladesh, the Andaman and Nicobar islands. They are predominantly tropical and only one, the Windmill Palm (*Trachycarpus takil*), ascends to temperate heights up to 2300 m in Kumaon, where it withstands snow. It has run wild in the summer resort of Mussoorie. The Date Palm occurs as far north as 35°N in Pakistan and bears fruits. The greatest number of palms are centred in the Asian tropics. Singapore island, which is roughly twice the size of Bombay island, has more palm species than the whole of Africa. The Date Palm (*Phoenix dactylifera*) occurs in semi-arid Sind, Rajasthan and the Punjab. The Mangrove Date Palm (*P. paludosa*) is common in the tidal forests of the Sunderbans, Andamans and Nicobars.

The biggest genus, *Calamus* (canes), are adapted to scrambling through forests by their sharp needle-like spines which are borne on the leaves. There is a remarkable palm *Nypa fruticans* growing among mangroves which is regarded as one of the seven earliest flowering plants in the fossil record, dating back 100 or 110 million years.

Some of the interesting or valuable indigenous palms of our area, including some exotics in cultivation, are briefly described.

Dwarf Ground Rattan (*Rhapis flabelliformis*) is a tufted cane from China commonly cultivated. Excellent walking-sticks are obtained from the stems.

Mazari Palm (*Nannorhops ritcheana*) is a low, tufted, gregarious shrub common in the arid regions of Sind, Baluchistan, Punjab and the North West Frontier Province (and common around the Khyber Pass). It is of great value to the inhabitants of these treeless, inhospitable areas as it serves as food, fibre, and fuel.

Trachycarpus martiniana is a magnificent palm in the Eastern Himalaya, Naga Hills, Manipur and Burma at 1500 m. Like *T. takil* it grows gregariously. Leaves form a handsome crown and are used for making hats.

Branching Palm (*Hyphaene dichotoma*) is endemic to the island of Diu in Gujarat from where it has spread to the adjacent mainland. It belongs to the tribe Borassinae to which Palmyra Palm belongs. The dichotomous

© K. C. Sahni

Branching palm. Note the repeatedly forking trunk and fan-shaped leaves. Diu island.

branching of the trunk is striking. The leaves are rounded and fan-shaped, with stalks armed with black spines. Male and female flowers occur on different trees, but occasionally hermaphrodite trees are found. The fruits are heart-shaped, about 8 × 6 cm, shining brown. It is gregarious on the coastal sands of Diu. Being endemic it has been listed as Endangered, so that its use as a firewood and fodder is halted.

Palmyra Palm (*Borassus flabellifer*) is a tall dioecious palm up to 30 m by 0·8 m in diameter from the dry parts of India and Sri Lanka. The trunk is black and swollen above the middle. The fruit is a drupe and varies in

Palmyra Palm

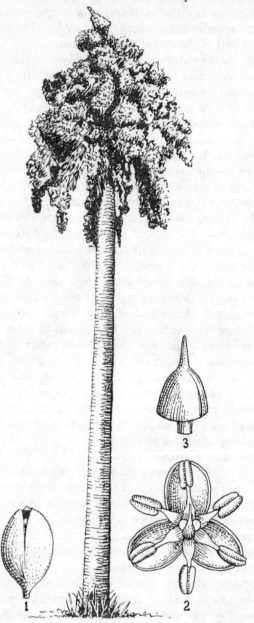

Talipot Palm bearing seed. The inflorescence bears 60 million flowers. It is the largest inflorescence in the world. (*Inset*) 1 Corolla just before opening. 2 Open corolla with stamens. 3 Longitudinal section through pistil—all enlarged.

colour from bright gold to brown. It gives shelter to birds at night, and to squirrels and monkeys during the day. Vast quantities of toddy are drawn from groves of this tree, just as the spathes begin to be formed. It is therefore also called the Toddy Palm. According to some, toddy resembles mild champagne, others associate its taste with cider. Sugar or Jaggery has been made from it for 4000 years. The fruits are edible, and fibre and cordage are obtained from many parts of the palm.

Fan-Palms (*Licula*) are recognized by their broad wedge-shaped leaflets. *L. peltata* is found in the hot valleys of Sikkim, Khasi and Naga Hills, Andamans and Burma.

The Talipot Palm (*Corypha umbraculifera*) of Malabar and Sri Lanka is one of the wonders of the plant kingdom. Its trunk, up to 24 m, is like a flagpole. Its inflorescence has been described earlier. Its leaves are also giant-sized, over 3 m wide, and will protect fifteen men against sun and rain. Ancient records were written on the leaves in Sri Lanka and S India. From the unopened leaves hats of the finest quality are made in the Philippines.

The Fish-tail Palm, also known as the Sago-palm, is called *Caryota urens* by botanists. A tall, fast-growing tree with twice pinnate leaves, its leaflets are lobed like the fins of a fish. It is characterized by a large drooping spadix up to 4 m long in the form of a horsetail. It is found in the humid forests of the subcontinent including Burma and Sri Lanka and is the source of the well-known kitul fibre, which is used in making brushes, fishing-lines and strong wiry ropes for tying wild elephants. The pith is excellent sago. The terminal bud too is edible and the leaves are relished by elephants.

Phoenix. About seven species are found in our area. Wild Date Palm (*P. sylvestris*) occurs in the subcontinent and Burma. It is usually up to 10 m tall with a dense hemispherical crown and feather-shaped leaves. The trunk is without root suckers at the base. The stem is characterized by persistent leaf-bases and it is a common sight to see other plants, mostly Banyan, growing in crevices. The initial host of the Great Banyan of Calcutta is said to be a wild date palm. The Date Palm (*P. dactylifera*) inhabits semi-arid and low-rainfall areas. It is distinguished from the wild date palm by having an open crown and the base of the trunk having a dense

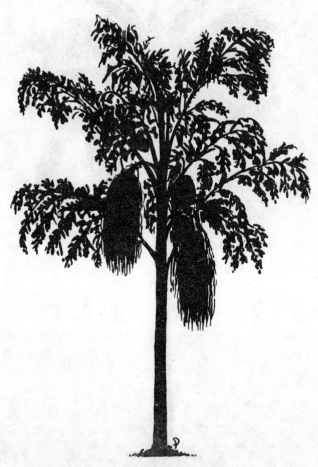

Fish-tail Palm

Date Palm

mass of root suckers. Its leaves are feathery and long, its flowers fragrant and its fruits sweet, fleshy and nutritious, though never equal to those obtained in North Africa and the Middle East. *P. acualis* is interesting because it is almost stemless, found in the plains and hills of northern India, in Burma, central India and in Kanara. In the Naga Hills it is associated with Khasi pine and in the western Himalayan foothills with chir and sal. The terminal bud is an excellent vegetable. *P. pusilla* of Sri Lanka and Tamil Nadu yields sago. The pulp of the seed is eaten. Mangrove Date Palm (*P. paludosa*) occurs in the mangrove swamps of Bengal, Andamans, Nicobars and Burma. It is gregarious, with tufted stems up to 6 m tall and 5 to 7·5 cm in diameter. Their upper portion is covered with long pointed leaf-bases, the lower portion being clear. Leaves are 1 to 3 m long, and leaflets 30 to 45 cm long. The fruit is 1·2 cm long, black in colour.

Areca catechu. The orange-coloured fruit of this species is the betel-nut. It is chewed with the leaf of *Piper betel*, the pan, and lime and is the origin of the splashes of dark red which decorate the streets of many Indian cities. It is a tall graceful tree with feathery foliage. The flowers are white and fragrant and fruits, up to 5 cm long, are olive-shaped with a soft fibrous covering. The seed is 2 cm in diameter and is reddish yellow. Introduced from Malaya from time immemorial it is found mostly in the Western Ghats, eastern India and Bangladesh. Supari collectors climb to the top of the trees, and hop from one tree to another collecting nuts, as in plantations the trees are closely spaced.

Pinanga, a genus of low palms found in the damp steamy atmosphere of evergreen forests in India and Burma, is often gregarious in swampy localities. *P. dicksonii* is common across the Ghats. It has a smooth, green stem 7 m × 3 cm; leaves of 1·5 to 2 m; leaflets 60 × 3 cm; with spike 20 cm long. The nut is eaten as a substitute for the betel-nut.

Bentinckia. *B. coddapanna* is a slender reed-like palm 6 to 9 m tall, found wild only in Kerala on steep cliffs. Being endemic and of botanical interest it is on the Endangered list of Indian flora. *B. nicobarica* is endemic in the Nicobars.

Coconut Palm (*Cocos nucifera*) is the most familiar palm. To the coastal people it is a vital source of income, food and shelter. Monkeys have been trained to harvest its fruits. The origin of the coconut is questionable, most probably it originated in the islands off the Pacific Coast of Panama.

Arenga. There are three species of this palm in our area, of which Sugar Palm (*A. saccharifera*) is important.

© K. C. Sahni

Betel-nut (Areca catechu) *plantation, native of Malaya, cultivated along the west coast from time immemorial. Goa coast, near Panaji.*

Coconut Palm

31 Spoonbill, Sunbird, Owlet, Bee-eater

Spoonbill (*Platalea leucorodia*)

Purplerumped Sunbird feeding young Plaintive Cuckoo
(*Cacomantis passerinus*)

Spotted Owlet (*Athene brama*)

Small Green Bee-eater (*Merops orientalis*)

32 Bonnet and Liontailed Macaques, Slender Loris, Palm Squirrel

Bonnet Macaque (*Macaca radiata*)

Liontailed Macaque (*Macaca silenus*)

Slender Loris (*Loris tardigradus*)

Three-striped Palm Squirrel (*Funambulus palmarum*)

It is wild in Assam, Martaban, Tenasserim and Pegu Yoma. A beautiful and magnificent palm 5 to 12 m tall. Crown oblong, dense, with large leaves 6 to 8 m long. It has up to 115 leaflets on each side. At the base of the petiole is found a beautiful black fibre, exactly like horsehair, known as Gomuta fibre, which is very strong and used for submarine telegraph cables. It is one of the handsomest of palms.

Nipa Palm (*Nypa fruticans*) is found in the mangrove swamps of Sunderbans, Burma, Andamans, Nicobars and Sri Lanka. The fruit is 30 cm in diameter and is a syncarp of many 1-seeded carpels of which the pericarp is fibrous and the endocarp is spongy. Such fruits float on sea-water and are carried far and wide. The leaves are durable for ten years and therefore prized for thatch. It is of value in stabilizing soil and functions as a breakwater. Wine, vinegar and sugar are obtained from it. It is a plant of interest to geologists and paleo-botanists as nuts of a fossil *Nypa* have been uncovered in large numbers in the Tertiary formations at the mouth of the Thames, where they once floated in great profusion as they do today in the rivers in the islands of the Indian and Pacific Oceans.

Canes (*Calamus, Plectocomia, Plectocomiopsis, Daemonorops*). These genera contain the important group of climbing palms, met with in our hot and humid forests. Their stems are of extraordinary value in the manufacture of cane furniture, which is highly popular all over the world, and fetching fancy prices. Enormous quantities of canes are extracted from the forests of northeast India, Bangladesh, Burma, Andaman and Nicobar Islands and a great industry has been built up in some of these areas. The uses to which canes or rattans are put are numerous. They are used for making fish-traps, blow-pipes, cane caps, baskets and furniture. Bridges up to 90 m long are built of canes in the Mishmi Hills and some of these are real feats of engineering. The more interesting and valuable of the canes are:

The Drinking Cane (*Calamus andamanicus*) is the largest cane in the evergreen and deciduous forests of the Andamans and Nicobars. It forms large loops on the ground and climbs over the tallest trees. The leaves are very large and prolonged into a slender whip-like tail which is armed with recurved sharp claws. The leaf rachis is armed with groups of black-tipped claws. These are adaptations to climb and cling to trees. The thick stems when cut are a source of safe drinking water. Sections 3 m long are cut and when held in a vertical position a sap trickles out for a time and then ceases. When the flow stops the lower 30 cm of the cane is cut

Nipa Palm (Nypa fruticans) *with graceful feathery foliage, growing among mangroves. Andamans.*

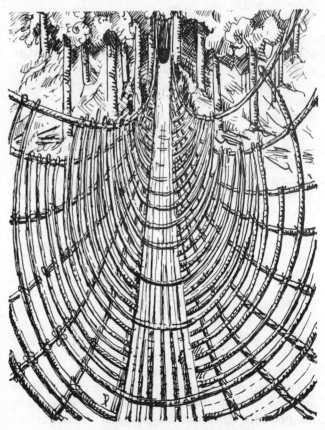

Cane bridge, 90 m long, Mishmi Hills.

433

and the trickle commences again. *C. tenuis*, a climbing palm growing in clumps from underground rhizomes, is distributed in the sub-Himalayan tracts from Dehra Dun to Assam. The leaves are feathery and prolonged into a whip-like tail. The fruits are covered with scales. The seeds are strung up as rosaries and worn by fakirs. Malacca Cane (*C. scipionum*) and Malaya Cane (*C. caesius*) have been introduced into Kerala. The former is used for making sticks and the latter for making chairs.

Plectocomia khasyana, from the Khasi Hills, is a climber with stems as thick as an arm and up to 20 m long. Its leaves are 3-ribbed with short spines. *Plectocomiopsis* is a climbing palm with a branched and spiked inflorescence.

Daemonorops kurzianus (from the Greek *daimon*, an evil spirit, and *rhops*, a low shrub). This is found in the Andamans and resembles the Drinking Cane, but is not so large. It exudes a red resin called Dragon's blood which is used in medicine. The true Dragon's Blood is from the Dragon's Blood Tree (*Dracaena draco*), endemic in the Canary Islands, which exudes a bright red gum.

The palms have the distinction of possessing the largest inflorescence, the largest seed and the largest leaf in the world. They therefore as a group truly deserve their scientific group-name Principes—the princes of the plant kingdom.

K.C.S.

PANGOLIN. Formerly classed in the order Edentata, meaning without teeth, along with the sloths and armadillos of America, pangolins or Scaly Anteaters are now grouped under Pholidota ('scaled animals'). The dorsal and lateral sides of a pangolin, and the outside of its limbs, are covered with large overlapping scales composed of modified agglutinated hairs. Coarse bristle-like hairs occur on the under-surface of the body and a few grow between the scales. Two species are found in the subcontinent: the Indian Pangolin, *Manis crassicaudata* is widespread from the base of the Himalayas to south India, and in Sri Lanka; the Chinese Pangolin, *M. pentadactyla*, ranges through Nepal and the eastern Himalayas to Burma and southern China.

Pangolins are nocturnal in habit, spending the day in their burrows, which are long tunnels ending into a large chamber. Burrows may be fairly deep (6 m) in loose soil. The entrance of the burrow is closed with earth when the animal is inside. It walks slowly with the back well arched and sometimes stands up on its hind feet with the body inclined forward.

The food of pangolins consists of various kinds of ants and termites. The termite mound is torn open by the powerful claws and the pangolin thrusts its long tongue, lubricated with saliva, into the passages and withdraws it with white ants adhering to it. Pangolins roll into a ball for defence and exhibit enormous muscular power that defies any ordinary attempt to unroll them. Probably, stronger carnivores can prey upon them.

In south India, young one are born in January, March and July. In Shekhawati region of Rajasthan, one was born in November, while another female from Las Belas delivered in January. Usually a single young one is produced, rarely two. The mother carries the young on her tail.

I.P.

PARAKEETS or PARROTS (family Psittacidae) are arboreal fruit- and grain-eating birds, chiefly green in colour. Their characteristic features are the stout, short, strongly hooked bill, and short legs with two toes directed forwards and two backwards, adapted for clinging (zygodactylous). The best known is the Roseringed Parakeet (*Psittacula krameri*), commonly kept as a pet. It is grass-green in colour with a red bill, long pointed tail, and a black and rose-pink collar in the male. Other common species are the large Indian or Alexandrine parakeet (*P. eupatria*), with a conspicuous maroon patch on the shoulder, the Bluewinged (*P. columboides*), and the Blossomheaded (*P. cyanocephala*). Their food consists of grains, seeds, fruits and

Blossomheaded parakeet (male) × ⅓

© E. Hanumantha Rao

flower-nectar. Large flocks descend on grain fields and orchards, causing serious damage since they destroy far more than they eat. They have thus earned an unsavoury reputation with the farmer. Parakeets roost communally. They nest in holes in tree-trunks and branches. Species like the Roseringed and Alexandrine are able to 'talk' indistinctly when trained.

The diminutive lorikeet (*Loriculus vernalis*) is a dainty sparrow-sized green parrot with a short square tail and a crimson rump. Its food consists largely of flower-nectar. It is noted for its habit of roosting upside down, bat-like, from branches, hence also called 'hanging parrot'.

PARASITES. A parasite is an organism which attaches to, and derives sustenance or support from, another usually larger organism, the host. Three types of parasites may be distinguished.

1. *Pathogenic*, or disease-producing, parasites gain their sustenance at the cost of deprivation or actual destruction of the host tissues, resulting in disease of the host.

2. *Saprophytic* parasites, which live on dead or decaying matter, and cause no disease or disability.

3. *Symbiotic*. When two organisms exist together in such a way that each supplies some support or sustenance to the other, they are said to be in symbiosis, e.g. the bacterium *Escherischia coli*, in the intestine of man and animals, derives its nutrition from the intestinal contents, and at the same time helps in digestion and preparation for absorption of the vitamin B complex. In the absence of these intestinal bacteria, a condition which can occur after treatment with antibiotics, absorption of the vitamin B complex is impaired, sometimes leading to symptoms of vitamin B deficiency.

PARASITIC PLANTS. The great majority of plants are independent, manufacturing their own food by photosynthesis, but parasitic plants develop an ability to obtain part or all of their food from tissues of other plants or animals. They pierce the tissues of the victim or host by small sucking organs, penetrating to storage or food-carrying tissues, and draw from it a supply of manufactured food for their requirements. In many cases the parasitic plants kill the host or create diseases which completely deform it, causing stunted growth and loss of production. They cause tremendous loss every year to our agricultural crops, fruits and forest trees.

'Total Parasites' do not contain the green colouring matter chlorophyll, which is necessary to manufacture the food they require. They are thus completely dependent on the host for nourishment. Many FUNGI, some bacteria and a few flowering plants are Total Parasites. Dodder or *Amarbel* (*Cuscuta*) is a well known example, which is a flowering plant, with bunches of small flowers borne on numerous yellow thread-like branches tightly twining around the host. It is common to see many trees, bushes and even smaller plants completely covered by yellow threads of this parasite. The seeds of dodder germinate later than those of the host plant, a very short anchorage root is formed, and the delicate yellow thread capable of standing erect, sways in wide circles in search of the host; as soon as it has clasped one, the root dies away, it twines around the host, developing sucking organs.

Broomrapes or *Boomphor* (*Balanophora, Orobanche, Cistanche*, etc.) of many types are also Total Parasites, and they obtain their food material from the roots of

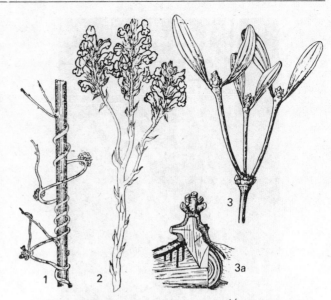

1 Dodder or *amarbel* (*Cuscuta reflexa*) × ⅓. Total parasite with yellow, thread[?] branches tightly twining round the stem of the host. [?]roomrape or *boomphor* (*Orobanche aegyptica*) × ⅙. Total parasite on the roots of the Mustard family; with colourless branched stem, scale-leaves, and blue-lilac flowers. 3 Mistletoe or *banda* (*Viscum orientale*) × ⅙. A partial parasite. 3a. Attachment of the parasite to the woody tissue of the host (after A. B. Randle).

the host plants. They mainly infect and cause great damage to the crops of Mustard and Tobacco families and to some extent even to the forest trees. All these parasites have colourless stems, bearing whitish yellow scale-leaves. The flowers are usually very beautiful, of purple, blue, red and even yellow colour, borne on the top of the stem. They produce an enormous number of tiny seeds which are viable and can germinate after several years.

Another example of this group of Total Parasites infecting the root of the host is a Malayan plant *Rafflesia*, which is one of the greatest peculiarities of the plant kingdom. The whole plant body is like a thread, producing a flower on the ground and in one species it is about 1 m in diameter weighing 7 or 8 kg and smells like putrid meat. The nearest Indian relative SAPRIA is found in the forests of Arunachal Pradesh.

'Partial Parasites' are other types of parasites which contain chlorophyll and obtain their food from the body of the host plants. They also cause great damage to fruits and forest trees. Mistletoe or Banda (*Loranthus, Viscum, Arceuthobium*, etc.) of which there are many types, belong to this group. These parasites are green in colour and develop normal fleshy leaves. They grow on the branches of the host and draw nourishment with the help of suckers. The seeds are surrounded by layers of sticky pulp, which sticks to the bills of birds, and while scraping it off their bills by rubbing against the branches, the seeds adhere there and germinate. The Himalayan

Himalayan Dwarf Mistletoe on Blue Pine bark. Note the minute plants which start as pimple-like growths.

Dwarf Mistletoe (*Arceuthobium minutissimum*) is the smallest dicotyledenous plant. It appears as a small pimple on the bark of Blue Pine. Each pimple bursts to open a minute plant not more than 5 mm in length. The fruits are berries, in which the seeds are embedded in a sticky pulp. When the berry ripens and bursts the seeds are shot out by explosive force. It does considerable damage to the Blue Pine forest and has been known to kill off numbers of fine trees. *A. oxycedri* attacks juniper in Lahul.

A fine example of this group of parasitic plants is the famous SANDAL tree, which looks quite normal from outside, but underground it sends suckers which establish contact with the roots of host plants and deprives them of their food material for its own use. Many other small Partial root Parasites (*Striga, Alectra, Centranthera*, etc.) all belong to the foxglove family, grow on the roots of several crop plants and many types of bushes.

S.D.

See ORNITHOPHILY.

PARASITIC WORMS. Among the multicellular or metazoan pathogenic parasites, worms are the largest and most widespread group. Most parasitic worms live in the intestinal canal of the host species, some are tissue parasites, and some pass through phases, spending their adult life as intestinal parasites, with one or more larval stages in intermediate hosts as tissue parasites, the tissues of the intermediate host being the food of the main host species.

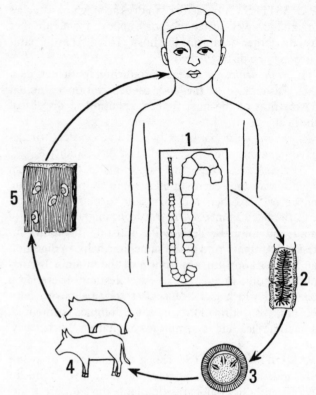

LIFE-CYCLE OF TAPEWORM (diagrammatic)

1 Parts of an adult worm in a human intestine; when fully grown a tapeworm can be many metres long. 2 A mature segment of the worm, containing a uterus loaded with eggs. 3 An egg, dropped with the faeces. 4 An intermediate host, such as a cow or pig, takes up the egg while grazing. 5 A piece of meat from the intermediate host containing larvas of tapeworm.

Flat tapeworms appear to be completely adapted to a parasitic existence, and except when there is massive infection, cause only minor degrees of disability to the host. Being completely parasitic, some worms adopt intricate and complicated life-cycles to ensure survival, one of the most interesting being that of the fish tapeworm, *Dibothriocephalus latus*, whose life-cycle contains four distinct phases, one free-living and the other three in three different host species. Other species of tapeworms have similar, though less complicated cycles. Adult tapeworms live in the human intestine, deriving their nutriment, through their skin, from the intestinal contents. They do not need any special organs except those of reproduction, and the mature segments of the worm are merely sacs containing male and female reproductive organs. Copulation occurs with other worms, or with other segments of the same worm, giving rise to large numbers of fertilized eggs, which are released with the faeces. After a brief free-living phase, the larvas are ingested by the intermediate host, ultimately coming to lie in the flesh of the animal which is then ingested by the definitive host. On digestion, larvas are released in the intestine where they mature. Larvas are destroyed by cooking, but survive in improperly cooked or pickled food. Tapeworms are common in humans, pet animals and in domestic cattle, and must be present in the wild carnivores in areas with high levels of worm infestation.

Other important pathogens include the hookworm (*Ankylostoma duodenale*, a round worm) which can cause severe anaemia in man, and the *Schistosoma* species and Filaria which are tissue parasites causing a variety of symptoms which depend upon the affected tissue. An interesting tapeworm, for which man is an intermediate host, is the *Taenia echinococcus*: the larval forms produce large cysts in various parts of the body and require to be removed by operation.

See also WORMS.

PARENTAL CARE.

In the vast majority of animals, the investment of the parents in their offspring ends with the fertilization of the eggs by sperm. The zygote is then left, sometimes simply floating in open waters, at other times deposited on the plant it will later feed on, without any further care. But in other animals, one or both of the parents may continue to protect and feed it.

The wasps, ants and bees exemplify a whole range of behaviours involving parental care. In all these groups parental care is restricted to the female sex. In more primitive wasps, such as mud wasps so common around our houses, the female constructs a mud nest, paralyzes a caterpillar on which she deposits a fertilized egg, and seals the caterpillar in her nest. On hatching, the larva feeds on the caterpillar as it grows. In other species of wasps, the mother does not provide the larva with prey just once and leave. Instead she repeatedly brings prey to the larva as it grows.

From such beginnings has developed the more elaborate parental and alloparental care of more advanced social insects. In such species a whole worker force takes care of the eggs and larvae (which are generally their brothers and sisters) while the queen confines herself to laying eggs. In an interesting variation of this, certain species of slave-making ants kidnap the brood of a slave species. When these slaves grow up, they in turn provide care for the brood of the master species. In fish, it is often the male that is involved in caring for the brood. This may be because the male fish must stay with the eggs spawned by the female to fertilize them, while the female can slip away, leaving the male to take care. Parental care in fish may involve guarding the eggs in a nest and carrying the fingerlings in the mouth. A number of species of frogs and crocodile also exhibit parental care.

Parental care is highly developed amongst birds. In the vast majority of birds, both the parents look after and actively feed the chicks. This seems to relate to the fact that bird chicks grow to adult size in a very

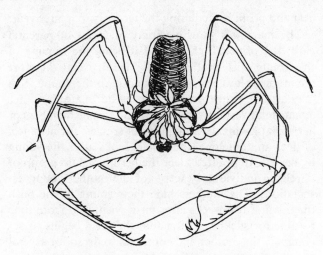

A female of the whip scorpion *Mastigoproctus giganteus* carries her newly hatched prenymphs in and around the brood chamber of her abdomen. (From Weygoldt, 1972)

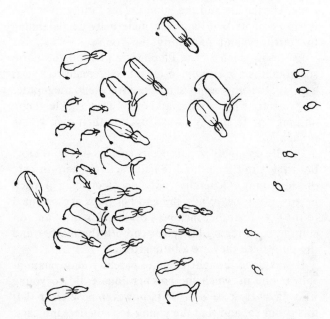

Large ungulates such as gaur and wild buffalo or elephants take a protective formation in which the young are flanked by the adults when threatened by predators like dhole.

short period of a few weeks, and therefore require a great deal of feeding, often involving over a hundred feeding trips a day. The cooperation of both parents is essential to successfully raise the chicks in such a case. In some species however the parental care is either restricted to the female, e.g. the Baya Weaver Bird, or the male, e.g. the Pheasant-tailed Jaçana.

We now know that in a number of species of babblers, bee-eaters and kingfishers parental care is not restricted to the parents but is also shared by other helpers at the nest, generally young of the previous year. The young benefit not only from helping their relatives, but also from acquiring their parental territory in a later year. Occasionally strangers may also help at the nest and profit by acquiring the territory in a later year.

The mammals show a variety of forms of parental care; the principal element of it being the nursing of the young with mother's milk. In all cases therefore the mother plays a major role in care of the young. She may also be helped in this task by her mate, by other female relatives or by both male and female members of her social group. In a majority of species, for instance, of deer and antelope, the female looks after the young by herself. In Spotted Deer the herd is a fluid group of unrelated individuals except for the mother-young associations. The mothers hide their young in the bush for the first few weeks after birth, visiting it from time to time for suckling. After that the fawn begins to accompany the mother and continues to do so for several months. No other member of the herd helps the mother in the task of caring for the young.

The basic unit of the Hoolock Gibbon society is the male, the female and the young. This family lives in a territory of its own, and the male helps in defending the territory and protecting the young.

In species such as the Elephant and Lion the basic social unit is a group of closely related females. In such a group, the other adult females actively help the mother in protecting the young, and may even suckle them, although preference is always given to their own young in suckling.

Finally, in highly social species such as the Wild Dog, both the adult males and females actively help in care of the young. Generally, only one female in a pack breeds at a time. While she and her pups are confined to the den, the other members of the pack hunt and on coming back regurgitate meat to both the mother and the pups once they are a little grown.

It was once thought that the parent-young relationship is one in which the parents donate aid to young who passively receive it. It is however now clear that this is not so, and that the young actively demand care, often at a level that the parents are not willing to provide. This conflict becomes particularly evident at the time of weaning of mammalian young. At one stage in their growth, the young demand continued suckling when the mother attempts to discontinue it. The mother then drives them off forcibly, leading to an often prolonged weaning conflict.

PARIAH KITE. The commonest bird of prey of our countryside, it is also a useful scavenger in towns and cities, taking kitchen scraps, offal or dead rats at garbage dumps but often also marauding poultry. Its large wings and forked tail enable remarkable gliding and manoeuvrability. Both sexes share the building of a large stick nest in a tree or on a building. Two or three eggs are laid.

PARROT FISHES of the family Scaridae have their small teeth fused together to form a beak like that of a parrot with which they browse on the corals or rooted aquatic weeds. Some of them have green colours also like common parrots and make attractive aquarium fishes.

PARTRIDGE. Partridges belong to the game bird family Phasianidae which also includes pheasants and junglefowl. They are squat, short-tailed ground-living birds of the size of a half-grown domestic hen. Several species are found, mostly in the plains, but some, e.g. Chukor, Snow Partridge, also high up in the Himalaya. The commonest and best known species both as a sporting bird and a table delicacy is the Grey Partridge (*Francolinus pondicerianus*) found throughout the drier parts of the subcontinent. It is greyish brown, mottled, barred and vermiculated with buff, rufous and black, blending in a beautifully camouflaging pattern. The challenging, spirited call of the male enlivens the open countryside. The female often weaves in a single high-pitched note, producing a perfectly timed duet. Taken young, grey partridges make interesting free-living pets. Cock birds are highly pugnacious and commonly kept by fanciers as fighting birds.

In the better-watered parts of N India and the peninsula southward, the place of the Grey is taken by the Black and Painted Partridge respectively. All three species are good runners and prefer to trust to their legs and lie low in thickets rather than fly to escape predators. They will burst out of cover with a whirr of wings only when beaten out.

See GAME BIRD.

D.J.P.

Grey partridge challenging × ¹/₃

PEAFOWL. The Peacock's world fame as nature's masterpiece of colourful splendour has obscured the original status of the species as a coveted game bird like the rest of the pheasant family. The large size (that of a domestic turkey) and the cock's 4-ft-long nuptial train, which he grows and sheds seasonally, are no hindrance to swift and dexterous movement through the jungle, on the ground and in flight, to evade pursuit. In this it is aided by its acute hearing and eyesight. The birds have a large repertoire of calls. The significance of most is not clear but the loud long-drawn cry *pai-aein, pai-aein* which usually heralds a thunderstorm is also a night-alarm, well understood by all jungle creatures as a signal that a predator is on the prowl.

The male's gorgeous plumage and spectacular courtship display has so powerfully impressed human sentimentality that the Peacock has become a symbol of beauty and splendour in India's folklore and mythology and an object of religious veneration. In consequence, although a truly wild denizen of deciduous jungles, peafowl have become semi-feral in areas like Rajasthan and Gujarat and enjoy a life of ease and security in the vicinity of human habitation where they often do considerable damage to field crops with impunity.

See plate 17 facing p. 240.

PEARLY NAUTILUS is also known as the Chambered Nautilus. It is a four-gilled cephalopod. Many may be familiar with its large, beautiful, discoidal shell coiled in the same plane. Its colour is porcellanous white outside with wavy bands of reddish brown. The interior is pearly. If a Nautilus shell is sawn across the middle lengthwise and the cross-section examined, it can be seen that the cavity of the shell is divided by a system of partitions (septa) into a series of chambers of gradually increasing size. The living animal occupies only the largest outer chamber of its shell, opening to the exterior. When the animal outgrows this accommodation it builds a larger chamber in front and moves onwards after sealing its former chamber with a septum. This is how a series of chambers with partitions are formed in the shell. Each partition is pierced by a small wire-like tube running through the middle all the way from the smallest initial chamber to the largest chamber lodging the animal. Through this tube a strand of living tissue called the siphuncle passes from the mantle of the animal to the initial chamber. Each empty chamber contains nitrogenous gas. The siphuncle helps to control the gas pressure of the empty chambers and renders the shell light and buoyant.

The Pearly Nautilus is unique not only for the peculiarity of its shell, which is entirely different from shells of other molluscs, but also for other reasons. It is the only cephalopod with four gills and the surviving

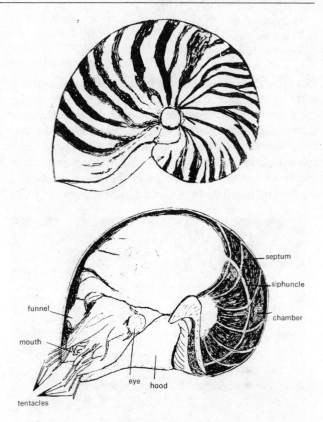

SHELL OF THE PEARLY NAUTILUS
and a cross-section showing the empty chambers divided from each other by septa, the siphuncle passing through them, and the animal in the last chamber. × ½ .

representative of an extinct order which was very large and varied in the earlier geological ages, 2000 fossil species having been found. The shalagram sacred to Hindus is derived from a fossil ancestor.

Although the large shells of the Pearly Nautilus are often found washed ashore on Indian beaches, the animal now lives in the deep waters of the Pacific, especially around the Philippines, New Guinea and Fiji. In many ways a Pearly Nautilus differs from other cephalopods. Unlike the squids and octopuses which have eight or ten sucker-clad arms, Nautilus has 60 to 90 prehensile tentacles given off from the lobes of the foot and arranged round its mouth. With these tentacles it can crawl over the sea bottom and can also swim after the manner of cuttlefish or squid. The tentacles also help in catching the crabs and shell-fish upon which it preys. The mouth of a Nautilus is armed with a beak which is solid and not horny as in other cephalopods. The animals are gregarious and go about in troops during the night in search of prey.

For protection from enemies the animal withdraws itself into the roomy chamber of its shell and closes the entrance with the dorsal lobe of its foot which functions as a thick and strong hood. The shell and hood fully protect it, and it lacks the ink-sac of other cephalopods.

The record size of the shell of a fully grown Nautilus is 27 cm across.

See also MOLLUSCS.

PELICANS are large aquatic birds (family Pelecanidae) with webbed feet and a stretchable skin-pouch under the enormous bill, used for scooping up fish. They are found on inland lakes and lagoons. Flocks fly in formation, soar on thermals, fish cooperatively and nest colonially. In India the White or Rosy pelican is a winter visitor but also breeds in the Rann of Kutch; the Spottedbilled or Grey is a resident species.

Grey Pelican × ¹/₆

Pelican braking to alight on nest

PERCH (suborder Percoidei) are a very large group of fishes, with 35 families in Indian waters, using most of the ecological niches in the aquatic kingdom. They have a single long dorsal fin, the front part of which has pointed spines and the hind part soft, jointed fin-rays. In some cases these parts are separate, indicating two fins. Although a large number are marine, a few freshwater and some estuarine species are also found in this group. For marine perch, see SEA PERCHES.

Among freshwater species, *Chanda nama* and *C. ranga* are the tiniest, measuring hardly 6 cm when full grown. Their body tissues being transparent and bones being visible, they are called 'glass fish' or the 'X-ray fish' and are admired as good aquarium fish. Eggs are laid in aquatic weeds and hatch out into larvae after 24 hours. They are commonly found in most of the rivers of India (see also MINNOWS). The next member of this group, in size, is *Toxotes* (see ARCHER FISH). Two more species of perch-like fishes occurring in fresh water are *Badis badis* and *Nandus nandus* of the family Nandidae. Both these are bottom-dwelling forms found mostly in perennial ponds and lakes of northeastern India.

Badis is a small, 8 cm long fish, known for its ability to change colour according to its surroundings, like a chameleon, from light green through purple and bluish black to dark stripes on the body, as camouflage to avoid being seen and caught. It is accepted as an aquarium fish though considered rather mischievous in a community tank. Being carnivorous in habit, it takes fish larvae, worms, *Daphnia*, *Cyclops* and other organic matter at the bottom. Eggs are laid on the underside of inverted tiles or any other broken earthenware and guarded by the male. *Nandus nandus* lives in similar habitat and has the same habits as the Badis; but it is different in colour, having unchangeable irregular black and white oblique stripes. It is slightly larger than Badis and is not welcome in home aquaria.

Among the estuarine perches, *Lates calcarifer* is undisputedly the most popular food fish in the Indian subcontinent and is also a game fish. It is known as beckti or begti in northeastern India. When grown to its full size of 1·7 m, it is bluish olive in colour but when about 2 cm long, it has black and white irregularly oblique stripes. When juvenile (20 cm) it has sometimes cloudy vertically disposed blotches. Adults are found near the mouths of creeks and breed early in the monsoon. Young ones frequent flooded salt-pans, low-lying puddles or ricefields in search of their favourite items of food such as prawns, crabs, small fish and other aquatic organisms and grow fast if food is plentiful. They are solitary in habit.

Another estuarine perch is the 'target perch', so called because of the slightly concentric dark bands on its body and a round blotch on the dorsal fin, similar to the bull's-eye of a shooting target. It is *Therepon jarbua*, the commonest of the four species of the genus and only 15 cm long when full grown. *T. theraps* has parallel bands, instead of concentric, and is found in offshore water. As a group they are also known as 'squeaking perches'. *T. jarbua* is very active both in the aquarium and in the field and subsists on prawns, larvae and the young of other fish, crabs, worms and other aquatic forms of life. It breeds in the estuaries and the young ones spread into smaller streams all round.

Scatophagus argus and *Sillago sihama*, two other estuarine perches, are both non-carnivorous. The first feeds largely on rooted aquatic plants, has a blunt mouth with browsing teeth, and an almost oblong body with large dark blue spots. Young ones are orbicular and more colourful, with reddish stripes added to the spots. They can be easily acclimatized to fresh water and utilized as beautiful aquarium fish. Known as 'scats', they have never been bred in aquarium tanks so far and fetch a good price abroad.

Sillago, lady-fish or sand-whiting, is a food fish on the west coast of India. It has an elongated sleek and slender body, tapering head and terminal mouth, well adapted for picking up polychaete worms from the bottom ooze, other benthic organisms and, at times, soft weeds and filamentous algae. Olive-green along the back, becoming light green on the sides and abdomen, it measures at most 28 cm at the end of the fourth year of its life. It breeds in estuarine waters, the spawning season extending from August to September. Being a prized fish in the market and non-predatory, its fingerlings are much valued in coastal aquaculture.

See plate 22 facing p. 305.

'PERCHING BIRDS', 'SONG BIRDS', or PASSERINE BIRDS.

Of the 8600 living species of birds grouped in 27 Orders, some 6000 species belong to the Order Passeriformes, which are completely terrestrial or largely arboreal. They are also loosely known as perching birds or passeres. They possess four unwebbed toes joined at the same level: the non-reversible hallux or hind toe, and three others in front. The articulation of the toes is by means of muscles and tendons that run the full length of the leg and which automatically pull in all the toes as soon as the legs are flexed. This mechanism enables the birds to perch securely even while asleep.

The bill in passeres has undergone great structural modification according to the nature of food eaten. The bill shape may be very similar within some families, such as swallows (Hirundinidae) and flycatchers (Muscicapidae) which have short triangular bills wide at the base, but in other closely related genera the bill exhibits wide adaptive specialization as seen in the Babblers (Timaliidae); e.g. Scimitar babblers (*Pomatorhinus* spp.) have long slender curved bills and Spotted babblers (*Pellorneum* spp.) have short straight bills. The bill shape, therefore, does not necessarily indicate phylogenetic relationship but is merely functional.

The majority of passeres can produce a great variety of notes. Some like the Greywinged Blackbird (*Turdus boulboul*) and the Shama (*Copsychus malabaricus*) have melodious voices and musical calls and can mimic the notes and songs of other birds. The Hill Myna (*Gracula religiosa*) can imitate human voice and speech to perfection. The organ responsible for producing voice in birds is known as the *syrinx*, and it is situated at the fork of the windpipe or trachea from where two bronchi originate. The syrinx varies in position and form and the muscles which operate it are known as syringeal or intrinsic muscles. Birds provided with more syringeal muscles are capable of producing a wider range of notes. Structural differences in the syrinx may be seen between a male that sings and its non-singing female. The Order Passeriformes has been divided into two primary groups on the basis of different arrangements of the intrinsic muscles: (i) Non-singing passerines Mesomyodian (Clamatores), which have the syringal muscles attached to the middle of the bronchial semirings, and (ii) Singing passerines, Acromyodian (Oscines) which have syringeal muscles attached to the extremities of the bronchial semirings. The syringeal morphology together with other anatomical characters such as scutellate and laminiplantar tarsus types, etc., have enabled the Order to be divided into five Suborders of which three, namely Eurylaimi, Tyranni and Oscines, are represented in the Indian subcontinent. They include 28 families, some 160 genera and 506 species. Lumping of some traditionally recognized full families such as Muscicapidae, Sylviidae, Turdidae, and Timaliidae into a single family Muscicapidae based on modern trends of ethological and anatomical research is considered controversial by some taxonomists. Whether the fringillids (finches) or the corvids (crows) are highest on the evolutionary ladder is also a matter of controversy.

The systematic arrangement of passerine families as basically proposed by Mayr and Greenway (1956) and Wetmore (1960), and accepted by the XI International Ornithological Congress (1954) is now generally adopted by the majority of international taxonomists. This sequence of the passerine Suborders and Families that concern the Indian subcontinent is followed, with minor adjustments, also by Ripley in *Synopsis of the Birds of India and Pakistan* (1982), and by Sálim Ali and Ripley in their *Handbook of the Birds of India and Pakistan* (1968-74).

Passeres in general are birds of small size, and in this subcontinent the smallest is the thumb-sized Tickell's Flowerpecker (*Dicaeum erythrorhynchos*). The largest is the Raven (*Corvus corax*), which is almost as big as the Pariah Kite. Plumages and colour-patterns vary greatly; some are drab earth-brown as the Wren-Warbler (*Prinia inornata*) and some brilliantly multicoloured like the metallic purple, green, crimson and yellow Sunbird (*Aethopyga siparaja*).

Some of the passerine families contain species that

Tree creeper × ¹/₂

are closely associated with man both aesthetically and economically. Many of our most popular 'cage' birds (finches, bulbuls, babblers, shama, hill myna, thrushes) make charming and engaging pets. Their food and feeding habits vary vastly; while some species live more or less entirely on seeds and vegetable matter (weaver birds, munias) and some on fruits and berries (bulbuls, orioles), others are purely insectivorous (flycatchers, drongos, shrikes). The vast majority, however, are omnivorous, and subsist on a mixed animal and vegetable diet. Depending on their food preferences they are of very great economic importance to man as foes or benefactors—as destroyers of cereal crops and orchard fruit or in keeping insect pests in check. Fruit-eating birds help in the dispersal of seeds of useful as well as harmful plants (mulberry, sandalwood, lantana, mistletoe), and nectar-eaters (sunbirds, leafbirds, white-eyes) in the pollination and fertilization of the specially adapted (ornithophilous) flowers they probe for food (*Loranthus*, *Gliricidia* and numerous others with tubular corollas). The omnivorous House Crow has become an unfailing commensal of Man in urban environments. It is a pilfering blackguard and ruthless persecutor of small garden birds, robbing their eggs and young. But side by side it is a full-time municipal scavenger helping to clean up a city's garbage and thus a useful adjunct to inefficient civic administrations.

A.K.M.

PEREGRINE. A large falcon (family Falconidae) of almost worldwide distribution in several geographical subspecies. It is a rare winter visitor to the Indian subcontinent from the Palearctic Region. Slaty grey above, white below closely barred with black in adult birds, spotted on the breast in immature birds. A prominent blackish cheek-stripe down sides of throat. In flight, its broad-shouldered bullet-like profile, pointed scythe-like wings, short tail and rapid powerful wing-beats make its identity unmistakable. Its natural prey is waterfowl, especially ducks (hence called 'duck hawk' in America). A prime favourite with falconers for dash and speed and spectacular 'dive-bombing' strike when flown at pigeons, partridges, pheasants and other large birds. Peregrine populations in western countries have suffered serious decline in recent years due to widespread use of chemical pesticides acting through the food chain. Success in artificial insemination and captive breeding offers hope of rehabilitation in denuded areas.

PERIYAR. This Sanctuary in Kerala was established in 1934 and is a rare instance of an environment where the establishment of a hydroelectric and irrigation project has left the surroundings not only undamaged but enhanced in certain respects. The artificial lake created by the dam has submerged much forested land, but nevertheless the waterspread has become an enchanting component of the Sanctuary. The area is 777 km² which includes the lake, the mount plateau, and grasslands and forests at altitudes of 800 to 2000 m. The average rainfall is approximately 2000 mm and supports impressive vegetation consisting of both evergreen and deciduous forest types interspersed with grasslands. The dominant evergreens include vellapine (*Vateria indica*), hali (*Palaquium elipticum*), semul, white cedar, rosewood etc. The dominant species in the deciduous forests are teak, *Terminalias*, and bijasal (*Pterocarpus marsupium*). In the cold season the variously coloured tender leaves, and the flowers of *Terminalias*, make the area unusually attractive.

There is abundant wildlife including elephant, gaur, sambar, barking deer, wild boar, Nilgiri langur and Malabar squirrel. It is a glorious sight to see otters gambolling in the lake.

Periyar has a large variety of birds and the most spec-

View of artificial lake at Periyar

tacular of the lot is the Great Indian Hornbill. Darters are almost always present perched on the stumps of trees sticking out of the water (remnants of the forest which was inundated when the lake was formed) and the forests ring with the piercing calls of grackles. There are a host of flycatchers and warblers and the myriad ecological niches of this rich habitat provide specialized sustenance to many species.

Nearly a lakh of visitors go to Periyar every year, and one of the great attractions is to view wildlife at close quarters from boats plying on the lake.

Periyar is included in Project Tiger (see WORLD WILDLIFE FUND).

PHEASANT. All true Indian Pheasants (family Phasianidae) are found only in the Himalaya and the adjacent NE hill tracts. 28 species and subspecies belonging to 9 genera occur within our limits. A few of these show marked habitat preferences: the Peacock-Pheasant and the Blood Pheasant are confined to the Eastern Himalaya while the Chir and the Koklas occupy only the western section. But by and large there is a notable absence of any clear-cut segregation; the habitats overlap and birds of several species usually share the same common tract. Thus the Kaleej and Monal Pheasants affect an extensive range from Afghanistan to Arunachal Pradesh. All the species are resident, but generally there is a regular altitudinal migration of populations seasonally. The birds move up after winter with ascending snow-line to remoter and less disturbed heights.

The majority of Indian pheasants are sexually dimorphic and the males of most spectacularly bright coloured. Though largely conforming to the domestic hen in size they vary markedly in shape. The dumpy Tragopans and Blood Pheasants look like oversized partridges, while the Chir Pheasant is adorned with a straight, pointed 20–inch tail. There is also a general conformity in habits among the numerous species. They affect steep hill slopes preferably with dense cover of scrub or deodar, oak, and rhododendron forest. Roosting in trees at night, they spend most of the day on the ground scratching for food and often digging for it through deep snow with their powerful beaks. They are polygynous birds, each cock holding a 'harem' of 3 or 4 hens. Coveys may be seen on the border of forests and on terraced slopes close to cover. Once alerted the birds are extremely shy and persistent skulkers, very difficult to flush out without the help of a trained dog. Their usual tendency is to run uphill through cover on their strong legs and then to shoot downhill at great speed. They lay their eggs on the ground; the chicks are born covered with down and able to run about almost immediately (nidifugous; precocial).

See GAME BIRDS.

D.J.P.

PHOTOSYNTHESIS. Photosynthesis is a process in which light energy from the sun is converted by plants into chemical energy in the form of carbohydrates. It occurs in the leaves in the presence of the green pigment, chlorophyll, and is associated with the photolysis of water and reduction of carbon dioxide (CO_2).

Historical. Plant tissue consists of 80 percent water and 20 percent dry matter. Of the latter, 14 percent is organic matter. How does the plant obtain this? Aristotle thought that plants get all their food from the soil. But a series of experiments in the last two centuries have shown that green plants synthesize their own food through complicated physical and chemical reactions.

Early in the 17th century Jan van Helmont, who believed water was the chief constituent of most living things, showed by simple measurements that the increase in the weight of a willow branch planted in a box over a period of 5 years was 77 kilograms, while the soil weight decreased by 57 grams only. The plant, obviously, got very little from the soil.

Priestley (1772) observed that an enclosed volume of air 'depleted' by a burning candle and unable to support the oxygen needs of a mouse, could be 'restored' by mint plants, so that it could now support the respiration of the animal. He concluded that green plants have the ability to 'purify' air by a process that is the reverse of respiration in animals. Seven years later, J. Ingenhousz reported that plants purify air only in the presence of light and the green part of the plant carries out this function. Jean Senebier showed in 1782 that carbon dioxide (CO_2) was required in the process.

A giant step towards the discovery of photosynthesis was Robert Meyer's recognition in 1842 that sunlight contributes the energy for the formation of photosynthetic products. Despite the efforts of several workers, however, the complex photochemical and biochemical reactions of photosynthesis remained a mystery till recently.

Biological Occurrence. A wide range of organisms have the capacity to carry out photosynthesis. These include not only the familiar green plants but also various types of algae, diatoms, dinoflagellates as well as some types of bacteria. It is interesting to mention that more than half of all photosynthesis on the surface of the earth is carried out in the sea by microscopic algae, diatoms and dinoflagellates.

Nature of photosynthesis. All photosynthetic organisms except bacteria use water and carbon dioxide. The biochemical nature of photosynthesis for these organisms is represented as below:

$$6CO_2 + 6H_2O \xrightarrow{\text{Light}} C_6H_{12}O_6 + 6O_2$$
$$\text{(glucose)}$$

Most of the photosynthetic bacteria are anaerobes, i.e. they neither produce nor use oxygen. Instead of water, these organisms use other compounds: for example, sulphur bacteria use hydrogen sulphide as shown below:

$$6CO_2 + 12H_2S \xrightarrow{\text{Light}} C_6H_{12}O_6 + 6H_2O + 12S$$

Mechanism of photosynthesis. Photosynthetic formation of glucose has two phases: the light reactions which are dependent on light-energy and the dark reactions which occur in the absence of light. This was suggested by the observation that when a green plant is subjected to intermittent illumination (flashes of light followed by dark intervals), the O_2 evolution was maximum.

Today it is known that the light-reactions of photosynthesis convert solar energy into high-energy molecules such as Adenosine triphosphate (ATP) and Nicotinamide adenine dinucleotide phosphate reduced (NADPH), whereas the dark reactions utilize the chemical energy of ATP and NADPH to bring about the reduction of CO_2 to carbohydrates. This however, should not mean that the dark reactions occur only in the dark. In green plants both sets of reactions take place in daytime. The mechanism of photosynthesis is described in greater detail below:

(i) *The photosynthetic pigments.* The light-trap system consists of pigment molecules that are firmly attached to the membranes of thylakoid in the chloroplast. The structure of a chloroplast, the nature of chlorophyll and their behaviour during photosynthesis are described under CHLOROPHYLL. In addition to chlorophylls, the chloroplast contains other pigments which participate in photosynthesis, such as carotenoids, phycocyanin and phycoerythrin.

(ii) *Plant photosystems.* The wave-lengths of light that are most efficient in promoting photosynthesis coincide closely with the wave-lengths of light absorbed maximally by plant pigments. However, in most plants, the photosynthetic efficiency of light wave-lengths above 680 nm drops sharply compared with their absorption. This deficit is called the *red drop.* (nm is short for nanometre, one-thousandth part of a micrometre.) Later, R. Emerson observed that if supplementary light of lower wave-lengths (650 nm) is present, the efficiency of longer wave-lengths (above 680 nm) is restored. This phenomenon, known as the Emerson-enhancement effect, yielded an important inference: two light-absorbing systems, one absorbing in the region 680 to 720 nm and the other at shorter wave-lengths, must co-operate to yield maximal photosynthetic efficiency.

The experimental evidence indeed supported the occurrence of two photosystems with somewhat different pigment composition. The pigment assemblies of the two photosystems are embedded in the membranes of thylakoid and appear to be physically distinct in the membrane structure. Each thylakoid may contain several hundred of these photosystems, although the detailed arrangement is unknown.

(iii) *Energy absorption during light reactions.* How is the solar energy trapped by these pigments converted to chemical energy? Visible light in the form of electromagnetic radiations of wave-length 400 to 700 nm acts as if it were composed of particles of light called 'photons'. Each photon carries a definite amount of energy called a 'quantum'. A pigment molecule is normally in the 'ground' state, a stable low-energy condition. In this condition, electrons within the molecule tend to occupy the orbitals with least energy (closest to the atomic nuclei). When the molecule absorbs a photon, it becomes excited, i.e. lifted from its normal ground state of lowest energy to an excited energy-rich but unstable state. The light-energy is said to have been converted into excitation energy. This is the primary event of conversion of solar energy to chemical energy. The organized photosystems described above exploit the excitation energy of the pigment molecules in a useful way to perform photosynthesis.

(iv) *Cyclic and non-cyclic photophosphorylation.* The preceding discussion raises some important questions: What is the function of two photosystems? How does the solar energy trapped by the pigment molecules lead to the formation of high-energy ATP and NADPH molecules in the light reactions?

The events that follow the arrival of solar energy are rather complex. The light energy drives a flow of electrons along a system of well-defined carrier molecules. Electrons move spontaneously along this carrier chain and the high-energy ATP and NADPH molecules are synthesized by a process called photophosphorylation.

(v) *The dark reactions.* The dark reactions utilize the chemical energy of ATP and NADPH formed in the light-reactions to bring about the reduction of CO_2 to carbohydrates. These reactions do not require light and are often called the C_3-pathway or carbon-cycle or calvin cycle (after an American physiologist Melvin Calvin, who won a Nobel prize for his contribution in discovering the dark reactions). This pathway is almost universally present in photosynthetic plants.

Photosynthesis and respiration. The processes of photosynthesis and respiration can be represented by the summary equation given below:

$$6CO_2 + H_2O \underset{\text{Respiration}}{\overset{\text{Photosynthesis}}{\rightleftarrows}} C_6H_{12}O_6 + 6O_2 \quad G = 680 \text{ Kcal}:$$

The difference of bond energy of reactants and products is 680 Kcal (Kcal = 1000 calories). This energy is stored during photosynthesis and liberated during respiration. Despite the fact that the photosynthesis and respiration appear to be the reverse of each other, experimental evidence proves that this is not so. Even the site of the two processes in the cell is different; respiration occurs in mitochondria, while photosynthesis involves chloroplasts.

What happens to the respiratory process during daytime when active photosynthesis is taking place? It has been observed that plants do respire during daytime as well, although the type of respiration is not mitochondrial. This 'light' respiration in green plants is called 'photorespiration'. It short-circuits photosynthesis and utilizes the intermediates of carbohydrates generated during the dark reactions of photosynthesis.

Photosynthesis and transpiration. The entry of CO_2 for photosynthetic fixation is through the stomata in the leaves. The rate of transpiration affects the stomatal aperture, thus altering the CO_2 levels in plant cells.

Photosynthesis and nutrition. Plants depend upon photosynthesis for their nutrition. Glucose, which is formed during photosynthesis, is converted to other forms of carbohydrates such as starch. These carbohydrates are a means for the storage of energy and also provide skeletons for most of the organic compounds that make up the plants.

Photosynthesis the basic source of energy. Solar energy is the immediate source of energy for green plants and other autotrophs (organisms able to feed on simple substances) and also the ultimate source of energy for nearly all heterotrophic organisms. Solar energy is also the basic source of well over 90 percent of the total energy used by man today because coal, petroleum and natural gas are all the decomposition products of biological material generated by photosynthesis.

See also SOLAR RADIATION.

K.K.N.

PIDDOCKS are boring bivalves. They are widely distributed in all warm waters. They are called stone-borers but in fact they do not bore hard stones but only chalk cliffs, peat and hardened mud and clay deposits found along the extreme low-tide mark. Their foot and shell are admirably suited to dig holes in which they live. The edges of their mantles are largely closed and their siphons, which are long and united, can communicate with the water outside the hole. The foot is strong, stout, truncate and disc-shaped. The shell is long, broad and round anteriorly and gradually tapering posteriorly. The umbo, which is near the anterior end, is strengthened by one or two accessory calcareous plates. Although brittle, the shell is hard and its exterior carries stiff radial ribs traversed by concentric striae resulting in a hard prickly surface. The valves are white in colour and gape at both ends.

For **boring holes in** soft rocks the action of the foot is supported by the file-like surface of the shell and its strong and sharp accessory plates. The piddock rasps the rock by adopting a rocking and twisting movement which grinds the toothed ends of the shell into the rock surface. The burrows of piddocks are about 30 cm deep while the animals themselves are less than 15 cm. They remain within the hole, extending their bifid siphon just a little beyond the mouth of the burrow.

Pholas orientalis and *Pholas bakeri* (called Angel Wing) are the common Indian large-sized piddocks. The former is more common along the eastern and southern shores and the latter along the west coast. *Pholas* is known for its luminescence, or light-giving capacity. The whole respiratory siphon glows brightly, particularly during summer nights.

There is a smaller piddock resembling a small *Pholas* called *Martesia*. This mollusc, while very young, gets attached to some floating timber. There it makes a hole just as a *Pholas* bores mud or peat by rocking or twisting its shell into the wood. It remains inside the hole with its siphons opening outside communicating with water. Remaining in the timber, it is carried over the surface of the ocean by the waves and feeds on tiny floating organisms.

There is an entirely different kind of wood-boring bivalve, more dangerous than *Martesia*, called *Teredo*, also known as the Ship-worm. It has a greatly reduced shell at its anterior end. It enters wood as a larva. It remains and grows, digging a circular burrow with the sharp edge of its shell. An adult Ship-worm has no resemblance to a bivalve at all, not even to a mollusc. It has a long, worm-like body, at one end of which are two small valves and at the other end the two siphons united. The animal bores using the shell end of the body. The valves of the shell are capable of being rocked on two points on their inner surface. The rocking movement grinds the wood and makes a narrow hole. The wood ground and converted into sawdust through this operation is not thrown out, but it is not known

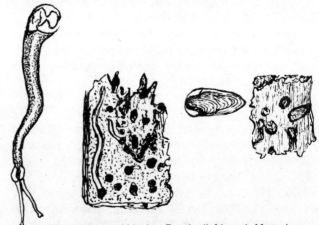

Wood boring piddocks, *Teredo* (left) and *Martesia*. The animals alongside the wood are about life size.

whether ship-worms feed upon the wood they excavate or not.

Anyway the damage the animal does to pilings and ships is very great. Before the days of steel, ships had their hulls sheathed with copper to keep out *Teredo*. Chemicals deter them but control is still a major problem. Cement concrete pilings have been substituted for wood at many landing stages.

See also MOLLUSCS.

PIGEON. The name is most commonly applied to the Blue Rock Pigeon, which in its feral form enjoys a worldwide distribution in urban environments. It is considered the main ancestor of the innumerable domestic varieties, some elegant, some bizarre, that have been artificially bred and are so highly esteemed by pigeon fanciers. It belongs to the family Columbidae which additionally includes the Green, Imperial and Snow pigeons, several species of dove, and the strangely beautiful ground-living *Caloenas nicobarica*, restricted to certain islands in the Andaman/Nicobar archipelago. The Blue Rock (*Columba livia*) and its congeners, as well as the doves (*Streptopelia* spp.) live almost entirely on fallen weedseeds, and cereals gleaned in harvested fields, while the staple food of the so-called 'fruit pigeons' (Green, Imperial, etc.) comprises drupes and berries in which wild figs (*Ficus* spp.) predominate. Pigeons' nests are loose sketchy platforms of a few crisscrossed sticks built in trees, on ledges or in holes of cliffs and buildings, or down shafts of old wells. Most species lay two pure white eggs, and both sexes incubate. In the early stages the nestlings are fed on a liquid secreted in the parents' crop known as 'pigeon's milk'; when the squab gets older this is mixed with predigested grain into a pap. Carrier pigeons produced by selective breeding from wild stock can be trained to fly long distances over predetermined courses. Apart from the popular sport of pigeon racing, this faculty has been utilized for carrying messages to distant parts. Before radio communication came into general use in India, a regular pigeon-post section was maintained by many police administrations for rapid communication between headquarters and remote outposts lacking roads and telegraphs and many awkward situations were averted by this means.

PIKAS. Pikas, variously called Rock Rabbits or Mouse Hares, resemble small guinea-pigs or large hamsters and are quite unlike rabbits, their nearest relatives. Pikas are about 20 cm long and adults weigh 200 to 300 grams. They are tailless and have mouse-like ears and hind feet. Upon close examination, however, they do have characteristics of the order Lagomorpha to which rabbits and hares belong: i.e. a second pair of upper incisors hidden behind the first pair, enamel on

© T. J. Roberts, *Mammals of Pakistan* (Benn)

Afghan or Collared Pika × ½

all sides of the incisors instead of only on the front as in rodents, lack of a baculum or penis bone, and many skeletal features typical of the order. Pikas have soft dense fur which is moulted twice a year. They have a gentle, inquisitive nature and make fine pets.

All species of pikas (genus *Ochotona*) are closely related and are difficult to distinguish from each other. The most recent studies indicate that there are 14 species, all found only in the mountains of Asia except one which has also spread to North America. Because they live in remote areas, most pikas have been little studied. In the Indian region there are thought to be five species, only two of which, *Ochotona rufescens* in Baluchistan, Afghanistan, and Iran; and *Ochotona roylei* in northern Pakistan, Kashmir, Nepal, and northern Burma, are widespread.

Most pikas pile up large heaps of green vegetation which sustains them through the cold winter. These hay piles and the area around them are defended by a single animal who makes loud whistles to announce ownership. In the spring, territories are not defended and vocalizations are used to warn other pikas of the approach of a predator. *Ochotona rufescens* is relatively silent.

In Baluchistan, pikas breed from late March through September. They may have from 4 to 10 young per litter and females can breed in the summer of their birth. Thus, under favourable conditions, pikas can become very abundant. In some places, they are a pest. They can kill apple trees by removing the bark in the winter and damage wheat, potatoes, and garden greens in the summer.

G.W.F.

PINE (*Pinus*). There are six species in our area (see CONIFERS), four are Himalayan (*P. roxburghii*, *P. wallichiana*, *P. bhutanica* and *P. gerardiana*); the fifth, *P. kesiya*, occurs in the Khasi, Naga and Manipur hills and Burma; the sixth, *P. merkusii*, of Burma, Sumatra, etc. is the only pine in the world that crosses the equator.

They are resinous trees and valuable for their timber when treated. They make good railway sleepers; and are also used for packing cases, furniture, etc. Chir, blue pine and khasi pine produce resins of commercial importance, while the chief value of chilgoza pine lies in its edible seeds. *P. roxburghii* (chir) occurs all along the Himalaya from Pakistan to Arunachal Pradesh at 450–2300 m in the outer ranges, where the full force of the monsoon is felt: In the Kashmir valley proper it is absent. It is characterized by needle-like long leaves in bundles of 3; cones 10-20 cm long, 7–13 cm broad with thick woody scales, the scales have a pyramid-shaped, pointed or curved beak.

P. wallichiana (blue pine), occurs all along the Himalaya from Pakistan to Arunachal at 1800–3700 m. It has five long needles, with a bluish tinge. The cones are banana-shaped with thin scales, closely adhering to the cone-axis like the scales of a fish.

P. bhutanica, also a five-needle pine, is new to science having been discovered in 1977 from Arunachal and in 1978 from Bhutan. It is allied to blue pine, differing in the anatomy of the needles.

P. gerardiana (chilgoza). It occurs from Bashahr (Himachal) westwards to Kashmir, Chitral and Baluchistan at 1800-3000 m in inner dry valleys where precipitation is in the form of snow with very scanty rain. It has three stiff needles. Its cones have very thick woody scales. It has cylindrical, edible seeds.

P. kesiya (Khasi pine) grows in the Khasi, Naga and Manipur hills and in Burma. It has three soft needles and cones 7·5 × 5 cm. It is a fast-growing pine.

P. merkusii (Merkus pine). This Burmese pine growing between 150 and 750 m is characterized by two needles and cones 5–7·5 cm long.

PIPITS AND WAGTAILS. The pipit and wagtail family (Motacillidae) comprises several species of slender,

Yellowheaded Wagtail × ¹/₂

sparrow-sized, chiefly ground-living birds of open country and marshland, with longish tails which are constantly 'bobbed'. Pipits (genus *Anthus*), of which our resident Paddyfield Pipit is typical, are sober-coloured brown birds rather like a lark or female house sparrow. Several species are winter visitors to India. They are difficult to identify without experience, but some have diagnostic calls and/or habitat preferences. Wagtails (genus *Motacilla*), of which our resident Large Pied is an example, are superficially similar to pipits but brighter and variously coloured in combinations of black, white, grey and yellow. Most species are winter migrants from Palearctic countries. Many, e.g. Yellow wagtails, are difficult to identify except in their distinctive summer (breeding) plumage. Pipits and wagtails feed on insects and worms on the ground. Nearly all are ground-nesters too.

PIROTAN is one among thirty or more coral and mangrove islands along the northern coast of the Saurashtra peninsula, between Jamnagar and Okha, in the Gulf of Kutch. The best known is perhaps Beyt island near Okha, visited for its temple.

At high tide, Pirotan some 14 km north of Jamnagar, is a horseshoe spit of sand enclosing a mudflat and a mangrove swamp, barely a kilometre across. The largest of these islands is Karumbhar, several square kilometres in area with fields cultivated during the SW monsoon. Chank and Bhydar are the two most prominent reef islands at the mouth of the Gulf. The remainder of the islands are separated from one another by narrow, winding creeks which form a maze of waterways, and are largely mangrove swamps. There are lighthouses on Pirotan, Karumbhar and Chank islands.

The exceptionally high range of tidal variation makes many of the islands awash by the tide waters, but at low tide it is possible to lead camels on to many of the islands for browsing on the mangroves. Huge areas of tidal mud and fringing coral reefs are exposed providing extensive and rich foraging area for the myriads of storks, ibis, spoonbills, herons and egrets, cormorants, pelicans and flamingos. Millions of shorebirds winter around these islands as do immense flocks of Lesser Flamingos. Storks, herons, egrets and cormorants nest on the larger mangroves.

White mangrove (*Avicennia marina*) forms the dominant forest tree. There are species of *Rhizophora* and *Choreopsis* in more sheltered locations. The terrestrial flora is composed of *Salvadora persica*, *Euphorbia neriifolia*, *Acacia senegal* and *Commiphora mukul*. The more stable sand dunes are colonized by xerophytic sedges and grasses. The excellent forest has been considerably exploited by professional charcoal-burners, camel graziers and fuel and fodder collectors.

The coral reefs forming the most extensive shelf coral

447

on the west coast have drawn science groups from the entire country for ease in studying and collecting marine life and a rich variety of marine algae. Careless human activity, largely the destruction of mangroves since 1947, has resulted in high siltation and pollution of the Gulf resulting in rapid death of the living reefs. In an attempt to restore the mangroves and the coral the Gujarat Government in 1981 notified the entire coastal area as a Marine Sanctuary with several of the islands designated as a Marine National Park. This action will extend protection to major Green Turtle-nesting islands like Bhydar.

PLANT ASSOCIATIONS. The general trend of plant successions, whatever their origin, is towards a characteristic climax type of vegetation, the nature of which is, to a great extent, determined by climate over large areas. The main climax regions, also known as *plant formations*, which are obvious as each has a distinct physiognomy, are the major natural vegetation types of the world exemplified by tropical rain-forest, coniferous forest, desert, tundra, etc. Each of these vegetation types predominates as climax over large areas which, even in different parts of the world, have obviously similar climates. A plant formation may thus be defined as one of the largest subdivisions of vegetation, usually of great geographical extent, composed of climax communities of the area that are similar in major physiognomic features and broad environmental parameters. The formation is thus a product as well as an indicator of climate whose inter-acting factors must be broadly equivalent throughout the range of similar vegetation. Although individual factors may differ, their biological effectiveness must be similar.

While the growth form of a formation is uniform and binding species are included among the dominants throughout, there are floristic variations that are normally to be expected. Again, because of the large areas involved, climate varies somewhat, and accordingly, species will assort themselves in different combinations of relative importance because they are not all identical in their responses or adaptations to environment. Because of the large extent of a formation, it may well include areas of different geological history and thus time is an additional factor for floristic variation. Therefore, within the range of a formation different climax communities are to be expected.

The climax vegetation of plant formations is, thus, not uniform, but made up of a number of well-defined smaller units called *associations* which are characterized by particular species which dominate them. An association is, therefore, one of the climax units of a formation, floristically and usually geographically distinct, but physiognomically similar and ecologically related to other units of the formation. It is a distinct unit of a formation characterized by two or more dominant species peculiar to it. If there is but a single dominant species, the unit is called a *consociation*.

Just as formations are made up of associations, so are further subdivisions recognizable. A subdivision of an association characterized by a particular group of dominant species (two or more dominant species but less than the total number of dominants in the association) and determined by minor climatic variations is termed as a *faciation*. Faciation may be further subdivided into *lociations* which are local units characterized by a specific local abundance and grouping of dominants of the faciation. A localized community within a consociation or an association which is characterized by a single subdominant species is called a *society*.

Units of the rank of association and consociation can be recognized not only in climax vegetation but at successional stages as well. Accordingly the developmental or successional equivalent of association is termed as *associes*, i.e. a seral community with two or more dominants. Similarly the seral equivalents of consociation, faciation, lociation and society are termed as *consocies*, *facies*, *locies* and *socies* respectively.

The system of classification just mentioned uses the association, a climax community, as the basic unit of vegetation and is developed by dividing and subdividing this major unit into lesser categories. Because climax communities are the basic units, the classification incorporates succession into its every interpretation of vegetation. Unfortunately, the term association is used in different systems of vegetation classification to express entirely different ideas. European ecologists use several systems of classification in which associations are basic units, which can be grouped into successively higher categories. Their associations are the simplest recognizable units and, therefore, the lowest ranking. This concept is, of course, in absolute contrast to the definition of association mentioned earlier, which is a community of the highest rank and divisible into numerous lesser categories. The two divergent uses of the term have become so well established and so widely used that there is little hope of clarification in the near future.

See also CLIMAX VEGETATION.

O.N.K.

PLANT DEFENCES. Like any other living organism plants protect themselves. Their enemies are grazing animals, seed-eating birds, butterflies and other insects that lay their eggs on or under leaves, gall-flies, fungi and many grubs and bacteria that attack roots, stems and leaves. Most plants multiply vigorously through their seeds and can survive extremes of heat or cold. They usually also survive attacks from animals, for if their leaves or buds are eaten they develop new ones, in the same way that a garden plant grows better after pruning. Given air, water and nutrients plants have the capacity to withstand most enemies, but they also have certain defence mechanisms.

Outgrowths on the surface of plants, such as prickles, thorns and spines, are common on roses, babul and cactus. Some of these may develop in response to dry climatic conditions as in some xerophytes such as cactus and euphorbias. In most plants spines develop, no matter what the conditions under which they grow. In a cactus, the flat stem performs the functions of a leaf, while the spines are leaves modified to reduce transpiration.

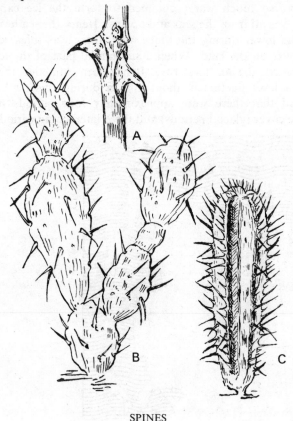

SPINES

A Rose × 1, B Cactus × 1/4, C Euphorbia × 1/2.

Some plants are protected by hairs (see NETTLES). Some protect themselves by physiological reaction (see SENSITIVE PLANT). Chemical substances such as formic acid are secreted in some plants, which cause local inflammation and swelling. Many plants, including mint, eucalyptus, nightshade and conifers (pine and cedar) yield aromatic oil, turpentine and tannins which deter animals from browsing.

An exceptionally hard surface, as in oleanders, also protects plants from browsers. Some plants have evolved special means of augmenting their nitrogen supply. One way is to have a partnership with nitrogen-fixing bacteria in the roots (see NITROGEN IN PLANTS).

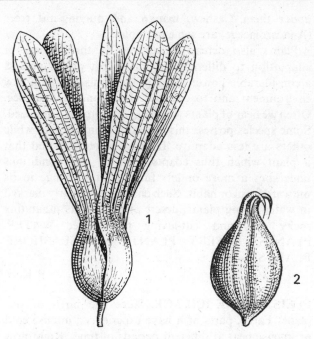

1 Winged seeds of sal (*Shorea robusta*) and 2 spined seeds of *Martynia annua* showing mechanism for dispersal and defence. Both × 1.

Members of several families of flowering plants capture and digest small insects as a source of nitrogen. Some insectivorous plants live in streams and ponds, others grow in waterlogged soils where most of the nitrogen is bound up in unavailable forms. Bladderwort is an insectivorous plant having small bladders scattered along the finely dissected leaves. The bladders have a hinge mechanism that swings inwards to admit tiny insects or small animals but does not swing outwards until the prey has been digested. Different kinds of pitcher plants (*Nepenthes*) have different odours and attract insects. The leaves of sundew (*Drosera*) are covered with stout spreading hairs that have a swollen, shining, sticky tip bearing protein-digesting enzymes. If any of the hairs are touched, the hairs near the margin turn towards the centre of the leaf, often trapping the insect (see INSECTIVOROUS PLANTS).

Certain plants are poisonous and cause dermatitis and irritation of the skin. The chemicals in plants include growth hormones, which may cause allergies, and secretions from glandular hairs as well as secretions following injury (balsams) which contain cutaneous irritants and sensitizers. The content and quality vary with the season and the area where the plant grows. Essential oils are notable substances responsible for smell, taste and fragrance and may be found in flowers (e.g. jasmine), fruits (citrus fruits), leaves (eucalyptus), bark (cinnamon or camphor) and wood (sandal). Oleoresins, resins, glucosides, terpenes, are also found. Certain plants cause dermatitis by contact, so that sensitive individuals do not touch them or avoid passing under them. Cashew, mango and purging-nut trees (Anacardiaceae) are some examples.

Plants also defend themselves by their power of adaptation to different habitats. Most plants possess a considerable power of adapting themselves to a new environment and to changed conditions of existence. Often we hear of plants which have become acclimatized. Some species possess this power in a high degree while others are less adaptive. It must be remembered that a plant which thus adapts itself to new conditions undergoes a more or less fundamental change in its outward form or habit. Such changes are clearly marked in water-loving plants, desert plants, plants inhabiting rocky sites and salt-loving plants (see WATER PLANTS, DESERT PLANTS and SEASHORE PLANTS).

R.K.G.

PLEISTOCENE ICE AGE. Since the birth of our planet Earth, parts of it have experienced intense cold or intense heat at different periods of time. Right now it is known that it is very cold in parts of Greenland and Antarctica, whereas it is quite hot in the deserts of Sahara (northern Africa), Saudi Arabia (West Asia), and Western Rajasthan in India. There have been cycles of climatic change due to which the same place in different successive times could have undergone varying types of climate.

Geologists have found that about one million years ago about 30 percent of the land was covered by an ice sheet. Now the cover extends over 10 percent only. Since this cover was over a part of the continents, it has been called Continental Glaciation, and the period popularly called the Great Ice Age. Wherefrom was so much water obtained to form the ice cap? It was all from the seas and oceans. Hence the sea level was lower during the times when there was thick ice cover on the land. When a substantial part of the ice melted, the sea level rose. It has been estimated that sea level fluctuated about 100 m during this period and that there were approximately four periods of ice cover (glacial periods) and three intervening periods

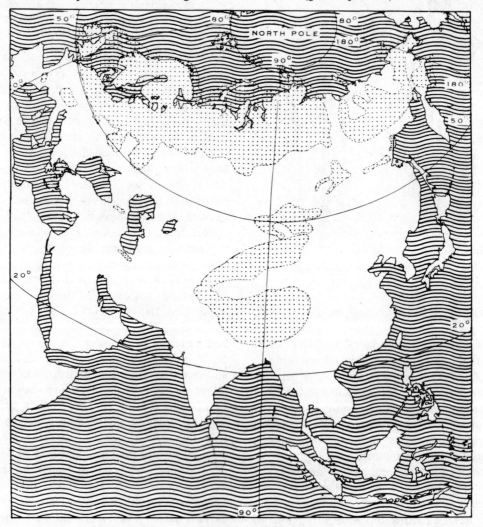

Approximate maximum extent of ice sheets in Eurasia about 30,000 years ago. Today only glaciers survive in mountains of Himalaya, Outer Mongolia, the Caucasus, Alps and Pyrenees.

of less ice cover (interglacial periods) over a period of a million years. Over Kashmir five periods of ice advance and four interglacial periods have been recognized.

What was the effect of these changes on land and sea? Vast areas in north America and the northern part of Eurasia were covered by ice and the moving ice mass carried along with it much rock debris and deposited it *en route*; these masses of debris are called moraines. Besides there were many glaciers over the mountains in other parts of the world as well (Rockies, Andes, Alps, Himalaya). The snow-line came down to a much lower level than what it is now. For example, the snow-line in the Himalaya is usually around 5000 m above mean sea level now, but during the Pleistocene it came down to 1700 m above mean sea level. This clearly indicates that a lot of ice has melted away since the Pleistocene: Another conspicuous change is along the coasts. Many exposed marine terraces indicate higher sea levels. Such terraces are present in parts of Gujarat and Tamil Nadu in India. Some areas have been drowned as some of the drowned valleys of the rivers along the west coast of India indicate. There are a few dried-up lakes with huge terraces in them like the Karewa in Kashmir. Similarly during this period there was a vast accumulation of river deposits, now seen as terraces at different levels along the Narmada and Tapti rivers in Central India.

Man had already come into this planet during Pleistocene times and he was a witness to these events, though he was in very few numbers scattered over different parts of the world, particularly in Asia and Africa. Some of his crude stone implements found within the deposits of this period indicate this. The animals that roamed over the land were very similar to some of our modern animals living in warm areas, with slight variations in size and a few parts of the body (rhinoceros, hippopotamus, elephants, horses etc.).

Is the Great Ice Age over or are we passing through an interglacial epoch? Scientists differ. Only time can tell.

R.V.

PLOVER. Plovers are short, thick-billed shore birds with long, pointed wings. They generally feed along the tide line, on coastal mudflats and other marshy areas usually in association with sandpipers, stilts and numerous other species, mostly migratory. Their food is mainly molluscs, crustaceans, insects and worms.

Plovers are strong fliers. On the slightest suspicion the flocks fly off *en masse*, turning and twisting in orderly formation to resettle some distance away.

The Little Ringed Plover (*Charadrius dubius*) is typical of the family. It is sandy brown above and white below, with a distinct black and white head, a prominent black collar round the neck and yellowish legs. The Red-wattled Lapwing (*titiri* in Hindi) is a resident plover, and perhaps the most common and widely known from its accusing 'Did you do it?' calls.

See plate 10 facing p. 129.

POISONOUS FISHES do not contain poison, as such, in their organs but consumption of some parts of their body, such as the liver, roe, intestine, skin, or brain tissue, creates some disorders in the human system and hence they are considered poisonous. By far the most notorious poisonous fishes found on the Indian coasts are the PUFFER fishes. Poison from these fishes is found in the skin, liver, ovaries and intestines. However, the flesh, if properly cleaned, is not poisonous, and is especially relished by the Japanese, who call the puffer fish *fugu*. Some kind of toxins are contained only in the roe of some fishes, the flesh and other organs being edible. This is so in a wide variety of fishes, such as sturgeon, garfish, trout, pike, and some minnows, catfish and perches. Indian fishes known to cause poisoning are three species of *Schizothorax* (freshwater fish found in Kashmir and Assam), the marine catfish *Plotosus anguilaris*, and sculpin (*Scorpaenichthys*) of the family Scorpaenidae. The concentration of the poison increases as the ovaries mature in the breeding season. Cooking is said to destroy the poison.

Fresh blood of freshwater eels (*Anguilla bicolor* and *A. bengalensis*), morays (*Muraena* and *Gymnothorax*) and snake eel (*Pisodonophis boro*), is also known to contain some poisonous substances. Similarly the brain tissue of rabbit fish (*Siganus gramin*), surgeon fish (*Acanthurus triostegus*), sergeant-major (*Abudefduf septemfasciatus*), goatfish (*Upeneus*) and groupers (*Epinephalus*) gives rise to hallucinations when eaten but is not fatal. Some fish become poisonous only when they feed on certain microscopic organisms (dinoflagellates).

As against poisonous fishes, venomous fishes do contain venom in their tissues which they inject into the blood circulatory system of the victim but its utility to the fishes themselves is not yet quite clear. The most venomous fish in the world, also found occasionally in our seas, is the Stone-fish (*Synanceia horrida*). A master of camouflage, it rests motionless on the sea bottom where it looks like a seaweed-covered stone. If trodden upon, the venom contained in the hollow spines of the dorsal fin enters the foot and is fatal.

More common poisonous forms are the Sting-rays (*Dasyatis*) of our coastal waters, both east and west. These have a whip-like tail, at the base of which are one or more saw-toothed spines. The sting-ray rests on the sea bottom covered with mud. When disturbed, the tail is lashed from side to side like a whip and the spine inflicts a very painful jagged wound.

The spines on the back and gill-covers of several fishes, such as the turkey (or lion) fish (*Pterois*), surgeon fish (*Acanthurus*), rabbit fishes (*Siganus = Teuthis*), trigger fish (*Balistes*), rock catfish (*Plotosus*), also contain some venom but although the wounds inflicted by the spines are quite painful, the venom is not very potent.

See also SHARKS & RAYS, DOCTOR FISH.

POLLINATION. The stamens are the male reproductive structures of the flower. A stamen in general shows a distinction into a slender axis called the filament which bears at its apex a 2-lobed anther. The anther at maturity has four chambers inside in which the pollen grains are found. By the splitting of the anther lobes or through openings formed at the apex, the grains are released. The pollen grains contain within them the male reproductive cell. The pistil is the female structure and has at its base an enlarged portion, the ovary, which contains the ovules. The ovules have within them the female reproductive cell, the egg. The ovary is topped by an elongated style whose apex is organized to function as the receptive surface for the pollen grains. This is the stigma. The pollen grains have to reach the stigmatic surface for bringing about sexual union. This transference of the pollen from the anther to the stigma is called pollination. In some bisexual flowers the anthers and the pistil come to maturity at the same time, in which case the pollen grains can germinate on the stigma of the same flower. This is self-pollination; which is not considered beneficial to the plant as it brings about weakening of the progeny in succeeding generations. During the course of evolution, plants have adopted themselves progressively to the transference of the pollen from one flower to another flower, preferably borne on another individual. The cross-pollination, as it is called, is of advantage to the plants because it results in new combinations of genes and the production of healthier and superior progeny.

Since the movement of the pollen over a distance is involved in cross-pollination, there is the need for an external agent to perform this transfer. In the case of plants like birches, willows, sedges and grasses which have very light, dry pollen, on release from the anther the grains are wafted away by wind and incidentally reach the stigma. Since this is a chance phenomenon, there is usually enormous pollen production in these plants and the stigmatic surfaces are also modified variously to catch the pollen. A good example is the 'silk' of the corn plant (*Zea mays*); the long filamentous structures forming a silky tuft at the opening of the sheath covering the female inflorescence are actually stigmas. Plants growing in water have the pollen carried by currents of water to the stigmatic surfaces. Here again, curious adaptations are seen in the floral structure to facilitate pollination.

The large majority of flowering plants, however,

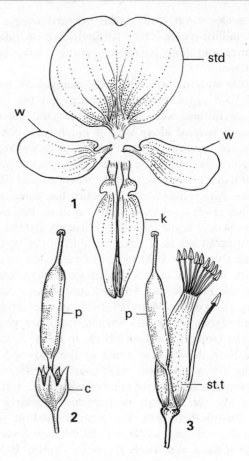

THE PEA FLOWER

1 Papilionaceous corolla with the petals spread out: std standard petal; w wings; k keel (from above, the two petals are fused along their anterior margin. 2 Calyx (c) of fused sepals and single-carpelled pistil (p). 3 Reproductive structures, androecium with stamens in two bundles (9,1), the staminal tube (st.t) opened out in diagram and the single-carpelled pistil (gynoecium) on stalk. *All figures diagrammatic.*

depend on insects for bringing about this transfer of pollen. In the course of evolution of the flowering plants, what began as an incidental visit of an insect to the flower to feed on the soft parts, gradually evolved into an intimate relationship between the insect and the flower. The rapid spread of the flowering plants in Tertiary times has been correlated with the profuse development of the insect fauna during the same period. This mutually beneficial association between the insect and the flower has resulted in an enormous diversity of the form and structure of the flower as well as of the insect. The insect during its visits to the flowers, attracted by the coloured petals or other parts and in search of nectar, comes in contact with the mature anthers and the pollen is carried on the body of the insect to the stigma of another flower. Many flowers show curious adaptations to facilitate insect visits. The pea flower, with its large standard petal (for attraction), the wings (for the insect to alight) and the two keel petals fused along one margin to form a boat-shaped structure containing the essential organs, is an example (Figs.1–3). The insect in its effort to reach the nectar brushes against the stigma or the anthers as the case may be. The bi-lipped flowers of the Mint family, the highly modified flowers of the Orchids are other examples. This relationship has progressed to such an extent that in some cases only a particular insect species can act successfully as a pollinating agent for a particular plant species. Both the insect and the flower are modified in their structure for this. If, by some chance, one of them, the flower or the insect, is absent, the other cannot survive. One of the orchid species (*Ophrys*) mimics the female of an insect species in coloration and body structure; the male of that insect indulges in pseudo-copulation, incidentally bringing about pollination. Bats and birds are also known to pollinate some flowers.

When the pollen reaches the stigma, if there is no barrier or incompatibility, it germinates producing a tube (pollen tube) which grows through the style and reaches the ovular region inside the ovary. The pollen tube enters the ovule through a narrow opening (micropyle) in the ovular coat (integument). The fusion of the male cell brought by the pollen tube with the egg within the embryo sac of the ovule results in the formation of a zygote, the beginning of a new generation. A second male cell brought by the pollen tube fuses with another nucleus in the embryo sac and initiates the development of a nutritive tissue (endosperm) for the developing embryo; this is a unique feature of the Angiosperms. The ovule, in due course, develops into the seed and concurrently the ovary is transformed into the fruit. The seeds are released when the fruit ripens and the seeds germinate to commence a new generation. There are many variations of the scheme and events described above, but essentially, sexually reproducing species go through the above succession of events in their life-cycle.

See also FLOWERS, FLOWERING PLANTS.

M.A.R.

POLLUTION. Like 'environment', pollution is a word which has acquired special significance in recent years. Formerly, when water was polluted by refuse of one kind or another, the damage was not too severe, and was seldom irreversible. Now, because of the creation of many synthetic chemicals which are not biodegradable, the situation has radically altered. When non-biodegradable effluents find their way into a river or the sea or into the air untold damage can be caused to various species of life, to the ecosystems of which they are a part, and to the environment in general.

It is for this reason that legislation has been passed to treat the effluents prior to their emission into air or their discharge into water. Two important Acts which have been passed for fighting pollution in India are the Water (Prevention and Control of Pollution) Act, 1974, and the Air (Prevention and Control of Pollution) Act, 1981.

To check pollution, action is required on a very wide front and this has to be the responsibility of the country at large. All citizens must cooperate to ensure that the country remains healthy and clean. The Department of Environment can only establish guidelines. Action has to be taken by everyone.

POMFRETS (Stromatidae) are the most popular edible fish in India and fetch the highest price in Bombay markets, their popularity stemming from the clean white appearance of the body, firmness of flesh and flavour. Three species, Silver pomfret (*Pampus argenteus*), Grey pomfret (*Pampus sinensis*) and *Parastromateus niger*, the Black pomfret, are commonly met with on the east and west coasts of India. The black pomfret has been placed into a separate family but it continues to be associated with the category of pomfrets in common parlance.

Silver pomfret, the commonest of the three, is somewhat rhomboid in shape, has a blunt nose, a single long dorsal fin and a caudal fin with long lobes, the lower caudal lobes being much extended and pointed. The lobes shorten as age progresses. Adults are silvery white with small silvery scales, but when about 5 to 8 cm they are black, and lightly striped when smaller. Though their average large size is about 1 kg and 30 cm long, the largest reach a length of 40 cm and weigh 2 kg. Being an offshore species, they normally prefer 20 to 40 fathom depth and remain in the middle column of water but nearer the bottom. Their first spawning period is calculated to be July and August, with a second in November and December, but where they spawn and their early life history are not known so far. Similarly, though they are known to take small pelagic protocordates like *Salpa* and *Doliolum*, more detailed study of their food has been rendered difficult because of a thick-walled stomach in which the food items are crushed to minute unidentifiable pieces.

The Grey pomfret is broader, heavier and larger than the silver pomfret and has shorter caudal lobes. It is greyish, as the name suggests. It is more common in the Gulf of Kutch than southwards and is much more in demand in the market. The Black pomfret is more elongated than the silver one and larger in size, the largest weighing 5 kg and being 60 cm long. Its colour is actually dark grey. Its tail region (caudal peduncle) is slender, hard and covered with thickened scales (scutes).

See plate 2 facing p. 33.

PONGAM OIL TREE, *Pongamia pinnata*, a tree with a spreading crown, wild in the Western Ghats and coastal areas of the Andaman and Nicobar islands. It has compound leaves, creamish-white flowers, woody, egg-shaped, beaked pods enclosing one, or rarely two seeds. The seeds yield an oil which is used as an illuminant and as a lubricant for gas and diesel engines.

Pongam Oil Tree. Leaves, inflorescence, and fruit × ²/₃.

POON TREE, *Calophyllum inophyllum*, a moderate-sized evergreen tree in coastal areas of South India,

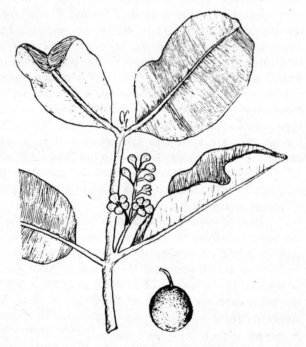

Poon Tree. Leaves, flowers and fruit, × ¹/₃ Characterized by shining leaves with closely parallel veins at right angles to the midrib.

Andaman and Nicobar islands, Burma and Sri Lanka, extending into islands in the Pacific. The genus is easy to spot in the forest provided one can obtain the leaves. This is difficult as the trees are very lofty, but leaves can be picked up from the forest floor. Their leathery texture, rounded tip, combined with numerous, closely parallel nerves distinguish this genus from all others. Fragrant white flowers appear in the cold season. The fruits are globose, yellow in colour and ripen in the summer. The wood is used for fishing-boats. A resinous gum exudes from the bark, called tacamahca in Tahiti where it is used as a perfume.

POONSPAR TREE, *Calophyllum elatum*, a lofty tree of great girth (45 m tall, 4·5 m girth), found in the evergreen forests of the Western Ghats and Sri Lanka. It is like the Poon but differs in its greater height, smaller leaves and bigger egg-shaped fruits with a pointed tip. It is called the Spar Tree because hundreds of years ago the Moors used to come to India to get this tree for masts of their dhows. The price paid was the number of rupees laid edge to edge along the length of the spar.

PORCUPINES. Porcupines (Hystricidae) are the largest Asiatic rodents characterized by long quills. Adult animals may weigh up to 18 kg. Three species occur in the subcontinent. The most common is the Crested Porcupine, *Hystrix indica*, which is distributed from Kashmir to Cape Comorin and Sri Lanka. In the west it occurs in the Middle East, up to Israel. *H. hodgsoni*, the Crestless Himalayan Porcupine, is found in Nepal, Assam, Burma, Thailand, Kampuchia and northwards in parts of China. The Brush-tailed Porcupine, *Atherurus macrourus*, found from Szechuan in China to Malaysia, also occurs in Assam (Khasi Hills).

H. indica favours rocky hillsides but is also adapted to live in burrows in the sand-dunes in the desert. It is strictly nocturnal, venturing out only well after dark.

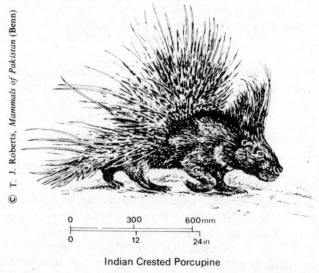

© T. J. Roberts, *Mammals of Pakistan* (Benn)

0	300	600mm
0	12	24in

Indian Crested Porcupine

Usually they live in pairs but sometimes singly. Their burrow has a number of openings and the tunnels are very long. The animal moves along regular runs or pathways. It is herbivorous, feeding on tree bark, roots, bulbs, tubers and fruits. At certain places it is very destructive to plantation and tuber crops. In the wild state, breeding of this rodent is reported during March and April but in the Jodhpur and Bikaner zoos, the female littered from March to December. Peak reproductive activity was observed during the monsoon and in December. The litter size varied from 1 to 3. The interval between two deliveries was 109 days.

Porcupine is well protected due to its spiny armour and it is known to tackle larger carnivores. However, it is preyed upon by panther, hyena and jackal.

I.P.

See also RODENTS.

PORTIA TREE, *Thespesia populnea*, is an umbrella-shaped tree in the coastal forests of south India, Andadamans, Bangladesh and Burma. It has heart-shaped leaves, their underside covered with scurfy scales. The Hibiscus-like flowers with a maroon centre open yellow and turn purple by nightfall, when they close. It is a tree often planted in avenues.

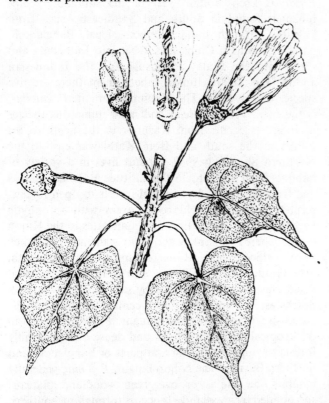

Portia Tree. Flowers (one with petals removed to show the inside), fruit and leaves (× ½). The poplar-like leaves explain the specific name. It is common in the Indian Ocean region, where the timber is known as Seychelles Rosewood, as its heartwood is reddish.

PRIMATES. Members of the order Primates are the most highly evolved of the mammalian fauna and are best identified by a combination of characteristics rather than by one or more unique or exclusive ones. Since all primates have evolved from arboreal ancestors, traits associated with arboreality are a common link. A feature of arboreality is prehensile hands and feet, with thumb and big toe opposable to the other digits. Primates have five digits on both hands and feet, each tipped with a flat nail, except for some families that have a claw on one or more digits. Arboreality has led to the reduction of olfactory abilities; the snout is reduced and the face relatively flattened. Both eyes look forward and have developed stereoscopic vision. Other evolutionary trends include an increase in the size and complexity of the brain, especially the cerebral hemispheres. The body is held more upright than other quadrupeds', and facultative bipedalism has developed. Postnatal development is prolonged. Primates have a complete bony rim surrounding the orbit. The dentition is greatly reduced from the primitive mammalian formula. The clavicles are well developed and the stomach is generally simple. Primates range in size from less than one kilogram in weight (e.g., South American marmosets) to massive animals weighing over 350 kg (e.g., Asian Orang-utans).

There are 15 species of primates in the Indian sub-continent. Two are lorises: the Slender loris, *Loris tardigradus*, and the Slow loris, *Nycticebus coucang*, of the family Lorisidae. The Hoolock or whitebrowed gibbon, *Hylobates hoolock*, of the family Hylobatidae, is the only ape found in the region covered by this encyclopedia. All other species belong to the family Cercopithicidae, which is subdivided into seven species of macaques (Cercopithecinae) and five species of langurs (Colobinae). The tail of these species may be as long or longer than the head and body combined, as in some macaques, such as *Macaca sinica*, and all of the langurs, *Presbytis* spp.; small and stunted as in *M. arctoides* and *M. nemestrina*; or altogether lacking as in *H. hoolock*.

Lorises

Loris E. Geoffroy, 1976, and *Nycticebus* E. Geoffroy, 1812, of the family Lorisidae, have dissimilar distributions. The Slender loris, *L. tardigradus*, is found in Sri Lanka and parts of southern India north to the Tapti river. It lives in tropical rain-forest, open woodland and swampy coastal forest, as well as evergreen forest up to 1850 m in elevation. In contrast, the Slow loris, *N. coucang*, lives in tropical rain-forest of northeast India (Assam, Nagaland), extending into Sylhet in Bangladesh, and east into the Malay peninsula and parts of Indonesia. Both lorises are characterized by their small size, large eyes for nocturnal vision, and hands and feet with well-developed thumbs and big

Slender Loris × ¹/₃

© M. Krishnan

toes that are prehensile and opposable to the other digits. The second toe is clawed, in contrast to flat nails on the other digits.

Lorises are nocturnal, sleeping during the day in shady, inaccessible places in tree hollows or on densely vegetated branches. They are mostly arboreal and seldom come to the ground. Both species are mainly insectivorous, supplementing their diet with fruit, shoots, young leaves, birds and eggs and lizards. They forage for prey slowly and stealthily, capturing it with a lightning grasp of both hands. They live singly or as monogamous pairs; and because of their habit of extensively scent-marking their space, may be territorial, but this is not established. Both species have a rich repertoire of vocalizations, *L. tardigradus* being the more vocal species.

Macaques

Macaca Lacépède, 1799, of the subfamily Cercopithecinae, is found in North Africa and South and Southeast Asia—from eastern Afghanistan through Tibet to China, Japan and Taiwan, south to India and Sri Lanka, east through parts of the Indonesian archipelago and the Philippines. In the Indian subcontinent there are seven species of *Macaca*, ranging from sea-level to altitudes of 3140 m. *M. arctoides*, the Stump-tailed macaque, lives in dense forest in the higher latitudes of northeast India (Assam, Nagaland). It is now rare in Assam and uncommon throughout much of its range, which extends into the Malay peninsula. The Assamese macaque, *M. assamensis*, also inhabits dense mountainous forest from Uttar Pradesh through Nepal, Sikkim, Bhutan and Assam to the Sundarbans and adjoining countries. The Rhesus macaque, *M. mulatta*, is the most common and widely distributed macaque, confined to the Himalayan regions of Pakistan, in India it occurs in Assam and north and central India as far south as the Godavari river in the east and the Tapti river in the west. It is exceptional among

the macaques by living in close association with humans around villages and temples. The Pig-tailed macaque, *M. nemestrina*, inhabits the same forests as *M. arctoides*, but is more arboreal and keeps strictly to dense evergreen forest. In contrast to the sociable *M. mulatta*, *M. nemestrina* is extremely wary of humans, because it is hunted for food and for medicinal purposes in many parts of its range. The Bonnet macaque, *M. radiata*, is the common monkey of southern India, occupying most of peninsular India north to the Krishna and Godavari rivers. It is more common in rural and suburban areas than in the interior of forests. The Lion-tailed macaque, *M. silenus*, is very rare, and its numbers are constantly being reduced with the steady deterioration in quality and quantity of its only habitat, the tropical rain-forests of the Western Ghats. Wild populations probably do not exceed 1000. It apparently has been totally eliminated in its northern range and is now confined to Kerala and scattered areas in the hills of southern Tamil Nadu. The Toque macaque, *M. sinica*, is confined to Sri Lanka, where it occupies a variety of habitats from lowland mixed deciduous and evergreen forests to urban settlements and temples.

Langurs

Presbytis Eschscholtz, 1821, of the subfamily Colobinae, is found in South and Southeast Asia—from Pakistan through India, Bhutan, Nepal, Bangladesh, parts of southern China and Tibet into Indochina and south into the Malay Peninsula and the Indonesian archipelago. In the Indian subcontinent there are five species of *Presbytis*. The Hanuman langur, *P. entellus*, is the most widely spread and most numerous species in India. It occurs from Kashmir in the north to Sri Lanka in the south and from Kathiawar east to the Northern Shan States. *P. entellus* lives in a variety of habitats from plains to 3660 m altitude in the Himalayas and from relatively dry tropical forests to open forests, scrub jungles, and arid rocky areas with xerophytic vegetation. Its association with man dates to the Hindu legend where the monkey god Hanuman and his monkey army helped Prince Rama recapture his bride Sita. The Golden langur, *P. geei*, lives in dense tropical deciduous forests dominated by sal, *Shorea robusta*, in northwest Assam and south-central Bhutan to the Sankosh river. The Capped langur, *P. pileatus*, lives in dry tropical deciduous forest and dense evergreen hilly forests throughout Assam and parts of Bangladesh and northern Burma. The Nilgiri langur, *P. johnii*, generally inhabits the sholas or evergreen woodland plateaus surrounded by grassland. It occurs throughout southern India in the Western Ghats south of Coorg, and in the hills of Nilgiri, Anaimalai, Brahmagiri and Palni. It is the most threatened of the Indian langurs, mostly due to habitat deterioration. The Purple-faced langur, *P. senex*, is confined to Sri Lanka, where it occupies a variety of habitats ranging from wet evergreen forests as well as parkland and dense primary montane cloud forests up to 2195 m in altitude.

Gibbons

Hylobates Illiger, 1811, of the family Hylobatidae, is found in Southeast Asia from eastern India east to Indochina, south through the Malay Peninsula into the Indonesian archipelago. Of the eight recognized species only *H. hoolock*, the Hoolock or whitebrowed gibbon, is found in India. *H. hoolock* once occurred throughout much of Assam, Burma, Bangladesh and parts of southern China, but its present distribution has been reduced considerably because of habitat destruction. Extant populations of *H. hoolock* are restricted to where evergreen and mixed evergreen forests still persist in hilly terrain of North Cachar, Garo, United Khasi-Jaintia and parts of Lakhimpur and Tirap districts of Assam and Arunachal. It also survives precariously in the Chittagong hills of Bangladesh bordering Arakan in Burma.

General Discussion

Macaques are generally smaller but more robust in shape than langurs. The smallest macaques are *Macaca radiata* (head-and-body length 35·0–65·0 cm, weight 2·5–8·9 kg), *M. sinica* (length 43·0–53·5 cm, weight 3·4–8·4 kg) and *M. mulatta* (length 47·0–63·5 cm, weight 3·0–10·9 kg). Intermediate sized macaques include *M. nemestrina* (length 43·2–60·0 cm, weight 4·4–14·5 kg), *M. silenus* (length 46·0–61·0 cm, weight unknown), and *M. assamensis* (length 44·0–68·0 cm, weight 4·6–13·0 kg). The largest and heaviest of the Indian macaques is *M. arctoides* (length 50·0–70·0 cm, weight 6·0–18·0 kg). Among the langurs *Presbytis senex* (head-and-body length 44·7–67·3 cm, weight 3·8–9·3 kg) and *P. geei* (length 48·8–72·0 cm, weight 9·5–12·0 kg) are the smallest. *P. johnii* (length 58·4–78·0 cm, weight 9·1–13·2 kg) and *P. pileatus* (length 49·8–76·2 cm, weight 9·5–14·0 kg) are the next largest. The largest and heaviest species of langur is *P. entellus* (length 51·0–108·0 cm, 7·5–20·9 kg). Gibbons, *Hylobates hoolock* (length 45·7–63·0 cm, 6·0–7·9 kg) are intermediate in size and weight, in contrast to the very small lorises. Of course, *Nycticebus coucang* (length 26·5–38·0 cm, weight 0·85–1·68 kg) is almost twice as large as *Loris tardigradus* (length 18·6–26·4 cm, weight 0·09–0·35 kg), which is the smallest Indian primate.

Among the macaques sexual dimorphism in body-size and weight is generally well marked, and in some cases it is considerable. In *M. arctoides*, *M. assamensis* and *M. sinica* males are considerably larger and heavier than females. They are less so in *M. mulatta* and *M. radiata*. In langurs males are also considerably larger and heavier than females but not to the same extent seen in macaques. The northern or Himalayan form of *P. entellus* has males much larger and heavier than seen in males and females from the Indian plains. Both species of lorises show some sexual dimorphism, with males being slightly heavier. The single species of Indian primate that has little or no sexual dimorphism in body-size and weight is *H. hoolock*. In general, the more arboreal species show little or no sexual dimorphism, and the more terrestrial species may show marked dimorphism between the sexes. One of the explanations advanced to explain such dimorphism is that males, with their greater size, are able to defend their group against predators which are more prevalent in open country than in closed forest canopy. Intrasexual competition between males for access to breeding females or to critical resources is also involved.

The habits of Indian primates range from almost exclusively arboreal to almost exclusively terrestrial, with many species between these two extremes depending on the habitat they live in. Among the arboreal species some live in the upper levels of the forest canopy, seldom if ever descending to the ground, and others live at lower levels. Perhaps the most arboreal species is *H. hoolock*, which almost never comes to the ground. Among the langurs *P. pileatus* and *P. geei* are mostly arboreal except for rare descents to drink ground water, especially in the dry season. *P. johnii* is frequently seen feeding, fighting and playing on the ground, but not as often as *P. senex*. The most terrestrial-living langur is *P. entellus*, which spends up to 80 percent of the time on the ground. In contrast, macaques are generally less dependent on trees than langurs. The most arboreal macaque is *M. silenus*, which normally remains in the upper canopy of the forest. In dense forest *M. radiata* is almost exclusively arboreal, but in less forested areas spends over 30 percent of the time on the ground. *M. nemestrina* is similar in habits. The other macaques are generally considered to be terrestrial, although they are capable of moving in trees, especially when alarmed.

Most Indian primates are vegetarians. Some species are exclusively so, while others are more omnivorous, supplementing their diets with occasional items of animal food such as insects, spiders, birds, bird eggs, crabs, molluscs, lizards, and even small mammals. In addition, several species are known to eat earth, as do *M. mulatta* and *P. johnii*, evidently to provide salts, minerals and trace elements; this is probably why *P. entellus* sometimes eats bones from cremation grounds. The range of plant foods that any one species will eat, which may include fruit, berries, leaves, flowers, buds, young shoots, seeds, stems and bark, and cultivated crops and vegetables, is more dependent upon what is edible and available in its area during a particular season. Certainly narrow preferences are occasionally seen in various groups, but this is the exception rather than the rule.

Primate groups have a variety of ways in which they

divide up the habitat so that each group has more or less exclusive use of the food resources in a particular area. In some species there is considerable overlap in space utilization, and in others there is virtually none. Space to which primate groups are usually attached are called either a home range or a territory. A primate group's home range is the space in which the group spends most of its time. The space remains constant even though it may partly overlap the home range of an adjacent group. It may vary in size from a few hectares to 20 or more km². In general large home ranges are correlated with poor food resources, and small ranges are rich in resources. Groups occupying overlapping home ranges generally avoid each other by giving loud spacing vocalizations, but occasionally there are attacks and aggressive fights between groups. For groups with large home ranges there will usually be one or more smaller areas, called core areas, used exclusively by that group. Groups spend most of their time in core areas, for these often contain critical food or water resources or preferred sleeping trees. Home ranges may overlap but core areas do not. When a home range is economically defensible by some combination of behaviours it is called a territory. Territories are characterized by having more or less discrete boundaries that may change through time. Territories are generally mutually exclusive. Thus, the space any one primate group occupies is part of a continuum, with broadly overlapping and loosely defined home ranges at one end and strict territoriality on the other. Most species fall somewhere in between these two extremes.

Gibbons, *H. hoolock*, are probably the most territorial of Indian primates. Their territories average about 22 ha in extent in the locations where they have been studied, but in other areas this may vary. Among the langurs *P. geei* and *P. pileatus* appear to have overlapping home ranges but to what extent is unknown. Home ranges of *P. senex* are relatively small, averaging about 3 to 6 ha depending on whether the group is bisexual or all males. *P. johnii* has home ranges of 6 to 260 ha in Tamil Nadu, but in Kerala, they are much smaller at 6 to 8 ha. Home-range overlap in *P. entellus* can be extensive. Also, the size varies enormously depending on group size and the quality and distribution of food resources. A similar situation prevails among the macaques, but very little information is available for most species. Home ranges of *M. radiata* vary enormously; depending on the area, they range from 40 to 520 ha with varying degrees of overlap. *P. silenus* has home ranges of about 200 ha in extent, and *M. mulatta*, the most extensively studied macaque, occupies ranges as small as 100 ha to as large as 1600 ha. It is of interest that urban populations in villages and temples are much more aggressive to each other than wild groups in forests.

Within their respective home range or territory primates live in groups of varying sizes and have varying degrees of social organization. The least common type of social organization is that of single or solitary-living individuals, in which a consort relationship occurs only for the short period when the female is receptive, as in lorises. The monogamous pair-bond, in which an adult male and adult female and up to four offspring live on a territory, is found only in *H. hoolock*. Macaques and Langurs live in larger groups of two general kinds, unisexual and bisexual. Unisexual groups may be all-male, as in most langurs but not macaques, or all female as in *P. johnii* and occasionally *P. entellus*. In contrast, bisexual groups, which are the most common, may be comprised of one or more adult males and a number of adult females and their offspring. These social groups may be stable for long periods, but it is not uncommon for periodic change in membership to occur through splitting, coalescence, desertion and birth and death. Grouping into societies has some obvious advantages, since it leads to more effective defence against predators, enhances mating opportunities, and also allows for the socialization of maturing young.

R.L.T.

See also HOOLOCKS, LANGURS, LORIS, MACAQUES. See plate 32 facing p. 433.

PRIMITIVE INSECTS. Three groups of primitive insects, wingless, with no eyes (rudimentary in some Collembola), poorly developed thorax and with internal mouth-parts withdrawn into the head, previously classified as orders of the class Insecta, are now treated separately as individual classes of the superclass Hexapoda. Their affinities and origins are still poorly understood and these minute organisms, confined to life in humid habitats, are considered to have diverged from the same stock that produced the insects, long ago, and have continued to evolve independently of them.

The Protura are very delicate, generally unpigmented hexapods with eyes and antennae wanting and the thorax only slightly developed, most species being less than 2 mm long. They appear to be quite common and occurring in all parts of the world, but because of their small size and soil-dwelling nature, only by special collecting apparatus can they be noticed. A little less than 500 species have so far been discovered and the Indian fauna has barely been scratched. The 'telsontails' occur in moist soil, leaf litter and moss, sometimes also under stones or bark. Their food consists of plant material and organic refuse in damp situations in forest or humid locations. Some described Indian species are *Eusentomon indicum*, *Acerentulus breviunguis* and *Proturentomon regale*.

The Diplura are mostly like the proturans, but possess bead-like antennae and cerci that are usually long and

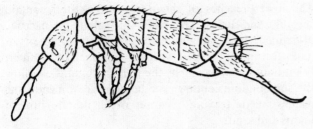

A springtail × 25

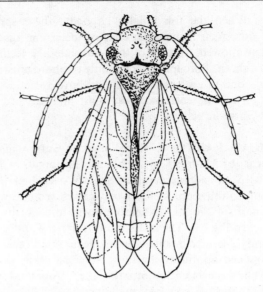

A psocid × 17

forceps-like. Of various lengths (3 to 50 mm), they also inhabit damp situations under logs and stones, in decaying wood, damp moss and the like, favouring tropical regions of the earth. They are basically herbivores, though a few are predacious. Some species may sometimes be mistaken for young earwigs (Dermaptera) on account of their forceps-like cerci. They often occur in small groups or colonies and sometimes occur, perhaps casually, in nests of ants and termites. *Campodea staphylinus*, *Burmajapyx oudemansi* and *Indjapyx indicus* are some known Indian species.

The Collembola, commonly called springtails, are small (2 to 3 mm) compact arthropods, varying in colour from white to almost black, or patterned attractively. Their bodies may be clothed with scales or even hairs and long sensory setae are usually present. Springtails, most of them, have a characteristic springing organ consisting of a spring (furcula) and catch (retinaculum). Some species also have a group of simple eyes. They are found in huge numbers on or near the soil surface especially in decaying vegetable matter, rotting logs, and the like. They inhabit ant and termite nests, fungi, moss, algae and lichens and are cosmopolitan, even inhabiting the Antarctic. *Smithuroides appendiculatus*, *Salina indica* and *Isotomurus palustris* are some known Indian species.

PSOCIDS. These are small (1 to 10 mm long) soft-bodied, winged or wingless insects, some of which, called 'book-lice', inhabit domestic situations, feeding on glue of book bindings and decaying vegetable and animal matter. Some are destructive to dried insect specimens preserved in museums. Only about 200 species of the 2000 to 3000 known Psocoptera are known from India, most of them found on vegetation where they feed on algae, fungi, lichen and debris, on foliage and tree bark. Many of them live in clusters of individuals as colonies and a few also frequent birds' nests. Some species, such as *Liposcelis divinatorius*, are also known to assume pest status, feeding on stored tea, cereal products, flour and other stored food. Many species occur under communal webs harbouring groups of nymphs and adults, and polymorphism is also exhibited in some families. Wings, when present, are membranous, sometimes darkly mottled and marked and held roof-wise over the abdomen. Of a primitive hemipteroid stock, they appear to be closely related to the ANIMAL LICE and some hairy species carry foreign matter, including fungal spores, which they disseminate.

PUFFER FISH and PORCUPINE FISH. Puffers (*Tetrodon*) have rounded bodies, with the scales replaced by minute prickles. The teeth are fused together to form a parrot-like beak consisting of two plates in each jaw. By swallowing water or air, these fishes can inflate their body like a balloon, thus feigning death but at the same time making it difficult for predators to swallow them. Swimming is accomplished by flapping the dorsal and anal fins like a fan.

Their internal organs are extremely poisonous, although the flesh is edible. In India, they are thrown away as inedible, though in Japan a special dish is prepared from them.

Porcupine fish (*Diodon*) are cousins of the Puffer fish, but have their body covered with long, sharp

Porcupine fish × ½

spines. When the fish inflates its body, these spines stand out erect, forming an effective deterrent against being swallowed. Instead of the four plate-like teeth of Puffer fish, the Porcupine fish has only two hard crushing plates—one in each jaw. Both the above varieties are common on the east and west coast and are usually caught in shore seines.

PULICAT LAKE is a large brackish-water lagoon northeast of Madras City. It has a waterspread of 350 km² with an average depth of slightly over 1 m. The opening to the sea is at the narrow southern end with the length of the lake extending north and northwest parallel to the sea. The lake mouth is narrow and shallow and is likely to be silted during the non-rainy season. It is opened up during the rains either naturally or by man. The freshwater flow into the lake is from land run-off and the inflow from two small rivers, which have been dammed and now contribute little to the lake's freshwater supply. While the lake is filled with fresh water during a good monsoon, it is hypersaline during the summer months, particularly in the northern areas. The microfauna and flora are tolerant of variations in the salinity. The area supports a large population of water birds seasonally, waders and flamingos predominating. The northern section now supports a breeding colony of pelicans.

PUNJAB. The word 'Punjab' is derived from the two Persian words *panj* (five) and *ab* (water) and is applied to the basins of the rivers Jhelum, Chenab, Ravi, Beas and Sutlej, all tributaries of the river Indus. Their drainage areas cut across the Indo-Pakistan international boundary. Bounded in the north by the state of Jammu and Kashmir, in the south by the state of Rajasthan, and in the east by the basin of the Ganga river system, the territory of Punjab lies in the northwest of the subcontinent. The total land area is 122,000 km² in India and 205,000 km² in Pakistan; and the estimated populations in India and Pakistan are 25 and 37 million respectively.

Punjab is largely an alluvial plain. The rivers rise in the Great Himalayan ranges and follow courses amid snow-clad mountains for a few hundred kilometres before entering the plains, providing varieties of scenery from the snow-peaked mountains to the semi-desert plains of shifting sands in their lower courses. Below the mountain stage lie the low Siwalik Hills (500–600m), and a narrow undulating foothill zone containing seasonal torrents or *chos*, many of which terminate in the plains below. In the plains the rivers gently meander in a south-southwesterly direction at an average gradient of one metre in a kilometre. The flat alluvial plains contain slightly raised banks between the river valleys.

The lower reaches of the rivers in Pakistan contain several sand-dunes. The rivers roll along in narrow, ill-defined, ever-changing channels. Their volume is at its lowest in winter, and with the approach of summer, the mountain snows melt, the rivers rise, and overflow the surrounding country. As the rainy season ends, the waters recede leaving expanses of fertile alluvium of loams and sand.

Geology and topography divide the Punjab into three main sections. The plains are composed entirely of the recent Indo-Ganga alluvium but at places contain beds of sedimentary rocks. The second or the foothill section which lies to the north of the plains is underlain by tertiary beds of sandstones of the Siwalik series. In the third, mountain section, rock-forms range in age from the Cambrian to the Cretaceous with quartzite, limestone and sandstone rocks predominating.

The inland subtropical location produces continental climates, ranging from the semi-arid to the subhumid, and generally given to large seasonal contrasts in temperature. Local conditions of temperature and precipitation in the mountains are affected by altitude, direction of mountain slopes, and the alignment of valleys. In the plains, summers are very hot (35°C to 43°C), and a few hot days register 45°C. The winters are fairly cold (January mean 13°C), with occasional frost. Precipitation is highest in the mountains. At Dharamsala in Himachal Pradesh it is 300 cm annually and decreases gradually to only 30 cm in the southwest parts of the Punjab plains in Pakistan. Most of the rainfall is received through the southwest monsoons between July to September. Winter rainfall is associated with the western cyclones occurring from December to March.

The Punjab plains have been largely cleared of their natural tropical and deciduous vegetation. Sissoo, dhak, pipal, jujube and babul are among the most common trees. In the southern part of the plains scrub and thorny bushes dominate, with babul and other trees. Along the streams, tamarisk trees grow. In the submontane parts (the Siwaliks) the sal is common, but deforestation has resulted in much bush vegetation. In the mountains above 2000 m altitude the deodar is the major tree. Valleys in the Himalaya have birches, and higher up pines, willows, oaks, yews, elms, and holly.

Since medieval times wildlife has become scarce, particularly in the plains with the increased use of land for agricultural purposes. Until the beginning of the 19th century lions and tigers were abundant. The areas between Delhi and Chandigarh in the Sutlej river basin were the favorite hunting grounds of the Mogul emperors.

R.T.

Indian Rock python, about × $^1/_{10}$

Python ingesting bandicoot rat × $^1/_5$

Python hatching

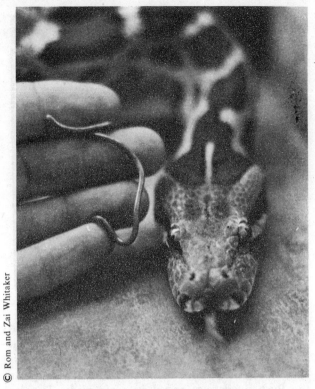

Python and worm snake × $^2/_3$

PYTHONS. Of the 16 species of pythons in the world two are found in our area. These are the Indian Rock Python, *Python molurus*, which grows to a maximum length of 6 m and is found throughout the subcontinent, and the Reticulated Python, of which the range is limited to the Nicobar Islands in the Bay of Bengal. The latter grows to 10 m and is the longest snake in the world. Pythons are nocturnal, resting during the day under a log or rock and hunting at night. They mainly eat warm-blooded animals such as rats, birds, civets, mongooses and other small forest animals, but also occasional monitor lizards. Large pythons have been known to take wild boar, a full-grown leopard, and even man! They rest for several days after a heavy meal and sometimes months elapse before they hunt again. In captivity they have been known to fast for two years without any ill effects. Pythons are not venomous, and prey is stealthily stalked and seized. Pythons coil tightly around the victim and do not crush it as is popularly believed, but squeeze it until suffocated.

Rock pythons breed between March and June. The female lays up to 100 eggs (depending on her size); these are the size of duck's eggs and the female lays them in a cave, hollow tree or hole. She stays with the eggs, and has even been observed contracting her body rhythmically, the mechanical heating providing temperature control for the eggs. Pythons are popular as pets, most being of a calm and gentle disposition. The odd individual will bite consistently and refuse to be tamed; such specimens are best released as they deteriorate rapidly in captivity. Python skins are very popular and pythons are killed for them throughout their range. They are extinct in several areas where once common. Some tribal groups kill pythons for the meat; but the main culprits are the efficient and deadly representatives of the snake-skin industry.

See also SNAKES.

461

QUEEN'S CRAPE MYRTLE, *Lagerstroemia speciosa*, is one of the most striking trees in the damp jungles of Bengal, Assam, Burma, southern India and Sri Lanka. It is not often that a tree with handsome flowers has valuable timber. The tree is covered with great clusters of mauve flowers with crinkled petals like crape paper and heavily ribbed calyces. Its trade name is Jarul. It has a strong timber and is ranked next in value after teak in Burma.

Queen's Crape Myrtle. Flowers, fruits and leaves × ¹/₄. The upright panicles may be a foot long. The tree is equally spectacular in winter, when the leaves turn coppery-red before falling.

QUAIL (family Phasianidae). Quails are rotund, short-tailed granivorous birds looking like miniature partridges, which in fact they are. Our two common species, the Grey Quail and the Rain Quail, along with the Bush quails, are prized as game birds. They are mostly resident, but in winter a vast influx of migrant Grey Quails from northern countries augments the local population. They spend their time hidden in crops, grassland or bushes in scattered pairs or coveys, and are seldom seen except when accidentally flushed or when put up by beaters or dogs for sport. They then fly low over the crops and soon dive into cover again. Bush quails have a disconcerting habit of squatting huddled together under a bush until almost stepped upon, the covey then exploding with a startling whirr, the individuals dispersing in all directions. Up to a few years ago large numbers of quail were netted by professional poachers and openly sold in markets and restaurants. The present game laws have curbed the practice somewhat, but their numbers had dropped dangerously and it may take years of protection to restore the populations.

See GAME BIRDS.

RABIES. A zoonotic disease (see ZOONOSIS), caused by the rabies virus, which attacks cells of the central nervous system. It is the most widespread infectious disease known, infecting, in nature, all warm-blooded animals. The virus is present in the salivary glands of the infected animal, and is transmitted from animal to animal, or from animal to man, by the bite, lick or scratch of a rabid animal. Human rabies is usually associated with dog bite, but it is important to note that *any animal*, not excluding man, suffering from rabies, can transmit the disease through its saliva. Therefore any exposure (bite, lick or scratch) to a suspect animal or human must be immediately washed and disinfected and expert advice sought on preventive vaccination. The vaccine treatment aims at raising the patient's immunity, but this immunity is ineffective once the virus has entered the nerve cell, hence the treatment, to be effective, must be started without delay. If, due to delay or lack of treatment, the virus succeeds in penetrating the nerve cells, onset of rabies and death is inevitable. The disease follows the same pattern in all animal species affected, but there may be differences in specific symptoms, e.g. the characteristic symptom of human disease, hydrophobia (fear of water; an inability to swallow water which sets up painful spasms of the throat muscles), may be absent in animals, and dogs have been known to drink water right up to the time of death.

A parasite which invariably kills the host is not normally an efficient parasite, and would soon die out itself. The rabies virus ensures its survival by extending its range of host species, and the cerebral irritation it causes makes the victim snap and bite at any moving object, which is frequently another animal, so that the parasite is readily transmitted to new victims. Eradi-cation of rabies in the wild appears almost impossible. But eradication of the disease from human communities requires only the determination to take the necessary measures. Most human cases are caused by the dog, pet or stray, or by other pets and domestic animals, which have been infected by a dog. To successfully eradicate the disease, it is necessary to destroy all stray dogs and to compulsorily vaccinate and license all pets, measures which have been applied with great success by several countries in recent times.

RAIN. Rain consists of liquid drops of water of various diameters falling to the ground from clouds above. By convention the diameter of a rain-drop should be half-a-millimetre or more; when the droplets are smaller the fall is called 'drizzle'.

The total volume of rain-water which falls during a given period over a given area of the ground will have a certain depth if spread over the same area, without any water being lost by evaporation or seepage or outflow. This depth of water is called the rainfall over that area for that period of time, whether it be an hour, a day, a month, a season, or a year.

The measurement is usually made by a rain-gauge. In its simplest form, a funnel with a truly circular bevelled rim is used to collect the rain and a narrow tube leads the collected water into a receiving vessel. At the end of the given period, usually 24 hours, the volume of water collected is measured in cubic centimetres; this volume divided by the area in square centimetres of the mouth of the funnel gives the rainfall in centimetres. There are other types of rain-gauges, the so-called self-recording rain-gauges. One such type records the intervals of time which have lapsed between the falling of successive small amounts, generally 0·1 millimetre of rain.

When the rainfalls at different meteorological stations are plotted on a blank geographical map, at the appropriate geographical locations of the stations, and smooth continuous lines are drawn connecting all places of equal rainfall, we obtain an isohyetal map. The lines of equal rainfall are called isohyets. The time interval to which the rainfall refers may be a day, a month, a season, or a year.

The annual rainfall averaged over the whole earth is around 80 cm. The annual rainfall averaged over the plains of India is around 100 cm.

See also CLIMATE.

RAIN TREE, *Samanea saman*, is a huge umbrella-shaped tree from Venezuela, cultivated in our area as an avenue tree. Its short massive trunk breaks up into large wide-spreading branches bearing feathery foliage that create an unsurpassed canopy of shade. The pinkish flowers in globose clusters are like shaving brushes. The fruits are succulent pods containing a sugary pulp which are relished by squirrels, cattle and horses. The pulp is fermentable and may be employed for production of alcohol. It gets its name from the fact that the leaflets fold together during a rain-storm.

RANGANATHITTU. There seem to be few places in India which have such a large congregation of breeding birds within a limited area as in Ranganathittu. This sanctuary is one of three islands on the river Cauvery 1·6 km from the historic city of Srirangapatna, and 14·4 km from Mysore. It is 750 m above sea level and the average temperature varies from 22·5 to 25°C and the annual rainfall is between 500 and 800 mm.

The areas surrounding the sanctuary are fertile fields under rice cultivation, with coconut trees irrigated by canals from Krishnaraj Sagar reservoir, upstream. During excessive rains the level of the river rises significantly at times; as a result of which the nests of many birds are submerged. There is need for better coordination between the Irrigation Department and the management of the sanctuary. With some foresight considerable damage to the bird life could be avoided.

Ranganathittu is beautiful at all times of the year; but it is of special interest to birdwatchers from June to the end of September. At this time the trees are studded with nests and the hatching time is generally in the middle of June when the surrounding lands are irrigated and there is plenty of food in the flooded areas. By the end of November the young birds are full-fledged and many species depart with the adults at this time.

The birds that can be commonly seen here are open-bill stork, white ibis, little egret, darter, spoonbill, pond heron, cattle egret, cormorant, stone curlews and river terns. The Indian mugger is often seen reclining on the exposed rock surfaces in the river.

Rowing-boats are available on hire and it is possible to make a very close approach to the trees where the birds are nesting. However, it is important to ensure that the birds are not disturbed by the presence of visitors. Considerable damage is done to the nests when the birds are frightened by human presence.

RANNS OF KUTCH. This is the name given to the extensive salt flats in the Kathiawar peninsula, southeast of Pakistan. The northern, larger portion (18,000 km²) is the Rann and the southeastern, smaller area (5000 km²) which is connected with the Rann by a narrow channel is known as the Little Rann. The Rann is believed to be a raised portion of the sea bed and was a navigable lake in Alexander's time (325 B.C.). It has subsequently become further raised and is now, except during the monsoon season, a desolate salt-encrusted wasteland. During the monsoon season the Rann is largely under brackish water, partly sea water driven in from the Gulf of Kutch by the monsoon winds and partly fresh water from the rain-fed streams. The Rann

comes to life as the water level goes down and the waters start to recede. An immense number of the raised mud-nests of flamingos can usually be seen at this period. Nearly a million flamingos are said to be present at this time, but the breeding conditions vary from year to year and in some years the birds do not breed in the area. The Rann is the only known flamingo breeding ground in India. Other than the Large Flamingo, the Lesser Flamingo and the Rosy Pelican also breed in the Rann. The Rann is also a staging area for birds migrating between India and the countries to the northwest. Very large numbers of birds move through the Rann in autumn and spring. The Little Rann is now the main stronghold of the Wild Ass. About 700 animals are believed to exist. Other than the Wild Ass the only notable large mammals are the Wolf and the Chinkara. The commonest reptiles are the Spiny-tailed Lizard and the Echis Viper.

RANTHAMBOR. Ranthambor is one of India's most spectacular wildlife sanctuaries. It proves that even when a comparatively small area is well protected it can abound with wildlife of many species.

Situated about 100 km SE of Ajmer, it was established as a sanctuary in 1959 and included in Project Tiger in 1972. It has an area of 392 km² and its altitude is between 200 and 500 m. It consists of numerous valleys and flat-topped hills of the Aravalli range. The river Banas, and its tributary Kandoli, flow through it. A large number of lakes make the landscape very beautiful and the magnificent fort invests the area with historical significance.

The forest is of the tropical dry deciduous variety and the main tree species are kardhas (*Anogeissus pendula*), palas, khair, ebony, ber and others.

There are very large herds of sambar and nilgai and a fair number of gazelles. The evening congregation of herbivores on the borders of the lake presents a magnificent sight. The other wild animals include tiger, panther, jungle cat, jackal, civet cat, striped hyena, sloth bear, wild boar, chinkara, rhesus and Hanuman monkeys, crocodiles and monitor lizards.

Because of the varied habitat there is rich birdlife, game species such as partridge, sandgrouse, peafowl, red spurfowl, and common green pigeon, and an interesting assortment of shore and water birds.

As a result of the total protection to the habitat since it became a part of Project Tiger the tiger population is on the increase and the Census taken in 1980 indicates that there are 25 tigers in the Reserve.

RAMIE, *Boehmeria nivea*, is a shrub from China cultivated in Bengal, Assam and elsewhere. It is characterized by rounded heart-shaped, coarsely toothed, triple-nerved leaves, which are covered with a snow-white felt beneath. The fibre from the bark is regarded as the longest (40–200 mm), toughest and most silky of all vegetable fibres. It is used in making bank notes, gas mantles and fishing lines.

RAPTORS or BIRDS OF PREY. Diurnal flesh-eating birds belonging to the order Falconiformes, represented in the subcontinent by three families, namely Accipitridae (hawks and vultures), Falconidae (falcons) and Pandionidae (osprey). They are all adapted for an animal diet, possessing a sharp hook-tipped bill for tearing flesh and (excepting vultures which have shorter toes and blunter claws) powerful curved claws for capturing and holding down prey. They either hunt living animals (hawks, falcons, eagles) or feed on carcasses and offal (kites, vultures) or fish (osprey). Falcons (peregrine, kestrel) are distinguished from hawks (shikra, goshawk) by narrower pointed streamlined wings as against broader and more rounded. Both the groups feed mainly on live animals—mammals, birds, reptiles of manageable size; also fish and insects; but their method of hunting prey differs, depending on their habitat and the shape of their wings. The long-winged falcons are usually found in open country and semi-desert. They capture their prey by stooping on it from aloft with closed wings in the manner of a dive-bomber plane at velocities close to 200 km per hour. The short-winged hawks live in wooded country where their broader and shorter wings make for easier manoeuvrability in pursuit of prey through shrubbery, tree-trunks and suchlike obstacles. Their normal method of hunting is to sit in ambush on a branch, hidden by foliage, and pounce unawares on prey on the ground or give swift chase, avoiding obstacles by lightning turns and twists. Falcons normally do not build nests of their own but appropriate the disused stick-nests of crows, eagles and suchlike birds on cliffs or trees for raising their families; hawks construct twig platforms usually up in trees.

For the sport of falconry both falcons and hawks (also some eagle species) are trained to fly from the sportsman's wrist and hunt birds (and mammals) often much larger and heavier than themselves. The Shaheen Falcon, which is our resident subspecies of the Peregrine, enjoys the highest esteem from falconers for

© M. Krishnan

Tawny Eagle × ¼

speed and audacity, followed closely by the broad-winged Goshawk and then by its smaller cousin the Sparrow-Hawk.

The vultures are a group of large highly beneficial carrion feeders which render great service to man as scavengers by speedily disposing of the carcasses of domestic livestock cast out in the precincts of villages and rural settlements which would otherwise breed pestilence. In recent times, however, presence of vultures around urban environments and their habit of soaring aloft and circling on thermals pose an increasing hazard to fast-flying jet aircraft, frequently causing mid-air collisions or 'strikes' resulting in expensive damage to aircraft and sometimes even serious crashes with loss of human life.

In addition to the Osprey, a migratory fish hawk of world-wide distribution, we have a number of eagles—some migratory, others resident—which live by active hunting or by piracy, i.e. robbing other raptors of their lawful prize, or even by scavenging. All birds of prey have suffered serious decline in their populations in recent years. This is believed to be due largely to the noxious effects of the free and indiscriminate use of toxic chemical pesticides and herbicides etc. acting through the birds' food-chain directly, or by inhibiting natural reproduction.

RAT SNAKE. The Rat Snake (*Ptyas mucosus*) is the best known and most often seen of all the larger snakes in our area. True to its name this large yellow, brown or black snake is mainly a rodent-eater, also preying on frogs, toads, birds, lizards and other snakes. It grows to over 3 m and is common wherever there are farm crops growing, particularly rice paddy. Rat snakes are fast-moving and quick to escape when a human approaches. Often mistaken for cobras, they are frequently killed.

Rat snakes lay up to 16 eggs and the young are about 30 cm long. When first caught, a specimen will

Bearded Vulture or Lämmergeier and Griffon Vultures. Both are about 120 cm from tip of bill to tip of tail.

Rat snake × ⅕

struggle and strike with vigour, making a low growling sound and inflating the body. They rapidly calm down and get used to handling.

See also SNAKES.

RATS. Rats are among man's most serious pests, both as destroyers of food and agricultural crops as well as carriers of diseases harmful to man. As such they have been intensively studied and much is known about their physiology, social behaviour, learning ability as well as many other facets of their biology.

They belong to the sub-order Muridae and are thought to have originated in southeast Asia where there are over 20 species, mostly forest-dwelling arboreal animals of little economic importance. It is the Roof Rat (*Rattus rattus*), so widespread throughout every Indian village, and the Brown Rat (*Rattus norvegicus*), the main pest animal in Europe and North America, which we mostly associate with the term 'rat'. In appearance all have long naked rather scaly tails and usually quite harsh dark brown body fur with relatively powerful incisors enabling them to gnaw holes through wooden doorways and even burrow through tile and cement floors of grain-stores.

Brown or Norway Rat, × ⅓

The Roof Rat is an agile climber which produces 5 to 7 litters a year after a gestation period of only 19 days with an average litter of between 5 and 6 in India and Pakistan. The Brown Rat prefers to live underground and is a good burrower and can produce litters of as many as 17 or 18.

G.W.F.

See also RODENTS.

RAVI. One of the five rivers that give the Punjab region its name, the Ravi rises in the snowy peaks of Bara Bangahal in Himachal Pradesh, and flows westnorthwest past Chamba before turning southwest at the boundary of Jammu and Kashmir. From there it flows generally southsouthwest making an international boundary between India and Pakistan before finally entering Pakistani territory. It flows past the historic city of Lahore and ends its 710 km course into the Chenab river.

In its mountain stage its basin lies between the two spectacular Himalayan ranges of Pir Panjal and Dhawaldhar. At the higher altitudes (2500 m and above) the topography is rugged but also contains several glacial-modified valleys with areas of fluvio-glacial deposits and moraines. Lower than this at altitudes of 1100 to 1500 m, synclinal valleys (the duns) are encountered followed by more or less continuous anticlinal ranges of lesser elevations (800–1200 m), known as the Siwaliks. These low-altitude ranges contain valleys in which fluvio-deposits, gravels, and unconsolidated materials have been dropped by the streams. Below this lies a generally flat plain with few topographic features along the river banks. Throughout its course of 450 km in the plains the river flows in a narrow valley, 3 to 5 km in width, with a generally meandering channel containing a few islands.

The Ravi is snow-fed and perennial, but its volume fluctuates seasonally. The discharge is smallest during the dry winter, whereas floods are common during the rainy season (July to September). Unlike the Ganga system, the Ravi passes through an area of decreasing rainfall, losing water on account of evaporation continually throughout its journey in the plains. Its mean annual flow is six times larger between July and September than over the dry period between October and December. Its long course in the semi-dry plains is ideally suited to irrigation. Here the soils are deep, of river-borne alluvium, and of silty and sandy loams. Under conditions of regulated water-supply by canals these highly fertile plains have become the principal granary of India and Pakistan.

Climate varies from the warm subtropical in the Siwaliks to the cold temperate above 2300 m. Rainfall is heaviest in the Himalayan section, between 100 and 250 cm annually, decreasing rapidly as the river moves into the plains. Lahore, in the middle of the plains section, gets only 60 cm. The plains are semi-arid, and their remoteness from the sea causes extreme seasonal contrasts in temperature. Temperatures occasionally fall below freezing point in January but shoot up to 42°C in summer. The rains arrive by the middle of July, bringing a welcome drop in temperature, and last until the end of August. Winters are generally sunny and dry, experiencing a few rainy days, created by winter cyclones, Temperatures follow altitudinal variations in the Himalayan section.

RAZOR SHELLS or Jack-knife shells have worldwide distribution. In all sandy bays they live in deep vertical burrows which they dig for themselves. The razor-fish is a fine example of how the habits of an organism modify its form and shape. Of all burrowing bivalves, such as soft-shelled clams and piddocks, the razor-fish is the deepest burrower. It is said that it can dig down to nearly 60 cm. To enable the animal to dig so deep, the foot becomes long, powerful and cylindrical and is shifted from the ventral to the anterior margin which is directed downwards. To obtain its food its posterior end, where the siphons open, must maintain contact with the water outside and the shell ends must gape. The enlarged end of the foot keeps both ends of the valves open or gaping. The length of the animal is only about a quarter of the burrow; and in order to ascend up or descend down the narrow hole rapidly and without resistance, the shell must be narrow and straight-margined. The streamlining is achieved by the foot shifting its position to the anterior end. Permanent living in a brine-filled burrow is likely to corrode a shell, so the shell has a thick horny surface. Living in a tube the animal should be free to move up and down and there should not be any tethering to a holdfast, so the razor-fish does not develop any byssus threads. Altogether, consistent with the mode of life, razor-fish, as their name suggests, have narrow, elongate, parallel-sided, smooth, equivalve shells. Their foot can expand into a sphere or contract into a sharp point as required by them. Their respiratory tube is also long. For most of the time the razor-fish remains at the upper part of its burrow, keeping the external orifice of its siphons at the mouth of the hole in contact with the water. The mouth of the siphon is fringed with hair-like filaments which are sensitive to touch and also to the slightest vibration in the water or ground. Water containing food particles such as animalcules and algae is sucked in through the siphon which circulates through the gills and is afterwards thrown out through the anal aperture of the siphon. At low tide the burrows of razor-fish can be seen on the mud-covered sand-banks as slot-like openings. Anybody walking near by can notice little jets of water squirted out here and there through these holes. The jets of water are those shot up by the razor-fish as they swiftly withdraw into the deeper part of their retreat, alerted by the vibration caused by the footsteps. Vibrations through any agency

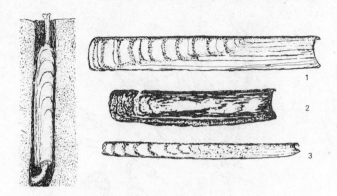

A razor-fish at the top of its burrow, × ¼ and some shells:
1 *Solen truncatus*, with purple and white bands, × ¼
2 *S. lamarckii*, × ⅓. 3 *S. linearis*, with purple and white bands, × ½

—a strong breeze, any abnormal sound, the movement of a fish at some distance—all are sensed by these animals, causing them to squirt water through their holes. Even the passing shadow of a low-flying gull will force the razors down to the bottom of their holes.

Like snails and slugs, razor-fish have the habit of amputating part of their feet to escape from dangerous attacks and are afterwards able to replace the lost part. Even if one succeeds in catching a razor-fish by its exposed siphon, it is difficult to pull the animal out. The foot of the animal sticks it fast to the burrow. If more force is used the animal may come out minus a portion of its foot. So to collect these molluscs, the fishermen adopt another method. They pour a strong solution of common salt into the holes and the animal, although habituated to sea-water, comes out by itself. Razor-fish are collected both for food and for baiting fishes.

In India several interesting species of razor-fish occur in the littoral mud-flats. The more common examples are: *Solen truncatus, S. lamarckii, S. linearis* and *Siliqua radiata*.

See also MOLLUSCS.

RED SANDERS, *Pterocarpus santalinus*, is a small tree, with compound leaves having 3 leaflets, yellow flowers and oblong winged pods. The bark exudes a reddish juice after a kukri-cut on the trunk. The fragrant heartwood is claret-purple to dull black and contains a dye called santolin. The wood is used in carving and also exported to Japan for making musical instruments. The timber with wavy grain fetches a fancy price as it produces a fine resonance. It is an endangered tree as it is endemic to a small area in Andhra Pradesh.

RED SILK-COTTON, *Bombax ceiba*, is a lofty deciduous tree with a buttressed base, widespread in the subcontinent including Sri Lanka and Burma. In the Himalaya it is found up to 1300 m. It is characterized

Red Silk-cotton. Leaf, bud, flower and unopened fruit × ¹/₄. Bursting pod with floss × ¹/₆. The flowers secrete a nectar attractive to birds and bees which effect pollination. The fallen flowers are eaten by deer.

by tiers of whorled branches. The young stems and branches are covered with stout conical prickles. The leaf is composed of lance-shaped leaflets arranged like the fingers of a hand. Its brilliant crimson, close-set, large flowers burst open from the dark buds during January to March. The fruit is a 15 cm pod that splits open on the tree in April-May and disgorges quantities of silky cotton in which the small seeds are embedded and are dispersed by wind. The floss is used for stuffing pillows and is excellent for making surgical dressings. The wood is used for making match-boxes and packing-cases.

REPRODUCTION. Asexual reproduction occurs in primitive forms of life. In simple organisms, the cell acquires more substance, enlarges and gives off buds or processes which break off to lead an independent existence. In more complex organisms there is a process of cell-division (mitosis). The cell first enlarges and then divides, a division which splits all its constituents, including its genes (see HEREDITY: GENETICS) into two equal parts. This halving or division is preceded by doubling of its genetic material which then divides into two, each daughter cell carrying a full complement of genetic material. This process can continue as long as the required nutrients are available in the environment (Fig. 2 in HEREDITY: GENETICS). Sexual re-

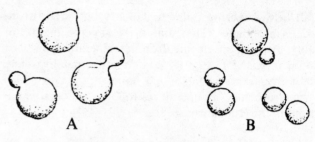

BUDDING OF YEAST CELLS
A Buds form as outgrowths on the surface of the cell.
B Buds separate to form new cells.

production consists essentially in the interchange of genetic material between living cells. A bacterial cell will adhere to another (conjugation), each will derive some genetic material from the other and then they drift apart, each cell having diversified its genetic structure. Many living forms use both asexual and sexual methods but all higher forms reproduce only by sexual means. Instead of two individuals conjugating for exchange of genetic material, the genetic material is mingled in the next generation. Each individual produces gametes (sex cells), the gamete bearing only half the chromosomes of the parent. This merges with another gamete, usually from another individual, to form a zygote with a full complement of genes which then grows into a new individual (Fig. 3 in HEREDITY: GENETICS). This new individual is different from both its parents though it has characters derived from both. Some of its differences may be more efficient, may fit the individual better to survive in the environment in which it is placed, and help to ensure survival of the species.

A further development is the separation of the sexes, each individual being either male or female. Each then forms gametes of one type and the combination of gametes from two individuals of different sexes is necessary for formation of a zygote which can develop into a new individual. Such a complicated mechanism slows down the process of reproduction, but it also offers greater chances of producing more efficient forms or variants since each generation differs from the one preceding it. The increase in variability gives the organism greater chances of survival under diverse or changing conditions.

To ensure that this complicated system works, all higher animals have a strongly developed instinct of mating between sexes, while higher plants develop colourful flowers and secrete nectar which attracts insects and birds. These, in passing from flower to flower carry the gametes from one plant to another. Many flowers are so constructed that the male gametes (pollen grains) are brushed off on the body of the insect, and are deposited in the next flower the insect visits to feed on its nectar.

REPTILES. Reptiles were the ruling animals on the planet for many millions of years during the Mesozoic Era. Today the only big dinosaur-like reptiles are some of the CROCODILIANS and MONITOR LIZARDS. There are about 6000 species of reptiles in the world of which about 540 are found in India. All are ectothermic, that is their body temperature varies according to the surrounding temperature. Almost all are covered with scales which vary in form from the armour of the crocodilians to the smooth shiny scales of the burrowing snakes.

The majority of Indian reptiles are oviparous: one or two eggs in the case of geckos, 20 for a cobra and 40 or more for a saltwater crocodile. A number of SNAKES like the vine snake, sand boas and most of the sea snakes bear living young, these are called ovoviviparous, referring to the fact that the eggs are incubated internally. Snakes lay their eggs in holes in the ground or trees, in crevices or caves. Lizards like chameleons and garden lizards dig holes for their eggs while monitors use termite mounds. Crocodiles, gharial and turtles have particular breeding seasons like the rest of our reptiles, and come ashore to dig holes in which to deposit their eggs.

In general Indian reptiles are carnivorous, snakes and crocodilians strictly so, while a number of TURTLES and a few agamid lizards include vegetable matter in their diets. Tortoises, the land-dwelling cousins of turtles, are almost totally vegetarian, grazing and browsing on a wide range of leaves, flowers and fruit.

See plate 3 facing p. 48, plate 9 facing p. 128, plate 16 facing p. 193 and plate 19 facing p. 256.

RESPIRATION. Respiration, or breathing in land animals, is the process by which oxygen is supplied to the animal body for its energy needs.
Oxygen is essential for life, and except anaerobic bacteria, no living organism can survive without free oxygen. Aquatic animals, whose requirement is relatively small, derive their oxygen from the air dissolved in water, whereas terrestrial animals derive it directly from the air they breathe. This oxygen is consumed for the activity of the animal, and waste gas in the form of carbon dioxide is thrown out.

In aquatic animals the gills constitute the respiratory system. This consists of a feather-like membrane, which, as the water passes over it, extracts the oxygen of the dissolved air, and throws out carbon dioxide. In land animals tracheal tubes, and in land vertebrates the lungs, take up oxygen from the air, and pass it to the blood, which carries it to all parts of the body. Some fishes respire through the vascular rectum where exchange of gases takes place with water which is alternately sucked in and squirted out through the anus. Annelids breathe through the skin, and molluscs through their mantle,

amphibia use both skin and lungs for breathing while higher vertebrates breathe only through the lungs. Lungs are spongy organs containing thousands of small air spaces, the alveoli. These are lined by a thin membrane surrounded by a rich network of blood vessels, the exchange of gases between the blood and alveolar air being made across the alveolar membrane.

The respiratory system, in addition to maintaining oxygen supply, also has certain subsidiary functions. Breathing can also help in temperature regulation, the panting of dogs in hot weather being a mechanism for discharge of excess heat.

The lungs of whales are probably hydrostatic as well as respiratory in function. Whales have capacious nasal chambers storing a large amount of air which would otherwise be forced out of the lungs by the enormous hydrostatic pressure during a deep plunge. The aperture from the nasal chamber to the lung can be closed off by a stopper-like arrangement, to permit the expulsion of stale air which comes out together with water vapour as a waterspout.

Chameleons have lungs which can swell up, perhaps to frighten predators. The inflatable lungs of a sea turtle act as floats, to help in maintaining its balance in turbulent water. Birds' lungs are connected with a system of air sacs which helps the bird, by adjusting the air contents, to alter the position of its centre of gravity and balance during flight. The flying muscles pump the reserve of fresh air from the air sacs to the lungs and also control respiratory movements. The faster the flight the greater is the automatic supply of air drawn through the lungs. Birds do not get out of breath or suffer from mountain sickness at high altitude. This is probably because the increase of wing-strokes in rarefied air brings in a compensatory supply of air. The frigate bird, which easily maintains a speed of 100 miles an hour, has the best-developed air sacs.

RHINOCEROSES. The first rhinoceroses lived about 50 million years ago. They were small, horse-like creatures with rather slender legs, hairy hides and no horns. The rhinoceroses soon became one of the dominant and most diverse groups of mammals. Among the many different forms which evolved was the largest land mammal ever to have walked the earth—the 6-metre-tall *Baluchitherium* which lived in Pakistan and Mongolia 20 million years ago.

Today only five rhinoceros species survive and they are all gravely threatened with extinction, by loss of habitat and poaching for their valuable horns. The Black Rhinoceros (*Diceros bicornis*) and the White Rhinoceros (*Ceratotherium simum*) live in Africa, whereas the Greater One-horned or Indian (*Rhinoceros unicornis*), the Lesser One-horned or Javan (*R. sondaicus*) and the Asian Two-horned or Sumatran (*Dicero-*

rhinus sumatrensis) rhinoceroses live in Asia.

Only the Greater One-horned or Indian Rhinoceros survives on the Indian subcontinent, although until early in this century there were a few Sumatran rhinos in the hills of northeastern India. Indian rhinos were once widely distributed on the flood plains of the Indus, the Ganga and the Brahmaputra rivers. Accurate representations of the Indian rhino found at Mohenjodaro indicate that the species occurred as far south and west as the present Sind province in 2000 B.C. The Mogul Emperor Babur hunted rhinos at Nowshera near Peshawar as late as 1519. The extent of the historical distribution of the Indian rhino to the east is complicated by confusion in the literature between the Indian and Javan rhinos. The Indian rhino has probably never been found far east of the present boundaries of India and Bangladesh, but there are a few records from northern Burma up until 15 or 20 years ago. Today the Indian rhino numbers fewer than 1500 individuals restricted almost entirely to eight small protected areas in Assam, West Bengal and Nepal: notably the Kaziranga National Park in Assam and the Royal Chitawan National Park in Nepal.

The Indian rhinoceros is the second largest of the five living species. Adult males weigh up to 2070 kg and stand up to 186 cm at the shoulder. Adult females are slightly smaller, weighing about 1600 kg and measuring about 160 cm at the shoulder. Both sexes have a single well-developed horn on the nose, which reaches a maximum length of about 60 cm but is normally between 15 and 45 cm long. In common with the other two Asian species of rhinoceros, both sexes have a pair of lower incisor tusks which reach lengths of up to 20 cm in adult males and are used in combat in preference to the horn. The chewing teeth (the molars and premolars) are high-crowned with a complex pattern of enamel which indicates a diet mainly of grass. The prehensile upper lip is used to gather in tall grasses and shrubs, but the tip can be folded under and opposed against the lower lip for cropping short grasses with the lips only. Two folds of skin encircle the body,

one behind the forelegs and one in front of the hindlegs. There are deep skin folds around the neck, particularly in adult males, and the skin of the rump is also folded and studded with tubercles.

The typical habitat of the Indian rhinoceros is the alluvial plain grasslands and woodlands of the often meandering rivers of northern India, where the grass may reach heights of 8 metres, where the *Bombax* trees flower in the spring time, and where abandoned river-beds and ox-bow lakes provide ample wallows and swampy feeding grounds. The rhinos' diet is varied. In Nepal 183 food species belonging to 57 botanical families have been recorded, but grass (50 species) made up between 70 and 90 percent of their diet according to the season. Other foods include shrubs, saplings, fruits and aquatic plants. Considerable seasonal variation in the availability of these foods results in regular movements of rhinos between types of vegetation. Most rhino movements are local and related to the availability of different foods, or flooding during the monsoon. As the rhinos have become more and more restricted to protected areas and their previous ranges have been cultivated, they have begun to raid the rice and maize crops of surrounding villages more frequently.

Although many rhinos may be seen grazing or wallowing together, they are generally solitary in habit and move independently of one another or in small groups. The most permanent associations are between cows and calves. For over three years they remain together, the mother only driving the youngster away shortly before the birth of a new calf. Some calves may stay with their mothers for up to five years. Physical contact is very important for a calf, which will often rub its head and flanks along its mother's body, sometimes climbing on to her back if she is lying down, or biting her ears and horn, or licking her skin. Frequently a calf frolics around the mother, sometimes picking up a stick in its mouth and charging back and forth with it like a young puppy. The mother rarely joins in the game, but if disturbed incessantly she will stand and suckle her calf from the side or between the hind legs. Nursing continues with decreasing frequency until the calf is over two years old.

Calves are always curious to initiate encounters with other rhinos, but mothers chase off strangers, sometimes quite fiercely. It is normally cows with calves that attack humans—and occasionally kill them—if surprised at close quarters. Oddly, however, cows sometimes leave their small calves unattended while they graze up to nearly a kilometre away. This only happens with calves younger than six months, and may explain how tigers can occasionally get near enough to kill calves. Each year about six calves are lost to tigers in Kaziranga and one or two in Chitawan. During the second year after the birth of her calf a cow comes into estrus again and is fertilized by a bull. The new calf is born after a

© M. Krishnan

Great Indian Rhinoceros cow with new-born calf, Kaziranga, Assam × ¹/₂₀ gestation period of about sixteen months. The older calf, driven away by its mother, suddenly becomes much more nervous. Young males in particular are subject to attacks by adult bulls and they respond to this risk by fleeing at the slightest hint of danger, standing alert at the crack of a twig or the approach of any animal. They also band together in groups of two or three, sometimes up to seven or eight, or with another cow-calf pair.

Aggressive interactions between rhinos, in particular between males, are frequent and sometimes result in the death of one of the combatants. Ten different vocal displays and visual displays such as the baring of the incisor tusks are used during encounters. Adult bulls fight among each other to determine a dominance order. The dominant bull in any one area will tolerate the presence of other bulls so long as they do not challenge him or attempt to mate with females in the area. The self-confident, dominant males display by squirting their strong-smelling urine in long jets behind them from their backward-pointing penises. They also drag their toes in the earth while walking, to make long parallel furrows on which the urine falls. These dominant males appear indifferent and even curious when disturbed by man, and may follow human scent instead of fleeing from it. They patrol their domains, visiting the numerous piles of dung which are typical of 'rhino country' and following the scent of any females in estrus. The mating chases of Indian rhinos are renowned for their length and noisiness: the female 'squeak pants' as the male charges after her for up to several kilometres. Although there is no proof, it is possible that this behaviour is a way for the female to advertise the fact that she is in estrus and to ensure that in the thick bush and grassland she is not mated by a bull of inferior strength.

Threats to the continued survival of the Indian rhino include poaching, encroachment by cultivators and stock-grazers, erosion as a result of annually increasing flood levels, and invasion of some of the Indian reserves by exotic plants. Ninety percent of the surviving rhinos live in two National Parks containing a total of only 500 km^2 of suitable habitat. Any catastrophe in Chitawan or Kaziranga, such as an epidemic disease or severe flooding, could drastically deplete the total rhino population. It is to guard against such an event that Indian rhinos have begun to be translocated to other protected areas within the former range of the species. As well as spreading out the population and insuring against a catastrophe in one reserve, removal of animals from the densely populated Kaziranga Park is helping to prevent overcrowding. Also, some of the smaller, isolated reserves in West Bengal and Assam may benefit from the introduction of new genes into their rhino populations.

W.A.L.

See UNGULATES.

RHIZOME. The rhizome is an underground, dorsiventral stem or branch which grows horizontally under the surface of the soil. Rhizomes are usually brownish in colour so that they are easily mistaken for roots. However, they are distinctly divided into nodes and internodes, bear scaly leaves at the nodes, and possess axillary as well as apical buds and thus can be easily distinguished from them. In several instances, adventitious roots may also develop at the lower surface of their nodes.

The more common types of rhizomes are rather fleshy due to the storage of food materials in them. Since an apical bud in them gives rise to the annual shoot which dies at the close of the season, leaving a scar, the rhizome continues its growth by a lateral bud. Common examples of such categories are ginger (*Zingiber officinale*) and turmeric (*Curcuma longa*) which are profusely branched. The rhizome of *Canna*, on the other hand, belongs to the other group since it shows poor branching. Rarely monopodial rhizomes may be found as in ferns like *Pteris*. A special type of rhizome is the root-stock of *Alocacia* which grows vertically rather than horizontally. The creeping sobole may be considered either as a thin rhizome or a runner.

From a single rhizome, several saplings can be obtained by dividing it into sections, each with at least one bud. For example, in certain orchids the rhizome is used to propagate the variety. Similarly, the rhizomes of several weeds get cut into sections during soil cultivation and each portion becomes potentially a new plant.

If we examine the internal tissue organization of rhizomes, it is seen that the cortex is usually parenchymatous, but it may consist of spongy or more compact tissues. In some genera it is made up of relatively thick-walled cells also.

In members of the Pineapple family (Bromeliaceae), the rhizome shows somewhat quantitative difference because here the periderm is often formed by the suberization of the outer cortex by the cell division. There are well developed intercellular air-spaces in the ground tissue as well. In some cane-like palms (e.g. *Bactris* and *Rhapis*), and climbing palms, creeping rhizomes are found. The rhizomes of *Rhapis* do not differ markedly from the aerial stems. Grass rhizomes are well developed and similar in structure to the culms. It is generally seen that sclerenchyma is less developed, and the ground tissue frequently serves for the storage of starch or other food reserves.

See also ROOT.

RIBBON FISHES of the family Trichiuridae, also known as hair-tails, are as the name suggests flattened, long ribbon-like forms with sharp teeth and long compressed jaws and a pointed tail. Two species, *Trichiurus lepturus* and *Lepturacanthus savala*, are more common among their six species. They occur on both coasts of

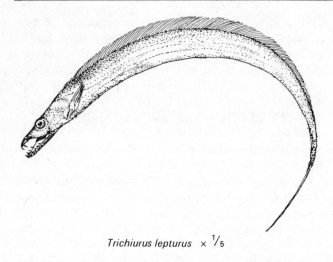

Trichiurus lepturus × ⅕

India and Sri Lanka, the west coast contributing a major part. They are pelagic and carnivorous. Being flat, they can be dried easily in the sun. The first species is thinner and grows up to 70 cm, while the other is comparatively thicker and longer (120 cm). Ribbon fishes together contribute about 65,000 tonnes annually to the country's fish supply and, as dried fish, are appreciated in rural areas.

RODENTS. Rats are amongst that group of mammals classified as Rodentia. This is by far the largest Order within the whole Class of Mammals and indeed rodents are not only the most numerous of all higher animals inhabiting the earth, but also account for about one third of all known mammal species. Rodents are for the most part rather small terrestrial species with short limbs having five digits and long slender tails. Most rodents walk on the soles of their feet, that is to say they are 'plantigrade'. But a detailed study of this Order reveals an amazing diversity of form and a development of abilities enabling them to utilize certain environments more successfully than other animal forms. They vary in size from the long-legged swamp-dwelling Capybara, a pig-sized rodent inhabiting South America which can weigh up to 50 kg; they vary in shape from the large-headed and bulky Crested Porcupine, to the long-tailed and graceful Flying Squirrel, and include some of the smallest known mammals, such as the tropical field-mice of both Africa and India which can weigh as little as 5 grams when adult.

What makes rodents so successful is that they have evolved into forms capable of occupying every part of the world's surface. There are species which can glide, species which live their whole lives in trees and others that can survive extremes of temperature by living their whole lives in underground burrows. Some species inhabiting our region can live and find food without hibernating, under a seven-foot-deep blanket of snow (the Murree Vole), others sleep out the long winter curled up together in an underground nest chamber for eight months of the year (Marmots). Still others can collect and store food underground for use in time of need, for example, the Migratory Hamster (*Cricetulus migratorius*) and the Lesser Bandicoot (*Bandicota bengalensis*). Others have developed extremely elongated hind legs with a reduced number of toes enabling them to traverse rapidly over soft sand-dunes and they can also burrow into and shelter comfortably in these unstable sand-dunes (e.g. Jerboas and certain *Gerbillus* species).

But the one unique feature, shared by all rodents, which enables them to be such successful feeders, is in the arrangement of their dentition. Rodents have only one pair of incisors in each jaw. These grow continuously from persistent pulp cavities throughout the rodent's life and are shaped as segments of a true arc. These incisors are only covered on the anterior surface by hard enamel, usually orange-coloured. The rest of the tooth comprises softer cement. In gnawing on food or other hard surfaces, the incisors therefore wear unevenly into a chisel-like cutting edge. Furthermore the incisors are separated from a reduced number of molars or grinding teeth by a long gap called the diastema and there are no canine teeth. Usually there are only four molars, and rodents are equipped with strongly developed masseter (or jaw) muscles. The arrangement and form of these masseter muscles have been found to fall into three distinct groups and this has been used to classify the Order. Even quite small rodents, with this arrangement can masticate the toughest of plant material, and gnaw through very hard surfaces such as the shells of nuts and can use their incisors to bite through soil as well as other protective materials without having to taste or swallow any of this until a potentially edible or palatable item is encountered. This partly explains why House-Mice and rats will gnaw such inedible items as candles (paraffin wax), and how many species can burrow through the hardest soil even though only equipped with comparatively weak and small fore-paws. Gnawing is also a necessary process to prevent incisors from growing too long and becoming less effective as cutting tools.

The classification of rodents according to the arrangement of the masseter muscles enables the relationship of seemingly unalike species to be better understood and helps to sort out this complicated assemblage of small animals. The first group, known as the Sciuromorphs, comprises about 365 species and is thought to include the most primitive families or the earliest to develop. These are the Mountain Beaver (*Alpodontia rufa*) found only in North America, as well as the many forms of Tree SQUIRRELS, Flying Squirrels and MARMOTS, representatives of which are found within within our area.

The next big group, known as the Myomorphs, is

by far the biggest and most successful, comprising over 1100 species. It includes the more familiar families such as mice (see MOUSE), RATS and mole-rats, as well as lesser-known and more specialized groups such as the Jumping Mice, JERBOAS, Dormice, Birch Mice and Voles, Hamsters and Lemmings.

The last group, known as the Hystricomorpha, comprises about 180 species which are mainly confined to the New World and includes the PORCUPINES, Guinea Pigs, Capybara and Chinchillas. The only representatives of this group found in our region are the Malaysian Brush-tailed Porcupine and the Indian Crested Porcupine.

Amongst the more notable Sciuromorphs found in our area are the Giant Tree Squirrel (*Ratufa* species). It is found in many varied forms according to the colour and pattern of its pelage and four species occur, from Sri Lanka and peninsular India up to Assam and Burma in the northeast. These are the largest known squirrels inhabiting mainly tropical forests in regions of higher rainfall. The much smaller Palm Squirrels (*Funambulus* species) are better adapted to drier conditions and open savanna country. Both these squirrels are diurnal in activity and subsist mostly upon berries, fruits, young buds and occasionally birds' eggs and insects. Flying Squirrels are nocturnal in activity and belong to several distinct genera. The Giant Flying Squirrels of the genus *Petaurista* measure up to three feet in length including the tail and can glide more than 150 metres' distance between trees. The much smaller Flying Squirrels of the genus *Hylopetes* inhabit Kashmir eastwards to Malaysia whilst the rare and little-known Woolly Flying Squirrel (*Eupetaurus cineres* seems to be adapted to live amongst rocks as much as trees. The MARMOTS of the Himalayas and Central Asia are also Sciuromorphs and are closely related to the Woodchuck of North America. Ground Squirrels are burrowing squirrels adapted to steppe areas and many species occur in both the prairies of North America where they are called Gophers and in central Russia where they are called Susliks.

Amongst the many Myomorphs inhabiting our region, the more primitive Birch Mouse (*Sicista betulis*) may be mentioned as inhabiting the drier inner Himalayan ranges and the beautiful little bushy-tailed Forest Dormouse (*Dryomys nitedula*) found in the juniper and holly-oak scrub forest of Baluchistan and northwestern Pakistan. Two genera of Jerboas, *Jaculus* and *Allactaga*, are found in the western part of Baluchistan. There are two more very big families within this sub-group, the Muridae and the Cricetidae. The Cricetidae (meaning squeaking ones) are thought to be of more primitive origins and these include the Voles with short tails and burrowing habits found in the northern regions, as well as the Hamsters having fur-lined cheek pouches

Indian Gerbil or Antelope Rat, nocturnal, × 1

Desert Jird or Indian Desert Gerbil, diurnal, × ½

enabling them to carry and store large quantities of food for use during the winter time. There are a large number of desert-adapted Gerbils and Jirds characterized by having hairy tails which terminate in a tuft of longer, usually darker, hairs. Amongst these, the nocturnal Indian Gerbil (*Tatera indica*) is widespread from the drier parts of Sri Lanka up to the northern foothills. They can often be so numerous as to cause considerable damage to farm crops. The jirds are highly colonial and usually diurnal in feeding activity with two species adapted to mountainous conditions in the west, and the Common Indian Jird (*Meriones hurrianae*) being especially abundant in Rajasthan where it has become an agricultural pest especially in young wheat or other green crops.

Turning to the family Muridae, this includes both the Large and Lesser Bandicoot Rats (*Bandicota indica* and *B. bengalensis*) as well as the household pests, the Roof Rat, House-Mouse and Brown or Norwegian

© T. J. Roberts

Soft-furred Field Rat, × ³/₄

Rat. The tiny field-mouse *Mus booduga* shuns human habitations and occurs in scrub-desert and cropland over most parts of India. The arboreal Long-tailed Climbing Mouse (*Vandeluria oleracea*) occurs from Sri Lanka up to Kumaon in better forested areas. Muridae also include the Woodmice (*Apodemus* sp.) found in palearctic regions of Russia and Europe as well as the northern Himalayas. The Soft-furred Field Rats or Metads (*Millardia* species) are unique to the subcontinent. The specialized Lesser Bamboo Rat (*Cannomys badius*) is a largely fossorial rat often classified in a separate family and occurs in Nepal and Assam. Altogether there are nearly 60 different species of rodents inhabiting the subcontinent.

Rodents can cope with extremes of climate and harsh conditions because of their ability to burrow, to feed only during the cool of the night, and to subsist without drinking any free water, but the success of this Order is not wholly explained by these adaptations. Most rodents are relatively defenceless against predators and they do in fact form the staple diet of many small carnivores and even snakes and birds of prey. However, the reproductive ability of most rodents is phenomenal. Most murids become sexually mature within 5 or 6 weeks and have a gestation period of only 20 days. They are capable of being impregnated within 2 or 3 days of parturition, post-partum estrus being normal, and thus can reproduce continuously and bear six to eight litters a year. When conditions are favourable and food supplies abundant they are capable of producing very large litters. *Bandicota bengalensis* females have been found with seventeen embryos. In times of food shortage or other physiological stress, pregnant females also have the faculty of being able to re-absorb their foetuses. Thus the maximum number of young which the mother is likely to be able to suckle and successfully rear are produced according to availability of food and shelter. This explains why, in ripening grain crops, populations of field rats or mice can sometimes seem to expand so suddenly as to reach almost plague proportions.

Mention has been made of the damage inflicted by rodents to farm crops as well as grain stores. Rodents in this region have been estimated to be responsible for well over 5 percent loss of the annual tonnage of cereals grown. Another authority estimated that the quantity of food grains destroyed by rodents in one year in India, would be enough to fill a freight train stretching for more than 3000 miles. Particularly heavy losses are suffered to rice where the Lesser Bandicoot is plentiful, but a variety of crops can be affected by different species. Bush Rats (*Golunda* sp.) damage coffee plantations and Metads (*Millardia* species) damage groundnuts in peninsular India. Rodents also act as reservoirs of diseases which can be passed on to man and they can directly infect man by contaminating his food. Besides bubonic plague, several forms of typhus and typhoid fever are rodent diseases which can be passed on to man. Not all rodents are harmful to man however, and this account would not be complete without mention of the valuable contributions made by such fur-bearing rodents as the Chinchilla, Beaver and Musk-rat. Laboratory strains of House-Mice and the Brown Rat have played an important part in biological research, not only in unravelling the mysteries of genetics but also in testing and developing many valuable drugs used by man.

See also BANDICOOT RATS, JERBOA, MARMOT, MOUSE, PORCUPINES, RATS, SQUIRRELS and VOLES.

ROLLER or 'BLUE JAY' (*Coracias benghalensis*). A rather sluggish pigeon-sized bird related to the kingfishers, largely rufous-brown at rest with brilliantly flashing dark and pale blue wings in flight. Commonly seen perched sedately on telegraph wires in open cultivated country, and light forest, flying lazily down to the ground now and again to pick up crawling prey—lizard, frog, grasshopper and suchlike. Has a spectacular mutual courtship display, both sexes flying around erratically, rolling from side to side, tumbling and nose-diving to the accompaniment of loud raucous screams, croaks and chuckles. Nests in tree-hollows, laying 3 or 4 roundish white eggs.

ROOSTING BEHAVIOUR. No animal is active for twentyfour hours of the day. To do this would require adaptation to move around when it is hot and cold, and bright and dark, and none of the animals has managed to stride such a wide range. Animals must remain inactive for many hours each day. This demands that they must evolve the timing, the locations and the pattern of behaviour for these periods of inactivity so as to maximize their chances of survival by avoiding predation, by enhancing physiological recuperation and by putting to best use any other opportunities that this bout of inactivity presents. All of this constitutes the roosting behaviour of an animal, and is in itself a subject full of fascinating variation.

The roosting times are of course related to the time of optimum activity for an animal. For day-active animals therefore night is a primary time for roosting, but they also roost in between bouts of activity during the daytime. The chital is essentially a diurnal animal, but at Bandipur it generally has a bout of rest in the forenoon as well as in mid-afternoon. When at rest during the day it sits down, but keeps its neck up. The resting site remains more or less the same from day to day. The herd so arranges itself that its members face outwards in every direction. And, as a ruminant, it uses this period for chewing the cud.

Other day-active animals may have a single bout of activity during the day and may roost continually for the rest of the time. The bigger vultures conform to this pattern. They take off when the day is warm enough for them to soar on warm updrafts, and come down when they discover a carcass. They generally roost in a tall tree wherever they have been feeding that afternoon, so that the roosting site keeps shifting from day to day. When roosting they try to get on as high a tree as possible for they cannot take-off quickly and easily and they roost with their bare necks tucked in to minimize loss of heat during the night.

Insectivorous bats hunt at night and roost during the day. They are particularly vulnerable while roosting, for their body temperature drops and they cannot move quickly if disturbed by a predator. They therefore choose deep, dark caves to afford them shelter from predators during the day. These caves have the additional advantage of high and constant temperature and humidity so that the bats exert little energy in keeping warm and suffer little water loss during their roosting time. But these caves with their narrow exits are also a trap, for at dusk hundreds of bats emerge through these crevices during half-an-hour or so. Owls and dusk-active hawks hang around these entrances to hunt the bats as they emerge.

For most animals the timing of onset and termination of roosting behaviour is related to intensity of light and temperature. In cold-blooded animals such as lizards the timing of roosting is determined by the period for which the body temperature cannot be maintained in the proper range. In the warm-blooded animals the timing is based on environmental cues provided by the light intensity. Different animals have the characteristic range of light intensity at which they terminate or commence roosting activity. Thus the house crows both wake up and sleep at much lower light intensities than the Indian myna, so that the crow is active over a much longer period of the day. With light intensity as a cue the animals are naturally active for longer periods during the summer. However some insectivorous bats at Madurai exhibit the remarkable behaviour of becoming active at the same time every evening regardless of the season. This, however, is exceptional.

The roosting sites tend to be sites of safety from predators. Thus they are caves for bats and swiftlets, rocky crevices for swifts and martins, open ground on which the bird is perfectly camouflaged for nightjars, tree-holes for owls, and high trees for peafowl, mynas, crows and many other species of birds. Remarkably enough, birds such as mynas prefer to roost on isolated trees in the cities or even in tall open buildings such as railway stations even though they do most of their feeding in the surrounding villages. Perhaps the much lower populations of predators like snakes is the reason for their preference for urban roosting sites.

The roosting posture is generally such as to enable the animal to conserve body-heat. So the limbs are withdrawn, the wings folded, the head tucked in.

Animals may roost in groups of various sizes. Many birds such as koels, magpie-robins and reed warblers roost solitarily. Others such as bulbuls and kingfishers roost in small groups. While yet others like mynas roost in groups which may exceed ten thousand in number. Occasionally birds of several species such as the Indian myna, house and jungle crows and the roseringed parakeet, may all roost together in thousands. The advantages of roosting in such groups may be warmth, safety from predators, or exchange of information about food sources. Bee-eaters roost huddled close together and their communal roosting may help them to conserve heat. An early-warning system must confer safety on birds sleeping in groups. Thus it has been shown that wagtails sleeping solitarily can easily be captured at night by hand, but it is impossible to do this for birds roosting in large groups. Communal roosting may also allow birds to derive information about patchy food sources from other birds. Thus a flock of parakeets may have fed on fruits of a tree which was nearing exhaustion. Next day they may be able to detect and follow the flight of another flock of parakeets which have discovered a very good fruiting tree the previous day and fly unhesitatingly towards it.

Birds roosting communally in large groups often exhibit spectacular social displays at the roosting sites. In many communally roosting Indian species such displays are particularly common just before the commencement of the breeding season. The function of such social displays is unclear.

ROOT. The plant organ, which originates from the radicle of the embryo, is devoid of leaves and flowers, and develops in a direction reverse to the stem, is known as the root. Its chief functions are to provide support to the plant and absorb moisture (water) and minerals from the substratum. In order to achieve these, the root grows inside the soil, away from light, under the influence of gravity, and slowly the germinating seed and seedlings get firmly entrenched.

There are two chief types of root development: (a)

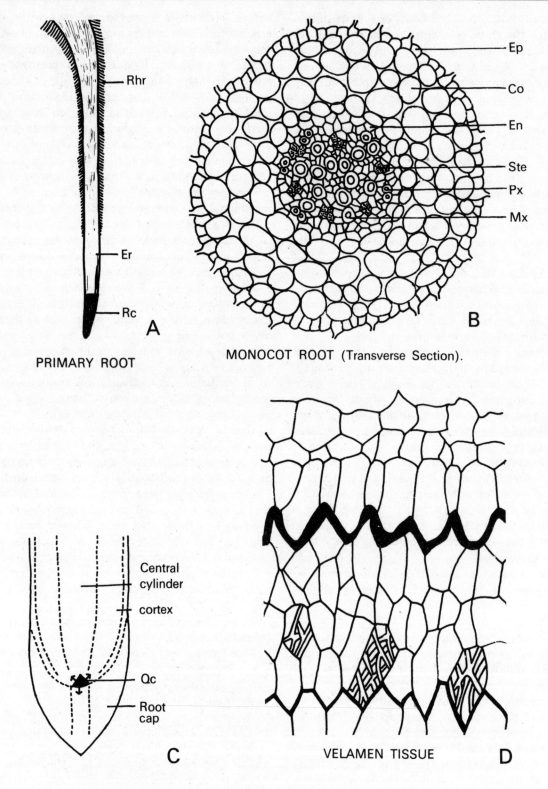

PRIMARY ROOT

MONOCOT ROOT (Transverse Section).

VELAMEN TISSUE

**Fig.1 APICAL ZONATION AND TISSUE DIFFERENTIATION
IN THE ROOT**

(Co cortex; En endodermis; Ep epidermis, Er elongation
region; Mx metaxylem; Px protoxylem; Qc quiescent centre;
Rc root cap; Rhr root hair region; Ste sieve-tube element).
A Different zones in a primary root, about × 5

B Transverse section of a monocot root, about × 25
C Organization of distal region of apical meristem in the
root, about × 25
D Velamen tissue in orchid roots, about × 100

The tap root system (e.g. radish—*Raphanus sativus*) and (b) fibrous *root system* (the common grasses, and bamboos). The former is the direct extension of the primary root of the seedlings. However, in the fibrous root system, at a very early stage, the primary organ is replaced by a number of roots produced at the base of the stem. All such roots which do not originate as branches of the primary root are known as *adventitious roots*. In the lower plants such as the bryophytes and the thalli of ferns, the function of roots is performed by thread-like structures—the rhizoids. In the members of the Hepaticeae these are made up of a single cell whereas in the mosses several cells constitute them.

At the tip of all roots occurs a group of cells known as the root cap, which protects the growing point as the root grows inside the soil by releasing lubrication. The cells of the root cap are produced by the growing point. Just below the root apex occurs the zone of cell elongation in which the cells are in a phase of continuous expansion and elongation. These cause the root to be pushed inside the soil with a marked force. This is followed by the root hair zone (Fig. 1A) which besides being largest of all is functionally the most important region of the organ. The hairs which develop as extensions of epidermal cells are the chief organs of water and mineral salt absorption owing to their proximity to the soil. Generally these have a short life-span and as soon as a hair dies, it is replaced by new ones.

The tissue distribution in the root differs from that in the stem (Fig. 1B) since here the mechanical elements are located in the centre. These have a broad cortex and its innermost layer is known as the endodermis. Except for some passage cells, all the constituents of this layer possess cell walls which get thickened by the deposition of corky substances: this process is called suberization. Inner to the endodermis occurs the pericycle whose outer layers are usually made up of parenchyma cells. The protoxylem is situated towards the outside—hence the exarch condition. Because of this orientation the xylem becomes star-shaped. In the roots of the monocotyledons, the number of arms of protoxylem is much larger than that of the dicotyledons. In the central region the amount of pith is also large.

But for a few exceptions, the secondary growth is restricted only to the roots of dicotyledonous plants. During this process the parenchymatous cells of the internal layer of phloem become active by the division of cells and thus the cambial layer is produced in the form of a ring. This gives rise to the secondary elements on both the outer and inner side. If the secondary growth is much more pronounced, as in trees, the entire pericycle becomes active and forms a corky layer towards the outside of the stele.

In the stem near the tip, buds develop from the external tissue layers. If this will happen in the root, buds will easily break apart, because the main root has to pass through several layers of soil. As such in the roots, branches originate more from the inner and lower sides of the growing regions: therefore they are known as endogenous in origin. During branch formation, the cells of pericycle become active and constitute a growing point like that of the main root just opposite the group of protoxylem elements. This new apex grows inside the cortex exactly in the same manner as the root grows in the soil. When it comes out after tearing away the parent root, its conducting elements have become fully formed. The profuseness of branching of roots depends upon the size and nature as well as the habitat of the plant. For example we see that in large trees the roots are highly ramified and thick to provide support and anchorage.

In between the stem and root is situated the hypocotyl region where the vascular tissue arrangement changes from root pattern to stem pattern. However, the conducting system remains continuous throughout without branching anywhere.

Usually the roots also become transformed to store food material. Man has learnt to use such of these which possess large amounts of carbohydrates, proteins, and fats, etc. Among the most common examples of this category are the carrot (*Daucus carota*), radish (*Raphanus sativus*), beet (*Beta vulgaris*). These are all tap roots. Sometimes adventitious roots may also undertake a similar function e.g. tuberous (sweet potato-*Ipomoea batatus*), fasciculated (*Dahlia*), nodulose (turmeric-*Curcuma amada*), moniliform (basil-*Basella rubra*). The rhizome of *Dahlia* and the roots of various orchids belong to this category (Fig. 2). In the climbers the roots are generally adventitious and continue to come out from all through the stem region. Similarly, in strawberry also the roots are adventitious. In some tropical orchids, which grow hanging in the air, a sponge-like tissue—the velamen—occurs. Its component cells absorb moisture and may also contain chlorophyll (Fig.1D).

In some members of the family Gramineae such as *Zea mays* roots appear from the nodes of the stem which provide additional support to the plant. The best example of such roots is seen in Banyan. These grow vertically and once they touch the soil, they grow inside it. In this manner they provide a solid support to the growing branches. In some marshy places which support mangrove vegetation, respiratory roots are seen, known as pneumatophores (Fig. 2). It is believed that these appear due to the fact that there is only a small quantity of oxygen present in the soil; simultaneously, some branches of roots grow upwards in the air. The oxygen becomes diffuse in the entire root system.

Besides, there are some roots which perform the function of carbon assimilation also. *Tinospora cordifolia*

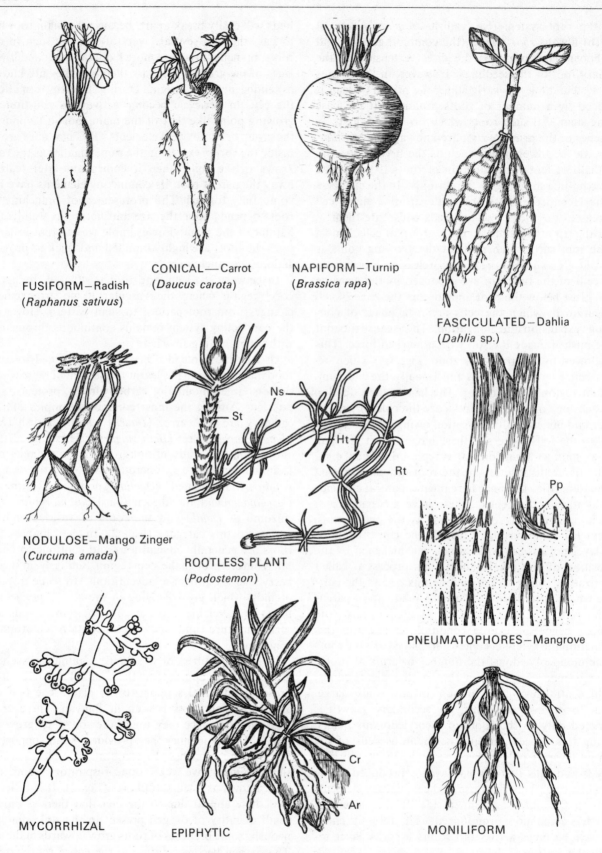

FUSIFORM—Radish
(*Raphanus sativus*)

CONICAL—Carrot
(*Daucus carota*)

NAPIFORM—Turnip
(*Brassica rapa*)

FASCICULATED—Dahlia
(*Dahlia* sp.)

NODULOSE—Mango Zinger
(*Curcuma amada*)

ROOTLESS PLANT
(*Podostemon*)

PNEUMATOPHORES—Mangrove

MYCORRHIZAL

EPIPHYTIC

MONILIFORM

Fig.2 COMMON MODIFICATIONS OF THE ROOTS

(Ar aerial root; Cr clinging root; Ht heptera; Ns new shoot; Pp pneumatophores; St shoot)

(giloy or gulancha) whose branches climb neighbouring trees and produce slender, hanging roots with abundant chlorophyll in its cells, is a good example of this. While growing in humus these become infested with fungal mycelia and constitute the mycorrhizal roots which absorb food solutions to be utilized both by the host plant and fungus (Fig. 2).

It has also been found that the roots grow and perform their function in a more or less conditioned atmosphere—the rhizosphere—in the soil. For example, the capacity for maintaining resistance against fungal infection or otherwise is based upon the specific milieu produced due to ecology which involves the release of enzymes by the host as well as the parasite (e.g. *Fusarium udum*), and the absorption of nutrients by the roots. There are some instances of plants with no roots. They are to be found among aquatic plants and among saprophytes (e.g. *Podostemon*).

A number of experiments have shown that individual root cells can divide and develop plants directly. In fact the first attempts for obtaining complete seedlings on a defined culture medium succeeded in carrot roots (phloem component) at the University of Cornell.

Research works over the past couple of decades have shown that at the root apices, just below the zone of root cap, there occurs a comparatively inactive group of cells—the *quiescent centre* (Fig. 1C). This is composed of almost semicircular cells surrounded on all sides by actively dividing initial cells. Its dimension varies in roots of different sizes, depending upon the number of constituents. Occasional divisions do occur in its centre and it may become active when the previously active initials are damaged, as for example by radiation. The component cells are thick-walled and resistant to damage due to their inactivity. They may be sites of auxin synthesis and the source of diploid cells below for replacement of polyploid and aneuploid cells that may get accumulated during somatic differentiation.

See also STEM, RHIZOME.

G.S.P. & U.R.

ROSEWOOD, *Dalbergia latifolia*, is a large deciduous tree of wide distribution attaining its best in the Western Ghats. The tree bears compound leaves with 5–7 leaflets, small whitish flowers and 1–3-seeded oblong, lance-shaped pods. Rosewood is characterized by a gold-brown to rose-purple or deep purple heartwood streaked with black. It ranks among the finest wood for furniture and is prized timber for making musical instruments, e.g. pianos, clarinets and guitars.

SAFFLOWER, *Carthamus tinctorius*, is a thistle-like herb, cultivated throughout the greater part of the sub-continent. It is characterized by toothed, spinous leaves, with orange-red flower-heads surrounded by bristly bracts. A dye extracted from the flowers is used for colouring wedding clothes; seeds yield a cooking oil, and are also used for paints and varnishes.

SAFFRON, *Crocus sativus*, a bulbous plant with linear, channeled leaves, dark, violet flowers with yellow anthers and brick-red stigmas. A native of S Europe, it has been in cultivation in the Kashmir valley for centuries. The saffron fields in Srinagar are a spectacular sight in autumn when millions of flowers come into bloom. For preparing saffron, the flowers are collected in the morning after the dew disappears, styles and stigmas are separated and dried in the sun. They are a favourite ingredient for flavouring and colouring confectionery, curries, rice and liquors.

SAL, *Shorea robusta*, is one of the most gregarious of trees, and is distributed from the Kangra Valley to Arunachal, Assam, Bangladesh and descending to the Godavari river through central India. Never quite leafless, it is characterized by shining leaves which are broad and ovate. Small creamish fragrant flowers appear in April. The ovoid fruits appear in June, and are topped by five propeller-like wings which help dispersal of seeds which lose their viability early. Sal has a high reputation for the excellence of its timber, is durable, not attacked by termites, and is in great demand for railway sleepers. A gum exuding from the stem is used in caulking boats and waterproofing roofs. It is called sal dammar. Sal butter, extracted from the seeds, is used in cooking and in the manufacture of chocolate.

SALT-WATER PLANTS. Mangroves are trees and bushes inhabiting tidal lands in the tropics. The vegetation community formed by mangrove plants is called a mangal. 58 principal salt-tolerant (halophilous) species are known from our area, distributed over 41 genera and 29 families.

The greatest concentration of mangals is in the Sundarbans of West Bengal and Bangladesh, with a total area of about 100,000 km² in the subcontinent. The essential conditions for the development of mangals are (1) a warm climate ranging from the semi-arid type of Kutch–Kathiawar to the wet type of Bengal, (2) periodic

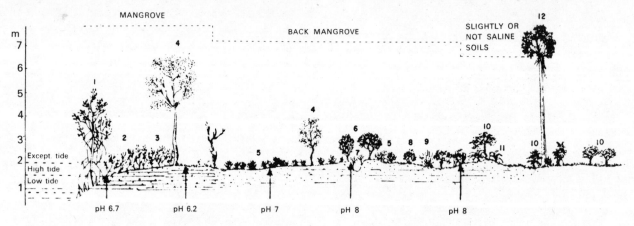

ZONATION IN PICHAVARAM MANGAL (After Blasco 1975)

1 *Rhizophora apiculata*	7	Sea holly, *Acanthus ilicifolius*
2 *Bruguiera cylindrica*	8	*Clerodendrum inerme*
3 *Ceriops decandra*	9	Dwarf Date Palm, *Phoenix pusilla*
4 *Avicennia*	10	*Thespesia populnea*, Portia tree
5 Saltworts: *Suaeda maritima* and *S. monoica*	11	*Prosopis juliflora*
6 *Excoecaria*	12	Palmyra palm, *Borassus*

480

The Sloth Bear (*Melursus ursinus*) works very hard at night to find enough fruit, honey, termites and beetles to feed itself.

The Red Panda (*Ailurus fulgens*), a smaller relative of the Giant Panda, is found only in Himalayan forests, sleeping in the branches by day and descending to the forest floor at night to eat leaves, roots and insects.

The Brown Bear (*Ursus arctos*), confined to the central and western Himalayas, wakes from long months of torpid sleep in spring, follows the melting snows up to graze on the new grass, and descends to lower levels only in summer, when fruit is abundant.

The Common Otter (*Lutra lutra*) in our area is only found in Himalayan regions and the hills of South India.

Danaid Eggfly *(Hypolimnas misippus)*
The female is a wonderful mimic of the Plain Tiger *(Danaus chrysippus)*

Common Indian Crow (*Euploea core*)
The commonest of the Crows, seen throughout the subcontinent below 2000 m.

Blue Pansy (*Precis orithyia*)
The Blue and Yellow Pansies are two of the most abundant and prettiest of Indian butterflies.

Yellow Pansy (*Precis hierta*)

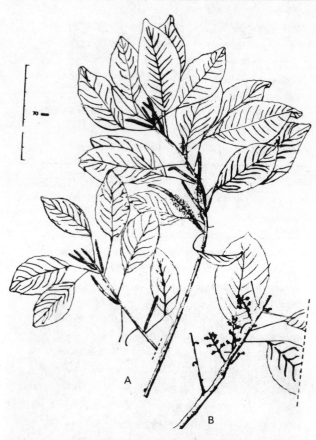

THE BLINDING TREE (*Excoecaria*)

A Male inflorescence. B Female inflorescence

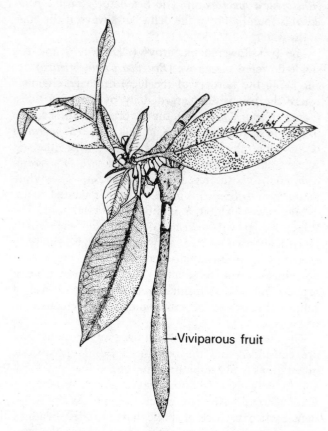

←Viviparous fruit

Rhizophora, or the twin-flowered mangrove × ¹/₃.

flooding by tides and (3) protected shores. Mangroves grow on soft muddy soils in that part of the tidal range where mud is deposited naturally. They form distinct zones in relation to the frequency of inundation by tidal waters and the depth of flooding, soil salinity, texture and structure.

A good example of complete zonation is that observed at Pichavaram near Parangipettai (Porto Novo) in the South Arcot district of Tamil Nadu. There is a narrow *Rhizophora* (twin-flowered mangrove) belt along the channels of the creek; on clayey, constantly wet terrain *Rhizophora apiculata* is the most common but *R. mucronata* (White Candle mangrove) is also abundant. Under *Rhizophora* species are seen Kankandan (*Bruguiera cylindrica*), Goran (*Ceriops decandra*), Madagascar mangrove (*Lumnitzera racemosa*), Goat's Horn mangrove (*Aegiceras corniculatum*) and a few scattered Kambala trees (*Sonneratia apetala*), rosewood climber (*Dalbergia spinosa*) and Panlata climber (*Derris trifoliata*).

Behind the *Rhizophora* belt is the zone of the White mangrove (*Avicennia marina*) with dwarf under-shrubs of the common Indian Saltwort (*Suaeda maritima*) and the stunted Blinding tree (*Excoecaria agallocha*), which is the only deciduous mangrove.

The third zone is that of the back mangrove consisting mainly of herbaceous and under-shrubby salt-marsh plants, halophytes like the common and the greater Indian Saltworts (*Suaeda maritima* and *S. monoica*), Seaside Parslane (*Sesuvium portulacastrum*), Heliotrope (*Heliotropium curassavicum*), English seaside grape (*Salicornia brachiata*), the grass *Aeluropus lagopoides* and the Sea Holly (*Acanthus ilicifolius*) among others. These grow on firm muddy soils with salt incrustation.

The Sundarbans area experiences scarcity of fresh water due to the shifting course of the Ganga. Consequently, species like Sunderi (*Heritiera fomes*) and a small palm *Nypa fruticans* are disappearing. All the tree species of the Indian mangroves are represented here but the plants vary from one island to the other. Dense mangals with Sunderi (*Heritiera*) are encountered only in the eastern part of the Gangetic delta. The dominant species is sometimes the Blinding tree (*Excoecaria*) or the Kambala tree (*Sonneratia apetala*). Others are Goran (*Ceriops decandra*), White mangrove (*Avicennia*), Single-flowered Calycine mangrove (*Bruguiera gymnorrhiza*), Pussur (*Xylocarpus granatum*), the crab oil plant (*X. moluccensis*), the Goat's Horn mangrove (*Aegiceras corniculatum*) and White Candle mangrove

(*Rhizophora mucronata*). The Sea-date (*Phoenix palu-dosa*) is found all over the delta but always at the edge of the water.

The twin-flowered mangrove (*Rhizophora*) and the single-flowered mangrove (*Bruguiera gymnorrhiza*) occur along the borders of the local drainage channels where the soil is submerged daily by the high tides. Goran (*Ceriops decandra*), on the other hand, occupies the highest terrain. The Kambala tree (*Sonneratia*) occurs along the gently sloping borders where the soil is constantly soaked in water. Wild rice (*Porteresia coarciata*) fixes the recent alluvial deposits. The White mangrove (*Avicennia*) is more near the inhabited zones.

Mangroves exhibit a number of adaptations. The White mangrove (*Avicennia marina*) and the Goat's Horn mangrove (*Aegiceras*) withstand high salinity in their sap because the salt-excretory glands on the leaf epiderms control the salinity. The twin-flowered mangrove (*Rhizophora*) maintains a low internal level of salinity by means of salt-eliminating mechanism in its roots.

The anaerobic mud in which the roots of mangroves are embedded and the periodic floodings necessitate adaptations to aid respiration and anchoring. An extensive radially disposed horizontal cable-root system provides a broad base upon which to grow. As the soil is deficient in oxygen, some species have air-filled breathing roots, 'pneumatophores', projecting above the soil. The cable roots and the pneumatophores produce anchor roots that anchor the plants to the muddy soil and also fine nutrition roots in the uppermost mud layer which is rich in oxygen. The pneumatophores may be thin and finger-like as in the White mangrove (*Avicennia*) or thickened and conical (Kambala tree, crab oil plant), knee-like (*Bruguiera* and Goran) or even convoluted, plate-like (Pussur). Mangroves have also numerous breathing holes (lenticels) covering the stem and roots. In *Rhizophora* there are long, arching, air-filled stilt-roots that arise from the lower trunk. Adult

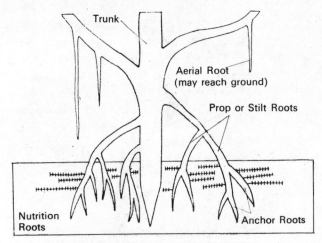

A sketch of the root system in *Rhizophora*. (After Semeniuk *et al.*)

Stilt roots of the mangrove **Rhizophora**

© G. Thanikaimoni

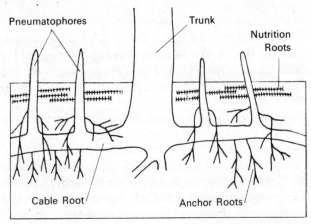

A sketch of the root system in the White mangrove, *Avicennia*. (After Semeniuk *et al.*)

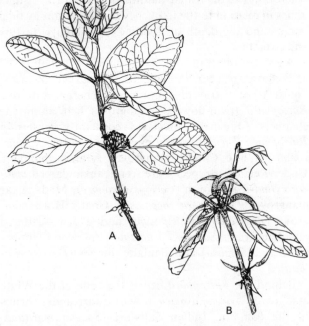

Goat's Horn mangrove, *Aegiceras*, × ⅓ A Flowering twig with young fruits. B Developing seedlings attached to the plant.

Rhizophora have in addition slender aerial roots descending from the lower branches which even on reaching soil do not seem to take root.

Many of the mangrove plants show vivipary, i.e. their fruits germinate while still attached to the parent plant. All of the mangrove species of the family Rhizophoraceae, the species of White mangrove (*Avicennia*) and Goat's Horn mangrove (*Aegiceras corniculatum*) exhibit this quality.

Excessive human interference has considerably affected the mangals. The twin-flowered mangrove (*Rhizophora*) has practically disappeared from the west coast. The mangals of Kerala have been almost eliminated and those of the Godavari delta and Kutch much disturbed because of their use as fuelwood or fodder. Mangroves have been used in a variety of ways as timber, firewood, charcoal and as a source of tannin. Mangals are also converted into salt-pans. However, their greatest value is as stabilizers of coasts. Wherever they have been disturbed, coastal erosion has been accentuated. As the mangrove swamps advance towards the sea, on the landward side mangals are replaced by terrestrial vegetation. Thus mangroves aid in reclamation of saline soils and creation of fertile land; their conservation is very necessary.

V.M.M-H.

See SEASHORE PLANTS, FOREST TYPES, SOILS.

SAMBHAR. The Sambhar lake is one of the largest inland saline depressions in India, situated 26°55′N, 75°11′E in the Jaipur and Nagaur districts of Rajasthan. The lake, virtually a hollow in the sand, normally covers an area of about 233 km². It is 35 km long and its width varies from 3 to 11 km. The dry bed of the lake lies at about 360 m a.s.1. The depth of water varies from 0·5 to 2 m. The area around the lake is covered with blown sand, except in the south and northeast where hillocks comprising schists and gneisses of Aravalli period occur. The lake is fed mainly by two flood rivers, Bandi and Mendha, and by several other small streams draining an area of about 7000 km². The lake dries out completely during summer. However, in recent years, due to heavy rains in the region, the water holds all through the year and the expanse of the lake has also been enlarged.

The origin of salt in the lake is much debated. Earlier it was maintained that salt is blown from the Rann of Kutch and is deposited in the lake. This theory has been severely criticized and discarded in favour of supposing that the salt lake is a remnant of the ancient transgressions of the sea. However, lack of calcium and magnesium in the lake water is difficult to explain if sea water had been the source of salt. The Geological Survey of India has recently elucidated the origin of salinity in Rajasthan. A single source has not been ascribed for the presence of salt accumulation but the extreme aridity and evaporation, coupled with centripetal drainage, are supposed to cause the prime inputs of salt from the catchment areas into the lake. On evaporation the brine of Sambhar lake shows three main components: NaCl 86%, Na_2SO_4 10%, and Na_2CO_3 4%.

The lake surface is devoid of any vegetation but halophytic plants like *Suaeda fruticosa*, *Salsola baryosma*, *Cressa cretica* are found on the shores. On higher grounds, around the lake, trees like *Tamarix articulata*, shrubs like *Tamarix dioica*, *Mimosa hamata*, and herbaceous cover of *Tephrosia* spp., *Cyperus* spp., *Portulaca oleracea*, *Aleuropes lagoprides*, *Chloris virgata* and *Sporobolus* sp. are supported by saline soil. About a kilometre away from the expanse of the lake, typical desert vegetation occurs.

Green and blue-green algae are, however, reported from the lake. Since Sambhar lake has not been drying during summers for the last few years, and due to excessive water, the salinity level of lake brine has considerably dropped. As a consequence, the growth of algae and availability of food has been enhanced. In turn, the monsoon-visiting flamingos, *Phoenicopterus ruber* and *P. minor*, are found in the lake in much larger numbers and all the year round, as compared to earlier years when the lake used to be dry during summer. Another spectacular change in Sambhar lake limnology due to its flooding is that the once abundantly found crustacean, *Artemia salina*, has totally disappeared! Sambhar lake with its vast fluctuations in brine salinity and evolutionary history is an open book of natural history.

I.P.

SAND-HOPPERS are amphipod crustaceans whose legs are modified both for walking and jumping. Their gills are at the base of their trunk appendages. The abdomen is fairly elongate and each of its first three segments bears a pair of swimmerets and each of the remaining segments a pair of appendages for springing. *Gammarus*

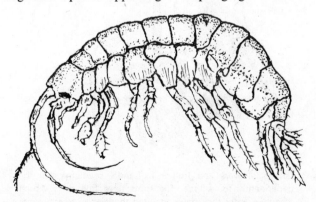

Sand-hopper *Gammarus* (amphipod) × 2

is a large-sized sand-hopper 3 to 5 cm long found in rock pools along both coasts of India. There are several smaller species of free-swimming sand-hoppers which during certain seasons form a major part of the plankton and serve as food for larger kinds of ocean life. The larger amphipods living on sandy shores are dug up for bait.

See also CRUSTACEANS.

SANDAL, *Santalum album,* a tree with dark-grey bark, drooping branchlets, small purple-brown flowers and black-purple fruits. It is famous for its strongly scented yellow heartwood used in carving. The heartwood also yields fragrant oil used in perfumery and soaps. Sandal is probably indigenous to SW India. It is certainly endemic in the Timor Islands. Some botanists consider that it was introduced to India from Timor in ancient times. Birds relish the fruit and are important dispersal agents and Sandal trees are found well inside forests in S India. There is evidence that it has been in India for 23 centuries as there are references to it or possibly to **RED SANDERS** in the Pali *Milinda-panha* (150 B.C.) and the *Mahabharata*. This is an un-

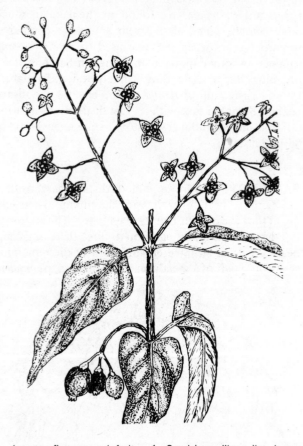

Leaves, flowers and fruit × 1. Sandal seedlings live independently for a time. The roots have fine hairs which disappear after they have attached themselves to the roots of a foster plant. So sandal is a semi-parasite. Its specific name refers to its white sapwood.

solved problem awaiting fossil findings and discovery of pollen grains deposited in sediments in the ancient past. India is now the main source of sandalwood.

SANDAN, *Ougeinia oojainensis,* is a middle-sized tree, common in the sal forests of Uttar Pradesh and widely distributed in the rest of the country. It has bark like crocodile skin, fissured horizontally and vertically, which when blazed reveals blood-red streaks and exudes a red gum similar to Kino. It is a spectacular sight when in bloom in May when leafless. The pink or lilac flowers are produced in dense clusters from the old wood and look like corals. The bark is used to stupefy fish.

SANDPIPER. Sandpipers are relatively longer legged and longer billed shore birds than plovers. They frequent our marshes and swampy areas during winter, returning to their northern breeding grounds in summer. Their prominent characteristic is the incessant tail and neck bobbing as they run about probing the mud for insects and worms.

The Spotted (*Tringa glareola*) is a typical example of a migratory sandpiper; it is greyish brown above, white below, with pale white spots on its upper parts.

SANDSTORMS. A sandstorm or a duststorm is an atmospheric phenomenon when large volumes of fine sand or dust are raised to heights of one or two kilometres above the ground by the action of strong gusty winds. The amount of sand/dust raised is so large that even objects 500 metres distant are rendered invisible. Sandstorms/Duststorms occur when the surface of the land has been dry for a couple of months and the soil has been loosened by alternations of heat and cold.

When a strong wind blows steadily for some time a lot of sand or dust is raised, but is not lifted above a hundred metres or so. However, there are a few occasions on hot summer afternoons, when some moisture is brought over the place by the upper winds at latitudes of two or three kilometres. Then the atmosphere becomes unstable, and strong upward and downward currents develop. The downward currents kick up the dust and carry the dust forward to be lifted upward by the ascending currents.

Duststorms occur on three or four days per year over the northwest, during the hot weather, mostly in the afternoons or evenings. They cease to form when the rains set in.

SAPRIA, *Sapria himalayana,* is a fantastic stemless and leafless parasite, found on the roots of a giant vine *Vitis* in the humid forests of Arunachal Pradesh and Manipur. The flowers are 36 cm across and the rosy pink flower-buds are of the size of a grapefruit. The flowers are a deep crimson with the upper surface covered with yellowish papillae emitting mephitic odours which attract beetles. It is threatened with extinction because it has only been sighted twice since its first discovery in the Mishmi Hills in 1836 and again a century later. It is now on the endangered list.

SARDINES (Clupeidae) include several species which occur on the east and west coast of India and also Sri Lanka. Of these, the main and most important species is the Oil Sardine, *Sardinella longiceps*, the rest of the species being termed as Lesser Sardines.

The Oil Sardine is a small, slightly compressed fish of about 16 to 18 cm in length. It is a resident of off-shore waters and migrates into the coastal belt in enormous numbers, forming huge shoals in the months of September to March, stretching from Quilon to Ratnagiri on the west coast. Sometimes the shoals move further south and to part of the east coast. Their migration is largely for feeding on the coastal plankton which is in abundance during this period of the year. Their average size at this time is 10 to 16 cm, and they are about a year in age.

The oil sardine matures when 14 to 15 cm long and has a longevity of about four years. They breed in the offshore region of the west coast of India during June to September. The eggs are pelagic (floating) and take about 24 hours to hatch. The larvae are transported with the ocean currents, with consequent high mortality, resulting in great fluctuations in yearly catches, from 120,000 to 300,000 tonnes. As its name suggests the fish is very oily and apart from being a popular food they are also used as a source of oil when in abundance, the residue being used as poultry and cattle feed.

Lesser Sardines are only slightly smaller than the oil sardines but they are not so oily or so numerous. They comprise *Sardinella gibbosa*, *S. fimbrieta* and several other species and though occasionally found on the west coast, are largely the east coast species. Their life-history is similar to that of the oil sardine.

SARISKA. Sariska was established as a sanctuary in August 1959 and was included in Project Tiger in 1972. It has an area of 210 km^2 and is approximately 700 m above sea level. It is 198 km from Delhi by road.

Scenically, Sariska is a very attractive area, as could have been expected from its location in the Aravalli range of hills. As is the case with most places in Rajasthan the landscape varies significantly in the different seasons of the year, being green in the rains, brown in winter and a drab grey in summer. Whatever the season the kardhas (*Anogeissus pendula*) which covers the undulating hillsides looks most attractive. The other common trees are khair, ber, ebony and palas.

The wildlife includes chital, sambar, fourhorned antelope, nilgai, rhesus and Hanuman monkeys, hyena, tiger, panther, jackal, jungle cat, and caracal. There is rich birdlife and the game species include partridge and spurfowl.

Sariska has been traditionally subjected to a great deal of overgrazing by cattle, being one of the few places in Rajasthan where fodder is available in the dry months. However, since the inception of Project Tiger the environment is effectively protected.

Sariska is an important centre from the tourist point of view and one of its attractions is viewing the tiger at night from machans. In 1979 there were thought to be 19 tigers in the reserve.

SCALLOPS are bivalve molluscs with handsome shells which are almost round and flat and neatly ribbed. At the upper edge they carry an ear or semicircular lobe on either side. The surface is invariably decorated with bright colours. The animal within the shell is more interesting than the shell itself. When young, it fixes itself to a rock or other solid object but in the adult stage it is free and moves through water in a zigzag course by rapidly opening and closing its valves. For this sort of swimming movement a foot is not necessary and it therefore becomes atrophied through disuse. Another unique feature of a scallop is that it has highly developed eyes borne on the tips of the short tentacles which hang down as a sort of fringe from the edge of the mantle. These eyes, iridescent and green in colour, are almost as complex as the eyes of higher animals, having a retina, cornea and lens.

Scallops generally prefer shallow waters where small seaweeds grow in plenty. Many large and attractive species of scallops abound along the Atlantic shores and there are a few handsome representatives here such as *Pecten tranquebaricus*, *P. senatorius* and *P. splendidulus*. They are all medium-sized measuring from 4 to

The two valves of *Pecten senatorius* opened out, × 1.

7 cm in height, triangular oval or round in shape, strongly ribbed radially, with prominent ears at the umbo and with various colour mottles or markings. These scallops are common along the western and eastern coasts of India and in the Pamban area.

In the deeper waters of the Bay of Bengal, off the Ganjam and Orissa coast, exists an exceedingly elegant scallop—*Amussium pleuronectes*, which is round, smooth, polished and painted red. It is known as the 'sun and moon shell' because its upper valve is of a livid hue and the lower one a dead white. This scallop

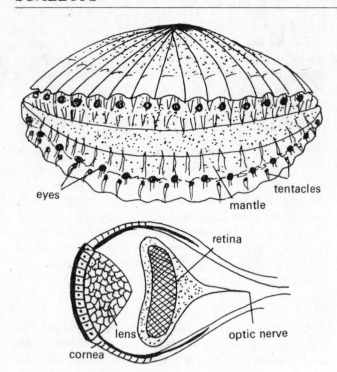

eyes

tentacles

mantle

retina

lens

cornea

optic nerve

The edge of a scallop's mantle, × 2 with a cross-section of one eye, × 120.

often gets entangled in fish-nets. Young ones and adults are found in large numbers from March to June every year. A fully grown specimen measures 8·3 cm in diameter. Along the edge of its mantle there are numerous eyes—as many as 100—of different sizes, round and staring like those of a doll. *Amussium* is a very active scallop. A living animal placed in a basin of sea water may be seen darting about with amazing quickness.

Closely allied to scallops are the Spondyles, popularly called Thorny Oysters. They are larger than scallops and possess spines or thorny growths on their shell. They are brilliantly coloured in yellow, orange and red. They are not free-living like scallops but the lower valves attach themselves to the substratum. Depending upon the nature of the surface of the objects upon which they rest, the shape of their valves also gets modified. With their mantle eyes the spondyles can spy their enemies. Noting a danger they tightly close their thick shell valves. The thorny outgrowths on the shell, coloured differently from the shell itself, also harmonize with the coralline growths among which they live.

The common Indian spondyles are: *Spondylus layardi*, *S. imperialis*, *S. hystrix* and *S. rubicundus*. They are edible.

See also MOLLUSCS.

SCIENTIFIC NAMES. The need for a single, unambiguous name for every plant or animal is explained under CLASSIFICATION and CLASSIFICATION AND

NAMING. A pachyderm is only thick-skin, an elephant; the epidermis is the outer skin, an epiphyte is a plant growing on another plant, phytopathology is the study of plant diseases: the meanings of all these words are clearer if one knows that *pachy-* implies thick, *dermis* skin, *epi-* upon, *phyton* a plant, *pathos* suffering, *logos* a word, hence *-ology* the thing studied. A few common elements, mostly Latin or Greek in origin, which often occur in combination in scientific names, are tabulated below for convenience of reference.

NUMBERS

1 mono-, uni-, haplos; 1st primus, protos
2 bi-, di-, duplus
3 ter-, tri-
4 quad-, tetra-
5 pent-, quin-
6 hex, sex
7 hepta-, sept-
8 octo
9 ennea, novem
10 deka, decem
100 hekaton, centum
1000 chillios, mille
0 zero
$^1/_2$ demi-, semi-, hemi-
all pan-, holos, catholicus
both amphi-
many poly-

PREFIXES

a-, an-, not
ante-, prae-, before
apo-, away from
bio-, life
carpo-, fruit
ceno-, caeno-, new
cyto-, cell
dino-, terrible
ecto-, out of
endo-, internal
epi-, upon
eu-, good, well, easy, true
ex-, out, out of
geo-, earth
holo-, sea, salt
hydro-, water
hyper-, above, beyond
hypo-, under
iso-, equal
macro-, long, large-scale
mega-, large
megalo-, great
meso-, middle
meta-, after, behind, change
micro-, small
neo-, new
paleo-, palaeo-, ancient, old
peri-, round, about
ortho-, straight

phil-, liking
phyllo-, leaf
phylo-, race, tribe
phyto-, plant
pre-, before, very
pseudo-, false
schizo-, split
super-, above
syn-, sym-, with
trans-, across
ultra-, beyond
xeno-, foreign
xero-, dry
xylo-, wood
zoo-, animal
zygo-, joining, pairing

SUFFIXES

-aceae, botanical Family
-ales, botanical Order
-cola, dweller, inhabitant
-ectomy, excision
-gamy, marriage
-graphy, writing
-idae, zoological Family
-inae, zoological Subfamily, botanical Subtribus
-itis, inflammation, disease
-logy, word, study of
-oid, resembling
-phagous, eating
-philous, loving
-saur, lizard

COLOURS

black. melanos, niger, sepia, ater
blue. caeruleus, viola, kyanos
colour. chroma
dark. fuscus
golden. aureus, chrysos
green. chloros, viridis
grey. canus, glaucus, griseus
red. erythros, flammeus, pyrrhos, rhodon, ruber, rufus
silver. argentum
white. albus, leukos, candidus, lacteus, niveus
yellow. flavus, xanthos

BOTANICAL

bark. cortex
berry. bacca, kokkos
bunch of grapes. racemus, botrys
branch. cyma, ramus, spadix, stolo, thallus
bud. blastos, gemma, germen
flower. flos, floris, anthos
fruit. fructus, karpos, pomum, drupa
herb. botane
leaf. folium, phyllon, pinna
petal. petalon
plant. phyton, -wort
root. radix, radicis, rhiza
seaweed. phykos
seed. gonos, semen, sperma

stem. rhachis, stipes, stirpa, culmus
tree. dendron
twig. virga

ZOOLOGICAL

arm. brachion
beak. rostrum, rhynchos
bird. avis, ornithos
bladder. kistis, cysto-
blood. haema, sanguis
bone. os
breast. mamma, mastos
chest. sternum, thorax
claw. chela
ear. auris, otos
eye. oculus, ophthalmos, opos
eyebrow. supercilium
feather. penna, pluma, pteron, ptilon
finger. dactylon, digitus
fish. piscis, ichthys
foot. pedis, podos
forefinger. index
hair(y). barba, capillus, coma, crinis, hispidus, pilus,
    pubes, thrix, villus, seta, hirsutus
head. kephale, caput
horn. cornu, keros
joint. arthron
knee. genus, gony
leg. skelos, crus
lung. pneumon, pulmo
mouth. boca, os, stoma
nail. unguis
nose. rhinos, rostrum, proboscis
nostril. naris
penis. lingam, phallos, stamen
scale. lepidos, squama
shoulder. humerus
skin. cutis, derma
stomach. gaster, stomachus
tail. cauda
throat. rumen
tongue. glossa, lingua
tooth. dentis, odontos
trunk. proboscis
voice. phone, vox
vulva. clitoris, stigma
wing. pteron, pteryx, ala
wrist. carpus

MISCELLANEOUS

air. aether, atmos, pneuma, spiritus, ventus
beautiful. formosus, ornatus, pulcher, speciosus, venustus
character. ethos
deep. abyssos, bathys, benthos, profundus
earth. ge, humus, terra, chersos, chthon
fire. flamma, focus, ignis, phlox, pyros
form. morphe, species
lake. lacus
man. homo, anthropos, andros, vir
marriage. hymen, conjugalis, gamos
marsh. paludis, stagnum

moon. luna, selene
nation. ethnos
old age. geros
river. fluvius, potamos, rivus
salt. sal; halos
sand. arena, sabulum
sea. halos, mare, oceanus, pelagos, pontus, thalassos, abyssos
shallow. superficialis
shell. cochlea, concha, testa
stone. lapis, lithos, petra, saxum, silex
sun. helios, sol
virgin. parthenos, virgo
woman. femina, gyne, gynaikos, parthenos
youth. hebos, juvenis, proles

SCORPIONS. Scorpions are Arachnids like spiders but of a different structure. They possess a pair of pedipalps bearing huge claws, four pairs of legs and a segmented abdomen the front half of which is broad and the hind half narrowed to a tail ending in a poison sting. In length they range from 13 to 180 mm. They inhabit warm countries and there are over 300 species. They are nocturnal and feed on insects, spiders and centipedes. Larger insects are paralyzed by the sting before being eaten. Scorpions are timid by nature and do not sting unless molested. During daytime they hide under stones or in pits made in sand. Their sense of sight is weak but the sense of touch is greatly developed in their pedipalps, and while walking they hold their pedipalps parallel to the ground to feel their way.

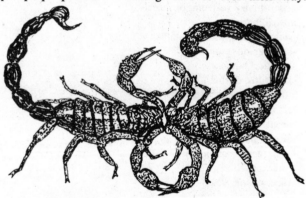

Scorpions greeting each other before mating, × 1

Before mating the partners face each other, curl up their tails, hold each other by their claws and playfully fight. The male then draws the female to a sheltered spot and mating takes place. Often the male is killed and eaten by the female after mating. Scorpions are viviparous. The newly-hatched young are carried by the mother on her back for about 10 days, i.e. until they are able to forage for themselves.

See Arachnids for diagram.

SCORPION FISHES (Scorpionidae) are usually bottom forms, some of them brilliantly coloured, and have sharp spines which inflict lacerating wounds (see POI-

Turkey or Lion fish, *Pterois*, × 1/3

SONOUS FISHES). They occur in most of the seas in the tropics and deep oceans. One of these fishes, the Turkey fish or the Lion fish, *Pterois russelli*, is very gorgeously coloured and has the habit of spreading its wide pectoral fins like a turkey and gulping its prey. Despite this beauty its spines are poisonous.

Another example of these Scorpion fishes is the Little Scorpion Fish, *Minous* of Indian seas. Many members of this genus live in a symbiotic manner with colonies of hydra-like animals covering their bodies. The fish receives protection from its enemies because, being covered with hydroid polyps, it looks like a weed-covered stone and is neglected; while the hydroid colony gets the advantage of movement for collecting food material.

SCORPIONFLIES. These are small to medium sized (2 to 30 mm) insects with a long, beak-like prolongation of the front of the head and black-marked membranous wings. They are usually found in cool, moist, shady habitats, generally in forest undergrowth beside rivulets. Males carry their terminal abdominal segments, with scorpion-sting-like structures, upwardly curved and thus resemble their namesakes. Both adults and larvae of Mecoptera are predacious on an assortment of mainly soft-bodied insects, the adults catching them on vegetation and the soil-dwelling, caterpillar-like larvae feeding underground. Around 400 species have so far been described, of which 30 are known from the Indian subcontinent. Those of *Boreus*, which are vegetarian, are also atypical in lacking wings. These, however, are not

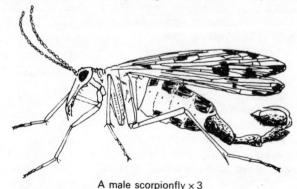

A male scorpionfly × 3

found in India, where species of *Neopanorpa* and *Panorpa* are common in the forests of the lower Himalaya. Assam and the Western Ghats. Some *Bittacus* occur mainly in warmer and drier areas on the plains.

SCORPION SPIDERS belong to the Order Pedipalpi and in all details including habits agree with WHIP SCORPIONS—the only difference is they have no tail. Genus *Admetus* is an example.

See also ARACHNIDS.

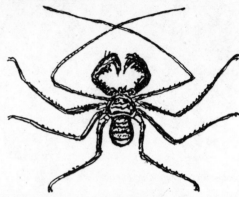

Scorpion spider (*Admetus*), × 1

SCREW-PINE, *Pandanus odoratissimus,* is a common coastal shrub with sword-like leaves, spiny on the margins and midrib. The male flowers are very fragrant, yielding keora oil, used in perfumery, and for flavouring sweets and syrups. The fruit is long, globose like a pineapple but drooping. The stems are cut into long pieces, one end beaten flat and used as whitewash brushes.

SEA-ANEMONES are sedentary coelenterates resembling hydras but stouter and larger. They are solitary and rarely form colonies. In internal structure they are far more advanced than hydras and *Obelias.* With their numerous spreading tentacles fringed with hair-like cilia and tinged white, lilac, yellow or orange and their habitat on mud banks and rock pools, sea-anemones form conspicuous objects of the seashore. Many of them however shroud their bodies with sand grains and shingle and when exposed during low tides they shrink to such insignificant masses that they are often overlooked.

A typical sea-anemone has a cylindrical body with an adhesive basal disc at the attached end and an oral disc with a slit-like mouth in its centre at the free end. The tentacles are hollow and their number varies from six to many arranged in circlets about the mouth. In one species, *Actinia equina,* there are 192 tentacles arranged in 6 circlets of 6, 6, 12, 24, 48 and 96. All tentacles bear stinging cells and are highly contractile. They can expand or contract considerably, capture prey and convey the food to the mouth. The animal can voluntarily shift its position by softly sliding its basal disc. The gastrovascular cavity of the sea-anemone is not simple. The slit-like mouth does not lead directly into the cavity but through a tube or gullet which is not circular but compressed from side to side. This gullet has two ciliated grooves one on either side. A constant stream of water enters into the body cavity through these grooves and supplies the animal with oxygen.

There are male and female sea-anemones producing either sperms or ova. The sperms are discharged into the gastric cavity from where they escape through the mouth. They are then carried partly by their own movement and partly by the action of their cilia into the cavity of a female where they fertilize the ova. The fertilized ovum divides and becomes an elongate ciliated larva. In this condition the larva leaves the parental cavity, swims for some time and settles on a suitable support where it develops into a fresh sea-anemone.

The distribution of sea-anemones is wide and many typical genera are found along almost all the sea coasts. They abound in tidal pools, and reefs have a rich representation of large and colourful anemones. There are many interesting forms and the tropical ones are the more brilliantly coloured.

Metridium is a typical sea-anemone found fixed to some solid object along the seashore where rock pools and mud banks abound. It has a dark brown trunk about 8 cm high and white, hollow tapering tentacles fringed with rich clusters of cilia. Delicately coloured *Sagartia,* about 2·5 to 3 cm high, is common on rocks. *Adamsia* is a stout reddish sea-anemone with grey tentacles found fixed on the back of empty gastropod

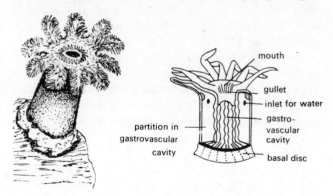

Sea-anemone (*Metridium*) and a longitudinal section, × 1½

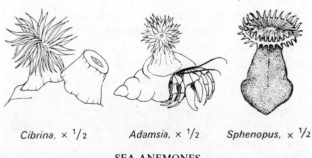

Cibrina, × 1/2   Adamsia, × 1/2   Sphenopus, × 1/2

SEA-ANEMONES

shells inhabited by hermit crabs. It has the benefit of being carried from place to place by the crab whereas the latter gets protection from predacious fishes which avoid the stinging weapons of the sea-anemone. Members of the genus allied to *Cibrina* measuring about 5 cm in diameter and 6 to 8 cm in height, bluish in colour with numerous slender tentacles and vertical rows of warty growths on the column, are found on mud-covered rocks. When they are exposed during low tide they contract and appear to be mere greyish lumps. These sea-anemones owe their green colour to the presence of algae in their cells. *Zoanthus* is another example where the original polyp sends out a horizontal branch (stolon) from which new polyps are formed by budding. *Sphenopus* is a somewhat flattened darkish sea-anemone 3 to 6 cm long, usually broad and triangular at one end and narrow and cylindrical at the other. Its outer layer is tough, coated with sand grains. In life the animal anchors in mud by the broader triangular end, exposing its tentacle-covered disc at the free end. Angry waves uproot these anemones and throw them to the shore as ugly lumps of jelly. But if returned to the sea they anchor again and expand vertically, exhibiting the disc and tentacles.

See COELENTERATE.

SEA CUCUMBERS are tough skinned, sausage-shaped echinoderms, in shape and architecture entirely different from the starfish and sea-urchin. They live buried in sand or under rocks below low-tide mark and also in deep water. They are plentiful in the Pacific Ocean and Indian waters. Especially around Krusadi Island and in the Ceylon pearl bank several genera occur. In size they range between 6 cm to 2 metres and more.

The body of a sea cucumber is roughly cylindrical and in some cases long and five-sided, not unlike that of a cucumber. Protruding from the mouth at the anterior end is a circlet of feathery retractive tentacles which are actually modified tube-feet. At the posterior end is the anus.

The body is soft and fleshy within, five-sided and covered by a thick transparent skin which has strong muscles enabling the animal to shrink and expand and move forwards. When it moves, a number of tube-feet with suckers protrude along the five sides and they aid locomotion. In those species lying parallel to the sandy surface a definite dorsal and ventral side can be recognized and in such species the tube-feet are confined to the ventral side only. Some species have sharp spicules embedded in the flesh. These are defensive and also assist in locomotion. Members of the genus *Synapta* have no tube-feet at all but a number of specialized spines with which they can successfully anchor to any soft body. Some live in the open sea and can float and swim propelled by the tentacles.

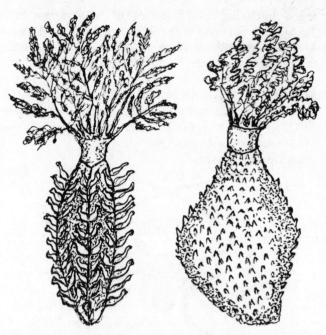

Two sea cucumbers, *Cucumaria* (*left*) and *Thyone* × ½

The ability of sea cucumbers to contract and escape from their enemies is marvellous. At the slightest molestation they squirt out water, shrink to an insignificant mass and recede into a hole in the rock.

The alimentary canal of a sea cucumber consists of the mouth, a cylindrical throat, a short gullet, a small muscular stomach and a very long, looped intestine ending in an anus.

The food consists of organic matter existing in the sand and mud in the sea bottom. Some species also feed on small sea organisms caught in their frill-like tentacles.

Sea cucumbers have great power of regenerating lost parts. Cases have been recorded where sea cucumbers which lost their tentacles, intestines and other organs have developed these parts again in two months.

Some large sea cucumbers are eaten by man, for example the *bêche-de-mer* or sea slug.

See ECHINODERMS, STARFISH and SEA-URCHINS.

SEA EAGLE. The Whitebellied Sea Eagle is usually seen in pairs. It affects almost the entire coastline of India from about Bombay south, up the east coast, and off Bangladesh in the Bay of Bengal, feeding on fish and sea snakes, and occasionally also raiding poultry. Two white eggs are laid in a massive stick-nest lined with green leaves high up in a tree or on a rocky offshore stack from October to June.

SEA LILIES, or Feather Stars, are echinoderms looking like shootlets of a fir tree. They are allies of starfish with a central pentangular disc and very long radiating arms. Although they can swim freely they prefer to remain fixed to rocks or weeds at the sea bottom. To

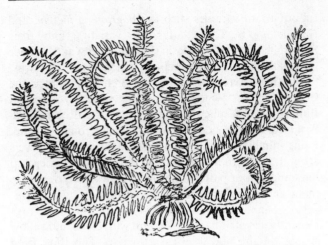

The sea lily *Antedon* × 1 (From Parker & Haswell).

suit this mode of life they have their mouth or oral side turned upwards and they fix their back to the mud by certain claw-like growths. Their anus is on the oral side. Each of the five arms is divided into two and in each branch there are rows of filamental growths giving a plume-like appearance. The arms are grooved longitudinally and are lined with tiny tube-feet. These tube-feet however are without ampullas and are not used for walking. They serve as respiratory organs.

Sea lilies feed on minute sea organisms which are wafted towards the mouth by the cilia fringing the grooves of the arms. The skeletons of sea lilies are of calcareous plates without spines.

The free-swimming larva of a feather star is a curious little, oval jelly which swims freely by cilia. During the swimming stage, which lasts for many days, lime plates are formed which arrange themselves in a cup form. A stalk is also formed. The jelly body becomes absorbed and the larva sinks to the bottom and gets fixed to some rock. Here it becomes transformed into a sea lily which grows into an adult.

Sea lilies are generally brilliantly coloured. Very many gigantic and beautiful species of stalked feather stars existed in the deep sea during the Palaeozoic era but most of them have vanished. Now there are only some 150 genera in the world distributed at varying depths. *Antedon* is the typical genus.

Two interesting species, *Tropiometra eucrinus* and *Lamprometra palmata*, are common on Indian shores: the former is a graceful purplish creature with 10 plume-like arms, the latter has 30 to 40 arms.

See ECHINODERMS and STARFISH.

SEA PERCHES constitute a varied group of marine fishes both large and small, covering several families occurring in seas of the entire subcontinent and even beyond. Only a few of these are, however, described here. In Indian seas the dominant group is the serranid perches or the groupers, also called 'rock cod' although having no relationship with the real cod. They are robust-bodied, with a large mouth beset with teeth, smaller behind and canine-like in front. Considerable confusion prevails about the scientific names of many sea perches but prevalent names are being used here.

Ephinephalus lanceolatus, *E. maculatus* and *E. tauvina* (Serranidae) are some of the large perches growing from 2 to 4 m and are very long-lived. The first one is very attractive, black-and-yellow coloured when young, and after 60 cm in length becoming dark black with yellow spots, a kind of speckled perch, though this name is given to the third species, which is pale chocolate-coloured with black nondescript spots. Living amidst submerged rocks and caves and being carnivorous they feed voraciously on small fish which also normally take shelter in the same rocky habitat. Prawns and lobsters also appear to be favoured food as perch have been seen to devour such material avidly even in the enclosed space of an aquarium.

Snappers (Lutjanidae) derive their name from their habit of snapping their prey, a habit that facilitates their capture by anglers on baited hooks. They are marked by a roundish dorsal profile, prominent eyes and a mouth with slightly shorter lower jaw. The abdominal line (ventral profile) is somewhat straight. Being carnivorous they always go after smaller fish. Some of them are copper-coloured and so are called *tambusa* or *tamb* in Bombay. One of them, *Lutianus johnii*, has a pale golden sheen all over, the upper half of the body being speckled with rich brown spots; large specimens measure 80 cm and weigh 8 kg. *L. annularis* is brilliant red but smaller in size (65 cm). *L. roseus* is cherry red in appearance and is caught in shoals, the catch making a dazzling red heap on the boat deck. The young of *L. argentimaculatus* enter estuaries and take shelter among loose stones of the marginal bunds in wait for their prey (small unwary fish). If any fisherman incautiously puts his hand on the stones the fish bites his finger, and hence the local name *chavri tamb* meaning 'biting snapper'.

Cardinal fishes belonging to the family Apogonidae are small perches found in coastal weeds, rock pools and coral reefs. Unlike other percoids, their dorsal fin is divided into two small fins. In Ambessidae, where except three freshwater forms the rest are marine, the dorsal fin was nearly divided; but in this family the division is complete. Being variously coloured to match the surroundings in their natural habitat, Cardinal fishes, being usually small, (less than 10 cm), are attractive aquarium fishes, taking copepods, *Mysis* crabs and other artificial feed.

Sciaenids (Sciaenidae), commonly known as Croakers because of the peculiar sound they make with the help of their air-bladder, and also called 'jew fishes', occur on both coasts of the Indian subcontinent including

Sri Lanka. There are 20 species. The dorsal fin is incompletely divided into a spinous part in front and a soft part behind; their nose is blunt and eyes prominent. Being largely demersal (bottom-dwelling) in nature, they subsist on prawns, crabs, squids and other fishes found at the bottom. Most of them inhabit coastal waters of about 40 fathoms depth but some species enter estuaries and backwaters for feeding. *Otolithoides brunneus*, *Protonibea diacanthus* and *Wak sina* are some of the large sciaenids of this group. The first is more elongated in body form, with a long dorsal fin, and is reported to grow to 2 m and weigh 45 kg. The same measurements are reported for *Wak sina* though *P. diacanthus* is known to grow only up to 1·4 m and 16 kg. Most of the other species do not grow beyond 30 cm and are referred to as Small Sciaenids or Dhoma. They occur almost throughout the year and spawn in April and May. Dhoma constitute a large part of trawler catch on the northwest coast of India. Because of their habit of making grunting sounds, fishermen listen to such sounds in quiet backwaters and dive to the bottom to catch them by hand or in small nets. The air-bladders of these fishes are dried and used for making isinglass, which is used in confectionery and for jellies.

The family Carangidae, which includes the Horse Mackerels, has several genera and species of varied types. Generally the fish are oblong, elevated, and flattened side to side. The central scales, especially in the tail region, are thickened, forming a hard keel. Most of them are fast swimmers inhabiting surface waters and having a tendency to school together to hunt other small fish for food. Many do not grow beyond 40 cm but some (e.g. *Megalaspis cordyla* and *Scombroides tala*) grow up to 1·5 m. The last-named is also called the Porthole fish because the circular blotches on its body resemble the portholes of a ship. Most of the smaller varieties of the genera *Selar*, *Citule* and *Caranx* are known as Horse Mackerel though they have no relationship with the true mackerel. Three species of *Trachynotus* look very similar to true pomfret but being different taxonomically are called pseudo-pomfrets or pumpano. As a group, the horse mackerels are not very popular as food but when cured with salt they are acceptable.

Drepane punctata, the Spotted Dory of the family Drepanidae, is a beautiful silver fish with head and body elevated and compressed side to side, and with black spots in parallel rows vertically disposed. Though not much valued as food, adults over 40 cm are outstanding because of their silvery colour and rounded shape. Sea-Bream (Sparidae), *Acanthopagrus berda*, is another perch of rocky environment which is popular as food, though not much is known about its life-history. Pomadasyidae is a heterogeneous family which contains the common food fish *Pomadasys maculatus* and also colour-ful aquarium varieties such as *Diagramma pictum*, and *Gaterin nigrus*. These latter two varieties, known as Squirrelfish, usually hide in coastal weeds and pounce on small prawns and copepods which come their way.

Another interesting family of perch-like fishes is the Chaetodontidae which includes the variously coloured Butterfly-fishes. They are usually short, deep-bodied, flattened side to side and with a small round tail fin. Being inhabitants of rocky coastal areas and coral reefs around southern India and Sri Lanka, the variegated lines and colour patterns on their bodies effectively help them as camouflage in their environment. The *Heniochus*, banner-fish, with one or two dorsal fin-rays elegantly elevated bannerlike, and with black and yellow stripes on the body, looks attractive. The Blue Ring *Pomocanthus annularis* has undulating blue stripes, as its name suggests. Its young ones are so differently coloured—dark blue body with white concentric rings—that for some time they were considered to be of a different species. They are also referred to as Blue-ring Angel-fish. The Emperor fish, *Pomocanthus imperator*, with its opercular spine, wavy yellow lines on a blue-golden body and a golden tail, is so outstanding and majestic that it deserves its name. Some of these fishes in their young larval stages are quite different in shape and colour and have bony shields on their head and spike-like structures jutting out from the opercular bones to protect them from their larger cousins. Chaetodonts, as a group, are some of the prettiest varieties of aquarium fishes from the tropical world and assume different names.

See also PERCH.

See plate 2 facing p. 33 and plate 22 facing p. 305.

SEA SNAKES. There are about 20 species of sea snakes found on the east and west coasts and coastal islands of our area. Although often seen in the deep sea, they are more commonly seen caught in offshore fishing-nets. Sea snakes have a highly potent neurotoxic venom, but fishermen handle them carelessly, scooping them out of nets and flinging them back into the sea. This is because they very rarely bite; in India there are at most a few deaths every year, always fishermen, never bathers. Antivenom serum for sea snake bite is available only in Japan and Australia. All sea snakes are characterized by flat, paddle-like tails for efficient swimming. They feed on fish, eels, and fish eggs. Fish are quickly paralyzed by the venom, and swallowed. Sea snakes, excepting the genus *Laticauda*, bear living young. *Laticauda* lays eggs on land and during the breeding season hundreds of females come ashore on little islands to deposit their eggs. This species, called sea krait, comes on land at other times as well, to bask, shed their skins and perhaps even feed.

From what little observations have been made on sea

Hook-nosed sea snake, average length about 60 cm.

snakes, they are active both day and night, and can stay under water for up to 5 hours at a time. They are equipped with glands to remove excess salt from their bodies, an adaptation in common with other marine animals.

Sea snakes are not exploited in our area and remain fairly common. In Hong Kong and Singapore they are killed for meat. Studies need to be done on their distribution, about which little is known except that *Enhydrina schistosa* is the commonest species.

See also SNAKES.

SEA-URCHINS are perhaps the most capricious and fantastic of all echinoderms. They look like prickly balls or small rolled-up hedgehogs. Their size varies from 4 to 25 cm. They are widely distributed in all shallow coastal waters in the cracks and crevices of rock pools or amidst soft seaweeds. The deep-water forms are gregarious, covering large areas of the bottom.

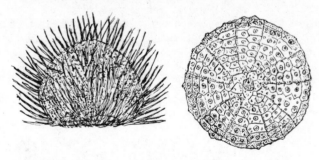

A sea-urchin with spines between which the tube-feet can be shot out, and (*right*) the arrangement of plates which cover them. The animal's mouth is below, unprotected. × ¹/₂

In appearance a sea-urchin is very different from an armed starfish; yet it has the same fundamental structure. Essentially it is a starfish and its globular form is the result of its five radial arms having bent upwards and united along their margins and tips. In starfish the skeletal plates are only loosely connected by fibrous tissues but in a sea-urchin the plates are fused to form a hard shell enclosing all the soft parts of the animal

including the tube-feet. The mouth however is uncovered. In life the plates are covered with slime outside and within and also between each plate. In a growing urchin these plates are continually thickened by a deposit of fresh limy matter formed by the action of slime and sea water. Additional plates are also added where the tips of the arms converge. Radiating upwards from the mouth along the arms are five bands of small holes through which the tube-feet can project. The pincers (pedicellarias) in a sea-urchin are more specialized than those of starfish in that they are stalked and have three jaws. The spines covering the entire body are brittle, long, sharp and movable. They aid the tube-feet in locomotion. It is a sight to see a sea-urchin moving on the tips of its spines, putting forth here and there its white elastic tube-feet. When some of the tube-feet shoot out and touch the ground, those already in contact with the ground are pulled back. In a short time the protruded ones are withdrawn and the retracted ones projected. Through this operation of the tube-feet coming out and going in the animal progresses.

The enemies of sea-urchins are large fishes.

There are three types of sea-urchins—the true Sea-urchins which are globular, Cake-urchins with almost circular and flattened bodies, and Heart-urchins shaped like a heart. Sea-urchins are coloured differently in purple, green, ochre and blue. Under each type there are several interesting forms along Indian coasts.

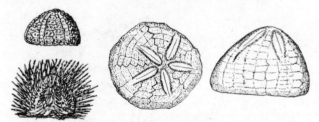

Left, side views of a *Temnopleura* shell and animal. *Right*, the heart-urchin *Echinolampas* from above and from the side. Both × ¹/₂

Temnopleura toreumaticus is commonly found on Madras beach. It is slightly conical in shape about 4 cm in diameter. The shell is greyish brown covered with somewhat long and reddish spines banded white. *Stomopneustes variolaris* is a big urchin with very long and thick spines mounted on tubercles and coloured purple. They are found at extreme low tide clinging to rocks.

Among cake-urchins several species of the genus *Echinodiscus* are common. They are recognized by the pair of oblique slits continuous with or near to the hind margin of the shell.

One of the commonest Indian heart-urchins is *Echinolampas*, having a slanting appearance and prominent petal markings. Living animals have long spines.

See ECHINODERMS and STARFISH.

SEAHORSE, PIPE FISH. With its armoured body, head like a horse, tail like a monkey, and a pouch like a kangaroo, the bizarre seahorse (*Hippocampus*) is unique among fishes. The head is set at right angles to the body, which is completely enclosed by cruciform, interlocking plates, whose edges often have spines or rounded knobs, quite unlike the scales of other fishes. The tail is prehensile and can be coiled around seaweed or twigs when the fish wishes to rest. It grows to 12 or 15 cm long.

The seahorse swims in an upright position, the transparent dorsal fin vibrating rapidly to allow the fish to glide slowly and gracefully. Its prey, consisting of crustaceans or small fish, is sucked into the tubular mouth.

The seahorse is known for its parental care. After mating, the female transfers her eggs into a brood pouch on the male's belly. In the breeding season this pouch becomes soft, inflated and richly supplied with blood vessels. The young develop inside the pouch and are ejected as tiny replicas of the parents, so that apparently it is the father which gives birth to the babies and not the mother!

The Pipe fish (*Syngnathus*) is closely related to the seahorse, but has a long, slender, tubular body encased in a series of jointed, bony rings. The brood pouch of the male consists of two elongated flaps of skin along the under side, where the eggs are incubated and young delivered in the same manner as the seahorse.

Both these species of fish are common in the coastal waters of India and are usually found among marine weeds and submerged rocks where they obtain their food.

SEASHORE PLANTS. The coastline of our region is about 7000 km long. The coastal sand and sand-dunes are characterized by mat-forming creepers like Red Adambo (*Ipomoea pes-caprae*), *Launaea sarmentosa*, the excellent sand-binder grass *Spinifex littoreus*, the sedges *Cyperus arenarius* and *C. pedunculatus*, the Seaside Bean (*Canavalia maritima*) associated with grasses like *Aeluropus lagopoides*, *Sporobolus virginicus* and the Manila grass (*Zoysia matrella*). The backshore zone, located a little further inland, is frequented by herbaceous plants like *Borreria articularis*, *Enicostemma hyssopifolium*, *Geniosporum tenuiflorum*, *Polycarpaea corymbosa*, *Scaevola plumieri*, Wild Indigo (*Tephrosia purpurea*) and the grass *Perotis indica*. The branching Doum Palm (*Hyphaene dichotoma*) is very characteristic of the coastal sands near the former Portuguese settlements at Diu and Daman.

Red Adambo, *Ipomoea pes-caprae* × ¹/₂.

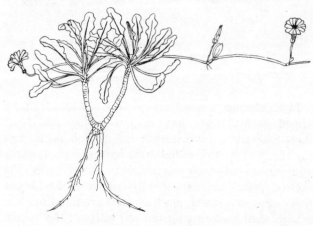

Launaea sarmentosa × ¹/₂

Above, Seahorse *Hippocampus* × ²/₃
Below, Pipe fish, *Syngnathus* × ²/₃

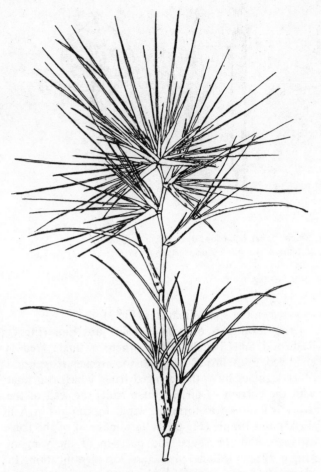

Sand-binder grass *Spinifex* × ¹/₂

On rocky strands the plants encountered are *Atriplex stocksii*, *Fagonia cretica*, *Limonium stocksii*, the Wall lizard (*Kickxia ramosissima*), Chawnlayi (*Portulaca quadrifida*) among others.

Coral strands, consisting of coral reef, sand, clay, and shell deposits, are confined to Krusadi and other islands in the Gulf of Mannar, and to the Lakshadweep, Amindivi, Andaman and Nicobar groups. The colonizers of the coral strand are *Pemphis acidula*, *Suriana maritima*, the Tahiti Town tree (*Cordia subcordata*), halophytes like Marsh Samphire (*Salicornia brachiata*), *Atriplex stocksii*, White mangrove (*Avicennia*), Madagascar mangrove (*Lumnitzera*) and the sedge *Cyperus pachyrrhizus*.

See SALT-WATER PLANTS.

SEASONS. Seasons represent the divisions of the year according to consistent changes, occurring year after year, in the character of the general weather at a place, in the course of the annual cycle of atmospheric events. European culture specifies four seasons: Winter, Spring, Summer and Autumn, which have reference to the life of plants—natural and cultivated. In fact the word 'season' has come from the French word *saison*, derived from the Latin word *satio*, meaning sowing. Winter is the period of dormancy, Spring that of sowing and germination, Summer of growth and Autumn of harvest. These terms are not quite appropriate to our conditions, or even to tropical conditions in general. In our area it is better to speak of Cold or Cool Weather, Hot Weather, Monsoon weather and Post-monsoon weather.

The climate and weather patterns associated with each season year after year is primarily caused by the apparent motion of the sun across the equator, to the extreme north—the Tropic of Cancer on 22 June and to the extreme south—the Tropic of Capricorn on 22 December. Such changes in the apparent position of the sun bring about differences in the solar heating of the different parts of the earth.

During the period June to August (summer) the days are longer in the Northern Hemisphere and most of the surface of the earth (in the Northern Hemisphere) is nearly normal to the sun's rays, and therefore the land and sea temperatures are higher. During the period December to February (winter), in the Northern Hemisphere, the conditions are reversed, the days are shorter and the land and sea surfaces are cooler. In the Southern Hemisphere, the seasons are just the opposite, it is winter there during June to August and summer during December to February.

See also SOLAR RADIATION, MONSOONS, CLIMATE.

SEED DISPERSAL. Fruits and seeds should be carried far and wide for propagation. Dispersal is an advantageous adaptation for safeguard against competition with other individuals or the mother plant to avoid overcrowding. A spacious distribution of seedlings provides adequate water, nutrition, light and air for proper growth and development. Dispersal is also a natural obligation for the survival of the species. In order to achieve wide distribution there are unique and manifold devices. Fruits and seeds are dispersed mainly by four processes:

1. *By wind*. Plants have taken wind as the best medium of dispersal for wide distribution. The chances of wastage of seeds are high by this process, but the loss of seeds is compensated by their number. In ORCHIDS the microscopic seeds are not only powdery but provided with wings, and when released in thousands are carried away by air.

More commonly wind dispersal is effected by wings or woolly growths on the fruits and seeds. There are not only woolly fruits but there are numerous examples where the seeds have a woolly growth as in Silk-cotton (Fig. 1), Yellow Silk-cotton, Poplar and Willow. Their seeds are taken far away by currents of air until they come to rest. Functionally the hairy growth is meant for dispersal but its source varies to a great extent. In silk-cotton the hairs are contributed by the carpel.

The achenes of Traveller's Joy are provided with persistent hairy styles which help the light fruits to be

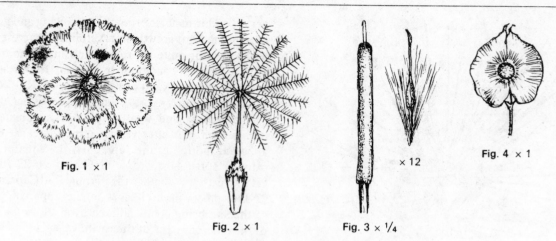

1 Seed of Indian Silk-cotton with floss by which it is carried to long distances.
2 Parachute adaptation in Dandelion with plumose pappus for dispersal by
wind. 3 A compact spike of Bullrush. On the right a tiny hairy fruit. 4 Fruit
of Elm, with papery wings for floating in air.

dispersed widely. The parachute mechanism in the cypselas of the Marigold family has made the family the most widespread in the world. Undesirable weeds like 'mile a minute climber' (*Mikania*) and *Parthenium* are thus spreading in our area and damaging forests and lands. The hairy crown or pappus shows manifold elaborations, typical of many genera. The pappus is bristly to finely hairy in Thoroughwort and Asters, and extremely plumose in Dandelion (Fig.2) and Thistles. The pappus helps the fruit to remain buoyant and able to be carried to far distant places. With a gentle current of air, the fruit is detached and the woolly crown floats off when the fruit comes to ground.

The caryopses of many grasses like Cogon grass, Reed, *Saccharum* and *Erianthus* have fine hairs at different parts of the spikelets which help easy dispersal by wind. In the Indian Botanic Garden, Calcutta, distribution of Cogon grass has followed the direction of the wind. Millions of hairy fruits of Bulrush are blown away when the spadix matures (Fig.3).

Other examples of winged fruits are Elm (Fig.4), Redwood and Switch Sorrel, where a single seed is encircled by a broad wing. Sissoo, some *Acacia* and other legumes have thin winged fruits which can float with the current of air. In these fruits the wall of the ovary is flattened to form the wing. In samaras of Ash (Fig.5) and Maple (Fig.6) the terminal end of the fruit expands, and the shape and position of the wing in Maple are important characters for identification. In Mountain Sorrel and Rhubarb also the fruits are membranously winged and dry fruits are easily carried away by wind. Most of the members of the Dipterocarpaceae are characterized by winged fruits derived from the enlargement of the calyx lobes. In *Hopea* and *Dipterocarpus* (Fig.7) two lobes are modified into wings but in Sal (Fig.8) all the five are enlarged. These help in propelling the seed and dispersing it far and wide. Sometimes immediate dispersal is essential; as a result Hol-

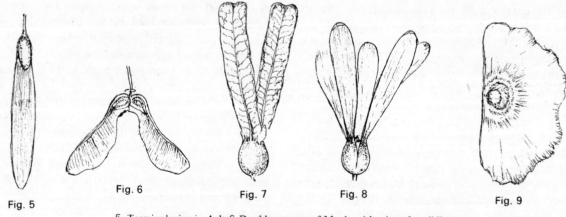

5 Terminal wing in Ash. 6 Double samara of Maple with wings for gliding.
7 Two-winged fruit of *Dipterocarpus*. 8 Five calyx lobes providing wings in
Sal. 9 Membranous winged seed of Canoe tree. All × 1

Papilionanthe teres. This species was known as *Vanda teres* till recently. It flowers from June to August and is widely distributed in eastern India, Bangladesh and Burma.

Dendrobium heterocarpum. This pale ochraceous yellow orchid is called the Golden Dendrobium. It flowers in April and is found in Kerala as well as the eastern Himalayas.

Dendrobium hookerianum. Distinguished from other yellow Dendrobs by the two maroon-purple blotches on the disc. It flowers in September.

Dendrobium lituiflorum. Found in the Khasi and Arakan hills, flowering during April-May.

© O.C. Edwards

The irregular pattern of the Sphinx moth *Deilaphila* sp., Sphingidae helps to camouflage it in foliage.

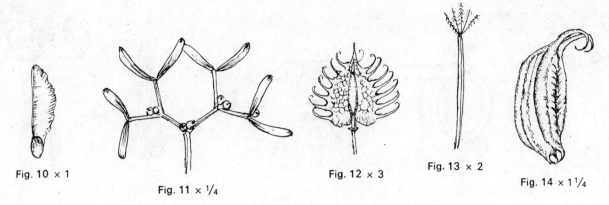

Fig. 10 × 1

Fig. 11 × ¼

Fig. 12 × 3

Fig. 13 × 2

Fig. 14 × 1¼

10 Winged naked seed of Pine. 11 A fruiting branch of parasitic Mistletoe.
12 Fruits with persistent hooked sepals in Dock. 13 Fruit of Burmarigold
with spinulose pappus. 14 Bihooked seed of Tiger's Nail.

long (*Dipterocarpus macrocarpus*) has failed to extend to the north across the very wide Brahmaputra river in Assam.

Several members of the family Bignoniaceae such as the Canoe Tree (Fig.9) and *Stereosperum* are packed with membraneous light winged seeds which are easily blown far away by the wind. Most of the CONIFERS like Pine (Fig.10), Deodar, Spruce and Fir have naked winged seeds which are dispersed by wind. Similarly seeds of Birch and Rhododendron are dispersed by wind.

From plants having porous dehiscence of capsules, like Poppies and Antirrhinums, the seeds escape in limited numbers as the fruits are swayed by high wind or get a jerk. Other special adaptations for wind dispersal are light inflated fruits, as in Balloon vine and *Kleinhovia*, which are carried easily by wind to distant places. A unique device is adopted by *Spinifex littoreus*, whose projecting long needles of the spikelets form a spherical ball which is drifted easily by wind along sandy sea coasts.

2. *By animals.* Fruits or seeds dispersed by animals or birds either stick to their body or are thrown away as a rejected part of the fruit. In Jack fruit and Fig, many berries like Date or Guava, or drupes such as Indian Jujube and Mango, the stony seeds are rejected or excreted after the pulp has been eaten. Birds, squirrels, jackals and human beings thus achieve dispersal of seeds. The seeds of Fig are relished by birds and bats and seeds coming unharmed through droppings germinate on roof-tops and trees. The Great Banyan of Calcutta is reported to have originated on a Date Palm. Mistletoe (Fig.11) and other members of the Loranthaceae family have berries with very sticky seeds. Birds get rid of such seeds by brushing their beak on the branches of other trees and the parasites get a new footing.

More commonly the fruits and seeds dispersed by animals have sticky glandular growth as in *Boerhavia* or *Plumbago*, or hooked hairs or barbed processes which cling easily to animal furs or clothing and are carried to distant places. Good examples of such fruits with hooked hairs are Cocklebur, *Cyathula*, and *Triumfetta*. Buffaloes move around muddy places where the Cocklebur predominates, and hundreds of fruits adhere to their hairy coats to be carried away. In the Himalaya the Dock (*Rumex nepalensis*) (Fig.12), is widespread as the fruits with hooked calyces are carried away by grazing sheep. Burmarigold (Fig.13), Lovethorn and Tiger's Nail (Fig.14) are examples in which the fruits are dispersed by animals. In Tiger's Nail the seed's two sharp hooked points stick to an animal body, and the spiny pappus in Burmarigold helps successful dispersal of the fruits in a similar way. Many aquatic birds and ducks carry minute fruits and seeds of aquatic weeds, or entire plants, with them and help their dispersal from one water source to another.

3. *By water.* Plants generally growing in water or near it develop fruit with a spongy or fibrous coat which helps it to float and get transported by water currents. The fibrous mesocarp of coconuts (Fig.15) helps them to float easily in water without being damaged even by the strongly saline water of the sea. Their abundance in islands, and along coasts, are the result of dispersal by water. The lighter fruits of the Nipa Palm and some mangroves are carried by tidal currents and remain floating till they get stuck in the mud. In most of the free-floating aquatics such as the Water Cabbage (Fig.16), Water Hyacinth, Large Duckweed, and *Salvinia*, the whole plants including fruits are spread by water dispersal. The buoyant seed-covers of Water-lily seeds keep them afloat long enough to be carried elsewhere before sinking. The spongy receptacle of the Lotus, with seeds embedded in it, can float until rotting frees the seeds.

4. *By explosive mechanism.* The capsular fruits of

497

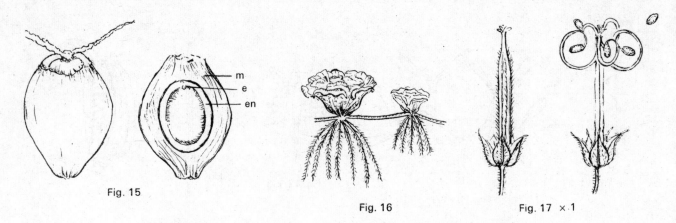

Fig. 15

Fig. 16

Fig. 17 × 1

15 Entire and cut-open Coconut showing m fibrous mesocarp, e position of embryo, en edible endosperm or kernel. 16 Floating Water Cabbage. 17 Mechanical dispersal of *Geranium* schizocarp.

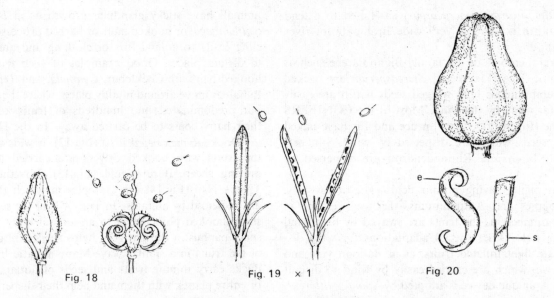

Fig. 18

Fig. 19 × 1

Fig. 20

18 Inrolling dehiscence of Balsam. 19 Capsule popping open. 20 Bulky capsule of Mahogany : e mechanical woody endocarp, s a winged light seed.

many plants burst and scatter seeds to a considerable distance. In Balsam (Fig.18), Wood Sorrel and *Geranium* (Fig.17) the capsules of the ripe fruits are extremely sensitive and split into segments at the slightest disturbance. In many Acanthaceae, like *Ruellia* (Fig.19). *Thunbergia* and *Barleria*, the entire capsule bursts longitudinally with great force and noise when ripe and seeds are thrown several feet away. In *Geranium* the styles curl suddenly upwards and in Wood Sorrel the seeds are ejected by sudden movement of the elastic aril. In Legumes like Camel's Foot Climber and Crab's-eye Creeper the seeds are scattered by sudden rupture and twisting of the valves. Excellent mechanical device along with winged seeds in Mahogany (Fig.20), *Soymida* and Toon have doubly insured themselves by combining winged seeds with an explosive mechanism.

See also FRUITS, FLOWERS.

U.C.B.

498

SEED DISPERSAL by BIRDS. Birds play a prominent role in dissemination of a variety of seeds. While some floating seeds and aquatic plants like *Lemna* may be transported accidentally in the plumage of ducks, e.g. when flushed from one tank to another, most plants provide a wide range of drupes and berries attractive to frugivorous birds who disperse the undigested seeds along with their faeces. Thrushes (Turdinae), bulbuls (Pycnonotidae), crows (Corvidae), ducks (Anatidae) and pigeons (Columbidae) appear to play a major role in seed dispersal. The spread of the *Lantana* weed over thousands of square kilometres of the Indian subcontinent in the last hundred and fifty years or so is chiefly due to frugivorous birds, particularly bulbuls. The spread of several species of *Loranthus* and *Viscum* plant-parasites (Loranthaceae) in Asia is the work chiefly of FLOWERPECKERS (Dicaeidae). The growth of pipal (*Ficus religiosa*) on buildings and as 'stranglers' of palm and other trees is the result of birds feeding on its fruits and broadcasting the seeds. The abundance of the sandalwood tree in S India and of mulberry in the Punjab and elsewhere is due largely to similar activities of frugivorous birds. Both are trees of great economic value. Most seeds pass through the alimentary canal of birds unharmed by the acidic digestive juices since they are ejected within a few minutes of swallowing. There is evidence to suggest that seeds sown this way actually germinate better than those obtained directly from fruits.

SEERFISH, being a member of the mackerel family (Scombridae), has a mackerel-like streamlined body, terminal mouth with teeth, two dorsal fins and 8 to 10 finlets behind the dorsal and anal fins. At the root of the forked tailfin a soft keel-like structure is present on either side. Their colour is shining steel-blue grey and the body is usually scaleless. Of the five species occurring in Indian waters, three, *Scomberomorus commersoni*, *S. gattatus* and *S. lineolatus*, are common. The first one grows to as much as 2 m and weighs 20 kg. The others are slightly smaller. They are essentially offshore pelagic species but during winter enter inshore waters following the shoals of sardines and mackerel which they attack and devour voraciously. Their normal food is small fish, squids etc. but they prey upon prawns and crabs also when close to the shore. They are considered as excellent food both in fresh and cured condition and contribute annually about 25,000 tonnes of good fish.

SELFISH BEHAVIOUR. Selfish behaviour may be defined as that act in which the initiator of the act gains something at the cost of the recipient of the act. Thus a Hanuman langur snatching a fruit from another member of the troop performs a selfish act. The individual that does the snatching improves, albeit ever so slightly, its own chances of survival or breeding at the cost of the individual from whom the fruit was snatched. Since natural selection favours those traits which enable an individual to maximize its ability to survive or reproduce we expect such selfish behaviour to be on the whole favoured by natural selection and therefore to be widely prevalent in the animal kingdom. Indeed this is the case. Animals continually compete for food, shelter, mates and other scarce resources in their own selfish interests.

The selfish behaviour may however be constrained in three major ways. When two males fight for a mate, the victorious male is often content to leave alone the male which accepts defeat; it rarely presses home its advantage to the point of killing the rival although it could do so. This is because if it were to escalate the fight to such a level the vanquished male may choose to fight to the bitter end, and the victorious male runs serious risks of injury even if it can succeed in killing its rival. The additional reproductive success it may achieve by killing its rival is unlikely to be worth the possible reduction in reproductive success through the injuries it sustains. This is why most fights in the animal kingdom are conventional; the males display their strengths to each other but rarely fight to the bitter end.

Selfish behaviour will also be constrained when the individuals that will lose are blood relatives. Such blood relatives share many genes in common, a trait which hurts blood relatives may be hurting other individuals which are carrying the same gene; hence genes coding for excessive selfishness towards blood relatives will not be favoured by natural selection. Therefore in social groups of relatives selfish behaviour will not be as rampant as in a group of animals not related to each other. The core of an elephant herd is a group of related females and calves. We therefore expect them to help each other and the calves, and this they do, protecting and even suckling each other's calves. But an elephant cow will suckle another female's calf only so long as she does not have a calf of her own. When she has her own calf she will actively drive away another calf trying to suckle her. She will also lash out with her trunk at a calf which is bullying her own calf.

A third context in which selfish behaviour will be constrained is when long-lived animals exchange social acts with known individuals. They may then learn to reciprocate by helpful and selfish behaviour. Chimpanzees beg meat from each other. It is then possible that a male which refuses to give meat would also find itself not being given any meat when he begs for it. If the cost of sharing some meat when it has made a small kill such as that of a gazelle fawn is small compared to the benefit of gaining meat every time any male makes a kill, the selfish behaviour of keeping a gazelle kill to itself will not evolve.

SENSES. The senses determine our experience of and our response to the world around us. These consist of the special senses of SIGHT, HEARING, TASTE and SMELL which are perceived by special sense organs while other general senses like Touch, Pain, Pressure, Temperature, and Vibration involve less clearly defined mechanisms. The senses are appreciated in unicellular animals by the cell membrane and in higher animals by the body wall or skin and by special organs. The organs of special senses are developed from modified nerves, which act as receptors for various sensations. These sensations received by receptors are transmitted as impulses along the nerves to the central nervous system or brain. Here they are interpreted according to inherited patterns modified by experience, to lead to appropriate action; for instance an amoeba will move away fast from an area of high temperatures and a deer in the forest will alert itself, ready to run, at the sound of a twig snapping, but may show no alarm at the sound of thunder during a rain storm.

SENSITIVE PLANT, *Mimosa pudica*, is a small prickly undershrub from tropical America naturalized in our area. It has very sensitive leaves, pinkish flowers and bristly pods. The leaflets fold immediately on being touched and droop down. The tender plants are relished by cattle and result in increased flow of milk.

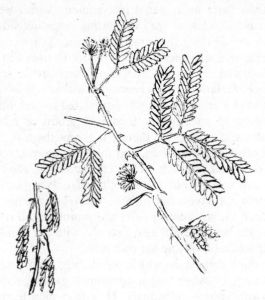

Sensitive Plant. Foliage, flowers and pods × 3/4. The shoot on the left shows folded, drooping leaflets which will unfold after half an hour.

SEXUAL DIMORPHISM. The invention of sex enabled living organisms to ensure generation of endless variety, but this does not necessarily call for two distinct sexes. The same variety could be produced if two identical sex cells or isogametes, as found in certain algae, fused together to produce the fertilized egg. But specialization in the function of the two sexes seems to have followed close upon the heels of sex itself. One sex, the males, produce gametes which are highly mobile, but contribute nothing except half the hereditary material to the fertilized egg. The other sex, the females, produce large gametes with much less mobility that provide all the nutrition plus half the hereditary material that the fertilized egg needs to develop. This specialization has two immediate consequences: firstly, young can be produced by females without any help from the males (parthenogenetically), but can never be produced by males alone. Secondly, all the eggs of a female can be fertilized by sperms from a single male, but a single male can potentially fertilize all the eggs of many, many females. Consequently females are a scarce resource for males who must compete for them, but males are not a scarce resource for females who

should only choose the best-quality male that comes along.

The divergence in the role of sex cells and the consequent divergence in the role of the two sexes has led to a number of differences in the morphology, physiology and behaviour of the two sexes. The term 'sexual dimorphism' encompasses this entire variety. The differences that merely relate to differences in the sexual organs and accessory organs such as mammary glands of mammals constitute the primary sexual characters and we shall not further concern ourselves with this component of sexual dimorphism here. What interests us more are the so-called secondary sexual characters; for instance, the much greater size of Hanuman langur males, the manes of male lions, the greater aggressive tendencies of tuskers.

The dispensibility of males has enabled the Echiuroid *Bonelia* to develop a most remarkable sexual dimorphism. The young stages of this marine worm float freely in sea-water. The adult stages are sedentary, and after a particular age the young have to settle down. Before settling the young are perfectly capable of developing either into males or females. Now they search for a female, and if they discover one they settle next to her and develop into males which are tiny in comparison with the female. If they fail to find a female, they settle down and develop into females themselves. Presumably, if such a female never gets a mate, she can reproduce parthenogenetically.

The most widespread differences among the two sexes however result from the fact that females are a scarce resource for males but not the other way around. Males must therefore compete with each other to gain access to females. The precise form of this competition depends on the social organization of the species, but very often involves a test of relative physical strength of the competing males. The most successful males in such a competition either take over long-term control of a group of females as in the Hanuman langur or lion, or gain possession of more desirable territories which enhance their access to females as in many species of lek birds such as grouse, or gain immediate access to a sexually receptive female as in the elephant.

It is notable that what counts is the relative and not the absolute capability of the males. Thus a male that is a little better off than others may gain a tremendous advantage. There will therefore be continuous selection pressure favouring an increase in the value of whichever character that is relevant in male-male competition. If this is the size of tusks, as may be the case in the elephant, males will continually evolve towards bigger and bigger tusk size. However this process cannot continue indefinitely, and at some size the tusks become so enormous a burden that they seriously reduce the chance of survival of the male with the big tusks. When

The large tusks of this elephant give him an advantage in mating

this reduction in survival more than balances any reproductive advantage he has, then the tusks will not evolve to any bigger size.

Such a process of runaway selection is expected to result in the males being considerably larger in physical size, and in the size of weaponry such as canine teeth, antlers, horns or tusks, than the females. Elephants, rhinos, deer, antelopes, goats, sheep and many species of monkeys show examples of such dimorphism.

But often the result of the competition may depend not on actual physical trial of strength at all, but on conventional or ritualized displays. In such a case any character which could initially have been related to the physiological vigour of the male may come to be preferred by females in their choice and will then be subject to the same runaway process of selection. The dazzling train of the peacock is a very likely result of such a process of selection, as are the brilliant plumages of many other birds.

The extent of sexual dimorphism would, in general, depend on the intensity of competition for mates. This is expected to be low in monogamous species where a single male mates with a single female at least for the duration of one breeding season. This of course is the case with many birds in which the sexual dimorphism is much less pronounced than in polygamous birds such as peafowl, junglefowl and ruffs and reeves. The sexual dimorphism is also absent in species like the wild dog or dhole. A pack of dhole is a group of closely

In the woodpecker, *Huia*, the male chips and the female probes for insects.

related cooperating individuals which compete little with each other. In such a species the males and females are practically indistinguishable. In polyandrous species like the jaçana where the role of the sexes is reversed, it is the females who compete for mates and are more conspicuous in coloration and more aggressive in behaviour.

Finally sexual dimorphism may derive not from sexual competition but from divergence in the ecological roles of the sexes. Thus in the woodpecker *Huia*, the females and males have beaks of strikingly different sizes and shapes and the two hunt in pairs, collaborating in getting more effectively at their prey.

SEXUAL SELECTION. The animal kingdom exhibits many examples of traits which seem to handicap an animal in its struggle for existence. Examples include morphological characters like the long train of the peacock which makes it difficult for it to fly off when startled, the elaborately branched antlers of barasingha which are apt to get tangled in vines, the enormous burden of the tusks of a big tusker; and behavioural traits such as the conspicuous courtship display of a Redwinged Bush Lark which renders it so conspicuous to a hawk and the tendency of the male wildebeest to always circle back to its territory when pursued by a hyena. In fact, it is known that adult males possessing such traits often do suffer higher rates of mortality than the females who lack them. How is it then that the evolutionary forces have not eliminated the genes coding for such traits in the continuing struggle for survival? The answer lies in the fact that while such males may run greater risks of mortality when compared to males without such handicaps, their greater reproductive success may more than make up for this handicap. For evolutionary fitness is measured in the currency of offspring produced, and the number of offspring fathered by a male will depend not only on his ability to survive, but also on his ability to mate. And characters such as tusks and antlers, and behaviours such as courtship displays and territoriality, confer on the male possessing them an advantage which may more than compensate for the increased risk of mortality that they entail. The process whereby an individual gains an advantage in mating either in competition with members of its own sex or through being given a higher preference by members of the other sex is known as sexual selection. Characters such as male weaponry and courtship displays are believed to have evolved through such a process of sexual selection.

Sexual selection owes its origin to the fact that the two sexes make very unequal investment in the zygote formed by the union of the sperm and the egg. The eggs are thousands or often millions of times larger than the sperms so that while any female can produce only a limited number of eggs, the males can produce practically a limitless number of sperms. Hence females have at their disposal a limitless supply of sperms to fertilize their eggs, but the males must compete for access for their sperms to the limited production of eggs. In consequence, while females are a scarce resource for the males, the males are not a scarce resource for females. Females are always guaranteed copulation, males are not and must struggle for it.

Bateson's classic experiments with the fruit-fly *Drosophila* were the first ones to demonstrate this most elegantly. He had males and females, all of whom carried genetic markers so that the paternity and maternity of any offspring produced could be determined with certainty. He then put together in a vial an equal number of unmated males and females and investigated their reproductive success. It was found that the number of offsprings produced by each female was pretty much the same, and even the few females which were sterile were courted vigorously by the males. A number of males, who attempted to court all right, however failed to achieve any copulations whatsoever, while a few successful males fathered a very large number of offspring. We of course know of many parallel examples from our own species. The maximum number of children produced by a female may at best approach a little over twenty, even the mythological Gandhari gave birth to only one hundred Kauravas. On the other hand Chengiz Khan fathered more than 500 sons, and Krishna had some 16,008 wives.

It is therefore evident that natural selection tending to maximize reproductive success will act very differently on the two sexes. For males copulations are scarce, so that many males may never achieve a single one; while for those males who do get to copulate, the reproductive success will go on increasing with the number of copulations they achieve. Males will therefore compete very vigorously for access to females, be aggressive towards other males in their attempts to gain such access to mates, and be quite indiscriminate in the choice of females they will mate with. For females on the other hand a few copulations are sufficient to realize their full reproductive potential. They would therefore not wish to merely acquire more and more copulations. The one way in which they could enhance their reproductive success would be to ensure that the mate they choose is a sexually competent mate who will complement her genes well, and who will produce offspring of a high quality which in turn will be successful in survival and reproduction. Females are therefore expected to be highly discriminatory in their choice of mate, to force the males to prove their ability to survive, to reproduce and to donate good care and only then to accept the highest quality available male as a mate.

Sexual selection will thus involve two elements, namely competition amongst males and choice by females. Which of these plays a dominating role will depend on the social organization of the species. In the Hanuman langur 10 to 50 adult females live in a troop which remains confined to a well-defined range. In such a case one or more adult males can attach themselves to the females and monopolize sexual access to them. The identity of the males who thus attach themselves to a group of females will be determined by fierce competition amongst the males, so that females will have little choice of their mates. Male-male competition will in this case lead to the development of larger size of the males, larger weapons such as canine teeth, and aggressive tendencies including killing of infants on newly taking over a troop.

Among baya weaver birds on the other hand, the males choose a nesting tree, and build and display at the nests. The females visit the nesting site and have a choice of mates. Sexual selection based on female choice has in this case led to the evolution of elaborate visual and auditory displays by the males.

There are however a few species such as jaçanas in which the role of the sexes has been reversed. In these birds the females lay a clutch, hand it over to a male to care for, and move on to another male for whom they will lay a second clutch and so on. In this case, it is now the males who are a scarce resource for the females and as expected it is the females that are brightly coloured and more aggressive than the males.

SHAHEEN FALCON—variously known as Sultan Falcon, Black Shaheen or Himalayan Peregrine—is the resident Indian subspecies of the Peregrine, black above, largely rusty red below cross-barred with black from belly down. The female is much larger than the male, weighing up to 900 g and held in highest esteem by falconers. It inhabits cliffs in the Himalayan foothills, and rugged hilly country in the peninsula.

See FALCONRY.

SHARKS and RAYS. The name 'shark' immediately brings to the mind of most people visions of horrid, voracious man-eaters roaming the tropical seas in search of human prey. This is only partially true because many sharks are hardly a metre or so long, and all are useful. Their liver yields oil rich in vitamin A, their fins are used to prepare delicious soups, their tough skin makes excellent leather, and the flesh is used for fish meal and manure and also as good human food when small. Their ferocity and man-hunting habit appear somewhat mythical. Experienced hunters have found them timid and easy to scare off, and they do not seem to have a special liking for human blood. In Australian waters, however, attacks on human beings even in comparatively shallow waters are reported and may be due to some individual frenzy.

Sharks and rays are more ancient than the true bony fishes. They have a cartilaginous (non-bony, gristly) skeleton, usually five pairs of independently opening gill-slits instead of a common gill-cover, and a tail with its upper lobe much longer than the lower. The skin is covered with minute placoid scales, which are enlarged in the mouth to form sharp cutting teeth. These occur in several rows, and a broken tooth is soon replaced by a new one. They have comparatively feeble eyesight but a strong sense of smell, especially for blood.

Sharks are rather streamlined and elongate in body while the rays are roundish and flattened on the underside. Rays normally stay on the bottom whereas sharks inhabit the wide upper strata of the sea. Both groups are solitary in habit and feed on fish and other aquatic animals. Some rays have flattened teeth for crushing hard shells on which they feed.

In sharks and rays, males can be distinguished by the finger-like 'claspers' they have near the anal fin as copulatory organs. Most of the sharks and rays are viviparous, giving birth to young ones; but some lay eggs which are encased in horny bag-like structures or egg cases with long processes which are popularly known as 'mermaid's purses'. These are oviviviparous, i.e. the eggs are fertilized within the mother's body and then encased by her own horny secretion. Being provided with the aforesaid protective cases and a large quantity of yolk, the eggs take two to seven months to hatch but the young hatchling is well developed to fend for itself. The egg cases are attached to submerged rocks or weeds and left to nature for their protection. Due to their horny cover, they are rarely eaten by other fish. Sharks breed almost throughout the year and there is no segregation of sexes. In an extreme case of the Nurse shark, where several embryos develop within the body, the first young to hatch out devours the others, so that only two young (one from each oviduct) are born. In other viviparous sharks, the embryo has a large yolk sac, a cord and a placenta like that of mammals, through which it receives nutrition from the mother and is well developed at birth to take care of itself. In some cases as many as 44 young ones are born at a time.

The dogfishes or cat sharks are small cousins of the sharks which, unlike sharks which constantly swim near the sea surface habitually rest on the sea bottom. They lay eggs in 'mermaid's purses' mentioned earlier. The largest shark is the Whale shark (*Rhinocodon typus*), which grows to 15 m. Fortunately, it is harmless to man, as it feeds on the minute plankton of the sea, filtering it from the water as it swims. The largest man-eater, *Carcharodon carcharias*, grows to more than ten metres long.

View of Whale Shark landed at Chowpatty beach. Bombay on 16th January 1940.

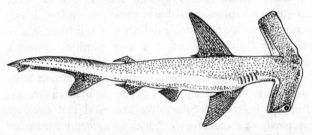

Hammerheaded shark, *Zygeana malleus* × $\frac{1}{16}$

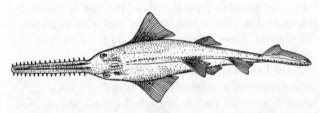

Sawfish, *Pristis perotteti*, which may have a body 5 metres long with a saw 2 m long

In the Hammerheaded sharks (*Sphyrna*), the eyes are situated at the extremities of the broad head, which is set transversely to the body like an artist's T-square.

The Rays, cartilaginous relatives of sharks, have the pectoral fins partly or wholly fused with the head Sawfish (*Pristis*) most resembles sharks as, although the pectoral fins are situated above the gill-slits, they do not extend on to the head. The tail has a typical shark shape, but the flat snout extends in front of the head like a sword and bears a row of teeth on each side. The fish uses this 'saw' to catch its food by rushing into a school of fish and swaying its head from side to side to smite its prey.

In the Guitar fish, also called shovel-nosed ray, (*Rhinobatis*), the pectoral fins continue forward along the sides of the head, but do not reach the snout. Their tail has the typical shark shape.

Sting rays (*Dasyatis*) have a flat, square or kite-shaped body since the pectoral fins meet in front of the

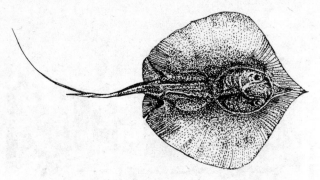

Sting Ray, *Dasyatis walga* × ¹/₈

head. There is no dorsal fin, and the upper surface of the body is coloured muddy brown, while the under-surface is white. The whip-like tail has one or more saw-toothed spines at its base; these are equipped with venom glands, and as the ray lashes its tail from side to side, the spines can inflict a very painful, festering wound.

The Butterfly rays (*Gymnura*) have a very broad body with short tail. The Electric ray (*Torpedo*) has a circular disc-like body with shark-like tail, and can produce electricity to give a mighty shock as a protective measure (see ELECTRIC FISHES).

In the Eagle ray (*Myliobatis*), part of the pectoral fin on the snout is separated from the rest of the fin by the eyes. The Cow-nosed ray uses the snout as a shovel for digging clams, which are crushed between flattened teeth and eaten.

In the Devil ray (*Mobula*) the front part of the pectoral fin is developed into two flexible 'horns'. Its body grows up to 6 m wide and it often leaps from the surface, landing with a thundering splash; the purpose of such acrobatics is not yet clearly understood.

SHRIKE or 'BUTCHER BIRD'. This family (Laniidae) contains some 97 species spread worldwide. Nine of these occur in the Indian subcontinent, ranging in size from bulbul to myna. Coloration is mainly grey in combination variously with black, white and rufous. Sexes alike. The name 'Butcher Bird' derives from their habit of maintaining a larder of prey (locusts, young birds, mice etc.) impaled on thorns for later consumption.

SIGHT. In the protozoa, the whole body reacts to light. A pigment spot considered to be specially reactive is present in some of these simple animals. Pigment spots are larger and well differentiated in some jellyfish and worms. Well developed eyes are found in molluscs and crustacea.

The development and the complexity of the eye varies in different groups of animals. Fishes, reptiles and invertebrates do not appear to differentiate

colour or shape but are quick to respond to movements. The compound eye of the arthropods is generally paired and the comparison of images from the two eyes gives a stereoscopic effect, thus permitting more accurate assessment of distance, speed and direction of the object which is seen. The attraction of bees to certain flowers appears to depend mainly on their scent, though appreciation of form and colour may also be a factor.

In vertebrates the components of the eye are well differentiated, with a lens which focuses the light rays to form an image on the sensitive surface of the retina. In higher animals there are specialized cells in the retinal membrane for appreciation of light (rods) and colours (cones). Birds of prey have highly developed eyes. In some species the eye weighs more than the brain. The retina of the diurnal bird contains few rods and a vast number of cones with coloured oil droplets which accentuate the appreciation of various colours in the environment. Birds of prey are also able to appreciate movements of small animals from a great distance. A kestrel can see the movement of a field mouse from a height at which most animals see nothing. Avian eyes contain a strongly pigmented membrane, the pecten, about whose functions there is considerable controversy. It is biggest in far-sighted diurnal hunters and it seems plausible that this membrane assists in perception of movements. It is suggested that shadows on the pleats of the pecten create photo-stimuli in the retina which help in selection of small moving objects at a distance. In the human eye a dimly seen object can be appreciated better by rapid blinking to produce a discontinuous image. The folds of the pecten may produce a similar effect in the bird's eye.

There is great variation in the field of vision. The compound eyes of most insects and crustaceans can be protruded on a stalk and moved in all directions to give all-round vision. The lizard, chameleon, and pigeon, with laterally placed eyes, have a visual field of 300°. Owls, like mammals, have their eyes close together and have binocular vision. The owls' eyes, however, cannot move in their sockets. This is compensated by an exceptionally mobile head which can be turned round on the neck almost 180° so that it faces backwards. The snake's eye has monocular vision but also has a remarkable ability of convergence which enables it to focus with both eyes on a object and obtain a stereoscopic image.

SILK-COTTON, see RED, WHITE and YELLOW.

SILVERFISH. A few species of Thysanura (about 600 known so far) are common associates of man, living in books etc., and escaping rapidly when disturbed, in a 'silver-streak'. Some other species are associated with other vertebrates, and many others live in nests of

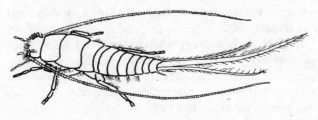

A silverfish × 8

ants or termites. These highly primitive, wingless, omnivorous (3 to 18 mm long) insects, with well-developed cerci, usually live a concealed life in soil, rotting wood, under stones or in forest leaf litter. A few principally vegetarian species are subterranean or cavernicolous. *Lepisma saccharina* and *Acrotelsa collaris* are almost cosmopolitan house insects, feeding on paper, binding of books, starched clothes, lace curtains, etc. Most species are brownish, grey or white in colour, some with silvery or metallic scales, exhibiting a sheen.

SIMILIPAL. The Similipal forest is situated centrally in the Mayurbhanj District of Orissa. 2750 km^2 of this forest has been established as a Tiger Reserve under Project Tiger and this contains an undisturbed core area of 300 km^2.

Similipal is an oval-shaped highland consisting of a rolling range of densely wooded hills and lush green valleys. It is a land of perennial water containing cascading rapids, deep pools and spectacular waterfalls. It is the meeting ground of the northern and southern flora of the country and among its animals there are examples of Himalayan as well as those inhabiting the southern reaches of the Nilgiri mountains. Its wildlife includes sambar, elephant, barking deer, fourhorned antelope, mouse deer, tiger, panther and a variety of lesser cats of which the leopard cat is the most commonly found.

Similipal owes its special natural character largely to the protection afforded to it by the rulers of Mayurbhanj prior to Independence. After the merger of this state with Orissa in 1950 there was large-scale destruction by local tribes as well as by more sophisticated hunters. Fortunately remnants of all species continue to exist and with the implementation of Project Tiger in 1973 faunal depletion was arrested and the wildlife staged a remarkable comeback. It is a splendid area for forest recreation and steps are being taken to promote tourism on a controlled basis.

SISSOO, *Dalbergia sissoo,* is a tree of the sub-Himalayan tracts of the subcontinent growing along riverbeds. It has alternate leaves, with 3–5 leaflets, small creamish flowers and pods shaped like a lance-blade. Its wood is strong and hard and in many ways like teak. It is the most extensively planted timber tree after

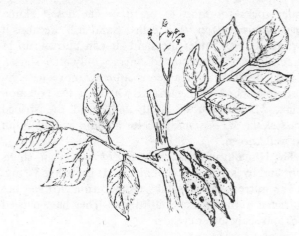

Sissoo. A twig showing leaves, flowers and pods × ⅓ .

teak. Fast-growing, adaptable and able to stand temperatures from below freezing to 50°C. It is therefore valuable for afforestation. It is planted in tea gardens to shade the bushes and is also a furniture wood and excellent fuelwood.

SKINKS. Skinks are generally small ground-dwelling lizards easily recognized by their glossy scales. They have enlarged symmetrical shields on the head, a broad flat tongue and movable eyelids. The small limbs and elongated body give them the appearance of snakes. Skinks are striped, cross-barred or spotted. They lack the ability to change colour but the males acquire red or orange hues during the breeding season. They are the commonest lizards of the world with over 600 species. Nearly half the species lay eggs, and the rest bear living young. They are generally insectivorous.

In India there are about 38 species of 13 genera, ranging in size from the 4-cm *Ristella travancorica* to the 45-cm *Mabuya tytleri*.

The limbless skink, *Barkudia insularis*, is found on Barkuda Island in Chilka Lake and Waltair. The yellow-bellied mole skink of Kashmir *Eumeces taeniolatus* is a secretive lizard which remains hidden in a burrow or

© Rom and Zai Whitaker

Common skink × ½

under rocks for most of the day. The desert skink, *Ophiomorus tridactylus*, is called 'sand fish' because it 'swims' its way under the sand. It can burrow up to 30 cm under the ground and emerges only after sunset.

Striped skinks of the genus *Mabuya* move with remarkable speed when pursued. Of these, the common skink, *M. carinata* of South India and the striped grass skink *M. dissimilis* of North India and Pakistan are well known.

The Himalayan skink, *Scincella himalayanum*, shuns light and lives in damp, cool places. It bears live young.

The burrowing cat skinks of the genus *Ristella* are restricted to the hills of South India. They have clawed toes and scaly lower eyelids.

See also LIZARDS.

SLUGS. True slugs are lung-breathing gastropods. They live on land and are almost worldwide in distribution. They are symmetrical animals with a well-defined head and an elongated body narrowing behind into a pointed tail. They have no spiral shell like snails. The common genus *Limax* found in India has only a small internal calcareous plate representing the shell. One characteristic of a land slug is that it has a hole on the right side of its body: this is the breathing aperture. The forepart of the body containing the lung, heart and other internal organs is covered over by the relatively small mantle which is sometimes called the shield. The head carries two pairs of tentacles. At the tips of the hinder pair are situated the eyes and the organs of smell. Both pairs of tentacles can be completely withdrawn into a hollow in the head. The mouth is at the lower part of the head and is a simple aperture. Land slugs are partly vegetarian and partly flesh-eating. Consistent with this feeding habit, their radula contains both large and small teeth—the marginal teeth are blade-like while the others are squarish. The radula is long and in *Limax maximus*, as many as 40,000 teeth are counted on it!

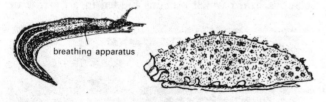

The land slugs *Limax* and *Onchidium* × $^2/_3$

Slugs are more active by night than by day, and they seek moisture. In India they are common in areas receiving regular annual rainfalls. During the dry season the slugs remain dormant under the earth, that is, they estivate. During this period of inactivity the slug contracts its body and covers it with a double coating of tenacious slime or **mucus which the animal** copiously

secretes. Unlike snails, slugs estivate separately and not gregariously.

Like Nassids or Dog-whelks (see WHELKS) slugs are sensitive to smell. Any putrid vegetable or decaying animal at once attracts them. They feed on these and also on earthworms, caterpillars, centipedes, etc. Certain species eat their own kind.

Slugs have many natural enemies including thrushes, toads, frogs, lizards and carnivorous beetles.

A land slug has both male and female organs capable of producing sperms and ova. They lay only a limited number of eggs. The black slug—*Arion*—lays 477 eggs in 48 days, i.e. at the rate of 10 eggs per day.

Belonging to the same sub-Order of land slugs, there is a curious slug—*Onchidium*—which lives by the sea coast near brackish marshes and mangrove swamps. Unlike land slugs it has no internal calcareous plate. Viewed from above it is oblong shaped and from the side it looks more or less convexly arched. The mantle is thick and leathery and covers the entire body. The back is covered by warty growths or tubercles of different sizes. In addition to the normal pair of eyes on the tentacles a number of other eyes are developed upon some of the tubercles on the back. These eyes are in structure similar to the eyes of SCALLOPS. The animal attains a size of 4·8 cm long and 2·9 cm broad and 2·5 cm high. It feeds on tiny seaweeds and is common on rocks especially near coral reefs. The animal is known to be subject to the attacks of certain predatory fish; but the mantle eyes enable the mollusc to become aware of the shadow of its approaching enemy and escape in time.

It has been noted in the case of one genus—*Prophysaon*—that when it is annoyed by being handled, an indented line appears at a point about two-thirds of the length from the head. This line deepens and eventually the tail is shaken completely off. If the animal is let loose before the actual dismemberment takes place, the deepening of the indented line stops forthwith and after some time completely closes up.

Sea slugs are perhaps the most curious and wonderful of all gastropods. They are all small shell-less shell-fish. They are widely distributed on the weed-covered rocks and reefs of all seashores. In cooler waters they are especially rich in number and variety. Sea slugs are not really slugs but naked gilled gastropods called Nudibranchiates. As an embryo the animal has a tiny coiled shell which is lost soon after the sea slug emerges. An adult sea slug is symmetrical, with a soft and elongate body like that of a true slug and the mouth and anus at opposite ends. It has no shell, no lung, or internal gills. It breathes through outgrowths on its back which are often elegant and brilliantly coloured in green, blue, crimson, red or yellow. There are fantastic forms in different genera, some are branching filaments, some club-shaped, some rosette-like and others mere folds.

Sea slugs feed on sea-anemones, corals, sponges and tiny molluscs. Many of them have no radula but grind what they eat by the horny teeth with which the inside of their stomach is lined. The liver in these animals is not a compact organ as in other molluscs but broken up into a number of glands. In the plumed and crested genera like *Aeolis*, the liver glands are contained in the plumes and crests.

Sea slugs are highly elastic. They can shrink to half a centimetre and expand to 8 cm. Most of them are far more richly coloured than any of the land slugs and some are of truly remarkable appearance. One and the same individual changes its colour scheme while expanding or contracting, while feeding or while gliding through different coloured seaweeds, sea-anemones and corals. Thus their ornamental and colourful appendages serve the purpose of protecting them from their enemies by harmonizing with their surroundings. The colours are not merely protective but have also a warning significance, because in some species the colours conceal the spicules on their skin or the stinging cells on their dorsal processes.

Along the Indian shores, especially in the waters of Pamban, Krusadi and Shingle Islands, several attractive and curious species of sea slugs are very common. They are found attached to the underside of stones and rocks in shallows abundant in seaweeds and corals. The more interesting among them are: the plumed types —*Bornella* and *Eubronchus*, the filamentose *Hervia* and the star-marked *Dendradoris* and *Trippa*.

See also MOLLUSCS.

SMALL CATS. Besides the big cats South-East Asia has a rich variety of small- to medium-sized cats, eleven of which occur in the subcontinent. Typically they are solitary in hunting, furtive and secretive in their habits, and seldom seen by man except through lucky chance encounters.

In the northern Himalayas, there is an isabelline race of the Lynx (*Felis lynx*) which is the size of a large dog (weighing up to 25 kg) and with long pointed ears terminating in black tassel-like hair tufts and an incongruously short black-tipped tail. Confined to the inner mountain ranges from alpine grassy slopes to tumbled rock talus they hunt by scent, preying upon snowcocks, marmots and hares. At the other extreme is the Clouded Leopard (*Neofelis nebulosa*), a long-tailed and very short-limbed cat (weighing up to 20 kg) which is adapted to an arboreal existence in the evergreen rain-forests of Assam, Bhutan, and Bangladesh. Descending in size, there are two more quite large and fearsome cats. One is the Caracal (*Felis caracal*), adapted to desert and low hill areas, extending from Rajasthan, Kutch, and the northern Punjab including the Salt Range westwards to the North West Frontier and Baluchistan. Slimmer in build than the Lynx, weighing about 16 kg, they have reddish buff fur with long legs and a medium-length tail. Incredibly agile in their final rush upon any quarry, they are capable of knocking over the Desert Hare even as it leaps away, or the Sandgrouse as it springs into the air. In the same humid forest regions of the north-east where the Clouded Leopard dwells, the Golden Cat (*Felis temmincki*) also occurs. They are of an unspotted golden red colour and have distinctive horizontal dark and light alternating stripes on their cheeks. Specimens are generally heavier in build than the Caracal, with a longer tail, and they are believed to prey upon chevrotains and young deer. They readily climb trees but are thought to be more at home in areas of tumbled rocks. Because of its rareness little is known about its habits.

Descending further in scale of size there are two cats which generally average considerably bigger and heavier than the domestic cat. The Fishing Cat (*Felis viverrina*) is a thick-set tabby-marked cat of savage disposition which swims freely both under and on the water, hunting waterfowl and any small mammals which it encounters. The Jungle Cat (*Felis chaus*) is a smaller, longer-legged greyish fawn cat which also has small tufts on its large upstanding ears and a few black rings on the tip of its rather short tail. Adapted to a wide variety of conditions from pure desert to low thorn scrub hilly country, they are partly diurnal in hunting.

There remain five truly small cats which are becoming increasingly rare and local in distribution. The beautifully patterned Leopard Cat (*Felis bengalensis*), perhaps the least rare, is confined to forest zones from the outer Himalayas down to the Ghats. They are largely arboreal in hunting and nocturnal. In south India and Sri Lanka the smallest representative of the family, known as the Rustyspotted Cat (*Felis rubiginosa*) occurs. About three-quarters of the size of a domestic cat (adults weigh a little over 1 kg), they have a fawn grey coat patterned with rusty brown bars and spots coalescing into horizontal lines. In the high steppic mountain areas of the North West Frontier and in the northern regions of Gilgit and Ladakh, the Steppe or Pallas's Cat (*Felis manul*) is found. In 1978 a fine reddish form of this species was captured from south Waziristan; the first authentic record of its survival south of the Tibetan plateau since about 1910. This small cat has a thick bushy tail, very low-set rounded ears and a broad skull framed by a ruff of long hair. Four faint vertical stripes traverse its body and there are two parallel radiating black stripes from the corners of its eyes. They prey upon pikas and chukor. In the extreme southwest of Baluchistan, in the Chaghai desert plateau, the rare and little known Sand Cat (*Felis margharita*) has recently been discovered. **With a broad skull and low-set ears** it looks superficially like Pallas's Cat but has the soles of its paws entirely covered by long greyish black fur. It lives in shifting sand-dune desert, preying largely upon lizards and gerbils.

In the desert regions of Sind and Rajasthan, the Desert Cat (*Felis lybica*) occurs sparingly. It is a small yellow-buff cat with prominent black spots over its body and vertical stripes on its forehead. Considered conspecific with the African Wild Cat it is believed to have provided the predominant ancestral stock from which domestic cats evolved.

Finally in the evergreen rain-forests of Assam and Bangladesh, another very elusive small cat, known as the Marbled Cat (*Felis marmorata*), survives. It has a long bushy tail, and is beautifully patterned in rufous buff with irregular black and darker brown patches. It is exclusively arboreal and like the Leopard Cat preys upon roosting birds, squirrels and even insects.

SMELL and TASTE. Smell or the ability to perceive different odours and scents appears to be a function of special nerve cells or receptors. Whereas sight can be understood in terms of intensity and colour of light, and hearing in terms of frequency and volume (decibels) of sound, we have not so far designed any methods for physically measuring the subtle blending of chemicals which causes the sensation of smell. The sense of smell therefore cannot be studied with any degree of precision. It is known however that smell plays a very important part in the life and behaviour of most animals, which can be attracted, or repelled by the smell, even in traces, emanating from a food source or from an enemy species. Sharks, in the Ganga, are said to get the scent of a dead body from miles downstream, and rapidly swim up to the site to get a meal. They possess a special organ, the vomerine organ, in the nasal cavity, which is supposed to help in picking up traces of the smell and guide them to the site. It is well-known that many animals, especially carnivores, use scent as a marker for individuals, and many animals will mark their own territory by urinating at the limits, or by the secretions of special scent glands. The scent of each animal is appreciated by others of the species, who recognize the territory marked in this way. Among insects, very powerful sexual attraction is exerted by certain secretions known as pheromones, which indicate that the individual is sexually active. Pheromones of some species have been identified and used to attract and entrap large numbers of the species concerned, and this is likely to become a useful method for control of insect pests, when their specific pheromones can be identified and artificially produced.

Taste and smell, both responses to chemical composition of the stimulus, appear to be interrelated in mammals. It is a common experience that certain medi-cines with unpleasant 'taste' actually have an offensive smell and can be easily swallowed if consumed quickly before the smell becomes apparent.

Taste is appreciated by special sense organs called taste-buds which contain highly sensitive nerve-cells. The hair-like structure of the bud appreciates the taste after the food is dissolved in the saliva. The nerve-cells of taste transmit the sensation by nerve fibres to the brain. Smell and taste are closely related in humans. Smell receptors occur in the nasal cavities and taste receptors around the beginning of the digestive tube. In fishes, such as carp and catfishes, the ability to taste extends over the entire ectodermal covering of the body, as far as the tail. In birds with cornified tongues taste buds are scarce. In birds like the parrot, which have fleshy tongues, they are on the tongue and along the sides of the upper half of the beak. Certain parrots have the most numerous taste organs of all birds.

Amphibians possess nerve endings in the skin which are sensitive to chemical stimuli but whether they interpret it as taste is not known.

Taste-buds reach their greatest development in mammals such as ruminants that have specialized grinding molars and retain forage within the mouth cavity for a prolonged period. The taste-buds in these animals are found on the surface of the cheek, and in papillas of the tongue. The numbers of taste-buds to each papilla varies greatly in different animals, sheep have 480, cows 176 and pigs 4760.

SNAILS form the largest group of molluscs with only one shell. There are many kinds, varying widely in habits and structure. There are the true lung-breathing land snails, freshwater snails and sea snails. All are widely distributed throughout the world.

Xestina mating, × ²/₃   *Helix* in motion, × ²/₃ .

LAND SNAILS

1. *Land snails* are familiar. They are marked by the slime-coated, trailing body which comes out of a roundish spiral shell. The animal moves so slowly that in English 'a snail's pace' means very slowly indeed. A snail has two pairs of tentacles on its head: a short pair in front and a long pair a little behind. The hinder pair has an eye at the tip of each tentacle. Snails can contract and completely withdraw these tentacles into a hollow in the head.

Land snails are vegetarians (plant-eating). Their mouth is a simple aperture below the head armed with an arched jaw which helps to grip the tender shoots. For grinding the vegetable matter they have a broad radula with rows of uniform squarish teeth. In length this radula does not exceed thrice its breadth. The throat of land snails widens inside into a crop which retains a quantity of masticated food before it is passed on to the stomach. Snails are well-known for the havoc they do to gardens.

Their habits are particularly interesting. As a rule they are moisture-loving animals. In rainy months (except when there are heavy downpours, which they shun) they are found in large numbers moving about over moist stones, moss-covered walls and plants of various kinds. During the hot dry season they close the lid of their shells tight and remain inactive, buried in loose earth or concealed under thick debris. In other words, they estivate. Snails estivate gregariously (many together) and not singly or separately. They are also known for their 'homing instinct'. After their search for food or partners they return to the place from which they started.

Land snails have a life-span of about 5 years. Their shell is completely developed with lip two years after birth; and their mating capacity commences from the third year. A land snail has both male and female organs which produce sperms and ova at about the same time in separate chambers and these are led by separate ducts to the common genital opening. For fertilization of the ovum, mating of two individuals is necessary. Compared with other molluscs snails lay few eggs. The typical land snail *Helix aspersa* has been stated to lay 40 to 100 eggs during its life. They are laid in cup-shaped hollows at the roots of grass with a little loose earth spread over them.

The more common species of land snails found in India belong to four genera: *Helix, Ariophanta, Xestina* and *Achatina*. *Helix* has a pale spiral shell growing to 4 cm in diameter and a reddish brown mouth. *Ariophanta* has a beautiful banded shell measuring 6 cm. Both are very common in western India, Nilgiris, Anaimalais and Nelliampathy Hills and they are eaten by hill tribes. *Xestina* has a regularly shaped and beautifully banded shell. It is abundant in Tamil Nadu. *Achatina* is an African snail with a tall-spired, large brownish shell 9 cm long and decorated with dark streaks. This snail has now infested all the rainy regions of India, particularly Kerala and Bengal, and does much damage to vegetation. The eggs of *Achatina* are 2·5 cm in diameter and yellow in colour.

Snails have many natural enemies. Carnivorous ground beetles, frogs, toads, lizards and birds all feed on them. One species *Helix vittata*—a white-shelled snail—has a bitter taste and is not eaten by birds.

2. *Pond and River Snails*. Like land snails, the pond and river snails are also lung-breathing. But in these molluscs the eyes are situated at the base of the tentacles, of which there is a single pair. The tentacles are capable of contraction but cannot be completely withdrawn into the hollow of the head. Two genera of Pond snails are common in our area. They are *Lymnaea* and *Planorbis*. The former has an elegant, conically spired shell. It is known for its worldwide distribution and unlimited capacity for adapting itself to physical extremes. One species is recorded from a tarn situated at over 5000 metres elevation in the Himalaya and another from a depth of nearly 250 m in the lake of Geneva. It survives also in polluted waters, sulphur springs, brackish water and even in the hot-water geysers of Iceland. Although it has reacquired the habit of living in water, *Lymnaea*'s lung is still functional, so that visits to the surface are necessary for the snail to obtain air.

Planorbis attached to the surface film, × 1.

In *Planorbis* the shell is flat and coiled anti-clockwise.

Both *Lymnaea* and *Planorbis* are found gliding upside down beneath the surface film of the pond where they live, the surface tension of water keeping them in position. These snails deposit their spawn in irregular gelatinous masses on the underside of water-plants and other debris. During the summer when the ponds dry up, they estivate buried in mud and remain there until the rains fill up the ponds again. The life-span of both these genera is 3 to 4 years.

Two other common genera of freshwater snails are the *Viviparus* and the *Pila*. Both have similar habits and habitat. In *Viviparus* the shell is thin, covered by an olive-green protective coating and coiled like a turban. As its very name implies, this snail is viviparous—the young develop within the parent and are born as

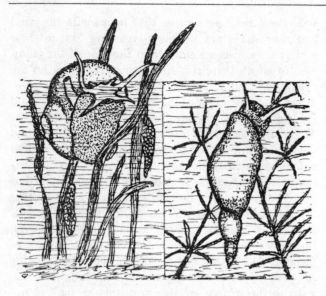

Pila with expanded foot, showing eyes and tentacles. Her spawn is on two leaves. Right, *Lymnaea*. Both × 1.

miniature adults. *Viviparus* is purely gill-breathing and remains in water.

Pila is the common Apple-snail, much larger than *Viviparus*, with a thicker and globular shell having a low spire. It is found in plenty in all ponds and water-clogged paddyfields. It has both a lung and a gill and can live in and out of water. It also possesses a fleshy siphon or tube by which it can suck air into the lung while actually under water. When, due to rains and floods, the pond waters get mixed with slime, mud and other debris, clogging the gills, this mollusc can come up periodically to the surface and suck air through the tube. During summer, when the ponds and fields are dried up, *Pila* withdraws into its shell, tightly closes it up with its lid and estivates embedded in mud until the return of favourable conditions. It lays its eggs in large clusters on the stems of water-plants under the surface.

3. *Sea snails.* The sea is also rich in several kinds of snails which vary very widely in shape, habits and other details. There are the Top and Turban Shells, Periwinkles and Violet snails.

Top shells and Turban shells are familiar univalve molluscs of all seashores, living on mud-covered rocks

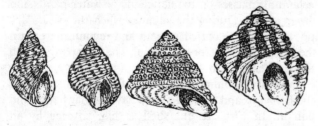

SEA SNAIL SHELLS

Left to right: two periwinkles (*Littorina* and *Planaxis*), a top shell (*Clanculus*) and a turban shell (*Turbo*), × 2/3.

of the intertidal region. In their internal structural details they agree with the primitive gastropods like LIMPETS. They possess a radula, carrying numerous tiny teeth for grinding sea plants. Their shells are fairly heavy, usually brightly coloured, ornamented outside and pearly within. Species like *Trochus radiatus* and *Clanculus depictus* have conically spiral shells carrying rows of small smooth tubercles over the whorls and coloured designs of red and greyish blue. They have a flat base and their shell-lids are thin and made of lime. *Turbo* is a large turban-shaped shell with rounded whorls, brownish or greenish in colour, rounded at the base and having a circular, thick, hemispherical operculum. The shells of *Turbo* have commercial value for the mother-of-pearl contained in them.

Closely allied to Top snails is an extremely curious species of snails called *Umbonium vestiarum* which occurs in plenty in all littoral mud-flats. Their shells are small, smooth, polished, flattened cones resembling coat buttons, whence their popular name Button shells. They are coloured differently—white, red, pink, chestnut, deep orange or any one of these colours mottled or streaked with a contrasting colour. They are beautiful objects, widely collected and exported to foreign countries for studding jewel boxes.

Periwinkles are represented abundantly on all the rocky shores of India by the genus *Littorina* and the allied *Planaxis*. They are a group of interesting sea snails. They can be seen slowly moving about on high coastal rocks and branches of mangrove trees feeding on tiny plant growths and algae. Their proboscis is short but the radula very long, well suited for chafing the vegetative food. Their shells are like pointed turbans, drab-coloured and quite in harmony with their environment—brown algae, greyish black rocks or dark brown branches of mangroves. The periwinkles living high up on rocks receive only a few sprays of water from high-tide waves but they do not suffer because they possess a rudimentary lung which enables them to breathe air and remain out of water for several hours. Periwinkles deposit their spawn on seaweed, rocks and stones. The eggs are enclosed in a glairy mass which is just able to retain its shape in the water—each egg has its own globe of jelly and is separated from the others by a very thin transparent membrane.

Then there are the Violet snails of the genus *Janthina*, which float on the surface of the sea in an inverted position with the help of a frothy, small boat-shaped float secreted by the foot. The female snail attaches her eggs to the underside of the float. The colour scheme of the shell has a protective significance. Birds cannot easily see it as the deep purple colour of the base of its shell harmonizes with the colour of the deep blue waters. Similarly, to predatory fishes the light blue spirals of the shell merge with the colour of the sky.

Communities of Violet snails and blue jellyfish (such as *Physalis*, the Portuguese man-of-war) float together on the surface of the ocean, the snails feeding on the jellyfish, and the jellyfish feeding on small fish that get entangled in their stinging tentacles.

Another interesting sea snail of pelagic habit is the Sea Butterfly—*Cavolina* belonging to the group Pteropoda. It is an extremely interesting mollusc in that its structure and habits amply justify the name 'sea butterfly' given to it. Its shell is not coiled but straight and usually elongated and its foot is modified into a pair of wing-like fins, arranged one on either side of the mouth, enabling it to swim with ease. These animals spend their whole life in the open sea in such vast swarms as to discolour the water for miles. They feed on plankton—minute floating sea organisms—and they themselves form the principal food of baleen whales.

See also MOLLUSCS.

SNAKES. Snakes are included in the order Squamata (snakes and lizards) of the Reptilia. They are vertebrates, and the long body is covered with scales, the numbers and size of which help to determine the different species. The skin is generally smooth, but some snakes such as the Sawscaled Viper and File Snake have rough scales.

Scales may differ to suit the habits of various species. Bronzeback tree snakes, for instance, have keeled belly scales. This assists the snake in climbing. The rough, saw-edged scales of sawscaled vipers are rubbed together to produce an intimidating, hissing sound to deter enemies.

A snake may be long, short, stocky or thin. It is divided into three distinct regions: the head, body and tail. The head may be either flat, pointed or rounded. Some snakes such as watersnakes have nostrils and eyes set high on the head, to enable them to see and breathe when the head surfaces. Some, like the pythons and pit vipers, have heat-sensitive 'pits' below or behind the nostrils which help to detect warm-blooded prey. A snake's eyes may be large as in the diurnal tree snakes and the Rat Snake, small as in the kraits, wolfsnakes and kukri snakes, covered or almost absent as in the shieldtails and worm snakes. Most snakes have round eye-pupils. Cat snakes and pit vipers have vertically elliptical pupils and only the vine snakes have a horizontally elliptical pupil.

Snakes have long, forked tongues which are actually organs of taste, feel and smell. A snake detects food, enemies or mates by darting out its tongue and transferring particles to Jacobson's organ in the roof of the mouth, which interprets the smell.

The head and body are divided by the neck which in some snakes is markedly thin and delicate. The tail may be short and stubby or thin and tapering. It is

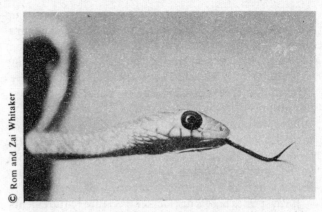

Head of rat snake showing sensory forked tongue, × ¹/₂

generally possible to differentiate between sexes because the female has a shorter and more rapidly tapering tail.

Snakes have evolved from lizards, and the vestiges of limbs can be seen in some species. Pythons are primitive snakes and have two prominent stumps on either side of the body. These are absent in the more evolved species.

That snakes lash with their tails is a fallacy. Sea snakes use their flattened, paddle-like tails for swimming. Female King Cobras use their tails and body to sweep up leaves to make a large mound nest in which the eggs are deposited. The Red Sand Boa, which is a thick-bodied burrowing snake, uses its tail in an interesting way to escape predation. The head and tail look exactly alike unless closely observed. When attacked, the tail is raised to fool the enemy, while the

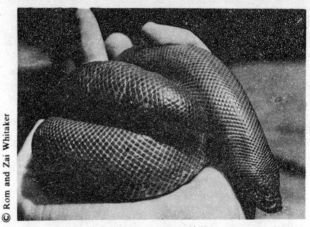

Red sand boa × ¹/₃

head is securely tucked under its coils for seeking escape. Young pit vipers use their tail as a lure. Curling it like a worm when a lizard or frog comes near they can lure the prey within striking range. Snakes such as pythons and pit vipers have prehensile tails and can cling to branches and climb with the help of the tail.

It was thought that snakes were totally deaf, since they have neither ear-drums nor ear openings. Recent

511

studies show that some air-borne sounds are picked up through the lungs. They also respond to surface vibrations; thus, they 'hear', and are quick to escape from footsteps, but the response to the snake-charmer's flute is probably only visual. The snake reacts to the movement of the flute and prodding by the snake-charmer, not his music.

Most snakes are near-sighted and depend partly on their sense of smell to avoid predators. Some of the more active, diurnal snakes such as King Cobras, rat snakes and tree snakes have good eyesight and may detect a human a hundred metres or more away.

Each species has its distinct temperature requirements. Some, like cobras and kraits, do well in hot, dry areas, spending the hottest part of the day in a relatively cool rat-hole, termite mound or crevice. Others, such as pit vipers, shieldtailed snakes and King Cobras, require cool, wet habitat and are found in evergreen jungle in the hills. They succumb quickly to drastic changes in temperature and humidity.

Snakes generally feed on rats, mice, frogs and small birds. There are several exceptions. Watersnakes eat

Vine snake $\times$ ²/₅

© Rom and Zai Whitaker

Shieldtailed snake $\times$ ¹/₂

© Rom and Zai Whitaker

Head of Vine snake $\times$ ¹/₂

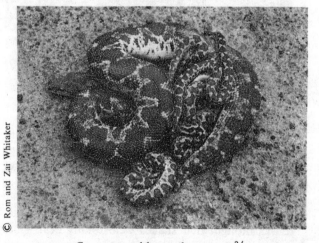

© Rom and Zai Whitaker

Common sand boa and young $\times$ ²/₅

fish and frogs, sea snakes fish and fish eggs. Kraits and king cobras are mainly snake-eaters. Smaller snakes feed on lizards, insects, earthworms and even scorpions. We do not know very much about the feeding habits of many snakes, especially some of the smaller forest snakes that are difficult to observe. Burrowers such as worm snakes and shieldtails are known to eat earthworms and insect larvae. Few snakes are insect-eaters. Even a newly born sand boa only 15 cm long can catch and swallow a mouse or a lizard. Tiny, new-born keelbacks will skilfully attack a baby frog or toad for their first meal.

Venomous snakes inject their prey on seizing it. Some snakes such as the harmless Vine Snake have fangs far back in the mouth and have a mild venom useful for killing their small prey. Constrictors such as pythons and sand boas catch and coil around prey, squeezing it to death before swallowing. Snakes prefer live prey but are sometimes seen eating carrion.

Snakes apparently find mates via a strong musk from the cloacal musk glands. When handling a freshly caught snake one often smells this distinctive musk on one's hands. Rat snakes, contrary to popular belief, do not mate with cobras. Male snakes have two sex organs called hemipenes. During mating, one is inserted into the female's cloaca. The male crawls on top of the female after a period of arousal procedures including rubbing and even biting in some species such as rat snakes. The vertical dance seen in two or more snakes is actually a ritual combat between males, and not mating. It is perhaps territorial combat and is observed often in rat snakes and kraits.

The majority of Indian snakes lay eggs, which generally hatch in 60 to 90 days; others bear living young.

Striped keelback laying eggs × 1

Striped keelback × ½

Some of the egg-layers are rat snakes, cobras, watersnakes, kraits and pythons. Vipers, sea snakes and sand boas, among others, bear living young. Eggs are laid in a sheltered hollow or crevice and the females of some species stay with the eggs for at least part of the incubation period. These include cobras, king cobras, kraits, checkered keelbacks and pythons. Eggs vary considerably in size; a big python may lay eggs the size of duck's eggs, while those of keelbacks and watersnakes can be only a centimetre in length. Snake eggs are always elongate, white, and with a leathery shell.

Emergence from the egg is aided by the egg tooth, a projection on the nose-tip that is later lost. Newly hatched snakes can be as small as earthworms and are preyed on by birds, mongooses, civets, monitor lizards, and many other predators. They eat tadpoles, baby mice and insects, and may double in length in the first month.

People are often curious as to why snakes shed their skins. Every few weeks or months snakes grow too big for their skins which are 'shed', or 'sloughed'. The thin, translucent outer skin peels off, inside out, often in one piece, and the snake emerges shiny and bright. Just prior to shedding it is dull-coloured and listless. During shedding, the eye-caps come off with the skin, resembling tiny contact lenses. Teeth, fangs and tongue-tips are replaced periodically as well, hence the Greek myth that snakes are immortal.

No studies have been done on the life-span of snakes in the wild. In captivity a Rock Python lived for 34 years, 2 months, a Rat Snake for 10 years, 7 months, and a Sawscaled Viper for 10 years, 3 months.

Snakes are surprisingly mobile animals. Many are excellent climbers and swimmers, while one species, the Flying Snake, flattens its ribs and glides through the air. The mode of locomotion we are most familiar with is what is popularly known as crawling. Protrusions on the surface of the ground are used by the snake to push itself forward. Some snakes, such as the Dog-faced Watersnake, are 'side-winders'—that is, they travel sideways on land. Sand boas, pythons and large vipers creep slowly while on the prowl in a characteristic 'caterpillar' crawl.

Even adult snakes are prey to a variety of animals. The wiry mongoose will attack and eat a snake if it is small enough; large monitors, birds of prey, water birds and wild pigs are also enemies. The adult snake's main enemy, however, is man. Since snakes are vulnerable animals they have several defence mechanisms to enable them to escape from predators. The most dramatic defence display is that of the Cobra. Hood spread, the Cobra hisses and turns around to display the bright hood markings. Contrary to popular belief, few snakes hiss. Some that do, include Russell's vipers, rat snakes, pythons and cobras. Some snakes poke with their tails when caught; others defecate, spray musk or regurgitate. Harmless snakes may rise up and spread a small, cobra-like 'hood'. Others raise the tail which may be coloured or shaped to resemble the head.

There are two factors which have contributed to a decline in many species of snakes in the last decade or so. The first is the destruction of specialized habitats. King Cobras, for instance, live in dense evergreen rain-forest, large tracts of which have been cleared to make way for tea and coffee plantations. They are big snakes and need plenty of thick jungle to hide in; thus the opening of evergreen forests is a very serious threat to them.

While the clearing of forest has been disastrous for forest snakes, the conversion of jungle to farm land, particularly paddy fields, has created ideal rat habitat resulting in dramatically increased populations of rat-eaters such as cobras and rat snakes. The large concentrations of these snakes in rice-growing parts of India (notably Bengal, Tamil Nadu, Kerala and Andhra Pradesh) did not go unnoticed by skin dealers. For several decades, tens of millions of snakes have been killed for skins for the export market. The loss of so many rodent predators has probably aggravated human food problems in India as rats continue to chew their way through close to half the country's rice production.

Snakes are eaten by several of the hill tribes of Northeast India and southern India. The oil and gall bladders of some snakes (pythons and rat snakes for example) are considered medicinal without justification.

Snakes are now protected in India by a ban on the export of skins, though people continue to kill them when encountered, whether venomous or not. In some parts of India (Shirala in Maharashtra and parts of Tamil Nadu and Andhra Pradesh) farmers will not kill snakes and recognize their valuable rat-killing services.

Several institutions are engaged full or part time with snake research. (a) Haffkine Institute, Bombay and Central Research Institute, Kasauli are the two anti-venom producers for which Haffkine keeps a collection of often several thousand snakes. (b) Madras Snake Park is primarily a public education, conservation and research organization. (c) The Bombay Natural History Society and Zoological Survey of India carry out mainly systematic work on snakes and (d) the Irula Snake Catchers Cooperative is the main producer of venom in South India.

Snakes and Man

There are over 2000 species of snakes in the world. Of these, about 230 are found on the Indian mainland and adjoining island groups. These vary from the tiny, primitive worm snakes to the giant King Cobra which

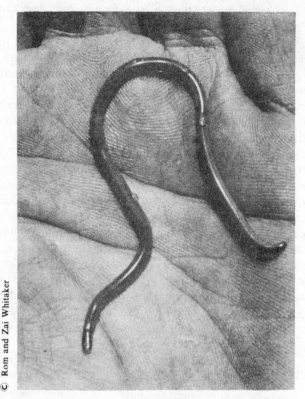

© Rom and Zai Whitaker

Worm snake × 1

grows to 5 m, and which is the only snake in the world known to build a nest. Most people fear all snakes, and believe that every snake is poisonous and interested in attacking man. In fact it is not so. Of the species in our area, for instance, 55 are venomous, yet the majority of these pose no threat to man as they are either rare and therefore hard to come across, or live in remote and isolated habitats, away from human dwellings, or are not given to biting. Some, like the pit vipers, have a low venom toxicity, which usually causes only a mild reaction.

Venomous snakes have hollow fangs, which function like hypodermic needles to inject venom when the snake bites. Snake venom is a complex mixture of toxins, proteins and enzymes produced by a pair of highly modified salivary glands, the parotid, in the 'cheeks' of the head of a venomous snake. Venoms vary in strength or toxicity. In India the range is from the venom of pit vipers which can cause serious pain and swelling to that of kraits, of which a fraction of a drop can be fatal to a man.

The venom of the sea snakes and elapids (snakes with short, fixed fangs such as cobras and kraits) is nerve-paralyzing (neurotoxic). Blood-destroying (haemotoxic) venom is characteristic of the vipers with their long, folding fangs. Some snakes, such as the Gaboon Viper of Africa, have a mixture of the two venom types.

The human body has natural defences against venom

and produces antibodies to combat a bite. The amount of venom a snake injects is what determines the severity of a bite, and this depends on the snake. Most bites are just defensive nips and fail to inject a dangerous amount of venom. The fangs carry the supply of venom from the venom glands through venom ducts. When the snake bites, muscles surrounding the venom gland are tensed, shooting a spurt of venom through the duct and out through the fang.

One of the enzymes is a 'spreading factor' called hyaluronidase which causes rapid absorption of the venom. Because venom is injected, the remedy too must be injected, preferably into a vein. Antivenom serum is made by injecting increasing doses of diluted venom into horses, until they become immune. Freeze-dried serum from immune horses is then packaged and supplied to hospitals throughout the country from Haffkine Institute, Bombay. This serum is effective for bites of the four common dangerous snakes found in our area: the Cobra, Krait, Sawscaled Viper and Russell's Viper—the Big Four. Haffkine serum need not be refrigerated. Its only disadvantage is that a small percentage of people are allergic to horse serum.

Snakes are not interested in biting humans. The highly efficient venom apparatus has evolved to meet the specialized feeding habits of each venomous species. Pit vipers, for instance, eat small birds and mice, to kill which their mild venom suffices. Kraits on the other hand have a highly toxic venom, the most toxic of any land snake in Asia. This is because they feed on other snakes, and cold-blooded prey needs a more toxic venom to incapacitate. Thus sea snakes, which feed on fish, have one of the most powerful venoms.

All over the world, elaborate stories have evolved about certain snakes that will chase and bite humans without provocation. Most popular shikar books include blood-chilling encounters with King Cobras. In actual fact, King Cobras, even when guarding their

Water snakes × ½

© Harry Miller

nests, will seek escape when encountered. This is true of all snakes. Unless injured or cornered, their instinct is to get away.

Cobras, kraits, sawscaled vipers and Russell's vipers, the Big Four, are common throughout most of our area. Bites from the Big Four are fairly common and most victims are barefoot rural people. It is important to be able to identify these dangerous snakes and others; and on no account should they be handled except by professionals. There have been tragic cases of people being fatally bitten by species thought to be harmless.

People believe that a venomous snake-bite is almost surely a fatal event. This is not true. Studies show that in India, 90 percent of venomous bites inject too small a quantity of venom to be fatal. The few serious bites can be treated with polyvalent anti-venom serum available from Haffkine Institute, Bombay.

See REPTILES, CAT SNAKES, COBRAS, ESTUARINE SNAKES, FLYING SNAKES, KING COBRA, KRAITS, KUKRI SNAKES, PYTHONS, RAT SNAKES, SEA SNAKES, TREE SNAKES, TRINKET SNAKES, VIPERS.

See plate 9 facing p. 128.

SNIPE. Snipe are very secretive and shy migratory marsh birds, dark brown above, streaked with black, rufous and buff. Their bill is long and slender with a sensitive tip and the eye prominently large and forwardly placed. They affect grassy marshland and wet paddyfields, probing in the mud for worms.

Snipe shooting is a popular but challenging sport, for the birds flush abruptly on being approached and fly off in lightning zigzags, completely baffling the shooter.

Our commonest Snipe is the Fantail (*Gallinago gallinago*).

SNOW. Snow is solid precipitation falling from the clouds above and occurs in the form of minute ice crystals at temperatures far below 0°Centigrade. Snow occurs as large snow flakes a few centimetres in size, when the temperatures are near 0°C.

Over high mountains, such as the Himalaya or the Alps, and over high latitudes polewards of sixty degrees, there are regions of perpetual snow. The boundary of an area of perpetual snow is called a snow-line. In Northern Scandinavia the snow-line is at a height of about 1200 metres, over the Alps at about 2400 metres, and over the Himalaya at about 4200 metres.

Snow-melt during the summer months contributes significantly to the flow of water in the Himalayan rivers.

A metre height of snow is roughly equivalent to ten centimetres of water.

SNOW LEOPARD. This is considered by many to be the most beautiful of the large cats, combining a lithe grace with a deeply luxuriant and handsomely marked pelage. The Snow Leopard (*Panthera uncia*) is an inhabitant of the highest inner ranges of the Himalayas and the rugged mountainous plateaus of Central Asia including Tibet and the Soviet Union. Unfortunately it has been heavily persecuted by local hunters, not only for the good price obtainable for its valuable fur but also because of the status earned by the successful hunter amongst his neighbours. The total world population has now been variously estimated at no more than 400 individuals or even less, and the IUCN has listed it as an endangered species.

Very similar in its ecology to the Panther or Leopard, Snow Leopards are known to be solitary animals, hunting over very extensive territories and seldom remaining for long in one area. Studies in Chitral and Nepal in the 1970s have revealed typical population densities of three individuals in an area of 3000 square kilometres. Like the Leopard, they tend to use well-worn trails, a habit which local hunters take advantage of in setting pitfall traps or poisoned spears. They prey mostly upon wild ungulates such as the Bharal in Nepal, or Ibex in Gilgit and Markhor in parts of Chitral. Unfortunately there are many recorded instances of hungry Snow Leopards coming into villages to seize domestic goats and sheep, and if they get into a penned flock they are capable of wantonly killing 7 or 8 animals before dragging one victim away for food. This trait, shared by many carnivores when confronted by penned domestic prey, has inevitably led to their being regarded as an enemy of the hill tribesmen who lose no opportunity to hunt and kill any Snow Leopard which appears in their vicinity. One hopeful sign for the future of this extremely rare animal is the reported fall in the market value of its fur subsequent to the adoption by both India and Pakistan of legislation banning trade in its skin as well as international restrictions on import of the pelt of skins of endangered animals.

Snow Leopards are smaller and lighter in build with relatively longer tails than panthers. Adults weigh up to 36 kg and measure 100–110 cm in head-and-body length with a tail of 84 cm. Their fur is a stone-grey colour with elongated black lozenge-shaped rings coalescing into solid black spots and lines in the spinal region. The long creamy white belly fur is interspersed with black spots and the face is also handsomely marked with solid black spots, the iris being green and the naked tip of the nose pinkish brown colour. Usually two young are produced in a litter and in Pakistan available information indicates that most cubs are born between May and August.

SOAP-NUT TREE, *Sapindus emarginatus*, mostly confined to peninsular India, is characterized by elliptic leaflets and globose fruits with a smooth black shell,

Soapnut tree. A shoot with foliage, inflorescence and fruits × ¹/₂

used for washing clothes and by jewellers for restoring brightness to ornaments. In the Himalaya an allied tree *S. mukorossi* is cultivated. Both are fast-growing, good fuel and valuable for afforestation.

SOCIAL HIERARCHY. Anyone who has even casually observed a troop of monkeys would have noticed that some members of the troop are more aggressive and dominant than others. If some food is offered to a relatively subordinate member of the troop it appears reluctant to accept the food and even if it does accept, it is very likely that a more dominant one would quickly snatch the food away. If one continues to watch the troop more carefully and tries to offer food to different members, a very interesting order or hierarchy in aggressiveness becomes evident. This is known as social hierarchy or dominance hierarchy and is present in a variety of group-living animals.

The form of social hierarchy can be highly variable. It can involve the dominance of one individual over all the others who are equal amongst themselves. Alternatively, and more commonly, there exist multiple ranks in a more or less linear order. In such a situation one animal, normally referred to as the alpha animal, dominates all the others in the group, the second or beta animal dominates everybody except the alpha animal and so on. It is this kind of situation that one is very likely to encounter while feeding a troop of monkeys. Occasionally a social hierarchy may deviate

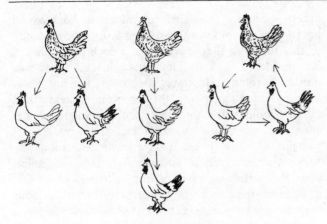

Three simple forms of dominance hierarchies. More complicated chains of hierarchies can arise by a combination of these basic types.

from a simple linear hierarchy and may be circular, i.e., A is dominant over B, B over C and C over A.

The rank of an individual in a social hierarchy is established after a series of fights during which the animals assess each other, and once the rank is established it remains stable, sometimes for very long periods of time. The juveniles in the troop of monkeys appear to be playing all the time without showing the kind of overt aggressiveness characteristic of the adults but it is during this play that they assess each other and a hierarchy gradually becomes established.

The domestic fowl illustrates vividly the way in which a social hierarchy comes into being. When a group of strange hens are brought together, as often happens in a poultry farm, they fight a great deal. At first all birds fight with everybody and the outcome of each fight decides whether a bird will fight with a given opponent again. Here, fighting mainly involves pecking at each other. In a very few days a 'peck-order' becomes established.

Social hierarchies of the kind described above are known to occur in many insects, some fishes and amphibians, many birds and most mammals. Often there are two separate hierarchies, one among the males and one among females. With few exceptions, males are dominant over females. In most cases that have been studied, there is a great deal of fighting before a hierarchy is established but subsequently the animals seem to accept their status so that there is now very little fighting. However, the animals seem to continuously keep reminding others of their status. Consider, for example, a troop of rhesus monkeys that seem to be resting at the end of the day. After all the fighting for food and mates is over, when one would least expect any expression of competition or dominance, look closely and you will find that a subordinate individual is grooming a dominant one which in turn is grooming a more dominant one. Animals have evolved a variety

of visual, chemical and acoustic ways of signalling their social status to their neighbours (see DOMINANCE BEHAVIOUR).

What then do the animals gain in arranging themselves in such a hierarchy? In answering this question one should probably think separately of the advantages to the dominant individuals and the advantages to the subordinate ones. In many instances the advantages to the dominant individuals seem very clear. They simply get the best of everything, food, mates, resting-place etc., without having to fight and demonstrate their superiority every time. When a flock of feeding pigeons were observed and were shot and cut open at the end of the day it was seen that the dominant ones had significantly more grain in their guts than the subordinate ones. The advantages of being dominant are very striking in rapidly growing young animals such as piglets and kittens that compete with each other for the anterior teats of their mothers, which yield more milk than the posterior teats. Such advantages would indirectly lead to greater genetic fitness. In other words, animals that are able to feed more are more likely to leave behind a larger number of offspring. This should be especially true when there is a severe shortage of food supply and the subordinate animals simply do not get enough to eat.

In addition, there can be advantages of being dominant which lead more directly to increased genetic fitness, because the dominant individuals end up mating with a disproportionately larger number of individuals of the opposite sex and thus leave behind more offspring. In the case of wild dogs, for example, there is a strong dominance hierarchy among the females and there is also cooperative hunting and communal feeding of the pups. Very often there is such a severe shortage of food supply that a pack can only successfully bring up the pups of one or two females and therefore only the very dominant bitches end up reproducing. This is often because the pups of subordinate bitches starve to death.

Recent studies suggest that in some primitive human tribes with polygyny and fierce forms of dominance hierarchy, there is an unmistakable advantage to the dominant males, who sire a very large proportion of the children born in the tribe.

The presence of social hierarchies is also often beneficial to the subordinate individual. Sometimes this is because an individual is much better off being subordinate in a group rather than being alone. The subordinate bitches at least survive and may occasionally bring up a pup but if they leave the pack and try to make it on their own they have no chance at all of even surviving, let alone reproducing. In other instances the presence of a hierarchy is beneficial to a subordinate individual because knowing its status it does not risk injury or death by fighting with a dominant individual but remains aloof and tries to steal a bite or a mate now and then.

R.G.

SOCIAL SYSTEMS. Many animals live in aggregations or groups consisting of several individuals of a species. Such groups may be called animal societies in analogy with human societies. The interaction between individuals in such animal societies is referred to as social behaviour. In recent times the study of social behaviour in animals has become a very popular new science called sociobiology. The aim of sociobiology is to understand the forces that bring animals together and that shape the form of social behaviour. Animals present a large variety of social systems to permit such an analysis. A survey of the animal kingdom reveals varying degrees of social behaviour. Edward O. Wilson has recognized colonial invertebrates, social insects, non-human mammals and human beings as the four pinnacles of social evolution.

Among the colonial invertebrates may be included the myxobacteria, the slime-moulds, siphonophores and corals. In myxobacteria and slime-moulds there is a free-swimming phase when individual bacteria or amoebae lead what can be termed a solitary life. Under certain conditions that are suggestive of deterioration of the habitat (such as starvation, lack of some crucial amino-acids etc.) the solitary individuals come together and a fraction of them make a 'supreme sacrifice' so that a certain other fraction can disperse and go on to new and perhaps better habitats. This is achieved by all the free-swimming single cells forming one giant multicellular aggregate. Out of this aggregate is formed a fruiting body with a base, a stalk, and a collection of spores at the tip. The cells that form the spores thus disperse to new habitats and once again begin to live as free-swimming forms. On the other hand the cells that formed the base and the stalk die after having made it possible for the spores to disperse. Since the different free-swimming cells that came together may not have been genetically identical, this act is no different from what we will call 'altruism' when we talk about higher organisms.

A number of colonial coelenterates such as the siphonophorans, the true jellyfish and corals, show such a curious form of organization that it is difficult to decide whether to call each unit a single organism or a society of very closely knit individuals. Typically each unit consists of a number of different kinds of individual members called zooids. Each kind of zooid performs a different function. Some are filled with gas and act as floats, some squirt jets of water and act as bellows, some are specialized in feeding and digestion, some others protect against external danger while still

others specialize in sexual reproduction. Each zooid has its own nervous system and acts fairly independently of the others so that it seems as if each zooid should be considered as a distinct organism and the whole unit as a colony. However, the zooids are physically interconnected, including their nervous systems, and all the zooids of a unit arise from a single zygote. Whatever we choose to call them, these bizarre organisms are sure to teach us a great deal about the forces that bring about and maintain social organization. From the few investigations carried out so far, it appears that such an organization provides four major kinds of advantage: namely, resistance to physical stress, liberation from a sessile condition, superior competitive abilities and defence against predators (see JELLYFISHES).

Insects show a remarkable degree of social organization, unparalleled among the higher invertebrates. The groups that have taken sociality to the extreme are ants, bees, wasps and termites. The following technical terms to define different levels of sociality have been developed with particular reference to the insects. *Eusocial* refers to the condition where there is cooperative brood care, reproductive division of labour with separate worker and reproductive castes and overlap between generations. When there is cooperative brood care and reproductive division of labour but no overlap between generations, the condition is referred to as *semisocial*. If there is only cooperative brood care, then it is called *quasisocial* and if none of these three traits is present the condition is referred to as *solitary*.

The social insects have received a great deal of attention by investigators interested in the origin and evolution of social behaviour in the animal kingdom. This is primarily because of two reasons. Firstly, the social insects provide a number of examples of the different levels of sociality described above. Often, the species showing different levels of sociality are fairly closely related so that they form attractive model systems for a comparative study that might help us to understand the forces responsible for the origin and maintenance of social behaviour. Secondly, it is only among the social insects that we see individuals totally giving up reproduction in favour of other members of the same species. Although the myxobacteria and slime-moulds also do this they do not have the complex multicellular body plan that makes the phenomenon of sterility as striking. In theory of course there is no real difference between the altruism of a slime-mould and that of a worker bee.

The act of a worker bee, giving up reproduction completely in order to aid its mother to produce more offspring, of forgoing its life by stinging a predator that might have otherwise harmed its colony, has attracted the attention of many a scientist not only because of its romantic analogy to certain acts of human beings but also because it presents one of the greatest difficulties for Darwin's theory of natural selection. The theory of natural selection argues that in the struggle for existence those kinds of individuals that manage to leave behind more progeny slowly come to replace other kinds of individuals who have been less successful in reproduction, for whatever reason. It follows then that any character that makes an organism more 'selfish' to increase its chances of reproduction should be favoured by natural selection. How then did the phenomenon of sterility and self-sacrifice evolve and be maintained for millions of years in hundreds of insect species? The crux of the problem lies in the fact that natural selection was believed to operate only at the level of individual organisms. Since the case of the social insects was clearly impossible to explain on the basis of such individual selection, Darwin himself and many others following him hinted that natural selection might also occasionally act at levels higher than the individual. For example, a colony of honey-bees, where some of the individuals sacrifice reproduction and work for the colony, might eventually produce a larger number of honey-bees than another colony in which everybody tried to reproduce and there was no one to bring sufficient food for the larvae. Such arguments of *group selection* did not gain popularity because the genetic mechanism underlying group selection was not at all clear.

The situation changed drastically when W. D. Hamilton recognized that altruism, especially towards close relatives, can be explained by a special form of group selection which has come to be known as *kin selection*. Hamilton urged that since relatives share a number of genes in common, an individual who sacrifices its own reproduction in favour of a relative does not lose everything in terms of genetic fitness (which may be defined as one's contribution to the gene pool of the next generation). This is because the offspring of the relative are bound to carry some of the genes of this individual. By this logic, Hamilton came up with the ingenious formula that in any social interaction, the cost to the actor should be less than the benefit to the recipient multiplied by the fraction of genes shared in common between the two individuals; cost and benefit measured in genetic fitness. It is easy to see that the greater the degree of relatedness the easier it is for social behaviour to evolve.

Hamilton's theory has achieved spectacular popularity because of its special application to the social insects. Most social insects (ants, bees and wasps but not termites) have a peculiar mode of reproduction. A female is capable of laying both fertilized and unfertilized eggs because she stores sperms received during copulation in an organ called the spermatheca and controls their release for fertilization. The fertilized diploid eggs develop into females while the unfertilized haploid eggs develop into males. As a result, the degree of relatedness between different relatives is not the same as in diploid organisms. In these so-called haplo-diploid organisms the mother would be related to her daughter by 1/2 but to her son by 1 because the son develops from her unfertilized egg and has no father at all! Similarly, a female would be related to her sister by three-fourths because the males have only a haploid set of chromosomes and therefore the father's contribution to each daughter is identical. On the other hand, a female is related to her 'brother' only by one-fourth. Thus a female is more closely related to her sister (three-fourths) than to her daughters (half) and sociality in the Hymenoptera (ants, bees and wasps) most often involves a female giving up reproduction on her own and helping her mother to raise more offspring (her sisters). As we have seen above, it is relatively easy for such behaviour to spread by natural selection because a female's sisters are more closely related to her than her daughters are. Haplo-diploidy is believed to be one of the most important factors that has led to the evolution of social behaviour in the Hymenoptera.

The hypothesis that haplo-diploidy has been a key factor in the evolution of hymenopteran sociality has an additional virtue in that it leads to a number of predictions that have been largely borne out by observations on the colonies of social insects. For example,

(i) eusociality should be more common among haplo-diploid organisms than among others,
(ii) males should be more selfish than females,
(iii) females should be more altruistic in their behaviour towards their sisters than towards their brothers,
(iv) workers should favour their own sons over their brothers.

It must be cautioned that a number of ecological factors favouring colonial over solitary existence must be operating hand in hand with haplo-diploidy to generate the extreme degrees of social organization and altruism witnessed in the Hymenoptera. The involvement of such other ecological factors is the reason that a number of haplo-diploid species have not attained eusociality and that the diploid termites have done so.

A group of animals that are considered not to have any elaborate social organization are the cold-blooded vertebrates comprising fishes, amphibians and reptiles. It is very likely that this notion is primarily by virtue of our ignorance regarding several aspects of the life of these animals. More recent studies appear to give the impression that schooling in fishes, territoriality and mating systems particularly of frogs and lizards are not far behind those of birds and mammals in their complexity of social organization. These groups of animals therefore represent social systems whose intricate details largely remain to be unravelled.

Birds' and mammals' are certainly the most highly evolved as well as the best studied social systems among the vertebrates. In many instances studies of these social systems have begun to throw light on the factors responsible for their social behaviour. Let us consider some instances of group life or sociality such as feeding flocks, colonial nesting and communal roosting in birds.

Feeding in large flocks is a widespread habit amongst many birds, particularly outside the breeding season. Birds such as hawks, falcons, kingfishers and warblers feed solitarily, babblers feed in small groups of 5 or 6, while the Rosy Pastor, cormorants, munias and rosefinches feed in large flocks. Feeding in flocks as compared to solitary feeding must have both its advantages and disadvantages. For example, feeding in flocks must often lead to competition for limited resources. On the other hand, feeding in flocks can sometimes enhance efficiency if it is easier for a group to detect and capture the food. Similarly, being in a flock can affect the chances of a bird's survival in the face of an attacking predator. This can also happen in both ways. For example, a large group of birds might be more efficient at detecting approaching predators or might simply be rendered more conspicuous and thus more vulnerable to predation. Being in a large group might also indirectly increase feeding efficiency because the birds on the average have to spend less time looking out for predators. In general, it may be said that the size of feeding groups can affect feeding efficiency and predator avoidance in rather complex ways depending upon the detailed circumstances involved. In many instances, observations of feeding birds have shown that the size of feeding flocks is adjusted to yield maximum feeding efficiency and escape from predators. It is a relatively simple exercise to show that the number of times a bird lifts its head to scan for predators is inversely correlated with the size of the group in which it is feeding.

Just as in feeding, birds have to make a choice between aggregations of different sizes during breeding activity. It is a familiar sight to see sparrows, crows and mynas nest solitarily while weaver birds, egrets, storks and vultures form large dense aggregations of nesting colonies. In this case, not only is the extent of crowding of nests determined by the costs and benefits in terms of food availability and efficiency of predator avoidance but there is an added factor. This factor brings about the almost simultaneous breeding of a large group of birds. The adaptive significance of this seems to be that if most of the chicks are produced simultaneously, predators will not be so likely to lurk around the nest-site on the chance of a one-day feast as they would be by the prospect of meals throughout a breeding season extending over several weeks.

Another familiar aggregation of birds is the communal roost. Most of us have observed the large noisy aggregations of mynas, crows and parakeets at about the time of dusk. Owls, koels, warblers, drongos and shrikes on the other hand roost singly or in very small groups. A roost is a place where the birds spend the night. Protection from predators by means of easy and quick detection is certainly an important factor favouring large communal roosts. Another possible subtle advantage that has more recently come to light is that a communal roost may not simply be a place for the birds to sleep but a congregation where there are plenty of opportunities to learn from other birds about the sources of food near by.

Now we move on to the mammals, which are the most highly evolved vertebrates. Their varied forms of social organization are significantly more complex. Since all mammals feed their young ones with milk, a mother-offspring association is the very least that one can expect in the social organization of any mammalian species. Apart from this, the presence of higher levels of social organization is very varied among mammals. The marsupials, insectivores, and hares are largely solitary. The bats and rodents are very diverse, while whales, dolphins, porpoises, ungulates, elephants, carnivores and primates are all highly social.

Among the bats and rodents, one can find species which are solitary, just with prolonged mating pairs, monogamous families, harems, or year round multi-male multi-female groups. Although the dolphins show considerable degrees of allo-maternal behaviour, school formation, communication, brain enlargement, imitating powers and acts of rescuing injured conspecifics, the stories one hears of their superior, near-human intelligence are largely the products of fertile imagination.

The ungulates as a group show strong tendencies towards herd formation. In some cases it is simply a selfish herd such as we see in Chital where every individual tries to get to the centre of the herd at the approach of a predator. Even here we see that the herds become smaller and more dispersed when grass is scarce. On the other hand we see the formation of harems in horses, zebras and camels.

Elephant society is very highly social and matriarchal. Several females and their calves always stay together and a considerable degree of altruism and allo-maternal care is seen. The males visit the herds only for mating and remain solitary at other times.

Most carnivores such as wild dogs, jackals and lions live in extended family groups and often hunt in groups. The males have a very important position in most carnivore societies unlike in the elephants. This is probably because many carnivores hunt in cooperative groups and sometimes chase their prey in relays so that they can bring down a large prey well beyond the capability of an individual animal.

Among the primates we find different kinds of social systems. The smaller lemurs and lorises are solitary. Some monkeys belonging to the families Indriidae and Hylobatidae live in parental family groups. The much-studied Hanuman langurs form two kinds of troops. The bisexual troops consists of one dominant male leader, several females and juveniles of both sexes. The males are driven away as they become adults. Males also live in separate all-male bands. These males are often subordinate to the males of bisexual troops, and therefore live at the periphery of the richer habitats occupied by the bisexual troops. Periodically the all-male bands attack the male of the bisexual troops and sometimes succeed in displacing him. In such an event the most dominant member of this band alone takes over all the females and kills the suckling infants, bringing their mothers into estrus much earlier. Among the gorillas there is a little more tolerance of males, so that in addition to the females there can be several males of different ages in a group. In the bonnet and rhesus monkeys we find true multi-male multi-female societies. Although there is a dominance hierarchy the subordinate individuals do seem to get occasional opportunities for reproduction.

Human societies show a number of properties reminiscent of the non-human primates. However, the advent of culture and a significant amount of cultural transmission of learned information somewhat obscures the biological relevance of human social behaviour. What is clear is that our unusually large brain volume provides for a high degree of behavioural plasticity and learning abilities which in turn have made it possible for us to practise norms of barter, reciprocal altruism, social bonding, sex and division of labour quite sophisticated by any animal standards. What the relative roles of our genetic heritage *via-à-vis* our cultural innovations are, is the next major area of research in sociobiology.

R.G

SOCIALIZATION. The higher mammalian societies exhibit a complex structure with considerable differentiation of behaviour with respect to sex, age and social status. Much of the behavioural repertoire in such species is learned and therefore develops through a process of continuing interaction amongst individuals of the social group. Socialization is a process whereby the young growing up in such a milieu gradually adjust their behaviour to the role they come to play in their social group.

At all stages during socialization, the interests of the young will conflict with those of all other individuals, including its own mother. At the earliest stages when the mammalian young is very vulnerable, the mother readily meets most of its demands. She lets it suckle and rest as much as it wants. As the young grows however it becomes less vulnerable and the mother does not have as much to lose by refusing its demands as earlier. For the mother would wish to conserve her energies for the production of future litters as well. She would also equally value different siblings from the current litter. Any young will however value its own welfare at a higher level than that of its present and future siblings and make higher demands on the mother in terms of her milk and attention, demands that the mother will not be willing to meet. From this conflict will arise attempts by mother and young to manipulate each other's behaviour—a conflict that is an essential element of the process of socialization.

This may most readily be observed in the behaviour of domestic dogs or the monkeys around villages or cities. In the Hanuman langur, for example, the mother is at first quite indulgent towards her infant, but as it reaches the age of 8 to 9 months she tries more and more to dissuade it from suckling and clinging to her by roughly pushing it away. The infant protests by screaming and pursuing her. The mother finally succeeds in weaning it, but only after much travail.

The young also has to adjust its behaviour towards other adults in the group. Among Hanuman langurs the other females try to carry the infant, perhaps in practice for their future role. The mother often allows them to do so because it frees her to attend better to her own feeding. But these other females handle the infant roughly, and it keeps squealing, but generally has no option except to lump it. In elephants and lions these 'aunts' may even nurse the calf and it attempts to get as much out of them as possible, but is often repelled.

The young also have to learn to adjust to their own siblings and other young of the same age. They practise this during social play which often involves testing each other's strength and trying to suppress the other young to a lower social position. The mothers may vigorously intervene in such a process and try to protect the interests of their own young. In this complex social environment the young grow and adjust themselves to the social role that they come to assume.

A number of experiments in which young are artificially deprived of social experience make it very clear that in a social species social experience is essential for the development of normal behaviour. For example, deprivation of social experience may result in females becoming completely incompetent as mothers and males completely incapable of copulation.

SOIL. The word 'soil' is derived from a Latin word *solum* meaning what is underneath, the ground. Knowledge of the soil has increased to such an extent that

A roadside cut showing the soil profile to a depth of 6m. The physical differences can be clearly seen here from top to bottom of the profile.

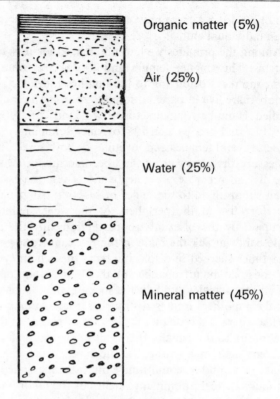

Fig. 1. Composition of surface soil. The constituents are shown according to the volume occupied under suitable conditions. The proportions of air and water change frequently.

there are different definitions of soil according to the area of interest and the use made of its properties. A fairly comprehensive definition of soil is: A three-dimensional heterogeneous body formed from natural material through the action of weathering and soil-forming agents. Under suitable physical, chemical and biological conditions it can provide anchorage and a medium for the growth of vegetation.

The soil develops from the rocks of the earth's crust through weathering processes. Weathering breaks up the rocks into loose material for soil development, called parent material. Soils can also be formed from material transported by water as in the Gangetic plains. Sometimes soil is so rich that it produces lush vegetation, on the other hand there are soils in which nothing can grow regardless of what is done to them. Most soils lie between these two limits. The composition of a typical soil is shown in Fig. 1.

The development of soil to a state at which it can support the growth of plants is dependent upon the climate, the parent material, relief, biosphere and the time elapsed during the process of soil formation and the development of physical, chemical and biological properties. At places, we can see an overriding influence of one factor over the others and the soil displays its properties accordingly. One such factor is climate. The differences between soils supporting extremely dense

vegetation in a rain-forest and a very sparse one in drier regions or deserts are dependent on the rainfall. The soils on valley bottoms are deep whereas on steep slopes, due to the movement of soil particles, the soils are shallow. As the soil acquires its characteristic properties, it develops features which separate it from soil of other localities. The soil on hill slopes in cooler regions would produce more brownish-black substance called humus whereas there is very little humus in soils in the plains: there most of the humus burns up during the hot summers.

Water plays a very important role in bringing about many physical and chemical changes in a soil. It not only carries soil particles from the top surface downwards, but also carries the products of biological and chemical activities to the lower depths. The succession of the horizons from top surface down to the depth where one can observe the parent material is called the 'soil profile' and is shown in Fig. 2.

The rock is now completely transformed due to physical, chemical and biological action and the minerals contained in the rocks also change to a decomposed state. So much degradation of the primary minerals takes place that one can find a developing complex mass of soil possessing particles of various size-fractions. These particle fractions are called sand, silt and clay. Depending upon the influence of several

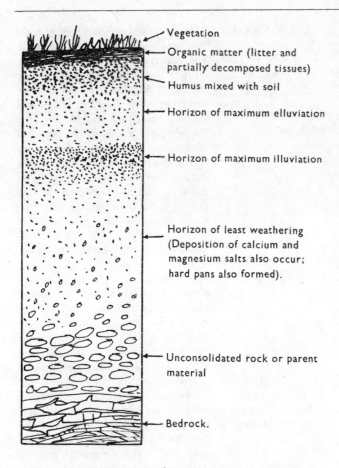

Fig. 2. A soil profile showing distinct features. In nature the thickness of horizons varies a great deal. (Elluviation= wash-out; Illuviation=wash-in process)

soil-forming factors, the mineral composition of rock being an extremely important one, the soil develops its physical and chemical behaviour. A complex mixture of soil particles below 0·002 mm derived from these minerals and humus in soil is known as the colloidal complex or clay humus complex of the soil. This complex is of great importance to plants since it controls the food supply to plant roots for absorption. In the absence of this complex, most of the food material of plants would be leached from the soil along with percolating water. Plants absorb food material from soil suspended in water. Scientists have discovered sixteen chemical elements which plants must take up in order to survive.

Many chemical elements such as nitrogen, phosphorus and sulphur are made available through the activities of micro-organisms. The most important contribution of microbes towards the nutrition of plants is to make nitrogen available by nitrogen fixation from the atmosphere. Therefore, the soils which lack a good population of these micro-organisms cannot feed the plant with this nutrient sufficiently.

Plants obtain food through their roots from the soil and water system in which the nutrient elements are dissolved. Plants can also absorb the nutrient elements directly through their leaves. This is a useful method of providing nutrient to the plants as it saves about 50 percent compared to feeding the plant through applying fertilizer to the soil. A soil which has a capacity to support vegetation at optimum growth level is called a fertile soil. A fertile soil has a balanced proportion of nutrient elements in it.

Due to natural factors such as high rainfall, the soil loses nutrient elements through percolating water. This process is called 'leaching'. The surface of the colloidal complex loses useful cations or bases and most of the exchange sites are replaced by H ions from the surrounding water. The soil thus develops acidity and poses a problem for productive use. Addition of lime improves this condition and the soil is gradually brought back to normal chemical condition. The chemical condition of soil depends on the concentration of active hydrogen ions in it. This concentration is referred to as the pH of soil, which is the negative logarithm of hydrogen ion concentration. Neutral soil has pH 7·00; acid soil has pH below 7.00 (e.g. 5.00); alkaline soil has pH above 7·00 (e.g. 9·50).

Sea water inundates coastal areas regularly and the soils of these areas develop excess of salts and are known as 'saline soils'. Also large areas are under irrigation. Non-judicious use of irrigation water over the years raises the ground-water level and this water comes to the surface during hot summers carrying dissolved salts with it. These salts accumulate on the soil surface in the form of a crust after the water has evaporated and do not allow plants to grow. These soils are also called 'saline soils'. If salinity is allowed to develop unchecked and if these soils have high sodium content, the soil deteriorates further to become alkaline. These soils are problem soils. There are mechanical and chemical means to improve these soils so that the plants can grow normally.

In order to systematize the study of soils, the soil characteristics have been classified and the important groups are shown in the Table below.

Soil Group	General Characteristics
INDIA	
1 Red Sandy	Derived from granites, quartzites, sandstones, etc., characterized by rich sandy materials. The clay is coated by a mixture of oxides of iron forming red, yellow or reddish yellow soil, red when dry and yellow when wet. Mostly slightly acid soils. The feel is generally gritty.
2 Red Loamy	Derived from gneisses, charnockites, diorites and other minerals, giving rise to high clay material. The soil has fine

3 Red and Yellow

grain, and lime concretions as nodules are also found. These soils are neutral to slightly alkaline in presence of lime. The ranges of colour vary from reddish yellow to yellowish brown, fine-grained and in some cases associated with laterites. Derived from micaceous quartzite schists, phyllites, etc.

4 Laterite

Generally reddish or yellowish red with massive and firm structure or loose aggregates of iron nodules. Used in brickmaking. Derived from basic rocks under high rainfall. These soils usually lack clay.

5 Podzols

Found on sandy quartzose and base-poor parent materials under high rainfall conditions.

6 Deep Black

Generally 120 to 200 cm deep, derived from the basalt rocks in the Deccan plateau. Also called black cotton soils. The clay content ranges from 40 to 60 percent and lime is present as nodules at lower depths: moderately alkaline soils.

7 Medium Black

Generally 50 to 120 cm in depth, formed from basaltic rocks etc., sticky soils, moderately rich in organic matter and well-drained. Lime is present in varying amounts. Gypsum deposits are also found in the peninsula.

8 Saline Alkali

Characterized by the occurrence of excessive soluble salts, can occur in many areas such as red, black and alluvial soils. In many cases they occur mostly on the coasts and low-lying areas. Vast areas in the Gangetic plain are covered by these soils. These soils may deteriorate to physically poor alkaline soils.

9 Peaty and Saline Peaty

Developed from brackish water sediments containing sulphides from acid alluvial soils. Located at the junction of hill streams of coastal hills and backwaters of sea. Slightly acid when ill drained, whereas can be very acid upon drainage; accumulated organic matter is also observed.

10 Alluvial

Formed by transportation by streams and rivers on the plains or along the coastal belt. Characteristic pale, pale grey and pale yellow to deep black soils having coarse to fine grains. Important formation is observed through the deposits of Yamuna, Ganga and their tributaries; fertile soil.

11 Coastal Alluvial

Found all along peninsular India between the sea and the hill ranges. Coarse to fine grain, usually deep, bright reddish brown and yellowish brown to grey; favourable soils growing mostly paddy.

12 Deltaic Alluvial

Represent varying sediments brought by rivers and deposited at the delta. The material deposited may be fine, medium to coarse. The soil at the mouth of the Ganga is swampy, accumulates a lot of organic matter and supports mangrove forest.

13 Desert

Found in the arid zones as sand and sand dunes, coarse-grained and derived from rocks of the nearby areas, coastal regions and the Indus valley. Composed of sand to 50 cm or more with shifting sand dunes at the top. Yellowish brown to very pale brown in colour. Sometimes have high salt content and high pH; very low in organic matter.

14 Terai

Found at the foot of Himalayan ranges, deposited through the water movement. Lime nodules are found in some areas; medium fine grain soils having gravels at lower depths. Continuous seepage of water gives rise to waterlogging and a lot of weed growth. Well drained soils are cultivated.

15 Hill (Brown Hill, Submontane and Mountain Meadow)

Found over a variety of rocks generally in high Himalayan regions with meadow soils growing beyond the tree-line. Derived from sandstone, micaceous sandy soils and shales. Originally growing conifers. Generally slightly acid soils on slopes above the tree-line, the soils are shallow, derived from sandstone and shales, skeletal in nature. Mainly grass grows, with some deposition of organic matter.

16 Skeleton

Shallow soils formed from sandstone, pale brown to dark brown up to 15 cm thickness, highly erodible. Coarse-grained due to silica, poor soils, support hardy vegetation, obtained both in colder and warmer regions.

17 Calcareous Alluvial

Developed on alluvium (10 to 40 percent lime) brought by the Gandak flowing towards the Ganga. Light-coloured good pale brown and yellow brown, coarse to moderately fine-grained, alkaline in nature.

PAKISTAN

Two broad soil regions can be observed in Pakistan apart from the humid slopes of Kashmir. (i) The Desert and (ii) the Dry Mediterranean mountain.

1 Desert

The desert includes alluvial soils; young soils; shallow soils of coarser nature; soils dominated by limestones; salt-affected soils in the plains. The soils of the plains are highly productive.

2 Dry Mediterranean

The mountainous region has young brown soils; yellowish brown soils; raw brown soils of deserts; shallow, very coarse soils and soils dominated by limestones.

The soils are highly fertile. Since the soils have developed from geologically young material, they are stony, shallow on slopes. The soils in parts of the plains suffer from waterlogging, salinity, alkalinity.

BANGLADESH — Alluvial and Tropical Mountain Skeleton soils

In the broad kaolinite-dominated alluvial soil region, the soils have undergone changes due to local factors such as presence of salts or flooding through river water. These soils are alluvial soils in which the profile has been influenced by inundation; yellowish brown soils which are mainly acidic in nature; salt-affected soils in mangrove forest; dark clay soils; soils rich in limestone throughout the profile; soils with hard clay pan. The plains of Bangladesh are frequently flooded by rivers, mainly the Brahmaputra and Meghna. Flash-flooding influences soils over a very extensive area.

The tropical mountainous regions have kaolinite-dominated and very coarse shallow soils.

SRI LANKA — Sri Lanka includes basically two types of soils, e.g. Kaolinite-dominated and Tropical Mountain. Yellowish brown soils are also found. Due to abundant vegetation in areas of heavy rainfall, organic soils are also found. The tropical mountains have yellowish brown and very coarse shallow soils.

Since the vast population of the world depends on soil and plants for its needs of food, fibre, fuel and timber, the soil has to produce ever more quantities of plant and plant products. This is possible, by improving the quality of plants by breeding to evolve more efficient utilizers of limited soil resources on the one hand and managing the soil by maintaining high fertility levels through the use of fertilizers on the other. Several technological advances have been made to treat the soil more efficiently and appropriately so that it not only gives high yields of plants but also does not deteriorate with time.

M.N.J.

SOLAR RADIATION. The surface of the sun continuously emits energy in the form of electromagnetic waves. This is called Solar Radiation. A fraction of this energy-flow is intercepted by the earth. This is the only external source of energy of any importance to the earth. All life and all manifestations of weather and climate depend on solar radiation.

The solar radiation that falls on a square centimetre of a surface exposed perpendicularly to the sun's beam, outside the earth's atmosphere, at the earth's mean distance from the sun, is called the Solar Constant. The latest value of this constant (measured with the help of man-made satellites) is 1·94 calories per square centimetre per minute. It is generally held that no significant variation of this Solar Constant occurs. The upper limit of such variations is about 0·1 per cent. However, there are large variations in the far ultra-violet regions and in the radio frequency regions of the solar spectrum, but the energies involved are extremely small, well below the 0·1 percent limit of the total solar energy received by the earth.

Besides electromagnetic waves, the sun also emits streams of charged particles at high speeds from the disturbed regions of the sun. No significant relation between the incidence of such streams and the weather and climate has yet been conclusively demonstrated, although their effects on the earth's magnetic field are large.

The spectrum of solar radiation outside the earth's atmosphere extends from the X-ray region, through the ultra-violet, through the visible and infra-red, to the radio-wave region. The visible region extends from a wavelength of about 0·4 thousandth of a millimetre to about 0·7 thousandth of a millimetre. About half of the total energy is within this visible region. The maximum occurs at a wavelength of about 0·474 thousandth of a millimetre, in the blue-green region—a region well absorbed by the CHLOROPHYLL of vegetation. About a third of the total solar radiation is in the infra-red.

The spectrum observed at the ground is sharply cut off in the near ultra-violet at a wavelength of about 0·29 thousandth of a millimetre, due to a complete absorption of shorter wavelengths by the gases present in the atmosphere. These shorter wavelengths contain about five percent of the energy. A small quantity of ozone present in the upper levels of the atmosphere is mainly responsible for absorbing these ultra-violet rays, which are detrimental to animal and plant life.

Taking the total radiation incident on the top of the atmosphere as 100 units, 30 units are diffusely reflected back into outer space without any change in wavelength. This is effected by the clouds (20 units), atmospheric gases (6 units), and the land and water on the earth's surface (4 units). This ratio, obtained by dividing the reflected radiation back into space by the total radiation, i.e. the number $30/100 = 0·3$, is called the albedo of the earth-atmosphere system.

About 20 units are absorbed by the constituents of the atmosphere. The remaining 50 units reach the earth's surface of land and water, where they are absorbed.

These land and water surfaces radiate heat in the infrared, in wavelengths with a maximum around ten-thousandths of a millimetre. (In the direct solar radiation there is practically no energy in the region of these wavelengths). The atmosphere absorbs a large fraction of the infra-red radiation emitted by the earth's surface. This is done by the clouds, water vapour and carbon-dioxide present in the atmosphere. After absorption, infra-red radiation at slightly longer wavelengths is re-radiated by the atmosphere, partly back to the earth's surface and partly to outer space. The heat radiation received back by the earth's surface from the atmosphere is in amount equal to about 75 units or 75 percent of the solar radiation incident at the top of the atmosphere, although in other wavelengths. Consequently the earth's surface receives 25 units more than what the sun provides at the top of the atmosphere, 50 units directly and 75 units as the back-radiation by the atmosphere. The effect due to the back-radiation in infra-red by the atmospheric gases, dust and clouds is called the 'greenhouse' effect. It is this greenhouse effect which keeps the earth warmer than it would have been if the earth had no atmosphere, or if the atmosphere had neither water vapour nor carbon-dioxide. Due to this effect, the earth's surface is warmer by about 15 °C than it would have been otherwise.

The heat radiated into outer space by the earth's atmosphere is in amount equivalent to about 70 units of our notation. This along with the 30 units directly reflected back into outer space, balances the total 100 units radiation received from the sun. The earth-atmosphere system as a whole is neither heated nor cooled over a long period of time. This is called the 'radiation balance' of the earth-atmosphere system.

As far as the earth's surface is concerned it receives 50 units of sun's radiation directly, and an equivalent of 75 units of radiation as the greenhouse effect. Because of its inherent temperature, the earth's surface radiates in the infra-red an amount very nearly equal to 100 units, that is as much as the total solar radiation; the remaining 25 units is imparted to the atmosphere through conduction and as the latent heat of evaporation of the water vapour evaporated from the oceans, rivers, lakes, etc. This latent heat is delivered to the atmosphere when and where the water vapour condenses into clouds and rain or snow. This is called the radiation balance of the earth's surface.

The surface of the earth is spherical and its axis of rotation is nearly perpendicular to the plane of the orbit in which it goes round the sun during the year. Consequently, the inclination of the solar radiation to the earth's surface is different at different latitudes. Near the poles, the rays have a grazing incidence, while near the equator the rays are incident nearly perpendicular to the surface of the earth. At intermediate latitudes the inclination varies between these values. Therefore, the equatorial regions get heated more than the polar regions. However, all these regions emit infra-red ra-

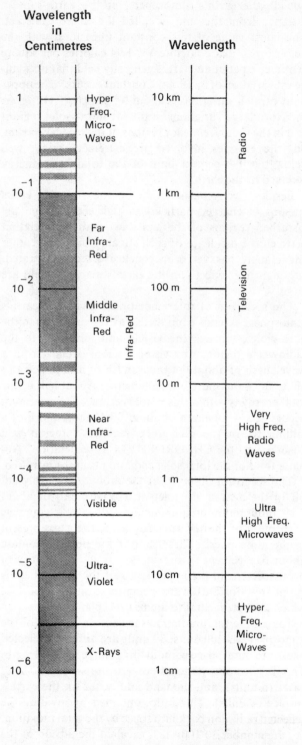

THE ELECTROMAGNETIC SPECTRUM FROM X-RAYS TO LONG RADIO WAVES

The shaded portions of the spectrum are absorbed by the earth's atmosphere totally or partially.

diation almost equally. Consequently there is excess heating over the equatorial regions and excess cooling over the polar regions. The ocean-atmosphere system transports the excess heat energy from the warm equatorial regions to the cold polar regions. The partial transport by the atmospheric air and water vapour manifests itself as the drama of weather—winds, clouds and rains. The partial transport by the oceans manifests itself as the poleward-moving warm ocean-currents and the equatorward-moving cold ocean-currents.

The earth's axis is not strictly perpendicular to the plane of the earth's orbit, it is now inclined at an angle of about 23·5 degrees to that perpendicular. Consequently there is an apparent motion of the sun across the equator north and south. On 22 June, the midday sun is vertically over places at latitudes 23·5 degrees north, and on 22 December the midday sun is vertically over places at latitude 23·5 degrees south. Thus the zone of maximum heating on the earth's surface varies with the time of the year. For the six months 22 March to 22 September, the northern hemisphere is heated more than the southern hemisphere; similarly for the other six months the southern hemisphere is heated more.

These changes in solar heating cause changes in the mean weather patterns and the mean climate in different parts of the globe during different parts of the year.

See SEASONS.

SPARROW. The House Sparrow (*Passer domesticus*) is probably the best-known bird of India because of its attachment to human surroundings. Though its normal food is seeds and insects, it has adapted itself to living on kitchen scraps and has multiplied enormously in urban environments.

The House Sparrow belongs to the Old World and its incautious introduction into Australia and America in order to control harmful agricultural insects resulted in the bird itself becoming a pest.

Sparrows go to bed early and their roosting trees usually resound with their ceaseless and intense twittering well before sundown. They are much given to dust-bathing and are very particular about maintaining their feathers in good condition.

Contrary to *domesticus*, the Tree Sparrow (*Passer montanus*) which replaces it in Assam and in the Himalaya, draws a line at the veranda and seldom enters dwellings. Another common sparrow in our area is the Yellowthroated (*Petronia xanthocollis*), which is a true countryside bird and never enters houses.

SPIDERS. The body of a spider consists of a cephalothorax (head and thorax united) and an abdomen, separated by a waist. The head carries eight simple eyes and along the front margin a pair each of jaws (chelicerae) with foldable poison fangs and pedipalps which are six-jointed. On either side of the thorax there are four seven-jointed legs, the size and disposition of which vary greatly in different genera. At the hind end of the abdomen are four or six spinning organs.

The spider has an alimentary canal with a sucking stomach. The heart is tubular. Respiration is by air tubes or book-lungs.

Silk Glands

The peculiarity of spiders is their ability to spin. Silk is part and parcel of their life. They are born in

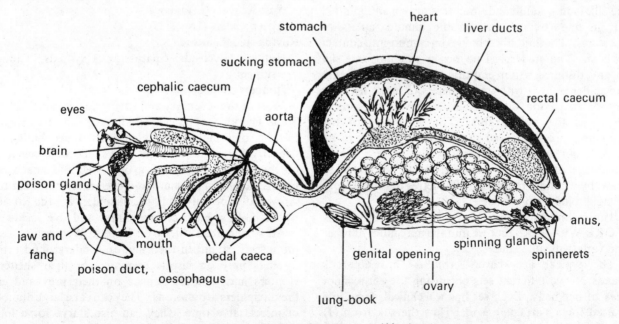

ANATOMY OF A FEMALE SPIDER, × 6 (After Leuckart)

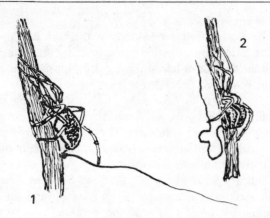

1 Spider paying out thread to the air, × 1
2 Spider with the hauling thread held by her leg, × 1

Araneus indicus, × 1 in part of her orb-web

silk, their webs and cocoons are of silk, they capture their prey with silk and they dangle and travel by silk threads. All spiders therefore are born with spinning organs and silk glands. The spinning organs are the four or six teat-like spinnerets at the hind end of their body. Each teat is not a single tap but carries numerous apertures connected with the various silk glands within. The spider can draw silk from the glands in whatever quantity or quality is required for a particular work. The animal by holding the spinnerets together can join threads, or by holding the tips a little apart manufacture broad bands of silk for swathing struggling victims. Besides the spinnerets, additional spinning organs such as a sieve-plate under the belly (cribellum) or a comb of hairs along one pair of legs (calamistrum), are possessed by some spiders (e.g. Eresids).

Nature has provided a spider with silk glands capable of yielding silk for as long as it lives. Within the glands the silk is in a fluid condition. It is drawn and emitted out in fine streams through the minute taplets or apertures. The fluid silk solidifies on coming into contact with air. The spider has no power to shoot out silk to any distance. After emitting a drop of liquid silk the spider pulls it with its spinnerets to make it into a thread. In paying out threads to the air the breeze serves as the pulling force. The sphincters of the spinnerets are weak and the escape of silk from its belly is not rigidly controlled by the spider. Taking advantage of this fact naturalists have been able to draw, by winding, more than 100 metres of thread from a single spider in one hour.

The spinning powers of spiders vary greatly. It is in ORB-WEAVERS that we find the maximum number and variety of silk glands.

The typical orb-weaver—*Araneus*—has 600 silk glands of five different sets occupying the entire floor area of her belly. Each set has a technical name and is used for a particular work. Thus the silk from (1) the *ambullaceal* set is used for boundary lines and spokes, (2) *aggregate* for viscid spirals, (3) *tubuliform* for cocoons, (4) *aciniform* for anchoring and (5) *piriform* for winding and enwrapping prey.

Life, including Sexual Habits

Spiders are widely distributed throughout temperate and tropical regions. They are ubiquitous and thrive in all environments. They vary greatly in size, shape and habits. There are Sedentary and Roving spiders. The Sedentary spiders include:

   Orb-weavers (Argyopids)
   Irregular Line Web-weavers (Pholcids, Theridiids etc.)
   Nest-weavers (Eresids)
   Water spiders (Lycosids)

Roving spiders include:

   Hunters (Lycosids, Sparassids, Oxyopids, Thomisids etc.)
   Jumpers (Attids)

Most of the sedentary and roving spiders are solitary. Some Theridiids, Pholcids and line-weavers are gregarious, i.e. they build their independent webs contiguous to each other. A few of them weave their webs generally on the periphery of an Argyopid web and poach the insects caught in its snare and even attack and eat the Argyopid. The only truly social spiders are the Eresids which build a huge common nest and live inside it.

All spiders feed chiefly on insects. Web-weavers eat the insects caught in their snares. Underground tube-dwellers prey on insects coming near their shelters. Hunters and jumpers pounce on their prey and eat them. Spiders are gluttons. They can eat a large number of insects at a time. They can also starve for a long period—some for nearly 18 months!

Blechnum spicant. *Blechnon* is the Greek name for a kind of fern. Ferns are commonly found growing in moist tropical habitats.

Langurs are the most common Indian primates.

During the mating season the croaking of frogs and toads is heard throughout the subcontinent.

© M. S. Hebbar

SPIDERS MATING

Left, *Tetragnatha mandibulata*, × 2/3
Right, *Heteropoda venatoria*, × 1/2

A Pholcid carrying her cocoon × 1

In the spider kingdom females are dominant. Males are generally pygmies and dull-coloured. In genera such as *Argyope* and *Nephila* the disparity is very great. A female *Nephila* is in length 20 times and in weight 1000 times the male. The mature males of web-weavers are nomadic, wandering in search of females. During the mating season they appear in large numbers along the borders of female webs, sometimes weaving rudimentary webs. Their only mission is to fertilize the female—and die during or soon after mating. In *Nephila* and allied genera the mating operation is fraught with considerable danger to the males. If the female is hungry even the most cautious approach of the male will result in his being eaten. If she is not hungry she may accept him but if he is not quick in escaping she is sure to devour him. The life of the males therefore is generally short. Females live longer—they have to lay eggs, protect them and nurse the spiderlings until the young can live by themselves.

Among the other sedentary spiders sexual relations are more pacific. Among roving spiders disparity in size is not great and mating is harmonious. A male rover leaps on the back of the female and after mating leaves in peace. Females of some Lycosids are aggressive. If the male lingers after the act, she may overpower and eat him. Among Attids, the male allures and woos his partner with his acrobatic dances, standing on his four hind legs and waving his fore-legs. Their mating is peaceful.

Most outdoor spiders live only for one year. Line-weavers within houses live for 5 years. Female orb-weavers live for 2 to 5 years and some trapdoor spiders live for even 25 years.

Cocoons and Spiderlings

The pregnant female weaves a silk matlet, lays her eggs on it and winds it into a cocoon. The size, shape and colour of cocoons differ in different genera.

Cocoons of Garden Spiders are nearly spherical, yellow or white and attached to one side of their webs. Cocoons of *Cyrtophora*, whose web resembles an umbrella, are ovoid and green, attached vertically above the web like a beaded chain. Lycosids and Pholcids have small globular cocoons which they carry by their spinnerets or jaws. The common Indian House Spider *Heteropoda venatoria* and other Sparassids have round, flat, disc-like cocoons which they carry beneath their belly. Others have cocoons of varying shapes which they conceal under patchy sheets in crevices of walls, bark of trees, axils of leaves, bunches of fruit, etc.

A cocoon of *Araneus* contains 600 eggs; of *Argyope* 1200 to 2000; line-weavers 100 and Attids only 2 to 4 eggs.

The eggs hatch producing spiderlings. Fresh spiderlings cannot feed or spin. They become active only after one or two moults (casting of the skin). The mother spiders are ready to avoid food and risk their own lives to protect the eggs. Some even carry the spiderlings on their backs until the young ones are able to live independently. When once independent, each spiderling sets off by air into the unknown. This occurs in spring or early autumn when the wind is gentle. The spiderlings sit on any raised convenient spot with head lowered and abdomen raised. In this position they emit silk which becomes threads. As the threads begin to float in the air the spiderlings leave their holds, cling to the thread and sail in the air. Although the direction of their flight will be determined by the wind, the spiderlings can land where they choose by hauling in the threads and rolling them into a ball. By this 'gossamer' method spiderlings cross wide rivers and oceans and are carried to considerable heights. Aeroplanes frequently run into gossamer.

Enemies of Spiders

Scorpions, lizards, birds, frogs, toads, insectivorous mammals—all feed on spiders and their cocoons. Solitary wasps and ichneumon flies particularly hunt them and store them in their cellars for the sustenance of their young. Within houses their cobwebs are brushed away by housewives. Digging, ploughing and tree-felling also destroy spiders.

Protective Adaptation and Mimicry

Sedentary spiders have their webs, nests or tubes which adequately shelter them from their enemies, besides enabling them to catch their prey. Roving spiders have strong and nimble legs armed with sharp spines. Some can make long and sudden leaps. Others can move zigzag like flies or sideways like crabs. Some mimic inedible insects such as ants and ladybirds or floral stamens and birds' droppings. Some are coloured in harmony with their surroundings and some can change their colour to suit the environment. Some have long and cylindrical bodies and long, thin legs (*Tetragnatha*) which they can stretch fore and aft and remain concealed along twigs and leaf-blades. *Argyope* and *Nephila* have warning colours. *Gasteracantha* imitates woody knots. Pholcids oscillate their webs and make spasmodic up-and-down movements of their bodies so rapidly that enemies leave them alone. Some sham death when disturbed, drop down and scurry away unnoticed.

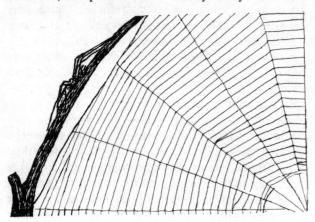

Tetragnatha, × 2/3 hiding along a twig and part of her web.

Uses of Spiders

Generally speaking spiders of this country are harmless to man except one or two species of the genus *Poecilotheria*. Spiders feed on mosquitoes, flies and cockroaches within our houses. Outside they are the principal agency that controls the bewildering multiplication of insects which destroy vegetation and crops.

There are hundreds of beautiful and interesting species of spiders in our area. The more common and outstanding ones are briefly described under ORB-WEAVERS, LINE-WEB WEAVERS, NEST-WEAVERS, HUNTING SPIDERS, WATER SPIDERS, JUMPING SPIDERS and TRAPDOOR SPIDERS.

SPONGES (Porifera) are the lowest type of slimy, multicellular animals. Their tissues are primitive and not differentiated as in other multicellular animals (Metazoa). They are therefore placed under an isolated division—Parazoa. All sponges except a single fresh-water family live in the sea attached to rocks, pilings, plants, etc. They grow much as plants do. Some are branched, some shaped like cups, some like gloves or domes. In size they range from 1 mm to a metre in diameter and 40 cm in thickness. Despite variation their essential structure is the same.

The form of a typical simple sponge is that of a vase or cylinder, closed and fixed at one end and with an opening called the osculum at the free end. The cylinder encloses a cavity—the gastric cavity or para-gaster. The wall of the cavity is of two layers—an outer layer of flat cells and an inner, lining layer of collared cells (cells with cilia or flagella). The cells in both these layers are but loosely connected and not fitted firm as in other multicelled animals.

Sponges feed on minute organic particles living or lifeless existing in the surrounding water which circulates in their body through a canal system which falls under three types.

On the outside of a sponge are a number of minute

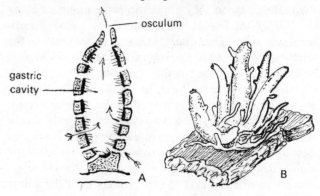

A. Ascon type of canal system
B. A small colony of *Leucosolenia*
(Both × 4), after Hegner, *Invertebrate Zoology*

pores. In simple sponges like *Leucosolenia*—a common seashore species attached to rocks—these pores open directly into the gastric cavity. Water enters into the cavity through the pores. The collared cells lining the cavity digest the food matter contained in the water and throw out the waste. The refuse water is finally pushed out through the osculum. The inflow of the water through the pores, its circulation within the cavity and its final ejection through the osculum are all due to the harmonious lashing of the vibratile cilia of the collared cells in the direction of the osculum. This simple canal system is called 'ascon'.

In the next higher type, which exists in the genus known by the name *Sycon*, the wall of the cavity is folded into a series of flagellated chambers. Outside water enters these chambers through special in-current

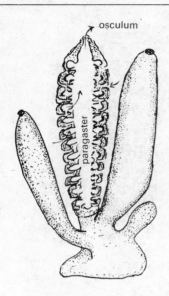

A sycon tuft with one cylinder bisected to show the cavity and sycon type of canal system, × 2

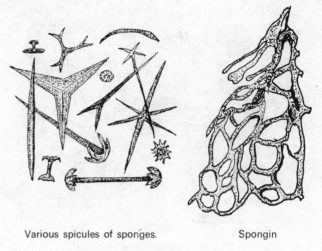

Various spicules of sponges.   Spongin

Both × 80

canals connecting the pores and the chambers. Digestion takes place in the chambers. The used water then flows into the cavity and thence out through the osculum.

In the third type, 'leucon' or *Rhagon*, the flagellated chambers become further folded into a number of smaller chambers each one connected to the pore by an incurrent canal and the gastric cavity by an excurrent canal. This elaborate system is found in all fibrous (bath and freshwater) sponges.

Spongin is a substance closely allied to silk. In the common bath sponge the skeleton is entirely of spongin threads which branch and form a compact but extensive supporting structure. The skeleton is not only supporting but also protective. The sharp spicules and the bitter taste of spongin keep predacious enemies off. Sponges are grouped according to their skeletal peculiarities.

In the cavities of large-sized sponges small shrimps, sea worms and molluscs enter and live for shelter, feeding on the waste matter. Species of *Cliona* bore the

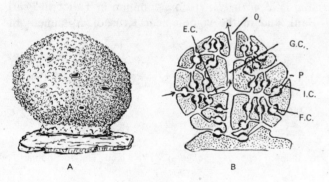

Ceratosa (A). A common bath sponge × 3
(B) A cross-section showing the leucon type of canal system, × 3. O osculum. G.C. gastric cavity. P pore. I.C. canal, incurrent. E.C. canal, excurrent. F.C. flagellated chamber.

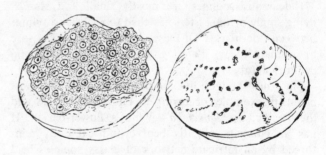

Cliona, a yellow sulphur sponge on a bivalve shell and (right) the holes made by the sponge on the shell, × 1/2

All sponges have a gelatinous layer of mesoglea between the two layers of cells. Embedded in this mesoglea is a skeletal framework. The skeleton consists of lime (calcareous), glass (siliceous) needles or spicules or horny fibre spongin. Calcareous spicules are generally single-, three- or four-rayed. The siliceous spicules are much more varied in shape and in one and the same kind of sponge there may be different forms of spicules each form having a special place in the skeleton of the various parts of the sponge body.

surface of bivalve shells and grow on them. Some get fixed on the back or legs of crabs. This association benefits both. The sponge is inedible and with it as a cover the mollusc and crab are protected from their enemies. The sponge is benefited by its being carried from place to place, enabling it to obtain adequate oxygenation and food.

Sponges reproduce asexually by budding or sexually by ova and sperms which unite to form a zygote. The zygote divides and becomes a spherical mass of cells successively called the blastula and the amphiblastula. Later it invaginates (or tucks), develops collared cells within, fixes itself to a holdfast and grows into an

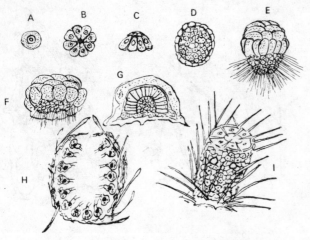

DEVELOPMENT OF A SPONGE

A zygote, B segmented ovum from above, C segmented ovum from the side, all × 100. D blastula, E amphiblastula, F invagination begins, all × 50. G larva attaches itself to a hold-fast, × 4. H young sponge with flagellated cells, × 2. I young fully formed sponge, × 1.

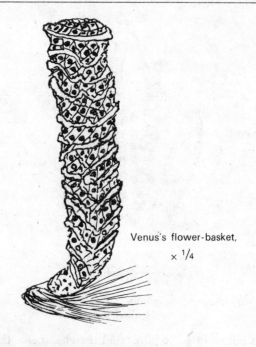

Venus's flower-basket, × 1/4

adult. The growth of the sponge in magnitude and direction is to a great extent dependent on the nature of the substrata.

There are nearly 3000 species of sponges distributed in all the seas between the tide-mark and the abyssal region. Many of them, both colourful and beautiful, occur in Indian waters especially along rocky coasts, gulfs and reefs.

Calcareous sponges are mostly small and simple, living singly or in clusters attached to rocks and pilings between tide-marks. Of these *Grantia* looks like a vase or miniature jar, *Poterion* a cup and *Leucosolenia* cylindrical tufts.

Siliceous sponges live in deep seas and the most beautiful example, occurring in Philippine waters and the Pacific, is *Euplectella*—Venus's flower-basket. It has an elongate, slightly bent cylindrical body reinforced by an intricate network of glassy spicules held together like a lattice and the fixed end fastened to the mud by a tuft of glassy fibre.

The most dominant among the sponges are those possessing fibrous threads. They have the widest distribution and include the largest number of species. They are often massive and brightly coloured in yellow, cream, orange, red, grey and green. The common bath sponge is Mediterranean but allied genera *Callyspongia, Adosia, Spirastrella* etc. are common in Tuticorin Pearl Bank and in the lagoon near Krusadi. Specimens of

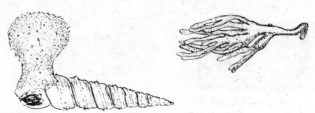

Sulphur sponge (*Suberite*) × 3/4, and Finger Sponge (*Chalina*) × 1/6

Grantia × 1   *Poterion*, Neptune's Cup × 1/8

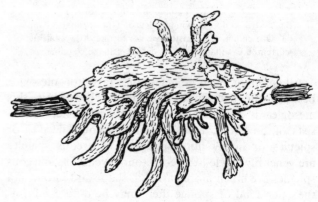

Freshwater sponge × 1, from Hegner's *Invertebrate Zoology*

brown finger-like *Chalina* are sometimes found washed ashore. The light brownish or yellow *Cliona* (sulphur sponge) and *Suberites* are common on dead shells.

Green freshwater sponges are plentiful in pools and ponds.

SQUIDS are two-gilled gastropods, free-swimming, very active and gregarious. The common species found around our shores are comparatively small, with body-lengths ranging from 10 to 30 cm, but the species found in Atlantic waters are larger and there are certain giant types measuring 10 to 15 metres. These large squids are the favourite food of sperm whales. Eye-witnesses say that when a whale meets a troop of speeding squids there starts a battle royal, the squids fighting 'tooth and nail' with their suckered arms and tentacles but eventually some of them being caught between the powerful jaws of the leviathan.

A squid can be at once recognized by its long torpedo-shaped body, very like a cuttlefish, except that its fins are triangular. There is a picture of a squid under MOLLUSCS. It has a distinct head region with a pair of prominent eyes. There are eight stumpy arms carrying disc-shaped suckers. There is an additional pair of very long tentacular arms whose extremities are swollen and sucker-clad. All these ten arms are arranged in a circle round the mouth of the animal. The tentacular arms can be withdrawn into a pouch on either side of the head. Squids seize their prey, mainly small fishes, by encircling them with their long tentacles and then eat them with the help of the sharp horny beaks set around their mouth. Like the octopus, the squid has also no external shell but it has an internal support called the 'pen' in the dorsal region. The pen is a thin horny structure almost transparent. A muscular mantle covers the body.

Squids are jet-propelled. They swim by ejecting a stream of water from the gill chamber through a siphon on the underside of their body. This shoots them backwards with astonishing speed, the squid's body being shaped to offer little resistance when propelled under water. Squids do not build individual shelters as octopuses do but are vagabonds, moving in schools from place to place in coastal waters. In calm weather they enter inshore waters in great shoals, and are found in plenty in deep embayments such as the Gulf of Kutch, parts of the Gulf of Mannar and in Palk Bay. During the southwest monsoon, when the southern waters are rough, they enter in great schools into the protected shallow waters around Rameshwaram partly to feed on the small fry that seek shelter in shallows from the violence of the waves and partly to lay their eggs on the thick growth of seaweeds abundant at that place.

As in all cephalopods, the sexes are separate. The ova are contained in small cylindrical cases measuring 7 to 10 × 1 cm and the number of ova is about 250 in each case. Hundreds of these cases are attached together like a bundle of sausages or young carrots. The total number of eggs range from 30,000 to 40,000. The eggs of squids develop in the same way as the eggs of CUTTLEFISHES. It is doubtful if all the ova in a capsule become fully developed due to limited accommodation in the capsule. Even if all of them come out as young adults many are eaten by other sea animals.

Squids are prized as food. They are fished by using suitable nets or by long jiggers (hooked poles 3 or 4 m long). Squids thus collected are split, dried in the sun and sold for food.

The common Indian species are *Loligo indica* and *Sepioteuthis arctipinnis*. Excluding the tentacular arms the body-length with the head and arms measures 16 cm in *Loligo*, 24 cm in the other.

See also MOLLUSCS.

SQUIRRELS. Squirrels belong to the family Sciuridae, a well-defined group within the order Rodentia. Over 50 genera have been included in this family out of which ten occur within the region covered by this encyclopedia, comprising twenty-two different species. All the family are characterized by grinding teeth which are low-crowned and cuspidate, and by a peculiar attachment of the masseter (main chewing) muscle which is inserted in the skull ahead of the zygomatic arch. Besides the familiar tree-dwelling and bushy-tailed squirrels, it is not generally realized that Marmots, Ground Hogs and the Ground Squirrels of the great natural prairie regions also belong to this family. Prairie Dogs and Susliks are Sciurids and are burrowing short-tailed rodents quite dissimilar in habits from tree squirrels. Twenty of the Indian species are, however, largely arboreal, and are exceedingly active, feeding on buds, fruits, nuts and other vegetable substances.

Squirrels have lived in captivity for over 10 years and have a potential longevity of 15 to 20 years. The females reach maturity between 6 and 8 months. Mating is promiscuous and the female when in season often mates with several males. Their love life lasts for a day. The gestation period varies from as short as 28 days in *Ratufa*, up to 42 days in *Funambulus*. During lactation the female spurns all attention from the male and often aggressively drives him away. She makes a nest in which the naked and helpless young are born and jealously guards and rears her offspring, unaided by the male. Palm Squirrels average three litters a year with an average of 3 young per litter, and Flying Squirrels in northern latitudes produce only one or two litters per year. Squirrels make large untidy nests of leaves, twigs and vegetable fibre in trees for rearing their young. The nest, known as a drey, is usually placed among the slender, upper branches hidden among the foliage of a tree or in the hollow of the tree-trunk. It may also be used for sleeping during the non-breeding season.

Not surprisingly because of the advantages offered by tall-tree evergreen forest, eleven out of the twenty tree squirrel species occur in Eastern India, in Assam and in the northeastern Himalaya. One species of Flying Squirrel, one species of Giant Squirrel and two species of Palm or Striped Squirrels are widespread in the peninsula.

Flying Squirrels are distinguished from other squirrels by the presence of an elastic membrane of skin between front and hind legs which acts as a parachute and permits the animal to glide from tree to tree, often over considerable distances on hill slopes. The method of 'flying' is to jump from a tree at a height to another at some distance and below. At the end of the glide the squirrel banks steeply and lands with all four feet on the tree-trunk. All Flying Squirrels are nocturnal and spend the day in tree hollows. Their presence in a forested area is often known only by their mewing call, which is distinctive.

The Common Giant Flying Squirrel (*Petaurista philippensis*) is found in forested areas throughout the peninsula. In the Himalaya three more species of Giant Squirrels and three smaller flying squirrels occur. Some of them, like Hodgson's Flying Squirrel (*Petaurista magnificus*), are splendidly coloured, with a bright yellow line down the middle of their back. The Woolly Flying Squirrel (*Eupetaurus cinereus*) of Kashmir is peculiar in that it only occurs in cold dry areas where tree growth is limited and where it is said to be capable of gliding from rock to rock. It is now an uncommon and rare species known only from a few localities in Gilgit and Chitral. The Small Travancore Flying Squirrel (*Petinomys fuscocapillus*) is peculiar to the forests of South Kerala, and is a rare and little-known species.

Giant Squirrels of the genus *Ratufa* are the largest known tree squirrels in the world, and exhibit a striking polymorphism in coat colour and pattern. Races of the Indian Giant Squirrel (*Ratufa indica*) are not uncommon in the forests of the peninsula. They vary in colour from the pale cream race (*R. i. dealbata*) of the Dangs in Gujarat through the Rufous Squirrels of the Konkan race (*R. i. indica*) to the almost Black Squirrels (*R. i. maxima*) of Kerala. A widespread race is the Central

© M. Krishnan

Indian Giant Squirrel × ¹/₄

© M. Krishnan

Three-striped Palm Squirrel eating a mango. × ¹/₂

Indian Giant Squirrel (*R. i. centralis*) of Madhya Pradesh, South Bihar, Orissa and the Eastern Ghats. Giant Squirrels make their presence known by their loud alarm-calls. They are partial to particular patches of forest and build large dreys for breeding and for roosting at night. The Grizzled Giant Squirrel (*R. macroura dandolena*) is rare and restricted to a small area of deciduous forest on the eastern slopes of the Western Ghats in Tamil Nadu. Other races of this species occur in Sri Lanka. The Malayan Giant Squirrel (*R. bicolor*) replaces the Indian Giant Squirrel in the Eastern Himalaya and the states of eastern India.

Striped or Palm Squirrels. These sprightly and diminutive Palm Squirrels comprise five species. The Five-striped (*Funambulus pennanti*) in the north and Three-striped (*Funambulus palmarum*) in the south, are both commensal with man in the peninsula. Thriving in all but the most densely populated areas of Indian cities, the Palm Squirrels are the commonest and best-known of Indian squirrels. They are most often seen hanging head down from a tree-trunk, and scolding shrilly at whatever has, at that moment, frightened them. They live largely on the vegetable food found in human vicinity, and occasionally raid birds' nests for eggs. Two to three young are reared at a time, in a nest of vegetable fibre built on trees or in the eaves of houses.

In the jungles of the southern peninsula three more species of Striped Squirrels occur: Layard's Striped Squirrel (*Funambulus layardi*), and the Dusky Striped Squirrel (*F. sublineatus*), both of which live in the forests of Kerala, and the Jungle Striped Squirrel (*F. sublineatus*) occurring throughout the Western Ghat forests. In the eastern Himalayas the Himalayan Striped Squirrel (*Callosciurus macclellandi*) replaces the Palm Squirrels. The eastern Himalayas are also the home of several very loud-voiced species of medium-sized squirrels like the Orange-bellied Himalayan Squirrel (*Dremomys lokriah*), the Red-cheeked Squirrel (*Dremomys rufigenis*) with its sharply pointed nose, and Perny's Squirrel (*D. pernyi*). Two species of *Callosciurus*, a genus which is well represented by species in Burma, also occurs in eastern India. Very little is known of the squirrels of Eastern India.

See also MARMOT.

J.C.D.

See plate 32 facing p. 433.

STARFISH, the most well-known and typical echinoderm, is common at low-tide mark, moving over sand or sides of rocks. Its shape is starlike with a central disc from which five tapering arms radiate. It has no head but can move in the direction of any arm. The upper surface of the animal is convex and the underside

flat. At the centre of the flat surface is a five-sided aperture containing the mouth. The corners of the aperture are produced as a groove along the mid-longitudinal line of each arm. Each groove is lined on either side by one or two rows of tube-feet. Each tube-foot is a short vertical tube with a sucker below and a small bladder (ampulla) above. The grooves with their numerous tube-feet on either side resemble miniature garden paths and hence are called ambulacral grooves. The arms and tube-feet are furnished with muscle fibres.

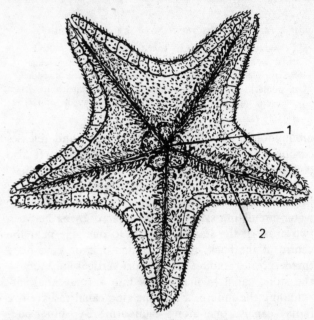

The underside of a starfish, showing the mouth (1) and the radial grooves (2) × 1.

The entire body of a starfish is protected by a tough, hard crust containing numerous limy (calcareous) plates called ossicles (small bones) embedded in soft flesh. This skeleton is not strictly rigid but somewhat flexible, for the ossicles are only loosely joined by connective tissues and muscles. The animal can therefore bend and twist its arms with ease. The ossicles are numerous, arranged regularly around the mouth region, in the ambulacral grooves and along the sides of the arms. Over the other parts they are scattered. Sharp spines, rigid and movable, project from the ossicles covering the back and sides of the arms. Surrounding the spines there are certain pincer-like structures called pedicellarias. In the living animal these pedicellarias continually twist and snap, picking out and cleaning dirt and debris off the exterior. They aid in keeping the respiratory and other pores from getting blocked and also in catching prey.

The starfish has an extensive internal cavity (coelom) which protects all the important internal organs. It is lined by a membrane and contains sea water mixed with

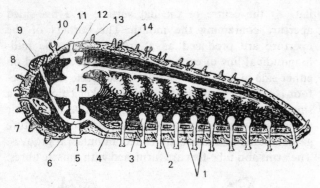

LENGTHWISE SECTION OF ONE ARM OF A STARFISH

1 Ossicles of the groove. 2 Tube-feet. 3 Ampulla. 4 Radial canal. 5 Mouth. 6 Ring canal. 7 Sex organ. 8 Stone canal. 9 Sieve-plate. 10 Pedicellaria. 11 Anus. 12 Rectal division. of stomach. 13 Spine. 14 Pyloric division of stomach and its extension into arm-bearing glands. 15 Cardiac division of stomach

albuminoids. This fluid takes in oxygen and releases carbon dioxide through respiratory openings. It also collects and excretes waste matter.

The locomotory apparatus and its working are unique in the starfish. The mechanism is hydraulic, run by the water-circulating system of the animal. Lying between two arms of the starfish, a little to one side near the centre of the back, is a perforated disc or sieve-plate (madreporite) leading into a short vertical limy tube—the stone canal—which opens into a ring canal surrounding the mouth. From the ring canal radiate five radial canals, one along each arm. By short side-branches the radial canals are connected to each tube-foot. Filtered sea water passes down through the stone canal into the ring canal and then into the radial canal. The bladders connected to the tube-feet also get filled with water. This distension at the upper parts of the tube-feet creates a vacuum at the lower ends, which act as suckers and hold fast. The distended bladders then contract, pushing the water down into the tube-feet. This removes the vacuum and releases the suckers. By the combined muscular action of the arms and the alternate distension and contraction of the bladders the animal walks on its tube-feet.

The mouth opens through a short throat into a capacious five-lobed sac—the cardiac division of the stomach. The walls of the sac have muscle fibres which can stretch or contract. If the animal finds an oyster or mussel too large for its mouth it thrusts out the cardiac portion of its stomach through the mouth and engulfs the prey. After digesting it the retractor muscles work and the stomach is drawn in. Next to the cardiac division is the pyloric division of the stomach which first projects as a short tube into each of the arms and then bifurcates into a pair of branches bearing a series of glandular pouches which secrete a digestive fluid. The pyloric division is followed by the intestinal or rectal

division which is also five-pouched. The rectum finally opens out through the anus on the back of the animal.

The food of starfish includes fish, molluscs, worms, crabs etc. and even garbage and decaying matter.

A starfish has nerve-cells scattered on its outer layer, a nerve-ring round the mouth, a nerve-cord along each arm supplying nerve-fibres to the tube-feet. At the tip of each arm is a pigmented mass which works as an eye.

There are separate male and female starfish. The sexual cells are produced by a pair of special organs at the base of each arm. The cells when mature are let out through pores on the back of the animal. Fertilization takes place in outside water. The zygote (fertilized egg) divides and forms a ciliated bilateral larva which for some time swims and later sinks to the sea bottom where it first becomes a tiny rosette and then changes into a small starfish with five knobs. It goes on growing for a year or two, strengthening the knobs into pointed rays, and eventually becomes an adult starfish.

Starfishes have the power to grow lost parts. Arms which get mutilated are redeveloped in a short time. Even if two or three arms and a part of the disc are destroyed the animal regenerates them in the course of a month.

Starfishes are great destroyers of shellfish. Since they consume garbage and decaying matter they are good scavengers of the seashore. Cod, haddocks and larger sea animals feed on starfishes.

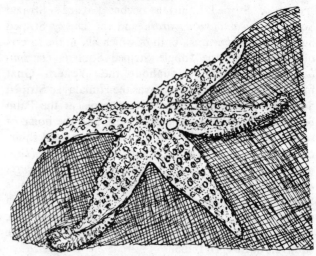

Typical starfish *Asterias* on a rock × 2/3. Note the anus near the middle.

There are more than 20 families of starfishes. In some the number of arms are many—about 40. In some the disc alone is prominent with short blunt arms. Mostly they are of reddish or copper colour but grey, ochre, scarlet and purple varieties are also common. They range in size from 8 to 80 cm. *Echinaster*, *Asterias*, *Astropecten* and *Anthenea* are common Indian genera. See ECHINODERMS.

STARGAZER (Labridae) has received its popular name because of its habit of remaining buried in the sand with only the eyes and mouth protruding above, lying all the time on the bottom of the sea as if to gaze at the stars above. This habit has resulted in certain structural modifications such as the mouth being almost dorsal in position and the eyes, instead of being on the sides, have shifted on to the top of the head. In this position it awaits the unwary prey to swim above it to be quickly sucked into the large mouth. This hiding habit is further assisted by the body also being spotted like grains of sand. Moreover, the fish has an electric organ just behind the eyes which is capable of administering shocks of painful intensity for stunning its prey.

STARLING or MYNA (family Sturnidae). All the 110 or so species of Starlings belong to the Old World. Two of the best known internationally are the common Starling (*Sturnus vulgaris*) and the Indian Myna (*Acridotheres tristis*), whose unwise introduction to other countries and the New World has resulted in many unfortunate ecological consequences. In our area the commonest migratory member of the family is the Rosy Pastor, one of the earliest winter immigrants. Large flocks keep together, raiding cereal crops and causing serious damage. But they feed regularly on nectar from the flowers of economic trees like semul (*Bombax*), and help in their cross-pollination, and also play a valuable role in destroying vast numbers of locusts. The commonest residents are the Indian, Greyheaded, Brahminy, Bank, and Pied mynas.

In common with the House Sparrow, the Indian Myna is greatly attracted to human habitations and thrives on kitchen scraps in addition to its natural food. The other species are less commensal with man and feed largely on fruits and insects.

STEM. It is the ascending organ of the plant which normally bears leaves, branches and flowers at its various regions. It branches exogenously and bears multicellular trichomes of different types. The stems carry out chiefly two types of functions. In the first instance, they provide support to the leaves and flowers and more importantly they allow conduction of water and food materials from one region of the plant to the other. At the young stages, owing to the presence of chlorophyll, photosynthesis also occurs in them. The characteristic stem of a flowering plant is cylindrical but it may and does show variations from angular to square. These may be herbaceous or woody. The region of the stem where the leaves originate is referred to as the *node* and the distance between successive nodes is known as an *internode*. There may be several nodes in a stem, each bearing one or more leaves. Each leaf has at its axis a dormant bud which may also produce branches, especially in such situations where the primary axis has been lost or destroyed due to injury. In the young stems, the outer leaves may become transformed into protective scales. When these fall off during spring, they leave their scars on stems and branches. Generally it is possible to estimate the age of a woody branch by counting the number of such marks.

The main axis of the stem may grow continuously by producing lateral branches. This situation is known as *racemose* or monopodial branching, and can be witnessed in the MAST tree. The *cymose* or sympodial branching differs from this owing to the fact that here the main axis ultimately becomes converted either into a flower or dies. In such a situation, the lateral bud takes over the main function each year.

The conducting and supporting tissues of the stem are spread over the plant body. These differentiate differently in the roots and stems of monocots and dicots. In order to understand the organization and structure of various tissues of a characteristic dicot stem, the description of *Helianthus annus*, the common sunflower, is being given below.

The transverse section of a young stem taken from the internodal region just below the apex shows a primary structure. The outermost layer of epidermal cells is organized in a regular manner, covered over by a thin cuticle. The cortical cells are made up of ordinary parenchyma. In the outer region of cortex is present collenchyma with its thick walls, chiefly at the point of attachment. It provides additional support to the stem. Following these occur, the *vascular tissues* and centrally the *pith*. The conducting tissue is recognizable as vascular bundles which possess on the outer side sclerenchymatous fibres as well. These are in the form of elongated cells and their walls are extremely thick owing to the deposition of lignin. They are devoid of protoplasm and are hard as well as elastic. Owing to these properties, the fibres help the plant to stand erect without allowing it to bend excessively due to the impact of wind. The phloem tissue composed of sieve elements, companion cells, and phloem parenchyma serves as conduit for the prepared food material. Just next to it is the *cambium* made up of meristematic cells that keep on dividing to produce elements both on the inner and outer sides and thus carry out secondary growth by the addition of vascular elements. The innermost region of the vascular bundle is *xylem*, comprising thick-walled and dead elements (vessels and fibres) except for parenchyma. This is chiefly responsible for the transportation of water from the soil to the tip of the plants with dissolved mineral salts and hormones. The medullary rays are usually made up of parenchymatous cells.

In monocotyledons, for example maize (*Zea mays*) or wheat (*Triticum aestivum*), the vascular bundles are irregularly scattered and are without cambium. These are, therefore, termed as *closed* in contrast to *open* in dicotyledons. The tissues which provide mechanical support or are situated towards the outer side in the cortical region and the vascular bundles are surrounded on all sides by sclerenchymatous fibres.

The growing point of the stem is situated just near the apical part and is made up of quickly dividing, unvacuolated (i.e. meristematic) cells and is termed the shoot apex (Fig. 1A). It is a vital, dynamic, everchanging growing system and is the region of initiation of leaves. The growth takes place just below the apical region where the new cells elongate by possessing vacuoles as well as expansion of the cell walls. This causes the stem to increase in height and undergo differentiation. Here the outside cells become transformed into epidermis below which, at a very early stage, the development of leaves is initiated and these cover the growing tip from all sides. The vegetative shoot apex varies in shape, size, cytological zonation and meristematic activity in different groups of plants (Fig. 1B). The origin of a leaf is initiated by periclinal divisions in a small group of cells in the peripheral zone of the shoot apex which cause the formation of a lateral prominence—the leaf primordium on the side of the shoot apex. In the reproductive stage of angiosperms, the floral apices replace the vegetative one, either directly or more frequently through the development of an inflorescence (Fig. 1C).

During cellular differentiation, at a very early stage the outer cells form cortex whereas still others located towards inside retain the capacity to divide and produce gradually the primary conducting elements of proto-

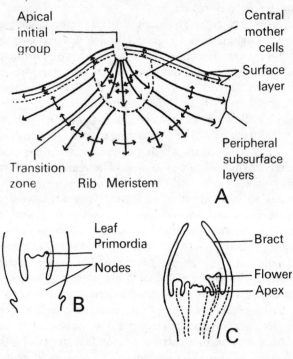

Fig. 1

A COMPARATIVE VIEW OF THE VEGETATIVE AND REPRODUCTIVE SHOOT APICES

A Zones and their mode of growth in shoot tip of *Gingko biloba* as seen in longitudinal sections, about × 200.
B *Hypericum uralum*, histology of shoot apex in longitudinal planes, about × 50.
C *Daucus carota*, transformation of apical meristem during shift from vegetative growth to the development of flowers, about × 75.

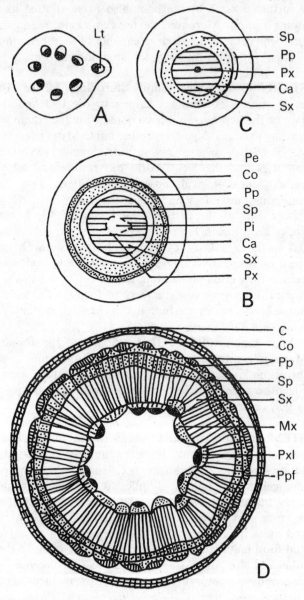

Fig. 2

Cross-sections at various stages of growth of the stem to show developmental stages beginning with the primary vascular tissue and ending with the first secondary increment of xylem and phloem. (C cork; Ca Cambium; Co cortex; Lt leaf trace; Mx metaxylem; Pe periderm; Pi pith; Pp primary phloem; Ppf primary phloem fibres; Px primary xylem; Pxl protoxylem; Sp secondary phloem; Sx secondary xylem).

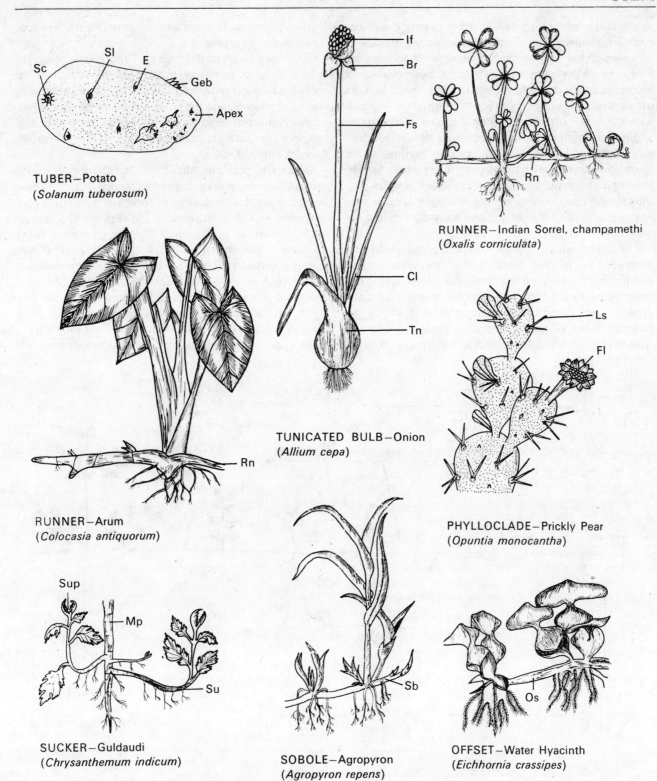

TUBER—Potato
(*Solanum tuberosum*)

RUNNER—Arum
(*Colocasia antiquorum*)

TUNICATED BULB—Onion
(*Allium cepa*)

RUNNER—Indian Sorrel, champamethi
(*Oxalis corniculata*)

PHYLLOCLADE—Prickly Pear
(*Opuntia monocantha*)

SUCKER—Guldaudi
(*Chrysanthemum indicum*)

SOBOLE—Agropyron
(*Agropyron repens*)

OFFSET—Water Hyacinth
(*Eichhornia crassipes*)

Fig. 3

COMMON MODIFICATIONS OF THE STEM FOR
FOOD STORAGE AND VEGETATIVE PROPAGATION

Br bract; Cl centric leaf; E eye; Fl flower; Fs flowering scape; Geb germinating eye bud; If inflorescence; Ls leaf spine; Mp mother plant; Os offset; Rn runner; Sb sobole; Sc scar of stem; Sl scar of scale leaf; Su sucker; Sup sucker plant; Tn tunic.

xylem and protophloem. These grow in length and become transformed into regular conducting tissues.

Although the secondary thickening invariably occurs in all the dicotyledonous plants, it is more evident in evergreen forms as well as those in the tropical regions of the world. During this process, in order to cope up with the needs of the increasing diameter of the trees, additional conducting and supporting tissues are formed. Some cells of the medullary rays become active again and experience divisions in various planes. Slowly they unite with the cambium of vascular bundles and thus form a circular cambial ring which produces secondary phloem towards outside and secondary xylem towards the inner side.

The xylem which is produced in larger quantity as compared to phloem towards the inner side, exerts continuous pressure on the cambium and phloem; the latter being in the form of a thin layer only. As such, gradually the outer tissues tend to break apart in the form of flakes of bark and cork, usually dead in the countries where marked climatic differences exist at various seasons. Distinct annual growth rings are seen in spring and autumn.

The cork is formed by one of the layers of cortex which behaves meristematically and starts cutting off cells both on the outer and inner sides, the former being called *rhytidome*, and the latter *secondary cortex*. Rhytidome comprises dead elements and the inner ones also acquire this fate gradually resulting in the formation of cork (Fig. 2A-D).

The stems perform other functions also besides these normal ones by undergoing modifications (Fig. 3). These are associated with storage of food and vegetative propagation as well as movement and support and may be underground [e.g. rhizome (*Zingiber officinale*), tuber (*Solanum tuberosum*), bulb (*Allium cepa*), corm (*Colocasia*)], sub-aerial [such as runner (*Oxalis corniculata*), sucker (*Mentha*), offset (*Eichhornia*), stolon (*Fragaria indica*)], or even aerial forms [like tendril (*Pisum sativum*), thorn (babul, *Bougainvillaea spectabilis*), phylloclade (*Optunia*), cladode (*Asparagus*), or bulbil (*Agave americana*)]. Certain devices have developed in the

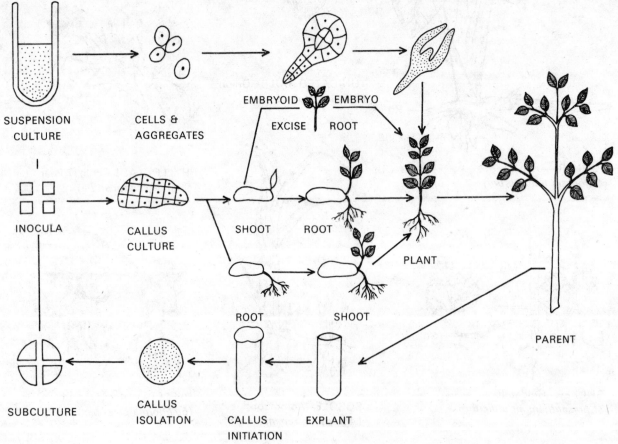

SUSPENSION CULTURE

CELLS & AGGREGATES

EMBRYOID   EMBRYO

EXCISE   ROOT

INOCULA

CALLUS CULTURE

SHOOT   ROOT

PLANT

ROOT

SHOOT

PARENT

SUBCULTURE

CALLUS ISOLATION

CALLUS INITIATION

EXPLANT

Fig. 4

TREE REGENERATION FROM SOMATIC CALLUS

Diagram showing alternative methods of regeneration of a tree, from somatic cells rather than from seed. Development in an artificial culture is much quicker than it would be from seeds germinating in nature and enables exact copies of the parent ('clonal images') to be propagated.

plants in course of evolution which directly or indirectly help them in protection. Most conspicuous of these are the armatures or defensive organs, which act as stationary weapons. These are thorns (babul, *Acacia nilotica*), spines (*Amaranthus spinosus*), prickles (rose *Rosa chinensis*, brinjal *Solanum melongena*) and hairs of different types such as stinging hairs (stinging nettle *Urtica dioica*), sticky glandular hairs (*Jatropha gossypifolia*), dense covering of hairs (Aak *Calotropis procera*). In terms of evolutionary history, it appears that the primitive stem (such as that of *Rhynia* and *Asteroxylon* among the fossil forms and *Psilotum* and *Tmesipteris* among the living ones) originated in the form of a simple, dichotomously branched appendage, from the thalloid form like the extinct liverworts and MOSSES. In the Silurian Period this situation continued for almost 100 million years. By the Devonian Period, complications and diversities developed which are distinctly manifested in the Carboniferous Period, particularly by those taxa which show arborescent habit, for example the representatives of Lepidodendrales and Calamitales which have been the chief source of coal deposits in the constituents of Northern Hemispheric land masses. It is surmised that *Heterangium* of the Devonian-Carboniferous strata showed first sign of secondary growth. In India and adjacent regions this feature is exhibited by taxa of the massive and woody form (which tended to be shrubby) of the Glossopteris flora (*Glossopteris*, *Gangamopteris*, *Palaeopteris*, etc.; see CONIFERS). *Vertebraria* shows a transition between stem and root comprising a creeping stem or rhizome of the massive type with manifold ramification. In the Mesozoic, the Cycadophyta dominated including *Ginkgo* leading to the origin of angiospermous habit—ranunculaceous on the one hand and magnoliaceous on the other, which is fully borne out by the oldest records of these in the Tertiary period.

During the past couple of decades, a lot of new information has been gathered which tends to suggest that various types of cells comprising the stem can be isolated and reared outside the plant in artificial culture media. The steps for such a formulation have been outlined in Fig. 4. This undoubtedly opens up new possibilities for acquiring knowledge pertaining to the various structures in relation to function of this organ as well as the differentiation of the various elements constituting it and there is no doubt that we will be able to supplement to a large extent the attempts to provide a global tree-cover, especially of fast-growing species, by new techniques of tissue culture.

Recent studies have also shown that the regular and organized differentiation for phloem on the outside and xylem on the inner side of the cambium is controlled by a number of factors, chiefly the ratio between IAA and kinins as well as the pressure exerted by the cortical elements on the cambial ring and its immediate derivative. Experiments with bark-peeling clearly suggest that if the pressure is released or nearly so on a temporary basis, the cambial initials produce only parenchymatous cells which become clumped together in the form of a mass of callus. However, if a pressure equivalent to 1·5 atmospheres is applied from outside with the help of a rubber band or some other source, regular differentiation of phloem and xylem sets in once again.

See also ROOT, BARK.

G.S.P. & U.R.

STICK and LEAF INSECTS. These extraordinarily mimetic insects, resembling sticks, grass blades or leaves, are usually fairly large and are dominantly tropical in distribution. Some individuals of the larger species may exceed a foot in length! The males and females are distinct in the same species, the male being smaller and slenderer than the female. These insects are probably very infrequent and rare and seldom noticed in the field. They are usually found on trees sitting quietly on the branches, trunks or leaves and also inhabit grassland and shorter shrubbery. They are very slow in movement and are mostly phytophagous. The males are usually winged while the females have reduced wings and are generally flightless. It is possible that most stick and leaf insects are nocturnal in habit, but they are seen active in the day too. More than 2500 species are known from all over the world, but the extent of the Indian fauna has perhaps not been seriously investigated.

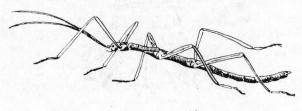

A stick insect × ¹/₂

Though stick and leaf insects are very difficult to spot, and make good use of their uncanny resemblance to sticks and leaves, if approached very closely, they drop to the ground and lie still for quite some time. Though males and females do copulate to produce eggs and further progeny, parthenogenesis is common in this group. The common Indian species, *Carausius morosus*, a stick insect widely used for laboratory studies, exhibits this and males are very rare. The nymphal stage lasts a few months and nymphs are of similar habit as the adults. Though rare, some species may be gregarious and in some years there may be an 'outbreak' of large numbers of individuals. The eggs are very hard, oval or barrel-shaped, and often sculptured intricately. They are extremely similar to seeds and can be mistaken for them. Though they have food in plenty as they are not very specific in their diets, and males do fly far, these insects are rare and the reason for this is not clear. Their enemies include ants, birds and other predators, as well as fly and wasp parasites. The common leaf insect in India, *Phyllium scythe*, inhabits thick forests.

See plate 26 facing p. 369.

STOMATOPODS include mainly *Mantis* or *Locust Shrimps* under the genus *Squilla*, and they are common

Mantis shrimp *Squilla* (stomatopod) × 1/3

along the east and west coasts. Unlike the decapods their carapace is short and does not cover the thoracic segments. The anterior three pairs of thoracic limbs are disposed forwards. The second pair is very large, sub-chelate and raptorial, giving the animal the appearance of a locust or praying mantis. The three pairs of posterior thoracic limbs and the abdominal appendages function as swimmerets. The tail is expanded and fin-like. The body is elongate and the abdomen broad. The animal grows to 20 cm. About eight species of *Squilla* have been recorded from Madras.

See also CRUSTACEANS.

STONEFLIES. A little more than 1500 species of these soft-bodied, membranous-winged, stream-frequenting insects (4 to 50 mm long) are known. Though fully winged, some adult males may be wingless, and brachypterous (stub-winged) species are also known. Stoneflies are mainly restricted to freshwater habitats, either swift-flowing streams or lake margins, generally in temperate zones or at high elevations in the tropics. They are an important insect diet of freshwater fish, especially trout, and, as they prefer pure water, rich in oxygen, they are good indicators of polluted water, which they avoid. The adults are poor fliers and are found resting on stones or vegetation near water; the larger species being commonly used as trout-bait by anglers. The nymphs

A stonefly × 5

are aquatic, living mainly on submerged stones or gravelly bottoms of streams or lakes. Most feed on algae, diatoms, lichen or mosses but perlid species are carnivorous, eating mayfly nymphs or larvae of chironomid flies. Some species are nocturnal in habit and attracted to artificial lights, but most are collected on vegetation or rocks along sides of streams. The Plecoptera are a primitive order of insects retaining many archaic features and perhaps only 100 Indian species are known so far. *Nemoura punjabensis* and *Rhabdiopteryx lunata* are common at higher altitudes on the northwest Himalaya.

542

STORK. The stork family (Ciconiidae) consists of long-legged marsh-inhabiting birds with partially bare tibia, longish necks and heavy pointed tapering bills. They lack a true voice-box (syrinx) and are mostly silent except for grunts and clattering of mandibles, especially when breeding. Our common resident species are: Painted Stork (*Mycteria leucocephala*), identified by its unfeathered waxy yellow face, long yellow bill, and the rose-pink about its shoulders and wings. The Open-billed Stork (*Anastomus oscitans*) is greyish white with black wings. The reddish black bill has a characteristic gap between the mandibles believed to help in opening up the larger snails—its favourite food. Possibly the least attractive of our storks is the Adjutant (*Leptoptilus dubius*), so called because of its martial stalking gait. It is black, grey and white with a massive yellowish bill and naked head and neck. A long naked reddish pouch hangs from its chest which its smaller congener (*L. javanicus*) lacks. Storks feed on fishes, frogs, aquatic arthopods and molluscs. The Adjutant Stork takes carrion in addition.

Painted Stork × ¹/₁₀

Openbilled Stork × ¹/₁₀

SUCKER FISH. Also called Remora, these fishes, about 30 cm when full grown, are capable of swimming,

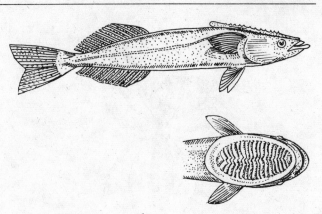

Remora or Sucker Fish × ¹/₄ , with its sucker enlarged × 2

but are usually found attached to sharks, turtles, dolphins, and even to ships. The first dorsal fin is modified into a suction apparatus; its rays become divided into two halves and are modified into transverse plates which are surrounded by a membranous fringe. Erection of the plates after the fish joins its host creates a vacuum which enables the fish to remain attached. The sucker fish cannot be dislodged until it wishes to get detached.

The fish is not a parasite and does not suck the blood of the host. It uses its host only for transport, and uses its own agility to avoid being caught and eaten by the shark. The sucker fish also helps itself to fragments of fish escaping out of the shark's mouth when the latter is feeding. Recently it has been suspected that the sucker fish may also be feeding on fish lice or other parasites living on the host's body and gills, as it is found to enter the mouth and gill-cavities of swordfish, sailfish and sunfish (*Mola mola*). Sucker fish are solitary in habit except during the breeding period and are oviparous.

Sucker fishes have been used to catch turtles. A long line is tied to the tail, and when a turtle is sighted, the sucker is released into the sea where its instinct makes it attach itself firmly to the turtle. The sucker fish together with the turtle is then pulled into the fishing-boat by the line.

SUN SPIDERS (Solifugae), known also as False Spiders, differ from all other arachnids in having a pair of armed (chelate) jaws, a body divided into a head, a 3-segmented thorax and a 10-segmented abdomen—which give them a close resemblance to insects. The pedipalps are long, leg-like and locomotory. The pedipalps and legs are hairy and with spines, giving the animal a dangerous appearance. There are two simple eyes on the head. They possess poison glands at the base of the jaws but are not poisonous to man. They breathe by air tubes like insects. Like spiders, Sun spiders are dimorphic—the male being much thinner

543

Sun spider, female *Galeodes* × $^1/_2$

and smaller than the female. Mating is fraught with danger to the males but their small size helps them to escape quickly from being devoured by the hungry females. Sun spiders are nimble and keen-sighted. They are fast and powerful hunters, overtaking and capturing any beetle, locust, moth or even a small lizard that comes in sight. Their formidable jaws crush the prey to small particles which they swallow. They are fearless of human beings—for they freely crawl over tea-shop tables to catch flies and suck up drops of water or tea adhering to mugs and glasses.

Sun spiders are widely distributed in the sandy regions of all warm countries.

Eggs are laid by the female in holes which she digs in sand or soil with her jaws. The eggs hatch in a fortnight. After another 15 days they begin to moult and become small sun spiders. The mother keeps guard over them until they can forage for themselves.

The term 'sun spiders' is a misnomer having regard to the fact that many of them are nocturnal. A few species however roam about in the bright sun and the Spaniards called them *aranhas del sol* or 'sun spiders', which name has come to stay.

The common genera found in our area are *Galeodes* and *Rhagodes*.

See also ARACHNIDS.

SUNBIRDS and SPIDERHUNTERS. Sunbirds (Nectariniidae) form a distinct group of very small birds of Africa and S Asia extending to the northern tip of Australia. Their small size, lively habits, especially the hovering while feeding, bright iridescent plumage of males, and their dependency on flower-nectar often invite comparison, even confusion, with humming-birds of the New World, to which however they are unrelated. Sunbirds are important flower pollinators and possess a highly adapted bill—long, thin and decurved—for flower-probing, and tubular suctorial nectar-eating tongue (see ORNITHOPHILY). Taxonomically sunbirds are closest to flowerpeckers.

Of the four genera of sunbirds the following two occur in both Asia and Africa: *Nectarinia* (containing the Purple, Purplerumped and the Small Sunbird, the last perhaps the smallest of Indian birds); *Anthreptes*, with a short somewhat straight bill and *Aethopyga*, with yellow rump-patch and two long projecting mid-tail feathers. The Spiderhunters, *Arachnothera*, occur only in Asia. They are much larger than sunbirds, both sexes without iridescent plumage. Both groups feed also on fruits, insects, spiders and small snails.

Sunbirds suspend their pouch-like nest of cobwebs and soft grasses from the tip of a branch at moderate heights. The spiderhunters' nest is a concave pad of skeleton leaves attached by its rim to the underside of a broad leaf (e.g. banana), with a side entrance-hole. Eggs in all species are streaked and blotched heavily.

See plate 24 facing p. 321 and plate 31 facing p. 432.

SUNDARBANS. The Sundarbans is located in the District of 24 Parganas in West Bengal, and within it the Sundarbans Tiger Reserve covers an area of 2500 km².

The Sundarbans is a Tidal Swamp Forest according to the classification of Champion and Seth. Human presence has been limited to honey collectors, wood collectors and people engaged in procuring minor forest produce. As a result it has remained a comparatively pristine area providing adequate cover for wildlife.

The tiger in the Sundarbans has been accused of undue man-eating propensities and one researcher attributed this to the saline environment. It seems more likely that honey collectors and others who suddenly come upon the animal invite a ferocious response.

The meeting place of water and land is usually an area of high productivity and so is the case with the Sundarbans. Diverse forms of life exist here including a variety of fish such as mud-skippers and climbing perch. There is a wide range of molluscs and crustacea and rather surprisingly high tides occur twice a day. Apart from the tiger the Sundarbans has large numbers of wild boar and chital. The Javan Rhinoceros, Swamp Deer and the Wild Buffalo which were once a feature of the place have unfortunately become extinct.

The core area of the Reserve (where no forestry operations are permitted) consists of 1330 km². The adjoining Pablakhali Wildlife Sanctuary has a nesting colony of water birds including Openbilled Storks, Little Cormorants and Large Egrets.

See plate 11 facing p. 144.

SUNFISH, *Mola mola*, is one of the two genera of round-tailed ocean sunfishes (Molidae) occurring in the

The wing-architecture of this Fruit Bat flying down is clearly visible in this mid-flight action-stopping flash photograph.

© A. J. T. Johnsingh

Hunting in packs, the Dhole is a formidable predator

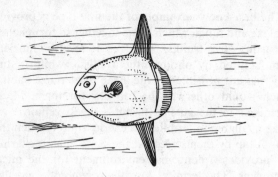

Sunfish, *Mola mola*, greatly reduced

Wiretailed swallow and young × ¹/₂

Indian seas, the other genus being *Ranzania*. These sunfishes are oval-bodied, with thick leathery skin and a small mouth with teeth fused together to form a single sharp-edged bony plate or beak in each jaw. Pelvic fins are absent and the dorsal and anal fins are extended vertically. The caudal fin is short and spread out along the entire rounded hind margin and hence the epithet 'round-tailed'. Their coloration is uniformly grey to olive brown with silvery reflections on the sides. It is reputed to grow a little over 3 m and weigh more than a tonne. Because of absence of caudal portion of the body and backward disposition of the dorsal fin, the fish appears to have an enormous head only and is sometimes referred to as a 'head fish'.

The sunfish is an inhabitat of the open oceans, often seen basking in the sun, lying on its side at the surface during calm weather, drifting more or less passively with ocean currents and feeding upon small fish, jellyfishes and crustaceans. With a semi-rounded body and no caudal portion, propulsion through water with speed must be a real problem and its dependence on ocean currents seems natural. Some people in the past attributed intelligence to this fish on account of its seemingly large head region, but it will allow a boat to come right up to it without making any effort to escape. It is found to have a very small brain despite its huge body.

The fish has a very thick skin having a layer of 5 to 7 cm thick gristly material, making harpooning difficult. The skin is said to be impervious even to a bullet from a gun. When accidentally captured in a net it is said to make grunting sounds.

Its outstanding feature is the enormous number of eggs (300 million) found in its ovary. The newly hatched larva is very tiny and elongated like a normal fish but later on it loses its tail portion and thereafter a short tail is developed on the rounded edge of the truncated body.

SWALLOW (family Hirundinidae). Includes Sand, Crag, and House martins and Swallows with sleek, streamlined, pointed wings, slightly to deeply forked tails, and short bills. Constantly flying, hawking insects.

Their nests are attached to buildings, rocks etc., built of mud, saliva, straw and feathers. Sand Martins burrow in earth banks.

SWALLOW-SHRIKE (family Artamidae), also called Wood-Swallow. A stoutly built grey-and-white bird with long pointed wings and stout finch-like bill. Affects open country. Captures insects on the wing in graceful swallow-like flight. Parties are commonly seen huddled together on telegraph wires. Sexes alike. Nests on palm trees, etc.

SWAMP DEER. The swamp deer is large, stags attaining a weight of 170 kg and more and a shoulder height of 125 cm. The characteristic antlers of a stag sweep upward for over half their length before branching; many have 12 tines, the basis for the animal's Hindi name *barasingha*.

In former times the deer ranged throughout the Indus, Ganga, and Brahmaputra river basins and as far south as the Godavari river in central India. Today it is extinct along the Indus, and survives primarily in a few reserves elsewhere. Being almost exclusively grazers and preferring the vicinity of water, swamp deer are found principally on grasslands near rivers, a habitat which has been almost completely converted to agriculture. The finest remaining deer populations are found in the Sukla Phanta reserve in Nepal, the Dudhwa National Park in U.P., and Kaziranga National Park in Assam, all three supporting a total of about 3000 animals. These northern deer are considered to be a different subspecies (*Cervus duvauceli duvauceli*) from the ones in central India (*C. d. branderi*), where a viable population of about 300 survives only in Kanha National Park.

Swamp deer are social animals, as many as 900 having been seen together, except that stags frequently wander alone and females become solitary when fawning. At the time of rut many may congregate at a favourite meadow. The time of rut varies, reaching a November peak in Nepal and a January one in central India. Several distinctive behaviour patterns are associated with the rut. At dawn and dusk, stags bugle, a mournful two-toned note, repeated 10 or more times. Stags also wallow in muddy depressions. Many stags use the same wallow whose location is traditional year after year.

A swamp deer stag does not collect a harem of females. Instead, many stags gather near the females and establish a rank order among themselves. The dominant stag has priority to an estrous female. Both intimidation and direct aggression are used to achieve or maintain rank. A stag may approach another with a stiff gait, his muzzle raised high, making himself impressive. Or he may lower his antlers and spar. Such fights are seldom serious, the animals merely testing their strength or reasserting themselves. True dominance battles are violent affairs, antlers clashing and sod flying.

After the rut, stags drop their antlers, beginning to do so in March in northern India and in May in central India. After a gestation period of 240 to 250 days, females have a single young. Females have their first young at the age of 3 years and one young per year thereafter.

See also DEER.

G.B.S.

SWIFT. Though superficially alike, the swift family (Apodidae) belongs to a different natural order from swallows. Swifts have all four toes pointing forward: therefore they cannot perch but can only cling to a vertical surface with their needle-sharp claws. Thus you would never see a swift sitting on a telegraph wire. Some of the species, e.g. the larger Spinetails and the Alpine Swift (*Chaetura* spp. and *Apus melba*), are among the world's fastest flying birds, credited with speeds of over 200 kmph. The birds spend the entire day in the air hawking insects, and evidently may also sleep on the wing. Their nests are untidy agglomerations ('villages') of globular 'pots' of straw and feathers stuck together and against ceilings in buildings or within rock fissures. The familiar black-and-white house swift (*Apus affinis*), with narrow bow-shaped wings, is common throughout the subcontinent.

SYMBIOSIS. Symbiosis is the association of two or more species of living organisms to their mutual advantage. Such associations may be classified into three types on the basis of the kind of advantages gained. Firstly, one partner provides food to the other partner which protects it; secondly, the parasites of one form the food of the other; and thirdly, the two partners facilitate food-finding by each other.

The best-known example of the first type is provided by algae and fungi, which come together to form lichens encrusting tree-trunks. The fungi receive nutrients manufactured by photosynthetic algae, while in turn they make it possible for the algae to colonize habitats which would otherwise be too dry for them. Microorganisms in the guts of termites digest cellulose and make available to termites an energy source which they cannot tap by themselves. At the same time, the termite gut provides a congenial environment for the microorganisms. The termites lose their symbiotic gut flora every time they moult and lose the lining of the gut. A new gut flora is then acquired by licking at the anus of other termites. In fact it has been suggested that termite sociality owes its origin to this necessity of acquiring fresh symbionts from another termite after every moult.

Several ant species have two notable symbiotic relations of the first kind, with fungi and with aphids. Many species of leaf-cutter ants cultivate fungal gardens in damp, humid chambers in their burrows. They provide leaf-cuttings for the fungi to grow upon, create an optimal physical environment for them, and protect them. In turn they feed on the fungi. Similarly, ants tend aphids. Aphids are insects which suck plant juice. This they convert into a sugary liquid, which is licked by the ants, that protect the aphids against predatory insects.

A number of instances of symbiosis involve one partner feeding on the ectoparasites of the other. The relationship may be a casual one, as with mynas and water-buffaloes. The birds do feed on ticks, lice and other ectoparasites on the buffaloes, but also profit by insects flushed by buffaloes while walking, as well as on many other food sources. The relationship is much more obligatory for the oxpecker birds of Africa, which attend various larger mammals such as rhinoceros.

A most intriguing symbiotic relationship of this type obtains amongst coral-reef fishes. Some species of such fish specialize as cleaners and obtain all their food from ectoparasites on the bodies and in the mouth and gill-chambers of other fish. These cleaners are highly sedentary small fish that live in cavities in coral reefs. The fish wishing to be cleaned visit them at their stations and often wait patiently in a queue to be cleaned. The cleaners as well as their customers employ specialized behavioural displays which imply an invitation to clean, and to stop cleaning. The cleaners are also distinctively marked. Notably enough, the customers of the cleaners do not eat the cleaners which enter their mouth cavities, although they eat other fish of similar size. There are however some fish species which mimic cleaners in their coloration and, having entered the mouth cavity of a fish wanting to be cleaned, bite pieces of flesh out of it.

The third kind of symbiosis relates to mutual help in feeding. The mixed feeding parties of insectivorous birds furnish an interesting example of this. These parties are particularly common in moister forests and may involve more than twenty species of birds in a given locality. In Western Ghats forests the drongos form the nuclear species of such mixed flocks, in which they are invariably present. Mynas, woodpeckers, nuthatches, chloropsises, and minivets are amongst the other components. All the partners presumably profit from better vigilance against predators, better flushing of their insect prey and more systematic search of the foraging area.

See DOCTOR FISHES.

TADOBA. In 1935 Tadoba was made into a game sanctuary and in 1955 it was converted into a national park because of the impressive wildlife which it contains. It has an area of 116 km² and there are plans for extending its size. Its altitude is approximately 200 m above sea level and it is about 90 km SSW of Nagpur.

The Tadoba lake situated almost in the centre of the forested area is an important ecological feature. The area is hilly with a gradual descent from the north to the south. The main tree species are teak, saj, salai (*Boswellia serrata*), axlewood, ebony, bijasal (*Pterocarpus marsupium*), karar (*Sterculia urens*), mahua, and jhingan.

Troops of the Hanuman langur, varying in size from elderly patriarchs to new-born young, can always be seen around the lake. The wildlife includes tiger, panther, jackal, dhole, civet cat, sloth bear, jungle cat, bison, spotted deer, sambar, nilgai, barking deer, four-horned antelope, mouse deer, flying squirrel and boar. There are many snakes.

TAHR. The tahr (genus *Hemitragus*) are closely allied to goats (genus *Capra*) even though their short horns are similar to those of serow and other goat-antelopes. During the Pleistocene, tahr ranged as far as Europe, but today they exist only in three widely separated mountain systems. The Himalayan tahr (*H. jemlahicus*) inhabits the southern flanks of the Himalayas from Kashmir to Bhutan where it 'revels in the steepest precipices', to quote one source. This species may be found as high as 5000 m in summer. Males have a distinctive dark and shaggy ruff and a long mantle of hair. Over 2000 km south of the Himalaya, in the lofty tablelands of Tamil Nadu and Kerala, is the home of the Nilgiri tahr (*H. hylocrius*). India is wholly responsible for this unique animal of which only about 2000 survive, many in the Eravikulam Reserve. Male Nilgiri tahr have a characteristic grizzled 'saddleback'. The third species, the Arabian tahr (*H. jayakari*), occurs only on the Arabian peninsula.

Himalayan and Nilgiri tahr are powerfully built animals, males weighing about 100 kg and females 60 kg. They are social, at times congregating into herds of up to 100. Males tend to separate from the females except during the rut which in Himalayan tahr takes place from November to January and in the Nilgiri tahr from June to September. One young is born after a gestation period of 180 days.

See MOUNTAIN SHEEP & GOATS.

G.B.S.

© Shekar Dattatri

Himalayan Tahr × ¹/₄

TAMARIND, *Tamarindus indica*, is native to Africa and was brought to India by the Arabs in ancient times. The name tamarind derives from the Arabic *tamr-hindi* which means 'date of India'. Its pleasant, acidic-tasting fruit was so popular that the plant's botanic and common names both point to its association with India. Senegal's capital city is named after the tree, whose local name is *dakar*. It is a handsome tree with a dome-shaped crown of graceful, airy leaves. Clusters of pale-yellow blossoms in summer turn into rust-coloured curved pods in late winter and contain shiny, brown seeds. Its strong, supple branches are not affected by

548

wind and it is known as a hurricane-resistant tree suitable for avenues. A valuable timber and choice fuel, it was a major fuel for producer-gas (gasogen) units that powered Indian trucks during World War II. The pulp from the pods is used to season chutneys and ice-cream.

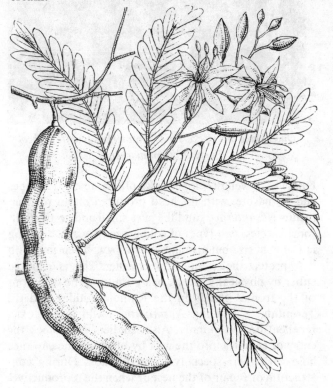

Tamarind. Leaves, flowers and pod × ³/₄

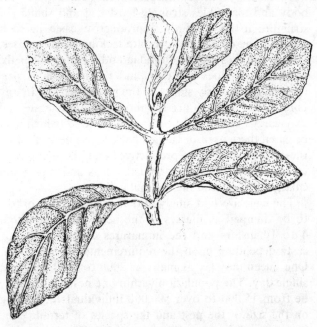

Teak. A branchlet with young leaves × ¹/₂. A tree 500 years old has been recorded from the Anaimalais.

TEAK, *Tectona grandis*, is a large tree, indigenous to peninsular India and Burma. Nobody is likely to fail to identify the teak tree, whose furrowed quadrangular branchlets, very large opposite leaves, with red dots beneath, when young, black when old, and the bladder-like fruits are sufficient to distinguish it from other trees. If the young leaves are rubbed between the hands, the palms will be stained red. Flowers are small and white in metre-long panicles. They appear from June to August and ripen from November to January. The first plantation in India was at Nilambur in 1844, where it attains the largest dimensions. The tree reaches maturity in 60 to 90 years. The timber is world-famous and its uses are well known. The hard knots which develop on trunks are prized for making tobacco pipes.

TEMPERATURE. Temperature is the condition which determines the flow of heat from one body (solid, liquid or gas) to another. It is measured by instruments called thermometers. The scales used are the Centigrade (also called Celsius) and the Fahrenheit. In the Centigrade scale, the freezing point of water is 0°C, and the boiling point of water is 100°C. In the Fahrenheit scale, the freezing point of water is 32°F and the boiling point of water is 212°F. In all current scientific literature, the Centigrade (Celsius) scale is used.

$$(°C \times {}^9/_5) + 32 = °F$$

Degrees C can be converted into Degrees F, by the formula:

On the Kerala coast the annual average temperature is about 27°C. In the northwestern plains summer temperatures soar up to about 47°C in the afternoons, while the winter temperatures drop to 0°C in the mornings.

As we go up in the free air, the temperature falls. The maximum rate of fall of temperature with height is ten degrees centigrade per kilometre increase in altitude. In general, the fall in temperature is about 7° or 8°C per kilometre. However, on moist and cloudy days, the fall in temperature with height is near 6°C per kilometre. This decrease of temperature continues to a height of about 16 or 17 kilometres over our area for most of the year. Above that height it begins to *increase* at the rate of five or six degrees C per kilometre.

Sea-surface temperature is less than that of adjoining land in summer, and greater than that of adjacent land in winter. Thus the seas act as a moderator of climate.

Over land there is a significant difference in the afternoon temperatures and the early morning temperatures, when the maxima and the minima occur respectively. National climatological tables provide the values of the monthly mean maximum temperatures, as well as the monthly mean minimum temperatures for different locations.

TERMITES. Termites, commonly called 'white ants', are highly organized social insects belonging to the order Isoptera. They are polymorphic, living in small to large communities either above or below ground. Their close association with crops and forests has attracted considerable attention and though they cause heavy losses to man, they are also beneficial in their rapid turnover of organic matter in the ecosystem. This is done by the breakdown of cellulose and returning it to the soil. They also bring subsoil to the surface and can be compared to the earthworms in this respect.

There are two basic types of habitats exploited by termites—soil and wood. The soil-inhabiting termites are common and build mounds above the soil surface or nest underground with no sign above. Wood-inhabiting termites are mostly arboreal, constructing their nests either outside the tree-trunks or branches or within them.

The Isoptera is a relatively small order of insects and related to the cockroaches and mantids. Over 2000 species are known from the world, where they are primarily tropical in distribution, with the greatest number found in Africa. More than 200 species are described from the Indian subcontinent but many more are still to be found. Termites are soft-bodied, pale insects which avoid sunlight and live cryptically, well protected from light. The winged forms are relatively large (6 to 18 mm) whereas the soldiers and workers are 3 to 15 mm long. Of the 270 or so species occurring in India, only 30 or 40 have been recorded as damaging crops or habitations and these belong to five of the nine known families: Kalotermitidae, Hodotermitidae, Rhinotermitidae, Stylotermitidae and Termitidae. Among them, the genera *Odontotermes*, *Macrotermes* and *Microtermes* cultivate fungus gardens in their nests. The genus *Odontotermes*, with 38 species in India, is our dominant termite group.

Termites have a fascinating biology. The individuals are differentiated into various castes or morphological forms which exhibit division of labour. The castes exhibit both anatomical and physiological specialization. Four distinct castes are recognizable: primary reproductives (king and queen), supplementary reproductives, soldiers (sterile males and females with heavily sclerotized heads), and workers (again sterile males and females). Also, a colony will contain immatures of all the castes at any particular time. The reproductives consist of winged males and females which leave the nest at a particular time each year to found new colonies. The primary function of both primary and supplementary reproductives is to keep the colony at optimum population by producing eggs (female) and spermatozoa (male). The soldier caste is very specialized with a well developed large head, and is of two types: (1) mandi-

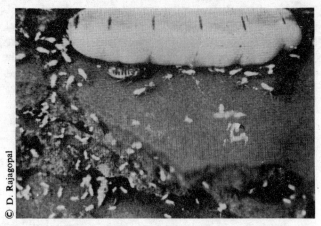

Queen, king and workers of Odontotermes *termite in royal chamber of termitarium* × 1

© D. Rajagopal

bulate with well developed jaws of characteristic shapes and (2) nasute, with the head produced into a rostrum (genus *Nasutitermes*) and the jaws small and vestigial. In some species, two types of soldiers may occur, differing not only in size but also in morphology. Their main job is to protect the colony from enemies. This is achieved either by physically biting, or by emitting the secretions of the frontal glands in the Rhinotermitidae and the Termitidae (including *Nasutitermes*) which are toxic, repellent or just gummy. Also, soldiers may block the entry of enemies into the nest by blocking the entrance hole with their specially formed heads. During construction or repair of the nest or when the reproductives fly out of the nest, the protective activities of the soldiers are most noticeable. Compared to the soldier caste, the workers are relatively unspecialized with a pale soft body and mandibles strongly hardened and similar in structure to those of the reproductive caste in each species. The compound eyes are lacking in most cases. In two families (Kalotermitidae and Termopsidae) the worker caste does not exist and the function is performed by the early stages of the reproductives. In any colony, the workers are the most numerous caste and they carry out chores like foraging for food (which may be eaten there itself or carried back to the nest), feeding the young, soldiers, reproductives and the nymphs of reproductives. They also tend the eggs and repair and enlarge the nest and the internal galleries.

The eggs are laid singly and carried by the workers to be dumped as masses in the nest. There are from 4 to 10 instars and the immatures develop into any caste depending upon the requirements of the colony. One queen may lay as many as 2000 to 3000 eggs in a single day. The population within one nest/colony may be from 15,000 to over 500,000 individuals depending on the size of the nest and the species of termite. The life of a colony may vary from 15 to 50 years under natural conditions.

New colonies are founded generally by the flight of numerous winged males and females from the nest at a particular time each year and, in some cases, at a certain hour of the day. In India the main time is just after the start of the monsoon. There may either be a single flight from one colony or the winged reproductives may fly out in several batches for a few weeks at irregular intervals. After a short, weak flight, the reproductives drop to ground, shed their wings and find the other sex with the help of an odour released by the female. The pair then forms a tandem one behind the other and they search for suitable nesting sites in soil or in wood. They excavate a 'nuptial chamber' once a suitable site is located and copulation takes place. The queen then starts laying eggs soon after a week and the small early brood is tended by the royal couple, until after several weeks when the queen's production of eggs increases and a nucleus of a strong colony is formed. The proportion of soldiers to workers in a colony may depend on the species concerned, and vary from 1 to over 15 percent. The winged reproductives are not produced until the colony is several years old. Both the original king and queen that formed the colony may live for a long time, from 15 to 50 years in the higher termites. In some species there may exist more than one pair of primary reproductives in the same colony. Colonies may also be founded by a group of nymphs going out of the colony and forming supplementary reproductives, as is often the case in the more primitive termites.

The food of termites consists of wood, either unaffected or decaying, grass, fungi, humus, bark, dead leaves and even herbivore dung. The termites that eat wood feed primarily on the cellulose contained in it which is converted into a form assimilable by the symbiotic flagellate protozoa that occur in the hind gut of the termite. In the Termitidae, which do not have a protozoan community in their digestive system, the cellulose, if fed upon, is digested by cellulase secreted by themselves or by that produced by bacterial flora in their guts. The species of termites that do not feed on wood have no protozoa at all or a very reduced number. Much of the food is eaten while foraging, but species which feed on grass or littered leaves, etc., carry it back to the nests. Foraging is carried out at night or in the day under sheetings of mud, which are conspicuous as 'trails' to our eyes.

Termites build very striking or cryptic nests either invisible to us or of complicated architecture and very large mounds. The nests are homeostatically regulated units with a constant relative humidity maintained within. Like the shell of an animal, the nest provides protection against weather and natural enemies. The simplest type of nest is found in the primitive forms like the Kalotermitidae and Termopsidae where the entire colony lives in a gallery system with chambers excavated in wood. The subterranean species construct a complex central nest from which underground galleries or covered 'trails' emerge and go to food sources above or below ground. The mounds or termitaria built by most Termitidae are the most complex of all termite nests. The outer layer of the termite mound is very hard and composed of hard earth or clay. Inside, it is much softer and comprises a network of galleries and chambers. The royal pair are often found in a special chamber and the eggs and young in the others. The gallery system of one colony may exploit all the food sources around the nest for one hectare or more and a single trail may extend for up to 75 or 100 m in length. Though most termitaria are found as mounds on the soil surface, the genus *Nasutitermes* builds carton nests (somewhat analogous to the tree ant, *Crematogaster*) on tree branches or even on poles or sawed-off tree-trunks.

In mound-building termites, we have either the fungus growers or the non-fungus growers. The fungus-growing termites are *Odontotermes*, *Macrotermes* and *Microtermes* and these cultivate fungus gardens within their nests for food. The non-fungus-growing termites construct small mounds (*Amitermes*, *Cubitermes*, *Nasutitermes* and *Trinervitermes*) and fill them with dry food.

Termite nests harbour an assortment of other organisms either within the nest (termitophiles) or in the outer wall (termitariophiles). The outer wall is often used for nesting by other termite species, birds or lizards. Within the nest and closely associated with the termites, several groups of animals occur, either feeding on the eggs or young termites or on the debris in the termitarium or involved in a symbiotic relationship with the host termite. Most common are beetles of many families and aphids. Also occur several flies, isopods, Collembola, **Thysanura**, **centipedes**, **shrews**, **geckos**, **rats**,

An Odontotermes *termite mound*

© D. Rajagopal

mongooses, cockroaches, scorpions and some other groups.

The emergence of winged termites after the first rains is so regular an event and the numbers of these frail nutritious creatures is so large, that many kinds of animals have taken the opportunity to congregate near the flying swarm and partake of it. Many birds, frogs, lizards, snakes, and even bats feed on them, while many orders of insects also utilize this event to satiate their hunger — the dragonflies conspicuously in the air and several beetles on the ground, mainly. Special animals like the anteaters and a few leptodactylid frogs exist primarily because the termites are their principal or only food.

See also EARTHWORMS.

TERN (family Laridae) Similar to the GULL but slimmer and daintier. Most species have a deeply forked tail, whence they are also called Sea-swallows. Terns frequent both coastal and inland waters and live chiefly on fish scooped up from the surface or caught by plunging from the air. While some species breed in the subcontinent, most are winter visitors. The Arctic Tern (not found here) is the champion of long-distance bird migration. It does a round trip annually of some 22,000 miles from the Arctic to the Antarctic and back.

TERRITORIALITY. Having gone to visit your friend you park your scooter in front of his house. As soon as you do this your friend's dog comes along and urinates on your scooter. This must have been a familiar experience for many of us. The dog treats the house and its surroundings as his *territory*. He keeps telling other dogs in his neighbourhood about his possession of this territory by marking various objects in his territory with his urine. Even if the dogs in the neighbourhood stray into his territory they are immediately told by the smell of his urine that this place belongs to somebody else. Since the smell keeps wearing out, the dog keeps marking his territory from time to time. The dog suddenly discovered your scooter in his territory with no trace of his unique scent mark and so he promptly amended the situation.

Such behaviour, which may be called territoriality, is found among a variety of animals. It is true that some animals, such as many micro-organisms and marine plankton for example, have no special reason to be attached to any space and spend their entire life drifting in any direction that the water currents might take them. However, barring these exceptions, most animals have certain rather fixed places where they live and rigid rules by which they share living-space with their neighbours. Thus animals in their lifetime come to associate with themselves a certain amount of land and this land is given different names depending on how closely the

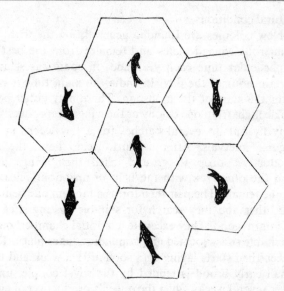

One of the most spectacular demonstrations of the formation of territories. When fish (*Tilapia mossambica*) are kept in a tank the males establish hexagonal territories. Each fish scoops out sand from his territory and pushes it out so that they result in hexagonal ridges of sand at the boundaries. Hexagons provide the maximum congruence of boundaries so that perfectly fitted hexagons leave no space between them.

animals associate themselves with it.

Consider a population of blackbuck which has been extensively studied in a place called Mudmal in Andhra Pradesh. The entire area traversed by an animal or the population in its lifetime is called the *total range* which may run to several tens of square kilometres. However, a population of blackbuck will associate itself more closely with a smaller area within this total range and use this area for purposes of feeding, rearing young, etc. This area is called the *home range* and has been estimated to be about 10 square kilometres. If one keeps following the movement of animals over time, it will become evident that the animals spend much more time within a certain portion of the home range and often learn the characters of the habitat within this smaller area, which is known as the *core area*. For example, the animals know precisely where to hide or in which direction to flee in the event of a predator arriving on the scene. Look at this population during the breeding season and you will see that some of the big males have established themselves as owners of small pieces of land of about ten hectares. These are the territorial males and this is their territory from which they actually chase away any intruding male and mate with any passing female.

The distinction between total range, home range, core area and territory may not be as clear in other animals but the important feature of what is defined as territory is that it is defended by its owner or owners against other members of the species by aggression or by advertising.

Territories are used by animals for different purposes. Bonnet monkeys in South India, rhesus monkeys in North India, the Hanuman langur of Mount Abu in Rajasthan, many squirrels and a variety of birds such as the common myna, the house sparrow, sunbirds and flowerpeckers, are known to defend territories for the sake of food supply. Snakes, crocodiles and ants, for example, defend territories for the opportunities available for building their nests.

A common function served by a territory is that it is the space utilized by the males for sexual display and the ensuing breeding activities. In such cases the territorial males advertise in a variety of ways. These advertisements serve two purposes. One is to inform other males that this space is already occupied and the other is to inform females that a potential mate is available. Often the form of the advertisement also tells the females the quality of the potential mate. A splendid example of this is the magpie-robin, where the male establishes a territory and sings; the song keeps other males away and brings females into his territory. The famous dancing of peafowl also appears to serve the same function. Territorial blackbucks are known to deposit their dung and urine in one particular spot in their territory and it is possible that the smell from these dung piles also serves the same functions. In these instances the strong males alone will be successful in establishing territories and therefore in being able to mate with females.

Males of many species of solitary bees and wasps also establish territories for the purpose of increasing their chances of mating, but in a much more direct manner. Eggs are laid under the ground where the entire development takes place and the adults emerge through holes in the ground. Males emerge first and after a lot of fighting the successful ones establish territories around the holes from which females are going to emerge. They stand guard at these holes and as soon as the female emerges they mate with her.

Territories need not always be a fixed place but can be moving. There is a very interesting case of a butterfly whose territory keeps moving every few minutes. The females are very fond of basking in the sun and therefore keep moving into little spots of sunlight that are scattered over the forest floor. So the males wake up early in the morning and, again after a great deal of fighting, occupy one of these sunny spots and keep moving as the sunny spot moves during the day. They then mate with the females that come to bask in the sunny spot.

R.G.

THAR. The Thar, or the Great Indian Desert, is spread over four Indian States (the Punjab, Haryana, Rajasthan and Gujarat) and two States of Pakistan (Punjab and Sind), covering an area of about 446,000 km². The average annual rainfall of the region varies from 100 to 500 mm with a coefficient of variation as high as 60–70 percent. This distribution of rainfall is also erratic, occurring mostly in the period July to September. During the summer, the daily maximum temperature is generally 40°C and 22°–28°C during the winter. The mean minimum temperature varies from 24° to 26°C during summer and 4° to 10°C during winter. The mean diurnal temperature variation ranges from 12° to 17°C. The relative humidity during summer varies from 36 to 50 percent and from 66 to 78 percent during monsoon during the morning hours. In the afternoon, however, there is an expected drop and the mean values range from 22 to 35 percent during summer and 48 to 60 percent in the monsoon season. The mean evaporation during summer exceeds 10 mm per day. The potential evapotranspiration values during summer (April to June) vary from 7 to 9 mm per day. The mean daily wind speed is recorded to be highest during summer and monsoon seasons, 8 to 20 km/h.

The origin of the Thar desert is a controversial subject. Whereas archaeologists and certain geologists consider it to be only 4000 to 10,000 years old, geomorphologists, pedologists and biologists have found evidence to suggest that aridity started to characterize the region much earlier. Recently, it has been found that the desert is not 'marching' towards the northeast, but the deterioration of the ecosystem is a local phenomenon. It is widely claimed that the destruction of biological potential in the Thar is fast intensifying. The escalation of human and livestock population in the desert are the major factors in this respect, resulting in degradation of soil fertility and vegetation, decline in productivity, inducing waterlogging, increase of salinity, less livestock production etc. Supplemented by climatic vagaries, these desertification factors are further

© B. L. Tak

60 percent of the Thar desert is affected by sand movement

© B.L. Tak

Livestock populations beyond the carrying capacity of the land result in degradation of the vegetation in the Thar to a point of no return

© B.L. Tak

Reforestation of the Thar can be a major process towards combating desertification

harming the desert ecosystem. However, the fast declining trend of productivity of arid land is being deterred by augmenting irrigation facilities. A massive effort is being made through the construction of the Rajasthan Canal, which is planned to transform about 11 percent of the barren areas of western Rajasthan into a vast granary. To bring water from the Sutlej river to Rajasthan, 204 km of feeder canal and 445 km of the Rajasthan canal are being constructed. The work started in 1958 and Phase I of the Project has been completed.

The desert is not a continuous stretch of sand. It is interspersed with hillocks, sandy as well as gravel plains, salt playas and lakes. Due to the presence of diversified habitats, the vegetation and animal life is quite rich in this arid land. The predominant trees are *Prosopis cineraria, Acacia nilotica*; and shrubs *Capparis decidua, Zizyphus nummularia, Calligonum polygonoides, Calotropis procera, Euphorbia caducipholia*. The monsoon herbaceous cover is composed of *Tephrosia purpurea,*

Tribulus terrestris, Crotalaria burhia, Aerva tomentosa, and a number of annual grasses. Perennial grasses of excellent fodder value are *Lasiurus sindicus, Cenchrus ciliaris, Panicum antidotale* and *Dichanthium annulatum.* Due to the impact of overgrazing, palatable and perennial species have been, in most parts of the desert, replaced by non-edible and annual species.

Since the Thar extends into the deserts of the Middle East and the Sahara, the majority of the biota show Saharo-Rajasthani affinities. The animals found in the Indian desert are well adapted to survive in the xeric environment. Locusts and their depredations are associated with arid zones. The seven-year cycle of the upsurge in their population and corresponding invasions in the Thar have now become occasional, chiefly due to excellent international cooperation in their control.

The desert region is holding two very important endangered wildlife species, the Lion in Gir forest and the Wild Ass, *Equus hemionus*, in the little Rann of Kutch. Certain wildlife species which are fast vanishing from other regions are however found in the Thar in fair numbers. The more important ones are the Great Indian Bustard, *Choriotis nigriceps*; the Blackbuck, *Antilope cervicapra* and the Indian Gazelle, *Gazella g. bennetti*. This has been possible because the transformation of the natural grasslands into croplands is relatively very slow in the desert, obviously for the lack of water; and due to the protection provided to them by a local community, the Bishnoi. Two winter migrants, the Houbara, *Chlamydotis undulata*, and the Imperial sandgrouse, *Pterocles orientalis*, visit the Thar in fairly large numbers. Besides, the desert holds a variegated fauna and deserves much more attention from naturalists.

I.P.

THREAD FINS (family Polynemidae) are represented in Indian waters by eight species. They derive their name from the thread-like free rays of the pectoral (shoulder) fin projecting out under the shoulder. They have a prominent snout, and a mouth on the lower side beset with small teeth. Adipose eyelids, two dorsal fins, pelvic fins below the shoulder; somewhat rounded body and large forked tail are other characteristics. Three species, *Polynemus tetradactylus*, *P. indicus* and *P. plebeius* are more common on the west coast while others which are of smaller size are numerous on the east. *P. tetradactylus*, because of some similarity between its head and body form with that of the Atlantic salmon is known as Indian salmon, though they have no taxonomic relationship. It has four free pectoral finrays and is a most popular food fish on both coasts, growing to a length of 1·2 m and weighing 10 kg. *P. indicus* has five free pectoral finrays and is the largest of the family, growing to 2 m and weighing more than

16 kg. It is more frequent near the bottom, feeding on small sciaenids (Dhoma), hemipterids, prawns, crabs and Bombay duck. *P. tetradactylus* has similar feeding habits but comes closer to the shore and even enters estuaries and prefers rocky environment where it obtains its favourite food more plentifully. It breeds near the shore and young ones also enter estuaries for feeding.

THRIPS. Mostly small (0·5 to 14 mm long) unnoticed insects with delicate, 'fringed' wings, brown/black or yellow in colour generally. Occur in colonies, which, if large, betray their presence. More than 2000 species (400+ Indian) have so far been described but there certainly remain an almost equal number to be collected and described. The Thysanoptera are dominantly tropical or subtropical in distribution, though the temperate regions do have a distinctive fauna, some being also found in the polar areas of the world. Among Indian insect groups, the thrips have had considerable attention paid to them recently and are pretty well known in our subcontinent.

There are two major suborders recognized: the Tubulifera and Terebrantia. The Terebrantia are plant- and animal-feeding and lay eggs inside plant tissue and pupate in soil. The females have a saw-like ovipositor and these thrips carry their wings side by side along the upper surface of the abdomen when resting. The predacious forms are mostly inquilines while the plant-feeding ones sometimes form galls. The Tubulifera, as they are called, have the last segment of the abdomen tubular and lack an ovipositor. Their wings are kept folded one over the other above the abdomen when resting. These are extremely diverse in form and habitat and many produce gall-like structures on the host plant. Eggs are laid on the plant surface and several species feed on fungi.

Besides rasping and sucking plant tissues, feeding on fungi and on other small insects and mites, thrips also inhabit flowers and are considered to be potential pollinators in some cases. Just one flower may hide several small Terebrantia which feed on pollen. Many species assume pest status and the common Indian

thrips that cause crop loss are *Thrips tabaci* (on cotton), *Scirtothrips dorsalis* (chilli), *Baliothrips biformis* (paddy) and *Rhipiphorothrips cruentatus* (groundnut). The common gall-forming thrips are *Arrhenothrips ramakrishnae* which cause the familiar leaf galls on *Mimusops elengi* (bulletwood) and *Gynaikothrips karnyi* on pepper.

THRUSH. Arboreal or terrestrial passerine birds belonging to the subfamily Turdinae, a division of the complex Flycatcher family Muscicapidae which includes robins (e.g. Magpie-Robin, Shama), chats (Pied Bush Chat, Redstart), dippers, ground and rock thrushes and many others. The other component subfamilies of Muscicapidae are the true flycatchers (Muscicapinae), warblers (Sylviinae) and babblers (Timaliinae) which include the laughing thrushes. However, the term Thrush or Blackbird in a particular sense is usually applied to members of the genus *Turdus*, about the size of a myna and predominantly black or blackish in colour. The genus is represented by a number of species in the Himalaya and by a single one (*Turdus merula*) with a number of subspecies in the peninsular hills. In the Indian subcontinent Blackbirds are essentially mountain-loving birds but they descend to lower altitudes in winter, some species then dispersing far and wide over the plains. Many others are true winter visitors, migrating long distances from northern lands. Thrushes include some of our finest avian songsters, and during spring and summer which are the breeding season, their native hills resound with their rich, fluty and long-sustained melody. Perhaps the most accomplished performer among them is the Greywinged Blackbird (*Turdus boulboul*) of the Himalaya, distinguished by a prominent pale grey wing-patch. It is a favourite cage bird in Kashmir and Punjab. Thrushes feed largely on drupes and berries, and invertebrate animals like snails and earthworms. The nest is a bulky cup of moss, rootlets and grass, usually intermixed with a quantity of mud. It is built in the fork of a small tree or shrub. Three to five eggs are laid, pale bluish- or greyish-green speckled with reddish brown.

THUNDERSTORMS. A thunderstorm is defined as the atmospheric phenomenon of one or more sudden electrical discharges, manifested by a flash of lightning and a sharp rumbling sound—thunder.

This phenomenon is usually associated with clouds of great vertical extent: five, ten or fifteen kilometres in thickness. They are called cumulo-nimbus clouds and are almost invariably associated with rain, often of great intensity. When ten or fifteen centimetres of rain falls within an hour or two, during a thunderstorm, it is called a cloud-burst. On some occasions over India and other tropical countries, the tops of the thunder-producing clouds may not reach a temperature of 0°C.

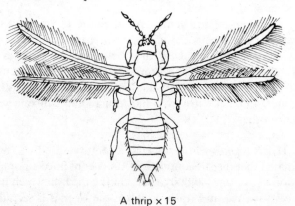

A thrip × 15

However, on most occasions the tops attain very low temperatures,—20°C or even less. The topmost portions of the cumulo-nimbus clouds consist of ice particles. They are blown towards one side by the prevailing wind at those heights, and the cloud top often acquires the shape of an anvil. They are called Anvil Clouds.

The mature thundercloud is cellular in structure, and is composed of several cloud cells. Some of these cells have strong updraughts with vertical velocities averaging 6 metres per second, and extreme values of about 30 metres per second. An adjacent cloud cell within the same thunderstorm complex may have a strong downdraft, although with smaller vertical velocities. When there are vertical upward velocities of 20 metres per second or more in a cumulo-nimbus cloud hailstones usually form within the cloud; they fall down as HAIL a few minutes later.

Charge separation in thunderclouds is such as to produce a positive charge in the upper part of the cloud and a negative charge in the lower part of the cloud. Occasionally there are pockets of positive and of negative charges embedded in the cloud. There are several theories about the mechanisms of the production and separation of charges. Some of the suggested mechanisms are (i) frictional rupture of raindrops into larger and smaller droplets, each carrying a different kind of charge, (ii) surface and volume interactions of ice particles, liquid water and even water vapour, all within the cloud, (iii) selective capture of ions (that is charged molecules of oxygen, water vapour, etc.) by water drops and ice particles. Most of these theories have some experimental support, but the relative importance of these mechanisms is as yet uncertain.

On the average, a thunderstorm cloud complex is twenty to fifty kilometres across. The downdraughts on striking the ground are deflected horizontally and manifest themselves as strong gusty winds. Such gusty winds may attain speeds of sixty kilometres per hour, and occasionally as much as a hundred kilometres per hour. They cause much damage to property and occasionally to life also—particularly to passenger and cargo boats on rivers. In Bengal the violent thunderstorms of April and May are called *kal-baisakhis*, 'the destroyers in the month of Baisakh'.

Thunderstorms usually develop during those summer afternoons when there is an abundant supply of water vapour in the lowest one or two kilometres of the atmosphere with comparatively dry air aloft. Over northwestern India and Kerala, ten to fifteen thunderstorms occur during each of the two months April and May, the pre-monsoon months, and a similar number during each of the two months September and October, the post-monsoon months. Over the rest of our area the frequency is less, about half-a-dozen during each of the hot months and one or two during each of the remaining months.

The highest frequency of thunderstorms for India occurs over Assam and Kerala, about sixty thunderstorms per annum. The most thundery part of the world is the island of Java—with about two hundred and twenty thunderstorms per annum.

It has been estimated that there are more than 40,000 thunderstorms occurring somewhere or other over the earth on each day; statistics indicate that there is a total of about one hundred lightning discharges each second.

TICKS and MITES. Ticks and their close allies Mites are more often than not mistaken for small spiders because they possess eight legs. They differ from spiders in structure and habits. Their body is round or an oval mass—the cephalothorax and the abdomen being completely fused without any distinct waist. Most of the members are microscopic forms parasitic on plants and living animals. Consistent with this mode of life the body assumes a round or oval shape—all belly without a distinct head or eyes: the mouth parts are modified for piercing and sucking. Their legs have two hooked claws, between which there may be suckers. With these claws they fasten upon the skin of animals so firmly that one can scarcely detach them without breaking their legs, as anyone who has de-ticked a dog knows. After holding fast to the skin, the ticks apply their piercing mouth-part to the body of the host and acquire a very considerable size by suction and their bellies get distended with blood like a blown-up balloon.

Ticks and mites are distributed all the world over. Unlike spiders they are detrimental to man and his economy. They are destructive to cultivation, cattle and poultry. They cause and transmit skin diseases and fever.

The so-called 'Harvest Bug' is a mite called *Sarcoptinae* which enters the human skin and causes itch or scabies.

The common 'Red Spider', smaller than a pinhead in size, found scurrying over pages of old books, decaying leaves, debris, etc. is not a spider but a tick, *Acarus telarius*, causing much damage to citrus and other plants.

Members of the genus *Argas*, found throughout the world, are always associated with fowls, transmitting a kind of disease. They are the carrier of 'tick-fever' in human beings. The Bont Tick, *Amblyomma*, causes a disease called 'heartwater' among cattle. All other mites and ticks are also harmful to man in one way or another.

TIGER (*Panthera tigris*). Half a century ago it was estimated that the subcontinent of India contained a population of perhaps 40,000 tigers. Unlike the lion which prefers savanna and open country, the tiger is a secretive animal depending upon extensive forests for its existence. Due to destruction of forests in India by 1972 a census then indicated that the total population had shrunk to probably less than 1800 animals. Throughout the latter part of the 19th century and the first half of the 20th, tigers have also been mercilessly hunted; to bag a tiger was a status symbol and indeed one Maharaja is reputed to have shot 1157 tigers during his lifetime.

In 1972 at the initiative of the World Wildlife Fund, the Government of India launched a massive programme to save the tiger. 18 million dollars were collected by public subscription from all over the world and the Conservation Project was initiated with special emphasis upon creating and managing nine reserves or national parks which contained remnant populations of this magnificent beast. It is hoped that the concern aroused and the action initiated will help to preserve this most spectacular representative of the country's fauna.

There are eight subspecies of tigers found mainly in underdeveloped and overpopulated countries, and comparative statistics over the decade 1970–79 indicate a favourable population trend in only three, namely the Siberian (*P. tigris altaica*), the nominate Indian (*P. tigris tigris*) and the Sumatran (*P. tigris sumatrae*). Of the remaining, the Chinese (*P. tigris amoyensis*), hitherto treated as an agricultural pest, is a doubtful population in spite of assurances by the Chinese. The remaining population of the Indo-Chinese (*P. tigris corbetti*), though officially listed at numbering 2000, is probably less after the napalm and defoliant bombing of the Vietnam war. The Caspian (*P. tigris virgata*) and the Balinese (*P. tigris balica*) are extinct, while the Javan (*P. tigris sondaica*) can be deemed to have lost the ability to survive. In India nine project areas containing 14 percent of India's tigers were launched in April 1973 for a period of six years, after which it was hoped that we would leapfrog to other problem areas.

The tiger population has varied in inverse proportion to the human increase. Dunbar Brander, a prominent forester-zoologist at the turn of the century, considered that at one time human survival was at stake in certain parts of India because of the depredations by marauding tigers. As we prepare to enter the 21st century with a conservative human population estimate of one billion, the pendulum seems to have swung the other way, and unless we can increase the size of tiger habitat areas and preserve their sanctity against intrusion, the process of attrition caused by genetic failures may not allow the tiger to enter the next century in spite of our good intentions.

A symptom of the present-day problem is the increasing incidence of the man-eater. When the tiger finds his habitat constricted by commercial and other biotic conditions, and his prey species exterminated by poachers and crop protectionists, he may take to preying

© Cyrus Adenwalla

Tiger × ¹/₁₂

upon humans. For the tiger does not fear man, though he shuns the unfamiliar human in the forest. But when by intrusion into his living space humans become familiar to tigers, they could well come to be accepted as a prey species. The problem then becomes a socio-ecological one.

The estimate of the IUCN scientists calls for a contiguous population of 300 to maintain a gene-pool in perpetuity, but such areas are only available in Manas and the Sundarbans, and as this magnificent predator fights his last battle we should be realistic, and while separating a Wildlife Service from the Forest Service, we should create Biosphere Reserves of the maximum area possible free from human intrusion to preserve pristine nature for future generations. For it has rightly been said that we have not inherited the universe from our parents, we have borrowed it from our children.

A great deal of literature by sport killers over the last two centuries has been devoted to the tiger, but these are chiefly accounts of their destruction, whereas behavioural and other observations are incidental, and it is surprising how little authentic information has emerged, for it relates to the unnatural reactions of hunted and harassed animals. The tiger is depicted as a savage and solitary beast always ready to take the offensive, and requiring little incentive to turn into an inveterate and congenital man-eater. It is only the recent observations of scientists and conservationists that have established that this powerful carnivore is a tolerant animal in areas where prey species are readily available, and subject to tiger protocol, whereby the killer feeds first, and when other feeders are not crowded by more powerful animals, any number of tigers can utilize the same kill. Communal hunting also takes place in favourable circumstances, and in areas of prey scarcity a tiger will readily share his kill with a tigress and cubs, in spite of the accepted premiss that the father is a menace to the young. In fact it may not be an exaggeration to state that a tiger's essentially solitary status is due to environmental conditions, and a restricted availability of suitable prey species rather than an intolerant temperament. It has been my experience that in the solitary cats the mother will give way to the feeding needs of her cub, and even carry prey to them, whereas in the communal feeding of the social lion the dictates of a power hierarchy ensure that the weaker may starve in times of scarcity. A condition which also pertains in human society.

Tigers, even males, are not rigidly territorial, and scent markings, apart from being a means of communication between the sexes, have a spacing value rather than a territorial intent. Tigers are normally silent animals and vocalizations, especially those of males, are another way of ensuring that chance encounters do not take place.

According to the prevailing system of social interaction the male and female come together for a brief period of courtship and mating which normally lasts three days and nights. The female tiger ovulates almost continuously when in estrus and copulations have been monitored as taking place every 21 minutes till conception takes place. Thereafter the pair separate, and the only stable relation is that of mother and cubs, which is thought to last two years. Though eight foetuses have been found, the average litter size is from two to four cubs. It is difficult to lay down a survival rate for the cubs as mortality is an imponderable factor. Certainly the optimistic population estimates postulated for Tiger Project areas do not appear to be justified and it is noteworthy that Schaller only calculated a 55 percent per annum increase for the lion under the very favourable conditions existing within the core area of the Serengeti National Park. The solitary tiger does not have the advantage of social security (in the shape of pride protection), and must be assumed to have a less favourable rate of increase.

The record length of a tiger is recorded at 11 ft 1 inch, but as the difference of measurement between pegs and over curves could be 5 to 7 inches, and trophy hunters normally used the more inaccurate method of measuring over curves, this record cannot be accepted in its entirety. The record weight is that of 705 lb for a Nepalese tiger, but as this figure has never been subsequently approached, it must be presumed to be that of an overfed tiger. Normally the Indian tiger measures between 9 and 9 ft 6 in, and weighs between 400 and 450 lb. The female is about a foot shorter and 100 or 150 lb less in weight. Siberians are supposed to weigh in excess of 700 lb, and have a maximum length of over 13 ft. The southern races are darker in colour and may have a maximum length of 9 ft. Paucity in numbers make it necessary to rely on old records.

The armament of the tiger is considerable, and the principal weapons of offence are four canines. Those in the upper jaw are larger and may measure up to three inches. In addition over half is encased in the jawbone. Two-inch retractile claws are used for holding prey, and a set of three carnassial teeth on each side is used for tearing and cutting flesh, for having no lateral movement the meat is swallowed whole. A set of incisors, and a tongue like a rasp, completes the essential equipment. Their life-span is short and 15 to 20 years can be considered the maximum.

A.S.

See plate 18 facing p. 241.

TIME-ENERGY BUDGETS. Every animal has a large number of different activities to perform and often a limited amount of time and energy to allocate to the different activities. An adult male bonnet monkey, for example, must eat enough to obtain sufficient nutrition to maintain itself, it must continually remain alert to any signs of danger, it must display and engage in fights with other males to establish its position in the social hierarchy to gain access to females, it must court and copulate with the females, it must groom itself and those higher up in the hierarchy and so forth.

The study of time-energy budgets has three aspects to it. The first involves making an exhaustive list of the various activities an animal performs and estimating the proportions of time spent in each of these. In the second phase we measure the amount of energy required to perform each activity. In the third phase one tries to relate the time-energy budgets to the ecology of the animal concerned and to understand how the time-energy budget of any animal has evolved through natural selection. The first phase of such a study has progressed a great deal and we have a number of examples of the time-budgets of animals. For instance, social insects such as ants, bees and wasps spend about one-third of their time sitting, one-third working and the remaining one-third patrolling or looking out for danger. Male orang-utans spend about 55 percent of their time feeding, 35 percent resting and 10 percent moving, while the female orang-utans spend 50, 35 and 15 percent respectively. Humming-birds devote 76–88 percent of their time sitting, 5–21 percent foraging, 0·5–1·8 percent fly-catching and 0·3–6·4 percent chasing other humming-birds from their territory. Feeding and connected activities occupy an overwhelmingly large fraction of the time of chital (80–90 percent), fighting (> 5 percent), displays, locomotion, grooming, play and sexual activities (< 5 percent). The exact allocation differs from one life-history stage to the other.

The second aspect, concerned with estimating the energetic costs of various activities, is still in a very primitive stage. The energetics of a number of animals has been studied by measuring the rate of oxygen consumption or by estimating the food consumed and excreted. Such studies yield information on the energy spent during rest and during some basic patterns of locomotion but we are far from estimating the energetic costs of such activities as sparring in deer. We know, for example, that a 500 kg bull spends 8000 Kcal/day if it is simply lying and 50 Kcal more per hour if it is standing and in addition 12 Kcal each time it stands and reclines.

Relating time-energy budgets to the ecology of an animal is an interesting exercise. Consider, for example, the time spent in feeding by an elephant and by a tiger. An elephant spends 12 to 14 hours a day in feeding and related activities while a tiger spends only an hour or two. This is primarily because an elephant has to spend very little time searching for food while a tiger spends most of its time searching for a suitable item of prey and once such an item has been located, the time actually spent in feeding is very small. Consider again the contrast between amphibians and reptiles on the one hand and birds and mammals on the other. Amphibians and reptiles are cold-blooded animals that do not maintain a constant body-temperature and become quite inactive when the environmental temperature falls. They therefore need much less energy than a bird or mammal of similar size that expends a good deal of energy in simply maintaining its body-temperature at a constant level in the face of environmental fluctuations.

Like any other character of a living organism, its time-energy budget would also be expected to have been so evolved as to maximize its genetic fitness. This is an aspect that we are just beginning to understand. Birds, for example, are known to spend more time in scanning for predators when outside or at the periphery of a flock than when in a flock where they get the benefit of alarm-calls from flock mates.

R.G.

TIT. Tits or Titmice (family Paridae) are vivacious, mostly sparrow-sized birds and great fun to watch as they hang upside down and probe under the leaves of trees for insects. Their merry calls are an added attraction. Many species are found in India, the commonest of which is the Grey Tit (*Parus major*), a resident species found in openly wooded country almost everywhere. The Himalaya are particularly rich in tit species.

TORNADOES. A tornado is a violent revolving cylinder of air, a few hundred metres across, and about two kilometres in height. This cylinder is rendered visible through the condensation of the moisture present in the air as cloud droplets. Part of the condensation is due to a rapid lifting of the moist air by the vertical currents in the cylinder as in any other cloud, and part through expansion into the low pressure within the revolving cylinder of air.

Outside Miranda House, University of Delhi, after the tornado on the night of 17 March 1978.

The high speed of rotation, in the range of 200 kilometres per hour to as much as 800 kilometres per hour, is attained through a coming together of slowly revolving air over a wide area. The diameter of this fast-rotating visible cloud varies from two hundred to four hundred metres, but occasionally can be as much as a kilometre. The geographical distance covered by a single tornado during its lifetime of five to ten minutes, is usually three or four kilometres, although on rare occasions it can cover a distance of about a hundred kilometres. The speed of movement of a tornado along its path of destruction is in the range of 20 to 80 kilometres per hour.

Besides the horizontal velocity of rotation and movement along the ground, the air in the revolving cloud has a large vertical velocity also. The air really has an upward spiralling motion. Estimates based on the weight of the objects lifted suggest upward velocities of 100–200 kilometres per hour.

The pressure inside the tornado cloud is very low; it can be as much as 30 percent less than the atmospheric pressure prevailing outside the tornado cloud. This decrease of pressure as the tornado cloud passes over a building has a devastating effect. The outward force exerted by the air inside a building over which a tornado passes can be as much as three tons per square metre of the ceiling and the side walls. It is a suction effect of the tornado. If the tornado grazes one side of a wall, the wall will be pushed sideways into the cloud.

The tornado passes over a particular object or a building in a minute or two. Within this short time, fantastic destruction is inflicted, buildings are destroyed. Individuals and vehicles on its path are lifted up by the vertical currents and left to fall to the ground safely or perched on tree tops or buildings. Trees are uprooted if directly on its path.

Tornadoes occur frequently in temperate latitudes, especially over regions where there are no high east–west mountain ranges. The USA alone reports about 200 tornadoes per year. They occur very rarely in our area in the hot weather and then only in the north. The tornadoes in Delhi and Orissa in 1978 were extremely rare events. The east–west alignment of the Great Himalayan Range acts as a deterrent to the formation of tornadoes in the subcontinent by preventing the inflow of very cold air from the north.

TRANSPIRATION. As for all living organisms, plants also need water for their existence. Terrestrial plants obtain their supply of water from the soil through the roots. The water is then translocated to the leaves and other parts through xylem tissue. While a plant absorbs considerable quantities of water, only a very small proportion is used up for its growth and metabolism. A very large proportion of the absorbed water just escapes in the form of water vapour. The process of loss of water from the aerial parts of the plant in the form of water vapour is known as 'transpiration'. This process differs from 'evaporation', which refers to the loss of water from any surface. Transpiration, thus, takes place from the living tissues of the plants and is controlled by them to a considerable extent.

Transpiration occurs mainly through the stomata (90 percent). The stomata are the microscopic pores which are interspersed in large numbers on the leaf surface. Stomata may also be present on the stem, petiole and even on the floral parts. The opening and closing of these pores is controlled by two specialized cells which border each pore and are called 'guard cells'. These guard cells swell when they take in water, thus opening the pore, and are influenced by the humidity of the air, the temperature, the water-content of the leaf and the intensity of light. At night transpiration almost ceases. The pores are microscopic in size (3 to 12 millionths of a metre across, 3–12 μm, and 10 to 40 μm in length) even when fully open. The number of pores or stomata on the leaf may vary from 1000 to 60,000 per square centimetre, depending on the species. In most of the land plants, more stomata are present on the lower than on the upper surface of the leaf. However, stomata may be present on both sides. In some water plants more stomata are found on the upper than on the lower side of the leaf.

In addition to transpiration through the stomata which is called 'stomal transpiration', some water is also lost in the form of water vapour directly from the surface of the leaf and even from the stem through lenticels. Lenticels are small openings in the corky tissue which covers the stem and twigs. The loss of water directly from the leaf surface is called 'cuticular transpiration' while that from the lenticels is called 'lenticular transpiration'. The loss of water by cuticular and lenticular transpiration is hardly about 10 percent of the water lost by stomatal transpiration.

The amount of water lost per unit area per unit time is called the 'rate of transpiration'. It is measured either with the help of instruments which are called potometers or by making use of some hygroscopic chemicals such as cobalt chloride and calcium chloride. A piece of blotting paper is dipped in cobalt chloride solution and then dried. The paper is blue when dry but turns pink and increases in weight when it absorbs moisture. The change in colour of the paper or increase in its weight is used as a measure of the rate of transpiration.

In order to prevent excessive loss of water, use is sometimes made of substances such as colourless plastics, silicon oils, low-viscosity wax in the form of a thin film or phenylmurcuric acetate as a foliar spray. These substances are called 'antitranspirants' and are used to economize water loss and to improve water-use efficiency of the plants.

The rate of transpiration is affected by light, humidity, temperature, wind velocity, availability of water in the soil, structural features of the plant and atmospheric pressure.

What role transpiration may be playing in the life of a plant is rather intriguing. On the one hand it is known that excessive loss of water by transpiration may cause water deficit and may result in wilting and even death of the plant. On the other hand, loss of water from the surface of the leaf sets in motion 'transpiration pull' which enables the water to rise even in very tall trees and also facilitates the uptake of some mineral salts along with the water stream. Transpiration may also prevent the plant from getting overheated due to direct sunlight and thus help in regulating its temperature.

Besides the loss of water vapour through stomata, cuticle and lenticels, in some plants like garden glory, balsam and grasses, water is lost in the form of droplets from the specialized structures at the leaf tips, known as hydathodes. The exudation of water droplets is called 'guttation'. However, it occurs only when the rate of absorption exceeds the rate of transpiration or when the plants are growing under conditions of high humidity.

K.K.N.

TRAPDOOR SPIDERS. There are several underground spiders of strictly nocturnal habits found under stones and hollows of trees. They fall under a special group called Mygalomorphs characterized by the formidable jaws directed forwards and the fangs, striking downwards like a pickaxe. Many of them build silk tubes, the mouths of which are in some cases closed by trapdoors. There are several of them in our area— they are of fairly large size, dark brown or sepia tinted, with stout hairy legs.

The one somewhat dangerous local species, *Poecilotheria* sp., is a mygalomorph spider. It is a large blackish,

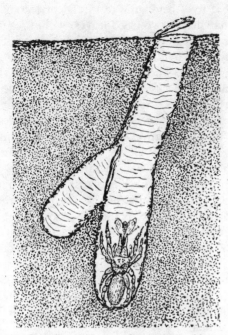

The silken tube-nest of a trapdoor spider. The spider, × 3/4, is at the bottom of the tube with a fly in its fangs.

hairy spider common in the hollows of jungle trees in South India. In Kerala it is called *Oorambuli* (Giant Hairy Spider), and in Tamil Nadu the Catleg Spider.

See also SPIDERS.

TREE SHREWS (family Tupaiidae). There are two species of tree shrews found in India. The Malay Tree Shrew (*Tupaia glis*) occurs in Assam, the Naga Hills and extending eastwards into the Himalayan foothills. It is also found in Burma and Malaysia. The Indian Tree Shrew (*Anathana ellioti*) is confined to the peninsula and southern tip of India. Thirteen other species are known, all confined to forested areas in southeast Asia.

In size and appearance these animals look like tree squirrels, with long bushy tails (15 to 20 cm in length) and the head and body of approximately equal length, but they differ from the squirrels in lacking any prominent long whiskers and having a long tapering pointed snout with moist rhinarium (nose-pad) like an insectivore. In colour the throat and chest of the Malay Tree Shrew is orange to buff whilst that of the Indian Tree Shrew is more greyish white. The rest of the body fur is grizzled blackish olive and ferruginous. Their exact phylogenetic relationship with other mammalian groups is still disputed. They show considerable anatomical differences from most insectivora except for the Elephant Shrews (Macroscelididae), and most zoologists prefer to classify them with primitive pro-simian ancestors of the primates.

In habits they are all diurnal, and though equipped with naked padded soles to their feet, and long and flexible digits with recurved claws, they spend as much time foraging upon the ground as in trees. Their diet consists partly of insects, soil-born crustacea and fruit, and probably birds' eggs and nestlings. All the Tree Shrews are forest-dwelling animals, typically associated with evergreen tropical rain-forest, though the Indian Tree Shrew occurs in dry deciduous forest also. They apparently can breed throughout the year and normally bear small litters of only 1 to 2 offspring, though the females sometimes have three pairs of mammae.

Though there are reports of females not tolerating each others' proximity and fighting any other Tree Shrews encountered, the writer has observed that the Malaysian Tree Shrew appears relatively unafraid of man and is remarkably active and agile, climbing up slender twigs, or probing and sniffing rapidly among the leaf litter and fallen branches of the forest floor.

TREE SNAKES. The snakes of genus *Dendrelaphis* are the typical thin-bodied, fast, harmless, diurnal snakes of India. These include the common bronzeback tree snakes (*D. tristis*) found throughout the country, green bronzebacks of the northeast and the Andamans, and five other species. Bronzebacks are generally shy and are usually only briefly seen as they dart up into a bush. They grow to about $1^{1}/_{2}$ m. Besides the flying snakes, the other well-known tree snakes are the harmless vine snakes, six species of long, slender, pointed-nosed snakes. The Common Vine Snake (*Ahaetulla nasuta*) is found throughout the subcontinent except the northwest, and grows to about 2 m. Other species are found in the hills and are all very well camouflaged. Whereas bronzebacks are conspicuous, fast and nervous, vine snakes blend in and move with a deliberate slowness. Both species feed largely on frogs and lizards, the larger vine snakes adding occasional birds to the menu.

Bronzebacks lay about 6 very elongated eggs. Vine snakes give birth to 6 or 8 living young. The sharp nose and bright green colour of the Vine Snake causes people to be afraid of it; in southern India it is falsely believed that it will poke your eyes out!

See also SNAKES, FLYING SNAKES.

© Rom and Zai Whitaker

Bronzeback tree snake × $^{1}/_{2}$

TREES. The subcontinent including Burma and Sri Lanka is the habitat of an immense variety of trees, bamboos, palms and tree-ferns: about 2000 species are known from this vast landmass. The plant life here is unique in the context of world flora. It has one of the richest flora in the world, with representatives of almost every family on the globe. This variety is due to the vastness of territory, embracing a range of climates and topography. There is every type of climate and habitat, from deserts—hot deserts of Sind and Rajasthan, and the cold deserts of Ladakh at 3600 to 5200 m—and the dry scrub of the Punjab to the tropical evergreen rain-forests of the Western Ghats and foothills of the Eastern Himalaya, and the coniferous and broad-leaved forests of the Himalaya. Sixteen FOREST TYPES are recognized in this area.

The Eastern Himalaya, adjacent hills and plains are some of the wettest areas in the world and consequently this is the most dramatic plant region on this planet. It has a climate which is typically tree-producing. Rainfall is also very heavy in the Western Ghats, where several trees common to the eastern region and Burma, e.g., Red Silk-Cotton, Champak, ironwood (*Mesua ferrea*), are found—an instance of disjunct distribution. In the Arunachal–Burma–China trijunction are located some of the deepest gorges on earth. Here we see conditions created by steep, uplifted mountains which could have created the first home of the forest trees of the Northern Hemisphere. The lower reaches here are regarded as the 'cradle of flowering plants'. It is a sanctuary of ancient and primitive flora and fauna which were spared the devastation of the ice ages. One such plant is a tree of the Magnolia family, *Magnolia pterocarpa* from Subansiri, which botanists consider as perhaps the most ancient species of living Angiosperms. The deciduous forests as a rule are valuable as they contain economically valuable trees. The tallest trees occur in the rain-forests, with some emergents standing head and shoulders above the rest, such as the Upas Tree up to 75 m. In the Thaungyin valley in Burma there is a forest with a mean height up to 60 m (200 ft), with a few Upas trees towering above it. These emergents have buttresses of 10 m or more up the trunk.

A tree by definition is a perennial plant with a single woody self-supporting stem or trunk capable of attaining at least 6 m height, usually unbranched for some distance above ground. There is a bewildering diversity that covers such widely different forms as conifers, broad-leaves, palms, bamboos and tree-ferns. In CONIFERS the leaves are needle-like and the reproductive structures are borne in cones—in form they are spire-like or pyramidal-shaped. Broad-leaved trees typically have leaves that are flattened and their reproductive structures are borne in flowers. Within the flowering plants broad-leaves are normally considered as being only those trees belonging to the Dicotyledons, that is those that have two seed-leaves within each seed. Another vital difference is that the female egg-containing body—the ovule—is enclosed in an ovary in all flowering plants, whereas in conifers it is not enclosed in an ovary but lies naked on the cone-scale. Apart from broad-leaves, tree-like forms also occur in the Monocotyledons which differ from broad-leaves in having typically fan-like leaves. Their main representatives are the PALMS which have unbranched trunks (except *Hyphaene*). Most have very large leaves which are few in number. Because of their giant size and often viciously spiny leaves, the palms are called 'the dinosaurs of the plant kingdom'.

Tree-ferns (see FERNS) have erect rhizomes, with generally unbranched trunks, topped by a crown of graceful feathery fronds forming a rosette at the apex. Reproduction is by minute spores borne on the back of fronds. They inhabit mist-enshrouded rain-forests of the E Himalaya and the montane forests of the Nilgiris and Sri Lanka.

The field naturalist can with practice recognize many trees by means of their field characters. They include bark features revealed by a kukri cut (a blaze) on the trunk, features of leaves, fruits, and 'spot characters', by which is meant the diagnostic characters of a species. For example:

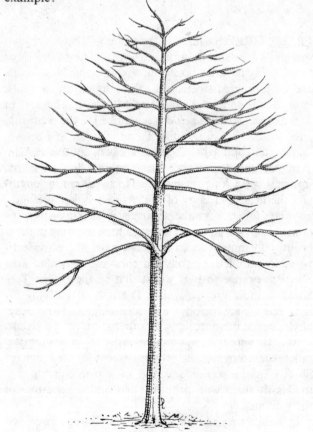

Whorled branches in Kadam

Mode of branching: Whorled branches are seen in Silk-Cottons, Kadam, drooping in Mast tree, dichotomous in Branching Palm, Screw-Pine. *Buttressed roots*: Silk-Cotton. *Stilt roots*: Screw-Pine. Mangroves.

Aerial roots: Banyan, *Spines* on trunk are confined to tropical trees and are generally absent in temperate trees—Silk-Cottons. *Bark* hard and resonant—White Dhup. Crocodile skin-like bark—Sandan. *Latex or*

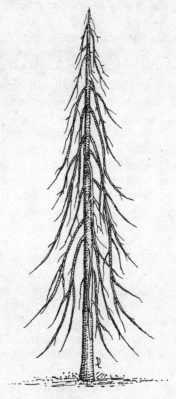

Drooping branches in Mast Tree

Dichotomous branches in Branching Palm, *Hyphaene dichotoma*

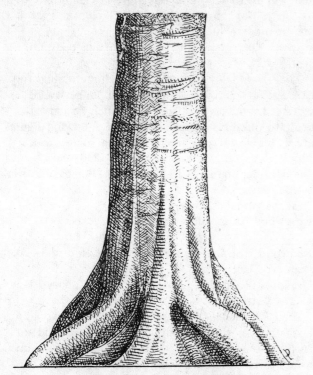

Buttressed root in Red Silk-cotton tree

Stilt roots in Screw-pine

Aerial roots in Banyan

sap: White latex turning muddy brown, Upas tree; grey turning black, Marking Nut, *Colour of slash or blaze*: Crimson marked with pink and white, Jhingan. Exudation of blood-red juice, Padauk. Blood-red streaks on a white ground, Sandan. *Leaves:* Very many parallel nerves, Poon, Poonspar. Young leaves purple or deep blue, *Memecylon.* Opposite, gland-dotted with prominent vein along leaf margin, Jaman. *Fruits/seeds*: Capsular fruits like woody figs, Crape Myrtles. Scarlet seeds with a black dot, Crab's-eye Creeper.

Flowering, fruiting, germination, growth: The spectacle of POLLINATION by wind, insects, birds and bats is dealt with elsewhere. Flowering is a prelude to fruiting; and just as a fruit is a ripened or mature ovary, so a seed is a matured fertilized ovule, containing the embryo plant. How they get dispersed is dealt with under SEED DISPERSAL.

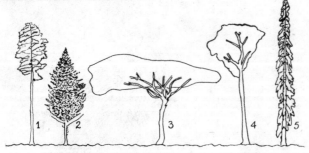

TYPICAL BOLE AND CROWN FORMS
1 Mature deodar. 2 Immature deodar in a rather open crop.
3 *Acacia planifrons.* 4 Gurjan. 5 *Abies pindrow.*
(Outlines traced from photographs)

When the seed germinates and the new plant begins to grow, several factors become operative. Soil, nutrients, moisture and temperature—all these are necessary if it is to survive. In the cotyledons, the seeds contain food reserves, which may be exhausted before successful establishment. Once established, the young tree grows: sal and oak seeds have considerable supplies of reserve food. The seedling, if kept moist, develops a root system and foliage. Unfavourable weather conditions may kill off seedlings. Growth of the seedling to the timber tree does not take place evenly but varies with conditions of temperature, moisture and food supplies. It is also linked with aspects of light—day-length and light-intensity. Reactivation of the cambium in spring is brought about through the production of hormones in the buds. There is an upsurge of activity in the apical meristems in the course of which stored food is used up and energy released, to attain cell-division. The reactivated vascular cambium loses no time in forming xylem and phloem which help transport of raw material from the root. There is intense chemical activity and PHOTOSYNTHESIS. As a result new leaves develop in the spring. A mature leaf has about 15 million cells.

Opening buds and new leaves are a refreshing sight. In the humid tropics a year's growth results in several metres of height growth (e.g. *Albizia falcataria*). In deserts, in some trees, the growth is very slow due to inadequate moisture and intense heat. The period of active height growth in sal and chir is 60 to 90 years. Chir growing in the open produces male and female cones in 10 to 15 years, and flowering continues after maturity. For purposes of timber, chir and deodar mature in 60 to 90 years, i.e., the period of active height growth. Thereafter, the height remains constant till death.

Longevity: Only for trees with definite annual rings can age at maturity and ultimate longevity be ascertained, and decay at the core often renders it difficult to obtain a direct count. The following figures are available for our trees including some exotics:
Pipal (*Ficus religiosa*), Sri Lanka—2200 years.
Deodar (section in F.R.I., Dehra Dun) Balcha, Garhwal—704 years.
Teak, Anaimalais—over 500 years.
Rosewood, Anamalais—about 600 years.
Bristle-cone Pine, N America—6000 years.
Redwood (*Sequoia sempervirens*), California—3000 years.
Dragon's Blood Tree (*Dracaena draco*), Canary Islands—2000 years.

Maximum height: On available records it would appear that in the subcontinent conifers considerably surpass broad-leaved trees in maximum height attained, as indicated by the following figures:
Deodar, Himachal Pradesh—73 m (240 ft)

E Himalayan Spruce (*Picea spinulosa*), Sikkim—67 m (227 ft)

W Himalayan Fir (*Abies pindrow*), Tehri Garhwal—65 m (215 ft)

Teak, S Malabar—58 m (192 ft)

Sal, Nepal—51 m (168 ft)

Blue gum, Nilgiris—71 m (234 ft)

Redwood (*Sequoia sempervirens*), California—120 m (400 ft)

Mountain Ash (*Eucalyptus regnans*), Australia—90 m (306 ft)

To S America goes the distinction of having a tree that grows at the highest altitude—*Polylepis* close to the snow-line in the equatorial Andes. *Betula utilis* grows up to 4570 m in the E Himalaya.

Rodents, insects and fungi destroy enormous quantities of seeds and timber. In addition, forest destruction has continued unabated during the last century due to a rising population and consequent demand for timber, fuel and competition for land to grow food crops. This is obvious to a passenger travelling by rail up the Gangetic plain and on into the Punjab and the North West Frontier in Pakistan. On his whole journey he will not see a vestige of a forest worthy of the name. Similarly, the traveller in parts of the Himalaya may see everywhere extensive views with range after range of hills completely denuded of tree growth, or with mere remnants of the original forest on the hilltops or in steep ravines. Nor is peninsular India much better off, and everywhere are to be seen hills not only denuded of trees which once covered them, but scoured of soil down to bed-rock and absolutely unproductive. Since the Stockholm Conference on Environment in 1972, in particular, and the spread of environmental consciousness, unplanned exploitation of trees has been halted to preserve our precious heritage of forests and wild life. Many National Parks and sanctuaries have been established in countries of the subcontinent. The modern forester is faced with a dilemma: the necessity to increase the sustained supply of wood from the forests as rapidly and as economically as possible and the need to protect the forest ecosystem to maintain all other benefits which are often not quantifiable in economic terms. His approach now is to achieve a balance between efficient production and safeguarding the environment, e.g., setting apart sizable natural areas in representative habitats, to preserve our ecosystems in pristine condition as bench-marks for conservation of the environment.

It is recognized that in addition to the production of wood, trees fulfil a variety of functions essential to Man's well-being. They protect the soil and so counter erosion, landslides, and reduce flooding; shelter and protect agriculture and influence local climate extremes; ensure clean water supplies and prevent pollution. In addition they provide a habitat for flora and fauna, a

© K.C. Sahni

A tree climber from foothill forests, Arunachal Pradesh. Note his cane cap with hornbill feather, also used as arrows with poisoned iron tips for hunting.

resource in itself. Many representative trees, ornamentals and interesting trees of our vast region have been described individually in this encyclopedia, such as the giant Gurjan of the Andamans, Flame Amherstia of Burma, Mangosteen from Sri Lanka, Ashoka, Durian, Neem, etc. Some have been treated under their families or groupings e.g., Bamboos, Palms, Conifers.

With every increase in population, more and more inroads are made into our forests, bringing in their wake soil erosion, floods, the advance of the deserts, and the denudation of the mountains. The world is faced with the problem of feeding a population of 7 or 8 billion by the year 2000. In India alone it is likely to touch about 913 million by the turn of the century. The world United Nations bodies such as the Food and Agricultural Organization, the World Bank, and the United Nations Development Programme are gearing themselves to meet this onslaught. Forestry has been integrated with Agriculture under the double-barrelled name Agro-Forestry, with the objective of raising massive plantations of food, fodder, fuel, fibre, fertilizer and fast-growing trees to provide a global tree cover.

To meet the threat of the population explosion and pollution of the environment, trees tolerant to pollution, fast-growing leguminous trees that are fodder resources of the future and help biological nitrogen-fixation are being identified and screened by botanists and foresters. In the future the thrust has to be on legumes (e.g. tamarind), conifers, eucalypts, casuarina, bamboos, etc. Air pollutants are reaching concentrations that are harmful to flora, fauna and humans. The production of energy by the burning of fossil fuels and their use to smelt metals have increased the concentration of air pollutants in the atmosphere. Trees which will flourish among the petrol and diesel fumes of city streets or in industrial areas are being identified and include many widespread trees such as mango, jaman, rain tree, poplar, chinar, sissoo and chir pine. Such trees act as biological filters or lungs and help to cleanse and cool the atmosphere. Trees are like Oxygen Banks, the Redwoods of California being the largest oxygen banks in the world. Legumes, conifers, eucalypts, casuarina, and bamboos are some of the important tree-groups that hold promise for quick afforestation.

Legumes help biological nitrogen-fixation and thus fertilize the soil. They are good fodders, fuels, ornamentals, luxury timbers and are fast-growing. As commercial nitrogen fertilizers are extremely expensive for our farmers, legumes are of special significance. Subabul (*Leucaena leucocephala*), an outstanding legume from Hawaii with high protein content, holds great promise and is being introduced on a vast scale. It has all the qualities of an ideal tree, providing fodder, fuel, fibre, fertilizer, and is fast-growing. *Albizia falcataria* of Indonesia—a 'miracle tree' introduced in Kerala, is recorded to have attained 7 m in one year and 13 to 18 m in three years, and is the fastest growing tree in the world. It is a promising tree for reforesting idle and denuded hill lands and is suitable as fuel and matchwood timber. Conifers have cylindrical straight stems, soft and easily worked wood, are useful in construction and produce the long fibres needed in the paper industry. Massive plantations of indigenous and exotic fast-growing tropical pines particularly from Mexico are under way in the subcontinent, including the hills of peninsular India. Conifers have an edge over eucalypts, being more useful. This is evidenced by the fact that eucalypts are no longer raised as plantation trees in their native Australia and New Zealand where Radiata Pine from California has been planted on a mass scale and is now the commonest tree in New Zealand. Eucacalypts will continue to be very useful in our semi-arid and low-rainfall areas where pines cannot succeed. Casuarina grows wild in the littoral zone of the Andamans and Nicobars. It is fast-growing, an excellent fuel and fertilizer tree, being the principal fuelwood on the Madras coast. During the last three decades large plantations have come up along the vast coastline.

Bamboos are of great importance in the economy of East Asia and are of outstanding value in the major wood-based industry of the subcontinent, viz., pulp and paper. Their strength, abundance, ease of propagation and fast growth make them invaluable for a range of uses. The Giant Bamboo of Burma (*Dendrocalamus giganteus*), the world's largest bamboo, is recorded to have grown 57 cm in one day.

For quick afforestation and to augment the depleted timber resources, scientists are now turning to vegetative propagation and tissue-culture techniques. Auxins and other growth-promotion hormones are being used in glasshouses under the best conditions of humidity and temperature to promote the rooting of cuttings of forest trees previously unknown to root by vegetati: e means, thus saving about three years spent on germination from seed to sapling stage. Under tissue-culture techniques scientists are producing test-tube trees in thousands. They take a tiny portion of a tree, put it in a test-tube with different chemicals and there it grows, leaves, roots and all. The plantlet is transferred to a nursery when it has grown to a viable size. By such techniques timber production has vastly accelerated. In addition seedlings are being scattered by helicopters on productive sites. To meet the challenges of the energy crisis, energy-producing forests of pine, eucalypts, casuarina, etc. have to be planted on a war footing in order to be less dependent on petroleum. The world's oil is fast dwindling but its trees remain a natural renewable resource of incomparable value.

K.C.S.

TRINKET SNAKES. There are ten species of trinket snakes in our area, the only well known one being the Common Trinket Snake (*Elaphe helena*). This is a harmless snake that grows to 1½ m. It is a bird- and rodent-eater that constricts its prey. Two of its close

© Rom and Zai Whitaker

Trinket snake × ½

New World relatives, the Red and Yellow Ratsnakes, are considered to be important rodent controllers in the southeastern United States. Common Trinkets are tan and chocolate brown with two prominent dark stripes on the latter half of the body. Trinkets are generally nocturnal but are sometimes seen moving on rainy or overcast days.

Trinkets lay 6 to 8 eggs and the young are tiny replicas of the adult. They are gentle snakes and quickly adjust to careful handling.

See also SNAKES.

TROGON. Trogons, of which we have three species in our region, are birds of the tropical forests. Their plumage is soft and ample, the wings short and rounded, and the tail long and graduated. Although they are brightly coloured birds they are often overlooked on account of their silent, sluggish and unobtrusive habits. The feet of trogons are unique among zygodactylic birds in that it is the second (inner) toe and not the fourth (outer) that is turned backwards. Trogons have very wide mouths like swifts and nightjars since their staple diet is flying insects. But they also take some fruits and berries. They nest in holes in forest trees.

TROUT as a group of Salmonid fishes, well known as excellent game fish, is not indigenous to India but was brought by British angling enthusiasts separately into Kashmir, Nilgiris, Kodai and Munnar range of Kerala. The Brown Trout (*Salmo trutta fario*) was first introduced into Kashmir waters in 1900 by importing eyed-eggs (eggs grown up to development of eyes) from Scotland as a present from the Duke of Bedford to the Maharaja of Kashmir. The first hatchery for breeding trout was established at Harwan near Srinagar and from there, after a few years, eyed-eggs were transplanted to many streams and lakes of Kashmir, Gilgit, Abbottabad, Chitral, Kangra, Kulu, Simla, Naini Tal and Shillong in the Indian subcontinent where suitable cold water prevailed. Another species, Rainbow Trout (*Salmo gairdneri irideus*) was introduced in 1912. Recently the American Brook Trout (*Salvelinus fontinalis*) from Canada, and a land-locked variety of Atlantic Salmon (*Salmo solar*) from north America, have also been transplanted into trout hatcheries in Kashmir.

Attempts to introduce Brown Trout in the Nilgiri hill streams were made several times after 1863 but did not succeed till importation of eggs and stock fish of Rainbow trout brought from Sri Lanka, Germany and New Zealand in 1909 proved fruitful. The same species has been thriving there ever since. In the Munnar High Range, introduction of trout succeeded only in 1932 and in the Kodai hills in 1943.

The culture of trout in India can be attributed mainly to its great popularity in angling, especially in fly-fishing. Trout normally inhabit cold waters ranging from 0°C to 20°C, though 10° to 12°C is considered optimum. In the Nilgiri hills the streams have a temperature range of 10° to 22°C at an altitude of about 2100 m, but the fish has been known to spread to lower altitudes of even 914 m in the same hill range, where the temperature is likely to exceed 22°C.

Normally trout is a fish of perennial mountain streams of clear cool water of high (above 6 p.p.m.) oxygen content and thrives in lakes also. As a habit it breeds in shallow slow-moving waters with gravelly bottom; but because of high mortality in natural streams and difficulty of collecting the young ones, the fish is usually bred artificially in hatcheries where ripe females are stripped of their eggs and fertilized by the semen of the males. After hatching of eggs and nurture of fry to fingerling stage on artificial feeds, the fingerlings are released into natural streams for further growth. Their natural food, and that of adults also, is usually small fish, shrimps, crabs, water-insect larvae and other aquatic organisms.

The trout, in general, has an elegant spindle-shaped body with medium sized dorsal fin and a small adipose fin. The colour varies according to species and subspecies. The Rainbow Trout is iridescent and blueish, with a broad purple longitudinal stripe along the middle part of the body and back and a few red spots mostly above the lateral line and on the tail whereas the Brown Trout is brownish blue and spotted all over. The Brown Trout in its original habitat (Western Europe) rarely grows above 80 to 100 cm and 10 to 15 kg in weight; while the Rainbow is at most 70 cm long and 7 kg in weight in its country of origin (America). The sizes depend on the quantity and quality of food and water. Both species mature after 3 or 4 years and produce 200 to 3000 eggs. In India they spawn in winter and the eggs hatch out in early spring after 25 to 50 days of incubation, depending on temperature and oxygen content of the waters. In cold countries they spawn after winter.

See also INDIAN TROUT.

TRUNK FISH (BOX FISH). These are diminutive relations of puffer fishes and are only about 10 cm long. The body of this fish is completely enclosed in hard, bony armour of hexagonal plates, leaving holes only for the mouth, tail and fins. In one species (*Ostracion*

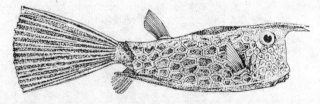

Ostracion cornutus × ¹/₂

cornutus), two long spines projecting forward from the forehead and looking like horns give it the name 'Cow fish'. As the body is not flexible, the fish swims by sculling movements of the dorsal and anal fins, with the tail acting as a rudder. The pectoral fins help to fan water into the gill-openings, and it is said that the fish can live outside water for 2 or 3 hours. These fishes manifest structurally very queer modifications of an otherwise agile and slippery fish tribe and consequently their movements are limited. Sometimes these fishes are washed ashore by strong tides, particularly on the western coast.

TUNA OR TUNNY, of the family Scombridae, are very widely distributed important fishes in the tropical and temperate regions of all the oceans of the world. Having a fusiform robust stature, they are powerful swimmers; every part of their body is streamlined: a bullet-shaped head, close-set jaws, the dorsal, pectoral and pelvic fins fitting into grooves and depressions to minimize resistance to water. Another unique feature of tunas is that their body temperature is higher than the surrounding water. Most of them are bright steel-blue on the back, becoming lighter silvery towards the belly. Some have stripes, some wavy markings and some spots on the sides. Similarly some species have long pectoral fins and some have short ones. Their migrations have been a matter of curiosity and speculation since ancient times but it has now been established that these rovers of the oceans migrate for breeding and feeding during certain seasons.

Several species of tuna visit the western coast of India during September to March when the smaller fishes such as sardines and mackerel also arrive in this area. Large species such as the Oceanic Skipjack, *Ketsunus pelamis* and Yellow-fin Tuna, *Thunnus albacora macropterus* grow to about a metre (weighing 25 kg) and 2 m (50 kg) respectively and are available near the off-shore islands of Lakshadweep. Smaller forms such as Frigate Mackerel (*Auxis* sp.), Bonito (*Sarda orientalis*), Little Tunny (*Euthynus affinis*) and the Northern Blue Fin (*Kishinoella* or *Thunnus tonggol*) approach coastal waters in search of live food. They disappear from the coastal waters at the end of the usual season. About

25,000 tonnes is annually landed in Indian ports, about a fifth of the entire Indian Ocean production.

Tuna, when canned, is very popular in Europe and the USA, where it is considered as the chicken of the sea.

TURTLES. About 220 species of turtles exist today. They inhabit land, the sea, rivers and bodies of fresh and brackish water in tropical and temperate regions. The Leatherback sea-turtle, largest of living turtles, has also been reported from arctic waters.

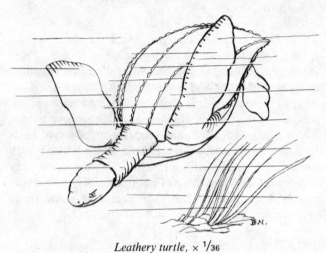

Leathery turtle, × ¹⁄₃₆

The term 'turtle' in this account will apply to any member—tortoises and terrapins included—of the order Testudines, one of the five orders of the class Reptilia.

The earliest true turtles which, unlike modern forms, possessed teeth in their palates, are known from the Triassic (about 200 million years ago). Of about 22 families then, twelve survive today. These are usually classified into two suborders—the more primitive Pleurodira, consisting of two families unrepresented in the Indian subcontinent, the members of which retract the head and neck sideways; and the Cryptodira, which do so in a vertical plane.

Six of the ten Cryptodire families are represented in the Indian region, here meaning the land and ocean south of the Himalaya, and extending westwards to Pakistan and eastwards to Burma, Thailand, Malaysia and Sumatra. They are the Emydidae, loosely, the fresh-water turtles; the Testudinidae, mainly land tortoises; the Platysternidae which has a lone species, the big-headed turtle; the Cheloniidae and Dermochelidae, both sea-turtle families; and the Trionychidae or soft-shelled turtles. Except for the last two families, in which leathery skin overlies the bones of the carapace (upper shell) and plastron (lower shell), turtles possess bony shells covered by horny shields or laminae. Typically, the larger of these shields number 13 on the carapace and 12 (six pairs) on the plastron. Those on the cara-

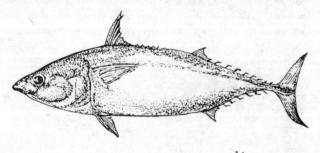

Little Tunny, *Euthynus affinis* × ¹⁄₅

Carapace of common terrapin × ¼

Plastron of common terrapin × ¼

pace are arranged in three rows: five vertebral shields along the midline, and rows of four costal shields on either side. Smaller shields called marginals, often numbering 12 on each side, usually with an anteriorly-placed nuchal shield, occur on the outer border, surrounding the vertebrals and costals. Carapace and plastron are often joined by a 'bridge' athwart the middle, leaving openings at the front for the head and forelegs and at the back for the hindlegs and tail. Most turtles can withdraw their head and limbs into the shell for protection, the exception being sea-turtles, which therefore occasionally lose limbs to predatory sharks.

Turtles have lungs and breathe air, but some aquatic species that have additional vascular organs near the throat and vent will also utilize dissolved oxygen directly from the water. Most turtles can remain for a long time without breathing, but even the most aquatic, the sea-turtles, can drown, as frequently happens when they are caught in trawl nets at sea.

All turtles are oviparous and lay eggs that are hatched by the sun's heat. Most species excavate a chamber in sand or mud to receive their eggs, others merely select a suitable crevice for this purpose. The body temperature of turtles, and therefore the level of activity, is largely determined by the ambient temperature. Hibernation in cold weather under mud and, in some sea-turtles, on the bed of the sea, occurs. Soft-shelled

turtles estivate under mud as their aquatic habitat dries up, and are set free with the onset of the monsoon.

Sex dimorphism may manifest itself as a concavity in the plastron of the male; size disparity between the sexes, the female often being the larger; a greater length and thickness in the tail of the male, with the opening of the vent being farther back than in the female.

The age of some turtles, especially land tortoises, can be reckoned by counting the annual growth of rings on individual shields. This method fails after the early years of its life, as the rings lose their definition.

About 55 kinds of turtles of 46 species representing 25 genera are found in the Indian region. The dearth of published literature subsequent to Malcolm Smith's classic *Fauna of British India* volume of 1931 may necessitate minor revisions in the account which follows.

The Emydidae

Half of India's turtle species are Emydids, distinguishable from the tortoises (family Testudinidae) by their more or less flattened limbs and webbed digits, as opposed to the relatively cylindrical limbs and separated digits of the tortoises.

Six species of the genus *Melanochelys* occur here. The commonest is *M. trijuga*, the Hardshelled terrapin, which has six races itself. At nightfall, particularly in wet weather and surroundings, these turtles often leave

Common terrapin × ⅕

Common terrapin hatchling × 1

569

the ponds or tanks which are their usual habitats by day, in order to forage on land. Plant matter is the food preferred. If handled when freshly caught, *M. trijuga* may produce a disagreeable-smelling scent-gland secretion; the odour may also serve as a sexual attractant— both males and females possess the glands. Turtles have a well-developed sense of smell.

In the genus *Cyclemys*, the rear margin of the shell has tooth-like projections (serrations). Two species, *C. mouhoti* and *C. dentata*, are found here. Both have an indistinct hinge across the undershell, thus enabling them to move the forepart of the plastron rather like a trapdoor.

Cuora amboinensis, the Malayan box turtle, which is common in ponds and swampy areas in southeast Asia, was first recorded in India as lately as 1979, at Great Nicobar and Trinkat islands. A well-defined hinge separates its undershell into two movable lobes which can 'box up' the turtle protectively. On being handled, the timid animal may squirt from its vent a fluid (not urine) from its anal sacs, which are accessory organs of respiration.

The two eggs laid by a 21 cm-long individual at Trinkat island were typical of the eggs of the Emydidae: ellipsoidal, white and brittle; one measured 51×29 mm. In all other turtle families found in India, the eggs are almost spherical. Sea-turtles lay soft, thin-shelled, parchment-textured eggs. Some Emydids lay over 30 eggs per clutch.

Head of Batagur turtle × ¹/₂

Batagur turtle of Bengal × ¹/₈

Tent turtle, egg tooth of hatchling × 1

Hardella thurgii × ¹/₁₀

Six species of the genus *Kachuga* occur in our area. They inhabit the rivers of northern and eastern India, are predominantly herbivorous and are much relished as food, except for *K. tecta tentoria*, one of the two races of the conical-shelled Tent or Dura turtle: its flesh is reportedly mildly poisonous if eaten. The more colourful species of *Kachuga*, being popular abroad as pets, were among those that were exported from India to Europe, the U.K. and the U.S.A. until the recent enactment of protective legislation.

K. kachuga lays its eggs in sand on the banks of the Ganga. The males apparently develop a red coloration on the top of the head, perhaps seasonally.

Batagur baska, an estuarine turtle whose shell reaches 590 mm in length, is probably the largest of the Emydidae. During the last century it was reported to abound at the mouth of the Hooghly river and that large numbers were brought to Calcutta for sale and consumption. The male in certain seasons assumes striking colours.

At 500 mm in shell length, the female of the riverine species *Hardella thurgi*, or Thurg's batagur, reportedly

Forest cane turtle × ²/₃

Travancore tortoise × ¹/₄

grows to a length almost three times that of the male. The turtle's flesh is esteemed in Bengal.

The carnivorous *Geoclemys hamiltoni* is elegantly spotted with yellow on its dark shell and soft parts. Its fully webbed digits proclaim the turtle's aquatic disposition.

Other emydids occurring to the east of India (i.e. in Bangladesh, Burma, Thailand, Malaysia or Sumatra) include the mollusc-eating *Malayemys subtrijuga; Hieremys annandalei*, individuals of which are deposited by local devotees at the Tortoise Temple in Bangkok (where, however, the turtles are starved); *Notochelys platynota*, a herbivorous, jungle-swamp-dwelling species that hisses like tortoises when alarmed; *Siebenrockiella crassicollis* which, as the last name implies, is thick-necked; and the two species of *Morenia*, one each from Bangladesh and Burma.

The Tortoises Family (Testudinidae)

Land tortoises are the most slow-moving of all the turtles, are predominantly herbivorous, and include the longest-lived among surviving backboned animals. One individual, probably from the Seychelles, and belonging to the genus *Geochelone*, was captured while an adult and lived 152 years in captivity before suffering an accidental death.

Geochelone is represented here by five species. *G. elegans*, the Starred tortoise, is the commonest and most strikingly patterned, each shield possessing yellow, ray-like streaks on a dark background. The turtle is most active during the rains, when it wanders all day, feeding and mating. To construct their underground egg-chamber, the females soften the earth to a workable, muddy consistency with water from their anal sacs. The eggs (about four in number) having been deposited, the turtle fills in the excavation and flattens and camouflages the nest-site by rising on all fours and repeatedly slamming the earth with her plastron.

In the dry hill regions of Baluchistan and even up to elevations of 3000 m the Four-toed or Afghan tortoise (*Testudo horsfieldi*) is still relatively common. The ex-

Starred tortoise × ¹/₅

tinct *Colossochelys atlas* of the Siwalik Hills had a shell over two metres long and is the most massive non-marine turtle known to science.

The Platysternidae

The family is represented by a single species, *Platysternon megacephalum*, the Big-headed turtle of Burma, Thailand and southeast Asia. It is incapable of withdrawing its large head under the shell. The tail is long, the claws sharp. Individuals of this species may climb trees and rocks.

The Soft-Shelled Turtles (Family Trionychidae)

The adults of most soft-shelled species found in our area, except those of the genus *Lissemys*, are capable of inflicting dangerous bites. Their long, flexible necks give access to most parts of their bodies; they can be held with safety only on the margin of the shell near the hindlegs.

The innocuous *Lissemys punctata*, the Flapshell turtle, can fast for extended periods, like many other reptiles.

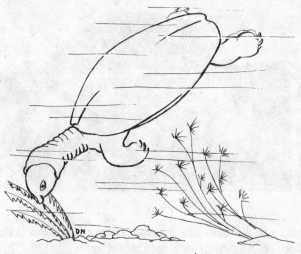

Soft-shelled turtle × ¹/₃

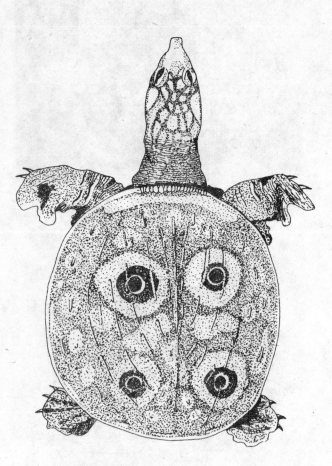

Peacock soft-shelled juvenile turtle × ¹/₂

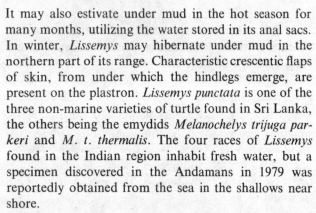

© Rom and Zai Whitaker

Common soft-shelled turtle × ¹/₄

It may also estivate under mud in the hot season for many months, utilizing the water stored in its anal sacs. In winter, *Lissemys* may hibernate under mud in the northern part of its range. Characteristic crescentic flaps of skin, from under which the hindlegs emerge, are present on the plastron. *Lissemys punctata* is one of the three non-marine varieties of turtle found in Sri Lanka, the others being the emydids *Melanochelys trijuga parkeri* and *M. t. thermalis*. The four races of *Lissemys* found in the Indian region inhabit fresh water, but a specimen discovered in the Andamans in 1979 was reportedly obtained from the sea in the shallows near shore.

The remaining soft-shelled turtles of our area, except *Dogania subplana* of Malaysia and Borneo, are large and riverine. These belong to the genera *Pelochelys*, *Trionyx* and *Chitra*. The upper shell of *Chitra indica* reaches a length of 800 mm but a 1914 report of two-metre-sized individuals cannot be discounted. These turtles were said to attack boats with blows.

The big soft-shell turtles of the genus *Trionyx* will bury themselves in the mud of their aquatic habitat, only the head and a part of the back being exposed. They catch prey—fish, frogs, molluscs, etc.—by suddenly extending their long neck, and sometimes by active swimming. Six species are known from our area.

One of the ten forms of the Lord Vishnu which is worshipped by Hindus is his turtle incarnation, Kurma Avtar, often depicted in carvings and paintings as a Trionychid. In parts of India, Bangladesh and Burma living soft-shelled turtles are venerated and fed in tanks near religious shrines and multiply there. One species, *Trionyx nigricans*, is known only from one such tank, near Chittagong.

The Sea-Turtles (Families Cheloniidae and Dermochelidae).

Sea-turtles possess paddle-shaped limbs (flippers) that may span 12 ft in *Dermochelys coriacea*, the leatherback. They come ashore voluntarily only briefly: to nest and, very rarely, to bask. The nesting female laboriously hauls herself on to a sandy beach where the eggs, usually 50 to 200 in number, are deposited in a roughly half-metre-deep chamber she excavates. Viscous tears that she sheds remove excess salt derived from the turtle's ability to drink sea water—a marine adaptation.

The nest site is levelled or camouflaged by the turtle which then returns to the sea where she may mate again. The female takes no further interest in her eggs or young, which hatch roughly two months later. Up to 8 clutches—representing a total of about 700 golf-ball-sized eggs—may be laid at 2-week intervals between clutches during a nesting year, but most species nest only every two or three years. Incubation temperatures that are too low (24–26 °C) or too high (30–32 °C) will, respectively, slow down or speed up hatching, and may result in all-male hatchlings (in the former case) or all-female (in the latter). Intermediate temperatures produce both males and females.

The hatchlings take two or more days to dig their way out to the sand surface; they emerge in a group usually at night, thus evading avian predators and lethal surface-sand temperatures. Braving ghost crabs and other enemies, the hatchlings rush to the sea, which they usually locate unerringly, even if invisible, by scampering towards the brightest horizon—a tendency which may also draw hatchlings landward to their doom at beaches such as Marina in Madras where disorienting lights exist. Predatory fish take a heavy toll at sea. The list of animals that prey on sea-turtle eggs or young reads like a representative cross-section from a *Who's Who* of scavengers and predators: dog, jackal, monitor lizards, wild and domestic pigs, leopard, hyena, estuarine crocodile, ghost crabs, hermit crabs, coconut crab, ants, rats, seagulls, crows, sharks and many more. Less than 0·1% of the hatchlings may survive to adulthood in nature. Yet it is mainly man's activities that have decimated many sea-turtle populations and necessitated urgent international action to reverse the trend towards extinction. The depletory activities include unrestricted egg collection; the killing of adults for commerce in meat or skin ('tortoiseshell', the beautiful horny shell-material, valuable in the curio industry, is obtained from the Hawksbill sea-turtle, *Eretmochelys imbricata*); commerce in calipee, the cartilaginous tissue which is the main ingredient of widely-relished turtle soup; the accidental drowning of sea-turtles in trawl nets; and land development that destroys nesting habitats.

Some species, including *Chelonia mydas* the Green turtle (named after its greenish fat), navigate with precision across hundreds of miles of open ocean between specific feeding grounds and breeding areas.

Sea-turtles are believed to nest invariably on, or near, the beach where they hatched. Renesting often occurs at sites lying within yards of nests the turtle had made six or more years earlier.

Of seven surviving species of sea-turtles, five, representing each of five existing genera, are known from the Indian Ocean. All except the aggressive Loggerhead turtle *Caretta caretta*, also nest along our shores, usually on sparsely-inhabited coasts and islands, where predator

Ridley laying eggs, 4 cm diameter

Ridley hatching

Ridley babies for release from the Madras Crocodile Bank hatchery × ¹/₇

pressure may be more sustainable. The Olive Ridley sea-turtle (*Lepidochelys olivacea*) nests on most of the mainland shores of India, but exceptionally vast numbers, 'arribadas' (Spanish for 'the Coming') nest each February in the Bhitar Kanika Wildlife Sanctuary in Orissa. The spectacular yearly arribada is one of the largest in the world, with over 200,000 nesters utilizing a 10 km stretch of beach in a period of about two weeks during which the nesting density is such that turtles accidentally dig up each others' eggs while laying their own.

The nesting of the Green Turtle (*Chelonia mydas*) in the environs of the modern-day Karachi city is another unique spectacle. Despite the throngs of people who enjoy the beaches at weekends, and have constructed concrete beach huts on the shore line, many hundreds still come ashore to lay their eggs between the months of July and November. The fact that competition for nest sites results in one female sometimes digging up a previously completed egg clutch probably indicates that there has not been much diminution in the local breeding population. The Sind Government in collaboration with the WWF has set up a protected turtle-hatching zone and the Sind Wildlife Management Board posts wardens along the beaches to protect this endangered species.

Females of at least some sea-turtle species have the ability to store viable sperm within their bodies for years. Copulation occurs in the sea near the nesting beach; to discourage males, which are promiscuous, the female may assume a 'refusal position': with body

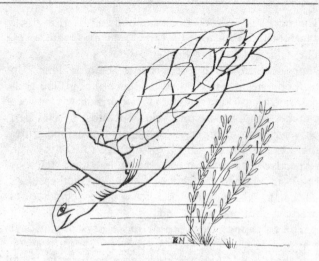

Hawksbill turtle × ⅙

held vertical in the water with the limbs outstretched, she faces the male.

Sea-turtles being large animals (the leatherback grows to over 2·5 m and 700 kg), and capable of forming

A Bengal spotted pond turtle market

Green sea turtles illegally slaughtered × ¹⁄₁₈

© Rom and Zai Whitaker

Young Hawksbill turtle

Turtles have survived and have changed little over the millennia during which many other reptile species have become extinct. It would therefore be a loss, aesthetically and economically speaking, were they to disappear through the agency of man who, since the sixteenth century, has often decimated and, in some instances, wiped out turtle populations the world over.

S.B.

See also REPTILES.
See plate 16 facing p. 193.

large populations if given the chance, are valuable as sources of tortoiseshell, oil, skin, eggs and meat (thousands are consumed annually in West Bengal and Tamil Nadu). The flesh of the Hawksbill turtle may be seasonally poisonous if consumed, a result of its diet: deaths occur intermittently among the coastal Tamilians who eat it.

TURUMTI or REDHEADED MERLIN (family Falconidae). An extremely active little falcon, mostly seen hunting in pairs. Distinguished by red head, longish tail, and blackbarred grey upperparts. Found in openly wooded country from Central Iran and southern Afghanistan to the Deccan. It is perhaps the only falcon known to build its own nest; other 'true' falcons only utilize disused nests of other birds. Feeds on rodents, birds, locusts and other insect pests. In the heyday of falconry was known as 'the Lady's hawk'.
See FALCONRY.

UTRASUM BEAD TREE, *Elaeocarpus sphaericus*, occurs in the evergreen forests of Nepal, Bihar, Bangladesh and Assam. It has oblong lance-shaped leaves, fragrant white drooping flowers with wedge-shaped petals. One of the anther lobes is tipped with 1 or 2 white bristles. The stones from the fruits of *Rudraksh*, as it is called in Hindi, are strung and made into rosaries and worn by priests. They are elegantly tubercled, marked with 5 vertical grooves, and when set in gold are made into necklaces. Freaks with fewer or more than 5 grooves fetch a fancy price.

UNCONFORMITIES. The existing rocks on the earth's crust are broken down by the mechanical and chemical action of water, ice, wind, and daily variations in the temperature. The loose fragments of varying sizes thus available are transported by moving water, ice, and wind and deposited into a basin. This process of deposition is known as sedimentation, and accumulated materials in the basin are called sediments. By hardening

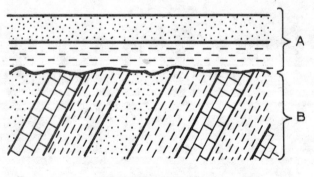

AN UNCONFORMITY

A Younger strata B Older strata

and consolidation of these sediments sedimentary rocks are formed.

In the geological history of a region the process of sedimentation may be continuous for some time. After which, for some reason or the other, there may be absence of deposition of sediments. Once again the sediments may start accumulating over the surfaces of the earlier sediments. The contact or the surface between the two sets of sediments or sedimentary rocks is known as an unconformity. The presence of an unconformity essentially indicates a period of non-deposition. During this period of non-deposition the earlier or older strata may be uplifted and subjected to the processes of weathering and erosion. During this uplift the strata may be tilted or even folded. Over these deformed formations a series of new and younger strata may be deposited. If the strata of a region represent the geological history of that place then the presence of an unconformity indicates that the geological record is missing for a period of time.

Unconformities are usually formed due to earth movements and they indicate the periods of mountain-building activity. Based on the study of unconformities the Earth's history has been divided into Eras, Periods, Epochs, and Ages. Eparchaean Unconformity between rocks of the Archaean era and younger rocks may be found in many places of peninsular India.

Some unconformities may be recognized while others require a detailed study of the rocks involved. If the rocks contain fossils their study will be of immense help in determining the period of unconformity. Much care has to be taken to distinguish unconformities from faults. Unconformities are useful in the search for mineral deposits. They also serve as reservoirs for oil and gas.

C.V.R.K.P.

UNGULATES or HOOFED MAMMALS. The term 'ungulates' is used here in the widest sense to include the two great orders of even-toed or 'cloven hoofed' animals in the order Artiodactyla, and the odd-toed mammals in the Perissodactyla. The latter order from fossil evidence evolved before the even-toed mammals and is therefore of more ancient origin.

Generally speaking these two orders of mammals show an inter-relationship in the development of their limbs with a reduced number of toes or digits and more specialized adaptation towards exploiting bulky leaf diets. The third digit has become greatly elongated and strengthened in the Perissodactyla and forms the axis of the body upon which the animal walks. However there is some variation in that one Family or group developed a single functional toe or hoof—the horses (Equidae)—whilst the other two Families of odd-toed ungulates comprising the Tapiridae and Rhinoscerotidae have three functional toes on the hind feet and in the case of Tapirs, four in the front feet but only three in the front feet of Rhinos. In all cases the first digit has been lost as is the case with the even-toed ungulates. All these animals walk on their toes rather than the whole sole of the foot, and these toes are often entirely encased in a horny sheath or hoof (in reality a modified claw or nail) which has special shock-absorbent qualities. Walking on the toes gives greater leverage and hence mobility (compare how a human athlete runs or dances up and down only on the fore part of his feet). In the ungulates the wrist and ankle bones are usually fused together and form a strong supporting column called the cannon bone. These hooved mammals, from the fossil record, first began to develop about 2·5 million years ago in the early Pleistocene era, and it appears that with the development of abundant leafy vegetation they spread outwards to exploit this potential food source often in less humid regions and evolved a modified leg and foot structure to traverse over a wider area of ground rather than to grasp trees. Greater mobility, coupled with speed and endurance in gait, enabled them not only to escape predators, but to harvest bulkier, less nutritious food and to colonize drier zones of vegetation, away from the dense steamy tropical forests where there was already much competition from other life forms. Large size enabled them to cope with bulky food of low nutritive value and some of the early prehistoric forms were undoubtedly giants, such as the extinct Giant Deer (*Megalocerus*) or the Long-horned Bison (*Bison latifrons*), and the Giant Rhinoceros such as *Elasmotherium*. These forms have long since died out, possibly because their very ungainliness made them vulnerable to predators, but gradually there evolved ungulate types with more complicated digestive systems enabling them to efficiently exploit such high fibrous diets as leaves afford, without a correspondingly enormous body size.

These more modern ungulates evolved multi-chambered stomachs as well as voluminous folded large intestines, frequently with a symbiotic relationship between a cultural flora of special bacteria which fed on the well macerated and masticated vegetable fibre in the stomach or large intestine, but in so doing broke down indigestible cellulose into nutrients which the hooved animal could itself digest. Even today no living animal has evolved digestive juices or enzymes capable of dissolving cellulose, from which all plant cell walls are constructed. Hence the total dependence of these large herbivores on the bacteria in their stomachs and intestines. The full extent of this development is believed to be possessed by the more recent and present-day successful members of this group and is represented by the Ruminants which have stomachs with three or four compartments called chambers. They re-cycle their food

to grind it down into more digestible particles by regurgitating from the first stomach compartment back into the mouth and chewing the 'cud'. Deer, camels, giraffes and sheep, goats and cattle all are ruminants. The odd-toed ungulates were the predecessors of the even-toed ungulates and were none of them cud-chewers.

The horses, which developed earlier and in an independent way, lack a multi-chambered stomach but have developed a multi-chambered enlarged lower gut (or caecum) wherein bacteria carry out the same digestive functions. The fossil record shows that these single-toed ungulates evolved originally as small dog-sized creatures inhabiting the North American continent (whence they died not many millions of years ago). The present-day survivors of this order include in our region the Tibetan Kiang (*Equus hemionus kiang*) and its close relative the Rann of Kutch Wild Ass (*Equus hemianus onager*). In Africa several species of zebra, and in the USSR the Wild Horse (*Equus przewalskii*) of Central Mongolia still survive. The Tapirs appear to be a very ancient stock which continued to exploit the humid swampy forested regions and depended on more succulent vegetation. Present-day survivors include two South American tapirs and the Malayan Tapir both of course occurring outside the region covered by this encyclopedia. The Rhinoscerotidae are now represented by only five living species, all sadly in grave danger of extinction. They are massive beasts with specially thickened skins and relatively short stout legs, and include the African Black and White Rhinoceroses, two smaller Southeast Asian species, and the Great Indian One-horned Rhinoceros (*Rhinoceros unicornus*), which now survives only precariously in a few Himalayan foothill sanctuaries in Nepal and Assam. The Asiatic or Sumatran Two-horned Rhinoceros (*Didermocerus sumatrensis*) is thought to survive very precariously in neighbouring Burma and Malaysia but probably became extinct in Bangladesh and Assam before World War II.

Coming back again to the order Artiodactyla or 'even-toed' ungulates, we have a much greater diversity of families and genera and many more surviving species. They all share a similarity of leg and foot structure but instead of supporting their bodies mainly on an enlarged third digit, their third and fourth digits are equally developed and form a cloven foot, usually entirely encased in horny hooves. The second and fifth digits however become largely non-functional or vestigial though their distal part is still often encased in horny sheaths equivalent to dew claws. Even within this order, however, there are wide variations in the way that the different families have evolved. Camels, Llamas and Alpacas represent the family Camelidae and instead of horny hooves have the third and fourth digits enlarged into flattened pads with a relatively small nail in front of each toe, and the first, second and fifth digits have entirely disappeared. The most primitive family of even-toed mammals is generally considered to be the Suidae or pigs which lack a compartmentalized or enlarged stomach and cannot subsist on bulky fibrous vegetable matter. They are largely forest dwellers adapted to dig for succulent roots and tubers and to subsist on insects, fallen fruit and even small mammals when available. The two species found in our region are the Wild Boar (*Sus scrofus*) and the Pigmy Hog (*Sus salvanius*). Wild Pigs have four separate developed digits on all feet but normally only the third and fourth come into contact with the ground. They also have canines in their upper jaws and incisors in both jaws. The South American Peccaries (a pig-like animal) comprise another family, as well as the Africa-dwelling Pigmy and Large Hippopotamuses (Hippopotamidae). Both these families have simple single-chambered stomachs, four toes on each foot and incisors in both jaws. The Camelidae as mentioned above include some South American domesticated and wild species, whilst Vicunas and Guanachos are New World representatives still with large wild populations, and in our region the one-humped Dromedary and the Bactrian two-humped camels are now only found domesticated though a remnant wild population of Mongolian Camels (*Camelus bactrianus*) survives in outer Mongolia. In this family, again the foot bones are united to form a single cannon bone and as mentioned above only two digits remain. However the digestive system is more advanced and camels are true cud-chewing ruminants with three-chambered stomachs.

One other family of cloven-hoofed mammals is also thought to represent a more primitive and earlier offshoot. This is the Tragulidae comprising Mouse Deer and Chevrotains. They, like the Pigs and Peccaries, have independent though non-functional second and fifth outer toes, and like Camelids have upper and lower incisors and ruminate, but their stomachs only have two well-developed compartments. The females have four mammae and no species bear any horns. In our region only one species, the Indian Chevrotain (*Tragulus meminna*), occurs.

The next family is much more widespread and is represented on the Indo-Pak subcontinent by eight different species. These are the true deer or Cervidae, generally characterized by branched antlers in the males, which are shed and re-grown annually from bony pedicles or bases; naked moist nose-pads; three-chambered stomachs; and long thin delicate legs. They are browsers as well as grazers, fleet of foot and shy in disposition.

Following from this family we come to the most highly developed and recently evolved family, namely the Bovidae, together with which two other small families are associated, the Giraffidae and Antilo-

capridae. Bovidae are the most highly developed ruminants with four stomach compartments, fused or entirely absent outer toes, permanent horns borne on hollow bony cores and usually carried by both sexes. They have no incisors in their upper jaws and usually no canines. Most forms are gregarious, associating in herds, but there is wide variation in food preferences as well as body form. The goats and sheep are thick-set animals adapted to climb and forage in steep montane conditions often in very arid regions. Goats are mainly browsers whereas sheep are grazers. Buffaloes, bison and cattle are also heavy-bodied, thick-set forms, whilst the long-legged slender gazelles and antelopes have a wide variety of horn shapes and sizes, from the ox-sized Giant Eland of South Africa to the tiny Dik-dik (*Madoqua*) antelope of South and East Africa. Giraffidae are represented only in Africa by the true giraffes and forest-dwelling Okapi and the Antilocapridae by one single species on the North American continent, viz. the Pronghorn Antelope.

In our region there are four gazelle-like species, the Chinkara or Ravine Deer (*Gazella gazella*), the Goitered Gazelle of the Iranian-Baluchistan border (*Gazella subgutturosa*), the Mongolian Gazelle (*Procapra gutturosa*) and the Tibetan Antelope or Chiru (*Pantholops hodgsoni*). Sometimes the Chiru is classified amongst the goats. Gazelles are the epitome of grace and speed, with their slender necks and long delicate legs. Females

© M. Krishnan

Chinkara buck, Rajasthan × $^1/_{15}$

Lone bull Buffalo at a wallow in Assam × ¹/₃₅

are generally hornless with two mammae in the inguinal region. Gazelles are particularly adapted to semi-desert regions and can subsist for many days without drinking free water. Amongst the ox-like representatives found in India are the wild water buffalo (*Bubalus bubalis*) still found in the foothill regions of Assam and parts of Orissa and Madhya Pradesh and living usually in swampy regions, the Tibetan Yak (*Bos grunniens*), the Gaur (*Bos gaurus*) and the Gayal (*Bos frontalis*). The Gayal is a smaller version of the Gaur, considered by some to be no more than a subspecies, still surviving precariously in the Chittagong hill tracts of Bangladesh and in Assam and northern Burma.

The wild sheep and goats are believed to have evolved from a more primitive goat-antelope group known as the Rupicaprinae, a subfamily which includes from our region the Goral (*Nemorhaedus goral*) of the Himalayas, the Serow (*Capricornis sumatraensis*), the Takin (*Budorcas taxicolor*), as well as the Rocky Mountain Goat (*Oreamnos*) of North America and the Chamois (*Rupicapra*) of the European Alps. Rupicaprinae are relatively small but stocky animals with small short horns and thin skulls.

Whilst we have seen that there is a considerable variety of cervids or deer in southeast Asia, it is surprising that outside of certain extreme northern mountain ranges, no deer are represented in the African subcontinent. But a bewildering variety of antelopes are typical of the vast plains and savanna of Africa. In India only three true antelopes survive. The beautiful Blackbuck (*Antelope cervicapra*), the ungainly cow-like Nilgai (*Boselaphus tragocamelus*) and the relatively tiny Chowsingha or Four-horned Antelope (*Tetracerus quadricornis*).

The Chowsingha is an inhabitant of dry deciduous forest mostly found in peninsular India. They have apparently never spread to Sri Lanka and are absent from Pakistan or Assam. The females are hornless but the males are unique among the Bovidae in their possession of two pairs of horns. These little animals stand about 60 cm high at the shoulder and subsist mainly upon grass. They are usually found singly and never associate in groups as is typical of all other Indian antelopes.

See ASIATIC WILD ASS, BHARAL, BISON, BLACKBUCK, BOAR, CHEVROTAIN, DEER, GORAL, IBEX, MOUNTAIN SHEEP & GOATS, NILGAI, RHINOCEROSES, URIAL, YAK.

UPAS TREE, *Antiaris toxicaria*, is a famous tree of legend and fairy-tale. A giant towering to 75 m, buttressed at the base, it inhabits the climax forests of the Western Ghats, Sri Lanka and the slopes of the Pegu Yoma. It has red, velvety fig-like fruits. Its fibrous bark is worn as cloth by jungle tribes. On blazing, milky latex oozes out which dries to a highly poisonous gum. Jungle-dwellers tip arrows and spears with this gum to hunt birds and animals. It is reported that all the Dutch soldiers, save one, were killed by poisoned arrows during an encounter with the tribes in the East Indies, that birds were said to fall dead from their perch upon its branches and travellers who went to sleep in its shade never woke up. Most of these stories are sensational flights of fancy. The juice does contain two glucosides, antiarin and strychnine, which have a powerful effect upon the heart and arrest its action.

URIAL. The Urial (*Ovis orientalis*), or Red Sheep as it is sometimes called, is thought to be the ancestor of domestic sheep stock though its appearance is very different from its domesticated descendants.

They are long-legged animals with harsh short reddish fur and no underwool. The belly and legs are whitish. Both sexes carry horns but in the male they curve out-

Urial or Red Sheep

579

wards often in a very symmetrical semicircle. Rams also have a prominent chest ruff of long black hair. Practically confined to Pakistan, in the region covered by this encyclopedia there are three recognizable subspecies occurring, one in Baluchistan, one confined to the Punjab Salt Range and a third found in the far northern regions of Chitral, Gilgit and Baltistan.

Normally they inhabit lower hills and spurs and avoid the precipitous crags so liked by Ibex and Markhor.

These wild sheep rest by day in some sheltered ravine and graze early in the morning and evening. Small bands of immature males and females will forage and rest together, whereas outside of the rutting season (during November) the old rams tend to congregate in separate flocks. The young lambs are usually born singly but twins do occur, and are dropped in early April and are able to run about and follow their mother within an hour of birth.

See UNGULATES.

VEDANTHANGAL. In 1936 the Collector of Chingleput District officially recognized the place as a sanctuary. The area of this waterbird sanctuary is 180 hectares and it is only 100 m above sea level. It is 80 km south of Madras.

The sanctuary consists of an irrigation tank and the requirements of irrigation have a priority, but over the years it has remained an excellent habitat for a large variety of birds. It fills up seasonally during the northeast monsoon in October and November.

Its main attraction for birds is the large number of elevated muddy platforms with trees, which include *Alangium*, *Acacia* and *Barringtonia*. The survival of

this spectacular sanctuary is due largely to the fact that the villagers in the adjoining region recognize the value of bird droppings (guano) as a fertilizer of great merit.

The sanctuary harbours a considerable segment of the waterbirds of South India and the bird species include grebes, moorhen, white ibis, night heron, spoonbill, openbilled stork, grey heron, pond heron, darter, shag, Little and Large cormorants. The migratory species include the garganey teal, shoveller, pintail, sandpiper, plover, wagtails and many more.

VIPERS. The two most venomous and notorious vipers here are Russell's and the Sawscaled Viper, which are found throughout our area. These are two of the Big Four dangerous snakes and have a haematoxic venom (typical of vipers) which can drastically affect blood clotting. Russell's vipers feed largely on mice and rats while sawscaled vipers, which grow to 50 cm in the south and 80 cm in the north, feed on mice, small birds, frogs and even scorpions. All Indian vipers bear living young. Russell's vipers have 20 to 40 young, sawscaled vipers 4 to 8. Russell's vipers grow to almost 2 m and have a beautifully marked yellow-and-brown skin; they are therefore heavily exploited by the skin

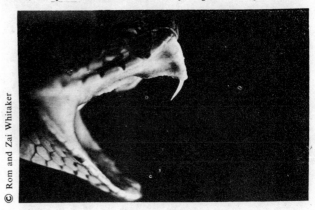

© Rom and Zai Whitaker

Russell's Viper fang

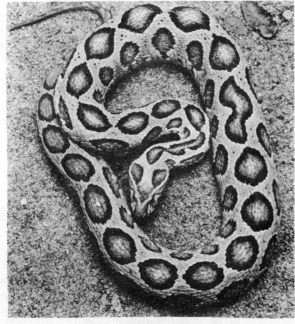

© Rom and Zai Whitaker

Russell's Viper × ²/₅

© Rom and Zai Whitaker

Sawscaled Viper × ¹/₂

Malabar Pit Viper × 1

© Rom and Zai Whitaker

Bamboo Pit Viper × 1/2

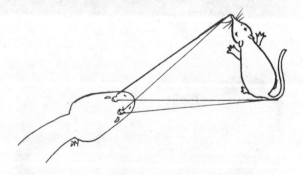

Directionality of heat sensor in pit viper

industry. Sawscaled vipers are protected by their size, but are collected in large numbers in some areas such as Ratnagiri for venom extraction in the process of making antivenom serum.

The large, brown Levantine Viper of eastern Europe

is found in parts of Kashmir where little is known about it. There are 16 species of pit vipers in India from two genera, *Agkistrodon* and *Trimeresurus*. Pit vipers are so called because they have a pair of pits between the nostril and the eye which are heat-sensitive, enabling pit vipers to locate their warm-blooded prey in the dark. Several interesting species are found in the southwestern ghats including the distinctive Hump-nosed Pit Viper and the Green Pit Viper with its shingle-like overlapping scales. All pit vipers have triangular-shaped heads, thin necks, and long, slender bodies. Patterns and colours vary considerably, even within the same species. Our pit vipers have haematoxic venoms. Bites can cause swelling and considerable pain; the few fatalities reported in the last hundred years probably involved exceptionally large specimens.

See also SNAKES. See plate 3 facing p. 48 and plate 9 facing p. 128.

VOLES. Voles are small rather secretive rodents belonging to the family Cricetidae. They are characterized by rather short tails and rounded blunt muzzles. Most voles live in the northern hemisphere and are palearctic in distribution. In the area covered by this encyclopedia there are several species adapted to the temperate conditions of the Himalayan ranges. In the northwestern part are two very large voles of strongly burrowing habits: True's Vole (*Hyperacrius fertilis*) and the Murree Vole (*H. wynnei*), both of which extend from Pakistan into Kashmir. The former occurs in alpine meadows above the tree-line whilst the latter is associated with mixed coniferous and deciduous forest in the lower or outer hill ranges and has become quite a serious pest of apple orchards, gnawing the bark at the base of the tree trunk.

In the inner hills at higher altitudes there are two more high-altitude voles with longer tails and less fossorial habits, Royle's Vole (*Alticola roylei*) and the Sikkim Vole (*Pitymys sikimensis*). The former occurs in the western part of the Himalayas from Pakistan to Kumaon, whilst the latter occurs in Sikkim. Baluchistan forms the easternmost range of the Mole Vole (*Ellobius fuscocapillus*) which is the most adapted of all to a fossorial life, having almost lost the power of vision and with tiny ears hidden in its short velvety body fur.

See also RODENTS.

VOLUTES and CHANKS. Volutes are deep-sea gastropods principally found in tropical seas. They are famous for their smooth and shining shells of bright and varied colours. They differ exceedingly in form and size: some are globular, others oval; some turreted and others with a small spire. They have no operculum. The animal inhabiting the shell has a distinct head and two short triangular tentacles with eyes at the base and a long proboscis or trunk. The foot is large. Under this large group India has some unique and beautiful representatives—the Melon shells, the Harp shells and the Olives.

The Melon shell must be familiar to many—a large, long ovoid, dome-shaped, smooth, yellowish orange shell generally carried by hermits and sadhus to receive alms, drinking water etc. The animal that owns this shell is *Melo indica*. It lives in waters 5 to 7 fathoms deep in Palk Bay and Pamban area. The shell measures over 20 cm and the animal is often larger than the shell. It prefers a bottom where sand and mud are mixed. It is carnivorous and its radula contains a small number of formidable teeth. In adult shells the spire is hidden by the inflated and overgrown body whorl. The inner lip of the shell mouth has strong folds or pleats. The animal—its foot, mantle and head—is striped like a tiger with yellow and black. When crawling the shell is engulfed in the folds of its gorgeous mantle.

Melo deposits her eggs in capsules, hundreds of them arranged in a honeycomb-like fashion, attached to a vertical stem, 30 cm in height and the whole structure resembling an elongated pineapple—the capsules representing the bracts. The animal carries this curious looking egg-mass with her until the young ones are free.

Harp shells (genus *Harpa*) are so called because the shells have a short spire and the longitudinal ribs on the body whorl look like the strings of a harp. The shell is thick-walled, large, exquisitely polished and the interspaces between the ribs are ornamented with crescent-shaped patches of golden yellow. The inner lip is smooth and polished. The foot is enormously developed—broad in front and pointed behind and with a lappet on either side. The expanded foot can be seen clearly when the animal moves. The head is small with thick, close-set, sharply pointed tentacles. The mantle lobes turn back over the shell, so protecting it from abrasion and scratches and accounting for its high polish.

The animal lives in rock pools among coral reefs in the Gulf of Mannar and Lakshadweep, feeding on sea organisms. Like certain snails and razor-fish, *Harpa* also has the habit of amputating part of its foot to escape from the grip of dangerous enemies. It presses the shell lip over the foot and cuts off the exposed part. The animal is capable of reproducing the lost part in due course.

Olives are all handsome volutes ranging from 1 to 5 cm in length. The shell has a conical, short spire and a long ovoid or cylindrical body whorl. The mouth opening is narrow and there is no operculum. The animal is larger than the shell and when it moves the folds of the mantle cover the shell to a great extent. The mantle is produced into a tentacle behind, which serves as a rudder while the animal moves. Olives are

all carnivorous and very active. They burrow in wet sand in quest of bivalves on which they feed. Certain species can also swim by expanding the lobes of their foot. Olives are famous for their glazed polish and fantastic colour design. Along the Indian shores and around Lakshadweep several beautiful species are common. Fishermen collect them from knee-deep water below low-tide mark by turning the sand over with their feet. Among them all *Oliva porphyrea* is the most attractive. It has a depressed spire and a prominent body whorl measuring 5 cm long. The surface colour is cream with black zigzag streaks described as a series of pictures of mountains done by a Chinese artist.

Sacred Chanks, belonging to the genus *Xancus*, are the most prized and popular volutes in India. The shells are heavy, pear-like in shape and snowy white in colour. The living shell is invariably covered by a golden brown periostracum. It has a narrow operculum. The chank (often called conch) is not merely a volute but a happy combination of the grace of a volute and the sturdiness of a whelk. In anatomical details, habits and larval development the Chanks resemble WHELKS; but in elegance, grace and weight they are unique. Chanks are found only in Indian waters and that too in the Gulf of Mannar, Palk Bay, coasts of Tinnevelly and Tuticorin and on the Kathiawar coast. The mouth of the shell is wide, ending in a hollow beak. The beak encloses the animal's siphon which can sense its prey, particularly sea worms. The inner margin of the aperture (the columella) is thickened and bears three or more prominent ridges.

Chanks are gregarious. They live in waters 8 to 10 fathoms deep and form distinct beds. Sandy bottoms where tube-worms abound are their favourite haunts. Some live in shallow waters also but there their growth is much reduced, the spire becomes short and the shell less heavy.

Chanks deposit their ova in capsules of papery consistency. These capsules are attached spirally one above the other along a twisted vertical support. The growing end of this egg column is wide. The lower end is narrow and is anchored to the sand or mud. Each column consists of 25 to 30 capsules. Each capsule has a slit which allows water to go inside. The larvas develop within the capsules until they are able to come out by themselves. The very young chank emerging from a capsule has a thin, spindle-shaped shell measuring less than a centimetre.

Chank shells are highly esteemed, especially the rare left-handed (sinistral) variety. They have a religious significance to Hindus and Buddhists. All Hindu temples and orthodox Hindu families own large chanks which are used for blowing during times of worship. The apex of the shell is lightly chipped and blowing through it produces a uniform sound, which is considered auspicious during festivals, pujas and ceremonial occasions.

In India, the Gulf of Mannar is famous for chank fishing. The peak season is between December and January. It is reported that 250,000 chanks of large size are taken from this area every year. The collection is done by experienced divers and the Fishery Department purchases the entire lot

See also MOLLUSCS.

WADER, SHORE BIRD. Names applied in general to aquatic or waterside birds belonging to the heterogeneous order Charadriiformes, represented in the Indian subcontinent by 11 families which include jaçanas, plovers, gulls, coursers, and a diverse assortment of other waterside birds. More specifically the names 'Wader' and 'Shore bird' are commonly applied to the family Charadriidae—plovers, curlews, snipes, sandpipers and suchlike birds that live around sea coasts, estuaries, and freshwater lakes and feed on aquatic invertebrates and vegetable matter in shallow water, or between tidemarks on the seashore, or on mudflats, or moist and inundated fields, etc. They mostly have long bare legs for wading in the shallows, and longish slender bills for probing in the wet mud.

The family contains both resident and migratory forms, some of the latter, e.g. Golden Plover, being among the longest distance non-stop trans-ocean fliers

known. The great majority of our migratory waders come from far northern Asia, chiefly the U.S.S.R., and from NE Europe. Ruffs ringed in India have been recovered on their breeding grounds in the Arctic Circle over 7000 km away, and many of the smaller sandpipers, e.g. Spotted Sandpiper, likewise travel astonishingly long distances. Waders are sober-coloured birds—usually sandy brown and white, often scalloped or otherwise patterned with blackish. But many species change into a colourful nuptial plumage on their breeding grounds in spring; in some cases this happens while the migrants are still in their winter quarters, before the summer exodus has commenced. This makes their identification confusing and difficult for the inexperienced bird-watcher. On account of the vast mixed concentrations and the open landscapes in which waders normally feed they hold a special fascination for the bird-watcher because there is always the exciting possibility of his picking out some rare and unexpected vagrant amongst the densely packed flocks of the commoner species. Among the more rewarding places in the subcontinent for wader watching in winter, are the Makran, Sind and Kathiawar coasts with their offshore islets in the west; and Point Calimere and the various brackish lagoons on the Coromandel coast, the low-lying flood plains of the Ganga in Bihar, the Chilka Lake in Orissa, and the Sundarbans of Bengal in the east.

Most waders lay their eggs on the ground in shallow saucer-like 'scrapes' on shingly or sandy river-beds, meadows, and the like. The normal clutch is of 2 to 4 cryptically coloured and patterned pegtop-shaped eggs. In most species both sexes take part in incubation. The hatchlings are clothed in camouflagingly coloured down and are capable of actively running about and hiding almost from the time they emerge from the shell (nidifugous).

Stone Curlew, young and egg

Ibisbill × 1/6

WARBLER. This is the name for a group of small, insectivorous birds forming one of the five subfamilies (Sylviinae) of the family Muscicapidae, with a wide distribution in the Eastern Hemisphere. Warblers are by and large nondescript and confusingly similar.

Within our limits we have 87 species belonging to 21 genera. Of these, the Longtailed Warblers (genus *Prinia*), the well-known Tailor Bird (*Orthotomus sutorius*) and the two little Fantail Warblers (genus *Cisticola*) are widespread residents. The Booted Warbler (*Hippolais caligata*), six species of Reed warblers (genus *Acrocephalus*), six species of Whitethroats (genus *Sylvia*) and two species of Grasshopper warblers (genus *Locustella*) are all widely spread winter visitors. The Whitethroat (*Sylvia communis*) is a very abundant passage migrant during September through western India. All these breed beyond the Himalaya, in temperate Eurasia. The Indian Great Reed Warbler (*Acrocephalus stentoreus*), however, is an abundant nester in the Kashmir Valley and also in mangroves along the West Coast.

The 20 species of Leaf warblers (genus *Phylloscopus*) breed in temperate Eurasia and along the Himalaya and winter in the subcontinent. They are largely subdued browns and greens above and sullied white or yellow below. Leaf warblers are very difficult to identify. A few species, however, have brighter green and yellow plumages. These can be confused with the 9 species of flycatcher-warblers (genera *Seicercus* and *Abroscopus*) found in mid-altitude Himalayan forests. The tiny Goldcrest *Regulus regulus* of high-altitude Himalayan coniferous forests resembles a small leaf warbler.

Several species of bush warblers (genera *Cettia* and *Bradypterus*) are all great skulkers inhabiting tangled vegetation along the Himalayan foothills with one wintering species confined to Pakistan and an endemic species to the central mountains of Sri Lanka. One species each of *Luciniola, Scotocerca, Graminicola, Chaetornis* and *Megalurus* are residents or local migrants with localized ranges in the plains, favouring marsh vegetation. The Thickbilled Warbler (*Phragmaticola aedon*) is a winter visitor from NE Asia.

L.K.

WARNING SIGNALS. Animals often give various signals when they suspect or detect the presence of a predator. Thus many species of deer such as chital have a conspicuous white rump-patch which they exhibit by raising their tails when alarmed. Many birds and mammals have shrill screeching alarm-calls, while the ants release a special chemical from their glands when their colony is raided by predators. These warning signals make known the presence of danger to neighbouring animals of their own or other species who may

Many species of deer raise their tails and display conspicuous rump patches when alarmed.

respond by freezing, fleeing, bunching or actively confronting the source of danger.

In a group of cooperating individuals, such as a herd of cow elephants and their calves, the warning signal serves to co-ordinate the defensive action of all the adult females. On a warning signal from any member of their herd, they rush together and form a solid phalanx confronting the source of danger with the calves tucked under their bellies and in between them. Some of the females may then make a charge while the calves retreat.

The function of the warning chemical signal is equal-

Many species of deer and antelope run away from the predator taking high running leaps which attract the attention of other members of the herd.

ly clear in the case of social insects such as ants and bees. In the honey-bee the worker bees will sting an attacking predator such as a bear. This releases a chemical which attracts other bees to sting at the same place. The release of the warning chemical by ants attracts other ants to the same place.

The function of the warning signals is however not so evident in those animals which do not cooperate in fighting off a predator. In such animals the individual that gives the warning signal may run a greater risk of being preyed upon. The question then is, why run such a risk at all? Why not simply keep quiet or slink away by itself?

That the warning signal in such species does increase the danger to the warner is suggested by the structure of these warning signals. Thus in birds the warning calls always tend to be a mixture of several frequencies and to begin and end slowly. Such signals are very difficult to localize. Furthermore, the warning calls are understood and responded to across a range of species; for instance, chital will respond to the warning calls of peafowl and Hanuman langur. All of this suggests that it is advantageous for any individual if all of its neighbours become aware of the predator. This should reduce the chance of successful hunting by a predator in that locality and thereby discourage the predator from visiting the locality again. This in turn would confer an advantage on the warning individual which is sufficient to offset the risk it runs when giving the warning signal.

At the same time, the warner may benefit not directly, but indirectly through benefiting its blood relative. Among Californian ground squirrels, for instance, a female is much more likely to warn when she is in her own territory and hence surrounded by blood relatives, than when she is away from her territory and surrounded by individuals not related to her.

Another possible function of the warning signal, such as tail-raising in deer, may be to force the predator to pursue its quarry when the predator is at a safe distance away rather than permit it to come closer before the pursuit starts. This would be advantageous to the deer which stand a better chance of survival if the pursuit starts from a greater distance.

Finally the warning signal may be used to turn the tables on the victim itself. Thus there are some slave-making ants. These ants have no workers of their own. The master species raids the colony of the slave species and captures their brood which then grows up to serve as the worker force for the master species. Now the slave species has a warning chemical which is released on alarm but breaks down into a non-alarming substance within minutes of release. The master species has a chemical which mimics the alarm chemical of the slave species but does not break down. During their raid the masters spray the slave workers and soldiers with this chemical. Since the effect of this chemical does not quickly wear down, it throws the defence forces of the slaves into total panic. This is to the advantage of the master species and the raiders make off with the slave brood.

WASPS. The order Hymenoptera perhaps includes the largest number of insects on earth, even though at the present time, the Coleoptera (beetles) possesses the maximum *described* species of insects. It includes what are commonly called the wasps, bees and ants. The latter two groups have been treated individually elsewhere and only what are generally called the wasps will be dealt with here. More than 100,000 species of wasps, bees and ants are now known to man throughout the world, but it is estimated that the final figure of all actually existing Hymenoptera, especially when the minute wasps are discovered and named, will probably exceed 500,000 species! Along with the flies (Diptera), the Hymenoptera are one of the most specialized insect orders. Two suborders are recognized: the primitive Symphyta which are plant-feeding insects, and the advanced Apocrita, which is the numerically dominant of the two suborders, and which includes both parasitic and predacious species leading to the bees that co-evolved to perfection with the flowering plants to feed only on pollen and nectar. Yet, as in all groups of insects, there are exceptions and some gall-making wasps do feed on plant tissues like their primitive relatives of the suborder Symphyta. Though some Hymenoptera are pests either as plant-feeders or indirectly, being hyper-parasites of beneficial parasitoids, probably no other insect order is more beneficial to man's interests.

The Hymenoptera are seen almost everywhere; digging in soil, in branches, using holes in man-made or natural structures, on all parts of plants, especially on flowering species, catching prey in flight, or even above or under water! Bees guarding their territories and performing the important job of pollination of many of our cultivated and wild plants have been discussed elsewhere, but many wasps also have similar habits. Social or subsocial behaviour first evolved in wasps and they are expert builders of nests too. Communication and defence are also carried out in the social groups of wasps, mainly by smell. Like the grasshoppers and crickets, many wasps also stridulate to produce sounds audible to them but many too weak for our ears to pick up. The reproductive strategies of the insects of this order is also interesting, males being haploid and in many groups being produced by parthenogenesis, whereas in most groups females are produced only by fertilization. Regulation of sexes has reached almost perfection in the wasps, which helps the species to determine the sex ratio at particular times depending on the resources and requirements.

Parental 'care' has been highly developed in this order of insects, so that only in the phytophagous Symphyta do the larvae depend on themselves for their food. The other groups of the Hymenoptera, like the Apocrita, are: (1) parasitic, and thus develop inside or outside the body of the host that the female chooses for them, (2) predatory, but being non-social, feeding on prey that the adults provide for them, or (3) social forms, that grow to maturity on food specially prepared by the colony.

The Symphyta lay their eggs either in leafy or woody tissues, or even in tunnels made by larvae of beetles that are filled with excreta. Like most Hymenoptera, the ovipositor or special egg-laying organ is well developed, in some being longer than the insect's body itself. Being a primitive section this order, the Symphyta larvae feed and develop mostly on conifers.

The Apocrita have three sections as discussed above, the most generalized among them being the parasitic forms. The adults of these forms are usually not associated with their host insects unless they need to find suitable species of hosts to paralyse and use their intricately designed ovipositors to lay the egg(s) either on or inside the body of the host insect. Many species of parasitic wasps find their hosts by first searching for the general habitat or environment that the host insect is usually found in, and then the host is located and parasitized. There may develop in the host, either a single parasitoid larva, or a whole mass of them, sometimes hundreds! Again, some parasitic wasp females are able to recognize if a particular host insect has already been parasitized by another female, and then she may or may not lay eggs, depending on the species to which she belongs. There are what are commonly termed 'hyperparasites', meaning wasps that lay eggs in the host body which already has been parasitized by another genus of wasp, but, instead of sharing the body fluids, its larvae attack the larvae of the species that had laid eggs earlier in the host body. In extreme cases this behaviour may go one step further so that in some cases we have species that are parasites, of parasites, of parasites, of the host insect!

The predatory forms of Hymenoptera that belong to another group and which later produced the highly evolved social group of wasps (wasps, bees and ants), search for their prey insect and paralyse it, laying egg(s) near the stunned prey. The most primitive of these predatory species sting a single spider, deposit a single egg on its body and the young feeds on this spider even though the prey individual may recover to resume normal life for some time. Other species may enter the nest of a single spider and paralyse it in its own 'lair' and deposit an egg. Other wasps make a cell and bring the paralysed victim into it and lay the egg there. Still others may provision the larval 'chamber' with more than one individual prey for the single young to feed and complete its development on. As in the bees, in wasps also, especially in the Sphecidae, progressive provisioning of the nest housing its larvae is carried out, the adult(s) continuously providing its/their larvae with fresh food when the already stored provision is almost totally devoured. The Vespoidea have species that either store paralysed hosts intact in the larval cells or even provide their young with masticated bodies of their prey.

The social vespoids go one step further in constructing many-celled nest structures and in the larvae being continually fed by the mother at first and then by a caste of sterile worker wasps. Some vespoids gather nectar and pollen from flowering plants, but most feed on animal matter and provide their young with the same, carefully prepared. Since so many wasp species are either parasitic or predacious on other insect groups,

© C.I.B.C., Bangalore

Larvae of parasitoid wasp in moth larva × 10

they are an important segment of life on earth that toil to produce that delicate balance that nature so assiduously maintains. Man has even looked upon the wasps to provide the biological control agents that he has partly successfully utilized to control the ravages of his crop plants by pest insects. The bees primarily, and also the wasps, are the important pollinators of plant life on earth and important for the rejuvenation of many flowering species.

The natural enemies of the Hymenoptera are mainly found in this order itself, though many Diptera (true flies) attack and feed on many wasps. Perhaps the higher Hymenoptera are themselves responsible for the regulation of their populations, and the main factors that decide on the rise or fall of numbers of several species of wasps are the sites that they use for nesting and the availability of their predators or prey. In social species of wasps, many bacterial, viral, protozoal and fungal diseases are rampant at times and some species are 'stylopised' by the curious beetle group, the Stylopoidea. Some parasitic beetles of the families Meloidae and Rhipiphoridae also take toll, as do many parasitic mites and mermithid worms, mainly of the social wasps.

The Tenthredinidae are the common family of the suborder Symphyta and contain some common pest species like *Athalia proxima* which attacks cruciferous vegetables. The females of this family have ovipositors with saw-like teeth on them, with the help of which they lay their eggs inside plant tissues. The larvae feed on leaves of plants of many kinds, including conifers and ferns, and they resemble the caterpillars of moths and butterflies to a surprising degree. Because of the saw-like ovipositors of the females, these insects are commonly termed 'sawflies'. Some species may have larvae that either mine leaves or bore stems. These larvae pupate in a silken cocoon beneath soil. Many adult sawflies are very attractively coloured.

More than 90 percent of all Hymenoptera belong to the large suborder Apocrita. Nearly a hundred families are recognized, most of them occurring in the Indian subcontinent. We will deal only with a few of the more obvious ones that are commonly encountered in nature The superfamily Trigonaloidea, with the single family Trigonalidae, may be mentioned, although their members are not the commonest of wasps, because they represent one of the more primitive families and one that perhaps is close to the point in the evolutionary history of the Order that gave rise to the two dominant groups of wasps, the 'aculeate' or predatory forms that are generally larger and the 'terebrant' or parasitic ones that contain some of the smallest known insects. The trigonalids are medium sized wasps that usually parasitize primary tachinid or ichneumonid parasitoids of sawfly or lepidopteran 'caterpillars'. The female has

an ingenious method of getting her larvae to enter the host's body. She lays eggs in slits on a maturing leaf, which when eaten by the caterpillar host get ingested conveniently so that the hatching larvae within may attack the larvae of ichneumonid wasp or tachinid fly parasitoids already developing inside the host's body.

The Ichneumonidae and Braconidae are two of the largest families of the Hymenoptera containing many very common and colourfully patterned wasps. The Ichneumonidae contain larger forms that parasitize larvae and pupae of most endopterygote insects and also adult and immature spiders and pseudoscorpions. Many species are very specific in their selection of host insects and other arthropods, but, as explained above, the kind of habitat plays an important role in their ultimate selection of host species. It may be generalized that most ichneumonids prefer humid environments and live in forests, but in drier, open areas and cultivation, they seem to prefer shady groves or orchards or the banks of streams lined with mixed vegetation.

© C.I.B.C., Bangalore

A braconid wasp laying egg in caterpillar inside stem with aid of ovipositor × 5

The more familiar genera found in India are *Rhyssa, Xanthopimpla, Theronia, Enicospilus, Netelia* and *Diplazon*, to mention just a few of the substantial Indian fauna of these useful insects.

The Braconidae are as large a family as the Ichneumonidae, and perhaps more diverse in form, size and habit. They do not possess the second 'recurrent vein' like the Ichneumonidae and this is what students of entomology get familiar with very soon. Members of this worldwide family parasitize immatures of many orders of insects, principally the Lepidoptera, Coleoptera, Diptera, Hymenoptera and Hemiptera. Some of the more common braconids in India belong to the genera *Apanteles, Microbracon, Chelonus* and *Aphidius*, the last containing species that specialize in parasitizing aphids which then become 'mummified'.

Aphidïïd wasp about to oviposit in aphid (× 10)

The next superfamily, the Evanioidea, comprises insects similar to the two preceding ones and sometimes has been placed in the Ichneumonoidea. It contains insects that parasitize cockroach oothecae (Evaniidae; even entering houses to search for them), larvae of wood-boring beetles and sawflies (Aulacidae), and larvae of bees and wasps nesting in wood, soil or in clay 'pots' (Gasteruptiidae).

The Proctotrupoidea is another large superfamily with the smaller parasitic wasps that resemble those of the Chalcidoidea. Many species parasitize eggs of other insects like beetles, moths, flies and wasps. The Scelionidae is one of the more common families including many species that are useful to man in specializing on grasshopper, cricket and locust eggs for survival. They also parasitize eggs of many bugs and moths and females of some species of scelionids settle on the abdomen of the host grasshopper female so that they are 'present' when she decides to lay her eggs in soil. The scelionids then lay their own eggs before the grasshopper removes her abdomen from soil, leaving no trace of where she deposited them. The common genera found in India are *Scelio*, *Telenomus*, *Macroteleia* and *Hadronotus*. The Cynipoidea contain the famous 'gall wasps', though many families in the Chalcidoidea also cause plant galls. But they also parasitize puparia of flies (Figitidae) and are hyperparasites of aphids through braconid primary parasites (Cynipidae).

The largest superfamily of the Hymenoptera is the Chalcidoidea or the small parasitic wasps belonging to more than 10 distinct families. Some species of this group are phytophagous and form plant galls. In build (size 0·2 to 30 mm) and habit they are also very diverse, but most of them are tiny wasps (1 to 3 mm) that are some of man's best friends in keeping his pest species in reasonable numbers and in maintaining nature's balance. Many chalcidoids are able to leap or jump with the help of a spur on the tibia of the middle leg. The Agaonidae are an interesting family containing the

familiar 'fig wasps'. All species live in 'fruits' of figs (*Ficus* spp.) and are important in performing the pollination of the particular host tree species. The female and male are very different-looking and our edible fig owes its popularity and survival to the species of these insects that are associated intimately with it. The Trichogrammatidae contains some of the smallest of all wasps and like some *Trichogramma* spp. that have been used in the biological control of pest species of moths, they are very useful egg parasites of many small insects. The Eulophidae is again a large and diverse family containing many wasps that are beneficial to man. They are parasites or hyperparasites of many Lepidoptera, Diptera, Hymenoptera and Coleoptera. The Mymaridae contains extremely tiny, fragile wasps that are parasites of eggs of many insects. The Chalcididae contains the larger forms and include wasps that parasitize other wasps, beetles, moths, butterflies, flies, etc. The common genera are *Antrocephalus*, *Brachymeria*, *Leucospis*, *Dirhinus* and the peculiarly formed *Uga*, which parasitizes larvae of ladybird beetles. The Eurytomidae are either gall-forming wasps or those that usually live in galls created by cecidomyiid flies or cynipid wasps. Wasps of the family Torymidae include the Podagrioninae that are specific parasitoids of oothecae of praying mantids, the Sycophaginae that are parasitoids of the fig-inhabiting Agaonidae, and other subfamilies that are, among other things, fly parasitoids. The Pteromalidae comprises a heterogeneous assemblage of wasps that includes the Eucharitinae that are parasitoids of ant immatures and other subfamilies that are parasitoids of immatures of the Lepidoptera, Coleoptera and Diptera among other orders. The Encyrtidae is perhaps the largest family of the Chalcidoidea and replete with many small wasps that are highly beneficial to man's interests. The spur on the middle leg of almost all species enables the wasps to jump and this is a characteristic very common in this family. The Encyrtinae are small insects that are either parasitoids or hyperparasites of scale insects of the Hemipterous superfamily Coccoidea, or others that attack nymphs of ticks, eggs or nymphs of bugs, puparia of flies or eggs and larvae of moths. The Eupelminae has the nominal genus *Eupelmus* which attacks insect eggs and other genera have species that parasitize larvae of Lepidoptera. The Aphelininae (most species being parasitoids of aphids, mealybugs and scales) and Thysaninae which also include hyperparasites of bug nymphs or eggs, are sometimes separated as distinct families. The former contains the species *Aphelinus mali* which is a common parasitoid of the apple woolly aphid. Some other genera of the Chalcidoidea that are frequently collected by the student of this order of insects in India are *Blastophaga*, *Euplectrus*, *Elasmus*, *Mymar*, *Eurytoma*, *Podagrion*, *Sycoryctes*, *Schizaspidia*, *Perilampus*,

Ormyrus, Cleonymus, Spalangia, Pachyneuron, Metaphycus, Aphycus, Comperiella, Azotus, Encarsia, Marietta, Ooencyrtus, Paralitomastix, Anastatus, Metapelma and *Psyllaephagus.*

The common metallic green wasps with brilliant red spots and marks that we see outside our windows inspecting them, but actually searching for the nests of vespoid and sphecoid wasps, belong to the family Chrysididae and the superfamily Chrysidoidea. When caught, we find these wasps rolling themselves up nicely, in the manner of a common millipede we have in India, and this is a trick to prevent the host wasp adults from tearing them apart. Several species of *Chrysis* and *Stilbum* occur in our country and are external parasitoids (or cleptoparasites) of the larvae of mud-dauber wasps especially.

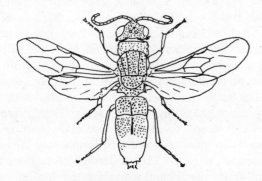

A jewel wasp (Chrysididae) × 5

The superfamily Bethyloidea comprises smaller wasps than the closely related Pompiloids and some species of this superfamily have wingless females. The Bethylidae are external parasitoids of larvae of small beetles and moths. This family has species that exhibit behaviour intermediate between the true 'parasitic' groups and the predatory groups that are fossorial (digger wasps). The common genera are *Goniozus* and *Perisierola* and some are associated with ants. The Sclerogibbidae are parasitoids of nymphs of the web-spinners (Embioptera). The Dryinidae contain peculiarly formed parasitoids that attack homopterous bugs, mainly Fulgoroidea and Cicadelloidea. Most are ectoparasites and live on the abdomen of their hosts. The females have their fore tarsi modified into clasping organs with which she captures the host leaf-hopper and stings it to lay an egg on its body. Then she drops the poor insect which recovers to lead a normal life for some time, allowing the parasitoid to feed on its body, and later to kill it. The adults may be dimorphic, many females being wingless, curiously resembling ants and often found close to ant trails.

The Pompiloidea are commonly called 'spider wasps' and belong to the fossorial section of this insect Order, possessing insects that are some of the largest living wasps. Most of the members belong to the family Pompilidae, which contains some primitive species that are truly parasitic. All species of this family are predators (or parasites) of spiders and exhibit interesting patterns of behaviour in searching for prey and then stinging it before dragging it to the burrow which is either dug before or after capturing the prey. Many species have orange or violet-black coloured wings which they flick while walking and they display a very colourful appearance. The common genera in India are *Macromeris, Pseudagenia, Cryptocheilus* and *Pompilus.*

The stout, black, hairy wasps that are commonly noticed hunting prey on the ground or flying (rather gliding) over field borders or irrigation channels belong to the family Scoliidae of the superfamily Scolioidea. All are parasitic and the Scoliids parasitize mainly the root grubs of Scarabaeid beetles. The species of *Campsomeris* are coloured yellow and orange and are generally slenderer than the *Scolia* type genera. The Mutillidae contain the popular 'velvet-ants' which comprise some of the most beautifully ornamented wasps. Females being generally wingless are found in roadside shrubs or on the surface of fields where they search for nests of sphecoid or vespoid wasps and bees to lay eggs in and let their larvae develop on those of their hosts. The males are very differently coloured from the females and once the male locates the female, he grabs her and lifts her off in flight when he mates with her before dropping her some distance away, thus accomplishing the duty of dispersal of his species in space. Many species of the genus *Mutilla*, which has now been divided into many other smaller genera, are common in India. The related Tiphiidae are duller in colour and females of one subfamily, Thynninae, are also wingless. Many species of *Tiphia* are useful parasitoids of pest species of Scarabaeidae.

The Vespoidea represent the apex of the evolutionary development of the wasps and contain many species that

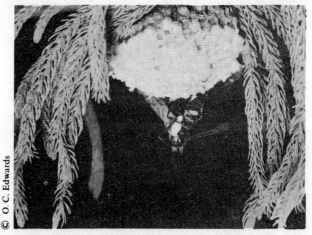

Vespa orientalis *feeding on larva of paper wasp,*
Polistes sp., × ½

© O. C. Edwards

591

are social. Many species are also good pollinators of several flowering plants and they are all predacious on caterpillars of moths and butterflies generally. Their nests are marvels of structural prowess of any kind of animal and may be made either of clay or paper-like material. Several species build their nests outside (or inside!) the homes of man and each species has a nest, shaped characteristically. In the higher vespids which exhibit social behaviour, many 'castes' may be present on the nest and though social life is not as elaborate as in ants, honey-bees or termites, it is substantially developed. The common genera in India are *Ropalidia*, *Vespula*, *Vespa* and *Polistes*. The related family Eumenidae consists of solitary wasps that also build nests and store them with larvae of moths, butterflies, sawflies or even beetles. Many genera like *Eumenes* have a slender 'waist' and their mud nests are a common sight

© Kumar Ghorpadé

Potter wasp, Eumenes conica, *sealing mud cell* × 1

on windows or walls of human habitations. The other common genera are *Rhynchium* and *Odynerus*.

Coming to the last superfamily of the wasps that will be dealt with here, the Sphecoidea are one of the largest groups and contain wasps that are thought to be the predecessors of what we now know as bees. They are also fossorial and prey on insects belonging to several orders like Hemiptera, Orthoptera, Lepidoptera, Hymenoptera, Diptera, Blattodea, Collembola, Thysanoptera and Coleoptera, besides on spiders. Most species are specific predators of a single genus or even a single species of prey and the adaptive behaviour of each sphecid species to its prey presents one of the most absorbing and intriguing aspects in the study of behaviour. The adult sphecids mainly feed on pollen, nectar, honeydew produced by homopterous bugs or on ripe fruits as well as on the exudates of plants. Only females take prey which they feed their larvae with. The more primitive sphecids take prey that are as large or larger than themselves to feed the larvae with one

prey individual. The more advanced forms store cells of their larvae with more than one prey individual and so take prey of smaller size. Some species practise progressive provisioning and others either leave the nest open while on the hunt for prey or close it and then hunt, opening it again on return to put the prey in. The Ampulicinae are the slender metallic green sphecids with red legs in many that are often found on trunks of trees or on house windows, as they store nests with cockroaches. The Sphecidae comprises many subfamilies that are themselves distinct enough or large enough to possess characteristic features of their own. The Astatinae hunt mainly heteropterous bugs, while the Sphecinae take lepidopterous larvae as well as grasshoppers, spiders and sawflies. Many species exhibit territoriality and the genus *Ammophila* is one of our common sphecids. *Sceliphron* contains the common mud-daubing sphecids that, along with the metallic *Chalybion*, provision their nests with spiders. Many species of Nyssoninae are inquilines in the nests of other wasps or bees but others like the Gorytini take cicadas or leaf-hoppers, while Stizini take flies and the familiar Bembicini store nests also mainly with flies and practise progressive provisioning. The Crabronini nest either in soil or in wood and take many orders of insects as prey and are fairly numerous in India. The slender, elongate Trypoxyloninae store nests with spiders and the Pemphredoninae consist of small sphecids that look like bees, visit flowers, and store nests (often made in wood) with small insects like collembolans, thrips and small bugs. The Cercerinae have a petiolated abdomen and the genus *Cerceris* has a number of Indian species that provision their larval cells with beetles, often weevils. The Larrinae contain some very beautiful sphecids that prey on crickets, mantids, cockroaches, tettigoniids and spiders, besides other things. Some other common Indian genera of the Sphecoidea are *Astata*, *Trypoxylon*, *Sphex*, *Liris*, *Ampulex*, *Gorytes*, *Philanthus*, *Bembix*, *Crabro*, *Stizus*, *Tachytes* and *Stigmus*.

See plate 26 facing p. 369.

WATER PLANTS. Water plants or hydrophytes may be defined as those that spend at least a part of their life-cycle in water, completely or partially submerged. They have been classified under various categories:

Floating types remain in contact with water and air only (examples are Duckweed, Water Chestnut or Singhara, Water Hyacinth, Water Lettuce and African payal, the last three being noxious weeds clogging waterways).

Suspended hydrophytes like Bladderworts and *Ceratophyllum* are rootless and in contact with water only.

Anchored submerged types remain in contact with soil and water either through entire life or part of it.

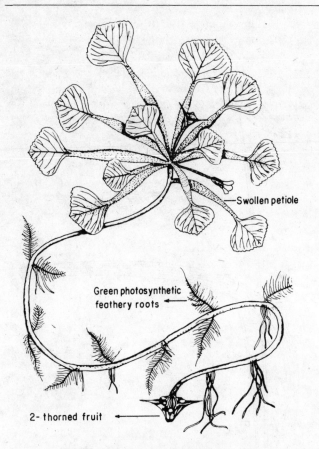

Water chestnut (Singhara), *Trapa* × ½

Labels on figure: Swollen petiole; Green photosynthetic feathery roots; 2- thorned fruit

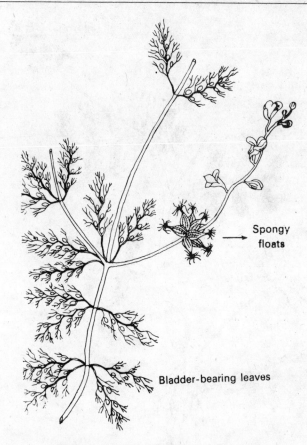

Bladderwort, *Utricularia stellaris* × ½

Labels on figure: Spongy floats; Bladder-bearing leaves

In *Hydrilla* (Jhangi) and Eel grass (*Vallisneria*), vegetative shoots are submerged but flowers may be at or above the water surface.

Anchored hydrophytes with floating leaves include the Indian Blue Water-lily (*Nymphaea*), and Arrowhead (*Sagittaria*).

Emergent amphibious types have their roots, lower parts of the stem and lower leaves submerged as in Talimkhana (*Hygrophila auriculata*), Didhen (*Aeschynomene indica*) or *Ludwigia*.

Wetland hydrophytes or halophytes are rooted in waterlogged soil at least during the early part of their life, as in certain sedges (*Cyperus* and *Scirpus* species), Khaki weed (*Alternanthera sessilis*), *Ammania baccifera*, *Polygonum plebeium*, *Eriocaulon*.

Certain plants may change their habit according to seasons. Thus Swamp Cabbage (*Ipomoea asarifolia*) belongs to the floating category when the water-level is high but turns into a halophyte in the dry season.

The submerged plants respire and absorb gases and salts dissolved in water over their entire surface. They do not transpire because of absence of stomata. Roots, if present, serve merely for anchorage. Woody tissue is lacking as it is not needed either for support or for conducting water from roots. The suspension in water is aided by large air-spaces in the tissues of all of the or-

gans, which also assure the supply of oxygen to the submerged parts.

Many water plants have floating leaves. In *Potamogeton* and *Aponogeton*, the lower surface in contact with water is deprived of stomata; however, unlike in the

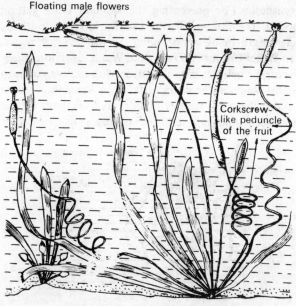

The Eel grass, *Vallisneria*, × ⅓

Labels on figure: Floating male flowers; Corkscrew-like peduncle of the fruit

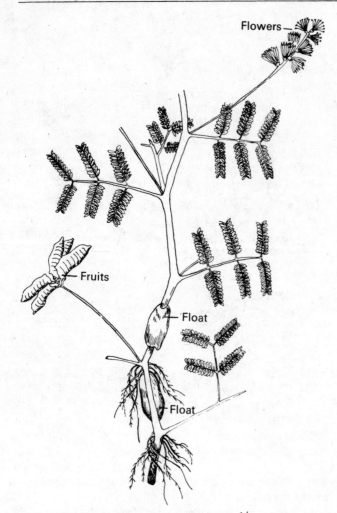

Flowers

Fruits

Float

Float

Neptunia oleracea (Lajalu), × ½

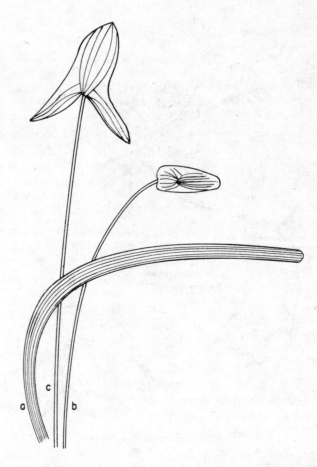

Arrowhead, *Sagittaria sagittaefolia* × ⅓. Leaf-blades developed (a) under water, (b) at the surface of the water, (c) in the air.

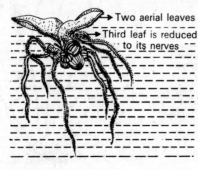

Two aerial leaves

Third leaf is reduced to its nerves

Water-fern (*Salvinea*) × 1

aerial leaves, they are numerous on the upper surface so that they can take in carbon dioxide needed for photosynthesis. For preventing the stomata from getting flooded, the floating leaves usually have long stalks, their length being proportional to the level of water.

In Water Chestnut (*Trapa*) and Water Hyacinth (*Eichhornia*) the petioles are swollen, serving as floats to keep the leaf blades at the water surface. In *Neptunia oleracea* (Lajalu) there are spongy bladder-like swellings on the internodes. *Ludwigia adscendens* floats by means of

white, spongy breathing roots formed in whorls at the nodes. In certain Bladderworts it is the flower-stalk that bears a whorl of spongy floats. In the Eel grass (*Vallisneria*), female flowers are borne on a slender peduncle which exposes the flower just at the water-level; later it twists in a spiral to bring the young fruits down to the mud for maturation.

Plants growing under flowing water have finely divided leaves, the advantage being that they expose a large surface for absorption of gases. In some *Myriophyllum* and Bladderworts, the leaf-blade is represented only by its nerves. In *Salvinia* (African payal), of the three leaves at each whorl, the submerged one is reduced to its nerves, functioning like a root. In the Arrowhead (*Sagittaria*), the leaves exposed to air are arrow-shaped, those floating are nearly cordate, whereas the submerged ones are ribbon-shaped. In *Limnophila aquatica*, the submerged leaves are finely laciniated whereas the floating leaves are not. In *Ceratophyllum demersum* it is the leafy branches that are modified as 'rhizoids' to perform the functions of a root. Yet another example of a rootless aquatic is the Bladderwort, in which leaves

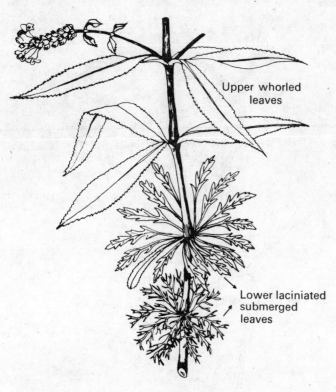

Limnophila aquaticus × 1

Upper whorled leaves

Lower laciniated submerged leaves

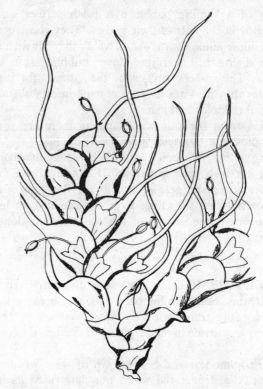

Podostemon subulatus growing in streams × 2¼

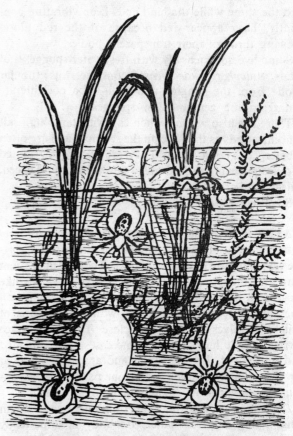

Water spiders, × ½ carrying bubbles of air to their dome-like nests. They may live 50 cm below the surface.

are modified to serve as roots. Besides, leaf segments produce small bladders which capture animalcules in this remarkable insectivorous water plant.

Two families of aquatic flowering plants deserve special mention. One is Lemnaceae, comprising the small or minute gregarious floating plants called Duck-weeds. The plant body lacks distinct stems and leaves; it may be rootless or with one to many hair-like roots. *Wolffia arrhiza* is amongst the smallest and simplest of flowering plants, resembling small grains.

The second remarkable aquatic family is the Podo-stemaceae, the members of which grow in rushing water and torrential hill streams. Their leaves are thin or almost nil and the stems short. The most important part of the vegetative body is formed by the roots, which are like green ribbons, resembling the thallus of algae and liverworts, sticking to the submerged rocks by hairs on their lower surface. Rich in chlorophyll, they carry out photosynthesis.

V.M.M-H.

See WATER-LILIES, INDIAN LOTUS.

WATER SPIDERS. Competition on land and the struggle for existence have been cited as contributory factors for certain Lycosids (Hunters) taking to a life in water. Such spiders are called Water spiders or Pirate spiders. The water spiders found in wells, ponds and rivers can live beneath the surface, breathing by means of a shining bubble of air which they carry entangled in the hair of their bodies. They construct a silken dome among waterplants which they fill with air like a diving bell, carrying down bubbles from the surface. The spiders wait inside their domes for prey. In water they feed not only on aquatic insects but also on small tadpoles and fish.

Apart from the true water spiders there are some Argyopids, e.g. *Orsinome marmorea*, which can remain within water for some hours. *Orsinome marmorea* builds her web near waterfalls. If disturbed she falls into the water and gets washed away. Reaching some rock or holdfast she clings to it and remains an inch or two below the surface of the water till the danger is over.

See also SPIDERS.

WATER WORMS. These are widely dispersed in rivers and streams, canals and ditches, probably because they can be easily transported by water systems or their cocoons are carried in mud clinging to the feet of wading birds.

The commonest and best known of these worms is *Tubifex*, the slender red worm popularly used as fish food. They are found in large aggregations in the slimy bottoms of ditches with their front ends thrust into the slime while the hind ends keep vibrating constantly. They appear red because of the red blood showing through their transparent bodies.

Some live commensally with freshwater sponges and snails. *Aulophorus tonkinensis* constructs its tubular abode from the stinging apparatus of a coelenterate and from the gemmules of sponges.

There are also worms which can withstand high salt concentrations as they live on the seashore and are often submerged by sea water. They are widely distributed and are probably drifted on to new sands by seaweed, flotsam and other debris.

See also WORMS.

WATER-FLEAS include several kinds of minute free-swimming crustaceans. Members of the genus *Cypris* are very common in stagnant pools. They are only 2 mm long and their carapace is formed of two valves like those of a bivalved mollusc with a hinge ligament on one side and adductor muscles within, by means of which the valves can open and close. The shell surface carries minute hairs and in some cases is ornamented and sculptured. Its antennules and antennae are used for swimming. Mandibles and four pairs of trunk appendages are present. Abdominal limbs are absent.

Daphnia is another water-flea of minute size, 1–2 mm long, living in fresh water. Its body is imperfectly segmented. The carapace is bivalved and reddish and the body laterally compressed except the head. A pair

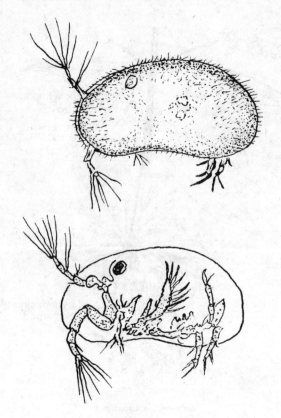

Cypris with and without shell, ×15

of sessile compound eyes are present. Antennules are large and the trunk appendages used for swimming by jerks. A prominent sharp caudal spine is a special feature. They constitute an important food item of small fishes in fresh water.

See also CRUSTACEANS.

WATER-LILIES. Represented by about 50 species in the world, the best examples in our area are the species of the genus *Nymphaea* well known in Indian, Greek and Roman mythology. In ancient Indian texts various species are referred to as Kumud, Kumudini and Nalini. Etymologically, the term *Nymphaea* owes its origin to the nature-goddess Nympha.

The Nymphaeas are perennial aquatic herbs with floating orbicular leaf-blades and flowers. Their fruit is spongy, with several seeds enclosed in a sac-like aril. Another member of the same family growing in shallow pools is the densely prickly aquatic herb Foxnut or Makhana (*Euryale ferox*). The circular leaf-blades are borne on heavily armed petioles. The undersurface of the leaf is also armed as in the Giant Water-Lily of the Amazon (*Victoria amazonica*) cultivated in India. The latter has large circular plate-shaped leaves, up to 2 m in diameter, with edges up-turned all round to prevent flooding; they can support the weight of a baby. The Water Snowflake (*Nymphoides indicum*) is often mistaken for a water-lily but in fact it is not a member of the *Nymphaea* family but belongs to the Gentian group.

See also INDIAN LOTUS.

WEAVER BIRD. Weaver Birds (subfamily Ploceinae), like the related House Sparrow, are characterized by a stout, conical, seed-crushing bill. The majority of species are African, only 5 occurring in Asia. Four of these, namely the common Baya (*Ploceus philippinus*), Finn's Baya (*P. megarhynchus*), the Blackthroated (*P. benghalensis*) and the Streaked (*P. manyar*) are Indian. They are seasonally dimorphic in plumage, the breeding males being largely yellow. The commonest and most widely spread is the Baya, best known for its cleverly woven, compact retort-shaped nests suspended in colonies from palm fronds or babul (*Acacia*) trees, usually over water. Its breeding biology, more or less common to the other three species as well, is of special interest. The sole nest-builder is the male. Several males start building in company on a selected tree. There are no females present in the early stages. When the nest is half completed a party of house-hunting females arrives who move from nest to nest as if to assess the workmanship. Some nests are accepted, others rejected. The satisfied female establishes possession and is accepted by the builder as his mate. The cock thereafter completes the nest, including the long downward-facing entrance tube, while the female lays her 2 or 3 white eggs within and begins to incubate. Thereupon the cock starts building a second nest nearby, which in course of time is similarly occupied by a second hen. In this way

a single cock may become possessor of 3, or even 4 nests and families in one and the same season. This type of polygamy is known as 'successive polygyny' in contrast to the 'harem' type of many game birds, in which a single cock holds mastery of a bevy of hens in the tradition of an oriental potentate.

© Salim Ali

Finn's Baya nest colony × ¹/₂

WEBSPINNERS. Webspinners are primitively social insects, usually living in colonies of 20 or more individuals, and are remarkable in that males, females, and nymphs spin the silken tunnels within which they live. The silk is produced by glands located in the basal segments of the fore tarsi. The slender, elongate (6 to 12 mm) insects are distributed mainly in the tropics and subtropics and though 800 species have been collected, it is estimated 2000 may actually exist. Owing

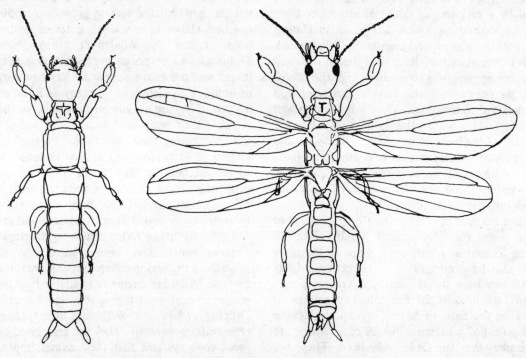

A pair of webspinners × 10

to their life activity being restricted to their silken labyrinths, except for dispersal, nonspecialists usually encounter only the winged males that come to lights in houses. Females are wingless and exhibit maternal care of eggs and young nymphs. They are confined to their nests, which may be in soil, under stones, under bark or lichen and in similar situations. Their food consists of bark, dead leaves, moss or lichen and is entirely of plant origin, though cannibalism is recorded. Webspinners are able to move extremely rapidly backwards into their tunnel, using their cerci as 'feelers' in their reverse movement, to escape from enemies. Probably 35 to 40 Indian species are known so far, *Oligotoma humbertiana* being common, having spread to many parts of the world through human agency.

WHALES and DOLPHINS. Amongst the class of Mammalia, several groups have adapted to find food and live for long periods in the water, but none so completely as the whales and dolphins which are grouped together in the order Cetacea. They are so adapted and modified in structure that they can reproduce and rear their young without ever coming on land. They are true mammals however, which means that they are warm-blooded, maintain a constant body temperature, must breathe air, and suckle their young with milk.

Some of the major structural modifications are in their long slim torpedo-like shape and the loss of hind limbs, whereas the front limbs have been transformed into flippers (though these still enclose some digital bones). Whales only have a scattering of hairs around the head or muzzle and their body temperatures are maintained by a very special thick subcutaneous layer of fat known as blubber which acts as an insulating layer in lieu of the hair of land mammals. Whales lack any glands in the skin, they have no external ear, and their breathing apparatus is so modified that the nostril is located far forward in the top of the head and directly connected to the lungs. Thus, when a cetacean comes to the surface to breathe its nostril is exposed without much of the body coming out of the water and air can be inhaled without mixing with any water from the mouth or throat cavity, as would be the case with other mammalian forms in which the trachea from the lungs opens into the throat or mouth.

Whales are not fish and can at once be distinguished from these lower forms of animal life by their tail flukes being located in a horizontal plane to the body axis. They also have no gills or scales on their body surface. All fish have the tail fin in a vertical plane.

Cetaceans are broadly divided into two groups of Sub-Orders on the basis of having or entirely lacking teeth. The so-called whalebone (or baleen) whales are grouped under the **Sub-Order Mysticeti.** They are nature's giants and have no teeth. Instead the mucous membrane of their palate is modified into two rows of vertical hanging horny plates known as baleen. These plates disintegrate or wear into separate fibres on their outer surface. These fibres form a compact hair-like mass which is used to entrap small marine organisms as the swimming whale sucks in vast quantities of water.

The surplus water is then expelled through the sides of the jaws by the enormous tongue which in baleen whales may weigh as much as several tons and be equipped with extremely powerful musculature. The Blue Whale (*Balaenoptera musculus*) is a Baleen Whale and has the distinction of being not only the biggest of the nine known species of toothless whale, but also it is the biggest animal known to have lived on our planet Earth, exceeding in length and weight the prehistoric reptiles such as the Dinosaur *Brachiosaurus*. The latter, it is estimated, weighed as much as 50 tons whereas an adult Blue Whale can weigh over 100 tons. *Diplodocus* measured in length as much as 85 feet but female Blue Whales are on record as exceeding this length by many feet. The new-born foetus measures up to 27 feet in length and will be 53 feet in length before it is weaned at about seven months of age.

The second great group is known as the Odontoceti and includes all the toothed whales. Some of these are specialized to feed upon squid, hunted near the bottom, whilst others catch all sorts of fish and hunt nearer the surface. Most are comparatively small in size varying from 6 to 10 feet in length, though a few species are much bigger notably the Killer Whale (*Orcinus orca*) which grows up to 25 feet and the Sperm Whale (*Physeter macrocephalus*), the largest of all toothed whales in which the female grows up to 50 feet. The toothed whales have a single external nostril or blowhole whereas the whalebone whales have a pair. Dolphins and porpoises are usually less than 8 feet in length and are characterized by having a large number of unmodified teeth with as many as 50 on each side of the jaw. They are usually gregarious in habits, travelling in schools or herds and maintaining social contact as well as locating their prey food by a highly developed system of sonar (echo-location). Dolphins are usually so named because their heads terminate in a narrow protracting snout or beak in which the teeth are located. This is called the rostrum. Porpoises are characterized by relatively rounded blunt foreheads and jaws.

There are living today about 90 different species of Cetacea which have been divided into 38 different genera. In the region covered by this encyclopedia there are two dolphins exclusively adapted to live in fresh water; in ecological terms, they are fluvial. The Indus Dolphin (*Platanista indi*), and the Ganges Dolphin (*Platanista gangetica*). They are associated with the two great river systems that their names imply. They are relatively small animals, quite similar to each other in external appearance, with pronounced long narrow beaks (rostra) bearing a number of sharp conical teeth. In the coastal waters several neritic species are found, such as the Finless Black Porpoise (*Neomeris phocaenoides*) and the Plumbeous Dolphin (*Sotalia plumbea*) and the Electra Dolphin (*Lagenorhynchus electra*). In deeper waters the Bottle-nosed Dolphin (*Tursiops aduncus*) and the various species of Dolphin such as the Cape Dolphin (*Delphinus capensis*) and Common Dolphin (*Delphinus delphis*) occur, often in huge schools numbering many thousands.

Amongst the giant baleen whales, the Blue Whale and the Humpback have both been authentically recorded in Indian waters and stranded specimens have been photographed, whilst the Fin Whale or Rorqual (*Balaenoptera physalus*) is perhaps more commonly sighted than the other two. In recent years there have been a number of strandings recorded of Bryde's Whale (*Balaenoptera edeni*) particularly along the Makran coast of Pakistan. This is the only member of the baleen whales which appears to spend its entire life in relatively warmer tropical and sub-tropical waters. All the other baleen whales have regular migratory movements, spending the arctic or antarctic summer in polar seas where an up-welling of the waters creates favourable conditions for the multiplication of vast quantities of small Euphasid shrimps known as krill. In northern Atlantic waters the baleen whales feed largely upon Copepods. Recent aerial photographs of feeding whales have revealed the technique of these animals; they cruise slowly just below the surface and through concentrated patches of zooplankton. A series of parallel grooves in the throat and ventral region allows this part of the whale's skin to expand into an enormous balloon-shaped pouch accommodating many tons of water and suspended food organisms. The mouth is then closed and the powerful and massive tongue compressed upwards against the fringed baleen plates which trap and sieve out the shrimps or copepods whilst the water is forced out through the sides of the jaws.

The so-called Right Whales have a different method of feeding. These mammals have enormous heads comprising up to one-third of their body-length and an upward arched jaw with baleen plates, each between 9 to 12 feet in length. These whales cruise on the surface with their mouths open when feeding. They are comparatively slow swimmers with blunt spade-shaped fore-flippers and are now confined to very small populations around the edge of the arctic and antarctic icefloes. They were hunted to the verge of extinction by the commercial whalers of the late nineteenth century.

Coming to the exploitation of whales by man the story is a sad one of greed and short-sightedness. After the development of the harpoon gun which could fire an iron harpoon attached to a length of coiled rope,

it became possible to pursue and kill the swifter-swimming whales from large diesel-propelled catcher boats. The subsequent development of factory ships and harpoons with explosive heads enabled greatly increased exploitation until by the mid 1950s the average annual catch in antarctic waters had reached 30,000 whales, most of which comprised Fin whales, the larger and more profitable Blue Whales having already begun to decline in the catches from the mid 1930s. The whales were hunted largely for the oil which could be extracted from their blubber, and all parts of their carcass. There was some attempt at international regulation of this industry before World War II but it was not until 1946 that an International Whaling Commission was proposed with the function of trying to establish catch quotas for the different whale species and to ensure future stocks. Up until the 1970s the Commission failed to impose realistic quotas and one by one the great whale species became so hard to locate as to be uneconomical to exploit. Today by the mid 80s decade most nations have ceased sending whaling fleets to the antarctic but the Soviet Union and Japan still refuse to agree to a total moratorium on whale-catching as urged by scientists and marine biologists.

Dolphins are still relatively plentiful and are for the most part too small in size to be worth exploiting for their oil. They are often in conflict, however, with commercial fishing fleets, largely because the dolphins themselves are attracted to schools of tuna and other commercially desirable fish and the boats often set their nets where fishermen observe concentrations of dolphins knowing that the area is likely to yield good catches. Consequently many thousands of dolphins get accidentally caught in these nets and are often drowned in the process or deliberately killed by the fishermen in order to free their nets.

It is only in recent decades that improvements in aquarium facilities and management techniques have enabled a variety of the smaller whales and porpoises to be kept in captivity. This has led to the surprising discovery that these small cetaceans are not only highly intelligent but that they can easily be taught to perform tricks and display definite signs of friendliness towards their keepers or trainers. Such behaviour by the dreaded Killer Whale has provided an entirely new understanding of the biology of this whale which is the largest living carnivorous animal, capable of attacking and tearing apart sea-lions and even baleen whales twice its own length.

Some of the larger whale species are thought to live up to 80 years but the smaller-toothed whales probably live no more than 15 or 20 years. The young are born after a gestation period of 6 to 11 months and normally only one at a time. The mother will assist it to the surface to draw its first breath, and in the first few days suckles it lying on her side so that it can breathe. Whales have only two mammae situated in longitudinal slits on either side of the ventral orifice. Their milk is known to be very rich, and young whales grow very rapidly. Recent studies have also revealed that whales communicate under water by a complicated repertoire of sounds and that migrating Humpback Whales (*Megaptera novaeangliae*) keep in contact with each other by unique and eerie-sounding songs.

When whales surface to breathe they can expel the breath from their lungs very rapidly. In colder oceans this warm humid breath condenses on contact with the air to form a fine mist or cloud. The shape and angle of this mist can be used in identifying the different larger species and is referred to as the whale's 'blow'. Most species surface to 'blow' several times in the space of a few minutes before diving to continue feeding. Sperm Whales, which feed at great depths, have been recorded as staying submerged for as long as 75 minutes, and the Bowhead or Right Whale regularly stays submerged for as long as 30 minutes. A frightened cetacean can easily swim two or three miles under water before surfacing to breathe.

See also DOLPHIN.

WHELKS are a large group of gastropods of various forms. They are all inhabitants of the sea and their shells are mostly shaped like a top or like a spindle or pear. They are all carnivorous and burrow in mud in search of the bivalves which they eat. They haunt littoral rocks, reefs and mud-flats, drill bivalve shells and rasp out the flesh with their radula. All whelks are characterized by a mouth situated at the end of a long proboscis which can be withdrawn into the cavity of the head. The radula is composed of only a few teeth in each row (total number only 220 to 250 whereas in a snail it is 30,000); but the teeth are large and often strongly serrated, well adapted to rasp shells or tear flesh. Like all carnivorous molluscs, whelks possess salivary glands opening into the throat which secrete sulphuric acid. This acid saliva softens the shell under attack and also aids digestion.

The sexes are separate in all whelks. Fertilization is internal. Eggs are laid in parchment-like capsules, in which is contained in addition to the eggs some quantity of albumin serving to nourish the embryos. The shape of the capsules varies in different genera; it is coin-like in typical whelks, flattened pouches in *Nassa*, grain-like in Purples and triangular in Murices. Sometimes the capsules are stalked, sometimes without stalks (sessile). Sometimes the capsules have lids which open to let the embryos out. Generally large numbers of these capsules (each capsule containing a varying number of eggs) are attached to a gristly thread or sheet which is fixed to a rock, seaweed or similar object. Often only a limited number—sometimes only one—of the embryos contained in each capsule becomes developed, the rest serving as nutriment for the survivors.

True Whelks belong to the family Buccinidae. Several kinds are abundant on the sand-covered mud-flats, in rock pools and among the stones on Indian coasts *Babylonia spirata* is the commonest and most beautiful of Indian whelks. It has a smooth white shell blotched with orange brown. The line joining two consecutive whorls is deeply sunk in a groove so that the spiral whorls look as though they are telescopically fitted. The shell mouth is broad and the outer lip is thin.

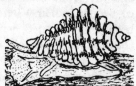

Whelks moving on a rock. Left, *Babylonia spirata*, moving; right, *Cantharus spiralis*, × 2/3.

The animal feeds on bivalves. When it moves gracefully over sand or rock its elongated flat foot, the prominent proboscis and head with tentacles, can all be seen clearly. Several species of *Cantharus* and *Pollia* are also abundant.

Apart from the true whelks, there are several families of carnivorous gastropods which possess whelk-like shells and live in similar places. The Rock-shells (Muricids) and the Mud-snails (Nassids) are called Dog Whelks. Both flourish in the intertidal region on both the shores of India, Muricids haunting the sides of oyster-covered rocks and Nassids trooping over mud-flats in thousands.

Muricid shells are often very beautiful and the growths on their shell have a protective function: they keep predatory fishes at a distance.

Muricids and their allies the Purples (*Purpura*) can be seen crawling in rock pools at half-tide, under blocks which rest on solid rocks at low-tide mark, in the

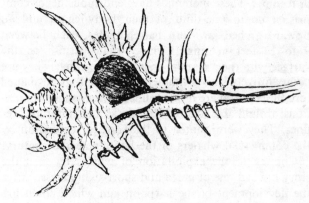

The shell of *Murex tribulus*, × 1.

crevices of flat-topped rocks and on oyster belts periodically washed by the rising tide. The frilled varieties of *Murex* are commonly found on isolated rocks at extreme low-tide mark.

During the breeding season Muricids appear in large assemblages near hollows of rocks. They lay their eggs in capsules measuring 1 cm long. The egg capsules of *Murex* look like three-sided grains with short stalks. They are deposited in clusters varying from 15 to 150 in number. Each capsule contains 20 eggs. The capsules of *Purpura* are like delicate pink grains of rice set on tiny stalks. They are not attached to one another but are set closely together in groups in sheltered nooks of the rocks. A single *Purpura* produces 245 capsules each containing 10 to 15 eggs.

In both Muricids and Purples there is a peculiar gland situated near the rectum and secreting a colourless fluid which turns to a dull crimson colour when exposed to the air. It is this dried secretion which was used by the ancient Greeks for dyeing, now popularly called 'Tyrian Purple'.

The other group of Dog whelks (Nassids) are all small molluscs with smooth spindle-shaped shells, lightly ribbed or sculptured. They look somewhat like snails. They are carnivorous, gregarious and their favourite haunts are mud-flats. Hence their popular name Mudsnails. Their long, broad foot and the prominent siphon can be seen as they move. They feed voraciously on living bivalves and also on any rotten and decaying matter. The sense of smell is greatly developed in Nassids and any heap of rubbish thrown on the beach soon attracts troops of different species of *Nassa*.

One genus of Nassids—*Bullia*—is of special interest. Unlike other Nassids it has an elongated auger-like shell. It crawls about rapidly on wet sand and attains its object of capturing a bivalve by a wide expansion of the foot on all sides. It slides over the sand instead of ploughing through it, the little lappets at the end of the tail probably serving as a rudder. *Bullia* also feeds on the little round-backed crabs scurrying over sand.

Some Nassids lay egg capsules shaped like flattened pouches with a short stalk and fasten them in rows to stems of seaweeds. Others on the other hand deposit solitary capsules which are shaped like small chattis or lotas.

Tritons are active univalve molluscs resembling whelks externally, internally and in feeding and breeding habits. Triton shells are characterized by the presence of varices (lengthwise thickening or ridges) on their whorls. Large Tritons found in other countries and used as Trumpet shells are not common in India but we have some exceedingly beautiful genera of Tritons—*Bursa, Ranella, Cymatium* and *Gyrenium*. They do not exceed 6 cm in length and they are handsomely ridged on both sides. Between the ridges they are beautifully noduled or tuber-

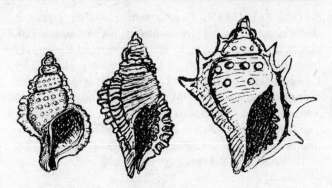

Some Triton shells. Left to right, *Gyrenium*, *Cymatium* and *Bursa*, all × 1.

culated. Some have long spines. They are found over a rough bottom in shallow waters. They feed on dead fish and other sea animals. Like Dog whelks they come out in troops to feed on any decaying matter. They are also called 'Frog shells' and 'Purse shells'.

See also Molluscs.

WHIP SCORPIONS (*Pedipalpi*) are arachnids inhabiting warm countries such as Arabia, India, Africa, tropical and central South America. They have a superficial resemblance both to scorpions and spiders. Their pedipalps are stout and end in powerful pincers whereas the first of the four pairs of legs are thin and long and function as feelers. At the posterior end of the segmented abdomen is a whip-like tail.

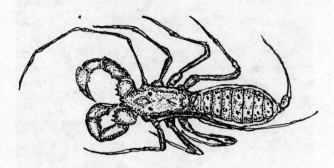

Whip scorpion (*Thelyphonus*) × ¹/₂

Like scorpions they are nocturnal in habits. During the day they conceal themselves in damp places, under stones, fallen logs, bark of trees and crevices. In view of their sinister appearance they are dreaded as venomous. In fact they are harmless and non-poisonous. They only emit a pungent and nasty acid secretion when handled. This secretion comes out from a pair of glands at the end of the posteriormost abdominal segment. Their food consists of small insects. The evil-smelling secretion protects them from their enemies.

Thelyphonus is a genus found in India. It has a long slender tail and its body is 7 to 8 cm long.

See also Arachnids.

WHITE DAMMAR, *Vateria indica,* is a large tree in the Western Ghats. It looks elegant with its bright-red, young leaves and fragrant white flowers which appear from January to March and is extensively planted by the road-sides in Kanara. The fruits are large 3-valved capsules ripening from May to July. The seeds yield a fat known as Malabar Tallow used in confectionery. The trunk on blazing exudes a resin called White Dammar, which is used in setting gold ornaments. The timber is in demand for making plywood.

WHITE DHUP, *Canarium strictum,* is a lofty tree in the evergreen forests of the East Himalayan foothills, Assam and the Western Ghats. The extremities are rusty-hairy and the tree looks spectacular when the new foliage in crimson-red comes into flush. Wounds in the bark exude a resin used as an incense to drive away mosquitoes and also used in varnishes. The timber produces excellent tea-boxes.

See also DHUP.

© K. C. Sahni

A White Dhup in evergreen forest in Arunachal Pradesh, height 38 m, girth at breast height 4·6 m. The bark is very hard and when struck with the back of a dao gives out a ringing sound.

WHITE SILK-COTTON (TRUE KAPOK) is a small to medium-sized tree known as *Ceiba pentandra* in botany. The specific name, which means 'five stamens', helps to distinguish it from the Red Silk-Cotton which

has 75 stamens. Introduced from the Amazon long back it is now common in S India, Sri Lanka, and in Burma. Young stems bear conical prickles, the branches arise in whorls and the adult trees are buttressed. The leaf has 5 to 9 lance-shaped, pointed leaflets arising from a long stalk. The flowers are dirty white with a milky smell and much smaller than Red Silk-Cotton. They appear in winter and the fruits ripen in late spring, disgorging lustrous white or pale yellow floss which is world-famous for use in lifebuoys and belts. The Red Silk-Cotton is also loosely called Kapok, which creates confusion and helps unscrupulous traders.

WHITE-EYE. The passerine family Zosteropidae, ranging from tropical Africa to New Zealand, North China and Japan, is represented in the Indian subcontinent by one species—*Zosterops palpebrosa.* It is a small warbler-like yellow-and-green bird with a conspicuous white ring round the eye, whence also called the 'spectacle bird'. It is commonly found foraging in flocks in hilly well-forested country uttering insect-like cheeps; also it has a jingling song. It feeds on arthropods, fruits, nectar, buds and seeds. Its semi-tubular tongue helps in nectar-feeding. The nest is a neat cobweb-plastered cup, slung between a horizontal twig fork. Two to four pale blue eggs are laid and both parents help in incubation and brood care.

WILLOWS, *Salix,* are well represented in the Himalaya. *S. tetrasperma* (Indian Willow) occurs throughout the greater part of the subcontinent along streams. It is a small tree, with lance-shaped leaves with flowers in male and female catkins. The wood is used in gunpowder and twigs in basketry. Willows are good fodder and check erosion by stabilizing hillsides. They are insect-pollinated, unlike Poplars which also belong to the willow family and are wind-pollinated. Because of the demand for cricket bats, the Cricket-bat Willow (*S. alba,* ssp. *coerulea*), the best wood for the purpose, was successfully introduced in Kashmir from the United Kingdom. The buyer takes care to select female trees only and rejects all male trees. It has a tapering, pyramidal crown with greenish-blue leaves and purple twigs. Leafless in winter, sprouting and flowering in spring, the fruits ripen 2 or 3 months after flowering. Minute seeds with silky hair are carried by the wind to long distances. *S. babylonica* (Weeping Willow), with graceful hanging branches, is popular in landscaping gardens. Introduced from the Levant, the young foliage has the fragrance of roses.

WINDS. The movement of air relative to the rotating surface of the earth is called wind. The air movement is essentially horizontal, the vertical component being of the order of a hundredth of the horizontal component. Hence by winds we mean the horizontal motion of the air. Its magnitude can be expressed as miles per hour, or kilometres per hour, or metres per second, or knots.

1 knot = 0·515 metre per second
      = 1·853 kilometres per hour
      = 1·152 miles per hour

Near the surface of the earth the winds are measured by different kinds of anemometers. Although on most occasions there is a significant wind from one direction or another, there are a few occasions when there is no wind. It is a 'calm'. In meteorology, it is conventional to specify the wind direction as the direction *from* which it blows. Thus a northeast wind means one which blows from the northeast to the southwest. Usually the wind directions are given in the eight points of the compass, N, NE, E, SE, etc.

The magnitudes of the surface winds (winds near the ground, usually not more than 10 metres above the ground), vary widely from place to place and from season to season and over different times of the same day. The India Meteorological Department has extensive data collected at about 400 stations distributed over the subcontinent, over a period of many years.

The surface winds are steady and strongest—15 to 20 knots—during the summer monsoon period. They are weakest during winter mornings. Associated with thunderstorms and duststorms, the speeds may increase to as much as 50 to 80 knots during brief spells of two to five minutes. Associated with severe cyclonic storms, the winds may become extremely violent, sometimes blowing a steady hundred knots for an hour or two. In all such cases of strong winds there are gusts, in which the speed further increases by 10 to 20 percent of the mean speed, for a few seconds. The strong winds due to CYCLONES affect coastal districts and the associated wind pressure and wind suction cause considerable damage to buildings and other structures. The winds generally increase with height, according to a logarithmic law, so that tall structures experience higher winds.

Winds are caused by horizontal gradients of atmospheric pressure on a rotating earth. They blow along the isobars (lines joining places with equal atmospheric pressure) with a small component towards the lower pressure to the left in the northern hemisphere. The ultimate cause of the differences is the variation in the heating of air-columns by direct and indirect solar radiation.

Winds at altitudes of 10 to 12 kilometres, where modern aircraft fly, can be thirty to forty knots most of the time, occasionally reaching 100 or even 150 knots over narrow bands encircling the globe, in a wavy pattern.

There are charts depicting the mean winds at different altitudes all over the globe, for the different seasons and months. The meteorological services of the respective geographical areas are the custodians of the details of wind structure.

WOLF. The Indian Wolf is recognized as a distinct subspecies *Canis lupus pallipes*, being smaller and shorter-haired than its larger cousins inhabiting the tundra and boreal forest zones of North America and Eurasia. There is however a much larger race *C. lupus chanco* inhabiting the inner Himalayan regions and Tibet.

Fifty years ago wolves were widespread in the drier open plains and desert zones of peninsular and northwestern India. The increase in human population and spread of cultivated areas has led to a rapid decline in their numbers, accelerated by man's incessant war against an animal which is a threat to his flocks of domestic stock. Wolves now only survive in the remoter, more sparsely populated regions such as in Rajasthan, Baluchistan, Gilgit and Ladakh in the north. In these areas they roam over large territories, often hunting in small family parties until the young become fully independent. Their principal prey is domestic sheep and goats but where wild game is abundant they can and do subsist upon hares, gazelle, and even gerbils, and desert locusts.

Wolves cannot tolerate constant human disturbance but in the areas where they do survive they are not particularly shy of mankind and will often stand and stare at the human intruder before loping off. They will hunt in late afternoon and early morning as well as throughout the hours of darkness, traversing distances up to 20 kilometres in a night.

By daytime they shelter in natural rock-caves or in burrows which they are capable of excavating themselves, and in the desert they will excavate a burrow in completely flat ground. The young are born in an underground nest-chamber and litter sizes vary from 3 to 9, the pups being blind and helpless at birth and taking about eight weeks to be weaned. The gestation period is around 68 days and females breed only once a year. In the northwestern part of India and in the plains of Pakistan most litters are produced in late winter and early spring. In the Himalayas cubs are born in spring or early summer. In captivity they have lived for 15 years. Recent studies have shown that wild wolves in North America form permanent pair-bonds. Certainly the male is attentive when the pups are young, helping to bring food, which is regurgitated by both parents in front of their offspring.

Adult male desert wolves weigh up to 24 kg and stand 71 cm at the shoulder, but the Tibetan race is much bigger with longer hair and may stand 80 cm at the shoulder. They have short bushy tails and a thick ruff of hair around their necks, being grizzled with grey and yellowish buff on their lower parts and predominantly black hairs on their back and the front of their limbs. Occasional individuals that are white or black have been recorded from Baluchistan and Ladakh and this same variation occurs in the sub-arctic regions. They have pointed upstanding ears, thickly fringed with white hair on their insides and black naked nosepad and lips.

The wolf will cross-breed with domestic dogs, and captive specimens are in fact easily tamed.

WOOD-LICE form a unique group of isopod crustaceans adapted to live on land. They have usually an oval body, convex above and almost flat below. The head is small and the abdomen short. The thorax is prominent with seven segments and seven pairs of walking legs. Abdominal appendages serve as breathing organs. *Oniscus* is the commonest example, 2 cm long, inhabiting damp places and living among decaying leaves. Its members are usually nocturnal and vegetarians. *Ligia* is a larger genus 2·5 to 3 cm long, found in crevices of stones by the seashore above high-tide mark.

See also ANIMAL LICE.

WOODPECKER. Insectivorous birds of well wooded country (family Picidae) with short legs, zygodactyl feet (two toes in front and two behind) and sharp curved claws for clinging to and climbing tree-trunks. They have a stiff pointed tail to serve as a tripod support in this activity. The chisel-shaped bill, specially adapted for digging in wood, and worm-like, barb-tipped extensile tongue capable of being shot out beyond the bill tip, enable the bird to skewer out hidden beetle larvae from the borings. The Indian subregion is particularly rich in woodpeckers. We have 15 genera and 32 species ranging from the crow-sized Great Black woodpecker (*Dryocopus javensis*) and the Himalayan Slaty *Mulleripicus pulverulentus* to the diminutive sparrow-sized Spotted and Rufous piculets *Picumnus innominatus* and *Sasia ochracea*. Woodpeckers are highly beneficial to forests in controlling timber pests. Enlightened forestry practices recognize their usefulness by leaving a few mature native trees standing for the birds to nest in amidst areas clear-felled for monoculture.

WORMS (Annelida). This comprises a group of segmented worms which consist of three classes:

1. The Oligochaeta—worms with few bristles on each segment.

2. The Polychaeta—worms with a distinct head often with tentacles and sensory structures and many bristles on each segment.

3. The Hirudinea—includes the leeches.

Each segment is an individual unit and has serially repeated organs except for the reproductive organs in Oligochaeta, Hirudinea and some Polychaetes which are localized to a few segments.

See EARTHWORMS, WATER WORMS, BRISTLEWORMS, PARASITIC WORMS and LEECHES.

WORMWOOD, *Artemisia absinthium*, is an aromatic herb, with a ribbed stem, egg-shaped leaves cut into blunt segments, and yellow, globose flower-heads. Its dried leaves and flowers repel bed bugs and protect garments from silverfish and other insects. It has recently been highly successful in combating malaria in China and is also used as a tonic. It is found from Kashmir northwards to northern Asia.

W.W.F. The World Wildlife Fund is an international charity. It was established in 1961 and is devoted to conserving nature in all its forms—fauna, flora, landscape, soil and water. The first President was Prince Bernhard of the Netherlands, and the Duke of Edinburgh took over in 1981. The Board of Trustees includes distinguished men and women from many parts of the world.

The headquarters of W.W.F. are at Gland in Switzerland, and there is a network of national organizations in 27 countries. These organizations interact with one another and establish priorities for conservation action throughout the world. The projects of the W.W.F. may be international, spanning many countries, or dealing with a species or ecosystem of worldwide significance, or national projects whose objectives are more limited.

By 1982, W.W.F. had raised 46 million US dollars and had channelled these funds into more than 2400 projects in 130 countries.

The Indian National Appeal of World Wildlife Fund was established in 1969 with Shri Fatehsinghrao, then

© Zafar Futehally

Tiger showing face markings that enable naturalists to identify individual animals

Gaekwad of Baroda, as the President. The World Wildlife Fund-India has engaged itself in many useful projects for conserving wild areas and wild species. Through its scheme of Nature Clubs it attempts to teach young people about conservation and the associated discipline of ecology.

The most important projects of W.W.F. in India relate to the tiger and the elephant. Project Tiger was instituted by W.W.F.-International and is now operated in association with W.W.F. by the Government of India. With regard to the elephant, an Asian Elephant Task Force has been established, largely on the initiative of W.W.F. and I.U.C.N., to survey and make conservation plans for saving the elephant.

YAK. The truly wild yak is probably a vanishing species, though it occurs in regions so remote and politically inaccessible that no reliable surveys have been carried out in recent times. In the subcontinent, they only survive in small numbers in the Changchenmo valley in northern Ladakh but their total range extends across Tibet into Kansu province of China.

The wild yak (*Bos grunniens*) is blackish brown all over, with perhaps a little white around the muzzle. An adult bull stands 170 cm at the shoulder and may weigh up to 545 kg. As in all the wild oxen, both sexes bear horns and those of the bull yak are smooth cylinders sweeping up like handlebars and measuring up to 75 cm in length. Their appearance is made more massive by the fact that their chest, shoulders and flanks are fringed by long hair hanging almost to the ground. Their tails are likewise covered in a thick and bushy tuft and this appendage is often used as a fly-whisk and badge of state by high dignitaries in the lamaseries. Yak have been domesticated since time immemorial and are widely used as pack animals in the northernmost reaches, from Chitral in the west through Ladakh, to Nepal and Sikkim in the east. There are also hybrid forms with domestic cattle locally called *zho*. Domestic yak are almost indistinguishable from wild yak except that they frequently are piebald and even dun-coloured, with less massive horns, whilst the *zho* are considerably smaller.

Amongst all the bovines this animal is perhaps best adapted not only to extreme cold but also to desert conditions. Yak can subsist on the scattered clumps of *Astragalus* and other thorny cushion plants and eat snow when water is unavailable. Unlike the domesticated oxen and buffaloes of the plains their faeces are voided in small rounded pellets with a very high dry content,

resembling closely those of a camel. Their voice is a deep grunt from which their scientific name is derived. Domesticated yaks are normally found at altitudes from 4270 up to 6100 metres and wild yak have seldom been observed below 5000 metres. They are gregarious, living in small herds, and in the wild are shy and wary beasts relying on an acutely developed sense of smell rather than keen eyesight to detect danger. The rutting season is late autumn, with calves being produced in April after a nine-month gestation period. Domesticated yak have a habit of grinding their teeth audibly, a trait probably connected with the fact that they are generally infested with intestinal tapeworms.

See also UNGULATES.

YAMUNA. Rich in religious tradition, folklore, and history, the river Yamuna is the principal tributary of the Ganga in north India. Rising from the Yamunotri glacier on the northwestern slope of Bandarpunch peak (6387 m) in the Himalaya, it flows for 1380 km in a course almost parallel to the Ganga, until meeting it at Allahabad. Giri and Tons are its major affluents in the hilly section and Sind, Chambal, Betwa, and Ken in the plains. Its total drainage area is over 30,000 km².

Physiographically, the Yamuna Basin is divisible into three major sections; the mountainous, the Siwalik or the submontane and the plains. In the mountainous Himalayan section, terrain is rugged, with deep gorges, and narrow U-shaped glaciated valleys, waterfalls, and river terraces. The Yamuna winds its way through this section for about 135 km. South of the river's origin, the area is dominated by several imposing granite peaks, like Chaur, Nag Tibba, Deoban, and Kharamba (all over 3000 m). The Tons, Pabar, and Giri are the main tributaries in this section. The river as do its tributaries, cuts through the Himalayan ridges in longitudinal valleys before entering the submontane or the Siwalik section. In general, the Himalayan topography is highly uneven, grand, and awe-inspiring with snowy peaks eternally in the background. Rocks are hard, dissected and folded (granites, limestone, predominating) but vary in geologic age.

After breaching the Mussoorie Range the Yamuna enters its Siwalik section. The Siwalik or submontane hills have a steep slope towards the south and a gentler profile to the north, enclosing small fertile valleys (the duns) between the Himalayan section and the submontane hills. Dehra Dun is a town located in such a fertile valley. At the foot of the Siwaliks is the terai area. The Siwaliks are forested or covered with tall grasses, and contain some wildlife including tigers.

Least interesting in topography but agriculturally the most productive and rich in human history is the flat, riverine plain section. The plain gradually slopes towards the east. Only undulations near the river-banks break the flat monotony of the Ganga-Yamuna doab. Flatness has induced considerable river meandering, resulting in occasional changes in river courses, cut-offs and the development of jheels (lakes). In this section the chief tributaries are the Chambal, Sind, Betwa and Ken, all of which issue from the Peninsular Plateau and cut deep gorges through hard rocks and have produced badland topography.

Soils in the mountainous section are varied, though generally coarse-grained and thin. In the valleys they are fine-grained and composed of river-deposited alluvium, and thick enough for crop growth. In the Siwaliks, alluvium is mixed with gravels and sand. The terai soils are generally ill-drained, swampy loams of mixed sand, gravel and silts. The plains section contains deep, river-borne alluvial and rich loamy silts (*khadar*) near the rivers. Earth borings near Delhi indicate that the river alluvium is over 1000 m deep. Away from the the river-course in the middle of the doab somewhat coarse or older alluvium exists whereas the younger, finer-grained material lies closer to the river. Hard rocks or kankar occur only occasionally in the plains.

Climatically, the river proceeds towards areas of increasing rainfall in the plains (from 55 cm to 140 cm annually). It is perennially fed from the glaciers, but is given to high seasonal fluctuations. There are clear-cut seasons: hot summer, wet summer, and dry winter. The incidence of winter rainfall increases towards the western parts of the basin. Summers are hot, even scorching, from mid-March to July, maxima rising from 20° to over 40°C. The Himalayan section remains considerably cooler, and acts as a haven for those seeking refuge from the uncomfortable heat of the plains. By mid-June the rains come and the temperatures fall gradually to 30°C.

Because of the Himalaya the river provides a continuous, if seasonally erratic, flow of water. The flat topography and soft alluvium provide excellent conditions for digging of canals and wells. Rainfall is insufficient for two good harvests. Irrigation, therefore, has become the mainstay of agriculture almost universally in the Yamuna plains.

R.T.

YELLOW SILK-COTTON, *Cochlospermum religiosum*, is a small deciduous tree of central peninsular India and Burma. It is a showy tree in the hot season with golden yellow flowers when leafless. The leaves are 5-lobed. The fruits are pear-shaped enclosing small seeds embedded in floss which is used for stuffing pillows. The bark yields a gum Kuteera which is used by cobblers. The specific name refers to its decorating temples.

ZOONOSIS. A disease which can be transmitted from animals to man e.g. RABIES, plague, various forms of encephalitis, ornithosis or psittacosis. Among zoonoses caused by metazoan parasites the most important are the WORM infestations, e.g. tapeworm, schistosomiasis, etc.

ZORAPTERANS. The Zoraptera comprise the smallest insect order with only 25 known species in a single genus. These extremely rare, peculiar insects were first discovered in 1913 and *Zorotypus ceylonicus* from Sri Lanka is our only known species. Most species have wingless individuals, which are blind and pale, as well as winged forms which are darker and possess eyes and ocelli. Rarely more than 2·5 mm long, with 9-segmented bead-like antennae, they are found gregariously, in the warmer parts of the world under bark, near termite

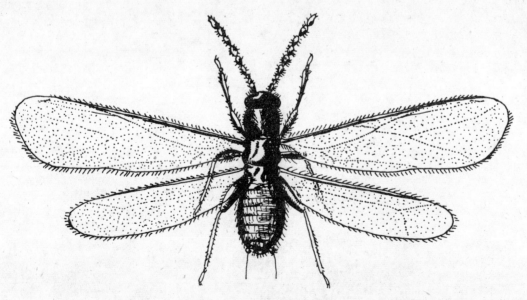

A zorapteran × 18

607

nests, or in rotting fallen logs and the like. Little is known of their food habits, but they appear to be scavengers, though mites and fungal bodies have been found in their gut. Their developmental period is extremely protracted and, though they sometimes occur in colonies of more than 100 individuals and are polymorphic (see INSECT EVOLUTION & BIOLOGY), no evidence of a social organization exists as in termites (Zorapterans also shed their wings after mating). Although there is considerable uncertainty regarding their nearest insect relatives, termites or psocids are though to be the most likely candidates.

Z.S.I. The Zoological Survey of India was established in 1916 and its collections cover the whole of our area. The Museum at Calcutta houses more than a million specimens belonging to all the animal groups from Protozoa to Mammalia. These collections are invaluable for research in the laboratory or in the field. Fourteen collecting stations are maintained in different ecological biotopes, such as the Desert Regional Station at Jodhpur and the High-altitude Station at Solan.

INDEX & GLOSSARY

Page numbers follow some of the longer headwords

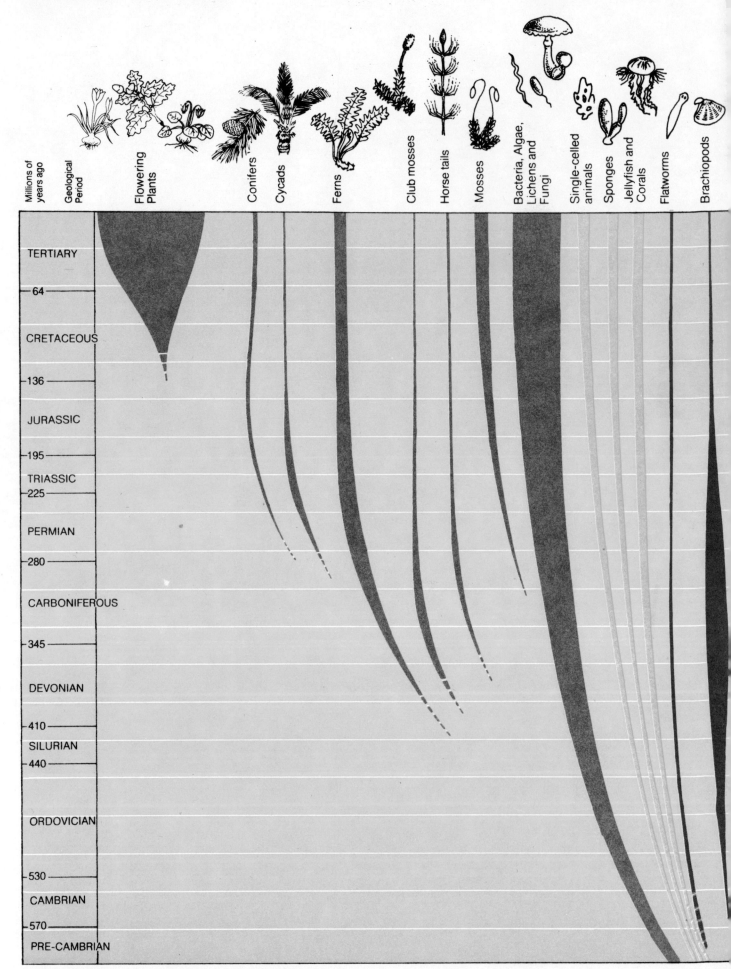

Simplified Tree of Life

The horizontal white lines are drawn at intervals of 30 million years. The widths of the

Millions of years ago

Geological Period

Flowering Plants

Conifers

Cycads

Ferns

Club mosses

Horse tails

Mosses

Bacteria, Algae, Lichens and Fungi

Single-celled animals

Sponges

Jellyfish and Corals

Flatworms

Brachiopods

TERTIARY

— 64 —

CRETACEOUS

— 136 —

JURASSIC

— 195 —

TRIASSIC
— 225 —

PERMIAN

— 280 —

CARBONIFEROUS

— 345 —

DEVONIAN

— 410 —

SILURIAN
— 440 —

ORDOVICIAN

— 530 —

CAMBRIAN

— 570 —

PRE-CAMBRIAN